Erik W. Grafarend
Friedrich W. Krumm
Map Projections

Charles Seale-Hayne Library

University of Plymouth

(01752) 588 588

Erik W. Grafarend
Friedrich W. Krumm

Map Projections

Cartographic Information Systems

With 230 Figures

 Springer

Professor Dr. Erik W. Grafarend
Universität Stuttgart
Institute of Geodesy
Geschwister-Scholl-Str. 24 D
70174 Stuttgart
Germany

E-mail: grafarend@gis.uni-stuttgart.de

Dr. Friedrich W. Krumm
Universität Stuttgart
Institute of Geodesy
Geschwister-Scholl-Str. 24 D
70174 Stuttgart
Germany

E-mail: krumm@gis.uni-stuttgart.de

Library of Congress Control Number: 2006929531

ISBN-10   3-540-36701-2   Springer Berlin Heidelberg New York
ISBN-13   978-3-540-36701-7   Springer Berlin Heidelberg New York

Springer is a part of Springer Science+Business Media
Springeronline.com
© Springer-Verlag Berlin Heidelberg 2006
Printed in Germany

The use of general descriptive names, registered names, trademarks, etc. in this publication does not imply, even in the absence of a specific statement, that such names are exempt from the relevant protective laws and regulations and therefore free for general use.

Cover design: E. Kirchner, Heidelberg
Design, layout, and software manuscript by Dr. Volker A. Weberruß, Im Lehenbach 18, 73650 Winterbach, Germany
Production: Almas Schimmel
Printing and binding: Stürtz AG, Würzburg

Printed on acid-free paper   32/3141/as 5 4 3 2 1 0

Preface

This book is dedicated to the Memory of US GS's J. P. Snyder (1926–1997), genius of inventing new Map Projections.

Our review of *Map Projections* has 21 chapters and 10 appendices. Let us point out the most essential details in advance in the following passages.

Foundations.

The first four chapters are of purely introductory nature. Chapter 1 and Chapter 2 are concerned with general mappings from *Riemann manifolds to Riemann manifolds* and with general mappings from *Riemann manifolds to Euclidean manifolds* and present the important *eigenspace analysis* of types *Cauchy–Green* and *Euler–Lagrange*. Chapter 3 introduces coordinates or parameters of a Riemann manifold, *Killing vectors of symmetry,* and oblique frames of reference for the sphere and for the ellipsoid-of-revolution. A special topic is the classification of surfaces of zero *Gaussian curvature* for ruled surfaces and for developable surfaces in Chapter 4.

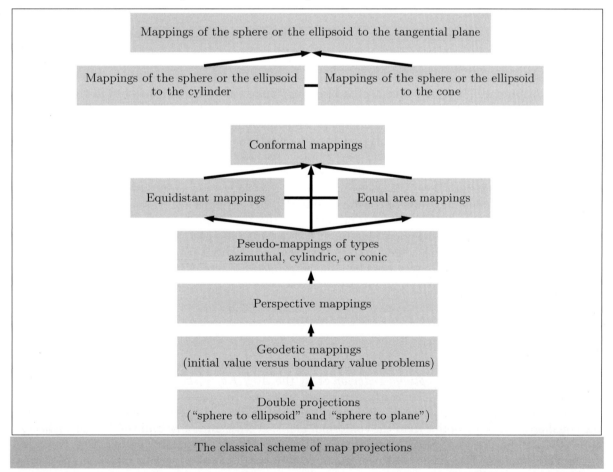

The classical scheme of map projections

Next, we intend to follow the classical scheme of map projections. Consult the formal scheme above for a first impression.

The standard map projections: tangential, cylindric, conic.

The Chapters 5–7 on mapping the *sphere to the tangential plane*, namely in the *polar aspect* (normal aspect) – for instance, the *Universal Polar Stereographic Projection* (UPS) – and the meta-azimuthal mapping in the *transverse* as well as the *oblique aspect*, follow. They range from equidistant mapping via conformal mapping to equal area mapping, finally to *normal perspective mappings*. Special cases are mappings of type "sphere to tangential plane" at maximal distance, at minimal distance, and at the equatorial plane (three cases). We treat the *line-of-sight*, the *line-of-contact*, and *minimal* versus *complete atlas*. The gnomonic projection, the orthographic projection, and the Lagrange projection follow. Finally, we ask the question: "what is the best projection in the class of polar and azimuthal projections of the sphere to the plane?" A special section on *pseudoazimuthal mappings*, namely the *Wiechel polar pseudoazimuthal mapping*, and another special section on *meta-azimuthal projections* (stereographic, transverse Lambert, oblique UPS and oblique Lambert) concludes the important chapter on various maps "sphere to plane".

Chapter 8 is the first chapter on mapping the *ellipsoid-of-revolution to the tangential plane*. We treat special mappings of type *equidistant*, *conformal*, and *equal area*, and of type *perspective*. Chapter 9 is the first chapter on *double projections*. First, we introduce the celebrated *Gauss double projection*. Alternatively, we introduce the *authalic equal area projection* of the ellipsoid to the sphere *and* from the sphere to the plane.

The four Chapters 10–13 are devoted to the mapping "sphere to cylinder", namely to the polar aspect, to the meta-cylindric projections of type *transverse* and of type *oblique*, and finally to the *pseudo-cylindrical mode*. Four examples, namely from mapping the sphere to a cylinder (polar aspect, transversal aspect, oblique aspect, pseudo-cylindrical equal area projections) in Chapters 10–13 document the power of these spherical projections. The resulting map projections are called (i) *Plate Carrée* (quadratische Plattkarte), (ii) *Mercator projection* (Gerardus Mercator 1512–1594), and (iii) *equal area Lambert projection*. A special feature of the Mercator projection is its property "mapping loxodromes (rhumblines, lines of constant azimuths) to a straight line crossing all meridians with a constant angle". The most popular map projection is the *Universal Transverse Mercator projection* (UTM) of the sphere to the cylinder, illustrated in Fig. 11.3. The pseudo-cylindrical equal area projections – they only exist – are widely used in the *sinusoidal* version (Cossin, Sanson–Flamsteed), in the *elliptic* version (Mollweide, very popular), in the *parabolic* version (Craster), and in the *rectilinear* version (Eckert II).

In Chapter 10, a special section is devoted to the question "what is the best cylindric projection when *best* is measured by the *Airy optimal criterion* or by the *Airy–Kavrajski optimal criterion*?" We have compared three mappings: (i) conformal, (ii) equal area, and (iii) distance preserving in the class of "equidistance on two parallel circles". We prove that the distance preserving maps are optimal and the equal area maps are better than the conformal maps, at least until a latitude of $\Phi = 56°$, when we apply the Airy optimal criterion. Alternatively, when we measure optimality by the Airy–Kavrajski optimal criterion, we find again that the optimum is with the distance preserving maps, but conformal maps produce exactly the same equal area maps, less optimal compared to distance preserving maps.

In contrast, Chapters 14–16 are a review in mapping an *ellipsoid-of-revolution to a cylinder*. We start with the *polar aspect* of type $\{x = A\Lambda, y = f(\Phi)\}$, specialize to normal equidistant, normal conformal, and normal equiareal, in general, to a rotationally symmetric figure (for example, the torus). The *transverse aspect* is applied to the *transverse Mercator projection* and the special *Gauss–Krueger coordinates* (UTM, GK) derived from the celebrated *Korn–Lichtenstein equations* subject to an integrability condition and an optimality condition for estimating the *factor of conformality* (dilatation factor) in a given quantity range $[-l_{\mathrm{E}}, +l_{\mathrm{E}}] \times [B_{\mathrm{S}}, B_{\mathrm{N}}] = [-3.5°, +3.5°] \times [80°\mathrm{S}, 84°\mathrm{N}]$ or $[-l_{\mathrm{E}}, +l_{\mathrm{E}}] \times [B_{\mathrm{S}}, B_{\mathrm{N}}] = [-2°, +2°] \times [80°\mathrm{S}, 80°\mathrm{N}]$, namely $\rho = 0.999, 578$ or $\rho = 0.999, 864$. Due to its practical importance, we have added three examples for the transverse Mercator projection and for the Gauss–Krueger coordinate system of type {Easting, Northing}, adding the meridian zone number. Another special topic is the *strip transformation* from one meridian strip system to another one, both for Gauss–Krueger coordinates and for UTM coordinates. We conclude with two detailed examples of strip transformation (Bessel ellipsoid, World Geodetic System 84). At the end, we present to you the *oblique aspect* of type *Oblique Mercator Projection* (UOM) of the ellipsoid-of-revolution, also called *rectified skew orthomorphic* by M. Hotine. J. P. Snyder calls it "Hotine Oblique Mercator Projection (HOM)". Landsat-type data are a satellite example.

Only in the *polar aspect*, we present in Chapter 17 the maps of the *sphere to the cone*. We use Fig. 17.1 as an illustration and the setup $\{a = \Lambda \sin \Phi_0, r = f(\Phi)\}$ in terms of polar coordinates. $n := \sin \Phi_0$ range from $n = 0$ for the cylinder to $n = 1$ for the azimuthal mapping. Thus, we are left with the rule $0 < n < 1$ for conic projections. The wide variety of conic projections were already known to Ptolemy as the equidistant and conformal version on the *circle-of-contact*. If we want a *point-like image* of the North Pole, the equidistant and conformal version on the circle-of-contact is our favorite. Another equidistant and conformal version on two parallels is the *de L'Isle mapping*. Various versions of conformal mapping range from the equidistant mappings on the circle-of-contact to the equidistant mappings on two parallels (*secant cone*, J. H. Lambert). The equal area mappings range from the case of an equidistant and conformal mapping on the circle-of-contact over the case of an equidistant and conformal mapping on the circle-of-contact and a point-like image of the North Pole to the case of equidistance and conformality on two parallels (*secant cone*, H. C. Albers).

Chapter 18 is an introduction into mapping the *sphere to the cone*, namely of type *pseudo-conic*. We specialize on the *Stab–Werner projection* and on the *Bonne projection*. Both types have the shape of the heart.

The polar aspect of mapping the *ellipsoid-of-revolution to the cone* is the key topic of Chapter 19. We review the line-of-contact and the principal stretches before we enter into special cases, namely of type *equidistant mappings* on the set of parallel circles of type *conformal* (variant equidistant on the circle-of-reference, variant equidistant on two parallel circles, *generalized Lambert conic projection*) and type *equal area* (variant equidistant and conformal on the reference circle, variant pointwise mapping of the central point and equidistant and conformal on the parallel circle, variant of an equidistant and conformal mapping on two parallel circles, *generalized Albers conic projection*).

Geodesics and *geodetic mappings*, in particular, the *geodesic circle*, the *Darboux frame*, and the *Riemann polar and normal coordinates*, are the topic of Chapter 20. We illustrate the *Lagrange* and the *Hamilton portrait* of a geodesic, introduce the *Legendre series*, the corresponding *Hamilton equations*, the notion of initial and boundary value problems, the Riemann polar and normal coordinates, *Lie series*, and specialize to the *Clairaut constant* and to the ellipsoid-of-revolution. Geodetic parallel coordinates refer to *Soldner coordinates*. Finally, we refer to *Fermi coordinates*. The deformation analysis of Riemann, Soldner, and Gauss–Krueger coordinates is presented.

Datum problems.

Datum problems, namely its analysis versus synthesis and its Cartesian approach versus curvilinear approach, are presented in Chapter 21. Examples reach from the transformation of conformal coordinates of type *Gauss–Krueger* and type *UTM* from a local datum (regional, national, European) to a global datum (WGS 84) of type *UM* (Universal Mercator).

Appendices.

Appendix A is entitled as "Law and order". It brings up *relation preserving maps*. We refer to Venn diagrams, Euler circles, power sets, Hesse diagrams, finally to *fibering*. The inversion of univariate, bivariate, in general, multivariate homogeneous polynomials is presented in Appendix B. In contrast, Appendix C reviews elliptic functions and elliptic integrals. Conformal mappings are the key subject of Appendix D. First, we treat the classical *Korn–Lichtenstein equations*. Second, we treat the celebrated *d'Alembert–Euler equations* (usually called *Cauchy–Riemann equations*) which generate both conformal mapping, (i) on the the basis of real algebra and (ii) on the basis of complex algebra. Lemma D.1 gives three alternative formulations of the Korn–Lichtenstein equations. The fundamental solutions of the d'Alembert–Euler equations subject to the *harmonicity condition* is reviewed in Lemma D.2 in terms of a *polynomial representation* (D.15)–(D.29). An alternative solution in terms of *matrix notation* based upon the *Kronecker–Zehfuss product* is provided by (D.30) and (D.31). Lemmas D.3 and D.4 review two solutions of the d'Alembert–Euler equations subject to the *integrability conditions* of harmonicity, by separation of variables this time. Two choices of solving the basic equations of the transverse Mercator projection are presented: $x = x(q\ p)$, $y = y(q\ p)$. We especially estimate (i) the boundary condition for the universal transverse Mercator projection modulo an unknown dilatation factor and (ii) we solve the already formulated boundary value problem with respect to the d'Alembert–Euler equations (Cauchy–Riemann equations). Finally, the unknown dilatation factor is optimally determined by optimizing the total distance distortion measure (Airy optimum) or the total areal distortion. Appendix E introduces the extrinsic terms *geodetic curvature*, *geodetic torsion*, and *normal curvature*, the notion of a geodesic circle, especially the *Newton form* of a geodesic in *Maupertuis gauge* on the sphere and on the ellipsoid-of-revolution. Mixed cylindrical maps of the ellipsoid-of-revolution of type *equiareal* based upon the *Lambert projection* and the sinusoidal *Sanson–Flamsteed projection*, especially as the horizontal weighted mean versus the vertical weighted mean, are the central topics of Appendix F. The *generalized Mollweide projection* and the *generalized Hammer projection* (generalized for the ellipsoid-of-revolution) are the key topics, especially of our studies in Appendix G and Appendix H. The *optimal* Mercator projection and the *optimal* polycylindric projection of type *conformal*, here developed on the ellipsoid-of-revolution, are applied to the many islands of the Indonesian Archipellagos in Appendix I. *Projection heights* in the geometry space are the topic of Appendix J. We treat the plane, the sphere, the ellipsoid-of-revolution, and the triaxial ellipsoid, and we review the solution algorithm for inverting Cartesian coordinates to projection heights. An example is the *Buchberger algorithm*. In detail, we review surface normal coordinates, for example, in the computation of the triaxial ellipsoids of type *Earth*, *Moon*, *Mars*, *Phobos*, *Amalthea*, *Io*, and *Mimas*.

We here would like to emphasize that our introduction into Map Projections is exclusively based upon *right handed coordinates*. In this orientation, we particularly got support from my German colleagues *J. Engels* (Stuttgart), *V. Schwarze* (Backnang), and *R. Syffus* (Munich). We here would like to note that the software manuscript was produced by *V. Weberruß* with expertise. To all our readers, we appreciate their care for the **Wonderful World of Map Projections**. We dedicate our work to *J. P. Snyder* (1926–1997), who worked for the US Geological Survey for a lifetime. We stay on the strong shoulders of great scientists, for example, C. F. Gauss, J. L. Lagrange, B. Riemann, E. Fermi, J. H. Lambert, and J. H. Soldner. May we remember their great works.

Erik W. Grafarend   *Friedrich W. Krumm*

Contents

1 From Riemann manifolds to Riemann manifolds

"It is vain to do with more what can be done with fewer."
(Entities should not be multiplied without necessity.)
William of Ockham (1285-1349)

Mappings from a left two-dimensional Riemann manifold to a right two-dimensional Riemann manifold, simultaneous diagonalization of two matrices, mappings (isoparametric, conformal, equiareal, isometric, equidistant), measures of deformation (Cauchy–Green deformation tensor, Euler–Lagrange deformation tensor, stretch, angular shear, areal distortion), decompositions (polar, singular value), equivalence theorems of conformal and equiareal mappings (conformeomorphism, areomorphism), Korn–Lichtenstein equations, optimal map projections.

There is no chance to map a curved surface (*left Riemann manifold*), which differs from a developable surface to a plane or to another curved surface (*right Riemann manifold*), without distortion or deformation. Such distortion or deformation measures are reviewed here as they have been developed in differential geometry, continuum mechanics, and mathematical cartography. The classification of various mappings from one Riemann manifold (called *left*) onto another Riemann manifold (called *right*) is conventionally based upon a comparison of the *metric*.

Example 1.1 (Classification).

The terms equidistant, equiareal, conformal, geodesic, loxodromic, concircular, and harmonic represent examples for such classifications.

End of Example.

In terms of the geometry of surfaces, this is taking reference to its *first fundamental form*, namely the *Gaussian differential invariant*. In particular, in order to derive certain invariant measures of such mappings outlined in the frontline examples and called *deformation measures*, a "canonical formalism" is applied. The simultaneous diagonalization of two symmetric matrices here is of focal interest. Such a diagonalization rests on the following Theorem 1.1.

Theorem 1.1 (Simultaneous diagonalization of two symmetric matrices).

If $A \in \mathbb{R}^{n \times n}$ is a symmetric matrix and $B \in \mathbb{R}^{n \times n}$ is a symmetric positive-definite matrix such that the product AB^{-1} exists, then there exists a non-singular matrix X such that both following matrices are diagonal matrices, where I_n is the n-dimensional unit matrix:

$$X^T A X = \mathrm{diag}(\lambda_1, \ldots, \lambda_n), \quad X^T B X = I_n = \mathrm{diag}(1, \ldots, 1). \tag{1.1}$$

End of Theorem.

According to our understanding, the theorem had been intuitively applied by C. F. Gauss when he developed his *theory of curvature* of parameterized surfaces (two-dimensional Riemann manifold). Here, the *second fundamental form* (Hesse matrix of second derivatives, symmetric matrix H) had been analyzed with respect to the first fundamental form (a product of Jacobi matrices of first derivatives, a symmetric and positive-definite matrix G). Equivalent to the simultaneous diagonalization of a symmetric matrix H and a symmetric and positive-definite matrix G is the *general eigenvalue problem*

$$|H - \lambda G| = 0, \tag{1.2}$$

which corresponds to the *special eigenvalue problem*

$$|HG^{-1} - \lambda I_n| = 0, \tag{1.3}$$

where HG^{-1} defines the *Gaussian curvature matrix*

$$-K = HG^{-1}. \tag{1.4}$$

In comparing two *Riemann manifolds* by a mapping from one (left) to the other (right), we here only concentrate on the corresponding metric, the first fundamental forms of two parameterized surfaces. A comparative analysis of the second and third fundamental forms of two parameterized surfaces related by a mapping is given elsewhere. F. Uhlig (1979) published a historical survey of the above theorem to which we refer. Generalizations to canonically factorize two symmetric matrices A and B which are only definite (which are needed for mappings between *pseudo-Riemann manifolds*) can be traced to J. F. Cardoso and A. Souloumiac (1996), M. T. Chu (1991a,b), R. W. Newcomb (1960), C. R. Rao and S. K. Mitra (1971), S. K. Mitra and C. R. Rao (1968), S. R. Searle (1982, pp. 312–316), W. Shougen and Z. Shuqin (1991), and F. Uhlig (1973, 1976, 1979). In mathematical cartography, the canonical formalism for the analysis of deformations has been introduced by N. A. Tissot (1881). Note that there exists a beautiful variational formulation of the simultaneous diagonalization of two symmetric matrices which motivates the notation of eigenvalues as *Lagrange multipliers* λ and which is expressed by Corollary 1.2.

Corollary 1.2 (Variational formulation, simultaneous diagonalization of two symmetric matrices).

If $A \in \mathbb{R}^{n \times n}$ is a symmetric matrix and $B \in \mathbb{R}^{n \times n}$ is a symmetric positive-definite matrix such that the product AB^{-1} exists, then there exist extremal (semi-)norm solutions of the *Lagrange function* $\left(\text{tr}\left[X^{T}AX\right]\right)^{1/2} =: \|X\|_A$, the A-weighted *Frobenius norm* of the non-singular matrix X subject to the constraint

$$\text{tr}\left[X^{T}BX - I_n\right] = 0 \,, \tag{1.5}$$

namely the constraint optimization

$$\|X\|_A^2 - \lambda \,\text{tr}\left[X^{T}BX - I_n\right] = \text{extr}_{X,\lambda} \,, \tag{1.6}$$

which is solved by the system of normal equations

$$(A - \lambda B)X = 0 \,, \tag{1.7}$$

subject to

$$X^{T}BX = I_n \,. \tag{1.8}$$

This is known as the *general eigenvalue–eigenvector problem*. The Lagrange multiplier λ is identified as eigenvalue.

End of Corollary.

Let here be given the left and right two-dimensional Riemann manifolds $\{\mathbb{M}_l^2, G_{MN}\}$ and $\{\mathbb{M}_r^2, g_{\mu\nu}\}$, with standard metric $G_{MN} = G_{NM}$ and $g_{\mu\nu} = g_{\nu\mu}$, respectively, both symmetric and positive-definite. A subset $U_l \subset \mathbb{M}_l^2$ and $U_r \subset \mathbb{M}_r^2$, respectively, is covered by the chart $V_l \subset \mathbb{E}^2 := \{\mathbb{R}^2, \delta_{IJ}\}$ and $V_r \subset \mathbb{E}^2 := \{\mathbb{R}^2, \delta_{ij}\}$, respectively, with respect to the standard canonical metric δ_{IJ} and δ_{ij}, respectively, of the left two-dimensional Euclidean space and the right two-dimensional Euclidean space. Such a chart is constituted by local coordinates $\{U, V\} \in S_\Omega \subset \mathbb{E}^2$ and $\{u, v\} \in S_\omega \subset \mathbb{E}^2$, respectively, over *open sets* S_Ω and S_ω. Figures 1.1 and 1.2 illustrate by a commutative diagram the mappings $\boldsymbol{\Phi}_l$, $\boldsymbol{\Phi}_r$ and $\overline{f}$, f. The left mapping $\boldsymbol{\Phi}_l$ maps a point from the left two-dimensional Riemann manifold (surface) to a point of the left chart, while $\boldsymbol{\Phi}_r$ maps a point from the right two-dimensional Riemann manifold (surface) to a point of the right chart. In contrast, the mapping $\overline{f}$ relates a point of the left two-dimensional Riemann manifold (surface) to a point of the right two-dimensional Riemann manifold (surface). Analogously, the mapping f maps a point of the left chart to a point of the right chart: $\overline{f}: \mathbb{M}_l^2 \to \mathbb{M}_r^2$, $f: V_l \to V_r = \boldsymbol{\Phi}_r \circ \overline{f} \circ \boldsymbol{\Phi}_l^{-1}$. All mappings are assumed to be a *diffeomorphism*: the mapping $\{dU, dV\} \to \{du, dv\}$ is bijective. Example 1.2 is the simple example of an isoparametric mapping of a point on an ellipsoid-of-revolution to a point on the sphere.

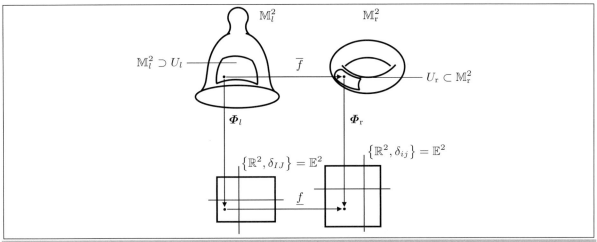

Fig. 1.1. Commutative diagram $(\overline{f}, \underline{f}, \boldsymbol{\Phi}_l, \boldsymbol{\Phi}_r)$; $\overline{f} : \mathbb{M}_l^2 \to \mathbb{M}_r^2$; $\underline{f} = \boldsymbol{\Phi}_r \circ \overline{f} \circ \boldsymbol{\Phi}_l^{-1}$.

Example 1.2 ($\mathbb{E}_{A_1, A_1, A_2}^2 \to \mathbb{S}_r^2$, isoparametric mapping).

As an example of the mapping $\overline{f} : \mathbb{M}_l^2 \longrightarrow \mathbb{M}_r^2$ and the commutative diagram $(\overline{f}, \underline{f}, \boldsymbol{\Phi}_l, \boldsymbol{\Phi}_r)$, think of an ellipsoid-of-revolution

$$\mathbb{E}_{A_1, A_1, A_2}^2 := \left\{ \boldsymbol{X} \in \mathbb{R}^3 \left| \frac{X^2 + Y^2}{A_1^2} + \frac{Z^2}{A_2^2} = 1, \ \mathbb{R}^+ \ni A_1 > A_2 \in \mathbb{R}^+ \right. \right\} \tag{1.9}$$

of semi-major axis A_1 and semi-minor axis A_2 as the left Riemann manifold $\mathbb{M}_l^2 = \mathbb{E}_{A_1, A_1, A_2}^2$, and think of a sphere

$$\mathbb{S}_r^2 := \left\{ \boldsymbol{x} \in \mathbb{R}^2 \left| x^2 + y^2 + z^2 = r^2, \ r \in \mathbb{R}^+ \right. \right\} \tag{1.10}$$

of radius r as the right Riemann manifold $\mathbb{M}_r^2 = \mathbb{S}_r^2$, $\overline{f}$ being the pointwise mapping of $\mathbb{E}_{A_1, A_1, A_2}^2$ to $\mathbb{S}_r^2$ one-to-one. $\underline{f}$ could be illustrated by a transformation of {ellipsoidal longitude Λ, ellipsoidal latitude Φ} onto {spherical longitude λ, spherical latitude ϕ} one-to-one. The mapping $\underline{f} = \text{id}$ is called isoparametric if $\{\Lambda = \lambda, \Phi = \phi\}$ or $\{U = u, V = v\}$ in general coordinates of the left Riemann manifold and the right Riemann manifold, respectively. Accordingly, in an isoparametric mapping, {ellipsoidal longitude, ellipsoidal latitude} and {spherical longitude, spherical latitude} are identical.

End of Example.

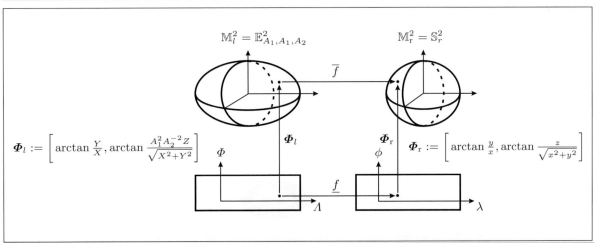

Fig. 1.2. Bijective mapping of an ellipsoid-of-revolution $\mathbb{E}_{A_1, A_1, A_2}^2$ to a sphere $\mathbb{S}_r^2$; $\overline{f} : \mathbb{E}_{A_1, A_1, A_2}^2 \to \mathbb{S}_r^2$; $\boldsymbol{\Phi}_l := [\Lambda, \Phi]$, $\boldsymbol{\Phi}_r := [\lambda, \phi]$; isoparametric mapping $\underline{f} = \text{id}$, namely $\{\Lambda, \Phi\} = \{\lambda, \phi\}$.

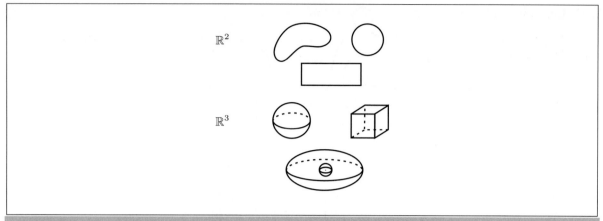

Fig. 1.3. Simply connected regions.

An *isoparametric mapping* of this type is illustrated by the commutative diagram of Fig. 1.2. We take notice that the differential mappings, conventionally called f_* and f^*, respectively, between the bell-shaped surface of revolution and the torus illustrated by Fig. 1.1 do *not* generate a diffeomorphism due to the different genus of the two surfaces. While Fig. 1.3 illustrates simply connected regions in $\mathbb{R}^2$ and $\mathbb{R}^3$, respectively, Fig. 1.4 demonstrates regions which are *not* simply connected. Those regions are characterized by closed curves which can be laid around the inner holes and which *cannot* be contracted to a point within the region. The holes are against contraction. The mapping $\overline{f} : \mathbb{M}_l^2 \to \mathbb{M}_r^2$ is usually called *deformation*. In addition, the mappings f_* (*pullback*) versus f^* (*pushforward*) of the left tangent space $T\mathbb{M}_l^2$ onto the right tangent space $T\mathbb{M}_r^2$, also called *pullback* (right derivative map, Jacobi map J_r), and of the right tangent space $T\mathbb{M}_r^2$ onto the left tangent map $T\mathbb{M}_l^2$, also called *pushforward* (left derivative map, Jacobi map J_l), are of focal interest for the following discussion. Indeed the pullback map f_* coincides with the mapping of the right cotangent space ${}^*T\mathbb{M}_r^2 \ni \{\mathrm{d}u, \mathrm{d}v\}$ onto the left cotangent space ${}^*T\mathbb{M}_l^2 \ni \{\mathrm{d}U, \mathrm{d}V\}$ as well as the pushforward map f^* with the mapping of the left cotangent space ${}^*T\mathbb{M}_l^2 \ni \{\mathrm{d}U, \mathrm{d}V\}$ onto the right cotangent space ${}^*T\mathbb{M}_r^2 \ni \{\mathrm{d}u, \mathrm{d}v\}$. This is illustrated by the relations

$$f_* : \begin{cases} T\mathbb{M}_l^2 \to T\mathbb{M}_r^2 \\ {}^*T\mathbb{M}_r^2 \to {}^*T\mathbb{M}_l^2 \end{cases} \quad \text{versus} \quad f^* : \begin{cases} {}^*T\mathbb{M}_l^2 \to {}^*T\mathbb{M}_r^2 \\ T\mathbb{M}_r^2 \to T\mathbb{M}_l^2 \end{cases}. \tag{1.11}$$

$$\text{(pullback)} \qquad\qquad\qquad \text{(pushforward)}$$

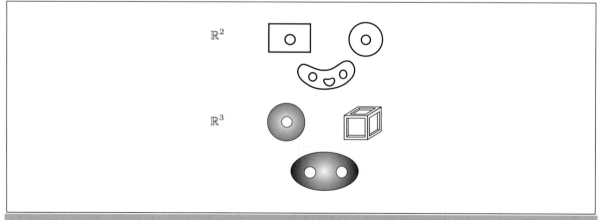

Fig. 1.4. Not simply connected regions.

1-1 Cauchy–Green deformation tensor

A first multiplicative measure of deformation: the Cauchy–Green deformation tensor, polar decomposition, singular value decomposition, Hammer retroazimuthal projection.

There are various local multiplicative and additive measures of deformation being derived from the infinitesimal distances dS^2 of $\mathbb{M}_l^2$ and ds^2 of $\mathbb{M}_r^2$, with

$$dS^2 = G_{MN}(U^L)dU^M dU^N \quad \text{versus} \quad ds^2 = g_{\mu\nu}(u^\lambda)du^\mu du^\nu. \tag{1.12}$$

The mapping of type deformation, $\overline{f} : \mathbb{M}_l^2 \to \mathbb{M}_r^2$, is represented locally by $\underline{f}$, in particular $U^M \to u^\mu$, the mapping of type inverse deformation, $\overline{f}^{-1} : \mathbb{M}_r^2 \to \mathbb{M}_l^2$, is represented locally by $\underline{f}^{-1}$, in particular $u^\mu \to U^M$, with $U^M \to u^\mu = f^\mu(U^M)$ and $u^\mu \to U^M = F^M(u^\mu)$. In the left and right tangent bundles $T\mathbb{M}_l^2 \times \mathbb{M}_l^2$ and $T\mathbb{M}_r^2 \times \mathbb{M}_r^2$, we represent locally the projections $\pi(T\mathbb{M}_l^2 \times \mathbb{M}_l^2) = T\mathbb{M}_l^2$ and $\pi(T\mathbb{M}_r^2 \times \mathbb{M}_r^2) = T\mathbb{M}_r^2$ by the *pullback map* and the *pushforward map*, in particular, by

$$f_* : \; dU^M = \frac{\partial U^M}{\partial u^\mu} du^\mu \quad \text{versus} \quad f^* : \; du^\mu = \frac{\partial u^\mu}{\partial U^M} dU^M . \tag{1.13}$$

$|\partial U^M/\partial u^\mu| > 0$ versus $|\partial u^\mu/\partial U^M| > 0$ *preserve the orientation* $\partial/\partial U \wedge \partial/\partial V$ and $\partial/\partial u \wedge \partial/\partial v$, respectively, of $\mathbb{M}_l^2$ and $\mathbb{M}_r^2$, respectively.

The first multiplicative measure of deformation has been introduced by A. L. Cauchy (1828) and G. Green (1839) reviewed in the sets of relations shown in Box 1.1, where the abbreviation Left CG indicates the *left Cauchy–Green deformation tensor* and the abbreviation Right CG indicates the *right Cauchy–Green deformation tensor*. With respect to the deformation gradients, the left and right Cauchy–Green tensors are represented in matrix algebra by

$$\mathsf{C}_l := \mathsf{J}_l^{\mathsf{T}} \mathsf{G}_r \mathsf{J}_l \quad \text{versus} \quad \mathsf{C}_r := \mathsf{J}_r^{\mathsf{T}} \mathsf{G}_l \mathsf{J}_r . \tag{1.14}$$

The set of *deformation gradients* is described by the two *Jacobi matrices* J_l and J_r, which obey the matrix relations

$$\mathsf{J}_l := \left\{ \frac{\partial u^\mu}{\partial U^M} \right\} = \mathsf{J}_r^{-1} \quad \text{versus} \quad \mathsf{J}_r := \left\{ \frac{\partial U^M}{\partial u^\mu} \right\} = \mathsf{J}_l^{-1} . \tag{1.15}$$

The abstract notation hopefully becomes more concrete when you work yourself through Example 1.3 where we compute the Cauchy–Green deformation tensor for an isoparametric mapping of a point on an ellipsoid-of-revolution to a point on a sphere.

Box 1.1 (Left and right Cauchy–Green deformation tensor).

Left CG:	Right CG:

$$ds^2 =$$

$$= g_{\mu\nu}\{f^\lambda(U^L)\} \frac{\partial u^\mu}{\partial U^M} \frac{\partial u^\nu}{\partial U^N} dU^M dU^N =$$

$$= c_{MN}(U^L) \, dU^M dU^N ,$$

$$c_{MN}(U^L) =$$

$$= g_{\mu\nu}(U^L) \frac{\partial u^\mu}{\partial U^M}(U^L) \frac{\partial u^\nu}{\partial U^N}(U^L) .$$

$$dS^2 =$$

$$= G_{MN}\{F^L(u^\lambda)\} \frac{\partial U^M}{\partial u^\mu} \frac{\partial U^N}{\partial u^\nu} du^\mu du^\nu =$$

$$= C_{\mu\nu}(u^\lambda)du^\mu du^\nu ,$$

$$C_{\mu\nu}(u^\lambda) =$$

$$= G_{MN}(u^\lambda) \frac{\partial U^M}{\partial u^\mu}(u^\lambda) \frac{\partial U^N}{\partial u^\nu}(u^\lambda) .$$

$$(1.16)$$

Example 1.3 (Cauchy-Green deformation tensor, $\overline{f} : \mathbb{E}^2_{A_1,A_1,A_2} \to \mathbb{S}^2_r$).

The embedding of an ellipsoid-of-revolution $\mathbb{M}^2_l = \mathbb{E}^2_{A_1,A_1,A_2}$ and a sphere $\mathbb{M}^2_r = \mathbb{S}^2_r$ into a three-dimensional Euclidean space $\{\mathbb{R}^3, \mathsf{I}_3\}$ with respect to a standard Euclidean metric I_3 (where I_3 is the 3×3 unit matrix) is governed by

$$\boldsymbol{X}(\Lambda, \Phi) = \boldsymbol{E}_1 \frac{A_1 \cos \Phi \cos \Lambda}{\sqrt{1 - E^2 \sin^2 \Phi}} + \boldsymbol{E}_2 \frac{A_1 \cos \Phi \sin \Lambda}{\sqrt{1 - E^2 \sin^2 \Phi}} + \boldsymbol{E}_3 \frac{A_1 (1 - E^2) \sin \Phi}{\sqrt{1 - E^2 \sin^2 \Phi}} =$$

$$= \begin{bmatrix} \boldsymbol{E}_1, \boldsymbol{E}_2, \boldsymbol{E}_3 \end{bmatrix} \frac{A_1}{\sqrt{1 - E^2 \sin^2 \Phi}} \begin{bmatrix} \cos \Phi \cos \Lambda \\ \cos \Phi \sin \Lambda \\ (1 - E^2) \sin \Phi \end{bmatrix} , \tag{1.17}$$

$$E^2 := \left(A_1^2 - A_2^2\right)/A_1^2 = 1 - \left(A_2^2/A_1^2\right) , \quad \left(A_2^2/A_1^2\right) = 1 - E^2 ,$$

and by

$$\boldsymbol{x}(\lambda, \phi) = \boldsymbol{e}_1 r \cos \phi \cos \lambda + \boldsymbol{e}_2 r \cos \phi \sin \lambda + \boldsymbol{e}_3 r \sin \phi =$$

$$= \begin{bmatrix} \boldsymbol{e}_1, \boldsymbol{e}_2, \boldsymbol{e}_3 \end{bmatrix} \begin{bmatrix} r \cos \phi \cos \lambda \\ r \cos \phi \sin \lambda \\ r \sin \phi \end{bmatrix} , \tag{1.18}$$

respectively. The coordinates (X, Y, Z) and (x, y, z) of the placement vectors $\boldsymbol{X}(\Lambda, \Phi) \in \mathbb{E}^2_{A_1,A_1,A_2}$ and $\boldsymbol{x}(\lambda, \phi) \in \mathbb{S}^2_r$ are expressed in the left and right orthonormal fixed frames $\{\boldsymbol{E}_1, \boldsymbol{E}_2, \boldsymbol{E}_3 | \mathcal{O}\}$ and $\{\boldsymbol{e}_1, \boldsymbol{e}_2, \boldsymbol{e}_3 | o\}$ at their origins $\mathcal{O}$ and o.

Next, we are going to construct the left tangent space $T\mathbb{M}^2_l$ as well as the right tangent space $T\mathbb{M}^2_r$, respectively. The vector field $\boldsymbol{X}(\Lambda, \Phi)$ is locally characterized by the *field of tangent vectors* $\left\{ \frac{\partial \boldsymbol{X}}{\partial \Lambda}, \frac{\partial \boldsymbol{X}}{\partial \Phi} \right\}$, the *Jacobi map* with respect to the "surface normal ellipsoidal longitude Λ" and the "surface normal ellipsoidal latitude Φ", namely

$$\left\{ \frac{\partial \boldsymbol{X}}{\partial \Lambda}, \frac{\partial \boldsymbol{X}}{\partial \Phi} \right\} = \begin{bmatrix} \boldsymbol{E}_1, \boldsymbol{E}_2, \boldsymbol{E}_3 \end{bmatrix} \begin{bmatrix} X_\Lambda & X_\Phi \\ Y_\Lambda & Y_\Phi \\ Z_\Lambda & Z_\Phi \end{bmatrix} =$$

$$= \begin{bmatrix} \boldsymbol{E}_1, \boldsymbol{E}_2, \boldsymbol{E}_3 \end{bmatrix} \begin{bmatrix} -\dfrac{A_1 \cos \Phi \sin \Lambda}{\sqrt{1 - E^2 \sin^2 \Phi}} & -\dfrac{A_1 (1 - E^2) \sin \Phi \cos \Lambda}{(1 - E^2 \sin^2 \Phi)^{3/2}} \\[3mm] +\dfrac{A_1 \cos \Phi \cos \Lambda}{\sqrt{1 - E^2 \sin^2 \Phi}} & -\dfrac{A_1 (1 - E^2) \sin \Phi \sin \Lambda}{(1 - E^2 \sin^2 \Phi)^{3/2}} \\[3mm] 0 & +\dfrac{A_1 (1 - E^2) \cos \Phi}{(1 - E^2 \sin^2 \Phi)^{3/2}} \end{bmatrix} , \tag{1.19}$$

as well as the vector field $\boldsymbol{x}(\lambda, \phi)$ is locally characterized by the *field of tangent vectors* $\left\{ \frac{\partial \boldsymbol{x}}{\partial \lambda}, \frac{\partial \boldsymbol{x}}{\partial \phi} \right\}$, the *Jacobi map* with respect to the "spherical longitude λ" and the "spherical latitude ϕ", namely

$$\left\{ \frac{\partial \boldsymbol{x}}{\partial \lambda}, \frac{\partial \boldsymbol{x}}{\partial \phi} \right\} = \begin{bmatrix} \boldsymbol{e}_1, \boldsymbol{e}_2, \boldsymbol{e}_3 \end{bmatrix} \begin{bmatrix} x_\lambda & x_\phi \\ y_\lambda & y_\phi \\ z_\lambda & z_\phi \end{bmatrix} =$$

$$= \begin{bmatrix} \boldsymbol{e}_1, \boldsymbol{e}_2, \boldsymbol{e}_3 \end{bmatrix} \begin{bmatrix} -r \cos \phi \sin \lambda & -r \sin \phi \cos \lambda \\ +r \cos \phi \cos \lambda & -r \sin \phi \sin \lambda \\ 0 & r \cos \phi \end{bmatrix} . \tag{1.20}$$

Next, we are going to identify the coordinates of the left metric tensor G_l and of the right metric tensor G_r, in particular, from the inner products

$$\left\langle \frac{\partial \boldsymbol{X}}{\partial \Lambda} \Big| \frac{\partial \boldsymbol{X}}{\partial \Lambda} \right\rangle = \frac{A_1^2 \cos^2 \Phi}{1 - E^2 \sin^2 \Phi} =: G_{11} \,, \qquad\qquad \left\langle \frac{\partial \boldsymbol{x}}{\partial \lambda} \Big| \frac{\partial \boldsymbol{x}}{\partial \lambda} \right\rangle = r^2 \cos^2 \phi =: g_{11} \,,$$

$$\left\langle \frac{\partial \boldsymbol{X}}{\partial \Lambda} \Big| \frac{\partial \boldsymbol{X}}{\partial \Phi} \right\rangle = \left\langle \frac{\partial \boldsymbol{X}}{\partial \Phi} \Big| \frac{\partial \boldsymbol{X}}{\partial \Lambda} \right\rangle =: G_{12} = 0 \,, \qquad\qquad \left\langle \frac{\partial \boldsymbol{x}}{\partial \lambda} \Big| \frac{\partial \boldsymbol{x}}{\partial \phi} \right\rangle = \left\langle \frac{\partial \boldsymbol{x}}{\partial \phi} \Big| \frac{\partial \boldsymbol{x}}{\partial \lambda} \right\rangle =: g_{12} = 0 \,,$$

$$\left\langle \frac{\partial \boldsymbol{X}}{\partial \Phi} \Big| \frac{\partial \boldsymbol{X}}{\partial \Phi} \right\rangle = \frac{A_1^2 (1-E^2)^2}{(1-E^2 \sin^2 \Phi)^3} =: G_{22} \,, \qquad\qquad \left\langle \frac{\partial \boldsymbol{x}}{\partial \phi} \Big| \frac{\partial \boldsymbol{x}}{\partial \phi} \right\rangle = r^2 =: g_{22} \,,$$

$$\mathrm{d}S^2 = \frac{A_1^2 \cos^2 \Phi}{1 - E^2 \sin^2 \Phi} \mathrm{d}\Lambda^2 + \frac{A_1^2 (1-E^2)^2}{(1-E^2 \sin^2 \Phi)^3} \mathrm{d}\Phi^2 \,, \qquad\qquad \mathrm{d}s^2 = r^2 \cos^2 \phi \, \mathrm{d}\lambda^2 + r^2 \, \mathrm{d}\phi^2 \,.$$

$$(1.21)$$

Resorting to this identification, we obtain the left metric tensor, i. e. G_l, and the right metric tensor, i. e. G_r, according to

$$\mathsf{G}_l := \begin{bmatrix} G_{11} & G_{12} \\ G_{12} & G_{22} \end{bmatrix} = \{G_{MN}\} = \qquad\qquad \mathsf{G}_r := \begin{bmatrix} g_{11} & g_{12} \\ g_{12} & g_{22} \end{bmatrix} = \{g_{\mu\nu}\} =$$

$$= \begin{bmatrix} \frac{A_1^2 \cos^2 \Phi}{1 - E^2 \sin^2 \Phi} & 0 \\ 0 & \frac{A_1^2 (1-E^2)^2}{(1-E^2 \sin^2 \Phi)^3} \end{bmatrix} \,, \qquad\qquad = \begin{bmatrix} r^2 \cos^2 \phi & 0 \\ 0 & r^2 \end{bmatrix} \,.$$

$$(1.22)$$

Finally, we implement the *isoparametric mapping* $\underline{f} = \mathrm{id}$. Applying the summation convention over repeated indices, this is realized by

$$U^M \to u^\mu = f^\mu(U^\mu) \,, \; u^\mu = \delta^\mu_M U^M \,, \; u^1 = U^1 \,, \; u^2 = U^2 \,, \; \lambda = \Lambda \,, \; \phi = \Phi \,, \; \mathsf{J}_l = \mathsf{I}_2 = \mathsf{J}_r \,, \qquad (1.23)$$

$$|\partial U^M / \partial u^\mu| = 1 > 0 \,, \qquad\qquad |\partial u^\mu / \partial U^M| = 1 > 0 \,,$$

$$f_* : \mathrm{d}U^M = \delta^M_\mu \mathrm{d}u^\mu \,, \qquad\qquad f^* : \mathrm{d}u^\mu = \delta^\mu_M \mathrm{d}U^M \,,$$

$$\begin{bmatrix} \mathrm{d}\Lambda \\ \mathrm{d}\Phi \end{bmatrix} = \begin{bmatrix} \mathrm{d}\lambda \\ \mathrm{d}\phi \end{bmatrix} \,, \qquad\qquad \begin{bmatrix} \mathrm{d}\lambda \\ \mathrm{d}\phi \end{bmatrix} = \begin{bmatrix} \mathrm{d}\Lambda \\ \mathrm{d}\Phi \end{bmatrix} \,.$$

$$(1.24)$$

Resorting to these relations and applying again the summation convention over repeated indices, we arrive at the left and right Cauchy–Green tensors, namely

$$c_{MN} = g_{\mu\nu} \frac{\partial u^\mu}{\partial U^M} \frac{\partial u^\nu}{\partial U^N} = g_{\mu\nu} \delta^\mu_M \delta^\nu_N \,, \qquad\qquad C_{\mu\nu} = G_{MN} \frac{\partial U^M}{\partial u^\mu} \frac{\partial U^N}{\partial u^\nu} = G_{MN} \delta^M_\mu \delta^N_\nu \,,$$

$$\mathsf{C}_l = \{c_{MN}\} = \mathsf{J}_l^\mathrm{T} \mathsf{G}_r \mathsf{J}_l = \begin{bmatrix} r^2 \cos^2 \Phi & 0 \\ 0 & r^2 \end{bmatrix}, \quad \mathsf{C}_r = \{C_{\mu\nu}\} = \mathsf{J}_r^\mathrm{T} \mathsf{G}_l \mathsf{J}_r = \begin{bmatrix} \frac{A_1^2 \cos^2 \phi}{1 - E^2 \sin^2 \phi} & 0 \\ 0 & \frac{A_1^2 (1-E^2)^2}{(1-E^2 \sin^2 \phi)^3} \end{bmatrix}, \quad (1.25)$$

$$\mathrm{d}s^2 = r^2 \cos^2 \Phi \, \mathrm{d}\Lambda^2 + r^2 \, \mathrm{d}\Phi^2 \,, \qquad\qquad \mathrm{d}S^2 = \frac{A_1^2 \cos^2 \phi}{1 - E^2 \sin^2 \phi} \mathrm{d}\lambda^2 + \frac{A_1^2 (1-E^2)^2}{(1-E^2 \sin^2 \phi)^3} \mathrm{d}\phi^2 \,.$$

By means of the *left Cauchy–Green tensor*, we have succeeded to represent the right metric or the metric of the right manifold $\mathbb{M}_r^2$ in the coordinates of the left manifold $\mathbb{M}_l^2$. Or we may say that we have *pulled back* $(\mathrm{d}\lambda, \mathrm{d}\phi) \in {}^*T_{\lambda,\phi}\mathbb{M}_r^2$ to $(\mathrm{d}\Lambda, \mathrm{d}\Phi) \in {}^*T_{\Lambda,\Phi}\mathbb{M}_l^2$, namely from the right cotangent space to the left cotangent space. By means of the *right Cauchy–Green tensor*, we have been able to represent the left metric or the metric of the left manifold $\mathbb{M}_l^2$ in the coordinates of the right manifold $\mathbb{M}_r^2$. Or we may say that we have *pushed forward* $(\mathrm{d}\Lambda, \mathrm{d}\Phi) \in {}^*T_{\Lambda,\Phi}\mathbb{M}_l^2$ to $(\mathrm{d}\lambda, \mathrm{d}\phi) \in {}^*T_{\lambda,\phi}\mathbb{M}_r^2$, namely from the left cotangent space to the right cotangent space.

End of Example.

There exists an intriguing representation of the matrix of deformation gradients J as well as of the matrix of Cauchy–Green deformation C, namely the *polar decomposition*. It is a generalization to matrices of the familiar polar representation of a complex number $z = r \exp i\phi$, $(r \geq 0)$ and is defined in Corollary 1.3.

Corollary 1.3 (Polar decomposition).

Let $\mathsf{J} \in \mathbb{R}^{n \times n}$. Then there exists a unique orthonormal matrix $\mathsf{R} \in SO(n)$ (called *rotation matrix*) and a unique symmetric positive-definite matrix S (called *stretch*) such that (1.26) holds and the expressions (1.27) are a polar decomposition of the matrix of Cauchy–Green deformation.

$$\mathsf{J} = \mathsf{RS} \; , \quad \mathsf{R}^*\mathsf{R} = \mathsf{I}_n \; , \quad \mathsf{S} = \mathsf{S}^* \; , \tag{1.26}$$

$$\mathsf{C}_l = \mathsf{J}_l^*\mathsf{G}_r\mathsf{J}_l = \mathsf{S}_l\mathsf{R}^*\mathsf{G}_r\mathsf{RS}_l \quad \text{versus} \quad \mathsf{S}_r\mathsf{R}^*\mathsf{G}_l\mathsf{RS}_r = \mathsf{J}_r^*\mathsf{G}_l\mathsf{J}_r = \mathsf{C}_r \; . \tag{1.27}$$

End of Corollary.

Question: "How can we compute the polar decomposition of the *Jacobi matrix*?" Answer: "An elegant way is the *singular value decomposition* defined in Corollary 1.4."

Corollary 1.4 (Polar decomposition by singular value decomposition).

Let the matrix $\mathsf{J} \in \mathbb{R}^{2 \times 2}$ have the singular value decomposition $\mathsf{J} = \mathsf{U\Sigma V}^*$, where the matrices $\mathsf{U} \in \mathbb{R}^{2 \times 2}$ and $\mathsf{V} \in \mathbb{R}^{2 \times 2}$ are orthonormal (unitary), i. e. $\mathsf{U}^*\mathsf{U} = \mathsf{I}_2$ and $\mathsf{V}^*\mathsf{V} = \mathsf{I}_2$, and where $\Sigma = \mathrm{diag}(\sigma_1, \sigma_2)$ in descending order $\sigma_1 \geq \sigma_2 \geq 0$ is the diagonal matrix of singular values $\{\sigma_1, \sigma_2\}$. If J has the polar decomposition $\mathsf{J} = \mathsf{RS}$, then $\mathsf{R} = \mathsf{UV}^*$ and $\mathsf{S} = \mathsf{V\Sigma V}^*$. $\lambda(\mathsf{J})$ and $\sigma(\mathsf{J})$ denote, respectively, the set of eigenvalues and the set of singular values of J. Then

the left eigenspace is spanned by the left eigencolumns $\boldsymbol{u}_1$ and $\boldsymbol{u}_2$ which are generated by

$$(\mathsf{JJ}^* - \lambda_i\mathsf{I}_2)\boldsymbol{u}_i = (\mathsf{JJ}^* - \sigma_i^2\mathsf{I}_2)\boldsymbol{u}_i = \mathbf{0} \; , \quad ||\boldsymbol{u}_1|| = ||\boldsymbol{u}_2|| = 1 \; ; \tag{1.28}$$

the right eigenspace is spanned by the right eigencolumns $\boldsymbol{v}_1$ and $\boldsymbol{v}_2$ generated by

$$(\mathsf{J}^*\mathsf{J} - \lambda_j\mathsf{I}_2)\boldsymbol{v}_j = (\mathsf{J}^*\mathsf{J} - \sigma_j^2\mathsf{I}_2)\boldsymbol{v}_j = \mathbf{0} \; , \quad ||\boldsymbol{v}_1|| = ||\boldsymbol{v}_2|| = 1 \; ; \tag{1.29}$$

the characteristic equation of the eigenvalues is determined by

$$|\mathsf{JJ}^* - \lambda\mathsf{I}_2| = 0 \; \text{ or } \; |\mathsf{J}^*\mathsf{J} - \lambda\mathsf{I}_2| = 0 \; , \tag{1.30}$$

which leads to $\lambda^2 - \lambda I + II = 0$, with the invariants

$$I := \mathrm{tr}\,[\mathsf{JJ}^*] = \mathrm{tr}\,[\mathsf{J}^*\mathsf{J}] \; , \quad II := (\det\,[\mathsf{J}])^2 = \det\,[\mathsf{JJ}^*] = \det\,[\mathsf{J}^*\mathsf{J}] \; ,$$
$$\lambda_1 = \sigma_1^2 = \tfrac{1}{2}\left(I + \sqrt{I^2 - 4II}\right) \; , \quad \lambda_2 = \sigma_2^2 = \tfrac{1}{2}\left(I - \sqrt{I^2 - 4II}\right) \; ; \tag{1.31}$$

the matrices S and R can be expressed as

$$\mathsf{S} = (\mathsf{J}^*\mathsf{J})^{1/2} = (\boldsymbol{v}_1, \boldsymbol{v}_2)\mathrm{diag}(\sigma_1, \sigma_2)(\boldsymbol{v}_1^*, \boldsymbol{v}_2^*) \; , \quad \mathsf{R} = \mathsf{JS}^{-1} = (\boldsymbol{u}_1, \boldsymbol{u}_2)(\boldsymbol{v}_1^*, \boldsymbol{v}_2^*) \; ; \tag{1.32}$$

J is normal if and only if $\mathsf{RS} = \mathsf{SR}$.

End of Corollary.

More details about the *polar decomposition* related to the *singular value decomposition* can be found in the classical text by N. J. Highham (1986), C. Kenney and A. J. Laub (1991), and T. C. T. Ting (1985). Example 1.4 is a numerical example for singular value decomposition and polar decomposition.

Example 1.4 (Singular value decomposition, polar decomposition).

Let there be given the Jacobi matrix J and the product matrices JJ^* and $\mathsf{J}^*\mathsf{J}$ such that the left and right characteristic equations of eigenvalues read

$$\mathsf{J} = \begin{bmatrix} 5 & 2 \\ -1 & 7 \end{bmatrix}, \quad \mathsf{JJ}^* = \begin{bmatrix} 29 & 9 \\ 9 & 50 \end{bmatrix}, \quad \mathsf{J}^*\mathsf{J} = \begin{bmatrix} 26 & 3 \\ 3 & 53 \end{bmatrix}, \tag{1.33}$$

$$|\mathsf{JJ}^* - \lambda\mathsf{I}_2| = \qquad\qquad |\mathsf{J}^*\mathsf{J} - \lambda\mathsf{I}_2| =$$

$$= \begin{vmatrix} 29 - \lambda & 9 \\ 9 & 50 - \lambda \end{vmatrix} = \qquad = \begin{vmatrix} 26 - \lambda & 3 \\ 3 & 53 - \lambda \end{vmatrix} =$$

$$= \lambda^2 - 79\lambda + 1369 = \qquad = \lambda^2 - 79\lambda + 1369 = \tag{1.34}$$

$$= 0, \qquad\qquad\qquad = 0,$$

$$I := \operatorname{tr}[\mathsf{JJ}^*] = \operatorname{tr}[\mathsf{J}^*\mathsf{J}] = 79, \quad II := \det[\mathsf{JJ}^*] = \det[\mathsf{J}^*\mathsf{J}] = 1369, \tag{1.35}$$

$$\lambda_1 = 53.329\,317, \quad \sigma_1 = \sqrt{\lambda_1} = 7.302\,692,$$
$$\lambda_2 = 25.670\,683, \quad \sigma_2 = \sqrt{\lambda_2} = 5.066\,624. \tag{1.36}$$

The left eigenspace is spanned by the left eigencolumns $(\boldsymbol{u}_1, \boldsymbol{u}_2)$, the right eigenspace by the right eigencolumns $(\boldsymbol{v}_1, \boldsymbol{v}_2)$, namely

$$(\mathsf{JJ}^* - \lambda_1\mathsf{I}_2)\boldsymbol{u}_1 = \boldsymbol{0}, \qquad (\mathsf{J}^*\mathsf{J} - \lambda_1\mathsf{I}_2)\boldsymbol{v}_1 = \boldsymbol{0},$$

$$(\mathsf{JJ}^* - \lambda_2\mathsf{I}_2)\boldsymbol{u}_2 = \boldsymbol{0}, \qquad (\mathsf{J}^*\mathsf{J} - \lambda_2\mathsf{I}_2)\boldsymbol{v}_2 = \boldsymbol{0}, \tag{1.37}$$

$$\text{or}$$

$$\begin{bmatrix} -24.329\,317 & 9 \\ 9 & -3.329\,317 \end{bmatrix}\begin{bmatrix} u_{11} \\ u_{21} \end{bmatrix} = \boldsymbol{0}, \qquad \begin{bmatrix} -27.329\,317 & 3 \\ 3 & -0.329\,317 \end{bmatrix}\begin{bmatrix} v_{11} \\ v_{21} \end{bmatrix} = \boldsymbol{0},$$

$$\begin{bmatrix} 3.329\,317 & 9 \\ 9 & 24.329\,317 \end{bmatrix}\begin{bmatrix} u_{12} \\ u_{22} \end{bmatrix} = \boldsymbol{0}, \qquad \begin{bmatrix} 0.329\,317 & 3 \\ 3 & 27.329\,317 \end{bmatrix}\begin{bmatrix} v_{12} \\ v_{22} \end{bmatrix} = \boldsymbol{0}. \tag{1.38}$$

Note that the matrices $\mathsf{JJ}^* - \lambda\mathsf{I}_2$ and $\mathsf{J}^*\mathsf{J} - \lambda\mathsf{I}_2$ have only rank one. Accordingly, in order to solve the homogenous linear equations uniquely, we need an additional constraint. Conventionally, this problem is solved by postulating *normalized eigencolumns*, namely

$$u_{11}^2 + u_{21}^2 = 1, \ u_{12}^2 + u_{22}^2 = 1, \quad v_{11}^2 + v_{21}^2 = 1, \ v_{12}^2 + v_{22}^2 = 1,$$

$$\|\boldsymbol{u}_1\| = \|\boldsymbol{u}_2\| = 1, \qquad\qquad \|\boldsymbol{v}_1\| = \|\boldsymbol{v}_2\| = 1. \tag{1.39}$$

The left eigencolumns, which are here denoted as $(\boldsymbol{u}_1, \boldsymbol{u}_2)$, are constructed from the following system of equations:

$$-24.329\,317u_{11} + 9u_{21} = 0, \quad +3.329\,317u_{12} + 9u_{22} = 0,$$

$$u_{11}^2 + u_{21}^2 = 1, \ u_{12}^2 + u_{22}^2 = 1. \tag{1.40}$$

This system of equations leads to two solutions. In the frame of the example to be considered here, we have chosen the following result:

$$u_{11} = +0.346\,946\,, \quad u_{12} = +0.937\,885\,,$$
$$u_{21} = +0.937\,885\,, \quad u_{22} = -0.346\,946\,. \tag{1.41}$$

The right eigencolumns, which are here denoted as (v_1, v_2), are constructed from the following system of equations:

$$-27.329 v_{11} + 3 v_{21} = 0\,, \quad +0.329 v_{12} + 3 v_{22} = 0\,,$$
$$v_{11}^2 + v_{21}^2 = 1\,, \quad v_{12}^2 + v_{22}^2 = 1\,. \tag{1.42}$$

This system of equations leads to two solutions. In the frame of the example to be considered here, we have chosen the following result:

$$v_{11} = +0.109\,117\,, \quad v_{12} = +0.994\,029\,,$$
$$v_{21} = +0.994\,029\,, \quad v_{22} = -0.109\,117\,. \tag{1.43}$$

In summary, the *left and right eigencolumns* are collected in the two following orthonormal matrices U and V:

$$\mathsf{U} = \begin{bmatrix} +0.346\,946 & +0.937\,665 \\ +0.937\,665 & -0.346\,946 \end{bmatrix}\,, \quad \mathsf{V} = \begin{bmatrix} +0.109\,117 & +0.994\,029 \\ +0.994\,029 & -0.109\,117 \end{bmatrix}\,. \tag{1.44}$$

The *polar decomposition* is now straightforward. According to the above considerations, we finally arrive at the result

$$\mathsf{R} = \mathsf{UV}^*\,, \quad \mathsf{S} = \mathsf{V\Sigma V}^*\,, \quad \mathsf{\Sigma} = \mathrm{diag}\,(\sigma_1, \sigma_2)\,, \tag{1.45}$$

$$\mathsf{R} = \begin{bmatrix} +0.970\,142 & +0.242\,536 \\ -0.242\,536 & +0.970\,142 \end{bmatrix}\,, \quad \mathsf{S} = \begin{bmatrix} +5.093\,248 & +0.242\,536 \\ +0.242\,536 & +7.276\,069 \end{bmatrix}\,. \tag{1.46}$$

Note that from this result immediately follows that R is an orthonormal matrix. Furthermore, note that S indeed is a symmetric matrix.

End of Example.

Before we consider a second multiplicative measure of deformation, please enjoy Fig. 1.5, which shows the *Hammer retroazimuthal projection*, illustrating special mapping equations of the sphere. The *ID card* of this special pseudo-azimuthal map projection is shown in Table 1.1.

Table 1.1. ID card of Hammer retroazimuthal projection of the sphere.

(i)	Classification	Retroazimuthal, modified azimuthal, neither conformal nor equal area.
(ii)	Graticule	Meridians: central meridian is straight, other meridians are curved. Parallels: curved. Poles of the sphere: curved lines. Symmetry: about the central meridians.
(iii)	Distortions	Distortions of area and shape
(iv)	Other features	The direction from any point to the center of the map is the angle that a straight line connecting the two points makes with a vertical line. This feature is the basis of the term "retroazimuthal". Scimitar-shaped boundary. Considerable overlapping when entire sphere is shown.
(v)	Usage	To determine the direction of a central point from a given location
(vi)	Origins	Presented by E. Hammer (1858–1925) in 1910. The author is the successor of E. Hammer in the Geodesy Chair of Stuttgart University (Germany). The map projection was independently presented by F. A. Reeves (1862 1945) and A. R. Hinks (1874–1945) of England in 1929.

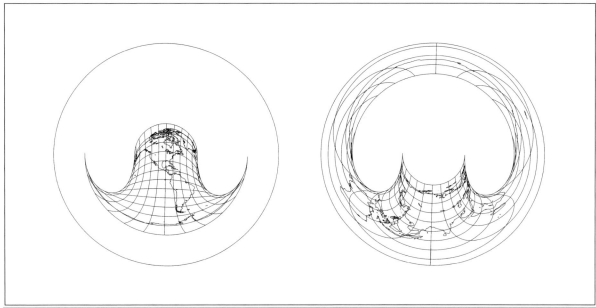

Fig. 1.5. Special map projection of the sphere, called *Hammer retroazimuthal projection*, centered near St.Louis (longitude 90° W, latitude 40° N), with shorelines, 15° graticule, two hemispheres, one of which appears backwards (they should be superimposed for the full map).

1-2 Stretch or length distortion

A second multiplicative measure of deformation: stretch or length distortion, Tissot portrait, simultaneous diagonalization of two matrices.

The *second multiplicative measure* of deformation is based upon the scale ratio, which is also called *stretch, dilatation factor*, or *length distortion*. One here distinguishes the left and right stretch:

left stretch: right stretch:

$$\Lambda^2 dS^2 = ds^2 \,, \quad \frac{ds^2}{dS^2} = \Lambda^2 =: \Lambda_l^2 \,, \qquad \lambda^2 ds^2 = dS^2 \,, \quad \Lambda_r^2 := \lambda^2 = \frac{dS^2}{ds^2} \,, \quad (1.47)$$

subject to duality $\Lambda^2 \lambda^2 = 1$.

Question: "What is the role of stretch $\{\Lambda^2, \lambda^2\}$ in the context of the pair of (symmetric, positive-definite) matrices $\{c_{MN}, G_{MN}\}$, $\{C_l, G_l\}$, and $\{C_{\mu\nu}, g_{\mu\nu}\}$, $\{C_r, G_r\}$, respectively?" **Answer:** "Due to a standard lemma of matrix algebra, both matrices can be simultaneously diagonalized, one matrix being the unit matrix."

We briefly outline the *simultaneous diagonalization* of the positive-definite, symmetric matrices $\{C_l, G_r\}$ and $\{C_r, G_l\}$, respectively, which is based upon a transformation called *"Kartenwechsel"*:

left "Kartenwechsel":   right "Kartenwechsel":

versus   (1.48)

$$T: \; V_l(U_{M_l^2}) \to \tilde{V}_l(U_{M_l^2}) \qquad\qquad \tau: \; V_r(U_{M_r^2}) \to \tilde{V}_r(U_{M_r^2}) \,.$$

The commutative diagram shown in Fig. 1.6 illustrates this "Kartenwechsel". Let us pay attention to Theorem 1.1 and Corollary 1.3, and let us present the various transformations in the Boxes 1.2 1.8.

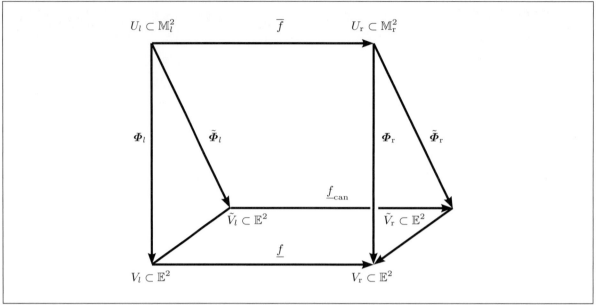

Fig. 1.6. Commutative diagram, canonical representation of pairs of metric tensors, "Kartenwechsel" T and τ, canonical mapping $\underline{f}_{\text{can}}$ from the left chart $\tilde{V}_l$ to the right chart $\tilde{V}_r$.

Box 1.2 (Left versus right Cauchy–Green deformation tensor).

Left CG:

$$\mathrm{d}s^2 = g_{\mu\nu}\big\{f^\lambda(U^L)\big\}\frac{\partial u^\mu}{\partial U^M}\frac{\partial u^\nu}{\partial U^N}\,\mathrm{d}U^M\mathrm{d}U^N =$$

$$= c_{MN}(U^L)\,\mathrm{d}U^M\mathrm{d}U^N \,,$$

$$c_{MN}(U^L) := g_{\mu\nu}(U^L)\frac{\partial u^\mu}{\partial U^M}(U^L)\frac{\partial u^\nu}{\partial U^N}(U^L)\,.$$

Right CG:

$$\mathrm{d}S^2 = G_{MN}\big\{F^L(u^\lambda)\big\}\frac{\partial U^M}{\partial u^\mu}\frac{\partial U^N}{\partial u^\nu}\,\mathrm{d}u^\mu\mathrm{d}u^\nu =$$

$$= C_{\mu\nu}(u^\lambda)du^\mu du^\nu \,,$$

$$C_{\mu\nu}(u^\lambda) := G_{MN}(u^\lambda)\frac{\partial U^M}{\partial u^\mu}(u^\lambda)\frac{\partial U^N}{\partial u^\nu}(u^\lambda)\,.$$

$$(1.49)$$

Box 1.3 (Left Tissot circle versus left Tissot ellipse, left Cauchy–Green deformation tensor: Ricci calculus).

Left Tissot circle $\mathbb{S}^1$:

$$\mathrm{d}S^2 = G_{MN}U_A^M U_B^N \mathrm{d}V^A \mathrm{d}V^B =$$

$$= \delta_{AB}\mathrm{d}V^A \mathrm{d}V^B =$$

$$= (\mathrm{d}V^1)^2 + (\mathrm{d}V^2)^2 = \Omega_1^2 + \Omega_2^2\,.$$

Left Tissot ellipse $\mathbb{E}^1_{\lambda_1,\lambda_2}$:

$$\mathrm{d}s^2 = g_{\mu\nu}u_M^\mu u_N^\nu U_A^M U_B^N \mathrm{d}V^A \mathrm{d}V^B =$$

$$= \Lambda_1^2(\mathrm{d}V^1)^2 + \Lambda_2^2(\mathrm{d}V^2)^2 =$$

$$= \Omega_1^2/\lambda_1^2 + \Omega_2^2/\lambda_2^2\,.$$

$$(1.50)$$

Box 1.4 (Left Tissot circle versus left Tissot ellipse, left Cauchy–Green deformation tensor: Cayley calculus).

Left Tissot circle $\mathbb{S}^1$:

$$\mathrm{d}S^2 = \boldsymbol{\Omega}^{\mathrm{T}}\mathsf{F}_l^{\mathrm{T}}\mathsf{G}_l\mathsf{F}_l\boldsymbol{\Omega} =$$

$$= \boldsymbol{\Omega}^{\mathrm{T}}\boldsymbol{\Omega} \Longleftrightarrow$$

$$\Longleftrightarrow \mathsf{F}_l^{\mathrm{T}}\mathsf{G}_l\mathsf{F}_l = \mathsf{I}\,.$$

Left Tissot ellipse $\mathbb{E}^1_{\lambda_1,\lambda_2}$:

$$\mathrm{d}s^2 = \boldsymbol{\Omega}^{\mathrm{T}}\mathsf{F}_l^{\mathrm{T}}\mathsf{C}_l\mathsf{F}_l\boldsymbol{\Omega} =$$

$$= \boldsymbol{\Omega}^{\mathrm{T}}\mathrm{diag}\left(\Lambda_1^2,\Lambda_2^2\right)\boldsymbol{\Omega} \Longleftrightarrow$$

$$\Longleftrightarrow \mathsf{F}_l^{\mathrm{T}}\mathsf{C}_l\mathsf{F}_l = \mathrm{diag}\left(\Lambda_1^2,\Lambda_2^2\right) = \mathrm{diag}\left(1/\lambda_1^2,1/\lambda_2^2\right)\,.$$

$$(1.51)$$

Box 1.5 (The right Tissot ellipse versus the right Tissot circle, right Cauchy–Green deformation tensor: Ricci calculus).

Right Tissot ellipse $\mathbb{E}^1_{\Lambda_1,\Lambda_2}$:

$$dS^2 = G_{MN}U^M_\mu U^N_\nu u^\mu_\alpha u^\nu_\beta dv^\alpha dv^\beta =$$
$$= (dv^1)^2/\Lambda_1^2 + (dv^2)^2/\Lambda_2^2 =$$
$$= \lambda_1^2\omega_1^2 + \lambda_2^2\omega_2^2 \ .$$

Right Tissot circle $\mathbb{S}^1$:

$$ds^2 = g_{\mu\nu}u^\mu_\alpha u^\nu_\beta dv^\alpha dv^\beta =$$
$$= \delta_{\alpha\beta}dv^\alpha dv^\beta =$$
$$= (dv^1)^2 + (dv^2)^2 = \omega_1^2 + \omega_2^2 \ .$$

(1.52)

Box 1.6 (The right Tissot ellipse versus the right Tissot circle, right Cauchy–Green deformation tensor: Cayley calculus).

Right Tissot ellipse $\mathbb{E}^1_{\Lambda_1,\Lambda_2}$:

$$dS^2 = \boldsymbol{\omega}^{\mathrm{T}}\mathsf{F}^{\mathrm{T}}_r\mathsf{C}_r\mathsf{F}_r\,\boldsymbol{\omega} =$$
$$= \boldsymbol{\omega}^{\mathrm{T}}\mathrm{diag}\left(\lambda_1^2,\lambda_2^2\right)\boldsymbol{\omega}$$
$$\Longleftrightarrow$$
$$\mathsf{F}^{\mathrm{T}}_r\mathsf{C}_r\mathsf{F}_r = \mathrm{diag}\left(\lambda_1^2,\lambda_2^2\right) = \mathrm{diag}\left(1/\Lambda_1^2,1/\Lambda_2^2\right) \ .$$

Right Tissot circle $\mathbb{S}^1$:

$$ds^2 = \boldsymbol{\omega}^{\mathrm{T}}\mathsf{F}^{\mathrm{T}}_r\mathsf{G}_r\mathsf{F}_r\boldsymbol{\omega} =$$
$$= \boldsymbol{\omega}^{\mathrm{T}}\boldsymbol{\omega}$$
$$\Longleftrightarrow$$
$$\mathsf{F}^{\mathrm{T}}_r\mathsf{G}_r\mathsf{F}_r = \mathsf{I} \ .$$

(1.53)

Box 1.7 (Left general eigenvalue problem and right general eigenvalue problem: Ricci calculus).

Left eigenvalue problem:

$$\Lambda^2 dS^2 = ds^2 \ ,$$
$$\Lambda^2 G_{MN}U^M_A U^N_B dV^A dV^B =$$
$$= g_{\mu\nu}u^\mu_M u^\nu_N U^M_A U^N_B dV^A dV^B$$
$$\Longleftrightarrow$$
$$\Lambda^2 G_{MN}U^N_B = c_{MN}U^N_B$$
$$\Longleftrightarrow$$
$$(c_{MN} - \Lambda^2 G_{MN})U^N_B = 0 \ ,$$
subject to
$$g_{\mu\nu}u^\mu_M u^\nu_N U^M_A U^N_B = \delta_{AB} \ .$$

Right eigenvalue problem:

$$\lambda^2 ds^2 = dS^2 \ ,$$
$$\lambda^2 g_{\mu\nu}u^\mu_\alpha u^\nu_\beta dv^\alpha dv^\beta =$$
$$= G_{MN}U^M_\mu U^N_\nu u^\mu_\alpha u^\nu_\beta dv^\alpha dv^\beta$$
$$\Longleftrightarrow$$
$$\lambda^2 g_{\mu\nu}u^\nu_\beta = C_{\mu\nu}u^\nu_\beta$$
$$\Longleftrightarrow$$
$$(C_{\mu\nu} - \lambda^2 g_{\mu\nu})u^\nu_\beta = 0 \ ,$$
subject to
$$G_{MN}U^M_\mu U^N_\nu u^\mu_\alpha u^\nu_\beta = \delta_{\mu\nu} \ .$$

(1.54)

Box 1.8 (Left general eigenvalue problem and right general eigenvalue problem: Cayley calculus).

Left eigenvalue problem:

$$\Lambda^2 dS^2 = ds^2 \ ,$$
$$\Lambda^2 d\boldsymbol{V}^{\mathrm{T}}\mathsf{F}^{\mathrm{T}}_l\mathsf{G}_l\mathsf{F}_l d\boldsymbol{V} =$$
$$= d\boldsymbol{V}^{\mathrm{T}}\mathsf{F}^{\mathrm{T}}_l\mathsf{C}_l\mathsf{F}_l d\boldsymbol{V}$$
$$\Longleftrightarrow$$
$$(\mathsf{C}_l - \Lambda^2\mathsf{G}_l)\mathsf{F}_l = 0 \ ,$$
subject to
$$\mathsf{F}^{\mathrm{T}}_l\mathsf{G}_l\mathsf{F}_l = \mathsf{I} \ .$$

Right eigenvalue problem:

$$\lambda^2 dS^2 = dS^2 \ ,$$
$$\lambda^2 d\boldsymbol{V}^{\mathrm{T}}\mathsf{F}^{\mathrm{T}}_r\mathsf{G}_r\mathsf{F}_r d\boldsymbol{V} =$$
$$= d\boldsymbol{V}^{\mathrm{T}}\mathsf{F}^{\mathrm{T}}_r\mathsf{C}_r\mathsf{F}_r d\boldsymbol{V}$$
$$\Longleftrightarrow$$
$$(\mathsf{C}_r - \lambda^2\mathsf{G}_r)\mathsf{F}_r = 0 \ ,$$
subject to
$$\mathsf{F}^{\mathrm{T}}_r\mathsf{G}_r\mathsf{F}_r = \mathsf{I} \ .$$

(1.55)

Certainly, we agree that the various transformations have to be checked by "paper and pencil", in particular, by means of Examples 1.2 and 1.3. In case that we are led to *"non-integrable differentials"* (namely differential forms), we have indicated this result by writing "đV" and "đv" according to the M. Planck notation. In this context, the *left* and *right Frobenius matrices*, F_l and F_r, have to be seen. They are used as matrices of *integrating factors* which transform *"imperfect differentials"* đV^A (namely đV^1, đV^2, or differential forms Ω_1, Ω_2) or đv^α (namely đv^1, đv^2, or differential forms ω_1, ω_2) to *"perfect differentials"* dU^A (namely dU^1, dU^2) or du^α (namely du^1, du^2). As a sample reference of the theory of differential forms and the *Frobenius Integration Theorem*, we direct the interested reader to J. A. de Azcarraga and J. M. Izquierdo (1995), M. P. do Carmo (1994), and H. Flanders (1970 p. 97). Indeed, we hope that the reader appreciates the triple notation index notation (Ricci calculus), matrix notation (Cayley calculus), and explicit notation (Leibniz–Newton calculus). Thus, we are led to the general eigenvalue problem as a result of simultaneous diagonalization of two positive-definite symmetric matrices $\{\mathsf{C}_l, \mathsf{G}_l\}$ or $\{\mathsf{C}_r, \mathsf{G}_r\}$, respectively. Compare with Lemma 1.5.

Lemma 1.5 (Left and right general eigenvalue problem of the Cauchy–Green deformation tensor).

For the pair of positive-definite symmetric matrices $\{\mathsf{C}_l, \mathsf{G}_l\}$ or $\{\mathsf{C}_r, \mathsf{G}_r\}$, respectively, a simultaneous diagonalization defined by

left diagonalization:

right diagonalization:

$$\mathsf{F}_l^{\mathrm{T}}\mathsf{C}_l\mathsf{F}_l = \mathrm{diag}\left(\Lambda_1^2, \Lambda_2^2\right) := \mathsf{D}_l\,, \qquad \mathsf{F}_r^{\mathrm{T}}\mathsf{C}_r\mathsf{F}_r = \mathrm{diag}\left(\lambda_1^2, \lambda_2^2\right) := \mathsf{D}_r\,, \qquad (1.56)$$

versus

$$\mathsf{F}_l^{\mathrm{T}}\mathsf{G}_l\mathsf{F}_l = \mathsf{I}_2 \qquad\qquad \mathsf{F}_r^{\mathrm{T}}\mathsf{G}_r\mathsf{F}_r = \mathsf{I}_2$$

is readily obtained from the following general eigenvalue–eigenvector problem of type *left eigenvalues* and *left principal stretches*:

$$\mathsf{C}_l\mathsf{F}_l - \mathsf{G}_l\mathsf{F}_l\mathsf{D}_l = 0$$

$$\Longleftrightarrow$$

$$(\mathsf{C}_l - \Lambda_i^2\mathsf{G}_l)\boldsymbol{f}_{li} = 0 \qquad\qquad (1.57)$$

$$\Longleftrightarrow$$

$$\left|\mathsf{C}_l - \Lambda^2\mathsf{G}_l\right| = 0\,,$$

$$\Lambda_{1,2}^2 = \Lambda_\pm^2 = \frac{1}{2}\left(\mathrm{tr}\left[\mathsf{C}_l\mathsf{G}_l^{-1}\right] \pm \sqrt{\left(\mathrm{tr}\left[\mathsf{C}_l\mathsf{G}_l^{-1}\right]\right)^2 - 4\det\left[\mathsf{C}_l\mathsf{G}_l^{-1}\right]}\right)\,,$$

subject to $\mathsf{F}_l^{\mathrm{T}}\mathsf{G}_l\mathsf{F}_l = \mathsf{I}_2$, and

$$\mathsf{C}_r\mathsf{F}_r - \mathsf{G}_r\mathsf{F}_r\mathsf{D}_r = 0$$

$$\Longleftrightarrow$$

$$(\mathsf{C}_r - \lambda_i^2\mathsf{G}_r)\boldsymbol{f}_{ri} = 0 \qquad\qquad (1.58)$$

$$\Longleftrightarrow$$

$$\left|\mathsf{C}_r - \lambda^2\mathsf{G}_r\right| = 0\,,$$

$$\lambda_{1,2}^2 = \lambda_\pm^2 = \frac{1}{2}\left(\mathrm{tr}\left[\mathsf{C}_r\mathsf{G}_r^{-1}\right] \pm \sqrt{\left(\mathrm{tr}\left[\mathsf{C}_r\mathsf{G}_r^{-1}\right]\right)^2 - 4\det\left[\mathsf{C}_r\mathsf{G}_r^{-1}\right]}\right)\,,$$

subject to $\mathsf{F}_r^{\mathrm{T}}\mathsf{G}_r\mathsf{F}_r = \mathsf{I}_2$, and

$$\Lambda_{1,2}^2 = 1/\lambda_{1,2}^2 \Longleftrightarrow 1/\Lambda_{1,2}^2 = \lambda_{1,2}^2\,. \qquad\qquad (1.59)$$

End of Lemma.

In order to visualize the eigenspace of the left and right Cauchy–Green deformation tensors C_l and C_r relative to the left and right metric tensors G_l and G_r, we are forced to compute in addition the *eigenvectors*, in particular, the *eigencolumns* (also called *eigendirections*) of the pairs $\{\mathsf{C}_l, \mathsf{G}_l\}$ and $\{\mathsf{C}_r, \mathsf{G}_r\}$, respectively. Compare with Lemma 1.6.

Lemma 1.6 (Left and right general eigenvectors, left and right principal stretch directions).

For the pair of positive-definite symmetric matrices $\{\mathsf{C}_l, \mathsf{G}_l\}$ and $\{\mathsf{C}_r, \mathsf{G}_r\}$, an explicit form of the left eigencolumns (also called *left principal stretch directions*) and of the right eigencolumns (also called *right principal stretch directions*) is

1st left eigencolumn, Λ_1:

$$\begin{bmatrix} F_{11} \\ F_{21} \end{bmatrix} = \frac{1}{\sqrt{\left(c_{22} - \Lambda_1^2 G_{22}\right)^2 G_{11} - 2\left(c_{12} - \Lambda_1^2 G_{12}\right)\left(c_{22} - \Lambda_1^2 G_{22}\right)G_{12} + \left(c_{12} - \Lambda_1^2 G_{12}\right)^2 G_{22}}} \times$$

$$\times \begin{bmatrix} c_{22} - \Lambda_1^2 G_{22} \\ -\left(c_{12} - \Lambda_1^2 G_{12}\right) \end{bmatrix},$$

(1.60)

2nd left eigencolumn, Λ_2:

$$\begin{bmatrix} F_{12} \\ F_{22} \end{bmatrix} = \frac{1}{\sqrt{\left(c_{11} - \Lambda_2^2 G_{11}\right)^2 G_{22} - 2\left(c_{11} - \Lambda_2^2 G_{11}\right)\left(c_{12} - \Lambda_2^2 G_{12}\right)G_{12} + \left(c_{12} - \Lambda_2^2 G_{12}\right)^2 G_{11}}} \times$$

$$\times \begin{bmatrix} -\left(c_{12} - \Lambda_2^2 G_{12}\right) \\ c_{11} - \Lambda_2^2 G_{11} \end{bmatrix},$$

1st right eigencolumn, λ_1:

$$\begin{bmatrix} f_{11} \\ f_{21} \end{bmatrix} = \frac{1}{\sqrt{\left(C_{22} - \lambda_1^2 g_{22}\right)^2 g_{11} - 2\left(C_{12} - \lambda_1^2 g_{12}\right)\left(C_{22} - \lambda_1^2 g_{22}\right)g_{12} + \left(C_{12} - \lambda_1^2 g_{12}\right)^2 g_{22}}} \times$$

$$\times \begin{bmatrix} C_{22} - \lambda_1^2 g_{22} \\ -\left(C_{12} - \lambda_1^2 g_{12}\right) \end{bmatrix},$$

(1.61)

2nd right eigencolumn, λ_2:

$$\begin{bmatrix} f_{12} \\ f_{22} \end{bmatrix} = \frac{1}{\sqrt{\left(C_{11} - \lambda_2^2 g_{11}\right)^2 g_{22} - 2\left(C_{11} - \lambda_2^2 g_{11}\right)\left(C_{12} - \lambda_2^2 g_{12}\right)g_{12} + \left(C_{12} - \lambda_2^2 g_{12}\right)^2 g_{11}}} \times$$

$$\times \begin{bmatrix} -\left(C_{12} - \lambda_2^2 g_{12}\right) \\ C_{11} - \lambda_2^2 g_{11} \end{bmatrix}.$$

End of Lemma.

A sketch of a proof is presented in the following. Note that there are four pairs of $\{F_{11}, F_{22}\}$ dependent on the sign choice $\{+, +\}$, $\{+, -\}$, $\{-, +\}$, and $\{-, -\}$. In Lemma 1.6, we have chosen the solution sign $\{F_{11}, F_{22}\} = \{+, +\}$. Furthermore, note that the proof for representing the right eigencolumns or right eigendirections runs analogeously. The dimension four of the solution space of eigencolumns or eigendirections has already been documented by J. M. Gere and W. Weaver (1965), for instance.

Proof (1st and 2nd left eigencolumns).

1st left eigencolumn, Λ_1:

$$
\begin{bmatrix} c_{11} - \Lambda_1^2 G_{11} & c_{12} - \Lambda_1^2 G_{12} \\ c_{12} - \Lambda_1^2 G_{12} & c_{22} - \Lambda_1^2 G_{22} \end{bmatrix} \begin{bmatrix} F_{11} \\ F_{21} \end{bmatrix} = \begin{bmatrix} 0 \\ 0 \end{bmatrix} ,
\tag{1.62}
$$

2nd identity:

$$
\left(c_{12} - \Lambda_1^2 G_{12} \right) F_{11} + \left(c_{22} - \Lambda_1^2 G_{22} \right) F_{21} = 0 \implies
$$

$$
\implies F_{21} = -\frac{c_{12} - \Lambda_1^2 G_{12}}{c_{22} - \Lambda_1^2 G_{22}} F_{11} \iff \begin{bmatrix} F_{11} \\ F_{21} \end{bmatrix} = F_{11} \begin{bmatrix} 1 \\ -\frac{c_{12} - \Lambda_1^2 G_{12}}{c_{22} - \Lambda_1^2 G_{22}} \end{bmatrix} .
\tag{1.63}
$$

2nd left eigencolumn, Λ_2:

$$
\begin{bmatrix} c_{11} - \Lambda_2^2 G_{11} & c_{12} - \Lambda_2^2 G_{12} \\ c_{12} - \Lambda_2^2 G_{12} & c_{22} - \Lambda_2^2 G_{22} \end{bmatrix} \begin{bmatrix} F_{21} \\ F_{22} \end{bmatrix} = \begin{bmatrix} 0 \\ 0 \end{bmatrix} ,
\tag{1.64}
$$

1st identity:

$$
\left(c_{11} - \Lambda_2^2 G_{11} \right) F_{21} + \left(c_{12} - \Lambda_2^2 G_{12} \right) F_{22} = 0 \implies
$$

$$
\implies F_{12} = -\frac{c_{12} - \Lambda_2^2 G_{12}}{c_{11} - \Lambda_2^2 G_{11}} F_{22} \iff \begin{bmatrix} F_{12} \\ F_{22} \end{bmatrix} = F_{22} \begin{bmatrix} -\frac{c_{12} - \Lambda_2^2 G_{12}}{c_{11} - \Lambda_2^2 G_{11}} \\ 1 \end{bmatrix} .
\tag{1.65}
$$

Left conditions:

$$
\mathsf{F}_l^{\mathsf{T}} \mathsf{G}_l \mathsf{F}_l = \mathsf{I}_2 \iff \begin{bmatrix} F_{11} & F_{21} \\ F_{12} & F_{22} \end{bmatrix} \begin{bmatrix} G_{11} & G_{12} \\ G_{12} & G_{22} \end{bmatrix} \begin{bmatrix} F_{11} & F_{12} \\ F_{21} & F_{22} \end{bmatrix} = \begin{bmatrix} 1 & 0 \\ 0 & 1 \end{bmatrix} .
\tag{1.66}
$$

1st and 2nd partitioning:

$$
\begin{bmatrix} F_{11}, F_{21} \end{bmatrix} \begin{bmatrix} G_{11} & G_{12} \\ G_{12} & G_{22} \end{bmatrix} \begin{bmatrix} F_{11} \\ F_{21} \end{bmatrix} = 1 , \quad \begin{bmatrix} F_{12}, F_{22} \end{bmatrix} \begin{bmatrix} G_{11} & G_{12} \\ G_{12} & G_{22} \end{bmatrix} \begin{bmatrix} F_{12} \\ F_{22} \end{bmatrix} = 1 .
\tag{1.67}
$$

2nd identity:

$$F_{11}^2 \left[1, -\frac{c_{12}-\Lambda_1^2 G_{12}}{c_{22}-\Lambda_1^2 G_{22}}\right] \begin{bmatrix} G_{11} & G_{12} \\ G_{12} & G_{22} \end{bmatrix} \begin{bmatrix} 1 \\ -\frac{c_{12}-\Lambda_1^2 G_{12}}{c_{22}-\Lambda_1^2 G_{22}} \end{bmatrix} = 1$$

$$\Longleftrightarrow F_{11}^2 \left[G_{11} - G_{12}\frac{c_{12}-\Lambda_1^2 G_{12}}{c_{22}-\Lambda_1^2 G_{22}}, G_{12} - G_{22}\frac{c_{12}-\Lambda_1^2 G_{12}}{c_{22}-\Lambda_1^2 G_{22}}\right] \begin{bmatrix} 1 \\ -\frac{c_{12}-\Lambda_1^2 G_{12}}{c_{22}-\Lambda_1^2 G_{22}} \end{bmatrix} = 1$$

$$\Longleftrightarrow F_{11}^2 \left[G_{11} - 2G_{12}\frac{c_{12}-\Lambda_1^2 G_{12}}{c_{22}-\Lambda_1^2 G_{22}} + G_{22}\frac{\left(c_{12}-\Lambda_1^2 G_{12}\right)^2}{\left(c_{22}-\Lambda_1^2 G_{22}\right)^2}\right] = 1$$

$$\Longrightarrow F_{11} = \pm \frac{c_{22} - \Lambda_1^2 G_{22}}{\sqrt{\left(c_{22} - \Lambda_1^2 G_{22}\right)^2 G_{11} - 2\left(c_{12} - \Lambda_1^2 G_{12}\right)\left(c_{22} - \Lambda_1^2 G_{22}\right)G_{12} + \left(c_{12} - \Lambda_1^2 G_{12}\right)^2 G_{22}}} \quad (1.68)$$

$$\Longleftrightarrow \begin{bmatrix} F_{11} \\ F_{21} \end{bmatrix} = F_{11} \begin{bmatrix} 1 \\ -\frac{c_{12}-\Lambda_1^2 G_{12}}{c_{22}-\Lambda_1^2 G_{22}} \end{bmatrix} =$$

$$= \pm \frac{1}{\sqrt{\left(c_{22} - \Lambda_1^2 G_{22}\right)^2 G_{11} - 2\left(c_{12} - \Lambda_1^2 G_{12}\right)\left(c_{22} - \Lambda_1^2 G_{22}\right)G_{12} + \left(c_{12} - \Lambda_1^2 G_{12}\right)^2 G_{22}}} \times$$

$$\times \begin{bmatrix} c_{22} - \Lambda_1^2 G_{22} \\ -\left(c_{12} - \Lambda_1^2 G_{12}\right) \end{bmatrix} \quad \text{q. e. d.}$$

1st identity:

$$F_{22}^2 \left[-\frac{c_{12}-\Lambda_2^2 G_{12}}{c_{11}-\Lambda_2^2 G_{11}}, 1\right] \begin{bmatrix} G_{11} & G_{12} \\ G_{12} & G_{22} \end{bmatrix} \begin{bmatrix} -\frac{c_{12}-\Lambda_2^2 G_{12}}{c_{11}-\Lambda_2^2 G_{11}} \\ 1 \end{bmatrix} = 1$$

$$\Longleftrightarrow F_{22}^2 \left[G_{12} - G_{11}\frac{c_{12}-\Lambda_2^2 G_{12}}{c_{11}-\Lambda_2^2 G_{11}}, G_{22} - G_{12}\frac{c_{12}-\Lambda_2^2 G_{12}}{c_{11}-\Lambda_2^2 G_{11}}\right] \begin{bmatrix} -\frac{c_{12}-\Lambda_2^2 G_{12}}{c_{11}-\Lambda_2^2 G_{11}} \\ 1 \end{bmatrix} = 1$$

$$\Longleftrightarrow F_{22}^2 \left[G_{22} - 2G_{12}\frac{c_{12}-\Lambda_2^2 G_{12}}{c_{11}-\Lambda_2^2 G_{11}} + G_{11}\frac{\left(c_{12}-\Lambda_2^2 G_{12}\right)^2}{\left(c_{11}-\Lambda_2^2 G_{11}\right)^2}\right] = 1$$

$$\Longrightarrow F_{22} = \pm \frac{c_{11} - \Lambda_2^2 G_{11}}{\sqrt{\left(c_{11} - \Lambda_2^2 G_{11}\right)^2 G_{22} - 2\left(c_{11} - \Lambda_2^2 G_{11}\right)\left(c_{12} - \Lambda_2^2 G_{12}\right)G_{12} + \left(c_{12} - \Lambda_2^2 G_{12}\right)^2 G_{11}}} \quad (1.69)$$

$$\Longleftrightarrow \begin{bmatrix} F_{12} \\ F_{22} \end{bmatrix} = F_{22} \begin{bmatrix} -\frac{c_{12}-\Lambda_2^2 G_{12}}{c_{11}-\Lambda_2^2 G_{11}} \\ 1 \end{bmatrix} =$$

$$= \pm \frac{1}{\sqrt{\left(c_{11} - \Lambda_2^2 G_{11}\right)^2 G_{22} - 2\left(c_{11} - \Lambda_2^2 G_{11}\right)\left(c_{12} - \Lambda_2^2 G_{12}\right)G_{12} + \left(c_{12} - \Lambda_2^2 G_{12}\right)^2 G_{11}}} \times$$

$$\times \begin{bmatrix} -\left(c_{12} - \Lambda_2^2 G_{12}\right) \\ c_{11} - \Lambda_2^2 G_{11} \end{bmatrix} \quad \text{q. e. d.}$$

End of Proof.

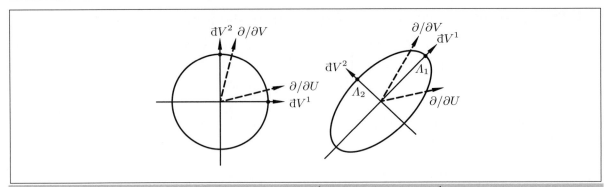

Fig. 1.7. Left Cauchy–Green tensor, left Tissot circle $\mathbb{S}^1$, left Tissot ellipse $\mathbb{E}^1_{\Lambda_1,\Lambda_2}$, the tangent vectors are $\partial/\partial U$ and $\partial/\partial V$.

The canonical forms of the metric, namely $\mathrm{d}S^2$ and $\mathrm{d}s^2$, have been interpreted as the following pairs:

$$
\begin{array}{ccc}
\text{left Tissot circle } \mathbb{S}^1 & & \text{right Tissot ellipse } \mathbb{E}^1_{\lambda_1,\lambda_2} \\
\text{versus} & \text{and} & \text{versus} \\
\text{left Tissot ellipse } \mathbb{E}^1_{\Lambda_1,\Lambda_2}, & & \text{right Tissot circle } \mathbb{S}^1.
\end{array}
\tag{1.70}
$$

Figure 1.7 illustrates the pair {left Cauchy–Green deformation tensor, left metric tensor} by means of the left Tissot circle $\mathbb{S}^1$ and the left Tissot ellipse $\mathbb{E}^1_{\Lambda_1,\Lambda_2}$ on the left tangent space $T\mathbb{M}^2_l$. In contrast, by means of Fig. 1.8, we aim at illustrating the pair {right Cauchy–Green deformation tensor, right metric tensor} by means of the right Tissot ellipse $\mathbb{E}^1_{\lambda_1,\lambda_2}$ and the right Tissot circle $\mathbb{S}^1$ on the right tangent space $T\mathbb{M}^2_r$. The left eigenvectors span canonically the left tangent space $T\mathbb{M}^2_l$, while the right eigenvectors span the right tangent space $T\mathbb{M}^2_r$, namely

$$
U^M_A \frac{\partial}{\partial U^M} \quad \text{versus} \quad u^\mu_\alpha \frac{\partial}{\partial u^\mu}, \qquad \mathsf{F}_l \frac{\partial}{\partial U} \quad \text{versus} \quad \mathsf{F}_r \frac{\partial}{\partial u}.
\tag{1.71}
$$

Indeed, they are generated from a dual holonomic base (coordinate base) $\{\mathrm{d}U^1, \mathrm{d}U^2\}$ versus $\{\mathrm{d}u^1, \mathrm{d}u^2\}$ to an anholonomic base $\{\mathrm{d}V^1, \mathrm{d}V^2\} = \{\Omega_1, \Omega_2\}$ versus $\{\mathrm{d}v^1, \mathrm{d}v^2\} = \{\omega_1, \omega_2\}$ by the transformations

$$
\begin{bmatrix} \mathrm{d}U^1 \\ \mathrm{d}U^2 \end{bmatrix} = \mathsf{F}_l \begin{bmatrix} \Omega_1 \\ \Omega_2 \end{bmatrix} \qquad \text{versus} \qquad \begin{bmatrix} \mathrm{d}u^1 \\ \mathrm{d}u^2 \end{bmatrix} = \mathsf{F}_r \begin{bmatrix} \omega_1 \\ \omega_2 \end{bmatrix}.
\tag{1.72}
$$

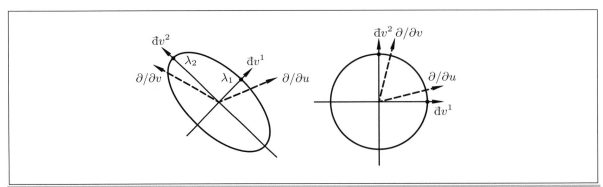

Fig. 1.8. Right Cauchy–Green tensor, right Tissot ellipse $\mathbb{E}^1_{\lambda_1,\lambda_2}$, right Tissot circle $\mathbb{S}^1$, the tangent vectors are $\partial/\partial u$ and $\partial/\partial v$.

1-3 Two examples: pseudo-cylindrical and orthogonal map projections

Two examples of deformation analysis: pseudo-cylindrical and orthogonal map projections (Cauchy–Green deformation tensor, its eigenspace, Tissot ellipses of distortion).

The general eigenspace analysis of the Cauchy–Green deformation tensor visualized by the Tissot ellipses of distortion is the heart of any map projection. It is for this reason that we present to you the *pseudo-cylindrical map projection* called *Eckert II* as Example 1.5 and the *orthogonal projection* of the northern hemisphere onto the equatorial plane as Example 1.6. We recommend to go through all details with "paper and pencil".

Example 1.5 (Pseudo-cylindric map projection of type Eckert II, left Cauchy–Green deformation tensor).

M. Eckert (1906) proposed six new pseudo-cylindrical map projections of the sphere which have some intrinsic properties. (i) The images of the central meridian and the pole have half the length of the equator, the line of zero latitude. (ii) The images of lines of equilatitude, called *parallel circles*, are parallel straight lines. Consult Fig. 1.9 for a more illustrative information. For instance, as a special pseudo-cylindrical projection, an equiareal mapping of the sphere onto a cylinder of type Eckert II, all meridians and parallel circles are mapped as straight lines. The mapping equations are given by

$$x = R \frac{2}{\sqrt{6\pi}} \Lambda \sqrt{4 - 3\sin|\Phi|} \,, \quad y = R\sqrt{\frac{2\pi}{3}} \left(2 - \sqrt{4 - 3\sin|\Phi|} \right) \operatorname{sign}\Phi \,,$$

$$\operatorname{sign}\Phi = \begin{bmatrix} +1 \ \forall \ \Phi \geq 0 \\ -1 \ \forall \ \Phi < 0 \end{bmatrix} . \tag{1.73}$$

End of Example.

We pose four problems. (i) Prove that the images of meridians and parallel circles are straight lines. Prove the half length condition between the images of the central meridian and the pole, respectively, and the equator. (ii) Derive the left Cauchy–Green deformation tensor. (iii) Solve the left general eigenvalue–eigenvector problem. Prove the condition of an equiareal mapping $\Lambda_1 \Lambda_2 = 1$. (iv) Prove that at $\{\Lambda = 0, \Phi = 0\}$ the special pseudo-cylindrical projection is not an isometry.

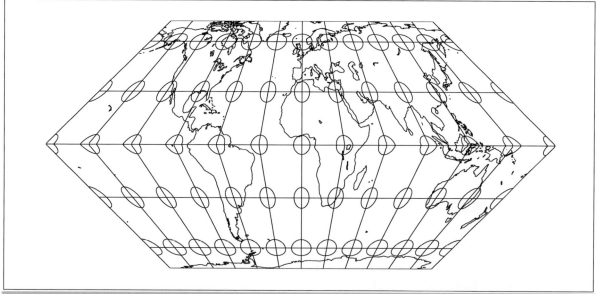

Fig. 1.9. Special pseudo-cylindrical projection of the sphere of type Eckert II (M. Eckert 1906), Tissot ellipses of distortion.

Solution (the first problem).

Let us rewrite the mapping equations in a more systematic form by introducing the two constants $c_1 := 2R/\sqrt{6\pi}$ and $c_2 := R\sqrt{2\pi/3}$ in Box 1.9 in order to analyze the graticule of "Eckert II". First, the geometrical shape of the image of the meridians is determined by removing the root $\sqrt{4 - 3\sin|\Phi|}$ from the second equation by substituting the root from the first equation. For $\Lambda = $ constant, we are led to the straight line $L^1(\Lambda = $ constant$)$. Second, the parallel circles are immediately fixed in shape by $\Phi = $ constant. x is a homogeneous linear form of longitude Λ and y is a constant. In summary, the meridians are tilted straights and the parallel circles are parallel straights. Third, let us compute the length of the circular equator $x(\Lambda = +\pi, \Phi = 0) - x(\Lambda = -\pi, \Phi = 0) = 8R\pi/\sqrt{6\pi} = 4\pi c_1$, the length of the central meridian $x(\Lambda = 0, \Phi = +\pi/2) - x(\Lambda = 0, \Phi = -\pi/2) = 4R\pi/\sqrt{6\pi} = 2\pi c_1$, and the length of the image of the pole $x(\Lambda = +\pi, |\Phi| = \pi/2) - x(\Lambda = -\pi, |\Phi| = \pi/2) = 4R\pi/\sqrt{6\pi} = 2\pi c_1$. Obviously, the length of image of the circular equator is twice the length of image of the central meridian or the pole.

End of Solution (the first problem).

Solution (the second problem).

In order to derive the left Cauchy–Green deformation tensor, according to Box 1.10, we depart from computing the left Jacobi matrix J_l. First, the partial derivatives $D_\Lambda x$, $D_\Phi x$, $D_\Lambda y$, and $D_\Phi y$ build up the left Jacobi matrix. Second, by means of the matrix product $C_l = J_l^* G_r J_l$, we are able to compute the left Cauchy–Green matrix for the right matrix of the metric $G_r = I_2$. Indeed, the chart $\{x, y\}$ is covered by Cartesian coordinates whose metric is simply given by $ds^2 = dx^2 + dy^2$. Though the special left Cauchy–Green matrix $C_l = J_l^* J_l$ looks simple, but is complicated in detail. The elements $\{c_{11}, c_{12} = c_{21}, c_{22}\}$ document these features.

End of Solution (the second problem).

Solution (the third problem).

Box 1.11 outlines the solution of the third problem, namely the laborious analytical computation of the left eigenvalues and the left eigencolumns. First, we refer to G_l as the matrix of the metric of the sphere $\mathbb{S}^2_R$ of radius R, and to C_l as the matrix of the left Cauchy–Green tensor, as computed in Box 1.10. The characteristic equation of the left general eigenvalue problem leads to the solution $\Lambda_{1,2}^2 = \Lambda_{+,-}^2$ as functions of the two *fundamental invariants* (i) $\text{tr}[C_l G_l^{-1}]$ and (ii) $\det[C_l G_l^{-1}]$. While the elements of the matrix $C_l G_l^{-1}$ evoke simple, its trace is complicated. In contrast, $\det[C_l G_l^{-1}] = 1$. Second, it is a straightforward proof that the product of eigenvalues squared is identical to the second invariant, i.e. $\Lambda_1^2 \Lambda_2^2 = \det[C_l G_l^{-1}]$. As proven, $\det[C_l G_l^{-1}] = 1$ (in consequence $\Lambda_1 \Lambda_2 = 1$) can be interpreted as the condition for an equiareal mapping. A detailed computation of the left eigenvalues $\{\Lambda_1, \Lambda_2\} = \{\Lambda_+, \Lambda_-\}$ is not useful due to the lengthy forms involved. Third, the same argument holds for the computed first eigencolumn, which is associated to $\Lambda_1 = \sqrt{\Lambda_1^2} \in \mathbb{R}^+$ and for the second eigencolumn, which is associated to $\Lambda_2 = \sqrt{\Lambda_2^2} \in \mathbb{R}^+$, and these are very lengthy. For practical use, a computation in a $\{\Lambda, \Phi\}$ lattice (for instance, $1° \times 1°$) is recommended.

End of Solution (the third problem).

Solution (the fourth problem).

Box 1.12 collects the details of the proof that the "Eckert II mapping" of the point $\{\Lambda, \Phi\} = \{0, 0\}$ is *not* an isometry. For an isometry, $\Lambda_1 = \Lambda_2 = 1$ is the postulate. If $\Lambda_1 = \Lambda_2$, then it holds that $(\text{tr}[C_l G_l^{-1}])^2 = 4\det[C_l G_l^{-1}]$. Since $(\text{tr}[C_l G_l^{-1}(\Lambda = 0, \Phi = 0)])^2 = [(64 + 9\pi^2)/24\pi]^2 \neq 4$ due to $\det[C_l G_l^{-1}] = 1$, it follows that $\Lambda_1 \neq \Lambda_2$.

End of Solution (the fourth problem).

Example 1.5 documents that for various map projections it is practically impossible to analytically compute the eigenspace which leads to the left and right Tissot ellipses. Numerically no problems appear when we have a computer at hand. For a large number of map projections, there is no problem to analytically compute the eigenspace. Such an example is considered after the boxes.

Box 1.9 (Eckert II, the first problem).

$$x = c_1 \Lambda \sqrt{4 - 3\sin|\Phi|} \;, \quad c_1 := \frac{2R}{\sqrt{6\pi}} \;,$$

$$y = c_2 \left(2 - \sqrt{4 - 3\sin|\Phi|}\right) \operatorname{sign}\phi \;, \quad c_2 := R\sqrt{\frac{2\pi}{3}} = \pi c_1 \;.$$

(1.74)

Meridians:

$$\sqrt{4 - 3\sin|\Phi|} = \frac{x}{c_1 \Lambda} \Rightarrow y = 2c_2 - \frac{c_2}{c_1}\frac{x}{\Lambda} = 2c_2 - \pi\frac{x}{\Lambda} \;,$$

$$\Lambda = \text{constant} \Rightarrow y = 2c_2 - c_3 x \;, \quad c_3 := \frac{\pi}{\Lambda} \;, \quad L^1(\Lambda = \text{constant}) := \left\{ x \in \mathbb{R}^2 \big| y = 2c_2 - c_3 x \right\} \;.$$

(1.75)

Parallel circles:

$$\Phi = \text{constant} \Rightarrow x = c_4 \Lambda \;, \; c_4 := c_1\sqrt{4 - 3\sin|\Phi|} \;, \; y = c_5 \;, \; c_5 := 2c_2 - c_2\sqrt{4 - 3\sin|\Phi|} \;,$$

$$L^1(\Phi = \text{constant}) := \left\{ x \in \mathbb{R}^2 \big| x = c_4 \Lambda, y = c_5 \right\} \;.$$

(1.76)

"Half":

(i) length of the circular equator:

$$x(\Lambda = +\pi, \Phi = 0) - x(\Lambda = -\pi, \Phi = 0) = 8R\pi/\sqrt{6\pi} \;,$$

(ii) length of the central meridian:

$$x(\Lambda = 0, \Phi = \pi/2) - x(\Lambda = 0, \Phi = -\pi/2) = 4R\pi/\sqrt{6\pi} \;,$$

(iii) length of the pole:

$$x(\Lambda = +\pi, |\Phi| = \pi/2) - x(\Lambda = -\pi, |\Phi| = \pi/2) = 4R\pi/\sqrt{6\pi} \;.$$

(1.77)

Box 1.10 (Eckert II, the second problem).

$$x = c_1 \Lambda \sqrt{4 - 3\sin|\Phi|} \;, \; c_1 := \frac{2R}{\sqrt{6\pi}} \;,$$

$$y = c_2 \left(2 - \sqrt{4 - 3\sin|\Phi|}\right) \operatorname{sign}\phi \;, \; c_2 := R\sqrt{\frac{2\pi}{3}} = \pi c_1 \;.$$

(1.78)

Left Jacobi matrix:

$$J_l := \begin{bmatrix} D_\Lambda x & D_\Phi x \\ D_\Lambda y & D_\Phi y \end{bmatrix} \;, \quad \begin{aligned} & D_\Lambda x = c_1\sqrt{4 - 3\sin|\Phi|} \;, \; D_\Phi x = -\frac{c_1}{2}\Lambda\frac{3\cos\Phi\operatorname{sign}\Phi}{\sqrt{4 - 3\sin|\Phi|}} \;, \\ & D_\Lambda y = 0 \;, \; D_\Phi y = \frac{3c_2}{2}\frac{\cos\Phi}{\sqrt{4 - 3\sin|\Phi|}} \;. \end{aligned}$$

(1.79)

Left Cauchy–Green matrix:

$$\mathsf{C}_l = \mathsf{J}_l^* \mathsf{G}_r \mathsf{J}_l \;, \; \mathsf{G}_r = \mathsf{I}_2 \Rightarrow \mathsf{C}_l = \mathsf{J}_l^* \mathsf{J}_l, \quad \begin{aligned} & c_{11} = \frac{2R^2}{3\pi}\left(4 - 3\sin|\Phi|\right) \;, \; c_{12} = -\frac{R^2}{\pi}\Lambda\cos\Phi\operatorname{sign}\Phi \;, \\ & c_{21} = c_{12} \;, \; c_{22} = \frac{3}{2}\frac{R^2}{\pi}\frac{\cos^2\Phi}{4 - 3\sin|\Phi|}\left(\Lambda^2 + \pi^2\right) \;. \end{aligned}$$

(1.80)

Box 1.11 (Eckert II, the third problem).

$$\mathsf{G}_l = R^2 \begin{bmatrix} \cos^2 \Phi & 0 \\ 0 & 1 \end{bmatrix} , \quad \mathsf{C}_l \quad \text{according to Box 1.10.} \tag{1.81}$$

Left general eigenvalue problem:

$$\left| \mathsf{C}_l - \Lambda^2 \mathsf{G}_l \right| = 0 \Leftrightarrow \Lambda^2_{1,2} = \Lambda^2_{+,-} = \frac{1}{2} \left(\text{tr} \left[\mathsf{C}_l \mathsf{G}_l^{-1} \right] \pm \sqrt{ \left(\text{tr} \left[\mathsf{C}_l \mathsf{G}_l^{-1} \right] \right)^2 - 4 \det \left[\mathsf{C}_l \mathsf{G}_l^{-1} \right] } \right) , \tag{1.82}$$

$$\left(\mathsf{C}_l \mathsf{G}_l^{-1} \right)_{11} = c_{11} \mathsf{G}_{11}^{-1} = + \frac{2}{3\pi} \frac{4 - 3 \sin |\Phi|}{\cos^2 \Phi} ,$$

$$\left(\mathsf{C}_l \mathsf{G}_l^{-1} \right)_{12} = c_{12} \mathsf{G}_{22}^{-1} = -\frac{1}{\pi} \Lambda \cos \Phi \, \text{sign} \Phi ,$$

$$\left(\mathsf{C}_l \mathsf{G}_l^{-1} \right)_{21} = c_{21} \mathsf{G}_{11}^{-1} = -\frac{1}{\pi} \Lambda \cos \Phi \, \text{sign} \Phi ,$$
$$\left(\mathsf{C}_l \mathsf{G}_l^{-1} \right)_{22} = c_{22} \mathsf{G}_{22}^{-1} = + \frac{3}{2\pi} \frac{\cos^2 \Phi}{4 - 3 \sin |\Phi|} \left(\Lambda^2 + \pi^2 \right) ,$$

$$\det \left[\mathsf{C}_l \mathsf{G}_l^{-1} \right] = 1 , \; \text{tr} \left[\mathsf{C}_l \mathsf{G}_l^{-1} \right] = \frac{2}{3\pi} \frac{4 - 3 \sin |\Phi|}{\cos^2 \Phi} + \frac{3}{2\pi} \frac{\cos^2 \Phi}{4 - 3 \sin |\Phi|} \left(\Lambda^2 + \pi^2 \right) . \tag{1.84}$$

$$\Lambda_1 \Lambda_2 = 1:$$

$$\Lambda_1^2 \Lambda_2^2 = \det \left[\mathsf{C}_l \mathsf{G}_l^{-1} \right] = 1 . \tag{1.85}$$

Left eigencolumns:

(i) $\quad \sqrt{} := \sqrt{ G_{11} (c_{22} - \Lambda_1^2 G_{22})^2 + G_{22} c_{12}^2 } \quad (G_{12} = 0) ,$

$$\begin{bmatrix} F_{11} \\ F_{22} \end{bmatrix} = \frac{1}{\sqrt{}} \begin{bmatrix} \dfrac{3}{2\pi} R^2 \dfrac{\cos^2 \Phi}{4 - 3 \sin |\Phi|} \left(\Lambda^2 + \pi^2 \right) - \Lambda_1^2 R^2 \\[2mm] \dfrac{1}{\pi} R^2 \Lambda \cos \Phi \, \text{sign} \Phi \end{bmatrix} ; \tag{1.86}$$

(ii) $\quad \sqrt{} := \sqrt{ G_{22} (c_{11} - \Lambda_2^2 G_{11})^2 + G_{11} c_{12}^2 } \quad (G_{12} = 0) ,$

$$\begin{bmatrix} F_{12} \\ F_{21} \end{bmatrix} = \frac{1}{\sqrt{}} \begin{bmatrix} \dfrac{1}{\pi} R^2 \Lambda \cos \Phi \, \text{sign} \Phi \\[2mm] \dfrac{2}{3\pi} R^2 (4 - 3 \sin |\Phi|) - \Lambda_2^2 R^2 \cos^2 \Phi \end{bmatrix} . \tag{1.87}$$

Box 1.12 (Eckert II, the fourth problem).

$$\Lambda_1 = \Lambda_2 \Leftrightarrow \left(\text{tr} \left[\mathsf{C}_l \mathsf{G}_l^{-1} \right] \right)^2 = 4 \det \left[\mathsf{C}_l \mathsf{G}_l^{-1} \right] , \tag{1.88}$$

$$\det \left[\mathsf{C}_l \mathsf{G}_l^{-1} \right] = 1 , \tag{1.89}$$

$$\left(\text{tr} \left[\mathsf{C}_l \mathsf{G}_l^{-1} (\Lambda = 0, \Phi = 0) \right] \right)^2 = \left(\frac{64 + 9\pi^2}{24\pi} \right)^2 \neq 4 \det \left[\mathsf{C}_l \mathsf{G}_l^{-1} \right] = 4 \Rightarrow \Lambda_1 \neq \Lambda_2 . \tag{1.90}$$

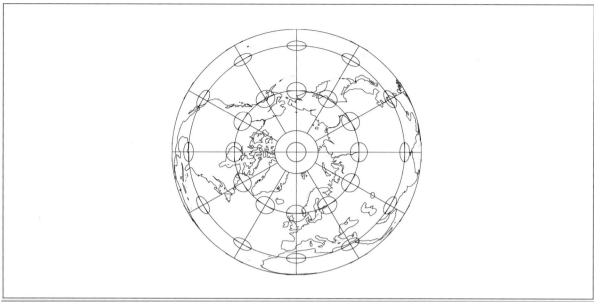

Fig. 1.10. Orthogonal projection of points of the sphere $\mathbb{S}^2_{R+}$ onto the tangent plane $\mathbb{P}^2_{\mathcal{O}}$ at the North Pole, shorelines, right Tissot ellipses of distorsion.

Example 1.6 (Orthogonal projection of points of the sphere onto the equatorial plane through the origin).

Let us assume that we make an orthogonal projection of points of the northern hemisphere onto the equatorial plane $\mathbb{P}^2_{\mathcal{O}}$ through the origin $\mathcal{O}$ of the plane $\mathbb{S}^2_{R+}$. Figure 1.10 and Figure 1.11 illustrate such an azimuthal projection by means of polar coordinate lines, shorelines, and right Tissot ellipses of distortion. The mapping equations are given by $x = X$, $y = Y$, $Z > 0$, $x = R\cos\varPhi\cos\varLambda$, $y = R\cos\varPhi\sin\varLambda$.

End of Example.

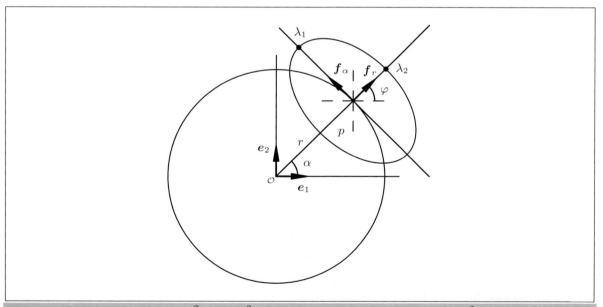

Fig. 1.11. Orthogonal projection $\mathbb{S}^2_R$ onto $\mathbb{P}^2_{\mathcal{O}}$, polar coordinates, right Tissot ellipse $\mathbb{E}^2_{\lambda_1,\lambda_2}$, right eigenvectors $\{\boldsymbol{f}_\alpha, \boldsymbol{f}_r \,|\, p\}$, right eigenvalues $\{\lambda_1, \lambda_2\}$, image of parallel circle.

We pose two problems. (i) Derive the right Cauchy–Green deformation tensor. (ii) Solve the right general eigenvalue–eigenvector problem.

Solution (the first problem).

By means of detailed derivations given in Boxes 1.13 and 1.14, we aim at an analytical analysis of the right Cauchy–Green deformation tensor in Cartesian coordinates $\{x, y\}$ and in polar coordinates $\{\alpha, r\}$, which cover the projection plane $\mathbb{P}_\mathcal{O}^2$. The right mapping equations $\{\Lambda(x, y), \Phi(x, y)\}$ and $\{\Lambda(\alpha), \Phi(r)\}$ are given first. They are constituted from the identities $x = X = R\cos\Phi\cos\Lambda$ and $y = Y = R\cos\Phi\sin\Lambda$, where $\{\Lambda, \Phi\}$ are the spherical coordinates. $\{\Lambda, \Phi\}$ or $\{$longitude, latitude$\}$ label a point in $\mathbb{S}_{R+}^2$. We use the symbol $+$ in order to allow only positive values $Z \in \mathbb{R}^+$, which are points in the northern hemisphere. Second, we compute the right Jacobi matrices $\mathsf{J}_\mathrm{r}(x, y)$ and $\mathsf{J}_\mathrm{r}(\alpha, r)$ in Cartesian coordinates $\{x, y\}$ and in polar coordinates $\{\alpha, r\}$. While $\mathsf{J}_\mathrm{r}(x, y)$ is a fully occupied matrix, $\mathsf{J}_\mathrm{r}(\alpha, r)$ is diagonal. Third, this difference continues when we are going to compute the right Cauchy–Green matrices $\mathsf{C}_\mathrm{r}(x, y)$ and $\mathsf{C}_\mathrm{r}(\alpha, r)$. Again, $\mathsf{C}_\mathrm{r}(x, y)$ is a fully occupied symmetric matrix, while $\mathsf{C}_\mathrm{r}(\alpha, r)$ is diagonal. Fourth, in Box 1.13, we represent the right Cauchy–Green deformation tensor as a tensor of second order in the Cartesian two-basis $\boldsymbol{e}_\mu \otimes \boldsymbol{e}_\nu$ for all $\{\mu, \nu\} = \{1, 2\}$. Note that $\mathbb{R}^2 = \mathrm{span}\{\boldsymbol{e}_1, \boldsymbol{e}_2\}$, where $\{\boldsymbol{e}_1, \boldsymbol{e}_2 \mid \mathcal{O}\}$ is an orthonormal two-leg at $\mathcal{O}$. Remarkably, $\mathsf{C}_\mathrm{r}(x, y)$ includes the components $\boldsymbol{e}_1 \otimes \boldsymbol{e}_2$, $\boldsymbol{e}_1 \otimes \boldsymbol{e}_2 + \boldsymbol{e}_2 \otimes \boldsymbol{e}_1$, and $\boldsymbol{e}_2 \otimes \boldsymbol{e}_2$. In contrast, the algebra of the right Cauchy–Green deformation tensor in Box 1.14, represented in polar coordinates, is slightly more complicated. A placement vector $\boldsymbol{x}(\alpha, r) \in \mathbb{P}_\mathcal{O}^2$ is locally described by the tangent space $T_{\boldsymbol{x}}\mathbb{M}_\mathrm{r}^2$ spanned by the tangent vectors $\boldsymbol{g}_1 = D_\alpha\boldsymbol{x}$ and $\boldsymbol{g}_1 = D_r\boldsymbol{x}$. In polar coordinates $\{\alpha, r\}$, the matrix of the right metric is given by $\mathsf{G}_\mathrm{r} = \mathrm{diag}(r^2, 1)$, a diagonal matrix. The first differential invariant of $\mathbb{M}_\mathrm{r}^2 \sim \mathbb{P}_\mathcal{O}^2$ is given by $(\mathrm{d}s)^2 = f(\mathrm{d}\alpha, \mathrm{d}r)$. The basis $\{\boldsymbol{g}^1, \boldsymbol{g}^2\}$, which is *dual* to $\{\boldsymbol{g}_1, \boldsymbol{g}_2\}$, also called *co-frame*, is computed next, namely by $\mathsf{G}^{-1}(\alpha, r)$. Due to the orthogonality of the two-leg $\{\boldsymbol{g}_1, \boldsymbol{g}_2 \mid p\}$, the co-frame amounts to $\boldsymbol{g}^1 = \boldsymbol{g}_1/g_{11}$ and $\boldsymbol{g}^2 = \boldsymbol{g}_2/g_{22}$, respectively. Question: "Why did we bother you with the notation of the co-frame $\{\boldsymbol{g}_1, \boldsymbol{g}_2 \mid p\}$?" Answer: "Often the moving frame $\{\boldsymbol{g}_1(\alpha, r), \boldsymbol{g}_2(\alpha, r)\}$ is called *covariant*, accordingly its dual $\{\boldsymbol{g}^1(\alpha, r), \boldsymbol{g}^2(\alpha, r)\}$ is called *contravariant*. The properly posed question can be answered immediately. The second-order tensor $\mathsf{C}_\mathrm{r}(\alpha, r)$ is represented in the contravariant or two-co-basis $\{\boldsymbol{g}^1 \otimes \boldsymbol{g}^1, \boldsymbol{g}^1 \otimes \boldsymbol{g}^2, \boldsymbol{g}^2 \otimes \boldsymbol{g}^1, \boldsymbol{g}^2 \otimes \boldsymbol{g}^2\}$, in general. Due to the diagonal structure of the right deformation tensor $\mathsf{C}_\mathrm{r}(r)$, contains only components $\boldsymbol{g}^1 \otimes \boldsymbol{g}^1$ and $\boldsymbol{g}^2 \otimes \boldsymbol{g}^2$, or $\boldsymbol{g}_1 \otimes \boldsymbol{g}_1$ and $\boldsymbol{g}_2 \otimes \boldsymbol{g}_2$, respectively."

End of Solution (the first problem).

Solution (the second problem).

The results on the right eigenspace analysis of the matrix pair $\{\mathsf{C}_\mathrm{r}, \mathsf{G}_\mathrm{r}\}$ are collected in Box 1.15 and Box 1.16, exclusively. In particular, we aim at computing the right eigenvalues, eigencolumns, and eigenvectors, namely in Box 1.15 in Cartesian coordinates $\{x, y\}$ along the fixed orthonormal frame $\{\boldsymbol{e}_1, \boldsymbol{e}_2\}$ and in Box 1.16 in polar coordinates $\{\alpha, r\}$ along the moving orthogonal frame $\{\boldsymbol{g}_1, \boldsymbol{g}_2 \mid p\}$. First, we solve the right general eigenvalue problem, both in Cartesian representation $\{\lambda_1(x, y), \lambda_2 = 1\}$ and in polar representation $\{\lambda_1(r), \lambda_2 = 1\}$. The characteristic equation $|\mathsf{C}_\mathrm{r} - \lambda^2\mathsf{G}_\mathrm{r}| = 0$ is solved in Box 1.16 if both C_r and G_r are diagonal. The determinantal identity is factorized directly into the right eigenvalues λ_1 and λ_2, a result we take advantage from in a following section. Second, we derive the simple structure of the eigencolumns $\{f_{11}(x, y), f_{21}(x, y)\}$ and $\{f_{12}(x, y), f_{22}(x, y)\}$ in case of Cartesian coordinates as well as of the eigencolumns $\{f_{11}(r), f_{21}(r)\}$ and $\{f_{12}(r), f_{22}(r)\}$ in polar coordinates. Third, let us derive the right eigenvectors. In Box 1.15, we succeed to represent the orthonormal right eigenvectors in the Cartesian basis $\{\boldsymbol{e}_1, \boldsymbol{e}_2 \mid p\}$. In contrast, in Box 1.16, we are able to compute the first right eigenvector as a tangent vector of the image of the parallel circle, while the second right eigenvector "radial" as a tangent vector of the image (straight line) of the meridian. Such a beautiful result is illustrated by Fig. 1.11.

End of Solution (the second problem).

Box 1.13 (Orthogonal projection $\mathbb{S}^2_{R+}$ onto $\mathbb{P}^2_{\mathcal{O}}$, Cartesian coordinates, the first problem).

$$x = r\cos\alpha\ , \quad y = r\sin\alpha\ , \quad \Lambda(x,y) = \arctan\frac{Y}{X} = \arctan\frac{y}{x} = \alpha\ ,$$

$$\Phi(x,y) = \arccos\frac{\sqrt{X^2+Y^2}}{R} = \arccos\frac{\sqrt{x^2+y^2}}{R} = \arccos\frac{r}{R}\ .$$

(1.91)

Right Jacobi matrix:

$$\mathsf{J_r} := \begin{bmatrix} D_x\Lambda & D_y\Lambda \\ D_x\Phi & D_y\Phi \end{bmatrix} = \frac{1}{\sqrt{x^2+y^2}}\begin{bmatrix} -\dfrac{y}{\sqrt{x^2+y^2}} & +\dfrac{x}{\sqrt{x^2+y^2}} \\ -\dfrac{x}{\sqrt{R^2-(x^2+y^2)}} & -\dfrac{y}{\sqrt{R^2-(x^2+y^2)}} \end{bmatrix}\ ,$$

$$D_x\Lambda = -\frac{y}{x^2+y^2}\ , \quad D_y\Lambda = +\frac{x}{x^2+y^2}\ ,$$

(1.92)

$$D_x\Phi = -\frac{x}{\sqrt{x^2+y^2}}\frac{1}{\sqrt{R^2-(x^2+y^2)}}\ , \quad D_y\Phi = -\frac{y}{\sqrt{x^2+y^2}}\frac{1}{\sqrt{R^2-(x^2+y^2)}}\ .$$

Right Cauchy–Green matrix:

$$\mathsf{C_r} := \mathsf{J_r^*G_lJ_r}\ , \quad \mathsf{G_l} = R^2\begin{bmatrix} \cos^2\Phi & 0 \\ 0 & 1 \end{bmatrix} = \begin{bmatrix} x^2+y^2 & 0 \\ 0 & R^2 \end{bmatrix}\ , \quad \mathsf{C_r} = \frac{1}{x^2+y^2}\times$$

$$\times\begin{bmatrix} \dfrac{y}{\sqrt{x^2+y^2}} & \dfrac{x}{\sqrt{R^2-(x^2+y^2)}} \\ -\dfrac{x}{\sqrt{x^2+y^2}} & \dfrac{y}{\sqrt{R^2-(x^2+y^2)}} \end{bmatrix}\begin{bmatrix} x^2+y^2 & 0 \\ 0 & R^2 \end{bmatrix}\begin{bmatrix} \dfrac{y}{\sqrt{x^2+y^2}} & -\dfrac{x}{\sqrt{x^2+y^2}} \\ \dfrac{x}{\sqrt{R^2-(x^2+y^2)}} & \dfrac{y}{\sqrt{R^2-(x^2+y^2)}} \end{bmatrix} =$$

(1.93)

$$= \frac{1}{x^2+y^2}\begin{bmatrix} y\sqrt{x^2+y^2} & x\dfrac{R^2}{\sqrt{R^2-(x^2+y^2)}} \\ -x\sqrt{x^2+y^2} & y\dfrac{R^2}{\sqrt{R^2-(x^2+y^2)}} \end{bmatrix}\begin{bmatrix} \dfrac{y}{\sqrt{x^2+y^2}} & -\dfrac{x}{\sqrt{x^2+y^2}} \\ \dfrac{x}{\sqrt{R^2-(x^2+y^2)}} & \dfrac{y}{\sqrt{R^2-(x^2+y^2)}} \end{bmatrix} =$$

$$= \frac{1}{R^2-(x^2+y^2)}\begin{bmatrix} R^2-y^2 & xy \\ xy & R^2-x^2 \end{bmatrix}\ .$$

Right Cauchy–Green tensor:

$$\mathsf{C_r} = \sum_{\mu,\nu=1}^{2} e^\mu\otimes e^\nu C_{\mu\nu} = \sum_{\mu,\nu=1}^{2} C_{\mu\nu}e^\mu\otimes e^\nu\ ,$$

$$\mathbb{R}^2 = \mathrm{span}\{e_1,e_2\} = \mathrm{span}\{e^1,e^2\}\ , \quad \langle e_\mu\,|\,e_\nu\rangle = \delta_{\mu\nu}\ , \quad \|e_\mu\|^2 = 1\ ,$$

$$\mathsf{C_r} = e_1\otimes e_1 C_{11} + \frac{1}{2}\big(e_1\otimes e_2 + e_2\otimes e_1\big)2C_{12} + e_2\otimes e_2 C_{22} =$$

(1.94)

$$= e_1\otimes e_1\frac{R^2-y^2}{R^2-(x^2+y^2)} + \frac{1}{2}\big(e_1\otimes e_2 + e_2\otimes e_1\big)\frac{2xy}{R^2-(x^2+y^2)} + e_2\otimes e_2\frac{R^2-x^2}{R^2-(x^2+y^2)}\ ,$$

$$\frac{1}{2}\big(e_\mu\otimes e_\nu + e_\nu\otimes e_\mu\big) =: e_\mu\vee e_\nu \quad (\text{symmetric product})\ ,$$

$$\mathsf{C_r} = e_1\vee e_1\frac{R^2-y^2}{R^2-(x^2+y^2)} + e_1\vee e_2\frac{2xy}{R^2-(x^2+y^2)} + e_2\vee e_2\frac{R^2-x^2}{R^2-(x^2+y^2)}\ .$$

Box 1.14 (Orthogonal projection $\mathbb{S}_{R+}^2$ onto $\mathbb{P}_{\mathcal{O}}^2$, polar coordinates, the first problem).

$$x = r\cos\alpha \,, \quad y = r\sin\alpha \,,$$

$$\Lambda(x,y) = \arctan\frac{y}{x} = \alpha \,,$$

$$\Phi(x,y) = \arccos\frac{\sqrt{x^2+y^2}}{R} = \arccos\frac{r}{R} \,.$$

(1.95)

Right Jacobi matrix:

$$\mathsf{J}_{\mathrm{r}} := \begin{bmatrix} D_\alpha\Lambda & D_r\Lambda \\ D_\alpha\Phi & D_r\Phi \end{bmatrix} = \begin{bmatrix} 1 & 0 \\ 0 & -\dfrac{1}{\sqrt{R^2-r^2}} \end{bmatrix} \,,$$

$$D_\alpha\Lambda = 1 \,, \quad D_r\Lambda = 0 \,,$$

(1.96)

$$D_\alpha\Phi = 0 \,, \quad D_r\Phi = -\frac{1}{\sqrt{1-r^2/R^2}}\frac{1}{R} = -\frac{1}{\sqrt{R^2-r^2}} \,.$$

Right Cauchy–Green matrix:

$$\mathsf{C}_{\mathrm{r}} := \mathsf{J}_{\mathrm{r}}^*\mathsf{G}_l\mathsf{J}_{\mathrm{r}} = \begin{bmatrix} r^2 & 0 \\ 0 & \dfrac{R^2}{R^2-r^2} \end{bmatrix} \,,$$

(1.97)

$$\mathsf{G}_l = R^2\begin{bmatrix} \cos^2\Phi & 0 \\ 0 & 1 \end{bmatrix} = \begin{bmatrix} r^2 & 0 \\ 0 & R^2 \end{bmatrix} \,.$$

Right Cauchy–Green tensor:

$$\boldsymbol{x}(\alpha,r) = \boldsymbol{e}_1 r\cos\alpha + \boldsymbol{e}_2 r\sin\alpha \,,$$

$$\boldsymbol{g}_1 := D_\alpha\boldsymbol{x} = -\boldsymbol{e}_1 r\sin\alpha + \boldsymbol{e}_2 r\cos\alpha \,, \quad \boldsymbol{g}_2 := D_r\boldsymbol{x} = +\boldsymbol{e}_1\cos\alpha + \boldsymbol{e}_2\sin\alpha \,,$$

$$g_{11} := \langle\boldsymbol{g}_1\,|\,\boldsymbol{g}_1\rangle = r^2 \,, \quad g_{12} := \langle\boldsymbol{g}_1\,|\,\boldsymbol{g}_2\rangle = 0 \,, \quad g_{22} := \langle\boldsymbol{g}_2\,|\,\boldsymbol{g}_2\rangle = 1 \,,$$

$$\mathsf{G}_{\mathrm{r}} = \begin{bmatrix} r^2 & 0 \\ 0 & 1 \end{bmatrix} \,,$$

$$(\mathrm{d}s)^2 = r^2(\mathrm{d}\alpha)^2 + (\mathrm{d}r)^2 \,,$$

(1.98)

$$\boldsymbol{g}^\mu = \sum_{\nu=1}^2 g^{\mu\nu}\boldsymbol{g}_\nu \,, \quad \boldsymbol{g}^1 = \frac{1}{g_{11}}\boldsymbol{g}_1 = \frac{1}{r^2}\boldsymbol{g}_1 \,, \quad \boldsymbol{g}^2 = \frac{1}{g_{22}}\boldsymbol{g}_2 = \boldsymbol{g}_2 \,,$$

$$\mathsf{C}_{\mathrm{r}} = \sum_{\mu,\nu=1}^2 \boldsymbol{g}^\mu \otimes \boldsymbol{g}^\nu C_{\mu\nu} =$$

$$-\boldsymbol{g}^1 \otimes \boldsymbol{g}^1 r^2 + \boldsymbol{g}^2 \otimes \boldsymbol{g}^2\frac{R^2}{R^2-r^2} = \boldsymbol{g}_1 \otimes \boldsymbol{g}_1 1 + \boldsymbol{g}_2 \otimes \boldsymbol{g}_2\frac{R^2}{R^2-r^2} \,.$$

Box 1.15 (Orthogonal projection $\mathbb{S}^2_{R+}$ onto $\mathbb{P}^2_O$, Cartesian coordinates, the second problem).

$$\mathsf{G}_r = \begin{bmatrix} 1 & 0 \\ 0 & 1 \end{bmatrix} , \quad \mathsf{C}_r \quad \text{according to Box 1.13.} \tag{1.99}$$

Right general eigenvalue problem:

$$\left| \mathsf{C}_r - \lambda^2 \mathsf{G}_r \right| = 0 \Leftrightarrow \lambda^2_{1,2} = \lambda^2_{+,-} = \frac{1}{2} \left(\text{tr}\left[\mathsf{C}_r \mathsf{G}_r^{-1}\right] \pm \sqrt{\left(\text{tr}\left[\mathsf{C}_r \mathsf{G}_r^{-1}\right]\right)^2 - 4\det\left[\mathsf{C}_r \mathsf{G}_r^{-1}\right]} \right) ,$$

$$\mathsf{C}_r = \frac{1}{R^2 - (x^2 + y^2)} \begin{bmatrix} R^2 - y^2 & xy \\ xy & R^2 - x^2 \end{bmatrix} , \quad \mathsf{G}_r = \mathsf{I}_2 ,$$

$$\text{tr}\left[\mathsf{C}_r \mathsf{G}_r^{-1}\right] = \text{tr}\left[\mathsf{C}_r\right] = \frac{2R^2 - (x^2 + y^2)}{R^2 - (x^2 + y^2)} ,$$

$$\det\left[\mathsf{C}_r \mathsf{G}_r^{-1}\right] = \det\left[\mathsf{C}_r\right] = \frac{\left(R^2 - x^2\right)\left(R^2 - y^2\right) - x^2 y^2}{\left[R^2 - (x^2 + y^2)\right]^2} = \frac{R^2}{R^2 - (x^2 + y^2)} , \tag{1.100}$$

$$\left(\text{tr}\left[\mathsf{C}_r \mathsf{G}_r^{-1}\right]\right)^2 - 4\det\left[\mathsf{C}_r \mathsf{G}_r^{-1}\right] = \frac{(x^2 + y^2)^2}{\left[R^2 - (x^2 + y^2)\right]^2} ,$$

$$\lambda_1^2 = \lambda_+^2 = \frac{R^2}{R^2 - (x^2 + y^2)} , \quad \lambda_1 = \lambda_+ = +\frac{R}{\sqrt{R^2 - (x^2 + y^2)}} , \quad \lambda_2^2 = \lambda_-^2 = 1 , \quad \lambda_2 = \lambda_- = +1 .$$

Right eigencolumns:

(i) $\quad \sqrt{} := \sqrt{g_{11}(C_{22} - \lambda_1^2 g_{22})^2 - 2g_{12}(C_{12} - \lambda_1^2 g_{12})(C_{22} - \lambda_1^2 g_{22}) + g_{22}(C_{12} - \lambda_1^2 g_{12})^2} =$

$$= \sqrt{(C_{22} - \lambda_1^2)^2 + C_{12}^2} = x\frac{\sqrt{x^2 + y^2}}{R^2 - (x^2 + y^2)} , \tag{1.101}$$

$$\begin{bmatrix} f_{11} \\ f_{21} \end{bmatrix} = \frac{1}{\sqrt{}} \begin{bmatrix} C_{22} - \lambda_1^2 \\ -C_{12} \end{bmatrix} = -\frac{1}{\sqrt{x^2 + y^2}} \begin{bmatrix} x \\ y \end{bmatrix} ;$$

(ii) $\quad \sqrt{} := \sqrt{g_{11}(C_{12} - \lambda_2^2 g_{12})^2 - 2g_{12}(C_{11} - \lambda_2^2 g_{11})(C_{12} - \lambda_2^2 g_{12}) + g_{22}(C_{11} - \lambda_2^2 g_{11})^2} =$

$$= \sqrt{(C_{11} - \lambda_2^2)^2 + C_{12}^2} = x\frac{\sqrt{x^2 + y^2}}{R^2 - (x^2 + y^2)} , \tag{1.102}$$

$$\begin{bmatrix} f_{12} \\ f_{22} \end{bmatrix} = \frac{1}{\sqrt{}} \begin{bmatrix} -C_{12} \\ C_{11} - \lambda_2^2 \end{bmatrix} = +\frac{1}{\sqrt{x^2 + y^2}} \begin{bmatrix} -y \\ x \end{bmatrix} .$$

Right eigenvectors:

1st eigenvector: $\quad \boldsymbol{f}_1 := \boldsymbol{e}_1 f_{11} + \boldsymbol{e}_2 f_{21} , \quad \boldsymbol{f}_1(x,y) = -\boldsymbol{e}_1 \frac{x}{\sqrt{x^2 + y^2}} - \boldsymbol{e}_2 \frac{y}{\sqrt{x^2 + y^2}} ;$

$$\tag{1.103}$$

2nd eigenvector: $\quad \boldsymbol{f}_2 := \boldsymbol{e}_1 f_{12} + \boldsymbol{e}_2 f_{22} , \quad \boldsymbol{f}_2(x,y) = -\boldsymbol{e}_1 \frac{y}{\sqrt{x^2 + y^2}} + \boldsymbol{e}_2 \frac{x}{\sqrt{x^2 + y^2}} .$

Notes:

$$\langle \boldsymbol{f}_1 \,|\, \boldsymbol{f}_2 \rangle = 0 \Rightarrow \measuredangle(\boldsymbol{f}_1, \boldsymbol{f}_2) = \pi/2 , \quad \left\| \boldsymbol{f}_1 \right\|_2 = \left\| \boldsymbol{f}_2 \right\|_2 = 1 . \tag{1.104}$$

Box 1.16 (Orthogonal projection $\mathbb{S}^2_{R+}$ onto $\mathbb{P}^2_{\mathbb{O}}$, polar coordinates, the second problem).

$$\mathsf{G}_r = \begin{bmatrix} r^2 & 0 \\ 0 & 1 \end{bmatrix}, \quad \mathsf{C}_r \quad \text{according to Box 1.14.} \tag{1.105}$$

Right general eigenvalue problem:

$$\left| \mathsf{C}_r - \lambda^2 \mathsf{G}_r \right| = 0 \Leftrightarrow \lambda^2_{1,2} = \lambda^2_{+,-} = \frac{1}{2}\left(\mathrm{tr}\left[\mathsf{C}_r\mathsf{G}_r^{-1}\right] \pm \sqrt{\left(\mathrm{tr}\left[\mathsf{C}_r\mathsf{G}_r^{-1}\right]\right)^2 - 4\det\left[\mathsf{C}_r\mathsf{G}_r^{-1}\right]} \right),$$

$$\mathsf{G}_r^{-1} = \begin{bmatrix} \dfrac{1}{r^2} & 0 \\ 0 & 1 \end{bmatrix}, \quad \mathsf{C}_r\mathsf{G}_r^{-1} = \begin{bmatrix} 1 & 0 \\ 0 & \dfrac{R^2}{R^2 - r^2} \end{bmatrix},$$

$$\mathrm{tr}\left[\mathsf{C}_r\mathsf{G}_r^{-1}\right] = \frac{2R^2 - r^2}{R^2 - r^2}, \quad \det\left[\mathsf{C}_r\mathsf{G}_r^{-1}\right] = \frac{R^2}{R^2 - r^2}, \tag{1.106}$$

$$\sqrt{\left(\mathrm{tr}\left[\mathsf{C}_r\mathsf{G}_r^{-1}\right]\right)^2 - 4\det\left[\mathsf{C}_r\mathsf{G}_r^{-1}\right]} = \frac{r^2}{R^2 - r^2},$$

$$\lambda^2_1 = \lambda^2_+ = \frac{R^2}{R^2 - r^2}, \quad \lambda_1 = \lambda_+ = +\frac{R}{\sqrt{R^2 - r^2}}, \quad \lambda^2_2 = \lambda^2_- = 1, \quad \lambda_2 = \lambda_- = +1.$$

Alternative solution, right general eigenvalue problem:

$$\mathsf{C}_r = \mathrm{diag}\left(r^2, \frac{R^2}{R^2 - r^2}\right), \quad \mathsf{G}_r = \mathrm{diag}\left(r^2, 1\right), \quad \left| \mathsf{C}_r - \lambda^2\mathsf{G}_r \right| = 0$$

$$\Leftrightarrow$$

$$\begin{vmatrix} r^2(1 - \lambda^2) & 0 \\ 0 & \dfrac{R^2}{R^2 - r^2} - \lambda^2 \end{vmatrix} = \begin{vmatrix} C_{11} - \lambda^2 g_{11} & 0 \\ 0 & C_{22} - \lambda^2 g_{22} \end{vmatrix} = 0$$

$$\Leftrightarrow \tag{1.107}$$

$$C_{11} - \lambda^2 g_{11} = r^2(1 - \lambda^2) = 0, \quad C_{22} - \lambda^2 g_{22} = \frac{R^2}{R^2 - r^2} - \lambda^2 = 0$$

$$\Rightarrow$$

$$\lambda^2_1 = \lambda^2_+ = \frac{R^2}{R^2 - r^2}, \quad \lambda_1 = \lambda_+ = +\frac{R}{\sqrt{R^2 - r^2}}, \quad \lambda^2_2 = \lambda^2_- = 1, \quad \lambda_2 = \lambda_- = +1.$$

Right eigencolumns:

$$\begin{bmatrix} f_{11} \\ f_{21} \end{bmatrix} = \frac{1}{\sqrt{g_{11}}}\begin{bmatrix} 1 \\ 0 \end{bmatrix} = \frac{1}{r}\begin{bmatrix} 1 \\ 0 \end{bmatrix}, \quad \begin{bmatrix} f_{12} \\ f_{22} \end{bmatrix} = \frac{1}{\sqrt{g_{22}}}\begin{bmatrix} 0 \\ 1 \end{bmatrix} = \begin{bmatrix} 0 \\ 1 \end{bmatrix}. \tag{1.108}$$

Right eigenvectors:

$$\boldsymbol{g}_1 := \boldsymbol{g}_\alpha = D_\alpha \boldsymbol{x} = -\boldsymbol{e}_1 r \sin\alpha + \boldsymbol{e}_2 r \cos\alpha, \quad \boldsymbol{g}_2 := \boldsymbol{g}_r = D_r \boldsymbol{x} = +\boldsymbol{e}_1 \cos\alpha + \boldsymbol{e}_2 \sin\alpha; \tag{1.109}$$

1st eigenvector: $\boldsymbol{f}_\alpha := \boldsymbol{g}_\alpha f_{11} + \boldsymbol{g}_r f_{21}, \quad \boldsymbol{f}_\alpha(r) = \boldsymbol{g}_\alpha \dfrac{1}{r} = -\boldsymbol{e}_1 \sin\alpha + \boldsymbol{e}_2 \cos\alpha$

(tangent vector of image of parallel circles) ;

$$\tag{1.110}$$

2nd eigenvector: $\boldsymbol{f}_r := \boldsymbol{g}_\alpha f_{12} + \boldsymbol{g}_r f_{22}, \quad \boldsymbol{f}_r(r) = \boldsymbol{g}_r$

(tangent vector of image of meridians) .

1-4 Euler–Lagrange deformation tensor

"Approach your problems from the right end and begin with the answers.
Then, one day, perhaps you will find the final question."
(The Hermit Clad in Crane Feathers, in R. van Gulik's The Chinese Maze Murders.)

A first additive measure of deformation: the Euler–Lagrange deformation tensor, relations between the Cauchy–Green and the Euler-Lagrange deformation tensor.

The first additive measure of deformation is based upon the scale differences $\mathrm{d}s^2 - \mathrm{d}S^2$ versus $\mathrm{d}S^2 - \mathrm{d}s^2$, which are represented by pullback $U^M \to u^\mu = f^\mu(U^M)$ or pushforward $u^\mu \to U^M = F^M(u^\mu)$, in particular, by

$$\mathrm{d}s^2 - \mathrm{d}S^2 = \mathrm{d}\boldsymbol{U}^{\mathrm{T}}\left(\mathsf{J}_l^{\mathrm{T}}\mathsf{G}_{\mathrm{r}}\mathsf{J}_l - \mathsf{G}_l\right)\mathrm{d}\boldsymbol{U} \quad \text{versus} \quad \mathrm{d}S^2 - \mathrm{d}s^2 = \mathrm{d}\boldsymbol{u}^{\mathrm{T}}\left(\mathsf{J}_{\mathrm{r}}^{\mathrm{T}}\mathsf{G}_l\mathsf{J}_{\mathrm{r}} - \mathsf{G}_{\mathrm{r}}\right)\mathrm{d}\boldsymbol{u} \,. \quad (1.111)$$

Accordingly, we are led to the *deformation measures* of Box 1.17, which have been introduced by L. Euler and J. L. Lagrange, called *strain*.

Box 1.17 (Left versus right Euler–Lagrange deformation tensor).

Left EL deformation tensor :   Right EL deformation tensor :

$$\mathrm{d}s^2 - \mathrm{d}S^2 = \mathrm{d}\boldsymbol{U}^{\mathrm{T}}\left(\mathsf{J}_l^{\mathrm{T}}\mathsf{G}_{\mathrm{r}}\mathsf{J}_l - \mathsf{G}_l\right)\mathrm{d}\boldsymbol{U} \qquad \mathrm{d}S^2 - \mathrm{d}s^2 = \mathrm{d}\boldsymbol{u}^{\mathrm{T}}\left(\mathsf{J}_{\mathrm{r}}^{\mathrm{T}}\mathsf{G}_l\mathsf{J}_{\mathrm{r}} - \mathsf{G}_{\mathrm{r}}\right)\mathrm{d}\boldsymbol{u}$$

$$\frac{1}{2}\left(\mathrm{d}s^2 - \mathrm{d}S^2\right) = +\mathrm{d}\boldsymbol{U}^{\mathrm{T}}\mathsf{E}_l\,\mathrm{d}\boldsymbol{U} \qquad \frac{1}{2}\left(\mathrm{d}S^2 - \mathrm{d}s^2\right) = -\mathrm{d}\boldsymbol{u}^{\mathrm{T}}\mathsf{E}_{\mathrm{r}}\,\mathrm{d}\boldsymbol{u} \qquad (1.112)$$

$$\mathsf{E}_l := \frac{1}{2}\left(\mathsf{J}_l^{\mathrm{T}}\mathsf{G}_{\mathrm{r}}\mathsf{J}_l - \mathsf{G}_l\right) \qquad \mathsf{E}_{\mathrm{r}} := \frac{1}{2}\left(\mathsf{G}_{\mathrm{r}} - \mathsf{J}_{\mathrm{r}}^{\mathrm{T}}\mathsf{G}_l\mathsf{J}_{\mathrm{r}}\right)$$

Question.

Question: "What is the role of strain in the context of the pair of matrices $\{\mathsf{E}_l, \mathsf{G}_l\}$ and $\{\mathsf{E}_{\mathrm{r}}, \mathsf{G}_{\mathrm{r}}\}$, respectively?" Answer: "$\{\mathsf{E}_l, \mathsf{E}_{\mathrm{r}}\}$ are symmetric matrices and $\{\mathsf{G}_l, \mathsf{G}_{\mathrm{r}}\}$ are symetric, positive-definite matrices. Thus, according to a standard lemma of matrix algebra, both matrices can be simultaneously diagonalized, one matrix being the unit matrix. With the reference to the general eigenvalue we experienced for the Cauchy–Green deformation tensor, we arrive at Lemma 1.7."

Lemma 1.7 (Left and right general eigenvalue problem of the Euler–Lagrange deformation tensor).

For the pair of symmetric matrices $\{\mathsf{E}_l, \mathsf{G}_l\}$ or $\{\mathsf{E}_{\mathrm{r}}, \mathsf{G}_{\mathrm{r}}\}$, where $\{\mathsf{G}_l, \mathsf{G}_{\mathrm{r}}\}$ are positive-definite matrices, a simultaneous diagonalization, namely

$$\mathsf{F}_l^{\mathrm{T}}\mathsf{E}_l\mathsf{F}_l = \mathrm{diag}\left(K_1, K_2\right), \ \mathsf{F}_l^{\mathrm{T}}\mathsf{G}_l\mathsf{F}_l = \mathsf{I}_2 \quad \text{versus} \quad \mathsf{F}_{\mathrm{r}}^{\mathrm{T}}\mathsf{E}_{\mathrm{r}}\mathsf{F}_{\mathrm{r}} = \mathrm{diag}\left(\kappa_1, \kappa_2\right), \ \mathsf{F}_{\mathrm{r}}^{\mathrm{T}}\mathsf{G}_{\mathrm{r}}\mathsf{F}_{\mathrm{r}} = \mathsf{I}_2, \quad (1.113)$$

is immediately obtained from the left and right general eigenvalue–eigenvector problems

$$\mathsf{E}_l\mathsf{F}_l - \mathsf{G}_l\mathsf{F}_l\,\mathrm{diag}\left(K_1, K_2\right) = 0 \qquad\qquad \mathsf{E}_{\mathrm{r}}\mathsf{F}_{\mathrm{r}} - \mathsf{G}_{\mathrm{r}}\mathsf{F}_{\mathrm{r}}\,\mathrm{diag}\left(\kappa_1, \kappa_2\right) = 0$$

$$\Leftrightarrow \qquad\qquad \Leftrightarrow$$

$$\left(\mathsf{E}_l - K_i\mathsf{G}_l\right)\boldsymbol{f}_{li} = 0 \quad (\forall i \in \{1, 2\}) \qquad \left(\mathsf{E}_{\mathrm{r}} - \kappa_i\mathsf{G}_{\mathrm{r}}\right)\boldsymbol{f}_{\mathrm{r}i} = 0 \quad (\forall i \in \{1, 2\})$$

$$\Leftrightarrow \qquad\qquad\qquad\qquad \text{and} \qquad\qquad \Leftrightarrow$$

$$\left|\mathsf{E}_l - K_i\mathsf{G}_l\right| = 0 \quad \left(\mathsf{F}_l^{\mathrm{T}}\mathsf{G}_l\mathsf{F}_l = \mathsf{I}_2\right), \qquad \left|\mathsf{E}_{\mathrm{r}} - \kappa_i\mathsf{G}_{\mathrm{r}}\right| = 0 \quad \left(\mathsf{F}_{\mathrm{r}}^{\mathrm{T}}\mathsf{G}_{\mathrm{r}}\mathsf{F}_{\mathrm{r}} = \mathsf{I}_2\right), \qquad (1.114)$$

$$K_{1\,2} = K_{+\,-} = \frac{1}{2}\mathrm{tr}\left[\mathsf{E}_l\mathsf{G}_l^{-1}\right]\pm \qquad \kappa_{1\,2} = \kappa_{+\,-} = \frac{1}{2}\mathrm{tr}\left[\mathsf{E}_{\mathrm{r}}\mathsf{G}_{\mathrm{r}}^{-1}\right]\pm$$

$$\pm\sqrt{\left(\mathrm{tr}\left[\mathsf{E}_l\mathsf{G}_l^{-1}\right]\right)^2 - 4\mathrm{det}\left[\mathsf{E}_l\mathsf{G}_l^{-1}\right]}\,, \qquad \pm\sqrt{\left(\mathrm{tr}\left[\mathsf{E}_{\mathrm{r}}\mathsf{G}_{\mathrm{r}}^{-1}\right]\right)^2 - 4\mathrm{det}\left[\mathsf{E}_{\mathrm{r}}\mathsf{G}_{\mathrm{r}}^{-1}\right]}\,.$$

End of Lemma.

In order to visualize the eigenspace of both the left and the right Euler–Lagrange deformation tensor E_l and E_r relative to the left and right metric tensors G_l and G_r, we are forced to compute in addition the left and right eigenvectors (namely the left and right eigencolumns, also called eigendirectories) of the pairs $\{E_l, G_l\}$ and $\{E_r, G_r\}$, respectively. Lemma 1.8 summarizes the results.

Lemma 1.8 (Left and right eigenvectors of the left and the right Euler–Lagrange deformation tensor).

For the pair of symmetric matrices $\{E_l, G_l\}$ or $\{E_r, G_r\}$, an explicit form of the left eigencolumns and the right eigencolumns is

1st left eigencolumns, K_1:

$$\begin{bmatrix} F_{11} \\ F_{21} \end{bmatrix} = \frac{1}{\sqrt{G_{11}(e_{22}-K_1 G_{22})^2 - 2G_{12}(e_{12}-K_1 G_{12})(e_{22}-K_1 G_{22}) + G_{22}(e_{12}-K_1 G_{12})^2}} \times$$

$$\times \begin{bmatrix} e_{22} - K_1 G_{22} \\ -(e_{12} - K_1 G_{12}) \end{bmatrix};$$

(1.115)

2nd left eigencolumns, K_2:

$$\begin{bmatrix} F_{12} \\ F_{22} \end{bmatrix} = \frac{1}{\sqrt{G_{22}(e_{11}-K_2 G_{11})^2 - 2G_{12}(e_{11}-K_2 G_{11})(e_{12}-K_2 G_{12}) + G_{11}(e_{12}-K_2 G_{12})^2}} \times$$

$$\times \begin{bmatrix} -(e_{12} - K_2 G_{12}) \\ e_{11} - K_2 G_{11} \end{bmatrix};$$

(1.116)

1st right eigencolumns, κ_1:

$$\begin{bmatrix} f_{11} \\ f_{21} \end{bmatrix} = \frac{1}{\sqrt{g_{11}(E_{22}-\kappa_1 g_{22})^2 - 2g_{12}(E_{12}-\kappa_1 g_{12})(E_{22}-\kappa_1 g_{22}) + g_{22}(E_{12}-\kappa_1 g_{12})^2}} \times$$

$$\times \begin{bmatrix} E_{22} - \kappa_1 g_{22} \\ -(E_{12} - \kappa_1 g_{12}) \end{bmatrix};$$

(1.117)

2nd right eigencolumns, κ_2:

$$\begin{bmatrix} f_{12} \\ f_{22} \end{bmatrix} = \frac{1}{\sqrt{g_{22}(E_{11}-\kappa_2 g_{11})^2 - 2g_{12}(E_{11}-\kappa_2 g_{11})(E_{12}-\kappa_2 g_{12}) + g_{11}(E_{12}-\kappa_2 g_{12})^2}} \times$$

$$\times \begin{bmatrix} -(E_{12} - \kappa_2 g_{12}) \\ E_{11} - \kappa_2 g_{11} \end{bmatrix}.$$

(1.118)

End of Lemma.

The proof of these relations follows the line of thought of the proof of Lemma 1.6. Accordingly, we skip any proof here.

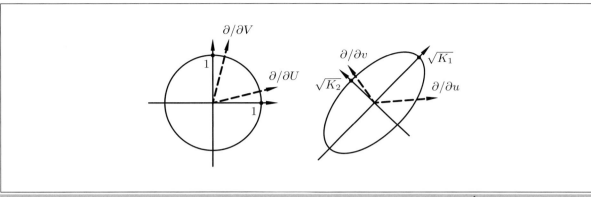

Fig. 1.12. Left Euler–Lagrange tensor, $K_1 > 0$, $K_2 > 0$, left Euler–Lagrange circle $\mathbb{S}^1$, left Euler–Lagrange ellipse $\mathbb{E}^1_{\sqrt{K_1},\sqrt{K_2}}$.

The canonical forms of the *scale difference* $(\mathrm{d}s)^2 - (\mathrm{d}S)^2$ and $(\mathrm{d}S)^2 - (\mathrm{d}s)^2$, respectively, have been interpreted as

left Euler–Lagrange circle $\mathbb{S}^1$ right Euler–Lagrange circle $\mathbb{S}^1$

versus   versus

left Euler–Lagrange ellipse   right Euler–Lagrange ellipse

$$\mathbb{E}^1_{\sqrt{K_1},\sqrt{K_2}} \left(K_i > 0 \,\forall\, i = 1, 2 \right), \quad \text{and} \quad \mathbb{E}^1_{\sqrt{\kappa_1},\sqrt{\kappa_2}} \left(\kappa_i > 0 \,\forall\, i = 1, 2 \right), \tag{1.119}$$

left Euler–Lagrange hyperbola   right Euler–Lagrange hyperbola

$$\mathbb{H}^1_{\sqrt{K_1},\sqrt{K_2}} \left(K_1 > 0, K_2 < 0 \right), \quad\quad\quad \mathbb{H}^1_{\sqrt{\kappa_1},\sqrt{\kappa_2}} \left(\kappa_1 > 0, \kappa_2 < 0 \right),$$

on the left tangent space $T_U \mathbb{M}^2_l$ and the right tangent space $T_u \mathbb{M}^2_r$, respectively. A deformation portrait with a positive eigenvalue $K(\mathsf{E}_l, \mathsf{G}_r)$ or $\kappa(\mathsf{E}_r, \mathsf{G}_l)$ is referred to as *extension*, with a negative eigenvalue $K(\mathsf{E}_l, \mathsf{G}_r)$ or $\kappa(\mathsf{E}_r, \mathsf{G}_l)$ as *compression*. Obviously, Cauchy–Green deformation and Euler–Lagrange deformation are related as outlined in Corollary 1.9. The four cases of the eigenspace analysis of the left and the right Euler–Lagrange deformation are illustrated in Figs. 1.12–1.15.

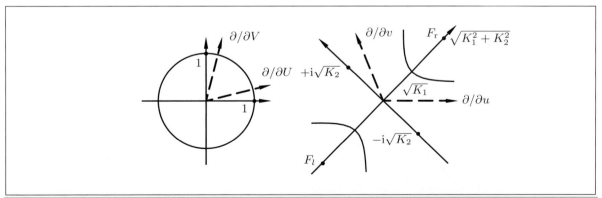

Fig. 1.13. Left Euler–Lagrange tensor, $K_1 > 0$, $K_2 < 0$, left Euler–Lagrange circle $\mathbb{S}^1$, left Euler–Lagrange hyperbola $\mathbb{H}^1_{\sqrt{K_1},\sqrt{K_2}}$, left and right focal points F_l and F_r.

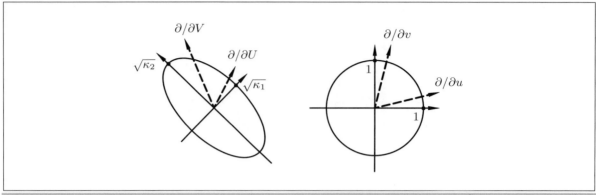

Fig. 1.14. Right Euler–Lagrange tensor, $\kappa_1 > 0$, $\kappa_2 > 0$, right Euler–Lagrange circle $\mathbb{S}^1$, right Euler–Lagrange ellipse $\mathbb{E}^1_{\sqrt{\kappa_1},\sqrt{\kappa_2}}$.

Corollary 1.9 (Relation between the Cauchy–Green and the Euler–Lagrange deformation tensor).

$$2\mathsf{E}_l = \mathsf{J}_l^* \mathsf{G}_r \mathsf{J}_l - \mathsf{G}_l = \mathsf{C}_l - \mathsf{G}_l \qquad\qquad 2\mathsf{E}_r = \mathsf{G}_r - \mathsf{J}_r^* \mathsf{G}_l \mathsf{J}_r = \mathsf{G}_r - \mathsf{C}_r$$

$$\text{versus} \qquad\qquad \text{versus} \qquad\qquad \text{versus}$$

$$\mathsf{C}_l = 2\mathsf{E}_l + \mathsf{G}_l\ , \qquad\qquad\qquad \mathsf{C}_r = \mathsf{G}_r - 2\mathsf{E}_r\ ; \qquad\qquad (1.120)$$

$$\mathsf{E}_l = \mathsf{J}_l^* \mathsf{E}_r \mathsf{J}_l \qquad\qquad \text{versus} \qquad\qquad \mathsf{E}_r = \mathsf{J}_r^* \mathsf{E}_l \mathsf{J}_r\ ;$$

$$2K_i = \Lambda_i^2\ \forall\ i = 1,2 \qquad\qquad \text{versus} \qquad\qquad 2\kappa_i = \lambda_i^2 - 1\ \forall\ i = 1,2\ .$$

End of Corollary.

Examples for the mapping between two Riemann manifolds are the following. C. F. Gauss (1822, 1844) presented his celebrated conformal mapping of the biaxial ellipsoid $\mathbb{E}^2_{A_1,A_1,A_2} = \mathbb{M}^2_l$ onto the sphere $\mathbb{S}^2_r = \mathbb{M}^2_r$, also called *double projection* due to a second conformal mapping of the sphere $\mathbb{S}^2_r$ onto the plane $\mathbb{R}^2$. M. Amalvict and E. Livieratos (1988) elaborated the isoparametric mapping of the triaxial ellipsoid $\mathbb{E}^2_{A_1,A_2,A_3} = \mathbb{M}^2_l$ onto the biaxial ellipsoid $\mathbb{E}^2_{A_1,A_1,A_2} = \mathbb{M}^2_r$. A. Dermanis, E. Livieratos, and S. Pertsinidou (1984) mapped the geoid onto the biaxial ellipsoid. While nearly all existing map projections are analyzed by means of the Cauchy–Green deformation tensor, A. Dermanis and E. Livieratos (1993) used the Euler–Lagrange deformation tensor for map projections, in particular, dilatation $\mathrm{tr}\left[\mathsf{E}_l \mathsf{G}_l^{-1}\right]$ or $\mathrm{tr}\left[\mathsf{E}_r \mathsf{G}_r^{-1}\right]$ and general shear $\left(\mathrm{tr}\left[\mathsf{E}_l \mathsf{G}_l^{-1}\right]\right)^2 - 4\det\left[\mathsf{E}_l \mathsf{G}_l^{-1}\right]$ or $\left(\mathrm{tr}\left[\mathsf{E}_r \mathsf{G}_r^{-1}\right]\right)^2 - 4\det\left[\mathsf{E}_r \mathsf{G}_r^{-1}\right]$. An elaborate example is discussed in Section 1-5. However, to give you some breathing time, please first enjoy the Berghaus star projection presented in Fig. 1.16.

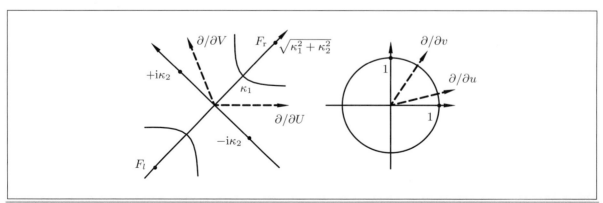

Fig. 1.15. Right Euler–Lagrange tensor, $\kappa_1 > 0$, $\kappa_2 < 0$, right Euler–Lagrange circle $\mathbb{S}^1$, right Euler–Lagrange hyperbola $\mathbb{H}^1_{\sqrt{\kappa_1},\sqrt{\kappa_2}}$, left and right focal points F_l and F_r.

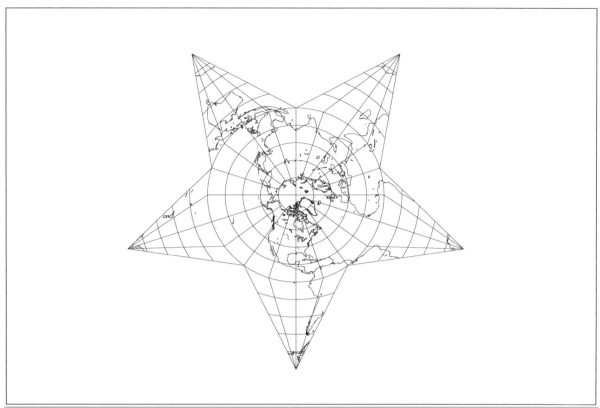

Fig. 1.16. Berghaus star projection, shorelines of a spherical Earth, 18° graticule, central meridian 90° W, "world map".

1-5 One example: orthogonal map projection

One example of deformation analysis (Euler–Lagrange deformation tensor, its eigenspace, ellipses and hyperbolae of distortion), orthogonal map projection, Hammer equiareal modified azimuthal projection.

The general eigenspace analysis of the *Euler–Lagrange deformation tensor analysis* visualized by ellipses and hyperbolae of distortion is close to the heart of any map projection. It is for this reason that we present to you as Example 1.7 the orthogonal projection of the northern hemisphere onto the equatorial plane. We recommend to go through all details with "paper and pencil".

Example 1.7 (Orthogonal projection of points of the sphere onto the equatorial plane through the origin).

Let us assume that we make an orthogonal projection of points of the northern hemisphere onto the equatorial plane $\mathbb{P}^2_{\mathcal{O}}$ through the origin $\mathcal{O}$ of the plane $\mathbb{S}^2_{R^+}$. For an illustration of such a map projection let us refer to Fig. 1.10. The direct mapping and inverse mapping equations are given by

$$x = X = R\cos\Phi\cos\Lambda, \qquad\qquad \Lambda(x,y) = \arctan(y/x),$$

$$y = Y = R\cos\Phi\sin\Lambda, \qquad \text{versus} \qquad \cos\Phi(x,y) = \frac{\sqrt{x^2+y^2}}{R}, \qquad (1.121)$$

$$\alpha = \Lambda, \quad r = \sqrt{X^2+Y^2} = R\cos\Phi, \qquad \Lambda = \alpha, \quad \cos\Phi = r/R.$$

End of Example.

We take advantage of Cartesian coordinates $\{x,y\}$ and polar coordinates $\{\alpha,r\}$ to cover $\mathbb{R}^2$. We pose two problems. (i) Derive the right Euler–Lagrange deformation tensor. (ii) Solve the right general eigenvalue–eigenvector problem.

We solve the two problems side-by-side in Box 1.18 in Cartesian coordinates and in Box 1.19 in polar coordinates. Given the right Euler–Lagrange matrix, by means of Box 1.20, we are giving the transform to the left Euler–Lagrange matrix.

- First, we transform the right Cauchy–Green matrix from Box 1.13 to Box 1.18. Second, we take advantage of Corollary 1.9 in order to compute the right Euler–Lagrange matrix $2E_r = I_2 - C_r$ in Cartesian coordinates as well as its right eigenvalues $2\kappa_i = \lambda_i^2 - 1$ from given right eigenvalues λ_i^2. In particular, we find $\kappa_1 \neq 0$ and $\kappa_2 = 0$. Third, we represent the right Euler–Lagrange tensor in the Cartesian base $e_1 \vee e_1$, $e_1 \vee e_2$, and $e_2 \vee e_2$, where $\vee$ denotes the symmetric product.
- Fourth, in contrast, we transfer the right Cauchy–Green matrix from Box 1.14 to Box 1.19. Fifth, we again use Corollary 1.9 in order to compute the right Euler–Lagrange matrix $2E_r = G_r - C_r$ ($G_r = \mathrm{diag}(r^2, 1)$) in polar coordinates as well as its right eigenvalues $2\kappa_i = \lambda_i^2 - 1$. Again, we find $\kappa_1 \neq 0$ and $\kappa_2 = 0$. Sixth, we represent the right Euler–Lagrange tensor in the polar base $g_2 \otimes g_2$.
- Seventh, Box 1.20 reviews the transformations of the right Euler–Lagrange tensor E_r to the left Euler–Lagrange tensor E_l by means of the left Jacobi matrix J_l transferred from Box 1.14 by $J_l = J_r^{-1}$. Eighth, we have computed the left eigenvalues of the left Euler–Lagrange deformation tensor. The degenerate distortion ellipse/hyperbola of the right Euler–Lagrange matrix is finally illustrated by Fig. 1.17.

Right Cauchy–Green matrix in Cartesian coordinates:

$$C_r = \frac{1}{R^2 - (x^2 + y^2)} \begin{bmatrix} R^2 - y^2 & xy \\ xy & R^2 - x^2 \end{bmatrix}. \tag{1.122}$$

Right Euler–Lagrange matrix in Cartesian coordinates:

$$2E_r = I_2 - C_r, \quad E_r = -\frac{1}{2} \frac{1}{R^2 - (x^2 + y^2)} \begin{bmatrix} x^2 & xy \\ xy & y^2 \end{bmatrix}. \tag{1.123}$$

Right eigenvalues:

$$2\kappa_i = \lambda_i^2 - 1 \; \forall \, i \in \{1, 2\},$$

$$\lambda_1^2 = \frac{R^2}{R^2 - (x^2 + y^2)}, \quad \lambda_2^2 = 1, \quad \kappa_1 = \frac{1}{2} \frac{x^2 + y^2}{R^2 - (x^2 + y^2)}, \quad \kappa_2 = 0. \tag{1.124}$$

Right Euler–Lagrange tensor:

$$E_r =$$

$$= -\frac{1}{2} e_1 \otimes e_1 \frac{x^2}{R^2 - (x^2 + y^2)} - \frac{1}{2} (e_1 \otimes e_2 + e_2 \otimes e_1) \frac{xy}{R^2 - (x^2 + y^2)} - \frac{1}{2} e_2 \otimes e_2 \frac{y^2}{R^2 - (x^2 + y^2)} =$$

$$= -\frac{1}{2} e_1 \vee e_1 \frac{x^2}{R^2 - (x^2 + y^2)} - e_1 \vee e_2 \frac{xy}{R^2 - (x^2 + y^2)} - \frac{1}{2} e_2 \vee e_2 \frac{y^2}{R^2 - (x^2 + y^2)} \tag{1.125}$$

subject to

$$e_\mu \vee e_\nu := \frac{1}{2} (e_\mu \otimes e_\nu + e_\nu \otimes e_\mu).$$

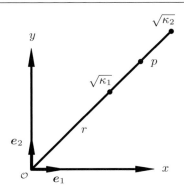

Fig. 1.17. Orthogonal projection $\mathbb{S}^2_{R+}$ onto $\mathbb{P}^2_{\mathcal{O}}$, degenerate Euler–Lagrange ellipse/hyperbola.

Box 1.19 (Orthogonal projection $\mathbb{S}^2_{R+}$ onto $\mathbb{P}^2_{\mathcal{O}}$, polar coordinates, the first and the second problem).

Right Cauchy–Green matrix in polar coordinates:

$$r^2 = x^2 + y^2 = X^2 + Y^2 = R^2 \cos^2 \Phi \,,$$

$$\mathsf{C}_r(r) = \begin{bmatrix} r^2 & 0 \\ 0 & \dfrac{R^2}{R^2 - r^2} \end{bmatrix} \,, \quad \mathsf{G}_r(r) = \begin{bmatrix} r^2 & 0 \\ 0 & 1 \end{bmatrix} \,. \tag{1.126}$$

Right Euler–Lagrange matrix in polar coordinates:

$$2\mathsf{E}_r = \mathsf{G}_r - \mathsf{C}_r \,, \quad \mathsf{E}_r = \frac{1}{2} \begin{bmatrix} 0 & 0 \\ 0 & -\dfrac{r^2}{R^2 - r^2} \end{bmatrix} \,. \tag{1.127}$$

Right eigenvalues:

$$2\kappa_i = \lambda_i^2 - 1 \,\,\forall\, i \in \{1,2\} \,, \quad \lambda_1^2 = \frac{R^2}{R^2 - r^2} \,, \quad \lambda_2^2 = 1 \,, \quad \kappa_1 = \frac{1}{2} \frac{r^2}{R^2 - r^2} > 0 \,, \quad \kappa_2 = 0 \,. \tag{1.128}$$

Right Euler–Lagrange tensor:

$$\mathsf{E}_r = -\frac{1}{2} \boldsymbol{g}_2 \otimes \boldsymbol{g}_2 \frac{r^2}{R^2 - r^2} \,. \tag{1.129}$$

Box 1.20 (Orthogonal projection $\mathbb{S}^2_{R+}$ onto $\mathbb{P}^2_{\mathcal{O}}$, polar coordinates, the transformations from the right Euler–Lagrange matrix to the left Euler–Lagrange matrix).

$$\mathsf{E}_r \rightarrow \mathsf{E}_l :$$

$$\mathsf{E}_l = \mathsf{J}_l^* \mathsf{E}_r \mathsf{J}_l \,, \quad r^2 = R^2 \cos^2 \Phi \,,$$

$$\mathsf{J}_l = \begin{bmatrix} 1 & 0 \\ 0 & -\sqrt{R^2 - r^2} \end{bmatrix} \,, \quad \mathsf{E}_l = \begin{bmatrix} 0 & 0 \\ 0 & -r^2/2 \end{bmatrix} = -\frac{1}{2} R^2 \begin{bmatrix} 0 & 0 \\ 0 & \cos^2 \Phi \end{bmatrix} \,. \tag{1.130}$$

Left eigenvalues:

$$2K_1 = \Lambda_1^2 - 1 = \frac{1}{\lambda_1^2} - 1 \,, \quad 2K_2 = \Lambda_2^2 - 1 = \frac{1}{\lambda_2^2} - 1 \,, \quad K_1 = -\frac{1}{2} \cos^2 \Phi \,, \quad K_2 = 0 \,. \tag{1.131}$$

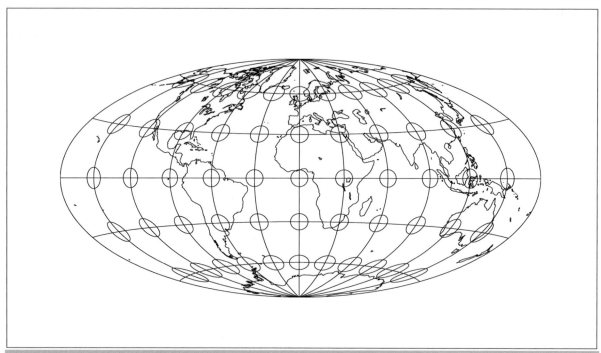

Fig. 1.18. Special map projection of the sphere: the Hammer equiareal modified azimuthal projection. This map projection is centered to the Greenwich meridian, with shorelines, 30° longitude, 15° latitude graticule, Tissot ellipses of distortion, "the world in one chart".

A map projection which is worth studying with all the machinery of deformation measures is the *Hammer equiareal modified azimuthal projection of the sphere* $\mathbb{S}^2_{R+}$ presented in Fig. 1.18. The ID card of this special map projection is shown in Table 1.2.

Table 1.2. ID card of Hammer equiareal modified azimuthal projection of the sphere.

(i)	Classification	Modified azimuthal, transverse, rescaled equiareal.
(ii)	Graticule	Meridians: central meridian is straight, other meridians are algebraic curves of fourth order. The limiting meridians form an ellipse. Parallels: curved. equator is straight, other parallels are algebraic curves of fourth order. Poles of the sphere: points. Symmetry: about the central meridians.
(iii)	Distortions	Product of principal stretches is one, equiareal, equidistant map of the equator.
(iv)	Direct mapping equations	$x = \dfrac{c_1 R \sqrt{2}\sqrt{1-c_4^2 \sin^2 \Phi}\sin(c_3 \Lambda)}{\sqrt{1+\sqrt{1-c_4^2 \sin^2 \Phi}\cos(c_3 \Lambda)}}$, $y = \dfrac{c_2 R \sqrt{2}c_4 \sin \Phi}{\sqrt{1+\sqrt{1-c_4^2 \sin^2 \Phi}\cos(c_3 \Lambda)}}$, $c_1 = 2$, $c_2 = 1$, $c_3 = \frac{1}{2}$, $c_4 = 1$, $c_1 c_2 c_3 c_4 = 1$.
(v)	Usage	Atlas cartography.
(vi)	Origins	Presented by E. Hammer (1858–1925) in 1892. The special Hammer projection has been generalized from the sphere $\mathbb{S}^2_{R+}$ to the ellipsoid-of-revolution $\mathbb{E}^2_{A_1,A_1,A_2}$ by E. Grafarend and R. Syffus (1997e).

1-6 Review: the deformation measures

Review: the family of twentytwo different deformation measures, compatibility conditions, integrability conditions, differential forms.

By means of Table 1.3, let us introduce a collection of various deformation measures, i. e. deformation tensors of the first kind based upon the reviews by D. B. Macvean (1968), K. N. Morman (1986), and E. Grafarend (1995). For the classification scheme various representation theorems of T. C. T. Ting (1985) are most useful. *Compatibility conditions* for Cauchy–Green deformation fields have been formulated by F. P. Duda and L. C. Martins (1995). They are needed for the problem to determine the mapping equations $U^K = f^K(u^k)$ or $u^k = f^k(U^K)$ from prescribed left or right Cauchy–Green deformation fields as tensor-valued functions. In the context of *exterior calculus*, these compatiblity conditions are classified as *integrability conditions*. The various deformation measures honor the works of E. Almansi (1911), A. Cauchy (1889,1890), J. Finger (1894a), G. Green (1839), H. Hencky (1928), Z. Karni and M. Reiner (1960), G. Piola (1836), and B. R. Seth (1964a,b). The inverse deformation matrices, namely E_5, E_6, E_{15}, E_{16}, E_{17}, and E_{18}, appear in the various forms of distortion energy. Logarithmic and root measures of deformation appear in special stress–strain relations, which very often are called *constitutive equations*. The measures E_3 and E_4 as well as E_{13} and E_{14} build up the special eigenvalue problems. They correspond to definitions of the curvature matrix $\mathsf{K} = -\mathsf{H}\mathsf{G}^{-1}$, in surface geometry built on the matrices of the *first differential form* $I \sim (\mathrm{d}g)^2 = g_{\mu\nu}\mathrm{d}u^\mu\mathrm{d}u^\nu$ as well as on the *second differential form* $II \sim (\mathrm{d}h)^2 = h_{\mu\nu}\mathrm{d}u^\mu\mathrm{d}u^\nu$, which is also called the *Hesse form*. Indeed, they establish the matrix pair $\{\mathsf{H}, \mathsf{G}\}$, where G is positive definite.

1-7 Angular shear

A second additive measure of deformation: angular shear (also called angular distortion), left and right surfaces, parameterized curves.

An alternative *additive measure* of deformation is *angular shear*, also called *angular distortion*. Assume that two parameterized curves in $\mathbb{M}_l^2$ as well as their images in $\mathbb{M}_r^2$ intersect at the point $\boldsymbol{U}_0$ as well as $\boldsymbol{u}_0$, respectively. Two vectors $\dot{\boldsymbol{U}}_1^M$ and $\dot{\boldsymbol{U}}_2^N$ as well as $\dot{\boldsymbol{u}}_1^\mu$ and $\dot{\boldsymbol{u}}_2^\nu$ being elements of the corresponding local tangent spaces $T_{\boldsymbol{U}_0}\mathbb{M}_l^2$ as well as $T_{\boldsymbol{U}_0}\mathbb{M}_r^2$,

$$\dot{\boldsymbol{U}}_1^M \in T_{\boldsymbol{U}_0}\mathbb{M}_l^2\,,\ \dot{\boldsymbol{U}}_2^N \in T_{\boldsymbol{U}_0}\mathbb{M}_l^2 \qquad \text{versus} \qquad \dot{\boldsymbol{u}}_1^\mu \in T_{\boldsymbol{U}_0}\mathbb{M}_r^2\,,\ \dot{\boldsymbol{u}}_2^\nu \in T_{\boldsymbol{U}_0}\mathbb{M}_r^2\,, \qquad (1.132)$$

include the angles Ψ_l and Ψ_r. (Note that prime differentiation is understood as differentiation with respect to arc length. In contrast, dot differentiation is understood as differentiation with respect to an arbitrary curve parameter, called "t_l" and "t_r", respectively.) As it is illustrated by Fig. 1.19, the left angle Ψ_l as well as the right angle Ψ_r are represented by the inner products

$$\cos\Psi_l = \langle \boldsymbol{U}_1' \mid \boldsymbol{U}_2' \rangle = \qquad\qquad \cos\Psi_r = \langle \boldsymbol{u}_1' \mid \boldsymbol{u}_2' \rangle =$$
$$\text{versus}$$
$$= \frac{G_{MN}\dot{\boldsymbol{U}}_1^M \dot{\boldsymbol{U}}_2^N}{\sqrt{G_{AB}\dot{\boldsymbol{U}}_1^A \dot{\boldsymbol{U}}_2^B}\sqrt{G_{\Gamma\Delta}\dot{\boldsymbol{U}}_1^\Gamma \dot{\boldsymbol{U}}_2^\Delta}} \qquad\qquad = \frac{g_{\mu\nu}\dot{\boldsymbol{u}}_1^\mu \dot{\boldsymbol{u}}_2^\nu}{\sqrt{g_{\alpha\beta}\dot{\boldsymbol{u}}_1^\alpha \dot{\boldsymbol{u}}_2^\beta}\sqrt{g_{\gamma\delta}\dot{\boldsymbol{u}}_1^\gamma \dot{\boldsymbol{u}}_2^\delta}}\,. \qquad (1.133)$$

The second additive measure of deformation is the *angular shear* or the *angle of shear* (Σ_l is of type "left" and Σ_r is of type "right", respectively)

$$\Sigma_l = \Sigma := \Psi_l - \Psi_r \qquad \text{versus} \qquad \Sigma_r = \sigma := \Psi_r - \Psi_l\,. \qquad (1.134)$$

The following Example 1.8 and the following Box 1.21 illustrate this second additive measure of deformation. In order to be simple, however, we have chosen the coordinate lines that are illustrated in Fig. 1.20.

Table 1.3. Various deformation tensors of the first kind.

Definitions	Author	Comments
$E_1 = C_l = S_l R^* G_r R S_l = J_l^* G_r J_l$	A. Cauchy (1889,1890) ("left Cauchy–Green")	if $G_r = I$, then $C_l = S_l^2 = J_l^* J_l$
$E_2 = C_r = S_r R^* G_l R S_r = J_r^* G_l J_r$	G. Green (1839) ("right Cauchy–Green")	if $G_l = I$, then $C_r = S_r^2 = J_r^* J_r$
$E_3 = C_l G_l^{-1}$	E. Grafarend (1995) ("left-right Cauchy–Green")	if $G_l = I$, then $E_3 = C_l$
$E_4 = C_r G_r^{-1}$	E. Grafarend (1995) ("right-left Cauchy–Green")	if $G_r = I$, then $E_4 = C_r$
$E_5 = G_l C_l^{-1}$	E. Grafarend (1995) ("inverse left-right Cauchy–Green")	J. Finger (1894a) if $G_l = I$, then $E_5 = E_3^{-1}$
$E_6 = G_r C_r^{-1}$	E. Grafarend (1995) ("inverse right-left Cauchy–Green")	G. Piola (1836), if $G_r = I$, then $E_6 = E_4^{-1}$
$E_7 = C_l^{m/2} \sim \{\Lambda_1^m, \Lambda_2^m\}$	B. R. Seth (1964a,b) $(m \in Z, m \neq 0)$	$m = 2 : E_7 = E_1$
$E_8 = \ln C_l \sim \{\ln \Lambda_1, \ln \Lambda_2\}$	H. Hencky (1928)	–
$E_9 = C_r^{m/2} \sim \{\lambda_1^m, \lambda_2^m\}$	B. R. Seth (1964a,b)	$m = 2 : E_9 = E_2$
$E_{10} = \ln C_r \sim \{\ln \lambda_1, \ln \lambda_2\}$	H. Hencky (1928)	–
$E_{11} = E_l = \frac{1}{2}(C_l - G_l)$	A. Cauchy (1889,1890) ("left Euler–Lagrange")	if $G_l = I$, then $E_l = \frac{1}{2}(C_l - I)$
$E_{12} = E_r = \frac{1}{2}(G_r - C_r)$	E. Almansi (1911) ("right Euler–Lagrange")	if $G_r = I$, then $E_r = \frac{1}{2}(I - C_r)$
$E_{13} = E_l G_l^{-1} = \frac{1}{2}(C_l G_l^{-1} - I)$	E. Grafarend (1995) ("left-right Euler–Lagrange")	if $G_l = I$, then $E_{13} = E_l$
$E_{14} = E_r G_r^{-1} = \frac{1}{2}(I - C_r G_r^{-1})$	E. Grafarend (1995) ("right-left Euler–Lagrange")	if $G_r = I$, then $E_{14} = E_r$
$E_{15} = \frac{1}{2}(C_l^{-1} - G_l^{-1})$	Z. Karni and M. Reiner (1960)	if $G_l = I$ then $E_{15} = \frac{1}{2}(C_l^{-1} - I)$
$E_{16} = \frac{1}{2}(G_r^{-1} - C_r^{-1})$	Z. Karni and M. Reiner (1960)	if $G_r = I$ then $E_{16} = \frac{1}{2}(I - C_r^{-1})$
$E_{17} = G_l E_l^{-1}$	E. Grafarend (1995) ("inverse left-right Euler–Lagrange")	if $G_l = I$, then $E_{17} = E_l^{-1}$
$E_{18} = G_r E_r^{-1}$	E. Grafarend (1995) ("inverse right-left Euler–Lagrange")	if $G_r = I$, then $E_{18} = E_r^{-1}$
$E_{19} = E_l^{m/2} \sim \{K_1^{m/2}, K_2^{m/2}\}$	B. R. Seth (1964a,b) $(m \in Z, m \neq 0)$	$m = 2 : E_{19} = E_{11}$
$E_{20} = \frac{1}{2} \ln E_l$	H. Hencky (1928)	–
$E_{21} = E_r^{m/2} \sim \{\kappa_1^{m/2}, \kappa_2^{m/2}\}$	B. R. Seth (1964a,b)	$m = 2 : E_{21} = E_{12}$
$E_{22} = \frac{1}{2} \ln E_r$	H. Hencky (1928)	–

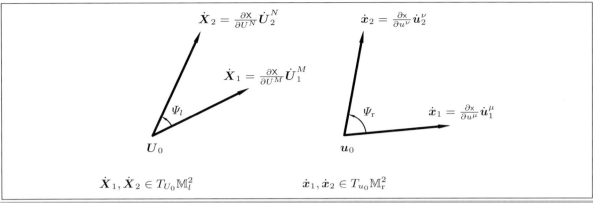

Fig. 1.19. Angular measure of deformation, left and right shear.

Example 1.8 (Angular shear or angular distortion, $\overline{f} : \mathbb{E}^2_{A_1\,A_1\,A_2} \to \mathbb{S}^2_r$).

Let us take reference to Example 1.3, where we analyze the isoparametric mapping $\underline{f} = \mathrm{id}$ from an ellipsoid-of-revolution $\mathbb{M}^2_l = \mathbb{E}^2_{A_1\,A_1\,A_2}$ to a sphere $\mathbb{M}^2_r = \mathbb{S}^2_r$. Here, we shall continue the analysis by computing angular shear or angular distortion of two parameterized curves in $\mathbb{M}^2_l = \mathbb{E}^2_{A_1\,A_1\,A_2}$ as well as their images in $\mathbb{M}^2_r = \mathbb{S}^2_r$.

Left surface, parameterized curves:

(i) parallel circles:

$$U^1 = \Lambda = t_l\,, \; U^2 = \Phi = \text{constant}\,;$$

(ii) meridians:

$$U^1 = \Lambda = \text{constant}\,, \; U^2 = \Phi = t_l\,.$$

Right surface, parameterized curves:

(i) parallel circles:

$$u^1 = \lambda = t_\mathrm{r}\,, \; u^2 = \phi = \text{constant}\,; \qquad (1.135)$$

(ii) meridians:

$$u^1 = \lambda = \text{constant}\,, \; u^2 = \phi = t_\mathrm{r}\,. \qquad (1.136)$$

$$\boldsymbol{U}_1(t_l) = \begin{bmatrix} \Lambda(t_l) \\ \Phi(t_l) \end{bmatrix} = \begin{bmatrix} t_l \\ \text{constant} \end{bmatrix}\,, \qquad \begin{bmatrix} t_\mathrm{r} \\ \text{constant} \end{bmatrix} = \begin{bmatrix} \lambda(t_\mathrm{r}) \\ \phi(t_\mathrm{r}) \end{bmatrix} = \boldsymbol{u}_1(t_\mathrm{r})\,,$$

$$\boldsymbol{U}_2(t_l) = \begin{bmatrix} \Lambda(t_l) \\ \Phi(t_l) \end{bmatrix} = \begin{bmatrix} \text{constant} \\ t_l \end{bmatrix}\,. \qquad \begin{bmatrix} \text{constant} \\ t_\mathrm{r} \end{bmatrix} = \begin{bmatrix} \lambda(t_\mathrm{r}) \\ \phi(t_\mathrm{r}) \end{bmatrix} = \boldsymbol{u}_2(t_\mathrm{r})\,. \qquad (1.137)$$

End of Example.

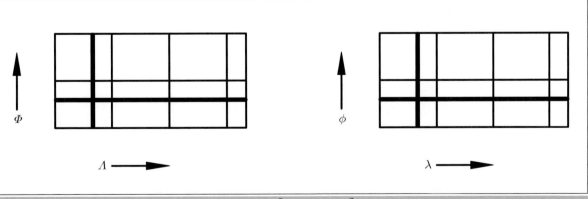

Fig. 1.20. Angular shear, isoparametric mapping $\mathbb{E}^2_{A_1,A_1,A_2} \to \mathbb{S}^2_r$, left and right parameterized curves of type {ellipsoidal parallel circle, ellipsoidal meridian} and {spherical parallel circle, spherical meridian}.

With these parameterized curves in $\mathbb{M}_l^2$ and $\mathbb{M}_r^2$, respectively, we enter Box 1.21. Here, we compute $\boldsymbol{\Phi}_l^{-1}$ and $\boldsymbol{\Phi}_r^{-1}$ in parameterized form, namely $(\Lambda, \Phi) \to \boldsymbol{X}(\Lambda, \Phi) \in \mathbb{R}^3$ and $(\lambda, \phi) \to \boldsymbol{x}(\lambda, \phi) \in \mathbb{R}^3$, respectively. The left and the right displacement field is used to derive the tangent vectors $\{\dot{\boldsymbol{X}}_1, \dot{\boldsymbol{X}}_2\}$ of type "left" and $\{\dot{\boldsymbol{x}}_1, \dot{\boldsymbol{x}}_2\}$ of type "right". The inner products vanish according to our test computations in Example 1.3. In consequence, $\Psi_l = \Psi_r = \pi/2$ and $\Sigma_l = \Sigma_r = 0$, i.e. no angular distortion appears.

Box 1.21 (Angular shear or angular distortion).

Left vector field:

$$\boldsymbol{X}(\Lambda, \Phi) = \boldsymbol{E}_1 \frac{A_1 \cos \Phi \cos \Lambda}{\sqrt{1 - E^2 \sin^2 \Phi}} +$$

$$+\boldsymbol{E}_2 \frac{A_1 \cos \Phi \sin \Lambda}{\sqrt{1 - E^2 \sin^2 \Phi}} + \boldsymbol{E}_3 \frac{(1 - E^2) A_1 \sin \Phi}{\sqrt{1 - E^2 \sin^2 \Phi}} \ .$$

Right vector field:

$$\boldsymbol{x}(\lambda, \phi) = \boldsymbol{e}_1 r \cos \phi \cos \lambda +$$

$$+\boldsymbol{e}_2 r \cos \phi \sin \lambda + \boldsymbol{e}_3 r \sin \phi \ .$$

(1.138)

Left displacement field:

$$\mathrm{d}\boldsymbol{X} = \frac{\partial \boldsymbol{X}}{\partial \Lambda} \frac{\mathrm{d}\Lambda}{\mathrm{d}t_l} \mathrm{d}t_l + \frac{\partial \boldsymbol{X}}{\partial \Phi} \frac{\mathrm{d}\Phi}{\mathrm{d}t_l} \mathrm{d}t_l \ .$$

Right displacement field:

$$\mathrm{d}\boldsymbol{x} = \frac{\partial \boldsymbol{x}}{\partial \lambda} \frac{\mathrm{d}\lambda}{\mathrm{d}t_r} \mathrm{d}t_r + \frac{\partial \boldsymbol{x}}{\partial \phi} \frac{\mathrm{d}\phi}{\mathrm{d}t_r} \mathrm{d}t_r \ .$$

(1.139)

1st left parameterized curve:

$$\dot{\Lambda} = 1 \ , \quad \dot{\Phi} = 0 \ , \quad \frac{\mathrm{d}\boldsymbol{X}}{\mathrm{d}t_l} = \frac{\partial \boldsymbol{X}}{\partial \Lambda} \ .$$

1st right parameterized curve:

$$\dot{\lambda} = 1 \ , \quad \dot{\phi} = 0 \ , \quad \frac{\mathrm{d}\boldsymbol{x}}{\mathrm{d}t_r} = \frac{\partial \boldsymbol{x}}{\partial \lambda} \ .$$

(1.140)

2nd left parameterized curve:

$$\dot{\Lambda} = 0 \ , \quad \dot{\Phi} = 1 \ , \quad \frac{\mathrm{d}\boldsymbol{X}}{\mathrm{d}t_l} = \frac{\partial \boldsymbol{X}}{\partial \Phi} \ .$$

2nd right parameterized curve:

$$\dot{\lambda} = 0 \ , \quad \dot{\phi} = 1 \ , \quad \frac{\mathrm{d}\boldsymbol{x}}{\mathrm{d}t_r} = \frac{\partial \boldsymbol{x}}{\partial \phi} \ .$$

(1.141)

Left angular shear:

$$\langle \dot{\boldsymbol{X}}_1 | \dot{\boldsymbol{X}}_2 \rangle = \left\langle \frac{\partial \boldsymbol{X}}{\partial \Lambda} \Big| \frac{\partial \boldsymbol{X}}{\partial \Phi} \right\rangle = 0 \ ,$$

$$\cos \Psi_l = 0 \Leftrightarrow \Psi_l = \pm \frac{\pi}{2} \ ,$$

$$\Sigma_l = \Psi_l - \Psi_r = 0 \ .$$

Right angular shear:

$$0 = \left\langle \frac{\partial \boldsymbol{x}}{\partial \lambda} \Big| \frac{\partial \boldsymbol{x}}{\partial \phi} \right\rangle = \langle \dot{\boldsymbol{x}}_1 | \dot{\boldsymbol{x}}_2 \rangle \ ,$$

$$\pm \frac{\pi}{2} = \Psi_r \Leftrightarrow \cos \Psi_r = 0 \ ,$$

$$\Sigma_r = \Psi_r - \Psi_l = 0 \ .$$

(1.142)

1-8 Relative angular shear

A third multiplicative measure of deformation: relative angular shear, Cauchy–Green deformation tensor, Euler–Lagrange deformation tensor.

The third multiplicative measure of deformation is the ratio Q_l and Q_r, respectively. This ratio is also called *relative angular shear*. In particular

$$Q_l \cos \Psi_l = \cos \Psi_r , \ Q_l = Q := \frac{\cos \Psi_r}{\cos \Psi_l} \quad \text{versus} \quad Q_r \cos \Psi_r = \cos \Psi_l , \ Q_r = q := \frac{\cos \Psi_l}{\cos \Psi_r} , \quad (1.143)$$

subject to duality $Qq = 1$. Note that additive angular shear and multiplicative angular shear are related by

$$\cos \Sigma_l = \qquad\qquad\qquad\qquad\qquad \cos \Sigma_r =$$

$$\text{versus}$$

$$= Q_l \cos^2 \Psi_l + \sqrt{1 - Q_l^2 \cos^2 \Psi_l} \sin \Psi_l \qquad = Q_r \cos^2 \Psi_r + \sqrt{1 - Q_r^2 \cos^2 \Psi_r} \sin \Psi_r \ .$$

(1.144)

In Box 1.22, we have collected various representations of angular shear, in particular, in terms of the Cauchy–Green deformation and Euler–Lagrange deformation tensors, as well as their eigenvalues.

Box 1.22 (Left and right angular shear).

$$\cos\Psi_l = \frac{\dot{U}_1^{\mathrm{T}}\mathsf{G}_l\dot{U}_2}{\|\dot{U}_1\|_{\mathsf{G}_l}\|\dot{U}_2\|_{\mathsf{G}_l}} = \qquad \cos\Psi_r = \frac{\dot{u}_1^{\mathrm{T}}\mathsf{G}_r\dot{u}_2}{\|\dot{u}_1\|_{\mathsf{G}_r}\|\dot{u}_2\|_{\mathsf{G}_r}} =$$

$$= \frac{\dot{u}_1^{\mathrm{T}}\mathsf{C}_r\dot{u}_2}{\|\dot{u}_1\|_{\mathsf{C}_r}\|\dot{u}_2\|_{\mathsf{C}_r}} , \qquad = \frac{\dot{U}_1^{\mathrm{T}}\mathsf{C}_l\dot{U}_2}{\|\dot{U}_1\|_{\mathsf{C}_l}\|\dot{U}_2\|_{\mathsf{C}_l}} , \tag{1.145}$$

$$Q_l := \frac{\cos\Psi_r}{\cos\Psi_l} = \frac{\dot{U}_1^{\mathrm{T}}\mathsf{C}_l\dot{U}_2}{\dot{U}_1^{\mathrm{T}}\mathsf{G}_l\dot{U}_2} \times \qquad Q_r := \frac{\cos\Psi_l}{\cos\Psi_r} = \frac{\dot{u}_1^{\mathrm{T}}\mathsf{C}_r\dot{u}_2}{\dot{u}_1^{\mathrm{T}}\mathsf{G}_r\dot{u}_2} \times$$

$$\times \frac{\|\dot{U}_1\|_{\mathsf{G}_l}\|\dot{U}_2\|_{\mathsf{G}_l}}{\|\dot{U}_1\|_{\mathsf{C}_l}\|\dot{U}_2\|_{\mathsf{C}_l}} = \qquad \times \frac{\|\dot{u}_1\|_{\mathsf{G}_r}\|\dot{u}_2\|_{\mathsf{G}_r}}{\|\dot{u}_1\|_{\mathsf{C}_r}\|\dot{u}_2\|_{\mathsf{C}_r}} = \tag{1.146}$$

$$= \frac{\dot{U}_1^{\mathrm{T}}\mathsf{C}_l\dot{U}_2}{\dot{U}_1^{\mathrm{T}}\mathsf{G}_l\dot{U}_2}\frac{1}{\Lambda(\dot{U}_1)\Lambda(\dot{U}_2)} , \qquad = \frac{\dot{u}_1^{\mathrm{T}}\mathsf{C}_r\dot{u}_2}{\dot{u}_1^{\mathrm{T}}\mathsf{G}_r\dot{u}_2}\frac{1}{\lambda(\dot{u}_1)\lambda(\dot{u}_2)} ,$$

$$Q_l = \frac{\dot{U}_1^{\mathrm{T}}(2\mathsf{E}_l+\mathsf{G}_l)\dot{U}_2}{\dot{U}_1^{\mathrm{T}}\mathsf{G}_l\dot{U}_2}\frac{\|\dot{U}_1\|_{\mathsf{G}_l}}{\sqrt{\dot{U}_1^{\mathrm{T}}(2\mathsf{E}_l+\mathsf{G}_l)\dot{U}_1}} \times \qquad Q_r = \frac{\dot{u}_1^{\mathrm{T}}(2\mathsf{E}_r+\mathsf{G}_r)\dot{u}_2}{\dot{u}_1^{\mathrm{T}}\mathsf{G}_r\dot{u}_2}\frac{\|\dot{u}_1\|_{\mathsf{G}_r}}{\sqrt{\dot{u}_1^{\mathrm{T}}(2\mathsf{E}_r+\mathsf{G}_r)\dot{u}_1}} \times$$

$$\times \frac{\|\dot{U}_2\|_{\mathsf{G}_l}}{\sqrt{\dot{U}_2^{\mathrm{T}}(2\mathsf{E}_l+\mathsf{G}_l)\dot{U}_2}} , \qquad \times \frac{\|\dot{u}_2\|_{\mathsf{G}_r}}{\sqrt{\dot{u}_2^{\mathrm{T}}(2\mathsf{E}_r+\mathsf{G}_r)\dot{u}_2}} , \tag{1.147}$$

$$Q_l = \frac{1+2(\dot{U}_1^{\mathrm{T}}\mathsf{E}_l\dot{U}_2)/(\dot{U}_1^{\mathrm{T}}\mathsf{G}_l\dot{U}_2)}{\sqrt{1+2(\dot{U}_1^{\mathrm{T}}\mathsf{E}_l\dot{U}_1)/(\dot{U}_1^{\mathrm{T}}\mathsf{G}_l\dot{U}_1)}\sqrt{1+2(\dot{U}_2^{\mathrm{T}}\mathsf{E}_l\dot{U}_2)/(\dot{U}_2^{\mathrm{T}}\mathsf{G}_l\dot{U}_2)}} ,$$

$$Q_r = \frac{1+2(\dot{u}_1^{\mathrm{T}}\mathsf{E}_r\dot{u}_2)/(\dot{u}_1^{\mathrm{T}}\mathsf{G}_r\dot{u}_2)}{\sqrt{1+2(\dot{u}_1^{\mathrm{T}}\mathsf{E}_r\dot{u}_1)/(\dot{u}_1^{\mathrm{T}}\mathsf{G}_r\dot{u}_1)}\sqrt{1+2(\dot{u}_2^{\mathrm{T}}\mathsf{E}_r\dot{u}_2)/(\dot{u}_2^{\mathrm{T}}\mathsf{G}_r\dot{u}_2)}} , \tag{1.148}$$

$$\cos\Psi_l = \frac{\dot{v}_1^{\mathrm{T}}\mathsf{F}_r^{\mathrm{T}}\mathsf{C}_r\mathsf{F}_r\dot{v}_2}{\|\dot{v}_1\|_{\mathsf{F}_r^{\mathrm{T}}\mathsf{C}_r\mathsf{F}_r}\|\dot{v}_2\|_{\mathsf{F}_r^{\mathrm{T}}\mathsf{C}_r\mathsf{F}_r}} = \qquad \cos\Psi_r = \frac{\dot{V}_1^{\mathrm{T}}\mathsf{F}_l^{\mathrm{T}}\mathsf{C}_l\mathsf{F}_l\dot{V}_2}{\|\dot{V}_1\|_{\mathsf{F}_l^{\mathrm{T}}\mathsf{C}_l\mathsf{F}_l}\|\dot{V}_2\|_{\mathsf{F}_l^{\mathrm{T}}\mathsf{C}_l\mathsf{F}_l}} =$$

$$= \frac{\dot{v}_1^{\mathrm{T}}\mathrm{diag}(\lambda_1^2,\lambda_2^2)\dot{v}_2}{\|\dot{v}_1\|_{\mathsf{D}_\lambda}\|\dot{v}_2\|_{\mathsf{D}_\lambda}} , \qquad = \frac{\dot{V}_1^{\mathrm{T}}\mathrm{diag}(\Lambda_1^2,\Lambda_2^2)\dot{V}_2}{\|\dot{V}_1\|_{\mathsf{D}_\Lambda}\|\dot{V}_2\|_{\mathsf{D}_\Lambda}} . \tag{1.149}$$

The following Example 1.9 and the following Box 1.23 illustrate this third multiplicative measure of deformation.

Example 1.9 (Relative angular shear).

Again, we refer to Example 1.3, and to Example 1.8 in addition, where the isoparametric mapping $\underline{f} = \mathrm{id}$ from an ellipsoid-of-revolution $\mathbb{M}_l^2 = \mathbb{E}_{A_1,A_1,A_2}^2$ to a sphere $\mathbb{M}_e^2 = \mathbb{S}_r^2$ with respect to the Cauchy–Green deformation tensor and the absolute angular shear has been analyzed. Here, we aim at relative angular shear. First, by means of Box 1.23, we are going to compute $\cos\Psi_l$ and $\cos\Psi_r$ from the two sets of left and right curves, namely from the left Cauchy–Green tensor and the right Cauchy–Green tensor. Second, we derive relative angular shear: $Q_l = Q_r = 1$.

End of Example.

Box 1.23 (Relative angular shear).

Left Cauchy–Green matrix:

$$\mathsf{C}_l = \begin{bmatrix} r^2 \cos^2 \Phi & 0 \\ 0 & r^2 \end{bmatrix}.$$

Right Cauchy–Green matrix:

$$\mathsf{C}_r = \begin{bmatrix} \dfrac{A_1 \cos^2 \phi}{1 - E^2 \sin^2 \phi} & 0 \\ 0 & \dfrac{A_1^2 \left(1 - E^2\right)^2}{\left(1 - E^2 \sin^2 \phi\right)^3} \end{bmatrix}. \tag{1.150}$$

Left angular shear:

$$\cos \Psi_l = \frac{\dot{\boldsymbol{u}}_1^{\mathrm{T}} \mathsf{C}_r \dot{\boldsymbol{u}}_2}{\|\dot{\boldsymbol{u}}_1\|_{\mathsf{C}_r} \|\dot{\boldsymbol{u}}_2\|_{\mathsf{C}_r}} ,$$

$$\cos \Psi_l \sim [1, 0]\, \mathsf{C}_r \begin{bmatrix} 0 \\ 1 \end{bmatrix} = 0 ,$$

$$\cos \Psi_l = 0 \Leftrightarrow \Psi_l = \pm \frac{\pi}{2} .$$

Right angular shear:

$$\cos \Psi_r = \frac{\dot{\boldsymbol{U}}_1^{\mathrm{T}} \mathsf{C}_l \dot{\boldsymbol{U}}_2}{\|\dot{\boldsymbol{U}}_1\|_{\mathsf{C}_l} \|\dot{\boldsymbol{U}}_2\|_{\mathsf{C}_l}} ,$$

$$\cos \Psi_r \sim [1, 0]\, \mathsf{C}_l \begin{bmatrix} 0 \\ 1 \end{bmatrix} = 0 ,$$

$$\cos \Psi_r = 0 \Leftrightarrow \Psi_r = \pm \frac{\pi}{2} . \tag{1.151}$$

Left relative angular shear:

$$Q_l := \frac{\cos \Psi_r}{\cos \Psi_l} = 1 .$$

Right relative angular shear:

$$Q_r := \frac{\cos \Psi_l}{\cos \Psi_r} = 1 . \tag{1.152}$$

In the following section, we consider the equivalence theorem for conformal mapping. However, in order to give you first some breathing time, please enjoy the Stab–Werner pseudo-conic projection that is presented in Fig. 1.21.

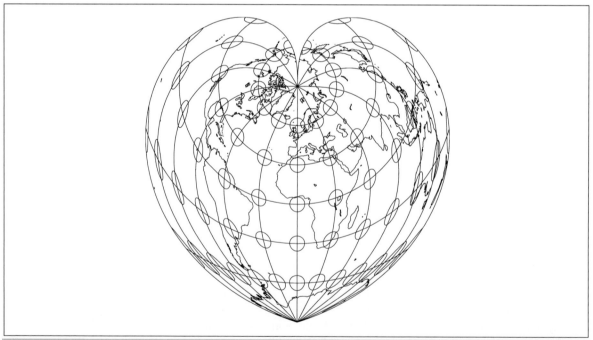

Fig. 1.21. Stab–Werner pseudo-conic projection, with shorelines of a spherical earth, equidistant mapping of the Greenwich meridian, Tissot ellipses of distortion, "cordiform mapping". (Johannes Werner: Libellus de quatuor terrarum orbis in plane figurationibus. Nova translativ primi libri geographiao. El. Ptolemai (Latin), Nenenberg 1514, "designed after instructions by Johann Stabius", first map by Petrus Aqianus, World Map of Ingolstadt).

1-9 Equivalence theorem of conformal mapping

"Experience proves that anyone who studied geometry is infinitely quicker to grasp difficult subjects than one who has not."
(Plato. The Republic Book 7, 375 B. C.)

The equivalence theorem of conformal mapping from the left to the right two-dimensional Riemann manifold (conformeomorphism), generalized Korn–Lichtenstein equations.

We shall define *conformeophism* as well as angular shear, and shall present the *equivalence theorem* that relates conformeomorphism to a special structure of the Cauchy–Green deformation tensor, the Euler–Lagrange deformation tensor, the left and right principal stretches (left and right eigenvalues) as well as dilatations, before we are led to the *generalized Korn–Lichtenstein equations* which govern any conformal mapping: compare with Definition 1.10 and Theorem 1.11. For a further motivation, we refer to Fig. 1.22, which presents an image of Lichtenstein's original publication "Zur Theorie der konformen Abbildung".

Definition 1.10 (Conformal mapping).

An orientation preserving diffeomorphism $\overline{f} : \mathbb{M}_l^2 \to \mathbb{M}_r^2$ is called *angle preserving* conformal mapping (conformeomorphism, inner product preserving) if $\Psi_l = \Psi_r$ and $\Sigma_l = \Sigma = 0 \Leftrightarrow \Sigma_r = \sigma = 0$ for all points of $\mathbb{M}_l^2$ and $\mathbb{M}_r^2$, respectively, holds.

End of Definition.

Theorem 1.11 (Conformeomorphism $\mathbb{M}_l^2 \to \mathbb{M}_r^2$, conformal mapping).

Let $\overline{f} : \mathbb{M}_l^2 \to \mathbb{M}_r^2$ be an orientation preserving conformal mapping. Then the following conditions (i)–(iv) are equivalent:

$$\text{(i)} \quad \Psi_l(\dot{U}_1, \dot{U}_2) = \Psi_r(\dot{u}_1, \dot{u}_2) \,, \tag{1.153}$$

for all tangent vectors $\{\dot{U}_1, \dot{U}_2\}$ and their images $\{\dot{u}_1, \dot{u}_2\}$, respectively;

$$\text{(ii)} \quad \mathsf{C}_l = \Lambda^2(\boldsymbol{U}_0)\mathsf{G}_l \,, \;\; \mathsf{C}_l\mathsf{G}_l^{-1} = \Lambda^2(\boldsymbol{U}_0)\mathsf{I}_2 \;\; \text{versus} \;\; \mathsf{C}_r = \lambda^2(\boldsymbol{u}_0)\mathsf{G}_r \,, \;\; \mathsf{C}_r\mathsf{G}_r^{-1} = \lambda^2(\boldsymbol{u}_0)\mathsf{I}_2 \,,$$
$$\mathsf{E}_l = K(\boldsymbol{U}_0)\mathsf{G}_l \,, \;\; \mathsf{E}_l\mathsf{G}_l^{-1} = K(\boldsymbol{U}_0)\mathsf{I}_2 \;\; \text{versus} \;\; \mathsf{E}_r = \kappa(\boldsymbol{u}_0)\mathsf{G}_r \,, \;\; \mathsf{E}_r\mathsf{G}_r^{-1} = \kappa(\boldsymbol{u}_0)\mathsf{I}_2 \,; \tag{1.154}$$

$$\text{(iii)} \quad K = (\Lambda^2 - 1)/2 \,, \;\; \Lambda^2 = 2K + 1 \;\; \text{versus} \;\; (\lambda^2 - 1)/2 = \kappa \,, \;\; 2\kappa + 1 = \lambda^2 \,,$$
$$\Lambda_1 = \Lambda_2 = \Lambda(\boldsymbol{U}_0) \qquad \text{versus} \qquad \lambda_1 = \lambda_2 = \lambda(\boldsymbol{u}_0) \,,$$
$$K_1 = K_2 = K(\boldsymbol{U}_0) \qquad \text{versus} \qquad \kappa_1 = \kappa_2 = \kappa(\boldsymbol{u}_0) \,,$$
$$\Lambda^2(\boldsymbol{U}_0) = \frac{1}{2}\text{tr}\left[\mathsf{C}_l\mathsf{G}_l^{-1}\right] \qquad \text{versus} \qquad \lambda^2(\boldsymbol{u}_0) = \frac{1}{2}\text{tr}\left[\mathsf{C}_r\mathsf{G}_r^{-1}\right] \,; \tag{1.155}$$

left dilatation: $\qquad\qquad\qquad$ right dilatation:
$$K = \frac{1}{2}\text{tr}\left[\mathsf{E}_l\mathsf{G}_l^{-1}\right] \qquad \text{versus} \qquad \kappa = \frac{1}{2}\text{tr}\left[\mathsf{E}_r\mathsf{G}_r^{-1}\right] \,,$$
$$\text{tr}\left[\mathsf{C}_l\mathsf{G}_l^{-1}\right] = 2\sqrt{\det\left[\mathsf{C}_l\mathsf{G}_l^{-1}\right]} \quad \text{versus} \quad \text{tr}\left[\mathsf{C}_r\mathsf{G}_r^{-1}\right] = 2\sqrt{\det\left[\mathsf{C}_r\mathsf{G}_r^{-1}\right]} \,, \tag{1.156}$$
$$\text{tr}\left[\mathsf{E}_l\mathsf{G}_l^{-1}\right] = 2\sqrt{\det\left[\mathsf{E}_l\mathsf{G}_l^{-1}\right]} \quad \text{versus} \quad \text{tr}\left[\mathsf{E}_r\mathsf{G}_r^{-1}\right] = 2\sqrt{\det\left[\mathsf{E}_r\mathsf{G}_r^{-1}\right]} \,;$$

(iv)   generalized Korn–Lichtenstein equations (special case: $g_{12} = 0$):
$$\begin{bmatrix} u_U \\ u_V \end{bmatrix} = \frac{1}{\sqrt{G_{11}G_{22} - G_{12}^2}} \sqrt{\frac{g_{11}}{g_{22}}} \begin{bmatrix} -G_{12} & G_{11} \\ -G_{22} & G_{12} \end{bmatrix} \begin{bmatrix} v_U \\ v_V \end{bmatrix} \,, \tag{1.157}$$

subject to the integrability conditions $u_{UV} = u_{VU}$ and $v_{UV} = v_{VU}$.

End of Theorem.

Przyczynek do teoryi podobnego odwzorowania. Odwzo-
rowanie podobne nieanalitycznej części powierzchni na
obszar płaski. − Zur Theorie der konformen Abbildung.
Konforme Abbildung nichtanalytischer, singularitätenfreier
Flächenstücke auf ebene Gebiete.

Mémoire

de M. *LÉON LICHTENSTEIN,*

présenté, dans la séance du 6 Mars 1916, par M. K. Żorawski m. c.

Es mögen x und y rechtwinklige Koordinaten eines Punktes in
der Ebene, X, Y, Z ebensolche Koordinaten eines Punktes im Raume,
c ein einfach zusammenhängendes, ganz im Endlichen liegendes Gebiet in der Ebene (x, y) bezeichnen. Es seien ferner $X(x, y)$, $Y(x, y)$,
$Z(x, y)$ in c erklärte reelle, nebst ihren partiellen Ableitungen erster
Ordnung stetige Funktionen. Es wird angenommen, daß die Funk-
tionaldeterminanten

(1) $$\frac{\partial(Y, Z)}{\partial(x, y)}, \qquad \frac{\partial(Z, X)}{\partial(x, y)}, \qquad \frac{\partial(X, Y)}{\partial(x, y)}$$

in c nicht gleichzeitig verschwinden. Durch die Gleichungen

(2) $$X = X(x, y), \qquad Y = Y(x, y), \qquad Z = Z(x, y)$$

wird ein ganz im Endlichen liegendes, singularitätenfreies Flächenstück C definiert.

Es sei $\omega(x, y)$ irgend eine partielle Ableitung erster Ordnung der
Funktionen (2). Wir nehmen zunächst an, daß $\omega(x, y)$ einer Un-
gleichheitsbeziehung

(3) $$|\omega(x + h, y + h') - \omega(x, y)| < a_0(|h| + |h'|)$$

$(a_0$ konstant$)$

genügt.

Fig. 1.22. Lichtenstein, L.: Zur Theorie der konformen Abbildung. Konforme Abbildung nichtanalytischer
singularitätenfreier Flächenstücke auf ebene Gebiete (Bull. Int. Acad. Sci. Cracovie, Chasse des Sciences,
Math. et Natur., Serie A, pp. 192 217, Cracovie 1916). Part one.

Konforme Abbildung   **193**

In einer in dem Anhänge zu den Abhandlungen der Kgl. Preußi-schen Akademie vom Jahre 1911 erschienenen Arbeit[1]) habe ich den folgenden Satz bewiesen: In der Umgebung eines jeden Punktes im Innern des Gebietes c läßt sich ein Flächenstück abgrenzen, das zusammenhängend und in den kleinsten Teilen ähnlich auf ein ebenes Flächenstück abgebildet werden kann.

Es sei

$$(4) \qquad ds^2 = E dx^2 + 2F dx dy + G dy^2$$

das Quadrat des Linienelementes der Fläche. Wir setzen

$$(5) \qquad a = \frac{G}{\sqrt{EG - F^2}}, \quad b = \frac{E}{\sqrt{EG - F^2}}, \quad d = \frac{F}{\sqrt{EG - F^2}},$$
$$ab - d^2 = 1.$$

Die Funktionen E, F, G, a, b, d sind stetig; ihre Differenzenquo-tienten sind nach (3) beschränkt[2]).

In der soeben zitierten Abhandlung wird eine in einem vor-geschriebenen Punkte $(\overline{x}, \overline{y})$ wie

$$(6) \quad \tfrac{1}{2}\log\left\{ b(\overline{x}, \overline{y})(x - \overline{x})^2 - 2d(\overline{x}, \overline{y})(x - \overline{x})(y - \overline{y}) + a(\overline{x}, \overline{y})(y - \overline{y})^2 \right\}$$

unstetige, den partiellen Differentialgleichungen

$$(7) \qquad \begin{cases} \dfrac{\partial v}{\partial x} = -d\dfrac{\partial u}{\partial x} - b\dfrac{\partial u}{\partial y}. \\[2mm] \dfrac{\partial v}{\partial y} = a\dfrac{\partial u}{\partial x} + d\dfrac{\partial u}{\partial y} \end{cases}$$

genügende Funktion $u(x, y)$ bestimmt. Durch Vermittlung der Funk-tion $u(x, y) + iv(x, y)$ wird ein zusammenhängender Teil des Flä-

[1]) L. Lichtenstein, Beweis des Satzes, daß jedes hinreichend kleine, im wesentlichen stetig gekrümmte, singularitätenfreie Flächenstück auf einen Teil einer Ebene zusammenhängend und in den kleinsten Teilen ähnlich abgebildet wer-den kann.

[2]) Nach bekannten Sätzen haben E, F, G, a, b, d in c, außer höchstens in einer Menge von Punkten vom Maße Null, beschränkte partielle Ableitungen erster Ordnung.

Fig. 1.23. Lichtenstein, L.: Zur Theorie der konformen Abbildung. Konforme Abbildung nichtanalytischer singularitätenfreier Flächenstücke auf ebene Gebiete (Bull. Int. Acad. Sci. Cracovie, Chasse des Sciences, Math. et Natur., Serie A, pp. 192–217, Cracovie 1916). Part two.

Before we present the sketches of proofs for the various conditions, it has to be noted that the generalized Korn–Lichtenstein equations, which govern conformal mapping $\mathbb{M}_l^2 \to \mathbb{M}_r^2$, suffer from the defect that they contain the unknown functions $g_{11}[u^\lambda(U^A)]$ and $g_{22}[u^\lambda(U^A)]$, and the reason is that the mapping functions $u^\lambda(U^A)$ have to be determined. In case of $\{\mathbb{M}_r^2, g_{\mu\nu}\} = \{\mathbb{R}^2, \delta_{\mu\nu}\}$, the corresponding Korn–Lichtenstein equations do not suffer since these functions do not appear. The stated problem is overcome by representing the right Riemann manifold $\mathbb{M}_r^2$ by *isometric coordinates* (also called *conformal coordinates* or *isothermal coordinates*) directly such that the quotient g_{22}/g_{11} is identical to one. This is exactly the procedure advocated by C. F. Gauss (1822, 1844) and applied to the conformal mapping of $\mathbb{E}^2$ onto $\mathbb{S}_r^2$. We shall come back to this point-of-view after the proof.

Proof (first part).

$$(i) \Rightarrow (ii).$$

$$\Psi_l = \Psi_r \to \cos \Psi_l = \cos \Psi_r \Leftrightarrow {\boldsymbol{U}'_1}^{\mathrm{T}} \mathsf{G}_l \boldsymbol{U}'_2 = {\boldsymbol{u}'_1}^{\mathrm{T}} \mathsf{G}_r \boldsymbol{u}'_2 \Leftrightarrow$$

$$\Leftrightarrow d\boldsymbol{u}_1^{\mathrm{T}} \mathsf{J}_r^{\mathrm{T}} \mathsf{G}_l \mathsf{J}_r d\boldsymbol{u}_2 = \frac{dS_1}{ds_1} d\boldsymbol{u}_1^{\mathrm{T}} \mathsf{G}_r d\boldsymbol{u}_2 \frac{dS_2}{ds_2} \Leftrightarrow d\boldsymbol{u}_1^{\mathrm{T}} \mathsf{C}_r d\boldsymbol{u}_2 = \lambda_1 d\boldsymbol{u}_1^{\mathrm{T}} \mathsf{G}_r d\boldsymbol{u}_2 \lambda_2 \Leftrightarrow \qquad (1.158)$$

$$\Leftrightarrow \lambda_1 = \lambda_2 = \lambda(\boldsymbol{u}_0) \, , \quad \mathsf{C}_r = \lambda^2(\boldsymbol{u}_0) \mathsf{G}_r \quad \text{q. e. d.}$$

$$\cos \Psi_r = \cos \Psi_l \Leftrightarrow {\boldsymbol{u}'_1}^{\mathrm{T}} \mathsf{G}_r \boldsymbol{u}'_2 = {\boldsymbol{U}'_1}^{\mathrm{T}} \mathsf{G}_l \boldsymbol{U}'_2 \Leftrightarrow d\boldsymbol{U}_1^{\mathrm{T}} \mathsf{J}_l^{\mathrm{T}} \mathsf{G}_r \mathsf{J}_l d\boldsymbol{U}_2 = \frac{ds_1}{dS_1} d\boldsymbol{U}_1^{\mathrm{T}} \mathsf{G}_l d\boldsymbol{U}_2 \frac{ds_2}{dS_2} \Leftrightarrow$$

$$\Leftrightarrow \Lambda_1 = \Lambda_2 = \Lambda(\boldsymbol{U}_0) \, , \quad \mathsf{C}_l = \Lambda^2(\boldsymbol{U}_0) \mathsf{G}_l \quad \text{q. e. d.} \qquad (1.159)$$

$$(i) \Leftarrow (ii).$$

$$\begin{bmatrix} \cos \Psi_l = {\boldsymbol{U}'_1}^{\mathrm{T}} \mathsf{G}_l \boldsymbol{U}'_2 = \frac{ds_1}{dS_1} \boldsymbol{u}_1^{\mathrm{T}} \mathsf{J}_r^{\mathrm{T}} \mathsf{G}_l \mathsf{J}_r d\boldsymbol{u}_2 \frac{ds_2}{dS_2} \\ \mathsf{J}_r^{\mathrm{T}} \mathsf{G}_l \mathsf{J}_r = \mathsf{C}_r = \lambda^2(\boldsymbol{u}_0) \mathsf{G}_r \, , \quad \lambda_1^{-1} = \lambda_2^{-1} = \lambda^{-1} \end{bmatrix} \Rightarrow \begin{bmatrix} \cos \Psi_l = {\boldsymbol{u}'_1}^{\mathrm{T}} \mathsf{G}_r \boldsymbol{u}'_2 = \cos \Psi_r \\ \text{orientation is preserved} \end{bmatrix} \Leftrightarrow \qquad (1.160)$$

$$\Leftrightarrow \Psi_l = \Psi_r \quad \text{q. e. d.}$$

End of Proof (first part).

Proof (second part).

$$(ii) \Rightarrow (iii).$$

Left eigenvalue problem:

$$\mathsf{C}_l = \Lambda^2(\boldsymbol{U}_0) \mathsf{G}_l \, , \quad \mathsf{E}_l = K(\boldsymbol{U}_0) \mathsf{G}_l \Leftrightarrow \begin{bmatrix} \Lambda^2(\boldsymbol{U}_0) = \Lambda_1^2 = \Lambda_2^2 \\ K(\boldsymbol{U}_0) = K_1^2 = K_2^2 \end{bmatrix} . \qquad (1.161)$$

Right eigenvalue problem:

$$\mathsf{C}_r = \lambda^2(\boldsymbol{u}_0) \mathsf{G}_r \, , \quad \mathsf{E}_r = \kappa(\boldsymbol{u}_0) \mathsf{G}_r \Leftrightarrow \begin{bmatrix} \lambda^2(\boldsymbol{u}_0) = \lambda_1^2 = \lambda_2^2 \\ \kappa(\boldsymbol{u}_0) = \kappa_1^2 = \kappa_2^2 \end{bmatrix} . \qquad (1.162)$$

$$(ii) \Leftarrow (iii).$$

$$\Lambda_1^2 = \Lambda_2^2 = \Lambda^2(\boldsymbol{U}_0), \quad \mathsf{F}_l^{\mathrm{T}^{-1}} \mathrm{diag}[\Lambda_1^2, \Lambda_2^2] \mathsf{F}_l^{-1} = \mathsf{C}_l, \quad \mathsf{F}_l^{\mathrm{T}^{-1}} \mathsf{F}_l^{-1} = \mathsf{G}_l \Rightarrow \mathsf{C}_l = \Lambda^2(\boldsymbol{U}_0) \mathsf{G}_l,$$

$$\lambda_1^2 = \lambda_2^2 = \lambda^2(\boldsymbol{u}_0), \quad \mathsf{F}_r^{\mathrm{T}^{-1}} \mathrm{diag}[\lambda_1^2, \lambda_2^2] \mathsf{F}_r^{-1} = \mathsf{C}_r, \quad \mathsf{F}_r^{\mathrm{T}^{-1}} \mathsf{F}_r^{-1} = \mathsf{G}_r \Rightarrow \mathsf{C}_r = \lambda^2(\boldsymbol{u}_0) \mathsf{G}_r. \qquad (1.163)$$

The statements for the quantities E_l, E_r, $\mathsf{E}_l \mathsf{G}_l^{-1}$, $\mathsf{E}_r \mathsf{G}_r^{-1}$, K, κ, Λ, and λ follow in the same way.

End of Proof (second part).

Proof (third part).

$$(ii) \Rightarrow (iv).$$

In order to derive a linear system of partial differential equations for $\{u_U, u_V, v_U, v_V\}$, we depart from the inverse right Cauchy–Green deformation tensor C_r since it contains just the above quoted partials. For an *inverse portrait* involving the partials $\{U_u, U_v, V_u, V_v\}$, in previous sections, we start from the inverse left Cauchy–Green deformation tensor, a procedure we are not following further.

1st step:

$$
C_r^{-1} = J_l G_l^{-1} J_l^T = \frac{G_r^{-1}}{\lambda^2} \Leftrightarrow
\begin{bmatrix}
\begin{bmatrix} u_U & u_V \\ v_U & v_V \end{bmatrix} G_l^{-1} \begin{bmatrix} u_U & v_U \\ u_V & v_V \end{bmatrix} = \frac{G_r^{-1}}{\lambda^2} \\[2mm]
\boldsymbol{x}_1 := \begin{bmatrix} u_U \\ u_V \end{bmatrix}, \quad \boldsymbol{x}_2 := \begin{bmatrix} v_U \\ v_V \end{bmatrix}
\end{bmatrix}
\Rightarrow
\begin{bmatrix}
(\alpha) \; \boldsymbol{x}_1^T G_l^{-1} \boldsymbol{x}_1 = +\dfrac{g_{22}}{\lambda^2} \\[2mm]
(\beta) \; \boldsymbol{x}_2^T G_l^{-1} \boldsymbol{x}_2 = +\dfrac{g_{11}}{\lambda^2} \\[2mm]
(\gamma) \; \boldsymbol{x}_1^T G_l^{-1} \boldsymbol{x}_2 = -\dfrac{g_{12}}{\lambda^2} \\[2mm]
(\delta) \; \boldsymbol{x}_2^T G_l^{-1} \boldsymbol{x}_1 = -\dfrac{g_{12}}{\lambda^2}
\end{bmatrix}. \quad (1.164)
$$

Without loss of generality – see through the remark that follows after the proof – let us here assume that the right two-dimensional Riemann manifold (i. e. the right parameterized surface) $\mathbb{M}_r^2$ is charted by *orthogonal parameters* (*orthogonal coordinates*) such that $g_{12} = 0$ holds. Such a parameterization of a surface can always be achieved though it might turn out to be a difficult numerical procedure.

2nd step:

$$
\begin{bmatrix}
\boldsymbol{x}_2^T G_l^{-1} \boldsymbol{x}_1 = 0 \quad (\delta) \\[2mm]
\text{"Ansatz" } \boldsymbol{x}_1 = G_l X \boldsymbol{x}_2 \; (X = \text{unknown matrix})
\end{bmatrix}
\Leftrightarrow \boldsymbol{x}_2^T G_l^{-1} \boldsymbol{x}_2 = 0 \; \forall \; \boldsymbol{x}_2 \in \mathbb{R}^{2\times 1} \quad (\varepsilon). \quad (1.165)
$$

A quadratic form over the field of real numbers can only be zero ("isotropic") if and only if X is antisymmetric, i. e. $X = -X^T$ (for a proof, we refer to A. Crumeyrolle (1990), Proposition 1.1.3):

$$
\text{"Ansatz" } X = Ax \; \forall \; A = -A^T, \quad A := \begin{bmatrix} 0 & 1 \\ -1 & 0 \end{bmatrix} \in \mathbb{R}^{2\times 2}, \quad x \in \mathbb{R}. \quad (1.166)
$$

3rd step:

$$
\begin{bmatrix}
\dfrac{1}{g_{22}} \boldsymbol{x}_1^T G_l^{-1} \boldsymbol{x}_1 = \dfrac{1}{g_{11}} \boldsymbol{x}_2^T G_l^{-1} \boldsymbol{x}_2 = \dfrac{1}{\lambda^2} \\[2mm]
\boldsymbol{x}_1 = G_l A x \boldsymbol{x}_2
\end{bmatrix} \Rightarrow
$$

$$
\Rightarrow \frac{1}{g_{22}} \boldsymbol{x}_1^T G_l^{-1} \boldsymbol{x}_1 = \frac{1}{g_{22}} \boldsymbol{x}_2^T A^T G_l A \boldsymbol{x}_2 x = \frac{1}{g_{11}} \boldsymbol{x}_2^T G_l^{-1} \boldsymbol{x}_2 \Leftrightarrow
$$

$$
\Leftrightarrow \frac{g_{11}}{g_{22}} A^T G_l A G_l x = I \Leftrightarrow \tag{1.167}
$$

$$
\Leftrightarrow \left[x = \frac{1}{\sqrt{G_{11} G_{22} - G_{12}^2}} \sqrt{\frac{g_{22}}{g_{11}}} \right] \Rightarrow
$$

$$
\boldsymbol{x}_1 = G_l A x \boldsymbol{x}_2
$$

$$
\Rightarrow \boldsymbol{x}_1 = G_l A \frac{1}{\sqrt{G_{11} G_{22} - G_{12}^2}} \sqrt{\frac{g_{22}}{g_{11}}} \boldsymbol{x}_2.
$$

The converse (iv) $\Rightarrow$ (ii) is obvious.

End of Proof (third part).

Here is the remark relating to G_l being diagonal, not unity, of course. An obvious generalization for solving (γ) and (δ) for $g_{12} \neq 0$ would be the

$$\text{``Ansatz''} \quad \boldsymbol{x}_1 = \mathsf{G}_l \mathsf{X} \boldsymbol{x}_2 \,, \quad \mathsf{X} = \begin{bmatrix} y & x \\ -x & -y \end{bmatrix} \,, \tag{1.168}$$

the superposition of a diagonal trace-free matrix $\mathrm{diag}\,[y, -y]$ and an antisymmetric matrix $\mathsf{A}x$. Indeed, we succeed in determining the unknowns x and y according to the above steps, but fail to arrive at linear relations between the partials $\{u_U, u_V, v_U, v_V\}$.

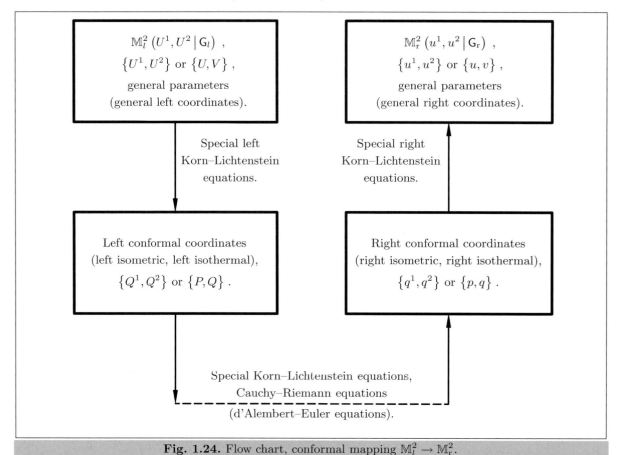

Fig. 1.24. Flow chart, conformal mapping $\mathbb{M}_l^2 \to \mathbb{M}_r^2$.

In practice, a different way in constructing a conformeomorphism $\mathbb{M}_l^2 \to \mathbb{M}_r^2$ has been chosen. In Fig. 1.24, the alternative path of generating a conformal mapping from a left curved surface to a right curved surface is outlined. First, the original coordinates $\{U^1, U^2\}$ or $\{U, V\}$, which parameterize the left surface, are transformed to alternative left *conformal coordinates* $\{P, Q\}$, which are also called *isometric* or *isothermal*. Indeed, the left differential invariant $I_l \sim \mathrm{d}S^2 = \Lambda^2(\mathrm{d}P^2 + \mathrm{d}Q^2)$ is described by identical metric coefficients $G_{PP} = G_{QQ} = \Lambda^2$ and $G_{PQ} = 0$. Second, the original coordinates $\{u^1, u^2\}$ or $\{u, v\}$, which parameterize the right surface, are transformed to alternative right conformal coordinates $\{p, q\}$, which are also called *isometric* or *isothermal*. Indeed, the right differential invariant $I_r \sim \mathrm{d}s^2 = \lambda^2(\mathrm{d}p^2 + \mathrm{d}q^2)$ is described by identical metric coefficients $g_{pp} = g_{qq} = \lambda^2$ and $g_{pq} = 0$. Third, the left conformal coordinates $\{P, Q\}$ are transformed to right conformal coordinates by solving the special Korn–Lichtenstein equations for $\mathbb{M}_l^2 \{P, Q \mid \mathsf{G}_l = \Lambda^2 \mathsf{I}_2\} \to \mathbb{M}_r^2 \{p, q \mid \mathsf{G}_r = \lambda^2 \mathsf{I}_2\}$, which are called *Cauchy–Riemann (d'Alembert–Euler) equations*, subject to an integrability condition. The integrability condition turns out to be the *vector-valued Laplace equation of harmonicity*, as stated in the following theorem and proven later on.

Theorem 1.12 (Conformeomorphism $\mathbb{M}_l^2 \to \mathbb{M}_r^2$, conformal mapping).

An orientation preserving conformal mapping $\mathbb{M}_l^2 \to \mathbb{M}_r^2$ can be constructed by three steps in solving special Korn–Lichtenstein equations.

$$1\text{st step or left step.}$$

The left Riemann manifold $\mathbb{M}_l^2\left(U^1, U^2 \mid \mathsf{G}_l\right)$, which is called *left surface*, is parameterized by *general left parameters* (general left coordinates) $\{U^1, U^2\}$ or $\{U, V\}$. The solution of the following special Korn–Lichtenstein equations (i), subject to the following integrability conditions of harmonicity (ii) and orientation conservation (iii), is needed.

(i) Special KL:

$$
\begin{bmatrix} P_U \\ P_V \end{bmatrix} = \frac{1}{\sqrt{G_{11}G_{22}-G_{12}^2}} \begin{bmatrix} -G_{12} & G_{11} \\ -G_{22} & G_{12} \end{bmatrix} \begin{bmatrix} Q_U \\ Q_V \end{bmatrix}, \quad \begin{bmatrix} P_U = \frac{1}{\sqrt{G_{11}G_{22}-G_{12}^2}}\left(-G_{12}Q_U + G_{11}Q_V\right) \\ P_V = \frac{1}{\sqrt{G_{11}G_{22}-G_{12}^2}}\left(-G_{22}Q_U + G_{12}Q_V\right) \end{bmatrix}. \tag{1.169}
$$

(ii) Left integrability:

$$P_{UV} = P_{VU} \quad \text{and} \quad Q_{UV} = Q_{VU} \tag{1.170}$$

or (in terms of the Laplace–Beltrami operator)

$$
\begin{bmatrix} \Delta_{UV}P := \left(\frac{G_{11}P_V - G_{12}P_U}{\sqrt{G_{11}G_{22}-G_{12}^2}}\right)_V + \left(\frac{G_{22}P_U - G_{12}P_V}{\sqrt{G_{11}G_{22}-G_{12}^2}}\right)_U = 0 \\ \Delta_{UV}Q := \left(\frac{G_{11}Q_V - G_{12}Q_U}{\sqrt{G_{11}G_{22}-G_{12}^2}}\right)_V + \left(\frac{G_{22}Q_U - G_{12}Q_V}{\sqrt{G_{11}G_{22}-G_{12}^2}}\right)_U = 0 \end{bmatrix}. \tag{1.171}
$$

(iii) Left orientation conservation:

$$\begin{vmatrix} P_U & P_V \\ Q_U & Q_V \end{vmatrix} = P_U Q_V - P_V Q_U > 0. \tag{1.172}$$

Note that the coordinates P and Q are the left conformal coordinates, which are also called *isometric* or *isothermal*.

$$2\text{nd step or right step.}$$

The right Riemann manifold $\mathbb{M}_r^2\left(u^1, u^2 \mid \mathsf{G}_r\right)$, which is called *right surface*, is parameterized by *general right parameters* (general right coordinates) $\{u^1, u^2\}$ or $\{u, v\}$. The solution of the following special Korn–Lichtenstein equations (i), subject to the following integrability conditions of harmonicity (ii) and orientation conservation (iii), is needed.

(i) Special KL:

$$
\begin{bmatrix} p_u \\ p_v \end{bmatrix} = \frac{1}{\sqrt{g_{11}g_{22}-g_{12}^2}} \begin{bmatrix} -g_{12} & g_{11} \\ -g_{22} & g_{12} \end{bmatrix} \begin{bmatrix} q_u \\ q_v \end{bmatrix}, \quad \begin{bmatrix} p_u = \frac{1}{\sqrt{g_{11}g_{22}-g_{12}^2}}\left(-g_{12}q_u + g_{11}q_v\right) \\ p_v = \frac{1}{\sqrt{g_{11}g_{22}-g_{12}^2}}\left(-g_{22}q_u + g_{12}q_v\right) \end{bmatrix}. \tag{1.173}
$$

(ii) Right integrability:

$$p_{uv} = p_{vu} \quad \text{and} \quad q_{uv} = q_{vu} \tag{1.174}$$

or (in terms of the Laplace–Beltrami operator)

$$\left[\begin{array}{l} \Delta_{uv} p := \left(\dfrac{g_{11} p_v - g_{12} p_u}{\sqrt{g_{11} g_{22} - g_{12}^2}} \right)_v + \left(\dfrac{g_{22} p_u - g_{12} p_v}{\sqrt{g_{11} g_{22} - g_{12}^2}} \right)_u = 0 \\[1.2em] \Delta_{uv} q := \left(\dfrac{g_{11} q_v - g_{12} q_u}{\sqrt{g_{11} g_{22} - g_{12}^2}} \right)_v + \left(\dfrac{g_{22} q_u - g_{12} q_v}{\sqrt{g_{11} g_{22} - g_{12}^2}} \right)_u = 0 \end{array} \right] . \tag{1.175}$$

(iii) Right orientation conservation:

$$\begin{vmatrix} p_u & p_v \\ q_u & q_v \end{vmatrix} = p_u q_v - p_v q_u > 0 . \tag{1.176}$$

3rd step (left–right).

The left Riemann manifold $\mathbb{M}_l^2 \left(P, Q \mid \Lambda^2 I_2 \right)$ which here is called *left surface* and is parameterized in left conformal coordinates $\{P, Q\}$, is orientation preserving conformally mapped onto the right Riemann manifold $\mathbb{M}_r^2 \left(p, q \mid \lambda^2 I_2 \right)$, which here is called *right surface* and is parameterized in right conformal coordinates $\{p, q\}$, if the following special Korn–Lichtenstein equations (i) (called Cauchy–Riemann (or d'Alembert–Euler) equations) subject to the following integrability conditions of harmonicity (ii) and orientation conservation (iii) are solved.

(i) Special KL (Cauchy–Riemann, d'Alembert–Euler):

$$\begin{bmatrix} p_P \\ p_Q \end{bmatrix} = \frac{1}{\sqrt{g_{11} g_{22} - g_{12}^2}} \begin{bmatrix} 0 & 1 \\ -1 & 0 \end{bmatrix} \begin{bmatrix} q_P \\ q_Q \end{bmatrix} , \quad p_P = q_Q \quad \text{and} \quad p_Q = -q_P . \tag{1.177}$$

(ii) Right integrability:

$$p_{PQ} = p_{QP} \quad \text{and} \quad q_{PQ} = q_{QP} \tag{1.178}$$

or (in terms of the Laplace–Beltrami operator)

$$\left[\begin{array}{l} \Delta_{PQ} p := p_{PP} + p_{QQ} = \left(\dfrac{\partial^2}{\partial P^2} + \dfrac{\partial^2}{\partial Q^2} \right) p(P, Q) = 0 \\[1em] \Delta_{PQ} q := q_{PP} + q_{QQ} = \left(\dfrac{\partial^2}{\partial P^2} + \dfrac{\partial^2}{\partial Q^2} \right) q(P, Q) = 0 \end{array} \right] . \tag{1.179}$$

(iii) Left–right orientation conservation:

$$\begin{vmatrix} p_P & p_Q \\ q_P & q_Q \end{vmatrix} = p_P q_Q - p_Q q_P = 0 . \tag{1.180}$$

The special Korn–Lichtenstein equations, which govern as Cauchy–Riemann (or d'Alembert–Euler) equations any harmonic, orientation preserving conformal mapping $\mathbb{M}_l^2(P, Q) \to \mathbb{M}_r^2(p, q)$, are uniquely solvable if a proper boundary value problem is formulated.

End of Theorem.

The proof of the operational theorem of conformal mapping $\mathbb{M}_l^2 \to \mathbb{M}_r^2$ rests upon the existence theorem of S. S. Cherne (1955), where it is shown that under rather mild certainty assumptions, namely $C^2{}^x$, conformal coordinates (isometric coordinates, isothermal coordinates) exist as solutions of the left or right Korn–Lichtenstein equations. Let us here also refer to the following authors. W. Blaschke and K. Leichtweiß (1973), D. G. L. Boulware, L. S. Brown and R. D. Peccei (1970), J. P. Bourguignon (1970), B. Y. Chen (1973), B. Y. Chen and K. Yano (1973), S. S. Chern (1967), S. S. Chern, P. Hartman and A. Wintner (1954), M. Do Carmo, M. Dajczer and F. Mercuri (1985), L. P. Eisenhart (1949), L. Euler (1755, 1777a), S. Ferrara, A. F. Grillo and R. Gatto (1972), A. Finzi (1922), C. F. Gauss (1822, 1844), H. Goenner, E. Grafarend and R. J. You (1994), C. G. J. Jacobi (1839), S. Heitz (1988), W. Klingenberg (1982), K. König and K. H. Weise (1951), A. Korn (1914), L. Krueger (1903, 1922), N. Kuiper (1949, 1950), R. S. Kulkarni (1969, 1972), R. S. Kulkarni and U. Pinkall (1988), J. Lafontaine (1988), J. L. Lagrange (1781), G. M. Lancaster (1969, 1973), L. Lichtenstein (1911, 1916), J. Liouville (1850), A. I. Markuschewitsch (1955), L. Mirsky (1960), C. W. Misner (1978), S. K. Mitra and C. R. Rao (1968), B. Moor and H. Zha (1991), J. D. Moore (1977), S. Nishikawa (1974), G. Ricci (1918), B. Riemann (1851), H. Samelson (1969), E. Schering (1857), H. Schmehl (1927), R. Schoen (1984), J. A. Schouten (1921), M. Spivak (1979), E. M. Stein and G. Weiss (1968), H. Weber (1867), H. Weyl (1918, 1921), T. Wray (1974), K. Yano (1970), A. I. Yanushauskas (1982), M. Zadro and A. Carminelli (1966), and J. Zund (1987).

Proof (sketch of the proof for the first step).

The special KL equations generate a conformal mapping, $\mathbb{M}_l\,(U, V \mid \mathsf{G}_l) \to \mathbb{M}_l\,(P, Q \mid \Lambda^2 \mathsf{l}_2)$, namely a conformal coordinate transformation from general left coordinates $\{U, V\}$ to left conformal coordinates $\{P, Q\}$. The left matrix of the metric, i.e. the matrix G_l, is transformed to the left matrix of the *conformally flat metric*, $\Lambda^2 \mathsf{l}_2$. Up to the *factor of conformality*, $\Lambda^2(P, Q)$, the transformed matrix of the metric is a unit matrix, l_2. Here, we only outline how the *integrability conditions* $P_{UV} = P_{VU}$ and $Q_{UV} = Q_{VU}$ are converted to the Laplace–Beltrami equation.

KL, 1st equation and 2nd equation, lead to

$$P_{UV} = \left(\frac{-G_{12}Q_U + G_{11}Q_V}{\sqrt{G_{11}G_{22} - G_{12}^2}} \right)_V \quad , \quad P_{VU} = \left(\frac{-G_{22}Q_U + G_{12}Q_V}{\sqrt{G_{11}G_{22} - G_{12}^2}} \right)_U \quad . \tag{1.181}$$

$$\text{1st: } P_{UV} = P_{VU} \Leftrightarrow \left(\frac{G_{11}Q_V - G_{12}Q_U}{\sqrt{G_{11}G_{22} - G_{12}^2}} \right)_V = -\left(\frac{G_{22}Q_U - G_{12}Q_V}{\sqrt{G_{11}G_{22} - G_{12}^2}} \right)_U \Rightarrow$$

$$\Rightarrow \left(\frac{G_{11}Q_V - G_{12}Q_U}{\sqrt{G_{11}G_{22} - G_{12}^2}} \right)_V + \left(\frac{G_{22}Q_U - G_{12}Q_V}{\sqrt{G_{11}G_{22} - G_{12}^2}} \right)_U = 0 \quad . \tag{1.182}$$

The KL matrix equation is inverted to

$$\begin{bmatrix} Q_U \\ Q_V \end{bmatrix} = \frac{1}{\sqrt{G_{11}G_{22} - G_{12}^2}} \begin{bmatrix} G_{12} & -G_{11} \\ G_{22} & -G_{12} \end{bmatrix} \begin{bmatrix} P_U \\ P_V \end{bmatrix}, \ Q_U = \frac{G_{12}P_U - G_{11}P_V}{\sqrt{G_{11}G_{22} - G_{12}^2}}, \ Q_V = \frac{G_{22}P_U - G_{12}P_V}{\sqrt{G_{11}G_{22} - G_{12}^2}} . \tag{1.183}$$

The inverted KL equations lead to

$$Q_{UV} = -\left(\frac{G_{11}P_V - G_{12}P_U}{\sqrt{G_{11}G_{22} - G_{12}^2}} \right)_V \quad , \quad Q_{VU} = \left(\frac{G_{22}P_U - G_{12}P_V}{\sqrt{G_{11}G_{22} - G_{12}^2}} \right)_U \quad . \tag{1.184}$$

$$\text{2nd: } Q_{UV} = Q_{VU} \Leftrightarrow -\left(\frac{G_{11}P_V - G_{12}P_U}{\sqrt{G_{11}G_{22} - G_{12}^2}} \right)_V = \left(\frac{G_{22}P_U - G_{12}P_V}{\sqrt{G_{11}G_{22} - G_{12}^2}} \right)_U \Rightarrow$$

$$\Rightarrow \left(\frac{G_{11}P_V - G_{12}P_U}{\sqrt{G_{11}G_{22} - G_{12}^2}} \right)_V + \left(\frac{G_{22}P_U - G_{12}P_V}{\sqrt{G_{11}G_{22} - G_{12}^2}} \right)_U = 0 \quad . \tag{1.185}$$

End of Proof (the first step)

Proof (sketch of the proof for the second step).

The special KL equations generate the conformal mapping $\mathbb{M}_r\left(u, v \mid \mathsf{G}_r\right) \to \mathbb{M}_r\left(p, q \mid \lambda^2 \mathsf{I}_2\right)$, a conformal coordinate transformation from general right coordinates $\{u, v\}$ to right conformal coordinates $\{p, q\}$. The right matrix of the metric, G_r, is transformed to the right matrix of the *conformally flat metric*, $\lambda^2 \mathsf{I}_2$. Up to the *factor of conformality*, $\lambda^2(p, q)$, the transformed matrix of the metric is a unit matrix, I_2. Here, we only outline how the *integrability conditions* $p_{uv} = p_{vu}$ and $q_{uv} = q_{vu}$ are converted to the Laplace–Beltrami equation.

KL, 1st equation and 2nd equation, lead to

$$p_{uv} = \left(\frac{-g_{12}q_u + g_{11}q_v}{\sqrt{g_{11}g_{22} - g_{12}^2}}\right)_v \quad , \quad p_{vu} = \left(\frac{-g_{22}q_u + g_{12}q_v}{\sqrt{g_{11}g_{22} - g_{12}^2}}\right)_u . \tag{1.186}$$

$$\text{1st: } p_{uv} = p_{vu} \Leftrightarrow \left(\frac{g_{11}q_v - g_{12}q_u}{\sqrt{g_{11}g_{22} - g_{12}^2}}\right)_v = -\left(\frac{g_{22}q_u - g_{12}q_v}{\sqrt{g_{11}g_{22} - g_{12}^2}}\right)_u \Rightarrow$$

$$\Rightarrow \left(\frac{g_{11}q_v - g_{12}q_u}{\sqrt{g_{11}g_{22} - g_{12}^2}}\right)_v + \left(\frac{g_{22}q_u - g_{12}q_v}{\sqrt{g_{11}g_{22} - g_{12}^2}}\right)_u = 0 . \tag{1.187}$$

The KL matrix equation is inverted to

$$\begin{bmatrix} q_u \\ q_v \end{bmatrix} = \frac{1}{\sqrt{g_{11}g_{22} - g_{12}^2}} \begin{bmatrix} g_{12} & -g_{11} \\ g_{22} & -g_{12} \end{bmatrix} \begin{bmatrix} p_u \\ p_v \end{bmatrix} , \quad q_u = \frac{g_{12}p_u - g_{11}p_v}{\sqrt{g_{11}g_{22} - g_{12}^2}} , \quad q_v = \frac{g_{22}p_u - g_{12}p_v}{\sqrt{g_{11}g_{22} - g_{12}^2}} . \tag{1.188}$$

The inverted KL equations lead to

$$q_{uv} = -\left(\frac{g_{11}p_v - g_{12}p_u}{\sqrt{g_{11}g_{22} - g_{12}^2}}\right)_v \quad , \quad q_{vu} = \left(\frac{g_{22}p_u - g_{12}p_v}{\sqrt{g_{11}g_{22} - g_{12}^2}}\right)_u . \tag{1.189}$$

$$\text{2nd: } q_{uv} = q_{vu} \Leftrightarrow -\left(\frac{g_{11}p_v - g_{12}p_u}{\sqrt{g_{11}g_{22} - g_{12}^2}}\right)_v = \left(\frac{g_{22}p_u - g_{12}p_v}{\sqrt{g_{11}g_{22} - g_{12}^2}}\right)_u \Rightarrow$$

$$\Rightarrow \left(\frac{g_{11}p_v - g_{12}p_u}{\sqrt{g_{11}g_{22} - g_{12}^2}}\right)_v + \left(\frac{g_{22}p_u - g_{12}p_v}{\sqrt{g_{11}g_{22} - g_{12}^2}}\right)_u = 0 . \tag{1.190}$$

End of Proof (the second step).

Proof (sketch of the proof for the third step).

The special KL equations generate a conformal mapping $\mathbb{M}_l^2\left(P, Q \mid \Lambda^2 \mathsf{I}_2\right) \to \mathbb{M}_r\left(p, q \mid \lambda^2 \mathsf{I}_2\right)$, namely a conformal transformation from left conformal (isometric, isothermal) coordinates $\{P, Q\}$ to right conformal (isometric, isothermal) coordinates $\{p, q\}$. The left matrix of the conformally flat metric, $\Lambda^2 \mathsf{I}_2$, is transformed to the right matrix of the conformally flat metric, $\lambda^2 \mathsf{I}_2$. Up to the factors of conformality, $\Lambda^2(P, Q)$ and $\lambda^2(p, q)$, the matrices of the left and right metrices are unit matrices, I_2. Here, we only outline how the *integrability conditions* $p_{PQ} = p_{QP}$ and $q_{PQ} = q_{QP}$ are converted to the special Laplace–Beltrami equation.

KL, 1st equation and 2nd equation, lead to the following relations.

$$\text{1st: } p_{PQ} = p_{QP} , \quad p_{PQ} = q_{QQ} , \quad p_{QP} = -q_{PP} ,$$

$$p_{PQ} = p_{QP} \Leftrightarrow q_{QQ} = -q_{PP} \Rightarrow q_{PP} + q_{QQ} = 0 . \tag{1.191}$$

$$\text{2nd: } q_{PQ} = q_{QP} , \quad q_{QP} = p_{PP} , \quad -q_{PQ} = p_{QQ} ,$$

$$q_{QP} = q_{PQ} \Leftrightarrow p_{PP} = -p_{QQ} \Rightarrow p_{PP} + p_{QQ} = 0 . \tag{1.192}$$

This concludes the proofs.

End of Proof (the second step).

Note that a more elegant proof of the Korn–Lichtenstein equations based upon *exterior calculus* has been presented by E. Grafarend and R. Syffus (1998d). In addition, the authors succeeded to generalize the fundamental differential equations which govern a conformeomorphism the number of dimensions being n (for $n = 3$, they coincide with the *Zund equations* (J. Zund (1987)) from $\mathbb{M}_l^3$ to $\mathbb{M}_r^3$), namely left (pseudo-)Riemann manifold $\mathbb{M}_l^n \to$ right (pseudo-)Riemann manifold $\mathbb{M}_r^n := \mathbb{E}^{r,s}$ ($r + s = n$). In general, conformal mappings from an arbitrary left (pseudo-)Riemann manifold $\mathbb{M}_l^n$ to an arbitrary right (pseudo-)Riemann manifold $\mathbb{M}_r^n$ do not exist. The dimension $n = 2$ is just an exception where conformal mappings always exist, though may be difficult to find. For instance, due to involved difficulties, the Philosphical Faculty of the University of Goettingen Georgia Augusta (dated 13 June 1857) set up the "Preisaufgabe" to find a conformal mapping of the triaxial ellipsoid which had already parameterized by C. F. Gauss in terms of "surface normal coordinates" applying the "Gauss map". Based upon the Jacobi's contribution on elliptic coordinates (C. G. J. Jacobi (1839)), which separate the Laplace–Beltrami equations of harmonicity, the "Preisschrift" of E. Schering (1857) was finally crowned, nevertheless leaving the numerical problem open as to how to construct a conformal map of the triaxial ellipsoid – up to now an open problem (W. Klingenberg (1982), H. Schmehl (1927), and B. Mueller (1991)). The case of dimension $n = 3$ is a special case to be treated. In contrast, for dimension $n > 3$, a general statement can be made: a conformeomorphism exists *if and only if* the *Weyl curvature tensor*, being a curvature element of the *Riemann curvature tensor*, vanishes. We have given in Table 1.4 a list of related, commented references. A typical example for the non-existence of a conformeomorphism is provided by the following example.

Example 1.10 (Non-existence of a conformeomorphism).

In general relativity, the solutions of the Einstein gravitational field equations (for instance, the Schwarzschild metric) generate a Weyl curvature different from zero. Accordingly, the space-time *pseudo-Riemann manifold* $\mathbb{M}_l^{3,1}$ (space–time) $\to \mathbb{M}_r^{3,1} := \{\mathbb{R}^{3,1}, \delta_{\mu\nu}^-\}$ does not allow a conformal mapping to the *pseudo-Euclidean manifold* $\{\mathbb{R}^{3,1}, \mathsf{I}_4^-\}$, where $\mathsf{I}_4^- := [\delta_{\mu\nu}^-] := \mathrm{diag}[1, 1, 1, -1]$. Note that details referred to those authors are listed in Table 1.4.

End of Example.

Physical aside.

There is another interesting perspective between the geometry of conformal mappings and the physical field equations, say of gravitostatics, electrostatics, and magnetostatics. It turns out as a result of *conformal field theory* that the *factor of conformality*, Λ^2 or λ^2, respectively, corresponds to the gravitational potential, the electric potential, and the magnetic potential, a notion being introduced by C. F. Gauss. A highlight has been the contribution of C. W. Misner (1978) who used the vector-valued four-dimensional Laplace–Beltrami equations (harmonic maps) as models of physical theories.

1-10 Two examples: Mercator Projection and Stereographic Projection

Two important examples for the equivalence theorem of conformal mapping, the conformal mapping from an ellipsoid-of-revolution to the sphere: Universal Mercator Projection (UMP), Universal Stereographic Projection (UPS).

The most famous examples for a conformal mapping of an ellipsoid-of-revolution $\mathbb{E}_{A_1,A_1,A_2}^2$ to a sphere $\mathbb{S}_r^2$ are the *Universal Mercator Projection (UMP)* and the *Universal Stereographic Projection (UPS)*, which we are going to present to you in Example 1.11 and Fig. 1.25, and in Example 1.12 and Fig. 1.26, respectively. For both examples, we pose four problems, namely (i) prove that left and right UMP as well as UPS fulfill the Korn–Lichtenstein equations subject to the integrability and orientation conditions, (ii) prove that the factor of left and right conformality has to fulfill a special Helmholtz differential equation derived from left and right Gaussian curvature, (iii) prove which coordinate line is mapped equidistantly, and (iv) derive a "simple conformal mapping" $\mathbb{E}_{A_1,A_1,A_2}^2 \to \mathbb{S}_r^2$.

Table 1.4. Conformal mapping $\mathbb{M}_l^n \to \mathbb{M}_r^n$, commented references.

W. Blaschke, K. Leichtweiß (1973)	Two-dimensional conformal mapping, Korn–Lichtenstein equations: $n = 2$.
J. P. Bourguignon (1970)	n-dimensional conformal mapping, Weyl curvature: $n \geq 3$.
L. P. Eisenhart (1949)	n-dimensional conformal mapping, Weyl curvature: $n = $ arbitrary.
A. Finzi (1922)	Three-dimensional conformal mapping, generalized Korn–Lichtenstein equations: $n = 3$.
C. F. Gauss (1822)	Classical contribution on two-dimensional conformal mapping: $n = 2$.
C. F. Gauss (1816–1827)	Classical contribution on conformal mapping of the ellipsoid-of-revolution.
E. Grafarend, R. Syffus (1998c)	n-dimensional conformal mapping, generalized Korn–Lichtenstein equations.
E. R. Hedrick, L. Ingold (1925a)	Analytic functions in three dimensions.
E. R. Hedrick, L. Ingold (1925b)	Laplace–Beltrami equations in three dimensions.
W. Klingenberg (1982)	Conformal mapping of the triaxial ellipsoid, elliptic coordinates.
R. S. Kulkarni (1969)	Curvature structures and conformal mapping.
R. S. Kulkarni (1972)	Conformally flat manifolds.
R. S. Kulkarni, U. Pinkall (eds.) (1988)	Conformal geometry.
J. Lafontaine (1988)	Conformal mapping "from the Riemann viewpoint".
J. Liouville (1850)	Three-dimensional conformal mapping.
G. Ricci (1918)	Conformal mapping.
J. A. Schouten (1921)	n-dimensional conformal mapping, Weyl curvature.
H. Weyl (1918)	Conformal mapping.
H. Weyl (1921)	Conformal mapping.
A. I. Yanushauskas (1982)	three-dimensional conformal mapping, generalized Korn–Lichtenstein equations.
J. Zund (1987)	three-dimensional conformal mapping, generalized Korn–Lichtenstein equations.

Example 1.11 (Conformal mapping of an ellipsoid-of-revolution $\mathbb{E}^2_{A_1,A_1,A_2}$ to a sphere $\mathbb{S}^2_r$: the Universal Mercator Projection (UMP) of type left $\mathbb{E}^2_{A_1,A_1,A_2}$ and right $\mathbb{S}^2_r$, the special Korn–Lichtenstein equations, and the Cauchy–Riemann equations (d'Alembert–Euler equations)).

Let us assume that we have found a solution of the left Korn–Lichtenstein equations of the ellipsoid-of-revolution $\mathbb{E}^2_{A_1,A_1,A_2}$ parameterized by the two coordinates $\{\Lambda, \Phi\}$ which conventionally are called $\{Gauss\ surface\ normal\ longitude,\ Gauss\ surface\ normal\ latitude\}$. Similarly, let us depart from a solution of the right Korn–Lichtenstein equations of the sphere $\mathbb{S}^2_r$ parameterized by the two coordinates $\{\lambda, \phi\}$ which are called $\{spherical\ longitude,\ spherical\ latitude\}$. Here, we follow the commutative diagram of Fig. 1.25 and identify the left conformal coordinates $\{P, Q\}$ with the Universal Mercator Projection (UMP) of $\mathbb{E}^2_{A_1,A_1,A_2}$, and the right conformal coordinates $\{p, q\}$ with the Universal Mercator Projection (UMP) of $\mathbb{S}^2_r$, which is outlined in Box 1.24. The ratios Q/A_1 and q/r are also called $\{ellipsoidal\ isometric\ latitude,\ spherical\ isometric\ latitude\}$ or ellipsoidal spherical $Lambert\ functions$ $Q = A_1 \operatorname{lam}\Phi$ and $q = r \operatorname{lam}\phi$, respectively. In addition, we adopt the left and right matrices of the metric $\{\mathsf{G}_l, \mathsf{G}_r\}$ of Example 1.3.

End of Example.

We pose four problems. (i) Do the left and right conformal maps that are parameterized by $\{P(\Lambda), Q(\Phi)\}$ and $\{p(\lambda), q(\phi)\}$ as "UMP left" and "UMP right" fulfil the Korn–Lichtenstein equations, the integrability conditions (vector-valued Laplace–Beltrami equations of harmonicity), and the condition "orientation preserving conformeomorphism"? (ii) Derive the left and right factors of conformality, $\Lambda^2 = \Lambda_1^2 = \Lambda_2^2$ and $\lambda^2 = \lambda_1^2 = \lambda_2^2$. Do the factors of conformality fulfil a special Helmholtz equation? (iii) Prove that under "UMP left" as well as "UMP right" both the equators of $\mathbb{E}^2_{A_1,A_1,A_2}$ and $\mathbb{S}^2_r$ are mapped equidistantly. Interpret this result as a boundary condition of the Korn–Lichtenstein equations. (iv) Derive a "simple conformal mapping" $\mathbb{E}^2_{A_1,A_1,A_2} \to \mathbb{S}^2_r$.

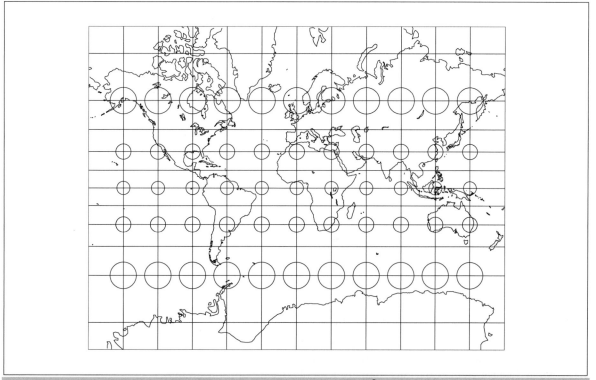

Fig. 1.25. Universal Mercator Projection (UMP) of the sphere $\mathbb{S}^2_r$ with shorelines and Tissot ellipses of distortion. Graticule: $30°$ in longitude, $15°$ in latitude. Domain: $\{-180° < \Lambda \leq +180°, -80° < \Phi < +80°\}$.

Box 1.24 (UMP of $\mathbb{E}^2_{A_1,A_1,A_2}$ versus UMP of $\mathbb{S}^2_r$).

Left conformal coordinates:

Right conformal coordinates:

$$P = A_1\Lambda \,, \quad Q = A_1 \ln\left(\tan\left(\frac{\pi}{4} + \frac{\Phi}{2}\right)\left[\frac{1 - E\sin\Phi}{1 + E\sin\Phi}\right]^{E/2}\right). \qquad q = r\ln\tan\left(\frac{\pi}{4} + \frac{\phi}{2}\right), \quad p = r\lambda \,. \quad (1.193)$$

Left matrix of the metric G_l:

Right matrix of the metric G_r:

$$\mathsf{G}_l = \begin{bmatrix} \dfrac{A_1^2\cos^2\Phi}{1 - E^2\sin^2\Phi} & 0 \\ 0 & \dfrac{A_1^2(1 - E^2)^2}{(1 - E^2\sin^2\Phi)^3} \end{bmatrix}. \qquad \begin{bmatrix} r^2\cos^2\phi & 0 \\ 0 & r^2 \end{bmatrix} = \mathsf{G}_r \,. \quad (1.194)$$

Left Jacobi matrix:

Right Jacobi matrix:

$$P_\Lambda = A_1 \,, \quad Q_\Lambda = P_\Phi = 0 \,, \quad Q_\Phi = \frac{A_1(1 - E^2)}{1 - E^2\sin^2\Phi}\frac{1}{\cos\Phi}\,. \qquad p_\lambda = r \,, \quad q_\lambda = p_\phi = 0 \,, \quad q_\phi = \frac{r}{\cos\phi}\,. \quad (1.195)$$

Solution (the first problem).

Start from the conformal map that is defined in Box 1.24. The three conditions to be fulfilled are given in Box 1.25 for "left UMP" and in Box 1.26 for "right UMP". First, we write down the left and right specified Korn–Lichtenstein equations, namely $\sqrt{G_{11}/G_{22}}$, $\sqrt{G_{22}/G_{11}}$ and $\sqrt{g_{11}/g_{22}}$, $\sqrt{g_{22}/g_{11}}$, respectively. Indeed, by transforming $\{Q_\Lambda, Q_\Phi, P_\Lambda, P_\Phi\}$ as well as $\{q_\lambda, q_\phi, p_\lambda, p_\phi\}$ left and right, KL 1st and KL 2nd are satified. Second, we specialize the left and right integrability conditions of the vector-valued Korn–Lichtenstein equations, namely the left and right Laplace–Beltrami equations, by $\{G_{11}, G_{22}, P_\Lambda, P_\Phi, Q_\Lambda, Q_\Phi\}$ of $\mathbb{E}^2_{A_1,A_1,A_2}$ as well as $\{g_{11}, g_{22}, p_\lambda, p_\phi, q_\lambda, q_\phi\}$ of $\mathbb{S}^2_r$. Indeed, we have succeeded that $\{P, Q\}$ are *left harmonic* and $\{p, q\}$ are *right harmonic*. Third, we prove left and right orientation by computing the left and right *Jacobians* which turn out positive.

End of Solution (the first problem).

Solution (the second problem).

In any textbook of Differential Geometry, you will find the representation of the Gaussian curvature of a surface in terms of conformal coordinates (isometric, isothermal). Let the left and the right matrix of the metric be equipped with a conformally flat structure $\{\mathsf{G}_l = \lambda_l^2 \mathsf{I}_2, \mathsf{G}_r = \lambda_r^2 \mathsf{I}_2\}$, which is generated by a left and a right conformal coordinate representation. Then the left and the right Gaussian curvature are given by $k_l = -(1/2\lambda_l^2)\Delta_l \ln\lambda_l^2 = -(1/\lambda_l^2)\Delta_l \ln\lambda_l$ and $k_r = -(1/2\lambda_r^2)\Delta_r \ln\lambda_r^2 = -(1/\lambda_r^2)\Delta_r \ln\lambda_r$ as well as $\Delta_l := D_{PP} + D_{QQ} = D_P^2 + D_Q^2$ and $D_p^2 + D_q^2 = D_{pp} + D_{qq} := \Delta_r$, where Δ_l and Δ_r represent the left and the right Laplace–Beltrami operator. Let us apply this result in solving the second problem. By means of Boxes 1.27, 1.28, and 1.29, we have outlined how to generate a conformally flat metric of an ellipsoid-of-revolution and of a sphere. It is the classical Gauss factorization.

End of Solution (the second problem).

Solution (the third problem).

By means of the left and the right mapping equations, i.e. by means of $\{P = A_1\Lambda, Q(\Phi = 0) = 0\}$ and $\{p = r\lambda, q(\phi = 0) = 0\}$, respectively, it is obvious that an equatorial arc of $\mathbb{E}^2_{A_1,A_1,A_2}$ and of $\mathbb{S}^2_r$ is mapped equidistantly. Similarly, the factor of conformality derived in Box 1.27 amounts to $\lambda_l^2(\Phi = 0) = 1$ for the left manifold and to $\lambda_r^2(\phi = 0) = 1$ for the right manifold. Such a configuration on the ellipsoidal as well as the spherical equator we call an *isometry*. Indeed, the postulate of an equidistant mapping of the left or right equator constitutes a boundary condition for the left and right Korn–Lichtenstein equations, namely to make their solution unique.

End of Solution (the third problem).

Box 1.25 (Left Korn–Lichtenstein equations, UMP of $\mathbb{E}^2_{A_1,A_1,A_2}$, harmonicity, orientation).

Left Korn–Lichtenstein equations:

$$\text{(1st)}\quad P_\Lambda = +\sqrt{\frac{G_{11}}{G_{22}}}\,Q_\Phi\ ,\quad P_\Phi = -\sqrt{\frac{G_{22}}{G_{11}}}\,Q_\Lambda \quad \text{(2nd)}\,,$$

$$\text{(1st)}\quad Q_\Lambda = -\sqrt{\frac{G_{11}}{G_{22}}}\,P_\Phi\ ,\quad Q_\Phi = +\sqrt{\frac{G_{22}}{G_{11}}}\,P_\Lambda \quad \text{(2nd)}\,;$$

(1.196)

$$\sqrt{\frac{G_{11}}{G_{22}}} = \cos\Phi\,\frac{1 - E^2\sin^2\Phi}{1 - E^2} \Leftrightarrow \sqrt{\frac{G_{22}}{G_{11}}} = \frac{1}{\cos\Phi}\,\frac{1 - E^2}{1 - E^2\sin^2\Phi}\ ;$$

(1.197)

$$\text{(UMP left)}\quad Q_\Phi = \frac{A_1(1 - E^2)}{1 - E^2\sin^2\Phi}\,\frac{1}{\cos\Phi}\ ,\quad P_\Lambda = A_1 \quad \text{(KL 2nd)}\,,$$

(1.198)

$$\text{(UMP left)}\quad Q_\Lambda = 0\ ,\quad P_\Phi = 0 \quad \text{(KL 1st)}\,.$$

Left integrability conditions:

$$\Delta_{\Lambda,\Phi}P = \left(\sqrt{\frac{G_{11}}{G_{22}}}\,P_\Phi\right)_\Phi + \left(\sqrt{\frac{G_{22}}{G_{11}}}\,P_\Lambda\right)_\Lambda = 0\ ,$$

$$\Delta_{\Lambda,\Phi}Q = \left(\sqrt{\frac{G_{11}}{G_{22}}}\,Q_\Phi\right)_\Phi + \left(\sqrt{\frac{G_{22}}{G_{11}}}\,Q_\Lambda\right)_\Lambda = 0\ ;$$

(1.199)

$$\text{(1st)}\quad \sqrt{\frac{G_{11}}{G_{22}}}\,P_\Phi = 0\ ,\quad \sqrt{\frac{G_{22}}{G_{11}}}\,P_\Lambda = \frac{A_1(1 - E^2)}{1 - E^2\sin^2\Phi}\,\frac{1}{\cos\Phi}$$

$$\Rightarrow$$

$$\Delta_{\Lambda,\Phi}P = \left(\sqrt{\frac{G_{11}}{G_{22}}}\,P_\Phi\right)_\Phi + \left(\sqrt{\frac{G_{22}}{G_{11}}}\,P_\Lambda\right)_\Lambda = \left(\frac{A_1(1 - E^2)}{1 - E^2\sin^2\Phi}\,\frac{1}{\cos\Phi}\right)_\Lambda = 0$$

(1.200)

q. e. d.

$$\text{(2nd)}\quad \sqrt{\frac{G_{11}}{G_{22}}}\,Q_\Phi = A_1\ ,\quad \sqrt{\frac{G_{22}}{G_{11}}}\,Q_\Lambda = 0$$

$$\Rightarrow$$

$$\Delta_{\Lambda,\Phi}Q = \left(\sqrt{\frac{G_{11}}{G_{22}}}\,Q_\Phi\right)_\Phi + \left(\sqrt{\frac{G_{22}}{G_{11}}}\,Q_\Lambda\right)_\Lambda = 0$$

(1.201)

q. e. d.

Left orientation:

$$\begin{vmatrix} P_\Lambda & P_\Phi \\ Q_\Lambda & Q_\Phi \end{vmatrix} = P_\Lambda Q_\Phi - P_\Phi Q_\Lambda = A_1^2\,\frac{1 - E^2}{1 - E^2\sin^2\Phi}\,\frac{1}{\cos\Phi} > 0$$

(1.202)

due to

$$-\pi/2 < \Phi < +\pi/2 \Rightarrow \cos\Phi > 0$$

(1.203)

q. e. d.

Box 1.26 (Right Korn–Lichtenstein equations, UMP of $\mathbb{S}_r^2$, harmonicity, orientation).

Right Korn–Lichtenstein equations:

$$(\text{1st}) \quad p_\lambda = +\sqrt{\frac{g_{11}}{g_{22}}}q_\phi \ , \quad p_\phi = -\sqrt{\frac{g_{22}}{g_{11}}}q_\lambda \quad (\text{2nd}) \ ,$$

$$(\text{1st}) \quad q_\lambda = -\sqrt{\frac{g_{11}}{g_{22}}}p_\phi \ , \quad q_\phi = +\sqrt{\frac{g_{22}}{g_{11}}}p_\lambda \quad (\text{2nd}) \ ; \tag{1.204}$$

$$\sqrt{\frac{g_{11}}{g_{22}}} = \cos\phi \ , \quad \sqrt{\frac{g_{22}}{g_{11}}} = \frac{1}{\cos\phi} \ ; \tag{1.205}$$

$$(\text{UMP right}) \quad q_\phi = \frac{r}{\cos\phi} \ , \quad p_\lambda = r \quad (\text{KL 2nd}) \ ,$$

$$(\text{UMP right}) \quad q_\lambda = 0 \ , \quad p_\phi = 0 \quad (\text{KL 1st}) \ . \tag{1.206}$$

Right integrability conditions:

$$\Delta_{\lambda,\phi}p = \left(\sqrt{\frac{g_{11}}{g_{22}}}p_\phi\right)_\phi + \left(\sqrt{\frac{g_{22}}{g_{11}}}p_\lambda\right)_\lambda = 0 \ ,$$

$$\Delta_{\lambda,\phi}q = \left(\sqrt{\frac{g_{11}}{g_{22}}}q_\phi\right)_\phi + \left(\sqrt{\frac{g_{22}}{g_{11}}}q_\lambda\right)_\lambda = 0 \ ; \tag{1.207}$$

$$(\text{1st}) \quad \sqrt{\frac{g_{11}}{g_{22}}}p_\phi = 0 \ , \quad \sqrt{\frac{g_{22}}{g_{11}}}p_\lambda = \frac{r}{\cos\phi}$$

$$\Rightarrow$$

$$\Delta_{\lambda,\phi}p = \left(\sqrt{\frac{g_{11}}{g_{22}}}p_\phi\right)_\phi + \left(\sqrt{\frac{g_{22}}{g_{11}}}p_\lambda\right)_\lambda = \left(\frac{r}{\cos\phi}\right)_\lambda = 0 \tag{1.208}$$

q. e. d.

$$(\text{2nd}) \quad \sqrt{\frac{g_{11}}{g_{22}}}q_\phi = r \ , \quad \sqrt{\frac{g_{22}}{g_{11}}}q_\lambda = 0$$

$$\Rightarrow$$

$$\Delta_{\lambda,\phi}q = \left(\sqrt{\frac{g_{11}}{g_{22}}}q_\phi\right)_\phi + \left(\sqrt{\frac{g_{22}}{g_{11}}}q_\lambda\right)_\lambda = 0 \tag{1.209}$$

q. e. d.

Right orientation:

$$\begin{vmatrix} p_\lambda & p_\phi \\ q_\lambda & q_\phi \end{vmatrix} = p_\lambda q_\phi - p_\phi q_\lambda = \frac{r^2}{\cos\phi} > 0 \tag{1.210}$$

due to

$$-\pi/2 < \Phi < +\pi/2 \Rightarrow \cos\Phi > 0 \tag{1.211}$$

q. e. d.

Box 1.27 (Conformally flat left and right manifolds, ellipsoid-of-revolution versus sphere).

<table>
<tr><td>

Left manifold,
$\{\Lambda, \Phi\}$ left coordinates:

</td><td>

Right manifold,
$\{\lambda, \phi\}$ right coordinates:

</td></tr>
</table>

$$\mathrm{d}S^2 = \frac{A_1^2 \cos^2 \Phi}{1 - E^2 \sin^2 \Phi} \mathrm{d}\Lambda^2 + \frac{A_1^2(1 - E^2)^2}{(1 - E^2 \sin^2 \Phi)^3} \mathrm{d}\Phi^2 . \qquad \mathrm{d}s^2 = r^2 \cos^2 \phi \, \mathrm{d}\lambda^2 + r^2 \mathrm{d}\phi^2 . \qquad (1.212)$$

Conformally flat left manifold,
$\{P, Q\}$ conformally
left coordinates
(isometric, isothermal):

Conformally flat right manifold,
$\{p, q\}$ conformally
right coordinates
(isometric, isothermal):

$$\mathrm{d}S^2 = \frac{\cos^2 \Phi}{1 - E^2 \sin^2 \Phi} \times \qquad\qquad \mathrm{d}s^2 = \cos^2 \phi \times$$

$$\times \left[A_1^2 \mathrm{d}\Lambda^2 + \frac{A_1^2(1 - E^2)^2}{(1 - E^2 \sin^2 \Phi)^2} \frac{1}{\cos^2 \Phi} \mathrm{d}\Phi^2 \right] . \qquad \times \left[r^2 \mathrm{d}\lambda^2 + \frac{r^2}{\cos^2 \phi} \mathrm{d}\phi^2 \right] . \qquad (1.213)$$

Left differential form:

Right differential form:

$$\mathrm{d}P := A_1 \mathrm{d}\Lambda , \qquad\qquad\qquad \mathrm{d}p := r \mathrm{d}\lambda ,$$

$$\mathrm{d}Q := A_1 \frac{1 - E^2}{1 - E^2 \sin^2 \Phi} \frac{1}{\cos \Phi} \mathrm{d}\Phi . \qquad \mathrm{d}q := r \frac{1}{\cos \phi} \mathrm{d}\phi . \qquad (1.214)$$

Left factor of conformality:

Right factor of conformality:

$$\lambda_l^2 := \frac{\cos^2 \Phi}{1 - E^2 \sin^2 \Phi} , \quad \frac{\mathrm{d}S^2}{\mathrm{d}s_l^2} = \lambda_l^2 , \qquad \cos^2 \phi =: \lambda_r^2 , \quad \lambda_r^2 = \frac{\mathrm{d}s^2}{\mathrm{d}s_r^2} ,$$

$$\mathrm{d}s_l^2 = \mathrm{d}P^2 + \mathrm{d}Q^2 , \qquad\qquad \mathrm{d}p^2 + \mathrm{d}q^2 = \mathrm{d}s_r^2 ,$$

$$\mathrm{d}S^2 = \lambda_l^2 \left(\mathrm{d}P^2 + \mathrm{d}Q^2 \right) , \qquad\qquad \lambda_r^2 \left(\mathrm{d}p^2 + \mathrm{d}q^2 \right) = \mathrm{d}s^2 , \qquad (1.215)$$

$$\mathrm{d}S^2 = \frac{\cos^2 \Phi}{1 - E^2 \sin^2 \Phi} \left(\mathrm{d}P^2 + \mathrm{d}Q^2 \right) . \qquad \cos^2 \phi \left(\mathrm{d}p^2 + \mathrm{d}q^2 \right) = \mathrm{d}s^2 .$$

Left Cauchy–Green matrix:

Right Cauchy–Green matrix:

$$\left| \mathsf{C}_l - \Lambda_l^2 \mathsf{G}_l \right| = 0 \Leftrightarrow \qquad\qquad \left| \mathsf{C}_r - \Lambda_r^2 \mathsf{G}_r \right| = 0 \Leftrightarrow$$

$$\Leftrightarrow \begin{vmatrix} A_1^2 - \Lambda_l^2 G_{11} & 0 \\ 0 & Q_\Phi^2 - \Lambda_l^2 G_{22} \end{vmatrix} = 0 \Leftrightarrow \qquad \Leftrightarrow \begin{vmatrix} r^2 - \Lambda_r^2 g_{11} & 0 \\ 0 & q_\phi^2 - \Lambda_r^2 g_{22} \end{vmatrix} = 0 \Leftrightarrow$$

$$\Leftrightarrow \begin{bmatrix} {}_l\Lambda_1^2 = \dfrac{A_1^2}{G_{11}} = \dfrac{1 - E^2 \sin^2 \Phi}{\cos^2 \Phi} \\[2mm] {}_l\Lambda_2^2 = \dfrac{Q_\Phi^2}{G_{22}} = \dfrac{1 - E^2 \sin^2 \Phi}{\cos^2 \Phi} \end{bmatrix} \Rightarrow \qquad \Leftrightarrow \begin{bmatrix} {}_r\Lambda_1^2 = \dfrac{r^2}{g_{11}} \\[2mm] {}_r\Lambda_2^2 = \dfrac{q_\phi^2}{g_{22}} = \dfrac{1}{\cos^2 \phi} \end{bmatrix} \Rightarrow \qquad (1.216)$$

$$\Rightarrow {}_l\Lambda_1^2 = {}_l\Lambda_2^2 = \Lambda_l^2 = \frac{1 - E^2 \sin^2 \Phi}{\cos^2 \Phi} \Leftrightarrow \qquad \Rightarrow {}_r\Lambda_1^2 = {}_r\Lambda_2^2 = \Lambda_r^2 = \frac{1}{\cos^2 \Phi} \Leftrightarrow$$

$$\Leftrightarrow \lambda_l^2 = \frac{\cos^2 \Phi}{1 - E^2 \sin^2 \Phi} . \qquad\qquad \Leftrightarrow \lambda_r^2 = \cos^2 \phi .$$

Box 1.28 (Representation of the factors of conformality in terms of conformal coordinates).

Left factor of conformality:

$$P = A_1 \Lambda \,, \quad Q = A_1 f(\Phi) \,, \quad f(\Phi) := \ln\left(\tan\left(\frac{\pi}{4} + \frac{\Phi}{2} \right) \left[\frac{1 - E\sin\Phi}{1 + E\sin\Phi} \right]^{E/2} \right) \,,$$

$$\lambda_l^2 = \frac{\cos^2\Phi}{1 - E^2\sin^2\Phi} \,, \quad \Lambda_l^2 = \frac{1 - E^2\sin^2\Phi}{\cos^2\Phi} \tag{1.217}$$

$$\Rightarrow$$

$$\lambda_l^2 = \frac{\cos^2 f^{-1}(Q/A_1)}{\sqrt{1 - E^2\sin^2 f^{-1}(Q/A_1)}} \,, \quad \Lambda_l^2 = \frac{1 - E^2\sin^2 f^{-1}(Q/A_1)}{\cos^2 f^{-1}(Q/A_1)} \,.$$

Right factor of conformality:

$$p = r\lambda \,, \quad q = r\ln\tan\left(\frac{\pi}{4} + \frac{\phi}{2} \right) = r\,\mathrm{artanh}\,\sin\phi \,,$$

$$\tanh(q/r) = \sin\phi \,, \quad \frac{1}{\cosh(q/r)} = \cos\phi \,,$$

$$\lambda_r^2 = \cos^2\phi \,, \quad \Lambda_r^2 = \frac{1}{\cos^2\phi} \tag{1.218}$$

$$\Rightarrow$$

$$\lambda_r^2 = \frac{1}{\cosh^2(q/r)} \,, \quad \Lambda_r^2 = \cosh^2(q/r) \,.$$

Box 1.29 (The differential equation which governs the factor of conformality).

Two versions of the special Helmholtz equations:

$$\text{(i)} \quad \Delta\ln\lambda^2 + 2k\lambda^2 = 0 \,, \quad \text{(ii)} \quad \Delta\lambda^2 + 2k\lambda^4 = 0 \,. \tag{1.219}$$

(k is the Gaussian curvature $k(p,q)$.)

Right differential equation of the factor of conformality ($\mathbb{S}_r^2$):

$$k_r = \frac{1}{r^2} = \text{constant} \,,$$

$$\Delta\ln\lambda_r^2 + \frac{2}{r^2}\lambda_r^2 = 0 \,, \quad \lambda_r^2 = \cosh^{-2}(q/r) \,, \quad \ln\lambda_r^2 = -2\ln\cosh(q/r) \,, \tag{1.220}$$

$$D_q\ln\lambda_r^2 = -\frac{2}{r}\tanh(q/r) \,, \quad \Delta_r\ln\lambda_r^2 = D_{qq}\ln\lambda_r^2 = -\frac{2}{r^2}\frac{1}{\cosh^2(q/r)} = -\frac{2}{r^2}\lambda_r^2$$

q. e. d.

Left differential equation of the factor of conformality ($\mathbb{E}_{A_1,A_1,A_2}^2$):

$$k_l = \frac{(1 - E^2\sin^2\phi)^2}{A_1^2(1 - E^2)} = \frac{1 - E^2\sin^2 f^{-1}(Q/A_1)}{A_1^2(1 - E^2)} \,,$$

$$\Delta\ln\lambda_l^2 + 2k(Q)\lambda_l^2 = 0 \,, \quad \lambda_l^2 = \frac{\cos^2 f^{-1}(Q/A_1)}{\sqrt{1 - E^2\sin^2 f^{-1}(Q/A_1)}} \,, \tag{1.221}$$

$$\ln\lambda_l^2 = 2\ln f^{-1}(Q/A_1) - \frac{1}{2}\ln\left[1 - E^2\sin^2 f^{-1}(Q/A_1) \right] \,,$$

$$\Delta_l\ln\lambda_l^2 = D_{QQ}\ln\lambda_l^2 = -2k(Q)\lambda_l^2$$

q. e. d.

In Box 1.27, we write down the metric forms "left dS^2" and "right ds^2" in the initial coordinates $\{\Lambda, \Phi\}$ and $\{\lambda, \phi\}$, respectively. Second, we factorize by (i) $\cos^2 \Phi / (1 - E^2 \sin^2 \Phi)$ and (ii) $\cos^2 \phi$. The first term $A_1 d\Lambda$ and $r d\lambda$, respectively, generates dP and dp, respectively. In contrast, the second term $([A_1(1-E^2)/(1-E^2 \sin^2 \Phi)] \cos \Phi) d\Phi$ and $(r/\cos \phi) d\phi$, respectively, generates dQ and dq, respectively. Indeed, the first factors $\cos^2 \Phi / (1 - E^2 \sin^2 \Phi)$ and $\cos^2 \phi$ produce the left and the right factor of conformality, called λ_l^2 and λ_r^2, respectively. They are reciprocal to Λ_l^2 and Λ_r^2, respectively. Third, by means of Box 1.28, we aim at representing the factors of conformality, λ_l^2 and λ_r^2, in terms of conformal (isometric, isothermal) latitude Q and q, respectively, namely $\lambda_l^2(Q)$ and $\lambda_r^2(q)$, respectively. Here, we have to invert the functions $Q/A_1 = f(\Phi)$ and $q/r = \ln \tan(\pi/4 + \phi/2) = \operatorname{artanh}(\sin \phi)$, also called the inverse *Lambert* or *Gudermann function*, lam or gd, respectively. $\phi = \operatorname{lam}(q/r) = \operatorname{gd}(q/r)$ or $\sin \phi = \tanh(q/r)$, $\cos \phi = 1/\cosh(q/r)$. While $\lambda_l^2(Q)$ and $\Lambda_l^2(Q)$ cannot be given in a closed form, $\lambda_r^2 = 1/\cosh^2(q/r)$ and $\Lambda_r^2 = \cosh^2(q/r)$ are available in a simple form. Fourth, by means of Box 1.29, we prove that λ_r^2 and λ_l^2, respectively, fulfill the conformal representation of the right and the left Gaussian curvature, here written in two versions as a special Helmholtz differential equation. For being simpler, we did first "right" followed by the more complex "left" computation. Indeed, for given Gaussian curvature $k_r = 1/r^2 = $ constant of the sphere $\mathbb{S}_r^2$ and $k_l = (1 - E^2 \sin^2 \Phi)^2/[A_1^2(1-E^2)]$ of the ellipsoid-of-revolution $\mathbb{E}_{A_1,A_1,A_2}^2$ finally transformed into $\{q, Q\}$ coordinates of type conformal (isometric, isothermal), we succeed to prove $\Delta \ln \lambda^2 + 2k\lambda^2 = 0$ of type "right" and "left".

Solution (the fourth problem).

A "simple conformal mapping" of $\mathbb{E}_{A_1,A_1,A_2}^2 \to \mathbb{S}_r^2$ is the isoparametric mapping characterized by

$$p = P, \; \lambda = \frac{A_1}{r}\Lambda, \; q = Q, \; A_1 \ln\left(\tan\left(\frac{\pi}{4} + \frac{\Phi}{2}\right)\left[\frac{1 - E \sin \Phi}{1 + E \sin \Phi}\right]^{E/2}\right) = r \ln \tan\left(\frac{\pi}{4} + \frac{\phi}{2}\right). \quad (1.222)$$

C. F. Gauss (1822, 1844) made some special proposals how to choose the radius r of $\mathbb{S}_r^2$ in an optimal way. Here, let us refer to Chapter 2, where the *Gauss projection* $\mathbb{E}_{A_1,A_1,A_2}^2 \to \mathbb{S}_r^2 \to \mathbb{P}^2$ is discussed in detail. Here, we conclude with a representation of the left as well as the right inverse mapping Φ_l^{-1} and Φ_r^{-1} in terms of conformal coordinates (isometric, isothermal) of Box 1.30, which specializes Φ_l^{-1} and Φ_r^{-1} of Box 1.21.

End of Solution (the fourth problem).

Box 1.30 (Representation of Φ_l^{-1} and Φ_r^{-1} in terms of conformal coordinates: $\mathbb{E}_{A_1,A_1,A_2}^2 \to \mathbb{S}_r^2$).

$$\Phi_l^{-1}: \; \boldsymbol{X}(\Lambda, \Phi) = \qquad\qquad \Phi_r^{-1}: \; \boldsymbol{x}(\lambda, \phi) =$$

$$= \boldsymbol{E}_1 \frac{A_1 \cos \Phi \cos \Lambda}{\sqrt{1 - E^2 \sin^2 \Phi}} + \qquad = \boldsymbol{e}_1 \, r \cos \phi \cos \lambda +$$

$$+ \boldsymbol{E}_2 \frac{A_1 \cos \Phi \sin \Lambda}{\sqrt{1 - E^2 \sin^2 \Phi}} + \qquad + \boldsymbol{e}_2 \, r \cos \phi \sin \lambda +$$

$$+ \boldsymbol{E}_3 \frac{A_1(1 - E^2) \sin \Phi}{\sqrt{1 - E^2 \sin^2 \Phi}} = \qquad + \boldsymbol{e}_3 \, r \sin \phi \qquad = \qquad\qquad (1.223)$$

$$= \boldsymbol{E}_1 \frac{A_1 \cos f^{-1}(Q/A_1) \cos(P/A_1)}{\sqrt{1 - E^2 \sin^2 f^{-1}(Q/A_1)}} + \qquad = \boldsymbol{e}_1 \, r \frac{\cos(p/r)}{\cosh(q/r)} +$$

$$+ \boldsymbol{E}_2 \frac{A_1 \cos f^{-1}(Q/A_1) \sin(P/A_1)}{\sqrt{1 - E^2 \sin^2 f^{-1}(Q/A_1)}} + \qquad + \boldsymbol{e}_2 \, r \frac{\sin(p/r)}{\cosh(q/r)} +$$

$$+ \boldsymbol{E}_3 \frac{A_1(1 - E^2) \sin f^{-1}(Q/A_1)}{\sqrt{1 - E^2 \sin^2 f^{-1}(Q/A_1)}}. \qquad + \boldsymbol{e}_3 \, r \tanh(q/r).$$

Isoparametric mapping: $p - P$ and $q - Q$.

Example 1.12 (Conformal mapping of an ellipsoid-of-revolution $\mathbb{E}^2_{A_1,A_1,A_2}$ to a sphere $\mathbb{S}^2_r$: the Universal Stereographic Projection (UPS) of type left $\mathbb{E}^2_{A_1,A_1,A_2}$ and right $\mathbb{S}^2_r$, special Korn–Lichtenstein equations, Cauchy–Riemann equations (d'Alembert–Euler equations)).

Let us assume that we have found a solution of the left Korn–Lichtenstein equations of the ellipsoid-of-revolution $\mathbb{E}^2_{A_1,A_1,A_2}$ parameterized by the two coordinates $\{\Lambda,\Phi\}$ which conventionally are called {*Gauss surface normal longitude, Gauss surface normal latitude*}. Similarly, let us depart from a solution of the right Korn–Lichtenstein equations of the sphere $\mathbb{S}^2_r$ parameterized by the two coordinates $\{\lambda,\phi\}$ which are called {*spherical longitude, spherical latitude*}. Here, we follow the commutative diagram of Fig. 1.26 and identify the left conformal coordinates $\{P,Q\}$ with the Universal Stereographic Projection (UPS) of $\mathbb{E}^2_{A_1,A_1,A_2}$, and the right conformal coordinates $\{p,q\}$ with the Universal Stereographic Projection (UPS) of $\mathbb{S}^2_r$, which is outlined in Box 1.31. In addition, we adopt the left and right matrices of the metric $\{\mathsf{G}_l,\mathsf{G}_r\}$ of Example 1.3.

End of Example.

We pose five problems. (i) Do the left and right conformal maps that are parameterized by $\{P(\Lambda,\Phi),Q(\Lambda,\Phi)\}$ and $\{p(\lambda,\phi),q(\lambda,\phi)\}$ as "UPS left" and "UPS right" fulfil the Korn–Lichtenstein equations, the integrability conditions (vector-valued Laplace–Beltrami equations of harmonicity, the condition "orientation preserving conformeomorphism"? (ii) Derive the left and right factors of conformality, $\Lambda^2 = \Lambda_1^2 = \Lambda_2^2$ and $\lambda^2 = \lambda_1^2 = \lambda_2^2$. Do the factors of conformality fulfil a special Helmholtz equation? (iii) Prove that under "UPS left" as well as "UPS right" both the ellipsoidal North Pole and the spherical North are mapped isometrically. (iv) Derive a "simple conformal mapping" $\mathbb{E}^2_{A_1,A_1,A_2} \to \mathbb{S}^2_r$. (v) Why is the conformal mapping "UPS" called *stereographic*?

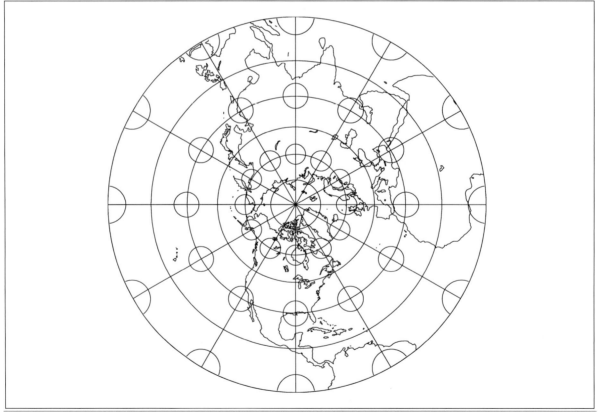

Fig. 1.26. Universal Polar Stereographic Projection of the sphere $\mathbb{S}^2_r$, shorelines of the northern hemisphere, Tissot ellipses of distortion. Graticule: $30°$ in longitude, $15°$ in latitude. Domain: $\{0 < \lambda < 2\pi, 0 < \phi < \pi/2\}$.

Box 1.31 (UPS of $\mathbb{E}^2_{A_1,A_1,A_2}$ versus UPS of $\mathbb{S}^2_r$).

<table>
<tr><td align="center">Left conformal coordinates
(isometric, isothermal):</td><td align="center">Right conformal coordinates
(isometric, isothermal):</td></tr>
</table>

$$P = \frac{2A_1}{\sqrt{1-E^2}}\left(\frac{1-E}{1+E}\right)^{E/2} \times$$

$$\times \tan\left(\frac{\pi}{4}-\frac{\Phi}{2}\right)\left(\frac{1+E\sin\Phi}{1-E\sin\Phi}\right)^{E/2}\cos\Lambda\,,$$

$$p = 2r\tan\left(\frac{\pi}{4}-\frac{\phi}{2}\right)\cos\lambda\,,$$

$$Q = \frac{2A_1}{\sqrt{1-E^2}}\left(\frac{1-E}{1+E}\right)^{E/2} \times$$

$$\times \tan\left(\frac{\pi}{4}-\frac{\Phi}{2}\right)\left(\frac{1+E\sin\Phi}{1-E\sin\Phi}\right)^{E/2}\sin\Lambda\,.$$

$$q = 2r\tan\left(\frac{\pi}{4}-\frac{\phi}{2}\right)\sin\lambda\,. \qquad (1.224)$$

<table>
<tr><td align="center">Left matrix of the metric G_l:</td><td align="center">Right matrix of the metric G_r:</td></tr>
</table>

$$\mathsf{G}_l = \begin{bmatrix} \dfrac{A_1^2\cos^2\Phi}{1-E^2\sin^2\Phi} & 0 \\[2ex] 0 & \dfrac{A_1^2(1-E^2)^2}{(1-E^2\sin^2\Phi)^3} \end{bmatrix}\,. \qquad \begin{bmatrix} r^2\cos^2\phi & 0 \\ 0 & r^2 \end{bmatrix} = \mathsf{G}_r\,. \qquad (1.225)$$

<table>
<tr><td align="center">Left Jacobi matrix:</td><td align="center">Right Jacobi matrix:</td></tr>
</table>

$$P = f(\Phi)\cos\Lambda\,,\ Q = f(\Phi)\sin\Lambda\,, \qquad\qquad p = g(\phi)\cos\lambda\,,\ q = g(\phi)\sin\lambda\,,$$

$$f(\Phi) := \frac{2A_1}{\sqrt{1-E^2}}\left(\frac{1-E}{1+E}\right)^{E/2} \times$$

$$\times\tan\left(\frac{\pi}{4}-\frac{\Phi}{2}\right)\left(\frac{1+E\sin\Phi}{1-E\sin\Phi}\right)^{E/2}\,,$$

$$g(\phi) := 2r\tan\left(\frac{\pi}{4}-\frac{\phi}{2}\right)\,,$$

$$\qquad\qquad (1.226)$$

$$P_\Lambda = -f(\Phi)\sin\Lambda\,,\ Q_\Lambda = +f(\Phi)\cos\Lambda\,, \qquad p_\lambda = -g(\phi)\sin\lambda\,,\ q_\lambda = +g(\phi)\cos\lambda\,,$$

$$P_\Phi = +f'(\Phi)\cos\Lambda\,,\ Q_\Phi = +f'(\Phi)\sin\Lambda\,, \qquad p_\phi = +g'(\phi)\cos\lambda\,,\ q_\phi = +g'(\phi)\sin\lambda\,,$$

$$f'(\Phi) = -\frac{1-E^2}{1-E^2\sin^2\Phi}\frac{f(\Phi)}{\cos\Phi}\,. \qquad\qquad g'(\phi) = -\frac{g(\phi)}{\cos\phi}\,.$$

<table><tr><td align="center">"Trigonometry":</td></tr></table>

$$\tan\frac{x}{2} = \frac{1-\cos x}{\sin x} : \tan\left(\frac{\pi}{4}-\frac{\phi}{2}\right) = \frac{1}{2}\tan\left(\frac{\pi}{2}-\phi\right) = \frac{1-\cos\left(\frac{\pi}{2}-\phi\right)}{\sin\left(\frac{\pi}{2}-\phi\right)} = \frac{1-\sin\phi}{\cos\phi}\,. \qquad (1.227)$$

<table><tr><td align="center">"Differential calculus":</td></tr></table>

$$y = \tan x\,,\ y' = \frac{1}{\cos^2 x} = 1+\tan^2 x :$$

$$\frac{\mathrm{d}}{\mathrm{d}\phi}\tan\left(\frac{\pi}{4}-\frac{\phi}{2}\right) = -\frac{1}{2}\left[1+\tan^2\left(\frac{\pi}{4}-\frac{\phi}{2}\right)\right] =$$

$$= -\frac{1}{2}\left[1+\tan^2\frac{1}{2}\left(\frac{\pi}{2}-\phi\right)\right] = -\frac{1}{2}\left[1+\frac{(1-\sin\phi)^2}{\cos^2\phi}\right] = \qquad (1.228)$$

$$= -\frac{1}{2}\frac{1}{\cos\phi}\frac{\cos^2\phi+1-2\sin\phi+\sin^2\phi}{\cos\phi} = -\frac{1}{\cos\phi}\frac{1-\sin\phi}{\cos\phi} = -\frac{1}{\cos\phi}\tan\left(\frac{\pi}{4}-\frac{\phi}{2}\right)\,.$$

Solution (the first problem).

Start from the conformal map defined in Box 1.31. The three conditions to be fulfilled are given in Box 1.32 for "left UPS" and in Box 1.33 for "right UPS". First, we specify the left and the right Korn–Lichtenstein equations, namely $\sqrt{G_{11}/G_{22}}$, $\sqrt{G_{22}/G_{11}}$ and $\sqrt{g_{11}/g_{22}}$, $\sqrt{g_{22}/g_{11}}$, respectively. Indeed, by transforming $\{Q_\Lambda, Q_\Phi, P_\Lambda, P_\Phi\}$ as well as $\{q_\lambda, q_\phi, p_\lambda, p_\phi\}$ left and right, KL 1st and KL 2nd are satified. Second, we analyze the left and the right Laplace–Beltrami equations as the integrability conditions of the left and the right Korn–Lichtenstein equations, namely by $\{G_{11}, G_{22}, P_\Lambda, P_\Phi, Q_\Lambda, Q_\Phi\}$ as well as $\{g_{11}, g_{22}, p_\lambda, p_\phi, q_\lambda, q_\phi\}$ of $\mathbb{E}^2_{A_1,A_1,A_2}$ and $\mathbb{S}^2_r$, respectively. Finally, we succeed to prove that $\{P, Q\}$ are *left harmonic coordinates* and $\{p, q\}$ are *right harmonic coordinates*. Third, we prove left and right orientation by computing the left and right *Jacobians* which are notably positive.

End of Solution (the first problem).

Solution (the second problem).

Again, we have to refer to standard textbooks of Differential Geometry, where you will find the representation of the Gaussian curvature of a surface in terms of conformal coordinates (isometric, isothermal). Let the left and the right matrix of the metric be equipped with a conformally flat structure $\{\mathsf{G}_l = \lambda_l^2 \mathsf{I}_2, \mathsf{G}_r = \lambda_r^2 \mathsf{I}_2\}$, which is generated by a left and a right conformal coordinate representation. Then the left Gaussian curvature and the right Gaussian curvature are provided by $k_l = -(1/2\lambda_l^2)\Delta_l \ln \lambda_l^2 = -(1/\lambda_l^2)\Delta_l \ln \lambda_l$ and $k_r = -(1/2\lambda_r^2)\Delta_r \ln \lambda_r^2 = -(1/\lambda_r^2)\Delta_r \ln \lambda_r$ as well as $\Delta_l := D_{PP} + D_{QQ} = D_P^2 + D_Q^2$ and $D_p^2 + D_q^2 = D_{pp} + D_{qq} := \Delta_r$, where Δ_l and Δ_r represent the left Laplace–Beltrami operator and the right Laplace–Beltrami operator. Let us apply this result in solving the second problem again.

- By means of Box 1.34, we have outlined how to generate a conformally flat metric of an ellipsoid-of-revolution and of a sphere. First, we depart from the arc lengths "left dS^2" given in "left coordinates" $\{\Lambda, \Phi\}$ as well as "right ds^2" given in "right coordinates" $\{\lambda, \phi\}$. Second, we compute the left and right Cauchy–Green matrices from "left $dP^2 + dQ^2$" and "right $dp^2 + dq^2$", the arc lengths squared of the projective plane covered by left conformal coordinates $\{P, Q\}$ and by right conformal coordinates $\{p, q\}$, respectively. In particular, we arrive at the two equations $dP^2 + dQ^2 = f^2(\Phi)d\Lambda^2 + f'^2(\Phi)d\Phi^2$ and $dp^2 + dq^2 = g^2(\phi)d\lambda^2 + g'^2(\phi)d\phi^2$, and the corresponding elements of the left Cauchy–Green matrix C_l and of the right Cauchy–Green matrix C_r. Third, we determine the left eigenvalues $\{_l\Lambda_1^2, _l\Lambda_2^2\}$ and the right eigenvalues $\{_r\Lambda_1^2, _r\Lambda_2^2\}$ in solving the left characteristic equation $|\mathsf{C}_l - \Lambda_l^2\mathsf{G}_l| = 0$ and the right characteristic equation $|\mathsf{C}_r - \Lambda_r^2\mathsf{G}_r| = 0$. In particular, we prove the identities "left $_l\Lambda_1^2 =_l \Lambda_2^2 = \Lambda_l^2$" and "right $_r\Lambda_1^2 =_r \Lambda_2^2 = \Lambda_r^2$", characteristic for a conformal mapping. Fourth, due to the duality relations $\Lambda_l^2\lambda_l^2 = 1$ $\Lambda_r^2\lambda_r^2 = 1$, we are able to compute λ_l^2 and λ_r^2, respectively, and the conformally flat metric of type "left $dS^2 = \lambda_l^2(dP^2 + dQ^2)$" and of type "right $ds^2 = \lambda_r^2(dp^2 + dq^2)$".

- Fifth, by means of Box 1.35, we aim at representing the factors of conformality, $\lambda_l^2(\phi)$ and $\lambda_r^2(\phi)$, in terms of left conformal coordinates $\{P, Q\}$ and right conformal coordinates $\{p, q\}$. We begin with transforming the right factor of conformality, $\lambda_r^2(\phi) \rightarrow \lambda_r^2(p, q)$, since it is available in closed form. In contrast, the transformation of the left factor of conformality, $\lambda_l^2(\Phi) \rightarrow \lambda_l^2(P, Q)$, is only symbolically written since $f^{-1}(\sqrt{P^2 + Q^2})$ is not available in closed form. Note the beautiful transformations $\{\sin\phi, \cos\phi\}$ and $\{\sin\lambda, \cos\lambda\}$ as functions of the "UPS coordinates" p and q.

- Sixth, Box 1.36 outlines that λ_r^2 and λ_l^2, respectively, fulfill the conformal representation of the right and the left Gaussian curvature, here written in two versions as a special Helmholtz differential equation. The simple representation is performed first, followed by the left representation. For a given Gaussian curvature $k_r = 1/r^2 = $ constant of the sphere $\mathbb{S}^2_r$ and $k_l = (1 - E^2\sin^2\Phi)^2/[A_1^2(1 - E^2)]$ of the ellipsoid-of-revolution $\mathbb{E}^2_{A_1,A_1,A_2}$ being transformed into $\{p, q\}$ and $\{P, Q\}$ left and right conformal coordinates, we succeed to prove $\Delta \ln \lambda^2 + 2k\lambda^2 = 0$ of type "right" and "left".

End of Solution (the second problem).

Box 1.32 (Left Korn–Lichtenstein equations, UPS of $\mathbb{E}^2_{A_1,A_1,A_2}$, harmonicity, orientation).

Left Korn–Lichtenstein equations:

$$\text{(1st)} \quad P_\Lambda = +\sqrt{\frac{G_{11}}{G_{22}}}Q_\Phi \, , \quad P_\Phi = -\sqrt{\frac{G_{22}}{G_{11}}}Q_\Lambda \quad \text{(2nd)} \, , \tag{1.229}$$

$$\text{(1st)} \quad Q_\Lambda = -\sqrt{\frac{G_{11}}{G_{22}}}P_\Phi \, , \quad Q_\Phi = +\sqrt{\frac{G_{22}}{G_{11}}}P_\Lambda \quad \text{(2nd)} \, ;$$

$$\sqrt{\frac{G_{11}}{G_{22}}} = \cos\Phi \frac{1-E^2\sin^2\Phi}{1-E^2} \Leftrightarrow \sqrt{\frac{G_{22}}{G_{11}}} = \frac{1}{\cos\Phi}\frac{1-E^2}{1-E^2\sin^2\Phi} \, ; \tag{1.230}$$

$$\text{(UPS left)} \quad \begin{cases} Q_\Phi = f'(\Phi)\sin\Lambda = -\dfrac{1-E^2}{1-E^2\sin^2\Phi}\dfrac{1}{\cos\Phi}f(\Phi)\sin\Lambda \\[2mm] P_\Lambda = -f(\Phi)\sin\Lambda \end{cases} \quad \text{(KL 2nd)} \, , \tag{1.231}$$

$$\text{(UPS left)} \quad \begin{cases} Q_\Lambda = f(\Phi)\cos\Lambda \\[2mm] P_\Phi = f'(\Phi)\cos\Lambda = -\dfrac{1-E^2}{1-E^2\sin^2\Phi}\dfrac{1}{\cos\Phi}f(\Phi)\cos\Lambda \end{cases} \quad \text{(KL 1st)} \, .$$

Left integrability conditions:

$$\Delta_{\Lambda,\Phi}P = \left(\sqrt{\frac{G_{11}}{G_{22}}}P_\Phi\right)_\Phi + \left(\sqrt{\frac{G_{22}}{G_{11}}}P_\Lambda\right)_\Lambda = 0 \, ,$$

$$\Delta_{\Lambda,\Phi}Q = \left(\sqrt{\frac{G_{11}}{G_{22}}}Q_\Phi\right)_\Phi + \left(\sqrt{\frac{G_{22}}{G_{11}}}Q_\Lambda\right)_\Lambda = 0 \, . \tag{1.232}$$

$$\text{(1st)} \quad \sqrt{\frac{G_{11}}{G_{22}}}P_\Phi = \cos\Phi\frac{1-E^2\sin^2\Phi}{1-E^2}\left(-\frac{1-E^2}{1-E^2\sin^2\Phi}\frac{1}{\cos\Phi}\right)f(\Phi)\cos\Lambda = -f(\Phi)\cos\Lambda \, ,$$

$$\sqrt{\frac{G_{22}}{G_{11}}}P_\Lambda = \frac{1-E^2}{1-E^2\sin^2\Phi}\frac{1}{\cos\Phi}\left[-f(\Phi)\sin\Lambda\right] \, ,$$

$$\Delta_{\Lambda,\Phi}P = -f'(\Phi)\cos\Lambda - \frac{1}{\cos\Phi}\frac{1-E^2\sin^2\Phi}{1-E^2}f(\Phi)\cos\Lambda = \tag{1.233}$$

$$= +\frac{1-E^2}{1-E^2\sin^2\Phi}\frac{1}{\cos\Phi}f(\Phi)\cos\Lambda - \frac{1-E^2}{1-E^2\sin^2\Phi}\frac{1}{\cos\Phi}f(\Phi)\cos\Lambda = 0 \quad \text{q.\,e.\,d.}$$

$$\text{(2nd)} \quad \sqrt{\frac{G_{11}}{G_{22}}}Q_\Phi = \cos\Phi\frac{1-E^2\sin^2\Phi}{1-E^2}\left(-\frac{1-E^2}{1-E^2\sin^2\Phi}\frac{1}{\cos\Phi}\right)f(\Phi)\sin\Lambda = -f(\Phi)\sin\Lambda \, ,$$

$$\sqrt{\frac{G_{22}}{G_{11}}}Q_\Lambda = \frac{1-E^2}{1-E^2\sin^2\Phi}\frac{1}{\cos\Phi}\left[+f(\Phi)\cos\Lambda\right] \, ,$$

$$\Delta_{\Lambda,\Phi}Q = -f'(\Phi)\sin\Lambda - \frac{1}{\cos\Phi}\frac{1-E^2\sin^2\Phi}{1-E^2}f(\Phi)\sin\Lambda = \tag{1.234}$$

$$= +\frac{1-E^2}{1-E^2\sin^2\Phi}\frac{1}{\cos\Phi}f(\Phi)\sin\Lambda - \frac{1-E^2}{1-E^2\sin^2\Phi}\frac{1}{\cos\Phi}f(\Phi)\sin\Lambda = 0 \quad \text{q.\,e.\,d.}$$

Left orientation:

$$\begin{vmatrix} P_\Lambda & P_\Phi \\ Q_\Lambda & Q_\Phi \end{vmatrix} = P_\Lambda Q_\Phi - P_\Phi Q_\Lambda = -ff'\left(\sin^2\Lambda + \cos^2\Lambda\right) = -f(\Phi)f'(\Phi) = \tag{1.235}$$

$$= \frac{1-E^2}{1-E^2\sin^2\Phi}\frac{1}{\cos\Phi}f^2(\Phi) > 0 \quad \text{due to} \quad -\pi/2 < \Phi < +\pi/2 \Rightarrow \cos\Phi > 0 \quad \text{q.\,e.\,d.}$$

Box 1.33 (Right Korn–Lichtenstein equations, UPS of $\mathbb{S}^2_r$, harmonicity, orientation).

Right Korn–Lichtenstein equations:

$$\text{(1st)} \quad p_\lambda = +\sqrt{\frac{g_{11}}{g_{22}}}\, q_\phi \;, \quad p_\phi = -\sqrt{\frac{g_{22}}{g_{11}}}\, q_\lambda \quad \text{(2nd)} \;,$$

$$\text{(1st)} \quad q_\lambda = -\sqrt{\frac{g_{11}}{g_{22}}}\, p_\phi \;, \quad q_\phi = +\sqrt{\frac{g_{22}}{g_{11}}}\, p_\lambda \quad \text{(2nd)} \;;$$

(1.236)

$$\sqrt{\frac{g_{11}}{g_{22}}} = \cos\phi \;, \quad \sqrt{\frac{g_{22}}{g_{11}}} = \frac{1}{\cos\phi} \;;$$

(1.237)

$$\text{(UPS right)} \quad q_\phi = g'(\phi)\sin\lambda \;, \quad p_\lambda = -g(\phi)\sin\lambda \;,$$

$$\text{(UPS right)} \quad q_\lambda = g(\phi)\cos\lambda \;, \quad p_\phi = +g'(\phi)\cos\lambda \;,$$

$$g'(\phi) = -\frac{g(\phi)}{\cos\phi}$$
$$\Rightarrow$$

(1.238)

$$\text{(KL 2nd)} \quad g'(\phi)\sin\lambda = -\frac{g(\phi)}{\cos\phi}\sin\lambda \quad \text{q.\,e.\,d.}$$

$$\text{(KL 1st)} \quad g(\phi)\cos\lambda = -\cos\phi\, g'(\phi)\cos\lambda \quad \text{q.\,e.\,d.}$$

Right integrability conditions:

$$\Delta_{\lambda,\phi}\, p = \left(\sqrt{\frac{g_{11}}{g_{22}}}\, p_\phi\right)_\phi + \left(\sqrt{\frac{g_{22}}{g_{11}}}\, p_\lambda\right)_\lambda = 0 \;,$$

$$\Delta_{\lambda,\phi}\, q = \left(\sqrt{\frac{g_{11}}{g_{22}}}\, q_\phi\right)_\phi + \left(\sqrt{\frac{g_{22}}{g_{11}}}\, q_\lambda\right)_\lambda = 0 \;.$$

(1.239)

$$\text{(1st)} \quad \sqrt{\frac{g_{11}}{g_{22}}}\, p_\phi = \cos\phi\, g'(\phi)\cos\lambda = -g(\phi)\cos\lambda \;,$$

$$\sqrt{\frac{g_{22}}{g_{11}}}\, p_\lambda = \frac{1}{\cos\phi}\,[-g(\phi)\sin\lambda] = -\frac{g(\phi)}{\cos\phi}\sin\lambda$$
$$\Rightarrow$$

(1.240)

$$\Delta_{\lambda,\phi}\, p = -g'(\phi)\cos\lambda - \frac{g(\phi)}{\cos\phi}\cos\lambda \;, \quad g'(\phi) = -\frac{g(\phi)}{\cos\phi} \Rightarrow \Delta_{\lambda,\phi}\, p = 0 \quad \text{q.\,e.\,d.}$$

$$\text{(2nd)} \quad \sqrt{\frac{g_{11}}{g_{22}}}\, q_\phi = \cos\phi\, g'(\phi)\sin\lambda = -g(\phi)\sin\lambda \;,$$

$$\sqrt{\frac{g_{22}}{g_{11}}}\, q_\lambda = \frac{1}{\cos\phi}\, g(\phi)\cos\lambda$$
$$\Rightarrow$$

(1.241)

$$\Delta_{\lambda,\phi}\, q = -g'(\phi)\sin\lambda - \frac{g(\phi)}{\cos\phi}\sin\lambda \;, \quad g'(\phi) = -\frac{g(\phi)}{\cos\phi} \Rightarrow \Delta_{\lambda,\phi}\, q = 0 \quad \text{q.\,e.\,d.}$$

Right orientation:

$$\begin{vmatrix} p_\lambda & p_\phi \\ q_\lambda & q_\phi \end{vmatrix} = p_\lambda q_\phi - p_\phi q_\lambda = -g(\phi)g'(\phi) = \frac{g^2(\phi)}{\cos\phi} > 0$$

(1.242)

due to

$$-\pi/2 < \phi < |\,\pi/2 \Rightarrow \cos\phi > 0 \quad \text{q.\,o.\,d.}$$

(1.243)

Box 1.34 (Conformally flat left and right manifold, ellipsoid-of-revolution versus sphere).

<table>
<tr><td>Left manifold,
$\{\Lambda, \Phi\}$ left coordinates:</td><td>Right manifold,
$\{\lambda, \phi\}$ right coordinates:</td></tr>
</table>

$$\mathrm{d}S^2 = \frac{A_1 \cos^2 \Phi}{1 - E^2 \sin^2 \Phi} \mathrm{d}\Lambda^2 + \frac{A_1^2 (1 - E^2)^2}{(1 - E^2 \sin^2 \Phi)^3} \mathrm{d}\Phi^2 \ . \qquad \mathrm{d}s^2 = r^2 \cos^2 \phi \, \mathrm{d}\lambda^2 + r^2 \mathrm{d}\phi^2 \ . \qquad (1.244)$$

Left Cauchy–Green matrix: Right Cauchy–Green matrix:

$$\begin{aligned}
\mathrm{d}P^2 + \mathrm{d}Q^2 &= \left(P_\Lambda^2 + Q_\Lambda^2\right) \mathrm{d}\Lambda^2 + \\
&+ 2\left(P_\Lambda P_\Phi + Q_\Lambda Q_\Phi\right) \mathrm{d}\Lambda \mathrm{d}\Phi + \\
&+ \left(P_\Phi^2 + Q_\Phi^2\right) \mathrm{d}\Phi^2 = \\
&= f^2(\Phi) \mathrm{d}\Lambda^2 + f'^{\,2}(\Phi) \mathrm{d}\Phi^2
\end{aligned} \qquad
\begin{aligned}
\mathrm{d}p^2 + \mathrm{d}q^2 &= \left(p_\lambda^2 + q_\lambda^2\right) \mathrm{d}\lambda^2 + \\
&+ 2\left(p_\lambda p_\phi + q_\lambda q_\phi\right) \mathrm{d}\lambda \mathrm{d}\phi + \\
&+ \left(p_\phi^2 + q_\phi^2\right) \mathrm{d}\phi^2 = \\
&= g^2(\phi) \mathrm{d}\lambda^2 + g'^{\,2}(\phi) \mathrm{d}\phi^2
\end{aligned} \qquad (1.245)$$

$$\Leftrightarrow \qquad\qquad\qquad \Leftrightarrow$$

$$_l c_{11} = f^2(\Phi) \ , \quad _l c_{12} = 0 \ , \quad _l c_{22} = f'^{\,2}(\Phi) \ ; \qquad _r c_{11} = g^2(\phi) \ , \quad _r c_{12} = 0 \ , \quad _r c_{22} = g'^{\,2}(\phi) \ ;$$

$$\left| \mathsf{C}_l - \Lambda_l^2 \mathsf{G}_l \right| = 0 \qquad\qquad\qquad \left| \mathsf{C}_r - \Lambda_r^2 \mathsf{G}_r \right| = 0$$

$$\Leftrightarrow \qquad\qquad\qquad \Leftrightarrow \qquad\qquad (1.246)$$

$$\begin{vmatrix} f^2(\Phi) - \Lambda_l^2 G_{11} & 0 \\ 0 & f'^{\,2}(\Phi) - \Lambda_l^2 G_{22} \end{vmatrix} = 0 \ ; \qquad \begin{vmatrix} g^2(\phi) - \Lambda_r^2 g_{11} & 0 \\ 0 & g'^{\,2}(\phi) - \Lambda_r^2 g_{22} \end{vmatrix} = 0 \ ;$$

$$_l \Lambda_1^2 = \frac{f^2(\Phi)}{G_{11}} = \frac{f^2(\Phi)}{A_1^2 \cos^2 \Phi}(1 - E^2 \sin^2 \Phi) \ , \qquad _r \Lambda_1^2 = \frac{g^2(\phi)}{g_{11}} = \frac{g^2(\phi)}{r^2 \cos^2 \phi} \ ,$$

$$_l \Lambda_2^2 = \frac{f'^{\,2}(\Phi)}{G_{22}} = \qquad\qquad\qquad _r \Lambda_2^2 = \frac{g'^{\,2}(\phi)}{g_{22}} =$$

$$= \frac{f'^{\,2}(\Phi)}{A_1^2 (1 - E^2)}(1 - E^2 \sin^2 \Phi) \ , \qquad\qquad = \frac{g'^{\,2}(\phi)}{r^2} \ , \qquad\qquad (1.247)$$

$$f'^{\,2}(\Phi) = \frac{(1 - E^2)^2}{(1 - E^2 \sin^2 \Phi)^2} \frac{f^2(\Phi)}{\cos^2 \Phi} \qquad\qquad g'^{\,2}(\phi) = \frac{g^2(\phi)}{\cos^2 \phi}$$

$$\Rightarrow \qquad\qquad\qquad \Rightarrow$$

$$_l \Lambda_1^2 = {}_l \Lambda_2^2 = \Lambda_l^2 \ ; \qquad\qquad\qquad _r \Lambda_1^2 = {}_r \Lambda_2^2 = \Lambda_r^2 \ ;$$

$$\frac{r^2 \cos \phi}{g^2(\phi)} = \frac{r^2 \cos^2 \phi}{4 r^2 \tan^2 \left(\frac{\pi}{4} - \frac{\phi}{2}\right)} = \frac{1}{4} \frac{\sin^2 \left(\frac{\pi}{2} - \phi\right)}{\tan^2 \left(\frac{\pi}{4} - \frac{\phi}{2}\right)} = \frac{\sin^2 \left(\frac{\pi}{4} - \frac{\phi}{2}\right) \cos^2 \left(\frac{\pi}{4} - \frac{\phi}{2}\right)}{\tan^2 \left(\frac{\pi}{4} - \frac{\phi}{2}\right)} =$$

$$= \cos^4 \left(\frac{\pi}{4} - \frac{\phi}{2}\right) \ ; \qquad\qquad (1.248)$$

$$\Lambda_l^2 = \frac{1 - E^2 \sin^2 \Phi}{A_1^2 \cos^2 \Phi} f^2(\Phi) \qquad\qquad\qquad \Lambda_r^2 = \frac{1}{\cos^4 \left(\frac{\pi}{4} - \frac{\phi}{2}\right)}$$

$$\Leftrightarrow \qquad\qquad\qquad \Leftrightarrow \qquad\qquad (1.249)$$

$$\lambda_l^2 = \frac{A_1^2 \cos^2 \Phi}{1 - E^2 \sin^2 \Phi} \frac{1}{f^2(\Phi)} \ . \qquad\qquad\qquad \lambda_r^2 = \cos^4 \left(\frac{\pi}{4} - \frac{\phi}{2}\right) \ .$$

Conformally flat left manifold: Conformally flat right manifold:

$$\mathrm{d}S^2 = \lambda_l^2 \left(\mathrm{d}P^2 + \mathrm{d}Q^2\right) \ . \qquad\qquad \mathrm{d}s^2 = \lambda_r^2 \left(\mathrm{d}p^2 + \mathrm{d}q^2\right) \ . \qquad (1.250)$$

Box 1.35 (Representation of the factors of conformality in terms of conformal coordinates).

Left factor of conformality:

$$P(\Lambda, \Phi) = f(\Phi) \cos \Lambda \;, \;\; Q(\Lambda, \Phi) = f(\Phi) \sin \Lambda \;,$$

$$\lambda_l^2 = \frac{A_1^2 \cos^2 \Phi}{1 - E^2 \sin^2 \Phi} \frac{1}{f^2(\Phi)} = \frac{A_1^2 \cos^2 f^{-1}\left(\sqrt{P^2 + Q^2}\right)}{1 - E^2 \sin^2 f^{-1}\left(\sqrt{P^2 + Q^2}\right)} \frac{1}{P^2 + Q^2} \;,$$

$$\Lambda_l^2 = \frac{1 - E^2 \sin^2 f^{-1}\left(\sqrt{P^2 + Q^2}\right)}{A_1^2 \cos^2 f^{-1}\left(\sqrt{P^2 + Q^2}\right)} \left(P^2 + Q^2\right) \;.$$

(1.251)

Right factor of conformality:

$$p(\lambda, \phi) = 2r \tan\left(\frac{\pi}{4} - \frac{\phi}{2}\right) \cos \lambda \;, \;\; q(\lambda, \phi) = 2r \tan\left(\frac{\pi}{4} - \frac{\phi}{2}\right) \sin \lambda \;, \;\; \tan(\alpha/2) = \sqrt{\frac{1 - \cos \alpha}{1 + \cos \alpha}}$$

$$\Leftrightarrow$$

$$\tan\left(\frac{\pi}{4} - \frac{\phi}{2}\right) = \tan\frac{1}{2}\left(\frac{\pi}{2} - \phi\right) = \sqrt{\frac{1 - \sin \alpha}{1 + \sin \alpha}} \;, \;\; 2r \tan\left(\frac{\pi}{4} - \frac{\phi}{2}\right) = 2r\sqrt{\frac{1 - \sin \alpha}{1 + \sin \alpha}} = \sqrt{p^2 + q^2}$$

$$\Rightarrow$$

$$\sin \phi = \frac{4r^2 - (p^2 + q^2)}{4r^2 + (p^2 + q^2)} \;, \;\; \cos \phi = \frac{4r\sqrt{p^2 + q^2}}{4r^2 + (p^2 + q^2)} \;,$$

(1.252)

$$\sin \lambda = \frac{\tan \lambda}{\sqrt{1 + \tan^2 \lambda}} = \frac{q}{\sqrt{p^2 + q^2}} \;, \;\; \cos \lambda = \frac{1}{\sqrt{1 + \tan^2 \lambda}} = \frac{p}{\sqrt{p^2 + q^2}} \;,$$

$$\lambda_r^2 = \cos^4\left(\frac{\pi}{4} - \frac{\phi}{2}\right) = \frac{1}{4}\left(1 + \sin \phi\right)^2 = \frac{16 r^4}{(4r^2 + p^2 + q^2)^2} \;,$$

$$\Lambda_r^2 = \frac{1}{\cos^4\left(\frac{\pi}{4} - \frac{\phi}{2}\right)} = \frac{4}{\left(1 + \sin \phi\right)^2} = \frac{\left(4r^2 + p^2 + q^2\right)^2}{16 r^4} \;.$$

Box 1.36 (The differential equation which governs the factor of conformality).

Two versions of the special Helmholtz equations (k is the Gaussian curvature $k(p, q)$):

$$\text{(i)} \quad \Delta \ln \lambda^2 + 2k\lambda^2 = 0 \;. \quad \text{(ii)} \quad \Delta \lambda^2 + 2k\lambda^4 = 0 \;.$$

(1.253)

Right differential equation of the factor of conformality ($\mathbb{S}_r^2$):

$$k_r = \frac{1}{r^2} = \text{constant} \;, \;\; \Delta \ln \lambda_r^2 + \frac{2}{r^2}\lambda_r^2 = 0 \;,$$

$$\lambda_r^2 = \frac{16 r^4}{(4r^2 + p^2 + q^2)^2} \;, \;\; \ln \lambda_r^2 = \ln 16 r^4 - 2 \ln\left(4r^2 + p^2 + q^2\right) \;,$$

$$D_p \ln \lambda_r^2 = -4\frac{p}{4r^2 + p^2 + q^2} \;, \;\; D_q \ln \lambda_r^2 = -4\frac{q}{4r^2 + p^2 + q^2} \;,$$

$$D_{pp} \ln \lambda_r^2 = D_p^2 \ln \lambda_r^2 = -4\frac{4r^2 + p^2 + q^2 - 2p^2}{(4r^2 + p^2 + q^2)^2} \;, \;\; D_{qq} \ln \lambda_r^2 = D_q^2 \ln \lambda_r^2 = -4\frac{4r^2 + p^2 + q^2 - 2q^2}{(4r^2 + p^2 + q^2)^2} \;,$$

(1.254)

$$D_{pp} \ln \lambda_r^2 = -\frac{4}{(4r^2 + p^2 + q^2)^2}\left(4r^2 - p^2 + q^2\right) \;, \;\; D_{qq} \ln \lambda_r^2 = -\frac{4}{(4r^2 + p^2 + q^2)^2}\left(4r^2 + p^2 - q^2\right) \;,$$

$$\Delta_r \ln \lambda_r^2 = -\frac{32 r^2}{(4r^2 + p^2 + q^2)^2} = -\frac{2}{r^2}\lambda_r^2 \quad \text{q.\,e.\,d.}$$

Continuation of Box.

Left differential equation of the factor of conformality ($\mathbb{E}^2_{A_1,A_1,A_2}$):

$$k_l = \frac{(1 - E^2 \sin^2 \phi)^2}{A_1^2(1 - E^2)} =$$

$$= \frac{1 - E^2 \sin^2 f^{-1}\left(\sqrt{P^2 + Q^2}\right)}{A_1^2(1 - E^2)} \ ,$$

$$\Delta \ln \lambda_l^2 + 2k(P,Q)\lambda_l^2 = 0 \ ,$$

$$\lambda_l^2 = \frac{A_1^2 \cos^2 f^{-1}\left(\sqrt{P^2 + Q^2}\right)}{1 - E^2 \sin^2 f^{-1}\left(\sqrt{P^2 + Q^2}\right)} \frac{1}{P^2 + Q^2} \ ,$$

(1.255)

$$\ln \lambda_l^2 = \ln A_1^2 + 2\ln \cos f^{-1}\left(\sqrt{P^2 + Q^2}\right) - \ln\left[1 - E^2 \sin^2 f^{-1}\left(\sqrt{P^2 + Q^2}\right)\right] - \ln\left(P^2 + Q^2\right) \ ,$$

$$\Delta_l \ln \lambda_l^2 = \left(D_P^2 + D_Q^2\right)\ln \lambda_l^2 = -2k(P,Q)\lambda_l^2$$

q. e. d.

Box 1.37 (Representation of Φ_l^{-1} and Φ_r^{-1} in terms of conformal coordinates: $\mathbb{E}^2_{A_1,A_1,A_2} \to \mathbb{S}^2_r$).

$$\Phi_l^{-1} : \ \boldsymbol{X}(\Lambda, \Phi) = \qquad\qquad \Phi_r^{-1} : \ \boldsymbol{x}(\lambda, \phi) =$$

$$= \boldsymbol{E}_1 \frac{A_1 \cos \Phi \cos \Lambda}{\sqrt{1 - E^2 \sin^2 \Phi}} + \qquad\qquad = \boldsymbol{e}_1 \, r \cos \phi \cos \lambda +$$

$$+ \boldsymbol{E}_2 \frac{A_1 \cos \Phi \sin \Lambda}{\sqrt{1 - E^2 \sin^2 \Phi}} + \qquad\qquad + \boldsymbol{e}_2 \, r \cos \phi \sin \lambda +$$

$$+ \boldsymbol{E}_3 \frac{A_1(1 - E^2)\sin \Phi}{\sqrt{1 - E^2 \sin^2 \Phi}} = \qquad\qquad + \boldsymbol{e}_3 \, r \sin \phi \quad =$$

(1.256)

$$= \boldsymbol{E}_1 \frac{A_1 \cos f^{-1}\left(\sqrt{P^2 + Q^2}\right)}{\sqrt{1 - E^2 \sin^2 f^{-1}\left(\sqrt{P^2 + Q^2}\right)}} \frac{P}{\sqrt{P^2 + Q^2}} + \qquad = \ \boldsymbol{e}_1 \, 4r^2 \frac{p}{4r^2 + p^2 + q^2} +$$

$$+ \boldsymbol{E}_2 \frac{A_1 \cos f^{-1}\left(\sqrt{P^2 + Q^2}\right)}{\sqrt{1 - E^2 \sin^2 f^{-1}\left(\sqrt{P^2 + Q^2}\right)}} \frac{Q}{\sqrt{P^2 + Q^2}} + \qquad + \boldsymbol{e}_2 \, 4r^2 \frac{q}{4r^2 + p^2 + q^2} +$$

$$+ \quad \boldsymbol{E}_3 \frac{A_1(1 - E^2)\sin f^{-1}\left(\sqrt{P^2 + Q^2}\right)}{\sqrt{1 - E^2 \sin^2 f^{-1}\left(\sqrt{P^2 + Q^2}\right)}} \cdot \qquad + \quad \boldsymbol{e}_3 \, r \frac{4r^2 - (p^2 + q^2)}{4r^2 + p^2 + q^2} \cdot$$

Isoparametric mapping:

$$p = P \ , \quad q = Q \ .$$

(1.257)

Solution (the third problem).

By means of the two mapping equations "left" $\{P = f(\Phi)\cos\Lambda, Q = f(\Phi)\sin\Lambda\}$ and of the two mapping equations "right" $\{p = g(\phi)\cos\lambda, q = g(\phi)\sin\lambda\}$, we are able to compute the factors of conformality $\{\Lambda_l^2, \lambda_l^2\}$ of type "left" and $\{\Lambda_r^2, \lambda_r^2\}$ of type "right". The detailed formulae are reviewed in Box 1.34. If we specify "left" $\Phi = \pi/2$ (ellipsoidal North Pole) or "right" $\phi = \pi/2$ (spherical North Pole), we are led to $_l\Lambda_1^2(\pi/2) = {}_l\Lambda_2^2(\pi/2) = \Lambda_l^2(\pi/2) = 1$ and $_r\Lambda_1^2(\pi/2) = {}_r\Lambda_2^2(\pi/2) = \Lambda_r^2(\pi/2) = 1$. Obviously, at the North Pole, "left UPS" and "right UPS" are an isometry. We shall see later that this is a built-in constraint for any UPS.

End of Solution (the third problem).

Solution (the fourth problem).

A "simple conformal mapping" of $\mathbb{E}_{A_1,A_1,A_2}^2 \to \mathbb{S}_r^2$ is the isoparametric mapping, which is conveniently characterized by

$$p = P, \; q = Q \; \text{ or } \;
\begin{aligned}
2r\tan\left(\tfrac{\pi}{4} - \tfrac{\phi}{2}\right)\cos\lambda &= \tfrac{2A_1}{\sqrt{1-E^2}} \tfrac{(1-E)^{E/2}}{(1+E)^{E/2}} \tan\left(\tfrac{\pi}{4} - \tfrac{\Phi}{2}\right)\left[\tfrac{1+E\sin\Phi}{1-E\sin\Phi}\right]^{E/2}\cos\Lambda\,, \\
2r\tan\left(\tfrac{\pi}{4} - \tfrac{\phi}{2}\right)\sin\lambda &= \tfrac{2A_1}{\sqrt{1-E^2}} \tfrac{(1-E)^{E/2}}{(1+E)^{E/2}} \tan\left(\tfrac{\pi}{4} - \tfrac{\Phi}{2}\right)\left[\tfrac{1+E\sin\Phi}{1-E\sin\Phi}\right]^{E/2}\sin\Lambda\,.
\end{aligned}
\tag{1.258}$$

Here, we conclude with a representation of the left as well as the right inverse mapping, namely $\Phi_l^{-1} : \{P, Q\} \to \mathbf{X}(P,Q)$ and $\Phi_r^{-1} : \{p, q\} \to \mathbf{x}(p,q)$ in terms of conformal coordinates (isometric, isothermal) of Box 1.37, which specializes Φ_l^{-1} and Φ_r^{-1} of Box 1.21.

End of Solution (the fourth problem).

Solution (the fifth problem).

By means of Fig. 1.27, we illustrate why "UPS" is called *stereographic*. The stereographic projection of the "left" ellipsoid-of-revolution $\mathbb{E}_{A_1,A_1,A_2}^2$ and the "right" sphere $\mathbb{S}_r^2$ is based upon three elements of *projective geometry* of type *central perspective*. First, we define the *perspective center*, here the ellipsoidal "left" South Pole S_l as well as the spherical "right" South Pole S_r. Second, we define the bundle of projection lines leaving S_l and S_r, respectively, and intersecting $\mathbb{E}_{A_1,A_1,A_2}^2$ at P_l and $\mathbb{S}_r^2$ at S_r. Third, we define the projective plane $\mathbb{P}_{N_l}^2$ and $\mathbb{P}_{N_r}^2$, respectively, namely the tangent planes $T_{N_l}\mathbb{E}_{A_1,A_1,A_2}^2$ at the "left" ellipsoidal North Pole and $T_{N_r}\mathbb{S}_r^2$ at the "right" spherical North Pole, respectively. The projection lines $S_l \to P_l$ intersect the projective plane at p_l, an element of the "left" tangent plane at the "left" North Pole, and the projection lines $S_r \to P_r$ intersect the projective plane at p_r, an element of the "right" tangent plane at the "right" North Pole. Note that we have collected the fundamental "left" and "right" ratios of projective geometry in Box 1.38. Their conversion to $\sqrt{P^2 + Q^2}$ "left" and $\sqrt{p^2 + q^2}$ "right" generates the map $\Phi \to f(\Phi)$ and $\phi \to g(\phi)$, respectively. The projective planes are covered by polar coordinates of type "left" $\{\sqrt{P^2 + Q^2}\cos\alpha_l, \sqrt{P^2 + Q^2}\sin\alpha_l\}$ and of type "right" $\{\sqrt{p^2 + q^2}\cos\alpha_r, \sqrt{p^2 + q^2}\sin\alpha_r\}$, respectively. $\{\sqrt{P^2 + Q^2}, \sqrt{p^2 + q^2}\}$ are the *radial coordinates*, $\{\alpha_l, \alpha_r\}$ are the "left and "right" *South azimuths*. The central perspective generates $\sqrt{P^2 + Q^2} = f(\Phi)$ versus $\sqrt{p^2 + q^2} = g(\phi)$ and $\alpha_l = \Lambda$ versus $\alpha_r = \lambda$. Indeed, "UPS" is *azimuth preserving*: the "left" azimuth is identified as ellipsoidal longitude, the "right" azimuth as spherical longitude, and

$$\begin{aligned}
P(\Lambda, \Phi) = f(\Phi)\cos\Lambda \quad &\text{versus} \quad p(\lambda, \phi) = g(\Phi)\cos\lambda\,, \\
Q(\Lambda, \Phi) = f(\Phi)\sin\Lambda \quad &\text{versus} \quad q(\lambda, \phi) = g(\Phi)\sin\lambda\,.
\end{aligned}
\tag{1.259}$$

End of Solution (the fifth problem).

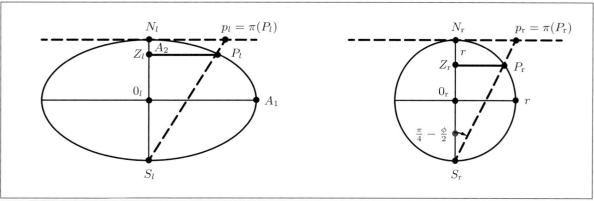

Fig. 1.27. Left: vertical section of the "left" ellipsoid-of-revolution $\mathbb{E}^2_{A_1,A_1,A_2}$, projective geometry of type central perspective (perspective center S_l, projection line $S_l P_l p_l$, projective plane $\mathbb{P}^2_{N_l}$). Right: vertical section of the "right" sphere $\mathbb{S}^2_r$, projective geometry of type central perspective (perspective center S_r, projection line $S_r P_r p_r$, projective plane $\mathbb{P}^2_{N_r}$).

Box 1.38 (Projective geometry of type central perspective).

Left projective ratio:

$$\frac{Z_l P_l}{N_l p_l} = \frac{S_l Z_l}{S_l N_l} , \quad \frac{\sqrt{X^2 + Y^2}}{\sqrt{P^2 + Q^2}} = \frac{A_2 + Z}{2A_2} .$$

Right projective ratio:

$$\frac{Z_r P_r}{N_r p_r} = \frac{S_r Z_r}{S_r N_r} , \quad \frac{\sqrt{x^2 + y^2}}{\sqrt{p^2 + q^2}} = \frac{r + z}{2r} . \quad (1.260)$$

$f(\Phi)$:

$$\frac{A_1 \cos \Phi}{\sqrt{1 - E^2 \sin^2 \Phi}} \frac{1}{\sqrt{P^2 + Q^2}} =$$

$$= \frac{1}{2} \left(1 + \frac{\sqrt{1 - E^2} \sin \Phi}{\sqrt{1 - E^2 \sin^2 \Phi}} \right) ,$$

$$\sqrt{P^2 + Q^2} = 2A_1 \frac{\cos \Phi}{\sqrt{1 - E^2 \sin^2 \Phi} + \sqrt{1 - E^2} \sin \Phi} ,$$

$$\sqrt{P^2 + Q^2} = \frac{2A_1}{\sqrt{1 - E^2}} \times$$

$$\times \frac{\cos \Phi}{\sqrt{1 - E^2 \sin^2 \Phi}/\sqrt{1 - E^2} + \sin \Phi} ,$$

$$\sqrt{P^2 + Q^2} = \frac{2A_1}{\sqrt{1 - E^2}} \times$$

$$\times \frac{\cos \Phi}{1 + \sin \Phi} \frac{1 + \sin \Phi}{\sqrt{1 - E^2 \sin^2 \Phi}/\sqrt{1 - E^2} + \sin \Phi} ,$$

$$\sqrt{P^2 + Q^2} = \frac{2A_1}{\sqrt{1 - E^2}} \tan \left(\frac{\pi}{4} - \frac{\phi}{2} \right) \times$$

$$\times \frac{\sqrt{(1 + E)(1 - E)}(1 + \sin \Phi)}{\sqrt{(1 + E \sin \Phi)(1 - E \sin \Phi)} + \sqrt{(1 + E)(1 - E)} \sin \Phi} =:$$

$$=: f(\Phi) .$$

$g(\phi)$:

$$\frac{r \cos \phi}{\sqrt{p^2 + q^2}} =$$

$$\tfrac{1}{2} (1 + \sin \phi) ,$$

$$\sqrt{p^2 + q^2} = 2r \frac{\sin \left(\frac{\pi}{2} - \phi \right)}{1 + \cos \left(\frac{\pi}{2} - \phi \right)} ,$$

$$\sqrt{p^2 + q^2} = 2r \tan \left(\frac{\pi}{4} - \frac{\phi}{2} \right)$$

$$:= g(\phi) . \quad (1.261)$$

According to the documents of Synesius (378–430), bishop of Ptolemaios, as well as of Prokius Diadochus (412–485), a philosopher in Athens, the *stereographic projection* originates from Hipparch (180–125 B. C.), astronomer in Nicaea (Bythinia). His planisphere shows the celestial sphere in a polar stereographic projection. For the use of terrestrial charts the *stereographic projection* has been used for the first time by Walter Lude (1507), canonicus in Lothringen. While his choice was polar projection, J. Stab and J. Werner (1514), respectively, used an arbitrary placement of the projection plane, finally Gemma Frisius (1540) its equatorial placement. The particular properties of the stereographic projection, namely *conformality* and the circular map of parallel circles of the sphere, has been recognized only later: Jordanius Nemorarius (1507) mentioned the circularity of transformal parallel circles. Gerhard Mercator (1587) invented conformality in his *Duisburg map* of the eastern and western half spheres in stereographic projection. At the bottom line of his map he writes: "…Etsi enim gradus a centro versus circumferentiam crescant, uti in gradibus aeqhimoctialibus vides, tamem latitudinis longitudinisque gradus in eadem a centro distantia eandem ad invicem proportionem servant quam in sphaera et quadranguli inter duos proximos parallelos dusque meridianos rectangulam figuram habent quemadmodum in sphaera, ita ut regiones undiquaque omnes motivam figuram obtineant sine omni tortuosa distractione." (Indeed though the distances grow from the center to the periphery as to be seen from the lines of constant aequinoctium, they preserve the lengths of longitude and latitude arcs in relative proportion with respect to the sphere. Quadrangles between to nearly parallels and two meridians are represented by a rectangular figure like on the sphere such that all areas keep their natural figure without distortions.) The name *stereographic projection* originates from the mathematician Aguilonius (1566–1617) of Belgium. Compare with Fig. 1.28, which gives an impression of a typical ancient map.

J. H. Lambert (1726–1777) was probably the first cartographer who compared different mappings and projections on a mathematical basis: in order to make the mapping of the sphere onto the plane locally similar ("in kleinsten Teilen ähnlich") he considered similar triangles on the sphere and the plane, which J. H. Lambert tested with respect to the stereographic projection as well as to the Mercator projection:

$$dx = a\frac{dQ}{\cos\Phi} + b\,d\Lambda$$

(spherical longitude Λ, spherical latitude Φ) ,

$$dy = b\frac{dQ}{\cos\Phi} + a\,d\Lambda$$

(1.262)

(righthand rectangular coordinates $\{\Lambda, \Phi\}$ of the plane) .

In support of J. L. Lagrange (1736–1813), he sets $d\Phi/\cos\Phi = dQ$, which leads to the famous differential equations for two-dimensional *conformal mapping*, namely

$$dx = -a\,dQ + b\,d\Lambda , \qquad y + ix = f(Q \pm i\Lambda) \begin{cases} + = \text{conformal} \\ \\ - = \text{anticonformal} \end{cases} , \qquad (1.263)$$
$$dy = +b\,dQ + a\,d\Lambda ,$$

with special reference to de Bougainville's "Traite du calcul integral" (Paris 1756, p. 140), who in turn gave reference to d'Alembert. It was only J. L. Lagrange (1779) who could work with the fundamental solution $y + ix = f(Q \pm i\Lambda)$. Meanwhile L. Euler (1777) had published the same result, finally leading to the notation of *d'Alembert-Euler equations* for two-dimensional *conformal mapping*. Additionally, note that the fundamental equations which govern *infinitesimal conformality* have been written as differential one-forms.

Fig. 1.28. "Ptolemeus Aegyptius", a detail of the star map by Albrecht Dürer (1471–1528). "Imagines coeli septentrionales cum duodecim imaginibus zodiaci" 1515, wood engraving, 42, 7×42, 7 cm, New York, Metropolitan Museum of Art, Harris Brisbane Dick Fund 1951.

1-11 Areal distortion

"It isn't that they can't see the solution. It is that they can't see the problem."
(G. K. Chesterton, The Scandal of Father Brown. The Point of a Pin.)

Fourth multiplicative and additive measures of deformation, dual deformation measures, areomorphism, equiareal mapping.

Up to now, all deformation measures have been built on the *first differential invariants I_l and I_r of surface geometry*, which are also called dS^2 and ds^2. Such an invariant "left" or "right" measures the infinitesimal distance between two points on the "left" or the "right" surface. A dual measure of a surface (two-dimensional Riemann manifold) immersed in $\mathbb{R}^3$ is the *infinitesimal surface element*. Indeed, the surface element "left versus right", (1.264), is *dual* to the infinitesimal distance element "left versus right", (1.265):

$$dS_l := \sqrt{\det[\mathsf{G}_l]}dU \wedge dV \quad \text{versus} \quad dS_r := \sqrt{\det[\mathsf{G}_r]}du \wedge dv , \tag{1.264}$$

$$dS^2 = G_{11}dU^2 + 2G_{12}dUdV + G_{22}dV^2 \quad \text{versus} \quad ds^2 = g_{11}du^2 + 2g_{12}dudv + g_{22}dv^2 . \tag{1.265}$$

In the context of the mapping $f : \mathbb{M}_l^2 \to \mathbb{M}_r^2$, we next define *areomorphism* as an *equiareal mapping* $\mathbb{M}_l^2 \to \mathbb{M}_r^2$: see Definition 1.13.

Definition 1.13 (Equiareal mapping).

An orientation preserving diffeomorphism $f : \mathbb{M}_l^2 \to \mathbb{M}_r^2$ is called *area preserving* and *equiareal* (vector product preserving, *areomorphism*) if

$$\sqrt{\det[\mathsf{G}_l]}dU \wedge dV = \sqrt{\det[\mathsf{G}_r]}du \wedge dv \tag{1.266}$$

or, equivalently,

$$\Phi_l^2 = \Phi^2 := \frac{\sqrt{\det[\mathsf{G}_r]}du \wedge dv}{\sqrt{\det[\mathsf{G}_l]}dU \wedge dV} = 1$$

$$\Leftrightarrow \tag{1.267}$$

$$1 = \frac{\sqrt{\det[\mathsf{G}_l]}dU \wedge dV}{\sqrt{\det[\mathsf{G}_r]}du \wedge dv} =: \Phi^2 = \Phi_r^2$$

or

$$S_{lr} := \sqrt{\det[\mathsf{G}_r]}du \wedge dv - \sqrt{\det[\mathsf{G}_l]}dU \wedge dV = 0$$

$$\Leftrightarrow \tag{1.268}$$

$$S_{rl} := \sqrt{\det[\mathsf{G}_l]}dU \wedge dV - \sqrt{\det[\mathsf{G}_r]}du \wedge dv = 0$$

for all points of $\mathbb{M}_l^2$ and $\mathbb{M}_r^2$, respectively, holds.

End of Definition.

Indeed, the left surface element $\sqrt{\det[\mathsf{G}_l]}dU \wedge dV$ as well as the right surface element $\sqrt{\det[\mathsf{G}_r]}du \wedge dv$ have enabled us to introduce *dual measures* to the left length element $dU^{\mathrm{T}}\mathsf{G}_l dU$ as well as to the right length element $du^{\mathrm{T}}\mathsf{G}_r du$. There exist representations of the multiplicative measure of areal distortion, $\{\Phi_l^2, \Phi_r^2\}$, and of the additive measure of areal distortion, $\{S_{lr}, S_{rl}\}$, in terms of the Cauchy–Green deformation tensor, the Euler Lagrange deformation tensor, and the principal stretches (left or right eigenvalues), which we collect in Box 1.39 and turn out to be useful in the equivalence theorem.

Box 1.39 (Areal distortion, representations of its multiplicative and additive deformation measures).

$$\mathbb{M}_l^2 \to \mathbb{M}_r^2:$$

(i)

$$\sqrt{\det[\mathsf{G}_l]}\,dU \wedge dV = \sqrt{\det[\mathsf{G}_r]}\,du \wedge dv \ ,$$

$$\sqrt{\det[\mathsf{G}_r]}\,du \wedge dv = \sqrt{\det[\mathsf{G}_l]}\,dU \wedge dV \ ;$$

$$\sqrt{\det[\mathsf{G}_l]}\,dU \wedge dV = \sqrt{\det[\mathsf{G}_r - 2\mathsf{E}_r]}\,du \wedge dv \ ,$$

$$\sqrt{\det[\mathsf{G}_r]}\,du \wedge dv = \sqrt{\det[\mathsf{G}_l - 2\mathsf{E}_l]}\,dU \wedge dV \ ;$$

$$\sqrt{\det[\mathsf{G}_l]}\,dU \wedge dV = \frac{1}{\det[\mathsf{F}_r]}\lambda_1\lambda_2 du \wedge dv \ ,$$

$$\sqrt{\det[\mathsf{G}_r]}\,du \wedge dv = \frac{1}{\det[\mathsf{F}_l]}\Lambda_1\Lambda_2 dU \wedge dV \ ;$$

$$\left\{\mathbb{M}_r^2, g_{\mu\nu}\right\} = \left\{\mathbb{R}^2, \delta_{\mu\nu}\right\} \Rightarrow \sqrt{\det[\mathsf{G}_l]}\,dU \wedge dV = \lambda_1\lambda_2 du \wedge dv \ .$$

(1.269)

(ii)

$$\Phi_l^2 = \sqrt{\det[\mathsf{C}_l\mathsf{G}_l^{-1}]} = \Lambda_1\Lambda_2 \ ,$$

$$\Phi_r^2 = \sqrt{\det[\mathsf{C}_r\mathsf{G}_r^{-1}]} = \lambda_1\lambda_2 \ .$$

(1.270)

(iii)

$$S_{lr} = \left(\sqrt{\det[\mathsf{C}_l]} - \sqrt{G_l}\right) dU \wedge dV \ ,$$

$$S_{rl} = \left(\sqrt{\det[\mathsf{C}_r]} - \sqrt{G_r}\right) du \wedge dv \ ;$$

$$S_{lr} = (\Lambda_1\Lambda_2 - 1)\frac{1}{\det[\mathsf{F}_l]}dU \wedge dV \ ,$$

$$S_{rl} = (\lambda_1\lambda_2 - 1)\frac{1}{\det[\mathsf{F}_r]}du \wedge dv \ ;$$

$$\left\{\mathbb{M}_r^2, g_{\mu\nu}\right\} = \left\{\mathbb{R}^2, \delta_{\mu\nu}\right\} \Rightarrow S_{rl} = (\lambda_1\lambda_2 - 1)\,du \wedge dv \ .$$

(1.271)

To give you again some breathing time, please enjoy Fig. 1.29, which presents the "quasicordiform" Bonne-pseudo-conic projection.

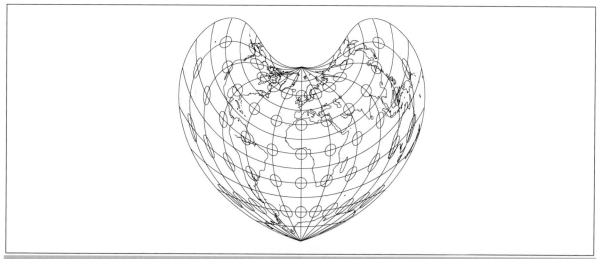

Fig. 1.29. Bonne-pseudo-conic projection, with shorelines of a spherical Earth, equidistant mapping of the line-of-contact of a circular cone, "quasicordiform". Tissot ellipses of distortion. (According to Rigobert Werner).

1-12 Equivalence theorem of equiareal mapping

The equivalence theorem of equiareal mapping from the left to the right two-dimensional Riemann manifold (areomorphism).

We have already defined areomorphism, namely areal distortion, in order to present here an equivalence theorem that relates areomorphism to a special partial differential equation whose solution guarantees an equiareal mapping. In particular, we make a "canonical statement" about the product of left and right principal stretches to be one. Furthermore, we specify the equiareal mapping for a right manifold $\{\mathbb{M}_r^2, g_{\mu\nu}\} = \{\mathbb{R}^2, \delta_{\mu\nu}\}$ to be Euclidean.

Theorem 1.14 (Areomorphism $\mathbb{M}_l^2 \to \mathbb{M}_r^2$, equiareal mapping).

Let $f : \mathbb{M}_l^2 \to \mathbb{M}_r^2$ be an orientation preserving equiareal mapping. Then the following conditions (i)–(iv) are equivalent.

$$\text{(i)}$$
$$\sqrt{\det[G_l]}\,dU \wedge dV = \sqrt{\det[G_r]}\,du \wedge dv \,.$$

$$\text{(ii)}$$
$$\det[C_r] = \det[G_r]\,,\ \det[C_l] = \det[G_l]\,,\ \det[G_r - 2E_r] = \det[G_r]\,,\ \det[G_l + 2E_l] = \det[G_l]\,.$$
$$\text{(iii)} \tag{1.272}$$
$$\Lambda_1\Lambda_2 = 1\,,\quad \lambda_1\lambda_2 = 1\,.$$

$$\text{(iv)}$$
$$U_u V_v - U_v V_u = \sqrt{\frac{\det[G_r]}{\det[G_l]}} = \sqrt{\frac{g_{11}g_{22} - g_{12}^2}{G_{11}G_{22} - G_{12}^2}}\,,\ u_U v_V - u_V v_U = \sqrt{\frac{\det[G_l]}{\det[G_r]}} = \sqrt{\frac{G_{11}G_{22} - G_{12}^2}{g_{11}g_{22} - g_{12}^2}}\,.$$

End of Theorem.

The proof is straightforward. For a better insight into the equivalence theorem of an equiareal mapping, we recommend a detailed study of the next example.

1-13 One example: mapping from an ellipsoid-of-revolution to the sphere

One example for the equivalence theorem of equiareal mapping: the equiareal mapping from an ellipsoid-of-revolution to the sphere.

A beautiful example for the equivalence theorem of equiareal mapping is the mapping of the ellipsoid-of-revolution $\mathbb{E}_{A_1,A_1,A_2}^2$ to the sphere $\mathbb{S}_r^2$, postulated by means of $\Lambda_1\Lambda_2 = 1$ to be area preserving. All notations are taken from Example 1.3. First, by means of Box 1.40, we set up the mapping equations $\mathbb{E}_{A_1,A_1,A_2}^2 \to \mathbb{S}_r^2$, namely by $\lambda = \Lambda, \phi = f(\Phi)$. Here, we compute the left Cauchy–Green matrix C_l as well as the left principal stretches $\{\Lambda_1, \Lambda_2\}$. Second, Box 1.41 illustrates the various steps to be taken in order to derive an equiareal map from the *canonical postulate* $\Lambda_1\Lambda_2 = 1$. As soon as we transfer the general form of the principal stretches $\{\Lambda_1, \Lambda_2\}$ into such a postulate, by means of separation of variables, we derive a first-order differential equation, which is directly solved by integration. Third, with respect to standard integrals and the boundary condition $\phi = f(\Phi = 0) = 0$, we find the classical formula for $\sin\phi$, where the mapping function $\phi = f(\Phi)$ is called *authalic latitude* (O. S. Adams (1921), p. 65; J. P. Snyder (1982), p. 19). Fourth, we solve the problem how to choose the radius of the sphere $\mathbb{S}_r^2$ when only the semi-major axis A_1 or the relative eccentricity $E^2 = (A_1^2 - A_2^2)/A_1^2$, $A_1 > A_2$ of $\mathbb{E}_{A_1,A_1,A_2}^2$ are given. A first choice is $A_1 = r$, a second choice, also called *optimal*, is the identity of the *left global surface element* S_l of $\mathbb{E}_{A_1,A_1,A_2}^2$ and of the *right global surface element* S_r of $\mathbb{S}_r^2$. As derived later, we give both area $(\mathbb{E}_{A_1,A_1,A_2}^2)$ as well as area $(\mathbb{S}_r^2)$ in closed form. Accordingly, we have succeeded to solve $r(A_1, E)$. Step five, based upon Box 1.42, summarizes the forward or direct equations of the special equiareal mapping, called *authalic*, of type $\lambda = \Lambda$ and $\sin\phi = \sin f(\Phi)$ for the optimal equiareal choice of the radius $r(A_1, E)$. In addition, we have computed the left and right principal stretches $\{\Lambda_1, \Lambda_2\}$ and $\{\lambda_1, \lambda_2\}$ for the *authalic mapping*.

Box 1.40 (Left Cauchy–Green matrix, left eigenspace: $\mathbb{E}^2_{A_1, A_1, A_2} \to \mathbb{S}^2_r$).

Left manifold ($\{\Lambda, \Phi\}$ coordinates):   Right manifold ($\{\lambda, \phi\}$ coordinates):

$$\mathrm{d}S^2 = \frac{A_1^2 \cos^2 \Phi}{1 - E^2 \sin^2 \Phi} \mathrm{d}\Lambda^2 + \frac{A_1^2 (1 - E^2)^2}{(1 - E^2 \sin^2 \Phi)^3} \mathrm{d}\Phi^2 . \qquad \mathrm{d}s^2 = r^2 \cos^2 \phi \, \mathrm{d}\lambda^2 + r^2 \mathrm{d}\phi^2 . \qquad (1.273)$$

"Ansatz":

$$\mathbb{E}^2_{A_1, A_1, A_2} \to \mathbb{S}^2_r \, ; \quad \lambda = \Lambda \, , \quad \phi = f(\Phi) . \qquad (1.274)$$

Left Cauchy–Green matrix:

$$\mathsf{C}_l = \mathsf{J}_l^\mathsf{T} \mathsf{G}_r \mathsf{J}_l = \begin{bmatrix} r^2 \cos^2 \phi & 0 \\ 0 & r^2 \, f'^{\,2}(\Phi) \end{bmatrix} ,$$

$$\mathsf{J}_l = \begin{bmatrix} D_\Lambda \lambda & D_\Lambda \phi \\ D_\Phi \lambda & D_\Phi \phi \end{bmatrix} = \begin{bmatrix} 1 & 0 \\ 0 & f'(\Phi) \end{bmatrix} , \quad \mathsf{G}_r = \begin{bmatrix} r^2 \cos^2 \phi & 0 \\ 0 & r^2 \end{bmatrix} . \qquad (1.275)$$

Left principal stretches, left eigenspace:

$$\left| \mathsf{C}_l - \Lambda_l^2 \mathsf{G}_l \right| = \begin{vmatrix} c_{11} - G_{11} \Lambda_l^2 & 0 \\ 0 & c_{22} - G_{22} \Lambda_l^2 \end{vmatrix} = 0 \quad \Leftrightarrow$$

$$\Leftrightarrow \begin{bmatrix} c_{11} - G_{11} \Lambda_l^2 = 0 \\ c_{22} - G_{22} \Lambda_l^2 = 0 \end{bmatrix} \Leftrightarrow \begin{bmatrix} \Lambda_1^2 = \dfrac{c_{11}}{G_{11}} = \dfrac{r^2 \cos^2 \phi}{A_1^2 \cos^2 \Phi} (1 - E^2 \sin^2 \Phi) \\ \Lambda_2^2 = \dfrac{c_{22}}{G_{22}} = \dfrac{r^2 f'^2(\Phi)}{A_1^2 (1 - E^2)^2} (1 - E^2 \sin^2 \Phi)^3 \end{bmatrix} . \qquad (1.276)$$

Box 1.41 (Equiareal mapping: $\mathbb{E}^2_{A_1, A_1, A_2} \to \mathbb{S}^2_r$, $\phi = f(\Phi)$).

Area preserving postulate:

$$\Lambda_1 \Lambda_2 = 1 \Leftrightarrow \frac{r \cos \phi}{A_1 \cos \Phi} \sqrt{1 - E^2 \sin^2 \Phi} \frac{r f'(\Phi)}{A_1 (1 - E^2)} (1 - E^2 \sin^2 \Phi)^{3/2} = 1 . \qquad (1.277)$$

Equation of variables:

$$r^2 \cos \phi \, \mathrm{d}\phi = \frac{\mathrm{d}\Phi}{(1 - E^2 \sin^2 \Phi)^2} A_1^2 (1 - E^2) \cos \Phi . \qquad (1.278)$$

Standard integrals:

$$\Delta := \pi/2 - \Phi \Rightarrow -\mathrm{d}\Delta = \mathrm{d}\Phi \, ,$$

$$\int \frac{\cos \Phi}{(1 - E^2 \sin^2 \Phi)^2} \mathrm{d}\Phi = -\int \frac{\sin \Delta}{(1 - E^2 \cos^2 \Delta)^2} \mathrm{d}\Delta =$$

$$= \frac{\cos \Delta}{2(1 - E^2 \cos^2 \Delta)} + \frac{1}{4E} \ln \frac{1 + E \cos \Delta}{1 - E \cos \Delta} + c_\mathrm{r} = \qquad (1.279)$$

$$= \frac{\sin \Phi}{2(1 - E^2 \sin^2 \Phi)} + \frac{1}{4E} \ln \frac{1 + E \sin \Phi}{1 - E \sin \Phi} + c_\mathrm{r} \, ,$$

$$\int \cos \phi \, \mathrm{d}\phi = \sin \phi + c_l \, .$$

Continuation of Box.

Boundary conditions:

$$\Phi = 0 \Leftrightarrow \phi = 0 \Rightarrow c_l = c_r = 0 . \tag{1.280}$$

Equiareal map of $\mathbb{E}^2_{A_1, A_1, A_2} \to \mathbb{S}^2_r$:

$$r^2 \sin \phi = A_1^2 (1 - E^2) \left[\frac{\sin \Phi}{2(1 - E^2 \sin^2 \Phi)} + \frac{1}{4E} \ln \frac{1 + E \sin \Phi}{1 - E \sin \Phi} \right] ;$$

case 1: $A_1 = r$; case 2: $\begin{cases} \text{left global surface element coincides} \\ \text{with right global surface element} \end{cases}$;

$$S_l = \text{area} \left(\mathbb{E}^2_{A_1, A_1, A_2} \right) = \text{area} \left(\mathbb{S}^2_r \right) = S_r ; \tag{1.281}$$

$$4\pi A_1^2 \left[\frac{1}{2} + \frac{1 - E^2}{2E} \ln \frac{1 + E}{1 - E} \right] = 4\pi r^2$$

$$\Rightarrow$$

$$r^2 = \frac{1}{2} A_1^2 \left[1 + \frac{1 - E^2}{2E} \ln \frac{1 + E}{1 - E} \right] .$$

Authalic latitude (O. S. Adams (1921), p. 65; J. P. Snyder (1982), p. 19):

$$\phi = f(\Phi) ,$$

$$\sin \phi = \sin f \left(\Phi \,|\, S_l = S_r \right) . \tag{1.282}$$

Box 1.42 (The authalic equiareal map: $\mathbb{E}^2_{A_1, A_1, A_2} \to \mathbb{S}^2_r$).

Authalic equiareal map:

$$\lambda = \Lambda ,$$

$$\sin \phi = (1 - E^2) \left[\frac{\sin \Phi}{1 - E^2 \sin^2 \Phi} + \frac{1}{2E} \frac{1 + E \sin \Phi}{1 - E \sin \Phi} \right] / \left[1 + \frac{1 - E^2}{2E} \ln \frac{1 + E}{1 - E} \right] . \tag{1.283}$$

Left and right principal stretches:

$$\Lambda_1 = \frac{r \cos \phi}{A_1 \cos \Phi} \sqrt{1 - E^2 \sin^2 \Phi} ,$$

$$\Lambda_2 = \frac{r}{A_1 (1 - E^2)} f'(\Phi)(1 - E^2 \sin^2 \Phi)^{3/2} ,$$

$$\phi = f(\Phi) \Rightarrow \phi' = f'(\Phi) = \frac{1}{r^2 \cos \phi} \frac{A_1^2 (1 - E^2)}{(1 - E^2 \sin^2 \Phi)^2} \cos \Phi , \tag{1.284}$$

$$\Lambda_1 = \lambda_1^{-1} = \frac{r \cos \phi}{A_1 \cos \Phi} \sqrt{1 - E^2 \sin^2 \Phi} ,$$

$$\Lambda_2 = \lambda_2^{-1} = \frac{A_1 \cos \Phi}{r \cos \phi} \frac{1}{\sqrt{1 - E^2 \sin^2 \Phi}} .$$

In the light of the *equivalence theorem* 1.14 (areomorphism), we are now prepared to solve the following problems. (i) Can we prove the first equivalence given by (1.285) and (ii) can we prove the third equivalence given by (1.286)?

$$\det[C_l] = \det[G_l] \text{ or } \det[C_r] = \det[G_r] \,, \tag{1.285}$$

$$u_U v_V - u_V u_U = \sqrt{\frac{\det[G_l]}{\det[G_r]}} = \sqrt{\frac{G_{11}G_{22} - G_{12}^2}{g_{11}g_{22} - g_{12}^2}} \,. \tag{1.286}$$

Solution (the first problem).

Start from the equiareal map of Box 1.41 in order to prove $\det[C_l] = \det[G_l]$, where the left Cauchy–Green matrix C_l as well as the left matrix G_l of the metric is given by means of Box 1.40. Here, again we collect all deviational items in Box 1.43. As soon as we implement $f'(\Phi)$ into the determinantal identity, the proof is closed.

End of Solution (the first problem).

Solution (the second problem).

By means of Box 1.44, let us work out the partial differential equation which governs an equiareal mapping. Note that we here specify $\{u = \lambda, v = \phi\}$ and $\{U = \Lambda, V = \Phi\}$ subject to the "Ansatz" $\{\lambda = \Lambda, \phi = f(\Phi)\}$. Indeed, we find $f(\Phi)$ as given already in Box 1.42.

End of Solution (the second problem).

Note that the second or canonical equivalence $\Lambda_1 \Lambda_2 = 1$ has already been used to construct the equiareal map $\mathbb{E}^2_{A_1,A_1,A_2} \to \mathbb{S}^2_r$.

Box 1.43 (Equiareal mapping: $\mathbb{E}^2_{A_1,A_1,A_2} \to \mathbb{S}^2_r$, $\det[C_l] = \det[G_l]$).

$$C_l = \begin{bmatrix} r^2 \cos^2 \phi & 0 \\ 0 & r^2 f'^2(\phi) \end{bmatrix} , \quad G_l = \begin{bmatrix} \dfrac{A_1^2 \cos^2 \Phi}{1 - E^2 \sin^2 \Phi} & 0 \\ 0 & \dfrac{A_1^2 (1 - E^2)^2}{(1 - E^2 \sin^2 \Phi)^3} \end{bmatrix} ,$$

$$\begin{bmatrix} \det[C_l] = r^4 \cos^2 \phi \, f'^2(\phi) \\ f'^2(\Phi) = \dfrac{1}{r^4 \cos^2 \phi} \dfrac{A_1^4 (1 - E^2)^2}{(1 - E^2 \sin^2 \Phi)^4} \cos^2 \Phi \end{bmatrix} \Rightarrow \det[C_l] = \dfrac{A_1^4 (1 - E^2)^2}{(1 - E^2 \sin^2 \Phi)^4} \cos^2 \Phi = \det[G_l]$$

(1.287)

q. e. d.

Box 1.44 (Equiareal mapping: $\mathbb{E}^2_{A_1,A_1,A_2} \to \mathbb{S}^2_r$, partial differential equation).

$$u_U v_V - u_V u_U = \sqrt{\frac{\det[G_l]}{\det[G_r]}} \,, \quad u = \lambda \,, \ v = \phi = f(\Phi) \,, \ U = \Lambda \,, \ V = \Phi \Leftrightarrow$$

$$\Leftrightarrow \lambda_\Lambda \phi_\Phi - \lambda_\Phi \phi_\Lambda = \sqrt{\frac{G_{11}G_{22}}{g_{11}g_{22}}} \Leftrightarrow f'^2(\Phi) = \frac{1}{r^2 \cos \phi} \frac{A_1^2 (1 - E^2)}{(1 - E^2 \sin^2 \Phi)^2} \cos \Phi \,, \tag{1.288}$$

$$f'(\Phi) : \quad \text{see Box 1.42}$$

q. e. d.

1-14 Review: the canonical criteria

"Where we cannot use the compass of mathematics or the torch of experience
. . . it is certain we cannot take a single step forward."
(Voltaire.)

Review: the canonical criteria for conformal equiareal, isometric, and equidistant mappings, optimal map projections, Gaussian curvatures.

Up to now, we have defined the conformal mapping (compare with Definition 1.10) as well as the equiareal mapping (compare with Definition 1.13) from the left two-dimensional Riemann manifold (here: left surface immersed into $\mathbb{R}^3$) to the right two-dimensional Riemann manifold (here: right surface immersed into $\mathbb{R}^3$). We demonstrated that under the action of the conformal map, angles were preserved. In contrast, an equiareal transformation preserves the surface element. However, what is to tell about *length preserving mappings* $\overline{f} : \mathbb{M}_l^2 \to \mathbb{M}_r^2$?

1-141 Isometry

Let us begin with the definition of an *isometry* and relate it in the form of an equivalence theorem to the other measures of deformation. In particular, we ask the question: When does an isometric mapping exist?

Definition 1.15 (Isometry).

An admissible mapping $\overline{f} : \mathbb{M}_l^2 \to \mathbb{M}_r^2$ is called *length preserving* or an *isometry* if for *any curve* in the left surface ("left curve": $c_l(t_l)$, $t_l \in I(c_l)$) the corresponding curve in the right surface ("right curve": $c_r(t_r)$, $t_r \in I(c_r)$) as its image $\overline{f} \circ c_l(t_l)$ has the identical length:

$$\int_{a_l}^{b_l} \dot{s}_l \mathrm{d}t_l = \int_{a_r}^{b_r} \dot{s}_r \mathrm{d}t_r . \tag{1.289}$$

Two Riemann manifolds $\mathbb{M}_l^2$ and $\mathbb{M}_r^2$, respectively, which are mapped on each other by means of an isometry are called *isometric*.

End of Definition.

Without any proof, we make the following equivalence statement. (Of course, we could make an equivalent statement for the right manifold $\mathbb{M}_r^2$.)

Theorem 1.16 (Isometry $\mathbb{M}_l^2 \to \mathbb{M}_r^2$).

An admissible mapping $\overline{f} : \mathbb{M}_l^2 \to \mathbb{M}_r^2$ is an isometry if and only if the following equivalent conditions are fulfilled.

(i) The coordinates of the left Cauchy–Green tensor C_l are identical to the coordinates of the left metric tensor G_l, i. e.

$$\mathsf{C}_l = \mathsf{G}_l . \tag{1.290}$$

(ii) The stretches Λ for any point $\boldsymbol{X} \in \mathbb{M}_l^1 \subset \mathbb{M}_l^2 \subset \mathbb{R}^3$ is independent of the directions of the tangent vector $\dot{\boldsymbol{X}}$, a constant to be one, i. e.

$$\Lambda(\dot{\boldsymbol{X}}) = 1 \,\forall\, \dot{\boldsymbol{X}} \neq 0 , \quad \dot{\boldsymbol{X}} \in T\mathbb{M}_l^1 \subset T\mathbb{M}_l^2 , \quad \dot{\boldsymbol{X}} = \sum_{I=1}^{3} \sum_{M=1}^{2} \boldsymbol{E}_I \frac{\partial X^I}{\partial U^M} \frac{\mathrm{d}U^M}{\mathrm{d}t_l} . \tag{1.291}$$

(iii) The left principal stretches for any point $\boldsymbol{X} \in \mathbb{M}_l^2$ are a constant to be one: $\Lambda_1 = \Lambda_2 = 1$.

End of Theorem.

If an isometric mapping $\overline{f} : \mathbb{M}_l^2 \to \mathbb{M}_r^2$ were existing for an arbitrary left and right two-dimensional Riemann manifold, we would have met an ideal situation. Let us therefore ask: when does an isometric mapping $\overline{f} : \mathbb{M}_l^2 \to \mathbb{M}_r^2$ exist? Unfortunately, we can only sketch the existence proof here which is based upon the intrinsic measure of curvature of a surface, namely Gaussian curvature, computed

$$k = \det[\mathsf{K}] = \frac{\det[\mathsf{H}]}{\det[\mathsf{G}]} \,,$$

$$\mathsf{K} := -\mathsf{H}\mathsf{G}^{-1} \in \mathbb{R}^{2 \times 2} \,.$$

(1.292)

The curvature matrix K of a surface is the negative product of the Hesse matrix H and the inverse of the Gauss matrix G defined as follows.

Box 1.45 (Curvature matrix of a surface).

$$\mathsf{G} = \sum_{I=1}^{3} \frac{\partial X^I}{\partial U^M} \frac{\partial X^I}{\partial U^N} = \begin{bmatrix} e & f \\ f & g \end{bmatrix} \,,$$

(1.293)

$$\mathsf{H} = \sum_{I=1}^{3} \frac{\partial^2 X^I}{\partial U^M \partial U^N} N^I = \begin{bmatrix} l & m \\ m & n \end{bmatrix} \,,$$

(1.294)

$$\mathsf{K} = \frac{1}{eg - f^2} = \begin{bmatrix} -gl + fm & fl - em \\ -gm + fn & fm - en \end{bmatrix} \,.$$

(1.295)

N^I denotes the coordinates of the surface normal vector $\boldsymbol{N} \in N\mathbb{M}_l^2$ with respect to the basis $\{\boldsymbol{E}_1, \boldsymbol{E}_2, \boldsymbol{E}_3, | \, \mathcal{O}\}$ fixed to the origin $\mathcal{O}$ and assumed to be orthonormal. $\boldsymbol{N} = \boldsymbol{E}_1 N^1 + \boldsymbol{E}_1 N^2 + \boldsymbol{E}_3 N^3$ and $X^I(U, V)$ are the representers of Φ_l^{-1}, which are also called *embedding functions* $\mathbb{M}_l^2 \subset \mathbb{R}^3$ if we exclude *self-intersections* and *singular points* (corners) of $\mathbb{M}_l^2$. The "Theorema Egregium" of C. F. Gauss states that the determinant of the curvature matrix, in short *Gaussian curvature*, depends only on (i) the metric coefficients e, f, g, (ii) their first derivatives $e_U, e_V, f_U, f_V, g_U, g_V$, and (iii) their second derivatives $e_{UU}, e_{UV}, e_{VV}, \ldots, g_{UU}, g_{UV}, g_{VV}$. The fundamental theorem of an isometric mapping can now be formulated as follows.

Theorem 1.17 (Isometric mapping).

If a left curvature is isometrically mapped to a right surface, then corresponding points $\boldsymbol{X} \in \mathbb{M}_l^2$ and $\boldsymbol{x} \in \mathbb{M}_r^2$ have identical Gaussian curvature.

End of Theorem.

A list of Gaussian curvatures for different surfaces is shown in Table 1.5. In consequence, there are no isometries (i) from ellipsoid to sphere, (ii) from ellipsoid or sphere to plane, cylinder, cone, any ruled surface (developable surfaces of Gaussian curvature zero).

Table 1.5. Gaussian curvatures for some surfaces.

Type of surface	Gaussian curvature
sphere $\mathbb{S}_R^2$	$k = \frac{1}{R^2} > 0$
ellipsoid-of-revolution $\mathbb{E}_{A_1, A_1, A_2}^2$	$k = \frac{1}{MN}$, $M := \frac{A_1(1 - E^2)}{(1 - E^2 \sin^2 \Phi)^{3/2}}$, $N := \frac{A_1}{\sqrt{1 - E^2 \sin^2 \Phi}}$
plane, cylinder, cone, ruled surface	$k = 0$

1-142 Equidistant mapping of submanifolds

Indeed, we are unable to produce an isometric landscape of the Earth, its Moon, the Sun, and planets, other celestial bodies, or the universe. In this situation, we have to look for a softer version of a length preserving mapping. Such an alternative concept is found by "dimension reduction". Only a one-dimensional *submanifold* $\mathbb{M}^1$ of the two-dimensional *Riemann manifold* $\mathbb{M}^2$ is mapped "length preserving". For instance, we map the left coordinate line "ellipsoidal equator" equidistantly to the right coordinate line "spherical equator", namely by the postulate $A_1\Lambda = r\lambda$. The arc length $A_1\Lambda$ of the ellipsoidal equator coincides with the arc length of the spherical equator $r\lambda$. A more precise definition is given in Definition 1.18.

Definition 1.18 (Equidistant mapping).

Let a particular mapping $\overline{f} : \mathbb{M}_l^2 \to \mathbb{M}_r^2$ of a left surface (left two-dimensional Riemann manifold) to a right surface (right two-dimensional Riemann manifold) be given. Beside the exceptional points, both parameterized surfaces $\mathbb{M}_l^2$ as well as $\mathbb{M}_r^2$ are covered by a set of coordinate lines $\{U = \text{constant}, V\}$, $\{U, V = \text{constant}\}$ as well as $\{u = \text{constant}, v\}$, $\{u, v = \text{constant}\}$, called *left curves* $c_l(t)$ (left one-dimensional submanifold) and *right curves* $c_r(t)$ (right one-dimensional submanifold). Under the mapping $\overline{f} \circ c_l(t) = c_r(t)$, the mapping

$$c_l(t) \to c_r(t) \quad \text{or} \quad \mathbb{R}^3 \supset \mathbb{M}_l^2 \supset \mathbb{M}_l^1 \stackrel{\text{equidistant}}{\longrightarrow} \mathbb{M}_r^1 \subset \mathbb{M}_r^2 \subset \mathbb{R}^3 \qquad (1.296)$$

is called *equidistant* if a finite section of a specific left curve $c_l(t)$ has the same length as a finite section of a corresponding right curve $c_r(t)$.

End of Definition.

Let us work out the equivalence theorem for an equidistant mapping from a left curve $c_l(t)$ to a right curve $c_r(t)$.

Theorem 1.19 (Equidistant mapping $\mathbb{R}^3 \supset \mathbb{M}_l^2 \supset \mathbb{M}_l^1 \longrightarrow \mathbb{M}_r^1 \subset \mathbb{M}_r^2 \subset \mathbb{R}^3$).

Let us assume that the left surface (left two-dimensional Riemann manifold) as well as the right surface (right two-dimensional Riemann manifold) has been parameterized by left coordinates $\{U, V\}$ and right coordinates $\{u, v\}$. If the directions of their left tangent vectors and their right tangent vectors coincide with the directions of the left principal stretches (left eigendirections, left eigenvectors) and of the right principal stretches (right eigendirections, right eigenvectors), then the following conditions of an equidistant mapping are equivalent.

(i) Equidistant mapping of a section of a specific left curve $c_l(t)$ to a
corresponding section of a specific right curve $c_r(t)$.

U coordinate line to u coordinate line: V coordinate line to v coordinate line:

$$\int_{a_l}^{b_l} \sqrt{G_{22}(t)}\,\dot{V}\mathrm{d}t = \int_{a_r}^{b_r} \sqrt{g_{22}(t)}\,\dot{v}\mathrm{d}t . \qquad \int_{a_l}^{b_l} \sqrt{G_{11}(t)}\,\dot{U}\mathrm{d}t = \int_{a_r}^{b_r} \sqrt{g_{11}(t)}\,\dot{u}\mathrm{d}t . \qquad (1.297)$$

(ii) Left or right Cauchy–Green matrix under an equidistant mapping $c_l(t) \to c_r(t)$.

U coordinate line to u coordinate line: V coordinate line to v coordinate line:

$$c_{22} = G_{22} \quad \text{or} \quad C_{22} = g_{22} . \qquad\qquad c_{11} = G_{11} \quad \text{or} \quad C_{11} = g_{11} . \qquad (1.298)$$

(iii) Left or right principal stretches under an equidistant mapping $c_l(t) \to c_r(t)$.

U coordinate line to u coordinate line: V coordinate line to v coordinate line:

$$\Lambda_2 = 1 \quad \text{or} \quad \lambda_2 = 1 . \qquad\qquad \Lambda_1 = 1 \quad \text{or} \quad \lambda_1 = 1 . \qquad (1.299)$$

End of Theorem.

The proof is straightforward. We refer to Example 1.11, where we solved the third problem: the ellipsoidal equator had been equidistantly mapped to the spherical equator, namely $r\lambda = A_1\Lambda$, such that $\Lambda_1(\Phi = 0) = 1$ or $\lambda_1(\phi = 0) = 1$.

1-143 Canonical criteria

By means of the various equivalence theorems, we are well-prepared to present to you, as beloved collectors items of Box 1.46, the canonical criteria or measures for a conformal, an equiareal, and an isometric mapping $\mathbb{M}_l^2 \to \mathbb{M}_r^2$ as well as for an equidistant mapping $c_l(t) \to c_r(t)$. These canonical measures are exclusively used to generate in following sections equidistant, conformal, and equiareal mappings of various surfaces like the ellipsoid-of-revolution to the sphere. Hilbert's invariant theory is finally used to generate scalar functions of the tensor-valued deformation measures. Box 1.47 reviews the two *fundamental Hilbert invariants* of the Cauchy–Green and Euler–Lagrange deformation tensors.

Box 1.46 (Canonical criteria for a conformal, equiareal, and isometric mapping $\mathbb{M}_l^2 \to \mathbb{M}_r^2$ as well as for an equidistant mapping $c_l(t) \to c_r(t)$).

Conformeomorphism:

$$\Lambda_1 = \Lambda_2 \quad \text{or} \quad \lambda_1 = \lambda_2 \,,$$
$$K_1 = K_2 \quad \text{or} \quad \kappa_1 = \kappa_2 \,, \tag{1.300}$$

for all points of $\mathbb{M}_l^2$ or $\mathbb{M}_r^2$, respectively.

Aeromorphism:

$$\Lambda_1\Lambda_2 = 1 \quad \text{or} \quad \lambda_1\lambda_2 = 1 \,,$$
$$K_1K_2 + \tfrac{1}{2}\left(K_1 + K_2\right) = 0 \quad \text{or} \quad \kappa_1\kappa_2 + \tfrac{1}{2}\left(\kappa_1 + \kappa_2\right) = 0 \,, \tag{1.301}$$

for all points of $\mathbb{M}_l^2$ or $\mathbb{M}_r^2$, respectively.

Isometry:

$$\Lambda_1 = \Lambda_2 = 1 \quad \text{or} \quad \lambda_1 = \lambda_2 = 1 \,,$$
$$K_1 = K_2 = 0 \quad \text{or} \quad \kappa_1 = \kappa_2 = 0 \,, \tag{1.302}$$

for all points of $\mathbb{M}_l^2$ or $\mathbb{M}_r^2$, respectively.

Equidistance:

$$\Lambda_1 = 1 \,, \quad \Lambda_2 = 1 \text{ or } \lambda_1 = 1 \,, \quad \lambda_2 = 1 \,,$$
$$K_1 = 0 \,, \quad K_2 = 0 \text{ or } \kappa_1 = 0 \,, \quad \kappa_2 = 0 \,, \tag{1.303}$$

for all points of $\mathbb{M}_l^2$ (left curve) and $\mathbb{M}_r^2$ (right curve) which are equidistantly mapped.

Box 1.47 (Canonical representation of Hilbert invariants derived from deformation measures).

$$I_1\left(\mathsf{C}_l\right) := \Lambda_1^2 + \Lambda_2^2 = \operatorname{tr}\left[\mathsf{C}_l\mathsf{G}_l^{-1}\right] \quad \text{versus} \quad i_1\left(\mathsf{C}_r\right) := \lambda_1^2 + \lambda_2^2 = \operatorname{tr}\left[\mathsf{C}_r\mathsf{G}_r^{-1}\right] \,,$$
$$I_2\left(\mathsf{C}_l\right) := \Lambda_1^2\Lambda_2^2 = \det\left[\mathsf{C}_l\mathsf{G}_l^{-1}\right] \quad \text{versus} \quad i_2\left(\mathsf{C}_r\right) := \lambda_1^2\lambda_2^2 = \det\left[\mathsf{C}_r\mathsf{G}_r^{-1}\right] \,, \tag{1.304}$$

or

$$I_1\left(\mathsf{E}_l\right) := K_1 + K_2 = \operatorname{tr}\left[\mathsf{E}_l\mathsf{G}_l^{-1}\right] \quad \text{versus} \quad i_1\left(\mathsf{E}_r\right) := \kappa_1 + \kappa_2 = \operatorname{tr}\left[\mathsf{E}_r\mathsf{G}_r^{-1}\right] \,,$$
$$I_2\left(\mathsf{E}_l\right) := K_1K_2 = \det\left[\mathsf{E}_l\mathsf{G}_l^{-1}\right] \quad \text{versus} \quad i_2\left(\mathsf{E}_r\right) := \kappa_1\kappa_2 = \det\left[\mathsf{E}_r\mathsf{G}_r^{-1}\right] \,. \tag{1.305}$$

$$\frac{1}{2}I_1 = \frac{1}{2}\left(\Lambda_1^2 + \Lambda_2^2\right) = \tfrac{1}{2}\mathrm{tr}\left[\mathsf{C}_l\mathsf{G}_l^{-1}\right],$$

$$\frac{1}{2}i_1 = \frac{1}{2}\left(\lambda_1^2 + \lambda_2^2\right) = \tfrac{1}{2}\mathrm{tr}\left[\mathsf{C}_r\mathsf{G}_r^{-1}\right] \tag{1.306}$$

represent the *average Cauchy–Green deformation, distortion energy density* of the first kind, also called *Cauchy–Green dilatation*. In contrast,

$$\ln\sqrt{I_2} = \frac{1}{2}\left(\ln\Lambda_1^2 + \ln\Lambda_2^2\right) = \ln\sqrt{\det\left[\mathsf{C}_l\mathsf{G}_l^{-1}\right]},$$

$$\ln\sqrt{i_2} = \frac{1}{2}\left(\ln\lambda_1^2 + \ln\lambda_2^2\right) = \ln\sqrt{\det\left[\mathsf{C}_r\mathsf{G}_r^{-1}\right]} \tag{1.307}$$

are the *geometric mean of Cauchy–Green deformation* or *distortion energy density* of the second kind. Note that similar *Hilbert invariants* can be formulated and interpreted for the Euler–Lagrange deformation tensor.

Physical aside.

Alternative measures of distortion energy density are introduced in continuum mechanics. By means of the *weighted Frobenius matrix norm* of Box 1.48, we have given quadratic forms of Cauchy–Green and Euler–Lagrange deformation density. The weight matrices W_l and W_r are *Hooke matrices*, also called *direct and inverse stiffness matrices*. Box 1.47 and Box 1.48 have reviewed local scalar-valued deformation measures, namely distortion densities of the first and the second kind. As soon as we have to map a certain part of the left surface as well as the right surface, we should consequently introduce global invariant distortion measures as summarized in Box 1.49, which constitute Cauchy–Green and Euler–Lagrange deformation energy. $\mathrm{d}S_l$ denotes the left surface element, while $\mathrm{d}S_r$ denotes the right surface element, for instance, $\mathrm{d}S_l = \sqrt{\det[\mathsf{G}_l]}\mathrm{d}U\mathrm{d}V$ and $\mathrm{d}S_r = \sqrt{\det[\mathsf{G}_r]}\mathrm{d}u\mathrm{d}v$, respectively. The vec operator is a mapping of a matrix as a two-dimensional array to a column as a one-dimensional array: under the operation $\mathrm{vec}[\mathsf{A}]$, the columns of the matrix A are stapled vertically one-by-one. An example is $\mathsf{A} \in \mathbb{R}^{2\times 2}$, $\mathrm{vec}[\mathsf{A}] = (a_{11}, a_{21}, a_{12}, a_{22})$.

Box 1.48 (Weighted matrix norms of Cauchy–Green and Euler–Lagrange deformations).

Cauchy–Green deformation:

$$\|\mathsf{C}_l\mathsf{G}_l^{-1}\|_{\mathsf{W}_l}^2 := \qquad\qquad \|\mathsf{C}_r\mathsf{G}_r^{-1}\|_{\mathsf{W}_r}^2 :=$$

$$\text{versus}$$

$$:= \mathrm{tr}\left[\left(\mathsf{C}_l\mathsf{G}_l^{-1}\right)^{\mathrm{T}}\mathsf{W}_l\left(\mathsf{C}_l\mathsf{G}_l^{-1}\right)\right] \qquad := \mathrm{tr}\left[\left(\mathsf{C}_r\mathsf{G}_r^{-1}\right)^{\mathrm{T}}\mathsf{W}_r\left(\mathsf{C}_r\mathsf{G}_r^{-1}\right)\right],$$

$$\|\mathsf{C}_l\mathsf{G}_l^{-1}\|_{\mathsf{W}_l}^2 = \qquad\qquad \|\mathsf{C}_r\mathsf{G}_r^{-1}\|_{\mathsf{W}_r}^2 = \tag{1.308}$$

$$\text{versus}$$

$$= \left(\mathrm{vec}\left[\mathsf{C}_l\mathsf{G}_l^{-1}\right]\right)^{\mathrm{T}}\mathsf{W}_l\left(\mathrm{vec}\left[\mathsf{C}_l\mathsf{G}_l^{-1}\right]\right) \qquad = \left(\mathrm{vec}\left[\mathsf{C}_r\mathsf{G}_r^{-1}\right]\right)^{\mathrm{T}}\mathsf{W}_r\left(\mathrm{vec}\left[\mathsf{C}_r\mathsf{G}_r^{-1}\right]\right).$$

Euler–Lagrange deformation:

$$\|\mathsf{C}_l\mathsf{G}_l^{-1}\|_{\mathsf{W}_l}^2 := \qquad\qquad \|\mathsf{E}_r\mathsf{G}_r^{-1}\|_{\mathsf{W}_r}^2 :=$$

$$\text{versus}$$

$$:= \mathrm{tr}\left[\left(\mathsf{E}_l\mathsf{G}_l^{-1}\right)^{\mathrm{T}}\mathsf{W}_l\left(\mathsf{E}_l\mathsf{G}_l^{-1}\right)\right] \qquad := \mathrm{tr}\left[\left(\mathsf{E}_r\mathsf{G}_r^{-1}\right)^{\mathrm{T}}\mathsf{W}_r\left(\mathsf{E}_r\mathsf{G}_r^{-1}\right)\right],$$

$$\|\mathsf{E}_l\mathsf{G}_l^{-1}\|_{\mathsf{W}_l}^2 = \qquad\qquad \|\mathsf{E}_r\mathsf{G}_r^{-1}\|_{\mathsf{W}_r}^2 = \tag{1.309}$$

$$\text{versus}$$

$$= \left(\mathrm{vec}\left[\mathsf{E}_l\mathsf{G}_l^{-1}\right]\right)^{\mathrm{T}}\mathsf{W}_l\left(\mathrm{vec}\left[\mathsf{E}_l\mathsf{G}_l^{-1}\right]\right) \qquad = \left(\mathrm{vec}\left[\mathsf{E}_r\mathsf{G}_r^{-1}\right]\right)^{\mathrm{T}}\mathsf{W}_r\left(\mathrm{vec}\left[\mathsf{E}_r\mathsf{G}_r^{-1}\right]\right).$$

Box 1.49 (Cauchy–Green distortion energy, Euler–Lagrange distortion energy).

(i) Cauchy–Green distortion energy:

(1st)
$$\frac{1}{2}\int dS_l\, \mathrm{tr}\left[\mathsf{C}_l\mathsf{G}_l^{-1}\right] =$$
$$= \frac{1}{2}\int dS_l\,\left(\Lambda_1^2 + \Lambda_2^2\right)$$
versus
$$\frac{1}{2}\int dS_r\, \mathrm{tr}\left[\mathsf{C}_r\mathsf{G}_r^{-1}\right] =$$
$$= \frac{1}{2}\int dS_r\,\left(\lambda_1^2 + \lambda_2^2\right)\ ;$$

(2nd)
$$\int dS_l\,\sqrt{\det\left[\mathsf{C}_l\mathsf{G}_l^{-1}\right]} =$$
$$= \int dS_l\,\Lambda_1\Lambda_2$$
versus
$$\int dS_r\, \mathrm{tr}\left[\mathsf{C}_r\mathsf{G}_r^{-1}\right] =$$
$$= \int dS_r\,\lambda_1\lambda_2\ ;$$
(1.310)

(3rd)
$$\int dS_l\,\left(\ln\Lambda_1^2 + \ln\Lambda_2^2\right)$$
versus
$$\int dS_r\,\left(\ln\lambda_1^2 + \ln\lambda_2^2\right)\ ;$$

(4th)
$$\int dS_l\, \mathrm{tr}\left[\left(\mathsf{C}_l\mathsf{G}_l^{-1}\right)^{\mathsf{T}}\mathsf{W}_l\left(\mathsf{C}_l\mathsf{G}_l^{-1}\right)\right] :=$$
$$:= |||\mathsf{C}_l\mathsf{G}_l^{-1}|||^2_{\mathsf{W}_l}$$
versus
$$\int dS_r\, \mathrm{tr}\left[\left(\mathsf{C}_r\mathsf{G}_r^{-1}\right)^{\mathsf{T}}\mathsf{W}_r\left(\mathsf{C}_r\mathsf{G}_r^{-1}\right)\right] :=$$
$$:= |||\mathsf{C}_r\mathsf{G}_r^{-1}|||^2_{\mathsf{W}_r}\ .$$

(ii) Euler–Lagrange distortion energy:

(1st)
$$\frac{1}{2}\int dS_l\, \mathrm{tr}\left[\mathsf{E}_l\mathsf{G}_l^{-1}\right] =$$
$$= \frac{1}{2}\int dS_l\,\left(K_1 + K_2\right)$$
versus
$$\frac{1}{2}\int dS_r\, \mathrm{tr}\left[\mathsf{E}_r\mathsf{G}_r^{-1}\right] =$$
$$= \frac{1}{2}\int dS_r\,\left(\kappa_1 + \kappa_2\right)\ ;$$

(2nd)
$$\int dS_l\,\sqrt{\det\left[\mathsf{E}_l\mathsf{G}_l^{-1}\right]} =$$
$$= \int dS_l\,\sqrt{K_1 K_2}$$
versus
$$\int dS_r\,\sqrt{\det\left[\mathsf{E}_r\mathsf{G}_r^{-1}\right]} =$$
$$= \int dS_r\,\sqrt{\kappa_1\kappa_2}\ ;$$
(1.311)

(3rd)
$$\frac{1}{2}\int dS_l(\ln K_1 + \ln K_2)$$
versus
$$\frac{1}{2}\int dS_r(\ln\kappa_1 + \ln\kappa_2)\ ;$$

(4th)
$$\frac{1}{2}\int dS_l\, \mathrm{tr}\left[\left(\mathsf{E}_l\mathsf{G}_l^{-1}\right)^{\mathsf{T}}\mathsf{W}_l\left(\mathsf{E}_l\mathsf{G}_l^{-1}\right)\right] :=$$
$$:= |||\mathsf{E}_l\mathsf{G}_l^{-1}|||^2_{\mathsf{W}_l}$$
versus
$$\frac{1}{2}\int dS_r\, \mathrm{tr}\left[\left(\mathsf{E}_r\mathsf{G}_r^{-1}\right)^{\mathsf{T}}\mathsf{W}_r\left(\mathsf{E}_r\mathsf{G}_r^{-1}\right)\right] :=$$
$$:= |||\mathsf{E}_r\mathsf{G}_r^{-1}|||^2_{\mathsf{W}_r}\ .$$

1-144 Optimal map projections

Optimal map projections relate to the invariant scalar measures of Cauchy–Green deformation. More than 1000 scientific contributions have been published on this topic. Harmonic maps, optimal Universal Mercator Projections (opt UMP) as well as optimal Universal Transverse Mercator (opt UTM) belong to this category. Let us only introduce here the *optimality conditions* as they are summarized in Box 1.50, Box 1.51, and Box 1.52. First, G. B. Airy (1861) and V. V. Kavrajski (1958) introduced local as well as global measures of $\overline{f} : \mathbb{M}_l^2 \to \mathbb{M}_r^2$ from isometry. Since for an isometry canonically $\Lambda_1 = \Lambda_2 = 1$ or $\lambda_1 = \lambda_2 = 1$ holds, $\{\Lambda_1 - 1, \Lambda_2 - 1\}$ or $\{\ln\Lambda_1, \ln\Lambda_2\}$ and $\{\lambda_1 - 1, \lambda_2 - 1\}$ or $\{\ln\lambda_1, \ln\lambda_2\}$ as "errors" ϵ_l and ϵ_r are measures of the local departure from isometry. When integrated over the part of the left or right surface to be mapped, we are led to the global measures of departure from isometry, namely I_A and I_{AK} of type "left" and "right". Second, we introduce local and global measures of $\overline{f} : \mathbb{M}_l^2 \to \mathbb{M}_r^2$ from an areomorphism or a conformeomorphism. Since for an equiareal mapping canonically $\Lambda_1\Lambda_2 = 1$ or $\lambda_1\lambda_2 = 1$ holds, $\{\Lambda_1\Lambda_2 - 1\}$ or $\{\lambda_1\lambda_2 - 1\}$ as "errors" ϵ_l and ϵ_r of type "areal" measure the local departure from an areomorphism. Similarly, for a conformal mapping canonically $\Lambda_1 = \Lambda_2$ or $\lambda_1 = \lambda_2$ holds. Accordingly, $\Lambda_1 - \Lambda_2$ or $\lambda_1 - \lambda_2$ as "errors" ϵ_l and ϵ_r as measures of type "conformal" describe the local departure from a conformeomorphism. When integrated over the part of the left or right surface to be mapped, we are led to global measures of departure from areomorphism or conformeomorphism, namely I_{areal} and I_{conf} of type "left" and "right". Examples are given in the following chapters.

Box 1.50 (Local measures for departure of the mapping $\mathbb{M}_l^2 \to \mathbb{M}_r^2$ from isometry).

(i) G. B. Airy (1861):

$$\epsilon_{lA}^2 := \frac{1}{2}\left[(\Lambda_1 - 1)^2 + (\Lambda_2 - 1)^2\right] \quad \text{versus} \quad \frac{1}{2}\left[(\lambda_1 - 1)^2 + (\lambda_2 - 1)^2\right] =: \epsilon_{rA}^2 . \tag{1.312}$$

(ii) V. V. Kavrajski (1958):

$$\epsilon_{lAK}^2 := \frac{1}{2}\left[(\ln \Lambda_1)^2 + (\ln \Lambda_2)^2\right] \quad \text{versus} \quad \frac{1}{2}\left[(\ln \lambda_1)^2 + (\ln \lambda_2)^2\right] =: \epsilon_{rAK}^2 . \tag{1.313}$$

Box 1.51 (Local measures for departure of the mapping $\mathbb{M}_l^2 \to \mathbb{M}_r^2$ from equiareal and conformal).

(i) Departure from an equiareal mapping:

$$\epsilon_{l\,\text{areal}}^2 := (\Lambda_1 \Lambda_2 - 1)^2 \quad \text{versus} \quad (\lambda_1 \lambda_2 - 1)^2 =: \epsilon_{r\,\text{areal}}^2 . \tag{1.314}$$

(ii) Departure from a conformal mapping:

$$\epsilon_{l\,\text{conf}}^2 := (\Lambda_1 - \Lambda_2)^2 \quad \text{versus} \quad (\lambda_1 - \lambda_2)^2 =: \epsilon_{r\,\text{conf}}^2 . \tag{1.315}$$

Box 1.52 (Global measures for departure of the mapping $\mathbb{M}_l^2 \to \mathbb{M}_r^2$ from isometry, areomorphism, and conformeomorphism).

(i) Isometry:

$$I_{lA} := \frac{1}{S_l} \int dS_l \epsilon_{lA}^2 \quad \text{versus} \quad \frac{1}{S_r} \int dS_r \epsilon_{rA}^2 =: I_{rA} ,$$

$$I_{lAK} := \frac{1}{S_l} \int dS_l \epsilon_{lAK}^2 \quad \text{versus} \quad \frac{1}{S_r} \int dS_r \epsilon_{rAK}^2 =: I_{rAK} . \tag{1.316}$$

(ii) Areomorphism:

$$I_{l\,\text{areal}} := \frac{1}{S_l} \int dS_l \epsilon_{l\,\text{areal}}^2 \quad \text{versus} \quad \frac{1}{S_r} \int dS_r \epsilon_{r\,\text{areal}}^2 =: I_{r\,\text{areal}} . \tag{1.317}$$

(iii) Conformeomorphism:

$$I_{l\,\text{conf}} := \frac{1}{S_l} \int dS_l \epsilon_{l\,\text{conf}}^2 \quad \text{versus} \quad \frac{1}{S_r} \int dS_r \epsilon_{r\,\text{conf}}^2 =: I_{r\,\text{conf}} . \tag{1.318}$$

1-145 Maximal angular distortion

The conformal mapping $\overline{f} : \mathbb{M}_l^2 \to \mathbb{M}_r^2$ had been previously defined by the *angular identity* $\Psi_l = \Psi_r$ or by *zero angular shear* $\sum_l = \Psi_l - \Psi_r = 0$ or $\sum_r = \Psi_r - \Psi_l = 0$. By means of the *canonical criteria* $\Lambda_1 = \Lambda_2$ or $\Lambda_1 - \Lambda_2 = 0$, we succeeded to formulate an equivalence for conformality. We shall concentrate here by means of a case study on the deviation of a general mapping $\overline{f} : \mathbb{M}_l^2 \to \mathbb{M}_r^2$ from conformality. In particular, we shall solve the *optimization problem* of *maximal angular shear* or of the largest deviation of such a general mapping from conformality. Fast first-hand information is offered by Lemma 1.20

Lemma 1.20 (Left and right general eigenvalue problem of the Cauchy–Green deformation tensor).

The angular distortion is maximal if $\Omega_l = 2\sum_l^+ = 2\arcsin\frac{\Lambda_1 - \Lambda_2}{\Lambda_1 + \Lambda_2}$ or $\Omega_r = 2\sum_r^+ = 2\arcsin\frac{\lambda_1 - \lambda_2}{\lambda_1 + \lambda_2}$.

End of Lemma.

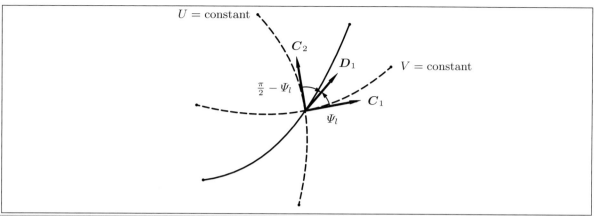

Fig. 1.30. Left angular shear $\sum_l := \Psi_l - \Psi_r$, left Gauss frame, left Cartan frame, left Darboux frame, angular shear parameter Ψ_l.

The general proof of such a lemma can be taken from C. Truesdell and R. Toupin (1960), pp. 257–266. here, we make the simplifying assumption $\{G_{12} = 0, c_{12} = 0\}$ and $\{g_{12} = 0, C_{12} = 0\}$. The off-diagonal elements of the left matrix of the metric G_l as well as of the left Cauchy–Green matrix C_l vanish. Or we may say that the coordinate lines "left" and their images "right" intersect at right angles. In consequence, the mapping equations are specified by $\{u(U), v(V)\}$. An analogue statement can be made for the special case $\{g_{12} = 0, C_{12} = 0\}$. First, we have to define the angular parameters Ψ_l and Ψ_r. According to Fig. 1.30 and Fig. 1.31, we refer the angle Ψ_l and Ψ_r, respectively, to the unit tangent vector $\boldsymbol{C}_1$ along the $V = $ constant coordinate line and to the unit tangent vector $\boldsymbol{D}_1$ of an arbitrary curve intersecting the coordinate line $V = $ constant, as well as to the unit tangent vector $\boldsymbol{c}_1$ along the $v = $ constant coordinate line and to the unit tangent vector $\boldsymbol{d}_1$ of an arbitrary curve intersecting the coordinate line $v = $ constant. Such an image curve is generated by mapping the original curve $\boldsymbol{C}(S) \in \mathbb{M}_l^1 \subset \mathbb{M}_l^2$ to $\boldsymbol{c}(s) \in \mathbb{M}_r^1 \subset \mathbb{M}_r^2$. Box 1.53 summarizes the related reference frames, namely

Gauss reference frame (3-leg):
$$\{\boldsymbol{G}_1, \boldsymbol{G}_2, \boldsymbol{G}_3 \,|\, U, V\}, \{\boldsymbol{g}_1, \boldsymbol{g}_2, \boldsymbol{g}_3 \,|\, u, v\};\tag{1.319}$$

Cartan reference frame (3-leg, orthonormal, repére mobile):
$$\{\boldsymbol{C}_1, \boldsymbol{C}_2, \boldsymbol{C}_3 \,|\, U, V\}, \{\boldsymbol{c}_1, \boldsymbol{c}_2, \boldsymbol{c}_3 \,|\, u, v\};\tag{1.320}$$

Darboux reference frame (3-leg, orthonormal):
$$\{\boldsymbol{D}_1, \boldsymbol{D}_2, \boldsymbol{D}_3 \,|\, U(S), V(S)\}, \{\boldsymbol{d}_1, \boldsymbol{d}_2, \boldsymbol{d}_3 \,|\, u(s), v(s)\}.\tag{1.321}$$

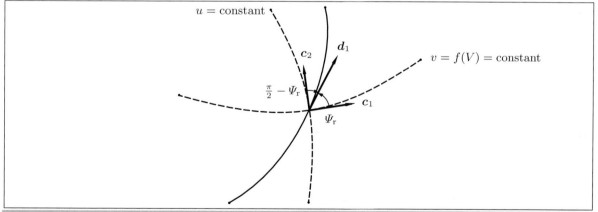

Fig. 1.31. Right angular shear $\sum_r := \Psi_r - \Psi_l$, right Gauss frame, right Cartan frame, right Darboux frame, angular shear parameter Ψ_r.

Box 1.53 (Reference frames (3-leg) of type Gauss, Cartan, and Darboux).

The left manifolds ($\mathbb{M}_l^1 \subset \mathbb{M}_l^2$, $G_{12} = 0$).

Gauss:

$$\boldsymbol{G}_1 := \frac{\partial \boldsymbol{X}(U,V)}{\partial U},$$

$$\boldsymbol{G}_2 := \frac{\partial \boldsymbol{X}(U,V)}{\partial V},$$

$$\boldsymbol{G}_3 := \frac{\boldsymbol{G}_1 \times \boldsymbol{G}_2}{\|\boldsymbol{G}_1 \times \boldsymbol{G}_2\|}.$$

Cartan:

$$\boldsymbol{C}_1 := \frac{\boldsymbol{G}_1}{\|\boldsymbol{G}_1\|} = \frac{\boldsymbol{G}_1}{\sqrt{G_{11}}},$$

$$\boldsymbol{C}_2 := \frac{\boldsymbol{G}_2}{\|\boldsymbol{G}_2\|} = \frac{\boldsymbol{G}_2}{\sqrt{G_{22}}},$$

$$\boldsymbol{C}_3 := \boldsymbol{C}_1 \times \boldsymbol{C}_2 =$$
$$= *(\boldsymbol{C}_1 \wedge \boldsymbol{C}_2) = \boldsymbol{G}_3.$$

Darboux:

$$\boldsymbol{D}_1 := \boldsymbol{X}' = \frac{\mathrm{d}\boldsymbol{X}}{\mathrm{d}S},$$

$$\boldsymbol{D}_2 := \boldsymbol{D}_3 \times \boldsymbol{D}_1 =$$
$$= *(\boldsymbol{D}_3 \wedge \boldsymbol{D}_1),$$

$$\boldsymbol{D}_3 = \boldsymbol{C}_3 = \boldsymbol{G}_3.$$

$$(1.322)$$

The right manifolds ($c_{12} = 0$).

Gauss:

$$\boldsymbol{g}_1 := \frac{\partial \boldsymbol{x}\big(u(U), v(V)\big)}{\partial U},$$

$$\boldsymbol{g}_2 := \frac{\partial \boldsymbol{x}\big(u(U), v(V)\big)}{\partial V},$$

$$\boldsymbol{g}_3 := \frac{\boldsymbol{g}_1 \times \boldsymbol{g}_2}{\|\boldsymbol{g}_1 \times \boldsymbol{g}_2\|}.$$

Cartan:

$$\boldsymbol{c}_1 := \frac{\boldsymbol{g}_1}{\|\boldsymbol{g}_1\|},$$

$$\boldsymbol{c}_2 := \frac{\boldsymbol{g}_2}{\|\boldsymbol{g}_2\|},$$

$$\boldsymbol{c}_3 := \boldsymbol{c}_1 \times \boldsymbol{c}_2 =$$
$$= *(\boldsymbol{c}_1 \wedge \boldsymbol{c}_2).$$

Darboux:

$$\boldsymbol{d}_1 := \boldsymbol{x}' = \frac{\mathrm{d}\boldsymbol{x}\big(u(s), v(s)\big)}{\mathrm{d}s},$$

$$\boldsymbol{d}_2 := \boldsymbol{d}_3 \times \boldsymbol{d}_1 =$$
$$= *(\boldsymbol{d}_3 \wedge \boldsymbol{d}_1),$$

$$\boldsymbol{d}_3 = \boldsymbol{c}_3 = \boldsymbol{g}_3.$$

$$(1.323)$$

Those forms of reference are needed to represent $\cos \Psi_l$ and $\cos \Psi_r$, the cosine of the angles between the tangent vector $\boldsymbol{C}_1$ and $\boldsymbol{c}_1$, respectively, and the tangent vector $\boldsymbol{D}_1$ and $\boldsymbol{d}_1$, respectively (also called "Cartan 1" and "Darboux 1") by means of the scalar product $\langle \boldsymbol{X}' \,|\, \boldsymbol{C}_1 \rangle$ and $\langle \boldsymbol{x}' \,|\, \boldsymbol{c}_1 \rangle$, respectively. Second, according to Box 1.54, we derive the basic relations $\cos \Psi_l = \sqrt{G_{11}} U'$ and $\sin \Psi_l = \sqrt{G_{22}} V'$ as well as $\cos \Psi_r = \sqrt{g_{11}} u'$ and $\sin \Psi_r = \sqrt{g_{22}} v'$. $\{U', V'\}$ and $\{u', v'\}$ express the derivative of the parameterized curve $C(S)$ and $c(s)$, respectively, with respect to the canonical curve parameters $\{\text{arc length } S, \text{arc length } s\}$. Third, outlined in Box 1.55, by means of the chain rule, we succeed to derive $\{U', V'\}$ and $\{u', v'\}$, respectively, in terms of the elements of the Jacobi matrices $[\partial \{U, V\}/\partial \{u, v\}]$ and $[\partial \{u, v\}/\partial \{U, V\}]$ and the stretches $\mathrm{d}s/\mathrm{d}S$ and $\mathrm{d}S/\mathrm{d}s$, respectively. In this way, we succeed to represent $\cos \Psi_l$ and $\sin \Psi_l$ and $\cos \Psi_r$ and $\sin \Psi_r$ in terms of the elements of the left and the right Cauchy–Green matrix C_l and C_r, respectively. Fourth, Box 1.56 leads us to the left and the right angular shear, $\sum_l$ and $\sum_r$, respectively. Our great results are presented in Corollary 1.21. The proof follows the lines of Box 1.56, namely the *addition theorem* $\tan(x - y) = (\tan x + \tan y)/(1 + \tan x \tan y)$. $\tan \sum_l (\Psi_l)$ as well as $\tan \sum_r (\Psi_r)$ establish the optimization crtiteria for *maximal angular distortion*. Fifth, the characteristic optimization problem $\sum_l (\psi_l) = \text{extr.}$ or $\sum_r (\psi_r) = \text{extr.}$ is dealt with in Box 1.57. Indeed, we find the two *stationary points* $\tan \Psi_l^\pm$ and $\tan \Psi_r^\pm$. These stationary solutions lead us to the extremal values of $\sum_l^\pm$ and $\sum_r^\pm$, the celebrated representations

$$\sin \sum_l^\pm = \pm \frac{\varLambda_1 - \varLambda_2}{\varLambda_1 + \varLambda_2} \quad \text{versus} \quad \sin \sum_r^\pm = \pm \frac{\lambda_1 - \lambda_2}{\lambda_1 + \lambda_2}. \tag{1.324}$$

From these extremal values of left and right angular shear $\sum_l^\pm$ and $\sum_r^\pm$, we derive the left and right *maximal angular distortion* $\varOmega_l$ and $\varOmega_r$, respectively, namely

$$\varOmega_l = 2 \arcsin \left| \frac{\varLambda_1 - \varLambda_2}{\varLambda_1 + \varLambda_2} \right| \quad \text{versus} \quad \varOmega_r = 2 \arcsin \left| \frac{\lambda_1 - \lambda_2}{\lambda_1 + \lambda_2} \right|, \tag{1.325}$$

based upon the symmetry $\sum_l^+ - = \sum_l^-$, $\sum_r^+ - = \sum_r^-$ and $\varOmega_l := \sum_l^+ - \sum_l^-$, $\varOmega_r := \sum_r^+ - \sum_r^-$. Indeed, $\varOmega_l$ and $\varOmega_r$ are the maximal data of angular distortion.

Box 1.54 (Angular parameters Ψ_l and Ψ_r).

"Left":

$$\boldsymbol{X}' := \frac{\partial \boldsymbol{X}}{\partial U}\frac{\mathrm{d}U}{\mathrm{d}S} + \frac{\partial \boldsymbol{X}}{\partial V}\frac{\mathrm{d}V}{\mathrm{d}S} \ ;$$

$$\cos\Psi_l = \|\boldsymbol{X}'\|\|\boldsymbol{C}_1\|\cos\Psi_l = \langle \boldsymbol{X}' \,|\, \boldsymbol{C}_1 \rangle \ ,$$

$$\cos\Psi_l = \sqrt{G_{11}}U' \ ;$$

$$\sin\Psi_l = \cos\left(\frac{\pi}{2} - \Psi_l\right) =$$

$$= \|\boldsymbol{X}'\|\|\boldsymbol{C}_2\|\cos\left(\frac{\pi}{2} - \Psi_l\right) = \langle \boldsymbol{X}' \,|\, \boldsymbol{C}_2 \rangle \ ,$$

$$\sin\Psi_l = \sqrt{G_{22}}V' \ .$$

"Right":

$$\boldsymbol{x}' := \frac{\partial \boldsymbol{x}}{\partial u}\frac{\mathrm{d}u}{\mathrm{d}s} + \frac{\partial \boldsymbol{x}}{\partial v}\frac{\mathrm{d}v}{\mathrm{d}s} \ ;$$

$$\cos\Psi_r = \|\boldsymbol{x}'\|\|\boldsymbol{c}_1\|\cos\Psi_r = \langle \boldsymbol{x}' \,|\, \boldsymbol{c}_1 \rangle \ ,$$

$$\cos\Psi_r = \sqrt{g_{11}}u' \ ; \tag{1.326}$$

$$\sin\Psi_r = \cos\left(\frac{\pi}{2} - \Psi_r\right) =$$

$$= \|\boldsymbol{x}'\|\|\boldsymbol{c}_2\|\cos\left(\frac{\pi}{2} - \Psi_r\right) = \langle \boldsymbol{x}' \,|\, \boldsymbol{c}_2 \rangle \ ,$$

$$\sin\Psi_r = \sqrt{g_{22}}v' \ .$$

Box 1.55 (Transformation of angular parameters Ψ_l and Ψ_r. Special case: $G_{12} = 0$, $c_{12} = 0$, $u(U), v(V)$ versus $U(u), V(v)$).

$$\cos\Psi_l = \sqrt{G_{11}}U' = \sqrt{G_{11}U'^2} \ ,$$

$$\sin\Psi_l = \sqrt{G_{22}}V' = \sqrt{G_{22}V'^2} \ .$$

$$\cos\Psi_r = \sqrt{g_{11}}u' = \sqrt{g_{11}u'^2} \ ,$$

$$\sin\Psi_r = \sqrt{g_{22}}v' = \sqrt{g_{22}v'^2} \ , \tag{1.327}$$

$$u' = \frac{\mathrm{d}u}{\mathrm{d}s} \ , \ v' = \frac{\mathrm{d}v}{\mathrm{d}s} \ .$$

$$U' = \frac{\mathrm{d}U}{\mathrm{d}u}\frac{\mathrm{d}u}{\mathrm{d}s}\frac{\mathrm{d}s}{\mathrm{d}S} \ , \ V' = \frac{\mathrm{d}V}{\mathrm{d}v}\frac{\mathrm{d}v}{\mathrm{d}s}\frac{\mathrm{d}s}{\mathrm{d}S}$$

$$u' = \frac{\mathrm{d}u}{\mathrm{d}U}\frac{\mathrm{d}U}{\mathrm{d}S}\frac{\mathrm{d}S}{\mathrm{d}s} \ , \ v' = \frac{\mathrm{d}v}{\mathrm{d}V}\frac{\mathrm{d}V}{\mathrm{d}S}\frac{\mathrm{d}S}{\mathrm{d}s}$$

$$\Rightarrow \qquad\qquad\qquad \Rightarrow$$

$$\cos\Psi_l = \sqrt{G_{11}\left(\frac{\mathrm{d}U}{\mathrm{d}u}\right)^2}u'\frac{\mathrm{d}s}{\mathrm{d}S} \ ,$$

$$\cos\Psi_r = \sqrt{g_{11}\left(\frac{\mathrm{d}u}{\mathrm{d}U}\right)^2}U'\frac{\mathrm{d}S}{\mathrm{d}s} \ , \tag{1.328}$$

$$\sin\Psi_l = \sqrt{G_{22}\left(\frac{\mathrm{d}V}{\mathrm{d}v}\right)^2}v'\frac{\mathrm{d}s}{\mathrm{d}S}$$

$$\sin\Psi_r = \sqrt{g_{22}\left(\frac{\mathrm{d}v}{\mathrm{d}V}\right)^2}V'\frac{\mathrm{d}S}{\mathrm{d}s}$$

$$\Rightarrow \qquad\qquad\qquad \Rightarrow$$

$$\cos\Psi_l = \sqrt{C_{11}}u'\frac{\mathrm{d}s}{\mathrm{d}S} \ , \ \sin\Psi_l = \sqrt{C_{22}}v'\frac{\mathrm{d}s}{\mathrm{d}S} \ .$$

$$\cos\Psi_r = \sqrt{c_{11}}U'\frac{\mathrm{d}S}{\mathrm{d}s} \ , \ \sin\Psi_r = \sqrt{c_{22}}V'\frac{\mathrm{d}S}{\mathrm{d}s} \ .$$

Corollary 1.21 (The canonical representation of left angular shear and right angular shear. Special case: $G_{12} = 0$, $c_{12} = 0$ and $g_{12} = 0$, $C_{12} = 0$).

Let $\sum_l := \Psi_l - \Psi_r$ and $\sum_r := \Psi_r - \Psi_l$, respectively, denote left and right angular shear, a measure of the deviation of the mapping $\mathbb{M}_l^2 \to \mathbb{M}_r^2$ from conformality. Then a canonical representation of the angular parameters Ψ_l and Ψ_r as well as the angular shear parameters $\sum_l$ and $\sum_r$ is

$$\tan\Psi_l = \frac{\lambda_2}{\lambda_1}\tan\Psi_r \quad \text{versus} \quad \tan\Psi_r = \frac{\Lambda_1}{\Lambda_2}\tan\Psi_l \ , \tag{1.329}$$

$$\tan\sum_l = (\Lambda_1 - \Lambda_2)\frac{\tan\Psi_l}{\Lambda_1 + \Lambda_2\tan^2\Psi_l} \quad \text{versus} \quad \tan\sum_r = (\lambda_1 - \lambda_2)\frac{\tan\Psi_r}{\lambda_1 + \lambda_2\tan^2\Psi_r} \ . \tag{1.330}$$

End of Corollary.

Box 1.56 (The canonical representation of left angular shear and right angular shear. Special case: $G_{12} = 0$, $c_{12} = 0$ and $g_{12} = 0$, $C_{12} = 0$).

Left and right angular parameters Ψ_l and Ψ_r:

$$\cos \Psi_l = \sqrt{C_{11}}\frac{du}{ds}\frac{ds}{dS} \quad \text{versus} \quad \sqrt{c_{11}}\frac{dU}{dS}\frac{dS}{ds} = \cos \Psi_r \ ,$$

$$\cos \Psi_l = \sqrt{C_{11}}\frac{du}{ds}\frac{1}{\lambda} \quad \text{versus} \quad \sqrt{c_{11}}\frac{dU}{dS}\frac{1}{\Lambda} = \cos \Psi_r \ ,$$

$$\sin \Psi_l = \sqrt{C_{22}}\frac{dv}{ds}\frac{ds}{dS} \quad \text{versus} \quad \sqrt{c_{22}}\frac{dV}{dS}\frac{dS}{ds} = \sin \Psi_r \ ,$$

$$\sin \Psi_l = \sqrt{C_{22}}\frac{dv}{ds}\frac{1}{\lambda} \quad \text{versus} \quad \sqrt{c_{22}}\frac{dV}{dS}\frac{1}{\Lambda} = \sin \Psi_r \ .$$

(1.331)

Left and right stretches, left and right principal stretches:

$$\lambda^2 := \frac{dS^2}{ds^2} \quad \text{versus} \quad \Lambda^2 := \frac{ds^2}{dS^2} \ ,$$

$$\lambda_1^2 = \frac{C_{11}}{g_{11}} \quad \text{versus} \quad \Lambda_1^2 = \frac{c_{11}}{G_{11}} \ ,$$

$$\lambda_2^2 = \frac{C_{22}}{g_{22}} \quad \text{versus} \quad \Lambda_2^2 = \frac{c_{22}}{G_{22}} \ ,$$

(1.332)

$$\cos \Psi_r = \sqrt{g_{11}}u' \qquad\qquad \sqrt{G_{11}}U' = \cos \Psi_l$$

$$\Rightarrow \qquad\qquad\qquad \Rightarrow$$

$$\cos \Psi_l = \cos \Psi_r\sqrt{\frac{C_{11}}{g_{11}}}\frac{1}{\lambda} \quad \text{versus} \quad \cos \Psi_l\sqrt{\frac{c_{11}}{G_{11}}}\frac{1}{\Lambda} = \cos \Psi_r \ ,$$

$$\sin \Psi_l = \sin \Psi_r\sqrt{\frac{C_{22}}{g_{22}}}\frac{1}{\lambda} \quad \text{versus} \quad \sin \Psi_l\sqrt{\frac{c_{22}}{G_{22}}}\frac{1}{\Lambda} = \sin \Psi_r \ ,$$

(1.333)

$$\cos \Psi_l = \cos \Psi_r\frac{\lambda_1}{\lambda} \quad \text{versus} \quad \cos \Psi_l\frac{\Lambda_1}{\Lambda} = \cos \Psi_r \ ,$$

$$\sin \Psi_l = \sin \Psi_r\frac{\lambda_2}{\lambda} \quad \text{versus} \quad \sin \Psi_l\frac{\Lambda_2}{\Lambda} = \sin \Psi_r \ ,$$

$$\tan \Psi_l = \frac{\lambda_2}{\lambda_1}\tan \Psi_r \quad \text{versus} \quad \frac{\Lambda_2}{\Lambda_1}\tan \Psi_l = \tan \Psi_r \ .$$

(1.334)

Left and right angular shear, left and right angular distortion:

$$\tan(\Psi_l - \Psi_r) = \tan\sum\nolimits_l \qquad \text{versus} \qquad \tan\sum\nolimits_r = \tan(\Psi_r - \Psi_l) \ ,$$

$$\tan\sum\nolimits_l = \frac{\tan \Psi_l - \tan \Psi_r}{1 + \tan \Psi_l \tan \Psi_r} \qquad \text{versus} \qquad \tan\sum\nolimits_r = \frac{\tan \Psi_r - \tan \Psi_l}{1 + \tan \Psi_r \tan \Psi_l} \ ,$$

$$\tan\sum\nolimits_l = \frac{\tan \Psi_l - \Lambda_2\Lambda_1^{-1}\tan \Psi_l}{1 + \Lambda_2\Lambda_1^{-1}\tan^2 \Psi_l} \qquad \text{versus} \qquad \tan\sum\nolimits_r = \frac{\tan \Psi_r - \lambda_2\lambda_1^{-1}\tan \Psi_r}{1 + \lambda_2\lambda_1^{-1}\tan^2 \Psi_r} \ ,$$

(1.335)

$$\tan\sum\nolimits_l = (\Lambda_1 - \Lambda_2)\frac{\tan \Psi_l}{\Lambda_1 + \Lambda_2\tan^2 \Psi_l} \quad \text{versus} \quad \tan\sum\nolimits_r = \frac{\tan \Psi_r}{\lambda_1 + \lambda_2\tan^2 \Psi_r}(\lambda_1 - \lambda_2) \ .$$

Box 1.57 (The optimization problem; extremal, left angular shear or right angular shear; maximal angular distortion).

Optimization problem:

$$\sum_l^\pm = \qquad\qquad\text{versus}\qquad\qquad \sum_r^\pm =$$
$$= \arg\left\{\sum_l \in [0, 2\pi] \,\big|\, \sum_l(\psi_l) = \text{extr.}\right\} \qquad = \arg\left\{\sum_r \in [0, 2\pi] \,\big|\, \sum_r(\psi_r) = \text{extr.}\right\}, \tag{1.336}$$

$$x := \Psi_l, \ f(x) := \sum_l(\Psi_l), \qquad\qquad x := \Psi_r, \ f(x) := \sum_r(\Psi_r),$$
$$a := \Lambda_1, \quad b := \Lambda_2. \qquad \tan f(x) := \frac{\tan x}{a + b\tan^2 x}. \qquad a := \lambda_1, \quad b := \lambda_2. \tag{1.337}$$

Stationary points:

$$(\tan x)' = 1 + \tan^2 x, \tag{1.338}$$

$$f'(x) = 0$$
$$\Leftrightarrow$$
$$(\tan f(x))' = (1 + \tan^2 f(x))f'(x) = 0,$$
$$(\tan f(x))' = \frac{1 + \tan^2 x}{(a + b\tan^2 x)^2}(a - b\tan^2 x),$$
$$(\tan f(x))' = 0$$
$$\Leftrightarrow$$
$$a - b\tan^2 x = 0$$
$$\Leftrightarrow$$
$$\tan x = \pm\sqrt{\tfrac{a}{b}}, \tag{1.339}$$

$$\tan\Psi_l^\pm = \pm\sqrt{\frac{\Lambda_1}{\Lambda_2}} \quad\text{versus}\quad \tan\Psi_r^\pm = \pm\sqrt{\frac{\lambda_1}{\lambda_2}}. \tag{1.340}$$

Extremal left or right angular shear:

$$\tan\sum_l^\pm = \pm\frac{1}{2}\frac{\Lambda_1 - \Lambda_2}{\sqrt{\Lambda_1\Lambda_2}} \quad\text{versus}\quad \tan\sum_r^\pm = \pm\frac{1}{2}\frac{\lambda_1 - \lambda_2}{\sqrt{\lambda_1\lambda_2}},$$
$$\sin x = \frac{\tan x}{\sqrt{1 + \tan^2 x}}, \tag{1.341}$$
$$\sin\sum_l^\pm = \pm\frac{\Lambda_1 - \Lambda_2}{\Lambda_1 + \Lambda_2} \quad\text{versus}\quad \sin\sum_r^\pm = \pm\frac{\lambda_1 - \lambda_2}{\lambda_1 + \lambda_2}.$$

Maximal angular distortion:

$$\Omega_l := \sum_l^+ - \sum_l^- = 2\sum_l^+ \quad\text{versus}\quad \Omega_r := \sum_r^+ - \sum_r^- = 2\sum_r^+,$$
$$\Omega_l = 2\arcsin\left|\frac{\Lambda_1 - \Lambda_2}{\Lambda_1 + \Lambda_2}\right| \quad\text{versus}\quad \Omega_r = \arcsin\left|\frac{\lambda_1 - \lambda_2}{\lambda_1 + \lambda_2}\right|. \tag{1.342}$$

1-15 Exercise: the Armadillo double projection

Exercise: the Armadillo double projection. First: sphere to torus. Second: torus to plane. The oblique orthogonal projection.

An excellent example of a mapping from a left two-dimensional Riemann manifold to a right two-dimensional Riemann manifold where we have to use all the power of the previous paragraphs is the *Armadillo map modified by Raisz*, which is illustrated in Fig. 1.32. First, points of the sphere $\mathbb{S}_R^2$ of radius R are mapped onto a specific torus $\mathbb{T}_{a,b}^2$. Second, subject to $a = b = R$, $\mathbb{T}_{a,b}^2$ is mapped as an *oblique orthogonal projection* onto a central plane $\mathbb{P}_\mathcal{O}^2$. Such a double projection is analytically presented in Box 1.58. The first mapping, namely $\mathbb{S}_R^2 \to \mathbb{T}_{a,b}^2$, is fixed by the postulate $\{\lambda = \Lambda/2, \phi = \Phi\}$, which cuts the spherical longitude Λ half to be gauged to the toroidal longitude λ. In contrast, spherical latitude Φ is set identical to the toroidal latitude ϕ. For generating the second mapping, namely $\mathbb{T}_{a,b}^2 \to \mathbb{P}_\mathcal{O}^2$, subject to $a = b = R$, we rotate around the 2 axis by $-\beta$ from $\{X, Y, Z\} \in \mathbb{R}^3$ to $\{X', Y', Z'\} \in \mathbb{R}^3$. In consequence, we experience an orthogonal projection of any point of the specific torus $\mathbb{T}_{a,b}^2$ onto the Y'–Z' plane such that $x = Y'$ and $y = Z'$. In this way, we have succeeded in parameterizing the double projection $\mathbb{S}_R^2 \to \mathbb{T}_{a,b}^2 \to \mathbb{P}_\mathcal{O}^2$ by $\{x(\Lambda, \Phi), y(\Lambda, \Phi)\}$. However, we pose the following problems. (i) Determine the left principal stretches $\{\Lambda_1, \Lambda_2\}$ from the direct mapping equations $x(\Lambda, \Phi)$ and $y(\Lambda, \Phi)$ subject to the matrix G_l of the metric, the right matrix G_r of the metric, the left Jacob matrix J_l, and the left Cauchy–Green matrix C_l viewed in Box 1.59. (ii) Prove that the Armadillo double projection is not equiareal. (iii) Prove that the images of the parallel circles of the sphere are ellipses. Determine their semi-major and semi-minor axes as well as the location of the center. (iv) Prove that the images of the meridians of the sphere are conic sections.

Solution (all problems).

Here are some ideas to solve the hard problems. For the second problem, we advise you to prove the inequality $\det[\mathsf{C}_l] \neq \det[\mathsf{G}_l]$. To solve the third problem, choose $\Phi = $ constant and eliminate Λ from the direct equations of the mapping, for instance, $\sin \Lambda/2 = x/[R(1 + \cos \Phi)]$ as well as $\cos \Lambda/2 = (R \cos \beta \sin \Phi - y)/[R(1 + \cos \Phi) \sin \beta]$. Next, add $\sin^2 \Lambda/2 + \cos^2 \Lambda/2 = 1$ and you are done. Similarly, to solve the fourth problem, choose $\Lambda = $ constant and eliminate Φ from the direct equations of the mapping, for instance, by $1 + \cos \Phi = x/[R \sin \Lambda/2]$ as $\Phi = (x - R \sin \Lambda/2)/(R \sin \Lambda/2)$ and $\cos^2 \Phi$ as well as by $y/R + (x \sin \beta)/(R \tan \Lambda/2) = \cos \beta \sin \Phi$, to be squared to $\cos^2 \beta \sin^2 \Phi$, leading to a quadratic form of type $ax^2 + bxy + cy^2 + d = 0$, indeed a conic section.

End of Solution (all problems).

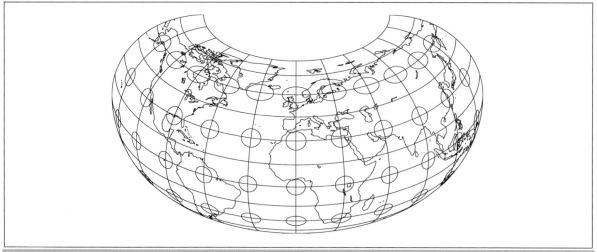

Fig. 1.32 *Armadillo projection modified by Raisz*: double projection, (i) sphere $\to$ torus, (ii) torus $\to$ plane, obliquity $\beta = 20°$, Tissot ellipses of distortion.

Box 1.58 (Armadillo double projection. 1st: sphere to torus. 2nd: torus to oblique plane).

Left manifold $\mathbb{S}^2_r$,
left coordinates
(spherical longitude Λ,
(spherical latitude Φ):

Right manifold $\mathbb{T}^2_{a,b} \sim \mathbb{S}^1_a \times \mathbb{S}^1_b$,
right coordinates
(toroidal longitude λ,
(toroidal latitude ϕ):

$$\mathbb{S}^2_r :=$$

$$\mathbb{T}^2_{a,b} :=$$

$$:= \left\{ X \in \mathbb{R}^3 \,\middle|\, X^2 + Y^2 + Z^2 - R^2 = 0, \right.$$

$$:= \left\{ x \in \mathbb{R}^3 \,\middle|\, \left(\sqrt{x^2 + y^2} - a \right)^2 + z^2 - b^2 = 0, \right.$$

$$\left. R \in \mathbb{R}^+, R > 0 \right\} ,$$

$$\left. a \in \mathbb{R}^+, b \in \mathbb{R}^+, b \le a \right\} ,$$

(1.343)

$$\Phi^{-1}_l : \boldsymbol{X}(\Lambda, \Phi) =$$

$$\Phi^{-1}_r : \boldsymbol{x}(\lambda, \phi) =$$

$$= \boldsymbol{E}_1 R \cos\Phi \cos\Lambda + \boldsymbol{E}_2 R \cos\Phi \sin\Lambda +$$

$$= \boldsymbol{e}_1 (a + b\cos\phi) \cos\lambda + \boldsymbol{e}_2 (a + b\cos\phi) \sin\lambda +$$

$$+ \boldsymbol{E}_3 R \sin\Phi .$$

$$+ \boldsymbol{e}_3 b \sin\phi .$$

1st mapping:

$$\begin{bmatrix} \lambda \\ \phi \end{bmatrix} = \begin{bmatrix} \Lambda/2 \\ \Phi \end{bmatrix} , \quad a = b = R .$$

(1.344)

2nd mapping (oblique orthogonal projection).

Left manifold $\mathbb{T}^2_{a,b}$:

Right manifold (oblique plane) $\mathbb{P}^2_{\mathcal{O}}$:

$$X = R(1 + \cos\Phi) \cos\Lambda/2 ,$$

$$x = Y' ,$$

$$Y = R(1 + \cos\Phi) \sin\Lambda/2 ,$$

$$y = Z' ;$$

(1.345)

$$Z = R \sin\Phi ;$$

$$\begin{bmatrix} X \\ Y \\ Z \end{bmatrix} = \mathsf{R}_2(\beta) \begin{bmatrix} X' \\ Y' \\ Z' \end{bmatrix}$$

$$\Leftrightarrow$$

$$\begin{bmatrix} X' \\ Y' \\ Z' \end{bmatrix} = \mathsf{R}_2(-\beta) \begin{bmatrix} X \\ Y \\ Z \end{bmatrix} ,$$

(1.346)

$$\mathsf{R}_2(\beta) = \begin{bmatrix} \cos\beta & 0 & -\sin\beta \\ 0 & 1 & 0 \\ \sin\beta & 0 & \sin\beta \end{bmatrix}$$

$$\Leftrightarrow$$

$$\begin{bmatrix} \cos\beta & 0 & \sin\beta \\ 0 & 1 & 0 \\ -\sin\beta & 0 & \cos\beta \end{bmatrix} = \mathsf{R}_2^{\mathrm{T}}(\beta) = \mathsf{R}_2^{\mathrm{T}}(-\beta) ;$$

$$x := Y = Y' , \; y := Z' = -\sin\beta X + \cos\beta Z ,$$

(1.347)

$$x = R(1 + \cos\Phi) \sin\Lambda/2 , \; y = -R(1 + \cos\Phi) \sin\beta \cos\Lambda/2 + R \cos\beta \sin\Phi .$$

Box 1.59 (Left principal stretches).

Left and right matrices of the metric:

$$\mathsf{G}_l := \begin{bmatrix} R^2 \cos^2 \Phi & 0 \\ 0 & R^2 \end{bmatrix} ,$$

$$\mathsf{G}_r := \begin{bmatrix} 1 & 0 \\ 0 & 1 \end{bmatrix} .$$

(1.348)

Left Jacobi matrix:

$$\mathsf{J}_l := \begin{bmatrix} D_\Lambda x & D_\Phi x \\ D_\Lambda y & D_\Phi y \end{bmatrix} =$$

$$= R \begin{bmatrix} \dfrac{1}{2}(1 + \cos \Phi) \cos \Lambda/2 & -\sin \Phi \sin \Lambda/2 \\ \dfrac{1}{2}(1 + \cos \Phi) \sin \beta \sin \Lambda/2 & \cos \beta \cos \Phi + \sin \beta \sin \Phi \cos \Lambda/2 \end{bmatrix} .$$

(1.349)

Left Cauchy–Green matrix:

$$\mathsf{C}_l := \mathsf{J}_l^{\mathrm{T}} \mathsf{G}_r \mathsf{J}_l = \begin{bmatrix} c_{11} & c_{12} \\ c_{12} & c_{22} \end{bmatrix} ,$$

$$c_{11} = x_\Lambda^2 + y_\Lambda^2 =$$

$$= \frac{1}{4} R^2 (1 + \cos \Phi)^2 \left(\cos^2 \Lambda/2 + \sin^2 \beta \sin^2 \Lambda/2 \right) =$$

$$= \frac{1}{4} R^2 (1 + \cos \Phi)^2 \left(1 - \cos^2 \beta \sin^2 \Lambda/2 \right) ,$$

$$c_{12} = x_\Lambda x_\Phi + y_\Lambda y_\Phi =$$

$$= \frac{1}{2} R^2 (1 + \cos \Phi) \sin \Lambda/2 \left[-\sin \Phi \cos \Lambda/2 + \sin \beta \left(\cos \beta \cos \Phi + \sin \Phi \sin \beta \cos \Lambda/2 \right) \right]$$

$$= \frac{1}{2} R^2 (1 + \cos \Phi) \sin \Lambda/2 \left(\sin \beta \cos \beta \cos \Phi - \sin \Phi \cos^2 \beta \cos \Lambda/2 \right)$$

$$= \frac{1}{2} R^2 (1 + \cos \Phi) \sin \Lambda/2 \cos \beta \left(\sin \beta \cos \Phi - \sin \Phi \cos \beta \cos \Lambda/2 \right) ,$$

$$c_{22} = x_\Phi^2 + y_\Phi^2 =$$

$$= R^2 \left[\sin^2 \Phi \sin^2 \Lambda/2 + (\cos \beta \cos \Phi + \sin \Phi \sin \beta \cos \Lambda/2)^2 \right] ,$$

(1.350)

$$\det [\mathsf{C}_l] = R^4 \frac{(1 + \cos \Phi)^2}{4} \left(\sin \beta \sin \Phi + \cos \Phi \cos \beta \cos \Lambda/2 \right)^2 ,$$

$$\det [\mathsf{G}_l] = R^4 \cos^2 \Phi \neq \det [\mathsf{C}_l] .$$

(1.351)

Left principal stretches:

$$\left| \mathsf{C}_l - \Lambda_l^2 \mathsf{G}_l \right| = 0 .$$

(1.352)

With this box, we finish the general consideration of mappings between Riemann manifolds. In the following chapter, we specialize the various rules for mappings between Riemann manifolds and Euclidean manifolds.

2 From Riemann manifolds to Euclidean manifolds

Mapping from a left two-dimensional Riemann manifold to a right two-dimensional Euclidean manifold, Cauchy–Green and Euler–Lagrange deformation tensors, equivalence theorem for equiareal mappings, conformeomorphism and areomorphism, Korn–Lichtenstein equations and Cauchy–Riemann equations, Mollweide projection, canonical criteria for (conformal, equiareal, isometric, equidistant) mappings, polar decomposition and simultaneous diagonalization for more than two matrices.

Let there be given the left two-dimensional Riemann manifold $\{\mathbb{M}_l^2, G_{MN}\}$ as well as the right two-dimensional Euclidean manifold $\{\mathbb{M}_r^2, g_{\mu\nu}\} = \{\mathbb{R}^2, \delta_{\mu\nu}\} = \mathbb{E}^2$. In many applications, the choice of $\{\mathbb{R}^2, \delta_{\mu\nu}\}$ is the "plane manifold", for instance, (i) the equatorial plane of the sphere or the ellipsoid, (ii) the meta-equatorial, also called *oblique equatorial plane* of the sphere or the ellipsoid, (iii) the plane generated by developing the cylinder, the cone, a ruled surface (namely surfaces which are "Gauss flat"), (iv) the tangent space $T_{U_0}\mathbb{M}_l^2$ of the left two-dimensional Riemann manifold fixed to the point $U_0 := \{U_0^1, U_0^2\}$ being covered by Cartesian coordinates. (Refer to all previous examples.) We shall not repeat the various *deformation measures* of type *multiplicative* and *additive* for the special case of the right two-dimensional Euclidean manifold $\{\mathbb{R}^2, \delta_{\mu\nu}\}$. Instead, we present to you (i) the left and right eigenspace analysis and synthesis of the Cauchy–Green deformation tensor, special case $\{\mathbb{M}_r^2, g_{\mu\nu}\} = \{\mathbb{R}^2, \delta_{\mu\nu}\}$, (ii) the left and right eigenspace analysis and synthesis of the Euler–Lagrange deformation tensor, special case $\{\mathbb{M}_r^2, g_{\mu\nu}\} = \{\mathbb{R}^2, \delta_{\mu\nu}\}$, (iii) conformeomorphism, conformal mapping, special case $\{\mathbb{M}_r^2, g_{\mu\nu}\} = \{\mathbb{R}^2, \delta_{\mu\nu}\}$; Korn–Lichtenstein equations, special case Cauchy–Riemann equations (d'Alembert–Euler equations).

2-1 Eigenspace analysis, Cauchy–Green deformation tensor

Left and right eigenspace analysis and synthesis of the Cauchy–Green deformation tensor, special case $\{\mathbb{M}_r^2, g_{\mu\nu}\} = \{\mathbb{R}^2, \delta_{\mu\nu}\}$.

First, let us confront you with Lemma 2.1, where we present detailed results of the left and right eigenspace analysis and synthesis of the Cauchy–Green deformation tensor for the special case of a right Euclidean manifold. Second, we focus on an interpretation of the results and additionally discuss a short example.

Lemma 2.1 (Left and right eigenspace analysis and synthesis of the Cauchy–Green deformation tensor, special case $\{\mathbb{M}_r^2, g_{\mu\nu}\} = \{\mathbb{R}^2, \delta_{\mu\nu}\}$).

(i) Synthesis.

For the matrix pair of positive-definite and symmetric matrices $\{C_l, G_l\}$ or $\{C_r, G_r\}$, a simultaneous diagonalization is (the right Frobenius matrix F_r is an orthonormal matrix)

$$C_l = J_l^T J_l\,,\quad F_l^T C_l F_l = \mathrm{diag}\left[\varLambda_1^2, \varLambda_2^2\right],\quad F_l^T G_l F_l = I \quad \text{versus} \quad F_r^T C_r F_r = \mathrm{diag}\left[\lambda_1^2, \lambda_2^2\right],\quad F_r^T F_r = I\,. \quad (2.1)$$

(ii) Analysis.

Left eigenvalues or left principal stretches:

$$\left|C_l - \varLambda_i^2 G_l\right| = 0\,,$$

$$\varLambda_{1,2}^2 = \varLambda_\pm^2 = \tfrac{1}{2}\left(\mathrm{tr}\left[C_l G_l^{-1}\right] \pm \sqrt{\left(\mathrm{tr}\left[C_l G_l^{-1}\right]\right)^2 - 4\det\left[C_l G_l^{-1}\right]}\right)\,. \quad (2.2)$$

Left eigencolumns:

$$\begin{bmatrix} F_{11} \\ F_{21} \end{bmatrix} = \frac{1}{\sqrt{G_{11}(c_{22} - \Lambda_1^2 G_{22})^2 - 2G_{12}(c_{12} - \Lambda_1^2 G_{12})(c_{22} - \Lambda_1^2 G_{22}) + G_{22}(c_{12} - \Lambda_1^2 G_{12})^2}} \times$$

$$\times \begin{bmatrix} +(c_{22} - \Lambda_1^2 G_{22}) \\ -(c_{12} - \Lambda_1^2 G_{12}) \end{bmatrix} \, ,$$

$$\begin{bmatrix} F_{12} \\ F_{22} \end{bmatrix} = \frac{1}{\sqrt{G_{22}(c_{11} - \Lambda_2^2 G_{11})^2 - 2G_{12}(c_{11} - \Lambda_2^2 G_{11})(c_{12} - \Lambda_2^2 G_{12}) + G_{11}(c_{12} - \Lambda_2^2 G_{12})^2}} \times$$

$$\times \begin{bmatrix} -(c_{12} - \Lambda_2^2 G_{12}) \\ +(c_{11} - \Lambda_2^2 G_{11}) \end{bmatrix} \, .$$

(2.3)

Right eigenvalues or right principal stretches
(the right general eigenvalue problem reduces to the right special eigenvalue problem):

$$\left| \mathsf{C_r} - \lambda_i^2 \mathsf{G_r} \right| = \left| \mathsf{C_r} - \lambda_i \mathsf{I}_2 \right| = 0 \; \forall \, i \in \{1, 2\} \, ,$$

$$\lambda_{1,2}^2 = \lambda_{\pm}^2 = \frac{1}{2} \left(\mathrm{tr}\, \left[\mathsf{C_r G_r^{-1}} \right] \pm \sqrt{\left(\mathrm{tr}\, \left[\mathsf{C_r G_r^{-1}} \right] \right)^2 - 4 \det \left[\mathsf{C_r G_r^{-1}} \right]} \right) =$$

$$= \frac{1}{2} \left(C_{11} + C_{22} \pm \sqrt{(C_{11} - C_{22})^2 + (2C_{12})^2} \right) \, .$$

(2.4)

Right eigencolumns:

$$\mathsf{F_r} = \begin{bmatrix} f_{11} & f_{12} \\ f_{21} & f_{22} \end{bmatrix} \begin{cases} \begin{bmatrix} f_{11} \\ f_{21} \end{bmatrix} = \dfrac{1}{\sqrt{(C_{22} - \lambda_1^2)^2 + C_{12}^2}} \begin{bmatrix} C_{22} - \lambda_1^2 \\ -C_{12} \end{bmatrix} \, , \\[2ex] \begin{bmatrix} f_{12} \\ f_{22} \end{bmatrix} = \dfrac{1}{\sqrt{(C_{11} - \lambda_2^2)^2 + C_{12}^2}} \begin{bmatrix} -C_{12} \\ C_{11} - \lambda_2^2 \end{bmatrix} \, . \end{cases}$$

(2.5)

Since the right Frobenius matrix $\mathsf{F_r}$ is an orthonormal matrix, it can be represented by

$$\mathsf{F_r} = \begin{bmatrix} \cos\varphi & \sin\varphi \\ -\sin\varphi & \cos\varphi \end{bmatrix} \; \forall \, \varphi \in [0, 2\pi] \, ,$$

(2.6)

$$\tan\varphi = \frac{C_{12}}{C_{11} - \lambda_-^2} \, , \quad \tan 2\varphi = \frac{2C_{12}}{C_{11} - C_{22}} \, .$$

End of Lemma.

The proof of Lemma 2.1 is straightforward from Lemma 1.6 as soon as we specialize $\mathsf{G_r} = \mathsf{I}_2$. Of special interest is the right eigenspace analysis. Here, the right Frobenius matrix $\mathsf{F_r}$ is orthonormal. As an orthonormal matrix (also called "proper rotation matrix"), it can be parameterized by a rotation angle φ. Such an angle of rotation orientates the right eigenvectors $\{\boldsymbol{f}_1, \boldsymbol{f}_2 \,|\, \mathcal{O}\}$ with respect to $\{\boldsymbol{e}_1, \boldsymbol{e}_2 \,|\, \mathcal{O}\}$, $\mathbb{R}^2 = \mathrm{span}\{\boldsymbol{e}_1, \boldsymbol{e}_2\}$. Indeed, the "$\tan 2\varphi$ identity" leads to an easy computation of the orientation of the right eigenvectors. We proceed to a short example.

Example 2.1 (Orthogonal projection of points of the sphere $\mathbb{S}^2_{R+}$ onto the equatorial plane $\mathbb{P}^2_{\mathcal{O}}$ through the origin $\mathcal{O}$).

In Example 1.6, we presented already to you the special map projection of the hemisphere $\mathbb{S}^2_{R+}$ onto the central equatorial plane $\mathbb{P}^2_{\mathcal{O}}$ by computing its characteristic right Cauchy–Green deformation tensor as well as its right eigenspace. Here, we aim at testing the right Frobenius matrix F_r on orthonormality. Let us transfer the right eigencolumns to build up

$$F_r = \begin{bmatrix} f_{11} & f_{12} \\ f_{21} & f_{22} \end{bmatrix} = -\frac{1}{\sqrt{x^2 + y^2}} \begin{bmatrix} x & y \\ y & -x \end{bmatrix}. \tag{2.7}$$

Is this Frobenius matrix of integrating factors an orthonormal matrix? Please test $F_r^* F_r = I_2$ to convince yourself. Here, we generate

$$F_r = \begin{bmatrix} \cos\varphi & \sin\varphi \\ -\sin\varphi & \cos\varphi \end{bmatrix} = -\frac{1}{\sqrt{x^2 + y^2}} \begin{bmatrix} x & y \\ y & -x \end{bmatrix}, \tag{2.8}$$

$$\tan\varphi = -\frac{y}{x}, \quad \tan 2\varphi = \frac{2\tan\alpha}{1 - \tan^2\alpha} = -\frac{2xy}{x^2 - y^2}, \tag{2.9}$$

$$C_{12} = \frac{xy}{R^2 - (x^2 + y^2)}, \quad C_{11} - C_{22} = \frac{x^2 - y^2}{R^2 - (x^2 + y^2)}, \tag{2.10}$$

$$\tan 2\varphi = \frac{2C_{12}}{C_{11} - C_{22}} = \frac{2xy}{x^2 - y^2}. \tag{2.11}$$

If $x = y$, then $\tan\varphi = -1$, $\tan 2\varphi \to \pm\infty$, $\varphi = \mp 45°$.

End of Example.

2-2 Eigenspace analysis, Euler–Lagrange deformation tensor

Left and right eigenspace analysis and synthesis of the Euler–Lagrange deformation tensor, special case $\{\mathbb{M}^2_r, g_{\mu\nu}\} = \{\mathbb{R}^2, \delta_{\mu\nu}\}$.

First, let us confront you with Lemma 2.2, where we present detailed results of the left and right eigenspace analysis and synthesis of the Euler–Lagrange deformation tensor for the special case of a right Euclidean manifold. Second, we focus on an interpretation of the results.

Lemma 2.2 (Left and right eigenspace analysis and synthesis of the Euler–Lagrange deformation tensor, special case $\{\mathbb{M}^2_r, g_{\mu\nu}\} = \{\mathbb{R}^2, \delta_{\mu\nu}\}$).

(i) Synthesis.

For the pair of symmetric matrices $\{E_l, G_l\}$ or $\{E_r, G_r\}$, where the matrices $\{G_l, G_r\}$ are positive definite, a simultaneous diagonalization is (the right Frobenius matrix F_r is an orthonormal matrix)

$$F_l^T E_l F_l = \operatorname{diag}[K_1, K_2], \quad F_l^T G_l F_l = I \text{ versus } F_r^T E_r F_r = \operatorname{diag}[\kappa_1, \kappa_2], \quad F_r^T F_r = I. \tag{2.12}$$

(ii) Analysis.

Left eigenvalues:

$$|E_l - K_i G_l| = 0, \quad K_{1,2} = K_\pm = \frac{1}{2}\left(\operatorname{tr}\left[E_l G_l^{-1}\right] \pm \sqrt{\left(\operatorname{tr}\left[E_l G_l^{-1}\right]\right)^2 - 4\det\left[E_l G_l^{-1}\right]} \right). \tag{2.13}$$

Left eigencolumns:

$$
\begin{bmatrix} F_{11} \\ F_{21} \end{bmatrix} = \frac{1}{\sqrt{G_{11}(e_{22} - K_1 G_{22})^2 - 2G_{12}(e_{12} - K_1 G_{12})(e_{22} - K_1 G_{22}) + G_{22}(e_{12} - K_1 G_{12})^2}} \times
$$

$$
\times \begin{bmatrix} e_{22} - K_1 G_{22} \\ -(e_{12} - K_1 G_{12}) \end{bmatrix} ,
$$

$$
\begin{bmatrix} F_{12} \\ F_{22} \end{bmatrix} = \frac{1}{\sqrt{G_{22}(e_{11} - K_2 G_{11})^2 - 2G_{12}(e_{11} - K_2 G_{11})(e_{12} - K_2 G_{12}) + G_{11}(e_{12} - K_2 G_{12})^2}} \times
$$

$$
\times \begin{bmatrix} -(e_{12} - K_2 G_{12}) \\ e_{11} - K_2 G_{11} \end{bmatrix} .
$$

(2.14)

Right eigenvalues
(the right general eigenvalue problem reduces to the right special eigenvalue problem):

$$
|E_r - \kappa_i I_r| = 0 ,
$$

$$
\kappa_{1,2} = \kappa_\pm = \frac{1}{2}\left(\mathrm{tr}\,[E_r] \pm \sqrt{(\mathrm{tr}\,[E_r])^2 - 4\det\,[E_r]} \right) =
$$

$$
= \frac{1}{2}\left(E_{11} + E_{22} \pm \sqrt{(E_{11} - E_{22})^2 + (2E_{12})^2} \right) .
$$

(2.15)

Right eigencolumns:

$$
F_r = \begin{bmatrix} f_{11} & f_{12} \\ f_{21} & f_{22} \end{bmatrix} \quad \begin{cases} \begin{bmatrix} f_{11} \\ f_{21} \end{bmatrix} = \dfrac{1}{\sqrt{(E_{22} - \kappa_1)^2 + E_{12}^2}} \begin{bmatrix} E_{22} - \kappa_1 \\ -E_{12} \end{bmatrix} , \\[3mm] \begin{bmatrix} f_{12} \\ f_{22} \end{bmatrix} = \dfrac{1}{\sqrt{(E_{11} - \kappa_2)^2 + E_{12}^2}} \begin{bmatrix} -E_{12} \\ E_{11} - \kappa_2 \end{bmatrix} . \end{cases}
$$

(2.16)

Since the right Frobenius matrix F_r is an orthonormal matrix, it can be represented by

$$
F_r = \begin{bmatrix} \cos\phi & \sin\phi \\ -\sin\phi & \cos\phi \end{bmatrix} \quad \forall\,\phi \in [0, 2\pi] ,
$$

(2.17)

$$
\tan\phi = \frac{E_{12}}{E_{11} - \kappa_-} , \quad \tan 2\phi = \frac{2E_{12}}{E_{11} - E_{22}} .
$$

End of Lemma.

Lemma 1.7 is the basis of the proof if we specialize $G_r = I_2$. Again, we emphasize that within the right eigenspace analysis the right Frobenius matrix is orthonormal. As an orthonormal matrix, i.e. $F_r \in SO(2) := \{F_r \in \mathbb{R}^{2\times 2} \mid F_r^T F_r = I_2 \text{ and } \det\,[F_r] = +1\}$, it can be properly parameterized by a rotation angle ϕ. Such an angle of rotation orientates the right eigenvectors $\{f_1, f_2 \mid \mathcal{O}\}$ with respect to $\{e_1, e_2 \mid \mathcal{O}\}$, $\mathbb{R}^2 = \mathrm{span}\{e_1, e_2\}$. Indeed, the "tan 2ϕ identity" leads to an easy computation of the orientation of the right eigenvectors.

2-3 The equivalence theorem for conformal mappings

The equivalence theorem for conformal mappings from the left two-dimensional Riemann manifold to the right two-dimensional Euclidean manifold (conformeomorphism), Korn–Lichtenstein equations and Cauchy–Riemann equations (d'Alembert–Euler equations).

The previous equivalence theorem for a conformeomorphism is specialized for the case of the two-dimensional right Euclidean manifold $\{\mathbb{M}_r^2, g_{\mu\nu}\} = \{\mathbb{R}^2, \delta_{\mu\nu}\} =: \mathbb{E}^2$. In many applications, the choice of $\{\mathbb{R}^2, \delta_{\mu\nu}\}$ is the planar manifold, for instance, the tangent space $T_{\boldsymbol{U}_0}\mathbb{M}_l^2$ of the left two-dimensional Riemann manifold fixed to the point $\boldsymbol{U}_0 = \{U_0^1, U_0^2\}$, being covered by Cartesian or polar coordinates. For an illustration of such a setup of a "planar manifold", go back to our previous examples.

2-31 Conformeomorphism

First, let us confront you with Lemma 2.3. The proof based upon Theorem 1.11 is straightforward. Examples are given in the following chapters.

Lemma 2.3 (Conformeomorphism, conformal mapping, special case $\{\mathbb{M}_r^2, g_{\mu\nu}\} = \{\mathbb{R}^2, \delta_{\mu\nu}\}$).

Let $\overline{f} : \mathbb{M}_l^2 \to \{\mathbb{R}^2, \delta_{\mu\nu}\}$ be an orientation preserving conformal mapping. Then the following conditions are equivalent.

$$\text{(i) } \Psi_l(\dot{\boldsymbol{U}}_1, \dot{\boldsymbol{U}}_2) = \Psi_r(\dot{\boldsymbol{u}}_1, \dot{\boldsymbol{u}}_2)$$

(2.18)

for all tangent vectors $\dot{\boldsymbol{U}}_1, \dot{\boldsymbol{U}}_2$ and their images $\dot{\boldsymbol{u}}_1, \dot{\boldsymbol{u}}_2$, respectively.

$$\text{(ii) } \mathsf{C}_l = \lambda^2(\boldsymbol{U}_0)\mathsf{G}_l \quad \text{versus} \quad \mathsf{C}_r = \lambda^2 \mathsf{I}_2 , \quad \mathsf{C}_r^{-1} = \mathsf{I}_2/\lambda^2 ,$$

$$C_{11} = C_{22} = \lambda^2 , \quad C_{12} = C_{21} = 0 , \quad C^{11} = C^{22} = \lambda^{-2} , \quad C^{12} = C^{21} = 0 ;$$

$$\mathsf{E}_l = K(\boldsymbol{U}_0)\mathsf{G}_l \quad \text{versus} \quad \mathsf{E}_r = \kappa \mathsf{I}_2 , \quad \mathsf{E}_r^{-1} = \mathsf{I}_2/\kappa ,$$

$$E_{11} = E_{22} = \kappa , \quad E_{12} = E_{21} = 0 , \quad E^{11} = E^{22} = \kappa^{-1} , \quad E^{12} = E^{21} = 0 .$$

(2.19)

$$\text{(iii) } \begin{bmatrix} K = (\Lambda^2 - 1)/2 \\ \Lambda^2 = 2K + 1 \end{bmatrix} \quad \text{versus} \quad \begin{bmatrix} (\lambda^2 - 1)/2 = \kappa \\ 2\kappa + 1 = \lambda^2 \end{bmatrix} ,$$

$$\Lambda_1 = \Lambda_2 = \Lambda(\boldsymbol{U}_0) \quad \text{versus} \quad \lambda_1 = \lambda_2 = \lambda(\boldsymbol{u}_0) ,$$

$$K_1 = K_2 = K(\boldsymbol{U}_0) \quad \text{versus} \quad \kappa_1 = \kappa_2 = \kappa(\boldsymbol{u}_0) ,$$

$$\Lambda^2(\boldsymbol{U}_0) = \text{tr}\left[\mathsf{C}_l \mathsf{G}_l^{-1}\right]/2 \quad \text{versus} \quad \lambda^2(\boldsymbol{u}_0) = \text{tr}\left[\mathsf{C}_r\right]/2 ;$$

(2.20)

$$\text{(left dilatation) } K = \text{tr}\left[\mathsf{E}_l \mathsf{G}_l^{-1}\right]/2 \quad \text{versus} \quad \text{(right dilatation) } \kappa = \text{tr}\left[\mathsf{E}_r\right]/2 ,$$

$$\text{tr}\left[\mathsf{C}_l \mathsf{G}_l^{-1}\right] = 2\sqrt{\det\left[\mathsf{C}_l \mathsf{G}_l^{-1}\right]} \quad \text{versus} \quad \text{tr}\left[\mathsf{C}_r \mathsf{G}_r^{-1}\right] = 2\sqrt{\det\left[\mathsf{C}_r\right]} ,$$

(2.21)

$$\text{tr}\left[\mathsf{E}_l \mathsf{G}_l^{-1}\right] = 2\sqrt{\det\left[\mathsf{E}_l \mathsf{G}_l^{-1}\right]} \quad \text{versus} \quad \text{tr}\left[\mathsf{E}_r\right] = 2\sqrt{\det\left[\mathsf{E}_r\right]} .$$

(iv) (Generalized Korn–Lichtenstein equations, Cauchy–Riemann equations, subject to the integrability conditions $u_{UV} = u_{VU}$ and $v_{UV} = v_{VU}$)

$$\begin{bmatrix} u_U \\ u_V \end{bmatrix} = \frac{1}{\sqrt{G_{11}G_{22} - G_{12}^2}} \begin{bmatrix} -G_{12} & G_{11} \\ -G_{22} & G_{12} \end{bmatrix} \begin{bmatrix} v_U \\ v_V \end{bmatrix} .$$

(2.22)

End of Lemma.

2-32 Higher-dimensional conformal mapping

In order to develop the theory of a higher-dimensional conformal diffeomorphism (in Gauss's words: "in kleinsten Teilen ähnlich"), we first derive the Korn–Lichtenstein equations of a two-dimensional conformal mapping $\mathbb{M}_l^2 \to \mathbb{M}_r^2 := \{\mathbb{R}^2, \delta_{\mu\nu}\} = \mathbb{E}^2$ by means of *exterior calculus*, namely by means of the *Hodge star operator*. With such an experience built up, second, we derive the *Zund equations* of a three-dimensional conformal mapping $\mathbb{M}_l^3 \to \mathbb{M}_r^3 := \{\mathbb{R}^3, \delta_{\mu\nu}\} = \mathbb{E}^3$ by means of exterior calculus taking advantage of the Hodge star operator in $\mathbb{R}^3$. Note that the Hodge star operator generalizes the *vector product*, also called *cross product* or *outer product*, to any dimension. Indeed, the classical vector product serves us only in $\mathbb{R}^3$. Box 2.1 summarizes the various steps to produce a conformal diffeomorphism $\mathbb{M}_l^2 \to \mathbb{M}_r^2 = \{\mathbb{R}^2, \delta_{\mu\nu}\} = \mathbb{E}^2$ in terms of exterior calculus. First, we introduce the left Jacobi map $\{dx, dy\} \to \{dU, dV\}$ and the right Jacobi map $\{dU, dV\} \to \{dx, dy\}$. Second, we compute the right Cauchy–Green matrix C_r subject to its conformal structure $\mathsf{C}_r = \lambda^2 \mathsf{I}_2$ and $\mathsf{C}_r^{-1} = \lambda^{-2} \mathsf{I}_2$. We are led to a representation of the conformal right Cauchy–Green matrix $\mathsf{C}_r = \mathsf{J}_r^T \mathsf{G}_l \mathsf{J}_r = \lambda^2 \mathsf{I}_2$ or $\mathsf{C}_r^{-1} = \mathsf{J}_l^T \mathsf{G}_l^{-1} \mathsf{J}_l = \lambda^{-2} \mathsf{I}_2$ in terms of the Jacobi matrices J_l and J_r. The rows of the left Jacobi matrix can be interpreted as "G_l^{-1} orthogonal", while the right Jacobi matrix can be interpreted as "G_l orthogonal". Third, this result of conformal geometry is used by the Hodge star operator. One-by-one, we define dx, x_1, x_2, and dy^*. Here, we make use of the two-dimensional permutation symbol $e_{LM} \in \mathbb{R}^{2 \times 2}$ ($L, M \in \{1, 2\}$). Fourth, we explicitly represent the exterior form $dx = dy^*$ of the Korn–Lichtenstein equations: compare with Lemma 2.4.

Lemma 2.4 (E. Grafarend and R. Syffus (1998d, p. 292), conformeomorphism $\mathbb{M}_l^2 \to \mathbb{M}_r^2 := \{\mathbb{R}^2, \delta_{\mu\nu}\}$, Korn–Lichtenstein equations).

The following formulations of the Korn–Lichtenstein equations producing a conformal diffeomorphism $\mathbb{M}_l^2 \to \mathbb{M}_r^2 := \{\mathbb{R}^2, \delta_{\mu\nu}\}$ are equivalent.

Formulation (i):

$$dx = *dy \ . \tag{2.23}$$

Formulation (ii):

$$\frac{\partial x}{\partial U^L} = e_{LM} \sqrt{\det [\mathsf{G}_l]} G^{MN} \frac{\partial y}{\partial U^N} \ . \tag{2.24}$$

Formulation (iii):

$$x_U = \frac{1}{\sqrt{|\mathsf{G}_l|}} \left(-G_{12} y_U + G_{11} y_V \right) \ , \quad x_V = \frac{1}{\sqrt{|\mathsf{G}_l|}} \left(-G_{22} y_U + G_{12} y_V \right) \ , \tag{2.25}$$

$$\mathsf{G}_l = [G_{MN}] = \begin{bmatrix} G_{11} & G_{12} \\ G_{12} & G_{22} \end{bmatrix} \quad \Leftrightarrow \quad \frac{1}{|\mathsf{G}_l|} \begin{bmatrix} G_{22} & -G_{12} \\ -G_{12} & G_{11} \end{bmatrix} = [G^{LM}] = \mathsf{G}_l^{-1} \ , \tag{2.26}$$

subject to the integrability conditions

$$\frac{\partial^2 x}{\partial U \partial V} = \frac{\partial^2 x}{\partial V \partial U} \ , \quad \frac{\partial^2 y}{\partial U \partial V} = \frac{\partial^2 y}{\partial V \partial U} \ . \tag{2.27}$$

End of Lemma.

Box 2.1 (Conformal diffeomorphism $\mathbb{M}_l^2 \to \mathbb{M}_r^2 = \{\mathbb{R}^2, \delta_{\mu\nu}\} = \mathbb{E}^2$, exterior calculus).

Diffeomorphism:

$$\begin{bmatrix} \mathrm{d}x \\ \mathrm{d}y \end{bmatrix} = \mathsf{J}_l \begin{bmatrix} \mathrm{d}U \\ \mathrm{d}V \end{bmatrix} \quad \text{or} \quad \begin{bmatrix} \mathrm{d}U \\ \mathrm{d}V \end{bmatrix} = \mathsf{J}_r \begin{bmatrix} \mathrm{d}x \\ \mathrm{d}y \end{bmatrix}$$

$$\Leftrightarrow$$

$$\mathsf{J}_l = \mathsf{J}_r^{-1}$$

$$\Leftrightarrow$$

$$\mathsf{J}_r = \mathsf{J}_l^{-1} \ . \tag{2.28}$$

Right Cauchy–Green matrix for a conformal diffeomorphism:

$$\mathsf{C}_r = \mathsf{J}_r^{\mathrm{T}} \mathsf{G}_l \mathsf{J}_r = \lambda^2 \mathsf{I}_2$$

$$\Leftrightarrow$$

$$\mathsf{C}_r^{-1} = \mathsf{J}_l \mathsf{G}_l^{-1} \mathsf{J}_l^{\mathrm{T}} = \lambda^{-2} \mathsf{I}_2 \ . \tag{2.29}$$

The rows of the left Jacobi matrix are G_l^{-1} orthogonal:

$$\mathrm{d}x = x_U \mathrm{d}U + x_V \mathrm{d}V = \sum_{M=1}^{2} x_M \mathrm{d}U^M \ , \quad x_1 := D_U x = x_U \ , \quad x_2 := D_V x = x_V \ . \tag{2.30}$$

Hodge star operator:

$$*\mathrm{d}y := \sum_{L,M,N=1}^{2} e_{LM} \sqrt{\det [\mathsf{G}_l]} G^{MN} y_N \mathrm{d}U^L \ ,$$

subject to

$$y_1 := D_U y = y_U \ , \quad y_2 := D_V y = y_V \ . \tag{2.31}$$

Permutation symbol:

$$e_{LM} = \begin{cases} +1 & \text{for an even permutation of the indices } L, M \in \{1, 2\} \\ -1 & \text{for an odd permutation of the indices } L, M \in \{1, 2\} \\ 0 & \text{otherwise} \end{cases} \ . \tag{2.32}$$

Korn–Lichtenstein equations in exterior calculus:

$$\mathrm{d}x = \sum_{M=1}^{2} x_M \mathrm{d}U^M = \sum_{L,M,N=1}^{2} e_{LM} \sqrt{\det [\mathsf{G}_l]} G^{MN} y_N \mathrm{d}U^L = \mathrm{d}y^*$$

$$\Leftrightarrow$$

$$\frac{\partial x}{\partial U^L} = e_{LM} \sqrt{\det [\mathsf{G}_l]} G^{MN} \frac{\partial y}{\partial U^N} \ , \quad \mathrm{d}x = \mathrm{d}y^* \ . \tag{2.33}$$

Box 2.2 summarizes the operational procedure for generating a conformal diffeomorphism, also called *conformeomorphism*, $\mathbb{M}_l^3 \to \mathbb{M}_r^3 = \{\mathbb{R}^3, \delta_{\mu\nu}\} = \mathbb{E}^3$, again in terms of exterior calculus. First, we introduce the differential one-forms, the differential two-forms, and the differential three-forms. Second, we apply the Hodge star operator (i) to $*\mathrm{d}x$ etc., (ii) to $*(\mathrm{d}y \wedge \mathrm{d}z)$ etc., and (iii) to $*(\mathrm{d}y \wedge \mathrm{d}y \wedge \mathrm{d}z)$. The columns $[x_1, x_2, x_3]^{\mathrm{T}}$, $[y_1, y_2, y_3]^{\mathrm{T}}$, and $[z_1, z_2, z_3]^{\mathrm{T}}$ may be considered orthogonal. Third, we represent the expression $*(\mathrm{d}y \wedge \mathrm{d}z)$ as an example explicitly. Again, the three-dimensional permutation symbol $e_{LM_1M_2} \in \mathbb{R}^{3\times3\times3}$ ($L, M_1, M_2 \in \{1, 2\}$) as a three-dimensional array is defined. Fourth, we explicitly compute the expression $\mathrm{d}x = *(\mathrm{d}y \wedge \mathrm{d}z)$, the *Zund equations* of a three-dimensional conformal mapping $\mathbb{M}_l^3 \to \mathbb{M}_r^3 = \mathbb{E}^3$: compare with Lemma 2.5.

Box 2.2 (Conformal diffeomorphism $\mathbb{M}_l^3 \to \mathbb{M}_r^3 = \{\mathbb{R}^3, \delta_{\mu\nu}\} = \mathbb{E}^3$, exterior calculus).

Differential frame:

$$
\begin{array}{l}
\mathrm{d}x = x_1 \mathrm{d}U \,+\, x_2 \mathrm{d}V \,+\, x_3 \mathrm{d}W \\
\text{(i)} \ \ \mathrm{d}y = y_1 \mathrm{d}U \,+\, y_2 \mathrm{d}V \,+\, y_3 \mathrm{d}W \\
\mathrm{d}z = z_1 \mathrm{d}U \,+\, z_2 \mathrm{d}V \,+\, z_3 \mathrm{d}W
\end{array} \left.\right\} \quad \text{(one-forms)} \ ,
$$

$$\text{(ii)} \ \mathrm{d}y \wedge \mathrm{d}z \,, \quad \mathrm{d}z \wedge \mathrm{d}x \,, \quad \mathrm{d}x \wedge \mathrm{d}y \,, \quad \text{(two-forms)} \ ,$$

$$\text{(iii)} \ \mathrm{d}x \wedge \mathrm{d}y \wedge \mathrm{d}z \qquad \text{(three-form)} \ .$$

(2.34)

Hodge star operator:

$$\text{(i)} \ *\mathrm{d}x = \mathrm{d}y \wedge \mathrm{d}z \,, \quad *\mathrm{d}y = \mathrm{d}z \wedge \mathrm{d}x \,, \quad *\mathrm{d}z = \mathrm{d}x \wedge \mathrm{d}y \,;$$

$$\text{(ii)} \ *(\mathrm{d}y \wedge \mathrm{d}z) = \mathrm{d}x \,, \quad *(\mathrm{d}z \wedge \mathrm{d}x) = \mathrm{d}y \,, \quad *(\mathrm{d}x \wedge \mathrm{d}y) = \mathrm{d}z \,;$$

$$\text{(iii)} \ *(\mathrm{d}x \wedge \mathrm{d}y \wedge \mathrm{d}z) = 1 \ .$$

(2.35)

Example:

$$\forall \, L, M_1, M_2, N_1, N_2 \in \{1, 2, 3\} :$$

$$*(\mathrm{d}y \wedge \mathrm{d}z) = \sum_{L,M_1,M_2,N_1,N_2=1}^{3} e_{LM_1M_2} \sqrt{|\mathsf{G}_l|} \, G^{M_1N_1} G^{M_2N_2} \frac{\partial y}{\partial U^{N_1}} \frac{\partial z}{\partial U^{N_2}} \mathrm{d}U^L \ .$$

(2.36)

Permutation symbol:

$$
e_{LM_1M_2} = \begin{cases}
+1 & \text{for an even permutation of the indices } L, M_1, M_2 \in \{1, 2, 3\} \\
-1 & \text{for an odd permutation of the indices } L, M_1, M_2 \in \{1, 2, 3\} \\
0 & \text{otherwise}
\end{cases} .
$$

(2.37)

Zund equations of a two-dimensional conformal diffeomorphism in exterior calculus:

$$\mathrm{d}x = \sum_{M=1}^{3} x_M \mathrm{d}U^M =$$

$$= \sum_{L,M_1,M_2,N_1,N_2=1}^{3} e_{LM_1M_2} \sqrt{|\mathsf{G}_l|} \, G^{M_1N_1} G^{M_2N_2} \frac{\partial y}{\partial U^{N_1}} \frac{\partial z}{\partial U^{N_2}} \mathrm{d}U^L = *(\mathrm{d}y \wedge \mathrm{d}z)$$

(2.38)

$$\Leftrightarrow$$

$$\frac{\partial x}{\partial U^L} = e_{LM_1M_2} \sqrt{\det [\mathsf{G}_l]} \, G^{M_1N_1} G^{M_2N_2} \frac{\partial y}{\partial U^{N_1}} \frac{\partial z}{\partial U^{N_2}} \ ,$$

$$\mathrm{d}x = *(\mathrm{d}y \wedge \mathrm{d}z) \ .$$

Lemma 2.5 (J. Zund (1987), E. Grafarend and R. Syffus (1998d, p. 292), the Zund equations of a three-dimensional conformeomorphism $\mathbb{M}_l^3 \to \mathbb{M}_r^3 = \{\mathbb{R}^3, \delta_{\mu\nu}\} = \mathbb{E}^3$).

Equivalent formulations of the equations producing a conformal mapping $\mathbb{M}_l^3 \to \mathbb{M}_r^3 = \mathbb{E}^3$ are provided by the following formulations.

Formulation (i):

$$\mathrm{d}x = *(\mathrm{d}y \wedge \mathrm{d}z) . \tag{2.39}$$

Formulation (ii):

$$\forall\ I, J_1, J_2, K_1, K_2 \in \{1, 2, 3\}:\ \frac{\partial x}{\partial U^I} = \tfrac{1}{2} e_{IJ_1J_2} \sqrt{|\mathsf{G}_l|} G^{J_1K_1} G^{J_2K_2} \frac{\partial y}{\partial U^{K_1}} \frac{\partial z}{\partial U^{K_2}} . \tag{2.40}$$

Formulation (iii):

$$\begin{aligned}
\frac{\partial x}{\partial U} = \tfrac{1}{2}\sqrt{|\mathsf{G}_l|}\ \Big[\ &\left(G^{21}G^{32} - G^{31}G^{12}\right)\frac{\partial y}{\partial U}\frac{\partial z}{\partial V} + \left(G^{21}G^{33} - G^{31}G^{23}\right)\frac{\partial y}{\partial U}\frac{\partial z}{\partial W} + \\
+ &\left(G^{22}G^{31} - G^{32}G^{21}\right)\frac{\partial y}{\partial V}\frac{\partial z}{\partial U} + \left(G^{22}G^{33} - G^{32}G^{23}\right)\frac{\partial y}{\partial V}\frac{\partial z}{\partial W} + \\
+ &\left(G^{23}G^{31} - G^{33}G^{21}\right)\frac{\partial y}{\partial W}\frac{\partial z}{\partial U} + \left(G^{23}G^{32} - G^{33}G^{22}\right)\frac{\partial y}{\partial W}\frac{\partial z}{\partial V}\ \Big]\ ,
\end{aligned} \tag{2.41}$$

$$\begin{aligned}
\frac{\partial x}{\partial V} = \tfrac{1}{2}\sqrt{|\mathsf{G}_l|}\ \Big[\ &\left(G^{31}G^{12} - G^{11}G^{32}\right)\frac{\partial y}{\partial U}\frac{\partial z}{\partial V} + \left(G^{31}G^{13} - G^{11}G^{33}\right)\frac{\partial y}{\partial U}\frac{\partial z}{\partial W} + \\
+ &\left(G^{32}G^{11} - G^{12}G^{31}\right)\frac{\partial y}{\partial V}\frac{\partial z}{\partial U} + \left(G^{32}G^{13} - G^{12}G^{33}\right)\frac{\partial y}{\partial V}\frac{\partial z}{\partial W} + \\
+ &\left(G^{33}G^{11} - G^{13}G^{31}\right)\frac{\partial y}{\partial W}\frac{\partial z}{\partial U} + \left(G^{33}G^{12} - G^{13}G^{32}\right)\frac{\partial y}{\partial W}\frac{\partial z}{\partial V}\ \Big]\ ,
\end{aligned} \tag{2.42}$$

$$\begin{aligned}
\frac{\partial x}{\partial W} = \tfrac{1}{2}\sqrt{|\mathsf{G}_l|}\ \Big[\ &\left(G^{11}G^{22} - G^{21}G^{12}\right)\frac{\partial y}{\partial U}\frac{\partial z}{\partial V} + \left(G^{11}G^{23} - G^{21}G^{13}\right)\frac{\partial y}{\partial U}\frac{\partial z}{\partial W} + \\
+ &\left(G^{12}G^{21} - G^{22}G^{11}\right)\frac{\partial y}{\partial V}\frac{\partial z}{\partial U} + \left(G^{12}G^{23} - G^{22}G^{13}\right)\frac{\partial y}{\partial V}\frac{\partial z}{\partial W} + \\
+ &\left(G^{13}G^{21} - G^{23}G^{11}\right)\frac{\partial y}{\partial W}\frac{\partial z}{\partial U} + \left(G^{13}G^{22} - G^{23}G^{12}\right)\frac{\partial y}{\partial W}\frac{\partial z}{\partial V}\ \Big]\ ,
\end{aligned} \tag{2.43}$$

subject to

$$\begin{aligned}
G^{11} &= \tfrac{1}{|\mathsf{G}_l|}\left(G_{22}G_{33} - G_{23}G_{32}\right) , & G^{12} &= \tfrac{1}{|\mathsf{G}_l|}\left(G_{13}G_{32} - G_{12}G_{33}\right) , \\
G^{13} &= \tfrac{1}{|\mathsf{G}_l|}\left(G_{12}G_{23} - G_{13}G_{22}\right) , & G^{22} &= \tfrac{1}{|\mathsf{G}_l|}\left(G_{11}G_{33} - G_{13}G_{31}\right) , \\
G^{23} &= \tfrac{1}{|\mathsf{G}_l|}\left(G_{12}G_{31} - G_{11}G_{32}\right) , & G^{33} &= \tfrac{1}{|\mathsf{G}_l|}\left(G_{11}G_{22} - G_{12}G_{21}\right) .
\end{aligned} \tag{2.44}$$

Formulation (iv):

$$\begin{aligned}
\frac{\partial x}{\partial U} &= \frac{1}{\sqrt{|\mathsf{G}_l|}}\left[G_{11}\left(\frac{\partial y}{\partial V}\frac{\partial z}{\partial W} - \frac{\partial y}{\partial W}\frac{\partial z}{\partial V}\right) + G_{12}\left(\frac{\partial y}{\partial W}\frac{\partial z}{\partial U} - \frac{\partial y}{\partial U}\frac{\partial z}{\partial W}\right) + G_{13}\left(\frac{\partial y}{\partial U}\frac{\partial z}{\partial V} - \frac{\partial y}{\partial V}\frac{\partial z}{\partial U}\right)\right], \\
\frac{\partial x}{\partial V} &= \frac{1}{\sqrt{|\mathsf{G}_l|}}\left[G_{12}\left(\frac{\partial y}{\partial V}\frac{\partial z}{\partial W} - \frac{\partial y}{\partial W}\frac{\partial z}{\partial V}\right) + G_{22}\left(\frac{\partial y}{\partial W}\frac{\partial z}{\partial U} - \frac{\partial y}{\partial U}\frac{\partial z}{\partial W}\right) + G_{23}\left(\frac{\partial y}{\partial U}\frac{\partial z}{\partial V} - \frac{\partial y}{\partial V}\frac{\partial z}{\partial U}\right)\right], \\
\frac{\partial x}{\partial W} &= \frac{1}{\sqrt{|\mathsf{G}_l|}}\left[G_{13}\left(\frac{\partial y}{\partial V}\frac{\partial z}{\partial W} - \frac{\partial y}{\partial W}\frac{\partial z}{\partial V}\right) + G_{23}\left(\frac{\partial y}{\partial W}\frac{\partial z}{\partial U} - \frac{\partial y}{\partial U}\frac{\partial z}{\partial W}\right) + G_{33}\left(\frac{\partial y}{\partial U}\frac{\partial z}{\partial V} - \frac{\partial y}{\partial V}\frac{\partial z}{\partial U}\right)\right],
\end{aligned} \tag{2.45}$$

subject to the integrability conditions $\frac{\partial^2 x}{\partial U \partial V} = \frac{\partial^2 x}{\partial V \partial U}, \frac{\partial^2 x}{\partial U \partial W} = \frac{\partial^2 x}{\partial W \partial U}, \frac{\partial^2 x}{\partial V \partial W} = \frac{\partial^2 x}{\partial W \partial V}$.

End of Lemma.

Question.

Question: "Why did we bother you with the three-dimensional conformal mapping of a three-dimensional Riemann manifold to a three-dimensional Euclidean manifold?" Answer: "One of the main reasons is the inability of the theory of complex manifolds to work conformally with odd-dimensional real manifolds. Only even-dimensional real manifolds $\mathbb{M}^{2n}(\mathbb{R})$ can be transformed to complex manifolds $\mathbb{M}^n(\mathbb{C})$."

Finally, Lemma 2.6 presents the partial differential equations of a conformeomorphism if it exists from a left n-dimensional (pseudo-)Riemann manifold $\mathbb{M}^n_l$ of signature l to a right n-dimensional (pseudo-)Riemann manifold $\mathbb{M}^n_r = \mathbb{E}^n$ of signature r.

Lemma 2.6 (E. Grafarend and R. Syffus (1998d, p. 293), conformeomorphism).

Equivalent formulations of the equations producing a conformal mapping $\mathbb{M}^n_l \to \mathbb{M}^n_r = \mathbb{E}^n$ are provided by the following formulations.

Formulation (i):

$$\mathrm{d}x^1 = *(\mathrm{d}x^2 \wedge \ldots \wedge \mathrm{d}x^n) \,. \tag{2.46}$$

Formulation (ii):

$$\forall\, L, M_1, \ldots, M_p, N_1, \ldots, N_p \in \{1, \ldots, n\}$$

$$(p = n - 1): \tag{2.47}$$

$$\frac{\partial x}{\partial U^L} = \frac{1}{p!} e_{LM_1 \ldots M_p} \sqrt{\det [\mathsf{G}_l]}\, G^{M_1 N_1} \ldots G^{M_p N_p} \frac{\partial x^2}{\partial U^{N_1}} \cdots \frac{\partial x^n}{\partial U^{N_p}} \,,$$

subject to the integrability conditions

$$\frac{\partial^2 x^1}{\partial U^L \partial U^N} = \frac{\partial^2 x^1}{\partial U^N \partial U^L} \,. \tag{2.48}$$

End of Lemma.

2-4 The equivalence theorem for equiareal mappings

The equivalence theorem for equiareal mappings from the left two-dimensional Riemann manifold to the right two-dimensional Euclidean manifold (areomorphism), Mollweide projection of the ellipsoid-of-revolution, principal stretches.

The previous equivalence theorem for an areomorphism is specialized for the case of the two-dimensional right Euclidean manifold $\{\mathbb{M}^2_r, g_{\mu\nu}\} = \{\mathbb{R}^2, \delta_{\mu\nu}\} =: \mathbb{E}^2$. In many applications, the choice of $\{\mathbb{R}^2, \delta_{\mu\nu}\}$ is the planar manifold, for instance, the tangent space $T_{U_0}\mathbb{M}^2_l$ of the left two-dimensional Riemann manifold fixed to the point $\boldsymbol{U}_0 = \{U^1_0, U^2_0\}$, being covered by Cartesian or polar coordinates. For an illustration of such a setup of a "planar manifold", go back to our previous examples. Here, we focus on the equivalence theorem, namely the differential equations which govern an equiareal mapping $\mathbb{M}^2_l \to \{\mathbb{R}^2, \delta_{\mu\nu}\}$.

Theorem 2.7 (Aeromorphism, $\mathbb{M}_l^2 \to \{\mathbb{R}^2, \delta_{\mu\nu}\}$, equiareal mapping).

Let $\overline{f} : \mathbb{M}_r^2 := \{\mathbb{R}^2, \delta_{\mu\nu}\} =: \mathbb{E}^2$ be an orientation preserving equiareal mapping. Then the following conditions are equivalent.

Condition (i):

$$\sqrt{\det[\mathsf{G}_l]}\, dU \wedge dV = du \wedge dv \ . \tag{2.49}$$

Condition (ii):

$$\det[\mathsf{C}_l] = 1 \ \text{ and } \ \det[\mathsf{C}_l\mathsf{G}_l^{-1}] = 1 \ ,$$
$$\det[\mathsf{I}_2 - 2\mathsf{E}_r] = 1 \ \text{ and } \ \det[2\mathsf{E}_l + \mathsf{G}_l] = \det[\mathsf{G}_l] \ . \tag{2.50}$$

Condition (iii):

$$\Lambda_1\Lambda_2 = 1 \ \text{ and } \ \lambda_1\lambda_2 = 1 \ . \tag{2.51}$$

Condition (iv):

$$U_u V_v - U_v V_u = 1/\sqrt{\det[\mathsf{G}_l]} =$$
$$= 1/\sqrt{G_{11}G_{22} - G_{12}^2} \ ,$$
$$u_U v_V - u_V v_U = \sqrt{\det[\mathsf{G}_l]} =$$
$$= \sqrt{G_{11}G_{22} - G_{12}^2} \ . \tag{2.52}$$

End of Theorem.

Here, we only have specialized Theorem 1.14 to $\mathbb{M}_r^2 := \{\mathbb{R}^2, \delta_{\mu\nu}\} =: \mathbb{E}^2$. One of the most popular equiareal mappings $\mathbb{E}_{A_1,A_1,A_2}^2 \to \{\mathbb{R}^2, \delta_{\mu\nu}\}$ is the *Mollweide projection of the ellipsoid-of-revolution* to the plane, which is presented in Example 2.2 and is illustrated in Fig. 2.1.

Example 2.2 (Mollweide projection of the ellipsoid-of-revolution, with reference to E. Grafarend and A. Heidenreich (1995)).

Let us assume that we have found a solution of the right characteristic equation, which generates an equiareal mapping of the ellipsoid-of-revolution $\mathbb{E}_{A_1,A_1,A_2}^2$ parameterized by the two coordinates $\{\Lambda, \Phi\}$ (called $\{$*Gauss surface normal longitude, Gauss surface normal latitude*$\}$) as outlined in Box 2.3, also called *generalized Mollweide projection*. Such a generalized Mollweide projection is classified as "pseudo-cylindric" and equiareal, mapping the circular equator equidistantly. Its mapping equations $x(\Lambda, \Phi)$ and $y(\Phi)$, where $\{x, y\}$ are Cartesian coordinates that cover $\{\mathbb{R}^2, \delta_{\mu\nu}\} = \mathbb{E}^2$, depend on $\cos t(\Phi)$ and $\sin t(\Phi)$. The auxiliary function $t(\Phi)$ is a solution of the *generalized Kepler equation* since for relative eccentricity $E^2 = (A_1^2 - A_2^2)/A_1^2 \to 0$ the generalized Kepler equations reduces to the Kepler equation. Such a Kepler equation is known from the *classical Mollweide projection* of the sphere or from solving the Kepler two-body problem in mechanics.

End of Example.

We pose two problems. (i) Prove that the generalized Mollweide projection of the ellipsoid-of-revolution is equiareal. For this purpose, observe the postulate $\det\left[\mathsf{C}_l\mathsf{G}_l^{-1}\right] = 1$. (ii) Determine the left principal stretches Λ_1 and Λ_2 by setting up the characteristic equations of the left eigenvalue problem that is presented in Box 2.4.

Solution (the first problem).

Here, we set up the test of an equiareal mapping to be based upon the postulate $\det\left[\mathsf{C}_l\mathsf{G}_l^{-1}\right] = 1$. First, by means of Box 2.5, we compute the left Jacobi matrix substituted by $D_\Lambda x$, $D_\Phi x$, $D_\Lambda y$, and $D_\Phi y$. Second, we set up the left Cauchy–Green matrix $\mathsf{C}_l = \mathsf{J}^*\mathsf{G}_r\mathsf{J}_l$ subject to $\mathsf{G}_r = \mathsf{I}_2$. We have to emphasize that C_l is not a diagonal matrix. Third, we adopt the left matrix of the metric G_l. Fourth, given the left Cauchy–Green matrix, C_l, and the left matrix of the metric, G_l, we derive the determinantal identity $\det\left[\mathsf{C}_l\mathsf{G}_l^{-1}\right] = 1$. By means of *implicit differentation* of the *generalized Kepler equation*, we compute (t'), $(t')^2$, $(t')^2\cos^4 t$, a^2b^2 and $1/G_{11}G_{22}$ in step five. Sixth, taking all individual terms into one, we have proven $\det\left[\mathsf{C}_l\mathsf{G}_l^{-1}\right] = 1$.

End of Solution (the first problem).

Solution (the second problem).

First, we set up the characteristic equations of the left general eigenvalue problem of Box 2.4 in order to compute the left principal stretches Λ_1 and Λ_2, respectively. Second, the solution of the left characteristic equation subject to the condition of an equiareal mapping, namely $\det\left[\mathsf{C}_l\mathsf{G}_l^{-1}\right] = 1$, accounts for computing the first left invariant $\operatorname{tr}\left[\mathsf{C}_l\mathsf{G}_l^{-1}\right]$. Indeed, a simple form of such an invariant is not available. Accordingly, we left $\operatorname{tr}\left[\mathsf{C}_l\mathsf{G}_l^{-1}\right]$ with a formula for $(t')^2$ and $1/G_{11}G_{22}$, respectively.

End of Solution (the second problem).

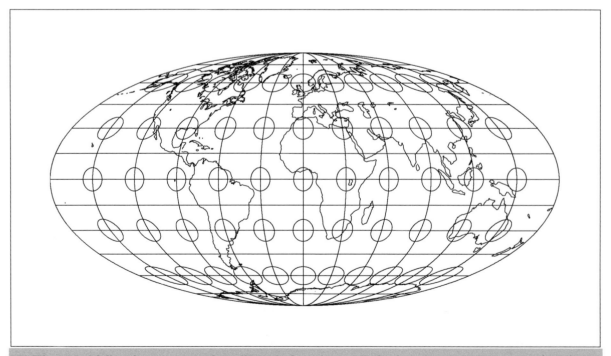

Fig. 2.1. Mollweide projection of an ellipsoid-of-revolution, E. Grafarend and A. Heidenreich (1995).

Box 2.3 (The Mollweide projection of $\mathbb{E}^2_{A_1,A_1,A_2}$; the pseudo-cylindric, equiareal, equidistant mapping of the circular equator).

Mapping equations:

$$x(\Lambda,\Phi) = a\Lambda\cos t(\Phi) \;,$$

$$y(\Lambda,\Phi) = b\sin t(\Phi) \;. \tag{2.53}$$

Generalized Kepler equations:

$$2t + \sin 2t = \pi\frac{\ln\frac{1+E\sin\Phi}{1-E\sin\phi} + \frac{2E\sin\Phi}{1-E^2\sin^2\Phi}}{\ln\frac{1+E}{1-E} + \ln\frac{2E}{1-E^2}} \;. \tag{2.54}$$

Scales:

$$a = A_1 \;,$$

$$b = \frac{A_1(1-E^2)}{\pi E}\left(\ln\frac{1+E}{1-E} + \frac{2E}{1-E^2}\right) \;. \tag{2.55}$$

Box 2.4 ([The left principal stretches, the left eigenvalues, and the generalized Mollweide projection of the ellipsoid-of-revolution).

Characteristic equation of the left general eigenvalue problem:

$$\Lambda^4 - \mathrm{tr}\left[\mathsf{C}_l\mathsf{G}_l^{-1}\right]\Lambda^2 + \det\left[\mathsf{C}_l\mathsf{G}_l^{-1}\right] = 0 \quad \text{subject to} \quad \det\left[\mathsf{C}_l\mathsf{G}_l^{-1}\right] = 1 \tag{2.56}$$

$$\Rightarrow$$

$$\Lambda^2_{1,2} = \frac{1}{2}\left[\mathrm{tr}\left[\mathsf{C}_l\mathsf{G}_l^{-1}\right] \pm \sqrt{\left(\mathrm{tr}\left[\mathsf{C}_l\mathsf{G}_l^{-1}\right]\right)^2 - 4}\right] \;. \tag{2.57}$$

Computation of the first invariant $\mathrm{tr}\left[\mathsf{C}_l\mathsf{G}_l^{-1}\right]$:

$$\mathrm{tr}\left[\mathsf{C}_l\mathsf{G}_l^{-1}\right] = \frac{a^2\cos^2 t}{G_{11}} + (t')^2\frac{a^2\Lambda^2\sin^2 t + b^2\cos^2 t}{G_{22}} \;,$$

$$\cos^4 t\,(t')^2 = \frac{E^2\cos^2\Phi}{(1-E^2\sin^2\Phi)^4}\frac{\pi^2}{\left(\ln\frac{1+E}{1-E} + \frac{2E}{1-E^2}\right)^2} \;,$$

$$\mathrm{tr}\left[\mathsf{C}_l\mathsf{G}_l^{-1}\right] = \frac{1}{G_{11}G_{22}}\left[a^2 G_{22}\cos^2 t + (t')^2 G_{11}\left(a^2\Lambda^2\sin^2 t + b^2\cos^2 t\right)\right] \;,$$

$$\frac{1}{G_{11}G_{22}} = \frac{(1-E^2\sin^2\Phi)^4}{A_1^4(1-E^2)^2\cos^2\Phi} \;,$$

$$G_{11} = \frac{A_1^2\cos^2\Phi}{1-E^2\sin^2\Phi} \;,$$

$$G_{22} = \frac{A_1^2(1-E^2)^2}{(1-E^2\sin^2\Phi)^3} \;. \tag{2.58}$$

Box 2.5 (Left Cauchy–Green matrix, generalized Mollweide projection of the ellipsoid-of-revolution).

Left Jacobi matrix:

$$\mathsf{J}_l := \begin{bmatrix} D_\Lambda x & D_\Phi x \\ D_\Lambda y & D_\Phi y \end{bmatrix} ,$$

$$D_\Lambda x = a\cos t , \quad D_\Phi x = D_t x D_\Phi t = -a\Lambda \sin t\, t' ,$$

$$D_\Lambda y = 0 , \quad D_\Phi y = D_t y D_\Phi t = +b\cos t\, t' .$$

(2.59)

Left Cauchy–Green matrix:

$$\mathsf{C}_l := \mathsf{J}_l^* \mathsf{G}_r \mathsf{J}_l , \quad \mathsf{G}_r = \mathsf{I}_2 \Rightarrow \mathsf{C}_l = \mathsf{J}_l^* \mathsf{J}_l ,$$

$$\mathsf{C}_l = \begin{bmatrix} a^2\cos^2 t & -a\Lambda^2 \cos t \sin t\, t' \\ -a\Lambda^2 \cos t \sin t\, t' & (a^2\Lambda^2 \sin^2 t + b^2\cos^2 t)(t')^2 \end{bmatrix} .$$

(2.60)

Left matrix of the metric:

$$\mathsf{G}_l = \begin{bmatrix} N^2(\Phi) & 0 \\ 0 & M^2(\Phi) \end{bmatrix} \quad (N(\Phi)\text{ and }M(\Phi)\text{: see Example 1.3}).$$

(2.61)

$$\det\!\left[\mathsf{C}_l \mathsf{G}_l^{-1}\right] = 1:$$

$$\det\!\left[\mathsf{C}_l \mathsf{G}_l^{-1}\right] = \frac{a^2\cos^2 t}{G_{11}} \frac{a^2\Lambda^2 \sin^2 t + b^2\cos^2 t}{G_{22}}(t')^2 - \frac{a^4\Lambda^2 \cos^2 t \sin^2 t}{G_{11}G_{22}}(t')^2 =$$

$$= \frac{\cos^4 t}{G_{11}G_{22}} a^2 b^2 (t')^2 .$$

(2.62)

$$(t'):$$

$$2(1+\cos 2t)\mathrm{d}t = \frac{\pi}{\ln\dfrac{1+E}{1-E} + \dfrac{2E}{1-E^2}} \times$$

$$\times \left[\frac{1-E\sin\Phi}{1+E\sin\Phi}\left(\frac{E\cos\Phi}{1-E\sin\Phi} + \frac{E\cos\Phi(1+E\sin\Phi)}{(1-E\sin\Phi)^2} \right) + \frac{2E\cos\Phi}{1-E^2\sin^2\Phi} + \frac{4E^3\sin^2\Phi\cos\Phi}{(1-E^2\sin^2\Phi)^2} \right]\mathrm{d}\Phi ,$$

$$1 + \cos 2t = 2\cos^2 t ,$$

(2.63)

$$\cos^2 t\,(t') = \frac{E\cos\Phi}{(1-E^2\sin^2\Phi)^2} \frac{\pi}{\ln\frac{1+E}{1-E} + \frac{2E}{1-E^2}} , \quad \cos^4 t\,(t')^2 = \frac{E^2\cos^2\Phi}{(1-E^2\sin^2\Phi)^4} \frac{\pi^2}{\left(\ln\frac{1+E}{1-E} + \frac{2E}{1-E^2}\right)^2} ,$$

$$\frac{1}{G_{11}G_{22}} = \frac{(1-E^2\sin^2\Phi)^4}{A_1^2\cos^2\Phi A_1^2(1-E)^2} , \quad a^2 b^2 = \frac{A_1^4(1-E^2)^2}{\pi^2 E^2}\left(\ln\frac{1+E}{1-E} + \frac{2E}{1-E^2}\right)^2 .$$

(6th) Determinantal identity:

$$\det\!\left[\mathsf{C}_l \mathsf{G}_l^{-1}\right] = 1 .$$

(2.64)

2-5 Canonical criteria for conformal, equiareal, and other mappings

Canonical criteria for conformal, equiareal, and isometric mappings as well as equidistant mappings $\mathbb{M}_l^2 \to \{\mathbb{R}^2, \delta_{\mu\nu}\}$, Hilbert invariants.

Question: "How can we generalize those canonical criteria for a conformal, an equiareal, or an isometric mapping $\mathbb{M}_l^2 \to \mathbb{M}_r^2 := \{\mathbb{R}^2, \delta_{\mu\nu}\} = \mathbb{E}^2$ if we restrict the right two-dimensional Riemann manifold to be two-dimensional Euclidean?" Answer: "Let us refer to Box 1.46 and Box 1.47 in order to formulate the answer. As it is outlined in Box 2.6, the fundamental four Hilbert invariants I_1 and I_2 or i_1 and i_2 become dependent, typically called "syzygetic", as soon as we are dealing with a conformal mapping $\mathbb{M}_l^2 \to \{\mathbb{R}^2, \delta_{\mu\nu}\}$."

Box 2.6 (Canonical representation of Hilbert invariants, $\mathbb{M}_l^2 \to \{\mathbb{R}^2, \delta_{\mu\nu}\}$).

$$I_1(\mathsf{C}_l) := \Lambda_1^2 + \Lambda_2^2 = \mathrm{tr}\left[\mathsf{C}_l \mathsf{G}_l^{-1}\right] \quad \text{versus} \quad i_1(\mathsf{C}_r) := \lambda_1^2 + \lambda_2^2 = \mathrm{tr}\left[\mathsf{C}_r\right] ,$$

$$I_2(\mathsf{C}_l) := \Lambda_1^2 \Lambda_2^2 = \det\left[\mathsf{C}_l \mathsf{G}_l^{-1}\right] \quad \text{versus} \quad i_2(\mathsf{C}_r) := \lambda_1^2 \lambda_2^2 = \det\left[\mathsf{C}_r\right] ,$$

$$\tag{2.65}$$

or

$$I_1(\mathsf{E}_l) := K_1 + K_2 = \mathrm{tr}\left[\mathsf{E}_l \mathsf{G}_l^{-1}\right] \quad \text{versus} \quad i_1(\mathsf{E}_r) := \kappa_1 + \kappa_2 = \mathrm{tr}\left[\mathsf{E}_r\right] ,$$

$$I_2(\mathsf{E}_l) := K_1 K_2 = \det\left[\mathsf{E}_l \mathsf{G}_l^{-1}\right] \quad \text{versus} \quad i_2(\mathsf{E}_r) := \kappa_1 \kappa_2 = \det\left[\mathsf{E}_r\right] .$$

$$\tag{2.66}$$

Special case: conformal mapping (syzygy).

$$I_1 = 2\sqrt{I_2} \quad \text{versus} \quad i_1 = 2\sqrt{i_2} .$$

$$\tag{2.67}$$

Note that for a general diffeomorphism, namely $\overline{f} : \{\mathbb{M}^2, G_{MN}\} \to \{\mathbb{R}^2, \delta_{\mu\nu}\}$, the first two Hilbert invariants $I_1(\mathsf{E}_l)$ and $i_1(\mathsf{E}_r)$ are also called *left* and *right dilatation*. They measure the *isotropic part* of a deformation, while the following *shear components* its *anisotropic part*:

$$\Gamma_1(\mathsf{C}_l) := C_{22} - C_{11} \quad \text{versus} \quad \gamma_1(\mathsf{C}_r) := c_{22} - c_{11} ,$$

$$\Gamma_1(\mathsf{E}_l) := E_{22} - E_{11} \quad \text{versus} \quad \gamma_1(\mathsf{E}_r) := e_{22} - e_{11} ,$$

$$\Gamma_2(\mathsf{C}_l) := 2C_{12} \quad \text{versus} \quad \gamma_2(\mathsf{C}_r) := 2c_{12} ,$$

$$\tag{2.68}$$

$$\Gamma_2(\mathsf{E}_l) := 2E_{12} \quad \text{versus} \quad \gamma_2(\mathsf{E}_r) := 2e_{12} .$$

2-6 Polar decomposition and simultaneous diagonalization of three matrices

Polar decomposition and simultaneous diagonalization of three matrices: $\{\mathsf{E}_l, \mathsf{C}_l, \mathsf{G}_l\}$ versus $\{\mathsf{E}_r, \mathsf{C}_r, \mathsf{G}_r\}$, stretch matrices.

A first remark has to be made towards the group theoretical representation of the left F_l and the right F_r matrix of eigenvectors. In case of $\{\mathbb{M}_r^2, g_{\mu\nu}\} = \{\mathbb{R}^2, \delta_{\mu\nu}\}$, we took advantage of the fact that the right matrix F_r of eigenvectors is an orthonormal matrix R. In the general case $\{\mathbb{M}_l^2, G_{MN}\} - \{\mathbb{M}_r^2, g_{\mu\nu}\}$, the left F_l and right the F_r matrix of eigenvectors enjoy the polar decomposition

$$\mathsf{F}_l = \mathsf{R}_1\mathsf{S}_1 \quad \text{versus} \quad \mathsf{F}_r = \mathsf{R}_3\mathsf{S}_3$$

$$\text{versus} \qquad\qquad \text{versus} \qquad , \tag{2.69}$$

$$\mathsf{F}_l = \mathsf{S}_2\mathsf{R}_2 \quad \text{versus} \quad \mathsf{F}_r = \mathsf{S}_4\mathsf{R}_4$$

where the matrices R_i are orthonormal, $\mathsf{R}_i^{-1} = \mathsf{R}_i^{\mathrm{T}}$, while the matrices S_i are by definition symmetric, $\mathsf{S}_i = \mathsf{S}_i^{\mathrm{T}}$. These symmetric matrices S_i are sometimes called *stretch matrices*. For more details including numerical examples, we refer to J. E. Marsden and T. J. R. Hughes (1983, pp. 51–55), R. W. Ogden (1984, pp. 92–94), J. C. Simo and R. L. Taylor (1991), and T. C. T. Ting (1985). Here, we conclude with a second remark relating again to the simultaneous diagonalization of two matrices, e., g. the pairs of Cauchy-Green deformation tensors $\{\mathsf{C}_l, \mathsf{G}_l\}$ or $\{\mathsf{C}_r, \mathsf{G}_r\}$ and the pairs of Euler-Lagrange deformation tensors $\{\mathsf{E}_l, \mathsf{G}_l\}$ or $\{\mathsf{E}_r, \mathsf{G}_r\}$, respectively. Of course, we could also aim at a simultaneous diagonalization of three matrices, e. g. the triplets

$$\{\mathsf{E}_l, \mathsf{C}_l, \mathsf{G}_l\} \text{ versus } \{\mathsf{E}_r, \mathsf{C}_r, \mathsf{G}_r\} , \tag{2.70}$$

in particular

$$\mathsf{U}_l^{\mathrm{T}}\mathsf{G}_l\mathsf{X}_l = \mathsf{S}_l^1 \Leftrightarrow \mathsf{G}_l = \mathsf{U}_l\mathsf{S}_l^1\mathsf{X}_l^{-1} \quad \text{versus} \quad \mathsf{G}_r = \mathsf{U}_r\mathsf{S}_r^1\mathsf{X}_r^{-1} \Leftrightarrow \mathsf{U}_r^{\mathrm{T}}\mathsf{G}_r\mathsf{X}_r = \mathsf{S}_r^1 , \tag{2.71}$$

$$\mathsf{X}_l^{\mathrm{T}}\mathsf{C}_l\mathsf{Y}_l = \mathsf{S}_l^2 \Leftrightarrow \mathsf{C}_l = \left(\mathsf{X}_l^{-1}\right)^{\mathrm{T}}\mathsf{S}_l^2\mathsf{Y}_l^{-1} \quad \text{versus} \quad \mathsf{C}_r = \left(\mathsf{X}_r^{-1}\right)^{\mathrm{T}}\mathsf{S}_r^2\mathsf{Y}_r^{-1} \Leftrightarrow \mathsf{X}_r^{\mathrm{T}}\mathsf{C}_r\mathsf{Y}_r = \mathsf{S}_r^2 , \tag{2.72}$$

$$\mathsf{Y}_l^{\mathrm{T}}\mathsf{E}_l\mathsf{V}_l = \mathsf{S}_l^3 \Leftrightarrow \mathsf{E}_l = \left(\mathsf{Y}_l^{-1}\right)^{\mathrm{T}}\mathsf{S}_l^3\mathsf{V}_l^{\mathrm{T}} \quad \text{versus} \quad \mathsf{E}_r = \left(\mathsf{Y}_r^{-1}\right)^{\mathrm{T}}\mathsf{S}_r^3\mathsf{V}_r^{\mathrm{T}} \Leftrightarrow \mathsf{Y}_r^{\mathrm{T}}\mathsf{E}_r\mathsf{V}_r = \mathsf{S}_r^3 , \tag{2.73}$$

where S^1, S^2, and S^3 are certain quasi-diagonal matrices, where V and U are unitary matrices, and non-singular matrices are X_l, Y_l and X_r, Y_r, respectively. But we are not able to diagonalize G_l and G_r, respectively, to unity. The diagonalization of G_l and G_r, respectively, to unit matrices is by all means recommendable since accordingly all other tensors, e.g. C_l and C_r, respectively, or E_l and E_r, alternatively, refer to unit vectors which span the local tangent space of $\mathbb{M}_l^2$ or $\mathbb{M}_r^2$, respectively. Before we proceed to the next chapter, let us here additionally note that a tree of generalization of the ordinary singular value decompositions has been developed by M. T. Chu (1991 a,b), B. de Moor and H. Zha (1991), H. Zha (1991), and others to which we refer.

3 Coordinates

Coordinates (direct, transverse, oblique aspects), coordinate transformations, charts (complete atlas, minimal atlas), homeomorphism, Killing vectors of symmetry, universal transverse Mercator projection, universal oblique Mercator projection.

Coordinates are in the heart of curves and surfaces, left and right, one- and two-dimensional Riemann manifolds. As parameters of curves and surfaces, they can be experimentally determined. For instance, by the satellite *Global Positioning System* ("global problem solver": GPS), we obtain {ellipsoidal longitude, ellipsoidal latitude, ellipsoidal height} as *Gauss surface normal coordinates* with respect to the *International Reference Ellipsoid* in order to coordinate points (so-called "bench marks") on the Earth's *topographic surface*. These coordinates are collected in *National Data Files*, each of the order of 10^{10} coordinate data. Here, we answer the following questions.

Question.

Question: "Why has the notion of the manifold and the chart, $\mathbb{U} \to \phi(\mathbb{U})$, been axiomatically introduced?" Question: "What are those coordinates as collectors items in mega data sets of the order of 10^{10} coordinate data?"

First, we present you a more careful explanation of the notion of a manifold, its chart, minimal and complete atlas, in particular, the change from one chart to another chart ("Kartenwechsel"). Second, we highlight the direct aspect, the transverse aspect, in general, the *oblique aspect* of a surface. Such a notion is needed (i) to understand the popular optimal *Universal Transverse Mercator Projection* (UTM) with respect to the International Reference Ellipsoid and (ii) to understand the so-called *Universal Oblique Mercator Projection* (UOM), also called "rectified skew orthomorphic" by M. Hotine (1947a–e) or *Hotine Oblique Mercator Projection* (HOM) by J. P. Snyder (1982, p. 76), which has been used for casting the *Heat Capacity Mapping Mission* (HCMM) imagery since 1978, particularly suitable for mapping *Landsat type data*.

3-1 Coordinates relating to manifolds

Coordinates relating to manifolds, in particular, differential manifolds, elements of topology (Hausdorff topological space, open and closed domains).

Let us here first consider manifolds and their charts, the complete atlas, and the minimal atlas. Let us here first define the term *chart* according to Definition 3.1.

Definition 3.1 (Manifold, chart).

An n-dimensional manifold $\mathbb{M}^n$ is locally an n-dimensional *Hausdorff topological space*. That is to any element of $\mathbb{M}^n$, called "point", there is a connected *open neighborhood* $\mathbb{U}$ and a *homeomorphism* $\Phi : \mathbb{U} \to \phi(\mathbb{U})$ to an open set $\phi(\mathbb{U})$ of $\mathbb{R}^n$. $\mathbb{M}^n$ is locally homeomorph to $\mathbb{R}^n$. Any homeomorphism ϕ is called a *chart* of $\mathbb{M}^n$. A one-dimensional manifold is called *curve*, a two-dimensional manifold with Riemann metric and without Cartan torsion is called *surface*, and a compact, connected manifold without boundary is called *closed manifold*.

End of Definition.

Indeed, we implemented many unknown notions from the *theory of morphism*, in particular, from topology, but we do not hope to lose you, the map maker. Therefore, just follow us to stroll along Hausdorff Street, Bonn (Germany) to meditate over Haussdorff's *axiom of separation* (T^2) within *Listing's topoploy*, shortly reviewed in the Appendix. What is more important here is the question that follows.

Question.

Question: "Why has the notion of the manifold and the chart, $\mathbb{U} \to \phi(\mathbb{U})$, been axiomatically introduced?" Answer: "The answer to our fundamental question is based on the "mathematical observation" that not all higher-dimensional curved surfaces can be embedded or immersed in a higher-dimensional Euclidean space. Or we do not know whether such an embedding or immersion exists."

For instance, since the twenties of the 20th century, we know that spacetime is a four-dimensional pseudo-Riemann manifold of signature "$+ + + -$" equipped with a pseudo-Riemann metric, Riemann curvature, zero Cartan torsion. But how to embed or immerse such a four-dimensional spacetime manifold in a pseudo-Euclidean space? In case we know nothing, we better work "intrinsically" with the manifold, neglecting the problem of embedding or immersion. Indeed, this is the majority vote procedure when dealing with spacetime. And, in addition, it would not be too helpful to think in terms of the following theorem: any analytical four-dimensional pseudo-Riemann manifold (analytical spacetime) can be immersed in a ten-dimensional pseudo-Euclidean space. Another example is the projective space $\mathbb{P}^n$, and this projective space cannot be embedded or immersed in an Euclidean space as the ambient space. Anyway, let us continue to explain *continuity* and *homeomorphism*, a situation similar to art, where many "... isms" exist. At least, the mathematical builders of the world are more careful in defining their "... isms", like *conformeomorphism* or *areomorphism*.

Definition 3.2 (Continuity, homeomorphism).

Let $\mathbb{X}$ and $\mathbb{Y}$ be two topological spaces and $f : \mathbb{X} \to \mathbb{Y}$ a mapping. f is called *sequence continuous* or *continous* in $x \in \mathbb{X}$ if f to x convergent sequences are continuously mapped to $f(x)$ convergent series. f is called *sequence continuous* or just *continuous* if f is continuous in any point x of $\mathbb{X}$. f is called *a topological map* or a *homeomorphism* if f is continuous, bijective, and f^{-1} is continous as well.

End of Definition.

In order to illustrate this in detail, let us here consider the topologies on $\mathbb{S}_r^1$ and $\phi(\mathbb{S}_r^1)$ as well as the topologies on $\mathbb{S}_r^2$ and $\phi(\mathbb{S}_r^2)$: compare with Example 3.1 and Fig. 3.1 as well as with Example 3.2 and Fig. 3.2, respectively.

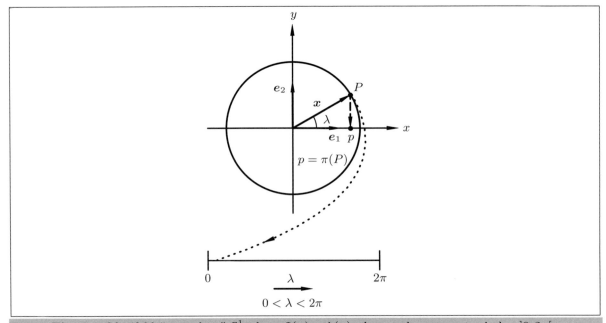

Fig. 3.1. Manifold "one-sphere" $\mathbb{S}_r^1$, chart $\Phi(\boldsymbol{x}) = \lambda(\boldsymbol{x})$: the angular parameter is $\lambda \in]0, 2\pi[$.

Example 3.1 (Circle $\mathbb{S}_r^1$, one-dimensional manifold, topology).

First, we present the topology on $\mathbb{S}_r^1$. Second, we present the topology on $\phi(\mathbb{S}_r^1)$. The "one-sphere" $\mathbb{S}_r^1$ (circle of radius r) is defined as the manifold

$$\mathbb{S}_r^1 := \left\{ \boldsymbol{x} \in \mathbb{R}^2 \,\middle|\, x^2 + y^2 = r^2, r \in \mathbb{R}^+, r > 0 \right\}, \quad \mathbb{U} := \mathbb{S}_r^1 / \{x = +r\}. \tag{3.1}$$

(i) Topology on $\mathbb{S}_r^1$.

The topology on $\mathbb{S}_r^1$ is defined by the *Euclidean metric*, namely the distance function of the ambient space $\mathbb{R}^3$, i. e.

$$\mathrm{d}(\boldsymbol{x}_1, \boldsymbol{x}_2) := \| \boldsymbol{x}_1 - \boldsymbol{x}_2 \|_2. \tag{3.2}$$

Along the orthonormal base $\{\boldsymbol{e}_1, \boldsymbol{e}_2 \mid \mathcal{O}\}$ attached to the origin $\mathcal{O}$, the center of the "one-sphere", we define a *Cartesian coordinate system* $\{x, y\}$ such that $\| \boldsymbol{x} \|_2^2 = x^2 + y^2 = r^2 > 0$. A point P of the "one-sphere" is orthogonally projected on the x axis such that $p = \pi(P)$. Refer to Fig. 3.1 for an illustration. The unit vector $\boldsymbol{e}_x := \boldsymbol{x} / \| \boldsymbol{x} \|_2$ and the unit base vector $\boldsymbol{e}_1$ include the angle $\lambda = \measuredangle(\boldsymbol{e}_x, \boldsymbol{e}_1)$, an element of the open interval $\lambda \in \{\mathbb{R} \mid 0 < \lambda < 2\pi\}$.

$$\Phi(\boldsymbol{x}) = \lambda(\boldsymbol{x}) \Leftrightarrow \Phi^{-1}(\boldsymbol{x}) : \boldsymbol{x}(\lambda) = \boldsymbol{e}_1 r \cos \lambda + \boldsymbol{e}_2 r \sin \lambda, \tag{3.3}$$

$$\Phi^{-1}(\boldsymbol{x}) = r \begin{bmatrix} \cos \lambda \\ \sin \lambda \end{bmatrix}, \tag{3.4}$$

$$\mathrm{d}(\boldsymbol{x}_1, \boldsymbol{x}_2) = r \sqrt{(\cos \lambda_1 - \cos \lambda_2)^2 + (\sin \lambda_1 - \sin \lambda_2)^2} =$$
$$= r\sqrt{2}\sqrt{1 - (\cos \lambda_1 \cos \lambda_2 + \sin \lambda_1 \sin \lambda_2)} = \tag{3.5}$$
$$= r\sqrt{2}\sqrt{1 - \cos(\lambda_1 - \lambda_2)}.$$

We apologize for our sloppy notation $\boldsymbol{x}(\lambda)$ meaning $\boldsymbol{x} = \boldsymbol{\kappa}(\lambda)$, but introduced for economical reason: save extra symbols.

$$\lambda(\boldsymbol{x}) = \begin{cases} \arctan(y/x) & \text{for } x > 0 \\ \arctan(y/x) + \pi & \text{for } x < 0 \\ (\pi/2)\,\mathrm{sgn}\, y & \text{for } x = 0 \text{ and } y \neq 0 \\ \text{undefined} & \text{for } x = 0 \text{ and } y = 0 \end{cases}. \tag{3.6}$$

Note that $\{\lambda = 0 \text{ or } 2\pi\}$ or, equivalently, $\{x = r, y = 0\}$ is the exceptional point which is not curved by the angular parameter λ!

(ii) Topology on $\Phi(\mathbb{S}_r^1)$.

The topology on $\Phi(\mathbb{S}_r^1)$ is defined by the Euclidean metric, namely the distance function

$$\mathrm{d}(\boldsymbol{y}_1, \boldsymbol{y}_2) := \| \boldsymbol{y}_1 - \boldsymbol{y}_2 \|_2 = |\lambda_1 - \lambda_2|. \tag{3.7}$$

End of Example.

Example 3.2 (Sphere $\mathbb{S}_r^2$, two-dimensional manifold, topology).

First, we present the topology on $\mathbb{S}_r^2$. Second, we present the topology on $\phi(\mathbb{S}_r^2)$. The "two-sphere" $\mathbb{S}_r^2$ (sphere of radius r) is defined as the manifold

$$\mathbb{S}_r^2 := \left\{ \boldsymbol{x} \in \mathbb{R}^3 \,\middle|\, x^2 + y^2 + z^2 = r^2, r \in \mathbb{R}^+, r > 0 \right\} . \tag{3.8}$$

(i) Topology on $\mathbb{S}_r^2$.

The topology on $\mathbb{S}_r^2$ is defined by the *Euclidean metric*, namely the distance function of the ambient space $\mathbb{R}^3$, i.e. $\mathrm{d}(\boldsymbol{x}_1, \boldsymbol{x}_2) := \| \boldsymbol{x}_1 - \boldsymbol{x}_2 \|_2$. Along the orthonormal base $\{ \boldsymbol{e}_1, \boldsymbol{e}_2, \boldsymbol{e}_3 \,|\, \mathcal{O} \}$ attached to the origin $\mathcal{O}$, the center of the "two-sphere", we define a *Cartesian coordinate system* $\{x, y, z\}$ in such a way that $\| \boldsymbol{x} \|_2^2 = x^2 + y^2 + z^2 = r^2 > 0$. A point P of the "two-sphere" is orthogonally projected on the (x, y) plane, which is also called the *equatorial plane*, such that $p = \pi(P)$. Refer to Fig. 3.2 for an illustration. The straight line p–$\mathcal{O}$ is oriented with respect to the unit vector $\boldsymbol{e}_1$ or the x axis by the angular parameter "spherical longitude" λ, an element of the open interval $\lambda \in \{ \mathbb{R} \,|\, 0 < \lambda < 2\pi \}$. In contrast, the straight line P–$\mathcal{O}$ is oriented with respect to the equatorial plane (x, y) by the angular parameter "spherical latitude" ϕ, an element of the open interval $\phi \in \{ \mathbb{R} \,|\, -\pi/2 < \phi < +\pi/2 \}$. Again, we emphasize the *open domain* $(\lambda, \phi) \in \{ \mathbb{R}^2 \,|\, 0 < \lambda < 2\pi, -\pi/2 < \phi < +\pi/2 \}$.

$$\Phi(\boldsymbol{x}) = \begin{bmatrix} \lambda(\boldsymbol{x}) \\ \phi(\boldsymbol{x}) \end{bmatrix} \Leftrightarrow \Phi^{-1}(\boldsymbol{x}) : \boldsymbol{x}(\lambda, \phi) = \boldsymbol{e}_1 r \cos\phi \cos\lambda + \boldsymbol{e}_2 r \cos\phi \sin\lambda + \boldsymbol{e}_3 r \sin\phi , \tag{3.9}$$

$$\Phi^{-1}(\boldsymbol{x}) = r \begin{bmatrix} \cos\phi \cos\lambda \\ \cos\phi \sin\lambda \\ \sin\phi \end{bmatrix} \quad \left(r = \sqrt{x^2 + y^2 + z^2} \right) , \tag{3.10}$$

$$\mathrm{d}(\boldsymbol{x}_1, \boldsymbol{x}_2) =$$

$$= r\sqrt{(\cos\phi_1 \cos\lambda_1 - \cos\phi_2 \cos\lambda_2)^2 + (\cos\phi_1 \sin\lambda_1 - \cos\phi_2 \sin\lambda_2)^2 + (\sin\phi_1 - \sin\phi_2)^2} =$$

$$= r\sqrt{2}\sqrt{1 - (\cos\phi_1 \cos\phi_2 \cos\lambda_1 \cos\lambda_2 + \cos\phi_1 \cos\phi_2 \sin\lambda_1 \sin\lambda_2 + \sin\phi_1 \sin\phi_2)} = \tag{3.11}$$

$$= r\sqrt{2}\sqrt{1 - \cos\phi_1 \cos\phi_2 \cos(\lambda_1 - \lambda_2) + \sin\phi_1 \sin\phi_2} = r\sqrt{2}\sqrt{1 - \cos\Psi} .$$

We here again apologize for our sloppy notation $\boldsymbol{x}(\lambda, \phi)$ meaning $\boldsymbol{x} = \boldsymbol{\kappa}(\lambda, \phi)$, but introduced for shorthand writing.

$$\lambda(\boldsymbol{x}) = \begin{cases} \arctan(y/x) & \text{for } x > 0 \\ \arctan(y/x) + \pi & \text{for } x < 0 \\ (\pi/2)\,\mathrm{sgn}\, y & \text{for } x = 0, y \neq 0 \\ \text{undefined} & \text{for } x = 0, y = 0 \end{cases} , \quad \phi(\boldsymbol{x}) = \begin{cases} \arctan \dfrac{z}{\sqrt{x^2 + y^2}} & \\ & \\ \text{undefined} & \text{for } x = y = z = 0 \end{cases} . \tag{3.12}$$

Note that $\{\lambda = 0$ or $2\pi, \phi = \pi/2$ or $-\pi/2\}$ is the exceptional point set, namely the *half meridian* South-Pole–North-Pole passing the point $x = r$, $y = 0$, $z = 0$, sometimes called *Greenwich Meridian*. Such a half meridian is not curved by the angular parameter set $\{\lambda, \phi\}$!

(ii) Topology on $\Phi(\mathbb{S}_r^2)$.

The topology on $\Phi(\mathbb{S}_r^2)$ is defined by the Euclidean metric, namely the distance function

$$\mathrm{d}(\boldsymbol{y}_1, \boldsymbol{y}_2) := \| \boldsymbol{y}_1 - \boldsymbol{y}_2 \|_2 = \left| (\lambda_1 - \lambda_2)^2 + (\phi_1 - \phi_2)^2 \right| . \tag{3.13}$$

End of Example.

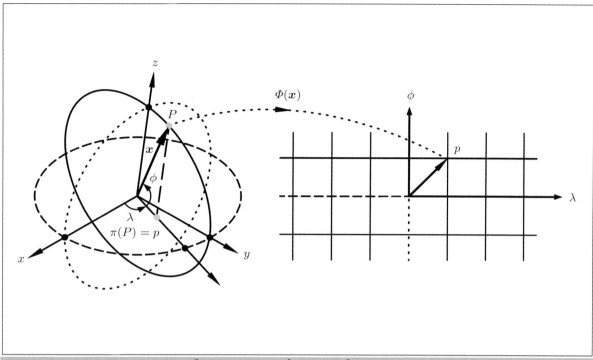

Fig. 3.2. Manifold "two-sphere" $\mathbb{S}_r^2$, chart $\varPhi(\boldsymbol{x}) = \big[\lambda(\boldsymbol{x}), \phi(\boldsymbol{x})\big]$: the angular parameters are $\lambda \in \,]0, 2\pi[$ and $\phi \in \,]-\pi/2, +\pi/2[$.

You may have wondered why did we introduce *open sets*, an *open domain of parameters* to coordinate a surface or a Riemann manifold of type $\mathbb{S}_r^1$ or $\mathbb{S}_r^2$, respectively. Actually, we have postulated that $\varPhi(\boldsymbol{x})$ should be "one-to-one". This is not guaranteed for the spherical *South Pole* or *North Pole* since $\lambda(x, y, z)$ for $x = 0$, $y = 0$, $z = \pm r$ as a mapping is "one-to-infinity", for instance. Further arguments are given as soon as we equip the manifold with a differential structure (differential topology). Indeed, you may have realized that in terms of open sets or an open domain of parameters, $\mathbb{S}_r^1$ or $\mathbb{S}_r^2$ is *not completely* covered. We are therefore forced to introduce more than one set of parameters, hoping that their union $\cup\, \mathbb{U}_i, i \in \{1, \ldots, I\}$ covers totally $\mathbb{S}_r^1$ and $\mathbb{S}_r^2$, and $\mathbb{M}^2$, in general.

Definition 3.3 (Atlas, complete atlas, minimal atlas).

An atlas of a manifold $\mathbb{M}^n$ of dimension n is a family of open sets $\mathbb{U}_i, i \in \{1, \ldots, I\}$, called *charts*, such that the two conditions (i) and (ii) hold: (i) $\cup_i \mathbb{U}_i(\boldsymbol{x}) = \mathbb{M}^n$, (ii) for each $i \in \{1, \ldots, I\}$ there is an open set $\mathbb{V}_i \subset \mathbb{E}^n$ and a bijective mapping $\varPhi_i : \mathbb{U}_i \to \mathbb{V}_i$ such that $\mathbb{V}_i$ is isomorphic with $\mathbb{E}^n := \{\mathbb{R}^n, \delta_{\mu\nu}\}$. Such an atlas is called "complete". Out of the choice of various charts whose union covers $\mathbb{M}^n$ completely, there is one called minimal atlas (which is sometimes also called maximal), where I is minimal.

End of Definition.

As an example think of a *Road Atlas* or a *Geographic Atlas* whose charts cover a part of or the whole surface of the Earth. In the first case, the atlas of the Earth would be *incomplete*. In the second case, *complete* but not minimal. The various notions of atlas, complete atlas, and minimal atlas are clarified by the examples that follow. Beforehand, however, let us give a short comment to the new notions, in particular, to the relation between "charts" and "coordinates". Indeed, the set of all charts enables us to associate to any point of $\mathbb{M}^n$ locally a set of coordinates. As coordinates of a point $\boldsymbol{x}$, we introduce the image $\varPhi(\boldsymbol{x})$ in $\{\mathbb{R}^n, \delta_{\mu\nu}\} =: \mathbb{E}^n$, most of the time equipped with an Euclidean metric or with a pseudo-Euclidean metric.

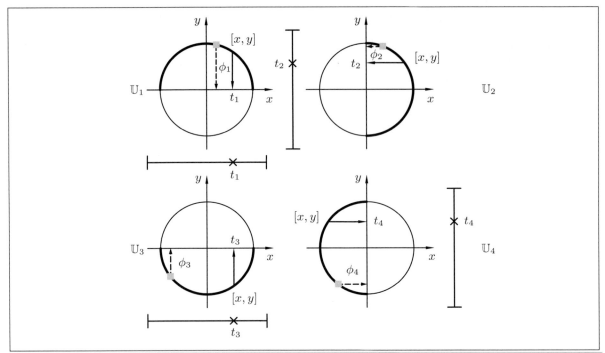

Fig. 3.3. "one-sphere" $\mathbb{S}_r^1$, complete atlas built on four charts.

Example 3.3 (Circle $\mathbb{S}_r^1$, complete atlas: $I = 4$).

A complete atlas of $\mathbb{S}_r^1$ is generated by four charts of the type

$$
\begin{aligned}
\mathbb{U}_1 &:= \{[x, y] \in \mathbb{S}_r^1 \,|\, y > 0\}\,, \quad \varPhi_1[x, y] := x = t_1\,, \\
\mathbb{U}_2 &:= \{[x, y] \in \mathbb{S}_r^1 \,|\, x > 0\}\,, \quad \varPhi_2[x, y] := y = t_2\,, \\
\mathbb{U}_3 &:= \{[x, y] \in \mathbb{S}_r^1 \,|\, y < 0\}\,, \quad \varPhi_3[x, y] := x = t_3\,, \\
\mathbb{U}_4 &:= \{[x, y] \in \mathbb{S}_r^1 \,|\, x < 0\}\,, \quad \varPhi_4[x, y] := y = t_4\,.
\end{aligned}
\tag{3.14}
$$

The sets $\mathbb{U}_i$ and their maps $\varPhi(\mathbb{U}_i) \in\,]-1, +1[$ are open with respect to the chosen topology.

$$I = 4:$$

$$
\begin{aligned}
\varPhi_1^{-1}(t_1) &= \left[t_1, +\sqrt{r^2 - t_1^2}\right] \sim \boldsymbol{x}_1(t_1) = \boldsymbol{e}_1 t_1 + \boldsymbol{e}_2 \sqrt{r^2 - t_1^2}\,, \\
\varPhi_2^{-1}(t_2) &= \left[+\sqrt{r^2 - t_2^2}, t_2\right] \sim \boldsymbol{x}_2(t_2) = +\boldsymbol{e}_1 \sqrt{r^2 - t_2^2} + \boldsymbol{e}_2 t_2\,, \\
\varPhi_3^{-1}(t_3) &= \left[t_3, -\sqrt{r^2 - t_3^2}\right] \sim \boldsymbol{x}_3(t_3) = \boldsymbol{e}_1 t_3 - \boldsymbol{e}_3 \sqrt{r^2 - t_3^2}\,, \\
\varPhi_4^{-1}(t_4) &= \left[-\sqrt{r^2 - t_4^2}, t_4\right] \sim \boldsymbol{x}_4(t_4) = -\boldsymbol{e}_1 \sqrt{r^2 - t_4^2} + \boldsymbol{e}_4 t_4\,.
\end{aligned}
\tag{3.15}
$$

Indeed, the union of the patches ("Umgebungsräume") $\mathbb{U}_1 \cup \mathbb{U}_2 \cup \mathbb{U}_3 \cup \mathbb{U}_4 = \mathbb{S}_r^1$, which is the "one-sphere" $\mathbb{S}_r^1$ is covered by the four charts $\varPhi_1 \in \mathbb{V}_1$, $\varPhi_2 \in \mathbb{V}_2$, $\varPhi_3 \in \mathbb{V}_3$, $\varPhi_4 \in \mathbb{V}_4$, and $\mathbb{V}_i :=\,]-1, +1[$, ($i \in \{1, 2, 3, 4\}$) completely. We have generated a complete atlas: consult Fig. 3.3 for animation.

End of Example.

Example 3.4 (Sphere $\mathbb{S}_r^2$, complete atlas: $I = 6$).

By means of an orthogonal projection $p = \pi(P)$, we already introduced a first coordinate set of the "two-sphere" $\mathbb{S}_r^2$ in terms of spherical longitude λ and spherical latitude ϕ. As local coordinates, $\{\lambda, \phi\}$ do not cover all points of the "two-sphere". As a set of exceptional points, we removed the South Pole, the North Pole, as well as the Greenwich Meridian, the meridian $\lambda = 0$. Here, we introduce a special union of six charts, which covers the "two-sphere" completely. Figure 3.4 illustrates those six charts. Their generators $\Phi_i = \Phi(\mathbb{U}_i)$ $(i = 1, 2, 3, 4, 5, 6)$ are the following:

$$\mathbb{U}_z^+ = \{[x, y, z] \in \mathbb{S}_r^2 \,\big|\, z = +\sqrt{r^2 - (x^2 + y^2)} > 0, x^2 + y^2 < r^2\}\,,$$

$$\Phi_1(x, y, z) := \begin{bmatrix} x \\ y \end{bmatrix} = \begin{bmatrix} u_1 \\ v_1 \end{bmatrix}\,,$$

$$\Phi_1^{-1}(u_1, v_1) = \left[u_1, v_1, +\sqrt{r^2 - (u_1^2 + v_1^2)}\right] \sim \boldsymbol{x}_1(u, v) = \boldsymbol{e}_1 u_1 + \boldsymbol{e}_2 v_1 + \boldsymbol{e}_3 \sqrt{r^2 - (u_1^2 + v_1^2)}\,,$$

$$\mathbb{U}_z^- = \{[x, y, z] \in \mathbb{S}_r^2 \,\big|\, z = -\sqrt{r^2 - (x^2 + y^2)} < 0, x^2 + y^2 < r^2\}\,,$$

$$\Phi_2(x, y, z) := \begin{bmatrix} x \\ y \end{bmatrix} = \begin{bmatrix} u_2 \\ v_2 \end{bmatrix}\,,$$

$$\Phi_2^{-1}(u_2, v_2) = \left[u_2, v_2, -\sqrt{r^2 - (u_2^2 + v_2^2)}\right] \sim \boldsymbol{x}_2(u, v) = \boldsymbol{e}_1 u_2 + \boldsymbol{e}_2 v_2 - \boldsymbol{e}_3 \sqrt{r^2 - (u_2^2 + v_2^2)}\,,$$

$$\mathbb{U}_y^+ = \{[x, y, z] \in \mathbb{S}_r^2 \,\big|\, y = +\sqrt{r^2 - (x^2 + z^2)} > 0, x^2 + z^2 < r^2\}\,,$$

$$\Phi_3(x, y, z) := \begin{bmatrix} x \\ z \end{bmatrix} = \begin{bmatrix} u_3 \\ v_3 \end{bmatrix}\,,$$

$$\Phi_3^{-1}(u_3, v_3) = \left[u_3, +\sqrt{r^2 - (u_3^2 + v_3^2)}, v_3\right] \sim \boldsymbol{x}_3(u, v) = \boldsymbol{e}_1 u_3 + \boldsymbol{e}_2 \sqrt{r^2 - (u_3^2 + v_3^2)} + \boldsymbol{e}_3 v_3\,,$$

$$\mathbb{U}_y^- = \{[x, y, z] \in \mathbb{S}_r^2 \,\big|\, y = -\sqrt{r^2 - (x^2 + z^2)} < 0, x^2 + z^2 < r^2\}\,, \tag{3.16}$$

$$\Phi_4(x, y, z) := \begin{bmatrix} x \\ z \end{bmatrix} = \begin{bmatrix} u_4 \\ v_4 \end{bmatrix}\,,$$

$$\Phi_4^{-1}(u_4, v_4) = \left[u_4, -\sqrt{r^2 - (u_4^2 + v_4^2)}, v_4\right] \sim \boldsymbol{x}_4(u, v) = \boldsymbol{e}_1 u_4 - \boldsymbol{e}_2 \sqrt{r^2 - (u_4^2 + v_4^2)} + \boldsymbol{e}_3 v_4\,,$$

$$\mathbb{U}_x^+ = \{[x, y, z] \in \mathbb{S}_r^2 \,\big|\, x = +\sqrt{r^2 - (y^2 + z^2)}, y^2 + z^2 < r^2\}\,,$$

$$\Phi_5(x, y, z) := \begin{bmatrix} y \\ z \end{bmatrix} = \begin{bmatrix} u_5 \\ v_5 \end{bmatrix}\,,$$

$$\Phi_5^{-1}(u_5, v_5) = \left[+\sqrt{r^2 - (u_5^2 + v_5^2)}, u_5, v_5\right] \sim \boldsymbol{x}_5(u, v) = +\boldsymbol{e}_1 \sqrt{r^2 - (u_5^2 + v_5^2)} + \boldsymbol{e}_2 u_5 + \boldsymbol{e}_3 v_5\,,$$

$$\mathbb{U}_x^- = \{[x, y, z] \in \mathbb{S}_r^2 \,\big|\, x = -\sqrt{r^2 - (y^2 + z^2)}, y^2 + z^2 < r^2\}\,,$$

$$\Phi_6(x, y, z) := \begin{bmatrix} y \\ z \end{bmatrix} = \begin{bmatrix} u_6 \\ v_6 \end{bmatrix}\,,$$

$$\Phi_6^{-1}(u_6, v_6) = \left[-\sqrt{r^2 - (u_6^2 + v_6^2)}, u_6, v_6\right] \sim \boldsymbol{x}_6(u, v) = -\boldsymbol{e}_1 \sqrt{r^2 - (u_6^2 + v_6^2)} + \boldsymbol{e}_2 u_6 + \boldsymbol{e}_3 v_6\,.$$

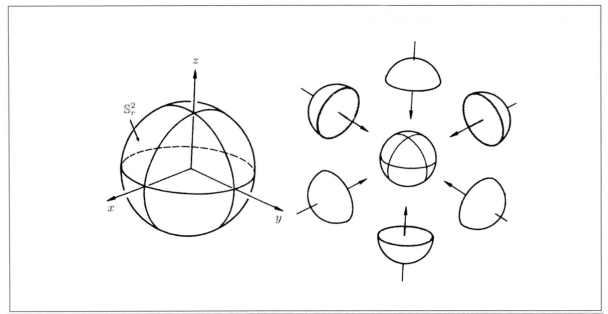

Fig. 3.4. "two-sphere" $\mathbb{S}_r^2$, complete atlas built on six charts.

The sets $\mathbb{U}_i$ and their images $\Phi(\mathbb{U}_i)$ are open with respect to the chosen topology. For instance, the set $\mathbb{U}_z^+$ and its image $\Phi_1(x, y, z)$:

$$\mathbb{U}_z^+ = \left\{ [x, y, z] \in \mathbb{S}_r^2 \,\middle|\, z = +\sqrt{r^2 - (x^2 + y^2)} > 0, x^2 + y^2 < r^2 \right\}, \quad \Phi_1(x, y, z) := \begin{bmatrix} x \\ y \end{bmatrix} = \begin{bmatrix} u \\ v \end{bmatrix}, \tag{3.17}$$

$$\Phi_1^{-1}(u, v) = \left[u, v, +\sqrt{r^2 - (u^2 + v^2)} \right] \sim \boldsymbol{x}_1(u, v) = \boldsymbol{e}_1 u + \boldsymbol{e}_2 v + \boldsymbol{e}_3 \sqrt{r^2 - (u^2 + v^2)}.$$

The terms $\Phi_1^{-1}(u, v)$ or $\boldsymbol{x}_1(u, v)$ determine an open set of the "two-sphere" over the (x, y) plane, namely $\{-r < u < +r, -r < v < +r\} =: \mathbb{V}_1$.

$$I = 6:$$

Again, the union of the patches ("Umgebungsräume") $\mathbb{U}_1 \cup \mathbb{U}_2 \cup \mathbb{U}_3 \cup \mathbb{U}_4 \cup \mathbb{U}_5 \cup \mathbb{U}_6 = \mathbb{S}_r^2$ is $\mathbb{S}_r^2$, completely covered by the six charts $\Phi_1 \in \mathbb{V}_1, \ldots, \Phi_6 \in \mathbb{V}_6$, and $\mathbb{V}_i := \{\,]-r, +r[, \,]-r, +r[\,\} \ni (u, v)$ ($i \in \{1, 2, 3, 4, 5, 6\}$). An illustration is offered by Fig. 3.4. In summary, we have generated the complete atlas of the "two-sphere" constructed by six charts. The choice of the open interval is motivated by the fact that the functions $\Phi_i(u, v)$ ($i \in \{1, 2, 3, 4, 5, 6\}$) at $u^2 + v^2 = r^2$ are singular when differentiated. Indeed, this result is documented by the following expressions:

$$\mathrm{d}\Phi_1^{-1}(u, v) \sim \begin{bmatrix} \text{partial derivative with respect to } u, \\ \left[1, 0, -u/\sqrt{r^2 - (u^2 + v^2)} \right] : \text{ singular at } u^2 + v^2 = r^2, \\ \text{partial derivative with respect to } v, \\ \left[0, 1, +v/\sqrt{r^2 - (u^2 + v^2)} \right] : \text{ singular at } u^2 + v^2 = r^2, \end{bmatrix}$$

$$\cdots \tag{3.18}$$

$$\mathrm{d}\Phi_6^{-1}(u, v) \sim \begin{bmatrix} \text{partial derivative with respect to } u, \\ \left[+u/\sqrt{r^2 - (u^2 + v^2)}, 1, 0 \right] : \text{ singular at } u^2 + v^2 = r^2, \\ \text{partial derivative with respect to } v, \\ \left[+v/\sqrt{r^2 - (u^2 + v^2)}, 0, 1 \right] : \text{ singular at } u^2 + v^2 = r^2. \end{bmatrix}$$

End of Example.

Example 3.5 (Circle $\mathbb{S}_r^1$, minimal atlas: $I = 2$).

Earlier, we generated a local coordinate system of the "one-sphere" $\mathbb{S}_r^1$ by the *orthogonal projection* $p = \pi_1(P)$ of a point P of the "one-sphere" $\mathbb{S}_r^1$ onto the x axis. Alternatively, we project the point P orthogonally by $q = \pi_2(P)$ onto the x' axis, chosen as the y axis. Again, we introduce an angular parameter by $\alpha = \measuredangle(e_x, e_2)$, an element of the *open interval* $\alpha \in \{\alpha \in \mathbb{R} \,|\, 0 < \alpha < 2\pi\}$.

1st chart, 1st parameter set: $\qquad$ 2nd chart, 2nd parameter set:

$$\Phi_1(\boldsymbol{x}) = \lambda(\boldsymbol{x}) \,, \qquad\qquad \Phi_2(\boldsymbol{x}) = \alpha(\boldsymbol{x}) \,,$$

$$\Phi_1^{-1}(\boldsymbol{x}) = r \begin{bmatrix} \cos\lambda \\ \sin\lambda \end{bmatrix} \sim \qquad\qquad \Phi_2^{-1}(\boldsymbol{x}) = r \begin{bmatrix} \cos\alpha \\ \sin\alpha \end{bmatrix} \sim \qquad (3.19)$$

$$\sim \boldsymbol{x}_1(\lambda) = \boldsymbol{e}_1 r\cos\lambda + \boldsymbol{e}_2 r\sin\lambda \,. \qquad \sim \boldsymbol{x}_2(\alpha) = \boldsymbol{e}_1 r\cos\alpha + \boldsymbol{e}_2 r\sin\alpha \,.$$

$I = 2$: $\{\Phi_1^{-1}(\mathbb{U})\} \cup \{\Phi_2^{-1}(\mathbb{U})\}$ covers the "one-sphere" completely in the sense of a *minimal atlas*. Formally, in Fig. 3.5 such a minimal atlas is illustrated.

End of Example.

Example 3.6 (Sphere $\mathbb{S}_r^2$, minimal atlas: $I = 2$).

Beforehand, a first coordinate system of the "two-sphere" $\mathbb{S}_r^2$ had been introduced by an orthogonal projection $p = \pi_1(P)$ of a point P of the "two-sphere" $\mathbb{S}_r^2$ onto the equatorial plane. Alternatively, let us make an orthogonal projection $q = \pi_2(P)$ of a point P of the "two-sphere" $\mathbb{S}_r^2$ onto the (x', y') plane, which coincides with the *Greenwich Meridian Plane* spanned by $\{e_2, e_3 \,|\, \mathcal{O}\}$. (The name *meridian* is derived from the word *noon*. Here, it coincides with the coordinate plane $\lambda = 0$.) Within the (x', y') plane spanned by $\{e_{1'}, e_{2'} \,|\, \mathcal{O}\} = \{e_2, e_3 \,|\, \mathcal{O}\}$, the point q is coordinated by the angular parameter α, namely $\alpha = \measuredangle(e_x, e_{1'})$, also called *meta-longitude*, an element of the open interval $\alpha \in \{\alpha \in \mathbb{R} \,|\, 0 < \alpha < 2\pi\}$. The elevation angle of the vector $\mathcal{O}$–P with respect to the (x', y') plane is the angular parameter β, also called *meta-latitude*, an element of the open interval $\beta \in \{\beta \in \mathbb{R} \,|\, -\pi/2 < \beta < +\pi/2\}$. The orientation of the *meta-equatorial plane* is conventionally denoted as *transverse*. Here, we only introduce the 1st and 2nd charts.

1st chart, 1st parameter set: $\qquad$ 2nd chart, 2nd parameter set:

$$\Phi_1(\boldsymbol{x}) = \begin{bmatrix} \lambda(\boldsymbol{x}) \\ \phi(\boldsymbol{x}) \end{bmatrix} \,, \qquad\qquad \Phi_2(\boldsymbol{x}) = \begin{bmatrix} \alpha(\boldsymbol{x}) \\ \beta(\boldsymbol{x}) \end{bmatrix} \,,$$

$$\Phi_1^{-1}(\boldsymbol{x}) = r \begin{bmatrix} \cos\lambda\cos\phi \\ \sin\lambda\cos\phi \\ \sin\phi \end{bmatrix} \sim \qquad \Phi_2^{-1}(\boldsymbol{x}) = r \begin{bmatrix} \cos\alpha\cos\beta \\ \sin\alpha\cos\beta \\ \sin\beta \end{bmatrix} \sim \qquad (3.20)$$

$$\sim \boldsymbol{x}_1(\lambda, \phi) = \qquad\qquad \sim \boldsymbol{x}_2(\alpha, \beta) =$$

$$= \boldsymbol{e}_1 r\cos\lambda\cos\phi + \boldsymbol{e}_2 r\sin\lambda\sin\phi + \boldsymbol{e}_3 r\sin\phi \,. \qquad = \boldsymbol{e}_1 r\cos\alpha\cos\beta + \boldsymbol{e}_2 r\sin\alpha\cos\beta + \boldsymbol{e}_3 r\sin\beta \,.$$

$I = 2$: $\{\Phi_1^{-1}(\mathbb{U})\} \cup \{\Phi_2^{-1}(\mathbb{U})\}$ covers the "two-sphere" as a *minimal atlas*. Let us identify the sets of exceptional points, both in the chart $\{\lambda, \phi\}$ and in the chart $\{\alpha, \beta\}$. In the left chart $\{\lambda, \phi\} \in \Phi_1(\boldsymbol{x})$, the North Pole, the South Pole, and the $\lambda = 0$ meridian define the set of *left exceptional points*. In the right chart $\{\alpha, \beta\} \in \Phi_2(\boldsymbol{x})$, the meta-North Pole ("West Pole"), the meta-South Pole ("East Pole"), and the $\alpha = 0$ meridian define the set of *right exceptional points*.

End of Example.

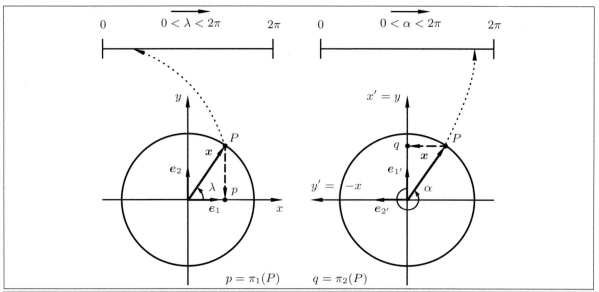

Fig. 3.5. "one-sphere" $\mathbb{S}_r^1$, minimal atlas of two charts, orthogonal projections $p = \pi_1(P)$, $q = \pi_2(P)$.

3-2 Killing vectors of symmetry

Killing vectors of symmetry for the surface-of-revolution and the sphere, transformation groups, first differential invariants, rotation group $\mathrm{R}_3(\Omega)$, special orthogonal groups $\mathrm{SO}(2)$ and $\mathrm{SO}(3)$.

In order to understand better the special aspects of a surface, called *transverse* and *oblique*, we have to analyse the special symmetries of the surface-of-revolution, in particular, the ellipsoid-of-revolution, and the sphere. Such a symmetry analysis is conventionally based on the *Killing vector of symmetry*, which we are going to compute here. As soon as we have identified at least one Killing vector of symmetry for the surface of revolution and the three Killing vectors of symmetry of the sphere, we discuss their impact on the definition of the transverse aspect as well as of the oblique aspect of a surface. We pose two questions.

Question.

Question 1: "Let a transformation group act on the coordinate transformation of a surface-of-revolution. Indeed, we make a coordinate transformation. What are the transformation groups (the coordinate transformations) which leave the *first differential invariant* $\mathrm{d}s^2$ of a surface-of-revolution *equivariant* or *form-invariant*?" Answer 1: "The transformation group, which leaves the first differential invariant $\mathrm{d}s^2$ (also called "arc length") equivariant is the one-dimensional rotation group $\mathrm{R}_3(\Omega)$, a rotation around the 3 axis of the ambient space $\{\mathbb{R}^3, \delta_{ij}\}$. The 3 axis establishes the Killing vector of symmetry.".

A proof of our answer is outlined in Box 3.1. First, we present a parameter representation of a surface-of-revolution, defined by $\{u, v\}$ in an equatorial frame of reference and defined by $\{u^*, v^*\}$ in a rotated equatorial frame of reference. Second, we follow the action of the rotation group $\mathrm{R}_3(\Omega) \in \mathrm{SO}(2)$. Third, we generate the forward and backward transformations $\{e_1, e_2, e_3 \mid \mathcal{O}\} \rightarrow \{e_{1*}, e_{2*}, e_{3*} \mid \mathcal{O}\}$ and $\{e_{1*}, e_{2*}, e_{3*} \mid \mathcal{O}\} \rightarrow \{e_1, e_2, e_3 \mid \mathcal{O}\}$ of orthonormal base vectors, which span the three-dimensional Euclidean ambient space. Fourth, we then fill in the backward transformation of bases into the first parameter representation of the surface-of-revolution and compare with the second one. In this way, we find the "Kartenwechsel" ("cha-cha-cha") $\{u^* = u - \Omega, v^* = v\}$. Fifth, we compute the first differential invariant $\mathrm{d}s^2$ of the surface-of-revolution, namely the matrix of the metric $\mathsf{G} = \mathrm{diag}\left[f^2, f'^2 + g'^2\right]$. Cha-cha-cha leads us via the Jacobi map J to the second representation $\mathrm{d}s^{*2}$ of the first differential invariant, which turns out to be equivariant or form-invariant. Indeed, we have shown that under the action of the rotation group $\mathrm{d}s^2 = \mathrm{d}s^{*2}$. Sixth, we identify e_3 or $[0, 0, 1]$ as the Killing vector of the symmetry of a surface-of-revolution.

Box 3.1 (Surface-of-revolution. Killing vector of symmetry, equivariance of the arc length under the action of the special orthogonal group $\mathrm{SO}(2)$).

Surface-of-revolution parameterized in an equatorial frame of reference:

$$\boldsymbol{x}(u,v) = \boldsymbol{e}_1 f(v) \cos u + \boldsymbol{e}_2 f(v) \sin u + \boldsymbol{e}_3 g(v) \,. \tag{3.21}$$

Surface-of-revolution parameterized in a rotated equatorial frame of reference:

$$\boldsymbol{x}(u^*,v^*) = \boldsymbol{e}_{1*} f(v^*) \cos u^* + \boldsymbol{e}_{2*} f(v^*) \sin u^* + \boldsymbol{e}_{3*} g(v^*) \,. \tag{3.22}$$

Action of the special orthogonal group $\mathrm{SO}(2)$:

$$\mathsf{R}_3(\varOmega) \in \mathrm{SO}(2) := \left\{ \mathsf{R}_3 \in \mathbb{R}^{3\times 3} \,\middle|\, \mathsf{R}_3^* \mathsf{R}_3 = \mathsf{I}_3, |\mathsf{R}_3| = 1 \right\} \,,$$

$$\begin{bmatrix} \boldsymbol{e}_{1*} \\ \boldsymbol{e}_{2*} \\ \boldsymbol{e}_{3*} \end{bmatrix} = \mathsf{R}_3(\varOmega) \begin{bmatrix} \boldsymbol{e}_1 \\ \boldsymbol{e}_2 \\ \boldsymbol{e}_3 \end{bmatrix} = \begin{bmatrix} \cos\varOmega & \sin\varOmega & 0 \\ -\sin\varOmega & \cos\varOmega & 0 \\ 0 & 0 & 1 \end{bmatrix} \begin{bmatrix} \boldsymbol{e}_1 \\ \boldsymbol{e}_2 \\ \boldsymbol{e}_3 \end{bmatrix} \,,$$

$$\begin{bmatrix} \boldsymbol{e}_1 \\ \boldsymbol{e}_2 \\ \boldsymbol{e}_3 \end{bmatrix} = \mathsf{R}_3^*(\varOmega) \begin{bmatrix} \boldsymbol{e}_{1*} \\ \boldsymbol{e}_{2*} \\ \boldsymbol{e}_{3*} \end{bmatrix} = \begin{bmatrix} \cos\varOmega & -\sin\varOmega & 0 \\ \sin\varOmega & \cos\varOmega & 0 \\ 0 & 0 & 1 \end{bmatrix} \begin{bmatrix} \boldsymbol{e}_{1*} \\ \boldsymbol{e}_{2*} \\ \boldsymbol{e}_{3*} \end{bmatrix} \,, \tag{3.23}$$

$$\boldsymbol{e}_1 = \boldsymbol{e}_{1*} \cos\varOmega - \boldsymbol{e}_{2*} \sin\varOmega \,, \quad \boldsymbol{e}_2 = \boldsymbol{e}_{1*} \sin\varOmega + \boldsymbol{e}_{2*} \cos\varOmega \,, \quad \boldsymbol{e}_3 = \boldsymbol{e}_{3*} \,.$$

Coordinate transformations:

$$\boldsymbol{x}(u,v) = f(v)\boldsymbol{e}_{1*}(\cos\varOmega \cos u + \sin\varOmega \sin u) + f(v)\boldsymbol{e}_{2*}(-\sin\varOmega \cos u + \cos\varOmega \sin u) + \boldsymbol{e}_{3*} g(v) \,,$$

$$\boldsymbol{x}(u,v) = f(v)\boldsymbol{e}_{1*} \cos(u-\varOmega) + f(v)\boldsymbol{e}_{2*} \sin(u-\varOmega) + \boldsymbol{e}_{3*} g(v) \,, \tag{3.24}$$

$$v = v^* \,, \; \boldsymbol{x}(u,v) = \boldsymbol{x}(u^*,v^*)$$

$$\Leftrightarrow$$

$$\cos u^* = \cos(u-\varOmega) \,, \; \sin u^* = \sin(u-\varOmega) \,, \; \tan u^* = \tan(u-\varOmega) \tag{3.25}$$

$$\Leftrightarrow$$

$$u^* = u - \varOmega \,.$$

Arc length (first differential invariant):

$$\mathrm{d}s^2 = [\mathrm{d}u, \mathrm{d}v] \, \mathsf{J}_{\boldsymbol{x}}^* \mathsf{J}_{\boldsymbol{x}} \begin{bmatrix} \mathrm{d}u \\ \mathrm{d}v \end{bmatrix} \,,$$

$$\mathsf{J}_{\boldsymbol{x}} = \begin{bmatrix} D_u x & D_v x \\ D_u y & D_v y \\ D_u z & D_v z \end{bmatrix} = \begin{bmatrix} -f \sin u & f' \cos u \\ f \cos u & f' \sin u \\ 0 & g' \end{bmatrix} \,, \quad \mathsf{G} := \mathsf{J}_{\boldsymbol{x}}^* \mathsf{J}_{\boldsymbol{x}} = \begin{bmatrix} f^2 & 0 \\ 0 & f'^2 + g'^2 \end{bmatrix} \,. \tag{3.26}$$

1st version: 2nd version:

$$\mathrm{d}s^2 = f^2 \mathrm{d}u^2 + \left(f'^2 + g'^2\right) \mathrm{d}v^2 \,. \qquad \mathrm{d}s^{*2} = f^{*2} \mathrm{d}u^{*2} + \left(f^{*'2} + g^{*'2}\right) \mathrm{d}v^{*2} \,.$$

$$u^* = u - \varOmega \,, \; v^* = v \Leftrightarrow \mathrm{d}u^{*2} = \mathrm{d}u^2 \,, \; \mathrm{d}v^{*2} = \mathrm{d}v^2 \,, \tag{3.27}$$

$$\mathrm{d}s^2 = f^2 \mathrm{d}u^2 + \left(f'^2 + g'^2\right) \mathrm{d}v^2 = f^2 \mathrm{d}u^{*2} + \left(f'^2 + g'^2\right) \mathrm{d}v^{*2} = \mathrm{d}s^{*2} \,.$$

Killing vector of symmetry (rotation axis):

$$\boldsymbol{e}_3 = [\boldsymbol{e}_1, \boldsymbol{e}_2, \boldsymbol{e}_3] \begin{bmatrix} 0 \\ 0 \\ 1 \end{bmatrix} \sim \begin{bmatrix} 0 \\ 0 \\ 1 \end{bmatrix} \,. \tag{3.28}$$

Question 2: "Let a transformation group act on the coordinate representation of a sphere. Or we may say, we make a coordinate transformation. What are the *transformation groups* (the coordinate transformations) which leave the *first differential invariant* ds^2 of a sphere *equivariant* or *form-invariant?*" Answer 2: "The transformation group, which leaves the first differential invariant ds^2 (also called "arc length") equivariant is the three-dimensional rotation group $R(\alpha, \beta, \gamma)$, a subsequent rotation around the 1 axis, the 2 axis, and the 3 axis of the ambient space $\{\mathbb{R}^3, \delta_{ij}\}$. The three axes establish the three Killing vectors of symmetry."

A proof of our answer is outlined in Box 3.2. First, we present a parameter representation of a sphere, defined by $\{u, v\}$ in an equatorial frame of reference and defined by $\{u^*, v^*\}$ in an oblique frame of reference generated by the three-dimensional orthogonal group SO(3). Second, the action of the transformation group SO(3) is parameterized by Cardan angles $\{\alpha, \beta, \gamma\}$, namely a rotation $R_1(\alpha)$ around the 1 axis, a rotation $R_2(\beta)$ around the 2 axis, and a rotation $R_3(\gamma)$ around the 3 axis. Third, we transform forward and backward the orthonormal system of base vectors $\{e_1, e_2, e_3 \mid \mathcal{O}\}$ and $\{e_{1^*}, e_{2^*}, e_{3^*} \mid \mathcal{O}\}$, which span the three-dimensional Euclidean space, the ambient space of the sphere $\mathbb{S}_r^2$. $\{e_1, e_2, e_3 \mid \mathcal{O}\}$ establish the conventional equatorial frame of reference, $\{e_{1^*}, e_{2^*}, e_{3^*} \mid \mathcal{O}\}$ at the origin the meta-equatorial reference frame. Fourth, the backward transformation is substituted into the parameter representation of the placement vector $e_1 r \cos v \cos u + e_2 r \cos v \sin u + e_3 r \sin v \in \mathbb{S}_r^2$, such that $e_{1^*} f_1(\alpha, \beta, \gamma \mid u, v) + e_{2^*} f_2(\alpha, \beta, \gamma \mid u, v) + e_{3^*} f_3(\alpha, \beta, \gamma \mid u, v)$ is a materialization of the "Kartenwechsel" ("cha-cha-cha"). In this way, we are led to $\tan u^* = f_2/f_1$ and $\sin v^* = f_3$, both functions of the parameters $\{\alpha, \beta, \gamma\} \in$ SO(3), of the longitude u, and the latitude v. Fifth, as soon as we substitute "cha-cha-cha", namely the diffeomorphism $\{du, dv\} \to \{du^*, dv^*\}$ by means of the Jacobi matrix J in the *first differential invariant* ds^{*2}, namely the matrix of the metric $G = \text{diag}[r^2 \cos^2 v, r^2]$, we are led to the first representation ds^2 of the first differential invariant, which is equivariant or form-invariant: $ds^2 = r^2 \cos^2 v du^2 + r^2 dv^2 = r^2 \cos^2 v^* du^{*2} + r^2 dv^{*2} = ds^{*2}$. Indeed, we have shown that under the action of the three-dimensional rotation group, namely $R(\alpha, \beta, \gamma) = R_1(\alpha) R_2(\beta) R_3(\gamma)$, $ds^2 = ds^{*2}$. Sixth, we accordingly identify the three Killing vectors $\{e_1, e_2, e_3\}$ or $[1, 0, 0]$, $[0, 1, 0]$, and $[0, 0, 1]$, respectively – the symmetry of the sphere $\mathbb{S}_r^2$.

Box 3.2 (Sphere. Killing vectors of symmetry, equivariance of the arc length under the action of the special orthogonal group SO(3)**).**

Sphere parameterized in an equatorial frame of reference:
$$x(u, v) = e_1 \cos v \cos u + e_2 \cos v \sin u + e_3 \sin v . \qquad (3.29)$$

Sphere parameterized in an oblique frame of reference:
$$x(u^*, v^*) = e_{1^*} \cos v^* \cos u^* + e_{2^*} \cos v^* \sin u^* + e_{3^*} \sin v^* . \qquad (3.30)$$

Action of the special orthogonal group SO(3):
$$R(\alpha, \beta, \gamma) \in SO(3) := \{R \in SO(3) \mid R^* R = I_3, |R| = 1\} ,$$

$$\begin{bmatrix} e_{1^*} \\ e_{2^*} \\ e_{3^*} \end{bmatrix} = R_1(\alpha) R_2(\beta) R_3(\gamma) \begin{bmatrix} e_1 \\ e_2 \\ e_3 \end{bmatrix}$$

$$\Leftrightarrow$$

$$\begin{bmatrix} e_1 \\ e_2 \\ e_3 \end{bmatrix} = R_3^*(\gamma) R_2^*(\beta) R_1^*(\alpha) \begin{bmatrix} e_{1^*} \\ e_{2^*} \\ e_{3^*} \end{bmatrix} , \qquad (3.31)$$

$$e_1 = e_{1^*}(\cos \gamma \cos \beta) - e_{2^*}(\sin \gamma \cos \alpha + \cos \gamma \sin \beta \sin \alpha) +$$
$$+ e_{3^*}(\sin \gamma \sin \alpha + \cos \gamma \sin \beta \cos \alpha) ,$$
$$e_2 = e_{1^*}(\sin \gamma \cos \beta) + e_{2^*}(\cos \gamma \cos \alpha + \sin \gamma \sin \beta \sin \alpha) +$$
$$+ e_{3^*}(- \cos \gamma \sin \alpha + \sin \gamma \sin \beta \cos \alpha) ,$$
$$e_3 = e_{1^*}(- \sin \beta) + e_{2^*}(\cos \beta \sin \alpha) + e_{3^*}(\cos \beta \cos \alpha) .$$

Continuation of Box.

Coordinate transformations:

$$\boldsymbol{x}(u,v) = \boldsymbol{x}(u^*, v^*)$$

$$\Leftrightarrow$$

$$\boldsymbol{e}_{1*} f_1\left(\alpha,\beta,\gamma\,|\,u,v\right) + \boldsymbol{e}_{2*} f_2\left(\alpha,\beta,\gamma\,|\,u,v\right) + \boldsymbol{e}_{3*} f_3\left(\alpha,\beta,\gamma\,|\,u,v\right) =$$

$$= \boldsymbol{e}_{1*} \cos v^* \cos u^* + \boldsymbol{e}_{2*} \cos v^* \sin u^* + \boldsymbol{e}_{3*} \sin v^* \,, \qquad (3.32)$$

$$\cos v^* \cos u^* = f_1\left(\alpha,\beta,\gamma\,|\,u,v\right) =$$

$$= \cos\gamma\cos\beta\cos v\cos u + \sin\gamma\cos\beta\cos v\sin u - \sin\beta\sin v \,,$$

$$\cos v^* \sin u^* = f_2\left(\alpha,\beta,\gamma\,|\,u,v\right) =$$

$$= -(\sin\gamma\cos\alpha + \cos\gamma\sin\beta\sin\alpha)\cos v\cos u +$$

$$+ (\cos\gamma\cos\alpha + \sin\gamma\sin\beta\sin\alpha)\cos v\sin u + \cos\beta\sin\alpha\sin v \,, \qquad (3.33)$$

$$\sin v^* = f_3\left(\alpha,\beta,\gamma\,|\,u,v\right) =$$

$$= (\sin\gamma\sin\alpha + \cos\gamma\sin\beta\cos\alpha)\cos v\cos u -$$

$$- (\cos\gamma\sin\alpha + \sin\gamma\sin\beta\cos\alpha)\cos v\sin u + \cos\beta\cos\alpha\sin v \,,$$

$$\tan u^* = f_2/f_1 \,, \quad \sin v^* = f_3 \,. \qquad (3.34)$$

Arc length (first differential invariant):

$$ds^2 = [du, dv] \begin{bmatrix} r^2\cos v & 0 \\ 0 & r^2 \end{bmatrix} \begin{bmatrix} du \\ dv \end{bmatrix} \quad (\text{``diffeomorphism''}) \,, \quad \begin{bmatrix} du^* \\ dv^* \end{bmatrix} = \mathsf{J} \begin{bmatrix} du \\ dv \end{bmatrix}, \quad \mathsf{J} := \begin{bmatrix} D_u u^* & D_v u^* \\ D_u v^* & D_v v^* \end{bmatrix},$$

$$d\tan u^* = (1 + \tan^2 u^*)du^* \Rightarrow du^* = \cos^2 u^* d\tan u^* \,, \qquad (3.35)$$

$$d\sin v^* = \cos v^* dv^* \Rightarrow dv^* = \frac{1}{\sqrt{1 - \sin^2 v^*}} d\sin v^* \,,$$

$$ds^2 = r^2\cos^2 vdu^2 + r^2 dv^2 = r^2\cos^2 v^* du^{*2} + r^2 dv^{*2} = ds^{*2} \,.$$

Killing vector of symmetry (rotation axis):

$$1 \text{ axis of symmetry: } \boldsymbol{e}_1 \sim \begin{bmatrix} 1 \\ 0 \\ 0 \end{bmatrix}, \quad 2 \text{ axis of symmetry: } \boldsymbol{e}_2 \sim \begin{bmatrix} 0 \\ 1 \\ 0 \end{bmatrix}, \quad 3 \text{ axis of symmetry: } \boldsymbol{e}_3 \sim \begin{bmatrix} 0 \\ 0 \\ 1 \end{bmatrix}. \quad (3.36)$$

Historical aside. W. Killing (1892) transformed the postulate of equivariance ("form invariance") of the first differential invariant under the action of a transformation group ("Lie group") into a system of partial differential equations, which are known as the *Killing equations* being subject to an integrability condition. An important historical reference on the theme *continuous groups of transformations* and *Killing's equations* is L. P. Eisenhart (1961, pp. 208–221). J. Zund succeeded to solve the Killing equations for the sphere (three Killing vectors) and for the ellipsoid-of-revolution (one Killing vector).

3-3 The oblique frame of reference of the sphere

The oblique frame of reference of the sphere (three Killing vectors): normal, oblique, and transverse aspects, Killing symmetry, designs of an oblique frame of reference of the sphere.

Let us confront you here with three aspects of the sphere, which are called *normal*, *oblique*, and *transverse*. These aspects form the basis of spherical coordinates of the sphere, taking into account the three Killing vectors of symmetry, typical for a spherical surface. The first oblique frame of reference of $\mathbb{S}_r^2$ is based upon the meta-North Pole, with the spherical coordinates $\{\lambda_0, \phi_0\}$ as design elements. Alternatively, the second oblique frame of reference of $\mathbb{S}_R^2$ refers to the centric oblique plane $\mathbb{P}_{\mathcal{O}}^2$, which intersects the sphere $\mathbb{S}_R^2$ and passes the origin $\mathcal{O}$. The oblique plane $\mathbb{P}_{\mathcal{O}}^2$ intersects $\mathbb{S}_R^2$ in a so-called *circular oblique equator*, also called *meta-equator*, which is oriented by two Kepler elements $\{\Omega, I\}$, namely the longitude Ω of the ascending node and the inclination I. Finally, the third frame of reference, which is called *transverse*, is defined as special oblique, namely by an inclination $I = \pi/2$. For all three frames of reference, we present to you the forward and backward transformation formulae, also called *direct* and *inverse*. Their derivation is technically done in a way which is suitable for other figures of reference which have less Killing symmetry, like the ellipsoid-of-revolution.

3-31 A first design of an oblique frame of reference of the sphere

The first design of an oblique frame of reference of the sphere $\mathbb{S}_r^2$ is taking reference to the following design aspects. (i) We make a choice about the three spherical coordinates $\{\lambda_0, \phi_0, r\}$ of the "new North Pole" (which is also called *meta-North Pole*) relative to the conventional equatorial frame of reference. We attach to the direction of the new North Pole a new equatorial frame of reference, which is called *meta-equatorial*, namely $\{e_{1^0}, e_{2^0}, e_{3^0} \mid x_0\}$, a set of orthonormal base vectors at the point $x(\lambda_0, \Phi_0, r) =: x_0$, such that $e_{3^0} := x_0/\parallel x_0 \parallel$. We connect the conventional equatorial frame of reference $\{e_1, e_2, e_3 \mid \mathcal{O}\}$ to the oblique or meta-equatorial frame of reference $\{e_{1^0}, e_{2^0}, e_{3^0} \mid \mathcal{O}\}$ at the origin $\mathcal{O}$, namely by parallel transport of $\{e_{1^0}, e_{2^0}, e_{3^0}\}$ from x_0 to $P_{\mathcal{O}}$ (in the Euclidean sense). (ii) We represent the coordinates of the placement vector x, both in the conventional equatorial frame of reference and in the oblique equatorial frame of reference. (iii) We finally derive the forward as well as backward equations of transformation between them.

Solution (the first problem).

The first problem can be solved (i) by representing the placement vector $x(\lambda_0, \phi_0, r)$ in terms of the chosen spherical coordinates $\{\lambda_0, \phi_0, r\}$, (ii) by computing the triplet of partial derivatives $\{D_{\lambda_0} x, D_{\phi_0} x, D_r x\}$, which are normalized by the Euclidean norms $\parallel D_{\lambda_0} x \parallel$, $\parallel D_{\phi_0} x \parallel$, and $\parallel D_r x \parallel$, leading to the triplet $-e_{\phi_0} := D_{\phi_0} x/\parallel D_{\phi_0} x \parallel$, $+e_{\lambda_0} := D_{\lambda_0} x/\parallel D_{\lambda_0} x \parallel$, and $+e_r := D_r x/\parallel D_r x \parallel$, and these three triplet terms are called *South*, *East*, and *Vertical*. Finally, we relate the base vectors $\{e_{1^0}, e_{2^0}, e_{3^0} \mid x_0\} := \{e_{\lambda_0}, e_{\phi_0}, e_r \mid x_0\}$ to $\{e_1, e_2, e_3 \mid \mathcal{O}\}$. This is outlined in Box 3.3. For geometrical details, consult Figs. 3.6 and 3.7.

End of Solution (the first problem).

Solution (the second problem).

The second problem, the representation of the placement vector x in the orthonormal equatorial frame of reference $\{e_1, e_2, e_3 \mid \mathcal{O}\}$ at the origin $\mathcal{O}$ as well as in the oblique frame of reference $\{e_{1^0}, e_{2^0}, e_{3^0} \mid \mathcal{O}\}$ (which is called *meta-equatorial*) at the origin $\mathcal{O}$ in terms of spherical coordinates $\{\lambda, \phi, r\}$ as well as in meta-spherical coordinates $\{\alpha, \beta, r\}$, is solved by forward and backward transformations. This is outlined in Boxes 3.3 and 3.4. For geometrical details, consult again Figs. 3.6 and 3.7. Here, we meet the particular problem to *parallel transport* the oblique frame of reference $\{e_{1^0}, e_{2^0}, e_{3^0} \mid x(\lambda_0, \phi_0, r)\}$ (which is defined at the point $x(\lambda_0, \phi_0, r)$) in the Euclidean sense from (λ_0, ϕ_0, r) to the origin $\mathcal{O}$ in order to generate the centric frame of reference $\{e_{1^0}, e_{2^0}, e_{3^0} \mid \mathcal{O}\}$.

End of Solution (the second problem).

Box 3.3 (Establishing an oblique frame of reference (meta-equatorial) of the sphere).

(i) Placement vector towards the meta-North Pole:

$$\boldsymbol{x}(\lambda_0, \phi_0, r) = \boldsymbol{e}_1 r \cos\phi_0 \cos\lambda_0 + \boldsymbol{e}_2 r \cos\phi_0 \sin\lambda_0 + \boldsymbol{e}_3 r \sin\phi_0 \ . \tag{3.37}$$

(ii) Jacobi map:

$$D_{\lambda_0}\boldsymbol{x} = -\boldsymbol{e}_1 r \cos\phi_0 \sin\lambda_0 + \boldsymbol{e}_2 r \cos\phi_0 \cos\lambda_0 \ ,$$

$$D_{\phi_0}\boldsymbol{x} = -\boldsymbol{e}_1 r \sin\phi_0 \cos\lambda_0 - \boldsymbol{e}_2 r \sin\phi_0 \sin\lambda_0 + \boldsymbol{e}_3 r \cos\phi_0 \ , \tag{3.38}$$

$$D_r\boldsymbol{x} = +\boldsymbol{e}_1 \cos\phi_0 \cos\lambda_0 + \boldsymbol{e}_2 \cos\phi_0 \sin\lambda_0 + \boldsymbol{e}_3 \sin\phi_0 \ .$$

(iii) Meta-equatorial (oblique) frame of reference:

$$\{\text{South}, \text{East}, \text{Vertical}\} :=$$

$$:= \left\{ \frac{-D_{\phi_0}\boldsymbol{x}}{\|D_{\phi_0}\boldsymbol{x}\|}, \frac{D_{\lambda_0}\boldsymbol{x}}{\|D_{\lambda_0}\boldsymbol{x}\|}, \frac{D_r\boldsymbol{x}}{\|D_r\boldsymbol{x}\|} \right\} =: \tag{3.39}$$

$$=: \left\{ \boldsymbol{e}_{1^0}, \boldsymbol{e}_{2^0}, \boldsymbol{e}_{3^0} \,\middle|\, [\lambda_0, \phi_0, r] \right\} \ ,$$

$$\boldsymbol{e}_{1^0} = \boldsymbol{e}_S = +\boldsymbol{e}_1 \sin\phi_0 \cos\lambda_0 + \boldsymbol{e}_2 \sin\phi_0 \sin\lambda_0 - \boldsymbol{e}_3 \cos\phi_0 \ ,$$

$$\boldsymbol{e}_{2^0} = \boldsymbol{e}_E = -\boldsymbol{e}_1 \sin\lambda_0 + \boldsymbol{e}_2 \cos\lambda_0 \ , \tag{3.40}$$

$$\boldsymbol{e}_{3^0} = \boldsymbol{e}_V = +\boldsymbol{e}_1 \cos\phi_0 \cos\lambda_0 + \boldsymbol{e}_2 \cos\phi_0 \sin\lambda_0 + \boldsymbol{e}_3 \sin\phi_0 \ .$$

$\{\boldsymbol{e}_S, \boldsymbol{e}_E, \boldsymbol{e}_V \,|\, \boldsymbol{x}_0\}$ is a moving frame (repére mobile) at $\boldsymbol{x}_0 := \boldsymbol{x}(\lambda_0, \phi_0, r)$:

$$\begin{bmatrix} \boldsymbol{e}_{1^0} \\ \boldsymbol{e}_{2^0} \\ \boldsymbol{e}_{3^0} \end{bmatrix} = \begin{bmatrix} \boldsymbol{e}_S \\ \boldsymbol{e}_E \\ \boldsymbol{e}_V \end{bmatrix} = \begin{bmatrix} \sin\phi_0\cos\lambda_0 & \sin\phi_0\sin\lambda_0 & -\cos\phi_0 \\ -\sin\lambda_0 & \cos\lambda_0 & 0 \\ \cos\phi_0\cos\lambda_0 & \cos\phi_0\sin\lambda_0 & +\sin\phi_0 \end{bmatrix} \begin{bmatrix} \boldsymbol{e}_1 \\ \boldsymbol{e}_2 \\ \boldsymbol{e}_3 \end{bmatrix} \ . \tag{3.41}$$

(iv) Parallel transport:

$$\{\boldsymbol{e}_S, \boldsymbol{e}_E, \boldsymbol{e}_V \,|\, \boldsymbol{x}(\lambda_0, \phi_0, r)\} = \{\boldsymbol{e}_S, \boldsymbol{e}_E, \boldsymbol{e}_V \,|\, \mathcal{O}\} \ . \tag{3.42}$$

Statement:

$$\begin{bmatrix} \boldsymbol{e}_{1^0} \\ \boldsymbol{e}_{2^0} \\ \boldsymbol{e}_{3^0} \end{bmatrix} = \begin{bmatrix} \boldsymbol{e}_S \\ \boldsymbol{e}_E \\ \boldsymbol{e}_V \end{bmatrix} = \mathsf{R}(\lambda_0, \phi_0, r) \begin{bmatrix} \boldsymbol{e}_1 \\ \boldsymbol{e}_2 \\ \boldsymbol{e}_3 \end{bmatrix} \ ,$$

$$\mathsf{R}(\lambda_0, \phi_0, r) := \mathsf{R}_2 (\pi/2 - \phi_0) \, \mathsf{R}_3(\lambda_0) \ , \tag{3.43}$$

$$\mathsf{R}_3(\lambda_0) := \begin{bmatrix} \cos\lambda_0 & \sin\phi_0 & 0 \\ -\sin\lambda_0 & \cos\lambda_0 & 0 \\ 0 & 0 & 1 \end{bmatrix} , \quad \mathsf{R}_2 (\pi/2 - \phi_0) := \begin{bmatrix} \sin\phi_0 & 0 & -\cos\phi_0 \\ 0 & 1 & 0 \\ \cos\phi_0 & 0 & \sin\phi_0 \end{bmatrix} \ .$$

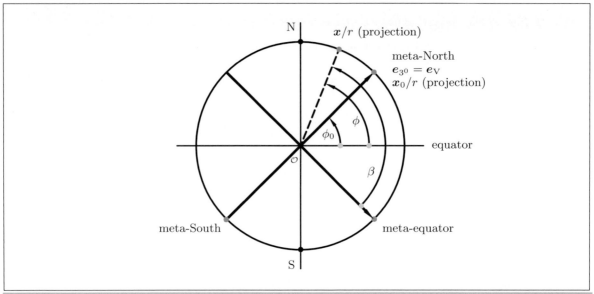

Fig. 3.6. Vertical section of $\mathbb{S}_r^2$, equatorial as well as meta-equatorial (oblique) frame of reference.

Now, we are well-prepared to solve the forward and backward transformation problems, which are also called the *direct* and the *inverse transformations*, which can be characterized as follows. (i) Direct transformation: given the longitude λ and the latitude ϕ of point $\boldsymbol{x} \in \mathbb{S}_r^2$ in the conventional equatorial frame of reference as well as the spherical coordinates $\{\lambda_0, \phi_0\}$ of the meta-North Pole, find the meta-longitude α and the meta-latitude β (alternatively, the meta-colatitude ψ) of an identical point in the meta-equatorial (oblique) frame of reference of the sphere $\mathbb{S}_r^2$. (ii) Inverse transformation: given the meta-longitude α and the meta-latitude β in the meta-equatorial (oblique) frame of reference as well as the spherical coordinates $\{\lambda_0, \phi_0\}$ of the meta-North Pole, find the longitude λ and the latitude ϕ of an identical point in the conventional equatorial frame of reference of the sphere $\mathbb{S}_r^2$.

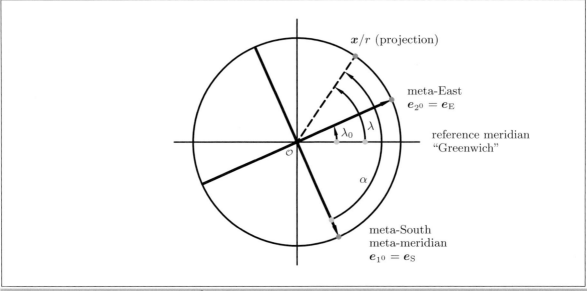

Fig. 3.7. Horizontal section of $\mathbb{S}_r^2$, equatorial as well as meta-equatorial (oblique) frame of reference.

Solution (the third problem, direct transformation).

Such a problem can be immediately solved as outlined in Boxes 3.3, 3.4, 3.5, and 3.6. First, we have parameterized the transformation of reference frames $\{e_{1^0}, e_{2^0}, e_{3^0} \mid \mathcal{O}\} \rightarrow \{e_1, e_2, e_3 \mid \mathcal{O}\}$ by means of the pole position $\{\lambda_0, \phi_0\}$. Second, the placement vector $x \in \mathbb{S}_r^2$ of a point of the reference sphere is represented in both the conventional equatorial frame of reference $\{e_1, e_2, e_3 \mid \mathcal{O}\}$ and in the meta-equatorial (oblique) frame of reference $\{e_{1^0}, e_{2^0}, e_{3^0} \mid \mathcal{O}\}$ at the origin $\mathcal{O}$. Third, we substitute $\{e_1, e_2, e_3 \mid \mathcal{O}\}$ in favor of $\{e_{1^0}, e_{2^0}, e_{3^0} \mid \mathcal{O}\}$ by means of the backward transformation of reference frames, our first setup. The final representation of the placement vector $x(\lambda, \phi; \lambda_0, \phi_0)$ is achieved in terms of (i) conventional equatorial coordinates $\{\lambda, \phi\}$ and (ii) equatorial coordinates $\{\lambda_0, \phi_0\}$ of the meta-North Pole. The corresponding two coordinate transformations $\alpha(\lambda, \phi; \lambda_0, \phi_0)$ and $\beta(\lambda, \phi; \lambda_0, \phi_0)$ are derived in Box 3.5. The three identities for (i) $\cos\alpha$, (ii) $\sin\alpha$, and (iii) $\sin\beta$ or $\cos\psi$ are derived by representing (i) $x^0 = r\cos\beta\cos\alpha$, (ii) $y^0 = r\cos\beta\sin\alpha$, and (iii) $z^0 = r\sin\beta = r\cos\psi$ in the oblique frame of reference. The first identity is also called *spherical sine lemma*, the second identity is also called *spherical sine–cosine lemma*, and the third identity is called *spherical side cosine lemma*. Indeed, we have derived the collective formulae of Spherical Trigonometry, however, in a way to be used for other reference surfaces, for instance, the ellipsoid-of-revolution – a surface with one Killing vector of symmetry. The direct transformation formulae $\{\lambda, \phi; \lambda_0, \phi_0\} \rightarrow \{\alpha, \beta\}$ are presented in Box 3.6. First, by dividing the second identity by the first identity, we arrive at $\tan\alpha = f(\lambda, \phi; \lambda_0, \phi_0)$. Alternatively, we may chose $\cos\alpha$ or $\sin\alpha$, which are additionally depending on $\cos\beta$. Second, we repeat the third identity $\sin\beta = \cos\psi$ either for the meta-latitude β or the meta-colatitude ψ, also called *meta-polar distance* – the space angle between the vectors x and x_0.

End of Solution (the third problem, direct transformation).

Solution (the third problem, inverse transformation).

Such a problem can be immediately solved as outlined in Box 3.7. First, we have parameterized the transformation of reference frames $\{e_1, e_2, e_3 \mid \mathcal{O}\} \rightarrow \{e_{1^0}, e_{2^0}, e_{3^0} \mid \mathcal{O}\}$ by means of the pole position $\{\lambda_0, \phi_0\}$. Second, the placement vector $x \in \mathbb{S}_r^2$ of a point of the reference sphere is represented in both the conventional equatorial frame of reference $\{e_1, e_2, e_3 \mid \mathcal{O}\}$ and in the meta-equatorial (oblique) frame of reference $\{e_{1^0}, e_{2^0}, e_{3^0} \mid \mathcal{O}\}$ at the origin $\mathcal{O}$. Third, we substitute $\{e_{1^0}, e_{2^0}, e_{3^0} \mid \mathcal{O}\}$ in favor of $\{e_1, e_2, e_3 \mid \mathcal{O}\}$ by means of the forward transformation of reference frames, our first setup. The final representation of the placement vector $x(\alpha, \beta; \lambda_0, \phi_0)$ is achieved in terms of (i) meta-equatorial coordinates $\{\alpha, \beta\}$ and (ii) equatorial coordinates $\{\lambda_0, \phi_0\}$ of the meta-North Pole. The corresponding two coordinate transformations $\lambda(\alpha, \beta; \lambda_0, \phi_0)$ and $\phi(\alpha, \beta; \lambda_0, \phi_0)$ are derived in Box 3.8. The three identities for (i) $\cos\lambda$, (ii) $\sin\lambda$, and (iii) $\sin\phi$ are based on representing (i) $x = r\cos\phi\cos\lambda$, (ii) $y = r\cos\phi\sin\lambda$, and (iii) $z = r\sin\phi$ in the oblique frame of reference as derived in the previous formulae. The inverse transformation formulae $\{\alpha, \beta; \lambda_0, \phi_0\} \rightarrow \{\lambda, \phi\}$ are presented in Box 3.9. First, by dividing the second identity by the first identity, we arrive at $\tan\lambda = f(\alpha, \beta; \lambda_0, \phi_0)$. Alternatively, we may take advantage of the elegant form $\tan(\lambda - \lambda_0)$, which is achieved as soon as we implement the addition theorem for $\tan(\alpha \pm \beta)$. Indeed, another simple derivation is the following. Take reference to the direct transformation formulae. Multiply, the third identity by $\cos\phi_0$, namely $\sin\beta\cos\phi_0$, as well as the first identity by $\sin\phi_0$, namely $\cos\beta\cos\alpha\sin\phi_0$, and sum up. In this way, you have found $\cos\phi\cos(\lambda - \lambda_0) = \sin\beta\cos\phi_0 + \cos\beta\cos\alpha\sin\phi_0$. From the second identity, you transfer $\cos\phi\sin(\lambda - \lambda_0) = \cos\beta\sin\alpha$ and divide to produce $\tan(\lambda - \lambda_0) = \sin(\lambda - \lambda_0)/\cos(\lambda - \lambda_0)$. Second, you transfer the third identity of the inverse transformation to gain $\sin\phi = g(\alpha, \beta; \lambda_0, \phi_0)$. Alternatively, you may multiply the first identity of the forward transformation by $\cos\phi_0$, namely $\cos\beta\cos\alpha\cos\phi_0$, and replace $\cos\beta\cos\phi_0\cos(\lambda - \lambda_0)$ by the third identity, i. e. $\sin\beta - \sin\phi\sin\phi_0 = \cos\psi - \sin\phi\sin\phi_0$. Finally, solve for $\sin\phi$.

End of Solution (the third problem, inverse transformation).

Box 3.4 (Representation of a placement vector $x \in \mathbb{S}_r^2$ in both the equatorial frame of reference indicated by $\{e_1, e_2, e_3 \,|\, \mathcal{O}\}$ and the meta-equatorial (oblique) frame of reference indicated by $\{e_{1^0}, e_{2^0}, e_{3^0} \,|\, \mathcal{O}\}$).

$$e_1 x + e_2 y + e_3 z = x = e_{1^0} x^0 + e_{2^0} y^0 + e_{3^0} z^0 \,,$$

$$e_1 \cos\phi\cos\lambda + e_2 \cos\phi\sin\lambda + e_3 \sin\phi = \frac{x}{r} = e_{1^0}\cos\beta\cos\alpha + e_{2^0}\cos\beta\sin\alpha + e_{3^0}\sin\beta \,. \tag{3.44}$$

Transformation $\{e_1, e_2, e_3 \,|\, \mathcal{O}\} \rightarrow \{e_{1^0}, e_{2^0}, e_{3^0} \,|\, \mathcal{O}\}$:

$$\begin{bmatrix} e_1 \\ e_2 \\ e_3 \end{bmatrix} = \begin{bmatrix} \sin\phi_0\cos\lambda_0 & -\sin\lambda_0 & \cos\phi_0\cos\lambda_0 \\ \sin\phi_0\sin\lambda_0 & \cos\lambda_0 & \cos\phi_0\sin\lambda_0 \\ -\cos\phi_0 & 0 & \sin\phi_0 \end{bmatrix} \begin{bmatrix} e_{1^0} \\ e_{2^0} \\ e_{3^0} \end{bmatrix} \,, \tag{3.45}$$

$$e_1 = +e_{1^0}\sin\phi_0\cos\lambda_0 - e_{2^0}\sin\lambda_0 + e_{3^0}\cos\phi_0\cos\lambda_0 \,,$$

$$e_2 = +e_{1^0}\sin\phi_0\sin\lambda_0 + e_{2^0}\cos\lambda_0 + e_{3^0}\cos\phi_0\sin\lambda_0 \,, \tag{3.46}$$

$$e_3 = -e_{1^0}\cos\phi_0 + e_{3^0}\sin\phi_0 \,,$$

$$x = e_1 x + e_2 y + e_3 z =$$

$$= e_{1^0}\big(x\sin\phi_0\cos\lambda_0 + y\sin\phi_0\sin\lambda_0 - z\cos\phi_0\big)+$$

$$+ e_{2^0}\big(-x\sin\lambda_0 + y\cos\lambda_0\big) + e_{3^0}\big(x\cos\phi_0\cos\lambda_0 + y\cos\phi_0\sin\lambda_0 + z\sin\phi_0\big) =$$

$$= r e_{1^0}\big(\cos\phi\cos\lambda\sin\phi_0\cos\lambda_0 + \cos\phi\sin\lambda\sin\phi_0\sin\lambda_0 - \sin\phi\cos\phi_0\big)+ \tag{3.47}$$

$$+ r e_{2^0}\big(-\cos\phi\cos\lambda\sin\lambda_0 + \cos\phi\sin\lambda\cos\lambda_0\big)+$$

$$+ r e_{3^0}\big(\cos\phi\cos\lambda\cos\phi_0\cos\lambda_0 + \cos\phi\sin\lambda\cos\phi_0\sin\lambda_0 + \sin\phi\sin\phi_0\big)$$

$$= r e_{1^0}\cos\beta\cos\alpha + r e_{2^0}\cos\beta\sin\alpha + r e_{3^0}\sin\beta \,.$$

Box 3.5 (From the equatorial frame of reference $\{e_1, e_2, e_3 \,|\, \mathcal{O}\}$ to the meta-equatorial (oblique) frame of reference $\{e_{1^0}, e_{2^0}, e_{3^0} \,|\, \mathcal{O}\}$: the direct transformation).

(i) The first identity (spherical sine lemma):

$$x^0 = r\cos\beta\cos\alpha \Rightarrow \cos\alpha = \frac{x^0}{r\cos\beta} \,,$$

$$\cos\alpha = \frac{1}{\cos\beta}\big(\cos\phi\cos\lambda\sin\phi_0\cos\lambda_0 + \cos\phi\sin\lambda\sin\phi_0\sin\lambda_0 - \sin\phi\cos\phi_0\big)\,, \tag{3.48}$$

$$\cos\alpha = \frac{1}{\cos\beta}\big(\cos\phi\sin\phi_0\cos(\lambda-\lambda_0) - \sin\phi\cos\phi_0\big)\,.$$

(ii) The second identity (spherical sine-cosine lemma):

$$y^0 = r\cos\beta\sin\alpha \Rightarrow \sin\alpha = \frac{y^0}{r\cos\beta} \,,$$

$$\sin\alpha = \frac{1}{\cos\beta}\big(-\cos\phi\cos\lambda\sin\lambda_0 + \cos\phi\sin\lambda\cos\lambda_0\big)\,, \tag{3.49}$$

$$\sin\alpha = \frac{1}{\cos\beta}\cos\phi\sin(\lambda-\lambda_0)\,.$$

(iii) The third identity (spherical side cosine lemma):

$$z^0 = r\sin\beta = r\cos\psi \Rightarrow \sin\beta = \cos\psi = \frac{z^0}{r} \,,$$

$$\sin\beta = \sin\psi = \cos\phi\cos\phi_0\cos\lambda\cos\lambda_0 + \cos\phi\cos\phi_0\sin\lambda\sin\lambda_0 + \sin\phi\sin\phi_0 \,, \tag{3.50}$$

$$\sin\beta = \sin\psi = \cos\phi\cos\phi_0\cos(\lambda-\lambda_0) + \sin\phi\sin\phi_0 \,.$$

Box 3.6 (The forward problem of transforming spherical frames of reference. Input variables: $\lambda, \phi, \lambda_0, \phi_0$. Output variables: α, β).

(i) The first and second identities:

$$\tan \alpha = \frac{\cos \phi \sin(\lambda - \lambda_0)}{\cos \phi \sin \phi_0 \cos(\lambda - \lambda_0) - \sin \phi \cos \phi_0} \; , \quad \tan \alpha = \frac{\sin(\lambda - \lambda_0)}{\sin \phi_0 \cos(\lambda - \lambda_0) - \tan \phi \cos \phi_0} \; . \quad (3.51)$$

Alternatives:

$$\sin \alpha = \frac{1}{\cos \beta} \cos \phi \sin(\lambda - \lambda_0) \; ,$$

$$\cos \alpha = \frac{1}{\cos \beta} \big(\cos \phi \sin \phi_0 \cos(\lambda - \lambda_0) - \sin \phi \cos \phi_0 \big) \; . \quad (3.52)$$

(ii) The third identity:

$$\sin \beta = \sin \psi = \cos \phi \cos \phi_0 \cos(\lambda - \lambda_0) + \sin \phi \sin \phi_0 \; . \quad (3.53)$$

Box 3.7 (Representation of a placement vector $\boldsymbol{x} \in \mathbb{S}_r^2$ in both the equatorial frame of reference $\{\boldsymbol{e}_1, \boldsymbol{e}_2, \boldsymbol{e}_3 \,|\, \mathcal{O}\}$ and the meta-equatorial (oblique) frame of reference $\{\boldsymbol{e}_{1^0}, \boldsymbol{e}_{2^0}, \boldsymbol{e}_{3^0} \,|\, \mathcal{O}\}$).

$$\boldsymbol{e}_{1^0} = +\boldsymbol{e}_1 \sin \phi_0 \cos \lambda_0 + \boldsymbol{e}_2 \sin \phi_0 \sin \lambda_0 - \boldsymbol{e}_3 \cos \phi_0 \; ,$$

$$\boldsymbol{e}_{2^0} = -\boldsymbol{e}_1 \sin \lambda_0 + \boldsymbol{e}_2 \cos \lambda_0 \; , \quad (3.54)$$

$$\boldsymbol{e}_{3^0} = +\boldsymbol{e}_1 \cos \phi_0 \cos \lambda_0 + \boldsymbol{e}_2 \cos \phi_0 \sin \lambda_0 + \boldsymbol{e}_3 \sin \phi_0 \; ,$$

$$\boldsymbol{x} = \boldsymbol{e}_1 x + \boldsymbol{e}_2 y + \boldsymbol{e}_3 z = \boldsymbol{e}_{1^0} x^0 + \boldsymbol{e}_{2^0} y^0 + \boldsymbol{e}_{3^0} z^0 =$$

$$= \boldsymbol{e}_1 \big(x^0 \sin \phi_0 \cos \lambda_0 - y^0 \sin \lambda_0 + z^0 \cos \phi_0 \cos \lambda_0 \big) +$$

$$+ \boldsymbol{e}_2 \big(x^0 \sin \phi_0 \sin \lambda_0 + y^0 \cos \lambda_0 + z^0 \cos \phi_0 \sin \lambda_0 \big) + \boldsymbol{e}_3 \big(-x^0 \cos \phi_0 + z^0 \sin \phi_0 \big) =$$

$$= r \boldsymbol{e}_1 \cos \phi \cos \lambda + r \boldsymbol{e}_2 \cos \phi \sin \lambda + r \boldsymbol{e}_3 \sin \phi = \quad (3.55)$$

$$= r \boldsymbol{e}_1 \big(\cos \beta \cos \alpha \sin \phi_0 \cos \lambda_0 - \cos \beta \sin \alpha \sin \lambda_0 + \sin \beta \cos \phi_0 \cos \lambda_0 \big) +$$

$$+ r \boldsymbol{e}_2 \big(\cos \beta \cos \alpha \sin \phi_0 \sin \lambda_0 + \cos \beta \sin \alpha \cos \lambda_0 + \sin \beta \cos \phi_0 \sin \lambda_0 \big) +$$

$$+ r \boldsymbol{e}_3 \big(-\cos \beta \cos \alpha \cos \phi_0 + \sin \beta \sin \phi_0 \big) \; .$$

Box 3.8 (From the meta-equatorial (oblique) frame of reference $\{\boldsymbol{e}_{1^0}, \boldsymbol{e}_{2^0}, \boldsymbol{e}_{3^0} \,|\, \mathcal{O}\}$ to the equatorial frame of reference $\{\boldsymbol{e}_1, \boldsymbol{e}_2, \boldsymbol{e}_3 \,|\, \mathcal{O}\}$: the inverse transformation).

(i) The first identity:

$$x = r \cos \phi \cos \lambda \Rightarrow \cos \lambda = \frac{x}{r \cos \phi} \; ,$$

$$\cos \lambda = \frac{1}{\cos \phi} \big(\cos \beta \cos \alpha \sin \phi_0 \cos \lambda_0 - \cos \beta \sin \alpha \sin \lambda_0 + \sin \beta \cos \phi_0 \cos \lambda_0 \big) \; . \quad (3.56)$$

(ii) The second identity:

$$y = r \cos \phi \sin \lambda \Rightarrow \sin \lambda = \frac{y}{r \cos \phi} \; ,$$

$$\sin \lambda = \frac{1}{\cos \phi} \big(\cos \beta \cos \alpha \sin \phi_0 \sin \lambda_0 + \cos \beta \sin \alpha \cos \lambda_0 + \sin \beta \cos \phi_0 \sin \lambda_0 \big) \; . \quad (3.57)$$

(iii) The third identity:

$$z = r \sin \phi \Rightarrow \sin \phi = \frac{z}{r} \; ,$$

$$\sin \phi = -\cos \beta \cos \alpha \cos \phi_0 + \sin \beta \sin \phi_0 \; . \quad (3.58)$$

Box 3.9 (The backward problem of transforming spherical frames of reference. Input variables: $\alpha, \beta, \lambda_0, \phi_0$. Output variables: λ, ϕ).

(i) The first and second identities:

$$\tan\lambda = \frac{\cos\beta\cos\alpha\sin\phi_0\sin\lambda_0 + \cos\beta\sin\alpha\cos\lambda_0 + \sin\beta\cos\phi_0\sin\lambda_0}{\cos\beta\cos\alpha\sin\phi_0\cos\lambda_0 - \cos\beta\sin\alpha\sin\lambda_0 + \sin\beta\cos\phi_0\cos\lambda_0} \, ,$$

$$\tan(\lambda - \lambda_0) = \frac{\tan\lambda - \tan\lambda_0}{1 + \tan\lambda\tan\lambda_0} \tag{3.59}$$

$$\Rightarrow$$

$$\tan(\lambda - \lambda_0) = \frac{\sin\alpha}{\tan\beta\cos\phi_0 + \cos\alpha\sin\phi_0} \, ,$$

$$\cos\lambda = \frac{1}{\cos\phi}\Big(\cos\beta\cos\alpha\cos\phi_0\sin\lambda_0 - \cos\beta\sin\alpha\sin\lambda_0 + \sin\beta\cos\phi_0\cos\lambda_0\Big) \, ,$$

$$\sin\lambda = \frac{1}{\cos\phi}\Big(\cos\beta\cos\alpha\sin\phi_0\sin\lambda_0 + \cos\beta\sin\alpha\cos\lambda_0 + \sin\beta\cos\phi_0\sin\lambda_0\Big) \, . \tag{3.60}$$

(ii) The third identity:

$$\sin\phi = -\cos\beta\cos\alpha\cos\phi_0 + \sin\beta\sin\phi_0 \, . \tag{3.61}$$

3-32 A second design of an oblique frame of reference of the sphere

The second design of an oblique frame of reference of the sphere $\mathbb{S}_R^2$ is taking reference to the following design aspects. (i) We intersect the sphere $\mathbb{S}_R^2$ by a central plane $\mathbb{P}_{\mathcal{O}}^2$, generating the oblique circular equator, also called *meta-equator*. We attach the oblique orthonormal frame of reference $\{\boldsymbol{E}_{1'}, \boldsymbol{E}_{2'}, \boldsymbol{E}_{3'} \,|\, \mathcal{O}\}$ at the origin $\mathcal{O}$ to the central plane $\mathbb{P}_{\mathcal{O}}^2$ such that $\{\boldsymbol{E}_{1'}, \boldsymbol{E}_{2'} \,|\, \mathcal{O}\}$ span the central plane as well as $\boldsymbol{E}_{3'}$, its unit normal vector. We connect the conventional equatorial frame of reference $\{\boldsymbol{E}_{1'}, \boldsymbol{E}_{2'}, \boldsymbol{E}_{3'} \,|\, \mathcal{O}\}$ by means of the Kepler elements Ω and I, also called *right ascension* of the *ascending node* Ω and *inclination* I. These Kepler elements constitute the Euler rotation matrix $\mathsf{R}(\Omega, I) := \mathsf{R}_1(I)\mathsf{R}_3(\Omega)$. (ii) Finally, based upon such a connection, the coordinates of the placement vector $\boldsymbol{x}$ have to be represented both in the conventional equatorial frame of reference and in the oblique equatorial frame of reference. (iii) The forward equations as well as the backward equations of transformation between them have to be derived.

Solution (the first, the second, and the third problem).

The three problems, in particular, can be solved (i) by representing the placement vector $\boldsymbol{X}(\Lambda, \Phi, R)$ in terms of the chosen spherical coordinates $\{\Lambda, \Phi, R\}$ with respect to the equatorial frame of reference $\{\boldsymbol{E}_1, \boldsymbol{E}_2, \boldsymbol{E}_3 \,|\, \mathcal{O}\}$, (ii) by transforming to an oblique frame of reference $\{\boldsymbol{E}_{1'}, \boldsymbol{E}_{2'}, \boldsymbol{E}_{3'} \,|\, \mathcal{O}\}$ by means of the Kepler elements (special Cardan angles) Ω and I, called *longitude* Ω of the *ascending node* and *inclination* I, and (iii) by introducing oblique spherical coordinates $\{A, B\}$, called *meta-longitude* A and *meta-latitude* B. The complete program is outlined in Boxes 3.10–3.15. In particular, we transform from the original equatorial frame of reference $\{\boldsymbol{E}_1, \boldsymbol{E}_2, \boldsymbol{E}_3 \,|\, \mathcal{O}\}$ to the oblique frame of reference, called *meta-equatorial*, by $\mathsf{R}_1(I)\mathsf{R}_3(\Omega)$. Indeed, we perform a first rotation by the Cardan angle Ω around the 3 axis (Z axis) and a second rotation by the Cardan angle I around the 1 axis (X' axis) in order to generate $[\boldsymbol{E}_{1'}, \boldsymbol{E}_{2'}, \boldsymbol{E}_{3'}]^* = \mathsf{R}_1(I)\mathsf{R}_3(\Omega)[\boldsymbol{E}_1, \boldsymbol{E}_2, \boldsymbol{E}_3]^*$. Such a procedure can be interpreted as follows: intersect $\mathbb{S}_R^2$ by a centric plane $\mathbb{P}_{\mathcal{O}}^2$ to produce a circular meta-equator, which is oriented by Ω and I. For geometrical details, consult Fig. 3.8.

End of Solution (the first, the second, and the third problem).

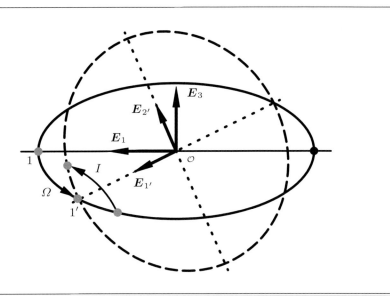

Fig. 3.8. The oblique plane $\mathbb{P}_{\mathcal{O}}^2$ intersecting the sphere $\mathbb{S}_R^2$, circular meta-equator.

Box 3.10 (Establishing an oblique frame of reference (meta-equatorial) of the sphere).

(i) The placement vector $\boldsymbol{X}$ represented in the conventional
as well as in the oblique frame of reference:

$$\boldsymbol{X}(\Lambda, \Phi, R) = \boldsymbol{E}_1 R \cos \Phi \cos \Lambda + \boldsymbol{E}_2 R \cos \Phi \sin \Lambda + \boldsymbol{E}_3 R \sin \Phi =$$

$$= \boldsymbol{E}_{1'} R \cos B \cos A + \boldsymbol{E}_{2'} R \cos B \sin A + \boldsymbol{E}_{3'} R \sin B = \boldsymbol{x}(A, B, R) \,.$$

(3.62)

(ii) The transformation of the frames of reference:

$$\begin{bmatrix} \boldsymbol{E}_{1'} \\ \boldsymbol{E}_{2'} \\ \boldsymbol{E}_{3'} \end{bmatrix} = \mathsf{R}_1(I)\mathsf{R}_3(\Omega) \begin{bmatrix} \boldsymbol{E}_1 \\ \boldsymbol{E}_2 \\ \boldsymbol{E}_3 \end{bmatrix} ,$$

$$\mathsf{R}_1(I)\mathsf{R}_3(\Omega) =$$

$$= \begin{bmatrix} \cos \Omega & \sin \Omega & 0 \\ -\sin \Omega \cos I & \cos \Omega \cos I & \sin I \\ \sin \Omega \sin I & -\cos \Omega \sin I & \cos I \end{bmatrix}$$

(3.63)

versus

$$\begin{bmatrix} \boldsymbol{E}_1 \\ \boldsymbol{E}_2 \\ \boldsymbol{E}_3 \end{bmatrix} = \mathsf{R}_3^*(\Omega)\mathsf{R}_1^*(I) \begin{bmatrix} \boldsymbol{E}_{1'} \\ \boldsymbol{E}_{2'} \\ \boldsymbol{E}_{3'} \end{bmatrix} ,$$

$$\mathsf{R}_3^*(\Omega)\mathsf{R}_1^*(I) =$$

$$= \begin{bmatrix} \cos \Omega & -\sin \Omega \cos I & \sin \Omega \sin I \\ \sin \Omega & \cos \Omega \cos I & -\cos \Omega \sin I \\ 0 & \sin I & \cos I \end{bmatrix} \,.$$

(3.64)

Box 3.11 (Representation of a placement vector $\boldsymbol{X} \in \mathbb{S}_R^2$ in both the equatorial frame of reference $\{\boldsymbol{E}_1, \boldsymbol{E}_2, \boldsymbol{E}_3 \,|\, \mathcal{O}\}$ and the meta-equatorial (oblique) frame of reference $\{\boldsymbol{E}_{1'}, \boldsymbol{E}_{2'}, \boldsymbol{E}_{3'} \,|\, \mathcal{O}\}$).

(i) The Cartesian representation:

$$\boldsymbol{E}_1 X + \boldsymbol{E}_2 Y + \boldsymbol{E}_3 Z = \boldsymbol{E}_{1'} X' + \boldsymbol{E}_{2'} Y' + \boldsymbol{E}_{3'} Z' \; . \tag{3.65}$$

(ii) The curvilinear representation $(\{\boldsymbol{E}_{1'}, \boldsymbol{E}_{2'}, \boldsymbol{E}_{3'} \,|\, \mathcal{O}\} \to \{\boldsymbol{E}_1, \boldsymbol{E}_2, \boldsymbol{E}_3 \,|\, \mathcal{O}\})$:

$$
\begin{aligned}
\boldsymbol{E}_1 &= \boldsymbol{E}_{1'} \cos \Omega - \boldsymbol{E}_{2'} \sin \Omega \cos I + \boldsymbol{E}_{3'} \sin \Omega \sin I \; , \\
\boldsymbol{E}_2 &= \boldsymbol{E}_{1'} \sin \Omega + \boldsymbol{E}_{2'} \cos \Omega \cos I - \boldsymbol{E}_{3'} \cos \Omega \sin I \; , \\
\boldsymbol{E}_3 &= \boldsymbol{E}_{2'} \sin I + \boldsymbol{E}_{3'} \cos I \; ,
\end{aligned}
\tag{3.66}
$$

$$
\begin{aligned}
\boldsymbol{X} &= \\
&= \boldsymbol{E}_1 X + \boldsymbol{E}_2 Y + \boldsymbol{E}_3 Z = \\
&= \boldsymbol{E}_{1'}\big(X \cos \Omega + Y \sin \Omega\big) + \\
&\quad + \boldsymbol{E}_{2'}\big(-X \sin \Omega \cos I + Y \cos \Omega \cos I + Z \sin I\big) + \\
&\quad + \boldsymbol{E}_{3'}\big(X \sin \Omega \sin I - Y \cos \Omega \sin I + Z \cos I\big) \; ,
\end{aligned}
\tag{3.67}
$$

$$
\begin{aligned}
\boldsymbol{X} &= \\
&= R\boldsymbol{E}_{1'}\big(\cos \Phi \cos \Lambda \cos \Omega + \cos \Phi \sin \Lambda \sin \Omega\big) + \\
&\quad + R\boldsymbol{E}_{2'}\big(-\cos \Phi \cos \Lambda \sin \Omega \cos I + \cos \Phi \sin \Lambda \cos \Omega \cos I + \sin \Phi \sin I\big) + \\
&\quad + R\boldsymbol{E}_{3'}\big(\cos \Phi \cos \Lambda \sin \Omega \sin I - \cos \Phi \sin \Lambda \cos \Omega \sin I + \sin \Phi \cos I\big) = \\
&= R\boldsymbol{E}_{1'} \cos B \cos A + R\boldsymbol{E}_{2'} \cos B \sin A + R\boldsymbol{E}_{3'} \sin B \; .
\end{aligned}
\tag{3.68}
$$

(iii) The curvilinear representation $(\{\boldsymbol{E}_1, \boldsymbol{E}_2, \boldsymbol{E}_3 \,|\, \mathcal{O}\} \to \{\boldsymbol{E}_{1'}, \boldsymbol{E}_{2'}, \boldsymbol{E}_{3'} \,|\, \mathcal{O}\})$:

$$
\begin{aligned}
\boldsymbol{E}_{1'} &= +\boldsymbol{E}_1 \cos \Omega + \boldsymbol{E}_2 \sin \Omega \; , \\
\boldsymbol{E}_{2'} &= -\boldsymbol{E}_1 \sin \Omega \cos I + \boldsymbol{E}_2 \cos \Omega \cos I + \boldsymbol{E}_3 \sin I \; , \\
\boldsymbol{E}_{3'} &= +\boldsymbol{E}_1 \sin \Omega \sin I - \boldsymbol{E}_2 \cos \Omega \sin I + \boldsymbol{E}_3 \cos I \; ,
\end{aligned}
\tag{3.69}
$$

$$
\begin{aligned}
\boldsymbol{X} &= \\
&= \boldsymbol{E}_{1'} X' + \boldsymbol{E}_{2'} Y' + \boldsymbol{E}_{3'} Z' = \\
&= \boldsymbol{E}_1\big(X' \cos \Omega - Y' \sin \Omega \cos I + Z' \sin \Omega \sin I\big) + \\
&\quad + \boldsymbol{E}_2\big(X' \sin \Omega + Y' \cos \Omega \cos I - Z' \cos \Omega \sin I\big) + \\
&\quad + \boldsymbol{E}_3\big(Y' \sin I + Z' \cos I\big) \; ,
\end{aligned}
\tag{3.70}
$$

$$
\begin{aligned}
\boldsymbol{X} &= \\
&= R\boldsymbol{E}_1\big(\cos B \cos A \cos \Omega - \cos B \sin A \sin \Omega \cos I + \sin B \sin \Omega \sin I\big) + \\
&\quad + R\boldsymbol{E}_2\big(\cos B \cos A \sin \Omega + \cos B \sin A \cos \Omega \cos I - \sin B \cos \Omega \sin I\big) + \\
&\quad + R\boldsymbol{E}_3\big(\cos B \sin A \sin I + \sin B \cos I\big) = \\
&= R\boldsymbol{E}_1 \cos \Phi \cos \Lambda + R\boldsymbol{E}_2 \cos \Phi \sin \Lambda + R\boldsymbol{E}_3 \sin \Phi \; .
\end{aligned}
\tag{3.71}
$$

Box 3.12 (From the equatorial frame of reference $\{E_1, E_2, E_3 \mid \mathcal{O}\}$ to the meta-equatorial (oblique) frame of reference $\{E_{1'}, E_{2'}, E_{3'} \mid \mathcal{O}\}$: the direct transformation).

(i) The first identity:

$$X' = R \cos B \cos A$$

$$\Rightarrow$$

$$\cos A = \frac{X'}{R \cos B},$$

$$\cos A = \frac{1}{\cos B}\left(\cos \Phi \cos \Lambda \cos \Omega + \cos \Phi \sin \Lambda \sin \Omega\right),$$

$$\cos A = \frac{\cos \Phi}{\cos B} \cos(\Lambda - \Omega).$$

(3.72)

(ii) The second identity:

$$Y' = R \cos B \sin A$$

$$\Rightarrow$$

$$\sin A = \frac{Y'}{R \cos B},$$

$$\sin A = \frac{1}{\cos B}\left(-\cos \Phi \cos \Lambda \sin \Omega \cos I + \cos \Phi \sin \Lambda \cos \Omega \cos I + \sin \Phi \sin I\right),$$

$$\sin A = \frac{1}{\cos B}\left(\cos \Phi \cos I \sin(\Lambda - \Omega) + \sin \Phi \sin I\right).$$

(3.73)

(iii) The third identity:

$$Z' = R \sin B$$

$$\Rightarrow$$

$$\sin B = \frac{Z'}{R},$$

$$\sin B = \cos \Phi \cos \Lambda \sin \Omega \sin I - \cos \Phi \sin \Lambda \cos \Omega \sin I + \sin \Phi \cos I,$$

$$\sin B = -\cos \Phi \sin I \sin(\Lambda - \Omega) + \sin \Phi \cos I.$$

(3.74)

Box 3.13 (The forward problem of transforming spherical frames of reference. Input variables: Λ, Φ, Ω, I. Output variables: A, B).

(i) The first and second identities:

$$\tan A = \frac{\cos I \sin(\Lambda - \Omega) + \tan \Phi \sin I}{\cos(\Lambda - \Omega)}.$$

(3.75)

Alternatives:

$$\sin A = \frac{\cos \Phi}{\cos B}\left(\cos I \sin(\Lambda - \Omega) + \tan \Phi \sin I\right),$$

$$\cos A = \frac{\cos \Phi}{\cos B} \cos(\Lambda - \Omega).$$

(3.76)

(ii) The third identity:

$$\sin B = -\cos \Phi \sin I \sin(\Lambda - \Omega) + \sin \Phi \cos I.$$

(3.77)

Box 3.14 (From the meta-equatorial (oblique) frame of reference $\{E_{1'}, E_{2'}, E_{3'} \mid \mathcal{O}\}$ to the equatorial frame of reference $\{E_1, E_2, E_3 \mid \mathcal{O}\}$: the inverse transformation).

(i) The first identity:

$$X = R\cos\Phi\cos\Lambda \Rightarrow \cos\Lambda = \frac{X}{R\cos\Phi} ,$$

$$\cos\Lambda = \frac{1}{\cos\Phi}\left(\cos B\cos A\cos\Omega - \cos B\sin A\sin\Omega\cos I + \sin B\sin\Omega\sin I\right) . \tag{3.78}$$

(ii) The second identity:

$$Y = R\cos\Phi\sin\Lambda \Rightarrow \sin\Lambda = \frac{Y}{R\cos\Phi} ,$$

$$\sin\Lambda = \frac{1}{\cos\Phi}\left(\cos B\cos A\sin\Omega + \cos B\sin A\cos\Omega\cos I - \sin B\cos\Omega\sin I\right) . \tag{3.79}$$

(iii) The third identity:

$$Z = R\sin\Phi \Rightarrow \sin\Phi = \frac{Z}{R} ,$$

$$\sin\Phi = \cos B\sin A\sin I + \sin B\cos I . \tag{3.80}$$

Box 3.15 (The backward problem of transforming spherical frames of reference. Input variables: A, B, Ω, I. Output variables: Λ, Φ).

(i) The first and second identities:

$$\tan\Lambda = \frac{\cos B\cos A\sin\Omega + \cos B\sin A\cos\Omega\cos I - \sin B\cos\Omega\sin I}{\cos B\cos A\cos\Omega - \cos B\sin A\sin\Omega\cos I + \sin B\sin\Omega\sin I} . \tag{3.81}$$

(ii) The third identity:

$$\sin\Phi = \cos B\sin A\sin I + \sin B\cos I . \tag{3.82}$$

3-33 The transverse frame of reference of the sphere: part one

In the framework of the *transverse aspect*, we are aiming at establishing a special oblique frame of reference by the inclination $I = 90°$. We have to deal with two problems depending on the input data.

Forward problem.

Input: Λ, Φ and $\Omega, I = \pi/2$.

Output: A, B.

Backward problem.

Input: A, B and $\Omega, I = \pi/2$.

Output: Λ, Φ.

Within the forward problem, which is solved in Box 3.16, we depart from (i) given spherical longitude Λ and spherical latitude Φ of a point in the sphere $\mathbb{S}_R^2$ and from (ii) given longitude of the ascending node Ω and inclination $I = \pi/2$ of the meta-equatorial plane in order to derive (iii) meta-longitude A and meta-latitude B of the homologous point in the sphere. Conversely, for solving the backward problem, which is outlined in Box 3.17, we inject (i) meta-longitude A and meta-latitude B of a point in the sphere $\mathbb{S}_R^2$ and (ii) longitude of the ascending node Ω and inclination $I = \pi/2$ of the meta-equatorial plane in order to derive spherical longitude Λ and spherical latitude Φ of the homologous point in the sphere. Consult Fig. 3.9, which is an illustration of the transverse aspect of the sphere.

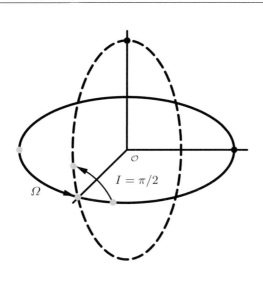

Fig. 3.9. The transverse aspect of the sphere.

Box 3.16 (The forward problem of transforming spherical frames of reference: the transverse aspect. Input variables: $\Lambda, \Phi, \Omega, I = \pi/2$. Output variables: A, B).

(i) Meta-longitude:

$$\tan A = \frac{\tan \Phi}{\cos(\Lambda - \Omega)} \; ,$$

$$\cos A = \frac{\cos \Phi}{\cos B} \cos(\Lambda - \Omega) =$$

$$= \frac{\cos \Phi \cos(\Lambda - \Omega)}{\sqrt{1 + \cos \Phi \sin(\Lambda - \Omega)} \sqrt{1 - \cos \Phi \sin(\Lambda - \Omega)}} \; , \tag{3.83}$$

$$\sin A = \frac{\sin \Phi}{\cos B} =$$

$$= \frac{\sin \Phi}{\sqrt{1 + \cos \Phi \sin(\Lambda - \Omega)} \sqrt{1 - \cos \Phi \sin(\Lambda - \Omega)}} \; .$$

(ii) Meta-latitude:

$$\sin B = -\cos \Phi \sin(\Lambda - \Omega) \; . \tag{3.84}$$

(iii) Substitutions:

$$\cos A = \frac{1}{\sqrt{1 + \tan^2 A}} \; , \quad \sin A = \frac{\tan A}{\sqrt{1 + \tan^2 A}} \; ,$$

$$\cos A = \frac{\cos(\Lambda - \Omega)}{\sqrt{\cos^2(\Lambda - \Omega) + \tan^2 \Phi}} \; , \quad \sin A = \frac{\tan \Phi}{\sqrt{\cos^2(\Lambda - \Omega) + \tan^2 \Phi}} \; , \tag{3.85}$$

$$\frac{1}{\cos B} = \frac{1}{\sqrt{1 - \cos^2 \Phi \sin^2(\Lambda - \Omega)}} = \frac{1}{\sqrt{1 - \cos \Phi \sin(\Lambda - \Omega)} \sqrt{1 - \cos \Phi \sin(\Lambda - \Omega)}} \; .$$

Box 3.17 (The backward problem of transforming spherical frames of reference: the transverse aspect. Input variables: $A, B, \Omega, I = \pi/2$. Output variables: Λ, Φ).

(i) Longitude:

$$\tan(\Lambda - \Omega) = -\frac{\tan B}{\cos A} \,. \tag{3.86}$$

(ii) Latitude:

$$\sin \Phi = \cos B \sin A \,. \tag{3.87}$$

(iii) Substitutions:

$$\sin(\Lambda - \Omega) = -\frac{\sin B}{\cos \Phi} \,,$$

$$\cos(\Lambda - \Omega) = +\frac{\tan \Phi}{\tan A} \,, \tag{3.88}$$

$$\tan(\Lambda - \Omega) = -\frac{\sin B}{\sin \Phi}\frac{\sin A}{\cos A} = -\frac{\tan B}{\cos A} \,.$$

3-34 The transverse frame of reference of the sphere: part two

The transverse case is a special case of an oblique frame of reference. Since it has gained great interest in map projections, we devote another special section to the transverse aspect. In short, for such a peculiar aspect, the meta-North Pole is chosen to be located in the conventional equator of the reference sphere $\mathbb{S}_r^2$. In short, the spherical latitude $\phi_0 = 0°$ of the meta-North Pole is fixed to zero. Accordingly, we specialize the forward and backward transformation formulae according to Boxes 3.6 and 3.9 to such a special pole configuration. In order to understand better the conventional choice of the transverse frame of reference, we additionally consider the following example.

Example 3.7 (On the transverse frame of refererence).

Let us choose $\lambda_0 = 270°$ and $\phi_0 = 0°$, namely a placement of the meta-North Pole in the West Pole. For such a configuration, the meta-equator (then called *transverse equator*) agrees with the Greenwich meridian of reference. $e_{1^0} = e_S$ is directed to the South, $e_{2^0} = e_E$ is directed to the East. Such an oblique frame of reference has not found the support of traditional map projectors. They prefer a transverse frame of reference $\{e_{1^*}, e_{2^*}, e_{3^*} | \mathcal{O}\}$, namely a right-handed orthonormal frame of reference that is oriented "East, North, Vertical" and relates to $\{e_{1^0}, e_{2^0}, e_{3^0} | \mathcal{O}\}$ by

$$e_{1^*} = +e_{2^0} = e_E \quad (\text{"Easting"}),$$

$$e_{2^*} = -e_{1^0} = e_S \quad (\text{"Northing"}), \tag{3.89}$$

$$e_{3^*} = +e_{3^0} = e_V \quad (\text{"Vertical"}).$$

In terms of this example, we may alternatively choose $\lambda^0 = \lambda_0 - 270° = 90° + \lambda_0$ or $\lambda_0 = 270° + \lambda^0$, and $\alpha = \alpha_S = 90° + \alpha_E$ or $\alpha_E = \alpha_S - 90° = 270° + \alpha_S$, such that $\{\lambda^0 = 0°, \phi_0 = 0°\}$ identifies the Greenwich meridian of reference. Indeed, we have shifted both λ_0 and α_E for $3\pi/2 \sim 270°$.

End of Example.

Note that the direct transformation $\{\lambda, \phi; \lambda^0, \phi_0\} \to \{\alpha_E, \beta\}$ can then be conveniently solved, namely with the result that is presented in Box 3.6. First, for the choice $\phi_0 = 0$, $\sin(\lambda - \lambda_0) = \cos(\lambda - \lambda^0)$, $\cos(\lambda - \lambda_0) = -\sin(\lambda - \lambda^0)$, and $\tan\alpha = \tan\alpha_S = -1/\tan\alpha_E = -\cot\alpha_E$, we have derived the transverse equations of reference. Second, if we substitute these identities into the representative formulae "from the equatorial frame of reference $\{e_1, e_2, e_3 | \mathcal{O}\}$ to the meta-equatorial ("oblique") frame of reference $\{e_{1^0}, e_{2^0}, e_{3^0} | \mathcal{O}\}$" that are collected in Box 3.5, we are directly led to the basic identities of transforming from the equatorial reference frame to the transverse reference frame.

3-35 Transformations between oblique frames of reference: first design, second design

Question: "How can the two oblique frames of reference e^0 and E', called *first design* and *second design*, respectively, be related?" Answer: "The two oblique frames of reference can be related to each other when we allow a third rotation in the Kepler orbital plane by means of $[\boldsymbol{E}_{1''}, \boldsymbol{E}_{2''}, \boldsymbol{E}_{3''}]^* = \mathsf{R}_3(\omega)[\boldsymbol{E}_{1'}, \boldsymbol{E}_{2'}, \boldsymbol{E}_{3'}]^*$."

Note that ω is called *longitude in the meta-equatorial plane*. Such an additional rotation may come as a surprise, but without such a longitude, the oblique frames of first and second kind cannot be identified without inconsistencies. Sometimes, the angular parameter ω is called *ambiguity*. As it is outlined in Box 3.18 and Box 3.19, the *identity postulate* (3.90) leads to trigonometric equations for $\{\omega, I, \Omega\}$ (given λ_0 and ϕ_0) and $\{\lambda_0, \phi_0\}$ (given ω, I, and Ω). Accordingly, we are able to transform forward and backward between the oblique frames of reference subject to $[\boldsymbol{E}_{1''}, \boldsymbol{E}_{2''}, \boldsymbol{E}_{3''}]^* = [e_{1^0}, e_{2^0}, e_{3^0}]^*$, $[\boldsymbol{E}_1, \boldsymbol{E}_2, \boldsymbol{E}_3]^* = [e_1, e_2, e_3]^*$. Compare with Fig. 3.10, which illustrates the commutative diagram for oblique frames of reference. The essential formulae for transforming $E'' \to e^0$ as well as $e^0 \to E''$ are collected in Lemma 3.4, Lemma 3.5, and Corollary 3.6. These transformation formulae are summarized and numerically tested in the following section.

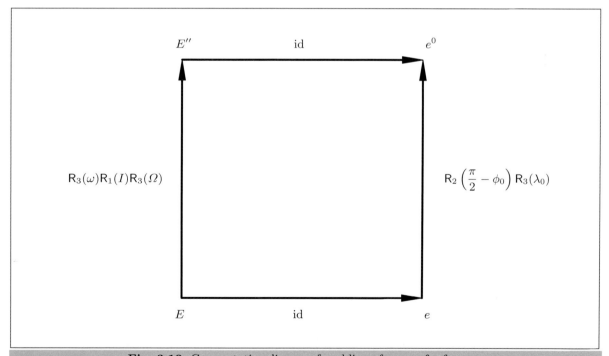

Fig. 3.10. Commutative diagram for oblique frames of reference.

$$\begin{bmatrix} \boldsymbol{E}_{1''} \\ \boldsymbol{E}_{2''} \\ \boldsymbol{E}_{3''} \end{bmatrix} =$$

$$= \mathsf{R}_3(\omega)\mathsf{R}_1(I)\mathsf{R}_3(\Omega) \begin{bmatrix} \boldsymbol{E}_1 \\ \boldsymbol{E}_2 \\ \boldsymbol{E}_3 \end{bmatrix} = \mathsf{R}_2\left(\tfrac{\pi}{2} - \phi_0\right)\mathsf{R}_3(\lambda_0) \begin{bmatrix} \boldsymbol{e}_1 \\ \boldsymbol{e}_2 \\ \boldsymbol{e}_3 \end{bmatrix} = \qquad (3.90)$$

$$= \begin{bmatrix} \boldsymbol{e}_{1^0} \\ \boldsymbol{e}_{2^0} \\ \boldsymbol{e}_{3^0} \end{bmatrix}$$

Box 3.18 (Transformation between oblique frames of reference: first design, second design).

(i) $E \to E''$

$$\begin{bmatrix} \boldsymbol{E}_{1''} \\ \boldsymbol{E}_{2''} \\ \boldsymbol{E}_{3''} \end{bmatrix} = \mathsf{R}_3(\omega) \begin{bmatrix} \boldsymbol{E}_{1'} \\ \boldsymbol{E}_{2'} \\ \boldsymbol{E}_{3'} \end{bmatrix} ,$$

$$\begin{bmatrix} \boldsymbol{E}_{1'} \\ \boldsymbol{E}_{2'} \\ \boldsymbol{E}_{3'} \end{bmatrix} = \mathsf{R}_1(I)\mathsf{R}_3(\Omega) \begin{bmatrix} \boldsymbol{E}_1 \\ \boldsymbol{E}_2 \\ \boldsymbol{E}_3 \end{bmatrix} , \qquad (3.91)$$

$$\begin{bmatrix} \boldsymbol{E}_{1''} \\ \boldsymbol{E}_{2''} \\ \boldsymbol{E}_{3''} \end{bmatrix} = \mathsf{R}_3(\omega)\mathsf{R}_1(I)\mathsf{R}_3(\Omega) \begin{bmatrix} \boldsymbol{E}_1 \\ \boldsymbol{E}_2 \\ \boldsymbol{E}_3 \end{bmatrix} . \qquad (3.92)$$

(ii) $e \to e^0$

$$\begin{bmatrix} \boldsymbol{e}_{1^0} \\ \boldsymbol{e}_{2^0} \\ \boldsymbol{e}_{3^0} \end{bmatrix} = \mathsf{R}_2\left(\frac{\pi}{2} - \phi_0\right)\mathsf{R}_3(\lambda_0) \begin{bmatrix} \boldsymbol{e}_1 \\ \boldsymbol{e}_2 \\ \boldsymbol{e}_3 \end{bmatrix} . \qquad (3.93)$$

(iii) $E = e$, $E'' = e^0$

$$\begin{bmatrix} \boldsymbol{E}_{1''} \\ \boldsymbol{E}_{2''} \\ \boldsymbol{E}_{3''} \end{bmatrix} = \mathsf{R}_3(\omega)\mathsf{R}_1(I)\mathsf{R}_3(\Omega) \begin{bmatrix} \boldsymbol{E}_1 \\ \boldsymbol{E}_2 \\ \boldsymbol{E}_3 \end{bmatrix} = \mathsf{R}_2\left(\frac{\pi}{2} - \phi_0\right)\mathsf{R}_3(\lambda_0) \begin{bmatrix} \boldsymbol{e}_1 \\ \boldsymbol{e}_2 \\ \boldsymbol{e}_3 \end{bmatrix} = \begin{bmatrix} \boldsymbol{e}_{1^0} \\ \boldsymbol{e}_{2^0} \\ \boldsymbol{e}_{3^0} \end{bmatrix}$$

$$\Leftrightarrow \qquad (3.94)$$

$$\mathsf{R}_3(\omega)\mathsf{R}_1(I)\mathsf{R}_3(\Omega) = \mathsf{R}_2\left(\frac{\pi}{2} - \phi_0\right)\mathsf{R}_3(\lambda_0) .$$

Box 3.19 (Individual equations of transforming oblique frames of reference).

Equation (i):

$$R_3(\omega)R_1(I)R_3(\Omega) =$$

$$= \begin{bmatrix} \cos\omega\cos\Omega - \sin\omega\sin\Omega\cos I & \cos\omega\sin\Omega + \sin\omega\cos\Omega\cos I & \sin\omega\sin I \\ -\sin\omega\cos\Omega - \cos\omega\sin\Omega\cos I & -\sin\omega\sin\Omega + \cos\omega\cos\Omega\cos I & \cos\omega\sin I \\ \sin\Omega\sin I & -\cos\Omega\sin I & \cos I \end{bmatrix} . \qquad (3.95)$$

Equation (ii):

$$R_2\left(\frac{\pi}{2} - \phi_0\right)R_3(\lambda_0) =$$

$$= \begin{bmatrix} \sin\phi_0\cos\lambda_0 & \sin\phi_0\sin\lambda_0 & -\cos\phi_0 \\ -\sin\lambda_0 & \cos\lambda_0 & 0 \\ \cos\phi_0\cos\lambda_0 & \cos\phi_0\sin\lambda_0 & \sin\phi_0 \end{bmatrix} . \qquad (3.96)$$

Equation (iii):

$$\cos\omega = 0 , \ \sin\omega = 1 \Rightarrow \omega = 90° . \qquad (3.97)$$

Equation (iv):

$$\cos\Omega = \sin\lambda_0 , \ \sin\Omega = -\cos\lambda_0 \Rightarrow \Omega = 270° + \lambda_0 , \ \lambda_0 = 90° + \Omega . \qquad (3.98)$$

Equation (v):

$$\cos I = \sin\phi_0 , \ \sin I = -\cos\phi_0 \Rightarrow I = 270° + \phi_0 , \ \phi_0 = 90° + I . \qquad (3.99)$$

Lemma 3.4 (The transformation of the first oblique frame of reference to the second oblique frame of reference: $\lambda_0, \phi_0 \rightarrow \omega, I, \Omega$).

If the first oblique frame of reference is given by defining a meta-North Pole $\{\lambda_0, \phi_0\}$, then the second oblique frame of reference is determined by the orbital Kepler elements $\omega = 90°$, $I = 270° + \phi_0$, and $\Omega = 270° + \lambda_0$.

End of Lemma.

Lemma 3.5 (The transformation of the second oblique frame of reference to the first oblique frame of reference: $\omega, I, \Omega \rightarrow \lambda_0, \phi_0$).

If the second oblique frame of reference is given by defining the orbital Kepler elements $\{\omega, I, \Omega\}$, then the first oblique frame of reference is determined by the meta-North Pole $\lambda_0 = 90° + \Omega$ and $\phi_0 = 90° + I$, subject to $\omega = 90°$.

End of Lemma.

For the transverse frame of reference, the inclination of the ascending node I is chosen ninety degrees, i.e. $I = 90°$. Accordingly, the transformation of reference frames leads us to Corollary 3.6.

Corollary 3.6 (Transformation of reference frames, transverse aspect, $I = 90°$).

If the second transverse frame of reference is given by defining the orbital Kepler elements as $\{\omega, I, \Omega\} = \{90°, 90°, \Omega\}$, then the first transverse frame of reference is determined by the meta-North Pole $\lambda_0 = 90° + \Omega$ and $\phi_0 = 0°$.

End of Corollary.

3-36 Numerical Examples

The following Table 3.1 and the following Table 3.2 show some selected examples of the two designs to be considered here.

Table 3.1. First design: compare with Section 3-31.

Oblique frame, meta-North Pole = Rio de Janeiro ($\lambda_0 = -43°12'$, $\phi_0 = -22°54'36''$).	Direct problem (3.51/3.53).	P = Stuttgart ($\lambda = 9°11'24''$, $\phi = 47°46'48''$) $\Rightarrow$ $\alpha = 147°41'30''.5$, $\beta = 5°07'56''.2$.
	Indirect problem (3.59/3.61).	$\alpha = -30°$, $\beta = -20°$ $\Rightarrow$ $\lambda = 173°26'06''.3$, $\phi = -38°03'29''.1$.
Transverse frame, meta-North Pole = West Pole ($\lambda_0 = 270°$, $\phi_0 = 0°$).	Direct problem (3.51/3.53).	P = Greenwich ($\lambda = 0°$, $\phi = 51°28'38''$) $\Rightarrow$ $\alpha = 141°28'38''$, $\beta = 0°$.
	Indirect problem (3.59/3.61).	$\alpha = -30°$, $\beta = -20°$ $\Rightarrow$ $\lambda = 143°56'51''.4$, $\phi = -54°28'07''.1$.

Table 3.2. Second design: compare with Section 3-32.

Meta-equator: inclination $I = 13°24'36''$, longitude of the ascending node $\Omega = 52°28'12''$.	Direct problem (3.75/3.77).	P = Stuttgart ($\lambda = 9°11'24''$, $\phi = 47°46'48''$) $\Rightarrow$ $A = -29°27'49''.7$, $B = 55°48'51''.1$.
	Indirect problem (3.81/3.82).	$A = -30°$, $B = -20°$ $\Rightarrow$ $\lambda = 27°34'19''.6$, $\phi = 47°46'48''$.
Transverse frame, meta-equator: inclination $I = 90°$, longitude of the ascending node $\Omega = 0°$.	Direct problem (3.75/3.77).	P = Greenwich ($\lambda = 0°$, $\phi = 51°28'38''$) $\Rightarrow$ $A = 51°28'38''$, $B = 0°$.
	Indirect problem (3.81/3.82).	$A = -30°$, $B = -20°$ $\Rightarrow$ $\lambda = 22°47'45''.2$, $\phi = 51°28'38''$.

3-4 The oblique frame of reference of the ellipsoid-of-revolution

The oblique frame of reference of the ellipsoid-of-revolution (one Killing vector), transverse aspect, direct transformation, indirect transformation, oblique quasi-spherical coordinates, quasi-spherical longitude, quasi-spherical latitude.

Indeed, for an ellipsoid-of-revolution, there exist also three aspects which are called *normal, oblique*, and *transverse*. Again, these aspects generate special ellipsoidal coordinates of the ellipsoid-of-revolution, taking into account *one Killing vector of symmetry*. The oblique frame of reference of $\mathbb{E}^2_{A_1\,A_2}$ is based upon the centric oblique plane $\mathbb{P}^2_{\mathcal{O}}$, which intersects the ellipsoid-of-revolution and passes the origin $\mathcal{O}$. Such an oblique plane $\mathbb{P}^2_{\mathcal{O}}$ intersects $\mathbb{E}^2_{A_1\,A_2}$ in an elliptic oblique equator, also called *meta-equator*, which is oriented by two Kepler elements $\{\Omega, I\}$, called the longitude Ω of the ascending node and the inclination I. The transverse frame of reference is obtained by choosing an inclination $I = \pi/2$.

3-41 The direct and inverse transformations of the normal frame to the oblique frame

Let us orientate a set of orthonormal base vectors $\{\boldsymbol{E}_1, \boldsymbol{E}_2, \boldsymbol{E}_3\}$ along the principal axes of the ellipsoid-of-revolution of semi-major axis A_1 and semi-minor axis A_2:

$$\mathbb{E}^2_{A_1\,A_2} := \left\{ \boldsymbol{X} \in \mathbb{R}^3 \,\big|\, (X^2 + Y^2)/A_1^2 + Z^2/A_2^2 = 1, A_1 > A_2, A_1 \in \mathbb{R}^+, A_2 \in \mathbb{R}^+ \right\} . \tag{3.100}$$

Against this frame of reference $\{\boldsymbol{E}_1, \boldsymbol{E}_2, \boldsymbol{E}_3 | \mathcal{O}\}$ at the origin $\mathcal{O}$, we introduce the oblique frame of reference $\{\boldsymbol{E}_{1'}, \boldsymbol{E}_{2'}, \boldsymbol{E}_{3'} | \mathcal{O}\}$ at the origin $\mathcal{O}$ built on an alternative set of orthonormal base vectors which are related by means of a rotation:

$$\begin{bmatrix} \boldsymbol{E}_{1'} \\ \boldsymbol{E}_{2'} \\ \boldsymbol{E}_{3'} \end{bmatrix} = \mathsf{R}_1(I)\mathsf{R}_3(\Omega) \begin{bmatrix} \boldsymbol{E}_1 \\ \boldsymbol{E}_2 \\ \boldsymbol{E}_3 \end{bmatrix} . \tag{3.101}$$

This rotation is illustrated by Fig. 3.8. The rotation around the 3 axis is denoted by Ω, the right ascension of the ascending node, while the rotation around the intermediate 1 axis is denoted by I, the inclination. R_1 and R_3 are orthonormal matrices such that the following relation holds:

$$\mathsf{R}_1(I)\mathsf{R}_3(\Omega) = \begin{bmatrix} 1 & 0 & 0 \\ 0 & \cos I & \sin I \\ 0 & -\sin I & \cos I \end{bmatrix} \begin{bmatrix} \cos \Omega & \sin \Omega & 0 \\ -\sin \Omega & \cos \Omega & 0 \\ 0 & 0 & 1 \end{bmatrix} = \begin{bmatrix} \cos \Omega & \sin \Omega & 0 \\ -\sin \Omega \cos I & \cos \Omega \cos I & \sin I \\ \sin \Omega \sin I & -\cos \Omega \sin I & \cos I \end{bmatrix} , \tag{3.102}$$

$$\mathsf{R}_1(I)\mathsf{R}_3(\Omega) \in \mathbb{R}^{3 \times 3} .$$

Accordingly, the following vector equation defines a representation of the placement vector $\boldsymbol{X}$ in the orthonormal bases $\{\boldsymbol{E}_1, \boldsymbol{E}_2, \boldsymbol{E}_3\}$ and $\{\boldsymbol{E}_{1'}, \boldsymbol{E}_{2'}, \boldsymbol{E}_{3'}\}$, respectively:

$$\boldsymbol{X} = \sum_{i=1}^{3} \boldsymbol{E}_i X^i = \boldsymbol{E}_1 X + \boldsymbol{E}_2 Y + \boldsymbol{E}_3 Z = \boldsymbol{E}_{1'} X' + \boldsymbol{E}_{2'} Y' + \boldsymbol{E}_{3'} Z' = \sum_{i'=1}^{3} \boldsymbol{E}_{i'} X^{i'} . \tag{3.103}$$

Note that the corresponding Cartesian coordinate transformations are dual to the following systems of coordinate transformations:

$$\begin{aligned} X^1 &= X^{1'} \cos \Omega - X^{2'} \sin \Omega \cos I + X^{3'} \sin \Omega \sin I =: X , \\ X^2 &= X^{1'} \sin \Omega + X^{2'} \cos \Omega \cos I - X^{3'} \cos \Omega \sin I =: Y , \\ X^3 &= X^{2'} \sin I + X^{3'} \cos I =: Z , \end{aligned} \tag{3.104}$$

or

$$\begin{aligned} X^{1'} &= +X^1 \cos \Omega + X^2 \sin \Omega =: X' , \\ X^{2'} &= -X^1 \sin \Omega \cos I + X^2 \cos \Omega \cos I + X^3 \sin I =: Y' , \\ X^{3'} &= +X^1 \sin \Omega \sin I - X^2 \cos \Omega \sin I + X^3 \cos I =: Z' . \end{aligned} \tag{3.105}$$

3-42 The intersection of the ellipsoid-of-revolution and the central oblique plane

In order to obtain an oblique equatorial plane, namely an oblique equator, we intersect the ellipsoid-of-revolution $\mathbb{E}^2_{A_1,A_2}$ of semi-major axis A_1 and semi-minor axis A_2 and the central oblique plane $\mathbb{L}^2_{\mathcal{O}}$ (two-dimensional linear manifold through the origin $\mathcal{O}$). Subsequently, the oblique equatorial plane as well as its normal enables us to establish an oblique *quasi-spherical coordinate system*. Our first result is summarized in Corollary 3.7.

Corollary 3.7 (The intersection of $\mathbb{E}^2_{A_1,A_2}$ and $\mathbb{L}^2_{\mathcal{O}}$).

The intersection of the ellipsoid-of-revolution $\mathbb{E}^2_{A_1,A_2}$ and the central oblique plane $\mathbb{L}^2_{\mathcal{O}}$ is the ellipse of semi-major axis $A_{1'} = A_1$ and semi-minor axis $A_{2'} = A_1\sqrt{1-E^2}/\sqrt{1-E^2\cos^2 I}$ with respect to the relative eccentricity $E^2 := (A_1^2 - A_2^2)/A_1^2$:

$$\mathbb{E}^1_{A_{1'},A_{2'}} :=$$

$$:= \left\{ \boldsymbol{x}' \in \mathbb{R}^2 \,\middle|\, \frac{x'}{A_{1'}^2} + \frac{y'}{A_{2'}^2} = 1, A_{1'} = A_1, A_{2'} = A_1\sqrt{1-E^2}/\sqrt{1-E^2\cos^2 I}, A_{1'} > A_{2'} \right\}. \tag{3.106}$$

End of Corollary.

For short, the proof of Corollary 3.7 has been given in J. Engels, E. Grafarend (1995, pp. 42–43). Compare with Fig. 3.8, which illustrates the oblique elliptic equator as well as the orthogonal projection of a point $\boldsymbol{X} \in \mathbb{E}^2_{A_1,A_2}$ onto the oblique equatorial plane, respectively. Note that in a following section, the oblique equator is used to establish the following elliptic cylinder:

$$\mathbb{C}^2_{A_{1'},A_{2'}} :=$$

$$:= \left\{ \boldsymbol{X}' \in \mathbb{R}^3 \,\middle|\, \frac{X'^2}{A_{1'}^2} + \frac{Y'^2}{A_{2'}^2} = 1, Z' \in \mathbb{R} \right\}. \tag{3.107}$$

We here note that the points of $\mathbb{E}^2_{A_1,A_2}$ are conformally mapped just to lay down the foundation of a cylindric map projection.

3-43 The oblique quasi-spherical coordinates

With respect to the oblique equatorial plane and its normal vector, namely the oblique orthonormal frame $\{\boldsymbol{E}_{1'}, \boldsymbol{E}_{2'}, \boldsymbol{E}_{3'}|\mathcal{O}\}$, let us introduce *oblique quasi-spherical coordinates* by means of

$$\begin{aligned} X' &= R(A,B)\cos A \cos B \,, \\ Y' &= R(A,B)\sin A \cos B \,, \\ Z' &= R(A,B)\sin B \,. \end{aligned} \tag{3.108}$$

A $(A \in [0, 2\pi[)$ usually is called *oblique quasi-spherical longitude*, B $(B \in [-\pi/2, +\pi/2])$ usually is called *oblique quasi-spherical latitude*, and $R(A,B)$ is the *oblique radius*, which in turn is a function of A and B. Corollary 3.8 gives the answer, how this radial function can be expressed.

Corollary 3.8 (The oblique radial function $R(A,B)$).

If a point $\boldsymbol{X} \in \mathbb{E}^2_{A_1,A_2}$ is given in terms of oblique quasi-spherical coordinates of type (3.108), its radial function is represented by

$$R(A,B) = \frac{A_1\sqrt{1-E^2}}{\sqrt{1-E^2[\cos^2 A \cos^2 B + (\sin A \cos B \cos I - \sin B \sin I)^2]}} \,, \tag{3.109}$$

where the angle I characterizes the inclination of the oblique equatorial plane.

End of Corollary.

Proof.

For the proof, we depart from the quadratic form which is characteristic for $\mathbb{E}^2_{A_1 \, A_2}$, namely we replace the Cartesian coordinates $\{X^1 \ X^2 \ X^3\} = \{X \ Y \ Z\}$ via (3.105) by the Cartesian coordinates $\{X^{1'} \ X^{2'} \ X^{3'}\} = \{X' \ Y' \ Z'\}$:

$$\frac{X^2 + Y^2}{A_1^2} + \frac{Z^2}{A_2^2} = 1$$

$$\Leftrightarrow$$

$$A_1^2 A_2^2 =$$

$$= \left(X^2 + Y^2\right) A_2^2 + Z^2 A_1^2 =$$

$$= A_2^2 X'^2 + \left(A_2^2 \cos^2 I + A_1^2 \sin^2 I\right) Y'^2 + \left(A_2^2 \sin^2 I + A_1^2 \cos^2 I\right) Z'^2 +$$

$$+ 2Y'Z' \left(A_1^2 - A_2^2\right) \sin I \cos I$$

(3.110)

$$E^2 := \frac{A_1^2 - A_2^2}{A_1^2}$$

$$\Leftrightarrow$$

$$E^2 = 1 - \frac{A_2^2}{A_1^2}$$

(3.111)

$$\Leftrightarrow$$

$$A_2^2 = A_1^2 \left(1 - E^2\right)$$

$$A_1^2 \left(1 - E^2\right) =$$

$$= \left(1 - E^2\right) X'^2 + \left[\left(1 - E^2\right) \cos^2 I + \sin^2 I\right] Y'^2 + \left[\left(1 - E^2\right) \sin^2 I + \cos^2 I\right] Z'^2 +$$

$$+ 2Y'Z'E^2 \sin I \cos I$$

and

$$A_1^2 \left(1 - E^2\right) =$$

$$= X'^2 + Y'^2 + Z'^2 - E^2 \left(X'^2 + Y'^2 \cos^2 I + Z'^2 \sin^2 I\right) +$$

$$+ 2Y'Z'E^2 \sin I \cos I$$

(3.112)

$$\frac{A_1^2 \left(1 - E^2\right)}{R^2} =$$

$$= 1 - E^2 \left(\cos^2 A \cos^2 B + \sin^2 A \cos^2 B \cos^2 I + \sin^2 B \sin^2 I - 2 \sin A \sin B \cos B \sin I \cos I\right)$$

$$= 1 - E^2 [\cos^2 A \cos^2 B + (\sin A \cos B \cos I - \sin B \sin I)^2]$$

(3.113)

$$\Rightarrow$$

$$(3 \ 109)$$

End of Proof.

Next, we have to compute the *arc length* of $\mathbb{E}^1_{A_{1'} \, A_{2'}}$, i.e. that part of the *oblique ecliptic equator* which ranges from the oblique quasi-spherical longitude zero to a fixed, but arbitrary value A.

3-44 The arc length of the oblique equator in oblique quasi-spherical coordinates

In order to compute the length of an arc in the oblique ecliptic equator $\mathbb{E}^1_{A_1, A_2}$, in terms of oblique quasi-spherical longitude, we are forced to represent the infinitesimal arc length by

$$dS = \sqrt{dX'^2 + dY'^2} \overset{B=0}{=} \sqrt{R^2(A) + R_1^2(A)}dA \qquad (3.114)$$

subject to

$$R_0(A\ B = 0) := R(A) := \frac{A_1\sqrt{1 - E^2}}{\sqrt{1 - E^2\left(1 - \cos^2 I \sin^2 A\right)}} \qquad (3.115)$$

$$R_1(A\ B = 0) := R_1(A) := \frac{1}{1!}\frac{dR(A)}{dA} = -\frac{A_1 E^2\sqrt{1 - E^2}\cos^2 I \sin A \cos A}{\left[1 - E^2\left(1 - \cos^2 I \sin^2 A\right)\right]^{3\ 2}} \qquad (3.116)$$

such that

$$S(A) = \int_{A=0}^{A}\sqrt{R^2(A^*) + R_1^2(A^*)}dA^* \qquad (3.117)$$

In the following steps, we perform the integration. (i) Series expansion of $R(A)$ according to (3.120) up to order E^6. (ii) Series expansion of $R_1(A) = dR\ dA$ according to (3.121) up to order E^6. (iii) Series expansion of $R^2(A) + R_1^2(A)$ according to (3.122) up to order E^6. (iv) Series expansion of $(R^2(A) + R_1^2(A))^{1\ 2}$ according to (3.125) up to order E^6.

Solution (the first step).

$$\frac{R(A)}{A_1\sqrt{1 - E^2}} = (1 - x)^{-1\ 2} = 1 + \frac{1}{2}x + \frac{1\cdot 3}{2\cdot 4}x^2 + \frac{1\cdot 3\cdot 5}{2\cdot 4\cdot 6}x^3 + O_-\left(x^4\right) \qquad (3.118)$$

subject to

$$x := -E^2\left(1 - \cos^2 I \sin^2 A\right) \qquad |x| < 1\ ; \qquad (3.119)$$

$$R(A) = A_1\sqrt{1 - E^2}\left[1 + \frac{1}{2}E^2\left(1 - \cos^2 I \sin^2 A\right) + \right.$$
$$\left. + \frac{1\cdot 3}{2\cdot 4}E^4\left(1 - \cos^2 I \sin^2 A\right)^2 + \frac{1\cdot 3\cdot 5}{2\cdot 4\cdot 6}E^6\left(1 - \cos^2 I \sin^2 A\right)^3 + O_-\left(E^8\right)\right] \qquad (3.120)$$

End of Solution (the first step).

Solution (the second step).

$$R_1(A) = \frac{dR}{dA} =$$
$$= A_1\sqrt{1 - E^2}\left[- E^2\cos^2 I \sin A \cos A - \right.$$
$$- \frac{1\cdot 3}{2}E^4\cos^2 I\left(1 - \cos^2 I \sin^2 A\right)\sin A \cos A - $$
$$\left. - \frac{1\cdot 3\cdot 5}{2\cdot 4}E^6\cos^2 I\left(1 - \cos^2 I \sin^2 A\right)\sin A \cos B - O_1\left(E^8\right)\right] \qquad (3.121)$$

End of Solution (the second step).

Solution (the third step).

$$R^2(A) + R_1^2(A) = A_1 \left(1 - E^2\right) \left(1 + E^2 \left(1 - \cos^2 I \cos^2 A\right) + \right.$$

$$+ E^4 \left[\left(1 - \cos^2 I \cos^2 A\right)^2 + \cos^4 I \sin^2 A \cos^2 A\right] + \tag{3.122}$$

$$+ E^6 \left[\left(1 - \cos^2 I \cos^2 A\right)^3 + 3\cos^4 I \left(1 - \cos^2 I \sin^2 A\right) \sin^2 A \cos^2 A\right] + O_1\left(E^8\right)\Big) .$$

End of Solution (the third step).

Solution (the fourth step).

$$\frac{dS}{dA} \frac{1}{A_1\sqrt{1 - E^2}} = (1 + x)^{+1/2} = 1 + \frac{1}{2}x - \frac{1 \cdot 1}{2 \cdot 4}x^2 + \frac{1 \cdot 1 \cdot 3}{2 \cdot 4 \cdot 6}x^3 - \frac{1 \cdot 1 \cdot 3 \cdot 5}{2 \cdot 4 \cdot 6 \cdot 8}x^4 + O_+\left(x^5\right), \tag{3.123}$$

subject to

$$x := E^2 \left(1 - \cos^2 I \sin^2 A\right) + E^4 \left[\left(1 - \cos^2 I \sin^2 A\right)^2 + \cos^4 I \sin^2 A \cos^2 A\right] +$$

$$+ E^6 \left[\left(1 - \cos^2 I \sin^2 A\right)^2 + 3\cos^4 I \left(1 - \cos^2 I \sin^2 A\right) \sin^2 A \cos^2 A\right] + O_+\left(E^8\right) ; \tag{3.124}$$

$$dS(B = 0) = A_1\sqrt{1 - E^2}\left(1 + \frac{1}{2}E^2 \left(1 - \cos^2 I \sin^2 A\right) + \right.$$

$$+ E^4 \left[\frac{3}{8}\left(1 - \cos^2 I \sin^2 A\right)^2 + \cos^4 I \sin^2 A \cos^2 A\right] + \tag{3.125}$$

$$+ E^6 \left[\frac{5}{16}\left(1 - \cos^2 I \sin^2 A\right)^3 + \frac{5}{4}\cos^4 I \left(1 - \cos^2 I \sin^2 A\right) \sin^2 A \cos^2 A\right] + O_+\left(E^8\right)\Big)dA .$$

End of Solution (the fourth step).

Note that all series (3.120), (3.121), (3.122), and (3.125) are uniformly convergent. Accordingly, we can interchange integration and summation within (3.117) when we substitute (3.125) as a series expansion. An alternative useful expansion of $S(A)$ in terms of powers of ΔA is provided by the following formulae:

$$S\left(A_0 + \Delta A\right) = S\left(A_0\right) + S_1\left(A_0\right)\Delta A + S_2\left(A_0\right)\left(\Delta A\right)^2 + O_S\left[\left(\Delta A\right)^3\right], \tag{3.126}$$

$$S_1\left(A\right) = \frac{1}{1!}\frac{dS}{dA} = A_1\sqrt{1 - E^2}\left(1 + \frac{1}{2}E^2 \left(1 - \cos^2 I \sin^2 A\right) + \right.$$

$$+ E^4 \left[\frac{3}{8}\left(1 - \cos^2 I \sin^2 A\right)^2 + \cos^4 I \sin^2 A \cos^2 A\right] + \tag{3.127}$$

$$+ E^6 \left[\frac{5}{16}\left(1 - \cos^2 I \sin^2 A\right)^3 + \frac{5}{4}\cos^4 I \left(1 - \cos^2 I \sin^2 A\right) \sin^2 A \cos^2 A\right] + O\left(E^8\right)\Big),$$

$$S_2\left(A\right) = \frac{1}{2!}\frac{d^2 S}{dA^2} = A_1\sqrt{1 - E^2}\left(-E^2 \cos^2 I \sin A \cos A - \right.$$

$$- E^4 \left[\frac{3}{4}\left(1 - \cos^2 I \sin^2 A\right)\cos^2 I \sin A \cos A - 2\cos^2 I \sin A \cos^3 A + \cos^4 I \sin^3 A \cos A\right] -$$

$$- E^6 \left[\frac{15}{8}\left(1 - \cos^2 I \sin^2 A\right)^2 \cos^2 I \sin A \cos A + \frac{5}{2}\cos^2 I \sin^3 A \cos A - \right. \tag{3.128}$$

$$- \frac{5}{2}\left(1 - \cos^2 I \sin^2 A\right)\cos^4 I \sin A \cos^3 A + \frac{5}{4}\left(1 - \cos^2 I \sin^2 A\right)\cos^4 I \sin^3 A \cos A\Big] + O\left(E^8\right)\Big).$$

Next, we have to work out how the oblique quasi-spherical longitude/latitude are related to the standard surface normal ellipsoidal longitude/latitude. This finally concludes the introduction of the oblique coordinate system of the ellipsoid-of-revolution. At first, we here aim at a transformation of oblique quasi-spherical longitude/latitude into surface normal ellipsoidal longitude/latitude to which we refer as *direct transformation*. Additionally, we here aim at a transformation of surface normal ellipsoidal longitude/latitude into oblique quasi-spherical longitude/latitude to which we refer as *inverse transformation*.

3-45 Direct transformation of oblique quasi-spherical longitude/latitude

The standard representation of a point $\boldsymbol{X}$ being an element of $\mathbb{E}^2_{A_1,A_2}$ is given in terms of surface normal ellipsoidal longitude/latitude $\{\Lambda, \Phi\} \in \{\mathbb{R}^2 | 0 \le \Lambda < 2\pi, -\pi/2 < \Phi < +\pi/2\}$, which excludes North Pole and South Pole of $\mathbb{E}^2_{A_1,A_2}$. Accordingly, $\{\Lambda, \Phi\}$ constitutes only a first chart of $\mathbb{E}^2_{A_1,A_2}$, i.e.

$$X^1 = X = \frac{A_1 \cos \Phi \cos \Lambda}{\sqrt{1 - E^2 \sin^2 \Phi}} ,$$

$$X^2 = Y = \frac{A_1 \cos \Phi \sin \Lambda}{\sqrt{1 - E^2 \sin^2 \Phi}} , \tag{3.129}$$

$$X^3 = Z = \frac{A_1 \left(1 - E^2\right) \sin \Phi}{\sqrt{1 - E^2 \sin^2 \Phi}} .$$

The ellipsoidal coordinates Λ and Φ are called *surface normal* since the surface normal of $\mathbb{E}^2_{A_1,A_2}$ enjoys the *spherical image*

$$\boldsymbol{N} = \boldsymbol{E}_1 \cos \Phi \cos \Lambda + \boldsymbol{E}_2 \cos \Phi \sin \Lambda + \boldsymbol{E}_3 \sin \Phi . \tag{3.130}$$

A minimal atlas of $\mathbb{E}^2_{A_1,A_2}$, which covers all points of $\mathbb{E}^2_{A_1,A_2}$, has to be based on two charts given by E. Grafarend and R. Syffus (1995). The direct mapping of type (3.129), namely $\{\Lambda, \Phi\} \to \{X, Y, Z\}$, has the inverse

$$\tan \Lambda = \frac{Y}{X} ,$$

$$\tan \Phi = \frac{1}{1 - E^2} \frac{Z}{\sqrt{X^2 + Y^2}} . \tag{3.131}$$

By means of (3.104), (3.108), and (3.109), one alternatively derives the direct mapping equations and inverse mapping equations

$$X = R(A, B) \left[\cos A \cos B \cos \Omega - \sin A \cos B \sin \Omega \cos I + \sin B \sin \Omega \sin I \right] ,$$

$$Y = R(A, B) \left[\cos A \cos B \sin \Omega + \sin A \cos B \cos \Omega \cos I - \sin B \cos \Omega \sin I \right] , \tag{3.132}$$

$$Z = R(A, B) \left[\sin A \cos B \sin I + \sin B \cos I \right] ,$$

$$\tan \Lambda = \frac{\cos A \cos B \sin \Omega + \sin A \cos B \cos \Omega \cos I - \sin B \cos \Omega \sin I}{\cos A \cos B \cos \Omega - \sin A \cos B \sin \Omega \cos I + \sin B \sin \Omega \sin I} ,$$

$$\tan \Phi = \frac{1}{1 - E^2} \frac{\sin A \cos B \sin I + \sin B \cos I}{\sqrt{\cos^2 A \cos^2 B + (\sin A \cos B \cos I - \sin B \sin I)^2}} . \tag{3.133}$$

Let us here additionally collect the result of the transformation $\{A, B\} \rightarrow \{\Lambda, \Phi\}$ by the following Corollary 3.9.

Corollary 3.9 (The change from one chart to another chart: cha-cha-cha, the oblique quasi-spherical longitude/latitude versus the surface normal ellipsoidal longitude/latitude).

Given the longitude of the ascending node Ω as well as the inclination I of the oblique equatorial plane, then the transformation of oblique quasi-spherical longitude/latitude in surface normal ellipsoidal longitude/latitude is represented by (3.133).

End of Corollary.

Next, let us assume that we know already a point $\{\Lambda_0, \Phi_0\}$, correspondingly $\{A_0, B_0\}$, in $\mathbb{E}^2_{A_1, A_2}$. Relative to such a fixed point, we are able to find the coordinates $\{\Lambda, \Phi\}$, correspondingly $\{A, B\}$, close to $\{\Lambda_0, \Phi_0\}$, correspondingly $\{A_0, B_0\}$, by a Taylor series expansion of the type presented in Boxes 3.20 and 3.21.

Box 3.20 (Taylor series expansion of the longitude function $\Lambda(A, B)$, Taylor polynomials).

$$\Delta\Lambda = \Lambda - \Lambda_0 =$$

$$= \frac{\partial\Lambda}{\partial A}(A_0, B_0)\,\Delta A + \frac{\partial\Lambda}{\partial B}(A_0, B_0)\,\Delta B+$$

$$+\frac{1}{2}\frac{\partial^2\Lambda}{\partial A^2}(A_0, B_0)\,(\Delta A)^2 + \frac{\partial^2\Lambda}{\partial A\partial B}(A_0, B_0)\,\Delta A\Delta B + \frac{1}{2}\frac{\partial^2\Lambda}{\partial B^2}(A_0, B_0)\,(\Delta B)^2 +$$

$$+\frac{1}{6}\frac{\partial^3\Lambda}{\partial A^3}(A_0, B_0)\,(\Delta A)^3 + \frac{1}{6}\frac{\partial^3\Lambda}{\partial B^3}(A_0, B_0)\,(\Delta B)^3 +$$

$$+\frac{1}{2}\frac{\partial^3\Lambda}{\partial A^2\partial B}(A_0, B_0)\,(\Delta A)^2\,\Delta B + \frac{1}{2}\frac{\partial^3\Lambda}{\partial A\partial B^2}(A_0, B_0)\,\Delta A\,(\Delta B)^2 +$$

$$+\mathrm{O}_\Lambda\left[(\Delta A)^4, (\Delta B)^4\right]\,,$$

(3.134)

$$\Delta\Lambda = \Lambda - \Lambda_0 =: l\,, \quad \Delta\Phi = \Phi - \Phi_0 =: b\,,$$

$$\Delta A = A - A_0 =: \alpha\,, \quad \Delta B = B - B_0 =: \beta\,.$$

(3.135)

Definition of partial derivatives:

$$l_{10} := \frac{\partial\Lambda}{\partial A}(A_0, B_0)\,, \quad l_{01} := \frac{\partial\Lambda}{\partial B}(A_0, B_0)\,,$$

$$l_{20} := \frac{1}{2}\frac{\partial^2\Lambda}{\partial A^2}(A_0, B_0)\,, \quad l_{11} := \frac{\partial^2\Lambda}{\partial A\partial B}(A_0, B_0)\,, \quad l_{02} := \frac{1}{2}\frac{\partial^2\Lambda}{\partial B^2}(A_0, B_0)\,,$$

$$l_{30} := \frac{1}{6}\frac{\partial^3\Lambda}{\partial A^3}(A_0, B_0)\,, \quad l_{03} := \frac{1}{6}\frac{\partial^3\Lambda}{\partial B^3}(A_0, B_0)\,,$$

$$l_{21} := \frac{1}{2}\frac{\partial^3\Lambda}{\partial A^2\partial B}(A_0, B_0)\,, \quad l_{12} := \frac{1}{2}\frac{\partial^3\Lambda}{\partial A\partial B^2}(A_0, B_0)\,.$$

(3.136)

Taylor series, powers of Taylor series:

$$l = l_{10}\alpha + l_{01}\beta + l_{20}\alpha^2 + l_{11}\alpha\beta + l_{02}\beta^2 + l_{30}\alpha^3 + l_{21}\alpha^2\beta + l_{12}\alpha\beta^2 + l_{03}\beta^3 + \mathrm{O}_1\left(\alpha^4, \beta^4\right)\,,$$

$$l^2 = l_{10}^2\alpha^2 + 2l_{10}l_{01}\alpha\beta + l_{01}^2\beta^2 + 2l_{10}l_{20}\alpha^3 +$$

$$+2\left(l_{10}l_{11} + l_{01}l_{20}\right)\alpha^2\beta + 2\left(l_{01}l_{20} + l_{10}l_{02}\right)\alpha\beta^2 + 2l_{01}l_{02}\beta^3 + \mathrm{O}_2\left(\alpha^4, \beta^4\right)\,,$$

$$l^3 = l_{10}^3\alpha^3 + 3l_{10}^2l_{01}\alpha^2\beta + 3l_{10}l_{01}^2\alpha\beta^2 + \mathrm{O}_3\left(\alpha^4, \beta^4\right)\,.$$

(3.137)

Box 3.21 (Taylor series expansion of the latitude function $\Phi(A, B)$, Taylor polynomials).

$$\Delta\Phi = \Phi - \Phi_0 =$$

$$= \frac{\partial\Phi}{\partial A}(A_0, B_0)\,\Delta A + \frac{\partial\Phi}{\partial B}(A_0, B_0)\,\Delta B +$$

$$+\frac{1}{2}\frac{\partial^2\Phi}{\partial A^2}(A_0, B_0)\,(\Delta A)^2 + \frac{\partial^2\Phi}{\partial A\partial B}(A_0, B_0)\,\Delta A\Delta B + \frac{1}{2}\frac{\partial^2\Phi}{\partial B^2}(A_0, B_0)\,(\Delta B)^2 +$$

$$+\frac{1}{6}\frac{\partial^3\Phi}{\partial A^3}(A_0, B_0)\,(\Delta A)^3 + \frac{1}{6}\frac{\partial^3\Phi}{\partial B^3}(A_0, B_0)\,(\Delta B)^3 +$$

$$+\frac{1}{2}\frac{\partial^3\Phi}{\partial A^2\partial B}(A_0, B_0)\,(\Delta A)^2\,\Delta B + \frac{1}{2}\frac{\partial^3\Phi}{\partial A\partial B^2}(A_0, B_0)\,\Delta A\,(\Delta B)^2 +$$

$$+\mathrm{O}_\Phi\left[(\Delta A)^4, (\Delta B)^4\right]\,,$$

(3.138)

$$\Delta\Lambda = \Lambda - \Lambda_0 =: l\,, \quad \Delta\Phi = \Phi - \Phi_0 =: b\,,$$

$$\Delta A = A - A_0 =: \alpha\,, \quad \Delta B = B - B_0 =: \beta\,.$$

(3.139)

Definition of partial derivatives:

$$b_{10} := \frac{\partial\Phi}{\partial A}(A_0, B_0)\,, \quad b_{01} := \frac{\partial\Phi}{\partial B}(A_0, B_0)\,,$$

$$b_{20} := \frac{1}{2}\frac{\partial^2\Phi}{\partial A^2}(A_0, B_0)\,, \quad b_{11} := \frac{\partial^2\Phi}{\partial A\partial B}(A_0, B_0)\,, \quad b_{02} := \frac{1}{2}\frac{\partial^2\Phi}{\partial B^2}(A_0, B_0)\,,$$

$$b_{30} := \frac{1}{6}\frac{\partial^3\Phi}{\partial A^3}(A_0, B_0)\,, \quad b_{03} := \frac{1}{6}\frac{\partial^3\Phi}{\partial B^3}(A_0, B_0)\,,$$

$$b_{21} := \frac{1}{2}\frac{\partial^3\Phi}{\partial A^2\partial B}(A_0, B_0)\,, \quad b_{12} := \frac{1}{2}\frac{\partial^3\Phi}{\partial A\partial B^2}(A_0, B_0)\,.$$

(3.140)

Taylor series, powers of Taylor series:

$$b = b_{10}\alpha + b_{01}\beta + b_{20}\alpha^2 + b_{11}\alpha\beta + b_{02}\beta^2 + b_{30}\alpha^3 + b_{21}\alpha^2\beta + b_{12}\alpha\beta^2 + b_{03}\beta^3 + \mathrm{O}_1\left(\alpha^4, \beta^4\right)\,,$$

$$b^2 = b_{10}^2\alpha^2 + 2b_{10}b_{01}\alpha\beta + b_{01}^2\beta^2 + 2b_{10}b_{20}\alpha^3 +$$

$$+2\left(b_{10}b_{11} + b_{01}b_{20}\right)\alpha^2\beta + 2\left(b_{01}b_{20} + b_{10}b_{02}\right)\alpha\beta^2 + 2b_{01}b_{02}\beta^3 + \mathrm{O}_2\left(\alpha^4, \beta^4\right)\,,$$

$$l^3 = b_{10}^3\alpha^3 + 3b_{10}^2 b_{01}\alpha^2\beta + 3b_{10}b_{01}^2\alpha\beta^2 + \mathrm{O}_3\left(\alpha^4, \beta^4\right)\,.$$

(3.141)

Note that all the partial derivatives that are quoted in these boxes can be computed by taking advantage of the identities $\mathrm{d}\tan x = (1 + \tan^2 x)\mathrm{d}x$ and $\mathrm{d}x = \mathrm{d}\tan x/(1 + \tan^2 x)$, which lead to the recursive scheme presented in Box 3.22.

Box 3.22 (Recursive relations for the partial derivatives l_{ij} and b_{ij} up to order three, $x \in \{\Lambda, \Phi\}$.

$$\frac{\partial x}{\partial A} = +\frac{1}{1 + \tan^2 x}\frac{\partial\tan x}{\partial A}\,, \quad \frac{\partial^2 x}{\partial A^2} = +\frac{2\tan x}{(1 + \tan^2 x)^2}\left(\frac{\partial\tan x}{\partial A}\right)^2 + \frac{1}{1 + \tan^2 x}\frac{\partial\tan x}{\partial A}\,,$$

$$\frac{\partial^3 x}{\partial A^3} = -2\frac{1 - 3\tan x}{(1 + \tan^2 x)^3}\left(\frac{\partial\tan x}{\partial A}\right)^3 - \frac{6\tan x}{(1 + \tan^2 x)^2}\left(\frac{\partial\tan x}{\partial A}\right)\left(\frac{\partial^2\tan x}{\partial A^2}\right) +$$

$$+\frac{1}{1 + \tan^2 x}\frac{\partial^3\tan x}{\partial A^3}\,,$$

(3.142)

$$\frac{\partial^2 x}{\partial A\partial B} = +\frac{1}{1 + \tan^2 x}\frac{\partial^2\tan x}{\partial A\partial B} - \frac{2\tan x}{(1 + \tan^2 x)^2}\left(\frac{\partial\tan x}{\partial B}\right)\left(\frac{\partial\tan x}{\partial A}\right)\,.$$

3-46 Inverse transformation of oblique quasi-spherical longitude/latitude

Let us depart from the representation (3.108) of oblique Cartesian coordinates $\{X', Y', Z'\}$ in terms of oblique quasi-spherical longitude/latitude $\{A, B\}$. The inverse map relates these oblique Cartesian coordinates to oblique quasi-spherical longitude/latitude:

$$\tan A = \frac{Y'}{X'} \,,$$
$$\tan B = \frac{Z'}{\sqrt{X'^2 + Y'^2}} \,. \tag{3.143}$$

As soon as we implement the transformation of normal Cartesian coordinates $\{X, Y, Z\}$ into oblique Cartesian coordinates $\{X', Y', Z'\}$ of type (3.104) and (3.105) as well as the surface normal ellipsoidal longitude/latitude $\{\Lambda, \Phi\}$ into normal Cartesian coordinates $\{X, Y, Z\}$ of type (3.129), we are led to

$$X' = X \cos \Omega + Y \sin \Omega =$$
$$= \frac{A_1}{\sqrt{1 - E^2 \sin^2 \Phi}} \cos \Phi \left(\cos \Lambda \cos \Omega + \sin \Lambda \sin \Omega\right) = \tag{3.144}$$
$$= \frac{A_1}{\sqrt{1 - E^2 \sin^2 \Phi}} \cos \Phi \cos(\Lambda - \Omega) \,,$$

$$Y' = -X \sin \Omega \cos I + Y \cos \Omega \cos I + Z \sin I =$$
$$= \frac{A_1}{\sqrt{1 - E^2 \sin^2 \Phi}} \left(- \cos \Phi \cos \Lambda \sin \Omega \cos I + \cos \Phi \sin \Lambda \cos \Omega \cos I + \left(1 - E^2\right) \sin \Phi \sin I\right) = \tag{3.145}$$
$$= \frac{A_1}{\sqrt{1 - E^2 \sin^2 \Phi}} \left(+ \cos \Phi \cos I \sin(\Lambda - \Omega) + \left(1 - E^2\right) \sin \Phi \sin I\right) \,,$$

$$Z' = X \sin \Omega \sin I - Y \cos \Omega \cos I + Z \cos I =$$
$$= \frac{A_1}{\sqrt{1 - E^2 \sin^2 \Phi}} \left(\cos \Phi \cos \Lambda \sin \Omega \sin I - \cos \Phi \sin \Lambda \cos \Omega \cos I + \left(1 - E^2\right) \sin \Phi \cos I\right) \,, \tag{3.146}$$

such that

$$\tan A = \frac{- \cos \Phi \cos I \sin(\Lambda - \Omega) + \left(1 - E^2\right) \sin \Phi \sin I}{\cos \Phi \cos(\Lambda - \Omega)} \,,$$
$$\tan B = \frac{\cos \Phi \cos \Lambda \sin \Omega \sin I - \cos \Phi \sin \Lambda \cos \Omega \cos I + \left(1 - E^2\right) \sin \Phi \cos I}{\sqrt{\cos^2 \Phi \cos^2(\Lambda - \Omega) + [\cos \Phi \cos I \sin(\Lambda - \Omega) + (1 - E^2) \sin \Phi \sin I]^2}} \,. \tag{3.147}$$

Let us here additionally collect the result of the transformation $\{\Lambda, \Phi\} \to \{A, B\}$ by the following Corollary 3.10.

Corollary 3.10 (The change from one chart to another chart: cha-cha-cha, the surface normal ellipsoidal longitude/latitude versus the oblique quasi-spherical longitude/latitude).

Given the longitude of the ascending node Ω as well as the inclination I of the oblique equatorial plane, then the transformation of surface normal ellipsoidal longitude/latitude into oblique quasi-spherical longitude/latitude is represented by (3.147).

End of Corollary.

It should be noted that the coordinates $\{A, B\}$ of type oblique quasi-spherical longitude/latitude are not orthogonal. Accordingly, the matrix of the metric of $\mathbb{E}^2_{A_1, A_2}$ in terms of these coordinates contains off-diagonal elements. Finally, we note that the terms up to order three of the corresponding Taylor series expansions can be determined by resorting to the partial derivatives of Box 3.23.

Box 3.23 (Partial derivatives up to order three).

$$(\tan \Lambda),_A = + \frac{\cos^2 B \cos I - \sin A \sin B \cos B \sin I}{(\cos A \cos B \cos \Omega - \sin A \cos B \sin \Omega \cos I + \sin B \sin \Omega \sin I)^2} ,$$

$$(\tan \Lambda),_B = - \frac{\cos A \sin I}{(\cos A \cos B \cos \Omega - \sin A \cos B \sin \Omega \cos I + \sin B \sin \Omega \sin I)^2} ;$$

(3.148)

$$N = N(A, B) := \cos^2 B \cos I - \sin A \sin B \cos B \sin I ,$$
$$M = M(A, B) := \cos A \sin I ,$$
$$D = D(A, B) := \cos A \cos B \cos \Omega - \sin A \cos B \sin \Omega \cos I + \sin B \sin \Omega \sin I ;$$

(3.149)

$$N,_A = - \cos A \sin B \cos B \sin I ,$$
$$N,_B = -2 \sin B \cos B - \sin A \cos^2 B \sin I + \sin A \sin^2 B \sin I ,$$
$$M,_A = - \sin A \sin I , \quad M,_B = 0 ,$$

(3.150)

$$N,_{AA} = + \sin A \sin B \cos B \sin I ,$$
$$N,_{AB} = - \cos A \cos^2 B \sin I + \cos A \sin^2 B \sin I = N,_{BA} ,$$
$$N,_{BB} = -2 \cos^2 B + 2 \sin^2 B + 2 \sin A \sin B \cos B \sin I + 2 \sin A \sin B \cos B \sin I ,$$
$$M,_{AA} = - \cos A \sin I , \quad M,_{AB} = 0 , \quad M,_{BA} = 0 , \quad M,_{BB} = 0 ,$$

(3.151)

$$D,_A = - \sin A \cos B \cos \Omega - \cos A \cos B \sin \Omega \cos I ,$$
$$D,_B = - \cos A \sin B \cos \Omega + \sin A \sin B \sin \Omega \cos I + \cos B \sin \Omega \sin I ,$$
$$D,_{AA} = - \cos A \cos B \cos \Omega + \sin A \cos B \sin \Omega \cos I ,$$
$$D,_{AB} = + \sin A \sin B \cos \Omega + \cos A \sin B \sin \Omega \cos I = D,_{DA} ,$$
$$D,_{BB} = - \cos A \cos B \cos \Omega + \sin A \cos B \sin \Omega \cos I - \sin B \sin \Omega \sin I ;$$

(3.152)

$$(\tan \Lambda),_A = \frac{N}{D^2} , \quad (\tan \Lambda),_B = \frac{M}{D^2} ,$$

$$(\tan \Lambda),_{AA} = \frac{D^2 N,_A - 2DD,_A N}{D^4} , \quad (\tan \Lambda),_{BB} = \frac{D^2 M,_B - 2D^2 D,_B M}{D^4} ,$$

$$(\tan \Lambda),_{AAA} = \frac{D^4 \left(2DD,_A N,_A + D^2 N,_{AA}\right) - 8D^4 D,_A^2 N}{D^8} ,$$

$$(\tan \Lambda),_{BBB} = \frac{D^4 \left(2DD,_B M,_B + D^2 M,_{BB}\right) - 4D^6 D,_B^2 M}{D^8} ,$$

(3.153)

$$(\tan \Lambda),_{AB} = \frac{D^2 N,_B - 2DD,_B N}{D^4} ,$$

$$(\tan \Lambda),_{ABB} = \frac{D^4 \left(2DD,_B N,_B + D^2 N,_{BB}\right) - 8D^4 D,_B^2 N}{D^8} ,$$

$$(\tan \Lambda),_{AB} = (\tan \Lambda),_{BA} , \quad (\tan \Lambda),_{ABB} = (\tan \Lambda),_{BAB} = (\tan \Lambda),_{BBA} .$$

In the following chapter, let us close a gap and introduce a classification scheme that is needed in the remaining chapters.

4 Surfaces of Gaussian curvature zero

Classification of surfaces of Gaussian curvature zero (Gauss flat, two-dimensional Riemann manifolds) in a two-dimensional Euclidean space, ruled surfaces, developable surfaces.

While in the first chapter we discuss the mapping of a left surface (two-dimensional Riemann manifold) to a right surface (two-dimensional Riemann manifold), the second chapter specializes the right surface to be a plane. In contrast, the third chapter answers the question of how to parameterize a surface (in general, a Riemann manifold) in order to cover all points of such differentiable manifolds completely by an atlas. Special attention is paid to the question of a minimal atlas. Here, we fill the gap between the first and the second chapter. In particular, we introduce a special ruled surface of Gaussian curvature zero which can be developed to a plane, a cylinder, a cone, or a "tangent developable". Such Gauss flat two-dimensional Riemann manifolds are fundamental for the classification of the right surface assumed to be developable. All following chapters are based upon this classification scheme. First, we clarify the notion of a *ruled surface*. Second, we specialize to *developable surfaces*, in short, "developables". The text is only explanatory and rich of illustrations. All proofs are referred to the literature.

4-1 Ruled surfaces

Ruled surfaces (circular cone of the sphere as a ruled surface, helicoid as a ruled surface, one-sheeted hyperboloid of revolution as a ruled surface, directrix).

Let us make familiar with a special surface, usually called *ruled surface*. First, we introduce a curve $\boldsymbol{x}(U) \in \{\mathbb{R}^3, \mathsf{I}_3\}$ in a three-dimensional Euclidean space, called the *directrix* of the surface. U is the parameter of the curve. Second, a ruler is moving along the directrix, generating the *ruling* of the surface. Alternatively, we may say that a ruled surface results from the motion of a straight line in space. Movements of this kind of surfaces or segments are found in many physical, in particular, mechanical applications. For instance, the motion of a robot arm generates a ruled surface. Example 4.1 together with Fig. 4.1 illustrates such *generators* of a ruled surface, here the circular cone of the sphere $\mathbb{S}_R^2$. In contrast, Fig. 4.2 presents the *helicoid* and the *one-sheeted hyberboloid of revolution* as alternative examples of a ruled surface.

Example 4.1 (Circular cone $\mathbb{C}_{R\cos\Phi_0}^2$ of the sphere $\mathbb{S}_R^2$).

Let us construct a circular cone of the sphere $\mathbb{S}_R^2$ as a ruled surface. First, we choose the parallel circle, also called *small circle*, of the parameterized sphere as the reference curve or directrix. Second, we attach locally to any point of the directrix a vector field which is generated by a ruler moving along the reference curve. Consult Box 4.1 and Fig. 4.1 for a more detailed analysis. In terms of spherical coordinates $\{\Lambda, \Phi, R\}$, we parameterize the "position vector" $\boldsymbol{X}(\Lambda, \Phi, R)$ with respect to an orthonormal frame of reference $\{\boldsymbol{E}_1, \boldsymbol{E}_2, \boldsymbol{E}_3 | \mathcal{O}\}$, spanning a three-dimensional Euclidean space $\mathbb{E}^3$, attached to the origin $\mathcal{O}$, which is the center of the sphere $\mathbb{S}_R^2$ of radius R. $\{\boldsymbol{C}_\Lambda, \boldsymbol{C}_\Phi | \Lambda, \Phi\}$ is the local frame of reference, i. e. Cartan's moving frame ("repére mobile"), attached to a point $\{\Lambda, \Phi\} \in \mathbb{S}_R^2$. As the reference curve, we have chosen the parallel circle $\Phi_0 = $ constant, namely the directrix $\boldsymbol{x}(U)$, where $U = \Lambda$ is the parameter of the reference curve. The generator or the ruler of the surface is the vector field $\boldsymbol{Y}(U) := \boldsymbol{C}_\Phi(U)$, the unit vector which is normal to $\boldsymbol{C}_\Lambda(\Lambda, \Phi_0)$, directed towards North. The linear manifold $V\boldsymbol{Y}(U)$, also called the *bundle of straight lines*, is forming the circular cone $\mathbb{C}_{R\cos\Phi_0}^2$ of radius $R\cos\Phi_0$ as soon as the ruler moves around the parallel circle. Finally, we have gained the parameterized ruled surface $\boldsymbol{X}(U, V)$. Its typical matrix of the metric, G, has been computed.

End of Example.

Additionally, let us more precisely define a ruled surface in Definition 4.1, which follows after Box 4.1 summarizing the vector definitions of Example 4.1.

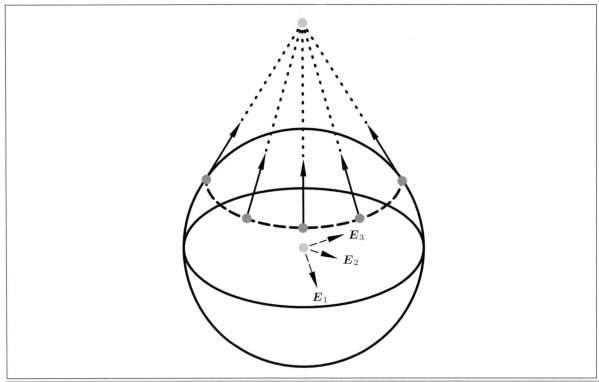

Fig. 4.1. Ruled surface of type circular cone $\mathbb{C}^2_{R\cos\Phi_0}$, directrix: parallel circle $\mathbb{S}^1_{R\cos\Phi_0}$, generating vector field $C_\Phi(\Lambda,\Phi_0)$.

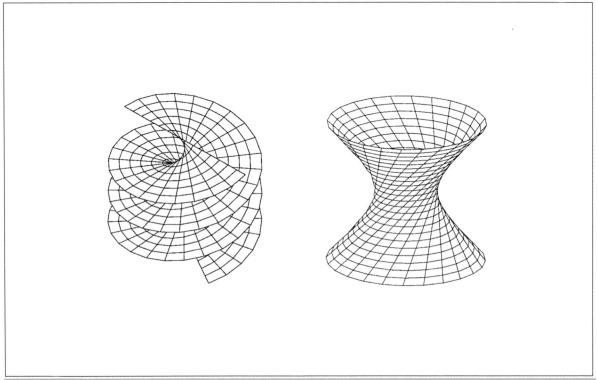

Fig. 4.2. Helicoid as ruled surface, one-sheeted hyberboloid of revolution as ruled surface.

Box 4.1 (Circular cone $\mathbb{C}^2_{R\cos\Phi_0}$ of the sphere $\mathbb{S}^2_R$: ruled surface).

Spherical coordinates and Cartan's frame of reference:

$$X(\Lambda,\Phi) = \begin{bmatrix} E_1, E_2, E_3 \end{bmatrix} \begin{bmatrix} R\cos\Phi\cos\Lambda \\ R\cos\Phi\sin\Lambda \\ R\sin\Phi \end{bmatrix} , \tag{4.1}$$

$$C_\Lambda := \frac{D_\Lambda X(\Lambda,\Phi)}{\|D_\Lambda X(\Lambda,\Phi)\|} = \begin{bmatrix} E_1, E_2, E_3 \end{bmatrix} \begin{bmatrix} -\sin\Lambda \\ +\cos\Lambda \\ 0 \end{bmatrix} ,$$

$$C_\Phi := \frac{D_\Phi X(\Lambda,\Phi)}{\|D_\Phi X(\Lambda,\Phi)\|} = \begin{bmatrix} E_1, E_2, E_3 \end{bmatrix} \begin{bmatrix} -\sin\Phi\cos\Lambda \\ -\sin\Phi\sin\Lambda \\ \cos\Phi \end{bmatrix} . \tag{4.2}$$

Directrix $x(U)$ (parallel circle, small circle, curve of constant latitude Φ_0, $U = \Lambda$):

$$x(U) := \begin{bmatrix} E_1, E_2, E_3 \end{bmatrix} \begin{bmatrix} R\cos\Phi_0\cos U \\ R\cos\Phi_0\sin U \\ R\sin\Phi_0 \end{bmatrix} . \tag{4.3}$$

Generator or ruler of the surface ($U = \Lambda$):

$$Y(U) := C_\Phi(\Phi_0) = \begin{bmatrix} E_1, E_2, E_3 \end{bmatrix} \begin{bmatrix} -\sin\Phi_0\cos U \\ -\sin\Phi_0\sin U \\ \cos\Phi_0 \end{bmatrix} . \tag{4.4}$$

Parameterized ruled surface:

$$X(U,V) = x(U) + V Y(U) ,$$

$$X(U,V) = \begin{bmatrix} E_1, E_2, E_3 \end{bmatrix} \begin{bmatrix} R\cos\Phi_0\cos U - V\sin\Phi_0\cos U \\ R\cos\Phi_0\sin U - V\sin\Phi_0\sin U \\ R\sin\Phi_0 + V\cos\Phi_0 \end{bmatrix} ,$$

$$x(U) = \begin{bmatrix} E_1, E_2, E_3 \end{bmatrix} \begin{bmatrix} R\cos\Phi_0\cos U \\ R\cos\Phi_0\sin U \\ R\sin\Phi_0 \end{bmatrix} \in \mathbb{S}^1_{R\cos\Phi_0} ,$$

$$Y(U) = \begin{bmatrix} E_1, E_2, E_3 \end{bmatrix} \begin{bmatrix} -\sin\Phi_0\cos U \\ -\sin\Phi_0\sin U \\ \cos\Phi_0 \end{bmatrix} =: C_\Phi . \tag{4.5}$$

Matrix of the metric:

$$G_{11} = E = \|D_U x\|^2 + V^2 \|D_U Y\|^2 = \|D_U X(U,V)\|^2 , \quad G_{11} = R^2\cos^2\Phi_0 + 1 ; \tag{4.6}$$

$$G_{12} = G_{21} = F = \langle D_U x(U)\,|\,Y(U)\rangle = \langle D_U X(U,V)\,|\,D_V X(U,V)\rangle , \quad G_{12} = G_{21} = 0 ; \tag{4.7}$$

$$G_{22} = G = \|Y(U)\|^2 = \|D_V X(U,V)\|^2 , \quad G_{22} = 1 ; \tag{4.8}$$

$$\mathsf{G} = \begin{bmatrix} 1 + R^2\cos^2\Phi_0 & 0 \\ 0 & 1 \end{bmatrix} . \tag{4.9}$$

Definition 4.1 (Ruled surface).

A surface is called *ruled surface* if there exists a parameterization of the continuity class $\mathbb{C}^2$ of type $\boldsymbol{X}(U\ V) = \boldsymbol{x}(U) + V\boldsymbol{Y}(U)$, where $\boldsymbol{x}(U)$ is a differentiable curve and $\boldsymbol{Y}(U)$ is a vector field along the curve $\boldsymbol{x}(U)$ which vanishes nowhere.

End of Definition.

The matrix of the metric, G, associated with a ruled surface is a typical element of this kind. Compare with Box 4.2, where we have computed the matrix G.

Box 4.2 (The matrix of the metric of a ruled surface).

$$G_{11} = E = \| D_U \boldsymbol{x} \|^2 + V^2 \| D_U \boldsymbol{Y} \|^2 = \| D_U \boldsymbol{X}(U\ V) \|^2 \tag{4.10}$$

$$G_{12} = G_{21} = F = \langle D_U \boldsymbol{x}(U) \,|\, \boldsymbol{Y}(U) \rangle = \langle D_U \boldsymbol{X}(U\ V) \,|\, D_V \boldsymbol{X}(U\ V) \rangle \tag{4.11}$$

$$G_{22} = G = \| \boldsymbol{Y}(U) \|^2 = \| D_V \boldsymbol{X}(U\ V) \|^2 \tag{4.12}$$

$$\mathsf{G} = \begin{bmatrix} E & F \\ F & G \end{bmatrix} = \begin{bmatrix} \| D_U \boldsymbol{x} \|^2 + V^2 \| D_U \boldsymbol{Y} \|^2 & \langle D_U \boldsymbol{x}(U) \,|\, \boldsymbol{Y}(U) \rangle \\ \langle D_U \boldsymbol{x}(U) \,|\, \boldsymbol{Y}(U) \rangle & \| \boldsymbol{Y}(U) \|^2 \end{bmatrix} \tag{4.13}$$

4-2 Developable surfaces

Developable surfaces (equivalence theorem for ruled surfaces, Gauss flat surfaces, tangent developable: developable helicoid.)

A ruled surface is called *developable* if it can be locally mapped to the plane, preserving the metric of the surface and the generating lines. One of the lines that lies in the plane and afterwards strips of the surface is developed on both sides of the plane, preserving both angles and lengths.

Theorem 4.2 (Equivalence theorem for ruled surfaces).

For a ruled surface, the following conditions are equivalent. (i) The surface is developable. (ii) The surface is Gauss flat: $k = 0$. (iii) Along each of the straight lines, the surface normales are parallel.

End of Theorem.

A ruled surface which satisfies one of the conditions (i), (ii), or (iii) is also called a *torse* or a *developable*. Every surface element without planar points which is Gauss flat ($k = 0$) is a ruled surface.

Theorem 4.3 (Torse, developable).

An open and dense subset of every torse consists of (i) planes, (ii) cylinders, (iii) cones, and (iv) tangent developables, namely ruled surfaces for which the vector $\boldsymbol{Y}$ is tangent to the directrix $\boldsymbol{x}$.

End of Theorem.

A detailed proof of Theorem 4.3 is given by W. S. Massey (1962) as well as by W. Kuehnel (2002, pp. 86–89). Here, we illustrate Gauss flat surfaces of type (i) plane, (ii) cylinder, (iii) cone, and (iv) tangent developable ("developable helicoid") in Figs. 4.3–4.6. In contrast, Fig. 4.7 illustrates a Gauss flat surface which is not a ruled surface based upon two segments of a cone.

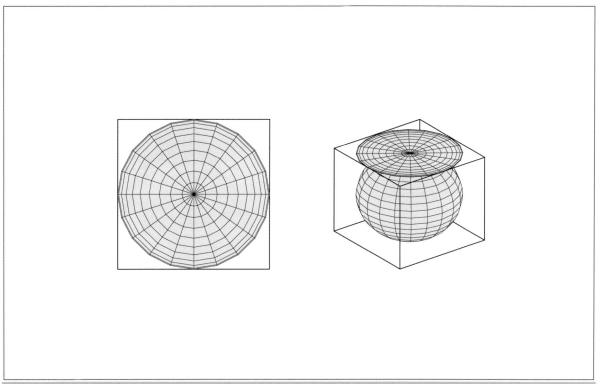

Fig. 4.3. Gauss flat surface ($k = 0$) of type plane. Here: tangent plane of the sphere attached to the North Pole: "developable".

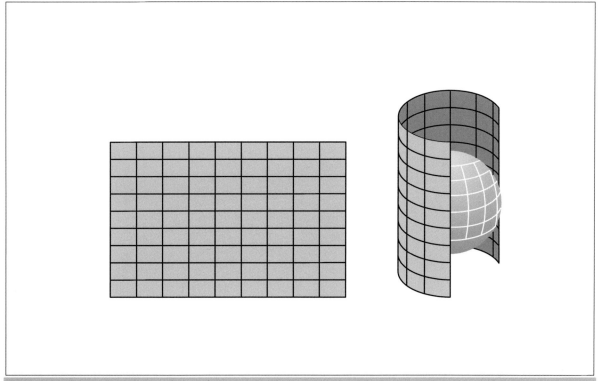

Fig. 4.4. Gauss flat surface ($k = 0$) of type cylinder. Here: cylinder wrapping the sphere, equator is the line-of-contact: "developable".

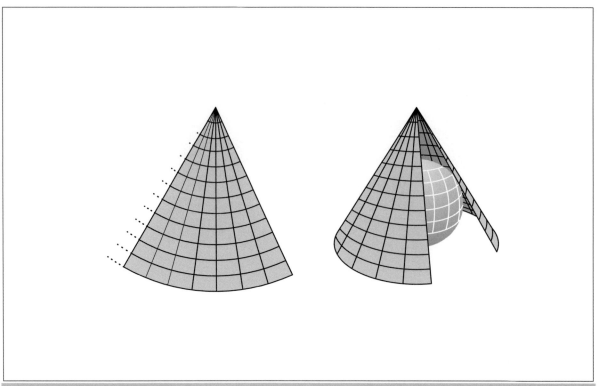

Fig. 4.5. Gauss flat surface ($k = 0$) of type conus. Here: cone wrapping the sphere, a special parallel circle is the line-of-contact: "developable".

Fig. 4.6. Tangent developable of a helix ("developable helix").

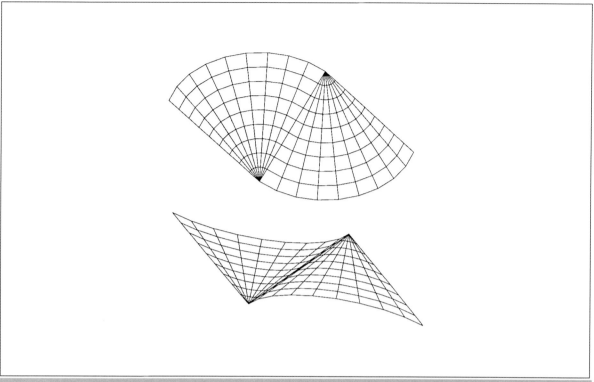

Fig. 4.7. Gauss flat surface ($k = 0$) which is not a ruled surface.

In the following chapter, focussing on the polar aspect, we study the mapping of the sphere to a tangential plane.

5 "Sphere to tangential plane": polar (normal) aspect

Mapping the sphere to a tangential plane: polar (normal) aspect. Equidistant, conformal, and equal area mappings. Normal perspective mappings. Pseudo-azimuthal mapping. Wiechel polar pseudo-azimuthal mapping. Northern tangential plane, equatorial plane, southern tangential plane. Gnomonic and orthographic projections. Lagrange projection.

For mapping local and regional areas, maps of a surface (for instance, a topographic surface $\mathbb{TOP}$, a reference figure of a celestial body like the sphere $\mathbb{S}^2_R$, a reference figure of a celestial body like the ellipsoid-of-revolution $\mathbb{E}^2_{A_1,A_2}$, or a reference figure of a celestial body like the triaxial ellipsoid $\mathbb{E}^2_{A_1,A_2,A_3}$) onto a tangential plane are without competition. In this introductory chapter, we focus on mapping the sphere to a tangential plane, which is located either at the North Pole or at the South Pole. Such a placement of the plane "we map onto" is conventionally called *polar aspect*. Since the spherical coordinate Λ coincides with the polar coordinate α of a point in the tangent plane, the mapping is called *azimuthal mapping*: $\alpha = \Lambda$. Later on, we generalize from the polar aspect to the transverse aspect, finally to the oblique aspect. For a first impression, consult Fig. 5.1.

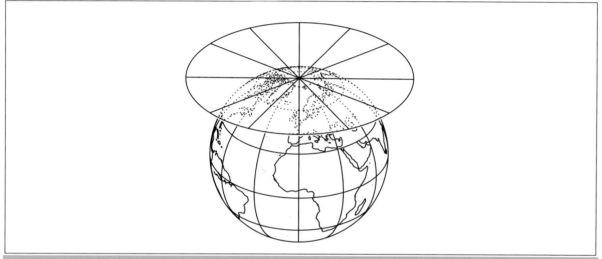

Fig. 5.1. Mapping the sphere to a tangential plane: polar aspect. Point-of-contact: North Pole. Parameters: $\Lambda_0 \in [0°, 360°]$, $\Phi_0 = 90°$.

A first set of maps is illustrated by the magic triangle that is depicted in Fig. 5.2. From the canonical postulates of principal stretches (i) $\Lambda_2 = 1$, (ii) $\Lambda_1 = \Lambda_2$, and (iii) $\Lambda_1 \Lambda_2 = 1$, we generate the differential equations which characterize (i) an *equidistant mapping*, (ii) a *conformal mapping* (conformeomorphism), and (iii) an *equiareal mapping* (areomorphism). These characteristic differential equations are uniquely solved with respect to a properly chosen initial value. The related maps are called (i) *Postel's map*, (ii) *Universal Polar Stereographic (UPS) map*, and (iii) *Lambert's map*. In addition, we produce a second set of maps called *normal perspective*. We identify the perspective center, the line-of-sight, and the line-of-contact, and we discuss the minimal and complete atlas. The guided tour through the world of azimuthal projective maps brings us to special maps, which are called (i) the *gnomonic projection*, (ii) the *orthographic projection*, and (iii) the *Lagrange projection*, and which are pointed out by Fig. 5.3. Finally, we answer the key question: What are the best polar azimuthal mappings of the sphere to the plane?

Historical aside.

Note that the gnomonic projection is believed to has been invented by Thales of Milet (1st half of 6th century B.C.), the equiareal azimuthal projection has been invented by J. H. Lambert (∗ 26 August 1728, Muelhausen, Elsass; † 25 September 1777, Berlin), and G. Postel used the equidistant azimuthal projection for a first map of France (1568, 1570).

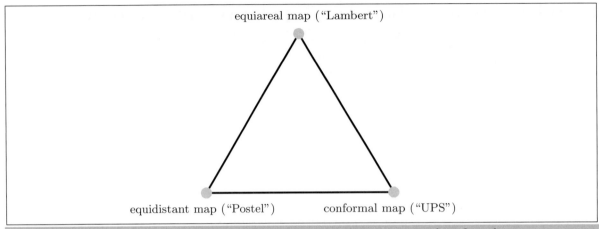

Fig. 5.2. The magic triangle: equiareal map, equidistant map, and conformal map.

The characteristics of the sphere $\mathbb{S}_R^2$ with radius R are reviewed by its ID card. Such an ID card is a list of (i) the embedding of the sphere $\mathbb{S}_R^2$ into a three-dimensional Euclidean space $\{\mathbb{R}^3, \mathsf{I}_3\}$ which is equipped with a canonical metric (the matrix of the metric is the unit matrix, namely $\mathsf{I}_3 = \mathrm{diag}[1, 1, 1]$); (ii) the Frobenius matrix F whose elements are called $\{a, b, c, d\}$ (the Frobenius matrix maps a two leg of tangent vectors to a two leg which is orthonormal and is also called *Cartan frame* of reference); (iii) the matrix $\mathsf{G} = \mathsf{J}^*\mathsf{J}$ of the metric of $\mathbb{S}_R^2$ whose elements are called $\{e, f, g\}$ (the letter G has been chosen in honour of C. F. Gauss, the matrix G builds the first fundamental form I: $\mathrm{d}s^2 = [\mathrm{d}\Lambda, \mathrm{d}\Phi]\mathsf{G}[\mathrm{d}\Lambda, \mathrm{d}\Phi]^*$); (iv) the matrix H of second derivatives, which is defined by $\left[\langle \boldsymbol{G}_3 | \partial^2 \boldsymbol{X} / \partial U^K \partial U^L \rangle\right]$ with respect to the surface normal vector $\boldsymbol{G}_3$ and the embedding function $\boldsymbol{X} = \boldsymbol{X}(U^1, U^2)$ or $\boldsymbol{X} = \boldsymbol{X}(\Lambda, \Phi)$ (the letter H has been chosen in honour of L. O. Hesse, the matrix H builds the second fundamental form II: $[\mathrm{d}\Lambda, \mathrm{d}\Phi]\mathsf{H}[\mathrm{d}\Lambda, \mathrm{d}\Phi]^*$, the elements of the Hesse matrix are denoted by $\{l, m, n\}$); (v) the Jacobi matrix J of the first derivatives of the embedding function, precisely $[D_\Lambda X, D_\Lambda Y, D_\Lambda Z,]$ and $[D_\Phi X, D_\Phi Y, D_\Phi Z,]$ (the letter J has been chosen in honour of C. G. J. Jacobi); the curvature matrix $\mathsf{K} = -\mathsf{H}\mathsf{G}^{-1}$, its negative trace taken half (denoted by the letter h), also called *mean curvature*, and its determinant (denoted by the letter k), also called *Gaussian curvature*; (vi) the *Christoffel symbols* of the second kind, which are named after E. B. Christoffel (∗ 10 November 1829, Monschau; † 15 March 1900, Strassburg), which are used to compute *geodesics*, and which are defined by

$$\begin{Bmatrix} M \\ K\,L \end{Bmatrix} = \frac{1}{2}G^{MN}(D_K G_{NL} + D_L G_{KN} - D_N G_{KL})\,. \tag{5.1}$$

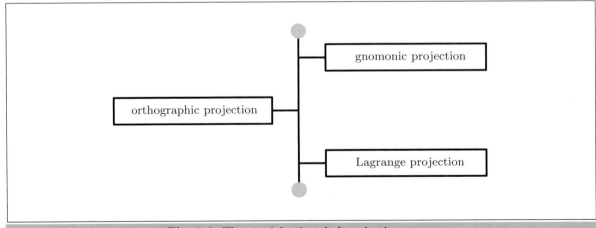

Fig. 5.3. The special azimuthal projective maps.

In Box 5.1, the ID card of the sphere $\mathbb{S}^2_R$ is summarized. According to the above considerations, F (with elements a, b, c, d) is the Frobenius matrix, G (with elements e, f, g) is the Gauss matrix, H (with elements l, m, n) is the Hesse matrix, J is the Jacobi matrix, and K is the curvature matrix, leading to the mean curvature h and to the Gaussian curvature k, and

$$
\mathsf{G} = \mathsf{J}^* \mathsf{J} , \quad \mathsf{H} = \left[\left\langle \boldsymbol{G}_3 \left| \frac{\partial^2 \boldsymbol{X}}{\partial U^K \partial U^L} \right. \right\rangle \right] = \left[\left\langle \boldsymbol{G}_3 | \boldsymbol{X}_{KL} \right\rangle \right] , \quad \mathsf{J} = \left[\frac{\partial X^J}{\partial U^K} \right] ,
$$

$$
\mathsf{K} = -\mathsf{H}\mathsf{G}^{-1} , \quad h = -\frac{1}{2}\mathrm{tr}[\mathsf{K}] , \quad k = \det[\mathsf{K}] .
$$

(5.2)

Box 5.1 (ID card of the sphere $\mathbb{S}^2_R$).

Spherical coordinates (1st chart: Λ, Φ):

$$
\{\Lambda, \Phi, R\} \to \{X, Y, Z\} :
$$

$$
\boldsymbol{X}(\Lambda, \Phi, R) = \boldsymbol{E}_1 R \cos \Phi \cos \Lambda + \boldsymbol{E}_2 R \cos \Phi \sin \Lambda + \boldsymbol{E}_3 R \sin \Phi \in \{\mathbb{R}^3, \mathsf{I}_3\} ;
$$

$$
\{X, Y, Z\} \to \{\Lambda, \Phi, R\} :
$$

(5.3)

$$
\Lambda(\boldsymbol{X}) = \arctan \frac{Y}{X} + 180° \left[-\frac{1}{2}\mathrm{sgn}Y - \frac{1}{2}\mathrm{sgn}Y\,\mathrm{sgn}X + 1 \right] , \quad \Phi(\boldsymbol{X}) = \arctan \frac{Z}{\sqrt{X^2 + Y^2}} ,
$$

$$
R = \sqrt{X^2 + Y^2 + Z^2} .
$$

Matrices F, G, H, J, K, and I (elements: a, b, c, d; e, f, g; l, m, n):

$$
\mathsf{F} = \begin{bmatrix} a & b \\ c & d \end{bmatrix} = \begin{bmatrix} \dfrac{1}{R \cos \Phi} & 0 \\ 0 & \dfrac{1}{R} \end{bmatrix} = \begin{bmatrix} \dfrac{1}{\sqrt{G_{11}}} & 0 \\ 0 & \dfrac{1}{\sqrt{G_{22}}} \end{bmatrix} \in \mathbb{R}^{2\times 2} ,
$$

(5.4)

$$
\mathsf{G} = \begin{bmatrix} e & f \\ f & g \end{bmatrix} = \begin{bmatrix} R^2 \cos^2 \Phi & 0 \\ 0 & R^2 \end{bmatrix} \in \mathbb{R}^{2\times 2} , \quad \mathsf{H} = \begin{bmatrix} l & m \\ m & n \end{bmatrix} = \begin{bmatrix} -R \cos^2 \Phi & 0 \\ 0 & -R \end{bmatrix} \in \mathbb{R}^{2\times 2} ,
$$

(5.5)

$$
\mathsf{J} = \begin{bmatrix} -R \cos \Phi \sin \Lambda & -R \sin \Phi \cos \Lambda \\ +R \cos \Phi \cos \Lambda & -R \sin \Phi \sin \Lambda \\ 0 & R \cos \Phi \end{bmatrix} \in \mathbb{R}^{3\times 2} , \quad \mathsf{K} = \begin{bmatrix} \dfrac{1}{R} & 0 \\ 0 & \dfrac{1}{R} \end{bmatrix} \in \mathbb{R}^{2\times 2} ,
$$

(5.6)

$$
h = -\frac{1}{R} , \quad k = \frac{1}{R^2} , \quad \mathsf{I} = \mathsf{I}_2 = \begin{bmatrix} 1 & 0 \\ 0 & 1 \end{bmatrix} \in \mathbb{R}^{2\times 2} .
$$

(5.7)

Christoffel symbols:

$$
\left\{ {1 \atop 1\,1} \right\} = \left\{ {1 \atop 2\,2} \right\} = \left\{ {2 \atop 1\,2} \right\} = \left\{ {2 \atop 2\,2} \right\} = 0 , \quad \left\{ {1 \atop 1\,2} \right\} = -\tan \Phi , \quad \left\{ {2 \atop 1\,1} \right\} = \sin \Phi \cos \Phi = \frac{1}{2}\sin 2\Phi .
$$

(5.8)

5-1 General mapping equations

Setting up general equations of the mapping "sphere to plane": the azimuthal projection in the normal aspect (polar aspect).

There are two basic postulates which govern the setup of general equations of mapping the sphere $\mathbb{S}^2_R$ of radius R to a tangential plane $T_{\boldsymbol{X}_0}\mathbb{S}^2_R$, which is attached to a point $\boldsymbol{X}_0 \in T\mathbb{S}^2_R$. Let the tangential plane be covered by polar coordinates $\{\alpha, r\}$. Then these postulates read as follows.

Postulate.

The polar coordinate α, which is also called *azimuth*, is identical to the spherical longitude, i.e. $\alpha = \Lambda$.

End of Postulate.

Postulate.

The polar coordinate r depends exclusively on the spherical latitude Φ or on the spherical colatitude $\Delta := \pi/2 - \Phi$, i.e. $r = \sqrt{x^2 + y^2} = f(\Delta) = f(\pi/2 - \Phi)$. If $\Phi = \pi/2$ or, equivalently, $\Delta = 0$, then $f(0) = 0$ holds.

End of Postulate.

In last consequence, the general equations of an azimuthal mapping are provided by the following vector equation:

$$\begin{bmatrix} x \\ y \end{bmatrix} = \begin{bmatrix} r \cos \alpha \\ r \sin \alpha \end{bmatrix} = \begin{bmatrix} f(\Delta) \cos \Lambda \\ f(\Delta) \sin \Lambda \end{bmatrix} . \tag{5.9}$$

Question.

Question: "How do the images of the coordinate line $\Lambda = $ constant and the coordinate line $\Phi = $ constant look like?" Answer ($y = x \tan \Lambda$ and $\Lambda = $ constant $= $ meridian): "The image of the meridian $\Lambda = $ constant under an azimuthal mapping is the radial straight line." Answer ($x^2 + y^2 = r^2 = f^2(\Delta)$ and $\Delta = $ constant $= $ parallel circle): "The image of the parallel circle $\Delta = $ constant (or $\Phi = $ constant) under an azimuthal mapping is the circle $\mathbb{S}_r^1$ of radius $r = f(\Delta)$. Such a mapping is called *concircular*."

Proof ($y = x \tan \Lambda$, $\Lambda = $ constant $= $ meridian).

Solve the first equation towards $f(\Delta) = x/\cos \Lambda$ and substitute $f(\Delta)$ in the second equation such that the following equation holds:

$$y = f(\Delta) \sin \Lambda = x \sin \Lambda / \cos \Lambda = x \tan \Lambda . \tag{5.10}$$

End of Proof ($y = x \tan \Lambda$, $\Lambda = $ constant $= $ meridian).

Proof ($x^2 + y^2 = r^2 = f^2(\Delta)$, $\Delta = $ constant $= $ parallel circle).

Compute the terms x^2 and y^2 and add the two:

$$x^2 + y^2 = f^2(\Delta) . \tag{5.11}$$

End of Proof ($x^2 + y^2 = r^2 = f^2(\Delta)$, $\Delta = $ constant $= $ parallel circle).

In summary, the images of the meridian and the parallel circle constitute the typical graticule of an azimuthal mapping, i.e.

$$\text{meridians } (\Lambda = \text{constant}) \quad \longrightarrow \quad \text{radial straight lines} ,$$

$$\text{parallel circles} \begin{pmatrix} \Delta = \text{constant} \\ \Phi = \text{constant} \end{pmatrix} \quad \longrightarrow \quad \text{equicentric circles} . \tag{5.12}$$

Box 5.2 shows a collection of formulae which describe the left Jacobi matrix J_l as well as the left Cauchy–Green matrix C_l for an azimuthal mapping $\mathbb{S}_R^2 \rightarrow \mathbb{P}_{\mathcal{O}}^2$. The left pair of matrices $\{C_l, G_l\}$ is canonically characterized by the left principal stretches Λ_1 and Λ_2 in their general form.

Box 5.2 ("Sphere to plane", distortion analysis, azimuthal projection, left principal stretches).

Parameterized mapping:

$$\alpha = \Lambda \,,$$

$$r = f(\Delta) \,,$$

$$x = r\cos\alpha = f(\Delta)\cos\Lambda \,,$$

$$y = r\sin\alpha = f(\Delta)\sin\Lambda \,. \tag{5.13}$$

Left Jacobi matrix:

$$J_l := \begin{bmatrix} D_\Lambda x & D_\Delta x \\ D_\Lambda y & D_\Delta y \end{bmatrix} = \begin{bmatrix} -f(\Delta)\sin\Lambda & f'(\Delta)\cos\Lambda \\ +f(\Delta)\cos\Lambda & f'(\Delta)\sin\Lambda \end{bmatrix} \,. \tag{5.14}$$

Left Cauchy–Green matrix $(G_r = I_2)$:

$$C_l = J_l^* G_r J_l = \begin{bmatrix} f^2(\Delta) & 0 \\ 0 & f'^2(\Delta) \end{bmatrix} \,. \tag{5.15}$$

Left principal stretches:

$$\Lambda_1 = +\sqrt{\frac{c_{11}}{G_{11}}} = \frac{f(\Delta)}{R\sin\Delta} \,,$$

$$\Lambda_2 = +\sqrt{\frac{c_{22}}{G_{22}}} = \frac{f'(\Delta)}{R} \,. \tag{5.16}$$

Left eigenvectors of the matrix pair $\{C_l, G_l\}$:

$$C_1 = \boldsymbol{E}_\Lambda = \frac{D_\Lambda \boldsymbol{X}}{\|D_\Lambda \boldsymbol{X}\|}$$

(Easting) ,

$$C_2 = \boldsymbol{E}_\Phi = \frac{D_\Phi \boldsymbol{X}}{\|D_\Phi \boldsymbol{X}\|}$$

(Northing) . $\tag{5.17}$

Next, we specialize the general azimuthal mapping to generate an equidistant mapping, a series of conformal mappings (stereographic projections) and an equiareal mapping.

5-2 Special mapping equations

Setting up special mappings "sphere to plane": azimuthal projections in the normal aspect (polar aspect). Equidistant Polar Mapping (EPM), Universal Polar Stereographic Projection (UPS). Conformal mapping, equiareal mapping, normal projective mapping.

5-21 Equidistant mapping (Postel projection)

Let us postulate an *equidistant mapping* of the family of meridians $\Lambda = $ constant, namely the mapping $r = f(\Delta) = R\Delta$. Indeed, $R \, \text{arc}(\pi/2 - \Phi) = r$ generates such a simple equidistant mapping, which we illustrate by means of Fig. 5.4. The corresponding distortion analysis is systematically presented in Box 5.3. The *EPM* (*Equidistant Polar Mapping*) is finally summarized in Lemma 5.1.

Box 5.3 (Equidistant mapping of the sphere to the tangential plane at the North Pole).

Parameterized mapping:

$$\alpha = \Lambda \, , \quad r = f(\Delta) = R\Delta \, ,$$

$$x = r\cos\alpha = R\Delta\cos\Lambda = R\left(\frac{\pi}{2} - \Phi\right)\cos\Lambda \, , \quad y = r\sin\alpha = R\Delta\sin\Lambda = R\left(\frac{\pi}{2} - \Phi\right)\sin\Lambda \, . \tag{5.18}$$

Left principal stretches:

$$\Lambda_1 = \frac{\Delta}{\sin\Delta} = \frac{\frac{\pi}{2} - \Phi}{\cos\Phi} \, , \quad \Lambda_2 = 1 \, . \tag{5.19}$$

Left eigenvectors:

$$\boldsymbol{C}_1\Lambda_1 = \boldsymbol{E}_\Lambda \frac{\Delta}{\sin\Delta} \quad (\text{Easting}) \, , \quad \boldsymbol{C}_2\Lambda_2 = \boldsymbol{E}_\Phi \quad (\text{Northing}) \, . \tag{5.20}$$

Parameterized inverse mapping:

$$\tan\Lambda = \frac{y}{x} \, , \quad \Delta = \sqrt{\frac{x^2 + y^2}{R^2}} \, . \tag{5.21}$$

Left maximal angular distortion:

$$\Omega_l = 2\arcsin\left|\frac{\Lambda_1 - \Lambda_2}{\Lambda_1 + \Lambda_2}\right| = 2\arcsin\left|\frac{\Delta - \sin\Delta}{\Delta + \sin\Delta}\right| \, . \tag{5.22}$$

Lemma 5.1 (EPM, equidistant mapping of the sphere to the tangential plane at the North Pole).

The equidistant mapping of the sphere to the tangential plane at the North Pole, in short, EPM (Equidistant Polar Mapping), is parameterized by

$$x = R\Delta\cos\Lambda = R\left(\frac{\pi}{2} - \Phi\right)\cos\Lambda \, , \quad y = R\Delta\sin\Lambda = R\left(\frac{\pi}{2} - \Phi\right)\sin\Lambda \, , \tag{5.23}$$

subject to the left Cauchy–Green eigenspace $\left\{\boldsymbol{E}_\Lambda\frac{\Delta}{\sin\Delta}, \boldsymbol{E}_\Phi\right\}$.

End of Lemma.

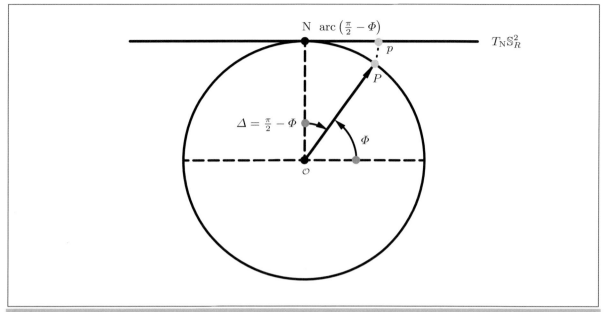

Fig. 5.4. A spherical vertical section, an equidistant mapping of the sphere to the tangential plane at the North Pole.

Historical aside. The equidistant mapping of the sphere to the tangential plane at the North Pole is associated with the name of G. Postel (1581), though it was already known to G. Mercator (1569). Both used it for mapping the polar regions. Nowadays, it is applied for plotting stars around the North Pole, for the World Map 1:2.5 Mio, and for charts in aerial navigation, remote sensing, and seismology.

In order to complete the considerations, we present to you Fig. 5.5, which shows a sample of a polar equidistant map of the sphere.

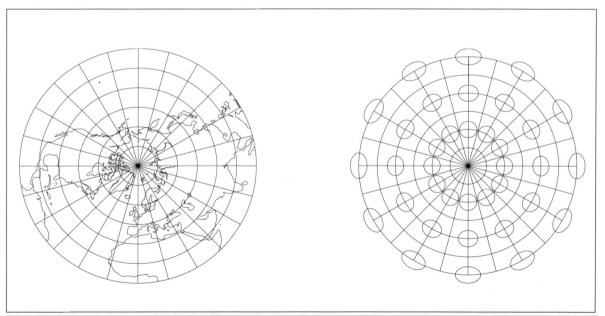

Fig. 5.5. An equidistant mapping of the sphere $\mathbb{S}^2_R$ onto the tangent space $T_N\mathbb{S}^2_R$, Tissot ellipses, polar aspect, graticule $15°$, shorelines.

5-22 Conformal mapping (stereographic projection, UPS)

Let us postulate a *conformal mapping* by means of the canonical measure of conformality, i. e. $\Lambda_1 = \Lambda_2$. Such a conformal mapping of the sphere to the tangential plane of the North Pole is illustrated by means of Fig. 5.6 that follows after the Boxes 5.4 and 5.5.

Question. Question 1: "How can we generate the conformal mapping equations?" Answer 1: "Following the procedure of Boxes 5.4 and 5.5, we here depart from the general representation of Λ_1 and Λ_2. By means of separation of variables, the relation $\Lambda_1 = \Lambda_2$ leads us to $\mathrm{d}f/f = \mathrm{d}\Delta/\sin\Delta$ as the characteristic differential equations. Integration of the left side as well as of the right side leads us to the indefinite mapping equation $f(\Delta) = c\tan\Delta/2$."

Question. Question 2: "How can we gauge the integration constant?" Answer 2: "The postulate $\lim_{\Delta\to 0}\Lambda_2(\Delta) = 1$ that is quoted in Box 5.5 establishes an isometry at the North Pole of the sphere. Indeed, the limit $\Delta \to 0$ of $\Lambda_2(\Delta) = c/(2R\cos^2\Delta/2)$ fixes c as $c = 2R$. Accordingly, the polar coordinate $r = f(\Delta) = 2R\tan\Delta/2$ leads to the parameterized conformal mapping $x = r(\Delta)\cos\Lambda$ and $y = r(\Delta)\sin\Lambda$."

This conformal mapping is called *UPS* (*Universal Polar Stereographic Projection*) for the following reason. Figure 5.6, which illustrates this stereographic projection, focuses on the peripheral angle $\Delta/2 = \pi/4 - \Phi/2$ at the South Pole. Obviously, a projection line departing from the perspective center intersects at $P \in \mathbb{S}_R^2$ and $p \in T_N\mathbb{S}_R^2$. Compare with Lemma 5.2, which summarizes the UPS (Universal Polar Stereographic Projection).

Lemma 5.2 (UPS, conformal mapping of the sphere to the tangential plane at the North Pole).

The conformal mapping of the sphere to the tangential plane at the North Pole, in short, UPS (Universal Polar Stereographic Projection), is parameterized by

$$x = 2R\tan\frac{\Delta}{2}\cos\Lambda =$$

$$= 2R\tan\left(\frac{\pi}{4} - \frac{\Phi}{2}\right)\cos\Lambda\,,$$

$$y = 2R\tan\frac{\Delta}{2}\sin\Lambda = \tag{5.24}$$

$$= 2R\tan\left(\frac{\pi}{4} - \frac{\Phi}{2}\right)\sin\Lambda\,,$$

subject to the left Cauchy–Green eigenspace

$$\text{left CG eigenspace} = \left\{\boldsymbol{E}_\Lambda\frac{1}{\cos^2\left(\frac{\pi}{4} - \frac{\Phi}{2}\right)}, \boldsymbol{E}_\Phi\frac{1}{\cos^2\left(\frac{\pi}{4} - \frac{\Phi}{2}\right)}\right\}\,. \tag{5.25}$$

End of Lemma.

In the case of UPS, the areal distortion increases fast with colatitude (polar distance Δ), namely $\Lambda_1\Lambda_2 - 1 = \cos^{-4}\left(\pi/4 - \Phi/2\right) - 1$ and $\Lambda_1\Lambda_2 - 1 \to \infty$ for $\Delta \to \pi$, and this is the reason for the application of UPS as outlined in the following historical aside.

Box 5.4 (Conformal mapping of the sphere to the tangential plane at the North Pole).

Postulate of a conformeomorphism:

$$\Lambda_1 = \Lambda_2 \ ,$$

$$\frac{f(\Delta)}{R \sin \Delta} = \frac{f'(\Delta)}{R} \Rightarrow \frac{\mathrm{d}f}{f} = \frac{\mathrm{d}\Delta}{\sin \Delta} \ . \tag{5.26}$$

Integration of the characteristic differential equation of a conformal mapping $\mathbb{S}_R^2 \to T_N \mathbb{S}_R^2$:

$$\int \frac{\mathrm{d}x}{\sin x} = \ln \left| \tan \frac{x}{2} \right| \ , \quad \int \frac{\mathrm{d}y}{y} = \ln y \ ,$$

$$\int \frac{\mathrm{d}f}{f} = \int \frac{\mathrm{d}\Delta}{\sin \Delta} \Leftrightarrow \ln f = \ln \left| \tan \frac{\Delta}{2} \right| + \ln c \ , \quad f(\Delta) = c \tan \frac{\Delta}{2} \, \forall \Delta \in \,]0, \pi[\ . \tag{5.27}$$

Parameterized conformal mapping:

$$\begin{bmatrix} x \\ y \end{bmatrix} = 2R \tan \frac{\Delta}{2} \begin{bmatrix} \cos \Lambda \\ \sin \Lambda \end{bmatrix} = 2R \tan \left(\frac{\pi}{4} - \frac{\Phi}{2} \right) \begin{bmatrix} \cos \Lambda \\ \sin \Lambda \end{bmatrix} \ . \tag{5.28}$$

Left principal stretches:

$$\Lambda_1 = \Lambda_2 = \frac{1}{\cos^2 \frac{\Delta}{2}} = \frac{1}{\cos^2 \left(\frac{\pi}{4} - \frac{\Phi}{2} \right)} \ . \tag{5.29}$$

Left eigenvectors:

$$\boldsymbol{C}_1 \Lambda_1 = \boldsymbol{E}_\Lambda \frac{1}{\cos^2 \left(\frac{\pi}{4} - \frac{\Phi}{2} \right)} \ (\text{``Easting''}) \ , \quad \boldsymbol{C}_2 \Lambda_2 = \boldsymbol{E}_\Phi \frac{1}{\cos^2 \left( \frac{\pi}{4} - \frac{\Phi}{2} \right)} \ (\text{``Northing''}) \ . \tag{5.30}$$

Left angular shear:

$$\sum\nolimits_l = \Psi_l - \Psi_r = 0 \ , \quad \Omega_l = 0 \ . \tag{5.31}$$

Parameterized inverse mapping:

$$\tan \Lambda = \frac{y}{x} \ , \quad \tan \frac{\Delta}{2} = \frac{1}{2R} \sqrt{x^2 + y^2} \ . \tag{5.32}$$

Box 5.5 (Distortion analysis at the North Pole).

Postulate of an isometry at the North Pole:

$$\lim_{\Delta \to 0} \Lambda_2(\Delta) = 1 \ . \tag{5.33}$$

Eigenspace analysis of the matrix pair $\{\mathsf{C}_l, \mathsf{G}_l\}$:

$$\Lambda_2 = \frac{f'(\Delta)}{R} \ , \ f'(\Delta) = \frac{\mathrm{d}f}{\mathrm{d}\Delta} = \frac{c}{2} \frac{1}{\cos^2 \frac{\Delta}{2}} \Rightarrow \Lambda_2 = \frac{c}{2R} \frac{1}{\cos^2 \frac{\Delta}{2}} \ ,$$

$$\Lambda_2 = \frac{c}{2R} \frac{1}{\cos^2 \frac{\Delta}{2}} \ , \ \lim_{\Delta \to 0} \Lambda_2(\Delta) = 1 \Rightarrow c = 2R \ , \tag{5.34}$$

$$\frac{\mathrm{d}f}{\mathrm{d}\Delta} = \frac{R}{\cos^2 \frac{\Delta}{2}} \ , \ r = f(\Delta) = 2R \tan \frac{\Delta}{2} \ .$$

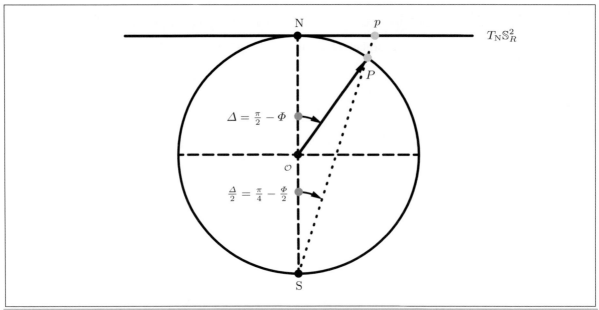

Fig. 5.6. A spherical vertical section, a conformal mapping of the sphere to the tangential plane at the North Pole, UPS (Universal Polar Stereographic Projection). This stereographic projection is highlighted by the radial coordinate $r = f(\Delta) = 2R \tan \Delta/2 = \mathrm{N}p$.

Historical aside. Note that the UPS (Universal Polar Stereographic Projection) is already found in Hipparch's works. Nowadays, it is applied for charts in aerial navigation, for the World Map of polar regions at latitudes larger than $+80°$, namely substituting the UTM (Universal Transverse Mercator Projection).

In order to complete the considerations, we present to you Fig. 5.7, which shows a sample of the UPS (Universal Polar Stereographic Projection) of the sphere.

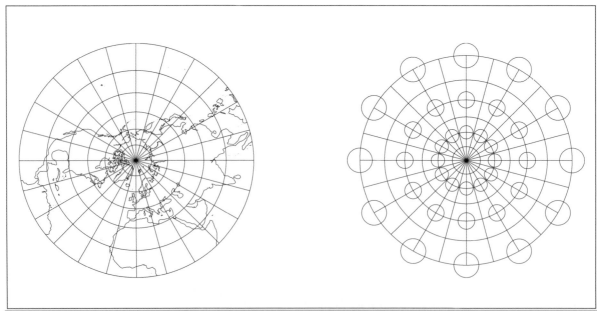

Fig. 5.7. Conformal map (UPS) of the sphere $\mathbb{S}_R^2$ onto the tangent space $T_{\mathrm{N}}\mathbb{S}_R^2$, Tissot ellipses, polar aspect, graticule $15°$, shorelines.

5-23 Equiareal mapping (Lambert projection)

Let us postulate an *equiareal mapping* by means of the canonical measure of areomorphism, i.e. $\Lambda_1 \Lambda_2 = 1$. Such an equiareal mapping of the sphere to the tangential plane of the North Pole is illustrated by means of Fig. 5.8 that follows after Box 5.6.

Question: "How can we construct the equiareal mapping equations?" Answer: "Following the procedure of Box 5.6, we here depart from the general representation of Λ_1 and Λ_2. The postulate of an equiareal mapping leads us to the characteristic differential equation, which we solve by separation of variables. We use the initial condition $r(0) = f(0) = 2R \sin \Delta/2$, namely the polar coordinate r as a function of the colatitude Δ, also called *polar distance*. The polar coordinate $\alpha = \Lambda$ is fixed by the postulate of an azimuthal projection. The parameterized equiareal mapping is finally used to compute the left principal stretches, namely $\Lambda_1 = 1/\cos \Delta/2$, $\Lambda_2 = \cos \Delta/2$. They build up the left eigenvectors along the East unit vector $\boldsymbol{E}_\Lambda$ and the North unit vector $\boldsymbol{E}_\Phi$ (the South unit vector is $\boldsymbol{E}_\Delta = -\boldsymbol{E}_\Phi$). These unit vectors are defined by $\boldsymbol{E}_\Lambda := D_\Lambda \boldsymbol{X}/\|D_\Lambda \boldsymbol{X}\|$ and $\boldsymbol{E}_\Phi := D_\Phi \boldsymbol{X}/\|D_\Phi \boldsymbol{X}\|$. In addition, we have computed the left maximal angular shear as well as the parameterized inverse mapping $\{\Lambda(x, y), \Phi(x, y)\}$."

The basic results of the equiareal azimuthal projection of the sphere to the tangential plane at the North Pole are collected in Lemma 5.3.

Lemma 5.3 (Equiareal azimuthal projection of the sphere to the tangential plane at the North Pole).

The equiareal mapping of the sphere to the tangential plane at the North Pole is parameterized by the two equations

$$x =$$
$$= 2R \sin \frac{\Delta}{2} \cos \Lambda =$$
$$= 2R \sin \left(\frac{\pi}{4} - \frac{\Phi}{2} \right) \cos \Lambda \,,$$
$$y =$$
$$= 2R \sin \frac{\Delta}{2} \sin \Lambda =$$
$$= 2R \sin \left(\frac{\pi}{4} - \frac{\Phi}{2} \right) \sin \Lambda \,,$$

(5.35)

subject to the left Cauchy–Green eigenspace

$$\text{left CG eigenspace} = \left\{ \boldsymbol{E}_\Lambda \frac{1}{\cos \left(\frac{\pi}{4} - \frac{\Phi}{2} \right)}, \boldsymbol{E}_\Phi \cos \left(\frac{\pi}{4} - \frac{\Phi}{2} \right) \right\} \,.$$

(5.36)

End of Lemma.

From the sketch that is shown in Fig. 5.8, we gain some geometric understanding of how to construct the normal equiareal mapping by a "pair of dividers and a ruler". The radial coordinate $r = Np$ coincides with the segment $NP = 2R \sin \Delta/2$, the peripheral point P within the vertical section constitutes a rectangular triangle NSP subject to $SN = 2R$.

Box 5.6 (Equiareal mapping of the sphere to the tangential plane at the North Pole).

Postulate of a areomorphism:

$$\Lambda_1 \Lambda_2 = 1 \ ,$$

$$\frac{f(\Delta)}{R^2 \sin \Delta} \frac{\mathrm{d}f(\Delta)}{\mathrm{d}\Delta} = 1 \Rightarrow f(\Delta)\mathrm{d}f(\Delta) = R^2 \sin \Delta \mathrm{d}\Delta \ . \tag{5.37}$$

Integration of the characteristic differential equation of an equiareal mapping $\mathbb{S}_R^2 \to T_\mathrm{N}\mathbb{S}_R^2$ subject to an initial condition:

$$\left[\begin{array}{l} \dfrac{f^2}{2} = -R^2 \cos \Delta + c \\ r(0) = f(0) = 0 \end{array} \right] \Rightarrow 0 = -R^2 + c \Rightarrow c = R^2 \ ,$$

$$\left[\begin{array}{l} f^2 = 2R^2(1 - \cos \Delta) = 4R^2 \sin^2 \dfrac{\Delta}{2} \\ \cos x = 1 - 2\sin^2 \dfrac{x}{2} \end{array} \right] \Rightarrow r = f(\Delta) = 2R \sin \dfrac{\Delta}{2} \ . \tag{5.38}$$

Parameterized equiareal mapping:

$$\begin{bmatrix} x \\ y \end{bmatrix} = 2R \sin \frac{\Delta}{2} \begin{bmatrix} \cos \Lambda \\ \sin \Lambda \end{bmatrix} = 2R \sin \left(\frac{\pi}{4} - \frac{\Phi}{2} \right) \begin{bmatrix} \cos \Lambda \\ \sin \Lambda \end{bmatrix} \ . \tag{5.39}$$

Left principal stretches:

$$f'(\Delta) = R \cos \frac{\Delta}{2} \ ,$$

$$\Lambda_1 = \frac{1}{\cos \frac{\Delta}{2}} = \frac{1}{\cos \left(\frac{\pi}{4} - \frac{\Phi}{2} \right)} \ , \quad \Lambda_2 = \cos \frac{\Delta}{2} = \cos \left(\frac{\pi}{4} - \frac{\Phi}{2} \right) \ ; \tag{5.40}$$

special value (isometry): $\Phi \to \dfrac{\pi}{2} : \quad \lim\limits_{\Phi \to \pi/2} \Lambda_1 = \lim\limits_{\Phi \to \pi/2} \Lambda_2 = 1 \ . \tag{5.41}$

Left eigenvectors:

$$\boldsymbol{C}_1 \Lambda_1 = \boldsymbol{E}_\Lambda \frac{1}{\cos \left(\frac{\pi}{4} - \frac{\Phi}{2} \right)} \qquad (\text{"Easting"}) \ ,$$

$$\boldsymbol{C}_2 \Lambda_2 = \boldsymbol{E}_\Phi \cos \left(\frac{\pi}{4} - \frac{\Phi}{2} \right) \qquad (\text{"Northing"}) \ . \tag{5.42}$$

Left maximal angular distortion:

$$\Omega_l = 2 \arcsin \left| \frac{\Lambda_1 - \Lambda_2}{\Lambda_1 + \Lambda_2} \right| = 2 \arcsin \frac{1 - \cos^2 \frac{\Delta}{2}}{1 + \cos^2 \frac{\Delta}{2}} \ . \tag{5.43}$$

Parameterized inverse mapping:

$$\tan \Lambda = \frac{y}{x} \ ,$$

$$\sin \frac{\Delta}{2} = \frac{1}{2R} \sqrt{x^2 + y^2} \ . \tag{5.44}$$

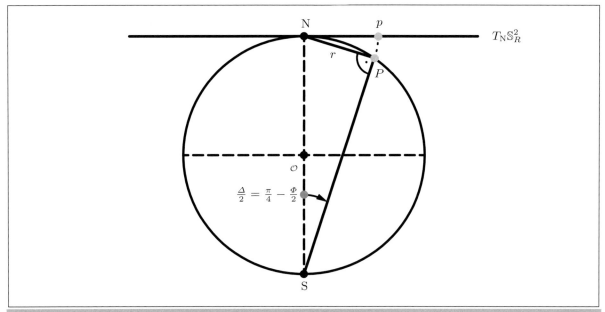

Fig. 5.8. Spherical vertical section, equiareal mapping of the sphere to the tangential plane at the North Pole, azimuthal projection.

Historical aside. The geometric construction to be considered here may have motivated J. H. Lambert (1772) to invent such an equiareal mapping of the sphere. Due to the postulate of an equiareal mapping, the equiareal azimuthal projection of the sphere is very popular in Geostatistics.

In order to complete the considerations, we present to you Fig. 5.9, which shows a sample of the polar equiareal projection of the sphere.

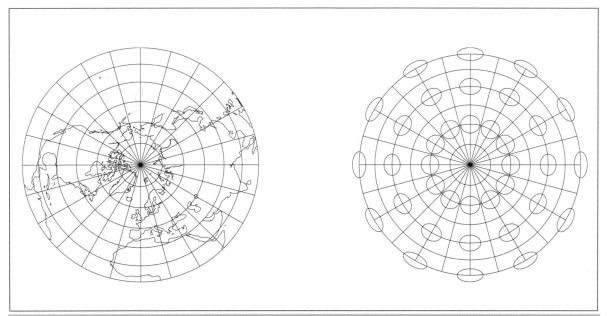

Fig. 5.9. Equiareal map of the sphere $\mathbb{S}_R^2$ onto the tangent space $T_N\mathbb{S}_R^2$, Tissot ellipses, polar aspect, graticule $15°$, shorelines.

5-24 Normal perspective mappings

The general normal perspective mapping of the sphere $\mathbb{S}_R^2$ of radius R to a tangential plane at the North Pole, the South Pole, or an equatorial plane is of focal interest in Mathematical Cartography, in Photogrammetry, in Machine Vision as well as in Aeronautics and Satellite Geodesy. Here, as soon as we have generated the general parameterized mappings of the perspective type, we introduce more specific projections: the *gnomonic projection*, the *orthographic projection*, and the *Lagrange projection*. Based on Figs. 5.10–5.12, we design the elements of a first perspective projection. At first, we locate the *perspective center* at $\mathcal{O}^*$ outside the sphere on the southern axis of symmetry North-Pole–South-Pole. The perspective center $\mathcal{O}^*$ is the origin of a bundle of projection lines, in particular, half straights. Second, we place the projection plane (i) at maximum distance from $\mathcal{O}^*$ at the North Pole to coincide with the tangential plane $T_N\mathbb{S}_R^2$, (ii) at the center $\mathcal{O}$ of the sphere $\mathbb{S}_R^2$ as the equatorial plane, and (iii) at minimum distance from $\mathcal{O}^*$ at the South Pole to coincide with the tangential plane $T_N\mathbb{S}_R^2$. Note that the projection lines intersect the sphere $\mathbb{S}_R^2$ at P, while the projection plane is intersected at p. The perspective center $\mathcal{O}^*$ is at distance D from the origin $\mathcal{O}$ of the sphere $\mathbb{S}_R^2$ or at height H above S, measured by S$\mathcal{O}^*$, such that $D = R + H$ holds.

Question: "How to find the polar coordinate $r = f(\Delta)$ in Figs. 5.10–5.12, where $\Delta = \pi/2 - \Phi$ is the spherical colatitude and Φ is the spherical latitude of the point $P \in \mathbb{S}_R^2$?" Answer: "Consult the sub-sections that follow, which compactly present the case studies for the individual geometrical situations."

Note that in all these cases the *perspective ratio* $r/QP = \mathcal{O}^*N/\mathcal{O}^*Q$ is the fundament for the answer to the well-posed question.

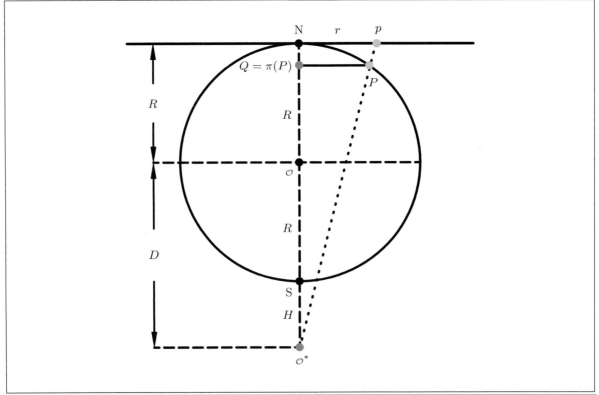

Fig. 5.10. Spherical vertical section, general normal perspective mapping of the sphere to the tangential plane at the North Pole, projection plane at maximal distance.

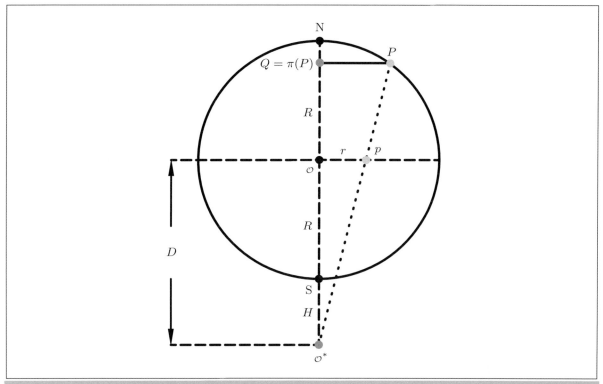

Fig. 5.11. Spherical vertical section, the general normal perspective mapping of the sphere to the equatorial plane. The geometrical details.

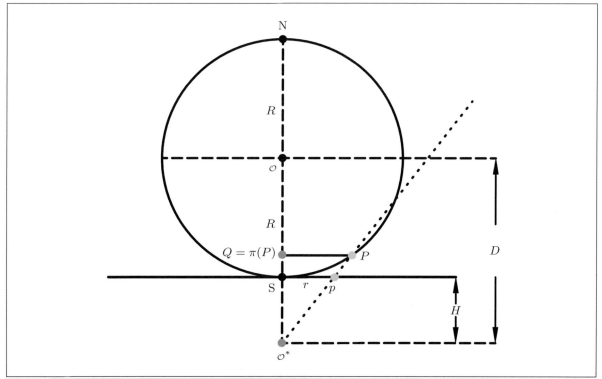

Fig. 5.12. Spherical vertical section, the general normal perspective mapping of the sphere to the tangential plane at the South Pole, projection plane at minimal distance. The geometrical details.

5-241 Case 1: northern tangential plane (tangential plane at maximal distance)

This situation is shown in Fig. 5.10. With reference to Boxes 5.7 and 5.8, we derive the general form of the parameterized mapping $r = f(\Delta)$. Note that $Q = \pi(P)$ is the point generated by an orthogonal projection of the point $P \in \mathbb{S}^2_R$ onto the axis of symmetry North-Pole–South-Pole. Let us refer to the following identities.

Identity (i):

$$QP = R\cos\Phi .$$

Identity (ii):

$$\mathcal{O}^*\mathrm{N} = R + D = 2R + H .$$

(5.45)

Identity (iii):

$$\mathcal{O}^*Q = \mathcal{O}^*\mathcal{O} + \mathcal{O}Q =$$

$$= D + R\sin\Phi = R(1 + \sin\Phi) + H .$$

Solving the perspective ratio for r, we are finally led to $r = f(\Delta)$. Such a representation of the radial function $f(\Delta)$ is supplemented by the computation of $f'(\Delta)$, a formula needed for the analysis of the left principal stretches.

Box 5.7 (Basics of the perspective ratio, northern tangential plane $T_\mathrm{N}\mathbb{S}^2_R$).

Basic ratio:

$$\frac{r}{QP} = \frac{\mathcal{O}^*\mathrm{N}}{\mathcal{O}^*Q} .$$

(5.46)

Explicit spherical representation of the basic ratio:

$$\frac{r}{R\cos\Phi} = \frac{R + D}{R\sin\Phi + D}$$

$$\Rightarrow$$

$$r = \frac{R + D}{R\sin\Phi + D}R\cos\Phi$$

(5.47)

$$\Rightarrow$$

$$r = \frac{R + D}{R\cos\Delta + D}R\sin\Delta =: f(\Delta) .$$

Derivative of the function $r = f(\Delta)$ with respect to colatitude (polar distance Δ):

$$f'(\Delta) = \frac{\mathrm{d}f}{\mathrm{d}\Delta} =$$

$$= R(R + D)\frac{R + D\cos\Delta}{(R\cos\Delta + D)^2} =$$

(5.48)

$$= R(R + D)\frac{R + D\sin\Phi}{(R\sin\Phi + D)^2} .$$

Box 5.8 (General normal perspective mapping of the sphere to the tangential plane at maximal distance).

Parameterized mapping (polar coordinates):

$$\alpha = \Lambda \,,$$

$$r = \frac{R+D}{R\sin\Phi + D}R\cos\Phi = \frac{1+\frac{R}{D}}{1+\frac{R}{D}\sin\Phi}R\cos\Phi \,. \tag{5.49}$$

Parameterized mapping (Cartesian coordinates):

$$\begin{bmatrix} x \\ y \end{bmatrix} = \begin{bmatrix} \dfrac{R+D}{R\sin\Phi + D}R\cos\Phi\cos\Lambda \\ \dfrac{R+D}{R\sin\Phi + D}R\cos\Phi\sin\Lambda \end{bmatrix} = \begin{bmatrix} \dfrac{1+\frac{R}{D}}{1+\frac{R}{D}\sin\Phi}R\cos\Phi\cos\Lambda \\ \dfrac{1+\frac{R}{D}}{1+\frac{R}{D}\sin\Phi}R\cos\Phi\sin\Lambda \end{bmatrix} \,,$$

$$\begin{bmatrix} x \\ y \end{bmatrix} = \frac{R+D}{R\sin\Phi + D}R\cos\Phi\begin{bmatrix} \cos\Lambda \\ \sin\Lambda \end{bmatrix} = \frac{1+\frac{R}{D}}{1+\frac{R}{D}\sin\Phi}R\cos\Phi\begin{bmatrix} \cos\Lambda \\ \sin\Lambda \end{bmatrix} \,. \tag{5.50}$$

Left principal stretches:

$$\Lambda_1 = \frac{f(\Delta)}{R\sin\Delta} \,, \quad \Lambda_2 = \frac{f'(\Delta)}{R} \,; \tag{5.51}$$

$$\Lambda_1 = \frac{R+D}{R\cos\Delta + D} = \frac{1+\frac{R}{D}}{1+\frac{R}{D}\cos\Delta} \,,$$

$$\Lambda_2 = \frac{R+D}{(R\cos\Delta + D)^2}(D\cos\Delta + R) = \left(1+\frac{R}{D}\right)\frac{\frac{R}{D}+\cos\Delta}{\left(1+\frac{R}{D}\cos\Delta\right)^2} \,, \tag{5.52}$$

or

$$\Lambda_1 = \frac{R+D}{R\sin\Phi + D} = \frac{1+\frac{R}{D}}{1+\frac{R}{D}\sin\Phi} \,,$$

$$\Lambda_2 = \frac{R+D}{(R\sin\Phi + D)^2}(D\sin\Phi + R) = \left(1+\frac{R}{D}\right)\frac{\frac{R}{D}+\sin\Phi}{\left(1+\frac{R}{D}\sin\Phi\right)^2} \,. \tag{5.53}$$

Special isometry:

$$\Delta \to 0 \quad \text{or} \quad \Phi \to \pi/2$$
$$\Rightarrow$$
$$\Lambda_1 = \Lambda_2 = 1 \,. \tag{5.54}$$

Box 5.8 is a summary of the general normal perspective mapping of the sphere $\mathbb{S}^2_R$ to the northern tangential plane, specifically of the parameterized mapping in both polar coordinates $\{\alpha, r\}$ and in Cartesian coordinates $\{x, y\}$, completed by the computation of the left principal stretches $\{\Lambda_1, \Lambda_2\}$. At the point of symmetry, namely $\Delta = 0$ or $\Phi = \pi/2$, we prove the isometry $\Lambda_1 = \Lambda_2$.

5-242 Case 2: equatorial plane of reference

This situation is shown in Fig. 5.11. With reference to Boxes 5.9 and 5.10, we derive the general form of the parameterized mapping $r = f(\Delta)$. It may be noticed newly that $Q = \pi(P)$ is the point generated by an orthogonal projection of the point $P \in \mathbb{S}_R^2$ onto the axis of symmetry North-Pole–South-Pole. Let us refer to the following identities.

$$\text{Identity (i):}$$

$$QP = R\cos\Phi \ .$$

$$\text{Identity (ii):}$$

$$\mathcal{O}^*\mathcal{O} = D = R + H \ . \tag{5.55}$$

$$\text{Identity (iii):}$$

$$\mathcal{O}^*Q = \mathcal{O}^*\mathcal{O} + \mathcal{O}Q = D + R\sin\Phi \ .$$

Solving the perspective ratio for r, we are finally led to $r = f(\Delta)$. There is the special case $\mathcal{O}^* = $ S, namely the identity of the perspective center $\mathcal{O}^*$ and the South Pole S, a case that is treated in all textbooks of Differential Geometry. Here, the distance D is identical to the radius R of the reference sphere $\mathbb{S}_R^2$. Indeed, for this special case, we probe $r = R\tan\Delta/2$. Finally, we compute $f'(\Delta)$, a formula going into the computation of the left principal stretches.

Box 5.9 (Basics of the perspective ratio, equatorial plane of reference).

Basic ratio:

$$\frac{r}{QP} = \frac{\mathcal{O}^*\mathcal{O}}{\mathcal{O}^*Q} \ . \tag{5.56}$$

Explicit spherical representation of the basic ratio:

$$\frac{r}{R\cos\Phi} = \frac{D}{R\sin\Phi + D}$$

$$\Rightarrow$$

$$r = \frac{D}{R\sin\Phi + D}R\cos\Phi \tag{5.57}$$

$$\Rightarrow$$

$$r = \frac{D}{R\cos\Delta + D}R\sin\Delta \ .$$

Special case $\mathcal{O}^* = $ S, $D = R$:

$$r = \frac{R\cos\Phi}{1 + \sin\Phi} = \frac{R\sin\Delta}{1 + \cos\Delta} \ , \quad r = R\tan\left(\frac{\pi}{4} - \frac{\Phi}{2}\right) = R\tan\frac{\Delta}{2} \ . \tag{5.58}$$

Derivative of the function $r = f(\Delta)$ with respect to colatitude (polar distance Δ):

$$f'(\Delta) = \frac{\mathrm{d}f}{\mathrm{d}\Delta} = DR\frac{R + D\cos\Delta}{(R\cos\Delta + D)^2} = DR\frac{R + D\sin\Phi}{(R\sin\Phi + D)^2} \ . \tag{5.59}$$

Box 5.10 (General normal perspective mapping of the sphere to the equatorial plane of reference).

Parameterized mapping (polar coordinates):

$$\alpha = \Lambda\,, \quad r = \frac{D}{R\sin\Phi + D}R\cos\Phi = \frac{R+H}{R(1+\sin\Phi)+H}\cos\Phi\,. \tag{5.60}$$

Special case $H = 0$:

$$r = R\tan\left(\frac{\pi}{4} - \frac{\Phi}{2}\right) = R\tan\frac{\Delta}{2}\,. \tag{5.61}$$

Parameterized mapping (Cartesian coordinates):

$$\begin{bmatrix} x \\ y \end{bmatrix} = \begin{bmatrix} \dfrac{D}{R\sin\Phi + D}R\cos\Phi\cos\Lambda \\[2mm] \dfrac{D}{R\sin\Phi + D}R\cos\Phi\sin\Lambda \end{bmatrix} = \begin{bmatrix} \dfrac{R+H}{R(1+\sin\Phi)+H}R\cos\Phi\cos\Lambda \\[2mm] \dfrac{R+H}{R(1+\sin\Phi)+H}R\cos\Phi\sin\Lambda \end{bmatrix}\,, \tag{5.62}$$

$$\begin{bmatrix} x \\ y \end{bmatrix} = \frac{D}{R\sin\Phi + D}R\cos\Phi\begin{bmatrix} \cos\Lambda \\ \sin\Lambda \end{bmatrix} = \frac{R+H}{R(1+\sin\Phi)+H}R\cos\Phi\begin{bmatrix} \cos\Lambda \\ \sin\Lambda \end{bmatrix}\,.$$

Special case $H = 0$:

$$\begin{bmatrix} x \\ y \end{bmatrix} = R\tan\left(\frac{\pi}{4} - \frac{\Phi}{2}\right)\begin{bmatrix} \cos\Lambda \\ \sin\Lambda \end{bmatrix} = R\tan\frac{\Delta}{2}\begin{bmatrix} \cos\Lambda \\ \sin\Lambda \end{bmatrix}\,. \tag{5.63}$$

Left principal stretches:

$$\Lambda_1 = \frac{f(\Delta)}{R\sin\Delta}\,, \quad \Lambda_2 = \frac{f'(\Delta)}{R}\,; \tag{5.64}$$

$$\Lambda_1 = \frac{D}{R\sin\Phi + D} = \frac{R+H}{R(1+\sin\Phi)+H}\,, \tag{5.65}$$

$$\Lambda_2 = D\frac{R+D\sin\Phi}{(R\sin\Phi+D)^2} = (R+H)\frac{R(1+\sin\Phi)+H\sin\Phi}{[R(1+\sin\Phi)+H]^2}\,.$$

Special case $H = 0$:

$$\Lambda_1 = \Lambda_2 = \frac{1}{1+\sin\Phi} = \frac{1}{1+\cos\Delta}\,, \quad \Lambda_1 = \Lambda_2 = \frac{1}{2}\frac{1}{\cos^2\frac{\Delta}{2}} = \frac{1}{2}\frac{1}{\cos^2\left(\frac{\pi}{4} - \frac{\Phi}{2}\right)}\,. \tag{5.66}$$

Box 5.10 is a summary of the general normal perspective mapping of the sphere $\mathbb{S}_R^2$ to the equatorial plane of reference, specifically of the parameterized mapping in both polar coordinates $\{\alpha, r\}$ and in Cartesian coordinates $\{x, y\}$, completed by the computation of the left principal stretches $\{\Lambda_1, \Lambda_2\}$. For the special case $\mathcal{O}^* = S$ or, equivalently, $D = R$ or $H = 0$, we prove conformality $\Lambda_1 = \Lambda_2$. For such a configuration of the southern perspective center, $\Phi = -\pi/2$ is singular: $\Lambda_1(-\pi/2) = \Lambda_2(\pi/2) \to \infty$.

5-243 Case 3: southern tangential plane (tangential plane at minimal distance)

This situation is shown in Fig. 5.12. According to Fig. 5.12, $Q = \pi(P)$ is the point generated by an orthogonal projection of the point $P \in \mathbb{S}_R^2$ onto the axis of symmetry North-Pole–South-Pole. Note that the southern projection plane is at distance D from the origin o or, alternatively, at spherical height H from o^*. Collected in Box 5.11, we present to you the basic identities

Identity (i):

$$QP = R\,|\cos\Phi| = R\,|\sin\Delta| \ .$$

Identity (ii):

$$o^*S = D - R = H \ . \qquad (5.67)$$

Identity (iii):

$$o^*Q = o^*o + oQ =$$

$$= D - R\,|\sin\Phi| = R(1 - |\sin\Phi|) + H \ .$$

Solving the perspective ratio for r, we are finally led to $r = f(\Delta)$. Such a representation of the radial function $f(\Delta)$ is supplemented by the computation of $f'(\Delta)$, a formula needed for the analysis of the left principal stretches.

Box 5.11 (Basics of the perspective ratio, tangential plane at minimal distance to o^*).

Basic ratio:

$$\frac{r}{QP} = \frac{o^*S}{o^*Q} \ . \qquad (5.68)$$

Explicit spherical representation of the basic ratio:

$$\frac{r}{R\cos\Phi} = \frac{H}{R + H - R\,|\sin\Phi|} = \frac{D - R}{D - R\,|\sin\Phi|}$$

$$\Rightarrow$$

$$r = \frac{HR\cos\Phi}{H + R(1 - |\sin\Phi|)} = \frac{D - R}{D - R\,|\sin\Phi|}R\cos\Phi \qquad (5.69)$$

$$\Rightarrow$$

$$r = \frac{D - R}{D - R\,|\cos\Delta|}R\,|\sin\Delta| =: f(\Delta) \ .$$

Derivative of the function $r = f(\Delta)$ with respect to colatitude (polar distance Δ):

$$f'(\Delta) = \frac{\mathrm{d}f}{\mathrm{d}\Delta} =$$

$$= R(D - R)\frac{D\,|\cos\Delta| - R}{(D - R\,|\cos\Delta|)^2} = \qquad (5.70)$$

$$= R(D - R)\frac{D\,|\sin\Phi| - R}{(D - R\,|\sin\Phi|)^2} \ .$$

By means of Box 5.12, we have collected the parameter equations which characterize the general normal perspective mapping of the sphere $\mathbb{S}_R^2$ to the southern tangential plane, specifically in terms of polar coordinates $\{\alpha, r\}$ and of Cartesian coordinates $\{x, y\}$, completed by the computation of the left principal stretches $\{\Lambda_1, \Lambda_2\}$. At the point of symmetry, namely $\Delta = \pi$ or $\Phi = -\pi/2$, we prove the isometry $\Lambda_1 = \Lambda_2 = 1$.

Box 5.12 (General normal perspective mapping of the sphere to the tangential plane at minimal distance).

Parameterized mapping (polar coordinates):

$$\alpha = \Lambda \,,$$

$$r = (D - R)\frac{R \cos \Phi}{D - R |\sin \Phi|} = \frac{H}{H + R(1 - |\sin \Phi|)} R \cos \Phi \,. \tag{5.71}$$

Parameterized mapping (Cartesian coordinates):

$$\begin{bmatrix} x \\ y \end{bmatrix} = \begin{bmatrix} (D - R)\dfrac{R \cos \Phi \cos \Lambda}{D - R |\sin \Phi|} \\ (D - R)\dfrac{R \cos \Phi \sin \Lambda}{D - R |\sin \Phi|} \end{bmatrix} \,,$$

$$\begin{bmatrix} x \\ y \end{bmatrix} = \frac{H}{H + R(1 - |\sin \Phi|)} R \cos \Phi \begin{bmatrix} \cos \Lambda \\ \sin \Lambda \end{bmatrix} \,. \tag{5.72}$$

Left principal stretches:

$$\Lambda_1 = \frac{f(\Delta)}{R \sin \Delta} \,,$$

$$\Lambda_2 = \frac{f'(\Delta)}{R} \,; \tag{5.73}$$

$$\Lambda_1 = \frac{D - R}{D - R |\cos \Delta|} = \frac{H}{H + R(1 - |\sin \Phi|)} \,,$$

$$\Lambda_2 = (D - R)\frac{D |\cos \Delta| - R}{(D - R |\cos \Delta|)^2} = H\frac{H - R(1 - |\sin \Phi|)}{[H + R(1 - |\sin \Phi|)]^2} \,. \tag{5.74}$$

Special isometry:

$$\Delta \to \pi \ (|\cos \Delta| \to 1) \quad \text{or} \quad \Phi \to -\pi/2 \ (|\sin \Phi| \to 1)$$

$$\Rightarrow \tag{5.75}$$

$$\Lambda_1 = \Lambda_2 = 1 \,.$$

5-244 *Line-of-sight, line-of-contact, minimal and complete atlas*

The *line-of-sight* as well as the *line-of-contact* for both the general normal perspective mapping to the tangential plane at the North Pole and to the tangential plane at the South Pole are illustrated in Fig. 5.13 and Fig. 5.14, respectively.

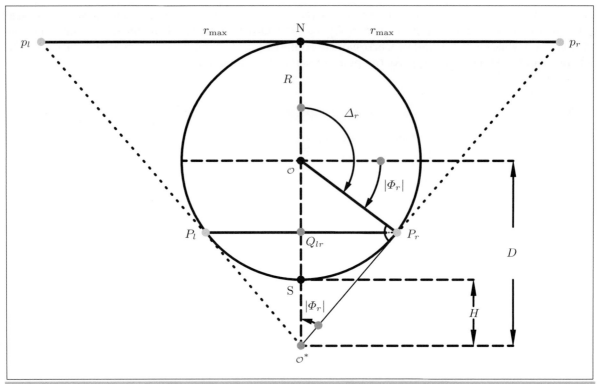

Fig. 5.13. Line-of-sight, normal perspective mapping of the sphere to the tangential plane at the North Pole, projection plane at maximal distance.

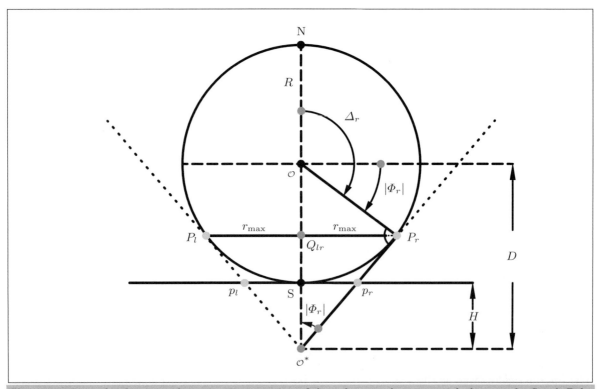

Fig. 5.14. Line-of-sight, normal perspective mapping of the sphere to the tangential plane at the South Pole, projection plane at minimal distance.

Question.

Question: "What is the line-of-sight or the line-of-contact and how can we compute the spherical latitude Φ_r of the line-of-contact or the maximal radial coordinate $r_{\max}$?" Answer 1: "The normal central projection $\mathcal{O}^* \to T_N \mathbb{S}_R^2$ or $\mathcal{O}^* \to T_S \mathbb{S}_R^2$ is restricted to points inside the circular cone $\mathbb{C}_{Q_{lr}P_r}^2$ or $\mathbb{C}_{P_l Q_{lr}}^2$. Indeed, the projection line, which contacts the sphere tangentially, restricts the domain of points of $\mathbb{S}_R^2$ which can be mapped to $T_N \mathbb{S}_R^2$ or $T_S \mathbb{S}_R^2$. The radius $Q_{lr}P_r$ or $P_l Q_{lr}$ determines the circular cone. Its related bundle of projection lines constitutes the characteristic circular cone-of-contact. The line-of-contact is the circle $\mathbb{S}_{R \cos \Phi_r}^1$ of radius $R \cos \Phi_r$. Its trace $P_l Q_{lr} P_r$ is illustrated in Fig. 5.13 and Fig. 5.14, respectively." Answer 2: "Let be given the distance $\mathcal{O}^* \mathcal{O}$ of the perspective center $\mathcal{O}^*$ and the origin $\mathcal{O}$ of $\mathbb{S}_R^2$, which is called D, or alternatively the spherical height H of the perspective center $\mathcal{O}^*$ relative to S. Then the critical spherical latitude Φ_r can be computed as outlined in Box 5.13. If $\mathcal{O}^*$ is placed south on the line NS, then the critical value is determined by $\sin |\Phi_r| = R/D$, regardless whether the projection plane is located at the North Pole or at the South Pole."

Box 5.13 (Data for the line-of-sight and the line-of-contact, critical spherical latitude, center of perspective under the South Pole).

Tangential plane at the North Pole   Tangential plane at the South Pole

$$\sin |\Phi_r| = \frac{R}{D} = \frac{R}{R+H} \qquad \text{versus} \qquad \sin |\Phi_r| = \frac{R}{D} = \frac{R}{R+H} \; ,$$

$$\tan |\Phi_r| = \frac{r_{\max}}{R+D} = \frac{r_{\max}}{2R+H} \qquad \text{versus} \qquad \tan |\Phi_r| = \frac{r_{\max}}{H} \; , \tag{5.76}$$

$$r_{\max} = (2R+H) \tan |\Phi_r| \qquad \text{versus} \qquad r_{\max} = H \tan |\Phi_r| \; ;$$

$$\tan x = \frac{\sin x}{\sqrt{1 - \sin^2 x}} \; , \quad \tan |\Phi_r| = \frac{R}{R+H} \frac{1}{\sqrt{1 - \frac{R}{(R+H)^2}}} = \frac{R}{\sqrt{(2R+H)H}} \; ; \tag{5.77}$$

$$r_{\max} = R\sqrt{1 + 2\frac{R}{H}} \qquad \text{versus} \qquad r_{\max} = \frac{R}{\sqrt{1 + 2\frac{R}{H}}} \; . \tag{5.78}$$

Let us compute the maximal extension of such a normal central perspective. According to the identities of Box 5.13, the maximal extension $r_{\max}$ is either $R\sqrt{1+x}$ for a projection plane at the North Pole or $R/\sqrt{1+x}$ for a projection plane at the South Pole and $x := 2R/H$. Figure 5.15 and Table 5.1 outline those functions in the domain $0 \le x \le 5$.

Example 5.1 (Numerical example I).

A first numerical example is $R/H = 3/2$ and $x = 3$, such that $\sqrt{1+x} = 2$, $1/\sqrt{1+x} = 1/2$, $r_{\max}(\text{North}) = 2R$, and $r_{\max}(\text{South}) = R/2$.

End of Example.

Example 5.2 (Numerical example II).

A second numerical example is $R/H = 40$ and $x = 80$, such that $\sqrt{1+x} = 9$, $1/\sqrt{1+x} = 1/9$, $r_{\max}(\text{North}) = 9R$, and $r_{\max}(\text{South}) = R/9$.

End of Example.

Obviously, by means of a normal central perspective from a southern perspective center to a projection plane at the North Pole, we can cover more points than on the northern hemisphere. In contrast, a normal central perspective from a southern perspective center to a projection plane at the South Pole, we can cover only few points of the southern hemisphere.

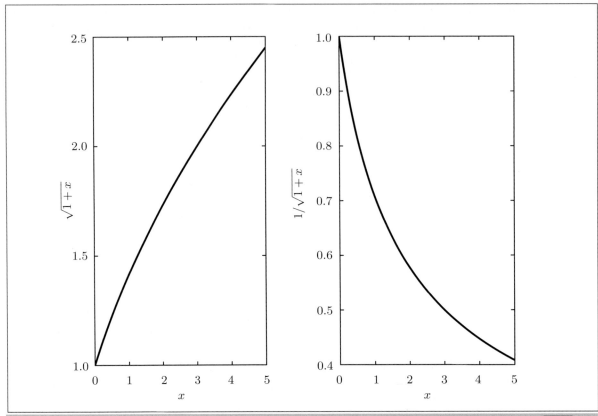

Fig. 5.15. Maximal extension of a normal central perspective of the sphere, the functions $\sqrt{1+x}$ and $1/\sqrt{1+x}$, domain $0 \leq x \leq 5$, $x := 2R/H$.

Such a discussion motivates the construction of a *minimal atlas* from the setup of a normal central perspective as follows. Consider the two charts (i) central perspective projection from a southern perspective center to a projection plane at the North Pole and (ii) central perspective projection from a northern perspective center to a projection plane at the South Pole. The union of the two charts covers the sphere $\mathbb{S}^2_R$ completely. The two charts constitute a *minimal atlas*. While Fig. 5.10 illustrates the first chart ($\mathcal{O}^*$ south on the NS line, projection plane at the North Pole), in contrast, Fig. 5.16 illustrates the second chart ($\mathcal{O}^*$ north on the SN line, projection plane at the South Pole).

Table 5.1. Maximal extension of a normal central perspective of the sphere, the functions $\sqrt{1+x}$ and $1/\sqrt{1+x}$, domain $0 \leq x \leq 5$, $x := 2R/H$.

x	$\sqrt{1+x}$	$1/\sqrt{1+x}$
0.0	1.000	1.000
0.5	1.225	0.816
1.0	1.414	0.707
1.5	1.581	0.632
2.0	1.732	0.577
2.5	1.871	0.535
3.0	2.000	0.500
3.5	2.121	0.471
4.0	2.236	0.447
4.5	2.345	0.426
5.0	2.449	0.408

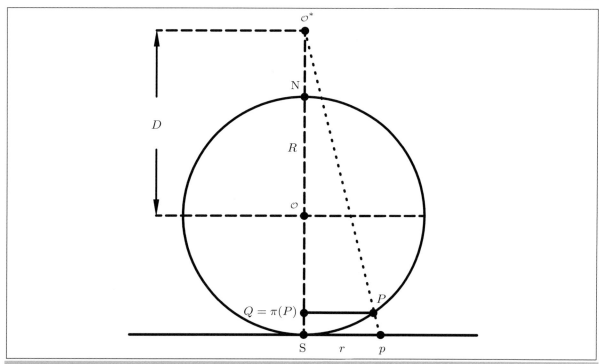

Fig. 5.16. Spherical vertical section, general normal perspective mapping of the sphere to the tangential plane at the South Pole, projection plane at maximal distance.

The normal perspective mapping, which corresponds to Fig. 5.12, but where the perspective center $\mathcal{O}^*$ is placed north on the SN line and the projection plane is identified as the tangent plane at the North Pole, is presented in Fig. 5.17.

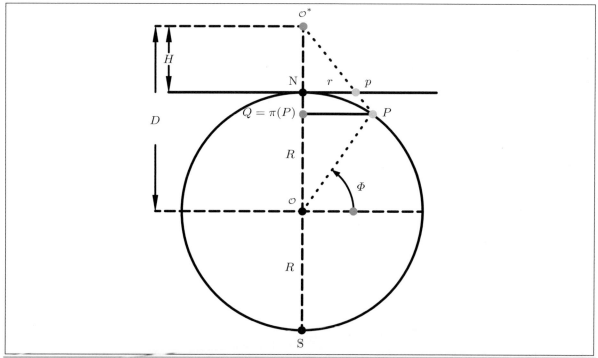

Fig. 5.17. Spherical vertical section, general normal perspective mapping of the sphere to the tangential plane which is at minimal distance to $\mathcal{O}^*$: $D = R + H$.

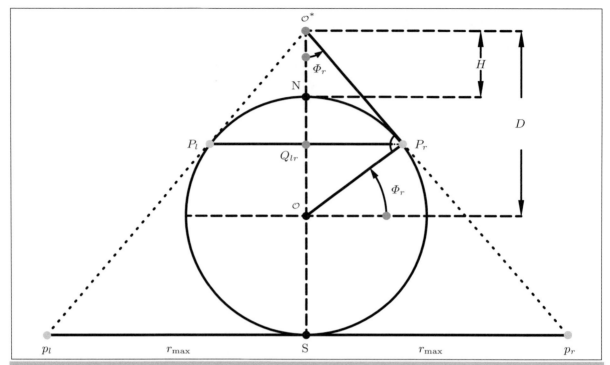

Fig. 5.18. Line-of-sight, normal perspective mapping of the sphere to the tangential plane at the South Pole, projection plane at maximal distance.

For later reference, Box 5.14 outlines the corresponding data for the line-of-sight and the line-of-contact: compare with Figs. 5.18 and 5.19. Note that the special case where the projection plane is placed in the center $\mathcal{O}$ of the sphere $\mathbb{S}_R^2$ is discussed at the end of this section.

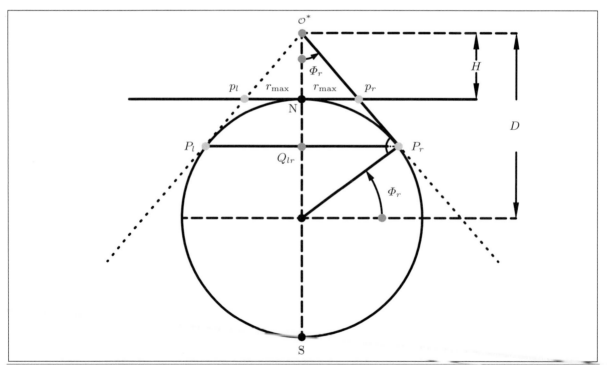

Fig. 5.19. Line-of-sight, normal perspective mapping of the sphere to the tangential plane at the North Pole, projection plane at minimal distance.

Box 5.14 (Data for the line-of-sight and the line-of-contact, critical spherical latitude, center of perspective over the North Pole).

Tangential plane at the South Pole   Tangential plane at the North Pole

$$\sin\Phi_r = \frac{R}{D} = \frac{R}{R+H} \qquad \text{versus} \qquad \sin\Phi_r = \frac{R}{D} = \frac{R}{R+H}\ ,$$

$$\tan\Phi_r = \frac{r_{\max}}{R+D} = \frac{r_{\max}}{2R+H} \qquad \text{versus} \qquad \tan\Phi_r = \frac{r_{\max}}{H}\ , \tag{5.79}$$

$$r_{\max} = (2R+H)\tan\Phi_r \qquad \text{versus} \qquad r_{\max} = H\tan\Phi_r\ ;$$

$$\tan x = \frac{\sin x}{\sqrt{1-\sin^2 x}}\ , \qquad \tan\Phi_r = \frac{R}{R+H}\frac{1}{\sqrt{1-\frac{R}{(R+H)^2}}} = \frac{R}{\sqrt{(2R+H)H}}\ ; \tag{5.80}$$

$$r_{\max} = R\sqrt{1+2\frac{R}{H}} \qquad \text{versus} \qquad r_{\max} = \frac{R}{\sqrt{1+2\frac{R}{H}}}\ . \tag{5.81}$$

5-245 The gnomonic projection

According to Fig. 5.20, the *gnomonic projection* is generated as a polar central perspective, where $\mathcal{O}^* = \mathcal{O}$ or $D = 0$ in the context of a normal general perspective mapping holds. In Box 5.15, the items of such a mapping of the sphere to a plane (namely, (i) the parameterized mapping, (ii) the left principal stretches of the left Cauchy–Green eigenspace, (iii) the left maximal angular shear, and (iv) the inverse parameterized mapping) are collected.

 Historical aside. Note that the gnomonic projection has been used in the antiquity for the construction of a sundial ("gnomon").

The basic results of the gnomonic projection or polar central perspective mapping of the sphere $\mathbb{S}_R^2$ are collected in Lemma 5.4.

Lemma 5.4 (Gnomonic projection of the sphere to the polar tangential plane).

The gnomonic projection of the sphere $\mathbb{S}_R^2$ to the tangential plane at the North Pole is parameterized by the two equations

$$x = R\cot\Phi\cos\Lambda\ ,$$
$$y = R\cot\Phi\sin\Lambda\ , \tag{5.82}$$

subject to the left Cauchy–Green eigenspace

$$\text{left CG eigenspace} = \left\{ \boldsymbol{E}_\Lambda \frac{1}{\sin\Phi},\, \boldsymbol{E}_\Phi \frac{1}{\sin^2\Phi} \right\}\ . \tag{5.83}$$

End of Lemma.

Box 5.15 (Gnomonic projection).

Parameterized central perspective mapping (polar coordinates):

$$\alpha = \Lambda \,,$$
$$r = R \tan \Delta = R \cot \Phi \,. \tag{5.84}$$

Parameterized central perspective mapping (Cartesian coordinates):

$$\begin{bmatrix} x \\ y \end{bmatrix} = R \cot \Phi \begin{bmatrix} \cos \Lambda \\ \sin \Lambda \end{bmatrix} \,. \tag{5.85}$$

Left principal stretches:

$$\Lambda_1 = \frac{1}{\sin \Phi} \,,$$
$$\Lambda_2 = \frac{1}{\sin^2 \Phi} \,. \tag{5.86}$$

Left eigenvectors:

$$\boldsymbol{C}_1 \Lambda_1 = \boldsymbol{E}_\Lambda \frac{1}{\sin \Phi} \quad \text{(Easting)} \,,$$
$$\boldsymbol{C}_2 \Lambda_2 = \boldsymbol{E}_\Phi \frac{1}{\sin^2 \Phi} \quad \text{(Northing)} \,. \tag{5.87}$$

Left maximal angular distortion:

$$\Omega_l = 2 \arcsin \left| \frac{\Lambda_1 - \Lambda_2}{\Lambda_1 + \Lambda_2} \right| = 2 \arcsin \left| \frac{1 - \sin \Phi}{1 + \sin \Phi} \right| \,. \tag{5.88}$$

Parameterized inverse mapping:

$$\tan \Lambda = \frac{y}{x} \,, \quad \tan \Phi = \frac{R}{\sqrt{x^2 + y^2}} \,. \tag{5.89}$$

Note that the northern gnomonic projection covers all points in the half open interval $\pi/2 \leq \Phi < 0$. The point $\Phi = 0$ moves to infinity. Accordingly, for a complete gnomonic atlas of the sphere, we need three charts: one northern, one southern, and one equatorial chart.

Question. Question: "What made the gnomonic projection particularly useful in marine, aerial, and space navigation?" Answer: "It is the property that the gnomonic projection is geodesic. Geodesics, namely great circles of the sphere, are mapped onto a straight line – a very important characteristic!"

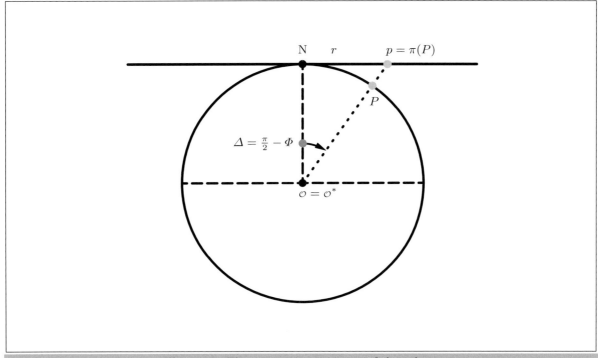

Fig. 5.20. The gnomonic projection of the sphere.

Last but not least, we present to you a nice sample of the polar gnomonic projection of the sphere in Fig. 5.21.

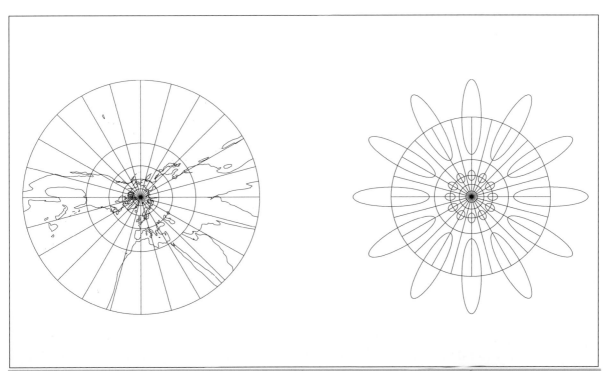

Fig. 5.21. Special perspective map of the sphere $\mathbb{S}_R^2$ onto the tangent space $T_N\mathbb{S}_R^2$: gnomonic projection, Tissot ellipses, polar aspect, graticule 15°, shorelines.

5-246 The orthographic projection

The *orthographic projection*, which is usually also called *parallel projection* or *orthogonal projection*, is generated as a parallel projection (orthogonal projection) of a point $P \in \mathbb{S}^2_R$ either on a polar tangent plane of $\mathbb{S}^2_R$ or on a plane parallel to the polar tangent plane through the origin $\mathcal{O}$: compare with Figs. 5.22 and 5.23. In the context of a general perspective mapping, we are able to generate an orthographic projection by moving the perspective center $\mathcal{O}^*$ to infinity, i.e. $D \to \infty$ or $R/D \to \infty$. In Figs. 5.22 and 5.23, such a parallel projection (orthogonal projection) is illustrated. In Box 5.16, the characteristics of such a projection (namely, (i) the parameterized mapping, (ii) the left principal stretches of the left Cauchy–Green eigenspace, (iii) the left maximal angular shear, and (iv) the inverse parameterized mapping) are collected.

Technical aside. Note that the orthographic projection is used for charting the Moon or the Earth, for example, on a TV screen.

The basic results of the orthographic projection (parallel projection, orthogonal projection) of the sphere $\mathbb{S}^2_R$ are collected in Lemma 5.5.

Lemma 5.5 (Orthographic projection of the sphere to the polar tangential plane or the equatorial plane).

The orthographic projection of the sphere $\mathbb{S}^2_R$ to the tangential plane or to the equatorial plane is parameterized by the two equations

$$x = R \cos \Phi \cos \Lambda = X \ ,$$

$$y = R \cos \Phi \sin \Lambda = Y \ , \tag{5.90}$$

subject to the left Cauchy–Green eigenspace

$$\text{left CG eigenspace} = \left\{ \boldsymbol{E}_\Lambda, \boldsymbol{E}_\Phi \frac{1}{\sin \Phi} \right\} \ . \tag{5.91}$$

End of Lemma.

Note that the northern orthographic projection covers all points of the northern hemisphere, while the southern orthographic projection covers all points of the southern hemisphere. Accordingly, the union of the two charts generated by a northern and a southern orthographic projection constitutes a minimal atlas of the sphere. Additionally, let us here emphasize that the left maximal angular shear of the gnomonic projection and the orthographic projection coincide.

Question. Question: "What makes the orthographic projection (parallel projection, orthogonal projection) particularly useful in Geographic Information Systems?" Answer: "It is the property, which is called *concircular*, that parallel circles of the sphere $\mathbb{S}^2_R$ are mapped onto circles of $T_N \mathbb{S}^2_R$, $T_S \mathbb{S}^2_R$, or $\mathbb{P}^2_{\mathcal{O}}$. By means of $r = R \cos \Phi$, they are *radius preserving* – an essential characteristic!"

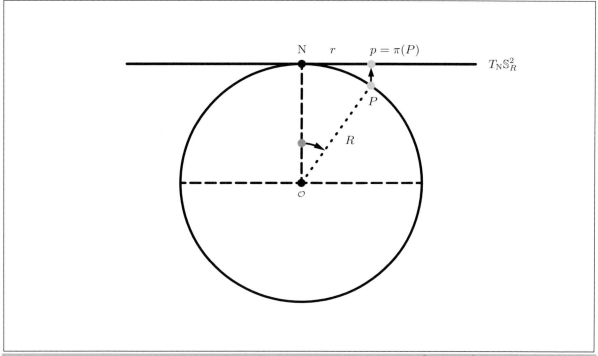

Fig. 5.22. The orthographic projection of the sphere: $\mathbb{S}_R^2 \to T_N\mathbb{S}_R^2$.

A sample of the polar orthographic projection (parallel projection, orthogonal projection) of the sphere is finally presented to you in Fig. 5.24.

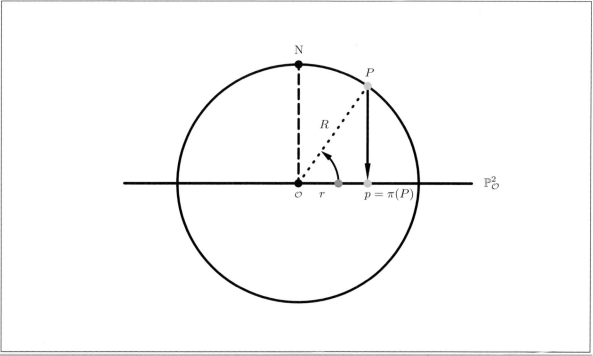

Fig. 5.23. The orthographic projection of the sphere: $\mathbb{S}_R^2 \to \mathbb{P}_\mathcal{O}^2$.

Box 5.16 (Orthographic projection).

Parameterized orthographic mapping (polar coordinates):

$$\alpha = \Lambda , \quad r = R\cos\Phi .$$

(5.92)

Parameterized orthographic mapping (Cartesian coordinates):

$$\begin{bmatrix} x \\ y \end{bmatrix} = R\cos\Phi \begin{bmatrix} \cos\Lambda \\ \sin\Lambda \end{bmatrix} = \begin{bmatrix} X \\ Y \end{bmatrix} .$$

(5.93)

Left principal stretches:

$$\Lambda_1 = 1 , \quad \Lambda_2 = \sin\Phi .$$

(5.94)

Left eigenvectors:

$$\boldsymbol{C}_1\Lambda_1 = \boldsymbol{E}_\Lambda \quad \text{(Easting)} , \quad \boldsymbol{C}_2\Lambda_2 = \boldsymbol{E}_\Phi\sin\Phi \quad \text{(Northing)} .$$

(5.95)

Left maximal angular distortion:

$$\Omega_l = 2\arcsin\left|\frac{\Lambda_1 - \Lambda_2}{\Lambda_1 + \Lambda_2}\right| = 2\arcsin\left|\frac{1 - \sin\Phi}{1 + \sin\Phi}\right| .$$

(5.96)

Parameterized inverse mapping:

$$\tan\Lambda = \frac{y}{x} , \quad \cos\Phi = \frac{\sqrt{x^2 + y^2}}{R} .$$

(5.97)

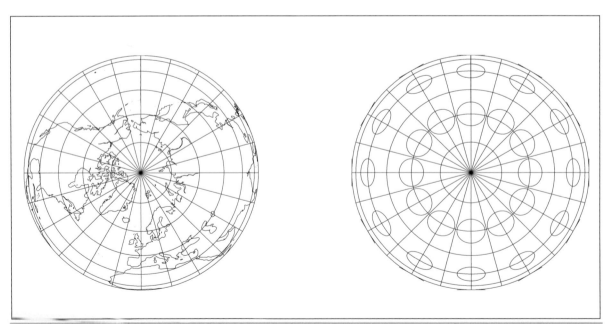

Fig. 5.24. Special perspective map of the sphere $\mathbb{S}^2_R$ onto the tangent space $T_N\mathbb{S}^2_R$: orthographic projection, Tissot ellipses, polar aspect, graticule $15°$, shorelines.

5-247 The Lagrange projection

The normal general perspective mapping of the sphere reduces to the Polar Stereographic Projection (UPS) if we specialize $H = 0$ or $D = R$. A special variant already mentioned is achieved if we choose the South Pole as the perspective center $\mathcal{O}^*$ (alternatively, the North Pole) and a projection plane to coincide with the equatorial frame $\mathbb{P}^2_{\mathcal{O}}$. In Figs. 5.25 and 5.26, such a central perspective mapping is illustrated. In Box 5.17, the characteristics of such a projection (namely, (i) the parameterized mapping, (ii) the left principal stretches of the left Cauchy–Green eigenspace, (iii) the left maximal angular shear, and (iv) the inverse parameterized mapping) are collected.

Historical aside. Such a central perspective mapping particularly is associated with the name of J. L. Lagrange (1736–1813). Note that his works on map projections are published in A. Wangerin, Über Kartenprojektionen, Abhandlungen von J. L. Lagrange and C. F. Gauss (Verlag W. Engelmann, Leipzig 1894).

The basic results of the Lagrange projection of the sphere $\mathbb{S}^2_R$ to the equatorial plane are collected in Lemma 5.6.

Lemma 5.6 (Special perspective mapping of the sphere: the Lagrange projection).

The Lagrange projection of the sphere $\mathbb{S}^2_R$ to the equatorial plane is parameterized by

$$
x = R \tan \frac{\Delta}{2} \cos \Lambda = R \tan \left(\frac{\pi}{4} - \frac{\Phi}{2} \right) \cos \Lambda \,,
$$

$$
y = R \tan \frac{\Delta}{2} \sin \Lambda = R \tan \left(\frac{\pi}{4} - \frac{\Phi}{2} \right) \sin \Lambda \,,
$$

(5.98)

subject to the left Cauchy–Green eigenspace

$$
\text{left CG eigenspace} = \left\{ \boldsymbol{E}_\Lambda \frac{1}{2 \cos^2 \frac{\Delta}{2}}, \boldsymbol{E}_\Phi \frac{1}{2 \cos^2 \frac{\Delta}{2}} \right\} \,.
$$

(5.99)

The Lagrange projection is conformal.

End of Lemma.

Note that the northern hemisphere is conformally mapped from the southern projective center $\mathrm{S} = \mathcal{O}^*$, while the southern hemisphere is conformally mapped from the northern projective center $\mathrm{N} = \mathcal{O}^*$, namely generating northern and southern points within a circle of radius R. The union of these two charts generates a *minimal atlas of conformal type*.

Question. Question: "What makes the Lagrange projection particularly useful when compared with the Universal Stereographic Projection (UPS)?" Answer: "It is the different factor of conformality $\Lambda_1 = \Lambda_2$: the left principal stretches of the Lagrange projection are half of the left principal stretches of the UPS: $\Lambda_1(\text{Lagrange}) = \Lambda_2(\text{Lagrange}) = \frac{1}{2}\Lambda_1(\text{UPS}) = \frac{1}{2}\Lambda_2(\text{UPS})$."

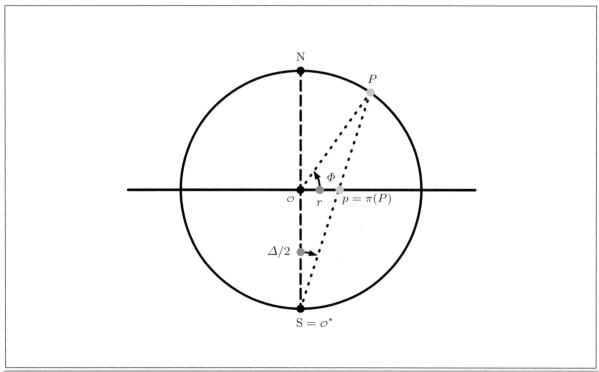

Fig. 5.25. Special perspective mapping: the Lagrange projection: the left chart. Together with the right chart, a minimal atlas of the sphere is constituted.

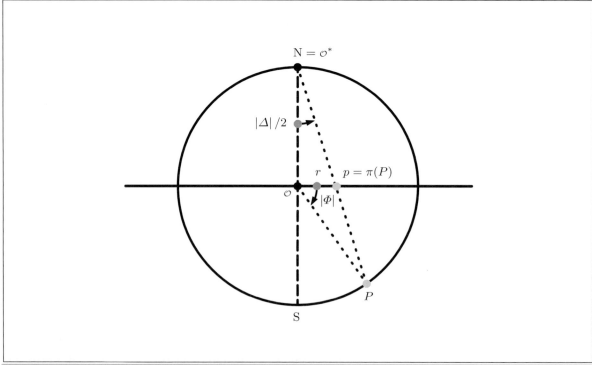

Fig. 5.26. Special perspective mapping; the Lagrange projection: the right chart. Together with the left chart, a minimal atlas of the sphere is constituted.

Box 5.17 (Lagrange projection).

Parameterized special perspective mapping (polar coordinates):

$$\alpha = \Lambda \,,$$

$$r = R \tan \frac{\Delta}{2} = R \tan \left(\frac{\pi}{4} - \frac{\Phi}{2} \right) \,. \tag{5.100}$$

Sine lemma (triangle $\mathcal{O}p\mathrm{S}$):

$$\frac{r}{\sin \frac{\Delta}{2}} = \frac{R}{\cos \frac{\Delta}{2}} \rightarrow r \,. \tag{5.101}$$

Parameterized special perspective mapping (Cartesian coordinates):

$$\begin{bmatrix} x \\ y \end{bmatrix} = R \tan \frac{\Delta}{2} \begin{bmatrix} \cos \Lambda \\ \sin \Lambda \end{bmatrix} = R \tan \left(\frac{\pi}{4} - \frac{\Phi}{2} \right) \begin{bmatrix} \cos \Lambda \\ \sin \Lambda \end{bmatrix} \,. \tag{5.102}$$

Case (i) ($\Phi = 0$):

$$\begin{bmatrix} x \\ y \end{bmatrix} = R \begin{bmatrix} \cos \Lambda \\ \sin \Lambda \end{bmatrix} \,. \tag{5.103}$$

Case (ii) ($\Phi = \pi/2$):

$$\begin{bmatrix} x \\ y \end{bmatrix} = \begin{bmatrix} 0 \\ 0 \end{bmatrix} \,. \tag{5.104}$$

Left principal stretches:

$$\Lambda_1 = \frac{f'(\Delta)}{R \sin \Delta} = \frac{\tan \frac{\Delta}{2}}{2 \sin \frac{\Delta}{2} \cos \frac{\Delta}{2}} = \frac{1}{2 \cos^2 \frac{\Delta}{2}} \,,$$

$$\Lambda_2 = \frac{f'(\Delta)}{R} = \frac{1}{2 \cos^2 \frac{\Delta}{2}} \,. \tag{5.105}$$

Conformality:

$$\Lambda_1 = \Lambda_2 \,. \tag{5.106}$$

Left eigenvectors:

$$\boldsymbol{C}_1 \Lambda_1 = \boldsymbol{E}_\Lambda \frac{1}{2 \cos^2 \frac{\Delta}{2}} \quad (\text{Easting}) \,,$$

$$\boldsymbol{C}_2 \Lambda_2 = \boldsymbol{E}_\Phi \frac{1}{2 \cos^2 \frac{\Delta}{2}} \quad (\text{Northing}) \,. \tag{5.107}$$

Continuation of Box.

Left maximal angular distortion:

$$\sum_l = \Psi_l - \Psi_r = 0 \;,$$

$$\Omega_l = 0 \;. \tag{5.108}$$

Parameterized inverse mapping:

$$\cos \Lambda = \frac{x}{\sqrt{x^2 + y^2}} \;,$$

$$\sin \Lambda = \frac{y}{\sqrt{x^2 + y^2}} \;, \tag{5.109}$$

$$\tan \frac{\Delta}{2} = \tan \left(\frac{\pi}{4} - \frac{\Phi}{2} \right) = \frac{1}{R} \sqrt{x^2 + y^2} \;,$$

$$2 \sin \frac{\Delta}{2} \cos \frac{\Delta}{2} = \frac{2 \tan \frac{\Delta}{2}}{1 + \tan^2 \frac{\Delta}{2}} = \sin \Delta = \cos \Phi \;,$$

$$\cos \Phi = \frac{2 \tan \frac{\Delta}{2}}{1 + \tan^2 \frac{\Delta}{2}} = \frac{2}{R} \frac{\sqrt{x^2 + y^2}}{1 + \frac{x^2 + y^2}{R^2}} = 2R \frac{\sqrt{x^2 + y^2}}{R^2 + x^2 + y^2} \;, \tag{5.110}$$

$$\cos^2 \frac{\Delta}{2} - \sin^2 \frac{\Delta}{2} = \frac{1}{1 + \tan^2 \frac{\Delta}{2}} - \frac{\tan^2 \frac{\Delta}{2}}{1 + \tan^2 \frac{\Delta}{2}} = \frac{1 - \tan^2 \frac{\Delta}{2}}{1 + \tan^2 \frac{\Delta}{2}} = \cos \Delta = \sin \Phi \;,$$

$$\sin \Phi = \frac{1 - \tan^2 \frac{\Delta}{2}}{1 + \tan^2 \frac{\Delta}{2}} = \frac{1 - \frac{x^2 + y^2}{R^2}}{1 + \frac{x^2 + y^2}{R^2}} = \frac{R^2 - (x^2 + y^2)}{R^2 + (x^2 + y^2)} \;,$$

$$\cos \Phi = 2R \frac{\sqrt{x^2 + y^2}}{R^2 + x^2 + y^2} \;,$$

$$\sin \Phi = \frac{R^2 - (x^2 + y^2)}{R^2 + (x^2 + y^2)} \;. \tag{5.111}$$

$$\mathbb{S}_R^2 \subset \mathbb{E}^3:$$

$$\boldsymbol{X}(\Lambda, \Phi, R) = [\boldsymbol{E}_1, \boldsymbol{E}_2, \boldsymbol{E}_3] \begin{bmatrix} X \\ Y \\ Z \end{bmatrix} \;, \tag{5.112}$$

$$X = R \cos \Phi \cos \Lambda = 2R \frac{x}{R^2 + (x^2 + y^2)} \;,$$

$$Y = R \cos \Phi \sin \Lambda = 2R \frac{y}{R^2 + (x^2 + y^2)} \;, \tag{5.113}$$

$$Z = R \sin \Phi = R \frac{R^2 - (x^2 + y^2)}{R^2 + (x^2 + y^2)} \;.$$

5-25 What are the best polar azimuthal projections of "sphere to plane"?

Most textbooks on map projections list those many azimuthal projections of the "sphere to plane" without taking any decision of which one may be the best. Indeed, for such a decision, we need an objective criterion, and we choose it according to Chapter 1 and Chapter 2, i. e. we choose the *distortion energy over a spherical cap* being covered by the chosen azimuthal projection of the sphere $\mathbb{S}_R^2$ to the tangent space $T_N\mathbb{S}_R^2$ or the plane $\mathbb{P}_O^2$. In order to prepare us for a rational decision of the best polar azimuthal projection "sphere to plane", in Table 5.2, we have tabulated a variety of values for the left principal stretches $\Lambda_1(\Delta)$ along the parallel circle and $\Lambda_2(\Delta)$ along the meridian for the area distortion $\Lambda_1(\Delta)\Lambda_2(\Delta)$ and the maximal angular shear $2\arcsin[|\Lambda_1(\Delta) - \Lambda_2(\Delta)| / (\Lambda_1(\Delta) + \Lambda_2(\Delta))]$ as functions of colatitude (polar distance Δ), namely for Δ given by $\Delta \in \{0°, 30°, 60°, 90°\}$ and six typical polar azimuthal projections.

Table 5.2. Distortion data of spherical mappings: "sphere to plane", azimuthal projections, normal aspect (polar, direct).

name	$\Delta = \pi/2 - \Phi$	Λ_1 (parallel circle)	Λ_2 (meridian)	$\Lambda_1\Lambda_2$ (area distortion)	$2\arcsin\frac{\|\Lambda_1(\Delta)-\Lambda_2(\Delta)\|}{(\Lambda_1(\Delta)+\Lambda_2(\Delta))}$ (max. ang. distortion)
equidistant	0°	1.000	1	1.000	0°00′
(Postel)	30°	1.047	1	1.047	2°38′
	60°	1.209	1	1.209	10°52′
	90°	1.571	1	1.571	25°39′
conformal	0°	1.000	1.000	1.000	0°
(UPS)	30°	1.072	1.072	1.149	0°
	60°	1.333	1.333	1.778	0°
	90°	2.000	2.000	4.000	0°
equiareal	0°	1.000	1.000	1.000	0°00′
	30°	1.035	0.966	1.000	3°58′
	60°	1.155	0.866	1.000	16°26′
	90°	1.414	0.707	1.000	38°57′
gnomonic	0°	1.000	1.000	1.000	0°00′
	30°	1.155	1.333	1.540	8°14′
	60°	2.000	4.000	8.000	38°57′
	90°	∞	∞	∞	180°00′
orthographic	0°	1	1.000	1.000	0°00′
	30°	1	0.866	0.866	8°14′
	60°	1	0.500	0.500	38°57′
	90°	1	0	0	180°00′
Lagrange	0°	0.500	0.500	0.250	0°
conformal	30°	0.536	0.536	0.287	0°
	60°	0.667	0.667	0.445	0°
	90°	1.000	1.000	1.000	0°

In addition, a collection of the distortion energy density $\mathrm{tr}[C_l G_l^{-1}]/2 = (\Lambda_1^2(\Delta) + \Lambda_2^2(\Delta))/2$, the arithmetic mean of the left principal stretches squared, is presented in Box 5.18. The distortion energy density has been given both as a function of colatitude Δ and latitude Φ. Next, by means of Box 5.19, we outline the computation of the total surface element S of a spherical cap between a parallel circle of latitude Φ (colatitude Δ) and $\Phi = \pi/2$ (North Pole). Finally, we are prepared to compute by means of Box 5.20 the distortion energy over a spherical cap, relatively to the six typical polar azimuthal projections. Note that all integral formulae were taken from W. Gröbner and N. Hofreiter (1973), in particular, 331.10 k (page 119) 331.11 k (page 120), and 333.8 b (page 130).

Box 5.18 (Distortion energy density $\mathrm{tr}[C_l G_l^{-1}]/2 = (\Lambda_1^2(\Delta) + \Lambda_2^2(\Delta))/2$ for various azimuthal map projections of the sphere, normal aspect (polar aspect)).

Equidistant (Postel):

$$\frac{1}{2}\left(\Lambda_1^2 + \Lambda_2^2\right) = \frac{1}{2}\frac{\sin^2\Delta + \Delta^2}{\sin^2\Delta} = \frac{1}{2}\frac{\cos^2\Phi + \left(\frac{\pi}{2} - \Phi\right)^2}{\cos^2\Phi} \ . \tag{5.114}$$

Conformal (UPS):

$$\frac{1}{2}\left(\Lambda_1^2 + \Lambda_2^2\right) = \frac{1}{\cos^4\frac{\Delta}{2}} = \frac{1}{\cos^4\left(\frac{\pi}{4} - \frac{\Phi}{2}\right)} \ . \tag{5.115}$$

Equiareal (Lambert):

$$\frac{1}{2}\left(\Lambda_1^2 + \Lambda_2^2\right) = \frac{1}{2}\frac{1 + \cos^4\frac{\Delta}{2}}{\cos^2\frac{\Delta}{2}} = \frac{1}{2}\frac{1 + \cos^4\left(\frac{\pi}{4} - \frac{\Phi}{2}\right)}{\cos^2\left(\frac{\pi}{4} - \frac{\Phi}{2}\right)} \ . \tag{5.116}$$

Gnomonic:

$$\frac{1}{2}\left(\Lambda_1^2 + \Lambda_2^2\right) = \frac{1}{2}\frac{\cos^2\Delta + 1}{\cos^4\Delta} = \frac{1}{2}\frac{\sin^2\Phi + 1}{\sin^4\Phi} \ . \tag{5.117}$$

Orthographic:

$$\frac{1}{2}\left(\Lambda_1^2 + \Lambda_2^2\right) = \frac{1}{2}\left(1 + \cos^2\Delta\right) = \frac{1}{2}\left(1 + \sin^2\Phi\right) \ . \tag{5.118}$$

Lagrange conformal:

$$\frac{1}{2}\left(\Lambda_1^2 + \Lambda_2^2\right) = \frac{1}{4}\frac{1}{\cos^4\frac{\Delta}{2}} = \frac{1}{4}\frac{1}{\cos^4\left(\frac{\pi}{4} - \frac{\Phi}{2}\right)} \ . \tag{5.119}$$

Box 5.19 (Total surface element S or the area of a spherical cap).

$$S = R^2\int_0^{2\pi}\mathrm{d}\Lambda\int_\Phi^{\pi/2}\mathrm{d}\Phi\cos\Phi = R^2\int_0^{2\pi}\mathrm{d}\Lambda\int_0^\Delta\mathrm{d}\Delta\sin\Delta \ , \tag{5.120}$$

subject to

$$\Delta := \frac{\pi}{2} - \Phi \quad \text{or} \quad \Phi = \frac{\pi}{2} - \Delta \ ,$$
$$\mathrm{d}\Phi = -\mathrm{d}\Delta \ ,$$
$$\int_\Phi^{\pi/2}\mathrm{d}\Phi = -\int_\Delta^0\mathrm{d}\Delta = +\int_0^\Delta\mathrm{d}\Delta \ ; \tag{5.121}$$

$$S = 2\pi R^2\left[+\sin\Phi\right]_\Phi^{\pi/2} = 2\pi R^2(1 - \sin\Phi) \ ,$$
$$S = 2\pi R^2\left[-\cos\Delta\right]_0^\Delta = 2\pi R^2(1 - \cos\Delta) \ . \tag{5.122}$$

Box 5.20 (Distortion energy of six polar azimuthal projections over a spherical cap).

General representation of the distortion energy:

$$J := \int dS \frac{1}{2} \operatorname{tr} \left[\mathsf{C}_l \mathsf{G}_l^{-1} \right] = \int dS \frac{1}{2} \left(\Lambda_1^2 + \Lambda_2^2 \right) \;, \quad J = 2\pi R^2 \int_0^\Delta \sin x \frac{1}{2} \left(\Lambda_1^2(x) + \Lambda_2^2(x) \right) dx \;. \quad (5.123)$$

(i) Equidistant (Postel) polar azimuthal projection:

$$J_1 := \frac{J}{2\pi R^2} = \frac{1}{2} \int_0^\Delta \frac{x^2}{\sin x} dx + \frac{1}{2} \int_0^\Delta \sin x \, dx \;. \quad (5.124)$$

1st integral:

$$\int_0^\Delta \frac{x^2}{\sin x} dx = \frac{\Delta^2}{2} + \lim_{K \to \infty} \sum_{k=1}^K (-1)^{k+1} \frac{2 \left(2^{2k-1} - 1 \right)}{(2+2k)(2k)!} B_{2k} \Delta^{2+2k} \;. \quad (5.125)$$

Bernoulli numbers:

$$B_0 = 1 \;, \; B_1 = -\frac{1}{2} \;, \; B_2 = +\frac{1}{6} \;, \; B_4 = -\frac{1}{30} \;, \; B_6 = +\frac{1}{42} \;, \; B_8 = -\frac{1}{30} \;,$$

$$B_{10} = +\frac{5}{66} \;, \; B_{12} = -\frac{691}{2730} \;, \; B_{14} = +\frac{7}{6} \;, \; B_{16} = -\frac{3617}{510} \;. \quad (5.126)$$

2nd integral:

$$\int_0^\Delta \sin x \, dx = [-\cos x]_0^\Delta = 1 - \cos \Delta \;. \quad (5.127)$$

J_1:

$$J_1 = \frac{1}{4} \Delta^2 + \lim_{K \to \infty} \sum_{k=1}^K (-1)^{k+1} \frac{\left(2^{2k-1} - 1 \right)}{(2+2k)(2k)!} B_{2k} \Delta^{2+2k} + \frac{1}{2} (1 - \cos \Delta) \;. \quad (5.128)$$

(ii) Conformal polar azimuthal projection (UPS):

$$J_2 := \frac{J}{2\pi R^2} = \int_0^\Delta \frac{\sin x}{\cos^4 \frac{x}{2}} dx = 2 \int_0^\Delta \frac{\sin \frac{x}{2}}{\cos^3 \frac{x}{2}} dx \;,$$

$$J_2/2 = [1/\cos^2 \tfrac{x}{2}]_0^\Delta = \frac{1}{\cos^2 \frac{\Delta}{2}} - 1 = \frac{1 - \cos^2 \frac{\Delta}{2}}{\cos^2 \frac{\Delta}{2}} \;, \quad J_2 = 2 \tan^2 \frac{\Delta}{2} \;. \quad (5.129)$$

(iii) Equiareal (Lambert) polar azimuthal projection:

$$J_3 := \frac{J}{2\pi R^2} = \int_0^\Delta \sin x \frac{1 + \cos^4 \frac{x}{2}}{\cos^2 \frac{x}{2}} dx = 2 \int_0^\Delta \sin \frac{x}{2} \frac{1 + \cos^4 \frac{x}{2}}{\cos \frac{x}{2}} dx =$$

$$= 2 \int_0^\Delta \tan \frac{x}{2} \left(1 + \cos^4 \frac{x}{2} \right) dx \;,$$

$$x/2 := y$$

$$\Rightarrow$$

$$\int \tan \frac{x}{2} \left(1 + \cos^4 \frac{x}{2} \right) dx = 2 \int \tan y \left(1 + \cos^4 y \right) dy = \quad (5.130)$$

$$= 2 \int \tan y \, dy + 2 \int \sin y \cos^3 y \, dy = -2 \ln \cos y - \frac{1}{2} \cos^4 y = -2 \ln \cos \frac{x}{2} - \frac{1}{2} \cos^4 \frac{x}{2} \;,$$

$$\left[\ln \cos \frac{x}{2} \right]_0^\Delta = \ln \cos \frac{\Delta}{2} \;, \quad \left[\cos^4 \frac{x}{2} \right]_0^\Delta = \cos^4 \frac{\Delta}{2} - 1 \;,$$

$$J_3 = 1 - \cos^4 \frac{\Delta}{2} - 4 \ln \cos \frac{\Delta}{2} \;.$$

Continuation of Box.

(iv) Gnomonic polar azimuthal projection:

$$J_4 := \frac{J}{2\pi R^2} = \frac{1}{2}\int_0^\Delta \sin x \frac{1+\cos^2 x}{\cos^4 x}\mathrm{d}x = \frac{1}{2}\int_0^\Delta \sin x \left(\frac{1}{\cos^4 x} + \frac{1}{\cos^2 x}\right)\mathrm{d}x \ ,$$

$$J_4 = \frac{1}{6}\left[1/\cos^3 x\right]_0^\Delta + \frac{1}{2}\left[1/\cos x\right]_0^\Delta \ ,$$

$$J_4 = \frac{1}{6}\left(\frac{1}{\cos^3 \Delta} - 1\right) + \frac{1}{2}\left(\frac{1}{\cos \Delta} - 1\right) \ ,$$

$$J_4 = \frac{1}{6}\frac{1-\cos^3 \Delta}{\cos^3 \Delta} + \frac{1}{2}\frac{1-\cos \Delta}{\cos \Delta} \ .$$

(5.131)

(v) Orthographic polar azimuthal projection:

$$J_5 := \frac{J}{2\pi R^2} = \frac{1}{2}\int_0^\Delta \sin x(1+\cos^2 x)\mathrm{d}x \ , \quad J_5 = \frac{1}{2}\int_0^\Delta \sin x\mathrm{d}x + \frac{1}{2}\int_0^\Delta \sin x \cos^2 x\mathrm{d}x \ ,$$

$$J_5 = \frac{1}{2}\left[-\cos x\right]_0^\Delta - \frac{1}{6}\left[\cos^3 x\right]_0^\Delta \ , \quad J_5 = \frac{1}{2}(1 - \cos\Delta) + \frac{1}{6}(1 - \cos^3 \Delta) \ .$$

(5.132)

(vi) Lagrange conformal polar azimuthal projection:

$$J_6 := \frac{J}{2\pi R^2} = \frac{1}{4}\int_0^\Delta \frac{\sin x}{\cos^4 \frac{x}{2}}\mathrm{d}x = \frac{1}{2}\int_0^\Delta \frac{\sin\frac{x}{2}}{\cos^3 \frac{x}{2}}\mathrm{d}x \ ,$$

$$\sin x = 2\sin\frac{x}{2}\cos\frac{x}{2} \ , \quad x/2 = y : \ \mathrm{d}x = 2\mathrm{d}y \ ,$$

$$J_6 = \int_0^{\Delta/2} \frac{\sin y}{\cos^3 y}\mathrm{d}y = \frac{1}{2}\left[1/\cos^2 y\right]_0^{\Delta/2} = \frac{1}{2}\left(\frac{1}{\cos^2 \frac{\Delta}{2}} - 1\right) \ ,$$

$$J_6 = \frac{1}{2}\tan^2\frac{\Delta}{2} \ .$$

(5.133)

The portrait of the distortion energy density and of the total distortion energy over a spherical cap $(0° \le \Delta \le 60°)$ is given by Fig. 5.27, Fig. 5.28, and Table 5.3 for six polar azimuthal projections of type (i) equidistant (Postel), (ii) conformal (UPS), (iii) equiareal (Lambert), (iv) gnomonic, (v) orthographic, and (vi) Lagrange conformal. Contact Appendix A in order to enjoy the ordering

$$J_6 < J_5 < J_1 < J_2 < J_4 < J_3 \text{ for } < 49°, 248502$$

and (5.134)

$$J_6 < J_5 < J_1 < J_2 < J_3 < J_4 \text{ for } > 49°, 248502 \ .$$

Denote for a moment the symbol $<$ by "better". Then we can make a most important qualitative statement about the six polar azimuthal projections based upon the ordering of the respective total distortion energies over a spherical cap, namely

conformal (Lagrange) $<$ orthographic $<$ equidistant (Postel) $<$ conformal (UPS) $<$
$<$ gnomonic $<$ equal area (Lambert) for $< 49°, 248502$

and (5.135)

conformal (Lagrange) $<$ orthographic $<$ equidistant (Postel) $<$ conformal (UPS) $<$
$<$ equal area (Lambert) $<$ gnomonic for $> 49°, 248502$.

Of course, in practice, decision makers for azimuthal map projections do not follow objective criteria: they prefer the equiareal (Lambert) projection.

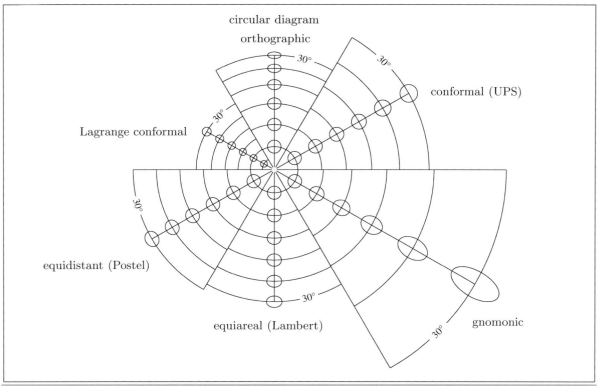

Fig. 5.27. Circular diagram, polar azimuthal projections.

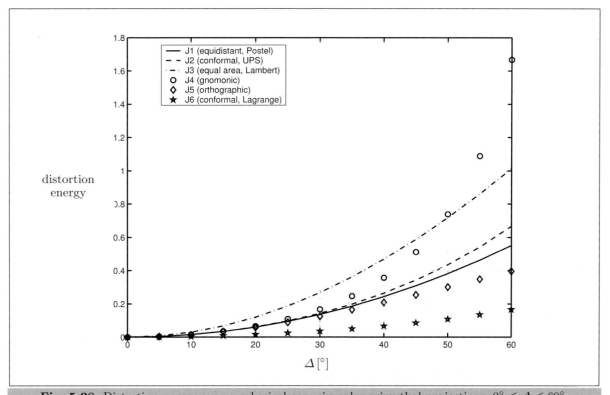

Fig. 5.28. Distortion energy over a spherical cap: six polar azimuthal projections, $0° \leq \Delta \leq 60°$.

Table 5.3. Distortion energy over a spherical cap: six polar azimuthal projections, $0° \leq \Delta \leq 60°$.

$\Delta[°]$	J_1	J_2	J_3	J_4	J_5	J_6
0	0.00000	0.00000	0.00000	0.00000	0.00000	0.00000
5	0.00381	0.00381	0.00761	0.00383	0.00380	0.00095
10	0.01523	0.01531	0.03038	0.01555	0.01508	0.00383
15	0.03427	0.03466	0.06815	0.03591	0.03350	0.00867
20	0.06093	0.06218	0.12063	0.06628	0.05853	0.01555
25	0.09521	0.09830	0.18746	0.10891	0.08944	0.02457
30	0.13713	0.14359	0.26816	0.16728	0.12540	0.03590
35	0.18670	0.19883	0.36222	0.24694	0.16548	0.04971
40	0.24397	0.26495	0.46908	0.35679	0.20872	0.06624
45	0.30899	0.34315	0.58814	0.51184	0.25419	0.08579
50	0.38184	0.43489	0.71882	0.73874	0.30101	0.10872
55	0.46264	0.54198	0.86056	1.08829	0.34843	0.13550
60	0.55155	0.66667	1.01286	1.66667	0.39583	0.16667

5-3 The pseudo-azimuthal projection

Setting up general equations of the mapping "sphere to plane": the pseudo-azimuthal projection in the normal aspect (polar aspect).

In a preceding section, we define polar azimuthal projections by the following two postulates. (i) The images of the circular meridians $\Lambda = $ constant under an azimuthal mapping are radial straight lines. (ii) The images of parallel circles $\Phi = $ constant or $\Delta = $ constant are concentric circles. Any deviation from these postulates generates *pseudo-azimuthal projections* or, in general, mappings of the sphere $\mathbb{S}^2_R$ of radius R to a polar tangential plane $T_N\mathbb{S}^2_R$ or to a plane $\mathbb{P}^2_O$ through the center o of the sphere $\mathbb{S}^2_R$. Here, we shall only consider general equations of a pseudo-azimuthal mapping of type

$$\begin{bmatrix} x(\Lambda, \Delta) \\ y(\Lambda, \Delta) \end{bmatrix} = r(\Delta) \begin{bmatrix} \cos \alpha(\Lambda, \Delta) \\ \sin \alpha(\Lambda, \Delta) \end{bmatrix}$$

or   (5.136)

$$\begin{bmatrix} x(\Lambda, \Delta) \\ y(\Lambda, \Delta) \end{bmatrix} = f(\Delta) \begin{bmatrix} \cos g(\Lambda, \Delta) \\ \sin g(\Lambda, \Delta) \end{bmatrix},$$

which are characterized by two functions, namely the *radial function* $f(\Delta)$ and the *azimuth function* $g(\Lambda, \Delta)$. The azimuth function $\alpha(\Lambda, \Delta) = g(\Lambda, \Delta)$ is *azimuth preserving* if $\alpha(\Lambda, \Delta) = \Delta$. Accordingly, in general, a pseudo-azimuthal mapping is not azimuth preserving, $\alpha(\Lambda, \Delta) \neq \Delta$.

Question: "How are the pseudo-azimuthal projections "sphere to plane" classified?" Answer: "By computing the left Cauchy–Green matrix as well as its left eigenspace."

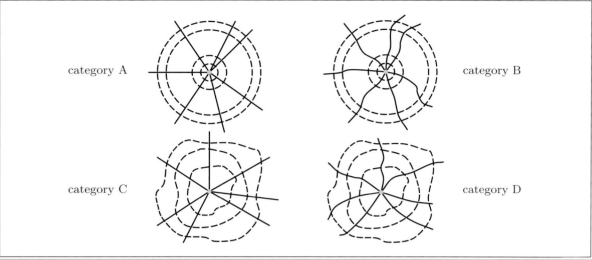

category A

category B

category C

category D

Fig. 5.29. Images of coordinate lines under four categories of mapping, special case: polar coordinates.

Let us bother you with the detailed analysis of distortion for pseudo-azimuthal mappings of type $x = f(\Delta) \cos g(\Lambda, \Delta)$ and $y = f(\Delta) \sin g(\Lambda, \Delta)$. In Box 5.21, we present the left Jacobi matrix, the left Cauchy–Green matrix and the left principal stretches. In order to supply you with a visual impression of what is going to happen when you switch from azimuthal to pseudo-azimuthal, in Fig. 5.29, we have made an attempt to highlight the images of the coordinate lines under the following four categories of mapping. (The various categories have been properly chosen by W. R. Tobler (1963a) as long as we intend to map "sphere to plane".)

$$
\begin{array}{llll}
\text{Category A:} & \text{Category B:} & \text{Category C:} & \text{Category D:} \\
\alpha = \Lambda\,, & \alpha = g(\Lambda, \Delta)\,, & \alpha = g(\Lambda)\,, & \alpha = g(\Lambda, \Delta)\,, \\
r = f(\Delta)\,. & r = f(\Delta)\,. & r = f(\Lambda, \Delta)\,. & r = f(\Lambda, \Delta)\,.
\end{array}
\qquad (5.137)
$$

Box 5.21 (Polar pseudo-azimuthal projections "sphere to plane", left principal stretches).

Parameterized mapping:

$$
\alpha = g(\Lambda, \Delta)\,, \quad r = f(\Delta)\,, \quad x = r \cos \alpha = f(\Delta) \cos g(\Lambda, \Delta)\,, \quad y = r \sin \alpha = f(\Delta) \sin g(\Lambda, \Delta)\,. \quad (5.138)
$$

Left Jacobi matrix:

$$
J_l = \begin{bmatrix} D_\Lambda x & D_\Delta x \\ D_\Lambda y & D_\Delta y \end{bmatrix} \begin{bmatrix} -f g_\Lambda \sin g & f' \cos g - f g_\Delta \sin g \\ +f g_\Lambda \cos g & f' \sin g + f g_\Delta \cos g \end{bmatrix}. \qquad (5.139)
$$

Left Cauchy–Green matrix:

$$
G_r = I_2\,, \quad C_l = J_l^* G_r J_l = J_l^* J_l = \begin{bmatrix} f^2 g_\Lambda^2 & f^2 g_\Lambda g_\Delta \\ f^2 g_\Lambda g_\Delta & f'^2 + f^2 g_\Delta^2 \end{bmatrix}. \qquad (5.140)
$$

Left principal stretches:

$$
|C_l - \Lambda^2 G_l| = 0 \Leftrightarrow \Lambda_{1,2}^2 = \Lambda_\pm^2 = \frac{1}{2} \left(\operatorname{tr} [C_l G_l^{-1}] \pm \sqrt{(\operatorname{tr} [C_l G_l^{-1}])^2 - 4 \det [C_l G_l^{-1}]} \right), \qquad (5.141)
$$

Continuation of Box.

$$\mathsf{G}_l = \begin{bmatrix} R^2 \sin^2 \Delta & 0 \\ 0 & R^2 \end{bmatrix} , \quad \mathsf{G}_l^{-1} = \begin{bmatrix} \dfrac{1}{R^2 \sin^2 \Delta} & 0 \\ 0 & \dfrac{1}{R^2} \end{bmatrix} ,$$

(5.142)

$$\mathsf{C}_l \mathsf{G}_l^{-1} = \begin{bmatrix} c_{11} G_{11}^{-1} & c_{12} G_{22}^{-1} \\ c_{12} G_{11}^{-1} & c_{22} G_{22}^{-1} \end{bmatrix} = \begin{bmatrix} \dfrac{f^2 g_\Lambda^2}{R^2 \sin^2 \Delta} & \dfrac{f^2 g_\Lambda g_\Delta}{R^2} \\ \dfrac{f^2 g_\Lambda g_\Delta}{R^2 \sin^2 \Delta} & \dfrac{f'^2 + f^2 g_\Delta^2}{R^2} \end{bmatrix} ,$$

$$\operatorname{tr}\left[\mathsf{C}_l \mathsf{G}_l^{-1}\right] = \frac{1}{R^2 \sin^2 \Delta} \left[f^2 g_\Lambda^2 + \left(f'^2 + f^2 g_\Delta^2 \right) \sin^2 \Delta \right] ,$$

$$\det\left[\mathsf{C}_l \mathsf{G}_l^{-1}\right] = \frac{1}{R^4 \sin^2 \Delta} \left[f^2 g_\Lambda^2 \left(f'^2 + f^2 g_\Delta^2 \right) - f^4 g_\Lambda^2 g_\Delta^2 \right] ,$$

$$\operatorname{tr}\left[\mathsf{C}_l \mathsf{G}_l^{-1}\right] = \frac{1}{R^2 \sin^2 \Delta} \left[f^2 \left(g_\Lambda^2 + g_\Delta^2 \sin^2 \Delta \right) + f'^2 \sin^2 \Delta \right] ,$$

$$\det\left[\mathsf{C}_l \mathsf{G}_l^{-1}\right] = \frac{1}{R^4 \sin^2 \Delta} f^2 f'^2 g_\Lambda^2 .$$

(5.143)

Let us comment on the left principal stretches that we have computed in Box 5.21. First, based on the parameterized mapping of category B, we have calculated the left Jacobi matrix, namely the partial derivatives of $x(\Lambda, \Delta)$ and $y(\Lambda, \Delta)$. Second, we succeeded to derive a simple form of the left Cauchy–Green matrix. Third, the general eigenvalue problem for the matrix pair $\{\mathsf{C}_l, \mathsf{G}_l\}$ leads to the characteristic equation $|\mathsf{C}_l - \Lambda^2 \mathsf{G}_l| = 0$. We did not explicitly compute the left principal stretches $\{\Lambda_1, \Lambda_2\}$. Instead, we took advantage of the invariant representation of the left eigenspace in terms of the Hilbert invariants $J_1 = \operatorname{tr}\left[\mathsf{C}_l \mathsf{G}_l^{-1}\right]$ and $J_2 = \det\left[\mathsf{C}_l \mathsf{G}_l^{-1}\right]$. J_1 as well as J_2 have been explicitly computed.

Question

Question: "Do conformal mappings of the pseudo-azimuthal type, category B, exist or do equiareal mappings of the pseudo-azimuthal type, category B, exist?" Answer: "No conformal mappings of the pseudo-azimuthal type, category B, exist, but equiareal indeed do."

This question may be asked with the left principal stretches Λ_1 and Λ_2 at hand. But how to prove this answer? Let us prove this answer in two steps.

Proof.

First, we prove the non-existence of a conformal pseudo-azimuthal mapping, category B. The canonical postulate of conformality, $\Lambda_1 = \Lambda_2$, is equivalent to (5.144). The sum of two positive numbers cannot be zero, in general. The special case $g_\Lambda = 1$ and $g_\Delta = 0$ transforms the pseudo-azimuthal mapping, category B, back to the azimuthal mapping, category A.

$$\operatorname{tr}\left[\mathsf{C}_l \mathsf{G}_l^{-1}\right] = 2\sqrt{\det\left[\mathsf{C}_l \mathsf{G}_l^{-1}\right]} ; \quad \text{here: } \left(f g_\Lambda - f' \sin \Delta\right)^2 + f^2 g_\Delta^2 \sin^2 \Delta = 0 .$$

(5.144)

Second, we characterize an equiareal pseudo-azimuthal mapping, category B. The canonical postulate of an equiareal mapping, $\Lambda_1 \Lambda_2 = 1$, is equivalent to (5.145). In consequence, we give an example of an equiareal pseudo-azimuthal mapping, category B. The special case $g_\Lambda = 1$ transforms the pseudo-azimuthal mapping, category B, back to the azimuthal mapping, category A.

$$\sqrt{\det\left[\mathsf{C}_l \mathsf{G}_l^{-1}\right]} = 1 ; \quad \text{here: } \frac{f f'}{R^2 \sin \Delta} g_\Lambda = 1 .$$

(5.145)

End of Proof.

5-4 The Wiechel polar pseudo-azimuthal projection

A special variant of Lambert's equiareal polar azimuthal projection: the Wiechel polar pseudo-azimuthal projection.

A special variant of Lambert's equiareal polar azimuthal projection has been given by H. Wiechel (1879). The direct equations for mapping the "sphere to plane" are presented in Box 5.22 and are illustrated in Fig. 5.30. Thanks to the azimuthal function $\alpha = g(\Lambda, \Delta) \neq \Lambda$ (in general, here we consider $\alpha = \Lambda + \Delta/2$), the *Wiechel map* is pseudo-azimuthal. A quick view to Wiechel's pseudo-azimuthal map of Fig. 5.30 motivates the following interpretation: we see the polar vortex at the North Pole directed to the Earth's rotation axis, namely e_3. Indeed, we compute the curl or vortex of the placement vector $x(\Lambda, \Delta) = e_1 x(\Lambda, \Delta) + e_2 y(\Lambda, \Delta)$:

$$\operatorname{curl} x(\Lambda, \Delta) = e_3 \left(D_\Lambda y - D_\Delta x \right) = e_3 \left[-R \cos(\Lambda + \Delta) + 2R \sin \frac{\Delta}{2} \cos \left(\Lambda + \frac{\Delta}{2} \right) \right] \neq 0. \quad (5.146)$$

Consult the original contribution of H. Wiechel (1879) for a deeper understanding. In particular, enjoy his arguments for "a rotational graticule". To become familiar with such a special pseudo-azimuthal mapping "sphere to plane", let us ask the following question.

Question

Question: "Is the Wiechel pseudo-azimuthal projection "sphere to plane" equiareal?" Answer: "Yes."

For the proof, follow the lines of the proof outlined in Box 5.22. First, we compute the left Jacobi matrix constituted by the partial derivatives $D_\Lambda x$, $D_\Delta x$, $D_\Lambda y$, and $D_\Delta y$. Second, we derive the left Cauchy–Green matrix by computing $C_l = J_l^* J_l$. Third, we derive the left principal stretches, the left eigenvalues of the matrix $C_l G_l^{-1}$, namely Λ_1 and Λ_2, from the trace $\operatorname{tr} \left[C_l G_l^{-1} \right]$ and the determinant $\det \left[C_l G_l^{-1} \right]$. Fourth, $\Lambda_1 \Lambda_2 = 1$ proves an equiareal mapping.

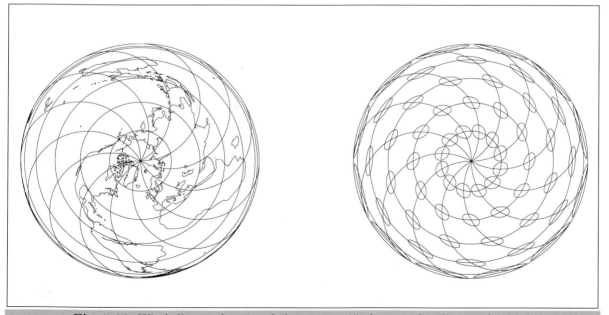

Fig. 5.30. Wiechel's pseudo-azimuthal projection "sphere to plane", normal aspect.

Box 5.22 (Wiechel's pseudo-azimuthal projection "sphere to plane").

Direct equations of Wiechel's map (polar coordinates):

$$\alpha = \Lambda + \frac{\Delta}{2} = \Lambda + \left(\frac{\pi}{4} - \frac{\Phi}{2} \right) =: g(\Lambda, \Delta) ,$$

$$r = 2R \sin \frac{\Delta}{2} = 2R \sin \left(\frac{\pi}{4} - \frac{\Phi}{2} \right) =: f(\Delta) .$$

(5.147)

Direct equations of Wiechel's map (Cartesian coordinates):

$$\begin{bmatrix} x \\ y \end{bmatrix} = 2R \sin \frac{\Delta}{2} \begin{bmatrix} \cos \left(\Lambda + \frac{\Delta}{2} \right) \\ \sin \left(\Lambda + \frac{\Delta}{2} \right) \end{bmatrix} , \quad \begin{bmatrix} x \\ y \end{bmatrix} = f(\Delta) \begin{bmatrix} \cos g(\Lambda, \Delta) \\ \sin g(\Lambda, \Delta) \end{bmatrix} .$$

(5.148)

Left Jacobi matrix:

$$J_l = \begin{bmatrix} D_\Lambda x & D_\Delta x \\ D_\Lambda y & D_\Delta y \end{bmatrix} \begin{bmatrix} -f \sin g & f' \cos g - f g_\Delta \sin g \\ +f \cos g & f' \sin g + f g_\Delta \cos g \end{bmatrix} , \quad f' = R \cos \frac{\Delta}{2} , \quad g_\Delta = \frac{1}{2} .$$

(5.149)

Left Cauchy–Green matrix:

$$G_r = I_2 , \quad C_l = J_l^* G_r J_l = J_l^* J_l = \begin{bmatrix} f^2 & \frac{1}{2} f^2 \\ \frac{1}{2} f^2 & f'^2 + f^2 g_\Delta^2 \end{bmatrix} = \begin{bmatrix} 4R^2 \sin^2 \frac{\Delta}{2} & 2R^2 \sin^2 \frac{\Delta}{2} \\ 2R^2 \sin^2 \frac{\Delta}{2} & R^2 \end{bmatrix} .$$

(5.150)

Left principal stretches:

$$\Lambda_{1,2}^2 = \Lambda_\pm^2 = \frac{1}{2} \left(\text{tr} \left[C_l G_l^{-1} \right] \pm \sqrt{\left(\text{tr} \left[C_l G_l^{-1} \right] \right)^2 - 4 \det \left[C_l G_l^{-1} \right]} \right) ,$$

(5.151)

$$C_l G_l^{-1} = \begin{bmatrix} \dfrac{1}{\cos^2 \frac{\Delta}{2}} & 2 \sin^2 \frac{\Delta}{2} \\ \dfrac{1}{2} \dfrac{1}{\cos^2 \frac{\Delta}{2}} & 1 \end{bmatrix} ,$$

$$\text{tr} \left[C_l G_l^{-1} \right] = \frac{1 + \cos^2 \frac{\Delta}{2}}{\cos^2 \frac{\Delta}{2}} , \quad \det \left[C_l G_l^{-1} \right] = 1 ,$$

(5.152)

$$\left(\text{tr} \left[C_l G_l^{-1} \right] \right)^2 - 4 \det \left[C_l G_l^{-1} \right] = \frac{1}{\cos^4 \frac{\Delta}{2}} \left[\left(1 + \cos^2 \frac{\Delta}{2} \right)^2 - 4 \cos^4 \frac{\Delta}{2} \right] ,$$

$$\Lambda_{1,2}^2 = \Lambda_\pm^2 = \frac{1}{2} \frac{1}{\cos^2 \frac{\Delta}{2}} \left[1 + \cos^2 \frac{\Delta}{2} \pm \sqrt{\left(1 + \cos^2 \frac{\Delta}{2} \right)^2 - 4 \cos^4 \frac{\Delta}{2}} \right]$$

(5.153)

$$\Rightarrow$$

$$\Lambda_1 \Lambda_2 = 1 .$$

With this box, we finish the discussion of the polar (normal) aspect of the mapping "sphere to tangential plane". In the chapter that follows, let us discuss the transverse aspect.

6 "Sphere to tangential plane": transverse aspect

Mapping the sphere to a tangential plane: meta-azimuthal projections in the transverse aspect. Equidistant, conformal (stereographic), and equal area (transverse Lambert) mappings.

Azimuthal projections may be classified by reference to the point-of-contact of the plotting surface with the Earth. While chapter 5 treated the case of a polar azimuthal projections (azimuthal projection in the polar aspect), this section concentrates on meta-azimuthal mappings of the Earth onto a plane in the transverse aspect, which are often also called equatorial: the point-of-contact (meta-North Pole) may be any point on the (conventional) equator of the reference sphere. According to chapter 3, its spherical coordinates, referring to the equatorial frame of reference, are specified through $\Lambda_0 \in [0°, 360°]$, $\Phi_0 = 0°$. In the special case $\Lambda_0 = 270°$, the meta-North Pole is located in the West Pole and the meta-equator (then called transverse equator) agrees with the Greenwich meridian of reference. For a first impression, consult Fig. 6.1.

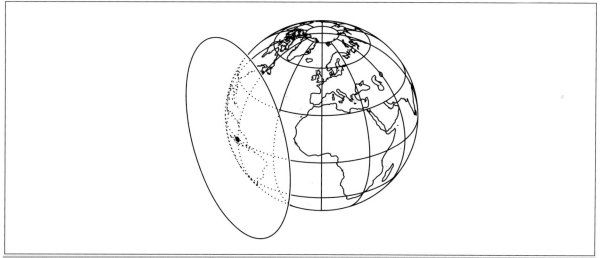

Fig. 6.1. Mapping the sphere to a tangential plane: transverse aspect. Point-of-contact: meta-North Pole at $\Lambda_0 = 300°$, $\Phi_0 = 0°$.

6-1 General mapping equations

Setting up general equations of the mapping "sphere to plane": the meta-azimuthal projections in the transverse aspect. Meta-longitude, meta-latitude.

The general equations for meta-azimuthal projections are based on the general equation (5.9) of Chapter 5, but spherical longitude Λ and spherical latitude Φ being replaced by their counterparts meta-longitude and meta-latitude. In order to distinguish the polar coordinate α in the plane from the meta-longitude α as introduced in Chapter 3, see (3.51) and (3.53), we here refer to A and B as the meta-coordinates *meta-longitude* and *meta-latitude*. In consequence, the general equations of a meta-azimuthal mapping in the transverse aspect are provided by the vector relation (6.1) taking into account the constraints (6.2):

$$\begin{bmatrix} x \\ y \end{bmatrix} = \begin{bmatrix} r \cos \alpha \\ r \sin \alpha \end{bmatrix} = \begin{bmatrix} f(B) \cos A \\ f(B) \sin A \end{bmatrix} , \tag{6.1}$$

$$\tan A = \frac{\sin(\Lambda - \Lambda_0)}{\tan \Phi} , \tag{6.2}$$

$$\sin B = \cos \Phi \cos(\Lambda - \Lambda_0) .$$

These equations result from (3.51) and (3.53) by setting the position of the meta-North Pole to $\Phi_0 = 0°$. As a matter of course, the polar coordinate α, usually called azimuth, is not anymore identical to the spherical longitude Λ. The images of (conventional) meridians and (conventional) parallels lose their typical behavior of being radial straight lines and equicentric circles. Since r equals $f(B)$, parallel circles $\Phi = $ constant are mapped as a function of longitude Λ and longitude Λ_0 of the meta-North Pole. Likewise, the image of a meridian $\Lambda = $ constant becomes a complicated curve satisfying the equation $y = -\sin(\Lambda - \Lambda_0)/\tan\Phi\, x$, i.e. y is a linear function of x but with a longitude and latitude dependent slope.

6-2 Special mapping equations

Setting up special equations of the mapping "sphere to plane": the meta-azimuthal projections in the transverse aspect. Equidistant mapping (transverse Postel projection), conformal mapping (transverse stereographic projection), equal area mapping (transverse Lambert projection).

6-21 Equidistant mapping (transverse Postel projection)

Let us formulate a transverse equidistant mapping of the sphere to a plane by the postulate that for the family of meta-meridians $A = $ constant the relations (6.3) hold true. The mapping equations and the corresponding distortion analysis are systematically presented in Box 6.1. A sketch of this mapping with the choice of $\Lambda_0 = 270°$ is given in Fig. 6.2.

$$r = R\operatorname{arc}(\pi/2 - B) \,,$$

$$\sin B = \cos\Phi\cos(\Lambda - \Lambda_0) \,. \tag{6.3}$$

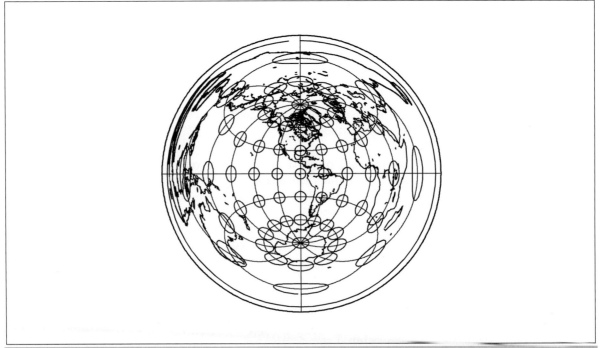

Fig. 6.2. Mapping the sphere to a tangential plane: transverse aspect, equidistant mapping. Point-of-contact: meta-North Pole at $\Lambda_0 = 270°$, $\Phi_0 = 0°$.

Box 6.1 (Transverse equidistant mapping of the sphere to a plane at the meta-North Pole. Parameters: $\Lambda_0 \in [0°, 360°]$, $\Phi_0 = 0$).

Parameterized mapping:

$$\alpha = A , \quad r = f(B) = R\operatorname{arc}(\pi/2 - B) , \tag{6.4}$$

$$x = r \cos \alpha = R(\pi/2 - B) \cos A , \quad y = r \sin \alpha = R(\pi/2 - B) \sin A ,$$

$$\tan A = \frac{\sin(\Lambda - \Lambda_0)}{-\tan \Phi} , \quad \sin B = \cos \Phi \cos(\Lambda - \Lambda_0) . \tag{6.5}$$

Left principal stretches:

$$\Lambda_1 = \frac{\pi/2 - B}{\cos B} , \quad \Lambda_2 = 1 . \tag{6.6}$$

Left eigenvectors:

$$\boldsymbol{C}_1 \Lambda_1 = \boldsymbol{E}_A \frac{\pi/2 - B}{\sin(\pi/2 - B)} , \quad \boldsymbol{C}_2 \Lambda_2 = \boldsymbol{E}_B . \tag{6.7}$$

Parameterized inverse mapping:

$$\tan A = \frac{y}{x} , \quad B = \frac{\pi}{2} - \sqrt{\frac{x^2 + y^2}{R^2}} ,$$

$$\tan(\Lambda - \Lambda_0) = \frac{\sin A}{\tan B} ,$$

$$\sin \Phi = -\cos B \cos A . \tag{6.8}$$

Left maximum angular distortion:

$$\Omega_l = 2 \arcsin \left| \frac{\Lambda_1 - \Lambda_2}{\Lambda_1 + \Lambda_2} \right| = 2 \arcsin \left| \frac{\frac{\pi}{2} - B - \cos B}{\frac{\pi}{2} - B + \cos B} \right| . \tag{6.9}$$

6-22 Conformal mapping (transverse stereographic projection, transverse UPS)

The *transverse conformal mapping* of the sphere to a tangential plane is easily derived with the knowledge of the preceding paragraph. We here conveniently rewrite the mapping equations of the *normal conformal mapping* of Section 5-22 in terms of the (meta-)coordinates meta-longitude A and meta-latitude B. Again, we take into account the relations (6.1) and (6.2) between meta coordinates, standard spherical coordinates (Λ, Φ), and the coordinates $\Lambda_0 \in [0°, 360°]$ and $\Phi_0 = 0°$ of the meta-North Pole. Then, the setup of the mapping equations is given by Lemma 6.1.

Lemma 6.1 (Transverse conformal mapping of the sphere to a tangential plane at the meta-North Pole $\Lambda_0 \in [0°, 360°]$, $\Phi_0 = 0°$).

$$x = 2R \tan\left(\frac{\pi}{4} - \frac{B}{2}\right) \cos A \ , \quad y = 2R \tan\left(\frac{\pi}{4} - \frac{B}{2}\right) \sin A \ ,$$

$$\tan A = \frac{\sin(\Lambda - \Lambda_0)}{-\tan \Phi} \ , \quad \sin B = \cos \Phi \cos(\Lambda - \Lambda_0) \ ,$$

(6.10)

subject to the principal stretches

$$\Lambda_1 = \Lambda_2 = \frac{1}{\cos^2\left(\frac{\pi}{4} - \frac{B}{2}\right)}$$

(6.11)

and the left Cauchy-Green eigenspace

$$\text{left CG eigenspace} = \left\{ \boldsymbol{E}_A \frac{1}{\cos^2\left(\frac{\pi}{4} - \frac{B}{2}\right)}, \boldsymbol{E}_B \frac{1}{\cos^2\left(\frac{\pi}{4} - \frac{B}{2}\right)} \right\} \ .$$

(6.12)

End of Lemma.

An idea of the appearance of the transverse conformal mapping to be considered here can be obtained from Fig. 6.3.

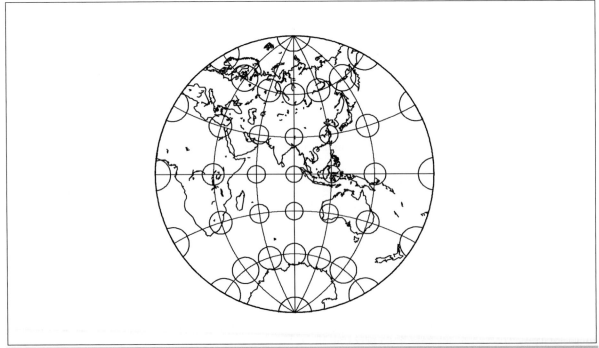

Fig. 6.3. Mapping the sphere to a tangential plane: transverse aspect, conformal mapping. Point-of-contact: meta-North Pole at $\Lambda_0 = 90°$, $\Phi_0 = 0°$.

6-23 Equal area mapping (transverse Lambert projection)

For displaying eastern and western hemispheres in atlas maps, the equatorial aspect of the well-known Lambert projection is widely used. In order to derive the mapping equations, we immediately start from equations (5.35) of Lemma 5.3 again substituting spherical longitude Λ and spherical latitude Φ by their counterparts meta-longitude A and meta-latitude B. We end up with the parameterization in Lemma 6.2. An illustration is given by Fig. 6.4. It is easily observed from Fig. 6.4 that meridians, except the central meridian, are complex curves unequally spaced at the equator. Spacing decreases with increasing distance from the central meridian. Parallels, except the equator which is a straight line, are as well complex curves. Distortions increase radially from the point-of-contact which is mapped isometrically, i. e. free from any distortion.

Lemma 6.2 (Transverse equal area mapping of the sphere to a tangential plane at the meta-North Pole $\Lambda_0 \in [0°, 360°]$, $\Phi_0 = 0°$).

$$x = 2R\sin\left(\frac{\pi}{4} - \frac{B}{2}\right)\cos A , \quad y = 2R\sin\left(\frac{\pi}{4} - \frac{B}{2}\right)\sin A \tag{6.13}$$

subject to

$$\tan A = \frac{\sin(\Lambda - \Lambda_0)}{-\tan\Phi} , \quad \sin B = \cos\Phi\cos(\Lambda - \Lambda_0) . \tag{6.14}$$

The left principal stretches and left Cauchy-Green eigenspace are specified through

$$\Lambda_1 = \frac{1}{\cos\left(\frac{\pi}{4} - \frac{B}{2}\right)} , \quad \Lambda_2 = \cos\left(\frac{\pi}{4} - \frac{B}{2}\right) ,$$

$$\text{left CG eigenspace} = \left\{ \boldsymbol{E}_A \frac{1}{\cos\left(\frac{\pi}{4} - \frac{B}{2}\right)}, \boldsymbol{E}_B \cos\left(\frac{\pi}{4} - \frac{B}{2}\right) \right\} . \tag{6.15}$$

End of Lemma.

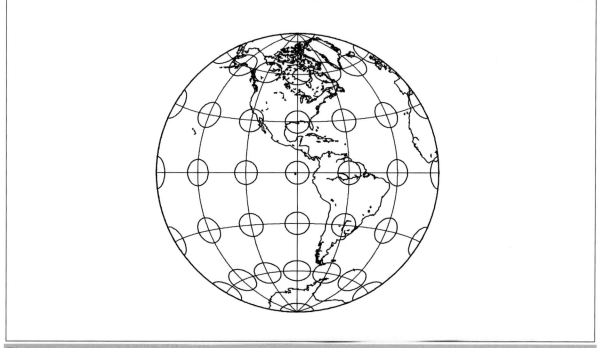

Fig. 6.4. Mapping the sphere to a tangential plane: transverse aspect, equal area mapping. Point-of-contact: meta-North Pole at $\Lambda_0 = 270°$, $\Phi_0 = 0°$.

Figure 6.4 finishes the discussion of the transverse aspect. In the following chapter, let us have a closer look at the oblique aspect.

7 "Sphere to tangential plane": oblique aspect

Mapping the sphere to a tangential plane: meta-azimuthal projections in the oblique aspect. Equidistant, conformal (oblique UPS), and equal area (oblique Lambert) mappings.

In this chapter, we generalize the concept of azimuthal projections and present the class of widely applied oblique azimuthal projection. The point-of-contact is not anymore restricted to be one of the poles or lying on the equator, it can be any point on the reference sphere, i.e. $\Lambda_0 \in [0°, 360°]$, $\Phi_0 \in [-90°, 90°]$. With this configuration, any region of interest can be mapped by an equidistant, conformal, or equal area projection. The latter one is in particular appropriate for regions which are approximately circular in extent. Figure 7.1 gives an impression of the geometrical situation for mappings of meta-azimuthal projections in the oblique aspect.

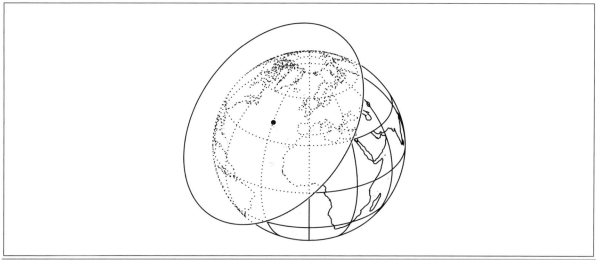

Fig. 7.1. Mapping the sphere to a tangential plane: oblique aspect. Point-of-contact: meta-North Pole at $\Lambda_0 = 330°$, $\Phi_0 = 40°$.

7-1 General mapping equations

Setting up general equations of the mapping sphere to plane: meta-azimuthal projections in the oblique aspect. Meta-longitude, meta-latitude.

The general equations for mapping the sphere to the plane using a meta-azimuthal projection in the oblique aspect involve the most general equations of meta-azimuthal mappings (7.1) in connection with the constraints (7.2) for oblique frames of references:

$$
\begin{bmatrix} x \\ y \end{bmatrix} = \begin{bmatrix} r \cos \alpha \\ r \sin \alpha \end{bmatrix} = \begin{bmatrix} f(B) \cos A \\ f(B) \sin A \end{bmatrix} , \tag{7.1}
$$

$$
\tan A = \frac{\cos \Phi \sin(\Lambda - \Lambda_0)}{\cos \Phi \sin \Phi_0 \cos(\Lambda - \Lambda_0) - \sin \Phi \cos \Phi_0} ,
$$

$$
\sin B = \cos \Phi \cos \Phi_0 \cos(\Lambda - \Lambda_0) + \sin \Phi \sin \Phi_0 . \tag{7.2}
$$

In order not to mix up the polar coordinate α in the plane and meta-longitude α as introduced in Chapter 3, see (3.51) and (3.53), we here refer to A and B as the meta-coordinates *meta-longitude* and *meta-latitude*. In contrast to previous sections, the latitude Φ_0 of the meta-North Pole is not restricted to $\Phi_0 = 90°$ (polar aspect) and $\Phi_0 = 0°$ (transverse aspect), respectively, but can take all values between $\Phi_0 = -90°$ and $\Phi_0 = 90°$, i.e. $\Lambda_0 \in [0°, 360°]$ and $\Phi_0 \in [-90°, 90°]$.

7-2 Special mapping equations

Setting up special equations of the mapping "sphere to plane": the meta-azimuthal projections in the oblique aspect. Equidistant mapping (oblique Postel projection), conformal mapping (oblique stereographic projection, UPS), equal area mapping (oblique Lambert projection).

7-21 Equidistant mapping (oblique Postel projection)

The oblique equidistant mapping of the sphere to a tangential plane is the generalization of equations derived earlier. The results are stated more precisely in Box 7.1. Figure 7.2 gives an impression of the famous oblique equidistant mapping of the sphere to a tangential plane with the meta-North Pole located in Stuttgart/Germany ($\Lambda_0 = 9°11'$, $\Phi_0 = 48°46'$).

Box 7.1 (Oblique equidistant mapping of the sphere to a plane at the meta-North Pole $\Lambda_0 \in [0°, 360°]$, $\Phi_0 \in [-90°, 90°]$).

Parameterized mapping:

$$\alpha = A , \quad r = f(B) = R\,\mathrm{arc}(\pi/2 - B) , \tag{7.3}$$

$$x = r \cos \alpha = R\,(\pi/2 - B) \cos A , \quad y = r \sin \alpha = R\,(\pi/2 - B) \sin A ,$$

$$\tan A = \frac{\cos \Phi \sin(\Lambda - \Lambda_0)}{\cos \Phi \sin \Phi_0 \cos(\Lambda - \Lambda_0) - \sin \Phi \cos \Phi_0} , \quad \sin B = \cos \Phi \cos \Phi_0 \cos(\Lambda - \Lambda_0) + \sin \Phi \sin \Phi_0 . \tag{7.4}$$

Left principal stretches:

$$\Lambda_1 = \frac{\pi/2 - B}{\cos B} , \quad \Lambda_2 = 1 . \tag{7.5}$$

Left eigenvectors:

$$\boldsymbol{C}_1 \Lambda_1 = \boldsymbol{E}_A \frac{\pi/2 - B}{\sin(\pi/2 - B)} , \quad \boldsymbol{C}_2 \Lambda_2 = \boldsymbol{E}_B . \tag{7.6}$$

Parameterized inverse mapping:

$$\tan A = \frac{y}{x} , \quad B = \frac{\pi}{2} - \sqrt{\frac{x^2 + y^2}{R^2}} , \tag{7.7}$$

$$\tan(\Lambda - \Lambda_0) = \frac{\sin A}{\tan B \cos \Phi_0 + \cos A \sin \Phi_0} , \quad \sin \Phi = -\cos B \cos A \cos \Phi_0 + \sin B \sin \Phi_0 .$$

Left maximum angular distortion:

$$\Omega_l = 2 \arcsin \left| \frac{\Lambda_1 - \Lambda_2}{\Lambda_1 + \Lambda_2} \right| = 2 \arcsin \left| \frac{\frac{\pi}{2} - B - \cos B}{\frac{\pi}{2} - B + \cos B} \right| . \tag{7.8}$$

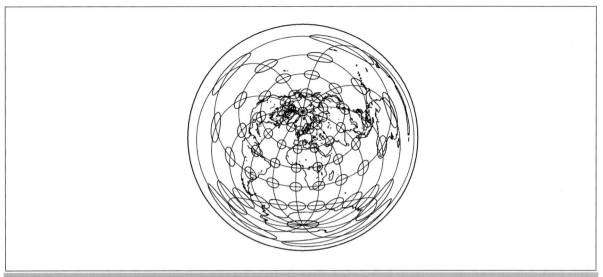

Fig. 7.2. Mapping the sphere to a tangential plane: oblique aspect, equidistant mapping. Point-of-contact: meta-North Pole at Stuttgart/Germany ($\Lambda_0 = 9°11'$, $\Phi_0 = 48°46'$).

7-22 Conformal mapping (oblique stereographic projection, oblique UPS)

The oblique conformal mapping of the sphere to a tangential plane is the generalization of equations derived earlier. The results are stated more precisely in Box 7.2. Figure 7.3 gives an impression of the famous oblique conformal mapping of the sphere to a tangential plane with the meta-North Pole located at Rio de Janeiro ($\Lambda_0 = -43°12'$, $\Phi_0 = -22°54'$).

Box 7.2 (Oblique conformal mapping of the sphere to a plane at the meta-North Pole $\Lambda_0 \in [0°, 360°]$, $\Phi_0 \in [-90°, 90°]$).

Parameterized mapping:

$$\alpha = A \ , \quad r = f(B) = 2R \tan \left(\frac{\pi}{4} - \frac{B}{2} \right) \ , \tag{7.9}$$

$$x = 2R \tan \left(\frac{\pi}{4} - \frac{B}{2} \right) \cos A \ , \quad y = 2R \tan \left(\frac{\pi}{4} - \frac{B}{2} \right) \sin A \ , \tag{7.10}$$

$$\tan A = \frac{\cos \Phi \sin(\Lambda - \Lambda_0)}{\cos \Phi \sin \Phi_0 \cos(\Lambda - \Lambda_0) - \sin \Phi \cos \Phi_0} \ , \quad \sin B = \cos \Phi \cos \Phi_0 \cos(\Lambda - \Lambda_0) + \sin \Phi \sin \Phi_0 \ .$$

Left principal stretches:

$$\Lambda_1 = \Lambda_2 = \frac{1}{\cos^2 \left(\frac{\pi}{4} - \frac{B}{2} \right)} \ . \tag{7.11}$$

Left eigenvectors:

$$\boldsymbol{C}_1 \Lambda_1 = \boldsymbol{E}_A \frac{1}{\cos^2 \left(\frac{\pi}{4} - \frac{B}{2} \right)} \ , \quad \boldsymbol{C}_2 \Lambda_2 = \boldsymbol{E}_B \frac{1}{\cos^2 \left(\frac{\pi}{4} - \frac{B}{2} \right)} \ . \tag{7.12}$$

Parameterized inverse mapping:

$$\tan A = \frac{y}{x} \ , \quad \tan \left(\frac{\pi}{4} - \frac{B}{2} \right) = \frac{1}{2R} \sqrt{x^2 + y^2} \ , \tag{7.13}$$

$$\tan(\Lambda - \Lambda_0) = \frac{\sin A}{\tan B \cos \Phi_0 + \cos A \sin \Phi_0} \ , \quad \sin \Phi = -\cos B \cos A \cos \Phi_0 + \sin B \sin \Phi_0 \ .$$

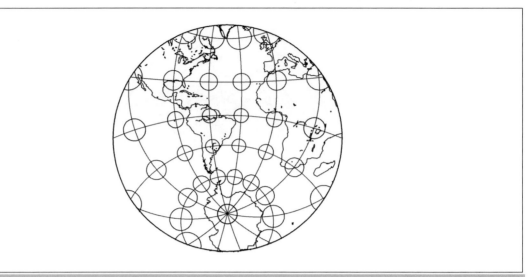

Fig. 7.3. Mapping the sphere to a tangential plane: oblique aspect, conformal mapping. Point-of-contact: meta-North Pole at Rio de Janeiro ($\Lambda_0 = -43°12'$, $\Phi_0 = -22°54'$).

7-23 Equal area mapping (oblique Lambert projection)

The oblique equal area mapping of the sphere to a tangential plane is the generalization of equations derived earlier. The results are stated more precisely in Box 7.3. Figure 7.4 gives an impression of the famous oblique conformal mapping of the sphere to a tangential plane with the meta-North Pole located at Perth ($\Lambda_0 = -115°52'$, $\Phi_0 = -31°57'$).

Box 7.3 (Oblique equal area mapping of the sphere to a plane at the meta-North Pole $\Lambda_0 \in [0°, 360°]$, $\Phi_0 \in [-90°, 90°]$).

Parameterized mapping:

$$\alpha = A, \quad r = f(B) = 2R \tan\left(\frac{\pi}{4} - \frac{B}{2}\right), \quad x = 2R \sin\left(\frac{\pi}{4} - \frac{B}{2}\right) \cos A, \quad y = 2R \sin\left(\frac{\pi}{4} - \frac{B}{2}\right) \sin A,$$

$$\tan A = \frac{\cos\Phi \sin(\Lambda - \Lambda_0)}{\cos\Phi \sin\Phi_0 \cos(\Lambda - \Lambda_0) - \sin\Phi \cos\Phi_0}, \quad \sin B = \cos\Phi \cos\Phi_0 \cos(\Lambda - \Lambda_0) + \sin\Phi \sin\Phi_0. \tag{7.14}$$

Left principal stretches:

$$\Lambda_1 = \frac{1}{\cos\left(\frac{\pi}{4} - \frac{B}{2}\right)}, \quad \Lambda_2 = \cos\left(\frac{\pi}{4} - \frac{B}{2}\right). \tag{7.15}$$

Left eigenvectors:

$$C_1\Lambda_1 = E_A \frac{1}{\cos\left(\frac{\pi}{4} - \frac{B}{2}\right)}, \quad C_2\Lambda_2 = E_B \cos\left(\frac{\pi}{4} - \frac{B}{2}\right). \tag{7.16}$$

Parameterized inverse mapping:

$$\tan A = \frac{y}{x}, \quad \sin\left(\frac{\pi}{4} - \frac{B}{2}\right) = \frac{1}{2R}\sqrt{x^2 + y^2},$$

$$\tan(\Lambda - \Lambda_0) = \frac{\sin A}{\tan B \cos\Phi_0 + \cos A \sin\Phi_0}, \quad \sin\Phi = -\cos B \cos A \cos\Phi_0 + \sin B \sin\Phi_0. \tag{7.17}$$

Left maximum angular distortion:

$$\Omega_l = 2\arcsin\left|\frac{\Lambda_1 - \Lambda_2}{\Lambda_1 + \Lambda_2}\right| = 2\arcsin\left|\frac{1 - \cos^2\left(\frac{\pi}{4} - \frac{B}{2}\right)}{1 + \cos^2\left(\frac{\pi}{4} - \frac{B}{2}\right)}\right|. \tag{7.18}$$

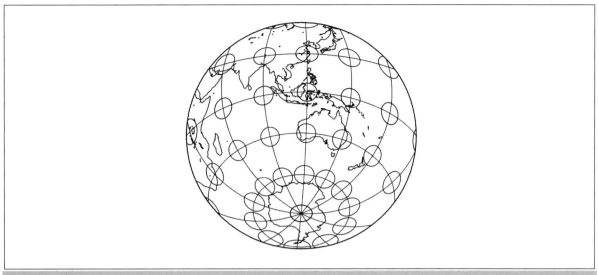

Fig. 7.4. Mapping the sphere to a tangential plane: oblique aspect, equal area mapping. Point-of-contact: meta-North Pole at Perth ($\Lambda_0 = -115°52'$, $\Phi_0 = -31°57'$).

Figure 7.4 finishes the discussion of the mappings "sphere to tangential plane". In the following chapter, let us have a closer look at the mapping "ellipsoid-of-revolution to tangential plane".

8 "Ellipsoid-of-revolution to tangential plane"

Mapping the ellipsoid-of-revolution to a tangential plane. Azimuthal projections in the normal aspect (polar aspect): equidistant, conformal, equiareal, and perspective mapping.

First and foremost, let us consider the ID card of the *ellipsoid-of-revolution* $\mathbb{E}^2_{A_1,A_2}$: see Box 8.1. As before, F (with elements a, b, c, d) is the Frobenius matrix, $G = J^*J$ (with elements e, f, g) is the Gauss matrix, $H = \left[\langle X_{KL} | G_3 \rangle\right]$ (with elements l, m, n) is the Hesse matrix, $J = \left[\partial X^J / \partial U^K\right]$ is the Jacobi matrix, and $K = -HG^{-1}$ is the curvature matrix, finally leading to the mean curvature $h = -\text{tr}[K]/2$ and to the Gaussian curvature $k = \det[K]$.

Box 8.1 (ID card of the ellipsoid-of-revolution $\mathbb{E}^2_{A_1,A_2}$).

Surface normal ellipsoidal coordinates (1st chart: Λ, Φ):

$$\{\Lambda, \Phi\} \in \mathbb{E}^2 / \{Z = \pm A_2\} := \left\{ X \in \mathbb{R}^3 \,\middle|\, (X^2 + Y^2)/A_1^2 + Z^2/A_2^2 = 1, A_1 > A_2, Z \neq \pm A_2 \right\} \,,$$

$$X := E_1 \frac{A_1 \cos \Phi \cos \Lambda}{\sqrt{1 - E^2 \sin^2 \Phi}} + E_2 \frac{A_1 \cos \Phi \sin \Lambda}{\sqrt{1 - E^2 \sin^2 \Phi}} + E_3 \frac{A_1 (1 - E^2) \sin \Phi}{\sqrt{1 - E^2 \sin^2 \Phi}} \,, \tag{8.1}$$

$$\Lambda(X) = \arctan \frac{Y}{X} + 180° \left[-\frac{1}{2}\text{sgn}Y - \frac{1}{2}\text{sgn}Y\,\text{sgn}X + 1 \right] \,, \quad \Phi(X) = \arctan \frac{1}{1 - E^2} \frac{Z}{\sqrt{X^2 + Y^2}} \,.$$

Matrices F, G, H, J, K , and I (elements $a, b, c, d; e, f, g; l, m, n$):

$$F = \begin{bmatrix} \dfrac{\sqrt{1 - E^2 \sin^2 \Phi}}{A_1 \cos \Phi} & 0 \\ 0 & \dfrac{(1 - E^2 \sin^2 \Phi)^{3/2}}{A_1(1 - E^2)} \end{bmatrix} = \begin{bmatrix} \dfrac{1}{\sqrt{G_{11}}} & 0 \\ 0 & \dfrac{1}{\sqrt{G_{22}}} \end{bmatrix} = \begin{bmatrix} \dfrac{1}{N \cos \Phi} & 0 \\ 0 & \dfrac{1}{M} \end{bmatrix} = \begin{bmatrix} a & b \\ c & d \end{bmatrix} \in \mathbb{R}^{2\times2} \,, \tag{8.2}$$

$$G = \begin{bmatrix} \dfrac{A_1^2 \cos^2 \Phi}{1 - E^2 \sin^2 \Phi} & 0 \\ 0 & \dfrac{A_1^2(1 - E^2)^2}{(1 - E^2 \sin^2 \Phi)^{3/2}} \end{bmatrix} = \begin{bmatrix} N^2 \cos^2 \Phi & 0 \\ 0 & M^2 \end{bmatrix} = \begin{bmatrix} e & f \\ f & g \end{bmatrix} \in \mathbb{R}^{2\times2} \,, \tag{8.3}$$

$$H = \begin{bmatrix} -\dfrac{A_1 \cos^2 \Phi}{\sqrt{1 - E^2 \sin^2 \Phi}} & 0 \\ 0 & -\dfrac{A_1(1 - E^2)}{(1 - E^2 \sin^2 \Phi)^{3/2}} \end{bmatrix} = \begin{bmatrix} l & m \\ m & n \end{bmatrix} \in \mathbb{R}^{2\times2} \,, \tag{8.4}$$

$$J = \begin{bmatrix} -\dfrac{A_1 \cos \Phi \sin \Lambda}{\sqrt{1 - E^2 \sin^2 \Phi}} & -\dfrac{A_1(1 - E^2)\sin \Phi \cos \Lambda}{(1 - E^2 \sin^2 \Phi)^{3/2}} \\ +\dfrac{A_1 \cos \Phi \cos \Lambda}{\sqrt{1 - E^2 \sin^2 \Phi}} & -\dfrac{A_1(1 - E^2)\sin \Phi \sin \Lambda}{(1 - E^2 \sin^2 \Phi)^{3/2}} \\ 0 & +\dfrac{A_1(1 - E^2)\cos \Phi}{(1 - E^2 \sin^2 \Phi)^{3/2}} \end{bmatrix} \in \mathbb{R}^{3\times2} \,, \tag{8.5}$$

$$K = \begin{bmatrix} \dfrac{\sqrt{1 - E^2 \sin^2 \Phi}}{A_1} & 0 \\ 0 & \dfrac{(1 - E^2 \sin^2 \Phi)^{3/2}}{A_1(1 - E^2)} \end{bmatrix} = \begin{bmatrix} \dfrac{1}{N} & 0 \\ 0 & \dfrac{1}{M} \end{bmatrix} \in \mathbb{R}^{2\times2} \,, \tag{8.6}$$

Continuation of Box.

$$I = I_2 = \begin{bmatrix} 1 & 0 \\ 0 & 1 \end{bmatrix} \in \mathbb{R}^{2 \times 2} \ . \tag{8.7}$$

1st curvature radius, normal curvature:

$$\kappa_1 = \frac{1}{A_1} \sqrt{1 - E^2 \sin^2 \Phi} \ , \quad \kappa_1^{-1} = \frac{A_1}{\sqrt{1 - E^2 \sin^2 \Phi}} =: N(\Phi) \ . \tag{8.8}$$

2nd curvature radius, meridianal curvature:

$$\kappa_2 = \frac{(1 - E^2 \sin^2 \Phi)^{3/2}}{A_1 (1 - E^2)} \ , \quad \kappa_2^{-1} = \frac{A_1 (1 - E^2)}{(1 - E^2 \sin^2 \Phi)^{3/2}} =: M(\Phi) \ . \tag{8.9}$$

Mean curvature, Gauss curvature:

$$h = -\frac{\sqrt{1 - E^2 \sin^2 \Phi}}{2 A_1} - \frac{(1 - E^2 \sin^2 \Phi)^{3/2}}{2 A_1 (1 - E^2)} = -\frac{1}{2} \frac{N + M}{NM} \ , \quad k = \frac{(1 - E^2 \sin^2 \Phi)^2}{A_1^2 (1 - E^2)} = \frac{1}{MN} \ . \tag{8.10}$$

Christoffel symbols of the 2nd kind:

$$\begin{Bmatrix} 1 \\ 1 \ 1 \end{Bmatrix} = \begin{Bmatrix} 1 \\ 2 \ 2 \end{Bmatrix} = \begin{Bmatrix} 2 \\ 1 \ 2 \end{Bmatrix} = 0 \ , \quad \begin{Bmatrix} 1 \\ 1 \ 2 \end{Bmatrix} = -\frac{(1 - E^2) \tan \Phi}{1 - E^2 \sin^2 \Phi} \ ,$$

$$\begin{Bmatrix} 2 \\ 1 \ 1 \end{Bmatrix} = \frac{1}{2} \sin 2\Phi \frac{1 - E^2 \sin^2 \Phi}{1 - E^2} \ , \quad \begin{Bmatrix} 2 \\ 2 \ 2 \end{Bmatrix} = 3 E^2 \sin \Phi \cos \Phi (1 - E^2 \sin^2 \Phi) \ ; \tag{8.11}$$

$$\begin{Bmatrix} M \\ K \ L \end{Bmatrix} := \frac{1}{2} G^{MN} (D_K G_{NL} + D_L G_{KN} - D_N G_{KL}) \in \mathbb{R}^{2 \times 2 \times 2} \ \forall \ K, L, M \in \{1, 2\} \ .$$

In addition to this ID card, we here present the central characteristics of the geodetic ellipsoidal system: see Table 8.1, "Geodetic Reference System 1980" (Bulletin Geodesique, 58, pp. 388–398, 1984) versus "World Geodetic Datum 2000" (Journal of Geodesy, 73, pp. 611–623, 1999).

Table 8.1. "Geodetic Reference System 1980" versus "World Geodetic Datum 2000".

	H. Moritz (1984)	E. Grafarend, A. Ardalan (1999) ("zero frequency tide geoid")
Semi-major axis A_1	6 378 137 m	6 378 136.602 ± 0.053 m
Semi-minor axis A_2	6 356 752.3141 m	6 356 751.860 ± 0.052 m
Relative eccentricity $E^2 = (A_1^2 - A_2^2)/A_1^2$	0.006 694 380 022 90	0.006 694 397 984 91
Absolute eccentricity $\epsilon = \sqrt{A_1^2 - A_2^2}$	521 854.0097 m	521 854.674 ± 0.015 m
Axis difference $A_1 - A_2$	21 384.686 m	21 384.742 m
Flattening $F = (A_1 - A_2)/A_1$	0.003 352 810 681 18	0.003 352 819 692 40
Inverse flattening $F^{-1} = A_1/(A_1 - A_2)$	298.257 222 101	298.256 420 489

8-1 General mapping equations

Setting up general equations of the mapping "ellipsoid-of-revolution to plane": azimuthal projections in the normal aspect (polar aspect).

There are again two basic postulates which govern the setup of general equations of mapping the ellipsoid-of-revolution $\mathbb{E}^2_{A_1,A_2}$ of semi-major axis A_1 and semi-minor axis A_2, which are characterized by $A_1 > A_2$, to a tangential plane $T\mathbb{E}^2_{A_1,A_2}$ attached to a point $\boldsymbol{X} \in T\mathbb{E}^2_{A_1,A_2}$. Let the tangential plane be covered by polar coordinates $\{\alpha, r\}$. Then the following postulates are valid.

Postulate.

The polar coordinate α, which is also called *azimuth*, is identical to the ellipsoidal longitude, i. e. $\alpha = \Lambda$.

End of Postulate.

Postulate.

The polar coordinate r depends only on the ellipsoidal latitude Φ or on the ellipsoidal colatitude $\Delta := \pi/2 - \Phi$, i. e. $r = \sqrt{x^2 + y^2} = f(\Delta) = f(\pi/2 - \Phi)$. If $\Phi = \pi/2$ or, equivalently, $\Delta = 0$, then $f(0) = 0$ holds.

End of Postulate.

In last consequence, the general equations of an azimuthal mapping are provided by the following vector equation:

$$\begin{bmatrix} x \\ y \end{bmatrix} = \begin{bmatrix} r \cos \alpha \\ r \sin \alpha \end{bmatrix} = \begin{bmatrix} f(\Delta) \cos \Lambda \\ f(\Delta) \sin \Lambda \end{bmatrix} . \tag{8.12}$$

Question. Question: "How can we identify the images of the special coordinate lines $\Lambda = $ constant and $\Phi = $ constant, respectively?" Answer ($y = x \tan \Lambda$, $\Lambda = $ constant : elliptic meridian): "The image of the elliptic meridian $\Lambda = $ constant under an azimuthal mapping is the radial straight line." Answer ($x^2 + y^2 = r^2 = f^2(\Delta)$, $\Delta = $ constant : parallel circle): "The image of the parallel circle $\Delta = $ constant (or $\Phi = $ constant) under an azimuthal mapping is the circle $\mathbb{S}^1_r$ of radius $r = f(\Delta)$. Such a mapping is called *concircular*."

Proof ($y = x \tan \Lambda$, $\Lambda = $ constant : elliptic meridian).

Solve the first equation towards $f(\Delta) = x/\cos \Lambda$ and substitute $f(\Delta)$ in the second equation such that $y = f(\Delta) \sin \Lambda = x \sin \Lambda / \cos \Lambda = x \tan \Lambda$ holds.

End of Proof ($y = x \tan \Lambda$, $\Lambda = $ constant : elliptic meridian).

Proof ($x^2 + y^2 = r^2 = f^2(\Delta)$, $\Delta = $ constant : parallel circle).

Compute the terms x^2 and y^2 and add the two: $x^2 + y^2 = f^2(\Delta)$.

End of Proof ($x^2 + y^2 = r^2 = f^2(\Delta)$, $\Delta = $ constant : parallel circle).

In summary, the images of the elliptic meridian and the parallel circle constitute the typical graticule of an azimuthal mapping, i. e.

$$\begin{array}{ccl} \text{meridians } (\Lambda = \text{constant}) & \longrightarrow & \text{radial straight lines} , \\ \text{parallel circles} \begin{pmatrix} \Delta = \text{constant} \\ \Phi = \text{constant} \end{pmatrix} & \longrightarrow & \text{equicentric circles} . \end{array} \tag{8.13}$$

Box 8.2 shows a collection of formulae which describe the left Jacobi matrix J_l as well as the left Cauchy–Green matrix C_l for an azimuthal mapping $\mathbb{E}^2_{A_1,A_2} \to \mathbb{P}^2_{\mathcal{O}}$. The left pair of matrices $\{\mathsf{C}_l, \mathsf{G}_l\}$ is canonically characterized by the left principal stretches Λ_1 and Λ_2 in their general form.

Box 8.2 ("Ellipsoid-of-revolution to plane", distortion analysis, azimuthal projection, left principal stretches).

Parameterized mapping:

$$\alpha = \Lambda \,,$$

$$r = f(\Delta) \,,$$

$$x = r \cos \alpha = f(\Delta) \cos \Lambda \,,$$

$$y = r \sin \alpha = f(\Delta) \sin \Lambda \,.$$

(8.14)

Left Jacobi matrix:

$$\mathsf{J}_l := \begin{bmatrix} D_\Lambda x & D_\Delta x \\ D_\Lambda y & D_\Delta y \end{bmatrix} = \begin{bmatrix} -f(\Delta) \sin \Lambda & f'(\Delta) \cos \Lambda \\ +f(\Delta) \cos \Lambda & f'(\Delta) \sin \Lambda \end{bmatrix} \,.$$

(8.15)

Left Cauchy–Green matrix ($\mathsf{G}_\mathrm{r} = \mathsf{I}_2$):

$$\mathsf{C}_l = \mathsf{J}_l^* \mathsf{G}_\mathrm{r} \mathsf{J}_l = \begin{bmatrix} f^2(\Delta) & 0 \\ 0 & f'^2(\Delta) \end{bmatrix} \,.$$

(8.16)

Left principal stretches:

$$\Lambda_1 = +\sqrt{\frac{c_{11}}{G_{11}}} = \frac{f(\Delta)\sqrt{1 - E^2 \cos^2 \Delta}}{A_1 \sin \Delta} \,,$$

$$\Lambda_2 = +\sqrt{\frac{c_{22}}{G_{22}}} = \frac{f'(\Delta)\left(1 - E^2 \cos^2 \Delta\right)^{3/2}}{A_1(1 - E^2)} \,.$$

(8.17)

Left eigenvectors of the matrix pair $\{\mathsf{C}_l, \mathsf{G}_l\}$:

$$C_1 = E_\Lambda = \frac{D_\Lambda X}{\| D_\Lambda X \|}$$

(Easting) ,

$$C_2 = E_\Phi = \frac{D_\Phi X}{\| D_\Phi X \|}$$

(Northing) .

(8.18)

Next, we specialize the general azimuthal mapping to generate an equidistant mapping, a series of conformal mappings (stereographic projections), and an equiareal mapping.

8-2 Special mapping equations

Setting up special mappings "ellipsoid-of-revolution to plane", equidistant mapping, conformal mapping, equiareal mapping.

8-21 Equidistant mapping

Let us postulate an *equidistant mapping* of the family of elliptic meridians $\Lambda =$ constant, namely $r = f(\Delta)$, by means of the canonical postulate of an equidistant mapping $\Lambda_2 = 1$. Figure 8.1 is an illustration of such a mapping, and Box 8.3 contains the mathematical details of the mapping equations $x = f(\Delta)\cos\Lambda$ and $y = f(\Delta)\sin\Lambda$, where the radial function is given as an elliptic integral of the second kind

$$f(\Delta^*) = A_1 E(\Delta^*, E) \,, \tag{8.19}$$

where Δ^* is the *circle reduced polar distance* and E is the *elliptic modulus*. Here, we address the reader to Appendix C, where some notes on elliptic functions and elliptic integrals of the first, second, and third kind are presented. At this point, we are left with the question of focal interest.

Question.

Question: "How can we prove the meridian arc length as an elliptic integral of the second kind?" Answer: "Let us work out this in the following passage in more detail."

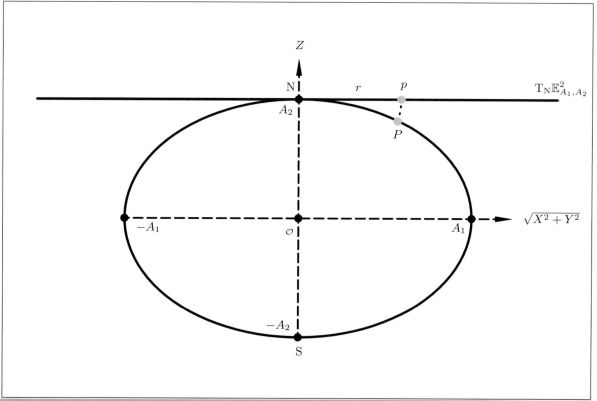

Fig. 8.1. Equidistant mapping of the ellipsoid-of-revolution to the tangential plane: normal aspect, meridian arc length $r = f(\Delta)$, $P \in \mathbb{E}^2_{A_1,A_2}$.

Box 8.3 (Equidistant mapping of the ellipsoid-of-revolution to the tangential plane at the North Pole).

Parameterized mapping:

$$\alpha = \Lambda , \ r = f(\Delta), \Delta := \pi/2 - \Phi ,$$
$$x = r\cos\alpha = f(\Delta)\cos\Lambda , \ y = r\sin\alpha = f(\Delta)\sin\Lambda . \tag{8.20}$$

Canonical postulate $\Lambda_2 = 1$,
equidistant mapping of the family of meridians:

$$\Lambda_2 = f'(\Delta)\frac{(1 - E^2\cos^2\Delta)^{3/2}}{A_1(1 - E^2)} = 1$$

$$\Leftrightarrow$$

$$\mathrm{d}f = A_1(1 - E^2)\frac{\mathrm{d}\Delta}{(1 - E^2\cos^2\Delta)^{3/2}} \tag{8.21}$$

$$\Rightarrow$$

$$f(\Delta) = A_1(1 - E^2)\int_0^\Delta \frac{\mathrm{d}\Delta}{(1 - E^2\cos^2\Delta)^{3/2}} \ .$$

Transformation of surface normal latitude Φ to reduced latitude Φ^*:

$$\tan\Phi^* = \sqrt{1 - E^2}\tan\Phi$$

$$\Leftrightarrow \tag{8.22}$$

$$\tan\Phi = \frac{1}{\sqrt{1 - E^2}}\tan\Phi^* \ .$$

Equidistant mapping of the family of meridians,
elliptic integral of the second kind:

$$f(\Delta) \to f(\Phi) ,$$

$$f(\Phi) = A_1(1 - E^2)\int_{\pi/2 - \Phi}^{\pi/2} \frac{\mathrm{d}\Phi'}{(1 - E^2\sin^2\Phi')^{3/2}} \ ;$$

$$f(\Phi) \to f(\Phi^*) ,$$

$$f(\Phi^*) = A_1\int_{\pi/2 - \Phi^*}^{\pi/2} \sqrt{1 - E^2\cos^2\Phi^*{}'}\mathrm{d}\Phi^*{}' \ ; \tag{8.23}$$

$$f(\Phi^*) \to f(\Delta^*) ,$$

$$f(\Delta^*) = A_1\int_0^{\Delta^*} \sqrt{1 - E^2\sin^2\Delta}\mathrm{d}\Delta =: A_1 E(\Delta^*, E) \ .$$

Elliptic integral of the second kind:

$$f(\Phi) = A_1 \boldsymbol{E}\left[\pi/2 - \arctan\left(\sqrt{1 - E^2}\tan\Phi\right), E\right] . \tag{8.24}$$

We depart from the representation of the meridian arc length as a function of the polar distance Δ, the complement of the surface normal latitude Φ. Let us transform the integral kernel (which is a function of Φ) to Φ^* (which is the circular reduced latitude). Such a polar coordinate is generated by projecting a meridianal point P vertically onto a circle $\mathbb{S}^1_{A_1}$ of radius A_1. Note that the geometrical situation is illustrated in Fig. 8.2 and Fig. 8.3. Furthermore, note that the relations $\Delta := \pi/2 - \Phi$ and $\Delta^* := \pi/2 - \Phi^*$ hold, and

$$f(\Delta) = A_1(1 - E^2) \int_0^\Delta \frac{\mathrm{d}\Delta'}{(1 - E^2 \cos^2 \Delta')^{3/2}}$$

$$\Leftrightarrow \tag{8.25}$$

$$f(\Delta^*) = A_1 \int_0^{\Delta^*} \sqrt{1 - E^2 \sin^2 \Delta}\, \mathrm{d}\Delta\ .$$

The transformation formulae $\Phi \to \Phi^*$ and $\Phi^* \to \Phi$, respectively, are summarized in Box 8.4, originating from $\sqrt{X^2 + Y^2} = A_1 \cos \Phi^*$ and $Z = A_2 \sin \Phi^*$, taking reference to the semi-major axis A_1 and the semi-minor axis A_2. Here, we refer to

$$\sin \Phi = \frac{\sin \Phi^*}{\sqrt{1 - E^2 \cos^2 \Phi^*}}$$

or

$$\cos \Delta = \frac{\cos \Delta^*}{\sqrt{1 - E^2 \sin^2 \Delta^*}}\ ,$$

and

$$\frac{A_1(1 - E^2)}{(1 - E^2 \cos^2 \Delta)^{3/2}} = \frac{A_1}{\sqrt{1 - E^2}}(1 - E^2 \sin^2 \Delta^*)^{3/2}\ ,$$

and $\tag{8.26}$

$$\sin \Delta = \frac{\sqrt{1 - E^2}}{\sqrt{1 - E^2 \sin^2 \Delta^*}} \sin \Delta^*$$

$$\Rightarrow$$

$$\cos \Delta\,\mathrm{d}\Delta = \frac{\sqrt{1 - E^2}}{(1 - E^2 \sin^2 \Delta^*)^{3/2}} \cos \Delta^*\,\mathrm{d}\Delta^*\ ,$$

$$\Rightarrow$$

$$\mathrm{d}\Delta = \frac{\cos \Delta^*}{\cos \Delta} \frac{\sqrt{1 - E^2}}{(1 - E^2 \sin^2 \Delta^*)^{3/2}}\mathrm{d}\Delta^* = \frac{\sqrt{1 - E^2}}{1 - E^2 \sin^2 \Delta^*}\mathrm{d}\Delta^*\ ,$$

in order to have derived

$$\frac{A_1(1 - E^2)}{(1 - E^2 \cos^2 \Delta)^{3/2}}\mathrm{d}\Delta = \frac{A_1}{\sqrt{1 - E^2}}(1 - E^2 \sin^2 \Delta^*)^{3/2}\frac{\sqrt{1 - E^2}}{1 - E^2 \sin^2 \Delta^*}\mathrm{d}\Delta^*\ ,$$

$$\tag{8.27}$$

$$\frac{A_1(1 - E^2)}{(1 - E^2 \cos^2 \Delta)^{3/2}}\mathrm{d}\Delta = A_1 \sqrt{1 - E^2 \sin^2 \Delta^*}\mathrm{d}\Delta^*\ .$$

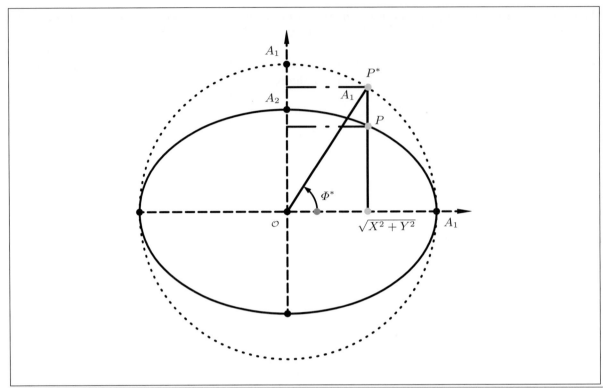

Fig. 8.2. Circle reduced latitude: $\sqrt{X^2 + Y^2} = A_1 \cos \Phi^*$.

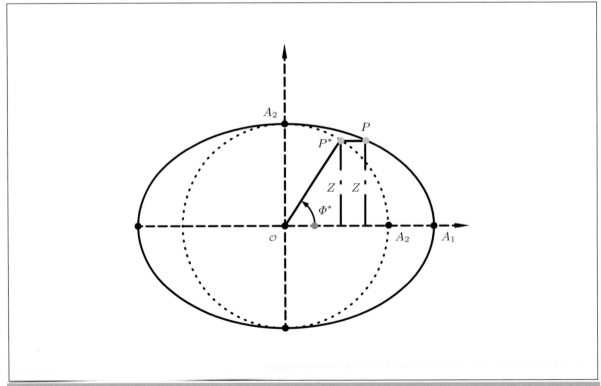

Fig. 8.3. Circle reduced latitude: $Z = A_2 \sin \Phi^*$.

Box 8.4 (Two ellipsoidal coordinate systems parameterizing the oblate ellipsoid-of-revolution).

Oblate ellipsoid-of-revolution:

$$\mathbb{E}_{A_1,A_1,A_2} := \left\{ \boldsymbol{X} \in \mathbb{R}^3 \,\middle|\, \frac{X^2 + Y^2}{A_1^2} + \frac{Z^2}{A_2^2} = 1, A_1 > A_2 \in \mathbb{R}^+ \right\} . \tag{8.28}$$

Ansatz 1 (surface normal coordinates):   Ansatz 2 (circle reduced coordinates):

$$X = \frac{A_1 \cos\Phi \cos\Lambda}{\sqrt{1 - E^2 \sin^2\Phi}} , \qquad\qquad X = A_1 \cos\Phi^* \cos\Lambda ,$$

$$Y = \frac{A_1 \cos\Phi \sin\Lambda}{\sqrt{1 - E^2 \sin^2\Phi}} , \qquad\qquad Y = A_1 \cos\Phi^* \sin\Lambda , \tag{8.29}$$

$$Z = \frac{A_1(1 - E^2)\sin\Phi}{\sqrt{1 - E^2 \sin^2\Phi}} , \qquad\qquad Z = A_2 \sin\Phi^* ,$$

subject to

$$E^2 := \frac{A_1^2 - A_2^2}{A_1^2} \qquad \text{and} \qquad \frac{A_2}{A_1} = \sqrt{1 - E^2} . \tag{8.30}$$

Direct and inverse transformation of surface normal latitude Φ to circle reduced latitude Φ^*:

$$\tan\Phi = \frac{1}{1 - E^2} \frac{Z}{\sqrt{X^2 + Y^2}} \qquad \text{versus} \quad \tan\Phi^* = \frac{A_1}{A_2} \frac{Z}{\sqrt{X^2 + Y^2}} = \frac{1}{\sqrt{1 - E^2}} \frac{Z}{\sqrt{X^2 + Y^2}} ,$$

$$\tan\Phi = \frac{1}{\sqrt{1 - E^2}} \tan\Phi^* \qquad \text{versus} \quad \tan\Phi^* = \sqrt{1 - E^2} \tan\Phi ,$$

$$\cos\Phi = \frac{\sqrt{1 - E^2}}{\sqrt{1 - E^2 \cos^2\Phi^*}} \cos\Phi^* \quad \text{versus} \quad \cos\Phi^* = \frac{1}{\sqrt{1 - E^2 \sin^2\Phi}} \cos\Phi , \tag{8.31}$$

$$\sin\Phi = \frac{1}{\sqrt{1 - E^2 \cos^2\Phi^*}} \sin\Phi^* \quad \text{versus} \quad \sin\Phi^* = \frac{\sqrt{1 - E^2}}{\sqrt{1 - E^2 \sin^2\Phi}} \sin\Phi .$$

In most practical cases, where we are aiming at an azimuthal projection of an equidistant type of the ellipsoid-of-revolution representing the Earth, the planets, or other celestial bodies, a series expansion of the meridian arc length has been a sufficient approximation. Accordingly, we are going to outline the series expansion of the meridian arc length as a function of surface normal latitude Φ or its complement, the polar distance Δ. In preparing such an series expansion, we have collected auxiliary formulae in Corollary 8.1 to Corollary 8.7. First, we expand $(1 + x)^y$ according to B. Taylor, just representing the meridian arc length by $x := -E^2 \cos^2\Delta$, $y = -3/2$, and $|x| > 1$. Second, we represent $(1 - E^2 \cos^2\Delta)^{3/2}$ in terms of powers $\{1, E^2 \cos^2\Delta, E^4 \cos^4\Delta, E^6 \cos^6\Delta, \dots\}$. Third, we transform the powers $\{\cos^2\Delta, \cos^4\Delta, \cos^6\Delta, \dots\}$ in terms of $\{1, \cos 2\Delta, \cos 4\Delta, \cos 6\Delta, \dots\}$. Fourth, an explicit version of the product sums is given in Corollary 8.4 to Corollary 8.6. Since the power series are *uniformly convergent*, we can term-wise integrate in order to achieve the meridian arc length in Corollary 8.7.

Corollary 8.1 (Power series $(1 + x)^y$, Taylor expansion).

$$(1 + x)^y =$$

$$= 1 + \frac{1}{1!}y[1 + x]_{x=0}^{y-1}x^1 + \frac{1}{2!}y(y-1)[1 + x]_{x=0}^{y-2}x^2 +$$

$$+ \frac{1}{3!}y(y-1)(y-2)[1 + x]_{x=0}^{y-3}x^3 + \mathrm{O}(x^4) \tag{8.32}$$

$$\forall \ |x| < 1 \ ,$$

$$(1 + x)^y := (1 - E^2 \cos^2 \varDelta)^{-3/2} \ , \quad x := -E^2 \cos^2 \varDelta \ , \quad y := -\frac{3}{2} \ .$$

End of Corollary.

Corollary 8.2 (Power series $(1 + x)^y$, reformulation 1).

$$(1 - E^2 \cos^2 \varDelta)^{-3/2} =$$

$$= 1 + \frac{3}{2}E^2 \cos^2 \varDelta + \frac{3 \cdot 5}{2 \cdot 4}E^4 \cos^4 \varDelta + \frac{3 \cdot 5 \cdot 7}{2 \cdot 4 \cdot 6}E^6 \cos^6 \varDelta + \mathrm{O}(E^8) \ . \tag{8.33}$$

End of Corollary.

Corollary 8.3 (Power series $(1 + x)^y$, cosine powers).

$$\cos^2 \varDelta = \frac{1}{2} + \frac{1}{2}\cos 2\varDelta \ ,$$

$$\cos^4 \varDelta = \frac{3}{8} + \frac{1}{2}\cos 2\varDelta + \frac{1}{8}\cos 4\varDelta \ , \tag{8.34}$$

$$\cos^6 \varDelta = \frac{5}{16} + \frac{15}{32}\cos 2\varDelta + \frac{3}{16}\cos 4\varDelta + \frac{1}{32}\cos 6\varDelta \ .$$

End of Corollary.

Corollary 8.4 (Power series $(1 + x)^y$, reformulation 2).

$$(1 - E^2 \cos^2 \varDelta)^{-3/2} =$$

$$= 1 + \frac{3}{2}E^2 \left(\frac{1}{2} + \frac{1}{2}\cos 2\varDelta\right) +$$

$$+ \frac{15}{8}E^4 \left(\frac{3}{8} + \frac{1}{2}\cos 2\varDelta + \frac{1}{8}\cos 4\varDelta\right) + \tag{8.35}$$

$$+ \frac{35}{16}E^6 \left(\frac{5}{16} + \frac{15}{32}\cos 2\varDelta + \frac{3}{16}\cos 4\varDelta + \frac{1}{32}\cos 6\varDelta\right) +$$

$$+ \mathrm{O}(E^8) \ .$$

End of Corollary.

Corollary 8.5 (Power series $(1+x)^y$, reformulation 3).

$$(1 - E^2 \cos^2 \Delta)^{-3/2} =$$

$$= 1 + \frac{3}{4}E^2 + \frac{45}{64}E^4 + \frac{175}{256}E^6 + O_1(E^8) +$$

$$+ \left(\frac{3}{4}E^2 + \frac{15}{16}E^4 + \frac{525}{512}E^6 + O_2(E^8) \right) \cos 2\Delta +$$

$$+ \left(\frac{15}{64}E^4 + \frac{105}{256}E^6 + O_3(E^8) \right) \cos 4\Delta + \left(\frac{35}{512}E^6 + O_4(E^8) \right) \cos 6\Delta + O_5(E^8) \,.$$

(8.36)

End of Corollary.

Corollary 8.6 (Power series $(1+x)^y$, multiplication with $(1-E^2)$).

$$(1 - E^2)(1 - E^2 \cos^2 \Delta)^{-3/2} =$$

$$= 1 - \frac{1}{4}E^2 - \frac{3}{64}E^4 - \frac{5}{256}E^6 + O_1(E^8) +$$

$$+ \left(\frac{3}{4}E^2 + \frac{3}{16}E^4 + \frac{45}{512}E^6 + O_2(E^8) \right) \cos 2\Delta +$$

$$+ \left(\frac{15}{64}E^4 + \frac{45}{256}E^6 + O_3(E^8) \right) \cos 4\Delta + \left(\frac{35}{512}E^6 + O_4(E^8) \right) \cos 6\Delta + O_5(E^8) \,.$$

(8.37)

End of Corollary.

Corollary 8.7 (Termwise integration of uniformly convergent series).

$$\int \cos nx \, dx = \frac{1}{n} \sin nx \ \forall \ n \in \mathbb{Z} \,,$$

$$\int_0^\Delta (1 - E^2)(1 - E^2 \cos^2 \Delta')^{-3/2} d\Delta' =$$

$$= \left(1 - \frac{1}{4}E^2 - \frac{3}{64}E^4 - \frac{5}{256}E^6 + O_1(E^8) \right) \Delta +$$

$$+ \left(\frac{3}{8}E^2 + \frac{3}{32}E^4 + \frac{45}{1024}E^6 + O_2(E^8) \right) \sin 2\Delta +$$

$$+ \left(\frac{15}{256}E^4 + \frac{45}{1024}E^6 + O_3(E^8) \right) \sin 4\Delta + \left(\frac{35}{3072}E^6 + O_4(E^8) \right) \sin 6\Delta + O_5(E^8) \,.$$

(8.38)

End of Corollary.

The hard work of the series expansion of the kernel representing the meridian arc length has finally led us to Lemma 8.8, where an elegant version of the meridian arc length up to the order $O(E^{12})$ has been achieved.

Lemma 8.8 (Meridian arc length, forward computation).

$$f(\Delta) = A_1 \int_0^\Delta (1 - E^2)(1 - E^2 \cos^2 \Delta')^{-3/2} d\Delta' = A_1(1 - E^2) \int_\Phi^{\pi/2} (1 - E^2 \sin^2 \Phi')^{-3/2} d\Phi'$$

$$\Rightarrow$$

$$f(\Phi) = A_1 \left[E_0 \left(\frac{\pi}{2} - \Phi \right) - E_2 \sin 2\Phi - E_4 \sin 4\Phi - E_6 \sin 6\Phi - E_8 \sin 8\Phi - \right. \tag{8.39}$$

$$\left. - E_{10} \sin 10\Phi + O(E^{12}) \right],$$

subject to

$$
\begin{aligned}
E_0 &= 1 - \frac{1}{4} E^2 - \frac{3}{64} E^4 - \frac{5}{256} E^6 - \frac{175}{16384} E^8 - \frac{441}{65536} E^{10}, \\
E_2 &= - \frac{3}{8} E^2 - \frac{3}{32} E^4 - \frac{45}{1024} E^6 - \frac{105}{4096} E^8 - \frac{2205}{131072} E^{10}, \\
E_4 &= \phantom{1 - \frac{3}{8} E^2} + \frac{15}{256} E^4 + \frac{45}{1024} E^6 + \frac{525}{16384} E^8 + \frac{1575}{65536} E^{10}, \\
E_6 &= \phantom{1 - \frac{3}{8} E^2 + \frac{15}{256} E^4} - \frac{35}{3072} E^6 - \frac{175}{12288} E^8 - \frac{3675}{262144} E^{10}, \\
E_8 &= \phantom{1 - \frac{3}{8} E^2 + \frac{15}{256} E^4 - \frac{35}{3072} E^6} + \frac{315}{131072} E^8 + \frac{2205}{524288} E^{10}, \\
E_{10} &= \phantom{1 - \frac{3}{8} E^2 + \frac{15}{256} E^4 - \frac{35}{3072} E^6 + \frac{315}{131072} E^8} - \frac{693}{1310720} E^{10}.
\end{aligned}
\tag{8.40}
$$

End of Lemma.

8-22 Conformal mapping

Let us postulate a *conformal mapping* of the ellipsoid-of-revolution onto a tangential plane at the North Pole by means of the canoncial measure of conformality, i. e. $\Lambda_1 = \Lambda_2$. Such a conformal mapping is illustrated by a vertical section of Fig. 8.4.

Question: "How can we generate the mapping equations of such a conformeomorphism?" Answer: "Let us work out this in the following passage in more detail."

The forward computation of the meridian arc length $r = f(\Delta)$ supplies us with the radial coordinate r of an equidistant mapping of a point of the ellipsoid-of-revolution to a corresponding point on the tangential plane at the North Pole: compare with Lemma 8.8. The central problem we are left with can be formulated as follows: given the radial coordinate r, find the surface normal ellipsoidal latitude Φ. Such a problem of generating the inverse function can be solved by series inversion. For details, we here have to direct you to *Appendix B*, where the standard series inversion of a homogeneous univariate polynomial is outlined, and where additional references of how to do it are given. Basic formulae are supplied by Lemma 8.9.

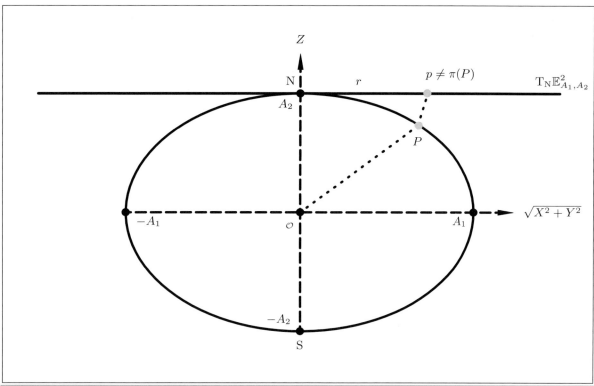

Fig. 8.4. Conformal mapping of the ellipsoid-of-revolution onto a tangential plane: normal aspect, $P \in \mathbb{E}^2_{A_1,A_2}$, $p \neq \pi(P)$, not UPS.

Lemma 8.9 (Meridian arc length, inverse computation).

A forward computation of the meridian arc length based upon a uniform series expansion is provided by formula (8.39). Its inverse function can be represented by

$$\Phi = \frac{\pi}{2} - \frac{r}{A_1 E_0} - F_2 \sin 2 \frac{r}{A_1 E_0} - F_4 \sin 4 \frac{r}{A_1 E_0} - F_6 \sin 6 \frac{r}{A_1 E_0} -$$
$$- F_8 \sin 8 \frac{r}{A_1 E_0} - F_{10} \sin 10 \frac{r}{A_1 E_0} + \mathrm{O}(E^{12}) , \tag{8.41}$$

subject to

$$E_0 = 1 - \frac{1}{4} E^2 - \frac{3}{64} E^4 - \frac{5}{256} E^6 - \frac{175}{16384} E^8 - \frac{441}{65536} E^{10} \tag{8.42}$$

and

$$
\begin{aligned}
F_2 &= \frac{3}{8} E^2 + \frac{3}{16} E^4 + \frac{213}{2048} E^6 + \frac{255}{4096} E^8 + \frac{166479}{655360} E^{10} , \\
F_4 &= \frac{21}{256} E^4 + \frac{21}{256} E^6 + \frac{533}{8192} E^8 - \frac{120563}{327680} E^{10} , \\
F_6 &= \frac{151}{6144} E^6 + \frac{155}{4096} E^8 + \frac{2767911}{9175040} E^{10} , \\
F_8 &= \frac{1097}{131072} E^8 - \frac{273697}{4587520} E^{10} .
\end{aligned} \tag{8.43}
$$

End of Lemma.

The *Equidistant Polar Mapping* (EPM) of the ellipsoid-of-revolution is summarized in Lemma 8.10, which is based upon the direct mapping equations, its left principal stretches, the left eigenvectors, the left maximal angular distortion, and the inverse mapping equations that are collected in Box 8.5.

Lemma 8.10 (Equidistant Polar Mapping (EPM), equidistant mapping of the ellipsoid-of-revolution to the tangential plane at the North Pole).

The *equidistant mapping* of the spheroid to the tangential plane at the North Pole of an oblate ellipsoid-of-revolution, in short, *Equidistant Polar Mapping (EPM)*, is parameterized by

$$x = f(\Delta) \cos \Lambda \, , \quad y = f(\Delta) \sin \Lambda \, , \tag{8.44}$$

subject to the *left Cauchy–Green eigenspace* $\{ \boldsymbol{E}_\Lambda \Lambda_1(\Delta), \boldsymbol{E}_\Phi \}$. The radial function $r = f(\Delta)$ that represents the meridian arc length from the North Pole to a point on the meridian $\Lambda = \text{constant}$ is given either in the form of an elliptic integral of the second kind or in the series expansion of Box 8.5.

End of Lemma.

Box 8.5 (Equidistant mapping of the ellipsoid-of-revolution to the tangential plane at the North Pole).

Parameterized mapping:

$$\alpha = \Lambda \, , \quad r = f(\Delta) \, , \quad \Delta := \pi/2 - \Phi \, , \quad f(\Delta) \to f(\Phi) \, ,$$
$$x = r \cos \alpha = f(\Phi) \cos \Lambda \, , \quad y = r \sin \alpha = f(\Phi) \sin \Lambda \, . \tag{8.45}$$

Series expansion, equidistant mapping of the family of meridians:

$$f(\Phi) = A_1 \left[E_0 \left(\frac{\pi}{2} - \Phi \right) - E_2 \sin 2\Phi - E_4 \sin 4\Phi - E_6 \sin 6\Phi - E_8 \sin 8\Phi - E_{10} \sin 10\Phi + \mathrm{O}(E^{12}) \right] . \tag{8.46}$$

Parameterized equidistant mapping:

$$x = A_1 E_0 \left(\frac{\pi}{2} - \Phi \right) \cos \Lambda -$$
$$- A_1 \left(E_2 \sin 2\Phi + E_4 \sin 4\Phi + E_6 \sin 6\Phi + E_8 \sin 8\Phi + E_{10} \sin 10\Phi + \mathrm{O}(E^{12}) \right) \cos \Lambda \, ,$$
$$y = A_1 E_0 \left(\frac{\pi}{2} - \Phi \right) \sin \Lambda -$$
$$- A_1 \left(E_2 \sin 2\Phi + E_4 \sin 4\Phi + E_6 \sin 6\Phi + E_8 \sin 8\Phi + E_{10} \sin 10\Phi + \mathrm{O}(E^{12}) \right) \sin \Lambda \, . \tag{8.47}$$

Left principal stretches and left eigenvectors:

$$\Lambda_1 = \frac{f(\Delta) \sqrt{1 - E^2 \cos^2 \Delta}}{A_1 \sin \Delta} = \frac{f(\Phi) \sqrt{1 - E^2 \sin^2 \Phi}}{A_1 \cos \Phi} \, , \quad \Lambda_2 = 1 \, , \tag{8.48}$$

$$\boldsymbol{C}_1 = \boldsymbol{E}_\Lambda = \frac{D_\Lambda \boldsymbol{X}}{\| D_\Lambda \boldsymbol{X} \|} \quad \text{("Easting")} \, , \quad \boldsymbol{C}_2 = \boldsymbol{E}_\Phi = \frac{D_\Phi \boldsymbol{X}}{\| D_\Phi \boldsymbol{X} \|} = -\boldsymbol{E}_\Delta \quad \text{("Northing")} \, ,$$
$$\text{(i)} \ \boldsymbol{C}_1 \Lambda_1 = \boldsymbol{E}_\Lambda \frac{f(\Phi) \sqrt{1 - E^2 \sin^2 \Phi}}{A_1 \cos \Phi} \, , \quad \text{(ii)} \ \boldsymbol{C}_2 \Lambda_2 = \boldsymbol{E}_\Phi = -\boldsymbol{E}_\Delta \, . \tag{8.49}$$

Left angular distortion:

$$d_l = 2 \arcsin \left| \frac{\Lambda_1 - \Lambda_2}{\Lambda_1 + \Lambda_2} \right| = 2 \arcsin \left| \frac{f(\Phi) \sqrt{1 - E^2 \sin^2 \Phi} - A_1 \cos \Phi}{f(\Phi) \sqrt{1 - E^2 \sin^2 \Phi} + A_1 \cos \Phi} \right| . \tag{8.50}$$

Parameterized inverse mapping, $\Lambda = \alpha$, $\tan \Lambda = y/x$ $(r = \sqrt{x^2 + y^2})$:

$$\Phi = \frac{\pi}{2} - \frac{r}{A_1 E_0} - F_2 \sin 2 \frac{r}{A_1 E_0} - F_4 \sin 4 \frac{r}{A_1 E_0} - F_6 \sin 6 \frac{r}{A_1 E_0} - F_8 \sin 8 \frac{r}{A_1 E_0} -$$
$$- F_{10} \sin 10 \frac{r}{A_1 E_0} + \mathrm{O}(E^{12}) \, . \tag{8.51}$$

Following the procedure that is outlined in Box 8.6, we are immediately able to generate the conformal mapping equations.

Box 8.6 (Conformal mapping of the ellipsoid-of-revolution to the tangential plane at the North Pole).

Postulate of conformeomorphism:

$$\Lambda_1 = \Lambda_2 \ ,$$

$$\frac{f(\Delta)\sqrt{1 - E^2 \cos^2 \Delta}}{A_1 \sin \Delta} = \frac{f'(\Delta)(1 - E^2 \cos^2 \Delta)^{3/2}}{A_1(1 - E^2)} \Rightarrow \frac{\mathrm{d}f}{f} = \frac{1 - E^2}{\sin \Delta(1 - E^2 \cos^2 \Delta)}\mathrm{d}\Delta \ . \tag{8.52}$$

Integration of the characteristic differential equations of a conformal mapping:

$$\ln f = \int \frac{1 - E^2}{\sin \Delta(1 - E^2 \cos^2 \Delta)}\mathrm{d}\Delta + \ln c \ . \tag{8.53}$$

Decomposition into rational partials:

$$\frac{1 - E^2}{\sin \Delta(1 - E^2 \cos^2 \Delta)} = \frac{1}{\sin \Delta} - \frac{E}{2}\left(\frac{E \sin \Delta}{1 + E \cos \Delta} + \frac{E \sin \Delta}{1 - E \cos \Delta}\right) \ ,$$

$$\int \frac{1 - E^2}{\sin \Delta(1 - E^2 \cos^2 \Delta)}\mathrm{d}\Delta = \ln \tan \frac{\Delta}{2} - \frac{E}{2}\ln \frac{1 - E \cos \Delta}{1 + E \cos \Delta} + \ln c =$$

$$= \mathrm{artanh}(\cos \Delta) - E\,\mathrm{artanh}(E \cos \Delta) + \ln c \tag{8.54}$$

$$\Rightarrow$$

$$f(\Delta) = c\left(\frac{1 + E \cos \Delta}{1 - E \cos \Delta}\right)^{E/2}\tan \frac{\Delta}{2} \ \forall \ \Delta \in [0, \pi[$$

or

$$f(\Delta) = c\exp\left[\mathrm{artanh}(\cos \Delta)\right]\exp\left[-E\,\mathrm{artanh}(E \cos \Delta)\right] \ .$$

Integration constant, postulate of isometry at the North Pole:

$$\lim_{\Delta \to 0} \Lambda_1(\Delta) = 1 \ ,$$

$$\lim_{\Delta \to 0} c\left(\frac{1 + E \cos \Delta}{1 - E \cos \Delta}\right)^{E/2}\tan \frac{\Delta}{2}\frac{\sqrt{1 - E^2 \cos^2 \Delta}}{A_1 \sin \Delta} = 1 \ , \tag{8.55}$$

$$\lim_{\Delta \to 0} \Lambda_1(\Delta) = c\left(\frac{1 + E}{1 - E}\right)^{E/2}\frac{\sqrt{1 - E^2}}{A_1}\lim_{\Delta \to 0}\frac{\tan(\Delta/2)}{\sin \Delta} = 1 \ .$$

L'Hospital's rule 0/0:

$$\lim_{\Delta \to 0}\frac{\tan(\Delta/2)}{\sin \Delta} = \lim_{\Delta \to 0}\frac{(\tan(\Delta/2))'}{(\sin \Delta)'} \ ,$$

$$(\tan(\Delta/2))' = \frac{1}{2}\frac{1}{\cos^2(\Delta/2)} = \frac{1}{1 + \cos \Delta} \ , \quad (\sin \Delta)' = \cos \Delta \ ,$$

$$\lim_{\Delta \to 0}\frac{\tan(\Delta/2)}{\sin \Delta} = \lim_{\Delta \to 0}\frac{1}{1 + \cos \Delta}\frac{1}{\cos \Delta} = \frac{1}{2} \ , \tag{8.56}$$

$$\lim_{\Delta \to 0}\Lambda_1(\Delta) = \frac{c}{2A_1}\left(\frac{1 + E}{1 - E}\right)^{E/2}\sqrt{1 - E^2} = 1$$

$$\Rightarrow$$

$$c = \frac{2A_1}{\sqrt{1 - E^2}}\left(\frac{1 - E}{1 + E}\right)^{E/2} \ .$$

Continuation of Box.

Parameterized conformal mapping:

$$\alpha = \Lambda , \quad r = f(\Delta) ,$$

$$f(\Delta) = \frac{2A_1}{\sqrt{1 - E^2}} \left(\frac{1 - E}{1 + E} \right)^{E/2} \left(\frac{1 + E \cos \Delta}{1 - E \cos \Delta} \right)^{E/2} \tan \frac{\Delta}{2} ,$$

$$f(\Delta) \to f(\Phi) , \qquad (8.57)$$

$$f(\Phi) = \frac{2A_1}{\sqrt{1 - E^2}} \left(\frac{1 - E}{1 + E} \right)^{E/2} \left(\frac{1 + E \sin \Phi}{1 - E \sin \Phi} \right)^{E/2} \tan \left(\frac{\pi}{4} - \frac{\Phi}{2} \right) ,$$

$$x = r \cos \alpha = f(\Phi) \cos \Lambda , \quad y = r \sin \alpha = f(\Phi) \sin \Lambda .$$

Left principal stretches and left eigenvectors:

$$\Lambda_1 = \Lambda_2 = \frac{f(\Phi) \sqrt{1 - E^2 \sin^2 \Phi}}{A_1 \cos \Phi} , \qquad (8.58)$$

$$\boldsymbol{E}_\Lambda = \frac{D_\Lambda \boldsymbol{X}}{\| D_\Lambda \boldsymbol{X} \|} \quad (\text{"Easting"}) , \quad \boldsymbol{E}_\Phi = \frac{D_\Phi \boldsymbol{X}}{\| D_\Phi \boldsymbol{X} \|} = -\boldsymbol{E}_\Lambda \quad (\text{"Northing"}) ,$$

$$\qquad (8.59)$$

$$(i) \; \boldsymbol{C}_1 \Lambda_1 = \boldsymbol{E}_\Lambda \frac{f(\Phi) \sqrt{1 - E^2 \sin^2 \Phi}}{A_1 \cos \Phi} , \quad (ii) \; \boldsymbol{C}_2 \Lambda_2 = \boldsymbol{E}_\Phi \frac{f(\Phi) \sqrt{1 - E^2 \sin^2 \Phi}}{A_1 \cos \Phi} .$$

Left angular shear:

$$\sum_l = \Psi_l - \Psi_r = 0 , \quad \Omega_l = 0 . \qquad (8.60)$$

Parameterized inverse mapping:

$$f(x) = \left(\frac{1 + x}{1 - x} \right)^{E/2} = f(0) + \frac{1}{1!} \; f'(x) \big|_{x=0} ,$$

$$E \cos \Delta = x \ll 1 ,$$

$$\qquad (8.61)$$

$$f(x) = \left(\frac{1 + x}{1 - x} \right)^{E/2} = 1 + \frac{1}{1!} \frac{E}{2} \left(\frac{1 + x}{1 - x} \right)^{E/2 - 1} \frac{(1 - x) - (1 + x)(-1)}{(1 - x)^2} \Bigg|_{x=0} x + O(2) =$$

$$= 1 + Ex + O(2) .$$

Alternative:

$$\operatorname{artanh} x = x + \frac{x^3}{3} + \frac{x^5}{5} + \frac{x^7}{7} + \frac{x^9}{9} + \frac{x^{11}}{11} + O(x^{13}) ,$$

$$E \cos \Delta = x \ll 1 ,$$

$$\qquad (8.62)$$

$$\operatorname{artanh}(E \cos \Delta) =$$

$$= E \cos \Delta + \frac{E^3}{3} \cos^3 \Delta + \frac{E^5}{5} \cos^5 \Lambda + \frac{E^7}{7} \cos^7 \Delta + \frac{E^9}{9} \cos^9 \Delta + \frac{E^{11}}{11} \cos^{11} \Delta + O(x^{13}) .$$

Question: "Is the conformal mapping of the ellipsoid-of-revolution to a tangential plane at the North Pole UPS?" Answer: "Let us work out this subject in the following passage in more detail."

Let us introduce the stereographic projection of the point $P \in \mathbb{E}^2_{A_1,A_2}$ of the ellipsoid-of-revolution $\mathbb{E}^2_{A_1,A_2}$ to the point $p = \pi(P)$, an element of the tangent space $T_N\mathbb{E}^2_{A_1,A_2}$ at the North Pole N. The South Pole S has been chosen as the perspective center, also called $\mathscr{O}^*$, the center of the projection. $Q = \pi(P)$ is the point on the z axis generated by an orthogonal projection. Consult Fig. 8.5 for further geometrical details. Naturally, $\angle NSp = \angle QSP$ denotes the characteristic parallactic angle of the central projection $p = \pi(P)$:

$$\tan \angle NSp = \tan \angle QSP \Leftrightarrow \frac{r}{2A_2} = \frac{\sqrt{X^2 + Y^2}}{A_2 + Z} \Rightarrow$$

$$r = \frac{2A_2}{A_2 + Z}\sqrt{X^2 + Y^2} = 2A_1 \cos\Phi \frac{A_2}{A_2\sqrt{1 - E^2 \sin^2 \Phi} + A_1(1 - E^2)\sin\Phi} \,, \tag{8.63}$$

$$f(\Phi) \to f(\Delta)\,, \qquad \begin{aligned} r = f(\Phi) &= \frac{2A_1 \cos\Phi}{\sqrt{1 - E^2 \sin^2 \Phi} + \sqrt{1 - E^2}\sin\Phi} \,, \\ r = f(\Delta) &= \frac{2A_1 \sin\Delta}{\sqrt{1 - E^2 \cos^2 \Delta} + \sqrt{1 - E^2}\cos\Delta} \,. \end{aligned} \tag{8.64}$$

The projective equations document a radial function $r \neq f(\Delta)$ which differs remarkably from the equations of an azimuthal conformal mapping. Definitely, the azimuthal conformal mapping of the ellipsoid-of-revolution is not UPS.

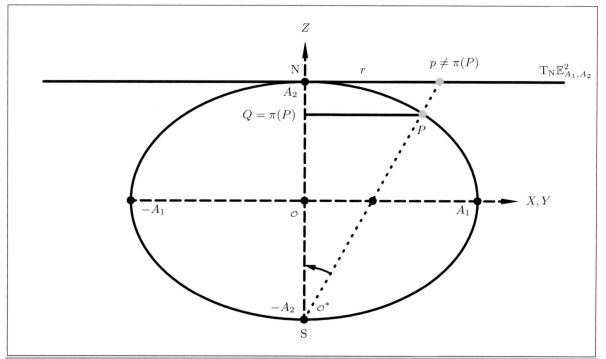

Fig. 8.5. Stereographic projection of $P \in \mathbb{E}^2_{A_1,A_2}$ to $p \in T_N\mathbb{E}^2_{A_1,A_2}$, perspective center S.

8-23 Equiareal mapping

Let us postulate an *equiareal mapping* of the ellipsoid-of-revolution onto a tangential plane at the North Pole by means of the measure $\Lambda_1\Lambda_2 = 1$. The details of such a mapping are collected in Box 8.7. At first, we have to start from the canonical postulate of an *equiareal mapping*, namely $\Lambda_1\Lambda_2 = 1$ or $f(\Delta)(1 - E^2\cos^2\Delta)^{1/2}f'(\Delta)(1 - E^2\cos^2\Delta)^{3/2}/(A_1^2\sin\Delta\,(1 - E^2)) = 1$, an equation solved for $f\,\mathrm{d}f = A_1^2(1 - E^2)\sin\Delta\,\mathrm{d}\Delta/(1 - E^2\cos^2\Delta)^2$. Direct integration leads to $f^2/2$ as an integral solved by "integration-by-parts". Four integrals lead us to the final integral $f^2/2$ as a function of (i) $\ln[(1 + E\cos\Delta)/(1 - E\cos\Delta)]$, (ii) $1/(1 - E\cos\Delta)$, and (iii) $1/(1 + E\cos\Delta)$. By the postulate $f(\Delta = 0) = 0$, we then gauge the integration constant c. In summary, we get the mapping equations $f(\Delta)$ and $f(\Phi)$, or $(\alpha = A, r = f(\Delta))$, or $(x = f(\Phi)\cos\Lambda, y = f(\Phi)\sin\Lambda)$. The *left principal stretches* and the *left eigenvectors* are collected in Box 8.7 by (8.73) and by (8.74). We finally conclude with the *left maximal angular distortion* (8.75).

> **Lemma 8.11 (Normal mapping: ellipsoid-of-revolution to plane, equiareal mapping).**

The equiareal mapping of the ellipsoid-of-revolution to the tangential plane at the North Pole is parameterized by

$$x = f(\Delta)\cos\Lambda ,$$

$$y = f(\Delta)\sin\Lambda ,$$
(8.65)

subject to the left Cauchy–Green eigenspace $\{\boldsymbol{E}_\Lambda\Lambda_1(\Phi), \boldsymbol{E}_\Phi\Lambda_2(\Phi)\}$. The radial function $r = f(\Phi)$ that represents an equiareal mapping is given as a *four terms integral* in a closed form.

> **End of Lemma.**

> **Box 8.7 (Equiareal mapping of the ellipsoid-of-revolution to the tangential plane at the North Pole).**
>
> Postulate of an areomorphism:
>
> $$\Lambda_1\Lambda_2 = 1 ,$$
>
> $$\frac{f(\Delta)\sqrt{1 - E^2\cos^2\Delta}}{A_1\sin\Delta}\,\frac{f'(\Delta)(1 - E^2\cos^2\Delta)^{3/2}}{A_1(1 - E^2)} = 1 \Rightarrow f\,\mathrm{d}f = A_1^2\frac{1 - E^2}{(1 - E^2\cos^2\Delta)^2}\sin\Delta\,\mathrm{d}\Delta . \tag{8.66}$$
>
> Integration of the characteristic differential equations of a conformal mapping
>
> $$\mathbb{E}^2_{A_1,A_2} \to T_N\mathbb{E}^2_{A_1,A_2}:$$
>
> $$\frac{1}{2}f^2 = A_1^2\int\frac{1 - E^2}{(1 - E^2\cos^2\Delta)^2}\sin\Delta\,\mathrm{d}\Delta + c . \tag{8.67}$$
>
> Decomposition into rational partials:
>
> $$y = E\cos\Delta$$
>
> $$\int\frac{\sin\Delta}{(1 - E^2\cos^2\Delta)^2}\,\mathrm{d}\Delta = -\frac{1}{E}\int\frac{\mathrm{d}(E\cos\Delta)}{(1 - E^2\cos^2\Delta)^2} = -\frac{1}{E}\int\frac{\mathrm{d}y}{(1 - y^2)^2} ,$$
>
> $$\frac{1}{(1 - y^2)^2} = \frac{A}{(1 - y)^2} + \frac{B}{(1 - y)} + \frac{C}{(1 + y)^2} + \frac{D}{(1 + y)} \Leftrightarrow A = B = C = D = \frac{1}{4} , \tag{8.68}$$
>
> $$\int\frac{\sin\Delta}{(1 - E^2\cos^2\Delta)^2}\,\mathrm{d}\Delta = -\frac{1}{4E}\int\left[\frac{1}{(1 - y)^2} + \frac{1}{(1 - y)} + \frac{1}{(1 + y)^2} + \frac{1}{(1 + y)}\right]\mathrm{d}y .$$

Continuation of Box.

Standard integrals:

$$\int \frac{dy}{ay+b} = \frac{1}{a}\ln|ay+b|\ ,$$

$$\int \frac{dy}{(1+y)^2} = -\frac{1}{1+y}\ ,\quad \int \frac{dy}{(1-y)^2} = +\frac{1}{1-y}\ ,$$

$$\int \frac{\sin\Delta}{(1-E^2\cos^2\Delta)^2}d\Delta = -\frac{1}{4E}\left[\ln\frac{1+y}{1-y} + \frac{1}{1-y} - \frac{1}{1+y}\right] =$$

$$= -\frac{1}{4E}\left[\ln\frac{1+E\cos\Delta}{1-E\cos\Delta} + \frac{1}{1-E\cos\Delta} - \frac{1}{1+E\cos\Delta}\right]\ . \tag{8.69}$$

Integration constant:

$$\frac{1}{2}f^2 = A_1^2\frac{1-E^2}{4E}\left[-\ln\frac{1+E\cos\Delta}{1-E\cos\Delta} - \frac{1}{1-E\cos\Delta} + \frac{1}{1+E\cos\Delta}\right] + c\ ,$$

$$f(\Delta=0) = 0 \Leftrightarrow f^2(\Delta=0) = 0 \Leftrightarrow c = A_1^2\frac{1-E^2}{4E}\left(\ln\frac{1+E}{1-E} + \frac{1}{1-E} - \frac{1}{1+E}\right)\ . \tag{8.70}$$

Parameterized mapping equations:

$$f(\Delta) \to f(\Phi)\ ,$$

$$f(\Delta) = A_1\sqrt{1-E^2}\sqrt{\frac{1}{1-E^2} + \frac{1}{2E}\ln\frac{1+E}{1-E} - \frac{\cos\Delta}{1-E^2\cos^2\Delta} - \frac{1}{2E}\ln\frac{1+E\cos\Delta}{1-E\cos\Delta}}\ , \tag{8.71}$$

$$f(\Phi) = A_1\sqrt{1-E^2}\sqrt{\frac{1}{1-E^2} + \frac{1}{2E}\ln\frac{1+E}{1-E} - \frac{\sin\Phi}{1-E^2\sin^2\Phi} - \frac{1}{2E}\ln\frac{1+E\sin\Phi}{1-E\sin\Phi}}\ ,$$

$$\alpha = \Lambda\ ,\ r = f(\Delta)\ \text{or}\ r = f(\Phi)\ ,\ x = f(\Phi)\cos\Lambda\ ,\ y = f(\Phi)\sin\Lambda\ . \tag{8.72}$$

Left principal stretches and left eigenvectors:

$$\Lambda_1 = \frac{f(\Phi)\sqrt{1-E^2\sin^2\Phi}}{A_1\cos\Phi}\ ,\quad \Lambda_2 = \frac{A_1\cos\Phi}{f(\Phi)\sqrt{1-E^2\sin^2\Phi}}\ , \tag{8.73}$$

$$\boldsymbol{C}_1 = \boldsymbol{E}_\Lambda = \frac{D_\Lambda\boldsymbol{X}}{\|D_\Lambda\boldsymbol{X}\|}\quad (\text{``Easting''})\ ,\quad \boldsymbol{C}_2 = \boldsymbol{E}_\Phi = \frac{D_\Phi\boldsymbol{X}}{\|D_\Phi\boldsymbol{X}\|}\quad (\text{``Northing''})\ ,$$

$$(\text{i})\ \boldsymbol{C}_1\Lambda_1 = \boldsymbol{E}_\Lambda\frac{f(\Phi)\sqrt{1-E^2\sin^2\Phi}}{A_1\cos\Phi}\ ,\quad (\text{ii})\ \boldsymbol{C}_2\Lambda_2 = \boldsymbol{E}_\Phi\frac{A_1\cos\Phi}{f(\Phi)\sqrt{1-E^2\sin^2\Phi}}\ . \tag{8.74}$$

Left maximal angular distortion:

$$\Omega_l = 2\arcsin\left|\frac{\Lambda_1-\Lambda_2}{\Lambda_1+\Lambda_2}\right| = 2\arcsin\left|\frac{\Lambda_1^2-1}{\Lambda_1^2+1}\right| =$$

$$= 2\arcsin\left|\frac{f^2(\Phi)(1-E^2\sin^2\Phi) - A_1^2\cos^2\Phi}{f^2(\Phi)(1-E^2\sin^2\Phi) + A_1^2\cos^2\Phi}\right|\ . \tag{8.75}$$

8-3 Perspective mapping equations

Setting up perspective mappings "ellipsoid-of-revolution to plane", the fundamental perspective graph, Space Photos.

In this section, we intend to present various *perspective mappings* from the ellipsoid-of-revolution to the tangential plane, placing the *perspective center* arbitrarily. Let the position P_c be on the top of the ellipsoid-of-revolution. Furthermore, let us use the orthogonal projection to locate the point $P_0 = p_0$ at *minimal distance* or *maximal distance*, namely $\| \boldsymbol{X}_c - \boldsymbol{X}_0 \| = \min$ or $\| \boldsymbol{X}_c - \boldsymbol{X}_0 \| = \max$. Alternatively, we can take advantage of an orthogonal projection of the ellipsoid-of-revolution to the sphere, which passes the center o of the ellipsoid-of-revolution. The three variants of the special perspective mappings "ellipsoid-of-revolution to plane" are illustrated by Figs. 8.8, 8.7, and 8.6.

Technical aside.

Note that the perspective mappings from the ellipsoid-of-revolution to the tangential plane are applied to map points-in-space to the tangential planes of the ellipsoid-of-revolution. Examples are visions from a tower or from an airplane and from an Earth satellite by eye or by a camera. A special example are images of TV cameras showing clouds – important information needed for weather reports.

For our introduction, we treat only the case of the mapping of minimal distance. The final mapping equations, given the coordinates of perspective center $\left(\Lambda_0, \Phi_0, H_0\right)$ to the plane which is located at minimal distance from the perspective center, are presented in Box 8.8 in terms of the coordinates $(x^*, y^*)_p$ in the tangential plane: see (8.76) and (8.77).

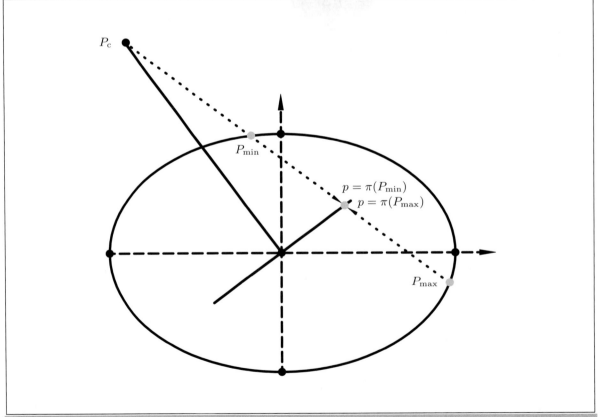

Fig. 8.6. Perspective mappings of a perspective center P_c to the plane which passes the center o of the ellipsoid-of-revolution $\mathbb{E}^2_{A_1, A_2}$.

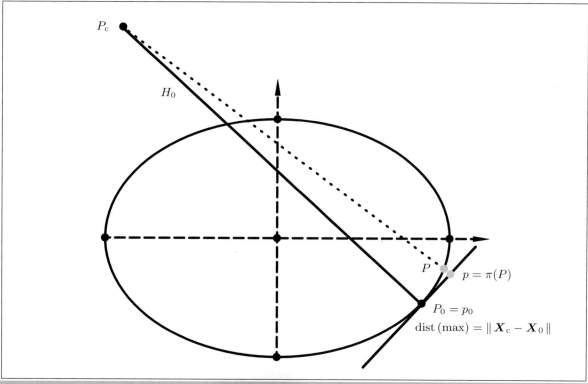

Fig. 8.7. Perspective mappings of a perspective center P_c to the plane which is at the maximal distance from an ellipsoid-of-revolution $\mathbb{E}^2_{A_1,A_2}$.

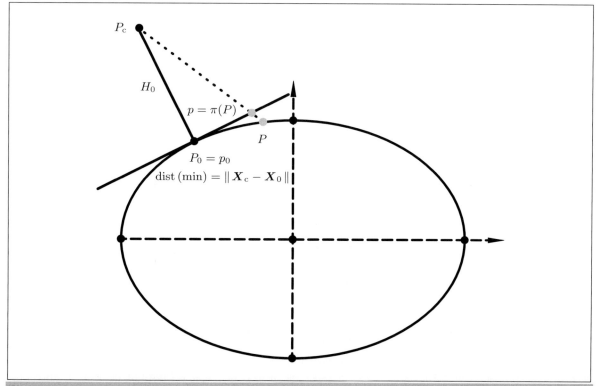

Fig. 8.8. Perspective mappings of a perspective center P_c to the plane which is at the minimal distance from an ellipsoid-of-revolution $\mathbb{E}^2_{A_1,A_2}$.

Box 8.8 (Perspective mapping equations, minimal distance, perspective center Λ_0, Φ_0, H_0).

South coordinates:

$$x^* = x^*(p) =$$

$$= H_0 \frac{-N \cos \Phi \sin \Phi_0 \cos(\Lambda - \Lambda_0) + N(1 - E^2) \sin \Phi \cos \Phi_0 + N_0 E^2 \sin \Phi_0 \cos \Phi_0}{N \cos \Phi \cos \Phi_0 \cos(\Lambda - \Lambda_0) + N(1 - E^2) \sin \Phi \sin \Phi_0 - N_0 + N_0 E^2 \sin^2 \Phi_0 - H_0} \, . \tag{8.76}$$

East coordinates:

$$y^* = y^*(p) =$$

$$= H_0 \frac{-N \cos \Phi \sin(\Lambda - \Lambda_0)}{N \cos \Phi \cos \Phi_0 \cos(\Lambda - \Lambda_0) + N(1 - E^2) \sin \Phi \sin \Phi_0 - N_0 + N_0 E^2 \sin^2 \Phi_0 - H_0} \, . \tag{8.77}$$

$$N := \frac{A_1}{\sqrt{1 - E^2 \sin^2 \Phi}} \, , \quad N_0 := \frac{A_1}{\sqrt{1 - E^2 \sin^2 \Phi_0}} \, . \tag{8.78}$$

Alternatives.

East coordinates:

$$x^{**} = y^* \, . \tag{8.79}$$

North coordinates:

$$y^{**} = -x^* \, . \tag{8.80}$$

Polar coordinates (South azimuth α, radial coordinate r):

$$\tan \alpha^* =$$

$$= \frac{N \cos \Phi \sin(\Lambda - \Lambda_0)}{N \cos \Phi \sin \Phi_0 \cos(\Lambda - \Lambda_0) - N(1 - E^2) \sin \Phi \cos \Phi_0 - N_0 E^2 \sin \Phi_0 \cos \Phi_0} \, , \tag{8.81}$$

$$r = \sqrt{x^{*\,2} + y^{*\,2}} = \sqrt{x^{**\,2} + y^{**\,2}} \, .$$

Alternative coordinates:

$$\alpha^{**} = 90° - \alpha^* \quad \text{(East azimuth)} \, . \tag{8.82}$$

At this point, you may enjoy our examples. Our first example, see Fig. 8.9, uses a *tilted perspective*, also called *Space Photo* projection: the eastern seaboard viewed from a point about 160 km above Newburgh, New York ($\Phi_0 = 41°30'$ northern latitude, $\Lambda_0 = 74°$ Western longitude, 1° graticule). Our second example, see Fig. 8.10, uses a *tilted perspective*, also called *Space Photo* projection: France and Central Europe viewed from a point about 640 km above central Spain ($\Phi_0 = 40°$ northern latitude, $\Lambda_0 = 5°$ western longitude, 2° graticule).

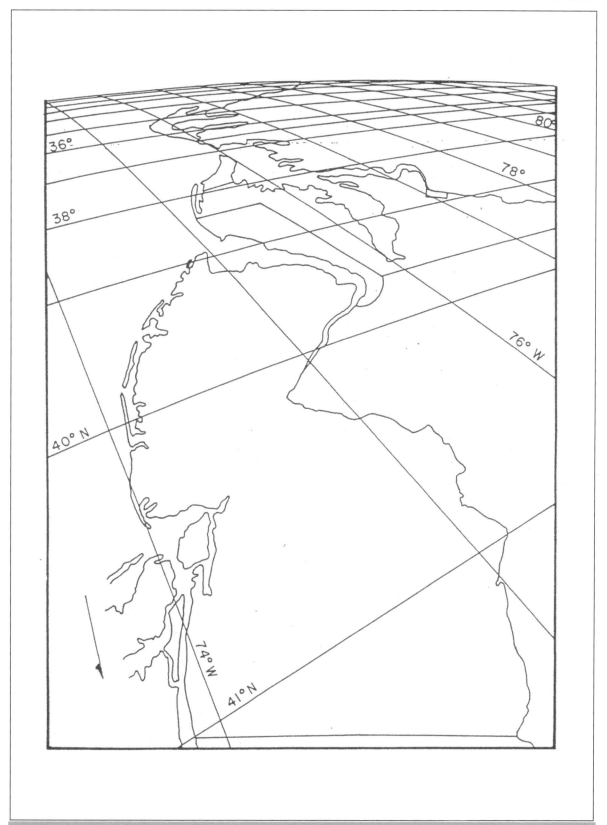

Fig. 8.9. A first example: Space Photo, J. P. Snyder. Reprinted with permission from "The perspective map projection of the Earth" by J. P. Snyder, The American Cartographer, vol. 8, no. 2, 1981, pp. 149–160.

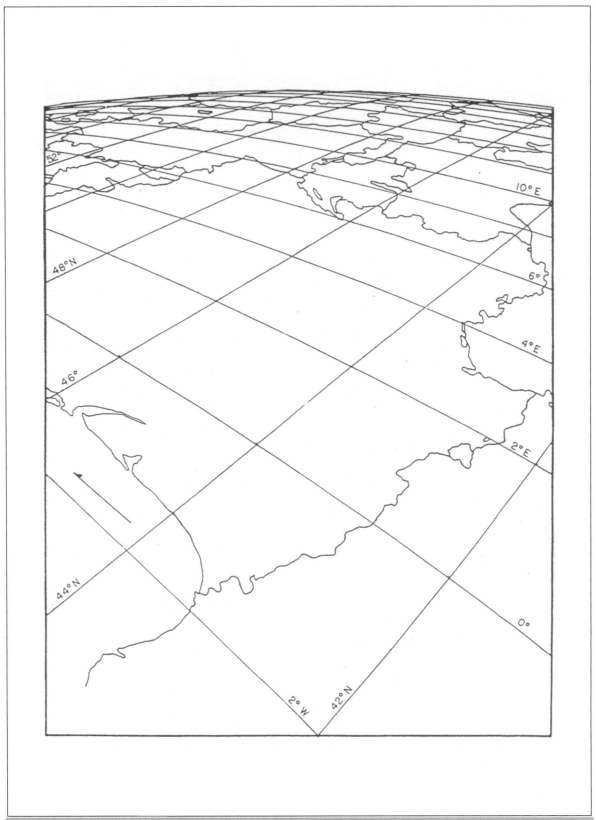

Fig. 8.10. A second example: Space Photo, J. P. Snyder. Reprinted with permission from "The perspective map projection of the Earth" by J. P. Snyder, The American Cartographer, vol. 8, no. 2, 1981, pp. 149–160.

8-31 The first derivation

The first derivation of the perspective equations is based upon the *fundamental perspective graph* denoted by $P_c P_0 P$ as illustrated by Fig. 8.11. Here, we take advantage of the basic equations which are based upon the so-called *normal intersection* in terms of the curve $P_0 P$, which coincides with the intersection line $\mathbb{E}^2_{A_1 \ A_2}$ and $\mathbb{P}_{P_c P_0 P}$. Note that δ is the angle of the cone in the triangle P_0, P_c, P at P_c. Furthermore, note that the point P_0 locates the point of minimal distance with respect to the point P_c and the tangent space $T_{P_0} \mathbb{E}^2_{A_1 \ A_2}$ at the point P_0. Moreover, note that $p = \pi(P)$ denotes the projection point, which is at minimal distance. In addition, $\boldsymbol{G}_3$ is the normal unit vector extending from P_0 to P_c. Here, we take advantage of the radial coordinate r, the first equation, the second equation, and the third equation, namely

$$r = \| P_0 - P \| , \quad P_0 = p_0$$
$$\text{(radial coordinate)} ,$$

(8.83)

$$\tan \delta = \frac{r}{h} \quad \text{or} \quad r = h \tan \delta$$
$$\text{(first equation)} ,$$

(8.84)

$$k^2 = g^2 + h^2 - 2gh \cos \delta \quad \text{or} \quad \cos \delta = \frac{g^2 + h^2 - k^2}{2gh}$$
$$\text{(second equation)} ,$$

(8.85)

$$\tan \delta = \frac{\pm\sqrt{1 - \cos^2 \delta}}{\cos \delta} = \frac{\pm\sqrt{4g^2 h^2 - (g^2 + h^2 - k^2)^2}}{g^2 + h^2 - k^2}$$
$$\text{(third equation)} .$$

(8.86)

The height $h = H_0$ of the perspective center P_c above the point P_0, which is nothing but an element of the ellipsoid-of-revolution, is given. The distance $g := \| \boldsymbol{X}_c - \boldsymbol{X}_P \|$ the distance $h := \| \boldsymbol{X}_c - \boldsymbol{X}_0 \|$, and the distance $k := \| \boldsymbol{X}_P - \boldsymbol{X}_0 \|$ are given. In summary, the above equations lead to a special formulation of r, namely

$$r = \frac{h}{g^2 + h^2 - k^2} \sqrt{4g^2 h^2 - (g^2 + h^2 - k^2)^2} .$$

(8.87)

In the passages that follow, we use the representation of the distances g, h, and k in surface normal ellipsoidal coordinates which are supported by $\mathbb{E}^2_{A_1 \ A_2}$.

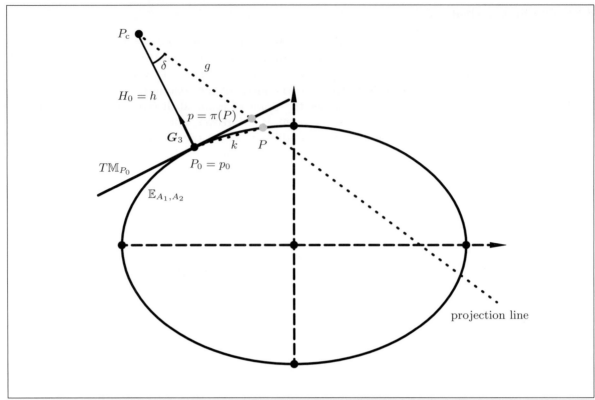

Fig. 8.11. Fundamental perspective graph. $P \in \mathbb{E}^2_{A_1,A_2}$, $P_0 P \in \mathbb{E}_2 \cup \mathbb{P}_{P_c P_0 P}$.

The point P_c $(\Lambda_c, \Phi_c, H_0 = h)$ is defined as follows:

$$\boldsymbol{X}_c =$$

$$= \boldsymbol{E}_1 \left[\frac{A_1}{\sqrt{1 - E^2 \sin^2 \Phi_0}} + H_0(\Lambda_c, \Phi_c) \right] \cos \Phi_0 \cos \Lambda_0 +$$

$$+ \boldsymbol{E}_2 \left[\frac{A_1}{\sqrt{1 - E^2 \sin^2 \Phi_0}} + H_0(\Lambda_c, \Phi_c) \right] \cos \Phi_0 \sin \Lambda_0 + \tag{8.88}$$

$$+ \boldsymbol{E}_3 \left[\frac{A_1(1 - E^2)}{\sqrt{1 - E^2 \sin^2 \Phi_0}} + H_0(\Lambda_c, \Phi_c) \right] \sin \Phi_0 \ .$$

Given the coordinates of the point $\boldsymbol{X}_c$, we derive h, g, and k as follows:

$$g := \sqrt{\left(X_c - X_P\right)^2 + \left(Y_c - Y_P\right)^2 + \left(Z_c - Z_P\right)^2} \ ,$$

$$h := \sqrt{\left(X_c - X_0\right)^2 + \left(Y_c - Y_0\right)^2 + \left(Z_c - Z_0\right)^2} \ , \tag{8.89}$$

$$k := \sqrt{\left(X_P - X_0\right)^2 + \left(Y_P - Y_0\right)^2 + \left(Z_P - Z_0\right)^2} \ .$$

The diverse differences are defined as follows:

$$X_c - X_P =$$

$$= \left[\frac{A_1}{\sqrt{1 - E^2 \sin^2 \Phi_0}} + H_0\left(\Lambda_c, \Phi_c\right) \right] \cos \Phi_0 \cos \Lambda_0 - \frac{A_1}{\sqrt{1 - E^2 \sin^2 \Phi}} \cos \Phi \cos \Lambda \, ,$$

$$Y_c - Y_P =$$

$$= \left[\frac{A_1}{\sqrt{1 - E^2 \sin^2 \Phi_0}} + H_0\left(\Lambda_c, \Phi_c\right) \right] \cos \Phi_0 \sin \Lambda_0 - \frac{A_1}{\sqrt{1 - E^2 \sin^2 \Phi}} \cos \Phi \sin \Lambda \, ,$$

$$(8.90)$$

$$Z_c - Z_P =$$

$$= \left[\frac{A_1(1 - E^2)}{\sqrt{1 - E^2 \sin^2 \Phi_0}} + H_0\left(\Lambda_c, \Phi_c\right) \right] \sin \Phi_0 - \frac{A_1(1 - E^2)}{\sqrt{1 - E^2 \sin^2 \Phi}} \sin \Phi \, ,$$

and

$$X_P - X_0 =$$

$$= \frac{A_1}{\sqrt{1 - E^2 \sin^2 \Phi}} \cos \Phi \cos \Lambda - \frac{A_1}{\sqrt{1 - E^2 \sin^2 \Phi_0}} \cos \Phi_0 \cos \Lambda_0 \, ,$$

$$Y_P - Y_0 =$$

$$= \frac{A_1}{\sqrt{1 - E^2 \sin^2 \Phi}} \cos \Phi \sin \Lambda - \frac{A_1}{\sqrt{1 - E^2 \sin^2 \Phi_0}} \cos \Phi_0 \sin \Lambda_0 \, ,$$

$$(8.91)$$

$$Z_P - Z_0 =$$

$$= \frac{A_1(1 - E^2)}{\sqrt{1 - E^2 \sin^2 \Phi}} \sin \Phi - \frac{A_1(1 - E^2)}{\sqrt{1 - E^2 \sin^2 \Phi_0}} \sin \Phi_0 \, .$$

Substituting g, h, and k into the basic formula for the radial coordinate r, r is obtained as follows:

$$r = \frac{\| \boldsymbol{X}_c - \boldsymbol{X}_0 \|}{\| \boldsymbol{X}_c - \boldsymbol{X}_P \|^2 + \| \boldsymbol{X}_c - \boldsymbol{X}_0 \|^2 - \| \boldsymbol{X}_P - \boldsymbol{X}_0 \|^2} \times$$

$$(8.92)$$

$$\times \sqrt{4 \| \boldsymbol{X}_c - \boldsymbol{X}_P \|^2 \| \boldsymbol{X}_c - \boldsymbol{X}_0 \|^2 - \left[\| \boldsymbol{X}_c - \boldsymbol{X}_P \|^2 + \| \boldsymbol{X}_c - \boldsymbol{X}_0 \|^2 - \| \boldsymbol{X}_P - \boldsymbol{X}_0 \|^2 \right]^2} \, .$$

Substitute the transformation of surface normal ellipsoidal coordinates $\{\Lambda, \Phi\}$, $\{\Lambda_0, \Phi_0, H_0(\Lambda_c, \Phi_c)\}$, and $\{\Lambda_0, \Phi_0, H = 0\}$ to the corresponding Cartesian coordinates, and you receive the new curvilinear representation of the radial coordinates.

Let us now take care of the *polar coordinate* and base our analysis on the transformation of reference frames, in particular, on the orthonormal Euclidean triad, corotating with the Earth, called $\{E_1, E_2, E_3\}$, and on the moving frame, called *South, East, Vertical*, an orthonormal triad in an astronomical orientation, namely $\{E_{1*}, E_{2*}, E_{3*}\}$: see Box 8.9. We here use the symbol of a star to identify the antipolar star orientation. Γ_{Gr} refers to the *gravity vector* at Greenwich, while Ω denotes the *global rotation vector* of the Earth. By contrast, E_{1*} refers to the South unit vector, E_{2*} refers to the East unit vector, and E_{3*} completes the orthonormal triad as the local vertical vector. The Euler rotation matrix $R_E(\Lambda_0, \Phi_0, 0)$ and the rotation matrices $R_3(\Lambda_0)$ and $R_2(\pi/2 - \Phi_0)$ are provided by (8.97). In Fig. 8.12, E_{1*} and E_{2*} are compactly illustrated.

Box 8.9 ($\{E_1, E_2, E_3\}$ and $\{E_{1*}, E_{2*}, E_{3*}\}$, Euler rotation matrix, Euler parameters).

Transformation of fixed and moving frame

($\{E_1, E_2, E_3\}$ versus $\{E_{1*}, E_{2*}, E_{3*}\}$):

$$
E_{1*} := -\frac{\partial X/\partial \Phi}{\|\partial X/\partial \Phi\|} \quad E_{2*} := +\frac{\partial X/\partial \Lambda}{\|\partial X/\partial \Lambda\|} \quad E_{3*} := +\frac{\partial X/\partial H}{\|\partial X/\partial H\|} \tag{8.93}
$$

(South) ,   (East) ,   (Vertical) ,

$$
\begin{bmatrix} E_{1*} \\ E_{2*} \\ E_{3*} \end{bmatrix} = R_E(\Lambda_0, \Phi_0, 0) \begin{bmatrix} E_1 \\ E_2 \\ E_3 \end{bmatrix} , \tag{8.94}
$$

$$
E_1 := E_2 \times E_3 ,
$$

$$
E_2 := \frac{-\Gamma_{\mathrm{Gr}} \times \Omega}{\|-\Gamma_{\mathrm{Gr}} \times \Omega\|} , \tag{8.95}
$$

$$
E_3 := \frac{\Omega}{\|\Omega\|} .
$$

Euler rotation matrix:

$$
R_E(\Lambda_0, \Phi_0, 0) := R_3(0) R_2(\pi/2 - \Phi_0) R_3(\Lambda_0) , \tag{8.96}
$$

$$
R_3(\Lambda_0) = \begin{bmatrix} \cos \Lambda_0 & \sin \Lambda_0 & 0 \\ -\sin \Lambda_0 & \cos \Lambda_0 & 0 \\ 0 & 0 & 1 \end{bmatrix} , \quad R_2(\pi/2 - \Phi_0) = \begin{bmatrix} \sin \Phi_0 & 0 & -\cos \Phi_0 \\ 0 & 1 & 0 \\ \cos \Phi_0 & 0 & \sin \Phi_0 \end{bmatrix} . \tag{8.97}
$$

Euler parameters:

$$
R_E(\Lambda_0, \Phi_0, 0) = \begin{bmatrix} \cos \Lambda_0 \sin \Phi_0 & \sin \Lambda_0 \sin \Phi_0 & -\cos \Phi_0 \\ -\sin \Lambda_0 & \cos \Lambda_0 & 0 \\ \cos \Lambda_0 \cos \Phi_0 & \sin \Lambda_0 \cos \Phi_0 & \sin \Phi_0 \end{bmatrix} . \tag{8.98}
$$

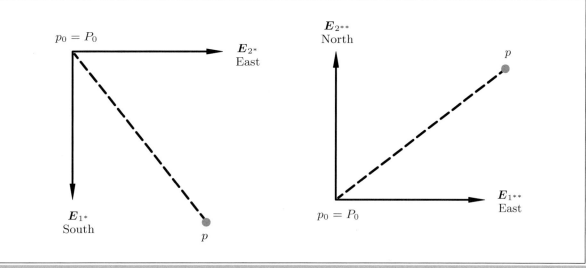

Fig. 8.12. Ellipsoidal horizontal plane at the point P_0.

In the frame that is located at the point P_0, let us here derive the spherical coordinates of the point P from the coordinates $\{\alpha, \beta, r\}$:

$$X_P^* - X_0^* = r \cos \beta \cos \alpha \ , \quad Y_P^* - Y_0^* = r \cos \beta \sin \alpha \ , \quad Z_P^* - Z_0^* = r \sin \beta \ ,$$

$$\tan \alpha = \frac{Y_P^* - Y_0^*}{X_P^* - X_0^*} = \frac{r_{21}\left(X_P - X_0\right) + r_{22}\left(Y_P - Y_0\right) + r_{23}\left(Z_P - Z_0\right)}{r_{11}\left(X_P - X_0\right) + r_{12}\left(Y_P - Y_0\right) + r_{13}\left(Z_P - Z_0\right)} \ , \tag{8.99}$$

$$\tan \beta = \frac{Z_P^* - Z_0^*}{\sqrt{\left(X_P^* - X_0^*\right)^2 + \left(Y_P^* - Y_0^*\right)^2}} \ .$$

Finally, we we transfrom the relative placement vector $\{X_P - X_0, Y_P - Y_0, Z_P - Z_0\}_{\boldsymbol{E}}$ to the relative placement vector $\{X_P^* - X_0^*, Y_P^* - Y_0^*, Z_P^* - Z_0^*\}_{\boldsymbol{E}^*}$:

$$\begin{bmatrix} X_P^* - X_0^* \\ Y_P^* - Y_0^* \\ Z_P^* - Z_0^* \end{bmatrix}_{\boldsymbol{E}^*} = \boldsymbol{R}_{\boldsymbol{E}}(\Lambda_0, \Phi_0, 0) \begin{bmatrix} X_P - X_0 \\ Y_P - Y_0 \\ Z_P - Z_0 \end{bmatrix}_{\boldsymbol{E}} \ , \tag{8.100}$$

$$\tan \alpha =$$
$$= \frac{-\sin \Phi_0\left(X_P - X_0\right) + \cos \Lambda_0\left(Y_P - Y_0\right)}{\sin \Phi_0 \cos \Lambda_0\left(X_P - X_0\right) + \sin \Phi_0 \cos \Lambda_0\left(Y_P - Y_0\right) - \cos \Phi_0\left(Z_P - Z_0\right)} \ , \tag{8.101}$$
$$\tan \beta \text{ analogous} \ .$$

The arctan leads to the orientation angle we need. But we have to pay attention to the *quadrant rule*. The mapping $\alpha \in [0, 2\pi] \to \tan \alpha$ is not injective. Therefore, we must apply the *quadrant rule*:

$$Y_P^* - Y_0^* \text{ positive, } X_P^* - X_0^* \text{ positive: 1st quadrant } 0 \leq \alpha < \pi/2 \ ,$$
$$Y_P^* - Y_0^* \text{ positive, } X_P^* - X_0^* \text{ negative: 2nd quadrant } \pi/2 \leq \alpha < 0 \ ,$$
$$Y_P^* - Y_0^* \text{ negative, } X_P^* - X_0^* \text{ negative: 3rd quadrant } \pi \leq \alpha < 3\pi/2 \ ,$$
$$Y_P^* - Y_0^* \text{ negative, } X_P^* - X_0^* \text{ positive: 4th quadrant } 3\pi/2 \leq \alpha < 2\pi \ . \tag{8.102}$$

8-32 The special case "sphere to tangential plane"

Let us here specialize to the mapping "sphere to tangential plane", namely to the case where P_0 is located at the North pole and P_c at the South pole, and we only treat the case where P_0 is at maximal distance from P_c. Consult Fig. 8.13 for more details. We here begin with identifying the points $\boldsymbol{X}_0$ and $\boldsymbol{X}_c$, respectively, by their coordinates $\{0, 0, Z\}$ and $\{0, 0, -Z\}$, respectively. $\Phi_0 = \pi/2$ and Λ_0 are not specified. g, h, and k are defined by (8.104).

$$\boldsymbol{X}_P \in \mathbb{S}_R^2 : \boldsymbol{X}_P = R\cos\Phi\cos\Lambda\boldsymbol{E}_1 + R\cos\Phi\sin\Lambda\boldsymbol{E}_2 + R\sin\Phi\boldsymbol{E}_3 \ , \tag{8.103}$$

$$
\begin{aligned}
g = \parallel \boldsymbol{X}_c - \boldsymbol{X}_P \parallel &= \sqrt{R^2\cos^2\Phi\cos^2\Lambda + R^2\cos^2\Phi\sin^2\Lambda + R^2(1+\sin\Phi)^2} = \\
&= R\sqrt{2}\sqrt{1+\sin\Phi} = R\sqrt{2}\sqrt{1+\cos\Delta} \ , \\
h &= H_0 = 2R \ ,
\end{aligned}
\tag{8.104}
$$

$$k = \parallel \boldsymbol{X}_0 - \boldsymbol{X}_P \parallel = R\sqrt{\cos^2\Phi + (1-\sin\Phi)^2} = R\sqrt{2}\sqrt{1-\sin\Phi} = R\sqrt{2}\sqrt{1-\cos\Delta} \ .$$

At this point we specialize $\alpha = \Lambda$ and $r = r(\Phi)$. Note that the analogous Φ representation is obtained by $4g^2h^2 = 32R^4(1+\sin\Phi) = 32R^4(1+\cos\Delta)$ and $(g^2+h^2-k^2)^2 = 16R^4(1+\sin\Phi)^2 = 16R^4(1+\cos\Delta)^2$.

$$\alpha = \Lambda \ ,$$

$$r = 2R\frac{\cos\Phi}{1+\sin\Phi} = 2R\frac{\sin\Delta}{1+\cos\Delta} = 2R\tan\Delta/2 = 2R\tan\left(\frac{\pi}{4}-\frac{\Phi}{2}\right) \ . \tag{8.105}$$

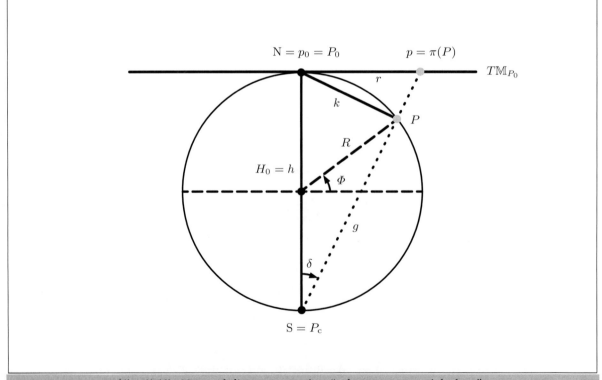

Fig. 8.13. Maximal distance mapping "sphere to tangential plane".

8-33 An alternative approach for a topographic point

We illustrate the perspective center P_c, the projection point P_0 on the ellipsoid-of-revolution, the topographic point $P(\Lambda, \Phi, H > 0)$, and the point $p = \pi(P)$ on the tangential plane through P_0 in Fig. 8.14. $\{\boldsymbol{E}_{1*}, \boldsymbol{E}_{2*}, \boldsymbol{E}_{3*}\}$ refers to the point P_0 as a triad in the local horizontal plane $p_0 p$ with reference to $P_0 = p_0$ and $p = \pi(P)$. $\left|\langle \boldsymbol{X}_p - \boldsymbol{X}_c | \boldsymbol{n} \rangle\right|$ is the length of the projection onto the normal vector $\boldsymbol{n}$, the local vertical at the point P_0 with respect to the local tangential plane. We may also write $\boldsymbol{n} = \boldsymbol{G}_3$. The reference frame is denoted by $\{\boldsymbol{E}_1, \boldsymbol{E}_2, \boldsymbol{E}_3\}$ and is oriented as described in the previous chapter. $H_0 = h$ is called the distance $P_0 P_c$ along the principal axis of the perspective mapping. Follow the illustration in Fig. 8.14. Here, we start from the fundamental equation, namely the ratio $\boldsymbol{X}_p - \boldsymbol{X}_c = \lambda(\boldsymbol{X}_P - \boldsymbol{X}_c)$, where $|\lambda|$ is the perspective factor:

$$|\lambda| = \frac{\|\boldsymbol{X}_p - \boldsymbol{X}_c\|}{\|\boldsymbol{X}_P - \boldsymbol{X}_c\|} = \frac{H_0}{\left|\langle \boldsymbol{X}_P - \boldsymbol{X}_c | \boldsymbol{n} \rangle\right|} ,$$

$$\lambda = -\frac{H_0}{\left|\langle \boldsymbol{X}_P - \boldsymbol{X}_c | \boldsymbol{n} \rangle\right|} ,$$

(8.106)

$$\boldsymbol{X}_p = \boldsymbol{X}_c - \frac{H_0}{\left|\langle \boldsymbol{X}_P - \boldsymbol{X}_c | \boldsymbol{n} \rangle\right|}(\boldsymbol{X}_P - \boldsymbol{X}_c) , \quad \boldsymbol{X}_c = \boldsymbol{X}_0 + H_0 \boldsymbol{n}$$

$$\Rightarrow$$

$$\boldsymbol{X}_p = \boldsymbol{X}_0 + H_0 \boldsymbol{n} - \frac{H_0(\boldsymbol{X}_P - \boldsymbol{X}_0 - H_0 \boldsymbol{n})}{\langle \boldsymbol{X}_P - \boldsymbol{X}_0 | \boldsymbol{n} \rangle - H_0} .$$

(8.107)

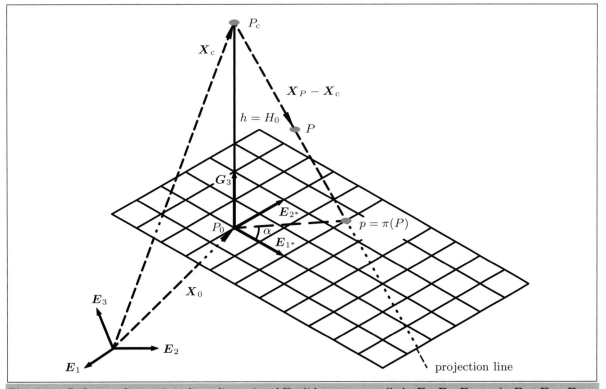

Fig. 8.14. Reference frames in a three-dimensional Euclidean space, called $\boldsymbol{E}_1, \boldsymbol{E}_2, \boldsymbol{E}_3$ and $\boldsymbol{E}_{1*}, \boldsymbol{E}_{2*}, \boldsymbol{E}_{3*}$, called *South, East, Vertical.* Special case: topographic point $P(\Lambda, \Phi, H > 0)$.

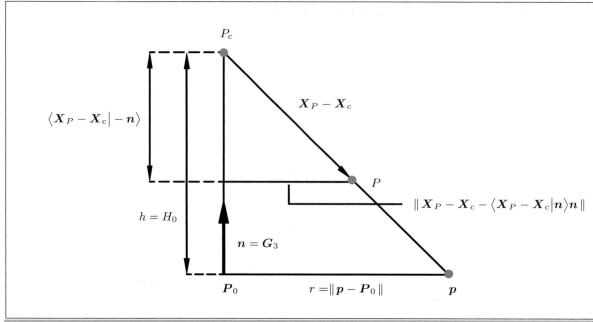

Fig. 8.15. The ratio $r/\parallel\boldsymbol{X}_P - \boldsymbol{X}_c - \langle\boldsymbol{X}_P - \boldsymbol{X}_c|\boldsymbol{n}\rangle\boldsymbol{n}\parallel = H_0/|\langle\boldsymbol{X}_P - \boldsymbol{X}_c|\boldsymbol{n}\rangle|$.

In the next phase, we compute the rectangular coordinates x_p^* and y_p^* as well as the angular parameters $r = \left(x_p^{*\,2} + y_p^{*\,2}\right)^{1/2}$ and $\alpha = \arctan y_p^*/x_p^*$, taking advantage of the ray condition. In Fig. 8.15, the principle situation is illustrated.

$$
\left.
\begin{aligned}
x_p^* &= \langle\boldsymbol{E}_{1*}|\boldsymbol{X}_p - \boldsymbol{X}_0\rangle \\
y_p^* &= \langle\boldsymbol{E}_{2*}|\boldsymbol{X}_p - \boldsymbol{X}_0\rangle
\end{aligned}
\right]
\Rightarrow
\left[
\begin{aligned}
x_p^* &= \frac{-H_0\langle\boldsymbol{E}_{1*}|\boldsymbol{X}_P - \boldsymbol{X}_0\rangle}{\langle\boldsymbol{X}_P - \boldsymbol{X}_0|\boldsymbol{n}\rangle - H_0} \\
y_p^* &= \frac{-H_0\langle\boldsymbol{E}_{2*}|\boldsymbol{X}_P - \boldsymbol{X}_0\rangle}{\langle\boldsymbol{X}_P - \boldsymbol{X}_0|\boldsymbol{n}\rangle - H_0}
\end{aligned}
\right. ,
\tag{8.108}
$$

$$
\alpha = \arctan\frac{\langle\boldsymbol{E}_{2*}|\boldsymbol{X}_p - \boldsymbol{X}_0\rangle}{\langle\boldsymbol{E}_{1*}|\boldsymbol{X}_p - \boldsymbol{X}_0\rangle} ,
\tag{8.109}
$$

$$
\frac{r}{\parallel\boldsymbol{X}_P - \boldsymbol{X}_c - \langle\boldsymbol{X}_P - \boldsymbol{X}_c|\boldsymbol{n}\rangle\boldsymbol{n}\parallel} = \frac{H_0}{|\langle\boldsymbol{X}_P - \boldsymbol{X}_c|\boldsymbol{n}\rangle|} = |\lambda|
$$
$$
\Rightarrow
\tag{8.110}
$$
$$
r = H_0\frac{\sqrt{\parallel\boldsymbol{X}_P - \boldsymbol{X}_c\parallel^2 - \langle\boldsymbol{X}_P - \boldsymbol{X}_c|\boldsymbol{n}\rangle^2}}{|\langle\boldsymbol{X}_P - \boldsymbol{X}_c|\boldsymbol{n}\rangle|} .
$$

In Box 8.10, the computational products are listed. As it is sketched in Box 8.10, the angular parameter r can be rewritten as (8.111).

$$
r = H_0\frac{\sqrt{\parallel\boldsymbol{X}_P - \boldsymbol{X}_0\parallel^2 - \langle\boldsymbol{X}_P - \boldsymbol{X}_0|\boldsymbol{n}\rangle^2}}{|\langle\boldsymbol{X}_P - \boldsymbol{X}_0|\boldsymbol{n}\rangle - H_0|} .
\tag{8.111}
$$

Box 8.10 (Computational products).

$$\boldsymbol{X}_c = \boldsymbol{X}_0 + H_0 \boldsymbol{n} \, ,$$

$$\| \boldsymbol{X}_P - \boldsymbol{X}_c \|^2 = \langle \boldsymbol{X}_P - \boldsymbol{X}_0 - H_0 \boldsymbol{n} | \boldsymbol{X}_P - \boldsymbol{X}_0 - H_0 \boldsymbol{n} \rangle =$$

$$= \| \boldsymbol{X}_P - \boldsymbol{X}_0 \|^2 - 2 H_0 \langle \boldsymbol{X}_P - \boldsymbol{X}_0 | \boldsymbol{n} \rangle + H_0^2 \, ,$$

$$\langle \boldsymbol{X}_P - \boldsymbol{X}_c | \boldsymbol{n} \rangle^2 = \langle \boldsymbol{X}_P - \boldsymbol{X}_0 - H_0 \boldsymbol{n} | \boldsymbol{n} \rangle^2 = \left(\langle \boldsymbol{X}_P - \boldsymbol{X}_0 | \boldsymbol{n} \rangle - H_0 \right)^2 =$$

$$= \left(\langle \boldsymbol{X}_P - \boldsymbol{X}_0 | \boldsymbol{n} \rangle \right)^2 - 2 H_0 \langle \boldsymbol{X}_P - \boldsymbol{X}_0 | \boldsymbol{n} \rangle + H_0^2$$

(8.112)

$$\Rightarrow$$

$$\| \boldsymbol{X}_P - \boldsymbol{X}_c \|^2 - \langle \boldsymbol{X}_P - \boldsymbol{X}_c | \boldsymbol{n} \rangle^2 = \| \boldsymbol{X}_P - \boldsymbol{X}_0 \|^2 - \langle \boldsymbol{X}_P - \boldsymbol{X}_0 | \boldsymbol{n} \rangle^2$$

$$\Rightarrow$$

$$r = H_0 \frac{\sqrt{\| \boldsymbol{X}_P - \boldsymbol{X}_0 \|^2 - \langle \boldsymbol{X}_P - \boldsymbol{X}_0 | \boldsymbol{n} \rangle^2}}{| \langle \boldsymbol{X}_P - \boldsymbol{X}_0 | \boldsymbol{n} \rangle - H_0 |} \, .$$

Finally, we have to represent the five vectors $\boldsymbol{n}$, $\boldsymbol{X}_P$, $\boldsymbol{X}_0$, $\boldsymbol{E}_{1*}$, and $\boldsymbol{E}_{2*}$ in the fixed reference frame $\{ \boldsymbol{E}_1 \ \boldsymbol{E}_2 \ \boldsymbol{E}_3 \}$ in order to be able to compute the projections onto $\boldsymbol{X}_P - \boldsymbol{X}_0$. In Boxes 8.11 and 8.12, the respective relations are collected.

Box 8.11 (Representation of the vectors $\boldsymbol{n}$, $\boldsymbol{X}_P$, $\boldsymbol{X}_0$, $\boldsymbol{E}_{1*}$, and $\boldsymbol{E}_{2*}$ in the fixed reference frame).

$$\boldsymbol{n} = [\boldsymbol{E}_1, \boldsymbol{E}_2, \boldsymbol{E}_3] \begin{bmatrix} \cos \Phi_0 \cos \Lambda_0 \\ \cos \Phi_0 \sin \Lambda_0 \\ \sin \Phi_0 \end{bmatrix} , \tag{8.113}$$

$$\boldsymbol{X}_0 = [\boldsymbol{E}_1, \boldsymbol{E}_2, \boldsymbol{E}_3] \begin{bmatrix} N_0 \cos \Phi_0 \cos \Lambda_0 \\ N_0 \cos \Phi_0 \sin \Lambda_0 \\ N_0 (1 - E^2) \sin \Phi_0 \end{bmatrix} ,$$

(8.114)

$$\boldsymbol{X}_P = [\boldsymbol{E}_1, \boldsymbol{E}_2, \boldsymbol{E}_3] \begin{bmatrix} (N + H) \cos \Phi \cos \Lambda \\ (N + H) \cos \Phi \sin \Lambda \\ ([N(1 - E^2) + H]) \sin \Phi \end{bmatrix} ,$$

$$\boldsymbol{E}_{1*} = [\boldsymbol{E}_1, \boldsymbol{E}_2, \boldsymbol{E}_3] \begin{bmatrix} \sin \Phi_0 \cos \Lambda_0 \\ \sin \Phi_0 \sin \Lambda_0 \\ - \cos \Phi_0 \end{bmatrix} \quad \text{(South)} ,$$

(8.115)

$$\boldsymbol{E}_{2*} = [\boldsymbol{E}_1, \boldsymbol{E}_2, \boldsymbol{E}_3] \begin{bmatrix} - \sin \Lambda_0 \\ \cos \Lambda_0 \\ 0 \end{bmatrix} \quad \text{(East)} .$$

Box 8.12 (Projections onto $\boldsymbol{X}_P - \boldsymbol{X}_0$).

$$\langle \boldsymbol{X}_P - \boldsymbol{X}_0 | \boldsymbol{n} \rangle =$$

$$= +(N + H)\cos\Phi\cos\Lambda\cos\Phi_0\cos\Lambda_0 + (N + H)\cos\Phi\sin\Lambda\cos\Phi_0\sin\Lambda_0 +$$

$$+ [N(1 - E^2) + H]\sin\Phi\sin\Phi_0 - N_0\cos^2\Phi_0\cos^2\Lambda_0 -$$

$$- N_0\cos^2\Phi_0\sin^2\Lambda_0 - N_0(1 - E^2)\sin^2\Phi_0 =$$

$$= +(N + H)\cos\Phi\cos\Phi_0\cos(\Lambda - \Lambda_0) +$$

$$+ [N(1 - E^2) + H]\sin\Phi\sin\Phi_0 - N_0 + N_0 E^2 \sin^2\Phi_0 \ ,$$

(8.116)

$$\langle \boldsymbol{E}_{1*} | \boldsymbol{X}_P - \boldsymbol{X}_0 \rangle =$$

$$= +(N + H)\cos\Phi\cos\Lambda\sin\Phi_0\cos\Lambda_0 + (N + H)\cos\Phi\sin\Lambda\sin\Phi_0\sin\Lambda_0 -$$

$$- [N(1 - E^2) + H]\sin\Phi\cos\Phi_0 - N_0\cos\Phi_0\cos\Lambda_0\sin\Phi_0\cos\Lambda_0 -$$

$$- N_0\cos\Phi_0\sin\Lambda_0\sin\Phi_0\sin\Lambda_0 + N_0(1 - E^2)\sin\Phi_0\cos\Phi_0 =$$

$$= +(N + H)\cos\Phi\cos\Phi_0\cos(\Lambda - \Lambda_0) -$$

$$- [N(1 - E^2) + H]\sin\Phi\cos\Phi_0 - N_0 E^2 \sin\Phi_0\cos\Phi_0 \ ,$$

(8.117)

$$\langle \boldsymbol{E}_{2*} | \boldsymbol{X}_P - \boldsymbol{X}_0 \rangle =$$

$$= -(N + H)\cos\Phi\cos\Lambda\sin\Lambda_0 + (N + H)\cos\Phi\sin\Lambda\cos\Lambda_0 +$$

$$+ N_0\cos\Phi_0\cos\Lambda_0\sin\Lambda_0 - N_0\cos\Phi_0\sin\Lambda_0\cos\Lambda_0 =$$

$$= +(N + H)\cos\Phi\sin(\Lambda - \Lambda_0) \ ,$$

(8.118)

$$\| \boldsymbol{X}_P - \boldsymbol{X}_0 \|^2 =$$

$$= +[(N + H)\cos\Phi\cos\Lambda - N_0\cos\Phi_0\cos\Lambda_0]^2 +$$

$$+ [(N + H)\cos\Phi\sin\Lambda - N_0\cos\Phi_0\sin\Lambda_0]^2 +$$

$$+ [(N(1 - E^2) + H)\sin\Phi - N_0(1 - E^2)\sin\Phi_0]^2 =$$

$$= +[N + H]^2 + E^2[-2N^2 + E^2 N^2 - 2HN]\sin^2\Phi -$$

$$- 2[N + H]N_0\cos\Phi\cos\Phi_0\cos(\Lambda - \Lambda_0) -$$

$$- 2[N(1 - E^2) + H]N_0(1 - E^2)\sin\Phi\sin\Phi_0 +$$

$$+ N_0^2 + E^2 N_0^2(-2 + E^2)\sin^2\Phi_0 \ .$$

(8.119)

The final formulae are given already before for $\{x_p^*, y_p^*\}$ or $\{x_p^{**}, y_p^{**}\} = \{y_p^*, -x_p^*\}$ and $\{\alpha^*, r\}$ for the South azimuth and the radial coordinate or $\{\alpha^{**}, r\}$ for the East azimuth $\alpha^{**} = 90° - \alpha^*$ and the radial coordinate.

In the following chapter, we study the mapping of the ellipsoid-of-revolution to the sphere and from the sphere to the plane.

9 "Ellipsoid-of-revolution to sphere and from sphere to plane"

Mapping the ellipsoid-of-revolution to sphere and from sphere to plane (the Gauss double projection, the "authalic" equal area projection): metric tensors, curvature tensors, principal stretches.

A special mapping, which was invented by C. F. Gauss (1822, 1844), is the double projection of the *ellipsoid-of-revolution to the sphere* and from the *sphere to the plane*. These are *conformal mappings*. A very efficient compiler version of the *Gauss double projection* was presented by M. Rosenmund (1903) (ROM mapping equations) and applied for mapping Switzerland and the Netherlands, for example. An alternative mapping, called "authalic", is equal area, first ellipsoid-of-revolution to sphere, and second sphere to plane.

9-1 General mapping equations "ellipsoid-of-revolution to plane"

Setting up general equations of the mapping "ellipsoid-of-revolution to plane": mapping equations, metric tensors, curvature tensors, differential forms.

Postulate.

The spherical longitude λ should be a linear function of the ellipsoidal longitude Λ: parallel circles of the ellipsoid-of-revolution should be transformed into parallel circles of the sphere.

End of Postulate.

Postulate.

The spherical latitude ϕ should only be a function of the ellipsoidal latitude Φ: meridians of the ellipsoid-of-revolution (lines of constant longitude) should be transformed into meridians of the sphere (ellipses of constant longitude).

End of Postulate.

9-11 The setup of the mapping equations "ellipsoid-of-revolution to plane"

$$\lambda = \lambda_0 + a(\Lambda - \Lambda_0), \quad \phi = f(\Phi). \tag{9.1}$$

Λ_0 is the ellipsoidal longitude of the reference point $P_0(\Lambda_0, \Phi_0)$, an element of the ellipsoid-of-revolution. First, let us compute the metric tensor (*first differential form*) of the ellipsoid-of-revolution and of the sphere. Second, let us compute the curvature tensor (*second differential form*) of the ellipsoid-of-revolution and the sphere. The mapping equations (9.2) ($\boldsymbol{X} = \Phi^{-1}(\boldsymbol{U})$ versus $\boldsymbol{x} = \phi^{-1}(\boldsymbol{u})$) form the basis of the computation of the first differential form and the second differential form of a surface. They lead to the inverse mapping equations (9.3).

$$\begin{bmatrix} X \\ Y \\ Z \end{bmatrix} = \frac{A_1}{\sqrt{1 - E^2 \sin^2 \Phi}} \begin{bmatrix} \cos \Phi \cos \Lambda \\ \cos \Phi \sin \Lambda \\ (1 - E^2) \sin \Phi \end{bmatrix} \quad \text{versus} \quad r \begin{bmatrix} \cos \phi \cos \lambda \\ \cos \phi \sin \lambda \\ \sin \phi \end{bmatrix} = \begin{bmatrix} x \\ y \\ z \end{bmatrix}, \tag{9.2}$$

$$\begin{bmatrix} U \\ V \end{bmatrix} = \begin{bmatrix} \Lambda \\ \Phi \end{bmatrix} = \begin{bmatrix} \arctan Y X^{-1} \\ \arctan \frac{1}{1 - E^2} \frac{Z}{\sqrt{X^2 + Y^2}} \end{bmatrix} \quad \text{versus} \quad \begin{bmatrix} u \\ v \end{bmatrix} = \begin{bmatrix} \lambda \\ \phi \end{bmatrix} = \begin{bmatrix} \arctan y x^{-1} \\ \arctan \frac{z}{\sqrt{x^2 + y^2}} \end{bmatrix}. \tag{9.3}$$

9-12 The metric tensor of the ellipsoid-of-revolution, the first differential form

First, let us here compute the first differential form of the surface of type *ellipsoid-of-revolution* as follows.

$$G_{KL} = \langle \boldsymbol{G}_K \,|\, \boldsymbol{G}_L \rangle = \sum_{J=1}^{3} \frac{\partial X^J}{\partial U^K} \frac{\partial X^J}{\partial U^L} \,, \tag{9.4}$$

$$\boldsymbol{G}_1 = \boldsymbol{G}_\Lambda := \frac{\partial \boldsymbol{X}}{\partial \Lambda} = \frac{A_1 \cos \Phi}{(1 - E^2 \sin^2 \Phi)^{1/2}} \left(-\sin \Lambda \, \boldsymbol{E}_1 + \cos \Lambda \, \boldsymbol{E}_2 \right) \,, \tag{9.5}$$

$$\boldsymbol{G}_2 = \boldsymbol{G}_\Phi := \frac{\partial \boldsymbol{X}}{\partial \Phi} = -\frac{A_1(1 - E^2)}{(1 - E^2 \sin^2 \Phi)^{3/2}} \left(\sin \Phi \cos \Lambda \, \boldsymbol{E}_1 + \sin \Phi \sin \Lambda \, \boldsymbol{E}_2 - \cos \Phi \, \boldsymbol{E}_3 \right) \,. \tag{9.6}$$

A_1 denotes the semi-major axis of the ellipsoid-of-revolution, A_2 denotes the semi-minor axis of the ellipsoid-of-revolution, and $E = \sqrt{A_1^2 - A_2^2}/A_1 = \sqrt{1 - A_2^2/A_1^2}$ defines the first *numerical eccentricity*. The basis vectors finally lead to the elements of the *metric tensor*.

$$E(\text{Gauss}) := \langle \boldsymbol{G}_\Lambda \,|\, \boldsymbol{G}_\Lambda \rangle := G_{\Lambda\Lambda} = G_{11} = \frac{A_1^2 \cos^2 \Phi}{1 - E^2 \sin^2 \Phi} \,,$$

$$F(\text{Gauss}) := \langle \boldsymbol{G}_\Lambda \,|\, \boldsymbol{G}_\Phi \rangle := G_{\Lambda\Phi} = G_{12} = G_{21} = G_{\Phi\Lambda} = 0 \,, \tag{9.7}$$

$$G(\text{Gauss}) := \langle \boldsymbol{G}_\Phi \,|\, \boldsymbol{G}_\Phi \rangle := G_{\Phi\Phi} = G_{22} = \frac{A_1^2 (1 - E^2)^2}{(1 - E^2 \sin^2 \Phi)^3} \,.$$

9-13 The curvature tensor of the ellipsoid-of-revolution, the second differential form

Second, let us here compute the second differential form of the surface of type *ellipsoid-of-revolution* as follows.

$$H_{KL} = \langle \boldsymbol{G}_3 \,|\, \partial^2 \boldsymbol{X} / \partial U^K \partial U^L \rangle = \frac{1}{\sqrt{\det [G_{KL}]}} [\boldsymbol{G}_{K,L}, \boldsymbol{G}_1, \boldsymbol{G}_2] \,. \tag{9.8}$$

The second differential form is related to the *determinantal form* of the ellipsoid-of-revolution. We shall compute the *surface normal vector* $\boldsymbol{G}_3$ and the *surface tangent vectors* $\boldsymbol{G}_1$ and $\boldsymbol{G}_2$. In Box 9.1, the various steps are collected. Subsequently, we shall collect the coordinates of the matrix H_{KL} which are derived from the second derivatives, see Box 9.2. In summary, we present the coordinates of the curvature tensor of the ellipsoid-of-revolution in (9.9).

$$L(\text{Gauss}) := \langle \boldsymbol{G}_3 \,|\, \partial \boldsymbol{G}_1 / \partial U^1 \rangle := H_{\Lambda\Lambda} = H_{11} = -\frac{A_1 \cos^2 \Phi}{(1 - E^2 \sin^2 \Phi)^{1/2}} \,,$$

$$M(\text{Gauss}) := \langle \boldsymbol{G}_3 \,|\, \partial \boldsymbol{G}_1 / \partial U^2 \rangle := H_{\Lambda\Phi} = H_{12} = H_{21} = H_{\Phi\Lambda} = 0 \,, \tag{9.9}$$

$$N(\text{Gauss}) := \langle \boldsymbol{G}_3 \,|\, \partial \boldsymbol{G}_2 / \partial U^2 \rangle := H_{\Phi\Phi} = H_{22} = -\frac{A_1(1 - E^2)}{(1 - E^2 \sin^2 \Phi)^{3/2}} \,.$$

Box 9.1 (The surface normal vector $\boldsymbol{G}_3$, the surface tangent vectors $\boldsymbol{G}_1$ and $\boldsymbol{G}_2$).

$$\boldsymbol{G}_3 = \frac{\boldsymbol{G}_1 \times \boldsymbol{G}_2}{\|\boldsymbol{G}_1 \times \boldsymbol{G}_2\|} = \cos\Phi\cos\Lambda\boldsymbol{E}_1 + \cos\Phi\sin\Lambda\boldsymbol{E}_2 + \sin\Phi\boldsymbol{E}_3 , \qquad (9.10)$$

$$\frac{\partial^2 \boldsymbol{X}}{\partial U^K \partial U^L} = \frac{\partial}{\partial U^L} G_K(U^1, U^2) , \quad \sqrt{\det[G_{KL}]} = \frac{A_1^2(1 - E^2)\cos\Phi}{(1 - E^2\sin^2\Phi)^2} , \qquad (9.11)$$

$$\frac{\partial \boldsymbol{G}_1}{\partial U^1} = \sum_{J=1}^{3} \frac{\partial^2 X^J}{\partial \Lambda^2} \boldsymbol{E}_J = -\frac{A_1\cos\Phi}{(1 - E^2\sin^2\Phi)^{1/2}}\big(\cos\Lambda\,\boldsymbol{E}_1 + \sin\Lambda\,\boldsymbol{E}_2\big) ,$$

$$\frac{\partial \boldsymbol{G}_1}{\partial U^2} = \sum_{J=1}^{3} \frac{\partial^2 X^J}{\partial \Lambda \partial \Phi} \boldsymbol{E}_J = \frac{A_1(1 - E^2)}{(1 - E^2\sin^2\Phi)^{3/2}}\big(\sin\Phi\sin\Lambda\,\boldsymbol{E}_1 - \sin\Phi\cos\Lambda\,\boldsymbol{E}_2\big) \qquad (9.12)$$

versus

$$\frac{\partial \boldsymbol{G}_2}{\partial U^1} = \sum_{J=1}^{3} \frac{\partial^2 X^J}{\partial \Phi \partial \Lambda} \boldsymbol{E}_J = \sum_{J=1}^{3} \frac{\partial^2 X^J}{\partial \Lambda \partial \Phi} \boldsymbol{E}_J , \quad \frac{\partial \boldsymbol{G}_2}{\partial U^2} = \sum_{J=1}^{3} \frac{\partial^2 X^J}{\partial \Phi^2} \boldsymbol{E}_J = \frac{A_1(1 - E^2)}{(1 - E^2\sin^2\Phi)^{5/2}} \times$$

$$\times\Big[-\cos\Phi\cos\Lambda(1 + 2E^2\sin^2\Phi)\boldsymbol{E}_1 - \cos\Phi\sin\Lambda(1 + 2E^2\sin^2\Phi)\boldsymbol{E}_2 - $$

$$- \sin\Phi(1 + 2E^2\sin^2\Phi - 3E^2)\boldsymbol{E}_3\Big] \qquad (9.13)$$

$$= -\frac{A_1(1 - E^2)(1 + 2E^2\sin^2\Phi)}{(1 - E^2\sin^2\Phi)^{5/2}}\Big[\cos\Phi\cos\Lambda\,\boldsymbol{E}_1 + \cos\Phi\sin\Lambda\,\boldsymbol{E}_2 + \Big(1 - \frac{3E^2}{1 + 2E^2\sin^2\Phi}\Big)\sin\Phi\,\boldsymbol{E}_3\Big].$$

Box 9.2 (The matrix H_{KL}).

$$H_{11} = \big\langle\boldsymbol{G}_3\,\big|\,\partial\boldsymbol{G}_1/\partial U^1\big\rangle = \big\langle\boldsymbol{G}_3\,\big|\,\partial\boldsymbol{G}_\Lambda/\partial\Lambda\big\rangle = -\frac{A_1\cos^2\Phi}{(1 - E^2\sin^2\Phi)^{1/2}} ,$$

$$H_{12} = \big\langle\boldsymbol{G}_3\,\big|\,\partial\boldsymbol{G}_1/\partial U^2\big\rangle = \big\langle\boldsymbol{G}_3\,\big|\,\partial\boldsymbol{G}_\Lambda/\partial\Phi\big\rangle =$$

$$= \frac{A_1(1 - E^2)}{(1 - E^2\sin^2\Phi)^{3/2}}\big(\sin\Phi\cos\Phi\sin\Lambda\cos\Lambda - \sin\Phi\cos\Phi\sin\Lambda\cos\Lambda\big) = 0 ,$$

$$H_{22} = \big\langle\boldsymbol{G}_3\,\big|\,\partial\boldsymbol{G}_2/\partial U^2\big\rangle = \big\langle\boldsymbol{G}_3\,\big|\,\partial\boldsymbol{G}_\Phi/\partial\Phi\big\rangle =$$

$$= -\frac{A_1(1 - E^2)(1 + 2E^2\sin^2\Phi)}{(1 - E^2\sin^2\Phi)^{5/2}}\Big(\cos^2\Phi + \sin^2\Phi\frac{1 + 2E^2\sin^2\Phi - 3E^2}{1 + 2E^2\sin^2\Phi}\Big) = \qquad (9.14)$$

$$= -\frac{A_1(1 - E^2)}{(1 - E^2\sin^2\Phi)^{5/2}}\big(\cos^2\Phi + 2E^2\sin^2\Phi\cos^2\Phi + \sin^2\Phi + 2E^2\sin^4\Phi - 3E^2\sin^2\Phi\big)$$

$$= -\frac{A_1(1 - E^2)}{(1 - E^2\sin^2\Phi)^{5/2}}\big[1 - E^2\big(3\sin^2\Phi - 2\sin^2\Phi\cos^2\Phi - 2\sin^4\Phi\big)\big] =$$

$$= -\frac{A_1(1 - E^2)}{(1 - E^2\sin^2\Phi)^{5/2}}\big[1 - E^2\big(3\sin^2\Phi - 2\sin^2\Phi\cos^2\Phi - 2\sin^2\Phi + 2\sin^2\Phi\cos^2\Phi\big)\big] =$$

$$= -\frac{A_1(1 - E^2)}{(1 - E^2\sin^2\Phi)^{5/2}}(1 - E^2\sin^2\Phi) = -\frac{A_1(1 - E^2)}{(1 - E^2\sin^2\Phi)^{3/2}} .$$

9-14 The metric tensor of the sphere, the first differential form

Third, we compute the first differential form of the surface of type *sphere*. In (9.15) and (9.16), r is the radius of the sphere. The basis vectors finally lead to the elements of the *spherical metric tensor*.

$$g_{kl} = \langle \boldsymbol{g}_k \,|\, \boldsymbol{g}_l \rangle = \sum_{j=1}^{3} \frac{\partial x^j}{\partial u^k} \frac{\partial x^j}{\partial u^l} \; ,$$

$$\boldsymbol{g}_1 = \boldsymbol{g}_\lambda := \frac{\partial \boldsymbol{x}}{\partial \lambda} = r \cos \phi \left(-\sin \lambda \, \boldsymbol{e}_1 + \cos \lambda \, \boldsymbol{e}_2 \right) \; ,$$

$$\boldsymbol{g}_2 = \boldsymbol{g}_\phi := \frac{\partial \boldsymbol{x}}{\partial \phi} = -r \left(\sin \phi \cos \lambda \, \boldsymbol{e}_1 + \sin \phi \sin \lambda \, \boldsymbol{e}_2 - \cos \phi \, \boldsymbol{e}_3 \right) \; ,$$

(9.15)

$$e(\text{Gauss}) := \langle \boldsymbol{g}_\lambda \,|\, \boldsymbol{g}_\lambda \rangle := g_{\lambda\lambda} = g_{11} = r^2 \cos^2 \phi \; ,$$

$$f(\text{Gauss}) := \langle \boldsymbol{g}_\lambda \,|\, \boldsymbol{g}_\phi \rangle := g_{\lambda\phi} = g_{12} = g_{21} = g_{\phi\lambda} = 0 \; ,$$

(9.16)

$$g(\text{Gauss}) := \langle \boldsymbol{g}_\phi \,|\, \boldsymbol{g}_\phi \rangle := g_{\phi\phi} = g_{22} = r^2 \; .$$

9-15 The curvature tensor of the sphere, the second differential form

Fourth, we compute the second differential form of the surface of type *sphere*. The second differential form is related to the *determinantal form* of the sphere. We compute first the surface normal vector and second the surface derivatives of the tangent vectors. In summary, we refer to the coordinates of the curvature tensor of the sphere.

$$h_{kl} = \langle \boldsymbol{g}_3 \,|\, \partial^2 \boldsymbol{x}/\partial u^k \partial u^l \rangle = \langle \boldsymbol{g}_3 \,|\, \partial^2 \boldsymbol{g}_k/\partial u^l \rangle = \frac{1}{\sqrt{\det [g_{kl}]}} \left[\boldsymbol{g}_{K,L}, \boldsymbol{g}_1, \boldsymbol{g}_2 \right] \; ,$$

$$\boldsymbol{g}_3 = \cos \phi \cos \lambda \, \boldsymbol{e}_1 + \cos \phi \sin \lambda \, \boldsymbol{e}_2 + \sin \phi \, \boldsymbol{e}_3 \; ,$$

(9.17)

$$\boldsymbol{g}_3 = \frac{\boldsymbol{g}_1 \times \boldsymbol{g}_2}{\| \boldsymbol{g}_1 \times \boldsymbol{g}_2 \|} \; , \quad \sqrt{\det [g_{kl}]} = r^2 \cos \phi \; ,$$

$$l(\text{Gauss}) := \langle \boldsymbol{g}_3 \,|\, \partial \boldsymbol{g}_1/\partial u^1 \rangle := h_{\lambda\lambda} = h_{11} = -r \cos^2 \phi \; ,$$

$$m(\text{Gauss}) := \langle \boldsymbol{g}_3 \,|\, \partial \boldsymbol{g}_1/\partial u^2 \rangle := h_{\lambda\phi} = h_{12} = h_{21} = h_{\phi\lambda} = 0 \; ,$$

(9.18)

$$n(\text{Gauss}) := \langle \boldsymbol{g}_3 \,|\, \partial \boldsymbol{g}_2/\partial u^2 \rangle := h_{\phi\phi} = h_{22} = -r \; .$$

Based upon the general mapping equations $\lambda = \lambda_0 + a(\Lambda - \Lambda_0)$ and $\phi = f(\Phi)$, let us here compute the deformation tensor of the first kind and the deformation tensor of the second kind.

9-16 Deformation of the first kind

We first consider the deformation of the first kind: the deformation tensor of the first kind is based upon the first fundamental form of differential geometry.

$$I := ds^2 = \sum_{k,l=1}^{2} g_{kl} du^k du^l = \sum_{K,L=1}^{2} c_{KL} dU^K dU^L , \quad c_{KL} := \sum_{k,l=1}^{2} g_{kl} \frac{\partial u^k}{\partial U^K} \frac{\partial u^l}{\partial U^L} . \tag{9.19}$$

The first invariant differential form $I := ds^2 = \sum_{K,L=1}^{2} c_{KL} dU^K dU^L$ is to be computed next. In Box 9.3, the various steps of computing the matrix c_{KL} are outlined.

Box 9.3 (The matrix c_{KL}).

$$\frac{\partial u^1}{\partial U^1} = \frac{\partial \lambda}{\partial \Lambda} = a \quad \text{and} \quad \frac{\partial u^1}{\partial U^2} = \frac{\partial \lambda}{\partial \Phi} = 0 ,$$

$$\frac{\partial u^2}{\partial U^1} = \frac{\partial \phi}{\partial \Lambda} = 0 \quad \text{and} \quad \frac{\partial u^2}{\partial U^2} = \frac{\partial \phi}{\partial \Phi} = f'(\Phi) \tag{9.20}$$

$$\Leftrightarrow$$

$$\frac{\partial u^k}{\partial U^K} = \begin{bmatrix} \partial\lambda/\partial\Lambda & \partial\lambda/\partial\Phi \\ \partial\phi/\partial\Lambda & \partial\phi/\partial\Phi \end{bmatrix} = \begin{bmatrix} a & 0 \\ 0 & f'(\Phi) \end{bmatrix} . \tag{9.21}$$

If $g_{12} = 0$, then

$$c_{11} = c_{\Lambda\Lambda} = \sum_{k,l=1}^{2} g_{kl} \frac{\partial u^k}{\partial \Lambda} \frac{\partial u^l}{\partial \Lambda} = g_{11} \left(\frac{\partial\lambda}{\partial\Lambda}\right)^2 + g_{22} \left(\frac{\partial\phi}{\partial\Lambda}\right)^2 ,$$

$$c_{12} = c_{\Lambda\Phi} = \sum_{k,l=1}^{2} g_{kl} \frac{\partial u^k}{\partial \Lambda} \frac{\partial u^l}{\partial \Phi} = g_{11} \left(\frac{\partial\lambda}{\partial\Lambda}\right)\left(\frac{\partial\lambda}{\partial\Phi}\right) + g_{22} \left(\frac{\partial\phi}{\partial\Lambda}\right)\left(\frac{\partial\phi}{\partial\Phi}\right) , \tag{9.22}$$

$$c_{22} = c_{\Phi\Phi} = \sum_{k,l=1}^{2} g_{kl} \frac{\partial u^k}{\partial \Phi} \frac{\partial u^l}{\partial \Phi} = g_{11} \left(\frac{\partial\lambda}{\partial\Phi}\right)^2 + g_{22} \left(\frac{\partial\phi}{\partial\Phi}\right)^2$$

$$\Rightarrow$$

$$c_{11} = c_{\Lambda\Lambda} = a^2 r^2 \cos^2\phi , \quad c_{12} = c_{\Lambda\Phi} = c_{21} = c_{\Phi\Lambda} = 0 , \quad c_{22} = c_{\Phi\Phi} = r^2 f'^2(\Phi) . \tag{9.23}$$

The matrix elements c_{KL} of the first Cauchy–Green deformation tensor are summarized according to (9.24). The principal stretches of the first kind amount to (9.25).

$$c_{KL} = \begin{bmatrix} a^2 r^2 \cos^2\phi & 0 \\ 0 & r^2 f'^2(\Phi) \end{bmatrix} , \tag{9.24}$$

$$\Lambda_1 = \sqrt{c_{11}/G_{11}} = \sqrt{\frac{a^2 r^2 \cos^2\phi}{A_1^2 \cos^2\Phi}(1 - E^2 \sin^2\Phi)} = \frac{ar\cos\phi}{N\cos\Phi} ,$$

$$\Lambda_2 = \sqrt{c_{22}/G_{22}} = \sqrt{\frac{r^2 f'^2(\Phi)}{A_1^2(1-E^2)^2}(1 - E^2 \sin^2\Phi)^3} = \frac{rf'(\Phi)}{M} = \frac{r}{M}\frac{d\phi}{d\Phi} . \tag{9.25}$$

The curvature tensors of the ellipsoid-of-revolution and the sphere, namely the Gauss curvature scalar and the trace as the alternative curvature scalar, are presented in Box 9.4. Note that N and M are the radii of principal type of the ellipsoid-of-revolution and that r is the curvature radius of the sphere.

Box 9.4 (Curvature tensors, Gauss's curvature tensors).

<table>
<tr><td align="center">Curvature tensor
(ellipsoid-of-revolution):</td><td align="center">Curvature tensor
(sphere):</td><td></td></tr>
</table>

$$\text{Grad } \boldsymbol{G}_3 = \qquad\qquad \text{grad } \boldsymbol{g}_3 =$$

$$= -\mathsf{H}\mathsf{G}^{-1} \begin{bmatrix} \boldsymbol{G}_1 \\ \boldsymbol{G}_2 \end{bmatrix} = \mathsf{K} \begin{bmatrix} \boldsymbol{G}_1 \\ \boldsymbol{G}_2 \end{bmatrix} . \qquad = -\mathsf{h}\mathsf{g}^{-1} \begin{bmatrix} \boldsymbol{g}_1 \\ \boldsymbol{g}_2 \end{bmatrix} = \mathsf{k} \begin{bmatrix} \boldsymbol{g}_1 \\ \boldsymbol{g}_2 \end{bmatrix} . \qquad (9.26)$$

Gauss's curvature tensor (ellipsoid-of-revolution):

Gauss's curvature tensor (sphere):

$$\mathsf{K} := -\mathsf{H}\mathsf{G}^{-1} , \qquad\qquad \mathsf{k} := \mathsf{h}\mathsf{g}^{-1} ,$$

$$\mathsf{K} := -\mathsf{H}\mathsf{G}^{-1} = \begin{bmatrix} 1/N & 0 \\ 0 & 1/M \end{bmatrix} , \qquad \mathsf{k} := -\mathsf{h}\mathsf{g}^{-1} = \begin{bmatrix} 1/r & 0 \\ 0 & 1/r \end{bmatrix} .$$

$$(9.27)$$

$$N := \frac{A_1}{(1 - E^2 \sin^2 \Phi)^{1/2}} ,$$

$$M := \frac{A_1(1 - E^2)}{(1 - E^2 \sin^2 \Phi)^{3/2}} .$$

Eigenvalues of the curvature tensor:

Eigenvalues of the curvature tensor:

$$K_1 := \frac{1}{N} , \quad K_2 := \frac{1}{M} , \qquad\qquad \kappa_1 = \kappa_2 = \frac{1}{r} ,$$

$$\text{Grad } \boldsymbol{G}_3 = \qquad\qquad \text{Grad } \boldsymbol{g}_3 = \qquad\qquad (9.28)$$

$$= \mathsf{K} \begin{bmatrix} \boldsymbol{G}_1 \\ \boldsymbol{G}_2 \end{bmatrix} = \begin{bmatrix} K_1 \boldsymbol{G}_1 \\ K_2 \boldsymbol{G}_2 \end{bmatrix} . \qquad = \mathsf{k} \begin{bmatrix} \boldsymbol{g}_1 \\ \boldsymbol{g}_2 \end{bmatrix} = \begin{bmatrix} \kappa_1 \boldsymbol{g}_1 \\ \kappa_2 \boldsymbol{g}_2 \end{bmatrix} .$$

Tangent space:

Tangent space:

$$\boldsymbol{G}_1 := \frac{\partial \boldsymbol{X}}{\partial \Lambda} , \quad \boldsymbol{G}_2 := \frac{\partial \boldsymbol{X}}{\partial \Phi} , \qquad\qquad \boldsymbol{g}_1 := \frac{\partial \boldsymbol{x}}{\partial \lambda} , \quad \boldsymbol{g}_2 := \frac{\partial \boldsymbol{x}}{\partial \phi} . \qquad (9.29)$$

9-17 Deformation of the second kind

We then consider the deformation of the second kind: the deformation tensor of the second kind is based upon the second fundamental form of differential geometry. We shall compute its representation. First, in the coordinate system $\{u, v\} = \{u^1, u^2\}$. Second, in the transformed coordinate system $u^k \rightarrow U^K = U^K(u^k)$. Or from the spherical coordinate system to the ellipsoidal coordinate system.

$$\mathrm{II} := \sum_{k,l=1}^{2} h_{kl} du^k du^l = \sum_{K,L=1}^{2} d_{KL} dU^K dU^L , \quad d_{KL} := \sum_{k,l=1}^{2} h_{kl} \frac{\partial u^k}{\partial U^K} \frac{\partial u^l}{\partial U^L} . \tag{9.30}$$

Box 9.5 (The matrix d_{KL}).

$$\frac{\partial u^1}{\partial U^1} = \frac{\partial \lambda}{\partial \Lambda} = a \quad \text{and} \quad \frac{\partial u^1}{\partial U^2} = \frac{\partial \lambda}{\partial \Phi} = 0 ,$$

$$\frac{\partial u^2}{\partial U^1} = \frac{\partial \phi}{\partial \Lambda} = 0 \quad \text{and} \quad \frac{\partial u^2}{\partial U^2} = \frac{\partial \phi}{\partial \Phi} = f'(\Phi) \tag{9.31}$$

$$\Leftrightarrow$$

$$\frac{\partial u^k}{\partial U^K} = \begin{bmatrix} \partial\lambda/\partial\Lambda & \partial\lambda/\partial\Phi \\ \partial\phi/\partial\Lambda & \partial\phi/\partial\Phi \end{bmatrix} = \begin{bmatrix} a & 0 \\ 0 & f'(\Phi) \end{bmatrix} . \tag{9.32}$$

If $h_{12} = 0$, then

$$d_{11} = d_{\Lambda\Lambda} = \sum_{k,l=1}^{2} h_{kl} \frac{\partial u^k}{\partial\Lambda} \frac{\partial u^l}{\partial\Lambda} = h_{11} \left(\frac{\partial\lambda}{\partial\Lambda}\right)^2 + h_{22} \left(\frac{\partial\phi}{\partial\Lambda}\right)^2 ,$$

$$d_{12} = d_{\Lambda\Phi} = \sum_{k,l=1}^{2} h_{kl} \frac{\partial u^k}{\partial\Lambda} \frac{\partial u^l}{\partial\Phi} = h_{11} \left(\frac{\partial\lambda}{\partial\Lambda}\right)\left(\frac{\partial\lambda}{\partial\Phi}\right) + h_{22} \left(\frac{\partial\phi}{\partial\Lambda}\right)\left(\frac{\partial\phi}{\partial\Phi}\right) , \tag{9.33}$$

$$d_{22} = d_{\Phi\Phi} = \sum_{k,l=1}^{2} h_{kl} \frac{\partial u^k}{\partial\Phi} \frac{\partial u^l}{\partial\Phi} = h_{11} \left(\frac{\partial\lambda}{\partial\Phi}\right)^2 + h_{22} \left(\frac{\partial\phi}{\partial\Phi}\right)^2$$

$$\Rightarrow$$

$$d_{11} = d_{\Lambda\Lambda} = -a^2 r \cos^2\phi , \quad d_{12} = d_{\Lambda\Phi} = 0 , \quad d_{22} = d_{\Phi\Phi} = -r f'^2(\Phi) . \tag{9.34}$$

In summary, let us here present the diverse coordinates of the *second deformation tensor*, namely its eigenvalues.

$$d_{KL} = \begin{bmatrix} -a^2 r \cos^2\phi & 0 \\ 0 & -r f'^2(\Phi) \end{bmatrix} , \tag{9.35}$$

$$d\Lambda_1 = \sqrt{d_{11}/H_{11}} = \sqrt{\frac{a^2 r \cos^2\phi}{A_1 \cos^2\Phi} (1 - E^2 \sin^2\Phi)^{1/2}} ,$$

$$d\Lambda_2 = \sqrt{d_{22}/H_{22}} = \sqrt{\frac{r f'^2(\Phi)}{A_1(1 - E^2)} (1 - E^2 \sin^2\Phi)^{3/2}} . \tag{9.36}$$

9-2 The conformal mappings "ellipsoid-of-revolution to plane"

The conformal mappings from the ellipsoid-of-revolution to the plane: the conditions of conformality, the standard integrals, spherical isometric latitude, ellipsoidal isometric latitude.

First, we postulate the condition of conformality $\Lambda_1 = \Lambda_2$ and subsequently we take advantage of the standard integrals that are collected in Box 9.6.

$$\Lambda_1 = \Lambda_2$$
$$\Leftrightarrow$$
$$\frac{ar\cos\phi}{A_1\cos\Phi}(1 - E^2\sin^2\Phi)^{1/2} = \frac{r}{A_1(1 - E^2)}\frac{\mathrm{d}\phi}{\mathrm{d}\Phi}(1 - E^2\sin^2\Phi)^{3/2} \tag{9.37}$$
$$\Leftrightarrow$$

$$\frac{\mathrm{d}\phi}{\cos\phi} = \frac{1 - E^2}{1 - E^2\sin^2\Phi}\frac{a}{\cos\Phi}\mathrm{d}\Phi . \tag{9.38}$$

We here note in passing that via the first standard integral, we introduce *spherical isometric latitude*. By integration-by-parts, we split the second standard integral into three parts, namely by introducing *ellipsoidal isometric latitude*.

Box 9.6 (The standard integrals).

First standard integral:

$$\int\frac{\mathrm{d}\phi}{\cos\phi} = I_{\text{first}} = \ln\tan\left(\frac{\pi}{4} + \frac{\phi}{2}\right) := q \tag{9.39}$$

("spherical isometric latitude").

Second standard integral:

$$\int\frac{1 - E^2}{1 - E^2\sin^2\Phi}\frac{\mathrm{d}\Phi}{\cos\Phi} = I_{\text{second}} . \tag{9.40}$$

$$\frac{1 - E^2}{1 - E^2\sin^2\Phi}\frac{1}{\cos\Phi} = \frac{1}{\cos\Phi} - \frac{E}{2}\left(\frac{E\cos\Phi}{1 + E\sin\Phi} - \frac{E\cos\Phi}{1 - E\sin\Phi}\right) , \tag{9.41}$$

$$\int\frac{1 - E^2}{1 - E^2\sin^2\Phi}\frac{\mathrm{d}\Phi}{\cos\Phi} =$$
$$= \ln\tan\left(\frac{\pi}{4} + \frac{\Phi}{2}\right) - \frac{E}{2}\ln\frac{1 + E\sin\Phi}{1 - E\sin\Phi} = \ln\left[\tan\left(\frac{\pi}{4} + \frac{\Phi}{2}\right)\left(\frac{1 - E\sin\Phi}{1 + E\sin\Phi}\right)^{E/2}\right] := Q \tag{9.42}$$

("ellipsoidal isometric latitude").

The combination of the first and the second standard integral leads us to the celebrated relation in terms of the integration constants c and k, namely

$$q = a(Q + k) , \quad k = \ln c , \quad c = \exp k ;$$
$$\ln\tan\left(\frac{\pi}{4} + \frac{\phi}{2}\right) - a\ln\tan\left(\frac{\pi}{4} + \frac{\Phi}{2}\right) - \frac{aE}{2}\ln\frac{1 + E\sin\Phi}{1 - E\sin\Phi} + a\ln c . \tag{9.43}$$

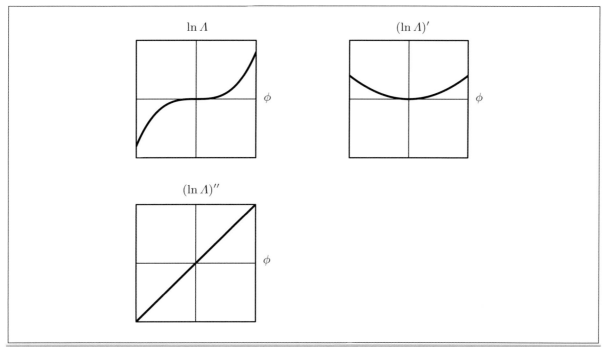

Fig. 9.1. The three postulates of the Gauss mapping "ellipsoid-of-revolution to sphere".

Let us fix the integration constants a, c, and r. In the so-called "fundamental point" $P_0(\Lambda_0, \Phi_0)$, we assume $r := N(\Phi_0)$, where the curvature form N_0 is defined by (9.44) ("first proposal of C. F. Gauss"). Around the "fundamental point" $P_0(\Lambda_0, \Phi_0)$, we assume the Taylor expansion that is defined by (9.45) ("second proposal of C. F. Gauss").

$$r = N_0 = \frac{A_1}{\sqrt{1 - E^2 \sin^2 \Phi}} \, , \tag{9.44}$$

$$\ln \Lambda = \ln \Lambda_0 + \left(\frac{\mathrm{d} \ln \Lambda}{\mathrm{d}\phi}\right)_{\phi_0} (\phi - \phi_0) + \frac{1}{2} \left(\frac{\mathrm{d}^2 \ln \Lambda}{\mathrm{d}\phi^2}\right)_{\phi_0} (\phi - \phi_0)^2 + \cdots . \tag{9.45}$$

The following three postulates specify the above relations. Note that we can summarize the three postulates in such a way that we postulate a *horizontal turning tangent* according to Fig. 9.1.

Postulate (first postulate).

$$\Lambda_0 = 1 \, . \tag{9.46}$$

In the so-called "fundamental point" $P_0(\Lambda_0, \Phi_0)$, we assume an equal lateral mapping.

$$\ln \Lambda_0 = 0 \, . \tag{9.47}$$

End of Postulate.

Postulate (second postulate).

$$(\ln \Lambda)' (\phi_0) = 0 \, . \tag{9.48}$$

End of Postulate.

Postulate (third postulate).

$$(\ln \Lambda)'' (\phi_0) = 0 \, . \tag{9.49}$$

End of Postulate.

Let us present the summary of the Gauss mapping "ellipsoid-of-revolution to sphere" based upon the stretch equation (9.50) in form of Lemma 9.1.

$$\frac{ar\cos\phi}{\frac{A_1\cos\Phi}{(1-E^2\sin^2\Phi)^{1/2}}} = \frac{ar\cos\phi}{N(\Phi)\cos\Phi} \,. \tag{9.50}$$

Lemma 9.1 (Gauss mapping "ellipsoid-of-revolution to sphere").

The *mapping equations* from the ellipsoid-of-revolution adequately parameterized by $\{\Lambda_e, \Phi_e\}$ to the sphere adequately parameterized by $\{\lambda_s, \phi_s\}$ of type *conformal* read

$$\lambda_s = \lambda_0^s + a(\Lambda_e - \Lambda_0^e) \,, \tag{9.51}$$

$$\tan\left(\frac{\pi}{4} + \frac{\phi_s}{2}\right) = c^a \left[\tan\left(\frac{\pi}{4} + \frac{\Phi_e}{2}\right)\right]^a \left(\frac{1 - E\sin\Phi_e}{1 + E\sin\Phi_e}\right)^{aE/2} \,, \tag{9.52}$$

and

$$a = \cos\Phi_0 \sqrt{\frac{N_0}{M_0} + \tan^2\Phi_0} \,,$$

$$c = \frac{\left[\tan\left(\frac{\pi}{4} + \frac{\phi_0}{2}\right)\right]^{1/a}}{\tan\left(\frac{\pi}{4} + \frac{\Phi_0}{2}\right)\left(\frac{1 - E\sin\Phi_0}{1 + E\sin\Phi_0}\right)^{E/2}} =$$

$$= \frac{\exp q_0/a}{\exp Q_0} \,, \tag{9.53}$$

$$\ln c = \frac{1}{a}q_0 - Q_0 \,,$$

$$\tan\phi_0 = \sqrt{\frac{N_0}{M_0}}\tan\Phi_0 \,.$$

Relative to the equidistant mapping of the "fundamental point", $P_0(\Lambda_0, \Phi_0) \to p_0 = p(\lambda_0, \phi_0)$, there hold the conditions

$$\Lambda_0 = 1 \,, \quad \Lambda_0' = 0 \,, \quad \Lambda_0'' = 0 \,. \tag{9.54}$$

The mean spherical radius reads

$$r = \sqrt{M_0 N_0} \,. \tag{9.55}$$

End of Lemma.

The following chains of calculations, on the one hand, supply us with proofs of the above relations, and on the other hand, supply us with additional relations needed to understand the above relations.

Proof (first postulate).

First postulate ($\Lambda_0 = 1$):

$$\Lambda_0 = \frac{ar \cos \phi_0}{N_0 \cos \Phi_0} \Rightarrow r = \frac{N_0 \cos \Phi_0}{a \cos \phi_0} . \tag{9.56}$$

End of Proof.

Proof (second postulate).

Second postulate ($\Lambda_0' = 0 \Leftrightarrow (\ln \Lambda)_0' = \Lambda_0'/\Lambda_0$):

$$\ln \Lambda = \ln ar + \ln \cos \phi - \ln[N(\Phi) \cos \Phi] ,$$

$$\frac{\mathrm{d} \ln \Lambda}{\mathrm{d}\phi} = -\frac{\sin \phi}{\cos \phi} - \frac{N'(\Phi) \cos \Phi - N(\Phi) \sin \Phi}{N(\Phi) \cos \Phi} \frac{\mathrm{d}\Phi}{\mathrm{d}\phi} ,$$

$$N(\Phi) = \frac{A_1}{(1 - E^2 \sin^2 \Phi)^{1/2}} , \quad N'(\Phi) = \frac{A_1 E^2 \sin \Phi \cos \Phi}{(1 - E^2 \sin^2 \Phi)^{3/2}} ,$$

$$\frac{\mathrm{d}\Phi}{\mathrm{d}\phi} = \frac{1 - E^2 \sin^2 \Phi}{1 - E^2} \frac{\cos \Phi}{a \cos \phi} = \frac{N(\Phi)}{M(\Phi)} \frac{\cos \Phi}{a \cos \phi} \tag{9.57}$$

$$\Rightarrow$$

$$\frac{\mathrm{d} \ln \Lambda}{\mathrm{d}\phi} = -\frac{\sin \phi}{\cos \phi} + \frac{\sin \Phi}{a \cos \phi} = -\tan \phi + \frac{\sin \Phi}{a \cos \phi}$$

$$\Rightarrow$$

$$(\ln \Lambda)'(\phi_0, \Phi_0) = 0 \Rightarrow a \sin \phi_0 = \sin \Phi_0 .$$

End of Proof.

Proof (third postulate).

Third postulate ($\Lambda_0'' = 0 \Leftrightarrow (\ln \Lambda)_0'' = 0$):

$$\frac{\mathrm{d}^2 \ln \Lambda}{\mathrm{d}\phi^2} = -\frac{1}{\cos^2 \phi} + \frac{a \cos \phi \cos \Phi \frac{\mathrm{d}\Phi}{\mathrm{d}\phi} + a \sin \phi \sin \Phi}{a^2 \cos^2 \phi} =$$

$$= -\frac{1}{\cos^2 \phi} \left[1 - \frac{1}{a} \sin \phi \sin \Phi - \frac{1}{a} \cos \phi \cos \Phi \frac{\mathrm{d}\Phi}{\mathrm{d}\phi} \right] ,$$

$$\frac{\mathrm{d}^2 \ln \Lambda}{\mathrm{d}\phi^2}(\phi_0, \Phi_0) = 0 \tag{9.58}$$

$$\Rightarrow$$

$$a \cos \phi_0 = \sqrt{\frac{1 - E^2 \sin^2 \Phi_0}{1 - E^2}} \cos \Phi_0 .$$

End of Proof.

Proof (lemma relations).

Intermediate results:

$$r = \frac{N_0 \cos \Phi_0}{a \cos \phi_0} \,, \tag{9.59}$$

$$a \sin \phi_0 = \sin \Phi_0 \,, \tag{9.60}$$

$$a \cos \phi_0 = a \sqrt{1 - \sin^2 \phi_0} = \sqrt{\frac{1 - E^2 \sin^2 \Phi_0}{1 - E^2}} \cos \Phi_0 = \sqrt{\frac{N_0}{M_0}} \cos \Phi_0 \,. \tag{9.61}$$

Action item: combine (9.59) and (9.61) and find

$$r = \frac{A_1 \cos \Phi_0}{a \cos \phi_0} \frac{1}{\sqrt{1 - E^2 \sin^2 \Phi_0}} = \frac{A_1 \sqrt{1 - E^2}}{1 - E^2 \sin^2 \Phi_0} = \sqrt{M_0 N_0} \,. \tag{9.62}$$

Action item: combine (9.60) and (9.61) and find

$$\tan \phi_0 = \sqrt{\frac{1 - E^2}{1 - E^2 \sin^2 \Phi_0}} \tan \Phi_0 = \sqrt{\frac{M_0}{N_0}} \tan \Phi_0 \,. \tag{9.63}$$

Action item: combine (9.60) and (9.61) and find

$$a = \sqrt{\sin^2 \Phi_0 + \frac{N_0}{M_0} \cos^2 \Phi_0} = \cos \Phi_0 \sqrt{\frac{N_0}{M_0} + \tan^2 \Phi_0} \,. \tag{9.64}$$

(If $N_0/M_0 = 1$, then $a = 1$.)

Solve the mapping equations at the initial fundamental point $P_0(\Lambda_0, \Phi_0)$
with respect to the integration constant c and find

$$c^a = \frac{\left[\tan\left(\frac{\pi}{4} + \frac{\phi_0}{2}\right)\right]}{\tan^a\left(\frac{\pi}{4} + \frac{\Phi_0}{2}\right)\left(\frac{1 - E \sin \Phi_0}{1 + E \sin \Phi_0}\right)^{aE/2}} \,,$$

$$\tag{9.65}$$

$$c = \frac{\left[\tan\left(\frac{\pi}{4} + \frac{\phi_0}{2}\right)\right]^{1/a}}{\tan\left(\frac{\pi}{4} + \frac{\Phi_0}{2}\right)\left(\frac{1 - E \sin \Phi_0}{1 + E \sin \Phi_0}\right)^{E/2}} \,.$$

End of Proof.

The principal stretches of the conformal mapping "ellipsoid-of-revolution to sphere" are explicitly given by (9.66).

$$\Lambda_1 = \Lambda_2 = \Lambda = \frac{a\sqrt{M_0 N_0}\cos\phi}{N(\Phi)\cos\Phi} \ , \quad a \text{ as given above}, \tag{9.66}$$

$$\phi = 2\arctan\left(c^a\left[\tan\left(\frac{\pi}{4} + \frac{\Phi}{2}\right)\right]^a \left(\frac{1 - E\sin\Phi}{1 + E\sin\Phi}\right)^{aE/2}\right) - \frac{\pi}{2} \ . \tag{9.67}$$

9-3 The equal area mappings "ellipsoid-of-revolution to plane"

The equal area mappings from the ellipsoid-of-revolution to the plane: the condition of equal area, the standard integrals, authalic latitude.

First, we postulate the condition of equal area $\Lambda_1\Lambda_2 = 1$ and subsequently we take advantage of the standard integrals that are collected in Box 9.7.

$$\Lambda_1\Lambda_2 = 1$$

$$\Leftrightarrow$$

$$\frac{ar\cos\phi}{A_1\cos\Phi}(1 - E^2\sin^2\Phi)^{1/2}\frac{r}{A_1(1 - E^2)}\frac{\mathrm{d}\phi}{\mathrm{d}\Phi}(1 - E^2\sin^2\Phi)^{3/2} = 1 \tag{9.68}$$

$$\Leftrightarrow$$

$$r^2\cos\phi\,\mathrm{d}\phi = \frac{A_1^2(1 - E^2)}{a}\cos\Phi\frac{\mathrm{d}\Phi}{(1 - E^2\sin^2\Phi)^2} \ . \tag{9.69}$$

Box 9.7 (The standard integrals).

First standard integral
(conic mapping: Lambert conformal):

$$r^2\sin\phi = \frac{A_1^2(1 - E^2)}{a}\int\cos\Phi\frac{\mathrm{d}\Phi}{(1 - E^2\sin^2\Phi)^2} + c \ ,$$

$$\Delta := \frac{\pi}{2} - \Phi \Rightarrow -\mathrm{d}\Delta = +\mathrm{d}\Phi \ , \quad c := 0 \ ,$$

$$-\int\frac{\sin\Delta}{(1 - E^2\cos^2\Delta)^2}\mathrm{d}\Delta = \frac{\cos\Delta}{2(1 - E^2\cos^2\Delta)} + \frac{1}{4E}\ln\frac{1 + E\cos\Delta}{1 - E\cos\Delta} = \tag{9.70}$$

$$= \frac{\sin\Phi}{2(1 - E^2\sin^2\Phi)} + \frac{1}{4E}\ln\frac{1 + E\sin\Phi}{1 - E\sin\Phi} \ .$$

Second standard integral
(equal area conic mapping):

$$ar^2\sin\phi = A_1^2(1 - E^2)\left[\frac{\sin\Phi}{2(1 - E^2\sin^2\Phi)} + \frac{1}{4E}\ln\frac{1 + E\sin\Phi}{1 - E\sin\Phi}\right] \ ,$$

$$\sin\phi = \frac{A_1^2(1 - E^2)}{ar^2}\left[\frac{\sin\Phi}{2(1 - E^2\sin^2\Phi)} + \frac{1}{4E}\ln\frac{1 + E\sin\Phi}{1 - E\sin\Phi}\right] \ . \tag{9.71}$$

The term *authalic latitude* ϕ has been introduced by O. S. Adams (1921, p. 65) or J. P. Snyder (1982, p. 19). In Box 9.8, the two interesting concepts are summarized.

Box 9.8 (The two interesting concepts).

First case:

$$A_1 = r , \quad a = 1 . \qquad (9.72)$$

Second case
("identical surface area"):

$$4\pi r^2 = 4\pi A_1^2 \left(\frac{1}{2} + \frac{1 - E^2}{4E} \ln \frac{1 + E}{1 - E} \right)$$

$$\Rightarrow$$

$$r^2 = \frac{1}{2} A_1^2 \left(1 + \frac{1 - E^2}{2E} \ln \frac{1 + E}{1 - E} \right) \qquad (9.73)$$

$$\Rightarrow$$

$$\sin \phi = \frac{(1 - E^2)}{a} \frac{\frac{\sin \Phi}{(1 - E^2 \sin^2 \Phi)} + \frac{1}{2E} \ln \frac{1 + E \sin \Phi}{1 - E \sin \Phi}}{1 + \frac{1 - E^2}{2E} \ln \frac{1 + E}{1 - E}} .$$

Example:

$$a = 1 . \qquad (9.74)$$

$$\Lambda_1 = \frac{a r \cos \phi}{A_1 \cos \Phi} (1 - E^2 \sin^2 \Phi)^{1/2} ,$$

$$\qquad (9.75)$$

$$\Lambda_2 = \frac{r \mathrm{d}\phi / \mathrm{d}\Phi}{A_1 (1 - E^2)} (1 - E^2 \sin^2 \Phi)^{3/2} ,$$

$$\frac{\mathrm{d}\phi}{\mathrm{d}\Phi} = \frac{1}{a \cos \phi} \frac{1}{r^2} \frac{A_1^2 (1 - E^2)}{(1 - E^2 \sin^2 \Phi)^2} \cos \Phi \qquad (9.76)$$

$$\Rightarrow$$

$$\Lambda_2 = \Lambda_1^{-1} =$$

$$= \frac{A_1 \cos \Phi}{a \cos \phi} \frac{1}{r} (1 - E^2 \sin^2 \Phi)^{-1/2} = \frac{1}{a} \frac{A_1 \cos \Phi}{r \cos \phi} \frac{1}{(1 - E^2 \sin^2 \Phi)^{1/2}} . \qquad (9.77)$$

We here also note the remarkable representations for $\sin \phi (\sin \Phi)$. Furthermore, note the remarkable representations for the cases $A_1 = r$ versus $4\pi r^2 = 4\pi A_1^2 (1/2 + [(1 - E^2)/4E] \ln[(1 + E)/(1 - E)])$ and Λ_1 and Λ_2, respectively.

The final step are the standard mapping procedures of mapping the sphere to the plane. In the following chapters, we study the mapping of the sphere to the cylinder.

10 "Sphere to cylinder": polar aspect

Mapping the sphere to a cylinder: polar aspect. Equidistant, conformal, and equal area mappings. Principle for constructing a cylindrical map projection. Optimal cylinder projections of the sphere, equidistant on two parallels.

In this chapter, we present a collection of most widely used map projections in the polar aspect in which meridians are shown as a set of equidistant parallel straight lines and parallel circles (parallels) by a system of parallel straight lines orthogonally crossing the images of the meridians. As a specialty, the poles are not displayed as points but straight lines as long as the equator. First, we derive the general mapping equations for both cases of (i) a tangent cylinder and (ii) a secant cylinder and describe the construction principle. The mapping equations and the equations for the left principal stretches involve a general latitude dependent function f, which is determined in a following section through the postulate of (i) an equidistant, (ii) a conformal, or (iii) an equal area mapping. The resulting map projection are the most simple Plate Carrée projection ("quadratische Plattkarte"), the famous conformal Mercator projection (presented by Gerardus Mercator (Latinized name of Gerhard Kremer, 1512–1594) of Flanders in 1569) and the equal area Lambert projection (presented by Johann Heinrich Lambert (1728–1777) of Alsace in 1772). While the Plate Carrée projection was mainly used for the representation of equatorial regions, the Mercator projection has found widespread use in (aero-)nautics and maps for displaying air and ocean currents. A special feature of this projection is that the loxodrome (rhumb line, line of constant azimuth) is displayed as a straight line crossing all meridians with a constant angle. The cylindrical Lambert projection, in contrast, has found only minimal usage, which is mainly due to the fact that the images of parallels lie very dense in medium and high latitudes. For a first impression, have a look at Fig. 10.1.

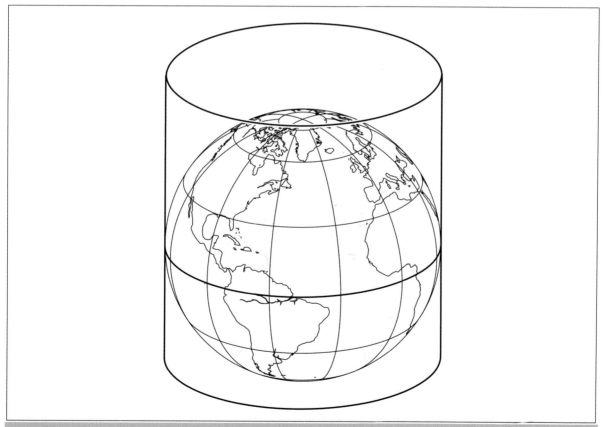

Fig. 10.1. Mapping the sphere to a (tangent) cylinder. Polar aspect. Line-of-contact: equator.

10-1 General mapping equations

Setting up general equations of the mapping "sphere to cylinder": projections in the polar aspect. Principle for constructing a cylindrical map projection.

There are two basic postulates which govern the setup of general equations of mapping the sphere $\mathbb{S}^2_R$ of radius R to a tangent or secant cylinder $\mathbb{C}^2_R$. First, the coordinate x depends only on the longitude Λ and the parallel circles $\Phi = \pm\Phi_0$ have to be mapped equidistantly, i.e. $x = R\Lambda\cos\Phi_0$. Second, the coordinate y is only a function of latitude Φ, i.e. $y = f(\Phi)$, compare with Fig. 10.2 for the case of a tangent cylinder. In case of the tangent variant, the cylinder is wrapping the sphere with the equator being the line-of-contact. In the second case of a secant cylinder, two parallel circles $\Phi = \pm\Phi_0$ are the lines-of-contact, compare with Fig. 10.3.

Box 10.1 ("Sphere to cylinder": distortion analysis, polar aspect, left principal stretches).

Parameterized mapping:

$$x = R\Lambda\cos\Phi_0 , \quad y = f(\Phi) . \tag{10.1}$$

Left Jacobi matrix:

$$\mathsf{J}_l := \begin{bmatrix} D_\Lambda x & D_\Phi x \\ D_\Lambda y & D_\Phi y \end{bmatrix} = \begin{bmatrix} R\cos\Phi_0 & 0 \\ 0 & f'(\Phi) \end{bmatrix} . \tag{10.2}$$

Left Cauchy–Green matrix ($\mathsf{G}_r = \mathsf{I}_2$):

$$\mathsf{C}_l = \mathsf{J}_l^* \mathsf{G}_r \mathsf{J}_l = \begin{bmatrix} R^2\cos^2\Phi_0 & 0 \\ 0 & f'^{\,2}(\Phi) \end{bmatrix} . \tag{10.3}$$

Left principal stretches:

$$\Lambda_1 = \sqrt{\frac{C_{11}}{G_{11}}} = \frac{\cos\Phi_0}{\cos\Phi} ,$$

$$\Lambda_2 = \sqrt{\frac{C_{22}}{G_{22}}} = \frac{f'(\Phi)}{R} . \tag{10.4}$$

Left eigenvectors of the matrix pair $\{\mathsf{C}_\lambda, \mathsf{G}_\lambda\}$:

$$\boldsymbol{C}_1 = \boldsymbol{E}_\Lambda = \frac{D_\Lambda \boldsymbol{X}}{\|D_\Lambda \boldsymbol{X}\|}$$

(Easting),

$$\boldsymbol{C}_2 = \boldsymbol{E}_\Phi = \frac{D_\Phi \boldsymbol{X}}{\|D_\Phi \boldsymbol{X}\|}$$

(Northing).

$$\tag{10.5}$$

Next, we specialize the general cylindrical mapping to generate an equidistant mapping, a conformal mapping, and an equal area mapping.

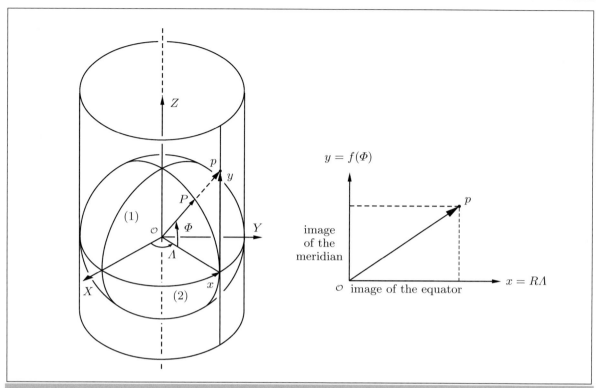

Fig. 10.2. Principle for constructing a cylindrical map projection in the polar aspect (tangent cylinder: $\Phi_0 = 0°$). Greenwich (1) and equator (2).

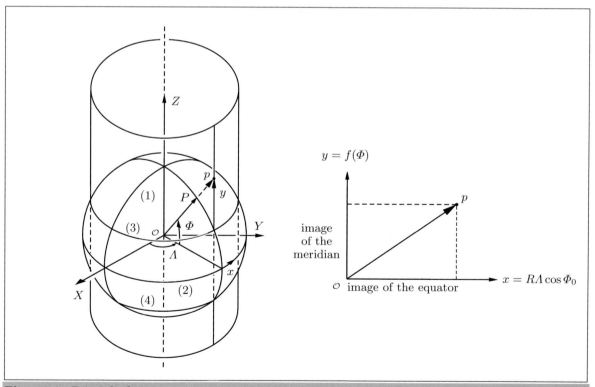

Fig. 10.3. Principle for constructing a cylindrical map projection in the polar aspect (secant cylinder: $\Phi_0 = \pm 30°$). Greenwich (1) and equator (2). $\Phi_0 = \text{const.}$ (3) and $-\Phi_0 = \text{const.}$ (4).

10-2 Special mapping equations

Setting up special equations of the mapping "sphere to cylinder". Equidistant mapping (Plate Carrée projection), conformal mapping (Mercator projection), equal area mapping (Lambert cylindrical equal area projection).

10-21 Equidistant mapping (Plate Carrée projection)

For the first mapping of the sphere to a cylinder, we postulate that all meridians shall be mapped equidistantly, namely

$$\Lambda_2 = 1 \Rightarrow \frac{f'(\Phi)}{R} = 1 \Rightarrow \mathrm{d}f = R\mathrm{d}\Phi \Rightarrow f(\Phi) = R\Phi + \text{const.} \tag{10.6}$$

The integration constant is determined from the additional constraint that for $\Phi = 0$ the coordinate y should vanish, $y = 0 \Rightarrow \text{const.} = 0$. We end up with the most simple mapping equations (10.7). The left principal stretches are provided by (10.8). For the parallel circle $\Phi = \pm\Phi_0$, we experience isometry, conformality $\Lambda_1 = \Lambda_2 = 1$, and no area distortion $\Lambda_1\Lambda_2 = 1$. Compare with Fig. 10.4.

$$\begin{bmatrix} x \\ y \end{bmatrix} = R \begin{bmatrix} \Lambda\cos\Phi_0 \\ \Phi \end{bmatrix}, \tag{10.7}$$

$$\Lambda_1 = \frac{\cos\Phi_0}{\cos\Phi}, \quad \Lambda_2 = 1. \tag{10.8}$$

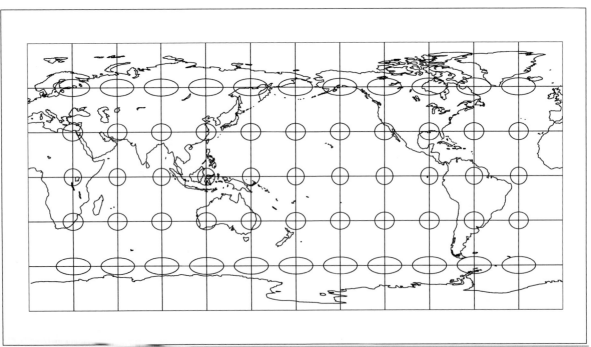

Fig. 10.4. Mapping the sphere to a cylinder: polar aspect, equidistant mapping, $\Phi_0 = 0°$: tangent cylinder (Plate Carrée projection, quadratische Plattkarte).

10-22 Conformal mapping (Mercator projection)

The requirement for conformality leads to the postulate (10.9). Again, the integration constant is determined from the additional constraint that for $\Phi = 0$ the coordinate y should vanish, namely $y = 0 \Rightarrow$ const. $= 0$. Therefore, the mapping equations are provided by (10.10). The left principal stretches are provided by (10.11). The parallel circle $\Phi = \pm\Phi_0$ is mapped free from any distortion. Compare with Fig. 10.5.

$$\Lambda_1 = \Lambda_2$$

$$\Rightarrow$$

$$\frac{\cos\Phi_0}{\cos\Phi} = \frac{1}{R}\frac{\mathrm{d}f}{\mathrm{d}\Phi} \Rightarrow \mathrm{d}f = R\frac{\cos\Phi_0\mathrm{d}\Phi}{\cos\Phi} \tag{10.9}$$

$$\Rightarrow$$

$$\int \mathrm{d}f = f(\Phi) = R\cos\Phi_0 \int \frac{\mathrm{d}\Phi}{\cos\Phi} = R\cos\Phi_0 \ln\cot\left(\frac{\pi}{4} - \frac{\Phi}{2}\right) + \text{const.} ,$$

$$\begin{bmatrix} x \\ y \end{bmatrix} = R\cos\Phi_0 \begin{bmatrix} \Lambda \\ \ln\cot\left(\frac{\pi}{4} - \frac{\Phi}{2}\right) \end{bmatrix} = R\cos\Phi_0 \begin{bmatrix} \Lambda \\ \ln\tan\left(\frac{\pi}{4} + \frac{\Phi}{2}\right) \end{bmatrix} , \tag{10.10}$$

$$\Lambda_1 = \Lambda_2 = \frac{\cos\Phi_0}{\cos\Phi} . \tag{10.11}$$

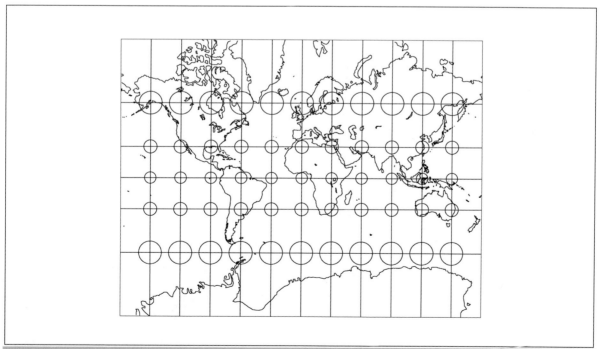

Fig. 10.5. Mapping the sphere to a cylinder: polar aspect, conformal mapping, $\Phi_0 = 0°\cdot$ tangent cylinder (Mercator projection).

10-23 Equal area mapping (Lambert projection)

$$\Lambda_1 \Lambda_2 = 1$$

$$\Rightarrow$$

$$\frac{\cos \Phi_0}{\cos \Phi} \frac{f'(\Phi)}{R} = 1$$

$$\Rightarrow$$

$$\mathrm{d}f = R \frac{\cos \Phi}{\cos \Phi_0} \mathrm{d}\Phi$$ (10.12)

$$\Rightarrow$$

$$\int \mathrm{d}f = f(\Phi) = R \frac{\sin \Phi}{\cos \Phi_0} + \text{const.}$$

As before, the integration constant is determined from the additional constraint that for $\Phi = 0$ the coordinate y should be zero, namely $y = 0 \Rightarrow \text{const.} = 0$. Therefore, the mapping equations are provided by (10.13). The left principal stretches are provided by (10.14). Compare with Fig. 10.6.

$$\begin{bmatrix} x \\ y \end{bmatrix} = R \begin{bmatrix} \Lambda \cos \Phi_0 \\ \frac{\sin \Phi}{\cos \Phi_0} \end{bmatrix} ,$$ (10.13)

$$\Lambda_1 = \frac{\cos \Phi_0}{\cos \Phi} , \quad \Lambda_2 = \frac{\cos \Phi}{\cos \Phi_0} .$$ (10.14)

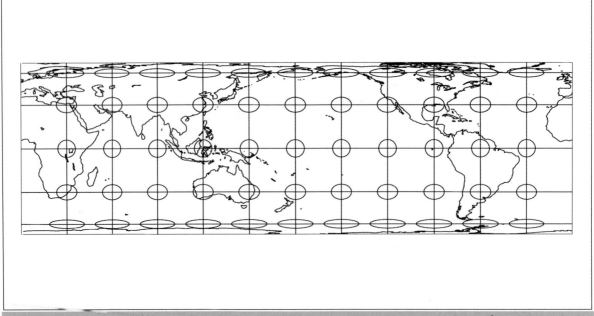

Fig. 10.6. Mapping the sphere to a cylinder; polar aspect, equal area mapping, $\Phi_0 = 0°$: tangent cylinder (normal Lambert cylindrical equal area projection).

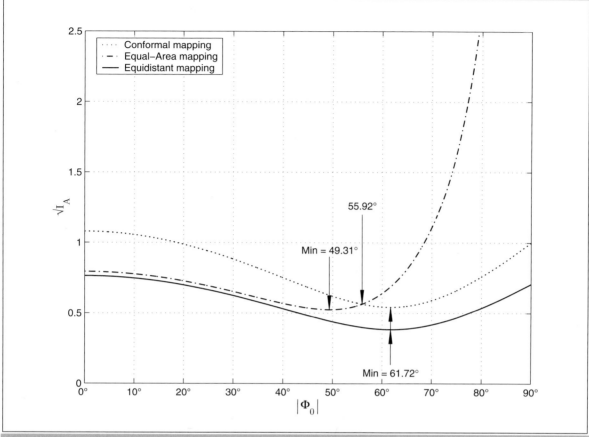

Fig. 10.7. The Airy optimum of three different mappings: (i) conformal maps, (ii) equiareal maps, and (iii) distance preserving maps.

10-3 Optimal cylinder projections

Optimal cylinder projections of the sphere of type *equidistant on two standard parallels*. Conformal cylindrical mapping, equal area cylindrical mapping, equidistant cylindrical mapping.

Many applications require a map projection the distortions of which do not excess a certain value in the mean. An example is given by Lemma 10.1.

Lemma 10.1 (Optimal cylinder projections of the sphere of type *equidistant on two parallel circles*).

If we compare (i) conformal maps, (ii) equiareal maps, and (iii) distance preserving maps in the class of *optimal cylinder projections* of the sphere, equidistant on two parallel circles, where the *equidistance* on two parallel circles is the *unknown parameter*, we find according to the *Airy optimal criterion* that the distance preserving maps are optimal and the equiareal maps are better than the conformal maps, at least up to a latitude of $\Phi = 56°$. According to the *criterion of Airy–Kavrajski*, again the distance preserving maps are optimal, but the conformal maps and the equiareal maps produce exactly equally good maps.

End of Lemma.

The two optima of type Airy and Airy–Kavrajski for the mapping of type cylinder projection of the sphere and equidistant on two parallel circles is illustrated by Fig. 10.7 and Fig. 10.8. Reference Papers are G. B. Airy (1861), N. Francula (1971), E. W. Grafarend (1995), E. W. Grafarend and A. Niermann (1984), E. W. Grafarend and R. Syffus (1998c), V. Hojovec and L. Jokl (1981), W. Jordan (1875, 1896), C. Kaltsikis (1980), V. V. Kavrajski (1958).

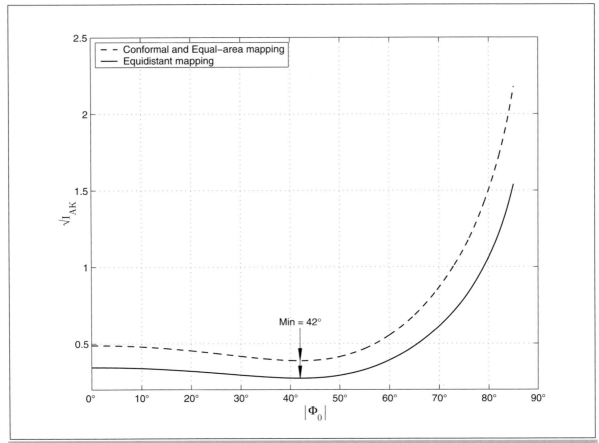

Fig. 10.8. The Airy–Kavrajski optimum of three different mappings: (i) conformal maps, (ii) equiareal maps, and (iii) distance preserving maps.

Let us finally prove our statements based upon (i) the various mapping equations of the sphere under the postulates of equidistant mappings on two parallels and type cylinder mappings and (ii) the corresponding principal stretches. The *Airy distortion energy* is based upon the integral (10.15), the global arithmetic mean of the surface integral of a spherical zone between the equator and the latitude circle Φ of the local measure respective global measure (10.16).

$$I_{\mathrm{A}} := \frac{1}{2S} \int\limits_{S} \mathrm{d}S[(\Lambda_1 - 1)^2 + (\Lambda_2 - 1)^2] \,, \tag{10.15}$$

$$\mathrm{d}S = 2\pi R^2 \cos\Phi \,\mathrm{d}\Phi \quad \text{versus} \quad S = 2\pi R^2 \sin\Phi \,. \tag{10.16}$$

Proof (conformal cylindrical mapping).

We start from the mapping equations of conformal type constrained to the equidistance postulate on two parallel circles. In addition, we enjoy the identity postulate of left principal stretches.

$$\begin{bmatrix} x \\ y \end{bmatrix} = R\cos\Phi_0 \begin{bmatrix} \Lambda \\ \ln\cot\left(\frac{\pi}{4} - \frac{\Phi}{2}\right) \end{bmatrix} = R\cos\Phi_0 \begin{bmatrix} \Lambda \\ \ln\tan\left(\frac{\pi}{4} + \frac{\Phi}{2}\right) \end{bmatrix} \,, \tag{10.17}$$

$$\Lambda_1 = \Lambda_2 = \frac{\cos\Phi_0}{\cos\Phi} \,. \tag{10.18}$$

Starting from the above relations, we obtain

$$\frac{(\Lambda_1 - 1)^2 + (\Lambda_2 - 1)^2}{2} = \left(\frac{\cos \Phi_0}{\cos \Phi} - 1\right)^2 = \frac{1}{\cos^2 \Phi}(\cos^2 \Phi_0 - 2 \cos \Phi \cos \Phi_0 + \cos^2 \Phi) , \quad (10.19)$$

$$I_A(\text{conformal}) = \frac{1}{\sin \Phi} \int_0^{\Phi} d\Phi^* \frac{1}{\cos \Phi^*}(\cos^2 \Phi_0 - 2 \cos \Phi^* \cos \Phi_0 + \cos^2 \Phi^*) \quad (10.20)$$

$$= \frac{1}{\sin \Phi}\left[\cos^2 \Phi_0 \ln \tan\left(\frac{\pi}{4} + \frac{\Phi}{2}\right) - 2\Phi \cos \Phi_0 + \sin \Phi\right] .$$

Auxillary integrals:

$$\int \frac{d\Phi}{\cos \Phi} = \ln \tan\left(\frac{\pi}{4} + \frac{\Phi}{2}\right) = \frac{1}{2} \ln \frac{1 + \sin \Phi}{1 - \sin \Phi} , \quad (10.21)$$

$$\int d\Phi = \Phi ,$$

$$\int d\Phi \cos \Phi = \sin \Phi . \quad (10.22)$$

In order to determine the unknown parameter Φ_0, we restrict the Airy distortion energy integral to the region between $\Phi = \pm 85°$.

$$I_A(\text{conformal}) = \min$$

$$\Leftrightarrow \quad (10.23)$$

$$dI_A/d\Phi_0 = 0 ,$$

$$-2 \sin \widehat{\Phi}_0 \cos \widehat{\Phi}_0 \ln \tan\left(\frac{\pi}{4} + \frac{\Phi}{2}\right) + 2\Phi \sin \widehat{\Phi}_0 = 0 ,$$

$$\sin \widehat{\Phi}_0 \neq 0 \quad (10.24)$$

$$\Rightarrow$$

$$\cos \widehat{\Phi}_0 = \frac{\Phi}{\ln \tan\left(\frac{\pi}{4} + \frac{\Phi}{2}\right)} , \quad (10.25)$$

$$\Phi = 85° , \quad \left|\widehat{\Phi}_0\right| = 61.72° , \quad \sqrt{I_A} = 0.5426 . \quad (10.26)$$

End of Proof.

Proof (equiareal cylindrical mapping).

Next, we deal with the mapping equations of equiareal type constrained to the equidistance postulate on two parallel circles. Again, we enjoy the condition of an equiareal mapping of type (10.28).

$$\begin{bmatrix} x \\ y \end{bmatrix} = R \begin{bmatrix} \Lambda \cos \Phi_0 \\ \sin \Phi / \cos \Phi_0 \end{bmatrix} , \tag{10.27}$$

$$\Lambda_1 = 1/\Lambda_2 = \cos \Phi_0 / \cos \Phi , $$

$$\Lambda_2 = 1/\Lambda_1 = \cos \Phi / \cos \Phi_0 . \tag{10.28}$$

Starting from the above relations, we obtain

$$\frac{(\Lambda_1 - 1)^2 + (\Lambda_2 - 1)^2}{2} =$$

$$= \frac{(\cos \Phi_0 / \cos \Phi - 1)^2 + (\cos \Phi / \cos \Phi_0 - 1)^2}{2} = \tag{10.29}$$

$$= \frac{1}{2} \frac{\cos^4 \Phi_0 - 2 \cos \Phi \cos^3 \Phi_0 + 2 \cos^2 \Phi \cos^2 \Phi_0 - 2 \cos^3 \Phi \cos \Phi_0 + \cos^4 \Phi}{\cos^2 \Phi_0 \cos^2 \Phi} ,$$

$$I_A(\text{equiareal}) = \frac{1}{2 \sin \Phi} \int_0^{\Phi} d\Phi^* \frac{1}{\cos \Phi^* \cos^2 \Phi_0} \times \tag{10.30}$$

$$\times \left(\cos^4 \Phi_0 - 2 \cos \Phi^* \cos^3 \Phi_0 + 2 \cos^2 \Phi^* \cos^2 \Phi_0 - 2 \cos^3 \Phi^* \cos \Phi_0 + \cos^4 \Phi^* \right) .$$

Auxillary integrals:

$$\int \frac{d\Phi^*}{\cos \Phi^*} = \ln \tan \left(\frac{\pi}{4} + \frac{\Phi^*}{2} \right) = \frac{1}{2} \ln \frac{1 + \cos \Phi^*}{1 - \cos \Phi^*} , \tag{10.31}$$

$$\int d\Phi^* = \Phi^* ,$$

$$\int d\Phi^* \cos^2 \Phi^* = \frac{\Phi^*}{2} + \frac{\sin 2\Phi^*}{4} = \frac{\Phi^*}{2} + \frac{1}{2} \sin \Phi^* \cos \Phi^* ,$$

$$\int d\Phi^* \cos \Phi^* = \sin \Phi^* , \tag{10.32}$$

$$\int d\Phi^* \cos^3 \Phi^* = \frac{1}{3} \sin \Phi^* (2 + \cos^2 \Phi^*) .$$

For the unknown parameter Φ_0, we shall compute the Airy distortion energy for the given region of $\Phi = \pm 85°$.

$$I_A(\text{equiareal}) = \min$$

$$\Leftrightarrow \tag{10.33}$$

$$dI_A/d\Phi_0 = 0 \ ,$$

$$I_A(\text{equiareal}) = I_A(\Phi_0) =$$

$$= \frac{1}{2\sin\Phi}\left[\cos^2\Phi_0\ln\tan\left(\frac{\pi}{4}+\frac{\Phi}{2}\right) - 2\Phi\cos\Phi_0 + 2\sin\Phi - \right. \tag{10.34}$$

$$\left. -\frac{1}{\cos\Phi_0}\left(\Phi+\frac{1}{2}\sin 2\Phi\right) + \frac{1}{3}\frac{\sin\Phi}{\cos^2\Phi_0}(2+\cos^2\Phi)\right] \ ,$$

$$dI_A/d\Phi_0 = 0$$

$$\Rightarrow$$

$$-2\sin\widehat{\Phi}_0\cos\widehat{\Phi}_0\ln\tan\left(\frac{\pi}{4}+\frac{\Phi}{2}\right) + 2\Phi\sin\widehat{\Phi}_0 -$$

$$-\frac{\sin\widehat{\Phi}_0}{\cos^2\widehat{\Phi}_0}\left(\Phi+\frac{1}{2}\sin 2\Phi\right) + \frac{2}{3}\frac{\sin\widehat{\Phi}_0}{\cos^3\widehat{\Phi}_0}\sin\Phi(2+\cos^2\Phi) = 0 \ , \tag{10.35}$$

$$\sin\widehat{\Phi}_0 \neq 0 \ , \quad \cos\widehat{\Phi}_0 \neq 0$$

$$\Rightarrow$$

$$-3\cos^4\widehat{\Phi}_0\ln\tan\left(\frac{\pi}{4}+\frac{\Phi}{2}\right) + 3\Phi\cos^3\widehat{\Phi}_0 - 3\cos\widehat{\Phi}_0\left(\frac{\Phi}{2}+\frac{1}{4}\sin 2\Phi\right) + (2+\cos^2\Phi)\sin\Phi = 0 \ .$$

The result is an algebraic equation of fourth order in terms of $\cos^4\widehat{\Phi}_0 = x$, namely
$$x^4 + ax^3 + bx + c = 0.$$

$$\cos^4\widehat{\Phi}_0 - \frac{\Phi\cos^3\widehat{\Phi}_0}{\ln\tan\left(\frac{\pi}{4}+\frac{\Phi}{2}\right)} + \frac{(2\Phi+\sin 2\Phi)\cos\widehat{\Phi}_0}{4\ln\tan\left(\frac{\pi}{4}+\frac{\Phi}{2}\right)} - \frac{(2+\cos^2\Phi)\sin\Phi}{3\ln\tan\left(\frac{\pi}{4}+\frac{\Phi}{2}\right)} = 0 \ , \tag{10.36}$$

$$a = -\frac{\Phi}{\ln\tan\left(\frac{\pi}{4}+\frac{\Phi}{2}\right)} \ , \quad b = +\frac{(2\Phi+\sin 2\Phi)}{4\ln\tan\left(\frac{\pi}{4}+\frac{\Phi}{2}\right)} \ , \tag{10.37}$$

$$c = -\frac{(2+\cos^2\Phi)\sin\Phi}{3\ln\tan\left(\frac{\pi}{4}+\frac{\Phi}{2}\right)} \ ,$$

$$\Phi = 85° \ , \quad \left|\widehat{\Phi}_0\right| = 49.31° \ , \quad \sqrt{I_A} = 0.5248 \ . \tag{10.38}$$

End of Proof.

Proof (distance preserving mapping).

Finally, we present the mapping equations of distance preserving type constrained to the postulate of an equidistance mapping on two parallel circles. We have to specify the principal stretches as (10.40).

$$
\begin{bmatrix} x \\ y \end{bmatrix} = R \begin{bmatrix} \varLambda \cos \varPhi_0 \\ \varPhi \end{bmatrix} ,
\tag{10.39}
$$

$$
\varLambda_1 = \cos \varPhi_0 / \cos \varPhi , \quad \varLambda_2 = 1 .
\tag{10.40}
$$

Starting from the above relations, we obtain

$$
\frac{(\varLambda_1 - 1)^2 + (\varLambda_2 - 1)^2}{2} = \frac{1}{2} (\cos \varPhi_0 / \cos \varPhi - 1)^2 ,
\tag{10.41}
$$

$$
I_A (\text{equidistant}) = \frac{1}{2} I_A (\text{conformal})
\tag{10.42}
$$

$$
\Rightarrow
$$

$$
\varPhi = 85° , \quad \left| \widehat{\varPhi_0} \right| = 61.72° , \quad \sqrt{I_A} = 0.3837 .
\tag{10.43}
$$

End of Proof.

In the following chapter, let us continue studying the mapping of the sphere to the cylinder, namely let us study the transverse aspect.

11 "Sphere to cylinder": transverse aspect

Mapping the sphere to a cylinder: meta-cylindrical projections in the transverse aspect. Equidistant, conformal, and equal area mappings.

Among cylindrical projections, mappings in the transverse aspect play the most important role. Although many worldwide adopted legal map projections use the ellipsoid-of-revolution as the reference figure for the Earth, the spherical variant forms the basis for the Universal Transverse Mercator (UTM) grid and projection. In the subsequent chapter, we first introduce the general concept of a cylindrical projection in the transverse aspect. Following this, three special map projections are presented: (i) the equidistant mapping (transverse Plate Carrée projection), (ii) the conformal mapping (transverse Mercator projection), and (iii) the equal area mapping (transverse Lambert projection). The transverse Mercator projection is especially appropriate for regions with a predominant North-South extent. As in previous chapters, the two possible cases of a tangent and a secant cylinder are treated simultaneously by introducing the meta-latitude $B = \pm B_1$ of a meta-parallel circle which is mapped equidistantly. For a first impression, have a look at Fig. 11.1.

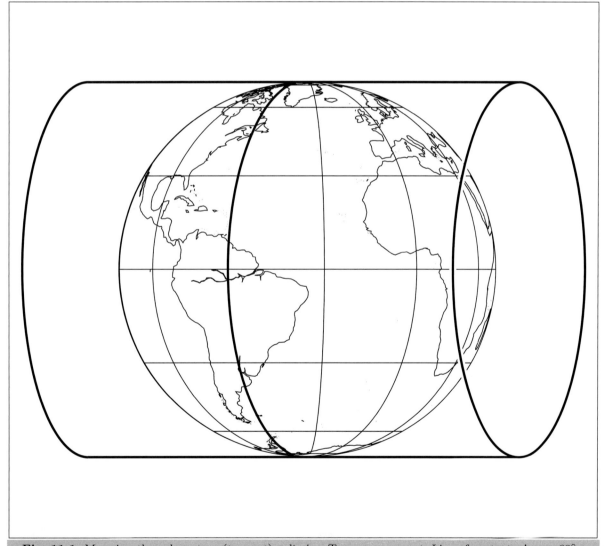

Fig. 11.1. Mapping the sphere to a (tangent) cylinder. Transverse aspect, Line-of-contact: $\Lambda_0 = -60°$.

11-1 General mapping equations

Setting up general equations of the mapping "sphere to cylinder": projections in the transverse aspect. Meta-spherical longitude, meta-spherical latitude.

The general equations for mapping the sphere to a cylinder in the transverse aspect are based on the general equation (10.1) of Chapter 10, but spherical longitude Λ and spherical latitude Φ being replaced by their counterparts meta-longitude and meta-latitude, which are indicated here by capital letters A and B. In order to treat simultaneously the transverse *tangent cylinder* and the transverse *secant cylinder*, we introduce B_0 as the meta-latitude of those meta-parallel circles $B = \pm B_0$ which shall be mapped equidistantly. In consequence, the general equations for this case are given by the very general vector relation (11.1), taking into account the constraints (3.51) and (3.53) for $\Phi_0 = 0°$, namely (11.2). For the distortion analysis, the left principal stretches result to (11.3).

$$\begin{bmatrix} x \\ y \end{bmatrix} = \begin{bmatrix} RA \cos B_0 \\ f(B) \end{bmatrix} , \tag{11.1}$$

$$\tan A = \frac{\sin(\Lambda - \Lambda_0)}{-\tan \Phi} , \quad \sin B = \cos \Phi \cos(\Lambda - \Lambda_0) , \tag{11.2}$$

$$\Lambda_1 = \frac{\cos B_0}{\cos B} , \quad \Lambda_2 = \frac{f'(B)}{R} . \tag{11.3}$$

The procedure of how to set up special equations of the mapping "sphere to cylinder" in the transverse aspect (transverse equidistant mapping, transverse conformal mapping, transverse equal area mapping) can be easily deduced from the preceding chapters. Far easier, in the mapping equations as well in the equations for the left principal stretches defined in Chapter 10, conventional coordinates *spherical longitude* Λ and *spherical latitudes* Φ and Φ_0 are simply replaced by their corresponding items *meta-spherical longitude* A and *meta-spherical latitudes* B and B_0. Transformations of conventional spherical coordinates to meta-spherical coordinates is performed by using (11.2).

11-2 Special mapping equations

Setting up special equations of the mapping "sphere to cylinder": meta-cylindrical projections in the *transverse aspect*. Equidistant mapping (transverse Plate Carrée projection), conformal mapping (transverse Mercator projection), equal area mapping (transverse Lambert cylindrical equal area projection).

11-21 Equidistant mapping (transverse Plate Carrée projection), see Fig. 11.2

$$\begin{bmatrix} x \\ y \end{bmatrix} = R \begin{bmatrix} A \cos B_0 \\ B \end{bmatrix} ,$$

$$\Lambda_1 - \frac{\cos B_0}{\cos B} , \quad \Lambda_2 = 1 . \tag{11.4}$$

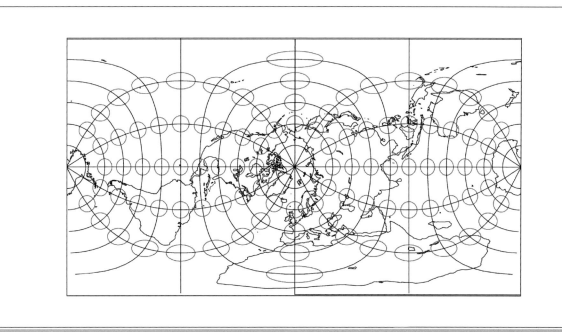

Fig. 11.2. Mapping the sphere to a cylinder: transverse aspect, equidistant mapping, $B_0 = 0°$, transverse Plate Carrée projection.

11-22 Conformal mapping (transverse Mercator projection), compare with Fig. 11.3

$$\begin{bmatrix} x \\ y \end{bmatrix} = R \cos B_0 \begin{bmatrix} A \\ \ln \cot \left(\frac{\pi}{4} - \frac{B}{2} \right) \end{bmatrix} = R \cos B_0 \begin{bmatrix} A \\ \ln \tan \left(\frac{\pi}{4} + \frac{B}{2} \right) \end{bmatrix} =$$

$$= R \cos B_0 \begin{bmatrix} A \\ \frac{1}{2} \ln \frac{1+\sin B}{1-\sin B} \end{bmatrix} = \tag{11.5}$$

$$= R \cos B_0 \begin{bmatrix} A \\ \operatorname{ar} \tanh(\sin B) \end{bmatrix} ,$$

$$\Lambda_1 = \Lambda_2 = \frac{\cos B_0}{\cos B} . \tag{11.6}$$

11-23 Equal area mapping (transverse Lambert projection), compare with Fig. 11.4

$$\begin{bmatrix} x \\ y \end{bmatrix} = R \begin{bmatrix} A \cos B_0 \\ \frac{\sin B}{\cos B_0} \end{bmatrix} , \tag{11.7}$$

$$\Lambda_1 = \frac{\cos B_0}{\cos B} , \quad \Lambda_2 = \frac{\cos B}{\cos B_0} . \tag{11.8}$$

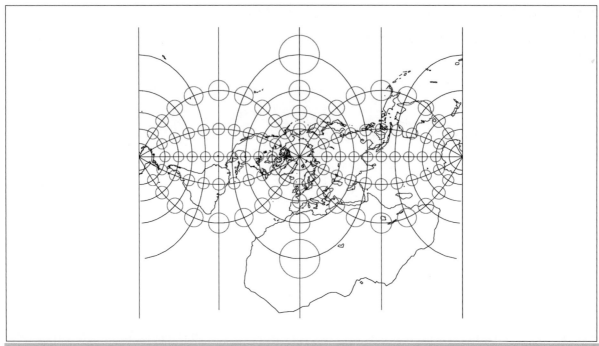

Fig. 11.3. Mapping the sphere to a cylinder: transverse aspect, conformal mapping, $B_0 = 0°$, transverse Mercator projection.

In the following chapter, let us continue studying the mapping of the sphere to the cylinder, namely let us study the oblique aspect.

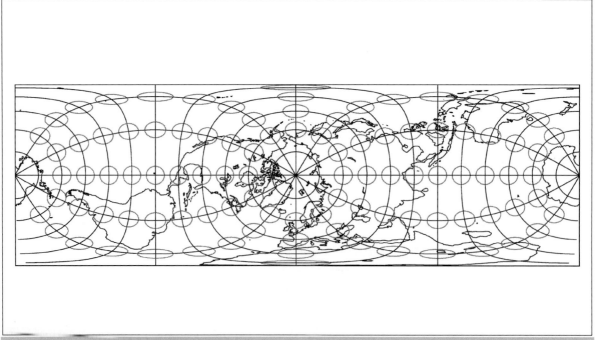

Fig. 11.4. Mapping the sphere to a cylinder: transverse aspect, equal area mapping, $B_0 = 0°$, transverse Lambert cylindrical equal area projection.

12 "Sphere to cylinder": oblique aspect

Mapping the sphere to a cylinder: meta-cylindrical projections in the oblique aspect. Equidistant, conformal, and equal area mappings.

Cylindrical projections in the oblique aspect are mainly used to display regions which have a predominant extent in the oblique direction, neither East-West nor North-South. In addition, they form the most general cylindrical projections because mapping equations for projections in the polar and the transverse aspect can easily be derived from it. This is done by setting the corresponding latitude of the meta-North Pole Φ_0 to a specific value: $\Phi_0 = 90°$ generates cylindrical projections in the polar aspect, $\Phi_0 = 0°$ result in cylindrical projections in the transverse aspect. As an introductory part, we present the equations for general cylindrical mappings together with the equations for the principal stretches, before derivations for specific cylindrical map projections of the sphere (oblique equidistant projection, oblique conformal projection and oblique equal area projection) are given. For a first impression, have a look at Fig. 12.1.

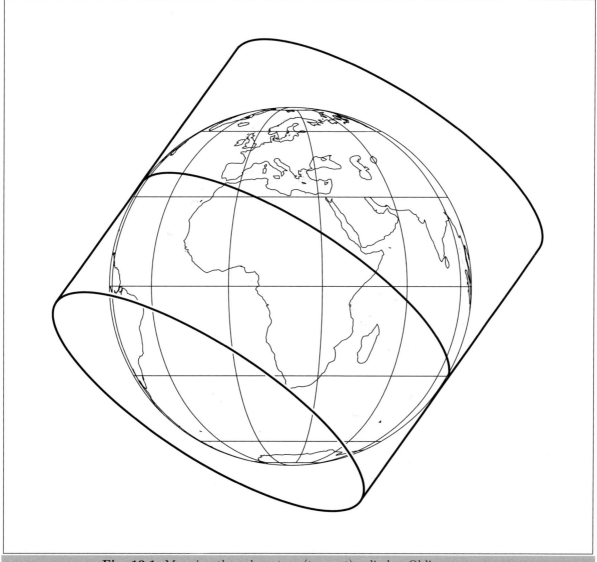

Fig. 12.1. Mapping the sphere to a (tangent) cylinder. Oblique aspect.

12-1 General mapping equations

Setting up general equations of the mapping "sphere to cylinder": projections in the oblique aspect. Meta-longitude, meta-latitude.

The general equations for mapping the sphere to a cylinder in the transverse aspect are based on the general equation (10.1) of Chapter 10, but spherical longitude Λ and spherical latitude Φ being replaced by their counterparts meta-longitude and meta-latitude, which are indicated here by capital letters A and B. In order to treat simultaneously both the transverse tangent cylinder and the transverse secant cylinder, we introduce B_0 as the meta-latitude of the meta-parallel circles $B = \pm B_0$ which shall be mapped equidistantly. In consequence, the general equations for this case are given by the very general vector relation (12.1), taking into account the constraints (3.51) and (3.53), namely (12.2). For the distortion analysis, the left principal stretches result to (12.3).

$$
\begin{bmatrix} x \\ y \end{bmatrix} = \begin{bmatrix} RA \cos B_0 \\ f(B) \end{bmatrix} , \tag{12.1}
$$

$$
\tan A = \frac{\cos \Phi \sin(\Lambda - \Lambda_0)}{\cos \Phi \sin \Phi_0 \cos(\Lambda - \Lambda_0) - \sin \Phi \cos \Phi_0} ,
$$
$$
\sin B = \cos \Phi \cos \Phi_0 \cos(\Lambda - \Lambda_0) + \sin \Phi \sin \Phi_0 , \tag{12.2}
$$

$$
\Lambda_1 = \frac{\cos B_0}{\cos B} , \quad \Lambda_2 = \frac{f'(B)}{R} . \tag{12.3}
$$

The procedure of how to set up special equations of the mapping sphere to cylinder in the oblique aspect (oblique equidistant mapping, oblique conformal mapping, oblique equal area mapping) can be easily deduced from the preceding chapters. Far easier, in the mapping equations as well in the equations for the left principal stretches defined in Chapter 10, conventional coordinates *spherical longitude* Λ and *spherical latitudes* Φ and Φ_0 are simply replaced by their corresponding items *meta-spherical longitude* A and *meta-spherical latitudes* B and B_0. Transformations of conventional spherical coordinates to meta-spherical coordinates is performed using (12.2).

12-2 Special mapping equations

Setting up special equations of the mapping "sphere to cylinder": meta-cylindrical projections in the *oblique aspect*. Equidistant mapping (oblique Plate Carrée projection), conformal mapping (oblique Mercator projection), equal area mapping (oblique Lambert cylindrical equal area projection).

12-21 Equidistant mapping (oblique Plate Carrée projection), compare with Fig. 12.2

$$
\begin{bmatrix} x \\ y \end{bmatrix} = R \begin{bmatrix} A \cos B_0 \\ B \end{bmatrix} , \tag{12.4}
$$

$$
\Lambda_1 = \frac{\cos B_0}{\cos B} , \quad \Lambda_2 = 1 .
$$

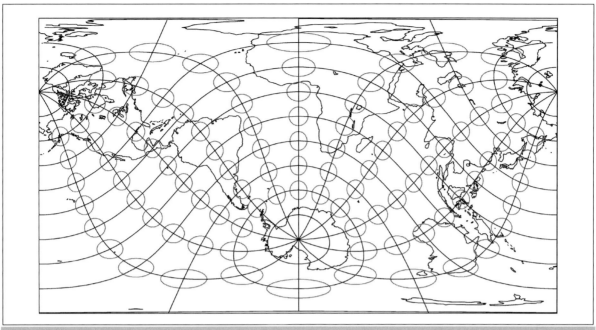

Fig. 12.2. Mapping the sphere to a cylinder: oblique aspect, equidistant mapping, $B_0 = 45°$, oblique Plate Carrée projection.

12-22 Conformal mapping (oblique Mercator projection), compare with Fig. 12.3

$$\begin{bmatrix} x \\ y \end{bmatrix} = R \cos B_0 \begin{bmatrix} A \\ \ln \cot\left(\frac{\pi}{4} - \frac{B}{2}\right) \end{bmatrix} = R \cos B_0 \begin{bmatrix} A \\ \ln \tan\left(\frac{\pi}{4} + \frac{B}{2}\right) \end{bmatrix} =$$

$$= R \cos B_0 \begin{bmatrix} A \\ \frac{1}{2} \ln \frac{1+\sin B}{1-\sin B} \end{bmatrix} = \tag{12.5}$$

$$= R \cos B_0 \begin{bmatrix} A \\ \operatorname{ar tanh}(\sin B) \end{bmatrix} ,$$

$$\Lambda_1 = \Lambda_2 = \frac{\cos B_0}{\cos B} . \tag{12.6}$$

12-23 Equal area mapping (oblique Lambert projection), compare with Fig. 12.4

$$\begin{bmatrix} x \\ y \end{bmatrix} = R \begin{bmatrix} A \cos B_0 \\ \frac{\sin B}{\cos B_0} \end{bmatrix} , \tag{12.7}$$

$$\Lambda_1 = \frac{\cos B_0}{\cos B} , \quad \Lambda_2 = \frac{\cos B}{\cos B_0} . \tag{12.8}$$

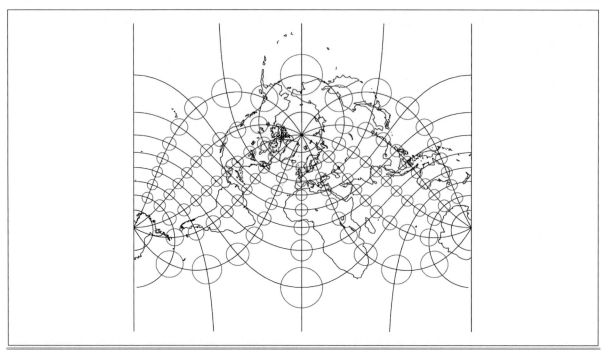

Fig. 12.3. Mapping the sphere to a cylinder: oblique aspect, conformal mapping, $B_0 = 45°$, oblique Mercator projection.

In the following chapter, let us continue studying the mapping of the sphere to the cylinder, namely let us study pseudo-cylindrical equal area projections.

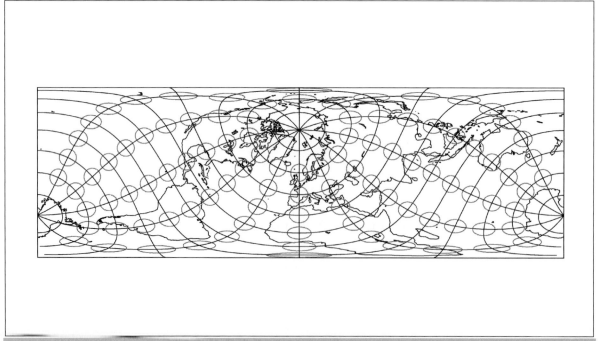

Fig. 12.4. Mapping the sphere to a cylinder: oblique aspect, equal area mapping, $B_0 - 60°$, oblique Lambert cylindrical equal area projection.

13 "Sphere to cylinder": pseudo-cylindrical projections

Mapping the sphere to a cylinder: pseudo-cylindrical projections. Sinusoidal pseudo-cylindrical mapping, elliptic pseudo-cylindrical mapping, parabolic pseudo-cylindrical mapping, rectilinear pseudo-cylindrical mapping. Jacobi matrix, Cauchy–Green matrix, principal stretches.

Pseudo-cylindrical projections have, in the normal aspect, straight parallel lines for parallels. The meridians are most often equally spaced along parallels, as they are on a cylindrical projection, but on which the meridians are curved. Meridians may be mapped as straight lines or general curves.

13-1 General mapping equations

General mapping equations and distortion measures for pseudo-cylindrical mappings of the sphere. Jacobi matrix, Cauchy–Green matrix, principal stretches.

The mapping equations are of the general form (13.1). The left Jacobi matrix is provided by (13.2) and the left Cauchy–Green matrix ($G_r = I_2$) by (13.3).

$$
\begin{aligned}
x &= x(\Lambda, \Phi) = R\Lambda \cos \Phi\, g(\Phi)\,, \\
y &= y(\Phi) = Rf(\Phi)\,,
\end{aligned}
\tag{13.1}
$$

$$
J_l := \begin{bmatrix} D_\Lambda x & D_\Phi x \\ D_\Lambda y & D_\Phi y \end{bmatrix} =
$$
$$
= R \begin{bmatrix} g(\Phi)\cos\Phi & -\Lambda[g(\Phi)\sin\Phi - g'(\Phi)\cos\Phi] \\ 0 & f'(\Phi) \end{bmatrix},
\tag{13.2}
$$

$$
C_l = J_l^* G_r J_l =
$$
$$
= R^2 \begin{bmatrix} g^2 \cos^2\Phi & -\Lambda g^2 \sin\Phi\cos\Phi + \Lambda g'g\cos^2\Phi \\ -\Lambda g^2 \sin\Phi\cos\Phi + \Lambda g'g\cos^2\Phi & \Lambda^2 g^2 \sin^2\Phi + f'^2 + \Lambda^2 {g'}^2 \cos^2\Phi - 2\Lambda^2 gg'\sin\Phi\cos\Phi \end{bmatrix}.
\tag{13.3}
$$

The left principal stretches are determined from the characteristic equation $\det[C_l - \Lambda_S^2 G_l] = 0$ and $G_l = \operatorname{diag}[R^2 \cos^2\Phi, R^2]$, which leads to the biquadratic equation (13.4), the solution of which is provided by (13.5).

$$
\Lambda_S^4 - \Lambda_S^2(\Lambda^2 g^2 \sin^2\Phi + f'^2 + g^2 + \Lambda^2 {g'}^2 \cos^2\Phi - 2\Lambda^2 gg'\sin\Phi\cos\Phi) + g^2 f'^2 = 0\,,
\tag{13.4}
$$

$$
\Lambda_S^2 = \tfrac{1}{2}(\Lambda^2 g^2 \sin^2\Phi + f'^2 + g^2 + \Lambda^2 {g'}^2 \cos^2\Phi - 2\Lambda^2 gg'\sin\Phi\cos\Phi)\pm
$$
$$
\pm\sqrt{\tfrac{1}{4}(\Lambda^2 g^2 \sin^2\Phi + f'^2 + g^2 + \Lambda^2 {g'}^2 \cos^2\Phi - 2\Lambda^2 gg'\sin\Phi\cos\Phi) - g^2 f'^2} =:
\tag{13.5}
$$

$$
=: a \pm b\,.
$$

The four roots are then given by (13.6). The postulate of "no area distortion", i.e.(13.7) now determines the relationship between the unknown functions f and g as (13.8).

$$(\Lambda_S)_{1,2} = \pm\sqrt{(\Lambda_S^2)_1} = \pm\sqrt{a+b} ,$$

$$(\Lambda_S)_{3,4} = \pm\sqrt{(\Lambda_S^2)_2} = \pm\sqrt{a-b} ,$$

$$(\Lambda_S)_{1,2}(\Lambda_S)_{3,4} =$$

$$= \sqrt{a+b}\sqrt{a-b} = a^2 - b^2 \overset{!}{=} 1 ,$$

$$f' = g^{-1} \Leftrightarrow g = f'^{-1} .$$
(13.8)

(13.6)

(13.7)

We therefore end up with the general mapping equations (13.9) and the left principal stretches (13.10). For the special case $f'(\Phi) = 1$, the left principal stretches can easily calculated as (13.11), which shows that on the equator, $\Phi = 0°$, we experience isometry (conformality).

$$x = x(\Lambda, \Phi) = R\Lambda \frac{\cos\Phi}{f'(\Phi)} = \frac{R^2\Lambda\cos\Phi}{\frac{dy}{d\Phi}} ,$$

$$y = y(\Phi) = Rf(\Phi) ,$$
(13.9)

$$\Lambda_S^2 =$$

$$= \frac{1}{2f'^4}(\Lambda^2 f'^2 \sin^2\Phi + f'^6 + f'^2 + \Lambda^2 f''^2 \cos^2\Phi + 2\Lambda^2 f'f'' \sin\Phi\cos\Phi)\pm$$

$$\pm\sqrt{\frac{1}{4f'^8}(\Lambda^2 f'^2 \sin^2\Phi + f'^6 + f'^2 + \Lambda^2 f''^2 \cos^2\Phi + 2\Lambda^2 f'f'' \sin\Phi\cos\Phi)^2 - 1} ,$$
(13.10)

$$(\Lambda_S)_{1,2} =$$

$$= \pm\frac{\sqrt{2}}{2}\sqrt{2 + \Lambda^2 \sin^2\Phi + \Lambda\sin\Phi\sqrt{4 + \Lambda^2 \sin^2\Phi}} ,$$

$$(\Lambda_S)_{3,4} =$$

$$= \pm\frac{\sqrt{2}}{2}\sqrt{2 + \Lambda^2 \sin^2\Phi - \Lambda\sin\Phi\sqrt{4 + \Lambda^2 \sin^2\Phi}} .$$
(13.11)

13-2 Special mapping equations

Special mapping equations for pseudo-cylindrical equal area mappings of the sphere. Sinusoidal pseudo-cylindrical mapping, elliptic pseudo-cylindrical mapping, parabolic pseudo-cylindrical mapping, rectilinear pseudo-cylindrical mapping.

The special mapping equations to be considered are the mapping equations of the sinusoidal pseudo-cylindrical mapping, the elliptic pseudo-cylindrical mapping, the parabolic pseudo-cylindrical mapping, and the rectilinear pseudo-cylindrical mapping. Let us study these special mapping equations in the sections that follow.

13-21 Sinusoidal pseudo-cylindrical mapping (J. Cossin 1570, N. Sanson 1650, J. Flamsteed 1646–1719), compare with Fig. 13.1

The mapping equations (13.12) are derived from (13.1) in connection with (13.8) and the special instruction $f(\Phi) = \Phi$. The left Jacobi matrix is given by (13.13) and the left Cauchy–Green matrix ($G_r = I_2$) by (13.14). The left principal stretches are defined by (13.15). The structure of the coordinate lines are defined by (13.16).

$$\begin{bmatrix} x \\ y \end{bmatrix} = R \begin{bmatrix} \Lambda \cos \Phi \\ \Phi \end{bmatrix} , \tag{13.12}$$

$$\mathsf{J}_l = R \begin{bmatrix} \cos \Phi & -\Lambda \sin \Phi \\ 0 & 1 \end{bmatrix} , \tag{13.13}$$

$$\mathsf{C}_l = \mathsf{J}_l^* \mathsf{G}_r \mathsf{J}_l = R^2 \begin{bmatrix} \cos^2 \Phi & -\Lambda \sin \Phi \cos \Phi \\ -\Lambda \sin \Phi \cos \Phi & \Lambda^2 \sin^2 \Phi + 1 \end{bmatrix} , \tag{13.14}$$

$$(\Lambda_S)_{1,2} = \pm \tfrac{\sqrt{2}}{2} \sqrt{2 + \Lambda^2 \sin^2 \Phi + \Lambda \sin \Phi \sqrt{4 + \Lambda^2 \sin^2 \Phi}} ,$$

$$(\Lambda_S)_{3,4} = \pm \tfrac{\sqrt{2}}{2} \sqrt{2 + \Lambda^2 \sin^2 \Phi - \Lambda \sin \Phi \sqrt{4 + \Lambda^2 \sin^2 \Phi}} , \tag{13.15}$$

$$\Phi = \frac{y}{R} \Rightarrow x = R\Lambda \cos \frac{y}{R} ,$$

$$\cos \Phi = \frac{x}{R\Lambda} \Rightarrow y = R \arccos \frac{x}{R\Lambda} . \tag{13.16}$$

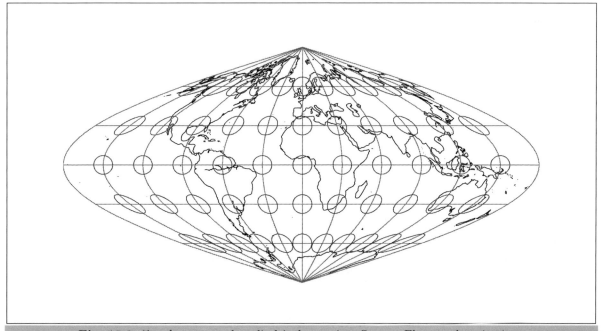

Fig. 13.1. Equal area pseudo-cylindrical mapping. Sanson–Flamsteed projection.

13-22 Elliptic pseudo-cylindrical mapping (C. B. Mollweide), compare with Fig. 13.2

Starting from the general equation of an ellipse, i.e. $x^2/a^2 + y^2/b^2 = 1$, with constant minor axis b and major axis $a = a(\Lambda)$ being a function of spherical longitude, we fix the size of b in such a way that a hemisphere $-\pi/2 \le \Lambda \le \pi/2$ is mapped onto a circle of the same area.

$$2\pi R^2 \qquad\qquad \pi r^2 = \pi b^2$$
$$\text{(area of the hemisphere)} \qquad \text{(area of a circle)}$$

(13.17)

$$\Rightarrow r = b = R\sqrt{2} \Rightarrow \frac{x^2}{a^2(\Lambda)} + \frac{y^2}{2R^2} = 1 \ .$$

Now the "Ansatz" (13.18) obviously fulfills the general ellipse equation. The choice (13.19) is motivated through the postulate of an equidistant mapping of the equator, $\Phi = t = 0$. In particular, we obtain $a(\pi/2) = b = R\sqrt{2}$!

$$x = a(\Lambda)\cos t \ , \quad y = b\sin t = R\sqrt{2}\sin t \ ,$$

(13.18)

$$t = t(\Phi) \ ,$$

$$a(\Lambda) = \frac{2\sqrt{2}}{\pi}R\Lambda \ .$$

(13.19)

The subsequent distortion analysis accompanied by the postulate of "no areal distortion" leads to the relationship (13.20) between the parameter t and spherical latitude Φ, which is solved by the separation-of-variables technique. The resulting equation (13.21) is a transcendental equation in t, the so-called *special Kepler equation*, well-known in satellite geodesy. It is best solved numerically, for example, by using the *Newton–Raphson method*.

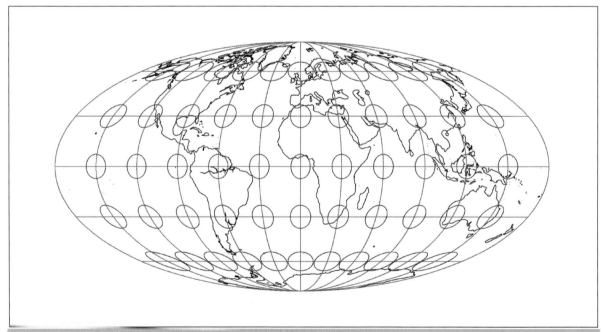

Fig. 13.2. Equal area pseudo-cylindrical mapping. Mollweide projection.

$$\cos \Phi = \pm \frac{4}{\pi} \frac{dt}{d\Phi} \cos^2 t \ , \tag{13.20}$$

$$\pi \sin \Phi = 2t + \sin 2t \ . \tag{13.21}$$

Table 13.1. The solution of the special Kepler equation.

Φ	t	Φ	t	Φ	t	Φ	t
0°	0	30°	0.415 85	60°	0.866 98	90°	$\pi/2$
10°	0.137 24	40°	0.559 74	70°	1.039 00		
20°	0.275 48	50°	0.709 10	80°	1.238 77		

The solution of the special Kepler equation is shown in Table 13.1. Thus, the final mapping equations are provided by (13.22). The left principal stretches are best determined by the numerical solution of the biquadratic characteristic equation based on the left Jacobi matrix (13.23) as well as the left Cauchy–Green deformation matrix (13.24) ($G_r = I_2$).

$$x = \frac{2\sqrt{2}}{\pi} R \Lambda \cos t \ , \quad y = R\sqrt{2} \sin t \ ,$$
$$2t + \sin 2t = \pi \sin \Phi \ , \tag{13.22}$$

$$J_l =$$
$$= \frac{R\sqrt{2}}{\pi} \begin{bmatrix} 2\cos t & -\frac{\pi \Lambda \tan t \cos \Phi}{2\cos t} \\ 0 & \frac{\pi^2 \cos \Phi}{4\cos t} \end{bmatrix} \ , \tag{13.23}$$

$$C_l = J_l^* G_r J_l =$$
$$= \frac{2R^2}{\pi^2} \begin{bmatrix} 4\cos^2 t & -\pi \Lambda \tan t \cos \Phi \\ -\pi \Lambda \tan t \cos \Phi & \frac{\pi^2 \cos^2 \Phi (\pi^2 + 4\Lambda^2 \tan^2 t)}{16\cos^2 t} \end{bmatrix} \ , \tag{13.24}$$

$$\det \left(C_l - \Lambda_S^2 G_l \right) =$$
$$= R^2 \begin{vmatrix} \frac{8}{\pi^2}\cos^2 t - \Lambda_S^2 \cos^2 \Phi & -\frac{2}{\pi}\Lambda \tan t \cos \Phi \\ -\frac{2}{\pi}\Lambda \tan t \cos \Phi & \frac{\cos^2 \Phi}{2\cos^2 t}(\Lambda^2 \tan^2 t + \frac{\pi^2}{4}) - \Lambda_S^2 \end{vmatrix} = \tag{13.25}$$
$$= 0 \ .$$

13-23 Parabolic pseudo-cylindrical mapping (J. E. E. Craster), compare with Fig. 13.3

This mapping is defined in such a way that the meridians except the central meridian, which is a straight line, are equally spaced parabolas. Parallels are unequally spaced straight lines, farthest apart near the equator. The mapping equations are defined by (13.26). The left Jacobi matrix is given by (13.27) and the left Cauchy–Green matrix is given by (13.28) ($G_r = I_2$). As can be seen from Fig. 13.3, map distortion is severe near outer meridians at high latitudes.

$$x =$$
$$= \sqrt{\tfrac{3}{\pi}} R \Lambda (2 \cos \tfrac{2\Phi}{3} - 1) \;,$$

$$y =$$
$$= \sqrt{3\pi} R \sin \tfrac{\Phi}{3} \;,$$

(13.26)

$$J_l =$$
$$= R \sqrt{\frac{3}{\pi}} \begin{bmatrix} 2 \cos \tfrac{2\Phi}{3} - 1 & -\tfrac{4}{3} \Lambda \sin \tfrac{2\Phi}{3} \\ 0 & \tfrac{\pi}{3} \cos \tfrac{\Phi}{3} \end{bmatrix} \;,$$

(13.27)

$$C_l = J_l^* G_r J_l =$$

$$= \frac{R^2}{3\pi} \begin{bmatrix} 9 \left(1 - 2 \cos \tfrac{2\Phi}{3}\right)^2 & 12\Lambda \sin \tfrac{2\Phi}{3} \left(1 - 2 \cos \tfrac{2\Phi}{3}\right) \\ 12\Lambda \sin \tfrac{2\Phi}{3} \left(1 - 2 \cos \tfrac{2\Phi}{3}\right) & 16\Lambda^2 \sin^2 \tfrac{2\Phi}{3} + \pi^2 \cos^2 \tfrac{\Phi}{3} \end{bmatrix} \;.$$

(13.28)

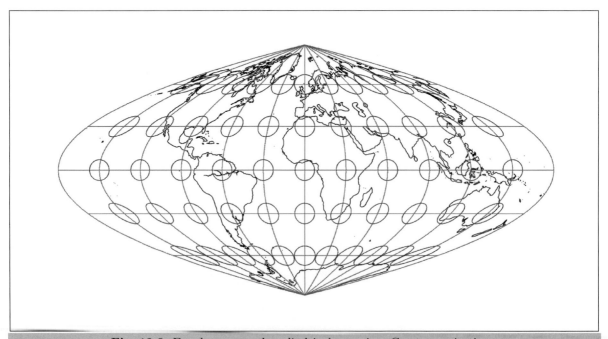

Fig. 13.3. Equal area pseudo-cylindrical mapping. Craster projection.

13-24 Rectilinear pseudo-cylindrical mapping (Eckert II), compare with Fig. 13.4

Here, the mapping instruction requires the meridians to be straight lines. This is generated by the mapping equations (13.29). (13.30) shows the left Jacobi matrix and (13.31) the left Cauchy–Green matrix ($G_r = I_2$). The structure of the meridian images is defined by (13.32). It is easily shown that the length of the poles and of the central meridian is half the length of the equator.

$$x = \frac{2R\Lambda}{\sqrt{6\pi}}\sqrt{4 - 3\sin|\Phi|} ,$$

$$y = R\sqrt{\frac{2\pi}{3}}\left(2 - \sqrt{4 - 3\sin|\Phi|}\right)\operatorname{sign}\Phi ,$$

$$\tag{13.29}$$

$$\mathsf{J}_l = R\begin{bmatrix} \frac{2}{\sqrt{6\pi}}\sqrt{4 - 3\sin|\Phi|} & -\frac{3\Lambda}{\sqrt{6\pi}}\frac{\cos\Phi\operatorname{sign}\Phi}{\sqrt{4-3\sin|\Phi|}} \\[2mm] 0 & \sqrt{\frac{3\pi}{2}}\frac{\cos\Phi}{\sqrt{4-3\sin|\Phi|}} \end{bmatrix} , \tag{13.30}$$

$$\mathsf{C}_l = \mathsf{J}_l^*\mathsf{G}_r\mathsf{J}_l = \frac{R^2}{\pi}\begin{bmatrix} \frac{2}{3}(4 - 3\sin|\Phi|) & -\Lambda\cos\Phi\operatorname{sign}\Phi \\[2mm] -\Lambda\cos\Phi\operatorname{sign}\Phi & \frac{3\cos^2\Phi(\Lambda^2+\pi^2)}{2(4-3\sin|\Phi|)} \end{bmatrix} , \tag{13.31}$$

$$\sqrt{4 - 3\sin|\Phi|} = \frac{\sqrt{6\pi}}{2R\Lambda}x$$
$$\Downarrow$$
$$y = \sqrt{\frac{2\pi}{3}}R\left(2 - \frac{\sqrt{6\pi}}{2R\Lambda}x\right)\operatorname{sign}\Phi = \left(-\pi\frac{x}{\Lambda} + R\sqrt{\frac{8\pi}{3}}\right)\operatorname{sign}\Phi . \tag{13.32}$$

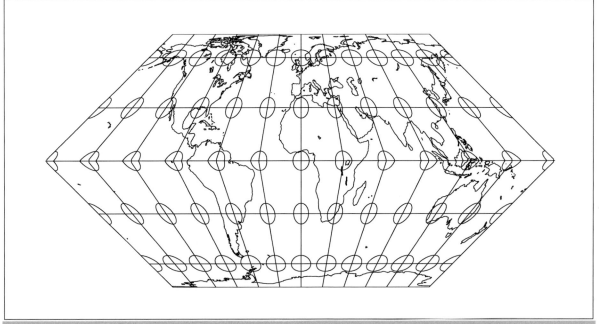

Fig. 13.4. Equal area pseudo-cylindrical mapping. Eckert II projection.

In the chapters that follow, let us study the mapping of the ellipsoid-of-revolution to the cylinder. Let us begin with the polar aspect.

14 "Ellipsoid-of-revolution to cylinder": polar aspect

Mapping the ellipsoid-of-revolution to a cylinder: polar aspect. Its generalization for general rotationally symmetric surfaces. Normal equidistant, normal conformal, and normal equiareal mappings. Cylindric mappings (equidistant) for a rotationally symmetric figure. Torus mapping.

At the beginning of this chapter, let us briefly refer to Chapter 8, where the data of the best fitting "ellipsoid-of-revolution to Earth" are derived in form of a table. Here, we specialize on the *mapping equations* and the *distortion measures* for mapping an ellipsoid-of-revolution $\mathbb{E}^2_{A_1,A_2}$ to a cylinder, equidistant on the equator. Section 14-1 concentrates on the structure of the mapping equations, while Section 14-2 gives special cylindric mappings of the ellipsoid-of-revolution, equidistant on the equator. At the end, we shortly review in Section 14-3 the general mapping equations of a rotationally symmetric figure different from an ellipsoid-of-revolution, namely the torus.

14-1 General mapping equations

General mapping equations of an ellipsoid-of-revolution to a cylinder: the polar aspect. Applications. Deformation tensor. Principal stretches.

The first postulate fixes the image coordinate y by the assumption of an exclusive dependence on the ellipsoidal latitude Φ. In contrast, the image coordinate x is only dependent on the longitude Λ, especially assuming that the equator is mapped equidistantly.

Postulate.

$$x = A_1 \Lambda , \quad y = f(\Phi) . \tag{14.1}$$

End of Postulate.

Assuming summation over repeated indices, we specialize the deformation tensor of first order c_{KL} according to (14.2). In detail, we note that (14.3) and (14.4) hold.

$$c_{KL} = g_{kl} \frac{\partial u^k}{\partial U^K} \frac{\partial u^l}{\partial U^l} = \delta_{kl} \frac{\partial u^k}{\partial U^K} \frac{\partial u^l}{\partial U^l} =$$
$$= \frac{\partial x^k}{\partial U^K} \frac{\partial x^l}{\partial U^l} , \tag{14.2}$$

$$c_{11} = \left(\frac{\partial x}{\partial \Lambda} \right)^2 + \left(\frac{\partial y}{\partial \Lambda} \right)^2 ,$$
$$c_{12} = \left(\frac{\partial x}{\partial \Lambda} \right) \left(\frac{\partial x}{\partial \Phi} \right) + \left(\frac{\partial y}{\partial \Lambda} \right) \left(\frac{\partial y}{\partial \Phi} \right) , \tag{14.3}$$
$$c_{22} = \left(\frac{\partial x}{\partial \Phi} \right)^2 + \left(\frac{\partial y}{\partial \Phi} \right)^2 ,$$

$$\frac{\partial x}{\partial \Lambda} = A_1 , \quad \frac{\partial x}{\partial \Phi} = \frac{\partial y}{\partial \Lambda} = 0 ,$$
$$\frac{\partial y}{\partial \Phi} = f'(\Phi) , \tag{14.4}$$
$$c_{KL} = \begin{bmatrix} A_1^2 & 0 \\ 0 & f'(\Phi) \end{bmatrix} .$$

At this point, let us finally review the principal stretches and let us finally give the general structure of the coordinate lines.

$$\Lambda_1 = \sqrt{c_{11}/G_{11}} = \frac{\sqrt{1 - E^2 \sin^2 \Phi}}{\cos \Phi} \ , \quad \Lambda_2 = \sqrt{c_{22}/G_{22}} = \frac{f'(\Phi)(1 - E^2 \sin^2 \Phi)^{3/2}}{A_1(1 - E^2)} \ , \quad (14.5)$$

$$x = A_1 \Lambda \ , \quad y = f(\Phi) \ . \tag{14.6}$$

14-2 Special mapping equations

Special mapping equations of cylindric mappings: normal equidistant, normal conformal, and normal equiareal mappings.

Next, we present special normal mappings of type *equidistant mapping*, *conformal mapping*, and *equiareal mapping* as second postulates.

14-21 Special normal cylindric mapping (equidistant: parallel circles, conformal: equator)

As it is shown in the following chain of relations, let us transfer the postulate of an equidistant mapping on the set of parallel circles.

$$\Lambda_2 = 1 \Rightarrow \frac{f'(\Phi)}{A_1(1 - E^2)}(1 - E^2 \sin^2 \Phi)^{3/2} = 1 \Leftrightarrow df = A_1(1 - E^2)\frac{d\Phi}{(1 - E^2 \sin^2 \Phi)^{3/2}} \ ,$$

$$\tag{14.7}$$

$$f(\Phi) = A_1(1 - E^2) \int_0^\Phi \frac{d\Phi'}{(1 - E^2 \sin^2 \Phi')^{3/2}} \ .$$

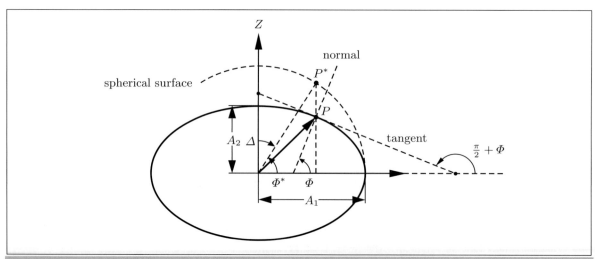

Fig. 14.1. Vertical section of the ellipsoid-of-revolution.

The integral $A_1(1-E^2) \int_0^\Phi \frac{\mathrm{d}\Phi'}{(1-E^2 \sin^2 \Phi')^{3/2}}$ is identified as the length of the meridian arc dependent on the *ellipsoidal latitude* Φ. Let us present the integral dependent on the *reduced latitude* Φ^*.

$$\tan \Phi^* = \frac{A_2}{A_1} \tan \Phi = \sqrt{1-E^2} \tan \Phi . \qquad (14.8)$$

Its derivation is based upon $\sqrt{X^2 + Y^2} = \sqrt{X^{*2} + Y^{*2}} = A_1 \cos \Phi^*$ and $Z^* = A_1 \sin \Phi^*$. A point P is characterized by the identical abszissa as a point P^* on the substitutional sphere of radius A.

$$\begin{bmatrix} X \\ Y \\ Z \end{bmatrix} = \begin{bmatrix} A_1 \cos \Lambda \cos \Phi^* \\ A_1 \cos \Lambda \cos \Phi^* \\ A_2 \sin \Phi^* \end{bmatrix} . \qquad (14.9)$$

We gain the parametric representation by $\Lambda = \Lambda^*$ and $\sqrt{X^2 + Y^2} = \sqrt{X^{*2} + Y^{*2}}$. The integral dependent of the reduced latitude then is obtained as follows.

$$\mathbb{E}^2_{A_1, A_2} := \left\{ (X, Y, Z) : \frac{X^2 + Y^2}{A_1^2} + \frac{Z^2}{A_2^2} = 1 \right\} \subset \mathbb{R}^3 , \qquad (14.10)$$

$$\frac{Z}{\sqrt{X^2 + Y^2}} = (1 - E^2) \tan \Phi = \frac{A_2}{A_1} \tan \Phi^* , \quad \frac{A_2}{A_1} = \sqrt{1 - E^2}$$

$$\Rightarrow$$

$$\tan \Phi^* = \sqrt{1 - E^2} \tan \Phi = \frac{A_2}{A_1} \tan \Phi \qquad (14.11)$$

$$\Rightarrow$$

$$f(\Phi) = A_1 (1 - E^2) \int_0^\Phi \frac{\mathrm{d}\Phi'}{(1 - E^2 \sin^2 \Phi')^{3/2}} = A_1 \int_0^{\Phi^*} \mathrm{d}\Phi^{*'} \sqrt{(1 - E^2 \cos^2 \Phi^{*'})} . \qquad (14.12)$$

Let us here also define the elliptic integral of the second kind (for example, consult Appendix C or I. S. Gradshteyn and I. M. Ryzhik (1983), namely page 905, formula 8.1113). The definition (14.13) leads for $f(\Phi)$ to the representation (14.14). The principal stretches are easily computed as (14.15).

$$f(\Phi) = A_1 (1 - E^2) \int_0^\Phi \frac{\mathrm{d}\Phi'}{(1 - E^2 \sin^2 \Phi')^{3/2}} , \quad f(\Phi^*) = A_1 \int_0^{\Phi^*} \mathrm{d}\Phi^{*'} \sqrt{1 - E^2 \cos^2 \Phi^{*'}} ,$$

$$f(\Delta^*) = A_1 \int_{\Delta^*}^{\pi/2} \mathrm{d}\Delta^{*'} \sqrt{1 - E^2 \sin^2 \Delta^{*'}} = A_1 \int_0^{\pi/2} \mathrm{d}\Delta^{*'} \sqrt{1 - E^2 \sin^2 \Delta^{*'}} - \qquad (14.13)$$

$$- A_1 \int_0^{\Delta^*} \mathrm{d}\Delta^{*'} \sqrt{1 - E^2 \sin^2 \Delta^{*'}} = A_1 \left[\boldsymbol{E}(\pi/2, E) - \boldsymbol{E}(\Delta^*, E) \right] ,$$

$$f(\Phi) = A_1 \left(\boldsymbol{E}(\pi/2, E) - \boldsymbol{E} \left[\frac{\pi}{2} - \arctan \left(\frac{A_2}{A_1} \tan \Phi \right), E \right] \right) , \qquad (14.14)$$

$$\Lambda_1 = \frac{\sqrt{1 - E^2 \sin^2 \Phi}}{\cos \Phi} , \quad \Lambda_2 = 1 . \qquad (14.15)$$

14-22 Special normal cylindric mapping (normal conformal, equidistant: equator)

First, let us here apply the postulate of conformal mapping. Similar as before, we obtain the following set of formulae.

$$\Lambda_1 = \Lambda_2 \Leftrightarrow \frac{\sqrt{1 - E^2 \sin^2 \Phi}}{\cos \Phi} = \frac{(1 - E^2 \sin^2 \Phi)^{3/2}}{A_1(1 - E^2)} f'(\Phi) \Rightarrow$$

$$\Rightarrow \mathrm{d}f = \frac{A_1(1 - E^2)}{\cos \Phi} \frac{1}{1 - E^2 \sin^2 \Phi} \mathrm{d}\Phi \Rightarrow f(\Phi) = A_1 \int_0^\Phi \frac{\mathrm{d}\Phi'}{\cos \Phi'} \frac{1 - E^2}{1 - E^2 \sin^2 \Phi'} \; . \tag{14.16}$$

The integral is called "isometric latitude". Applying "integration-by-parts", we obtain the following set of formulae.

$$f(\Phi) = A_1 \int_0^\Phi \mathrm{d}\Phi' \left[\frac{1}{\cos \Phi'} - \frac{E}{2} \left(\frac{E \cos \Phi'}{1 + E \sin \Phi'} + \frac{E \cos \Phi'}{1 - E \sin \Phi'} \right) \right] , \tag{14.17}$$

$$f(\Phi) = A_1 \ln \tan \left(\frac{\pi}{4} + \frac{\Phi}{2} \right) - \frac{A_1 E}{2} \ln \frac{1 + E \sin \Phi}{1 - E \sin \Phi} \; . \tag{14.18}$$

At this point, let us present the mapping equations as well as the principal stretches. They are easily computed as follows.

$$\begin{bmatrix} x \\ y \end{bmatrix} = \begin{bmatrix} A_1 \Lambda \\ A_1 \ln \left[\tan \left(\frac{\pi}{4} + \frac{\Phi}{2} \right) \left(\frac{1 - E \sin \Phi}{1 + E \sin \Phi} \right)^{E/2} \right] \end{bmatrix} , \tag{14.19}$$

$$\Lambda_1 = \Lambda_2 = \frac{\sqrt{1 - E^2 \sin^2 \Phi}}{\cos \Phi} \; . \tag{14.20}$$

14-23 Special normal cylindric mapping (normal equiareal, equidistant: equator)

Second, let us here apply the postulate of equiareal mapping. In doing so, we obtain the following chain of relations.

$$\Lambda_1 \Lambda_2 = 1 \Leftrightarrow \frac{\sqrt{1 - E^2 \sin^2 \Phi}}{\cos \Phi} \frac{f'(\Phi)}{A_1(1 - E^2)} (1 - E^2 \sin^2 \Phi)^{3/2} = 1 \Rightarrow$$

$$\Rightarrow \mathrm{d}f = A_1(1 - E^2) \frac{\cos \Phi}{(1 - E^2 \sin^2 \Phi)^2} \mathrm{d}\Phi \Rightarrow f(\Phi) = A_1(1 - E^2) \int_0^\Phi \mathrm{d}\Phi' \frac{\cos \Phi'}{(1 - E^2 \sin^2 \Phi')^2} , \tag{14.21}$$

$$f(\Phi) = \frac{A_1(1 - E^2)}{4E} \left(\ln \frac{1 + E \sin \Phi}{1 - E \sin \Phi} + \frac{2E \sin \Phi}{1 - E^2 \sin^2 \Phi} \right) \; . \tag{14.22}$$

Proof.

"Integration-by-parts":

$$\int_0^x \frac{\cos x' \, dx'}{(1 - E^2 \sin^2 x')^2} = \frac{1}{E} \int_0^x \frac{d(E \sin x')}{(1 - E^2 \sin^2 x')^2} = \frac{1}{E} \int_0^{E \sin x} \frac{dy}{(1 - y^2)^2} \,,$$

$$\frac{1}{(1 - y^2)^2} = \frac{1}{(1 + y)^2 (1 - y)^2} = \frac{A}{(1 - y)^2} + \frac{B}{1 - y} + \frac{C}{(1 + y)^2} + \frac{D}{1 + y} \,,$$

$$A(1 + y)^2 + B(1 + y)^2 (1 - y) + C(1 - y)^2 + D(1 - y)^2 (1 + y) = 1$$

$$\Rightarrow$$

$$A(1 + 2y + y^2) + B(1 + y - y^2 - y^3) + C(1 - 2y + y^2) + D(1 - y - y^2 + y^3) = 1 \,;$$

(14.23)

(i)
$$y^3 : -B + D = 0 \,,$$

(ii)
$$y^2 : A - B + C - D = 0 \,,$$

(iii)
$$y : 2A + B - 2C - D = 0 \,,$$

(iv)
$$1 : A + B + C + D = 1 \,;$$

(14.24)

$$-(\text{ii}) + (\text{iv}) : 2(B + D) = 1 \,, \quad (\text{i}) : -B + D = 0 \Rightarrow B = D = \frac{1}{4} \,,$$

$$B = D = \frac{1}{4} \,, \quad (\text{ii}) : A + C = \frac{1}{2} \,, \quad (\text{iii}) : 2(A - C) = 0 \Rightarrow A = C = \frac{1}{4} \,;$$

(14.25)

$$\frac{1}{E} \int_0^{E \sin x} \frac{dy}{(1 - y^2)^2} = \frac{1}{4E} \int_0^{E \sin x} \left[\frac{1}{(1 - y)^2} + \frac{1}{(1 + y)^2} + \frac{1}{1 - y} + \frac{1}{1 + y} \right] dy \,.$$

(14.26)

Standard Integrals:

$$\int \frac{dy}{ay + b} = \frac{1}{a} \ln |ay + b| \,, \quad \int \frac{dy}{(+y + 1)^2} = -\frac{1}{+y + 1} \,, \quad \int \frac{dy}{(-y + 1)^2} = \frac{1}{-y + 1} \,,$$

(14.27)

$$\frac{1}{4E} \left[\ln \left(\frac{1 + y}{1 - y} \right) + \frac{1}{1 - y} - \frac{1}{1 + y} \right] \Bigg|_0^{E \sin x} = \frac{1}{4E} \left[\ln \left(\frac{1 + E \sin x}{1 - E \sin x} \right) + \frac{2E \sin x}{1 - E^2 \sin^2 x} \right] \,.$$

(14.28)

End of Proof.

To this end, we review the mapping equations and the principal stretches. They are easily computed as follows.

$$
\begin{bmatrix} x \\ y \end{bmatrix} = \begin{bmatrix} A_1 \Lambda \\ \frac{A_1(1-E^2)}{4E} \left[\ln \left(\frac{1+E \sin \Phi}{1-E \sin \Phi} \right) + \frac{2E \sin \Phi}{1-E^2 \sin^2 \Phi} \right] \end{bmatrix} , \tag{14.29}
$$

$$
\Lambda_1 = \frac{\sqrt{1 - E^2 \sin^2 \Phi}}{\cos \Phi} , \quad \Lambda_2 = \frac{\cos \Phi}{\sqrt{1 - E^2 \sin^2 \Phi}} . \tag{14.30}
$$

14-24 Summary (cylindric mapping equations)

For the convenience of the reader, the central formulae that specify the mapping equations and the principal stretches are summarized in the following Box 14.1.

Box 14.1 (Summary).

Type 1 (equidistant on the set of parallel circles):

$$
x = A_1 \Lambda , \quad y = f(\Phi) , \tag{14.31}
$$

$$
f(\Phi) = A_1 \left(E(\pi/2, E) - E \left[\frac{\pi}{2} - \arctan \left(\frac{A_2}{A_1} \tan \Phi, E \right) \right] \right) , \tag{14.32}
$$

$$
\Lambda_1 = \frac{\sqrt{1 - E^2 \sin^2 \Phi}}{\cos \Phi} , \quad \Lambda_2 = 1 . \tag{14.33}
$$

"Elliptic integral".

Type 2 (normal conformal):

$$
x = A_1 \Lambda , \quad y = f(\Phi) , \tag{14.34}
$$

$$
f(\Phi) = A_1 \ln \left[\tan \left(\frac{\pi}{4} + \frac{\Phi}{2} \right) \left(\frac{1 - E \sin \Phi}{1 + E \sin \Phi} \right)^{E/2} \right] , \tag{14.35}
$$

$$
\Lambda_1 = \Lambda_2 = \frac{\sqrt{1 - E^2 \sin^2 \Phi}}{\cos \Phi} . \tag{14.36}
$$

Type 3 (normal equiareal):

$$
x = A_1 \Lambda , \quad y = f(\Phi) , \tag{14.37}
$$

$$
f(\Phi) = \frac{A_1(1 - E^2)}{4E} \left[\ln \left(\frac{1 + E \sin \Phi}{1 - E \sin \Phi} \right) + \frac{2E \sin \Phi}{1 - E^2 \sin^2 \Phi} \right] , \tag{14.38}
$$

$$
\Lambda_1 = \frac{\sqrt{1 - E^2 \sin^2 \Phi}}{\cos \Phi} , \quad \Lambda_2 = \frac{\cos \Phi}{\sqrt{1 - E^2 \sin^2 \Phi}} . \tag{14.39}
$$

14-3 General cylindric mappings (equidistant, rotational-symmetric figure)

General mapping equations and distortion measures of cylindric mappings of type *equidistant mappings* in case of a rotationally symmetric figure.

Let us here review the structure of the general mapping equations of a rotationally symmetric figure mapped onto a cylinder: in Box 14.2, we collect the parameterization of a rotationally symmetric figure, the left coordinates of the metric tensor, the right coordinates of the metric tensor, the left Cauchy–Green matrix, and the left principal stretches. Following this, we present special *cylindric mappings* of a *rotationally symmetric figure* which are equidistant on the equator. In addition, let us assume that the image coordinate y depends only on the *latitude* Φ, while the image coordinate x depends on the *longitude* Λ under the constraint that the equator is mapped *equidistanly*. $x(\Lambda)$ and $y(\Phi)$ are the result. Finally, special cylindric mappings onto the rotationally symmetric figure are presented. As an example, we present the *torus*.

Box 14.2 (Rotationally symmetric figure mapped onto a cylinder).

Parameterization of a rotationally symmetric figure:

$$\{\Lambda, \Phi\} \to \{X, Y, Z\} \, ,$$

$$\boldsymbol{X}(\Lambda, \Phi) = \boldsymbol{E}_1 F(\Phi) \cos \Lambda + \boldsymbol{E}_2 F(\Phi) \sin \Lambda + \boldsymbol{E}_3 G(\Phi) \, .$$

(14.40)

Inverse parameterization:

$$\{X, Y, Z\} \to \{\Lambda, \Phi\} \, ,$$

$$\Lambda(\boldsymbol{X}) = \arctan Y X^{-1} \, , \quad \Phi(\boldsymbol{X}) : \text{ the general form is not representable} \, .$$

(14.41)

Coordinates of the left metric tensor (rotationally symmetric figure):

$$\mathsf{G}_l = \begin{bmatrix} F^2(\Phi) & 0 \\ 0 & F'^{\,2}(\Phi) + G'^{\,2}(\Phi) \end{bmatrix} \, .$$

(14.42)

Coordinates of the right metric tensor (cylinder):

$$\mathsf{G}_\mathrm{r} = \mathsf{I}_2 \, .$$

(14.43)

Parameterized mapping:

$$x = F(0)\Lambda \, , \quad y = f(\Phi) \, .$$

(14.44)

Left Jacobi matrix:

$$\mathsf{J}_l = \begin{bmatrix} D_\Lambda x & D_\Phi x \\ D_\Lambda y & D_\Phi y \end{bmatrix} = \begin{bmatrix} F(0) & 0 \\ 0 & f'(\Phi) \end{bmatrix} \, .$$

(14.45)

Left Cauchy–Green matrix:

$$\mathsf{C}_l = \mathsf{J}_l^* \mathsf{G}_\mathrm{r} \mathsf{J}_l = \begin{bmatrix} F^2(0) & 0 \\ 0 & f'^{\,2}(\Phi) \end{bmatrix} \, .$$

(14.46)

Continuation of Box.

Left principal stretches:

$$\Lambda_1 = \sqrt{c_{11}/G_{11}} = \frac{F(0)}{F(\Phi)} \ , \quad \Lambda_2 = \sqrt{c_{22}/G_2} = \frac{f'(\Phi)}{\sqrt{F'^{\,2}(\Phi) + G'^{\,2}(\Phi)}} \ . \tag{14.47}$$

Structure of the coordinate lines:

$$\text{(i)}: x = F(0)\Lambda \ . \tag{14.48}$$

(Straight line through the origin for $\Lambda = $ const.)

$$\text{(ii)}: y = f(\Phi) \ . \tag{14.49}$$

(Straight line through the origin for $\Phi = $ const.)

14-31 Special normal cylindric mapping (equidistant: equator, set of parallel circles)

We start off by the postulate of an equidistant mapping on the set of parallel circles. This leads to the following result.

$$\Lambda_2 = 1 \Rightarrow \frac{f'(\Phi)}{\sqrt{F'^{\,2}(\Phi) + G'^{\,2}(\Phi)}} = 1 \Leftrightarrow \mathrm{d}f = \sqrt{F'^{\,2}(\Phi) + G'^{\,2}(\Phi)}\mathrm{d}\Phi \ ,$$

$$f(\Phi) = \int_0^\Phi \sqrt{F'^{\,2}(\tilde{\Phi}) + G'^{\,2}(\tilde{\Phi})}\mathrm{d}\tilde{\Phi} + \text{const.} \ , \quad f(0) = 0 \Rightarrow \text{const.} = 0 \ , \tag{14.50}$$

$$\begin{bmatrix} x \\ y \end{bmatrix} = \begin{bmatrix} F(0)\Lambda \\ \int_0^\Phi \sqrt{F'^{\,2}(\tilde{\Phi}) + G'^{\,2}(\tilde{\Phi})}\mathrm{d}\tilde{\Phi} \end{bmatrix} \ , \quad \Lambda_1 = \frac{F(0)}{F(\Phi)} \ , \quad \Lambda_2 = 1 \ . \tag{14.51}$$

14-32 Special normal conformal cylindric mapping (equidistant: equator)

Alternatively, let us start off by the postulate of a conformal mapping. Similar as before, this leads to the following result.

$$\Lambda_1 = \Lambda_2 \Rightarrow \frac{F(0)}{F(\Phi)} = \frac{f'(\Phi)}{\sqrt{F'^{\,2}(\Phi) + G'^{\,2}(\Phi)}} \Leftrightarrow \mathrm{d}f = F(0)\frac{\sqrt{F'^{\,2}(\Phi) + G'^{\,2}(\Phi)}}{F(\Phi)}\mathrm{d}\Phi \ ,$$

$$f(\Phi) = \int_0^\Phi F(0)\frac{\sqrt{F'^{\,2}(\tilde{\Phi}) + G'^{\,2}(\tilde{\Phi})}}{F(\tilde{\Phi})}\mathrm{d}\tilde{\Phi} + \text{const.} \ , \quad f(0) = 0 \Rightarrow \text{const.} = 0 \ , \tag{14.52}$$

$$\begin{bmatrix} x \\ y \end{bmatrix} = F(0)\begin{bmatrix} \Lambda \\ \int_0^\Phi \frac{\sqrt{F'^{\,2}(\tilde{\Phi}) + G'^{\,2}(\tilde{\Phi})}}{F(\tilde{\Phi})}\mathrm{d}\tilde{\Phi} \end{bmatrix} \ , \quad \Lambda_1 = \Lambda_2 = \frac{F(0)}{F(\Phi)} \ . \tag{14.53}$$

14-33 Special normal equiareal cylindric mapping (equidistant + conformal: equator)

Here, let us depart from the the postulate of an equiareal mapping. Step by step, this leads to the following result.

$$
\Lambda_1 \Lambda_2 = 1 \Rightarrow \frac{F(0)}{F(\Phi)} \frac{f'(\Phi)}{\sqrt{F'^{\,2}(\Phi) + G'^{\,2}(\Phi)}} = 1 \Leftrightarrow \mathrm{d}f = \frac{F(\Phi)}{F(0)} \sqrt{F'^{\,2}(\Phi) + G'^{\,2}(\Phi)} \, \mathrm{d}\Phi \,,
$$

$$
f(0) = 0 \Rightarrow f(\Phi) = \frac{1}{F(0)} \int_0^\Phi F(\tilde{\Phi}) \sqrt{F'^{\,2}(\tilde{\Phi}) + G'^{\,2}(\tilde{\Phi})} \, \mathrm{d}\tilde{\Phi} + \mathrm{const.} \,, \tag{14.54}
$$

$$
f(0) = 0 \Rightarrow \mathrm{const.} = 0 \,.
$$

The following formulae define the general mapping equations and the left principal stretches for an equiareal cylindric mapping.

$$
\begin{bmatrix} x \\ y \end{bmatrix} = \begin{bmatrix} F(0)\Lambda \\ \frac{1}{F(0)} \int_0^\Phi F(\tilde{\Phi}) \sqrt{F'^{\,2}(\tilde{\Phi}) + G'^{\,2}(\tilde{\Phi})} \mathrm{d}\tilde{\Phi} \end{bmatrix} \,, \quad \Lambda_1 = \frac{F(0)}{F(\Phi)} \,, \quad \Lambda_2 = \frac{F(\Phi)}{F(0)} = \frac{1}{\Lambda_1} \,. \tag{14.55}
$$

14-34 An example (mapping the torus)

The torus is the *product manifold* $\mathbb{S}_A^1 \times \mathbb{S}_B^1$, especially for the parameter range $A > B$, $0 < U < 2\pi$, and $0 < V < 2\pi$.

$$
\begin{bmatrix} X \\ Y \\ Z \end{bmatrix} = \begin{bmatrix} (A + B \cos V) \cos U \\ (A + B \cos V) \sin U \\ B \sin V \end{bmatrix} = \begin{bmatrix} (A + B \cos \Phi) \cos \Lambda \\ (A + B \cos \Phi) \sin \Lambda \\ B \sin \Phi \end{bmatrix} \,. \tag{14.56}
$$

The torus is the special surface which is generated by rotating a circle of radius B relative to a circle of radius $A > B$ around the center of the circle.

$$
F(\Phi) = A + B \cos \Phi \,, \quad G(\Phi) = B \sin \Phi \,,
$$
$$
U = \Lambda = \arctan \tfrac{Y}{X} \,, \quad V = \Phi = \arctan \frac{Z}{\sqrt{X^2 + Y^2} - A} \,. \tag{14.57}
$$

Let us summarize in Box 14.3 the (left) tangent vectors and the (left) coordinates of the metric tensor of the torus.

Box 14.3 (A special rotational figure: the torus).

Left tangent vectors:

$$
\boldsymbol{G}_\Lambda := \frac{\partial \boldsymbol{X}}{\partial \Lambda} = -\boldsymbol{E}_1 (A + B \cos \Phi) \sin \Lambda + \boldsymbol{E}_2 (A + B \cos \Phi) \cos \Lambda \,,
$$

$$
\boldsymbol{G}_\Phi := \frac{\partial \boldsymbol{X}}{\partial \Phi} = -\boldsymbol{E}_1 B \sin \Phi \cos \Lambda - \boldsymbol{E}_2 B \sin \Phi \sin \Lambda + \boldsymbol{E}_3 B \cos \Phi \,, \tag{14.58}
$$

$$
F'(\Phi) = -B \sin \Phi \,, \quad G'(\Phi) = B \cos \Phi \,. \tag{14.59}
$$

Coordinates of the metric tensor:

$$
\boldsymbol{G}_l = \begin{bmatrix} (A + B \cos \Phi)^2 & 0 \\ 0 & B^2 \end{bmatrix} \,. \tag{14.60}
$$

The mapping equations are provided by the following formulae. As "equator", let us define the coordinate line $\Phi = 0$ in the X, Y plane.

$$\begin{bmatrix} X \\ Y \\ Z \end{bmatrix}_{\Phi=0} = \begin{bmatrix} (A+B)\cos\Lambda \\ (A+B)\sin\Lambda \\ 0 \end{bmatrix} , \tag{14.61}$$

$$(X^2 + Y^2)_{\Phi=0} = (A+B)^2 , \quad \sqrt{(X^2+Y^2)_{\Phi=0}} = A+B , \quad F(0) = A+B , \tag{14.62}$$

$$x = (A+B)\Lambda , \quad y = f(\Phi) . \tag{14.63}$$

Most notable, we could have alternatively chosen the "equator" as $\Phi = \pi$. From this, we conclude the special case $F(\pi) = A - B$. In addition, we refer to Fig. 14.2 illustrating the geometry of the torus, namely its *vertical section*. As a case study, we present the special forms of the deformation tensor for the torus as well as its left principal stretches.

$$\mathsf{C}_l = \begin{bmatrix} F^2(0) & 0 \\ 0 & f'^{\,2}(\Phi) \end{bmatrix} = \begin{bmatrix} (A+B)^2 & 0 \\ 0 & f'^{\,2}(\Phi) \end{bmatrix} . \tag{14.64}$$

$$\Lambda_1 = \frac{F(0)}{F(\Phi)} = \frac{A+B}{A+B\cos\Phi} , \quad \Lambda_2 = \frac{f'(\Phi)}{\sqrt{F'^{\,2}(\Phi) + G'^{\,2}(\Phi)}} = \frac{f'(\Phi)}{B} . \tag{14.65}$$

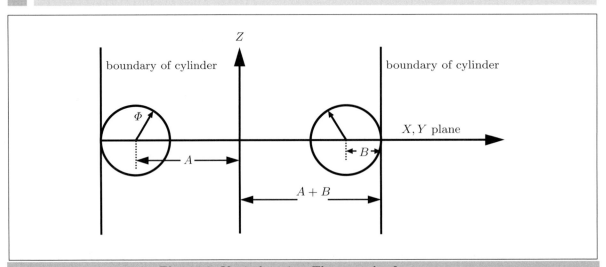

Fig. 14.2. Vertical section. The example of a torus.

The special case *normal cylindric mapping*, equidistant on the equator and the set of parallel circles, the special case *normal conformal cylindric mapping*, equidistant on the equator, and the special *normal equiareal cylindric mapping*, equidistant on the equator are summarized in Box 14.4.

Box 14.4 (Summary).

Case 1
(normal cylindric mapping, equidistant on the equator and the set of parallel circles):

$$f(\Phi) = \int_0^\Phi B\mathrm{d}\Phi' = B\Phi, \quad \begin{bmatrix} x \\ y \end{bmatrix} = \begin{bmatrix} (A+B)\Lambda \\ B\Phi \end{bmatrix}. \tag{14.66}$$

Case 2
(normal conformal cylindric mapping, equidistant on the equator):

$$f(\Phi) = (A+B) \int_0^\Phi \frac{B}{A + B\cos\Phi'} \mathrm{d}\Phi' =$$

$$= (A+B) \int_0^\Phi \frac{\mathrm{d}\Phi'}{AB^{-1} + \cos\Phi'} = \tag{14.67}$$

$$= \frac{2B(A+B)}{\sqrt{A^2-B^2}} \arctan\frac{\tan\dfrac{\Phi}{2}\sqrt{A^2-B^2}}{A+B}.$$

Mapping equations and principal stretches:

$$\begin{bmatrix} x \\ y \end{bmatrix} = \begin{bmatrix} (A+B)\Lambda \\ \dfrac{2B(A+B)}{\sqrt{A^2-B^2}} \arctan\dfrac{\tan\dfrac{\Phi}{2}\sqrt{A^2-B^2}}{A+B} \end{bmatrix}, \quad \Lambda_1 = \Lambda_2 = \frac{F(0)}{F(\Phi)} = \frac{A+B}{A+B\cos\Phi}. \tag{14.68}$$

Case 3
(normal equiareal cylindric mapping, equidistant on the equator):

$$f(\Phi) = \frac{B}{A+B} \int_0^\Phi \left(A + B\cos\Phi'\right) \mathrm{d}\Phi' =$$

$$= \frac{AB}{A+B} \int_0^\Phi \mathrm{d}\Phi' + \frac{B^2}{A+B} \int_0^\Phi \mathrm{d}\Phi' \cos\Phi' = \frac{AB}{A+B}\Phi + \frac{B^2}{A+B}\sin\Phi. \tag{14.69}$$

Mapping equations and principal stretches:

$$\begin{bmatrix} x \\ y \end{bmatrix} = \begin{bmatrix} (A+B)\Lambda \\ \dfrac{AB}{A+B}\Phi + \dfrac{B^2}{A+B}\sin\Phi \end{bmatrix}, \quad \Lambda_1 = \frac{F(0)}{F(\Phi)} = \frac{A+B}{A+B\cos\Phi}, \quad \Lambda_2 = \frac{1}{\Lambda_1} = \frac{A+B\cos\Phi}{A+B}. \tag{14.70}$$

Let us now "switch" from the polar aspect of the mapping "ellipsoid-of-revolution to cylinder" to the transverse aspect of the mapping "ellipsoid-of-revolution to cylinder".

15 "Ellipsoid-of-revolution to cylinder": transverse aspect

Mapping the ellipsoid-of-revolution to a cylinder: transverse aspect. Transverse Mercator projection, Gauss–Krueger/UTM coordinates. Korn–Lichtenstein equations, Laplace–Beltrami equations.

Conventionally, conformal coordinates, also called *conformal charts*, representing the surface of the Earth or any other Planet as an *ellipsoid-of-revolution*, also called the *Geodetic Reference Figure*, are generated by a two-step procedure. First, *conformal coordinates* (isometric coordinates, isothermal coordinates) of type *UMP* (*Universal Mercator Projection*, compare with Example 15.1) or of type *UPS* (*Universal Polar Stereographic Projection*, compare with Example 15.2) are derived from geodetic coordinates such as surface normal ellipsoidal longitude/ellipsoidal latitude. UMP is classified as a conformal mapping on a circular cylinder, while UPS refers to a conformal mapping onto a polar tangential plane with respect to an ellipsoid-of-revolution, an azimuthal mapping. The conformal coordinates of type UMP or UPS, respectively, are consequently complexified, just describing the two-dimensional Riemann manifold of type of ellipsoid-of-revolution as one-dimensional *complex manifold*. Namely, the *real-valued conformal coordinates* x and y of type UMP or UPS, respectively, are transformed into the *complex-valued conformal coordinate* $z = x + iy$. Second, the conformal coordinates $(x, y) \sim z$ of type UMP or UPS, respectively, are transformed into another set of conformal coordinates, called *Gauss–Krueger* or *UTM*, by means of holomorphic functions $w(z)$ ($w := u + iv \in \mathbb{C}$) with respect to complex algebra and complex analysis. Indeed, holomorphic functions directly fulfill the *d'Alembert–Euler equations* (*Cauchy–Riemann equations*) of conformal mapping as outlined by E. Grafarend (1995), for instance. Consult Figs. 15.1 and 15.2 for a first impression.

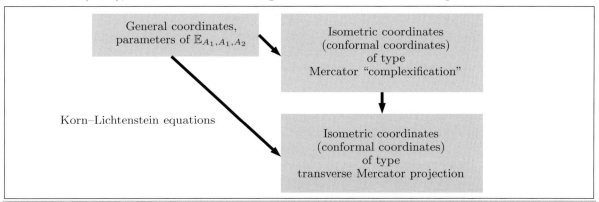

Fig. 15.1. Change from one conformal chart to another conformal chart (c:c: Cha-Cha-Cha) according to a proposal by C. F. Gauss (1822, 1844). First conformal coordinates: Mercator projection. Second conformal coordinates: transverse Mercator projection. Ellipsoid-of-revolution $\mathbb{E}_{A_1, A_1, A_2}$.

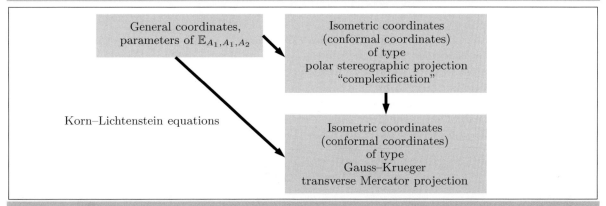

Fig. 15.2. Change from one conformal chart to another conformal chart (c:c: Cha-Cha-Cha) according to a proposal by L. Krueger (1922). First conformal coordinates: polar stereographic projection. Second conformal coordinates: transverse Mercator projection. Ellipsoid-of-revolution $\mathbb{E}_{A_1, A_1, A_2}$.

This two-step procedure has at least two basic disadvantages. On the one hand, it is in general difficult to set up a first set of conformal coordinates. For instance, due to the involved difficulties the Philosophical Faculty of the University of Goettingen Georgia Augusta dated 13th June 1857 set up the "Preisaufgabe" to find a conformal mapping to the triaxial ellipsoid. Based upon Jacobi's contribution on elliptic coordinates (C. G. J. Jacobi 1839) the "Preisschrift" of E. Schering (1857) was finally crowned, nevertheless leaving the numerical problem open as how to construct a conformal map of the triaxial ellipsoid of type UTM. For an excellent survey, we refer to W. Klingenberg (1982), H. Schmehl (1927), recently, to B. Mueller (1991). There is another disadvantage of the two-step procedure. The equivalence between two-dimensional real-valued Riemann manifolds and one-dimensional complex-valued manifolds holds only for analytic Riemann manifolds. In E. Grafarend (1995), we give two counterexamples of surfaces of revolution which are from the differentiability class $\mathbb{C}^\infty$, but which are not analytical. Accordingly, the theory of holomorphic functions does not apply. Finally, one encounters great difficulties in generalizing the theory of conformal mappings to higher-dimensional (pseudo-) Riemann manifolds. Only for even-dimensional (pseudo-)Riemann manifolds of analytic type, multi-dimensional complex analysis can be established. We experience a total failure for odd-dimensional (pseudo-)Riemann manifolds as they appear in the theory of refraction, Newton mechanics, or plumb line computation, to list just a few conformally flat three-dimensional Riemann manifolds. The theory of conformal mapping took quite a different direction when A. Korn (1914) and L. Lichtenstein (1911, 1916) set up their general differential equations for two-dimensional Riemann manifolds, which govern conformality. They allow the straightforward transformation of ellipsoidal coordinates of type surface normal longitude L and latitude B into conformal coordinates of type Gauss-Krueger or UTM (x, y) without any intermediate conformal coordinate system of type UMP or UPS! Accordingly, our objective here is a proof of our statement!

Section 15-1.

Section 15-1 offers a review of the *Korn–Lichtenstein equations* of *conformal mapping* subject to the *integrability conditions* which are vectorial *Laplace–Beltrami equations* on a curved surface, here with the metric of the ellipsoid-of-revolution. Two examples, namely UMP and UPS, are chosen to show that the mapping equations $x(L, B)$ and $y(L, B)$ fulfill the Korn–Lichtenstein equations as well as the Laplace–Beltrami equations. In addition, we present in Appendix D a fresh derivation of the Korn–Lichtenstein equations of conformal mapping for a (pseudo-)Riemann manifold of arbitrary dimension. The standard Korn–Lichtenstein equations of conformal mapping for a (pseudo-)Riemann manifold of arbitrary dimension extend initial results of higher-dimensional manifolds, for instance, by J. Zund (1987). The standard equations of type Korn–Lichtenstein which generate a conformal mapping of a two-dimensional Riemann manifold can be taken from standard textbooks like W. Blaschke and K. Leichtweiss (1973) or S. Heitz (1988).

Section 15-2.

Section 15-2 aims at a solution of *partial differential equations* of type *Laplace–Beltrami* (second order) as well as *Korn–Lichtenstein* (first order) in the function space of bivariate polynomials $x(l, b)$ and $y(l, b)$ subject to the definitions (15.1). The coefficients constraints are collected in Corollary 15.1 and Corollary 15.2. Note that the solution space is different from that of type separation of variables known to geodesists from the analysis of the three-dimensional Laplace–Beltrami equation of the gravitational potential field.

$$l := L - L_0 \,,$$
$$b := B - B_0 \,.$$
(15.1)

Section 15-3.

Section 15-3, in contrast, outlines the constraints to the general solution of the Korn–Lichtenstein equations subject to the *integrability conditions* of type *Laplace–Beltrami*, which lead directly to the conformal coordinates of type *Gauss–Krueger* or *UTM*. Such a solution is generated by the equidistant mapping of the *meridian of reference* L_0, for UTM up to a dilatation factor, as the proper constraint ($x(0, b) = 0$ and $y(0, b)$ given). The highlight is the theorem which gives the solution of the partial differential equations for the conformal mapping in terms of a conformal set of bivariate polynomials. Throughout, we use a right-handed coordinate system, namely x "Easting" and y "Northing". Box 15.4 and Box 15.5 contain the non-vanishing polynomial coefficients in a closed form.

Section 15-4.

Section 15-4 introduces by four corollaries the left *Cauchy–Green tensor* and the *dilatation factor* for both the *UTM reference frame* as well as the *Gauss–Krueger reference frame* with the values (15.2) and (15.3) based upon the geometry of the "Geodetic Reference System 1980" (H.Moritz 1984). Such a result was achieved by (i) minimizing the total distance distortion or (ii) minimizing the total areal distortion with the identical result.

UTM:

$$[-l_E, +l_E] \times [B_S, B_N] = [-3.5°, +3.5°] \times [80°S, 84°N] ,$$

$$\rho = 0.999\,578$$

(scale reduction factor $1 : 2\,370$) ,

(15.2)

Gauss–Krueger:

$$[-l_E, +l_E] \times [B_S, B_N] = [-2°, +2°] \times [80°S, 80°N] ,$$

$$\rho = 0.999\,864$$

(scale reduction factor $1 : 7\,353$) .

(15.3)

(The symbols S, N, E, and W as indices denote South, North, East, and West.)

Section 15-5.

Examples are the subject of Section 15-5. In particular, compare with Figs. 15.6–15.16 dealing with the *transverse Mercator projection*.

Section 15-6.

Strip transformations of conformal coordinates of type *Gauss–Krueger* as well as of type *UTM* are finally the subject of Section 15-6.

Appendix

In Appendix D, we outline the theory of the *Cauchy–Green deformation tensor* and its related general *eigenvalue–eigenvector problem*, in particular, its conformal structure, which leads us to three forms of the related *Korn–Lichtenstein equations*.

15-1 The equations governing conformal mapping

The equations governing conformal mapping and their fundamental solution. The Korn–Lichtenstein equations, the Laplace–Beltrami equations.

Here, we are concerned with a conformal mapping of the biaxial ellipsoid $\mathbb{E}_{A_1,A_1,A_2}$ (ellipsoid-of-revolution, spheroid, semi-major axis A_1, semi-minor axis A_2) embedded in a three-dimensional Euclidean manifold $\mathbb{E}^3 = \{\mathbb{R}^3, \delta_{ij}\}$ with a standard canonical metric δ_{ij}, the Kronecker delta of ones in the diagonal, of zeros in the off-diagonal, namely by means of (15.4), introducing surface normal ellipsoidal longitude L and surface normal ellipsoidal latitude B.

$$X^1 = \frac{A_1 \cos B \cos L}{\sqrt{1 - E^2 \sin^2 B}} , \quad X^2 = \frac{A_1 \cos B \sin L}{\sqrt{1 - E^2 \sin^2 B}} , \quad X^3 = \frac{A_1(1 - E^2) \sin B}{\sqrt{1 - E^2 \sin^2 B}} . \tag{15.4}$$

$E^2 := (A_1^2 - A_2^2)/(A_1^2) = 1 - A_2^2/A_1^2$ denotes the first *numerical eccentricity* squared. According to $[L, B] \in [-\pi, +\pi] \times [-\pi/2, +\pi/2]$, we exclude from the domain $[L, B]$ North Pole and South Pole. Thus, $[L, B]$ constitute only a first chart of $\mathbb{E}^2_{A_1,A_1,A_2}$: a minimal atlas of $\mathbb{E}^2_{A_1,A_1,A_2}$ based upon two charts, which covers all points of the ellipsoid-of-revolution, is given in all detail by E. Grafarend and R. Syffus (1995).

Conformal coordinates $\{x, y\}$ (isometric coordinates, isothermal coordinates) are constructed from the surface normal ellipsoidal coordinates $\{L, B\}$ as solutions of the Korn–Lichtenstein equations (conformal change from one chart to another chart: c: Cha-Cha-Cha)

$$\begin{bmatrix} x_L \\ x_B \end{bmatrix} = \frac{1}{\sqrt{G_{11}G_{22} - G_{12}^2}} \begin{bmatrix} -G_{12} & G_{11} \\ -G_{22} & G_{12} \end{bmatrix} \begin{bmatrix} y_L \\ y_B \end{bmatrix} , \tag{15.5}$$

subject to the integrability conditions

$$x_{LB} = x_{BL} , \quad y_{LB} = y_{BL} \tag{15.6}$$

or

$$\Delta_{LB}x := \left(\frac{G_{11}x_B - G_{12}x_L}{\sqrt{G_{11}G_{22} - G_{12}^2}} \right)_B + \left(\frac{G_{22}x_L - G_{12}x_B}{\sqrt{G_{11}G_{22} - G_{12}^2}} \right)_L = 0 ,$$

$$\Delta_{LB}y := \left(\frac{G_{11}y_B - G_{12}y_L}{\sqrt{G_{11}G_{22} - G_{12}^2}} \right)_B + \left(\frac{G_{22}y_L - G_{12}y_B}{\sqrt{G_{11}G_{22} - G_{12}^2}} \right)_L = 0 , \tag{15.7}$$

$$\begin{vmatrix} x_L & x_B \\ y_L & y_B \end{vmatrix} = (x_L y_B - x_B y_L) > 0 \tag{15.8}$$

(orientation conserving conformeomorphism) .

$\Delta_{LB}x = 0$ and $\Delta_{LB}y = 0$, respectively, are called *vectorial Laplace-Beltrami equations*. The matrix of the metric of the first fundamental form of $\mathbb{E}^2_{A_1,A_1,A_2}$ is defined by

$$G_{MN} = \begin{bmatrix} G_{11} & G_{12} \\ G_{12} & G_{22} \end{bmatrix} \forall M, N \in \{1, 2\} . \tag{15.9}$$

A derivation of the Korn–Lichtenstein equations is given in Appendix D. Here, we are interested in some examples of the Korn–Lichtenstein equations (15.5) subject to the integrability conditions (15.7) and the condition of orientation conservation (15.8).

Example 15.1 (Universal Mercator Projection (UMP)).

$$x = A_1 L \,,$$

$$y = A_1 \ln \left[\tan \left(\frac{\pi}{4} + \frac{B}{2} \right) \left(\frac{1 - E \sin B}{1 + E \sin B} \right)^{E/2} \right] \,. \tag{15.10}$$

The matrix of the metric of the ellipsoid-of-revolution $\mathbb{E}^2_{A_1, A_1, A_2}$ is represented by

$$G_{MN} = \begin{bmatrix} G_{11} & G_{12} \\ G_{12} & G_{22} \end{bmatrix} = \begin{bmatrix} \frac{A_1^2 \cos^2 B}{1 - E^2 \sin^2 B} & 0 \\ 0 & \frac{A_1^2 (1 - E^2)^2}{(1 - E^2 \sin^2 B)^3} \end{bmatrix} \,. \tag{15.11}$$

The mapping equations of type UMP imply

$$x_L = A_1 \,, \quad x_B = 0 \,, \quad y_L = 0 \,, \quad y_B = \frac{A_1 (1 - E^2)}{(1 - E^2 \sin^2 B) \cos B} \,. \tag{15.12}$$

Korn–Lichtenstein equations:

$$x_L = \sqrt{\frac{G_{11}}{G_{22}}} y_B \,, \quad x_B = -\sqrt{\frac{G_{22}}{G_{11}}} y_L \,, \quad y_L = -\sqrt{\frac{G_{11}}{G_{22}}} x_B \,, \quad y_B = -\sqrt{\frac{G_{22}}{G_{11}}} x_L \,,$$

$$\sqrt{\frac{G_{11}}{G_{22}}} = \frac{1 - E^2 \sin^2 B}{1 - E^2} \cos B \Rightarrow y_B = \frac{A_1 (1 - E^2)}{(1 - E^2 \sin^2 B) \cos B} \,. \tag{15.13}$$

Integrability conditions:

$$\Delta_{LB} x = \left(\sqrt{\frac{G_{11}}{G_{22}}} x_B \right)_B + \left(\sqrt{\frac{G_{22}}{G_{11}}} x_L \right)_L = 0 \,, \quad \Delta_{LB} y = \left(\sqrt{\frac{G_{11}}{G_{22}}} y_B \right)_B + \left(\sqrt{\frac{G_{22}}{G_{11}}} y_L \right)_L = 0 \,, \tag{15.14}$$

$$\sqrt{\frac{G_{11}}{G_{22}}} x_B = 0 \,, \quad \sqrt{\frac{G_{22}}{G_{11}}} x_L = \frac{A_1 (1 - E^2)}{(1 - E^2 \sin^2 B) \cos B} \,, \quad \left(\sqrt{\frac{G_{22}}{G_{11}}} x_L \right)_L = 0 \,,$$

$$\sqrt{\frac{G_{11}}{G_{22}}} y_B = A_1 \,, \quad \sqrt{\frac{G_{22}}{G_{11}}} y_L = 0 \,, \quad \left(\sqrt{\frac{G_{11}}{G_{22}}} y_B \right)_B = 0 \,. \tag{15.15}$$

Orientation preserving conformeomorphism:

$$\begin{vmatrix} x_L & x_B \\ y_L & y_B \end{vmatrix} = (x_L y_B - x_B y_L) = \frac{A_1^2 (1 - E^2)}{(1 - E^2 \sin^2 B) \cos B} > 0 \,, \tag{15.16}$$

due to $-\pi/2 < B < +\pi/2 \to \cos B > 0$.

End of Example.

The UMP solution of the Korn–Lichtenstein equations subject to the vectorial Laplace–Beltrami equations as integrability conditions and the condition of orientation conservation is based upon the constraint of the following type: map the equator equidistantly, for instance, $x(B = 0) = A_1 \Lambda$.

Example 15.2 (Universal Polar Stereographic Projection (UPS)).

$$x = \frac{2A_1}{\sqrt{1-E^2}} \left(\frac{1-E}{1+E}\right)^{E/2} \tan\left(\frac{\pi}{4} - \frac{B}{2}\right) \left(\frac{1+E\sin B}{1-E\sin B}\right)^{E/2} \cos L ,$$

$$y = \frac{2A_1}{\sqrt{1-E^2}} \left(\frac{1-E}{1+E}\right)^{E/2} \tan\left(\frac{\pi}{4} - \frac{B}{2}\right) \left(\frac{1+E\sin B}{1-E\sin B}\right)^{E/2} \sin L .$$

(15.17)

The matrix of the metric of the ellipsoid-of-revolution $\mathbb{E}^2_{A_1,A_1,A_2}$ is represented by

$$G_{MN} = \begin{bmatrix} G_{11} & G_{12} \\ G_{12} & G_{22} \end{bmatrix} = \begin{bmatrix} \frac{A_1^2 \cos^2 B}{1-E^2 \sin^2 B} & 0 \\ 0 & \frac{A_1^2(1-E^2)^2}{(1-E^2\sin^2 B)^3} \end{bmatrix} .$$

(15.18)

The mapping equations of type UPS imply

$$x_L = -f(B)\sin L , \quad x_B = f'(B)\cos L ,$$

$$y_L = f(B)\cos L , \quad y_B = f'(B)\sin L ,$$

(15.19)

subject to

$$f(B) :=$$

$$:= \frac{2A_1}{\sqrt{1-E^2}} \left(\frac{1-E}{1+E}\right)^{E/2} \tan\left(\frac{\pi}{4} - \frac{B}{2}\right) \left(\frac{1+E\sin B}{1-E\sin B}\right)^{E/2} ,$$

$$f'(B) :=$$

$$:= -\frac{2A_1}{\sqrt{1-E^2}} \left(\frac{1-E}{1+E}\right)^{E/2} \frac{\tan\left(\frac{\pi}{4} - \frac{B}{2}\right)}{\cos B} \left(\frac{1-E^2}{1-E^2\sin^2 B}\right) \left(\frac{1+E\sin B}{1-E\sin B}\right)^{E/2} =$$

$$= -\frac{1-E^2}{\cos B(1-E^2\sin^2 B)} f(B) .$$

(15.20)

Korn–Lichtenstein equations:

$$x_L = \sqrt{\frac{G_{11}}{G_{22}}} y_B , \quad x_B = -\sqrt{\frac{G_{22}}{G_{11}}} y_L , \quad y_L = -\sqrt{\frac{G_{11}}{G_{22}}} x_B , \quad y_B = \sqrt{\frac{G_{22}}{G_{11}}} x_L ,$$

(15.21)

$$\sqrt{\frac{G_{11}}{G_{22}}} = \cos B \frac{1-E^2\sin^2 B}{1-E^2}$$

$$\Rightarrow$$

$$y_B = -\frac{1-E^2}{\cos B(1-E^2\sin^2 B)} f(B)\sin L = f'(B)\sin L ,$$

$$y_L = -\frac{\cos B(1-E^2\sin^2 B)}{1-E^2} f'(B)\cos L = f(B)\cos L .$$

(15.22)

End of Example.

15-2 A fundamental solution for the Korn–Lichtenstein equations

A fundamental solution for the Korn–Lichtenstein equations of conformal mapping. The ellipsoidal Korn–Lichtenstein equations, the ellipsoidal Laplace–Beltrami equations.

For the biaxial ellipsoid $\mathbb{E}^2_{A_1\,A_1\,A_2}$, we shall construct a fundamental solution for the Korn–Lichtenstein equations of conformal mapping (15.5) subject to the vectorial Laplace–Beltrami equations (15.7). The condition of orientation conservation (15.8) is automatically fulfilled.

$$x_L = \sqrt{G_{11}\,G_{22}}\,y_B \qquad y_L = -\sqrt{G_{11}\,G_{22}}\,x_B$$
$$\text{or}$$
$$x_B = -\sqrt{G_{22}\,G_{11}}\,y_L \qquad y_B = \sqrt{G_{22}\,G_{11}}\,x_L \tag{15.23}$$

$$\left(\sqrt{G_{22}\,G_{11}}\,x_L\right)_L + \left(\sqrt{G_{11}\,G_{22}}\,x_B\right)_B = 0$$
$$\left(\sqrt{G_{22}\,G_{11}}\,y_L\right)_L + \left(\sqrt{G_{11}\,G_{22}}\,y_B\right)_B = 0 \tag{15.24}$$

$$x_L y_B - x_B y_L = \sqrt{G_{22}\,G_{11}}\,x_L^2 + \sqrt{G_{11}\,G_{22}}\,x_B^2 > 0 \tag{15.25}$$

$$\sqrt{G_{22}\,G_{11}} \in \mathbb{R}^+ \qquad \sqrt{G_{11}\,G_{22}} \in \mathbb{R}^+ \tag{15.26}$$

Here, we are interested in a local solution of the ellipsoidal Korn–Lichtenstein equations around a point $\{L_0\ B_0\}$ such that the relations $L = L_0 + l$ and $B = B_0 + b$ hold. A polynomial setup of the local solution of the ellipsoidal Korn–Lichtenstein equations subject to the ellipsoidal vectorial Laplace–Beltrami equations,

$$y_l = -\sqrt{G_{11}\,G_{22}}\,x_b \qquad y_b = \sqrt{G_{22}\,G_{11}}\,x_l \tag{15.27}$$

$$\left(\sqrt{G_{22}\,G_{11}}\,x_l\right)_l + \left(\sqrt{G_{11}\,G_{22}}\,x_b\right)_b = 0$$
$$\left(\sqrt{G_{22}\,G_{11}}\,y_l\right)_l + \left(\sqrt{G_{11}\,G_{22}}\,y_b\right)_b = 0 \tag{15.28}$$

is

$$x(l\ b) = x_0 + x_{10}l + x_{01}b + x_{20}l^2 + x_{11}lb + x_{02}b^2 + x_{30}l^3 + x_{21}l^2b + x_{12}lb^2 + x_{03}b^3 +$$
$$+ O(4)$$

$$y(l\ b) = y_0 + y_{10}l + y_{01}b + y_{20}l^2 + y_{11}lb + y_{02}b^2 + y_{30}l^3 + y_{21}l^2b + y_{12}lb^2 + y_{03}b^3 +$$
$$+ O(4) \tag{15.29}$$

or

$$x(l\ b) = \sum_{n=0}^{\infty} P_n(l\ b)$$

$$y(l\ b) = \sum_{n=0}^{\infty} Q_n(l\ b) \tag{15.30}$$

with

$$P_0(l, b) := x_0 \,,$$

$$P_1(l, b) := x_{10}l + x_{01}b = \sum_{\alpha+\beta=1} x_{\alpha\beta}l^\alpha b^\beta \,,$$

$$P_2(l, b) := x_{20}l^2 + x_{11}lb + x_{02}b^2 = \sum_{\alpha+\beta=2} x_{\alpha\beta}l^\alpha b^\beta \,, \tag{15.31}$$

$$\vdots$$

$$P_n(l, b) := \sum_{\alpha+\beta=n} x_{\alpha\beta}l^\alpha b^\beta \,,$$

and

$$Q_0(l, b) := y_0 \,,$$

$$Q_1(l, b) := y_{10}l + y_{01}b = \sum_{\alpha+\beta=1} y_{\alpha\beta}l^\alpha b^\beta \,,$$

$$Q_2(l, b) := y_{20}l^2 + y_{11}lb + y_{02}b^2 = \sum_{\alpha+\beta=2} y_{\alpha\beta}l^\alpha b^\beta \,, \tag{15.32}$$

$$\vdots$$

$$Q_n(l, b) := \sum_{\alpha+\beta=n} y_{\alpha\beta}l^\alpha b^\beta \,,$$

subject to the Taylor expansion

$$r := \sqrt{G_{11}/G_{22}} = \cos B \frac{1 - E^2 \sin^2 B}{1 - E^2} = r_0 + r_1 b + r_2 b^2 + r_3 b^3 + \mathrm{O}(4) \,, \tag{15.33}$$

namely

$$r_0 := \tfrac{1}{0!} r^{(0)}(B_0) = r(B_0) \,,$$

$$r_1 := \tfrac{1}{1!} r^{(1)}(B_0) = r'(B_0) \,,$$

$$\vdots \tag{15.34}$$

$$r_n := \tfrac{1}{n!} r^{(n)}(B_0) = \frac{1}{n(n-1)\cdot\cdots\cdot2\cdot1} r^{(n)}(B_0) \,,$$

and *vice versa*

$$s := \sqrt{G_{22}/G_{11}} = \frac{1}{\cos B} \frac{1 - E^2}{1 - E^2 \sin^2 B} = s_0 + s_1 b + s_2 b^2 + s_3 b^3 + \mathrm{O}(4) =$$

$$= r_0^{-1} - r_0^{-2} r_1 b + \left(r_0^{-3} r_1^2 - r_0^{-2} r_2 \right) b^2 + \left(-r_0^{-4} r_1^3 + 2 r_0^{-3} r_1 r_2 - r_0^{-2} r_3 \right) b^2 + \mathrm{O}(4) \,, \tag{15.35}$$

namely

$$s_0 := \tfrac{1}{0!} s^{(0)}(B_0) = s(B_0) \,,$$

$$s_1 := \tfrac{1}{1!} s^{(1)}(B_0) = s'(B_0) \,,$$

$$\vdots \tag{15.36}$$

$$s_n := \tfrac{1}{n!} s^{(n)}(B_0) = \frac{1}{n(n-1)\cdot\cdots\cdot2\cdot1} s^{(n)}(B_0) \,,$$

given in detail by the coefficients of Box 15.1.

Box 15.1 (Taylor expansion of $r(B)$ and $s(B)$).

Taylor expansion of $r(B)$ up to order three:

$$r := \sqrt{G_{11}/G_{22}} =$$

$$= \cos B \frac{1 - E^2 \sin^2 B}{1 - E^2} = \tag{15.37}$$

$$= \sum_{n=0}^{N} \frac{1}{n!} r^{(n)}(B_0) b^n = \sum_{n=0}^{N} r_n b^n \ ,$$

$$r_0 = + \frac{\cos B_0 (1 - E^2 \sin^2 B_0)}{1 - E^2} \ ,$$

$$r_1 = - \frac{\sin B_0 (1 + 2E^2 - 3E^2 \sin^2 B_0)}{1 - E^2} \ ,$$

$$r_2 = - \frac{\cos B_0 (1 + 2E^2 - 9E^2 \sin^2 B_0)}{2(1 - E^2)} \ , \tag{15.38}$$

$$r_3 = + \frac{\sin B_0 (1 + 20E^2 - 27E^2 \sin^2 B_0)}{6(1 - E^2)} \ .$$

Taylor expansion of $s(B)$ up to order three:

$$s := \sqrt{G_{22}/G_{11}} =$$

$$= \frac{1 - E^2}{\cos B (1 - E^2 \sin^2 B)} = \tag{15.39}$$

$$= \sum_{n=0}^{N} \frac{1}{n!} s^{(n)}(B_0) b^n = \sum_{n=0}^{N} s_n b^n \ ,$$

$$s_0 = \frac{1 - E^2}{\cos B_0 (1 - E^2 \sin^2 B_0)} \ ,$$

$$s_1 = \frac{(1 - E^2) \sin B_0 (1 + 2E^2 - 3E^2 \sin^2 B_0)}{\cos^2 B_0 (1 - E^2 \sin^2 B_0)^2} \ ,$$

$$s_2 = \frac{(1 - E^2)}{2 \cos^3 B_0 (1 - E^2 \sin^2 B_0)^3} \times$$

$$\times \left[1 + 2E^2 + \sin^2 B_0 (1 - 4E^2 + 6E^4) - E^2 \sin^4 B_0 (2 + 13E^2) + 9E^4 \sin^6 B_0 \right] \ , \tag{15.40}$$

$$s_3 = \frac{(1 - E^2) \sin B_0}{6 \cos^4 B_0 (1 - E^2 \sin^2 B_0)^4} \times$$

$$\times \left[5 + 4E^2 + 24E^4 + \sin^2 B_0 (1 - 17E^2 - 80E^4 + 24E^6) - E^2 \sin^4 B_0 (5 - 91E^2 + 68E^4) - \right.$$

$$\left. - E^4 \sin^6 B_0 (17 - 65E^2) - 27E^6 \sin^8 B_0 \right] \ .$$

First, let us here consider the ellipsoidal vectorial Laplace–Beltrami equations which are defined by (15.28), namely

$$\Delta_{lb}x = \left(\sqrt{G_{22}/G_{11}}\,x_l\right)_l + \left(\sqrt{G_{11}/G_{22}}\,x_b\right)_b = 0 \ ,$$

$$\Delta_{lb}y = \left(\sqrt{G_{22}/G_{11}}\,y_l\right)_l + \left(\sqrt{G_{11}/G_{22}}\,y_b\right)_b = 0 \ ,$$

(15.41)

$$sx_{ll} + (rx_b)_b = sx_{ll} + r_b x_b + r x_{bb} = 0 \ , \tag{15.42}$$

$$sy_{ll} + (ry_b)_b = sy_{ll} + r_b y_b + r y_{bb} = 0 \ , \tag{15.43}$$

$$x(l,b) = x_0 + x_{10}l + x_{01}b + x_{20}l^2 + x_{11}lb + x_{02}b^2 + x_{30}l^3 + x_{21}l^2b + x_{12}lb^2 + x_{03}b^3 +$$
$$+x_{40}l^4 + x_{31}l^3b + x_{22}l^2b^2 + x_{13}lb^3 + x_{04}b^4 + O(5) \ , \tag{15.44}$$

$$x_l(l,b) = x_{10} + 2x_{20}l + x_{11}b + 3x_{30}l^2 + 2x_{21}lb + x_{12}b^2 +$$
$$+4x_{40}l^3 + 3x_{31}l^2b + 2x_{22}lb^2 + x_{13}b^3 + O(4) \ , \tag{15.45}$$

$$x_{ll}(l,b) = 2x_{20} + 6x_{30}l + 2x_{21}b + 12x_{40}l^2 +$$
$$+6x_{31}lb + 2x_{22}b^2 + O(3) \ , \tag{15.46}$$

$$sx_{ll}(l,b) = (s_0 + s_1b + s_2b^2 + O(3))x_{ll} =$$
$$= 2s_0x_{20} + 6s_0x_{30}l + 2s_0x_{21}b + 2s_1x_{20}b + 12s_0x_{40}l^2 + 6s_0x_{31}lb +$$
$$+6s_1x_{30}lb + 2s_0x_{22}b^2 + 2s_1x_{21}b^2 + 2s_2x_{20}b^2 + O(3) \ , \tag{15.47}$$

$$x_b(l,b) = x_{01} + x_{11}l + 2x_{02}b + x_{21}l^2 + 2x_{12}lb + 3x_{03}b^2 +$$
$$+x_{31}l^3 + 2x_{22}l^2b + 3x_{13}lb^2 + 4x_{04}b^3 + O(4) \ , \tag{15.48}$$

$$x_{bb}(l,b) = 2x_{02} + 2x_{12}l + 6x_{03}b +$$
$$+2x_{22}l^2 + 6x_{13}lb + 12x_{04}b^2 + O(3) \ , \tag{15.49}$$

$$r_bx_b(l,b) = (r_1 + 2r_2b + 3r_3b^2 + O(3))x_b =$$
$$r_1x_{01} + r_1x_{11}l + 2r_1x_{02}b + 2r_2x_{01}b + r_1x_{21}l^2 + 2r_1x_{12}lb +$$
$$+2r_2x_{11}lb + 3r_1x_{03}b^2 + 4r_2x_{02}b^2 + 3r_3x_{01}b^2 + O(3) \ , \tag{15.50}$$

$$rx_{bb}(l,b) = (r_0 + r_1b + r_2b^2 + O(3))x_{bb} =$$
$$2r_0x_{02} + 2r_0x_{12}l + 6r_0x_{03}b + 2r_1x_{02}b + 2r_0x_{22}l^2 + 6r_0x_{13}lb +$$
$$2r_1x_{12}lb + 12r_0x_{04}b^2 + 6r_1x_{03}b^2 + 2r_2x_{02}b^2 + O(3) \ . \tag{15.51}$$

While (15.47), (15.50), and (15.51) represent the polynomial solution of (15.42), namely for $x(l,b)$, a corresponding solution for (15.43) could be found as soon as we replace x and y, namely for the polynomial solution $y(l,b)$. Let us write down the $n-1$ constraints for $n+1$ polynomials given by the zero identiy of the sum of the three terms of (15.47) (sx_{ll}, first term), (15.50) ($r_b x_b$, second term), and (15.51) ($r x_{bb}$, third term).

Corollary 15.1 (Laplace–Beltrami equations solved in the function space of bivariate polynomials).

If a polynomial (15.29)–(15.32) of degree n fulfills the Laplace–Beltrami equations (15.42) and (15.43), then there are $n-1$ coefficient constraints, namely

$$n = 2:$$

$$2s_0 x_{20} + 2r_0 x_{02} + r_1 x_{01} = 0 \,, \tag{15.52}$$

$$2s_0 y_{20} + 2r_0 y_{02} + r_1 y_{01} = 0 \,; \tag{15.53}$$

$$n = 3:$$

$$6s_0 x_{30} + 2r_0 x_{12} + r_1 x_{11} = 0 \,, \tag{15.54}$$

$$6s_0 y_{30} + 2r_0 y_{12} + r_1 y_{11} = 0 \,, \tag{15.55}$$

$$s_0 x_{21} + s_1 x_{20} + 3r_0 x_{03} + 2r_1 x_{02} + r_2 x_{01} = 0 \,, \tag{15.56}$$

$$s_0 y_{21} + s_1 y_{20} + 3r_0 y_{03} + 2r_1 y_{02} + r_2 y_{01} = 0 \,; \tag{15.57}$$

$$n = 4:$$

$$12s_0 x_{40} + 2r_0 x_{22} + r_1 x_{21} = 0 \,, \tag{15.58}$$

$$12s_0 y_{40} + 2r_0 y_{22} + r_1 y_{21} = 0 \,, \tag{15.59}$$

$$3s_0 x_{31} + 3s_1 x_{30} + 3r_0 x_{13} + 2r_1 x_{12} + r_2 x_{11} = 0 \,, \tag{15.60}$$

$$3s_0 y_{31} + 3s_1 y_{30} + 3r_0 y_{13} + 2r_1 y_{12} + r_2 y_{11} = 0 \,, \tag{15.61}$$

$$2s_0 x_{22} + 2s_1 x_{21} + 2s_2 x_{20} + 12r_0 x_{04} + 9r_1 x_{03} + 6r_2 x_{02} + 3r_3 x_{01} = 0 \,, \tag{15.62}$$

$$2s_0 y_{22} + 2s_1 y_{21} + 2s_2 y_{20} + 12r_0 y_{04} + 9r_1 y_{03} + 6r_2 y_{02} + 3r_3 y_{01} = 0 \,; \tag{15.63}$$

and in general

$$sx_{ll} + (rx_b)_b = \sum_{n=2}^{\infty} \sum_{i=0}^{n-2} \sum_{j=0}^{i} \Big[(j+1)[(i-j+1)r_{j+1} x_{n-i-2,i-j+1} +$$

$$+(j+2)r_{i-j} x_{n-i-2,j+2}] + (n-i)(n-i-1)s_i x_{n-i,i-j} \Big] l^{n-i-2} b^i = 0 \,,$$

$$sy_{ll} + (ry_b)_b = \sum_{n=2}^{\infty} \sum_{i=0}^{n-2} \sum_{j=0}^{i} \Big[(j+1)[(i-j+1)r_{j+1} y_{n-i-2,i-j+1} +$$

$$\tag{15.64}$$

$$+(j+2)r_{i-j} y_{n-i-2,j+2}] + (n-i)(n-i-1)s_i y_{n-i,i-j} \Big] l^{n-i-2} b^i = 0 \,.$$

End of Corollary.

Finally, we here have to constrain the general solution $x(l,b)$ of the Laplace–Beltrami equation (compare with (15.29)) to the ellipsoidal Korn–Lichtenstein equation $y_l = -\sqrt{G_{11}/G_{22}}x_b$ (compare with (15.27)), in particular

$$y_l = -r(b)x_b = -(r_0 + r_1 b + r_2 b^2 + r_3 b^3 + \mathrm{O}(4))x_b \;, \tag{15.65}$$

$$y_l = y_{10} + 2y_{20}l + y_{11}b + 3y_{30}l^2 + 2y_{21}lb + y_{12}b^2 + 4y_{40}l^3 + 3y_{31}l^2 b + 2y_{22}lb^2 + y_{13}b^3 + \mathrm{O}(4) =$$
$$= -r_0 x_{01} - r_0 x_{11}l - 2r_0 x_{02}b - r_1 x_{01}b - r_0 x_{21}l^2 - 2r_0 x_{12}lb - r_1 x_{11}lb -$$
$$-3r_0 x_{03}b^2 - 2r_1 x_{02}b^2 - r_2 x_{01}b^2 - r_0 x_{31}l^3 - 2r_0 x_{22}l^2 b - r_1 x_{21}l^2 b - 3r_0 x_{13}lb^2 -$$
$$-2r_1 x_{12}lb^2 - r_2 x_{11}lb^2 - 4r_0 x_{04}b^3 - 3r_1 x_{03}b^3 - 2r_2 x_{02}b^3 - r_3 x_{01}b^3 + \mathrm{O}(4) \;, \tag{15.66}$$

Alternatively, we here have to constrain the general solution to the ellipsoidal Korn–Lichtenstein equation $y_b = -\sqrt{G_{22}/G_{11}}x_l$ (compare with (15.27)), in particular

$$y_b = s(b)x_l = (s_0 + s_1 b + s_2 b^2 + s_3 b^3 + \mathrm{O}(4))x_l \;, \tag{15.67}$$

$$y_b = y_{01} + y_{11}l + 2y_{02}b + y_{21}l^2 + 2y_{12}lb + 3y_{03}b^2 + y_{31}l^3 + 2y_{22}l^2 b + 3y_{13}lb^2 + 4y_{04}b^3 + \mathrm{O}(4) =$$
$$= s_0 x_{10} + 2s_0 x_{20}l + s_0 x_{11}b + s_1 x_{10}b + 3s_0 x_{30}l^2 + 2s_0 x_{21}lb + 2s_1 x_{20}lb +$$
$$+s_0 x_{12}b^2 + s_1 x_{11}b^2 + s_2 x_{10}b^2 + 4s_0 x_{40}l^3 + 3s_0 x_{31}l^2 b + 3s_1 x_{30}l^2 b + 2s_0 x_{22}lb^2 +$$
$$+2s_1 x_{21}lb^2 + 2s_2 x_{20}lb^2 + s_0 x_{13}b^3 + s_1 x_{12}b^3 + s_2 x_{11}b^3 + s_3 x_{10}b^3 + \mathrm{O}(4) \;. \tag{15.68}$$

Corollary 15.2 (Korn–Lichtenstein equations solved in the function space of bivariate polynomials).

If a polynomial (15.29)–(15.32) of degree n fulfills the Korn–Lichtenstein equations (15.27) with respect to an ellipsoid-of-revolution and subject to the $n-1$ constraints given by (15.52)–(15.64), then the following mixed coefficient relations hold.

$$n = 1:$$
$$y_{10} = -r_0 x_{01}, \quad y_{01} = s_0 x_{10} \;. \tag{15.69}$$

$$n = 2:$$
$$2y_{20} = -r_0 x_{11}, \quad y_{11} = -2r_0 x_{02} - r_1 x_{01}, \quad y_{11} = 2s_0 x_{20}, \quad 2y_{02} = s_0 x_{11} + s_1 x_{10} \;. \tag{15.70}$$

$$n = 3:$$
$$3y_{30} = -r_0 x_{21}, \quad 2y_{21} = -2r_0 x_{12} - r_1 x_{11}, \quad y_{12} = -3r_0 x_{03} - 2r_1 x_{02} - r_2 x_{01}, \tag{15.71}$$
$$y_{21} = 3s_0 x_{30}, \quad 2y_{12} = 2s_0 x_{23} + 2s_1 x_{20}, \;, \quad 3y_{03} = s_0 x_{12} + s_1 x_{11} + s_2 x_{10} \;.$$

$$n = 4:$$
$$4y_{40} = -r_0 x_{31}, \quad 3y_{31} = -2r_0 x_{22} - r_1 x_{21}, \quad 2y_{22} = -3r_0 x_{13} - 2r_1 x_{12} - r_2 x_{11}, \tag{15.72}$$
$$y_{13} = -4r_0 x_{04} - 3r_1 x_{03} - 2r_2 x_{02} - r_3 x_{01}, \quad y_{31} = 4s_0 x_{40}, \quad 2y_{22} = 3s_0 x_{31} + 3s_1 x_{30},$$
$$3y_{13} = 2s_0 x_{22} + 2s_1 x_{21} + 2s_2 x_{20}, \quad 4y_{04} = s_0 x_{13} + s_1 x_{12} + s_2 x_{11} + s_3 x_{10} \;.$$

In general:

$$y_l = \sum_{n=1}^{\infty} \sum_{i=0}^{n-1} (n-i)y_{n-i,i}l^{n-i-1}b^i = -\sum_{n=1}^{\infty} \sum_{i=0}^{n-1} \sum_{j=0}^{i} (i-j+1)r_j x_{n-1-i,i-j+1}l^{n-i-1}b^i =$$
$$= -r(b)x_b \;, \tag{15.73}$$

$$y_b = \sum_{n=1}^{\infty} \sum_{i=0}^{n-1} (n-i)y_{n-i-1,i+1}l^{n-i-1}b^i = \sum_{n=1}^{\infty} \sum_{i=0}^{n-1} \sum_{j=0}^{i} (n-i)s_j x_{n-i,i-j}l^{n-i-1}b^i =$$
$$= s(b)x_l \;.$$

End of Corollary.

15-3 Constraints to the Korn–Lichtenstein equations (Gauss–Krueger/UTM mappings)

The constraints to the Korn–Lichtenstein equations generating the Gauss–Krueger conformal mapping or the UTM conformal mapping.

The *equidistant mapping* of a meridian of reference L_0 immediately establishes the proper constraint to the Korn–Lichtenstein equations which leads to the standard *Gauss–Krueger* conformal mapping or *universal transverse Mercator projection* conformal mapping. The arc length of the coordinate line $L_0 = \text{const.}$, namely the *meridian*, between latitude B_0 and B is computed by (15.74) as soon as we set up uniformly convergent Taylor series of type (15.75) and integrate term-wise.

$$y(0, b) = \int_{B_0}^{B} \sqrt{G_{22}(B^*)} dB^* = \int_{B_0}^{B} M(B^*) dB^* = \sum_{n=1}^{\infty} y_{0n} b^n \,, \tag{15.74}$$

$$\sqrt{G_{22}(B)} = M(B) = \frac{A_1(1 - E^2)}{(1 - E^2 \sin^2 B)^{3/2}} = \sum_{n=1}^{\infty} \frac{1}{n!} G_{22}^{(n)}(B_0) b^n \,. \tag{15.75}$$

Box 15.2, which follows subsequently, contains a list of resulting coefficients y_{0n}, which establish the setup of the constraints defined in Definition 15.3.

Definition 15.3 (Constraints to the Korn–Lichtenstein equations of conformal mapping).

Let there be given the ellipsoidal Korn–Lichtenstein equations (15.76), subject to the integrability condition, the Laplace–Beltrami equations (15.77), which generate a conformal mapping via a polynomial representation of type (15.29)–(15.32) and the coefficient constraints given by (15.69)–(15.73).

$$y_l = -\sqrt{G_{11}/G_{22}} x_b \,, \quad y_b = \sqrt{G_{22}/G_{11}} x_l \,, \tag{15.76}$$

$$sx_{ll} + (rx_b)_b = 0 \,, \quad sy_{ll} + (ry_b)_b = 0 \,. \tag{15.77}$$

The *equidistant mapping* of the meridian of reference L_0 establishes by means of constraints of type (15.78) the *conformal mapping* of type *Gauss–Krueger* or *UTM*.

$$x(0, b) = 0 \,, \quad y(0, b) = \sum_{n=1}^{\infty} y_{0n} b^n \,. \tag{15.78}$$

End of Definition.

Box 15.2 (The equidistant mapping of the meridian of reference L_0, $y(0, b) = \sum_{n=1}^{\infty} y_{0n} b^n$, coefficients $y_{01}, \ldots, y_{04}$).

$$y_{01} = \sqrt{G_{22}}\Big|_{B_0} = \frac{A_1(1 - E^2)}{(1 - E^2 \sin^2 B_0)^{3/2}} \,,$$

$$y_{02} = \frac{1}{2}[\sqrt{G_{22}}]' = \frac{1}{4} G_{22}' / \sqrt{G_{22}}\Big|_{B_0} = \frac{3}{2} \frac{A_1 E^2(1 - E^2) \cos B_0 \sin B_0}{(1 - E^2 \sin^2 B_0)^{5/2}} \,,$$

$$y_{03} = \frac{1}{24} [2G_{22}G_{22}'' - G_{22}'^2]/G_{22}^{3/2}\Big|_{B_0}$$

$$= \frac{1}{2} \frac{A_1 E^2(1 - E^2)}{(1 - E^2 \sin^2 B_0)^{7/2}}(1 - 2\sin^2 B_0 + 4E^2 \sin^2 B_0 - 3E^2 \sin^2 B_0) \,,$$

$$y_{04} = \frac{1}{192} [4G_{22}^2 G_{22}''' - 6G_{22}G_{22}'G_{22}'' + 3\,G_{22}'^3]/G_{22}^{5/2}\Big|_{B_0}$$

$$= \frac{1}{8} \frac{A_1 E^2(1 - E^2)}{(1 - E^2 \sin^2 B_0)^{9/2}} \cos B_0 \sin B_0 (4 - 15E^2 + 22E^2 \sin^2 B_0 - 20E^4 \sin^2 B_0 + 9E^4 \sin^4 B_0)$$

$$\tag{15.79}$$

Let us now give the solution of the Korn–Lichtenstein equations with respect to the ellipsoid-of-revolution and subject to the integrability condition of the type of the vectorial Laplace–Beltrami equation in the function space of bivariate polynomials of type (15.29)–(15.32) and restricted to the coefficient constraints given by (15.69)–(15.73). The quoted result is collected in the following Box 15.3.

Box 15.3 (Vanishing and non-vanishing polynomial coefficients x_{ij} and y_{ij}: $n = 1 \ldots n = 4$).

$$n = 1:$$

$$x_{01} = 0\,, \qquad y_{01} \text{ given}\,,$$
$$[(15.69)] \quad x_{10} = \tfrac{1}{s_0} y_{01}\,. \qquad y_{10} = 0\,. \tag{15.80}$$

$$n = 2:$$

$$x_{02} = 0\,, \qquad\qquad y_{02} \text{ given}\,,$$
$$[(15.70)] \quad x_{20} = 0\,, \qquad [(15.53)] \quad y_{20} = -\tfrac{1}{2s_0}(2r_0 y_{02} + r_1 y_{01})\,, \tag{15.81}$$
$$[(15.52)] \quad x_{11} = \tfrac{1}{s_0}(2y_{02} - s_1 x_{10})\,. \quad [(15.70)] \quad y_{11} = 0\,.$$

$$n = 3:$$

$$x_{03} = 0\,, \qquad\qquad y_{03} \text{ given}\,,$$
$$[(15.54)] \quad x_{30} = -\tfrac{1}{6s_0}(2r_0 x_{12} + r_1 x_{11})\,, \qquad [(15.71)] \quad y_{30} = 0\,,$$
$$[(15.55)] \quad x_{21} = 0\,, \qquad\qquad [(15.71)] \quad y_{21} = 2s_0 x_{30}\,, \tag{15.82}$$
$$[(15.71)] \quad x_{12} = \tfrac{1}{s_0}(3y_{03} - s_1 x_{11} - s_2 x_{10})\,. \quad [(15.71)] \quad y_{12} = 0\,.$$

$$n = 4:$$

$$x_{04} = 0\,, \qquad\qquad y_{04} \text{ given}\,,$$
$$[(15.58)] \quad x_{40} = 0\,, \qquad\qquad [(15.72)] \quad y_{40} = -\tfrac{1}{4}r_0 x_{31}\,,$$
$$[(15.72)] \quad x_{31} = \tfrac{1}{3s_0}(2y_{22} - 3s_1 x_{30})\,, \quad [(15.72)] \quad y_{31} = 0\,, \tag{15.83}$$
$$[(15.72)] \quad x_{22} = 0\,, \qquad\qquad [(15.72)] \quad y_{22} = -\tfrac{3}{2}r_0 x_{13} - r_1 x_{12} - \tfrac{1}{2}r_2 x_{11}\,,$$
$$[(15.72)] \quad x_{13} = \tfrac{1}{s_0}(4y_{04} - s_1 x_{12} - s_2 x_{11} - s_3 x_{10})\,. \quad [(15.72)] \quad y_{13} = 0\,.$$

Theorem 15.4 (The solution of the Korn–Lichtenstein equations of conformal mapping which generates directly Gauss–Krueger or UTM conformal coordinates).

The equidistant mapping of the meridian of reference L_0, which is the constraint fixing the general solution (15.84) of the Korn–Lichtenstein equations (15.85) subject to the integrability conditions, the Laplace–Beltrami equations given by (15.86), leads us to the solution (15.87) in the function space of bivariate polynomials.

$$x(l, b) = 0\,, \quad y(0, b) = \sum_{n=1}^{\infty} y_{0n} b^n\,, \tag{15.84}$$

$$x_l - \sqrt{G_{11}/G_{22}}\, y_b = 0\,, \quad x_b - \sqrt{G_{22}/G_{11}}\, y_l = 0\,, \tag{15.85}$$

$$\Delta_{LB}
\begin{bmatrix} r(l, b) \\ y(l, b) \end{bmatrix} = 0\,, \tag{15.86}$$

$$x(l,b) =$$

$$= x_{10}l + x_{11}lb + x_{30}l^3 + x_{12}lb^2 + x_{31}l^3b + x_{13}lb^3 + x_{50}l^5 + x_{32}l^3b^2 +$$

$$+ x_{14}lb^4 + O(6) \ ,$$

$$(15.87)$$

$$y(l,b) =$$

$$= y_{01}b + y_{20}l^2 + y_{02}b^2 + y_{21}l^2b + y_{03}b^3 + y_{40}l^4 + y_{22}l^2b^2 + y_{04}b^4 + y_{41}l^4b + y_{23}l^2b^3 +$$

$$+ y_{05}b^5 + O(6) \ .$$

Box 15.4 and Box 15.5 are a collection of the coefficients $x_{10}, \ldots, y_{05}$.

End of Theorem.

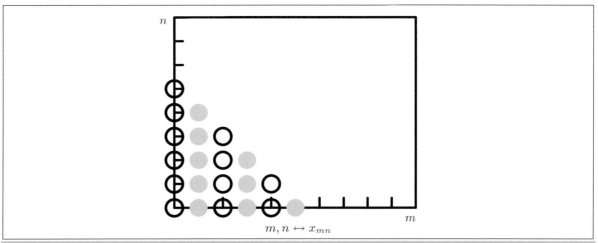

Fig. 15.3. Monomial diagram. The ideal J of conformal bivariate polynomials of type Gauss–Krueger/UTM. The solid dots illustrate monomials in J, those not in J are open circles, $x(l,b)$, according to D. Cox, J. Little, and D. O'Shea (1996).

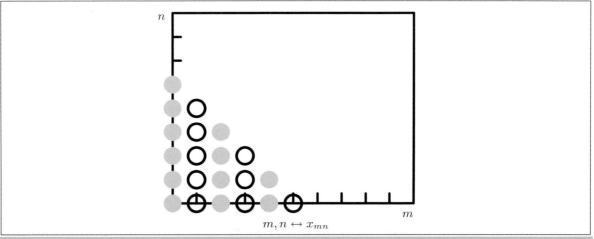

Fig. 15.4. Monomial diagram. The ideal J of conformal bivariate polynomials of type Gauss–Krueger/UTM. The solid dots illustrate monomials in J, those not in J are open circles, $y(l,b)$, according to D. Cox, J. Little, and D. O'Shea (1996).

Box 15.4 (A representation of the non-vanishing coefficients in a polynomial setup of a conformal mapping of type Gauss–Krueger or UTM).

$$x(l, b) =$$

$$= x_{10}l + x_{11}lb + x_{30}l^3 + x_{12}lb^2 + x_{31}l^3b + x_{13}lb^3 + x_{50}l^5 + x_{32}l^3b^2 +$$

$$+ x_{14}lb^4 + O(6) ,$$

(15.88)

$$x_{10} = \frac{A_1 \cos B_0}{(1 - E^2 \sin^2 B_0)^{1/2}} ,$$

$$x_{11} = \frac{-A_1(1 - E^2) \sin B_0}{(1 - E^2 \sin^2 B_0)^{3/2}} ,$$

$$x_{30} = \frac{A_1 \cos B_0(1 - 2\sin^2 B_0 + E^2 \sin^4 B_0)}{6(1 - E^2)(1 - E^2 \sin^2 B_0)^{1/2}} ,$$

$$x_{12} = \frac{-A_1(1 - E^2) \cos B_0(1 + 2E^2 \sin^2 B_0)}{2(1 - E^2 \sin^2 B_0)^{5/2}} ,$$

$$x_{31} = \frac{-A_1 \sin B_0}{6(1 - E^2)(1 - E^2 \sin^2 B_0)^{3/2}} \times$$

$$\times [5 - E^2 - 6\sin^2 B_0(1 + E^2) + 3E^2 \sin^4 B_0(3 + E^2) - 4E^4 \sin^6 B_0] ,$$

$$x_{13} = \frac{A_1(1 - E^2) \sin B_0}{6(1 - E^2 \sin^2 B_0)^{7/2}} \times$$

$$\times [1 - 9E^2 + 2E^2 \sin^2 B_0(5 - 3E^2) + 4E^4 \sin^4 B_0] ,$$

(15.89)

$$x_{30} = \frac{A_1 \cos B_0}{120(1 - E^2)^3(1 - E^2 \sin^2 B_0)^{1/2}} \times$$

$$\times [5 - E^2 - 4\sin^2 B_0(7 + 4E^2) + 2\sin^4 B_0(12 + 43E^2 + 13E^4) -$$

$$- 4E^2 \sin^6 B_0(18 + 25E^2 + 3E^4) + E^4 \sin^8 B_0(77 + 39E^2) - 28E^6 \sin^{10} B_0] ,$$

$$x_{32} = \frac{-A_1 \cos B_0}{120(1 - E^2)(1 - E^2 \sin^2 B_0)^{5/2}} \times$$

$$\times [5 - E^2 - 2\sin^2 B_0(9 + 4E^2 + E^4) + 15E^2 \sin^4 B_0(3 + E^2) -$$

$$- 2E^4 \sin^6 B_0(23 + 3E^2) + 16E^6 \sin^8 B_0] ,$$

$$x_{14} = \frac{A_1(1 - E^2) \cos B_0}{24(1 - E^2 \sin^2 B_0)^{9/2}} \times$$

$$\times [1 - 9E^2 + 36E^2 \sin^2 B_0(1 - 2E^2) +$$

$$+ 12E^4 \sin^4 B_0(5 - 2E^2) + 8E^6 \sin^6 B_0] .$$

Box 15.5 (A representation of the non-vanishing coefficients in a polynomial setup of a conformal mapping of type Gauss–Krueger or UTM).

$$y(l,b) =$$

$$= y_{01}b + y_{20}l^2 + y_{02}b^2 + y_{21}l^2b + y_{03}b^3 + y_{40}l^4 + y_{22}l^2b^2 + y_{04}b^4 + y_{41}l^4b + y_{23}l^2b^3 + \quad (15.90)$$

$$+ y_{05}b^5 + O(6) ,$$

$$y_{01} = \frac{A_1(1-E^2)}{(1-E^2\sin^2 B_0)^{3/2}} ,$$

$$y_{20} = \frac{A_1\cos B_0\sin B_0}{2(1-E^2\sin^2 B_0)^{1/2}} ,$$

$$y_{02} = \frac{3A_1E^2(1-E^2)\cos B_0\sin B_0}{2(1-E^2\sin^2 B_0)^{5/2}} ,$$

$$y_{21} = \frac{A_1(1-2\sin^2 B_0+E^2\sin^4 B_0)}{2(1-E^2\sin^2 B_0)^{3/2}} ,$$

$$y_{03} = \frac{A_1E^2(1-E^2)}{2(1-E^2\sin^2 B_0)^{7/2}} \times$$

$$\times[1-2\sin^2 B_0(1-2E^2)-3E^2\sin^2 B_0] ,$$

$$y_{40} = \frac{A_1\cos B_0\sin B_0}{24(1-E^2)(1-E^2\sin^2 B_0)^{1/2}} \times$$

$$\times[5-E^2-6\sin^2 B_0(1+E^2)+3E^3\sin^4 B_0(3+E^2)-4E^4\sin^5 B_0] ,$$

$$y_{22} = \frac{-A_1\cos B_0\sin B_0}{4(1-E^2\sin^2 B_0)^{5/2}} \times$$

$$\times[4-3E^2-2E^2\sin^2 B_0+E^4\sin^4 B_0] , \qquad (15.91)$$

$$y_{04} = \frac{-A_1E^2(1-E^2)\cos B_0\sin B_0}{8(1-E^2\sin^2 B_0)^{9/2}} \times$$

$$\times[4-15E^2+2E^2\sin^2 B_0(11-10E^2)+9E^2\sin^4 B_0] ,$$

$$y_{41} = \frac{A_1}{24(1-E^2)^2(1-E^2\sin^2 B_0)^{3/2}} \times$$

$$\times[5-E^2-4\sin^2 B_0(7+4E^2)+2\sin^4 B_0(12+43E^2+13E^4)-4E^2\sin^6 B_0(18+25E^2+3E^4)+$$

$$+E^4\sin^8 B_0(77+39E^2)-28E^6\sin^{10} B_0] ,$$

$$y_{23} = \frac{-A_1}{12(1-E^2\sin^2 B_0)^{7/2}} \times$$

$$\times[4-3E^2-4\sin^2 B_0(2-4E^2+3E^4)-2E^2\sin^4 B_0(2-5E^2)-$$

$$-4E^4\sin^6 B_0+E^6\sin^8 B_0] ,$$

$$y_{05} = \frac{-A_1E^2(1-E^2)}{40(1-E^2\sin^2 B_0)^{11/2}} \times$$

$$\times[4-15E^2-4\sin^2 B_0(2-32E^2+45E^4)-2E^2\sin^4 B_0(38-181E^2+60E^4)-$$

$$-4E^4\sin^6 B_0(41-34E^2)-27E^5\sin^5 B_0]$$

15-4 Principal distortions and various optimal designs (UTM mappings)

Principal distortions and various optimal designs of the Universal Transverse Mercator Projection (UTM) with respect to the dilatation factor.

By means of the general eigenvalue problem, we can constitute the principal distortions. At first, we compute the left Cauchy–Green tensor for the universal transverse Mercator projection modulo an unknown dilatation parameter according to Corollary 15.5.

Corollary 15.5 ($\mathbb{E}^2_{A_1,A_1,A_2}$, left Cauchy–Green tensor, Universal Transverse Mercator Projection (UTM) modulo an unknown dilatation parameter).

The solution of the boundary value problem subject to the integrability conditions of type Box 15.4 and Box 15.5 constitute the Universal Transverse Mercator Projection (UTM) modulo an unknown dilatation parameter ρ, namely (15.92), in the function space of bivariate polynomials.

$$x(l,b) =$$
$$= \rho\big(x_{10}l + x_{11}lb + x_{30}l^3 + x_{12}lb^2 + x_{31}l^3b + x_{13}lb^3 + x_{50}l^5 + x_{32}l^3b^2 + x_{14}lb^4 +$$
$$+O(6)\big) \ ,$$

$$y(l,b) =$$
$$= \rho\big(y_{01}b + y_{20}l^2 + y_{02}b^2 + y_{21}l^2b + y_{03}b^3 + y_{40}l^4 + y_{22}l^2b^2 + y_{04}b^4 + y_{41}l^4b + y_{23}l^2b^3 + y_{05}b^5 +$$
$$+O(6)\big) \ .$$

(15.92)

The coordinates of the left Cauchy–Green deformation tensor $\mathbf{C}_l$ are represented by

$$c_{11} := x_l^2 + y_l^2 \ ,$$
$$c_{12} := c_{21} := x_l x_b + y_l y_b = 0 \ ,$$
$$c_{22} := x_b^2 + y_b^2 \ ,$$

(15.93)

or

$$x_l = \rho\big(x_{10} + x_{11}b + 3x_{30}l^2 + x_{12}b^2 + O_{lx}(3)\big) \ ,$$
$$y_l = \rho\big(2y_{20}l + 2y_{21}lb + O_{ly}(3)\big) \ ,$$
$$x_b = \rho\big(x_{11}l + 2x_{12}lb + O_{bx}(3)\big) \ ,$$
$$y_b = \rho\big(y_{01} + 2y_{02}b + y_{21}l^2 + 3y_{03}b^2 + O_{by}(3)\big) \ ,$$

(15.94)

and

$$c_{11} = \rho^2\big(x_{10}^2 + (4y_{20} + 6x_{10}x_{30})l^2 + 2x_{10}x_{11}b + x_{11}^2b^2 + O_l(3)\big) \ ,$$
$$c_{22} = \rho^2\big(y_{01}^2 + (x_{11} + 2y_{01}y_{21})l^2 + 4y_{01}y_{02}b + (4y_{02}^2 + 6y_{01}y_{03})b^2 + O_b(3)\big) \ .$$

(15.95)

End of Corollary.

The proof of Corollary 15.5 4 is lengthy, namely for $c_{12} = c_{21} = 0$. Instead, we refer to the solution of the general eigenvalue problem in Corollary 15.6.

Corollary 15.6 ($\mathbb{E}^2_{A_1, A_1, A_2}$, principal distortions, Universal Transverse Mercator Projection (UTM) modulo an unknown dilatation parameter).

Under the mapping equations (15.92), which constitute the Universal Transverse Mercator Projection (UTM) modulo an unknown dilatation parameter ρ, the *principal distortion* or *factor of conformality*, after a lengthy computation, amounts to

$$\Lambda^2 := \Lambda_1^2 = \Lambda_2^2 = \frac{c_{11}}{G_{11}} = \frac{c_{22}}{G_{22}} \tag{15.96}$$

or

$$\Lambda^2 = \rho^2 \left(1 + \cos^2 B \left(1 + \frac{E^2}{1 - E^2} \cos^2 B \right) l^2 + \mathrm{O}_{\Lambda^2}(l^4) \right) . \tag{15.97}$$

End of Corollary.

In summarizing, we get the squared factor of conformality proportional to the order of squared l^2. In the following few passages, we determine the unknown dilation factor either by the postulate of *minimal total distance distortion* (Airy optimality) or by the postulate of *minimal total areal distortion*. Results are collected in two corollaries, two examples (UTM and Gauss–Krueger conformal coordinate systems) and five graphical illustrations.

Corollary 15.7 (Dilatation factor for an optimal transversal Mercator projection, minimal total distance distortion, Airy optimum).

(i)

For a conformal map of the half-symmetric strip $[-l_{\mathrm{E}}, +l_{\mathrm{E}}] \times [B_{\mathrm{S}}, B_{\mathrm{N}}]$ of type Universal Transverse Mercator Projection (UTM), the unknown dilatation factor ρ is optimally designed under the postulate of minimal total distance distortion if (15.98) accurate to the order $\mathrm{O}(E^4)$ holds.

$$\rho = 1 - \frac{1}{6} l_{\mathrm{E}}^2 \left[\frac{\sin B_{\mathrm{N}} + E^2 \sin B_{\mathrm{N}} - \frac{1}{3} \sin^3 B_{\mathrm{N}} - \frac{E^2}{5} \sin^5 B_{\mathrm{N}}}{\sin B_{\mathrm{N}} + \frac{2}{3} E^2 \sin^3 B_{\mathrm{N}} - \sin B_{\mathrm{S}} - \frac{2}{3} E^2 \sin^3 B_{\mathrm{S}}} + \right.$$

$$\left. + \frac{- \sin B_{\mathrm{S}} - E^2 \sin B_{\mathrm{S}} + \frac{1}{3} \sin^3 B_{\mathrm{S}} + \frac{E^2}{5} \sin^5 B_{\mathrm{S}}}{\sin B_{\mathrm{N}} + \frac{2}{3} E^2 \sin^3 B_{\mathrm{N}} - \sin B_{\mathrm{S}} - \frac{2}{3} E^2 \sin^3 B_{\mathrm{S}}} + \cdots \right] . \tag{15.98}$$

(ii)

For the symmetric strip $[-l_{\mathrm{E}}, +l_{\mathrm{E}}] \times [-B_{\mathrm{N}}, B_{\mathrm{N}}]$, we specialize

$$\rho = 1 - \frac{1}{6} l_{\mathrm{E}}^2 \frac{(1 + E^2) \sin B_{\mathrm{N}} - \frac{1}{3} \sin^3 B_{\mathrm{N}} - \frac{1}{5} E^2 \sin^5 B_{\mathrm{N}}}{\sin B_{\mathrm{N}} + \frac{2}{3} E^2 \sin^2 B_{\mathrm{N}}} . \tag{15.99}$$

(iii)

If $B_{\mathrm{N}} - B_{\mathrm{S}} = \pi/2$ up to $\mathrm{O}(E^4)$, ρ amounts to

$$\rho(\pi/2) = 1 - \frac{1}{9} \left(1 + \frac{8}{15} E^2 \right) l_{\mathrm{E}}^2 . \tag{15.100}$$

End of Corollary.

Example 15.3 ($[-l_E, +l_E] \times [B_S, B_N] = [-3.5°, +3.5°] \times [80°\,S, 84°\,N]$).

The classical UTM conformal coordinate system is chosen for a strip of 6° width with 1° overlays and between $B_S = -80°$ of southern latitude and $B_N = +84°$ of northern latitude. Once we refer to the Geodetic Reference System 1980 (H. Moritz, 1984), $E^2 = 0.006\,694\,380\,022\,90$, in particular, with l_E given by $l_E = 3.5° = 0.061\,086\,5\,\text{rad}$, the dilatation parameter amounts to

$$\rho = 0.999\,578 \quad \text{(scale reduction factor 1 : 2\,370)} . \tag{15.101}$$

End of Example.

Example 15.4 ($[-l_E, +l_E] \times [B_S, B_N] = [-2°, +2°] \times [80°\,S, 80°\,N]$).

The classical Gauss–Krueger conformal coordinate system is chosen for a strip of 3° width with 0.5° overlays and between $B_S = -80°$ of southern latitude and $B_N = +80°$ of northern latitude. Once we refer to the Geodetic Reference System 1980, $E^2 = 0.006\,694\,380\,022\,90$, in particular, with l_E given by $l_E = 2° = 0.034\,906\,5\,\text{rad}$, the dilatation parameter amounts to

$$\rho = 0.999\,864 \quad \text{(scale reduction factor 1 : 7\,353)} . \tag{15.102}$$

End of Example.

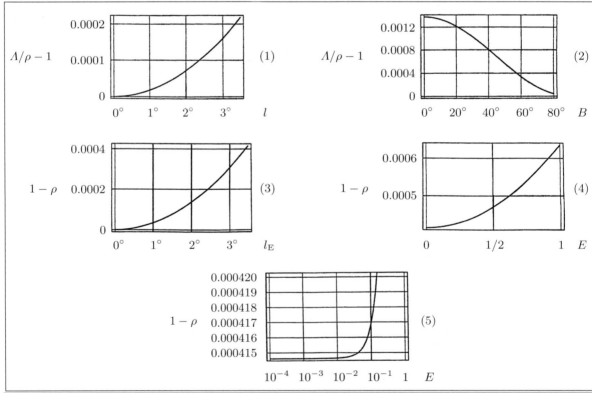

Fig. 15.5. (1) The ratio of scale factors $\Lambda/\rho(l)$ as a function of l, $B = 70°$ ($\to$ (15.97)). (2) The ratio of scale factors $\Lambda/\rho(B)$ as a function of B, $l = 3°$ ($\to$ (15.97)). (3) Dilatation factor $\rho(l_E)$ as a function of $l_E = 0.006\,694\,380\,022\,90$ ($\to$ (15.100)). (4) $\rho(l_E)$ as a function of eccentricity E, $l_E = 3.5°$, first illustration ($\to$ (15.100)). (5) $\rho(l_E)$ as a function of eccentricity E $l_E = 3.5°$, second illustration ($\to$ (15.100)),

For the proof, we start from the formula $\Lambda^2(l,b)$ as a representation of formula (15.97), namely the *principal distortion* as a function of the longitudinal difference $L - L_0 =: l$ and the latitude B. The *criterion of optimality* for the first design of the transverse Mercator projection modulo an unknown dilatation factor ρ is the *minimal total distance distortion* over a meridian strip $[l_W, l_E] \times [B_S, B_N]$ between a longitudinal extension L_W and L_E and a latitudinal extension B_S and B_N (namely the symbols S, N, E, and W as indices denote South, North, East, and West), in particular, the *G. B. Airy* (1861) *distortion measure* (15.103) with respect to the principal distortions Λ_1 and Λ_2 and the spheroidal surface element, locally (15.104) and globally (15.105). The G. B. Airy distortion minimization subject to $\Lambda_1 = \Lambda_2 = \Lambda$, the criterion for conformality, leads directly to the representations (15.98)–(15.100).

$$
I_{lA} :=
$$
$$
:= \frac{1}{2S} \int dS\big[(\Lambda_1 - 1)^2 + (\Lambda_2 - 1)^2\big] = \min ,
$$
(15.103)

$$
dS =
$$
$$
= A_1^2(1 - E^2)\cos B\, dl\, dB(1 - E^2 \sin^2 B)^2 ,
$$
(15.104)

$$
S =
$$
$$
= 2A_1^2(1 - E^2)l_E \times
$$
$$
\times[\sin B_N + (2/3)E^2 \sin^3 B_N - (\sin B_S + (2/3)E^2 \sin^3 B_S)+
$$
$$
+O_N(E^4) + O_S(E^4)] .
$$
(15.105)

As an alternative, we could base a second design of the transverse Mercator projection modulo an unknown dilatation factor ρ on a *minimal total areal distortion* over a meridian strip $[l_W, l_E] \times [B_S, B_N]$ between a longitudinal extension L_W and L_E and a latitudinal extension B_S and B_N, in particular, the distortion measure (15.106) with respect to principal distortions Λ_1 and Λ_2, and the spheroidal surface element. But surprisingly, it does not differ from the one of the previous corollary, the optimal Airy design, see Corollary 15.8, the proof of which we want to leave to the reader.

$$
I_l(\text{areal}) :=
$$
$$
:= \frac{1}{S} \int dS(\Lambda_1 \Lambda_2 - 1) .
$$
(15.106)

Corollary 15.8 (The dilatation factor for an optimal universal transverse Mercator projection, the zero total distortion).

For a conformal map of the half-symmetric strip $[-l_E, l_E] \times [B_S, B_N]$ of type universal transverse Mercator projection, the postulate of *minimal total distance distortion* (Airy optimum) and the postulate of *minimal total areal distortion* lead to the same unknown dilatation factor ρ ($\to$ (15.98)–(15.100)). The total areal distortion amounts to zero!

End of Corollary.

For this chapter and the other chapters, please consult the following publications. G. B. Airy (1861), M. Amalvict and E. Livieratos (1988), E. Beltrami (1869), W. Blaschke and K. Leichtweiß (1973), K. Bretterbauer (1980), M. do Carmo, M. Dajczer, and F. Mercuri (1985), A. Cauchy (1823, 1828), A. R. Clarke and F. R. Helmert (1911), J. H. Cole (1943), D. Cox, J. Little, and D. O'Shea (1996), A. Dermanis and E. Livieratos (1983, 1993), A. Dermanis, E. Livieratos, and S. Pertsinidou (1984), J. Engels and E. Grafarend (1995), L. Euler (1755, 1770), A. Finzi (1922), C. F. Gauss (1813, 1816–1827, 1822, 1844), H. Glasmacher, K. Krack (1984), H. Goenner, E. Grafarend, and R. J. You (1994), E. Grafarend (1995), E. Grafarend and R. Syffus (1995, 1998c), G. Green (1841), E. R. Hedrick and L. Ingold (1925a,b), S. Heitz (1988), M. Hotine (1946, 1947), C. G. J. Jacobi (1839), C. Kaltsikis (1980), V. V. Kavrajski (1958), W. Klingenberg (1982), R. König and K. H. Weise (1951), A. Korn (1914), L. Krueger (1903, 1912, 1914, 1922), R. S. Kulkarni (1969, 1972), R. S. Kulkarni and U. Pinkall (eds. 1988), Laborde (1928), J. Lafontaine (1988a,b), J. L. Lagrange (1781), L. P. Lee (1944, 1976), L. Lichtenstein (1911, 1916), R. Lilienthal (1902–1927), J. Liouville (1850), E. Livieratos (1987), C. F. van Loan (1976), D. H. Maling (1960, 1973), H. Maurer (1935), A. I. Markuschewitsch (1955), O. M. Miller (1941), C. W. Misner (1978), S. K. Mitra and C. R. Rao (1968), B. de Moor and H. Zha (1991), B. Mueller (1991), G. Ricci (1918), P. Richardus and R. K. Adler (1972a,b), C. F. B. Riemann (1851), M. Rosenmund (1903), E. Schering (1857), H. Schmehl (1927), J. A. Schouten (1921), J. P. Snyder (1979a–c, 1982), K. Spallek (1980), E. M. Stein and G. Weiss (1968), T. C. T. Ting (1985), N. A. Tissot (1881), F. Uhlig (1976, 1979), H. Weber (1867), T. Wray (1974), K. Yano (1970), A. I. Yanushaushas (1982), M. Zadro and A. Carminelli (1966), H. Zha (1991) and J. Zund (1987).

15-5 Examples (Gauss–Krueger/UTM coordinates)

Various interesting Examples. Mapping of the transverse Mercator projection. Gauss–Krueger/UTM coordinates. Strip system, meridian strip system of Germany.

There has been the result that the regular transverse Mercator projection of the sphere is simple and its mathematical version does not cause any problem. The picture changes if we move to the transverse Mercator projection of the ellipsoid-of-revolution.

It relates to the *elliptical transverse cylinder*. It is *conformal*. Its *central meridian* and each meridian 90° apart from it are *straight lines*. Its equator is a straight line, other meridians and parallels are *complex curves*. Scale is true along the central meridian or along two straight lines in the map equidistant from and parallel to the central meridian, constant along any straight line on the map parallel to the central meridian. Scale becomes *infinite* 90° from the reference meridian. It is used extensively for *quadrangle maps* at scales from 1 : 25000 to 1 : 250000.

We recall the representation of Transverse Mercator coordinates for the ellipsoid-of-revolution in the following form.

$$x(l, b) =$$
$$= x_{10}l + x_{11}lb + x_{30}l^3 + x_{12}lb^2 + x_{31}l^3b + x_{13}lb^3 + x_{50}l^5 + x_{32}l^3b^2 + x_{14}lb^4 +$$
$$+ O(6)$$
$$\text{(Easting)} ,$$

$$\tag{15.107}$$

$$y(l, b) =$$
$$= y_{01}b + y_{20}l^2 + y_{02}b^2 + y_{21}l^2b + y_{03}b^3 + y_{40}l^4 + y_{22}l^2b^2 + y_{04}b^4 + y_{41}l^4b + y_{23}l^2b^3 +$$
$$+ O(6)$$
$$\text{(Northing)} .$$

$$\tag{15.108}$$

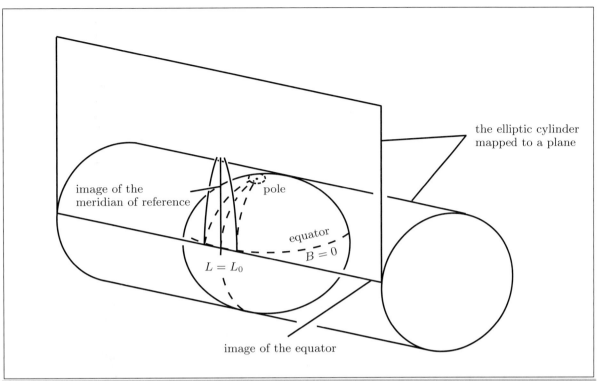

Fig. 15.6. Mapping of the transverse Mercator projection, the surface of the elliptical cylinder and the line-of-contact L_0.

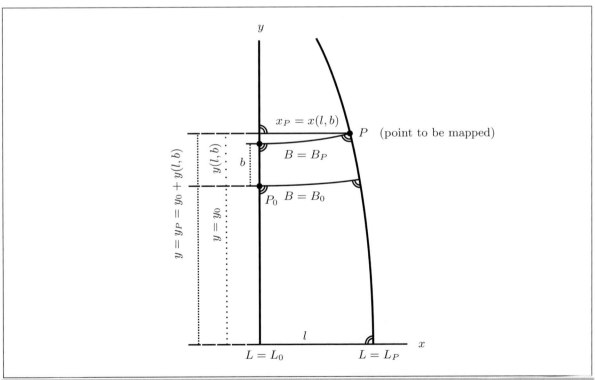

Fig. 15.7. Mapping of the transverse Mercator projection, central meridian $L = L_0$. Mapping of the point P with respect to the point $P_0(L_0, B_0)$ of reference, $l := L - L_0$, $b := B - B_0$.

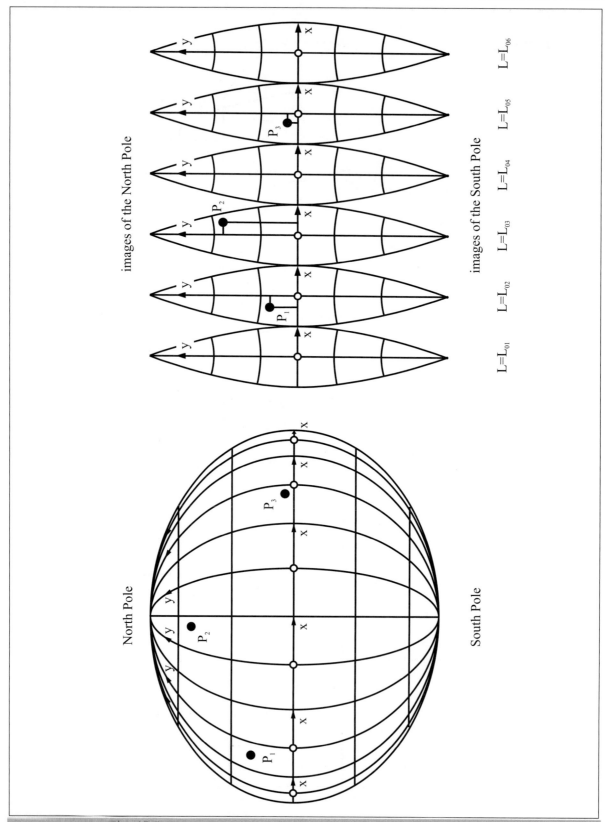

Fig. 15.8. Mapping of the transverse Mercator projection, strip system

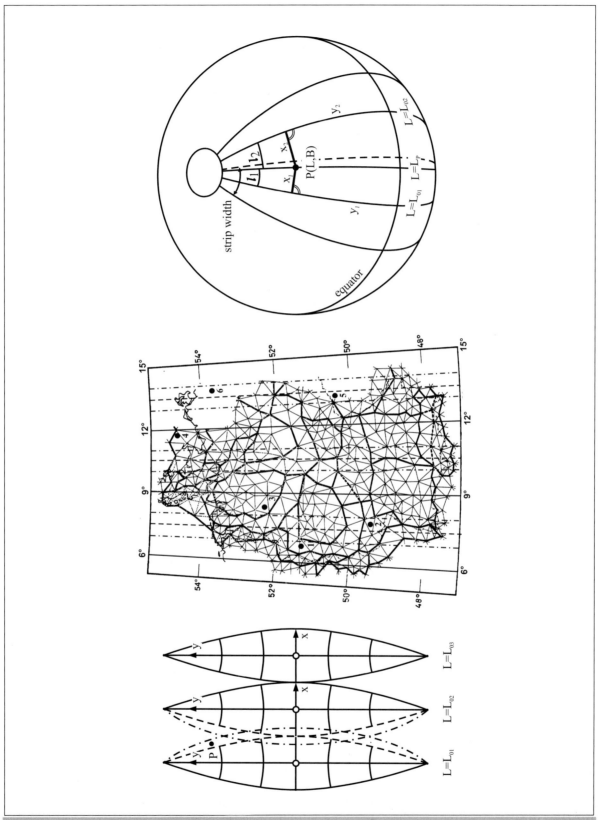

Fig. 15.9. Mapping of the transverse Mercator projection, meridian strip system of Germany: reference meridians $L_0 = \{6°, 9°, 12°, 15°\}$, overlapping range $\pm 0.5°$.

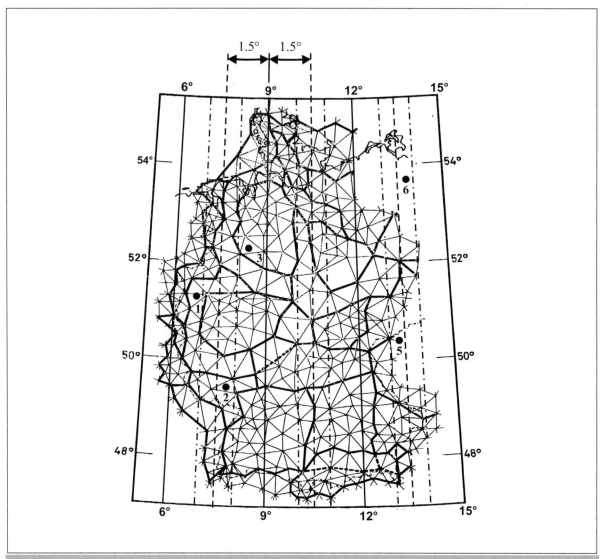

Fig. 15.10. Mapping of the transverse Mercator projection, meridian strip system of Germany, strip width ±1.5°, Gauss–Krueger coordinates.

Important!

West and East of the *reference meridian*, we choose a strip of ±1.5° in longitude for a Gauss–Krueger strip system, for instance, according to Example 15.5 and Example 15.6 in the strips 6°, 9°, and 12° for Germany. In contrast, ETRS 89 is given in UTM coordinates requiring a strip of ±3° width, a 6° wide strip reference system.

In order to avoid negative coordinates which are located *West of the reference meridian* and not to lose reference to the reference meridian, Easting coordinates as well as Northing coordinates of the Gauss–Krueger strip system are to changed in the following way.

Important!

(i) x: add 10^6 times the meridian number $L_0/3°$, (ii) x: add the number $500000\,\mathrm{m}$, and (iii) y: define $y_0 + y(l, b)$ as the number reflecting the distance of a point from the equator. y_0 is the length of the meridian arc from the equator to the ellipsoidal latitude B_0.

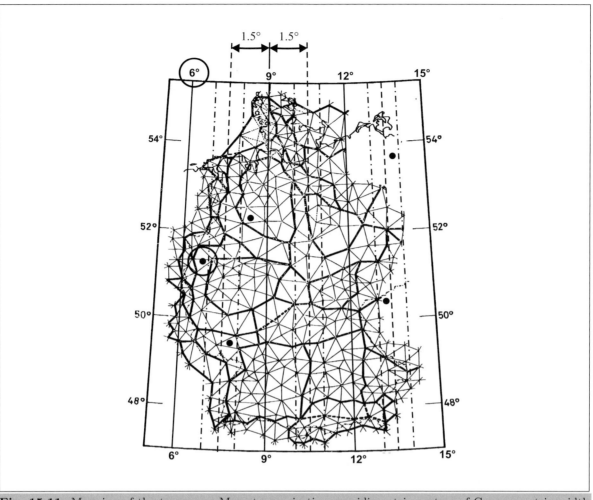

Fig. 15.11. Mapping of the transverse Mercator projection, meridian strip system of Germany, strip width ±1.5°, Gauss–Krueger coordinates, point 1 is indicated by the dot–circle.

Example 15.5 (Gauss–Krueger coordinates. Easting versus Northing coordinates).

The point 1, which is located on the *Bessel ellipsoid*, is described by ellipsoidal normal coordinates $L = 6.8°$ and $B = 51.2°$. Compare with Fig. 15.11.

Important!

$$x_1(\text{Gauss–Krueger}) = 55\,909.151 \text{ m}, \quad y_1(\text{Gauss–Krueger}) = 5\,674\,057.263 \text{ m}.$$

Easting: $2\,555\,909.151$ m, Northing: $5\,674\,057.263$ m.

Important!

The point 1 is located $55\,909.151$ m East of the *reference meridian* with the *meridian number* 2, that is $L_0 = 6°$ and $5\,674\,057.263$ m *North of the equator*.

End of Example.

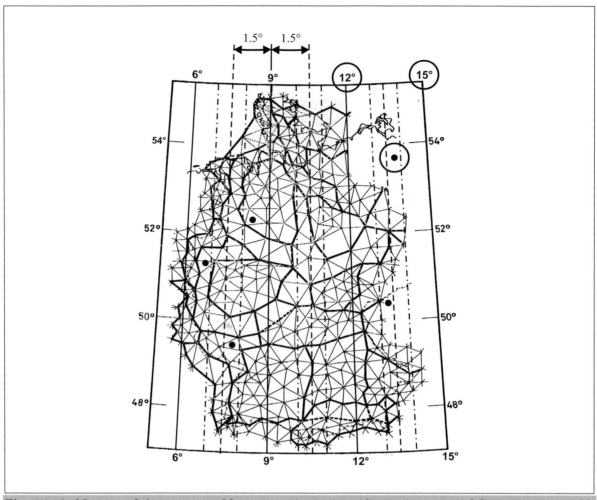

Fig. 15.12. Mapping of the transverse Mercator projection, meridian strip system of Germany, strip width ±1.5°, Gauss–Krueger coordinates, point 6 (dot–circle) in two separate strips.

Example 15.6 (Gauss–Krueger coordinates. Easting versus Northing coordinates).

The point 6, which is located on the *Bessel ellipsoid*, is described by ellipsoidal normal coordinates $L = 13.75°$ and $B = 53.8°$. Compare with Fig. 15.12.

First meridan strip.

x_1(Gauss–Krueger) $= 115\,287.428$ m, y_1(Gauss–Krueger) $= 5\,964\,460.428$ m.

Easting: $4\,615\,287.428$ m, Northing: $5\,964\,460.428$ m.

The point 6 is located $115\,287.428$ m East of the *reference meridian* with the *meridian number 4*, that is $L_0 = 12°$ and $5\,964\,460.428$ m *North of the equator*.

Second meridan strip.

$$x_2(\text{Gauss–Krueger}) = -82\,350\,056\,\text{m}, \ y_2(\text{Gauss–Krueger}) = 5\,963\,764\,424\,\text{m}.$$

Easting: $5\,417\,649\,944\,\text{m}$, Northing: $5\,963\,764\,424\,\text{m}$.

The point 6 is located $82\,350\,056\,\text{m}$ West of the *reference meridian* with the *meridian number* 5, that is $L_0 = 15°$ and $5\,963\,464\,424\,\text{m}$ *North of the equator.*

End of Example.

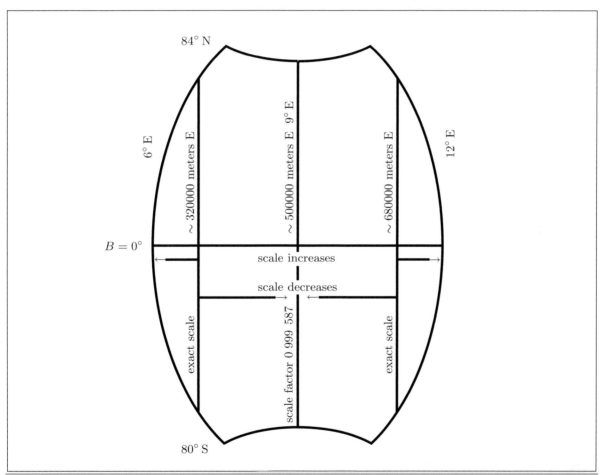

Fig. 15.13. Mapping of the universal transverse Mercator projection, strip width $\pm 3°$, scale factor $0\,999\,587$ (Geodetic Reference System 1980, H. Moritz 1984).

If we choose a *Universal Transverse Mercator Projection* as our reference system, we have to acknowledge that the central meridian is not equidistantly mapped. Instead, two meridians *West and East of the central meridian* are mapped equidistantly.

We compute the scale factor of the reference meridian by minimizing the *Airy distortion measure* in a given strip with respect to the ellipsoid-of-revolution in a preceding section. There is an International Agreement to use the scale factor 0.999 587.

Important!

In practice, we use relative to the *reference meridian* the *zone number* $(L_0+3°)/6°+30$. In use is also the reference *Easting/Northing* with respect to the meridians $L_0 = -177° = 177°\,\text{W}$, $L_0 = -171°, \cdots, L_0 = -3°, L_0 = 3°, \ldots, L_0 = 171° = 171°\,\text{E}$. For instance, Germany is located between the zone 32 and the zone 33. The easterly coordinate is equipped with an offset number 500000 m, called "false Easting". In contrast, Northings are changed by an offset of 10 000 000 m on the southern hemisphere $(y < 0)$ to avoid negative coordinates, an artifical effect, called "false Northing". The strip overlap is chosen as $0.5°$.

Example 15.7 (UTM coordinates. Easting versus Northing coordinates).

The point 1 is described relative to the Geodetic Reference System 80 (GRS 80), the surface normal cooordinates by $L = 6.8°$ and $B = 51.2°$. Compare with Figs. 15.14–15.16.

Important!

First meridan strip (scale factor 0.999 578).

$$x_1(\text{UTM}) = -153\,697.036\,\text{m}, \ y_1(\text{UTM}) = 5\,674\,241.346\,\text{m}.$$

Easting: 346 302.964 m, Northing: 5 674 241.346 m, zone 32.

Important!

The point 1 is located 153 697.036 m West of the *reference meridian* of the zone 32, that is $L_0 = 9°$ and 5 674 241.346 m *North of the equator*.

Important!

Second meridan strip (scale factor 0.999 578).

$$x_2(\text{UTM}) = 265\,448.926\,\text{m}, \ y_2(\text{UTM}) = 5\,678\,805.917\,\text{m}.$$

Easting: 765 448.926 m, Northing: 5 678 805.917 m, zone 31.

Important!

The point 1 is located 265 448.926 m East of the *reference meridian* of the zone 31, that is $L_0 = 3°$ and 5 678 805.917 m *North of the equator*.

End of Example.

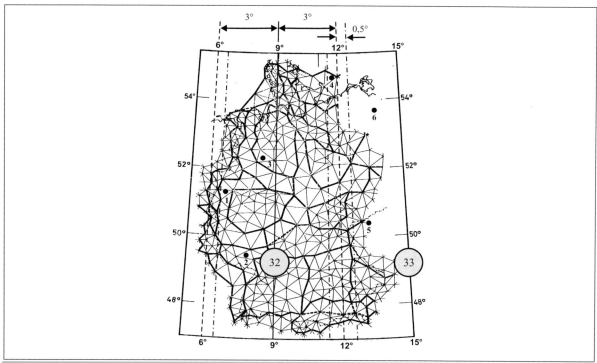

Fig. 15.14. Mapping of the universal transverse Mercator projection, strip width ±3°, scale factor 0.999 578, point 1 (dot–circle), zone 32 and zone 33.

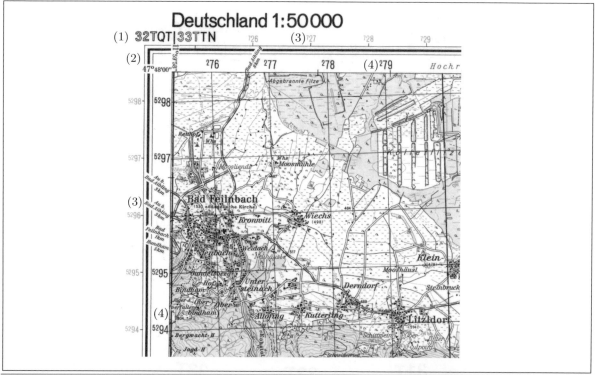

Fig. 15.15. Mapping of the universal transverse Mercator projection, strip width ±3°, scale factor 0.999 578, point 1 (dot–circle), zone 32 and zone 33, map scale 1 : 50000 (Germany). (1) Zone n, UTM grid. (2) Ellipsoidal normal coordinates of the underlying ellipsoid. (3) East and North values with respect to the central meridian, zone 32. (4) East and North values with respect to the central meridian, zone 33.

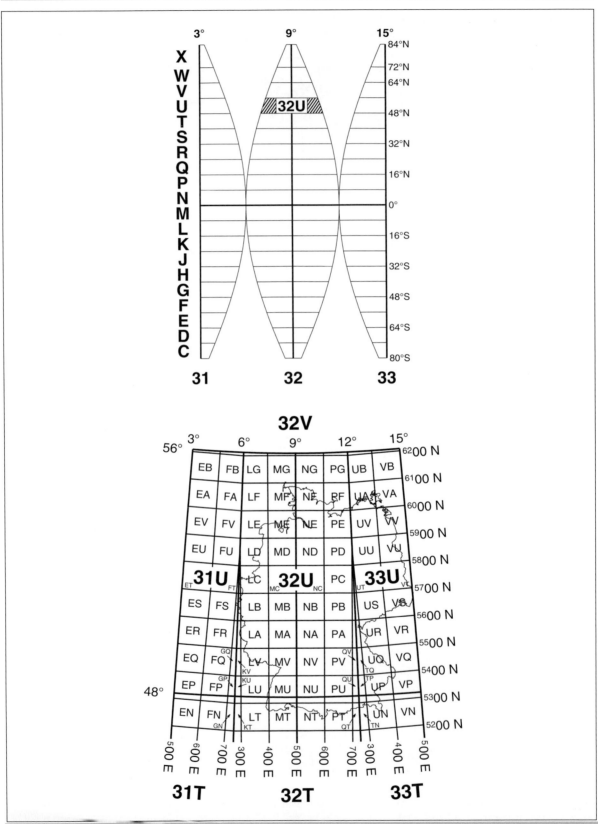

Fig. 15.16. Mapping of the transverse Mercator projection, zones 31, 32, and 33, design of identity zones.

The summary concerning the Gauss–Krueger coordinates is presented in Box 15.6, the summary concerning the UTM coordinates is presented in Box 15.7, and a comparison is shown in Box 15.8.

Box 15.6 (Summary: Gauss–Krueger coordinates).

(1)

Choose L_0, B_0 such that $l = L - L_0 < l_{\max} = 2°$, $b = B - B_0 < 1°$, and L_0 multiple of $3°$.

False Easting: $x(l, b) + \frac{L_0}{3°} \times 10^6 + 5 \times 10^5$, Northing: $y_0 + y(l, b)$.

y_0 is the length of meridian from the equator to the latitude point B_0: $B_0 = B$ is permitted.

(2)

Choose L, B out of False Easting, Northing .

Choose y_0 such that $y = $ Northing $- y_0 < 100\,\mathrm{km}$ and
$$x = \left| \text{False Easting} - 10^6 \times \text{meridian number} - 5 \times 10^5 \right| < l_{\max} \times R_0 \times \cos B / \rho,$$
$$\rho := 180°/\pi, \ R_0 = 6\,380\,000\,\mathrm{m}.$$

Longitude: $L = L_0 + l(x, y)$, latitude: $B = B_0 + b(x, y)$.

B_0 is the latitude of the meridian arc y_0, $y_0 = $ Northing is admissible, $L_0 = 3° \times$ meridian number.

Box 15.7 (Summary: UTM coordinates).

(1)

Choose L_0, B_0 such that $l = L - L_0 < l_{\max} = 3.5°$, $b = B - B_0 < 1°$, and $L_0 + 3°$ multiple of $6°$.

False Easting: $x(l, b) + 5 \times 10^5$, False Northing: $y_0 + y(l, b) + \begin{cases} 0 & \text{if Northing } > 0, \\ 10^7 & \text{if Northing } < 0. \end{cases}$

y_0 is the length of meridian from the equator to the latitude point B_0: $B_0 = B$ is permitted.

(2)

Choose L, B out of False Easting, False Northing subject to scale $0.999\,578$.

Choose y_0 such that $y = $ False Northing $- y_0 < 100\,\mathrm{km}$ and
$$x = \left| \text{False Easting} - 5 \times 10^5 \right| < l_{\max} \times R_0 \times \cos B / \rho,$$
$$\rho := 180°/\pi, \ R_0 = 6380000\,\mathrm{m}.$$

Longitude: $L = (\text{zone} - 30) \times 6° - 3° + l(x, y)$, latitude: $B = B_0 + b(x, y)$.

B_0 is the latitude of the meridian arc y_0, $y_0 = $ Northing is admissible.

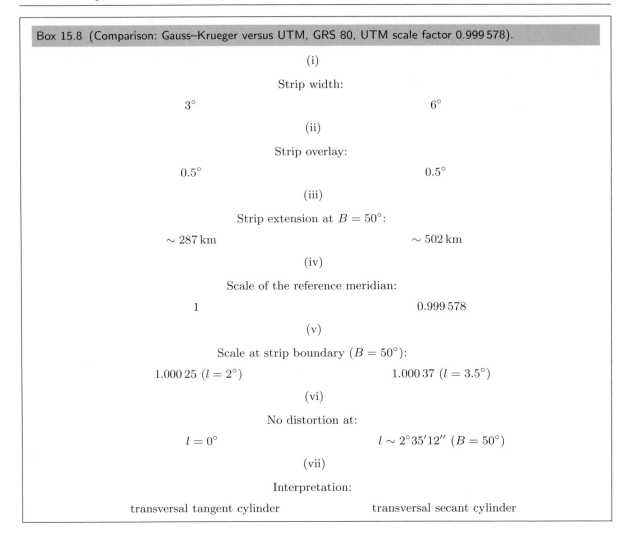

Box 15.8 (Comparison: Gauss–Krueger versus UTM, GRS 80, UTM scale factor 0.999 578).

(i)

Strip width:

3°   6°

(ii)

Strip overlay:

0.5°   0.5°

(iii)

Strip extension at $B = 50°$:

$\sim 287\,\mathrm{km}$ $\sim 502\,\mathrm{km}$

(iv)

Scale of the reference meridian:

1   0.999 578

(v)

Scale at strip boundary $(B = 50°)$:

1.000 25 $(l = 2°)$ 1.000 37 $(l = 3.5°)$

(vi)

No distortion at:

$l = 0°$ $l \sim 2°35'12''$ $(B = 50°)$

(vii)

Interpretation:

transversal tangent cylinder   transversal secant cylinder

15-6 Strip transformation of conformal coordinates (Gauss–Krueger/UTM mappings)

Strip transformation of conformal coordinates of type Gauss–Krueger and of type UTM. Conformal polynomial, inverse conformal polynomial.

Due to the increasing demand of connectivity of *geodetic charts* of the Earth surface, namely caused by digital cartography in transport systems ("vehicles"), *strip transformations* of *conformal coordinates* have gained new interest, namely under the postulate of efficiency and speed of computation. Accordingly, we derive here a set of new formulae for the strip transformation of conformal coordinates of type Gauss–Krueger and of type Universal Transverse Mercator Projection (UTM) with an optimal dilatation factor different from one.

Section 15-61 has its objective in the derivation of transformation formulae of conformal coordinates $\{x_1, y_1\}$ of a strip of ellipsoidal longitude L_{01} to conformal coordinates $\{x_2, y_2\}$ of a strip of ellipsoidal longitude L_{02}. A two-step-approach is proposed which generates the solution (15.123) and (15.124) of the strip transformation problem. Section 15-612 focuses on two examples of strip transformations relating to (i) the *Bessel reference ellipsoid* and (ii) the *World Geodetic Reference System 1984* (WGS84). In particular, we compare the strip transformation results with those produced by a direct transformation of ellipsoidal longitude/latitude of a point on the reference ellipsoid (ellipsoid-of-revolution) into conformal coordinates in the first and second strip.

15-61 Two-step-approach to strip transformations

Here, we outline the two-step-approach which leads us by inversion technology of bivariate homogeneous polynomials to the strip transformation $x_2 = X(x_1, y_1)$ and $y_2 = Y(x_1, y_1)$ of conformal coordinates $\{x_1, y_1\}$ of the first L_{01}-strip into conformal coordinates $\{x_2, y_2\}$ of the second L_{02}-strip, namely for conformal coordinates of type Gauss–Krueger (GK) and UTM.

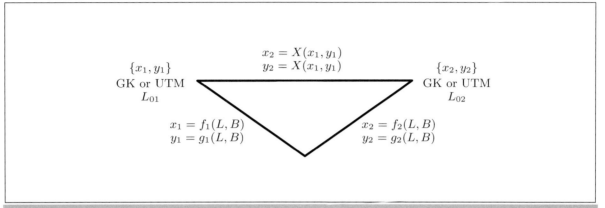

Fig. 15.17. Commutative diagram for a strip transformation of conformal coordinates of type Gauss–Krueger or of type UTM.

Assume the conformal coordinates $\{x_1, y_1\}$ in the first Gauss–Krueger or UTM strip system of ellipsoidal longitude L_{01} to be given. We also refer to L_{01} as the ellipsoidal longitude of the meridian of reference which is mapped equidistantly (or up to an optimal dilatation factor) under a conformal mapping of Gauss–Krueger type (or of UTM type). The minimal distance mapping of a topographic point on the Earth surface onto the ellipsoid-of-revolution $\mathbb{E}^2_{A_1, A_2}$ of semi-major axis A_1 and semi-minor axis A_2 as outlined by E. Grafarend and P. Lohse (1991) identifies the point {ellipsoidal longitude, ellipsoidal latitude} $= \{L, B\}$ of surface normal type on $\mathbb{E}^2_{A_1, A_2}$. The problem of a strip transformation may be formulated as following: given the conformal coordinates $\{x_1, y_1\}$ with respect to a first strip system L_{01} of a point $\{L, B\}$ on $\mathbb{E}^2_{A_1, A_2}$, find its conformal coordinates $\{x_2, y_2\}$ with respect to a second strip system L_{02}. An illustration of the involved transformations is presented in the commutative diagram of Figs. 15.17 and 15.18. The transformation $x_2(x_1, y_1)$ and $y_2(x_1, y_1)$ to which we refer as the strip transformation of conformal coordinates of type Gauss–Krueger or of type UTM is generated as following.

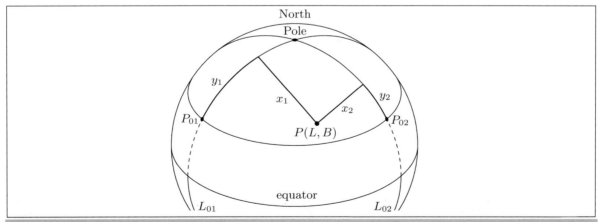

Fig. 15.18. Oblique orthogonal projection of an ellipsoid-of-revolution $\mathbb{E}^2_{A_1, A_2}$, semi-major axis A_1, semi-minor axis A_2; meridian of reference L_{01} and L_{02}, respectively, reference points $\{L_{01}, B_{01} = B_0\}$ and $\{L_{02}, B_{02} = B_0\}$, respectively; L_{01}-strip, L_{02}-strip; a point $P(L, B)$ on $\mathbb{E}^2_{A_1, A_2}$.

15-611 The first step: polynomial representation of conformal coordinates in the first strip and bivariate series inversion

The standard polynomial representation of conformal coordinates of type Gauss–Krueger or UTM in the L_{01}-strip is given by (15.109) and (15.110) subject to the longitude/latitude differences $l_1 := L - L_{01}$ and $b_1 := B - B_{01}$ with respect to the longitude L_{01} of the reference meridian and the latitude B_{01} of the reference point $\{L_{01}, B_{01}\}$ of series expansion.

Easting:

$$x_1 = \rho \left(x_{10} l_1 + x_{11} l_1 b_1 + x_{30} l_1^3 + x_{12} l_1 b_1^2 + O_{4x} \right) .$$

(15.109)

Northing:

$$y_1 = \rho \left(y_0 + y_{01} b_1 + y_{20} l_1^2 + y_{02} b_1^2 + y_{03} b_1^3 + O_{4y} \right) .$$

(15.110)

y_0 denotes the length of the meridian arc from zero ellipsoidal latitude to the ellipsoidal latitude B_{01} of the reference point $\{L_{01}, B_{01}\}$. The dilatation factor ρ amounts to one for a classical Gauss–Krueger conformal mapping. Optimal alternative values for the dilatation factor depending on the width of the strip, namely for UTM, are given in Box 15.9. The coefficients $\{x_{ij}, y_{ij}\}$ of the conformal polynomial of type (15.109) and (15.110) of order five are derived in E. Grafarend (1995, p. 457–459), for instance, and listed in Boxes 15.4 and 15.5. The length y_0 of the meridian arc from the equator to the reference point is computed from (15.114) in Box 15.10.

Box 15.9 (Optimal dilatation factor for a Universal Transverse Mercator mapping of an ellipsoid-of-revolution $\mathbb{E}^2_{A_1, A_2}$ according to E. Grafarend (1995 p. 459–461), $l_E := L - L_0$ eastern longitude difference, B_S and B_N southern latitude and northern latitude).

Strip width $[-l_E, l_E] \times [B_S, B_N]$:	Optimal dilatation factor:
$[-3.5°, +3.5°] \times [80°, 84°]$,	0.999 578 ,
$[-2°, +2°] \times [80°, 80°]$.	0.999 864 .

(15.111)

As outlined by E. Grafarend, T. Krarup and R. Syffus (1996, p. 279–284), the inversion of the bivariate homogeneous conformal polynomial (15.109) and (15.110) leads us to the bivariate homogenous polynomial (15.112) and (15.113) with coefficients $\{l_{ij}, b_{ij}\}$ summarized in Box 15.11.

$$l_1 = L - L_{01} =$$

$$= l_{10} \frac{x_1}{\rho} + l_{11} \frac{x_1}{\rho} \left(\frac{y_1}{\rho} - y_0 \right) + l_{30} \left(\frac{x_1}{\rho} \right)^3 +$$

$$+ l_{12} \frac{x_1}{\rho} \left(\frac{y_1}{\rho} - y_0 \right)^2 + O_{4l} ,$$

(15.112)

$$b_1 = B - B_{01} =$$

$$= b_{01} \left(\frac{y_1}{\rho} - y_0 \right) + b_{20} \left(\frac{x_1}{\rho} \right)^2 + b_{02} \left(\frac{y_1}{\rho} - y_0 \right)^2 +$$

$$+ b_{21} \left(\frac{x_1}{\rho} \right)^2 \left(\frac{y_1}{\rho} - y_0 \right) + b_{03} \left(\frac{y_1}{\rho} - y_0 \right)^3 + O_{4b} .$$

(15.113)

Box 15.10 (Meridian arc $y_0 = M(B_0)$).

$$y_0 = A_1(1 - E^2) \int_0^{B_0} \frac{dB}{(1 - E^2 \sin^2 B)^{3/2}} = A_1 \left[B_0 \left(1 - \frac{1}{4} E^2 - \frac{3}{64} E^4 - \frac{5}{256} E^6 - \frac{175}{16384} E^8 \right) - \right.$$

$$- \frac{3}{8} E^2 \left(1 + \frac{1}{4} E^2 + \frac{15}{128} E^4 - \frac{35}{512} E^6 \right) \sin 2B_0 + \frac{15}{256} E^4 \left(1 + \frac{3}{4} E^2 + \frac{35}{64} E^4 \right) \sin 4B_0 - \qquad (15.114)$$

$$\left. - \frac{35}{3072} E^6 \left(1 + \frac{5}{4} E^2 \right) \sin 6B_0 + \frac{315}{131072} E^8 \sin 8B_0 \right] + O(E^{10}) .$$

Box 15.11 (Inverse conformal polynomial $l(x, y), b(x, y)$, coefficients l_{ij}, b_{ij}).

$$l_{10} = \frac{(1 - E^2 \sin^2 B_0)^{1/2}}{A_1 \cos B_0} , \quad l_{11} = \frac{(1 - E^2) \tan B_0 (1 - E^2 \sin^2 B_0)}{A_1^2 (1 - E^2) \cos B_0} ,$$

$$l_{30} = \frac{-(1 - E^2 \sin^2 B_0)^{3/2}}{6 A_1^3 (1 - E^2) \cos^3 B_0} (1 + \sin^2 B_0 - 3E^2 \sin^2 B_0 + E^2 \sin^4 B_0) ,$$

$$l_{12} = \frac{(1 - E^2 \sin^2 B_0)^{3/2}}{2 A_1^3 (1 - E^2) \cos^3 B_0} (1 + \sin^2 B_0 - 3E^2 \sin^2 B_0 + E^2 \sin^4 B_0) ,$$

$$l_{31} = \frac{- \tan B_0 (1 - E^2 \sin^2 B_0)^2}{6 A_1^4 (1 - E^2)^2 \cos^3 B_0} \times$$

$$\times (5 - 9E^2 + \sin^2 B_0 - 4E^2 \sin^2 B_0 + 15E^4 \sin^2 B_0 + E^2 \sin^4 B_0 - 13E^4 \sin^4 B_0 + 4E^4 \sin^6 B_0) ,$$

$$l_{13} = \frac{\tan B_0 (1 - E^2 \sin^2 B_0)^2}{6 A_1^4 (1 - E^2)^2 \cos^3 B_0} \times$$

$$\times (5 - 9E^2 + \sin^2 B_0 - 4E^2 \sin^2 B_0 + 15E^4 \sin^2 B_0 + E^2 \sin^4 B_0 - 13E^4 \sin^4 B_0 + 4E^4 \sin^6 B_0) ,$$

$$l_{50} = \frac{(1 - E^2 \sin^2 B_0)^{5/2}}{120 A_1^5 (1 - E^2)^3 \cos^5 B_0} \times$$

$$\times (5 - 9E^2 + 18 \sin^2 B_0 - 64E^2 \sin^2 B_0 + 90E^4 \sin^2 B_0 + \sin^4 B_0 - E^2 \sin^4 B_0 - \qquad (15.115)$$

$$- 31E^4 \sin^4 B_0 - 105E^6 \sin^4 B_0 + 2E^2 \sin^6 B_0 + 20E^4 \sin^6 B_0 + 162E^6 \sin^6 B_0 -$$

$$- 7E^4 \sin^8 B_0 - 109E^6 \sin^8 B_0 + 28E^6 \sin^{10} B_0) ,$$

$$l_{32} = \frac{-(1 - E^2 \sin^2 B_0)^{5/2}}{12 A_1^5 (1 - E^2)^3 \cos^5 B_0} \times$$

$$\times (5 - 9E^2 + 18 \sin^2 B_0 - 64E^2 \sin^2 B_0 + 90E^4 \sin^2 B_0 + \sin^4 B_0 - E^2 \sin^4 B_0 -$$

$$- 31E^4 \sin^4 B_0 - 105E^6 \sin^4 B_0 + 2E^2 \sin^6 B_0 + 20E^4 \sin^6 B_0 + 162E^6 \sin^6 B_0 -$$

$$- 7E^4 \sin^8 B_0 - 109E^6 \sin^8 B_0 + 28E^6 \sin^{10} B_0) ,$$

$$l_{14} = \frac{(1 - E^2 \sin^2 B_0)^{5/2}}{24 A_1^5 (1 - E^2)^3 \cos^5 B_0} \times$$

$$\times (5 - 9E^2 + 18 \sin^2 B_0 - 64E^2 \sin^2 B_0 + 90E^4 \sin^2 B_0 + \sin^4 B_0 - E^2 \sin^4 B_0 -$$

$$- 31E^4 \sin^4 B_0 - 105E^6 \sin^4 B_0 + 2E^2 \sin^6 B_0 + 20E^4 \sin^6 B_0 + 162E^6 \sin^6 B_0 -$$

$$- 7E^4 \sin^8 B_0 - 109E^6 \sin^8 B_0 + 28E^6 \sin^{10} B_0) ;$$

Continuation of Box.

$$b_{01} = \frac{(1 - E^2 \sin^2 B_0)^{3/2}}{A_1(1 - E^2)} \,,$$

$$b_{02} = \frac{-3E^2 \cos B_0 \sin B_0 (1 - E^2 \sin^2 B_0)^2}{2A_1^2(1 - E^2)^2} \,,$$

$$b_{20} = \frac{-\tan B_0 (1 - E^2 \sin^2 B_0)^2}{2A_1^2(1 - E^2)} \,,$$

$$b_{21} = \frac{-(1 - E^2 \sin^2 B_0)^{5/2}(1 - 5E^2 \sin^2 B_0 + 4E^2 \sin^4 B_0)}{2A_1^3(1 - E^2)^2 \cos^2 B_0} \,,$$

$$b_{03} = \frac{-E^2(1 - E^2 \sin^2 B_0)^{5/2}}{2A_1^3(1 - E^2)^3} \times$$

$$\times (1 - 2 \sin^2 B_0 - 5E^2 \sin^2 B_0 + 6E^2 \sin^4 B_0) \,,$$

$$b_{40} = \frac{\tan B_0 (1 - E^2 \sin^2 B_0)^3}{24A_1^4(1 - E^2)^3 \cos^2 B_0} \times$$

$$\times (5 - 9E^2 - 2 \sin^2 B_0 - 7E^2 \sin^2 B_0 + 21E^4 \sin^2 B_0 + 10E^2 \sin^4 B_0 - 22E^4 \sin^4 B_0 + 4E^4 \sin^6 B_0) \,,$$

$$b_{22} = \frac{-\tan B_0 (1 - E^2 \sin^2 B_0)^3}{4A_1^4(1 - E^2)^3 \cos^2 B_0} \times$$

$$\times (2 - 15E^2 + 19E^2 \sin^2 B_0 + 35E^4 \sin^2 B_0 -$$

$$-8E^2 \sin^4 B_0 - 61E^4 \sin^4 B_0 + 28E^4 \sin^6 B_0) \,,$$

$$b_{04} = \frac{E^2 \cos B_0 \sin B_0 (1 - E^2 \sin^2 B_0)^3}{8A_1^4(1 - E^2)^4} \times$$

$$\times (4 + 15E^2 - 38E^2 \sin^2 B_0 - 35E^4 \sin^2 B_0 + 54E^4 \sin^4 B_0) \,,$$

$$b_{41} = \frac{(1 - E^2 \sin^2 B_0)^{7/2}}{24A_1^5(1 - E^2)^4 \cos^4 B_0} \times$$

$$\times (5 - 9E^2 + 4 \sin^2 B_0 - 74E^2 \sin^2 B_0 + 126E^4 \sin^2 B_0 + 88E^2 \sin^4 B_0 - 83E^4 \sin^4 B_0 - 189E^6 \sin^4 B_0 -$$

$$-32E^2 \sin^6 B_0 - 80E^4 \sin^6 B_0 + 368E^6 \sin^6 B_0 + 64E^4 \sin^8 B_0 - 228E^6 \sin^8 B_0 + 40E^6 \sin^{10} B_0) \,,$$

$$b_{23} = \frac{-(1 - E^2 \sin^2 B_0)^{7/2}}{12A_1^5(1 - E^2)^4 \cos^4 B_0} \times$$

$$\times (2 - 15E^2 + 4 \sin^2 B_0 + 13E^2 \sin^2 B_0 + 210E^4 \sin^2 B_0 - 32E^2 \sin^4 B_0 - 536E^4 \sin^4 B_0 - 315E^6 \sin^4 B_0 +$$

$$+16E^2 \sin^6 B_0 + 520E^4 \sin^6 B_0 + 881E^6 \sin^6 B_0 - 176E^4 \sin^8 B_0 - 852E^6 \sin^8 B_0 + 280E^6 \sin^{10} B_0) \,,$$

$$b_{05} = \frac{E^2(1 - E^2 \sin^2 B_0)^{7/2}}{40A_1^5(1 - E^2)^5} \times$$

$$\times (4 + 15E^2 - 8 \sin^2 B_0 - 172E^2 \sin^2 B_0 - 210E^4 \sin^2 B_0 + 184E^2 \sin^4 B_0 + 872E^4 \sin^4 B_0 +$$

$$+315E^6 \sin^4 B_0 - 704E^4 \sin^6 B_0 - 944E^6 \sin^6 B_0 + 648E^6 \sin^8 B_0) \,.$$

(15.116)

15-612 The second step: polynomial representation of conformal coordinates in the second strip
  replaced by the conformal coordinates in the first strip

The standard polynomial representation of conformal coordinates of type Gauss–Krueger or UTM in the L_{02}-strip is given by (15.117) and (15.118) subject to the longitude/latitude differences $l_2 := L - L_{02}$ and $b_2 := B - B_{02}$ with respect to the longitude L_{02} of the reference meridian and the latitude B_{02} of the reference point $\{L_{02}, B_{02}\}$ of series expansion.

$$\text{Easting:} \quad x_1 = \rho\left(x_{10}l_2 + x_{11}l_2b_2 + x_{30}l_2^3 + x_{12}l_2b_2^2 + O_{4x}\right) . \tag{15.117}$$

$$\text{Northing:} \quad y_1 = \rho\left(y_0 + y_{01}b_2 + y_{20}l_2^2 + y_{02}b_2^2 + y_{21}l_2^2b_2 + y_{03}b_2^3 + O_{4y}\right) . \tag{15.118}$$

y_0 denotes the length of the meridian arc from zero ellipsoidal latitude to the ellipsoidal latitude B_{02} of the reference point $\{L_{02}, B_{02}\}$ chosen by the identity $B_{01} = B_{02} = B_0$ for operational reasons. The coefficients $\{x_{ij}, y_{ij}\}$ can be taken from Boxes 15.4 and 15.5. In addition, the optimal dilatation factor ρ has been set identical in the L_{01}-strip and the L_{02}-strip of the same strip width. The longitude/latitude differences $\{l_1, b_1\}$ as well as $\{l_2, b_2\}$ are related by $l_2 = (L_{01} - L_{02}) + l_1$ and $b_2 = (B_{01} - B_{02}) + b_1$ being derived from the invariance $L = L_{01} + l_1 = L_{02} + l_2$ and $B = B_{01} + b_1 = B_{02} + b_2$. Now we are on duty to replace $\{l_2, b_2\}$ by means of $l_2 = (L_{01} - L_{02}) + l_1$ and $b_2 = (B_{01} - B_{02}) + b_1$ by $\{(L_{01} - L_{02}) + l_1, b_1\}$ within (15.117) and (15.118), which leads us to

$$x_2 = \rho\big[x_{10}(L_{01} - L_{02}) + x_{10}l_1 + x_{11}(L_{01} - L_{02})b_1 + x_{11}l_1b_1 + x_{30}(L_{01} - L_{02})^3 +$$
$$+ 3x_{30}(L_{01} - L_{02})^2 l_1 + 3x_{30}(L_{01} - L_{02})l_1^2 + x_{30}l_1^3 + x_{12}(L_{01} - L_{02})b_1^2 + x_{12}l_1b_1^2 + O_{4x}\big], \tag{15.119}$$

$$y_2 = \rho\big[y_0 + y_{01}b_1 + y_{20}(L_{01} - L_{02})^2 + 2y_{20}(L_{01} - L_{02})l_1 + y_{20}l_1^2 + y_{02}b_1^2 +$$
$$+ y_{21}(L_{01} - L_{02})^2 b_1 + 2y_{21}(L_{01} - L_{02})l_1b_1 + y_{21}l_1^2b_1 + y_{03}b_1^3 + O_{4y}\big]. \tag{15.120}$$

Obviously, the conformal coordinates $\{x_2, y_2\}$ in the second strip L_{02} depend on the difference $L_{01} - L_{02}$ of the chosen L_{01}-strip, respectively. Finally, we have to replace $\{l_1, b_1\}$ within $\{x_2, y_2\}$ by the bivariate homogeneous polynomial $\{l_1(x_1, y_1), b(x_1, y_1)\}$ given by (15.112) and (15.113) and coefficients $\{l_{ij}, b_{ij}\}$ of Box 15.11. In this way, we have achieved a solution of the strip transformation problem presented in the form

$$x_2 = \rho\Bigg(x_{10}(L_{01} - L_{02}) + x_{10}\left[l_{10}\frac{x_1}{\rho} + l_{11}\frac{x_1}{\rho}\left(\frac{y_1}{\rho} - y_0\right) + O_{3l}\right] +$$
$$+ x_{11}(L_{01} - L_{02})\left[b_{01}\left(\frac{y_1}{\rho} - y_0\right) + b_{20}\left(\frac{x_1}{\rho}\right)^2 + b_{02}\left(\frac{y_1}{\rho} - y_0\right)^2 + O_{3b}\right] +$$
$$+ x_{11}\left[l_{10}\frac{x_1}{\rho} + l_{11}\frac{x_1}{\rho}\left(\frac{y_1}{\rho} - y_0\right) + O_{3l}\right] \times$$
$$\times \left[b_{01}\left(\frac{y_1}{\rho} - y_0\right) + b_{20}\left(\frac{x_1}{\rho}\right)^2 + b_{02}\left(\frac{y_1}{\rho} - y_0\right)^2 + O_{3b}\right] + O_{3x}\Bigg), \tag{15.121}$$

$$y_2 = \rho\Bigg(y_0 + y_{01}\left[b_{01}\left(\frac{y_1}{\rho} - y_0\right) + b_{20}\left(\frac{x_1}{\rho}\right)^2 + b_{02}\left(\frac{y_1}{\rho} - y_0\right)^2 + O_{3b}\right] +$$
$$+ y_{20}(L_{01} - L_{02})^2 + 2y_{20}(L_{01} - L_{02})\left[l_{10}\frac{x_1}{\rho} + l_{11}\frac{x_1}{\rho}\left(\frac{y_1}{\rho} - y_0\right) + O_{3l}\right] +$$
$$+ y_{20}\left[l_{10}\frac{x_1}{\rho} + l_{11}\frac{x_1}{\rho}\left(\frac{y_1}{\rho} - y_0\right) + O_{3l}\right]^2 +$$
$$+ y_{02}\left[b_{01}\left(\frac{y_1}{\rho} - y_0\right) + b_{20}\left(\frac{x_1}{\rho}\right)^2 + b_{02}\left(\frac{y_1}{\rho} - y_0\right)^2 + O_{3b}\right]^2 + O_{3y}\Bigg) . \tag{15.122}$$

In general:

$$
x_2 = x_1 + \rho \left[s_{00} + s_{10} \frac{x_1}{\rho} + s_{01} \left(\frac{y_1}{\rho} - y_0 \right) + s_{20} \left(\frac{x_1}{\rho} \right)^2 + \right.
$$
$$
\left. + s_{11} \frac{x_1}{\rho} \left(\frac{y_1}{\rho} - y_0 \right) + s_{02} \left(\frac{y_1}{\rho} - y_0 \right)^2 + O_{3s} \right] ,
$$

(15.123)

$$
y_2 = y_1 + \rho \left[t_{00} + t_{10} \frac{x_1}{\rho} + t_{01} \left(\frac{y_1}{\rho} - y_0 \right) + t_{20} \left(\frac{x_1}{\rho} \right)^2 + \right.
$$
$$
\left. + t_{11} \frac{x_1}{\rho} \left(\frac{y_1}{\rho} - y_0 \right) + t_{02} \left(\frac{y_1}{\rho} - y_0 \right)^2 + O_{3t} \right] .
$$

(15.124)

The coefficients $\{s_{ij}, t_{ij}\}$ are collected in Box 15.12 and are computed by MATHEMATICA up to order five as a polynomial in terms of $L_{01} - L_{02}$. Fig. 15.19 outlines the transformation steps of various orders. It should be noted that the explicit formula manipulation of the coefficients $\{s_{ij}, t_{ij}\}$ in Box 15.12 was restricted by the postulate $x_1 = x_2$ and $y_1 = y_2$ if $L_{01} - L_{02} = 0$ holds!

Box 15.12 (Conformal polynomial $x_2 = X(x_1, y_1), y_2 = Y(x_1, y_1)$, coefficients s_{ij}, t_{ij}).

$$
s_{00} = \frac{A_1(L_{01} - L_{02}) \cos B_0}{(1 - E^2 \sin^2 B_0)^{1/2}} + \frac{A_1(L_{01} - L_{02})^3 \cos B_0(1 - 2\sin^2 B_0 + E^2 \sin^4 B_0)}{6(1 - E^2)(1 - E^2 \sin^2 B_0)^{1/2}} +
$$
$$
+ A_1(L_{01} - L_{02})^5 \cos B_0 \left[5 - E^2 - 28\sin^2 B_0 - 16E^2 \sin^2 B_0 + 24\sin^4 B_0 + 86E^2 \sin^4 B_0 + \right.
$$
$$
+ 26E^4 \sin^4 B_0 - 72E^2 \sin^6 B_0 - 100E^4 \sin^6 B_0 - 12E^6 \sin^6 B_0 + 77E^4 \sin^8 B_0 + 39E^6 \sin^8 B_0 -
$$
$$
\left. - 28E^6 \sin^{10} B_0 \right] \left[120(1 - E^2)^3(1 - E^2 \sin^2 B_0)^{1/2} \right]^{-1} ,
$$
$$
s_{10} = \frac{(L_{01} - L_{02})^2(1 - 2\sin^2 B_0 + E^2 \sin^4 B_0)}{2(1 - E^2)} +
$$
$$
+ (L_{01} - L_{02})^4 \left[5 - E^2 - 28\sin^2 B_0 - 16E^2 \sin^2 B_0 + 24\sin^4 B_0 + 86E^2 \sin^4 B_0 + \right.
$$
$$
+ 26E^4 \sin^4 B_0 - 72E^2 \sin^6 B_0 - 100E^4 \sin^6 B_0 - 12E^6 \sin^6 B_0 + 77E^4 \sin^8 B_0 + 39E^6 \sin^8 B_0 -
$$
$$
\left. - 28E^6 \sin^{10} B_0 \right] \left[24(1 - E^2)^3 \right]^{-1} ,
$$
$$
s_{01} = -(L_{01} - L_{02}) \sin B_0 -
$$
$$
- (L_{01} - L_{02})^3 \sin B_0 \left[5 - E^2 - 6\sin^2 B_0 - 6E^2 \sin^2 B_0 + 9E^2 \sin^4 B_0 + \right.
$$

(15.125)

$$
\left. + 3E^4 \sin^4 B_0 - 4E^4 \sin^6 B_0 \right] \left[6(1 - E^2)^2 \right]^{-1} ,
$$
$$
s_{20} = \frac{(L_{01} - L_{02}) \cos B_0(1 - E^2 \sin^2 B_0)^{3/2}}{2A_1(1 - E^2)} +
$$
$$
+ (L_{01} - L_{02})^3 \cos B_0(1 - E^2 \sin^2 B_0)^{3/2} \left[5 - E^2 - 18\sin^2 B_0 - 18E^2 \sin^2 B_0 + 45E^2 \sin^4 B_0 + \right.
$$
$$
\left. + 15E^4 \sin^4 B_0 - 28E^4 \sin^6 B_0 \right] \left[12A_1(1 - E^2)^3 \right]^{-1} ,
$$
$$
s_{11} = -\frac{2(L_{01} - L_{02})^2 \cos B_0 \sin B_0(1 - E^2 \sin^2 B_0)^{5/2}}{A_1(1 - E^2)^2} +
$$
$$
+ (L_{01} - L_{02})^4 \tan B_0(1 - E^2 \sin^2 B_0)^{1/2} \left[5 - E^2 - 28\sin^2 B_0 - 16E^2 \sin^2 B_0 + 24\sin^4 B_0 + \right.
$$
$$
+ 86E^2 \sin^4 B_0 + 26E^4 \sin^4 B_0 - 72E^2 \sin^6 B_0 - 100E^4 \sin^6 B_0 - 12E^6 \sin^6 B_0 +
$$
$$
\left. + 77E^4 \sin^8 B_0 + 39E^6 \sin^8 B_0 - 28E^6 \sin^{10} B_0 \right] \left[24A_1(1 - E^2)^3 \right]^{1} ,
$$

Continuation of Box.

$$s_{02} = -\frac{(L_{01} - L_{02})\cos B_0(1 - E^2\sin^2 B_0)^{3/2}}{2A_1(1 - E^2)} -$$

$$-(L_{01} - L_{02})^3\cos B_0(1 - E^2\sin^2 B_0)^{3/2}\left[5 - E^2 - 18\sin^2 B_0 - 18E^2\sin^2 B_0 + 45E^2\sin^4 B_0 +\right.$$

$$\left.+15E^4\sin^4 B_0 - 28E^4\sin^6 B_0\right]\left[12A_1(1 - E^2)^3\right]^{-1};$$

$$t_{00} = \frac{A_1(L_{01} - L_{02})^2\cos B_0\sin B_0}{2(1 - E^2\sin^2 B_0)^{1/2}} +$$

$$+A_1(L_{01} - L_{02})^4\cos B_0\sin B_0\left[5 - E^2 - 6\sin^2 B_0 - 6E^2\sin^2 B_0 + 9E^2\sin^4 B_0 +\right.$$

$$\left.+3E^4\sin^4 B_0 - 4E^4\sin^6 B_0\right]\left[24(1 - E^2)^2(1 - E^2\sin^2 B_0)^{1/2}\right]^{-1},$$

$$t_{10} = (L_{01} - L_{02})\sin B_0 +$$

$$+(L_{01} - L_{02})^3\sin B_0\left[5 - E^2 - 6\sin^2 B_0 - 6E^2\sin^2 B_0 + 9E^2\sin^4 B_0 +\right.$$

$$\left.+3E^4\sin^4 B_0 - 4E^4\sin^6 B_0\right]\left[6(1 - E^2)^2\right]^{-1},$$

$$t_{01} = \frac{(L_{01} - L_{02})^2(1 - 2\sin^2 B_0 + E^2\sin^4 B_0)}{2(1 - E^2)} +$$

$$+(L_{01} - L_{02})^4\left[5 - E^2 - 28\sin^2 B_0 - 16E^2\sin^2 B_0 + 24\sin^4 B_0 + 86E^2\sin^4 B_0 +\right.$$

$$+26E^4\sin^4 B_0 - 72E^2\sin^6 B_0 - 100E^4\sin^6 B_0 - 12E^6\sin^6 B_0 + 77E^4\sin^8 B_0 + 39E^6\sin^8 B_0 -$$

$$\left.-28E^6\sin^{10} B_0\right]\left[24(1 - E^2)^3\right]^{-1},$$ (15.126)

$$t_{20} = \frac{(L_{01} - L_{02})^2\cos B_0\sin B_0(1 - E^2\sin^2 B_0)^{5/2}}{A_1(1 - E^2)^2} -$$

$$-(L_{01} - L_{02})^4\tan B_0(1 - E^2\sin^2 B_0)^{1/2}\left[5 - E^2 - 28\sin^2 B_0 - 16E^2\sin^2 B_0 + 24\sin^4 B_0 +\right.$$

$$+86E^2\sin^4 B_0 + 26E^4\sin^4 B_0 - 72E^2\sin^6 B_0 - 100E^4\sin^6 B_0 - 12E^6\sin^6 B_0 +$$

$$\left.+77E^4\sin^8 B_0 + 39E^6\sin^8 B_0 - 28E^6\sin^{10} B_0\right]\left[48A_1(1 - E^2)^3\right]^{-1},$$

$$t_{11} = \frac{(L_{01} - L_{02})\cos B_0(1 - E^2\sin^2 B_0)^{3/2}}{A_1(1 - E^2)} +$$

$$+(L_{01} - L_{02})^3\cos B_0(1 - E^2\sin^2 B_0)^{3/2}\left[5 - E^2 - 18\sin^2 B_0 - 18E^2\sin^2 B_0 + 45E^2\sin^4 B_0 +\right.$$

$$\left.+15E^4\sin^4 B_0 - 28E^4\sin^6 B_0\right]\left[6A_1(1 - E^2)^3\right]^{-1},$$

$$t_{02} = -\frac{(L_{01} - L_{02})^2\cos B_0\sin B_0(1 - E^2\sin^2 B_0)^{5/2}}{A_1(1 - E^2)^2} -$$

$$-(L_{01} - L_{02})^4 E^2\cos B_0\sin B_0(1 - E^2\sin^2 B_0)^{1/2}\left[5 - E^2 - 28\sin^2 B_0 - 16E^2\sin^2 B_0 + 24\sin^4 B_0 +\right.$$

$$+86E^2\sin^4 B_0 + 26E^4\sin^4 B_0 - 72E^2\sin^6 B_0 - 100E^4\sin^6 B_0 - 12E^6\sin^6 B_0 +$$

$$\left.+77E^4\sin^8 B_0 + 39E^6\sin^8 B_0 - 28E^6\sin^{10} B_0\right]\left[16A_1(1 - E^2)^4\right]^{-1}.$$

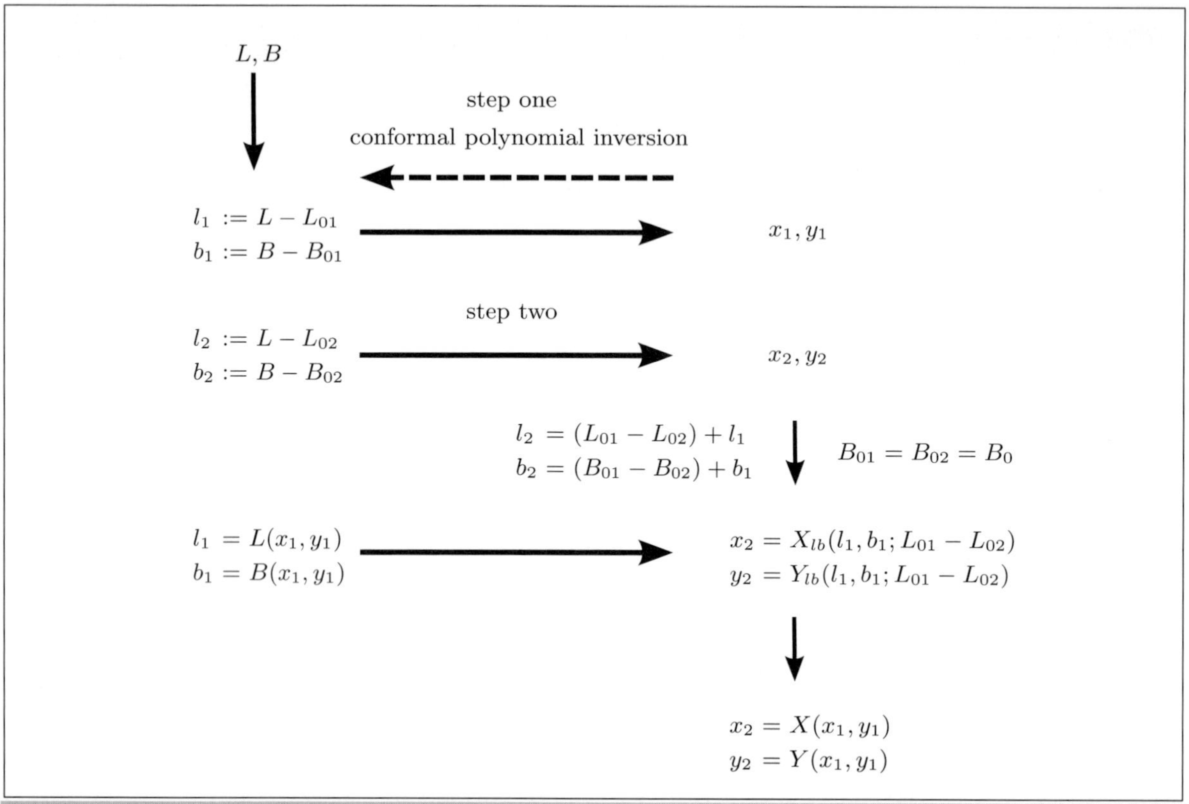

Fig. 15.19. Flow chart of the two step approach for generating the strip transformation $x_2 = X(x_1, y_1)$ and $y_2 = Y(x_1, y_1)$.

15-62 Two examples of strip transformations

Let us consider two examples of a strip transformation of conformal coordinates of Gauss–Krueger type with a dilatation factor $\rho = 1$ (A. Schoedlbauer 1981c, 1982c,a) and a strip transformation of conformal coordinates of UTM type with $\rho = 0.999\,578$. Due to Example 15.8, we start from the coordinates ellipsoidal longitude/latitude of TP I.O. Bonstetten ($\{L, B\} = \{10°42'59.''3215, 48°26'45.''4355\}$) on the Bessel ellipsoid of semi-major axis $A = 6377\,397.155$ m and reciprocal flattening $f^{-1} = 299.152\,812\,85$. The first and second strip has been fixed with $L_{01} = 9°$ and $L_{02} = 12°$, respectively. The ellipsoidal latitude of the reference point was chosen to $B_0 = 48°$. In addition, we compared the direct transformation $\{L, B\} \rightarrow \{x_1, y_1\}$ and $\{L, B\} \rightarrow \{x_2, y_2\}$ as illustrated by the commutative diagram of Fig. 15.17, leading to differences in the submillimeter range. By contrast, Example 15.9 gives the strip transformation of a point $\{L, B\} = \{12.01°, 49°\}$ on the ellipsoid referring to WGS84 ($A = 6378\,137$ m, $f^{-1} = 298.257\,223\,563$) with a first reference meridian of $L_{01} = 9°$ and a second of $L_{01} = 15°$. The differences again in the comparison of both ways of calculating the UTM coordinates have been in the submillimeter range the closer B_0 is chosen to the point $\{L, B\}$.

Important! The strip transformation of conformal coordinates of type Gauss–Krueger ($\rho = 1$) or UTM ($\rho = 0.999\,578$) for a strip $[-l_E, l_E] \times [B_S, B_N] = [-3.5°, 3.5°] \times [80°S, 84°N]$) represented by $x_2 = X(x_1, y_1)$ and $y_2 = Y(x_1, y_1)$ is derived in terms of a bivariate polynomial up to order five. $\{x_1, y_1\}$ represent the conformal coordinates in the first strip of ellipsoidal longitude L_{01}, while $\{x_2, y_2\}$ represent those conformal coordinates in the second strip of ellipsoidal longitude L_{02}. $X(x_1, y_1)$ and $Y(x_1, y_1)$ are power series in terms of $L_{01} - L_{02}$ given by (15.123), (15.124), and Box 15.12. Two examples (Bessel ellipsoid, World Geodetic Reference System 1984 (WSGS84)) document the numerical stability of the derived strip transformation.

Example 15.8 (Bessel ellipsoid, strip transformation $x_2(x_1, y_1)$ and $y_2(x_1, y_1)$ of conformal coordinates of Gauss–Krueger (GK) type versus direct transformations $\{L, B\} \rightarrow \{x_1, y_1\}$ with respect to $L_{01} = 9°$ and $\{L, B\} \rightarrow \{x_2, y_2\}$ with respect to $L_{02} = 12°$, TP 1.O. Bonstetten ($L = 10°42'59.''3215$, $B = 48°26'45.''4355$)).

$$\{L, B\} \rightarrow \{x_1, y_1\}:$$

$$L_{01} = 9° \ ,$$

$$\text{dilatation factor } \rho = 1 \ ,$$

$$B_0 = 48° \ ,$$

$$x_1 = 126\,967.248\,33\,\text{m} \ , \ y_1 = 51\,005.569\,24\,\text{m} \ ,$$

$$y_0(B_0 = 48°) = 5317\,885.232\,32\,\text{m} \ .$$

Conventional Gauss–Krueger coordinates:

$$\text{Northing } y_{\text{GK}} = y_0 + y_1 = 5368\,890.801\,55\,\text{m} \ ,$$

$$\text{False Easting } x_{\text{GK}} = \tfrac{L_{01}}{3°} \times 10^6\,\text{m} + 500\,000\,\text{m} + x_1 = 3626\,967.248\,33\,\text{m} \ .$$

$$\{L, B\} \rightarrow \{x_2, y_2\}:$$

$$L_{02} = 12° \ ,$$

$$\text{dilatation factor } \rho = 1 \ ,$$

$$B_0 = 48°,$$

$$x_2 = -94\,942.371\,14\,\text{m} \ , \ y_2 = 50\,378.015\,51\,\text{m} \ .$$

Conventional Gauss–Krueger coordinates:

$$\text{Northing } y_{\text{GK}} = 5368\,263.247\,82\,\text{m} \ ,$$

$$\text{False Easting } x_{\text{GK}} = \tfrac{L_{02}}{3°} \times 10^6\,\text{m} + 500\,000\,\text{m} + x_2 = 4405\,057.628\,86\,\text{m} \ .$$

$$\{x_1, y_1\} \rightarrow \{x_2, y_2\}:$$

$$B_0 = 48° \ ,$$

$$x_2 = -94\,942.371\,10\,\text{m} \ , \ y_2 = 50\,378.015\,51\,\text{m} \ ,$$

End of Example.

Example 15.9 (WGS84 reference ellipsoid, strip transformation $x_2(x_1, y_1)$ and $y_2(x_1, y_1)$ of conformal coordinates of UTM type versus direct transformations $\{L, B\} \to \{x_1, y_1\}$ with respect to $L_{01} = 9°$ and $\{L, B\} \to \{x_2, y_2\}$ with respect to $L_{02} = 15°$, $B = 49°$, and $L = 12°0'36''$).

$$\{L, B\} \to \{x_1, y_1\}:$$

$$L_{01} = 9°,$$

dilatation factor $\rho = 0.999\,578$,

$$B_0 = 48° \begin{cases} x_1 = \rho \times 220\,233.080\,33\,\text{m}, \\ y_1 = \rho \times 115\,567.839\,91\,\text{m}, \\ y_0 = \rho \times 5318\,427.595\,49\,\text{m}, \end{cases}$$

$$B_0 = 48.8° \begin{cases} x_1 = \rho \times 220\,233.080\,33\,\text{m}, \\ y_1 = \rho \times 26\,609.363\,14\,\text{m}, \\ y_0 = \rho \times 5407\,386.072\,26\,\text{m}. \end{cases}$$

Conventional UTM coordinates:

Northing $y_{\text{UTM}} = y_0 + y_1 = 5431\,702.289\,33\,\text{m}$,

False Easting $x_{\text{UTM}} = 500\,000\,\text{m} + x_1 = 720\,140.142\,00\,\text{m}$, zone $\frac{L_{01}+3°}{6°} + 30 = 32$.

$$\{L, B\} \to \{x_2, y_2\}:$$

$$L_{02} = 15°,$$

dilatation factor $\rho = 0.999\,578$,

$$B_0 = 48° \begin{cases} x_2 = -\rho \times 218\,769.922\,64\,\text{m}, \\ y_2 = \rho \times 115\,509.967\,94\,\text{m}, \end{cases}$$

$$B_0 = 48.8° \begin{cases} x_2 = -\rho \times 218\,769.922\,64\,\text{m}, \\ y_2 = \rho \times 26\,551.491\,17\,\text{m}. \end{cases}$$

Conventional UTM coordinates:

Northing $y_{\text{UTM}} = 5431\,644.441\,78\,\text{m}$,

False Easting $x_{\text{UTM}} = 281\,322.398\,27\,\text{m}$, zone $\frac{L_{02}+3°}{6°} + 30 = 33$,

$$\{x_1, y_1\} \to \{x_2, y_2\}:$$

$$B_0 = 48° \{x_2 = -\rho \times 218\,769.919\,79\,\text{m}, \quad y_2 = \rho \times 115\,509.968\,26\,\text{m},$$

$$B_0 = 48.8° \{x_2 = -\rho \times 218\,769.921\,93\,\text{m}, \quad y_2 = \rho \times 26\,551.491\,46\,\text{m}.$$

End of Example.

Within the world of map projections, the Oblique Mercator projection (UOM) plays an important role. In the next chapter, let us have a closer look at the Oblique Mercator Projection (UOM).

16 "Ellipsoid-of-revolution to cylinder": oblique aspect

Mapping the ellipsoid-of-revolution to a cylinder: oblique aspect. Oblique Mercator Projection (UOM), rectified skew orthomorphic projections. Korn–Lichtenstein equations, Laplace–Beltrami equations.

In the world of *conformal mappings* of the Earth or other celestial bodies, the *Mercator projection* plays a central role. The Mercator projection of the sphere $\mathbb{S}_r^2$ or of the ellipsoid-of-revolution $\mathbb{E}_{A_1,A_2}^2$ beside *conformality* is characterized by the *equidistant mapping* of the equator. In contrast, the transverse Mercator projection is conformal and maps the transverse meta-equator, the meridian of reference, equidistantly. Accordingly, the Mercator projection is very well suited for regions which extend East–West around the equator, while the transverse Mercator projection fits well to those regions which have a South–North extension. Obviously, several geographical regions are centered along lines which are neither equatorial, parallel circles, or meridians, but may be taken as central intersection of a plane and the reference figure of the Earth or other celestial bodies, the ellipsoid-of-revolution (spheroid). For geodetic applications, conformality is desired in such cases, the *Universal Oblique Mercator Projection* (UOM) is the projection which should be chosen. A study of the conformal projection of the ellipsoid-of-revolution by M. Hotine (1946, 1947) is the basis of the ellipsoidal oblique Mercator projection, which M. Hotine called the "rectified skew orthomorphic", mainly applied in the United States (e. g. for Alaska), for Malaysia, and for Borneo (M. Hotine 1947), for the sphere by Laborde (1928) for Madagaskar, by M. Rosenmund (1903) for Switzerland and by J. H. Cole (1943) for Italy, namely in the context of the celebrated Gauss double projection (conformal mapping of the ellipsoid-of-revolution to the sphere and of the sphere to the plane). According to J. P. Snyder (1982 p. 76), the *Hotine Oblique Mercator Projection* (HOM) is the most suitable projection available for mapping Landsat type data. HOM has also been used to cast the Heat Capacity Mapping Mission (HCMM) imagery since 1978. Note that our interest in the Oblique Mercator was raised by the personally obscure procedure to derive the mapping equations which should be based on similar concepts known for Normal Mercator and Transverse Mercator. The mapping equations should guarantee that the elliptic meta-equator should be mapped equidistantly. Accordingly, we derive here the general mapping equations $x(L,B)$ and $y(L,B)$ for conformal coordinates (isometric coordinates, isothermal coordinates) as a function of ellipsoidal longitude L and ellipsoidal latitude B, which map the line-of-intersection (an ellipse) of an inclined central plane and the ellipsoid-of-revolution equidistantly.

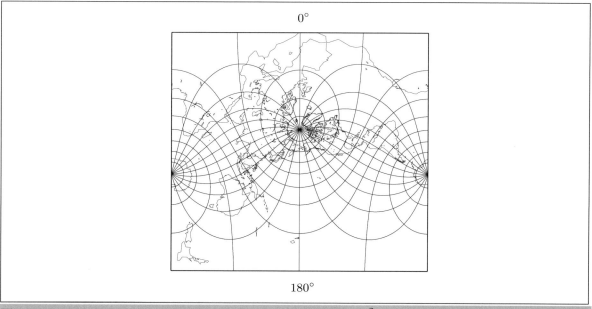

Fig. 16.1. Universal Oblique Mercator Projection of the sphere $\mathbb{S}_r^2$, meta-pole coordinates $L_0 = 180°$ and $B_0 = -30°$. Compare with Fig. 16.2.

Section 16-1.

In particular, in Section 16-1, we review the fundamental equations which govern conformal mapping of a two-dimensional Riemann manifold, namely (i) the *Korn–Lichtenstein equations*, (ii) the *Laplace–Beltrami equations* (the integrability conditions of the Korn–Lichtenstein equations), and (iii) the condition *preserving the orientation* of a conformeomorphism, for the ellipsoid-of-revolution $\mathbb{E}^2_{A_1,A_2}$ parameterized by ellipsoidal longitude L and ellipsoidal latitude B. Two examples for the solution of the fundamental equations (i), (ii), and (iii) are given, namely (1) the *Universal Mercator Projection* (UMP), and (2) the *Universal Polar Stereographic Projection* (UPS). If the equations (i), (ii), and (iii) of a conformeomorphism are specialized to UMP or UPS as input conformal coordinates, the equations for output conformal coordinates of another type are obtained as (α) the d'Alembert–Euler equations (the Cauchy–Riemann equations), (β) the Laplace–Beltrami equations (the integrability conditions of the d'Alembert–Euler equations), (γ) the condition *preserving the orientation* of a conformeomorphism. A fundamental solution of the equations (α), (β), and (γ) is given in the class of homogeneous polynomials and interpreted with respect to the two-dimensional conformal group $C_6(2)$ constituted by six parameters (2 for translation, 1 for rotation, 1 for dilatation, 2 for special conformal) embedded in the two-dimensional conformal group $C_\infty(2)$, which is described by infinite set of parameters.

Section 16-2, Section 16-3.

Section 16-2 introduces the oblique reference frame of $\mathbb{E}^2_{A_1,A_2}$, in particular, the oblique meta-equator $\mathbb{E}^2_{a',b'}$ which is parameterized by reduced meta-longitude α. Section 16-3 determines the unknown coefficients of the fundamental solution for the equations (α), (β), and (γ) which govern conformeomorphism by an equidistant map of the oblique meta-equator. In such a way, a boundary value problem for the d'Alembert–Euler equations (Cauchy–Riemann equations) is defined and solved. Finally, we show that special cases of the universal oblique Mercator projection for $\mathbb{E}^2_{A_1,A_2}$ are normal Mercator and transverse Mercator. In addition, we shortly outline the local reduction of the universal oblique Mercator projection of $\mathbb{E}^2_{A_1,A_2}$ towards $\mathbb{S}^2_r$ given in Box 16.1 and as an example plotted in Fig. 16.1.

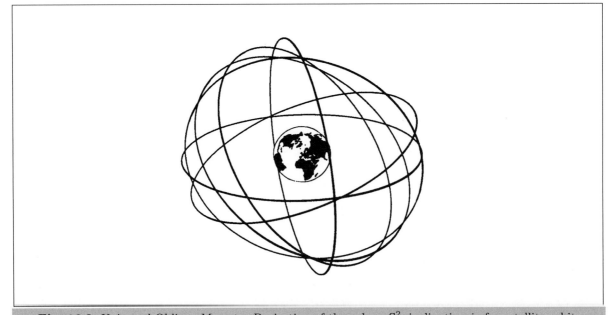

Fig. 16.2. Universal Oblique Mercator Projection of the sphere $\mathbb{S}^2_r$, inclination i of a satellite orbit.

Box 16.1 (The universal oblique Mercator projection of the sphere $\mathbb{S}_r^2$. α, β : meta-longitude, meta-latitude. L, B : longitude, latitude. Ω, i : longitude, inclination of the oblique meta-equator).

$$x = r\alpha =$$

$$= r \arctan\left[\cos i \tan(L - \Omega) + \sin i \tan B / \cos(L - \Omega)\right] , \qquad (16.1)$$

$$y = r \ln \tan\left(\frac{\pi}{4} + \frac{\beta}{2}\right) =$$

$$= r \operatorname{ar} \tanh(\sin \beta) = \qquad (16.2)$$

$$= r \operatorname{ar} \tanh\left[\cos i \sin B - \sin i \cos B \sin(L - \Omega)\right] ,$$

$$\tan\frac{x}{r} =$$

$$= \cos i \tan(L - \Omega) + \sin i \tan B / \cos(L - \Omega) , \qquad (16.3)$$

$$\tanh\frac{y}{r} =$$

$$= \cos i \sin B - \sin i \cos B \sin(L - \Omega) . \qquad (16.4)$$

16-1 The equations governing conformal mapping

The equations governing conformal mapping and their fundamental solution. Korn–Lichtenstein equations and Laplace–Beltrami equations. Universal Mercator Projection (UMP) and Universal Polar Stereographic Projection (UPS).

We are concerned with a conformal mapping of the biaxial ellipsoid $\mathbb{E}^2_{A_1,A_2}$ ("ellipsoid-of-revolution"), "spheroid", semi-major axis A_1, semi-minor axis A_2) embedded in a three-dimensional Euclid manifold $\mathbb{E}^3 = \{\mathbb{R}^3, \delta_{ij}\}$ with standard "canonical" metric $\{\delta_{ij}\}$, the Kronecker delta of ones in the diagonal, of zeros in the off-diagonal, namely by means of

$$x^1 = \frac{A_1 \cos B \cos L}{\sqrt{1 - E^2 \sin^2 B}} ,$$

$$x^2 = \frac{A_1 \cos B \sin L}{\sqrt{1 - E^2 \sin^2 B}} , \qquad (16.5)$$

$$x^3 = \frac{A_1(1 - E^2) \sin B}{\sqrt{1 - E^2 \sin^2 B}} ,$$

introducing "surface normal" elipsoidal longitude L as well as "surface normal" ellipsoidal latitude B, where $E^2 := (A_1^2 - A_2^2)/A_1^2 = 1 - A_2^2/A_1^2$ denotes the first relative eccentricity. According to the relation $L, B \in \{0 \le L \le 2\pi, -\pi/2 < B < \pi/2\}$, we exclude from the domain $\{L, B\}$ North and South Pole. Thus, $\{L, B\}$ constitutes only a first chart of $\mathbb{E}^2_{A_1,A_2}$. A minimal atlas of $\mathbb{E}^2_{A_1,A_2}$ based on two charts and which covers all points of the ellipsoid-of-revolution is given by E. Grafarend and R. Syffus (1994), in great detail.

Conformal coordinates x and y (isometric coordinates, isothermal coordinates) are constructed from the "surface normal" ellipsoidal coordinates L and B as solutions of the Korn–Lichtenstein equations (conformal change from one chart to another chart, c:cha-cha-cha)

$$\begin{bmatrix} x_L \\ x_B \end{bmatrix} = \frac{1}{\sqrt{G_{11}G_{22} - G_{12}^2}} \begin{bmatrix} -G_{12} & G_{11} \\ -G_{22} & G_{12} \end{bmatrix} \begin{bmatrix} y_L \\ y_B \end{bmatrix} ,$$

(16.6)

subject to the integrability conditions $x_{LB} = x_{BL}$ and $y_{LB} = y_{BL}$ or

$$\triangle_{LB}x := \left(\frac{G_{11}x_B - G_{12}x_L}{\sqrt{G_{11}G_{22} - G_{12}^2}} \right)_B + \left(\frac{G_{22}x_L - G_{12}x_B}{\sqrt{G_{11}G_{22} - G_{12}^2}} \right)_L = 0 ,$$

$$\triangle_{LB}y := \left(\frac{G_{11}y_B - G_{12}y_L}{\sqrt{G_{11}G_{22} - G_{12}^2}} \right)_B + \left(\frac{G_{22}y_L - G_{12}y_B}{\sqrt{G_{11}G_{22} - G_{12}^2}} \right)_L = 0 ,$$

(16.7)

and

$$\begin{vmatrix} x_L & y_L \\ x_B & y_B \end{vmatrix} > 0$$

(16.8)

(orientation preserving conformeomorphism) ,

$$\{g_{\mu\nu}\} := \begin{bmatrix} G_{11} & G_{12} \\ G_{12} & G_{22} \end{bmatrix} \ \forall\, \mu, \nu \in \{1, 2\}$$

(16.9)

(metric of the first fundamental form of $\mathbb{E}^2_{A_1, A_2}$) .

$\triangle_{LB}x = 0$ and $\triangle_{LB}y = 0$, respectively, are called the *vectorial Laplace–Beltrami equations*. We here note that a Jacobi map (16.6) can be made unique by a proper boundary condition, e. g. the equidistant map of a particular coordinate line. Examples are equidistant mappings of the circular equator (Mercator projection) or of the elliptic meridian (transverse Mercator projection). Furthermore, we here note that only few solutions of the Korn–Lichtenstein equations (16.6) subject to the integrability condition (16.7) (vectorial Laplace–Beltrami equations) and the condition of orientation preservation are known. We list two in the following.

Universal Mercator Projection (UMP):

$$x = A_1 L =: p_{\text{UMP}} ,$$

(16.10)

$$y = A_1 \ln \left[\tan\left(\frac{\pi}{4} + \frac{B}{2} \right) \left(\frac{1 - E \sin B}{1 + E \sin B} \right)^{E/2} \right] =: q_{\text{UMP}} .$$

Universal Polar Stereographic Projection (UPS):

$$x = \frac{2A_1}{\sqrt{1 - E^2}} \left(\frac{1 - E}{1 + E} \right)^{E/2} \tan\left(\frac{\pi}{4} - \frac{B}{2} \right) \left(\frac{1 + E \sin B}{1 - E \sin B} \right)^{E/2} \cos L =: p_{\text{UPS}} ,$$

(16.11)

$$y = \frac{2A_1}{\sqrt{1 - E^2}} \left(\frac{1 - E}{1 + E} \right)^{E/2} \tan\left(\frac{\pi}{4} - \frac{B}{2} \right) \left(\frac{1 + E \sin B}{1 - E \sin B} \right)^{E/2} \sin L =: q_{\text{UPS}} .$$

Once one system of conformal coordinates is established, we can use it as the input for another system of conformal coordinates (conformal change from one conformal chart to another conformal chart, c:c:cha-cha-cha). Accordingly, the Korn–Lichtenstein equations reduce to the d'Alembert–Euler equations (16.12) (more known as the Cauchy–Riemann equations) subject to the integrability conditions (16.13) or (16.14), which is automatically *orientation preserving* according to (16.15). Here, we have denoted $\{p, q\}$ as being generated by (16.10) (UMP) or by (16.11) (UPS).

$$x_p = y_q \ , \quad x_q = -y_p \ , \tag{16.12}$$

$$x_{pq} = x_{qp} \ , \quad y_{pq} = y_{qp} \ , \tag{16.13}$$

$$\Delta_{LB} x := x_{pp} + x_{qq} = 0 \ , \quad \Delta_{LB} y := y_{pp} + y_{qq} = 0 \ , \tag{16.14}$$

$$x_p y_q - x_q y_p = x_p^2 + y_p^2 = y_q^2 + x_q^2 > 0 \ . \tag{16.15}$$

A fundamental solution of the d'Alembert–Euler equations (16.12) (Cauchy–Riemann equations) subject to the integrability conditions (16.14) in the class of polynomials is provided by (16.16), or in matrix notation, based on the Kronecker–Zehfuss product $\otimes$ and transposition T, provided by (16.17).

$$x = \alpha_0 + \alpha_1 q + \beta_1 p + \alpha_2 (q^2 - p^2) + \beta_2 2pq +$$

$$+ \sum_{r=3}^{N} \alpha_r \sum_{s=0}^{r/2} (-1)^s \binom{r}{2s} q^{r-2s} p^{2s} + \sum_{r=3}^{N} \beta_r \sum_{s=1}^{(r+1)/2} (-1)^{s+1} \binom{r}{2s-1} q^{r-2s+1} p^{2s-1} \ ,$$

$$y = \beta_0 + \beta_1 q - \alpha_1 q + \beta_2 (q^2 - p^2) + \alpha_2 2pq +$$

$$+ \sum_{r=3}^{N} \beta_r \sum_{s=0}^{r/2} (-1)^s \binom{r}{2s} q^{r-2s} p^{2s} \sum_{r=3}^{N} \alpha_r \sum_{s=1}^{(r+1)/2} (-1)^{s+1} \binom{r}{2s-1} q^{r-2s+1} p^{2s-1} \ , \tag{16.16}$$

$$\begin{bmatrix} x \\ y \end{bmatrix} = \begin{bmatrix} \alpha_0 \\ \beta_0 \end{bmatrix} + (\beta_1 I_2 + \alpha_1 A) \begin{bmatrix} p \\ q \end{bmatrix} +$$

$$+ \left[\mathrm{vec} \begin{bmatrix} -\alpha_2 & \beta_2 \\ \beta_2 & \alpha_2 \end{bmatrix}, \mathrm{vec} \begin{bmatrix} -\beta_2 & -\alpha_2 \\ -\alpha_2 & \beta_2 \end{bmatrix} \right]^{\mathrm{T}} \begin{bmatrix} p \\ q \end{bmatrix} \otimes \begin{bmatrix} p \\ q \end{bmatrix} + O_3 \ , \tag{16.17}$$

identifying the conformal transformation group, namely of type translation (parameters α_0 and β_0), of type rotation (parameter α_1), of type dilatation (parameter β_1), and of type special-conformal (parameters α_2 and β_2) up to order three (O_3), actually the six-parameter subalgebra $C_6(2)$ of the infinite dimensional algebra $C_\infty(2)$ in two dimensions $\{q, p\} \in \mathbb{R}^2$. Note that the rotation parameter α_1 operates on the antisymmetric matrix (16.18), while the matrices (16.19), which generate the special conformal transformation, are traceless and symmetric.

$$A := \begin{bmatrix} 0 & 1 \\ -1 & 0 \end{bmatrix} \ , \tag{16.18}$$

$$H^1 := \begin{bmatrix} -\alpha_2 & \beta_2 \\ \beta_2 & \alpha_2 \end{bmatrix} \ , \quad H^2 := \begin{bmatrix} -\beta_2 & -\alpha_2 \\ -\alpha_2 & \beta_2 \end{bmatrix} \ . \tag{16.19}$$

There remains the task to determine the coefficients α_0, β_0, α_1, β_1, α_2, β_2 etc. by means of properly chosen boundary condition.

16-2 The oblique reference frame

Oblique reference frame and normal reference frame, central oblique plane, circle-reduced meta-longitude and circle-reduced meta-pole.

In the following discussion, let us orientate a set of orthonormal base vectors $\{e_1, e_2, e_3\}$ along the principal axes of $\mathbb{E}^2_{A_1, A_2} := \{x \in \mathbb{R}^3 \mid [(x^1)^2 + (x^2)^2]A_1^{-2} + (x^3)^2 A_2^{-2} = 1, A_1 \in \mathbb{R}^+, A_2 \in \mathbb{R}^+\}$. Against this frame of reference $\{e_1, e_2, e_3, \mathcal{O}\}$ (consisting of the base vectors e_i, and the origin $\mathcal{O}$), we introduce the oblique one $\{e_{1'}, e_{2'}, e_{3'}, \mathcal{O}\}$ by means of (16.20) illustrated by Figure 16.3.

$$\begin{bmatrix} e_{1'} \\ e_{2'} \\ e_{3'} \end{bmatrix} = R_1(i) R_3(\Omega) \begin{bmatrix} e_1 \\ e_2 \\ e_3 \end{bmatrix} . \tag{16.20}$$

The rotation around the 3 axis, we have denoted by Ω, the "right ascension of the ascending node", while the rotation around the intermediate 1 axis by i, the "inclination". $R_1(i)$ and $R_3(\Omega)$ are orthonormal matrices such that (16.21) holds.

$$R_1(i) R_3(\Omega) = \begin{bmatrix} \cos \Omega & \sin \Omega & 0 \\ -\sin \Omega \cos i & +\cos \Omega \cos i & \sin i \\ +\sin \Omega \sin i & -\cos \Omega \sin i & \cos i \end{bmatrix} \in \mathbb{R}^{3 \times 3} . \tag{16.21}$$

Accordingly, (16.22) is a representation of the placement vector x in the orthonormal bases $\{e_1, e_2, e_3, \mathcal{O}\}$ and $\{e_{1'}, e_{2'}, e_{3'}, \mathcal{O}\}$, respectively. Note that (16.23) and (16.24) hold.

$$x = \sum_{i=1}^{3} e_i x^i = \sum_{i'=1}^{3} e_{i'} x^{i'} , \tag{16.22}$$

$$x^1 = x^{1'} \cos \Omega - x^{2'} \sin \Omega \cos i + x^{3'} \sin \Omega \sin i ,$$

$$x^2 = x^{1'} \sin \Omega + x^{2'} \cos \Omega \cos i - x^{3'} \cos \Omega \sin i , \tag{16.23}$$

$$x^3 = x^{2'} \sin i + x^{3'} \cos i ,$$

$$x^{1'} = +x^1 \cos \Omega + x^2 \sin \Omega =: x' ,$$

$$x^{2'} = -x^1 \sin \Omega \cos i + x^2 \cos \Omega \cos i + x^3 \sin i =: y' , \tag{16.24}$$

$$x^{3'} = +x^1 \sin \Omega \sin i - x^2 \cos \Omega \sin i + x^3 \cos i =: z' .$$

Corollary 16.1 (Intersection of $\mathbb{E}^2_{A_1, A_2}$ and $\mathbb{L}^2_{\mathcal{O}}$).

The intersection of the ellipsoid-of-revolution $\mathbb{E}^2_{A_1, A_2}$ and the central oblique plane $\mathbb{L}^2_{\mathcal{O}}$ (two-dimensional linear manifold through the origin $\mathcal{O}$) is the ellipse (16.25) of semi-major axis $A'_1 = A_1$ and semi-minor axis $A'_2 = A_1 \sqrt{1 - E^2} / \sqrt{1 - E^2 \cos^2 i}$.

$$\mathbb{E}^1_{A'_1, A'_2} :=$$

$$:= \left\{ x \in \mathbb{R}^2 \mid \frac{x'^2}{A_1'^2} + \frac{y'^2}{A_2'^2} = 1, A'_1 = A_1, A'_2 = A_1 \frac{\sqrt{1 - E^2}}{\sqrt{1 - E^2 \cos^2 i}}, A'_1 > A'_2 \right\} . \tag{16.25}$$

End of Corollary.

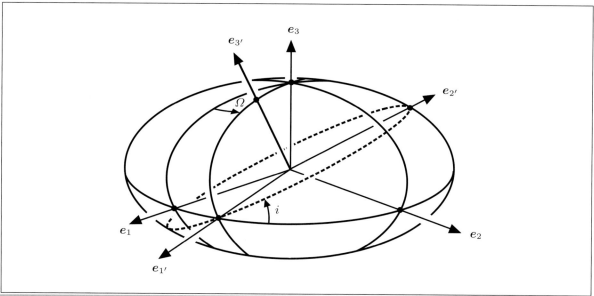

Fig. 16.3. Oblique reference frame $\{e_{1'}, e_{2'}, e_{3'}, o\}$ with respect to the normal reference frame $\{e_1, e_2, e_3, o\}$ along the principal axes of $\mathbb{E}^2_{A_1,A_2} := \{x \in \mathbb{R}^3 \,|\, [(x^1)^2 + (x^2)^2]A_1^{-2} + (x^3)^2 A_2^{-2} = 1, A_1 \in \mathbb{R}^+, A_2 \in \mathbb{R}^+\}$.

Proof.

$$[(x^1)^2 + (x^2)^2]A_1^{-2} + (16.23)$$

$$\Rightarrow \tag{16.26}$$

$$[(x^1)^2 + (x^2)^2]A_1^{-2} = A_1^{-2}[x'^2 + y'^2 \cos^2 i + z'^2 \sin^2 i - 2y'z' \sin i \cos i] \,,$$

$$[(x^3)^2]A_2^{-2} + (16.23)$$

$$\Rightarrow \tag{16.27}$$

$$[(x^3)^2]A_2^{-2} = A_2^{-2}[y'^2 \sin^2 i + z'^2 \cos^2 i] \,,$$

if $x' = 0$, then

$$[(x^1)^2 + (x^2)^2]A_1^{-2} + [(x^3)^2]A_2^{-2} = \frac{x'^2}{A_1^2} + \left(\frac{\cos^2 i}{A_1^2} + \frac{\sin^2 i}{A_2^2} \right) y' = 1 \,,$$

$$A_2^2 = A_1^2(1 - E^2)$$

$$\Rightarrow \tag{16.28}$$

$$\frac{x'^2}{A_1^2} + \frac{y'^2}{A_1^2(1 - E^2)}[(1 - E^2)\cos^2 i + \sin^2 i] = \frac{x'^2}{A_1^2} + \frac{y'^2}{A_1^2(1 - E^2)}(1 - E^2 \cos^2 i) = 1 \,.$$

End of Proof.

In the plane $\{x', y'\} \in \{x' \in \mathbb{R}^2 \,|\, Ax' + By' + C = 0\}$, we introduce circle-reduced meta-longitude α in order to parameterize $\mathbb{E}^1_{A'_1, A'_2}$, namely by (16.29), illustrated by Fig. 16.4.

$$x' = A'_1 \cos\alpha = A'_1 \sin\alpha^* \,, \quad \alpha^* = \tfrac{\pi}{2} - \alpha \,,$$

$$y' = A'_2 \sin\alpha = A'_2 \cos\alpha^* \,, \quad \alpha = \tfrac{\pi}{2} - \alpha^* \,. \tag{16.29}$$

In terms of circle-reduced metalongitude α or of circular reduced meta pole distance $\alpha^* - \pi/2 - \alpha$, we are able to represent the arc length of $\mathbb{E}^1_{A'_1, A'_2}$ as an elliptic integral of the second kind.

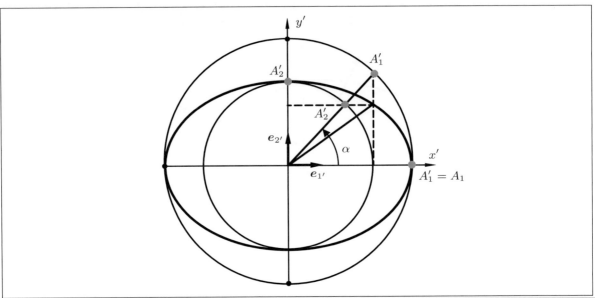

Fig. 16.4. Oblique reference frame $e_{1'}, e_{2'}, e_{3'}, \mathcal{O}$, intersection of $\mathbb{E}^2_{A_1,A_2}$ and $\mathbb{L}^2_{\mathcal{O}}$: the ellipse $\mathbb{E}^1_{A'_1,A'_2}$ as given by (16.25), circle-reduced meta-longitude α.

Corollary 16.2 (Arc length of $\mathbb{E}^1_{A'_1,A'_2}$).

The arc length $s(\alpha)$ of $\mathbb{E}^1_{A'_1 A'_2}$ can be represented by (16.30) with respect to the elliptic integral of the second kind $\boldsymbol{E}(\cdot; E')$ and the first relative eccentricity $E' := (1 - A'^2_2/A'^2_1)^{1\ 2}$. A series expansion of $s(\alpha)$ up to order E'^{12} is provided by (16.31).

$$s(\alpha) = A'_1 \left[\boldsymbol{E}\left(\pi/2; E'\right) - \boldsymbol{E}\left(\alpha^*; E'\right) \right] , \tag{16.30}$$

$$s(\alpha) = A'_1 \alpha \left[1 - \frac{1}{2\cdot 2} E'^2 - \frac{1\cdot 1}{2\cdot 4}\frac{3\cdot 1}{4\cdot 2} E'^4 - \frac{1\cdot 1\cdot 3}{2\cdot 4\cdot 6}\frac{5\cdot 3\cdot 1}{6\cdot 4\cdot 2} E'^6 - \right.$$
$$- \frac{1\cdot 1\cdot 3\cdot 5}{2\cdot 4\cdot 6\cdot 8}\frac{7\cdot 5\cdot 3\cdot 1}{8\cdot 6\cdot 4\cdot 2} E'^8 - \frac{1\cdot 1\cdot 3\cdot 5\cdot 7}{2\cdot 4\cdot 6\cdot 8\cdot 10}\frac{9\cdot 7\cdot 5\cdot 3\cdot 1}{10\cdot 8\cdot 6\cdot 4\cdot 2} E'^{10} - \mathrm{O}(E'^{12}) \right] -$$
$$- A'_1 \cos\alpha \sin\alpha \left[\frac{1}{2}\frac{1}{2} E'^2 + \frac{1\cdot 1}{2\cdot 4}\frac{3\cdot 1}{4\cdot 2} E'^4 + \frac{1\cdot 1\cdot 3}{2\cdot 4\cdot 6}\frac{5\cdot 3\cdot 1}{6\cdot 4\cdot 2} E'^6 + \right.$$
$$\left. + \frac{1\cdot 1\cdot 3\cdot 5}{2\cdot 4\cdot 6\cdot 8}\frac{7\cdot 5\cdot 3\cdot 1}{8\cdot 6\cdot 4\cdot 2} E'^8 + \frac{1\cdot 1\cdot 3\cdot 5\cdot 7}{2\cdot 4\cdot 6\cdot 8\cdot 10}\frac{9\cdot 7\cdot 5\cdot 3\cdot 1}{10\cdot 8\cdot 6\cdot 4\cdot 2} E'^{10} + \mathrm{O}(E'^{12}) \right] -$$
$$- A'_1 \cos^3\alpha \sin\alpha \left[\frac{1\cdot 1}{2\cdot 4}\frac{1}{4} E'^4 + \frac{1\cdot 1\cdot 3}{2\cdot 4\cdot 6}\frac{5\cdot 1}{6\cdot 4} E'^6 + \frac{1\cdot 1\cdot 3\cdot 5}{2\cdot 4\cdot 6\cdot 8}\frac{7\cdot 5\cdot 1}{8\cdot 6\cdot 4} E'^8 + \right.$$
$$\left. + \frac{1\cdot 1\cdot 3\cdot 5\cdot 7}{2\cdot 4\cdot 6\cdot 8\cdot 10}\frac{9\cdot 7\cdot 5\cdot 1}{10\cdot 8\cdot 6\cdot 4} E'^{10} + \mathrm{O}(E'^{12}) \right] - \tag{16.31}$$
$$- A'_1 \cos^5\alpha \sin\alpha \left[\frac{1\cdot 1\cdot 3}{2\cdot 4\cdot 6}\frac{1}{6} E'^6 + \frac{1\cdot 1\cdot 3\cdot 5}{2\cdot 4\cdot 6\cdot 8}\frac{7\cdot 1}{8\cdot 6} E'^8 + \frac{1\cdot 1\cdot 3\cdot 5\cdot 7}{2\cdot 4\cdot 6\cdot 8\cdot 10}\frac{9\cdot 7\cdot 1}{10\cdot 8\cdot 6} E'^{10} + \mathrm{O}(E'^{12}) \right] -$$
$$- A'_1 \cos^7\alpha \sin\alpha \left[\frac{1\cdot 1\cdot 3\cdot 5}{2\cdot 4\cdot 6\cdot 8}\frac{1}{8} E'^8 + \frac{1\cdot 1\cdot 3\cdot 5\cdot 7}{2\cdot 4\cdot 6\cdot 8\cdot 10}\frac{9\cdot 1}{10\cdot 8} E'^{10} + \mathrm{O}(E'^{12}) \right] .$$

An alternative expansion in terms of powers of $\Delta\alpha$ is

$$s(\alpha_0 + \Delta\alpha) = s(\alpha_0) + s_1(\alpha_0)\Delta\alpha + s_2(\alpha_0)\Delta\alpha^2 + \mathrm{O}(\Delta\alpha^3) , \tag{16.32}$$

$$s_1(\alpha) := \frac{\mathrm{d}s}{\mathrm{d}u} = A'_1\sqrt{1 - E'^2\cos^2\alpha} , \quad s_2(\alpha) := \frac{1}{2!}\frac{\mathrm{d}^2s}{\mathrm{d}\alpha^2} = \frac{1}{2}\frac{A'_1\ E'^2\sin\alpha\cos\alpha}{\sqrt{1 - E'^2\cos^2\alpha}} \tag{16.33}$$

End of Corollary.

Proof.

$$\mathbb{E}^1_{A'_1, A'_2} :$$

$$ds = \sqrt{dx'^2 + dy'^2} = A'_1 \sqrt{1 - E'^2 \cos^2 \alpha}\, d\alpha = -A'_1 \sqrt{1 - E'^2 \sin^2 \alpha^*}\, d\alpha^*$$

$$\Rightarrow$$

$$s(\alpha) = A'_1 \int_0^\alpha \sqrt{1 - E'^2 \cos^2 \alpha'}\, d\alpha' =$$

$$= -A'_1 \int_{\pi/2}^{\alpha^*} \sqrt{1 - E'^2 \sin^2 \alpha^*}\, d\alpha^* = A'_1 \int_{\alpha^*}^{\pi/2} \sqrt{1 - E'^2 \sin^2 \alpha^*}\, d\alpha^* =$$

$$= A'_1 \int_0^{\pi/2} \sqrt{1 - E'^2 \sin^2 \alpha^*}\, d\alpha^* - A'_1 \int_0^{\alpha^*} \sqrt{1 - E'^2 \sin^2 \alpha^*}\, d\alpha^* =$$

$$= A'_1 \left[\boldsymbol{E}\left(\pi/2; E'\right) - \boldsymbol{E}\left(\alpha^*; E'\right) \right] \ \forall \ E'^2 = \frac{E^2 \sin^2 i}{1 - E^2 \cos^2 i} .$$

(16.34)

End of Proof.

Proof.

$$(1 - x)^{1/2} = 1 - \frac{1}{2}x - \frac{1 \cdot 1}{2 \cdot 4}x^2 - \frac{1 \cdot 1 \cdot 3}{2 \cdot 4 \cdot 6}x^3 - \frac{1 \cdot 1 \cdot 3 \cdot 5}{2 \cdot 4 \cdot 6 \cdot 8}x^4 - \cdots \ \forall \ |x| \leq 1 ,$$

(16.35)

$$\sqrt{1 - E'^2 \cos^2 \alpha'} = 1 - \frac{1}{2} E'^2 \cos^2 \alpha' - \frac{1 \cdot 1}{2 \cdot 4} E'^4 \cos^4 \alpha' - \frac{1 \cdot 1 \cdot 3}{2 \cdot 4 \cdot 6} E'^6 \cos^6 \alpha' -$$

$$- \frac{1 \cdot 1 \cdot 3 \cdot 5}{2 \cdot 4 \cdot 6 \cdot 8} E'^8 \cos^8 \alpha' - \frac{1 \cdot 1 \cdot 3 \cdot 5 \cdot 7}{2 \cdot 4 \cdot 6 \cdot 8 \cdot 10} E'^{10} \cos^{10} \alpha' - \mathrm{O}(E'^{12}) \ \forall E' < 1 .$$

(16.36)

These series are uniformly convergent. Accordingly, in the arc length integral, we can interchange integration and summation and are directly led to (16.31).

End of Proof.

The proof for (16.32) is now straightforward. In Corollary 16.3, the relation of meta-longitude α to longitude L and latitude B is summarized.

Corollary 16.3 (Cha-cha-cha: meta-longitude α versus longitude L and latitude B).

$$\tan(L - \Omega) = \sqrt{1 - E'^2} \cos i \tan \alpha ,$$

(16.37)

$$\tan B = \frac{A'_2 \sin i}{(1 - E^2)} \frac{\sin \alpha}{\sqrt{A'^2_1 \cos^2 \alpha + A'^2_2 \cos^2 i \sin^2 \alpha}}$$

(16.38)

versus

$$\tan \alpha = \frac{\sqrt{1 - E^2 \cos^2 i}}{\sqrt{1 - E^2}} \frac{1}{\cos(L - \Omega)} \left[(1 - E^2) \sin i \tan B + \cos i \sin(L - \Omega) \right] .$$

(16.39)

End of Corollary.

Proof.

$$x^{3'} = z' = 0$$

$$\Rightarrow$$

$$\tan L = \frac{x^2}{x^1} = \frac{x^{1'} \sin \Omega + x^{2'} \cos \Omega \sin i}{x^{1'} \cos \Omega - x^{2'} \sin \Omega \cos i} = \frac{A_1' \cos \alpha \sin \Omega + A_2' \sin \alpha \cos \Omega \cos i}{A_1' \cos \alpha \cos \Omega - A_2' \sin \alpha \sin \Omega \cos i}$$

$$\Rightarrow$$

$$(A_1' \cos \alpha \cos \Omega - A_2' \sin \alpha \sin \Omega \cos i) \sin L = (A_1' \cos \alpha \sin \Omega + A_2' \sin \alpha \cos \Omega \cos i) \cos L$$

$$\Rightarrow \tag{16.40}$$

$$A_1' \cos \alpha (\cos \Omega \sin L - \sin \Omega \cos L) = A_2' \sin \alpha \cos i (\cos \Omega \cos L + \sin \Omega \sin L)$$

$$\Rightarrow$$

$$A_1' \cos \alpha \sin(L - \Omega) = A_2' \sin \alpha \cos i \cos(L - \Omega)$$

$$\Rightarrow$$

$$\tan(L - \Omega) = \frac{A_2'}{A_1'} \cos i \tan \alpha = \sqrt{1 - E'^2} \cos i \tan \alpha \ .$$

End of Proof.

Proof.

$$x^{3'} = z' = 0$$

$$\Rightarrow$$

$$\tan B = \frac{1}{1 - E^2} \frac{x^3}{\sqrt{(x^1)^2 + (x^2)^2}} = \tag{16.41}$$

$$= \frac{1}{1 - E^2} \left[x^{2'} \sin i \right] \left[\left(x^{1'} \cos \Omega - x^{2'} \sin \Omega \cos i \right)^2 + \left(x^{1'} \sin \Omega + x^{2'} \cos \Omega \cos i \right)^2 \right]^{-1/2} \ ,$$

$$\sqrt{(x^{1'})^2 + (x^{2'})^2 \cos i} = \sqrt{A_1'^2 \cos^2 \alpha + A_2'^2 \sin^2 \alpha \cos i} \ ,$$

$$x^{2'} \sin i = A_2' E \sin \alpha \sin i$$

$$\Rightarrow \tag{16.42}$$

$$\tan B = \frac{A_2' \sin i}{(1 - E^2)} \frac{\sin \alpha}{\sqrt{A_1'^2 \cos^2 \alpha + A_2'^2 \cos^2 i \sin^2 \alpha}} \ .$$

End of Proof.

Proof.

$$x^{3'} = z' = 0$$

$$\Rightarrow$$

$$\tan \alpha = \frac{y'}{x'} \frac{A_1'}{A_2'} =$$

$$= \frac{y'}{x'} \frac{1}{\sqrt{1 - E'^2}} =$$

$$= \frac{\sqrt{1 - E^2 \cos^2 i}}{\sqrt{1 - E^2}} \left(\frac{-x^1 \sin \Omega \cos i + x^2 \cos \Omega \cos i + x^3 \sin i}{x^1 \cos \Omega + x^2 \sin \Omega} \right) = \qquad (16.43)$$

$$= \frac{\sqrt{1 - E^2 \cos^2 i}}{\sqrt{1 - E^2}} \left(\frac{\cos B \sin(L - \Omega) \cos i + (1 - E^2) \sin B \sin i}{\cos B \cos(L - \Omega)} \right)$$

$$\Rightarrow$$

$$\tan \alpha = \frac{\sqrt{1 - E^2 \cos^2 i}}{\sqrt{1 - E^2}} \frac{1}{\cos(L - \Omega)} \left[(1 - E^2) \sin i \tan B + \cos i \sin(L - \Omega) \right] .$$

End of Proof.

16-3 The equations of the oblique Mercator projection

Universal oblique Mercator projection. D'Alembert–Euler equations (Cauchy–Riemann equations), oblique elliptic meta-equator.

The fundamental solution (16.16) of the d'Alembert–Euler equations (Cauchy–Riemann equations) here are specified by $\{p, q\}_{\mathrm{UMP}}$ of type (16.10) and by the boundary condition of an equidistant mapping of the oblique elliptic meta-equator illustrated by Figs. 16.3 and 16.4. In particular, we depart from (16.16) and (16.10), conventionally written as (16.44), here only given up to degree three.

$$\Delta x := x - \alpha_0 =$$
$$= \alpha_1 \Delta q + \beta_1 \Delta l + \alpha_2 (\Delta q^2 - \Delta l^2) + \beta_2 2 \Delta q \Delta l + \mathrm{O}_{x3} ,$$
$$\Delta y := y - \beta_0 = \qquad (16.44)$$
$$= \beta_1 \Delta q - \alpha_1 \Delta l + \beta_2 (\Delta q^2 - \Delta l^2) - \alpha_2 2 \Delta q \Delta l + \mathrm{O}_{y3} .$$

We are left with the problem to determine the unknown coefficients α_1, β_1, α_2, β_2 etc. by a properly chosen boundary condition we outline as follows.

Definition 16.4 (Universal oblique Mercator projection).

A conformal mapping of the ellipsoid-of-revolution $\mathbb{E}^2_{A_1, A_2}$ is called *Universal oblique Mercator projection* if its oblique elliptic meta-equator $\mathbb{E}^1_{A_1', A_2'}$ for $A_1' = A_1$ and $A_2' = A_1 (1 - E^2)/\sqrt{1 - E^2 \cos^2 i}$ is mapped equidistantly as a straight line.

End of Definition.

Theorem 16.5 (Universal oblique Mercator projection).

The boundary condition of the equidistantly mapped elliptic meta-equator $\mathbb{E}^1_{A'_1, A'_2}$,

$$\Delta x(\text{meta-equator}) = \Delta s(\Delta \alpha) , \quad \Delta y(\text{meta-equator}) = 0 , \tag{16.45}$$

with respect to first power series $s(\alpha)$,

$$\Delta s(\Delta \alpha) = s_1 \Delta \alpha + s_2 \Delta \alpha^2 + O_{s3} , \quad \Delta \alpha := \alpha - \alpha_0 , \tag{16.46}$$

the second power series $B(\alpha)$ and $L(\alpha)$,

$$\Delta b(\Delta \alpha) = b_1 \Delta \alpha + b_2 \Delta \alpha^2 + O_{b3} ,$$
$$\Delta b := B - B_0 , \tag{16.47}$$

$$\Delta l(\Delta \alpha) = l_1 \Delta \alpha + l_2 \Delta \alpha^2 + O_{l3} ,$$
$$\Delta l := L - L_0 , \tag{16.48}$$

$$\Delta b^2(\Delta \alpha) = b_1^2 \Delta \alpha^2 + O_{b3}^2 , \quad \Delta l^2(\Delta \alpha) = l_1^2 \Delta \alpha^2 + O_{l3}^2 , \tag{16.49}$$

and the third power series $q_{\text{UMP}}(B)$,

$$\Delta q = q_1 \Delta b + q_2 \Delta b^2 + O_{q3} , \quad \Delta q := q_{\text{UMP}}(B) - q_{\text{UMP}}(B_0) , \tag{16.50}$$

leads to the parameters of the second order universal oblique Mercator projection,

$$\Delta x =$$
$$= \alpha_1 q_1 \Delta b + \beta_1 \Delta l + (\alpha_1 q_2 + \alpha_2 q_1^2) \Delta b^2 + 2\beta_2 q_1 \Delta b \Delta l - \alpha_2 \Delta l^2 + O_{x3} ,$$
$$\Delta y =$$
$$= \beta_1 q_1 \Delta b - \alpha_1 \Delta l + (\beta_1 q_2 + \beta_2 q_1^2) \Delta b^2 - 2\alpha_2 q_1 \Delta b \Delta l - \beta_2 \Delta l^2 + O_{y3} , \tag{16.51}$$

namely

$$\alpha_1 = \frac{q_1 b_1 s_1}{q_1^2 b_1^2 + l_1^2} , \quad \beta_1 = \frac{l_1 s_1}{q_1^2 b_1^2 + l_1^2} , \tag{16.52}$$

$$\alpha_2 = \frac{1}{(q_1^2 b_1^2 + l_1^2)^3} \times$$
$$\times \left(s_2(q_1^2 b_1^2 - l_1^2)(q_1^2 b_1^2 + l_1^2) + s_1[(q_1 b_2 + q_2 b_1^2)(3l_1^2 - q_1^2 b_1^2)q_1 b_1 - l_1 l_2(3q_1^2 b_1^2 - l_1^2)] \right) , \tag{16.53}$$

$$\beta_2 = \frac{1}{(q_1^2 b_1^2 + l_1^2)^3} \times$$
$$\times \left(2q_1 b_1 l_1(q_1^2 b_1^2 + l_1^2)s_2 + s_1[(q_1 b_2 + q_2 b_1^2)(-3q_1^2 b_1^2 + l_1^2)l_1 + (-3l_1^2 + q_1^2 b_1^2)q_1 b_1 l_1] \right) . \tag{16.54}$$

End of Theorem.

The coefficients $\{q_1, q_2\}$, $\{s_1, s_2\}$, $\{b_1, b_2\}$, and $\{l_1, l_2\}$ are collected in the following Boxes 16.2–16.5.

Box 16.2 (Isometric latitude $q(b)$ as a function of latitude b).

Power series expansion $\Delta q = \sum_{r=1}^{N} q_r \Delta b^r$ up to order $N = 2$

(higher-order terms are given by J. Engels, E. Grafarend (1995)):

$$q_1 := \frac{1 - E^2}{\cos B_0 (1 - E^2 \sin^2 B_0)} ,$$

$$q_2 := \frac{\sin B_0}{2 \cos^2 B_0 (1 - E^2 \sin^2 B_0)^2} [1 + E^2(1 - 3\sin^2 B_0) + E^4(-2 + 3\sin^2 B_0)] .$$

(16.55)

Box 16.3 (Arc length of the oblique meta-equator).

Power series expansion $\Delta s = \sum_{r=1}^{N} s_r \Delta \alpha^r$ up to order $N = 2$:

$$s_1(\alpha_0) := A_1 \sqrt{1 - E'^2 \cos^2 \alpha_0} ,$$

$$s_2(\alpha_0) := \frac{1}{2} \frac{A_1' \, E'^2 \sin \alpha_0 \cos \alpha_0}{\sqrt{1 - E'^2 \cos^2 \alpha_0}} .$$

(16.56)

Box 16.4 (Latitude $B(\alpha)$ as a function of meta-longitude α).

Power series expansion $\Delta b = \sum_{r=1}^{N} b_r \Delta \alpha^r$ up to order $N = 2$:

$$b_1 := -\frac{A_2' A_1'^2 (1 - E^2) \sin i \cos \alpha_0}{[A_1'^2(1 - E^2)^2 + E^2 A_2'^2 \sin^2 i \sin^2 \alpha_0]} [A_1'^2 + (A_2'^2 \cos^2 i - A_1'^2) \sin^2 \alpha_0]^{-1/2} ,$$

$$2b_2 := -\frac{A_2' A_1'^2 (1 - E^2) \sin i}{[A_1'^2(1 - E^2)^2 + E^2 A_2'^2 \sin^2 i \sin^2 \alpha_0]^2} [A_1'^2 + (A_2'^2 \cos^2 i - A_1'^2) \sin^2 \alpha_0]^{-3/2} \times$$

$$\times \bigg([A_1'^2(1 - E^2)^2 + E^2 A_2'^2 \sin^2 i \sin^2 \alpha_0][A_1'^2 + (A_2'^2 \cos^2 i - A_1'^2) \sin^2 \alpha_0] \sin \alpha_0 +$$

$$+ 2[A_1'^2 + (A_2'^2 \cos^2 i - A_1'^2) \sin^2 \alpha_0] E^2 A_2'^2 \sin^2 i \sin \alpha_0 \cos^2 \alpha_0 +$$

$$+ [A_1'^2(1 - E^2)^2 + E^2 A_2'^2 \sin^2 i \sin^2 \alpha_0](A_2'^2 \cos^2 i - A_1'^2) \sin \alpha_0 \cos^2 \alpha_0 \bigg) .$$

(16.57)

Box 16.5 (Longitude $L(\alpha)$ as a function of meta-longitude α).

Power series expansion $\Delta l = \sum_{r=1}^{N} l_r \Delta \alpha^r$ up to order $N = 2$:

$$l_1 := +\frac{A_1'^2 A_2'^2 \cos i}{A_1'^2 \cos^2 \alpha_0 + A_2'^2 \cos^2 i \sin^2 \alpha_0} ,$$

$$2l_2 := -2 \frac{A_1'^2 A_2'^2 \cos i (A_2'^2 \cos^2 i - A_1'^2) \sin \alpha_0 \cos \alpha_0}{(A_1'^2 \cos^2 \alpha_0 + A_2'^2 \cos^2 i \sin^2 \alpha_0)^2} .$$

(16.58)

The relations (16.44) together with the relations (16.50) lead to the relations (16.51). Let us prove the other central relations here.

Proof: (16.47): b_1.

$$\tan B = \frac{A_2'}{1 - E^2} \sin i \frac{\sin \alpha}{\sqrt{A_1'^2 + (A_2'^2 \cos^2 i - A_1'^2) \sin^2 \alpha}} \ ,$$

$$\frac{\mathrm{d} \tan B}{\mathrm{d}\alpha} = \frac{1}{\cos^2 B} \frac{\mathrm{d}B}{\mathrm{d}\alpha} = \frac{A_2'}{1 - E^2} \sin i \frac{A_1'^2 \cos \alpha}{[A_1'^2 + (A_2'^2 \cos^2 i - A_1'^2) \sin^2 \alpha]^{3/2}} \ ,$$

$$\frac{\mathrm{d}B}{\mathrm{d}\alpha} = \cos^2 B \frac{\mathrm{d} \tan B}{\mathrm{d}\alpha} = \frac{1}{1 + \tan^2 B} \frac{\mathrm{d} \tan B}{\mathrm{d}\alpha} = \tag{16.59}$$

$$= [A_2' A_1'^2 (1 - E^2) \sin i \cos \alpha][A_1'^2 + (A_2'^2 \cos^2 i - A_1'^2) \sin^2 \alpha]^{-1/2} \times$$

$$\times [A_1'^2 (1 - E^2)^2 + E^2 A_2'^2 \sin^2 i \sin^2 \alpha]^{-1}$$

$$\Rightarrow$$

$$b_1 := \frac{\mathrm{d}B}{\mathrm{d}\alpha}(\alpha_0) \ .$$

End of Proof.

Proof: (16.48): l_1.

$$\tan(L - \Omega) = \frac{A_2'}{A_1'} \cos i \tan \alpha \ , \tag{16.60}$$

$$\frac{\mathrm{d} \tan(L - \Omega)}{\mathrm{d}\alpha} = \frac{1}{\cos^2(L - \Omega)} \frac{\mathrm{d}L}{\mathrm{d}\alpha} = \frac{A_2'}{A_1'} \frac{\cos i}{\cos^2 \alpha} \ ,$$

$$\cos^2(L - \Omega) = \frac{1}{1 + \tan^2(L - \Omega)} = \frac{1}{1 + \frac{A_2'^2 \cos^2 i}{A_1'^2} \tan^2 \alpha}$$

$$\Rightarrow$$

$$\frac{\mathrm{d}L}{\mathrm{d}\alpha} = \frac{A_2'}{A_1'} \frac{\cos i}{\cos^2 \alpha} \frac{1}{1 + \frac{A_2'^2 \cos^2 i}{A_1'^2} \tan^2 \alpha} = \tag{16.61}$$

$$= \frac{A_1' A_2' \cos i}{A_1'^2 \cos^2 \alpha + A_2'^2 \cos^2 i \sin^2 \alpha}$$

$$\Rightarrow$$

$$l_1 := \frac{\mathrm{d}L}{\mathrm{d}\alpha}(\alpha_0) \ .$$

End of Proof.

Proof: (16.47): b_2.

$$\frac{d^2 B}{d\alpha^2} = A'_2 A'_1{}^2 (1 - E^2) \sin i \times$$

$$\times \left(- \sin \alpha [A'_1{}^2 (1 - E^2)^2 + E^2 A'_2{}^2 \sin^2 i \sin^2 \alpha]^{-1} [A'_1{}^2 + (A'_2{}^2 \cos^2 i - A'_1{}^2) \sin^2 \alpha]^{-1/2} - \right.$$

$$-2E^2 A'_2{}^2 \sin^2 i \sin \alpha \cos^2 \alpha \times$$

$$\times [A'_1{}^2 (1 - E^2)^2 + E^2 A'_2{}^2 \sin^2 i \sin^2 \alpha]^{-2} [A'_1{}^2 + (A'_2{}^2 \cos^2 i - A'_1{}^2) \sin^2 \alpha]^{-1/2} -$$

$$-(A'_2{}^2 \cos^2 i - A'_1{}^2) \sin \alpha \cos^2 \alpha \times$$

$$\left. \times [A'_1{}^2 (1 - E^2)^2 + E^2 A'_2{}^2 \sin^2 i \sin^2 \alpha]^{-1} [A'_1{}^2 + (A'_2{}^2 \cos^2 i - A'_1{}^2) \sin^2 \alpha]^{-3/2} \right)$$

(16.62)

$$\Rightarrow$$

$$2b_2 := \frac{d^2 B}{d\alpha^2}(\alpha_0) \ .$$

End of Proof.

Proof: (16.48): l_2.

$$\frac{d^2 L}{d\alpha^2} = -2 \frac{A'_1 A'_2 \cos i (A'_2{}^2 \cos^2 i - A'_1{}^2) \sin \alpha \cos \alpha}{(A'_1{}^2 \cos^2 \alpha + A'_2{}^2 \cos^2 i \sin^2 \alpha)^2} \ ,$$

$$2l_2 := \frac{d^2 L}{d\alpha^2}(\alpha_0) \ .$$

(16.63)

End of Proof.

Proof: (16.52)–(16.54).

In a first step, (16.44) is specified by

$$\Delta x(\text{meta-equator}) =$$
$$= \alpha_1 \Delta q + \beta_1 \Delta l + \alpha_2 (\Delta q^2 - \Delta l^2) + \beta_2 2 \Delta q \Delta l + O_{x3} \ ,$$
$$\Delta y(\text{meta-equator}) =$$
$$= \beta_1 \Delta q - \alpha_1 \Delta l + \beta_2 (\Delta q^2 - \Delta l^2) - \alpha_2 2 \Delta q \Delta l + O_{y3} = 0 \ .$$

(16.64)

Implementation of (16.50) constitutes the second step:

$$\Delta x(\text{meta-equator}) =$$
$$= \alpha_1 q_1 \Delta b + \alpha_1 q_2 \Delta b^2 + \beta_1 \Delta l + \alpha_2 (q_1^2 \Delta b^2 - \Delta l^2) + \beta_2 2 q_1 \Delta b \Delta l + O_{x3} \ ,$$
$$\Delta y(\text{meta-equator}) =$$
$$= \beta_1 q_1 \Delta b + \beta_1 q_2 \Delta b^2 - \alpha_1 \Delta l + \beta_2 (q_1^2 \Delta b^2 - \Delta l^2) - \alpha_2 2 q_1 \Delta b \Delta l + O_{y3} = 0 \ .$$

(16.65)

In a third step, the boundary condition in the above form is represented in the meta-longitude dependence by means of (16.47)–(16.49) (where in a fourth step we identify (16.45) by (16.46)):

$$\Delta x(\text{meta-equator}) =$$

$$= \alpha_1(q_1 b_1 \Delta\alpha + q_1 b_2 \Delta\alpha^2 + q_2 b_1^2 \Delta\alpha^2) + \beta_1(l_1 \Delta\alpha + l_2 \Delta\alpha^2) +$$

$$+ \alpha_2(q_1^2 b_1^2 \Delta\alpha^2 - l_1^2 \Delta\alpha^2) + \beta_2 2 q_1 b_1 l_1 \Delta\alpha^2 + O_{x3} =$$

$$= s_1 \Delta\alpha + s_2 \Delta\alpha^2 \; ,$$

$$(16.66)$$

$$\Delta y(\text{meta-equator}) =$$

$$-\alpha_1(l_1 \Delta\alpha + l_2 \Delta\alpha^2) + \beta_1(q_1 b_1 \Delta\alpha + q_1 b_2 \Delta\alpha^2 + q_2 b_1^2 \Delta\alpha^2) -$$

$$-\alpha_2 2 q_1 b_1 l_1 \Delta\alpha^2 + \beta_2(q_1^2 b_1^2 \Delta\alpha^2 - l_1^2 \Delta\alpha^2) + O_{y3} = 0 \; .$$

A comparison of the coefficients of the two polynomials $\Delta x(\Delta\alpha)$ and $\Delta y(\Delta\alpha)$ constitutes the fifth step:

$$\Delta x(\text{meta-equator})$$

$$\Rightarrow$$

$$\Delta\alpha : q_1 b_1 \alpha_1 + l_1 \beta_1 = s_1 \; ,$$

$$\Delta\alpha^2 : (q_1 b_2 + q_2 b_1^2)\alpha_1 + l_2 \beta_1 + (q_1^2 b_1^2 - l_1^2)\alpha_2 + 2 q_1 b_1 l_1 \beta_2 = s_2 \; .$$

$$(16.67)$$

$$\Delta y(\text{meta-equator})$$

$$\Rightarrow$$

$$\Delta\alpha : -l_1 \alpha_1 + q_1 b_1 \beta_1 = 0 \; ,$$

$$\Delta\alpha^2 : -l_2 \alpha_1 + (q_1 b_2 + q_2 b_1^2)\beta_1 - 2 q_1 b_1 l_1 \alpha_2 + (q_1^2 b_1^2 - l_1^2)\beta_2 = 0 \; .$$

$$(16.68)$$

A matrix version of the above equations is

$$\begin{bmatrix} q_1 b_1 & l_1 & 0 & 0 \\ q_1 b_2 + q_2 b_1^2 & l_2 & q_1^2 b_1^2 - l_1^2 & 2 q_1 b_1 l_1 \\ -l_1 & q_1 b_1 & 0 & 0 \\ -l_2 & q_1 b_2 + q_2 b_1^2 & -2 q_1 b_1 l_1 & q_1^2 b_1^2 - l_1^2 \end{bmatrix} \begin{bmatrix} \alpha_1 \\ \beta_1 \\ \alpha_2 \\ \beta_2 \end{bmatrix} = \begin{bmatrix} s_1 \\ s_2 \\ 0 \\ 0 \end{bmatrix} \; .$$

$$(16.69)$$

1st row, 3rd row

$$\Rightarrow$$

$$q_1 b_1 \alpha_1 + l_1 \beta_1 = s_1 \,, \quad -l_1 \alpha_1 + q_1 b_1 \beta_1 = 0$$

$$\Rightarrow$$

$$(q_1^2 b_1^2 l_1^{-1} + l_1) \beta_1 = s_1 \,, \quad \alpha_1 = q_1 b_1 l_1^{-1} \beta_1$$

(16.70)

$$\Rightarrow$$

$$\beta_1 = l_1 s_1 (q_1^2 b_1^2 + l_1^2)^{-1} \,,$$

$$\alpha_1 = q_1 b_1 s_1 (q_1^2 b_1^2 + l_1^2)^{-1} \,.$$

2nd row, 3rd row

$$\Rightarrow$$

$$(q_1 b_2 + q_2 b_1^2) \alpha_1 + l_2 \beta_1 + (q_1^2 b_1^2 - l_1^2) \alpha_2 - 2 q_1 b_1 l_1 \beta_2 = s_2 \,,$$

$$-l_2 \alpha_1 + (q_1 b_2 + q_2 b_1^2) \beta_1 - 2 q_1 b_1 l_1 \alpha_2 + (q_1^2 b_1^2 - l_1^2) \beta_2 = 0$$

$$\Rightarrow$$

$$(q_1 b_2 + q_2 b_1^2) q_1 b_1 s_1 (q_1^2 b_1^2 + l_1^2)^{-1} + l_2 l_1 s_1 (q_1^2 b_1^2 + l_1^2)^{-1} - s_1 =$$

(16.71)

$$= (l_1^2 - q_1^2 b_1^2) \alpha_2 - 2 q_1 b_1 l_1 \beta_2 \,,$$

$$(q_1 b_2 + q_2 b_1^2) l_1 s_1 (q_1^2 b_1^2 + l_1^2)^{-1} - l_2 q_1 b_1 s_1 (q_1^2 b_1^2 + l_1^2)^{-1} =$$

$$= 2 q_1 b_1 l_1 \alpha_2 + (l_1^2 - q_1^2 b_1^2) \beta_2 \,,$$

$$\begin{bmatrix} q_1^2 b_1^2 - l_1^2 & 2 q_1 b_1 l_1 \\ -2 q_1 b_1 l_1 & q_1^2 b_1^2 - l_1^2 \end{bmatrix} \begin{bmatrix} \alpha_2 \\ \beta_2 \end{bmatrix} =$$

$$= \begin{bmatrix} s_2 - (q_1^2 b_1^2 + l_1^2)^{-1} l_1 l_2 s_1 - (q_1 b_2 + q_2 b_1^2)(q_1^2 b_1^2 + l_1^2)^{-1} q_1 b_1 s_1 \\ (q_1^2 b_1^2 + l_1^2)^{-1} l_2 q_1 b_1 s_1 - (q_1^2 b_1^2 + l_1^2)^{-1} (q_1 b_2 + q_2 b_1^2) l_1 s_1 \end{bmatrix}$$

$$\Rightarrow$$

$$\alpha_2 = \frac{1}{(q_1^2 b_1^2 + l_1^2)^3} \times$$

(16.72)

$$\times \left(s_2 (q_1^2 b_1^2 - l_1^2)(q_1^2 b_1^2 + l_1^2) + s_1 [(q_1 b_2 + q_2 b_1^2)(3 l_1^2 - q_1^2 b_1^2) q_1 b_1 - l_1 l_2 (3 q_1^2 b_1^2 - l_1^2)] \right) ,$$

$$\beta_2 = \frac{1}{(q_1^2 b_1^2 + l_1^2)^3} \times$$

$$\times \left(2 q_1 b_1 l_1 (q_1^2 b_1^2 + l_1^2) s_2 + s_1 [(q_1 b_2 + q_2 b_1^2)(-3 q_1^2 b_1^2 + l_1^2) l_1 + (-3 l_1^2 + q_1^2 b_1^2) q_1 b_1 l_1] \right) .$$

End of Proof.

The equations (16.51), which represent locally the oblique Mercator projection, reduce (i) to the equations of the standard Mercator projection of $\mathbb{E}^2_{A_1,A_2}$ for zero inclination, see Box 16.6, or (ii) to the equations of the transverse Mercator projection of $\mathbb{E}^2_{A_1,A_2}$ for ninety degrees inclination, see Box 16.7, or (iii) to the equations of the oblique Mercator projection of $\mathbb{S}^2_r$ for zero relative eccentricity $E = 0$, compare with Box 16.1 presented already before.

Box 16.6 (The equations of the standard Mercator projection of $\mathbb{E}^2_{A_1,A_2}$ for zero inclination).

$$i = 0$$
$$\Rightarrow \tag{16.73}$$
$$E' = 0 \,, \quad A'_1 = A_1 \,, \quad A'_2 = A_1 \,,$$

$$\tan(L - \Omega) = \tan\alpha$$
$$\Rightarrow \tag{16.74}$$
$$\alpha = L - \Omega \,,$$

$$\tan B = 0$$
$$\Rightarrow \tag{16.75}$$
$$b_j = 0 \ \forall \, j = 1, 2, \dots \,,$$

$$s_1(\alpha) = A'_1 = a \,, \quad s_2(\alpha) = 0 \,, \quad s_3(\alpha) = 0 \,, \dots \,, \tag{16.76}$$

$$\alpha_1 = 0 \,, \quad \alpha_2 = 0 \,, \quad \beta_1 = s_1 = a \,, \quad \beta_2 = 0$$
$$\Rightarrow \tag{16.77}$$
$$\Delta x = a\Delta l \,, \quad \Delta y = a\Delta q = aq_1\Delta b + aq_2\Delta b^2 + \mathrm{O}_3 \,.$$

Box 16.7 (The equations of the transverse Mercator projection of $\mathbb{E}^2_{A_1,A_2}$ for ninety degrees inclination).

$$i = \pi/2$$
$$\Rightarrow \tag{16.78}$$
$$E' = E \,, \quad A'_1 = A_1 \,, \quad A'_2 = A_1\sqrt{1 - E^2} \,,$$

$$\tan(L - \Omega) = 0$$
$$\Rightarrow \tag{16.79}$$
$$L = \Omega \,,$$

$$\tan B = \tan\alpha/\sqrt{1 - E^2} \,, \tag{16.80}$$

$$s_1(\alpha) = A_1\sqrt{1 - E^2\cos^2\alpha} \,, \quad s_2(\alpha) = \frac{1}{2}\frac{A_1 E^2 \sin\alpha\cos\alpha}{\sqrt{1 - E^2\cos^2\alpha}} \,, \tag{16.81}$$

$$l_1 = 0 \,, \quad l_2 = 0 \,, \dots \,, \tag{16.82}$$

$$b_1 = \frac{dB}{d\alpha}(\alpha_0) = \sqrt{1 - E^2}/(1 - E^2\cos^2\alpha_0) = \frac{\cos^2 D_0}{\sqrt{1 - E^2}}[1 + (1 - E^2)\tan^2 B_0] \,, \quad b_2 = \cdots \,. \tag{16.83}$$

The oblique Mercator projection is particularly well suited for long and narrow countries. As an example, Fig. 16.5 shows the HOM of Italy. Note that i and Ω of the intersecting ellipse (which appears in the map as x axis) were fitted to the location of this country.

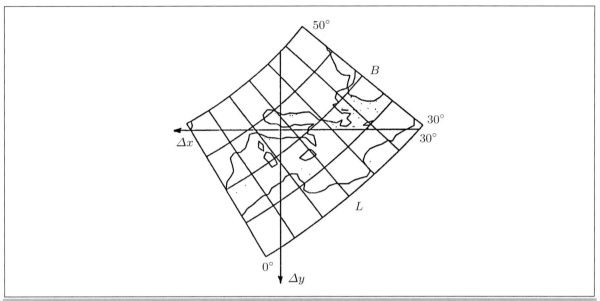

Fig. 16.5. Oblique Mercator projection of $\mathbb{E}^2_{A_1,A_2}$. Inclination $i = 125.02°$, right ascension of the ascending node $\Omega = 53.01°$.

We close this chapter on the rectified skew orthomorphic projection or HOM by referring to K. Bretterbauer (1980), J. H. Cole (1943), E. Grafarend (1995), E. Grafarend and R. Syffus (1995), and J. Engels and E. Grafarend (1995).

17 "Sphere to cone": polar aspect

Mapping the sphere to a cone: polar aspect. Equidistant, conformal, and equal area mappings. Ptolemy, de L'Isle, Lambert, and Albers projections. Point-like North Pole. Tangent cones, secant cones, and circles-of-contact.

For mapping regional areas of medium latitude, conic mappings are particularly adequate (compare with Fig. 17.1). The characteristic feature of conic mappings is that in the polar aspect meridians are represented by straight lines which intersect in one point, the *apex*. Parallels are mapped onto arcs of equicentric circles with the apex as the central point. As with cylindrical mappings, there exist two cases: first, the cone touches the sphere along a parallel circle (compare with Fig. 17.2, top) and, second, it intersects the sphere along two parallels (compare with Fig. 17.2, bottom). Both cases are driven by the opening angle $\Theta \in (0, \pi/2)$, which is the vertex angle made by a cross section through the apex and center of the base (compare with Fig. 17.3).

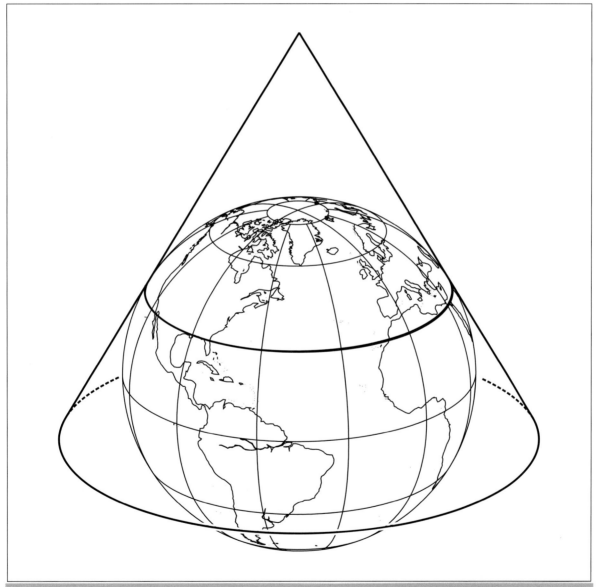

Fig. 17.1. Mapping the sphere to a (tangent) cone. Polar aspect. Line-of-contact: $\Phi_0 = 30°$.

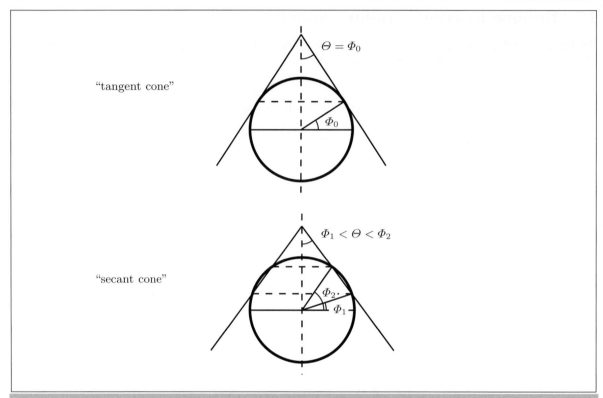

Fig. 17.2. Tangent cone and secant cone.

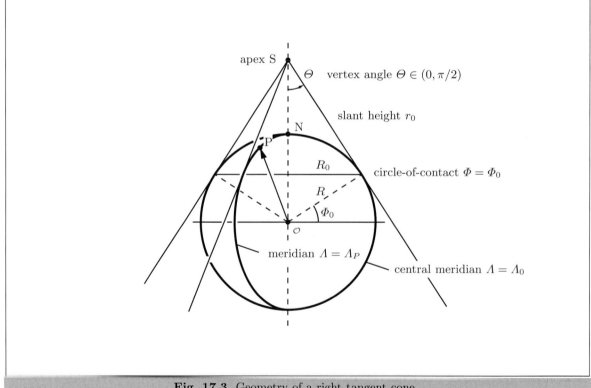

Fig. 17.3. Geometry of a right tangent cone.

17-1 General mapping equations

Setting up general equations of the mapping "sphere to cone": projections in the polar aspect. Jacobi matrix, Cauchy–Green matrix, principal stretches.

The axis of the cone coincides with the polar axis of the Earth, i.e. the straight line passing through the North Pole N and the center o of the sphere. The main construction principals are that first two points of equal spherical latitude Φ have the same distance r_0 from the map center, which is the image of the apex. Second, the cone is sliced along the image of that meridian which is diametrically opposed to the image of the central meridian (compare with Fig. 17.4). Third, the cone can be developed into the plane. The circle-of-contact is that parallel circle $\Phi = \Phi_0$ where the cone touches the sphere. If necessary, the cone is shifted along the polar axis until the touching position is reached. The radius R_0 of the circle-of-contact is given by $R_0 = R \cos \Phi_0$. The slant height r_0, which is the radius of the map image of the circle-of-contact, is $r_0 = R_0 / \sin \Phi_0 = R \cot \Phi_0$.

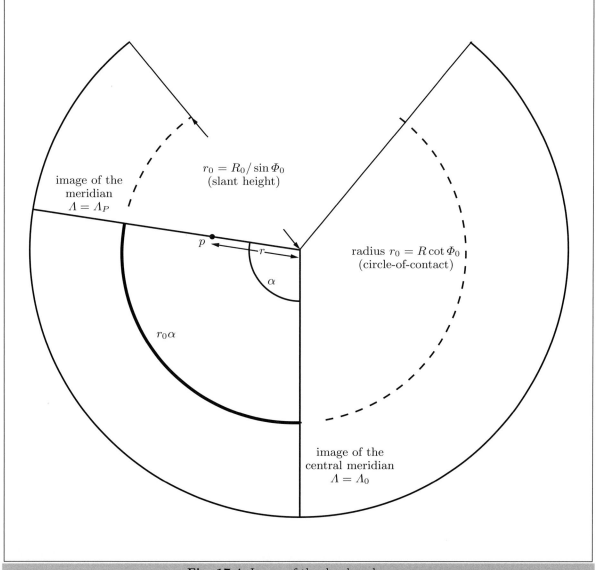

Fig. 17.4. Image of the developed cone.

We know that there are two fundamental rules how to map longitudes Λ and latitudes Φ. The angle (first polar coordinate) $\alpha = \alpha(\Lambda)$ of the image p of a spherical point $P(\Lambda_P, \Phi_P)$ shall only depend on its spherical longitude $\Lambda = \Lambda_P$. In particular, corresponding arcs on the circle-of-contact and their images shall coincide, and this is expressed by (17.1).

$$R_0\Lambda = r_0\alpha \,,\; R_0 = R\cos\Phi_0 \Rightarrow R\Lambda\cos\Phi_0 = r_0\alpha = R_0\frac{\alpha}{\sin\Phi_0} = R\frac{\cos\Phi_0}{\sin\Phi_0}\alpha$$

$$\Rightarrow$$

$$\alpha = \Lambda\sin\Phi_0\,. \tag{17.1}$$

The term $n := \sin\Phi_0$ is called the cone constant, $0 < n < 1$. For $n = 0$, a cylindrical, for $n = 1$, an azimuthal mapping is generated. The second rule concerns to the second polar coordinate r which shall depend only on the latitude $\Phi = \Phi_P$, i.e. $r = f(\pi/2 - \Phi)$. We therefore obtain the general mapping equations for conical mappings (17.2) with the left Jacobi matrix (17.3) and the left Cauchy–Green matrix (17.4) ($\mathsf{G}_r = \operatorname{diag}[r^2, 1] = \operatorname{diag}[f^2, 1]$). For the reason that both C_l and G_r are diagonal matrices, the left principal stretches are easily computed as follows (17.5).

$$\begin{bmatrix} \alpha \\ r \end{bmatrix} = \begin{bmatrix} n\Lambda \\ f(\Phi) \end{bmatrix}\,, \tag{17.2}$$

$$\mathsf{J}_l = \begin{bmatrix} n & 0 \\ 0 & f' \end{bmatrix}\,, \tag{17.3}$$

$$\mathsf{C}_l = \mathsf{J}_l^*\mathsf{G}_r\mathsf{J}_l = \begin{bmatrix} n^2 f^2 & 0 \\ 0 & f'^2 \end{bmatrix}\,. \tag{17.4}$$

$$\Lambda_1 = \sqrt{\frac{C_{11}}{G_{11}}} = \frac{nf}{R\cos\Phi}\,,\quad \Lambda_2 = \sqrt{\frac{C_{22}}{G_{22}}} = \frac{f'}{R}\,. \tag{17.5}$$

17-2 Special mapping equations

Setting up special equations of the mapping "sphere to cone". Equidistant, conformal, and equal area mappings. Ptolemy, de L'Isle, Lambert, and Albers projections. Point-like North Pole.

17-21 Equidistant mapping (de L'Isle projection)

The general mapping equations for this type of mappings are derived from the postulate (17.6) such that (17.7) holds. The integration constant c has to be determined from the additional requirement that the image of the North Pole is a point or a circular arc. Setting $c = 0$, a point-like image of the North Pole is attained.

$$\Lambda_2 = \frac{f'\left(\frac{\pi}{2}-\Phi\right)}{R} = 1 \Leftrightarrow f'\left(\frac{\pi}{2} - \Phi\right) = R \Rightarrow f\left(\frac{\pi}{2} - \Phi\right) = R\left(\frac{\pi}{2} - \Phi\right) + c\,, \tag{17.6}$$

$$\begin{bmatrix} \alpha \\ r \end{bmatrix} = \begin{bmatrix} n\Lambda \\ R\left(\frac{\pi}{2} - \Phi\right) + c \end{bmatrix}\,,$$

$$\Lambda_1 = \frac{n[R(\frac{\pi}{2} - \Phi) + c]}{R\cos\Phi}\,,\quad \Lambda_2 = 1\,. \tag{17.7}$$

17-211 Equidistance and conformality on the circle-of-contact (C. Ptolemy, 85–150 AD), compare with Fig. 17.5

We require the circle-of-contact to be mapped equidistantly and thus state (17.8) from which – together with the cone constant $n = \sin \Phi_0$ – the integration constant c is determined as (17.9).

$$\Lambda_1|_{\Phi=\Phi_0} = \frac{n[R(\frac{\pi}{2} - \Phi_0) + c]}{R \cos \Phi_0} = 1 , \qquad (17.8)$$

$$c = R(\frac{\cos \Phi_0}{n} - \frac{\pi}{2} + \Phi_0) = R(\cot \Phi_0 - \frac{\pi}{2} + \Phi_0) . \qquad (17.9)$$

Since $c \neq 0$, the image of the North Pole is a circular arc. The final mapping equations now result to (17.10) or (17.11) with the left principal stretches (17.12). The circle-of-contact, $\Phi = \Phi_0$, is mapped equidistantly and conformally, i.e. $\Lambda_1|_{\Phi=\Phi_0} = \Lambda_2|_{\Phi=\Phi_0} = 1$.

$$\begin{bmatrix} \alpha \\ r \end{bmatrix} = \begin{bmatrix} \Lambda \sin \Phi_0 \\ R(\Phi_0 - \Phi + \cot \Phi_0) \end{bmatrix} , \qquad (17.10)$$

$$\begin{bmatrix} x \\ y \end{bmatrix} = R(\Phi_0 - \Phi + \cot \Phi_0) \begin{bmatrix} \cos(\Lambda \sin \Phi_0) \\ \sin(\Lambda \sin \Phi_0) \end{bmatrix} , \qquad (17.11)$$

$$\Lambda_1 = \frac{\sin \Phi_0 (\Phi_0 - \Phi + \cot \Phi_0)}{\cos \Phi} , \qquad (17.12)$$

$$\Lambda_2 = 1 .$$

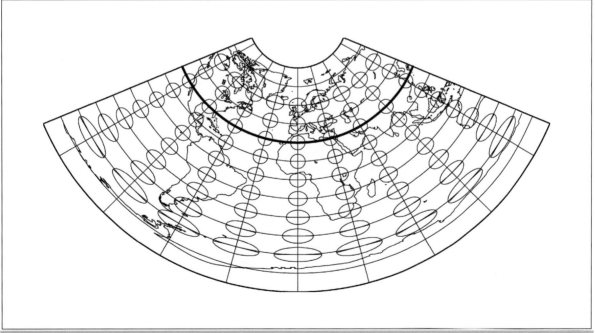

Fig. 17.5. Mapping the sphere to a cone. Polar aspect, equidistant mapping of the set of meridians, equidistant and conformal on the standard parallel $\Phi = \Phi_0 = 30°$ (Ptolemy projection).

17-212 Equidistance and conformality on the circle-of-contact, point-like image of the North Pole, compare with Fig. 17.6

As a special case of the Ptolemy projection the equidistant mapping with point-like pole is obtained by setting the integration constant c to zero. The mapping equations (17.14) or (17.15) and the left principal stretches (17.16) are easily derived from equations (17.7).

$$\Lambda_1|_{\Phi=\Phi_0,c=0} = \frac{nR(\pi/2-\Phi_0)}{R\cos\Phi_0} = 1$$

$$\Rightarrow \tag{17.13}$$

$$n = \frac{\cos\Phi_0}{\pi/2-\Phi_0} \; ,$$

$$\begin{bmatrix} \alpha \\ r \end{bmatrix} = \begin{bmatrix} \Lambda\dfrac{\cos\Phi_0}{\pi/2-\Phi_0} \\ R(\pi/2-\Phi) \end{bmatrix} \; , \tag{17.14}$$

$$\begin{bmatrix} x \\ y \end{bmatrix} = R(\frac{\pi}{2}-\Phi) \begin{bmatrix} \cos\left(\Lambda\dfrac{\cos\Phi_0}{\pi/2-\Phi_0}\right) \\ \sin\left(\Lambda\dfrac{\cos\Phi_0}{\pi/2-\Phi_0}\right) \end{bmatrix} \; , \tag{17.15}$$

$$\Lambda_1 = \frac{\cos\Phi}{\cos\Phi_0}\frac{\pi/2-\Phi_0}{\pi/2-\Phi} \; , \quad \Lambda_2 = 1 \; . \tag{17.16}$$

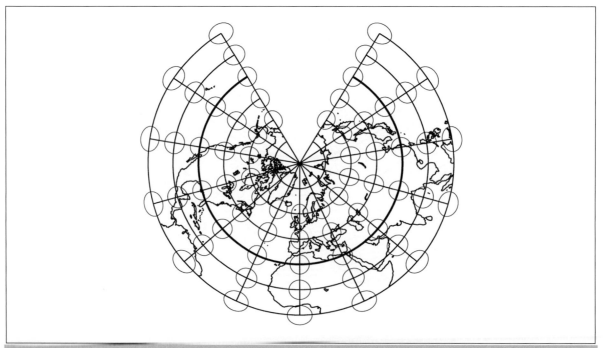

Fig. 17.6. Mapping the sphere to a cone. Polar aspect, equidistant mapping of the set of meridians, equidistant and conformal on the standard parallel $\Phi = \Phi_0 = 30°$, point-like North Pole.

*17-213 Equidistance and conformality on two parallels (secant cone, J. N. de L'Isle 1745),
   compare with Fig. 17.7*

If instead of one parallel two parallel circles are required to be mapped equidistantly, this approach leads to a secant cone, the so-called *de L'Isle projection*, named after the French astronomer Joseph Nicolas de L'Isle. We start from (17.7) and demand that (17.17) is satisfied for the two parallel circles $\Phi = \Phi_1$ and $\Phi = \Phi_2$. We obviously receive two equations for the two unknowns $n := \sin \Phi_0$ (cone constant!) and c, the result of which is (17.18). We end up with the mapping equations (17.19) or (17.20) with the left principal stretches (17.21). For $\Phi = \Phi_1$ or $\Phi = \Phi_2$, we even experience conformality (isometry), $\Lambda_1 = \Lambda_2 = 1$.

$$\Lambda_1|_{\Phi=\Phi_1} = \frac{n[R(\frac{\pi}{2} - \Phi_1) + c]}{R \cos \Phi_1} = \Lambda_1|_{\Phi=\Phi_2} = \frac{n[R(\frac{\pi}{2} - \Phi_2) + c]}{R \cos \Phi_2} = 1 , \tag{17.17}$$

$$\sin \Phi_0 = n = \frac{\cos \Phi_1 - \cos \Phi_2}{\Phi_2 - \Phi_1} , \quad c = R \frac{(\frac{\pi}{2} - \Phi_1) \cos \Phi_2 - (\frac{\pi}{2} - \Phi_2) \cos \Phi_1}{\cos \Phi_1 - \cos \Phi_2} , \tag{17.18}$$

$$\begin{bmatrix} \alpha \\ r \end{bmatrix} = \begin{bmatrix} \dfrac{\cos \Phi_1 - \cos \Phi_2}{\Phi_2 - \Phi_1} \Lambda \\ R\left(-\Phi + \dfrac{\Phi_1 \cos \Phi_2 - \Phi_2 \cos \Phi_1}{\cos \Phi_2 - \cos \Phi_1}\right) \end{bmatrix} , \tag{17.19}$$

$$\begin{bmatrix} x \\ y \end{bmatrix} = R\left(-\Phi + \frac{\Phi_1 \cos \Phi_2 - \Phi_2 \cos \Phi_1}{\cos \Phi_2 - \cos \Phi_1}\right) \begin{bmatrix} \cos\left(\dfrac{\cos \Phi_1 - \cos \Phi_2}{\Phi_2 - \Phi_1} \Lambda\right) \\ \sin\left(\dfrac{\cos \Phi_1 - \cos \Phi_2}{\Phi_2 - \Phi_1} \Lambda\right) \end{bmatrix} , \tag{17.20}$$

$$\Lambda_1 = \frac{\Phi_2 \cos \Phi_1 - \Phi_1 \cos \Phi_2 + \Phi(\cos \Phi_2 - \cos \Phi_1)}{(\Phi_2 - \Phi_1) \cos \Phi} , \quad \Lambda_2 = 1 . \tag{17.21}$$

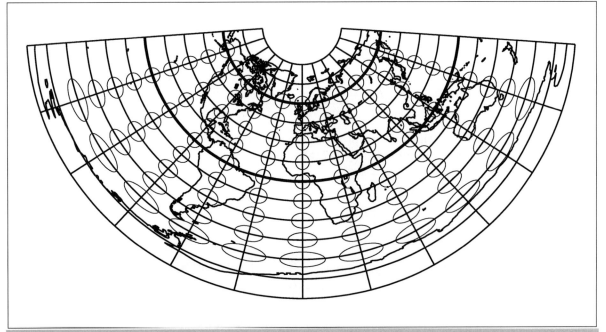

Fig. 17.7. Mapping the sphere to a cone. Polar aspect, equidistant mapping of the set of meridians, equidistant and conformal on two parallels $\Phi = \Phi_1 = 0°$ and $\Phi = \Phi_2 = 60°$ (de L'Isle projection).

17-22 Conformal mapping (Lambert projection)

The general mapping equations for this type of mappings are derived from the identity (17.22). The mapping equations are obtained as (17.25). The left principal stretches are obtained as (17.26).

$$\Lambda_1 = \sqrt{\frac{C_{11}}{G_{11}}} = \frac{nf}{R\cos\Phi} = \Lambda_2 = \sqrt{\frac{C_{22}}{G_{22}}} = \frac{f'}{R} \tag{17.22}$$

$$\Downarrow$$

$$\frac{f'}{f} = \frac{n}{\cos\Phi} \Rightarrow \int \frac{\mathrm{d}f}{f} = n \int \frac{\mathrm{d}\Phi}{\cos\Phi} \tag{17.23}$$

$$\Downarrow$$

$$\ln f = n \ln \tan\left(\frac{\pi}{4} - \frac{\Phi}{2}\right) + \ln c , \tag{17.24}$$

$$\begin{bmatrix} \alpha \\ r \end{bmatrix} = \begin{bmatrix} n\Lambda \\ c\tan\left(\frac{\pi}{4} - \frac{\Phi}{2}\right)^n \end{bmatrix} , \tag{17.25}$$

$$\Lambda_1 = \Lambda_2 = \frac{cn\tan\left(\frac{\pi}{4} - \frac{\Phi}{2}\right)^n}{R\cos\Phi} . \tag{17.26}$$

17-221 Equidistance on the circle-of-contact, compare with Fig. 17.8

The constant n is defined using the parallel circle $\Phi = \Phi_0$ which shall be mapped equidistantly, i. e. through the cone constant $n = \sin\Phi_0$. It follows from (17.26) that (17.27) holds.

$$\Lambda_1|_{\Phi=\Phi_0} = \Lambda_2|_{\Phi=\Phi_0} = \frac{cn\tan\left(\frac{\pi}{4} - \frac{\Phi_0}{2}\right)^n}{R\cos\Phi_0} = 1$$

$$\Leftrightarrow \tag{17.27}$$

$$c = \frac{R\cos\Phi_0}{n\tan\left(\frac{\pi}{4} - \frac{\Phi_0}{2}\right)^n} = \frac{R\cot\Phi_0}{\tan\left(\frac{\pi}{4} - \frac{\Phi_0}{2}\right)^n} .$$

The mapping equations for this kind of projection are therefore defined through (17.28) or (17.29). The left principal stretches are provided by (17.30).

$$\begin{bmatrix} \alpha \\ r \end{bmatrix} = \begin{bmatrix} \Lambda\sin\Phi_0 \\ R\cot\Phi_0 \left(\dfrac{\tan\left(\frac{\pi}{4} - \frac{\Phi}{2}\right)}{\tan\left(\frac{\pi}{4} - \frac{\Phi_0}{2}\right)}\right)^n \end{bmatrix} , \tag{17.28}$$

$$\begin{bmatrix} x \\ y \end{bmatrix} = R\cot\Phi_0 \left(\frac{\tan\left(\frac{\pi}{4} - \frac{\Phi}{2}\right)}{\tan\left(\frac{\pi}{4} - \frac{\Phi_0}{2}\right)}\right)^n \begin{bmatrix} \cos(\Lambda\sin\Phi_0) \\ \sin(\Lambda\sin\Phi_0) \end{bmatrix} , \tag{17.29}$$

$$\Lambda_1 = \Lambda_2 = \frac{\cos\Phi_0}{\cos\Phi} \left(\frac{\tan\left(\frac{\pi}{4} - \frac{\Phi}{2}\right)}{\tan\left(\frac{\pi}{4} - \frac{\Phi_0}{2}\right)}\right)^n . \tag{17.30}$$

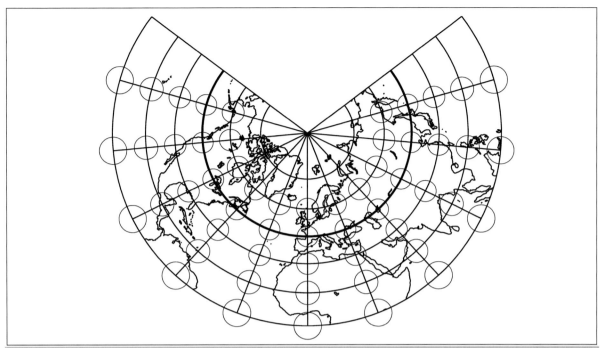

Fig. 17.8. Mapping the sphere to a cone. Polar aspect, conformal mapping, equidistant on the standard parallel $\Phi = \Phi_0 = 45°$.

17-222 Equidistance on two parallels (secant cone, J. H. Lambert 1772), compare with Fig. 17.9

The basic idea is to determine the cone constant $n = \sin\Phi_0$ from an equidistant mapping of two standard parallel circles $\Phi = \Phi_1$ and $\Phi = \Phi_2$. Starting from (17.31), we immediately arrive at (17.32), from which the integration constant c according to (17.33) is computed as a function of the unknown cone constant n. Since c can also be determined via $\Lambda_1|_{\Phi=\Phi_2} = \Lambda_2|_{\Phi=\Phi_2} := 1$, the equality (17.34) is used to compute n according to (17.35).

$$\Lambda_1 = \Lambda_2 = \frac{cn\tan\left(\frac{\pi}{4} - \frac{\Phi}{2}\right)^n}{R\cos\Phi} , \tag{17.31}$$

$$\Lambda_1|_{\Phi=\Phi_1} = \Lambda_2|_{\Phi=\Phi_1} = \frac{cn\tan\left(\frac{\pi}{4} - \frac{\Phi_1}{2}\right)^n}{R\cos\Phi_1} = 1 , \tag{17.32}$$

$$c = \frac{R\cos\Phi_1}{n\tan\left(\frac{\pi}{4} - \frac{\Phi_1}{2}\right)^n} , \tag{17.33}$$

$$\frac{R\cos\Phi_1}{n\tan\left(\frac{\pi}{4} - \frac{\Phi_1}{2}\right)^n} = \frac{R\cos\Phi_2}{n\tan\left(\frac{\pi}{4} - \frac{\Phi_2}{2}\right)^n} , \tag{17.34}$$

$$n = \frac{\ln\cos\Phi_1 - \ln\cos\Phi_2}{\ln\tan\left(\frac{\pi}{4} - \frac{\Phi_1}{2}\right) - \ln\tan\left(\frac{\pi}{4} - \frac{\Phi_2}{2}\right)} . \tag{17.35}$$

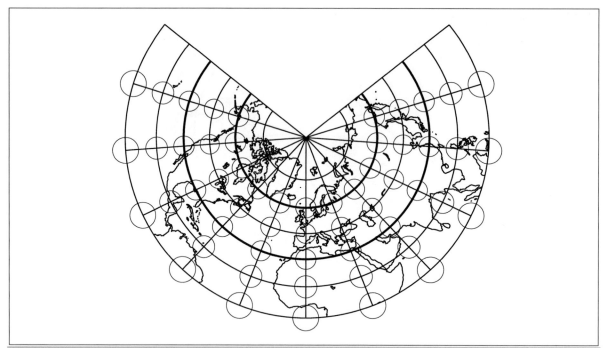

Fig. 17.9. Mapping the sphere to a cone. Polar aspect, conformal mapping, equidistant on two standard parallels $\Phi = \Phi_0 = 30°$ and $\Phi = \Phi_0 = 60°$ (Lambert projection).

The resulting mapping equations are given by (17.36) or (17.37). The left left principal stretches are given by (17.38).

$$
\begin{bmatrix} \alpha \\ r \end{bmatrix} =
$$

$$
= \begin{bmatrix} n\Lambda \\ R\dfrac{\cos\Phi_1}{n} \left[\dfrac{\tan\left(\frac{\pi}{4} - \frac{\Phi}{2}\right)}{\tan\left(\frac{\pi}{4} - \frac{\Phi_1}{2}\right)} \right]^n \end{bmatrix} = \begin{bmatrix} n\Lambda \\ R\dfrac{\cos\Phi_2}{n} \left[\dfrac{\tan\left(\frac{\pi}{4} - \frac{\Phi}{2}\right)}{\tan\left(\frac{\pi}{4} - \frac{\Phi_2}{2}\right)} \right]^n \end{bmatrix} , \tag{17.36}
$$

$$
\begin{bmatrix} x \\ y \end{bmatrix} =
$$

$$
= R\frac{\cos\Phi_1}{n} \left[\frac{\tan\left(\frac{\pi}{4} - \frac{\Phi}{2}\right)}{\tan\left(\frac{\pi}{4} - \frac{\Phi_1}{2}\right)} \right]^n \begin{bmatrix} \cos n\Lambda \\ \sin n\Lambda \end{bmatrix} = R\frac{\cos\Phi_2}{n} \left[\frac{\tan\left(\frac{\pi}{4} - \frac{\Phi}{2}\right)}{\tan\left(\frac{\pi}{4} - \frac{\Phi_2}{2}\right)} \right]^n \begin{bmatrix} \cos n\Lambda \\ \sin n\Lambda \end{bmatrix} , \tag{17.37}
$$

$$
\Lambda_1 = \Lambda_2 =
$$

$$
= \frac{\cos\Phi_1}{\cos\Phi} \left[\frac{\tan\left(\frac{\pi}{4} - \frac{\Phi}{2}\right)}{\tan\left(\frac{\pi}{4} - \frac{\Phi_1}{2}\right)} \right]^n = \frac{\cos\Phi_2}{\cos\Phi} \left[\frac{\tan\left(\frac{\pi}{4} - \frac{\Phi}{2}\right)}{\tan\left(\frac{\pi}{4} - \frac{\Phi_2}{2}\right)} \right]^n . \tag{17.38}
$$

It is worthwhile noting that this famous map (*Lambert map*, also called *conical orthomorphic mapping*) has interesting limiting forms. First, if one of the poles is selected as a single standard parallel, the cone is a plane and a stereographic azimuthal projection is generated. If the equator or two parallels $\Phi = \Phi_1$ and $\Phi = -\Phi_1$ are chosen as the standard parallels, the cone becomes a cylinder and the Mercator projection results.

17-23 Equal area mapping (Albers projection)

The general mapping equations for this type of mappings are derived from the requirement that the product of the principal stretches equals unity, i.e.

$$\Lambda_1 \Lambda_2 = \frac{nf}{R\cos\Phi}\frac{f'}{R} = 1 \Rightarrow ff' = \frac{R^2\cos\Phi}{n} \Rightarrow \int f \mathrm{d}f = \frac{R^2}{n}\int \cos\Phi \mathrm{d}\Phi$$

$$\Downarrow \tag{17.39}$$

$$\frac{1}{2}f^2 = -\frac{R^2}{n}\sin\Phi + \frac{1}{2}c \Rightarrow f = \sqrt{-\frac{2R^2}{n}\sin\Phi + c}\ .$$

For the root to be real for all Φ, the integration constant c should fulfill the inequality $c \geq 2\frac{R^2}{n}$. The general mapping equations thus are given by (17.40) or (17.41), and the general left principal stretches are given by (17.42).

$$\begin{bmatrix} \alpha \\ r \end{bmatrix} = \begin{bmatrix} n\Lambda \\ \sqrt{-\frac{2R^2}{n}\sin\Phi + c} \end{bmatrix} , \tag{17.40}$$

$$\begin{bmatrix} x \\ y \end{bmatrix} = \sqrt{-\frac{2R^2}{n}\sin\Phi + c}\begin{bmatrix} \cos(n\Lambda) \\ \sin(n\Lambda) \end{bmatrix} , \tag{17.41}$$

$$\Lambda_1 = \frac{n\sqrt{-\frac{2R^2}{n}\sin\Phi + c}}{R\cos\Phi} , \quad \Lambda_2 = \frac{R\cos\Phi}{n\sqrt{-\frac{2R^2}{n}\sin\Phi + c}} . \tag{17.42}$$

17-231 Equidistance and conformality on the circle-of-contact, compare with Fig. 17.10

For the reason to map the standard parallel (circle-of-contact) $\Phi = \Phi_0$ equidistantly, we claim that (17.43) holds, with the consequence that – together with the cone constant $n = \sin\Phi_0$ – (17.44) is immediately obtained.

$$\Lambda_1|_{\Phi=\Phi_0} = \frac{n\sqrt{-\frac{2R^2}{n}\sin\Phi_0 + c}}{R\cos\Phi_0} = 1 , \tag{17.43}$$

$$c = R^2(2 + \cot^2\Phi_0) . \tag{17.44}$$

The mapping equations therefore are provided by (17.45) or (17.46). The left principal stretches are provided by (17.47).

$$\begin{bmatrix} \alpha \\ r \end{bmatrix} = \begin{bmatrix} n\Lambda \\ R\sqrt{-\frac{2}{n}\sin\Phi + \cot^2\Phi_0 + 2} \end{bmatrix} , \tag{17.45}$$

$$\begin{bmatrix} x \\ y \end{bmatrix} = R\sqrt{-\frac{2}{n}\sin\Phi + \cot^2\Phi_0 + 2}\begin{bmatrix} \cos(n\Lambda) \\ \sin(n\Lambda) \end{bmatrix} , \tag{17.46}$$

$$\Lambda_1 = \frac{\sqrt{2n\sin\Phi + n^2 + 1}}{\cos\Phi} , \quad \Lambda_2 = \frac{\cos\Phi}{\sqrt{-2n\sin\Phi + n^2 + 1}} . \tag{17.47}$$

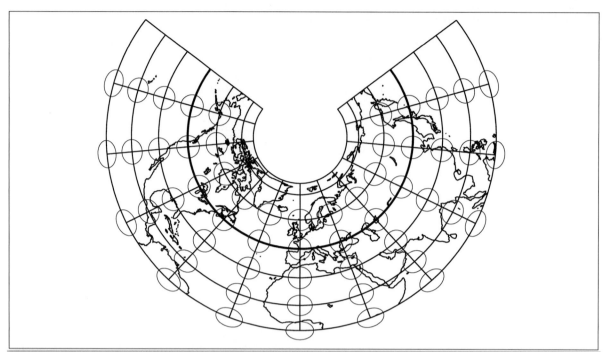

Fig. 17.10. Mapping the sphere to a cone. Polar aspect, equal area mapping, conformal on the standard parallel $\Phi = \Phi_0 = 45°$.

17-232 Equidistance and conformality on the circle-of-contact, point-like image of the North Pole, compare with Fig. 17.11

Starting from the general mapping equations, in (17.40) the postulate of a point-like image of the pole is achieved by setting $r|_{\Phi=90°} := 0$, which is equivalent to assigning $c = 2R^2 n^{-1}$. We therefore obtain after some trigonometric conversions the general mapping equations and general left principal stretches that are defined by (17.48) and (17.49). The further requirement that the parallel circle $\Phi = \Phi_1$ shall be mapped equidistantly now determines the cone constant $n = \sin \Phi_0$. From the postulate (17.50), we get the value (17.51).

$$\begin{bmatrix} \alpha \\ r \end{bmatrix} = \begin{bmatrix} n\Lambda \\ \frac{2R}{\sqrt{n}} \sin\left(\frac{\pi}{4} - \frac{\Phi}{2}\right) \end{bmatrix} , \tag{17.48}$$

$$\Lambda_1 = \frac{2\sqrt{n} \sin\left(\frac{\pi}{4} - \frac{\Phi}{2}\right)}{\cos \Phi} = \frac{\sqrt{n}}{\cos\left(\frac{\pi}{4} - \frac{\Phi}{2}\right)} ,$$

$$\Lambda_2 = \frac{\cos\left(\frac{\pi}{4} - \frac{\Phi}{2}\right)}{\sqrt{n}} , \tag{17.49}$$

$$\Lambda_1|_{\Phi=\Phi_1} = \frac{\sqrt{n}}{\cos\left(\frac{\pi}{4} - \frac{\Phi_1}{2}\right)} = 1 , \tag{17.50}$$

$$n = \cos^2\left(\frac{\pi}{4} - \frac{\Phi_1}{2}\right) . \tag{17.51}$$

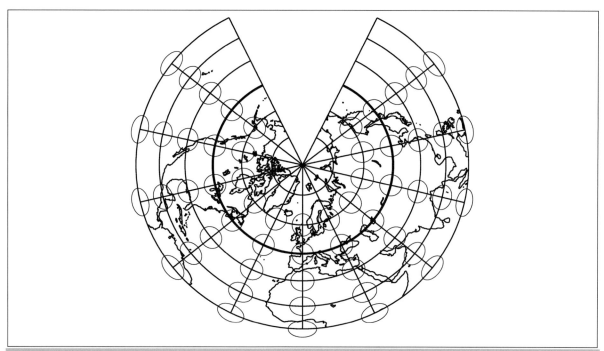

Fig. 17.11. Mapping the sphere to a cone. Polar aspect, equal area mapping, equidistant and conformal on the standard parallel $\Phi = \Phi_0 = 45°$, point-like North Pole.

The final mapping equations thus are given by (17.52), and the left principal stretches are given by (17.53). It is easily seen that for the standard parallel $\Phi = \Phi_1$ conformality and isometry is guaranteed.

$$
\begin{bmatrix} \alpha \\ r \end{bmatrix} = \begin{bmatrix} \cos^2\left(\dfrac{\pi}{4} - \dfrac{\Phi_1}{2}\right) \Lambda \\[2mm] \dfrac{2R}{\cos\left(\frac{\pi}{4} - \frac{\Phi_1}{2}\right)} \sin\left(\dfrac{\pi}{4} - \dfrac{\Phi}{2}\right) \end{bmatrix} , \quad \begin{bmatrix} x \\ y \end{bmatrix} = 2R\frac{\sin\left(\frac{\pi}{4} - \frac{\Phi}{2}\right)}{\cos\left(\frac{\pi}{4} - \frac{\Phi_1}{2}\right)} \begin{bmatrix} \cos\left(\cos^2\left(\dfrac{\pi}{4} - \dfrac{\Phi_1}{2}\right)\Lambda\right) \\[2mm] \sin\left(\cos^2\left(\dfrac{\pi}{4} - \dfrac{\Phi_1}{2}\right)\Lambda\right) \end{bmatrix} , \quad (17.52)
$$

$$
\Lambda_1 = \frac{\cos\left(\frac{\pi}{4} - \frac{\Phi_1}{2}\right)}{\cos\left(\frac{\pi}{4} - \frac{\Phi}{2}\right)} , \quad \Lambda_2 = \frac{\cos\left(\frac{\pi}{4} - \frac{\Phi}{2}\right)}{\cos\left(\frac{\pi}{4} - \frac{\Phi_1}{2}\right)} . \tag{17.53}
$$

*17-233 Equidistance and conformality on two parallels (secant cone, H. C. Albers),
    compare with Fig. 17.12*

This famous projection which was introduced by Heinrich Christian Albers (1773–1833) in 1805 has interesting limiting forms. If one of the poles is defined to be the single standard parallel, then the Lambert azimuthal equal area projection in the polar aspect (compare with Section 5-23) is generated: the cone becomes a plane. If, on the other hand, the equator is used as the single standard parallel, the cylindrical equal area projection (Lambert projection, compare with Section 10-23) is obtained. In order to derive the mapping equation, we again start from equations (17.43) and claim that for an equidistant mapping of the standard parallel $\Phi = \Phi_1$, we have

$$
\Lambda_1|_{\Phi = \Phi_1} = \frac{n\sqrt{-\frac{2R^2}{n}\sin\Phi_1 + c}}{R\cos\Phi_1} = 1 . \tag{17.54}
$$

We solve this equation in order to determine the integration constant c and obtain (17.55). Since c can also be computed from $\Lambda_1|_{\Phi=\Phi_2} := 1$, we obtain (17.56) and (17.57). Resubstituting this result into (17.56) gives the final form of c, namely (17.58).

$$c = R^2 \frac{\cos^2 \Phi_1 + 2n \sin \Phi_1}{n^2} \;, \tag{17.55}$$

$$R^2 \frac{\cos^2 \Phi_1 + 2n \sin \Phi_1}{n^2} = R^2 \frac{\cos^2 \Phi_2 + 2n \sin \Phi_2}{n^2}$$
$$\Rightarrow$$
$$\cos^2 \Phi_1 + 2n \sin \Phi_1 = \cos^2 \Phi_2 + 2n \sin \Phi_2 \;, \tag{17.56}$$

$$n = \frac{\cos^2 \Phi_1 - \cos^2 \Phi_2}{2(\sin \Phi_2 - \sin \Phi_1)} = \frac{1 - \sin^2 \Phi_1 - 1 + \sin^2 \Phi_2}{2(\sin \Phi_2 - \sin \Phi_1)} = \frac{\sin^2 \Phi_2 - \sin^2 \Phi_1}{2(\sin \Phi_2 - \sin \Phi_1)} =$$
$$= \frac{(\sin \Phi_2 - \sin \Phi_1)(\sin \Phi_2 + \sin \Phi_1)}{2(\sin \Phi_2 - \sin \Phi_1)} = \tag{17.57}$$
$$= \frac{1}{2}(\sin \Phi_1 + \sin \Phi_2) \;,$$

$$c = 4R^2 \frac{1 + \sin \Phi_1 \sin \Phi_2}{(\sin \Phi_1 + \sin \Phi_2)^2} \;. \tag{17.58}$$

As a matter of course, the cone constant and the integration constant are symmetric in Φ_1 and Φ_2. The final mapping equations are given by (17.59) or (17.60), and the final left principal stretches are provided by (17.61). Indeed, the requirements $\Lambda_1|_{\Phi=\Phi_1} = \Lambda_2|_{\Phi=\Phi_1} = \Lambda_1|_{\Phi=\Phi_2} = \Lambda_2|_{\Phi=\Phi_2} = 1$ are met and the inverse mapping equations are defined through (17.62).

$$\begin{bmatrix} \alpha \\ r \end{bmatrix} =$$
$$= \begin{bmatrix} n\Lambda \\ \frac{R}{n}\sqrt{\cos^2 \Phi_1 + 2n \sin \Phi_1 - 2n \sin \Phi} \end{bmatrix} \;, \tag{17.59}$$

$$\begin{bmatrix} x \\ y \end{bmatrix} =$$
$$= \frac{R}{n}\sqrt{\cos^2 \Phi_1 + 2n \sin \Phi_1 - 2n \sin \Phi} \begin{bmatrix} \cos(n\Lambda) \\ \sin(n\Lambda) \end{bmatrix} \;, \tag{17.60}$$

$$\Lambda_1 = \frac{\sqrt{\cos^2 \Phi_1 + 2n \sin \Phi_1 - 2n \sin \Phi}}{\cos \Phi} = \frac{\sqrt{1 + \sin \Phi_1 \sin \Phi_2 - (\sin \Phi_1 + \sin \Phi_2) \sin \Phi}}{\cos \Phi} \;,$$
$$\Lambda_2 = \Lambda_1^{-1} \;. \tag{17.61}$$

$$\Lambda = \frac{\alpha}{n} \;, \quad \Phi = \arcsin \frac{1}{2n}\left[\cos^2 \Phi_1 + 2n \sin \Phi_1 - \left(\frac{nr}{R}\right)^2\right] \;. \tag{17.62}$$

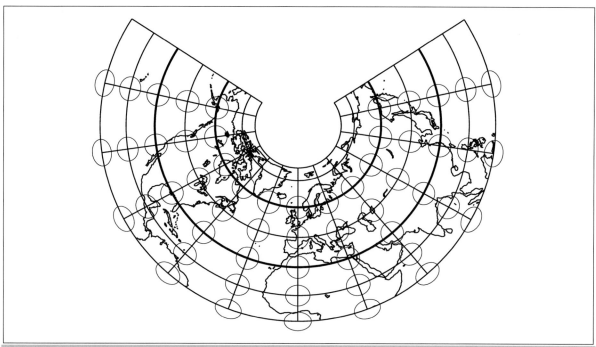

Fig. 17.12. Mapping the sphere to a cone. Polar aspect, equal area mapping, equidistant and conformal on two standard parallel $\Phi = \Phi_1 = 30°$ and $\Phi = \Phi_2 = 60°$ (Albers projection).

With these formulae, we close the discussion of the polar aspect of the mappings "sphere to cone". In the chapter that follows, let us discuss the pseudo-conic aspect.

18 "Sphere to cone": pseudo-conic projections

Mapping the sphere to a cone: pseudo-conic projections. The Stab–Werner mapping and the Bonne mapping. Tissot indicatrix.

First, let us develop the general setup of *pseudo-conic projections* from the *sphere to a cone*. Second, let us present special pseudo-conic mappings like the *Stab–Werner mapping* and the *Bonne mapping* including illustrations.

18-1 General setup and distortion measures of pseudo-conic projections

Conic projections, polyconic projections. Mapping equations, deformation tensor. Lemma of Vieta and postulate of equal area mapping.

In general, *pseudo-conic projections* are based upon the setting (18.1) if we use spherical longitude Λ and spherical co-latitude $\Delta = \pi/2 - \Phi$ and polar coordinates α and r. The next extension leaves us with the polyconic projections of type (18.2). Our analysis is based upon Lemma 18.1.

$$\alpha = \alpha(\Lambda, \Delta) = \Lambda \cos \Delta , \quad r = r(\Delta) = f(\Delta) , \tag{18.1}$$

$$\alpha(\Lambda, \Delta) = g(\Delta)\Lambda \cos \Delta , \quad r = r(\Delta) = f(\Delta) . \tag{18.2}$$

Lemma 18.1 (Equiareal mapping).

A general mapping of any surface to the plane is equiareal or area preserving if and only if

$$\det\left[\mathsf{C}_l\right]/\det\left[\mathsf{G}_l\right] = 1 . \tag{18.3}$$

C_l and G_l are the left Cauchy–Green matrix and the left metric matrix, respectively.

End of Lemma.

Proof.

$$\det\left[\mathsf{C}_l - \Lambda_\mathrm{S}^2 \mathsf{G}_l\right] = 0$$

$$\Leftrightarrow \tag{18.4}$$

$$\Lambda_\mathrm{S}^4 - \Lambda_\mathrm{S}^2 \mathrm{tr}\left[\mathsf{C}_l \mathsf{G}_l^{-1}\right] + \det\left[\mathsf{C}_l \mathsf{G}_l^{-1}\right] = 0 ,$$

$$\left(\Lambda_\mathrm{S}^2\right)^+ \left(\Lambda_\mathrm{S}^2\right)^- = 1$$

(canonical postulate of equal area mapping)

$$\Leftrightarrow \tag{18.5}$$

$$\det\left[\mathsf{C}_l\right]/\det\left[\mathsf{G}_l\right] = 1$$

(Lemma of Vieta:
the product of the solutions of a quadratic equation equals the absolute term) .

End of Proof.

From the general form of the *deformation tensor*, the metric tensor, and the postulate of equal area mapping, we derive the general structure of equal area pseudo-conic mappings of the sphere in Box 18.1. As a side remark, we use the result that *only pseudo-conic projections of type equal area exist*.

Box 18.1 (General structure of equal area pseudo-conic mappings of the sphere).

Mapping equations:

$$\alpha = g(\Delta)\Lambda \cos \Delta = h(\Delta)\Lambda , \quad r = f(\Delta) . \tag{18.6}$$

Left Cauchy–Green matrix:

$$\mathsf{C}_l = \mathsf{J}_l^{\mathsf{T}} \mathsf{G}_r \mathsf{J}_l = \begin{bmatrix} f^2 h^2 & f^2 hh'\Lambda \\ f^2 hh'\Lambda & f^2 h'^2 \Lambda^2 + f'^2 \end{bmatrix} , \tag{18.7}$$

$$\det[\mathsf{C}_l] = f^2 h^2 (f^2 h'^2 \Lambda^2 + f'^2) - f^4 h^2 h'^2 \Lambda^2 = f^4 h^2 h'^2 \Lambda^2 + f^2 h^2 f'^2 - f^4 h^2 h'^2 \Lambda^2 = f^2 f'^2 h^2 . \tag{18.8}$$

Left Jacobi matrix:

$$\mathsf{J}_l = \begin{bmatrix} D_\Lambda \alpha & D_\Delta \alpha \\ D_\Lambda r & D_\Delta r \end{bmatrix} = \begin{bmatrix} h(\Delta) & \Lambda h'(\Delta) \\ 0 & f'(\Delta) \end{bmatrix} . \tag{18.9}$$

Right metric tensor:

$$\mathsf{G}_r = \begin{bmatrix} r^2 & 0 \\ 0 & 1 \end{bmatrix} = \begin{bmatrix} f^2 & 0 \\ 0 & 1 \end{bmatrix} . \tag{18.10}$$

Left metric tensor:

$$\mathsf{G}_l = \begin{bmatrix} R^2 \sin^2 \Delta & 0 \\ 0 & R^2 \end{bmatrix} , \quad \det[\mathsf{G}_l] = R^4 \sin^2 \Delta . \tag{18.11}$$

Postulate of an equal area mapping:

$$\det[\mathsf{C}_l] = \det[\mathsf{G}_l] \Rightarrow ff'g \cos \Delta = +R^2 \sin \Delta$$

(only the $+$ sign is here correct due to the orientation constance) $\qquad$ (18.12)

$$\Leftrightarrow$$

$$g = 2R^2 \frac{\tan \Delta}{(f^2)'} = R^2 \frac{\tan \Delta}{ff'} . \tag{18.13}$$

General structure
(equal area mapping: pseudo-conic):

$$\alpha = \alpha(\Lambda, \Delta) = g(\Delta)\Lambda \cos \Delta = 2R^2 \frac{\sin \Delta}{[f^2(\Delta)]'} \Lambda = R^2 \frac{\sin \Delta}{ff'} \Lambda ,$$
$$r = r(\Delta) = f(\Delta) . \tag{18.14}$$

Proof.

$$(\Lambda_S^2)^+ =$$

$$= \frac{1}{2}\left[\operatorname{tr}\left[C_l G_l^{-1}\right] + \sqrt{\left(\operatorname{tr}\left[C_l G_l^{-1}\right]\right)^2 - 4\det\left[C_l G_l^{-1}\right]}\,\right],$$

(18.15)

$$(\Lambda_S^2)^- =$$

$$= \frac{1}{2}\left[\operatorname{tr}\left[C_l G_l^{-1}\right] - \sqrt{\left(\operatorname{tr}\left[C_l G_l^{-1}\right]\right)^2 - 4\det\left[C_l G_l^{-1}\right]}\,\right],$$

$$\det\left[C_l\right] =$$

$$= \frac{1}{4}\left[(f^2)'\right]^2 h^2 =$$

$$= \frac{1}{4}\left[(f^2)'\right]^2 \cos^2\Delta\, g^2(\Delta) =$$

(18.16)

$$= \frac{1}{4}\left[(f^2)'\right]^2 \cos^2\Delta\, 4R^4\frac{\tan^2\Delta}{\left[(f^2)'\right]^2} =$$

$$= R^4 \sin^2\Delta\,.$$

End of Proof.

Proof.

$$\operatorname{tr}\left[C_l G_l^{-1}\right] =$$

$$= c_{11}G_{11}^{-1} + c_{22}G_{22}^{-1} =$$

$$= f^2 h^2\left(R^2\sin^2\Delta\right)^{-1} + \left(f^2 h'^2\Lambda^2 + f'^2\right)\left(R^2\right)^{-1} =$$

$$= \frac{1}{R^2\sin^2\Delta}f^2 g^2\cos^2\Delta + \frac{1}{R^2}\left[f^2\left(g'\cos\Delta - g\sin\Delta\right)^2\Lambda^2 + f'^2\right] =$$

(18.17)

$$= 4R^2\frac{f^2}{\left[(f^2)'\right]^2} +$$

$$+ \frac{1}{R^2}\left(f^2\left[2R^2\frac{\cos\Delta(1+\tan^2\Delta)}{(f^2)'} - 2R^2\frac{\sin\Delta}{\left[(f^2)'\right]^2}(f^2)'' - 2R^2\frac{\sin\Delta\tan\Delta}{(f^2)'}\right]^2\Lambda^2 + f'^2\right),$$

$$\det\left[G_l\right] = R^4\sin^2\Delta\,.$$

(18.18)

End of Proof.

18-2 Special pseudo-conic projections based upon the sphere

The Stab–Werner mapping and the Bonne mapping. The mapping equations and the principal stretches. Tissot indicatrix.

We use the setup ("Ansatz")

$$r(\Delta) = f(\Delta) = a\Delta + b \,, \quad f'(\Delta) = a \,. \tag{18.19}$$

18-21 Stab–Werner mapping

In the framework of the Stab–Werner mapping, let us take advantage of the following two postulates.

Postulate.

The North Pole should be mapped to a point.

$$b = 0 : \ r(\Delta = 0) = 0 \,. \tag{18.20}$$

End of Postulate.

Postulate.

An arc on the meridian should be mapped equidistantly.

$$a = R : \ r(\Delta) = R\Delta \,. \tag{18.21}$$

End of Postulate.

$$\alpha(\Lambda, \Delta = 0) = \lim_{\Delta \longrightarrow 0} = \frac{\sin \Delta}{\Delta}\Lambda = \lim_{\Delta \longrightarrow 0} \frac{\cos \Delta}{1}\Lambda = \Lambda \,, \tag{18.22}$$

$$\alpha(\Lambda, \Delta) = \frac{\sin \Delta}{\Delta}\Lambda = \frac{\cos \Phi}{\frac{\pi}{2} - \Phi}\Lambda \,, \quad r(\Delta) = R\Delta = R\left(\frac{\pi}{2} - \Phi\right) \tag{18.23}$$

(direct mapping equations),

$$\Lambda = \frac{r}{R\sin\frac{r}{R}}\alpha \,, \quad \Phi = \frac{\pi}{2} - \frac{r}{R} \tag{18.24}$$

(inverse mapping equations).

At this point, we collect the Stab–Werner mapping equations and analyze the principal stretches. In particular, we observe that the principal stretch components are not directed along the coordinate lines $\Delta = \text{const.}/\Phi = \text{const.}$ and $\Lambda = \text{const.}$ because C_l is not a diagonal matrix, in general. We here additionally note that Johannes Werner (1514) in his work "Libellus de quatuor terrarum orbis in plano figurationibus, Nova translatio primi libri geographiae El. Ptolemai: Neuenberg (Latin)" remarks: "computed assisted by Johann Stablus". Furthermore, note that the first published map is due to Petrus Aqianus, World Map of Ingolstadt (1530). Moreover, note that the term *cardioform* is translated in the *form of a heart*.

Mapping equations (Stab–Werner):

$$x = r \cos \alpha =$$

$$= R \left(\frac{\pi}{2} - \Phi \right) \cos \left(\frac{\cos \Phi}{\frac{\pi}{2} - \Phi} \Lambda \right) , \tag{18.25}$$

$$y = r \sin \alpha =$$

$$= R \left(\frac{\pi}{2} - \Phi \right) \sin \left(\frac{\cos \Phi}{\frac{\pi}{2} - \Phi} \Lambda \right) .$$

Principal stretches (Stab–Werner):

$$\left(\Lambda_{\mathrm{S}}^2 \right)_1 = \frac{1}{2} \Bigg[2 +$$

$$+ \Lambda^2 \left(\cos \Delta - \frac{\sin \Delta}{\Delta} \right)^2 + \sqrt{4 \Lambda^2 \left(\cos \Delta - \frac{\sin \Delta}{\Delta} \right)^2 + \Lambda^4 \left(\cos \Delta - \frac{\sin \Delta}{\Delta} \right)^4} \Bigg] , \tag{18.26}$$

$$\left(\Lambda_{\mathrm{S}}^2 \right)_2 = \frac{1}{2} \Bigg[2 +$$

$$+ \Lambda^2 \left(\cos \Delta - \frac{\sin \Delta}{\Delta} \right)^2 - \sqrt{4 \Lambda^2 \left(\cos \Delta - \frac{\sin \Delta}{\Delta} \right)^2 + \Lambda^4 \left(\cos \Delta - \frac{\sin \Delta}{\Delta} \right)^4} \Bigg] .$$

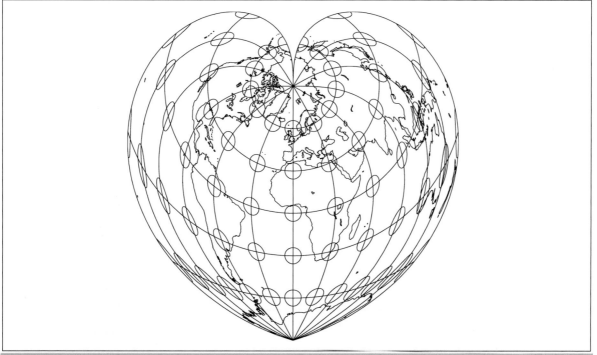

Fig. 18.1. Stab–Werner mapping, pseudo-conic projection, Tissot ellipses of distortion.

18-22 Bonne mapping

We agree upon the postulate that the line-of-contact of the cone shall be mapped equidistantly. We start the theory of the *Bonne mapping* from the following two postulates, from which follows (18.29). Compare with Fig. 18.2.

Postulate.

$$r(\Delta) = a\Delta + b \,.\tag{18.27}$$

End of Postulate.

Postulate.

$$r(\Delta_0) = R\tan\Delta_0 \,, \quad a = R \,.\tag{18.28}$$

End of Postulate.

$$b = R(\tan\Delta_0 - \Delta_0) \,.\tag{18.29}$$

Next, we summarize the mapping equations of type Bonne, the inverse mapping equations, and the principal stretches. Note that $r(\Delta = 0) = R(\tan\Delta_0 - \Delta_0)$: "Pointwise mapping of the North Pole, but not at the coordinate origin $\{x = 0, y = 0\}$"! Note that Rigobert Bonne's work can be read in "Ptolemaeus Geographia" (Francesco Berlinghieri, Florenz 1482). Furthermore, note that very often $\Phi_0 = 50°\,\mathrm{N}$ is chosen.

Mapping equations:

$$\alpha = \alpha(\Lambda, \Delta) = \frac{\sin\Delta}{\Delta - \Delta_0 + \tan\Delta_0}\Lambda = \frac{\cos\Phi}{\Phi_0 - \Phi + \cot\Phi_0}\Lambda \,,\tag{18.30}$$

$$r = r(\Delta) = R(\Delta - \Delta_0) + R\tan\Delta_0 = R(\Phi_0 - \Phi) + R\cot\Phi_0 \,.$$

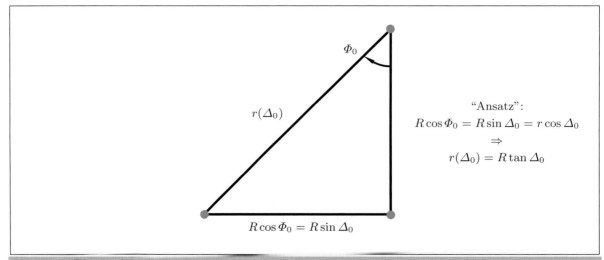

Fig. 18.2. Bonne mapping, pseudo-conic projection, parallel circle and conic center.

Direct mapping equations:

$$
\begin{bmatrix} x \\ y \end{bmatrix} = R(\Phi_0 - \Phi + \cot \Phi_0) \begin{bmatrix} \cos\left(\dfrac{\cos \Phi}{\Phi_0 - \Phi + \cot \Phi_0} \Lambda \right) \\[2ex] \sin\left(\dfrac{\cos \Phi}{\Phi_0 - \Phi + \cot \Phi_0} \Lambda \right) \end{bmatrix} . \tag{18.31}
$$

Inverse mapping equations:

$$
\Lambda = \frac{\frac{r}{R}\alpha}{\cos\left(\cot \Phi_0 + \Phi_0 - \frac{r}{R} \right)} , \tag{18.32}
$$

$$
\Phi = \cot \Phi_0 + \Phi_0 - \frac{r}{R} .
$$

Principal stretches:

$$
(\Lambda_{\mathrm{S}}^2)_1 = \frac{1}{2}\left[2 + \Lambda^2\left(\cos \Delta - \frac{\sin \Delta}{\Delta + \tan \Delta_0 - \Delta_0} \right)^2 + \right.
$$

$$
\left. + \sqrt{4\Lambda^2\left(\cos \Delta - \frac{\sin \Delta}{\Delta + \tan \Delta_0 - \Delta_0} \right)^2 + \Lambda^4\left(\cos \Delta - \frac{\sin \Delta}{\Delta + \tan \Delta_0 - \Delta_0} \right)^4} \right], \tag{18.33}
$$

$$
(\Lambda_{\mathrm{S}}^2)_2 = \frac{1}{2}\left[2 + \Lambda^2\left(\cos \Delta - \frac{\sin \Delta}{\Delta + \tan \Delta_0 - \Delta_0} \right)^2 - \right.
$$

$$
\left. - \sqrt{4\Lambda^2\left(\cos \Delta - \frac{\sin \Delta}{\Delta + \tan \Delta_0 - \Delta_0} \right)^2 + \Lambda^4\left(\cos \Delta - \frac{\sin \Delta}{\Delta + \tan \Delta_0 - \Delta_0} \right)^4} \right].
$$

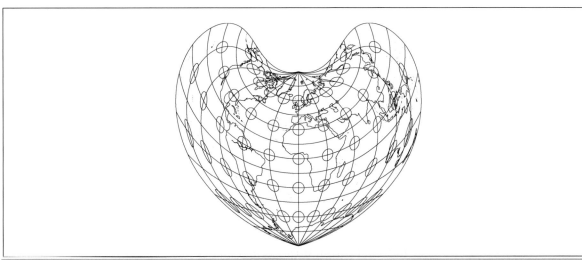

Fig. 18.3. Bonne mapping, pseudo-conic projection, Tissot ellipses of distortion.

In the 16th century and the 17th century, the pseudo-conic, the equal area, and the cordiform (heart shaped) *Stab–Werner projections* (J. Stab and J. Werner, $\sim$ 1514) was frequently used for *world maps* and some *continental maps*. The mapping equations are specified trough polar coordinates α and r or Cartesian coordinates x and y as follows. (Λ and Φ describe spherical longitude and spherical latitude.)

$$\alpha = \Lambda \frac{\cos\Phi}{\pi/2 - \Phi} \ , \quad r = R(\pi/2 - \Phi) \ , \tag{18.34}$$

$$x = R(\pi/2 - \Phi)\cos\left(\Lambda \frac{\cos\Phi}{\pi/2 - \Phi}\right) \ , \quad y = R(\pi/2 - \Phi)\sin\left(\Lambda \frac{\cos\Phi}{\pi/2 - \Phi}\right) \ . \tag{18.35}$$

(i) Prove analytically that the Stab–Werner projection is equal area and (ii) determine the numerical values of the elements of the Tissot indicatrix (Tissot ellipse: minor distortions and major distortions, and coordinates of the corresponding eigendirections in the map) for the point *Aachen, Germany* ($\Lambda = 6°06'$ E, $\Phi = 50°46'$ N).

Solution.

From the general eigenvalue problem in Lemma 1.7 or from (18.26), the minor and the major distortions are easily calculated as (18.36), and $\det D_l = \Lambda_{\min}\Lambda_{\max} = 1$. The matrix F_l of eigenvectors, fulfilling the requirements $F_l^T C_l F_l = D_l^2$ and $F_l^T G_l F_l = I_2$ ("left diagonalization"), results to (18.37). These eigenvectors refer to the base vectors of the left tangential space TM_l^2. In order to properly plot the Tissot ellipses of distortion, the eigendirections in the right tangential space TM_r^2 are needed.

$$D_l :=$$

$$:= \text{diag}\,[\Lambda_{\min}, \Lambda_{\max}] = \begin{bmatrix} 0.992\,10 & 0 \\ 0 & 1.007\,97 \end{bmatrix} \ , \tag{18.36}$$

$$F_l =$$

$$= \begin{bmatrix} 1.122\,42 & 1.113\,55 \\ -0.704\,30 & 0.709\,91 \end{bmatrix} \ . \tag{18.37}$$

With the help of the left Jacobi matrix

$$J_l = \begin{bmatrix} D_\Lambda x & D_\Phi x \\ D_\Lambda y & D_\Phi y \end{bmatrix}_{\Lambda = 6°06'\,\text{E},\ \Phi = 50°46'\,\text{N}} = \begin{bmatrix} -0.062\,10 & -0.996\,73 \\ 0.629\,42 & -0.082\,38 \end{bmatrix} \ , \tag{18.38}$$

the transformation left-to-right is easily performed as

$$F_r = J_l F_l D_l^{-1} = J_l F_l D_r = \begin{bmatrix} 0.770\,59 & 0.637\,33 \\ -0.637\,33 & 0.770\,59 \end{bmatrix} \ , \tag{18.39}$$

and

$$F_r^T C_r F_r = D_r^2 \ , \quad F_r^T G_r F_r = F_r^T F_r = I_2 \ . \tag{18.40}$$

Indeed, F_r is an orthonormal matrix.

End of Exercise.

Exercise 18.2 (Bonne projection).

Until recently, the pseudo-conic equal area *Bonne projection* (R. Bonne, 1727–1795) was frequently used for *atlas maps of continents*. The mapping equations are specified through polar coordinates α and r or Cartesian coordinates x and y as follows. (Λ and Φ describe spherical longitude and spherical latitude. $\Phi_0 = 60°$ N is the parallel circle which is mapped isometrically, i. e. without any distortion.)

$$\alpha = \Lambda \frac{\cos\Phi}{\Phi_0 - \Phi + \cot\Phi_0} , \quad r = R(\Phi_0 - \Phi) + R\cot\Phi_0 , \tag{18.41}$$

$$x = R(\Phi_0 - \Phi + \cot\Phi_0) \cos\left(\Lambda\frac{\cos\Phi}{\Phi_0 - \Phi + \cot\Phi_0}\right) ,$$
$$y = R(\Phi_0 - \Phi + \cot\Phi_0) \sin\left(\Lambda\frac{\cos\Phi}{\Phi_0 - \Phi + \cot\Phi_0}\right) . \tag{18.42}$$

(i) Prove analytically that the Bonne projection is equal area and (ii) determine the numerical values of the elements of the Tissot indicatrix (Tissot ellipse: minor distortions and major distortions, and coordinates of the corresponding eigendirections in the map) for the point *Alexandria, Egypt* ($\Lambda = 29°55'$ E, $\Phi = 31°13'$ N).

Solution.

In this case, the minor and the major distortions amount ot (18.43), and $\det\mathsf{D}_l = \Lambda_{\min}\Lambda_{\max} = 1$. The matrix F_l of eigendirections from the left diagonalization is provided by (18.44). As before, $\mathsf{F}_l^{\mathsf{T}}\mathsf{C}_l\mathsf{F}_l = \mathsf{D}_l^2$ and $\mathsf{F}_l^{\mathsf{T}}\mathsf{G}_l\mathsf{F}_l = \mathsf{I}_2$ are satisfied.

$$\mathsf{D}_l :=$$

$$:= \operatorname{diag}\left[\Lambda_{\min}, \Lambda_{\max}\right] = \begin{bmatrix} 0.931\,07 & 0 \\ 0 & 1.074\,03 \end{bmatrix} , \tag{18.43}$$

$$\mathsf{F}_l =$$

$$= \begin{bmatrix} 0.855\,79 & 0.796\,80 \\ -0.681\,43 & 0.731\,88 \end{bmatrix} . \tag{18.44}$$

The Jacobi matrix J_l reads

$$\mathsf{J}_l = \begin{bmatrix} D_\Lambda x & D_\Phi x \\ D_\Lambda y & D_\Phi y \end{bmatrix}_{\Lambda=29°55'\,\mathrm{E},\ \Phi=31°13'\,\mathrm{N},\ \Phi_0=60°\,\mathrm{N}} = \begin{bmatrix} -0.343\,70 & -0.973\,14 \\ 0.783\,11 & -0.270\,98 \end{bmatrix} . \tag{18.45}$$

Finally, the orthonormal matrix of (right) eigendirections results to

$$\mathsf{F}_\mathrm{r} = \begin{bmatrix} 0.918\,11 & 0.396\,32 \\ -0.396\,32 & 0.918\,11 \end{bmatrix} . \tag{18.46}$$

$\mathsf{F}_\mathrm{r}^{\mathsf{T}}\mathsf{C}_\mathrm{r}\mathsf{F}_\mathrm{r} = \mathsf{D}_\mathrm{r}^2$ and $\mathsf{F}_\mathrm{r}^{\mathsf{T}}\mathsf{G}_\mathrm{r}\mathsf{F}_\mathrm{r} = \mathsf{\Gamma}_\mathrm{r}^{\mathsf{T}}\mathsf{F}_\mathrm{r} = \mathsf{I}_2$, which enables us to orientate the Tissot ellipse in the map.

End of Exercise.

The above two exercises close the discussion of the mappings "sphere to cone". In the next chapter, let us study the mappings "ellipsoid-of-revolution to cone".

19 "Ellipsoid-of-revolution to cone": polar aspect

Mapping the ellipsoid-of-revolution $\mathbb{E}^2_{A_1,A_2}$ to a cone: polar aspect. Lambert conformal conic mapping and Albers equal area conic mapping.

Section 19-1, Section 19-2.

First, in Section 19-1, we review the general equations of a *conic mapping to the ellipsoid-of-revolution*, the polar aspect only. Second, in Section 19-2, we treat a special set of conic mappings, namely three types and special aspects. Indeed, the detailed computations are rather elaborate.

19-1 General mapping equations of the ellipsoid-of-revolution to the cone

Deformation tensor of first order, the meridian radius, the radius of curvature in the prime vertical, the principal stretches.

The first postulate fixes the surface normal ellipsoidal coordinate as follows. The opening angle half, representing the latitude Φ_0 of the cone, agrees to the latitude of the circle-of-contact. For practical reasons, we use the polar distance $\Delta_0 = \pi/2 - \Phi_0$. We refer to Fig. 19.1.

Line-of-contact:

$$\Delta_0 = \pi/2 - \Phi_0 \,. \tag{19.1}$$

For the deformation tensor of first order, we specialize $C_l = J_l^T G_r J_l$. In detail, we note

$$\begin{bmatrix} \alpha \\ r \end{bmatrix} = \begin{bmatrix} n\Lambda \\ f(\Delta) \end{bmatrix} \,, \tag{19.2}$$

$$J_l = \begin{bmatrix} D_\Lambda \alpha & D_\Delta \alpha \\ D_\Lambda r & D_\Delta r \end{bmatrix} = \begin{bmatrix} n & 0 \\ 0 & f'(\Delta) \end{bmatrix} \,, \quad G_r = \begin{bmatrix} r^2 & 0 \\ 0 & 1 \end{bmatrix} = \begin{bmatrix} f^2(\Delta) & 0 \\ 0 & 1 \end{bmatrix} \,, \tag{19.3}$$

$$c_{11} = n^2 f^2(\Delta) \,, \quad c_{12} = 0 \,, \quad c_{21} = 0 \,, \quad c_{22} = f'^2(\Delta) \,, \tag{19.4}$$

$$M = \frac{1}{\kappa_1} = \frac{A_1(1 - E^2)}{(1 - E^2 \sin^2 \Phi)^{3/2}} \quad \text{(meridian radius)} \,,$$

$$N = \frac{1}{\kappa_2} = \frac{A_1}{(1 - E^2 \sin^2 \Phi)^{1/2}} \quad \text{(radius of curvature in the prime vertical)} \,, \tag{19.5}$$

$$G_l = \begin{bmatrix} N^2 \cos^2 \Phi & 0 \\ 0 & M^2 \end{bmatrix} = \begin{bmatrix} N^2 \sin^2 \Delta & 0 \\ 0 & M^2 \end{bmatrix} \,, \tag{19.6}$$

and finally, we note the principal stretches

$$\Lambda_1 = \sqrt{c_{11}/G_{11}} = \frac{nf(\Delta)}{N \sin \Delta} \,, \quad \Lambda_2 = \sqrt{c_{22}/G_{22}} = \frac{f'(\Delta)}{M} \,. \tag{19.7}$$

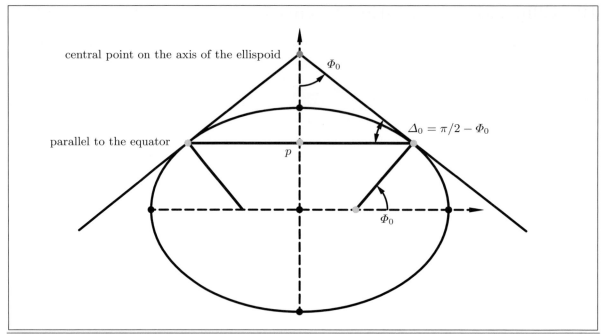

Fig. 19.1. Mapping the ellipsoid-of-revolution to the cone, polar aspect, line-of-contact.

19-2 Special conic projections based upon the ellipsoid-of-revolution

Normal mappings of type *equidistant*, *conformal*, and *equal area*. The mapping equations and the principal stretches. Lambert mapping and Albers mapping.

In this section, we present special normal mappings of type *equidistant*, *conformal*, and *equal area* as second postulates.

19-21 Special conic projections of type equidistant on the set of parallel circles

Let us transfer the postulate of an equidistant mapping on the set of parallel circles. As it is shown in (19.9), we get a typical *elliptic integral* of the second kind.

$$\Lambda_2 = 1 \quad \text{and} \quad \Lambda_2 = \frac{f'(\Delta)}{M}$$

$$\Rightarrow$$

$$f'(\Delta) = M \tag{19.8}$$

$$\Leftrightarrow$$

$$f'(\Delta) = \frac{A(1 - E^2)}{(1 - E^2 \cos^2 \Delta)^{3/2}} \, ,$$

$$f(\Delta) = A(1 - E^2) \int (1 - E^2 \cos^2 \Delta)^{-3/2} \, \mathrm{d}\Delta$$

$$= A(1 - E^2) \int_0^\Delta (1 - E^2 \cos^2 \Delta^*)^{-3/2} \, \mathrm{d}\Delta^* \, . \tag{19.9}$$

19-22 Special conic projections of type conformal

Here, we depart from the postulate of conformality. After integration-by-parts and after application of the addition theorem, we are finally led to (19.15).

$$\Lambda_1 = \Lambda_2 \Rightarrow \frac{nf(\Delta)}{N\sin\Delta} = \frac{f'(\Delta)}{M} \Rightarrow \frac{f'}{f} = \frac{nM}{N\sin\Delta}$$

$$\Leftrightarrow$$ (19.10)

$$\int \frac{\mathrm{d}f}{f} = n \int \frac{(1-E^2)\mathrm{d}\Delta}{(1-E^2\cos^2\Delta)\sin\Delta} \Rightarrow \ln f = n(1-E^2) \int \frac{\mathrm{d}\Delta}{(1-E^2\cos^2\Delta)\sin\Delta} \; .$$

Here, let us substitute $u := E\cos\Delta$:

$$\sin^2\Delta = \frac{E^2-u^2}{E^2} \; , \quad \frac{\mathrm{d}u}{\mathrm{d}\Delta} = E(-\sin\Delta) \; , \quad \ln f = -n(1-E^2)E \int \frac{\mathrm{d}u}{(1-u^2)(E^2-u^2)} \; . \quad (19.11)$$

By integration-by-parts, we find in detail:

$$\int \frac{\mathrm{d}u}{(1-u^2)(E^2-u^2)} = \int \frac{A\mathrm{d}u}{1-u^2} + \int \frac{B\mathrm{d}u}{E^2-u^2} \; , \quad (19.12)$$

$$A(E^2-u^2) + B(1-u^2) = 1 \; , \quad AE^2 + B - u^2(A+B) = 1 \; ,$$

$$AE^2 + B = 1 \; , \quad A+B = 0$$

$$\Rightarrow$$

$$A = -\frac{1}{1-E^2} \; , \quad B = \frac{1}{1-E^2}$$

$$\Rightarrow$$ (19.13)

$$\int \frac{\mathrm{d}u}{(1-u^2)(E^2-u^2)} = -\frac{1}{1-E^2}\left[\int \frac{\mathrm{d}u}{1-u^2} - \int \frac{\mathrm{d}u}{E^2-u^2}\right] =$$

$$= -\frac{1}{1-E^2}\left[\frac{1}{2}\ln\frac{1+u}{1-u} - \frac{1}{2E}\ln\frac{E+u}{E-u}\right] =$$

$$= -\frac{1}{2(1-E^2)}\left[\ln\frac{1+E\cos\Delta}{1-E\cos\Delta} + \frac{1}{E}\ln\frac{1-\cos\Delta}{1+\cos\Delta}\right] \; .$$

We take advantage of the addition theorem:

$$\int \frac{\mathrm{d}\Delta}{(1-E^2\cos^2\Delta)\sin\Delta} = -\frac{1}{2(1-E^2)}\left[\ln\frac{1+E\cos\Delta}{1-E\cos\Delta} + \frac{1}{E}\ln\tan^2\frac{\Delta}{2}\right] + \ln c''$$

$$\Rightarrow$$ (19.14)

$$\ln f = \frac{nE}{2}\ln\left[\frac{1+E\cos\Delta}{1-E\cos\Delta}\tan^{2/E}\frac{\Delta}{2}\right] + \ln c' = \ln\left[\left(\frac{1+E\cos\Delta}{1-E\cos\Delta}\right)^{En/2}\tan^n\frac{\Delta}{2}c\right]$$

$$\Rightarrow$$

$$f(\Lambda) := c\left[\left(\frac{1+E\cos\Delta}{1-E\cos\Delta}\right)^{E/2}\tan\frac{\Delta}{2}\right]^n \; . \quad (19.15)$$

Let us here also summarize the general form of the conformal mapping equations as well as the principal stretches.

$$
\begin{bmatrix} \alpha \\ r \end{bmatrix} = \begin{bmatrix} n\Lambda \\ c \left[\left(\frac{1+E\cos\Delta}{1-E\cos\Delta} \right)^{E/2} \tan\frac{\Delta}{2} \right]^n \end{bmatrix} , \tag{19.16}
$$

$$
\begin{bmatrix} \alpha \\ r \end{bmatrix} = \begin{bmatrix} n\Lambda \\ c \left[\left(\frac{1+E\sin\Phi}{1-E\sin\Phi} \right)^{E/2} \tan\left(\frac{\pi}{4}-\frac{\Phi}{2}\right) \right]^n \end{bmatrix} , \tag{19.17}
$$

$$
\Lambda_1 = \Lambda_2 = \frac{cn}{N\sin\Delta} \left[\left(\frac{1+E\sin\Phi}{1-E\sin\Phi} \right)^{E/2} \tan\left(\frac{\pi}{4}-\frac{\Phi}{2}\right) \right]^n . \tag{19.18}
$$

We here distinguish between two cases.

$$\Delta = 0 : r = f(0)$$

(the central point is mapped to a point)

and

$$\Delta = \pi/2 : r = f(\pi/2) = c$$

(the parallel circle-of-reference $\Delta = \pi/2$ or $\Phi = 0$ is mapped to a circle of radius c).

There are two variants of conformal mappings.

19-221 Conformal mapping: the variant of type equidistant on the parallel circle-of-reference

We first consider the variant of type *equidistant on the parallel circle-of-reference*. In this context, let us fix the projection constant n by

$$
n := \sin\Phi_0 = \cos\Delta_0 . \tag{19.19}
$$

The radius of the parallel circle p is computed as follows.

Input:

$$
\frac{X^2 + Y^2}{A_1^2} + \frac{Z^2}{A_2^2} = 1 \tag{19.20}
$$

(equation of the ellipsoid-of-revolution) .

Output:

$$
p^2 := X^2 + Y^2 , \quad A_2^2 = A_1^2(1 - E^2) , \quad Z = \frac{(1-E^2)A_1\sin\Phi}{\sqrt{1 - E^2\sin^2\Phi}} \tag{19.21}
$$

$$\Rightarrow$$

$$
p = \frac{A_1\cos\Phi}{\sqrt{1 - E^2\sin^2\Phi}} . \tag{19.22}
$$

We use the postulate of an equidistant mapping on the parallel circle-of-reference Φ_0.

$$\frac{A_1 \cos \Phi_0}{\sqrt{1 - E^2 \sin^2 \Phi_0}} \Lambda = f(\Phi_0)\alpha$$

$$\Rightarrow$$

$$\frac{A_1 \sin \Delta_0}{\sqrt{1 - E^2 \cos^2 \Delta_0}} \Lambda = c \left[\left(\frac{1 + E \cos \Delta_0}{1 - E \cos \Delta_0} \right)^{E/2} \tan \frac{\Delta_0}{2} \right]^n \cos \Delta_0 \Lambda \qquad (19.23)$$

$$\Rightarrow$$

$$c = \frac{A_1 \tan \Delta_0}{\sqrt{1 - E^2 \cos^2 \Delta_0}} \left[\left(\frac{1 + E \cos \Delta_0}{1 - E \cos \Delta_0} \right)^{E/2} \tan \frac{\Delta_0}{2} \right]^{-n}$$

$$\Rightarrow$$

$$\begin{bmatrix} \alpha \\ r \end{bmatrix} = \begin{bmatrix} n\Lambda \\ f(\Delta) \end{bmatrix},$$

$$f(\Delta) = \frac{A_1 \tan \Delta_0}{\sqrt{1 - E^2 \cos^2 \Delta_0}} \left[\frac{\tan \frac{\Delta}{2}}{\tan \frac{\Delta_0}{2}} \left(\frac{1 - E \cos \Delta_0}{1 - E \cos \Delta} \frac{1 + E \cos \Delta}{1 + E \cos \Delta_0} \right)^{E/2} \right]^n \qquad (19.24)$$

$$= \frac{A_1}{\tan \Phi_0 \sqrt{1 - E^2 \sin^2 \Phi_0}} \left[\frac{\tan \left(\frac{\pi}{4} - \frac{\Phi}{2} \right)}{\tan \left(\frac{\pi}{4} - \frac{\Phi_0}{2} \right)} \left(\frac{1 + E \sin \Phi}{1 + E \sin \Phi_0} \frac{\sqrt{1 - E^2 \sin^2 \Phi_0}}{\sqrt{1 - E^2 \sin^2 \Phi}} \right)^{E} \right]^n .$$

19-222 Conformal mapping: the variant of type equidistant on two parallel circles (Lambert conformal mapping)

We then consider the variant of type *equidistant on two parallel circles*. In this context, the projection constant n is determined byt he postulate of an equidistant mapping on two parallel circles fixed by $\Delta_1 = \frac{\pi}{2} - \Phi_1$ and $\Delta_2 = \frac{\pi}{2} - \Phi_2$.

$$\Lambda_1(\Delta_1) = \Lambda_2(\Delta_2) = 1 \Rightarrow \begin{cases} \dfrac{cn}{N_1 \sin \Delta_1} \left[\left(\dfrac{1 + E \sin \Phi_1}{1 - E \sin \Phi_1} \right)^{E/2} \tan \left(\dfrac{\pi}{4} - \dfrac{\Phi_1}{2} \right) \right]^n \\ = \\ \dfrac{cn}{N_2 \sin \Delta_2} \left[\left(\dfrac{1 + E \sin \Phi_2}{1 - E \sin \Phi_2} \right)^{E/2} \tan \left(\dfrac{\pi}{4} - \dfrac{\Phi_2}{2} \right) \right]^n , \end{cases} \qquad (19.25)$$

$$\frac{\sin \Delta_2 (1 - E^2 \cos^2 \Delta_1)^{1/2}}{\sin \Delta_1 (1 - E^2 \cos^2 \Delta_2)^{1/2}} = \left[\frac{\left(\frac{1 + E \cos \Delta_2}{1 - E \cos \Delta_2} \right)^{E/2} \tan \frac{\Delta_2}{2}}{\left(\frac{1 + E \cos \Delta_1}{1 - E \cos \Delta_1} \right)^{E/2} \tan \frac{\Delta_1}{2}} \right]^n \qquad (19.26)$$

$$\Rightarrow$$

$$n = \frac{1}{2} \frac{\ln[(1 - E^2 \cos^2 \Delta_1) \sin^2 \Delta_2] - \ln[(1 - E^2 \cos^2 \Delta_2) \sin^2 \Delta_1]}{\ln \frac{\tan \frac{\Delta_2}{2}}{\tan \frac{\Delta_1}{2}} + \frac{E}{2} \ln \frac{(1 + E \cos \Delta_2)(1 - E \cos \Delta_1)}{(1 - E \cos \Delta_2)(1 + E \cos \Delta_1)}} . \qquad (19.27)$$

The general form of the conformal, conic mapping takes the special form (19.28) subject to the equidistant mapping of the parallel circle (19.29), a formula from which we derive the constant c and finally the radial component r according to (19.30) and (19.31).

$$r = c \left[\left(\frac{1 + E \cos \Delta}{1 - E \cos \Delta} \right)^{E/2} \tan \frac{\Delta}{2} \right]^n , \qquad (19.28)$$

$$\frac{A_1 \sin \Delta_1}{\sqrt{1 - E^2 \cos^2 \Delta_1}} = cn \left[\left(\frac{1 + E \cos \Delta_1}{1 - E \cos \Delta_1} \right)^{E/2} \tan \frac{\Delta_1}{2} \right]^n , \qquad (19.29)$$

$$c = \frac{A_1 \sin \Delta_1}{n\sqrt{1 - E^2 \cos^2 \Delta_1} \left[\left(\frac{1+E \cos \Delta_1}{1-E \cos \Delta_1} \right)^{E/2} \tan \frac{\Delta_1}{2} \right]^n} \qquad (19.30)$$

$$\Rightarrow$$

$$r = \frac{A_1 \sin \Delta_1}{n\sqrt{1 - E^2 \cos^2 \Delta_1}} \left[\frac{\left(\frac{1+E \cos \Delta}{1-E \cos \Delta} \right)^{E/2} \tan \frac{\Delta}{2}}{\left(\frac{1+E \cos \Delta_1}{1-E \cos \Delta_1} \right)^{E/2} \tan \frac{\Delta_1}{2}} \right]^n . \qquad (19.31)$$

19-23 Special conic projections of type equal area

We here depart from the postulate of equal area. Fixing the integration constant, we finally arrive at the general form of the mapping equations, namely (19.35).

$$\Lambda_1 \Lambda_2 = 1 \Rightarrow \frac{nf(\Delta)}{N \sin \Delta} \frac{f'(\Delta)}{M} = 1$$

$$\Rightarrow$$

$$f df = \frac{MN \sin \Delta}{n} , \quad M := \frac{A_1(1 - E^2)}{(1 - E^2 \cos^2 \Delta)^{3/2}} , \quad N := \frac{A_1}{(1 - E^2 \cos^2 \Delta)^{1/2}} \qquad (19.32)$$

$$\Rightarrow$$

$$\int f df = \int \frac{A_1^2(1 - E^2) \sin \Delta}{(1 - E^2 \cos^2 \Delta)^2 n} d\Delta \Rightarrow \frac{1}{2} f^2 = \frac{A_1^2(1 - E^2)}{n} \int \frac{\sin \Delta}{(1 - E^2 \cos^2 \Delta)^2} d\Delta .$$

Here, let us substitute $u := E \cos \Delta$:

$$-\int \frac{\sin \Delta}{(1 - E^2 \cos^2 \Delta)^2} d\Delta = \int \frac{1}{E(1 - u^2)^2} du =$$

$$= \frac{u}{2E(1 - u^2)} + \frac{1}{4E} \ln \frac{1 + u}{1 - u} + c' = \frac{\cos \Delta}{2(1 - E^2 \cos^2 \Delta)} + \frac{1}{4E} \ln \frac{1 + E \cos \Delta}{1 - E \cos \Delta} + c'$$

$$\Rightarrow$$

$$\frac{1}{2} f^2 = \frac{1}{2} c^2 - \frac{A_1^2(1 - E^2)}{2n} \left(\frac{\cos \Delta}{1 - E^2 \cos^2 \Delta} + \frac{1}{2E} \ln \frac{1 + E \cos \Delta}{1 - E \cos \Delta} \right) \qquad (19.33)$$

$$\Rightarrow$$

$$f = \left[c^2 - \frac{A_1^2(1 - E^2)}{n} \left(\frac{\cos \Delta}{1 - E^2 \cos^2 \Delta} + \frac{1}{2E} \ln \frac{1 + E \cos \Delta}{1 - E \cos \Delta} \right) \right]^{1/2} .$$

We fix the integration constant:

$$\Delta = \frac{\pi}{2} :$$

$$r = f(\pi/2) = c .$$

(19.34)

We summarize the general form of the mapping equations:

$$\begin{bmatrix} \alpha \\ r \end{bmatrix} = \begin{bmatrix} n\Lambda \\ \left[c^2 - \dfrac{A_1^2(1-E^2)}{n} \left(\dfrac{\cos \Delta}{1 - E^2 \cos^2 \Delta} + \dfrac{1}{2E} \ln \dfrac{1 + E \cos \Delta}{1 - E \cos \Delta} \right) \right]^{1/2} \end{bmatrix} .$$

(19.35)

19-231 Equiareal mapping: the variant of type equidistant and conformal on the reference circle

The projection constant n is fixed by the postulate of an equidistant and conformal mapping on the reference circle Φ_0.

Condition on Φ_0:

$$\frac{A_1 \sin \Delta_0 \Lambda}{(1 - E^2 \cos^2 \Delta_0)^{1/2}} = f(\Delta_0) n\Lambda ,$$

$$n = \cos \Delta_0 ,$$

$$\frac{A_1^2 \sin^2 \Delta_0}{1 - E^2 \cos^2 \Delta_0} = \left[c^2 - \frac{A_1^2(1-E^2)}{n} \left(\frac{\cos \Delta_0}{1 - E^2 \cos^2 \Delta_0} + \frac{1}{2E} \ln \frac{1 + E \cos \Delta_0}{1 - E \cos \Delta_0} \right) \right] n^2 \qquad (19.36)$$

$$\Rightarrow$$

$$c^2 = \frac{A_1^2(1 + \tan^2 \Delta_0 - E^2)}{1 - E^2 \cos^2 \Delta_0} + \frac{A_1^2(1 - E^2)}{2E \cos \Delta_0} \ln \frac{1 + E \cos \Delta_0}{1 - E \cos \Delta_0} ,$$

$$\frac{1}{\cos^2 \Delta_0} = 1 + \tan^2 \Delta_0$$

$$\Rightarrow$$

$$c^2 = A_1^2 \left[\frac{1}{\cos^2 \Delta_0} + \frac{1 - E^2}{2E \cos \Delta_0} \ln \frac{1 + E \cos \Delta_0}{1 - E \cos \Delta_0} \right]$$

(19.37)

$$\Rightarrow$$

$$r = f(\Delta) =$$

$$= A_1 \left[\frac{1}{\cos^2 \Delta_0} - \frac{(1 - E^2) \cos \Delta}{\cos \Delta_0 (1 - E^2 \cos^2 \Delta)} + \frac{1 - E^2}{2E \cos \Delta_0} \ln \left(\frac{1 + E \cos \Delta_0}{1 - E \cos \Delta_0} \frac{1 - E \cos \Delta}{1 + E \cos \Delta} \right) \right]^{1/2} .$$

We summarize the general form of the mapping equations:

$$
\begin{bmatrix} \alpha \\ r \end{bmatrix} =
$$

$$
= \begin{bmatrix} \cos \Delta_0 \Lambda \\ A_1 \left[\dfrac{1}{\cos^2 \Delta_0} - \dfrac{(1-E^2)\cos \Delta}{\cos \Delta_0 (1-E^2 \cos^2 \Delta)} + \dfrac{1-E^2}{2E \cos \Delta_0} \ln \left(\dfrac{1+E\cos \Delta_0}{1-E\cos \Delta_0} \dfrac{1-E\cos \Delta}{1+E\cos \Delta} \right) \right]^{1/2} \end{bmatrix}, \tag{19.38}
$$

$$
\begin{bmatrix} \alpha \\ r \end{bmatrix} =
$$

$$
= \begin{bmatrix} \sin \Phi_0 \Lambda \\ A_1 \left[\dfrac{1}{\sin^2 \Phi_0} - \dfrac{(1-E^2)\sin \Phi}{\sin \Phi_0 (1-E^2 \sin^2 \Phi)} + \dfrac{1-E^2}{2E \sin \Phi_0} \ln \left(\dfrac{1+E\sin \Phi_0}{1-E\sin \Phi_0} \dfrac{1-E\sin \Phi}{1+E\sin \Phi} \right) \right]^{1/2} \end{bmatrix}. \tag{19.39}
$$

19-232 Equiareal mapping: the variant of a pointwise mapping of the central point, equidistant and conformal on the parallel circle

We here apply two postulates. First, we map the central point pointwise ($\Delta = 0$: $r = f(0) = 0$). Second, let us apply the equidistant mapping on the parallel circle ($\Delta_1 = \frac{\pi}{2} - \Phi_1$).

The first postulate:

$$
\Delta = 0 : r = f(0) = 0
$$

$$
\Rightarrow
$$

$$
f(0) = c^2 - \frac{A_1^2 (1-E^2)}{n} \left(\frac{1}{1-E^2} + \frac{1}{2E} \ln \frac{1+E}{1-E} \right) = 0
$$

$$
\Rightarrow
$$

$$
c^2 = \frac{A_1^2}{n} \left(1 + \frac{1-E^2}{2E} \ln \frac{1+E}{1-E} \right) \Rightarrow c = \frac{A_1}{\sqrt{n}} \left(1 + \frac{1-E^2}{2E} \ln \frac{1+E}{1-E} \right)^{1/2} \tag{19.40}
$$

$$
\Rightarrow
$$

$$
f(\Delta) = \frac{A_1}{\sqrt{n}} \left[1 - \frac{(1-E^2)\cos \Delta}{1-E^2 \cos^2 \Delta} + \frac{1-E^2}{2E} \left(\ln \frac{1+E}{1-E} - \ln \frac{1+E\cos \Delta}{1-E\cos \Delta} \right) \right]^{1/2}.
$$

A first form of the mapping equations is given by

$$
\begin{bmatrix} \alpha \\ r \end{bmatrix} = \begin{bmatrix} n\Lambda \\ \dfrac{A_1}{\sqrt{n}} \left[1 - \dfrac{(1-E^2)\sin \Phi}{1-E^2 \sin^2 \Phi} + \dfrac{1-E^2}{2E} \left(\ln \dfrac{1+E}{1-E} - \ln \dfrac{1+E\sin \Phi}{1-E\sin \Phi} \right) \right]^{1/2} \end{bmatrix}. \tag{19.41}
$$

The second postulate:

$$\Delta_1 = \frac{\pi}{2} - \Phi_1 : \frac{A_1 \sin \Delta_1}{(1 - E^2 \cos^2 \Delta_1)^{1/2}} = f(\Delta_1) n$$

$$\Rightarrow$$

$$\frac{\sin \Delta_1}{(1 - E^2 \cos^2 \Delta_1)^{1/2}} = \tag{19.42}$$

$$= \sqrt{n} \left[1 - \frac{(1 - E^2) \cos \Delta_1}{1 - E^2 \cos^2 \Delta_1} + \frac{1 - E^2}{2E} \left(\ln \frac{1 + E}{1 - E} - \ln \frac{1 + E \cos \Delta_1}{1 - E \cos \Delta_1} \right) \right]^{1/2}$$

$$\Rightarrow$$

$$n = \frac{1}{1 - E^2 \cos^2 \Delta_1} \frac{\sin^2 \Delta_1}{1 - \frac{(1-E^2) \cos \Delta_1}{1 - E^2 \cos^2 \Delta_1} + \frac{1-E^2}{2E} \left(\ln \frac{1+E}{1-E} - \ln \frac{1+E \cos \Delta_1}{1-E \cos \Delta_1} \right)} . \tag{19.43}$$

19-233 Equiareal mapping: the variant of an equidistant and conformal mapping on two parallel circles (Albers equal area conic mapping)

In contrast, we here use the two postulates of equidistant mapping on two parallel circles Φ_1 and Φ_2. We finally arrive at the relations (19.46)–(19.49).

$$\frac{A_1 \sin \Delta_i \Lambda}{(1 - E^2 \cos^2 \Delta_i)^{1/2}} = f(\Delta_i) n \Lambda \; \forall i \in \{1, 2\} , \tag{19.44}$$

$$f(\Delta_i) = \left[c^2 - \frac{A_1^2 (1 - E^2)}{n} \left(\frac{\cos \Delta_i}{1 - E^2 \cos^2 \Delta_i} + \frac{1}{2E} \ln \frac{1 + E \cos \Delta_i}{1 - E \cos \Delta_i} \right) \right]^{1/2}$$

$$\Rightarrow$$

$$\frac{A_1^2 \sin^2 \Delta_i}{n^2 (1 - E^2 \cos^2 \Delta_i)} = c^2 - \frac{A_1^2 (1 - E^2)}{n} \left(\frac{\cos \Delta_i}{1 - E^2 \cos^2 \Delta_i} + \frac{1}{2E} \ln \frac{1 + E \cos \Delta_i}{1 - E \cos \Delta_i} \right) \tag{19.45}$$

$$\Rightarrow$$

$$c = \frac{A_1}{\sqrt{n}} \left[\frac{\sin^2 \Delta_i}{n(1 - E^2 \cos^2 \Delta_i)} + (1 - E^2) \left(\frac{\cos \Delta_i}{1 - E^2 \cos^2 \Delta_i} + \frac{1}{2E} \ln \frac{1 + E \cos \Delta_i}{1 - E \cos \Delta_i} \right) \right]^{1/2} ,$$

$$c(\Delta_1) = c(\Delta_2) .$$

Let us substitute the two functions $h(\Delta_i)$ and $g(\Delta_i)$:

$$h(\Delta_i) = h_i := \frac{\sin^2 \Delta_i}{1 - E^2 \cos^2 \Delta_i} ,$$

$$g(\Delta_i) = g_i := (1 - E^2) \left(\frac{\cos \Delta_i}{1 - E^2 \cos^2 \Delta_i} + \frac{1}{2E} \ln \frac{1 + E \cos \Delta_i}{1 - E \cos \Delta_i} \right) . \tag{19.46}$$

For c, we then arrive at

$$c = \frac{A_1}{\sqrt{n}} \left[\frac{h_i}{n} + g_i \right]^{1/2} .$$

(19.47)

For n, we then arrive at

$$\frac{A_1}{\sqrt{n}} \left[\frac{h_1}{n} + g_1 \right]^{1/2} = \frac{A_1}{\sqrt{n}} \left[\frac{h_2}{n} + g_2 \right]^{1/2}$$

$$\Rightarrow$$

(19.48)

$$n = \frac{h_1 - h_2}{g_2 - g_1}$$

$$\Rightarrow$$

$$n = \frac{\frac{\sin^2 \Delta_1}{1 - E^2 \cos^2 \Delta_1} - \frac{\sin^2 \Delta_2}{1 - E^2 \cos^2 \Delta_2}}{(1 - E^2) \left[\frac{\cos \Delta_2}{1 - E^2 \cos^2 \Delta_2} - \frac{\cos \Delta_1}{1 - E^2 \cos^2 \Delta_1} + \frac{1}{2E} \ln \left(\frac{1 + E \cos \Delta_2}{1 - E \cos \Delta_2} \frac{1 - E \cos \Delta_1}{1 + E \cos \Delta_1} \right) \right]} .$$

(19.49)

Important!

In this section, we review mappings of the ellipsoid-of-revolution onto the circular cone. They range from equidistant mappings on the set of parallel circles (they lead to typical elliptic integrals of the second kind) to conformal mappings (summarized by (19.16)–(19.18), of type equidistant on one circle-of-reference and of type equidistant on two parallel circles: the celebrated *Lambert conformal conic mapping*), and finally to the equal area mappings of type equidistant and conformal on the reference circle as given by (19.38) and (19.39), of type of a pointwise mapping of the central point, equidistant and conformal on the parallel circle, and of type of an equidistant and conformal mapping on two parallel circles (the celebrated *Albers equal area conic mapping*). The Lambert conformal conic mapping and the Albers conformal conic mapping were, of course, developed on the sphere instead of the ellipsoid-of-revolution.

With this summary, we close this chapter. In the chapter that follows, let us have a more detailed look at geodesics and geodetic mappings.

20 Geodetic mapping

Geodesics, geodetic mapping. Riemann, Soldner, and Fermi coordinates on the ellipsoid-of-revolution, initial values, boundary values. Initial value problems versus boundary value problems.

A *global length preserving mapping* of a geodetic reference surface such as the *sphere* or such as the *ellipsoid-of-revolution* (spheroid) onto the *plane* (the chart) does *not* exist. Thus, as a compromise, *equidistant mappings* of certain coordinate lines like the equator or the central meridian of a UTM/Gauss–Krueger strip system have been proposed. Of focal interest are *geodetic mappings*: a mapping of a surface (two-dimensional Riemann manifold) is called *geodetic* if *geodesics* on the given surface (in particular, shortest geodesics like the "great circles" on the sphere) are mapped onto straight lines in the plane (the chart). In the plane (the chart), straight lines are geodesics, of course. According to a fundamental lemma of E. Beltrami (1866), a geodetic mapping of a surface exists if and only if the surface is characterized by *constant Gaussian curvature*. Thus, a geodetic mapping of the sphere *does* exist, for example, (the *gnomonic projection*). Compare with Fig. 20.1.

Important!

> E. Beltrami (1866): a geodetic mapping of a surface exists if and only if the surface is characterized by constant Gaussian curvature.

Unfortunately, the ellipsoid-of-revolution (spheroid) is not of *constant Gaussian curvature*; to the contrary, its Gaussian curvature depends on ellipsoidal latitude. In this situation, B. Riemann (1851) has proposed to use instead a geodetic mapping with respect to one central point only: with respect to one particular point P of contact, a tangential plane $T_P\mathbb{M}^2$ of the surface (two-dimensional Riemann manifold $\mathbb{M}^2$) is chosen to map P-passing geodesics equidistantly onto the tangential plane $T_P\mathbb{M}^2$. In the tangential plane $T_P\mathbb{M}^2$ at point P, either polar coordinates $\{\alpha, r\}$ or normal coordinates $\{x, y\} = \{r\cos\alpha, r\sin\alpha\}$ are used where α is the *azimuth* of the geodesic passing $P \in T_P\mathbb{M}^2$ and r is its *length*. These *Riemann coordinates* (polar or normal) represent *length preserving mappings* with respect to the central point $P \in T_P\mathbb{M}^2$.

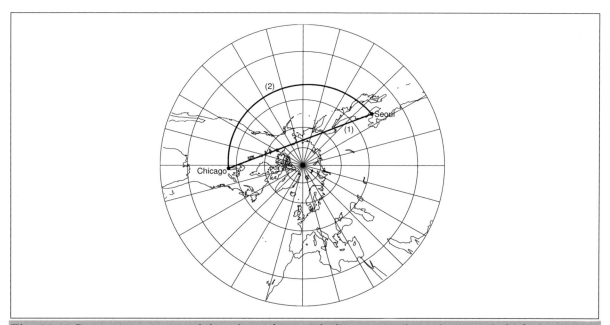

Fig. 20.1. Gnomonic projection of the sphere, the straight lines are geodesics (great circles), the loxodromes (rhumblines) are circular. Great circle (1) and rhumb line (2).

Section 20-1.

The elaborate presentation of *Riemann polar/normal coordinates* starts in Section 20-1 by the setup of a *minimal atlas* of the biaxial ellipsoid, namely in terms of {ellipsoidal longitude, ellipsoidal latitude} and {meta-longitude, meta-latitude}. Box 20.1 and Box 20.2 collect all fundamental elements of surface geometry of $\mathbb{E}^2_{A_1,A_2}$ (two-dimensional ellipsoid-of-revolution, semi-major axis A_1, semi-minor axis A_2). The *Darboux frame* of a one-dimensional submanifold in the two-dimensional manifold $\mathbb{E}^2_{A_1,A_2}$ is reviewed, in particular, by Corollary 20.3, the representation of geodetic curvature, geodetic torsion, and normal curvature in terms of elements of the first and second fundamental form as well as of Christoffel symbols. First, we define the *geodesic*. Second, we define the *geodesic circle* following A. Fialkow (1939), J. A. Schouten (1954), W. O. Vogel (1970, 1973) and K. Yano (1940a–d, 1942) enriched by two examples. Corollary 20.2 states that a curve is a *geodesic* if and only if it fulfills a system of *second order* ordinary differential equations (20.42). In contrast, a curve is a *geodesic circle* if and only if it fulfills a system of *third order* ordinary differential equations (20.43). Proofs are presented in Appendix E-1 and E-2. Finally, we define *Riemann polar/normal coordinates* and by Definition 20.4 the Riemann mapping.

Section 20-2.

Section 20-2 concentrates on the computation of Riemann polar/normal coordinates. First, by solving the two *second order* ordinary differential equations of a geodesic in the *Lagrange portrait*, namely by means of the *Legendre recurrence* ("Legendre series"), in particular, initial value problem versus boundary value problem, by the technique of *standard series inversion*. Second, the three *first order* ordinary differential equations of a geodesic in the *Hamilton portrait* ("phase space") subject to the A. C. Clairaut constant for a rotational symmetric surface like the ellipsoid-of-revolution are solved by means of the *Lie recurrence* ("Lie series"), in particular, initial value problem versus boundary value problem, by the technique of *standard series inversion*.

Section 20-3.

Section 20-3 treats the elaborate *Soldner coordinates* or *geodetic parallel coordinates*. These coordinates compete with Gauss–Krüger coordinates and Riemann normal coordinates. The way of construction is illustrated by Fig. 20.4. As the first problem of computing such a *geodetic projection*, we treat the case (1). In contrast, the second problem in computing Soldner coordinates may be summarized by (2). As an example, we introduce the *Soldner map* centered at the Tübingen Observatory.

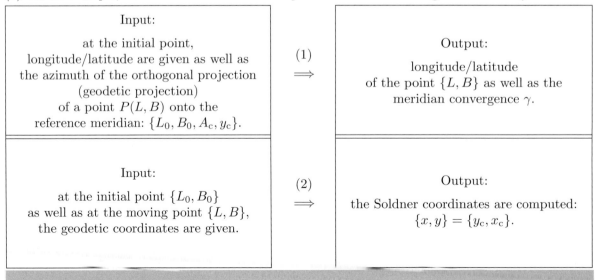

Input: at the initial point, longitude/latitude are given as well as the azimuth of the orthogonal projection (geodetic projection) of a point $P(L,B)$ onto the reference meridian: $\{L_0, B_0, A_c, y_c\}$.	(1) $\Longrightarrow$	Output: longitude/latitude of the point $\{L, B\}$ as well as the meridian convergence γ.
Input: at the initial point $\{L_0, B_0\}$ as well as at the moving point $\{L, B\}$, the geodetic coordinates are given.	(2) $\Longrightarrow$	Output: the Soldner coordinates are computed: $\{x, y\} = \{y_c, x_c\}$.

Section 20-4.

Section 20-4 focuses on the celebrated *Fermi coordinates* which extend the notion of a geodetic projection. An initial point $\{L_0, B_0\}$ is chosen. A moving point $\{L, B\}$ is projected at right angles onto the point P_{F}, which is fixed. The two-step solution from $\{L_0, B_0, u, v, u_{\mathrm{F}}, v_{\mathrm{F}}\}$ to $\{L, B, A_{PF}^c\}$ is given. $\{u, v\}$ are the Fermi coordinates of the moving point and $\{u_{\mathrm{F}}, v_{\mathrm{F}}\}$ are the Fermi coordinates of the geodetic projections $\{L, B\}$ onto $\{u_{\mathrm{F}}, v_{\mathrm{F}}\}$. The geodetic projections are fixed by identifying the coordinates of the point $\{u_{\mathrm{F}}, v_{\mathrm{F}}\}$.

Section 20-5.

Section 20-5 reviews all the details of *Riemann coordinates*, compares them with the Soldner coordinates, and additionally compares them with the Gauss–Krueger coordinates. First, we introduce the left deformation analysis or distortion analysis of the Riemann mapping, namely by outlining the *additive measure of deformation*, called the left *Cauchy–Green deformation tensor*. Solving the general eigenvalue/eigenvector problem for the pair $\{\mathsf{C}_l, \mathsf{G}_l\}$ of symmetric matrices, G_l positive-definite, we succeed to compute and illustrate the *principal distortions* of the Riemann mapping. We conclude with a global distortion analysis generated by charting the ellipsoid-of-revolution $\mathbb{E}^2_{A_1, A_2}$ by means of conformal Gauss–Krueger coordinates, parallel Soldner coordinates and normal Riemann coordinates summarized by the *Airy measure* of total deformation or total distortion for a symmetric strip $[-l_{\mathrm{E}}, +l_{\mathrm{E}}] \times [-b_{\mathrm{N}}, +b_{\mathrm{N}}]$ relative to a point $\{L_0, B_0\}$. A special highlight is Table 20.5, comparing those three coordinate systems in favor of *normal Riemann coordinates*. Assume a celestial body like the Earth can be globally modeled by an ellipsoid-of-revolution. Then in terms of the Airy measure of total deformation on a symmetric strip $[-l_{\mathrm{E}}, +l_{\mathrm{E}}] \times [-b_{\mathrm{N}}, +b_{\mathrm{N}}]$ given by Table 20.5, normal coordinates produce the minimal Airy global distortion when compared to parallel Soldner coordinates and conformal Gauss–Krueger coordinates. Let us therefore push forward the geodetic application of normal Riemann coordinates.

20-1 Geodesic, geodesic circle, Darboux frame, Riemann coordinates

Riemann polar/normal coordinates. Frobenius matrix, Gauss matrix, Hesse matrix, Christoffel symbols. Surface fundamental forms.

Let there be given the *ellipsoid-of-revolution* $\mathbb{E}^2_{A_1, A_2}$ (biaxial ellipsoid, spheroid, with semi-major A_1, with semi-minor axis A_2, and with relative eccentricity $E^2 := (A_1^2 - A_2^2)/A_1^2$). It is embedded in $\mathbb{E}^3 := \{\mathbb{R}^3, \delta_{IJ}\}$, the three-dimensional Euclidean space of canonical metric $I = \{\delta_{IJ}\}$ of *Kronecker type*. The Latin indices I and J are elements of $\{1, 2, 3\}$.

$$\mathbb{E}^2_{A_1, A_2} := \{\boldsymbol{X} \in \mathbb{R}^3 | (X^2 + Y^2)/A_1^2 + Z^2/A_2^2 = 1\}. \tag{20.1}$$

The ellipsoid-of-revolution $\mathbb{E}^2_{A_1, A_2}$ is *globally covered* by the union of the *two charts* $\{L, B\}$ and $\{U, V\}$ constituted by {ellipsoidal longitude, ellipsoidal latitude} and {meta-longitude, meta-latitude}, in particular (20.2), for *open sets* (20.3) illustrated in Fig. 20.2.

$$
\begin{aligned}
\boldsymbol{X}_I = &+\boldsymbol{E}_1 \frac{A_1 \cos B \cos L}{\sqrt{1 - E^2 \sin^2 B}} & & & \boldsymbol{X}_{II} = &+\boldsymbol{E}_{1'} \frac{A_1 \cos V \cos U}{\sqrt{1 - E^2 \sin^2 U \cos^2 V}} \\
&+\boldsymbol{E}_2 \frac{A_1 \cos B \sin L}{\sqrt{1 - E^2 \sin^2 B}} & &\text{versus} & &+\boldsymbol{E}_{2'} \frac{A_1 (1 - E^2) \cos V \sin U}{\sqrt{1 - E^2 \sin^2 U \cos^2 V}} + \\
&+\boldsymbol{E}_3 \frac{A_1 (1 - E^2) \sin B}{\sqrt{1 - E^2 \sin^2 B}} & & & &+\boldsymbol{E}_{3'} \frac{A_1 \sin V}{\sqrt{1 - E^2 \sin^2 U \cos^2 V}},
\end{aligned}
\tag{20.2}
$$

$$
\begin{aligned}
0 < L < 2\pi, \quad -\tfrac{\pi}{2} < B < +\tfrac{\pi}{2}, \\
0 < U < 2\pi, \quad -\tfrac{\pi}{2} < V < +\tfrac{\pi}{2},
\end{aligned}
\tag{20.3}
$$

The placement vector $\boldsymbol{X} \in \{\mathbb{R}^3, \delta_{IJ}\}$ is represented either in the *orthonormal triad* $\{\boldsymbol{E}_1, \boldsymbol{E}_2, \boldsymbol{E}_3\}$, which is oriented along the ordered principal axes of the ellipsoid-of-revolution, or is represented in the *transverse orthonormal triad* $\{\boldsymbol{E}_{1'}, \boldsymbol{E}_{2'}, \boldsymbol{E}_{3'}\} = \{-\boldsymbol{E}_1, \boldsymbol{E}_3, \boldsymbol{E}_2\}$. $\boldsymbol{X}_I \in \boldsymbol{X}_I \cup \boldsymbol{X}_{II} \ni \boldsymbol{X}_{II}$ constitute the *minimal atlas* of $\mathbb{E}^2_{A_1,A_2}$. *Inverse formulae* $\{L, B\}(X, Y, Z)$ and $\{U, V\}(X, Y, Z)$ are given by E. Grafarend and P. Lohse (1991). Note that in both charts (local coordinates $\{L, B\}$ and $\{U, V\}$, respectively) the surface normal vector $\boldsymbol{G}_3$ is represented by (20.4), motivating why the local coordinates $\{L, B\}$ and $\{U, V\}$, respectively, are called *surface normal coordinates*.

$$\boldsymbol{G}_3 = +\boldsymbol{E}_1 \cos B \cos L + \boldsymbol{E}_2 \cos B \sin L + \boldsymbol{E}_3 \sin B$$

versus   (20.4)

$$\boldsymbol{G}'_3 = +\boldsymbol{E}_{1'} \cos V \cos U + \boldsymbol{E}_{2'} \cos V \sin U + \boldsymbol{E}_{3'} \sin V \ .$$

The embedding $\mathbb{E}^2_{A_1,A_2} \subset \{\mathbb{R}^3, \delta_{IJ}\}$ is characterized by the *mapping equations*

$$\tan L = \frac{Y}{X} \quad \text{versus} \quad \tan U = -\frac{Z}{(1 - E^2)X} \ , \tag{20.5}$$

$$\tan B = \frac{Z}{(1 - E^2)\sqrt{X^2 + Y^2}} \quad \text{versus} \quad \tan V = \frac{(1 - E^2)Y}{\sqrt{(1 - E^2)X^2 + Z^2}} \ . \tag{20.6}$$

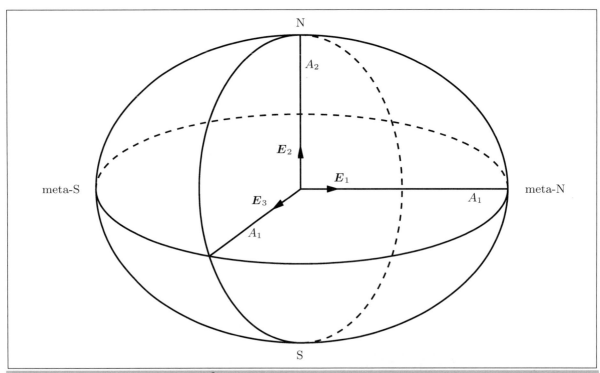

Fig. 20.2. The minimal atlas of $\mathbb{E}^2_{A_1,A_2}$. First chart: "surface normal" ellipsoidal longitude, ellipsoidal latitude $\{0 < L < 2\pi, -\pi/2 < B < +\pi/2\}$. Second chart: "surface normal" meta-longitude, meta-latitude $\{0 < U < 2\pi, -\pi/2 < V < +\pi/2\}$. Half ellipse $L = 0$, South Pole $B = -\pi/2$, North Pole $B = +\pi/2$ excluded in the first chart. Half circle $U = 0$, meta-South Pole $V = -\pi/2$, meta-North Pole $V = +\pi/2$ excluded in the second chart. $\{\boldsymbol{E}_1, \boldsymbol{E}_2\}$ span the equator plane, $\{\boldsymbol{E}_{1'}, \boldsymbol{E}_{2'}\} - \{\boldsymbol{E}_1, \boldsymbol{E}_3\}$ span the meta-equator plane.

From these mapping equations, we derive *cha-cha-cha* ("change from one chart to another chart")

$$\tan L = -\frac{\tan V}{\cos U} \quad \text{versus} \quad \tan U = -\frac{\tan B}{\cos L} \ , \tag{20.7}$$

$$\tan B = \frac{\tan U}{\sqrt{1 + \tan^2 V / \cos^2 U}} \quad \text{versus} \quad \tan V = \frac{\tan L}{\sqrt{1 + \tan^2 B / \cos^2 L}}$$

$$= \frac{\sin U}{\sqrt{\cos^2 U + \tan^2 V}} \qquad\qquad = \frac{\sin L}{\sqrt{\cos^2 L + \tan^2 B}} \ , \tag{20.8}$$

in particular, the diffeomorphism

$$\begin{bmatrix} dU \\ dV \end{bmatrix} =$$

$$= \begin{bmatrix} -\frac{\sin L \tan B}{\cos^2 L + \tan^2 B} & -\frac{\cos L}{\cos^2 B \cos^2 L + \sin^2 B} \\ \frac{\cos L}{(\cos^2 L + \tan^2 B)^{1/2}} & \frac{-\sin L \tan B}{(\cos^2 L + \tan^2 B)^{1/2}} \end{bmatrix} \begin{bmatrix} dL \\ dB \end{bmatrix} \ , \tag{20.9}$$

$$\begin{bmatrix} dL \\ dB \end{bmatrix} =$$

$$= \begin{bmatrix} \frac{-\tan V \sin U}{\cos^2 U + \tan^2 V} & \frac{-\cos U}{\cos^2 V (\cos^2 U + \tan^2 V)} \\ \frac{\cos U}{(\cos^2 U + \tan^2 V)^{1/2}} & \frac{-\sin U \tan V}{(\cos^2 U + \tan^2 V)^{1/2}} \end{bmatrix} \begin{bmatrix} dU \\ dV \end{bmatrix} \ . \tag{20.10}$$

Box 20.1 and Box 20.2 summarize the surface geometry of $\mathbb{E}^2_{A_1, A_2} \subset \{\mathbb{R}^3, \delta_{IJ}\}$, in particular, the matrices F, G, H, and J of type Frobenius, Gauß, Hesse, and Jacobi as well as the curvature matrix $\mathsf{K} := -\mathsf{H}\mathsf{G}^{-1}$, especially the *surface fundamental forms* $\{I, II, III\}$.

Let $C : [0, \infty] \to \{\mathbb{E}^2_{A_1, A_2}, G_{KL}\}$ be a smooth curve which is parameterized by arc length. Denote by $\{D_1, D_2, D_3\}$ its *Darboux frame* defined by (20.11). Its derivational equations are given by (20.12), introducing the *antisymmetric connection matrix* Ω (κ_g, κ_n, τ_g) containing *geodetic curvature* κ_g, *normal curvature* κ_n, and *geodetic torsion* τ_g. In terms of the first fundamental form, in particular $\{G_{KL}\}$, the second fundamental form, in particular $\{H_{KL}\}$, the Riemann connection, in particular the Christoffel symbols $\{^M_{KL}\}$, the curvature measures of the curve C as a submanifold of $\{\mathbb{E}^2_{A_1, A_2}, G_{KL}\}$ can be represented by Corollary 20.1.

$$D_1 := \frac{d\boldsymbol{X}\{U^k(S)\}}{dS} = \frac{\partial \boldsymbol{X}}{\partial U^K} U'^K = \boldsymbol{G}_K U'^K \ ,$$

$$D_2 := *(D_3 \wedge D_1) = D_3 \times D_1 \ , \tag{20.11}$$

$$D_3 := \boldsymbol{G}_3 \ ,$$

$$\boldsymbol{D}'_1 = +\kappa_g \boldsymbol{D}_2 + \kappa_n \boldsymbol{D}_3 \ ,$$

$$\boldsymbol{D}'_2 = -\kappa_g \boldsymbol{D}_1 + \tau_g \boldsymbol{D}_3 \ ,$$

$$\boldsymbol{D}'_3 = -\kappa_n \boldsymbol{D}_1 - \tau_g \boldsymbol{D}_2 \ , \tag{20.12}$$

$$\begin{bmatrix} \boldsymbol{D}'_1 \\ \boldsymbol{D}'_2 \\ \boldsymbol{D}'_3 \end{bmatrix} = \begin{bmatrix} 0 & \kappa_g & \kappa_n \\ -\kappa_g & 0 & \tau_g \\ -\kappa_n & \tau_g & 0 \end{bmatrix} \begin{bmatrix} \boldsymbol{D}_1 \\ \boldsymbol{D}_2 \\ \boldsymbol{D}_3 \end{bmatrix} \ , \quad \boldsymbol{D}' = \Omega \boldsymbol{D} \ , \quad \Omega = \begin{bmatrix} 0 & \kappa_g & \kappa_n \\ -\kappa_g & 0 & \tau_g \\ -\kappa_n & -\tau_g & 0 \end{bmatrix} \ .$$

Box 20.1 (Surface geometry of $\mathbb{E}^2_{A_1,A_2}$).

Matrices

(Frobenius matrix F (elements a, b, c, d), Gauss matrix $\mathsf{G} = \mathsf{J}^{\mathrm{T}}\mathsf{J}$, (elements e, f, g),
Hesse matrix $\mathsf{H} = \{\langle \boldsymbol{X},_{KL}, \boldsymbol{G}_3 \rangle\}$ (elements l, m, n), curvature matrix K,
Jacobi matrix $\mathsf{J} = \{\partial X^J / \partial U^K\}$):

$$\mathsf{F} = \{F_{KL}^1\} = \begin{bmatrix} \dfrac{\sqrt{1-E^2\sin^2 B}}{A_1\cos B} & 0 \\ 0 & \dfrac{(1-E^2\sin^2 B)^3{}^2}{A_1(1-E^2)} \end{bmatrix}, \quad \mathsf{F} \in \mathbb{R}^{2\times 2}, \tag{20.13}$$

$$\mathsf{G} = \{G_{KL}^1\} = \begin{bmatrix} \dfrac{A_1^2\cos^2 B}{1-E^2\sin^2 B} & 0 \\ 0 & \dfrac{A_1^2(1-E^2)^2}{(1-E^2\sin^2 B)^3} \end{bmatrix}, \quad \mathsf{G} \in \mathbb{R}^{2\times 2}, \quad \mathsf{G} = \mathsf{J}^{\mathrm{T}}\mathsf{J}, \tag{20.14}$$

$$\mathsf{H} = \{H_{KL}^1\} = \begin{bmatrix} -\dfrac{A_1\cos^2 B}{\sqrt{1-E^2\sin^2 B}} & 0 \\ 0 & -\dfrac{A_1(1-E^2)}{(1-E^2\sin^2 B)^3{}^2} \end{bmatrix}, \quad \mathsf{H} \in \mathbb{R}^{2\times 2}, \tag{20.15}$$

$$\mathsf{K} = \{K_{KL}^1\} = \begin{bmatrix} \dfrac{\sqrt{1-E^2\sin^2 B}}{A_1} & 0 \\ 0 & \dfrac{(1-E^2\sin^2 B)^3{}^2}{A_1(1-E^2)} \end{bmatrix} = -\mathsf{H}\mathsf{G}^{-1}, \quad \mathsf{K} \in \mathbb{R}^{2\times 2}, \tag{20.16}$$

$$h = -\frac{\mathrm{tr}[K]}{2} = -\frac{\sqrt{1-E^2\sin^2 B}(2-E^2(1+\sin^2 B))}{2A_1(1-E^2)}, \tag{20.17}$$

$$k = \det[K] = \frac{(1-E^2\sin^2 B)^2}{A_1^2(1-E^2)}, \tag{20.18}$$

$$\mathsf{J} = \{J_{KL}^1\} = \frac{\partial(X,Y)}{\partial(L,B)} = \begin{bmatrix} -\dfrac{A_1\cos B\sin L}{\sqrt{1-E^2\sin^2 B}} & -\dfrac{A_1(1-E^2)\sin B\cos L}{(1-E^2\sin^2 B)^3{}^2} \\ \dfrac{A_1\cos B\cos L}{\sqrt{1-E^2\sin^2 B}} & -\dfrac{A_1(1-E^2)\sin B\sin L}{(1-E^2\sin^2 B)^3{}^2} \\ 0 & +\dfrac{A_1(1-E^2)\cos B}{(1-E^2\sin^2 B)^3{}^2} \end{bmatrix}, \quad \mathsf{J} \in \mathbb{R}^{3\times 2}. \tag{20.19}$$

Eigenvalues:

1st eigenvalue of K: $\kappa_1 = \sqrt{1-E^2\sin^2 B}/A_1$;

$\kappa_1^{-1} =: N(B) = A_1/\sqrt{1-E^2\sin^2 B}$ (1st curvature radius);

2nd eigenvalue of K: $\kappa_2 = (1-E^2\sin^2 B)^{3/2}/A_1(1-E^2)$;

$\kappa_2^{-1} =: M(B) = A_1(1-E^2)/(1-E^2\sin^2 B)^{3/2}$ (2nd curvature radius).

Christoffel symbols $\left\{ {}^{M}_{KL} \right\}$:

$$\left\{ {}^1_{11} \right\}(L,B) = \left\{ {}^1_{22} \right\}(L,B) = \left\{ {}^2_{12} \right\}(L,B) = 0, \quad \left\{ {}^1_{12} \right\}(L,B) = -\frac{-\tan B(1-E^2)}{1-E^2\sin^2 B},$$

$$\left\{ {}^2_{11} \right\}(L,B) = \frac{\sin B\cos B(1-E^2\sin^2 B)}{1-E^2}, \quad \left\{ {}^2_{22} \right\}(L,B) = 3E^2\sin B\cos B(1-E^2\sin^2 B). \tag{20.20}$$

Box 20.2 (Surface geometry of $\mathbb{E}^2_{A_1,A_2}$).

Matrices

(Frobenius matrix F (elements a, b, c, d), Gauss matrix $\mathsf{G} = \mathsf{J}^{\mathrm{T}}\mathsf{J}$, (elements e, f, g),
Hesse matrix $\mathsf{H} = \{\langle \boldsymbol{X},_{KL}, \boldsymbol{G}_3\rangle\}$ (elements l, m, n), curvature matrix K,
Jacobi matrix $\mathsf{J} = \{\partial X^J/\partial U^K\}$):

$$\mathsf{F} = \{F^2_{KL}\} = \frac{(1-E^2\sin^2 U\cos^2 V)^{1\ 2}}{A_1(1-\sin^2 U\cos^2 V)^{1\ 2}} \begin{bmatrix} \dfrac{-\sin U\sin V}{\cos V} & \dfrac{\cos U(1-E^2\sin^2 U\cos^2 V)}{(1-E^2)\cos V} \\[2ex] -\cos U & \dfrac{-\sin U\sin V(1-E^2\sin^2 U\cos^2 V)}{(1-E^2)} \end{bmatrix} , \tag{20.21}$$

$$\mathsf{G} = \{G^2_{KL}\} = \\ = \begin{bmatrix} G_{11} & G_{12} \\ G_{21} & G_{22} \end{bmatrix}, \quad \begin{cases} G_{11} = \dfrac{A_1^2\cos^2 V(1-2E^2(1-\sin^2 U\sin^2 V)+E^4(1-\sin^2 U\sin^2 V(1+\sin^2 U\cos^2 V)))}{(1-E^2\sin^2 U\cos^2 V)^3} , \\[2ex] G_{12} = G_{21} = \dfrac{A_1^2 E^2\cos U\sin U\cos V\sin V(2-E^2(1+\cos^2 V\sin^2 U))}{(1-E^2\sin^2 U\cos^2 V)^3} , \\[2ex] G_{22} = \dfrac{A_1^2(1-2E^2\sin^2 U+E^4\sin^2 U(1-\cos^2 V\cos^2 U))}{(1-E^2\sin^2\cos^2 V)^3} , \end{cases} \tag{20.22}$$

$$\mathsf{H} = \{H^2_{KL}\} = \begin{bmatrix} \dfrac{-A_1\cos^2 V(1-E^2(1-\sin^2 U\sin^2 V))}{(1-E^2\sin^2 U\cos^2 V)^{3\ 2}} & \dfrac{-A_1 E^2\sin U\cos U\sin V\cos V}{(1-E^2\sin^2 U\cos^2 V)^{3\ 2}} \\[2ex] \text{symmetric} & \dfrac{-A_1(1-E^2\sin^2 U)}{(1-E^2\sin^2 U\cos^2 V)^{3\ 2}} \end{bmatrix} , \tag{20.23}$$

$$\mathsf{K} = \{K^2_{KL}\} = \\ = \begin{bmatrix} \dfrac{(1-E^2\sin^2 U)(1-E^2\sin^2 U\cos^2 V)^{1\ 2}}{A_1(1-E^2)} & \dfrac{-\cos U\cos V\sin U\sin V E^2(1-E^2\sin^2 U\cos^2 V)^{1\ 2}}{A_1(1-E^2)} \\[2ex] \dfrac{-\cos U\sin U\sin V E^2(1-E^2\sin^2 U\cos^2 V)^{1\ 2}}{A_1(1-E^2)\cos V} & \dfrac{(1-E^2(1-\sin^2 U\sin^2 V))(1-E^2\sin^2 U\cos^2 V)^{1\ 2}}{A_1(1-E^2)} \end{bmatrix} , \tag{20.24}$$

$$h = -\frac{\operatorname{tr}[K]}{2} = -\frac{\sqrt{1-E^2\sin^2 U\cos^2 V}(2-E^2(1+\sin^2 U\cos^2 V))}{2A_1(1-E^2)} , \quad k = \det[K] = \frac{(1-E^2\sin^2 U\cos^2 V)^2}{A_1^2(1-E^2)} , \tag{20.25}$$

$$\mathsf{J} = \{J^2_{KL}\} = \begin{bmatrix} \dfrac{A_1\cos V\sin U(1-E^2\cos^2 V)}{(1-E^2\sin^2 U\cos^2 V)^{3\ 2}} & \dfrac{-A_1\cos U\sin V}{(1-E^2\sin^2 U\cos^2 V)^{3\ 2}} \\[2ex] \dfrac{A_1(1-E^2)\cos V\cos U}{(1-E^2\sin^2 U\cos^2 V)^{3\ 2}} & \dfrac{-A_1(1-E^2)\sin U\sin V}{(1-E^2\sin^2 U\cos^2 V)^{3\ 2}} \\[2ex] \dfrac{A_1 E^2\sin U\cos U\sin V\cos^2 V}{(1-E^2\sin^2 U\cos^2 V)^{3\ 2}} & \dfrac{A_1\cos V(1-E^2\sin^2 U)}{(1-E^2\sin^2 U\cos^2 V)^{3\ 2}} \end{bmatrix} . \tag{20.26}$$

Eigenvalues:

1st eigenvalue of K: $\kappa_1 = \sqrt{1-E^2\sin^2 U\cos^2 V}/A_1$;

2nd eigenvalue of K: $\kappa_2 = (1-E^2\sin^2 U\cos^2 V)^{3/2}/A_1(1-E^2)$.

Christoffel symbols $\{{}^M_{KL}\}$:

$$\{{}^1_{11}\}(U,V) = \frac{E^2\sin U\cos U\cos^2 V(3-E^2(3-\sin^2 U\sin^2 V))}{(1-E^2\sin^2 U\cos^2 V)(1-E^2)} ,$$

$$\{{}^1_{12}\}(U,V) = -\frac{\sin V(1-E^2-E^4\sin^2 U\cos^2 U\cos^2 V)}{(1-E^2\sin^2 U\cos^2 V)(1-E^2)\cos V} , \quad \{{}^1_{22}\}(U,V) = \frac{E^2\sin U\cos U(1-E^2\sin^2 U)}{(1-E^2\sin^2 U\cos^2 V)(1-E^2)} ,$$

$$\{{}^2_{11}\}(U,V) = \frac{\sin V\cos V(1-E^2(1+2\sin^2 U\cos^2 V)+E^4\sin^2 U\cos^2 V(2-\sin^2 U\sin^2 V))}{(1-E^2\sin^2 U\cos^2 V)(1-E^2)} \tag{20.27}$$

$$\{{}^2_{12}\}(U,V) = \frac{E^2\sin U\cos U\cos^2 V(1-E^2(1+\sin^2 U\sin^2 V))}{(1-E^2\sin^2 U\cos^2 V)(1-E^2)} , \quad \{{}^2_{22}\}(U,V) = \frac{-E^2\sin^2 U\sin V\cos V(3-E^2(?+\sin^2 U))}{(1-E^2\sin^2 U\cos^2 V)(1-E^2)} .$$

Corollary 20.1 $(\kappa_{\mathrm{g}}, \kappa_{\mathrm{n}}, \tau_{\mathrm{g}})$.

$$\kappa_{\mathrm{g}}^2 = G_{M_1 M_2}[U''^{M_1} + U'^{K_1} U'^{L_1}\{{}^{M_1}_{K_1 L_1}\}][U''^{M_2} + U'^{K_2} U'^{L_2}\{{}^{M_2}_{K_2 L_2}\}] , \tag{20.28}$$

$$\kappa_{\mathrm{n}} = H_{KL} U'^K U'^L , \tag{20.29}$$

$$\tau_{\mathrm{g}} = [H_{KL} U''^K U'^L + \{{}^N_{KL}\} H_{NM} U'^K U'^L U'^M] \times$$
$$\times [G_{M_1 M_2}[U''^{M_1} + U'^{K_1} U'^{L_1}\{{}^{M_1}_{K_1 L_1}\}][U''^{M_2} + U'^{K_2} U'^{L_2}\{{}^{M_2}_{K_2 L_2}\}]]^{-1 2} . \tag{20.30}$$

End of Corollary.

The curve C is called *geodesic* if $\kappa_{\mathrm{g}} = 0$ and the curve C is called a *geodesic circle* if $\kappa_{\mathrm{g}} = \text{const.}$, $\kappa_{\mathrm{n}} = \text{const.}$, and $\tau_{\mathrm{g}} = 0$. Compare with Examples 20.1 and 20.2.

Example 20.1 (Geodesic as a submanifold in $\{\mathbb{E}^2_{A_1\ A_2}, G_{KL}\}$, $L = c = \text{const}$: "meridian").

$$\boldsymbol{X} = +\boldsymbol{E}_1 \frac{A_1 \cos T \cos c}{\sqrt{1 - E^2 \sin^2 T}} + \boldsymbol{E}_2 \frac{A_1 \cos T \sin c}{\sqrt{1 - E^2 \sin^2 T}} + \boldsymbol{E}_3 \frac{A_1(1 - E^2) \sin T}{\sqrt{1 - E^2 \sin^2 T}} , \tag{20.31}$$

$$\boldsymbol{D}_1 = -\boldsymbol{E}_1 \sin T \cos c - \boldsymbol{E}_2 \sin T \sin c + \boldsymbol{E}_3 \cos T =: \dot{\boldsymbol{X}}/\|\dot{\boldsymbol{X}}\| ,$$

$$\boldsymbol{D}_2 = \boldsymbol{E}_1 \sin c - \boldsymbol{E}_2 \cos c , \tag{20.32}$$

$$\boldsymbol{D}_3 = \boldsymbol{E}_1 \cos T \cos c + \boldsymbol{E}_2 \cos T \sin c + \boldsymbol{E}_3 \sin T ,$$

$$\boldsymbol{D} = \begin{bmatrix} -\sin T \cos c & -\sin T \sin c & \cos T \\ \sin c & -\cos c & 0 \\ \cos T \cos c & \cos T \sin c & \sin T \end{bmatrix} \boldsymbol{E} , \tag{20.33}$$

$$\boldsymbol{D} = \mathsf{R}\boldsymbol{E} , \quad \boldsymbol{D}' = \mathsf{R}'\boldsymbol{E} = \mathsf{R}'\mathsf{R}^{\mathrm{T}}\boldsymbol{D} = \boldsymbol{\Omega}\boldsymbol{D} \ \forall \ \mathsf{R} \in \mathrm{SO}(3) , \tag{20.34}$$

$$\mathsf{R}' = \begin{bmatrix} -T' \cos T \cos c & -T' \cos T \sin c & -T' \sin T \\ 0 & 0 & 0 \\ -T' \sin T \cos c & -T' \sin T \sin c & +T' \cos T \end{bmatrix} , \quad \boldsymbol{\Omega} := \mathsf{R}'\mathsf{R}^{\mathrm{T}} = \begin{bmatrix} 0 & 0 & -T' \\ 0 & 0 & 0 \\ +T' & 0 & 0 \end{bmatrix} , \tag{20.35}$$

$$\kappa_{\mathrm{g}} = 0 , \ \kappa_{\mathrm{n}} = -T' , \ \tau_{\mathrm{g}} = 0 ,$$
$$\|\dot{\boldsymbol{X}}\| = \frac{dS}{dT} = \frac{A_1(1 - E^2)}{(1 - E^2 \sin^2 T)^{3\ 2}} , \quad T' := \frac{\mathrm{d}T}{\mathrm{d}S} = \frac{(1 - E^2 \sin^2 T)^{3\ 2}}{A_1(1 - E^2)} , \tag{20.36}$$

$$\kappa_{\mathrm{g}} = 0 , \ \kappa_{\mathrm{n}} = -\frac{(1 - E^2 \sin^2 T)^{3\ 2}}{A_1(1 - E^2)} , \ \tau_{\mathrm{g}} = 0 . \tag{20.37}$$

End of Example

Obviously, the meridian $L = \text{const.}$ is a *geodesic*.

Example 20.2 (Geodesic as a submanifold in $\{\mathbb{E}^2_{A_1\,A_2}, G_{KL}\}$, $B = c = $ const: "parallel circle").

$$\boldsymbol{X} = +\boldsymbol{E}_1 \frac{A_1 \cos c \cos T}{\sqrt{1 - E^2 \sin^2 c}} + \boldsymbol{E}_2 \frac{A_1 \cos c \sin T}{\sqrt{1 - E^2 \sin^2 c}} + \boldsymbol{E}_3 \frac{A_1(1 - E^2) \sin c}{\sqrt{1 - E^2 \sin^2 c}} \,, \tag{20.38}$$

$$\boldsymbol{D}_1 = -\boldsymbol{E}_1 \sin T + \boldsymbol{E}_2 \cos T =: \dot{\boldsymbol{X}}/\|\dot{\boldsymbol{X}}\| \,,$$

$$\boldsymbol{D}_2 = -\boldsymbol{E}_1 \sin c \cos T - \boldsymbol{E}_2 \sin c \sin T + \boldsymbol{E}_3 \cos c \,, \tag{20.39}$$

$$\boldsymbol{D}_3 = +\boldsymbol{E}_1 \cos c \cos T + \boldsymbol{E}_2 \cos c \sin T + \boldsymbol{E}_3 \sin c \,,$$

$$\boldsymbol{D} = \begin{bmatrix} -\sin T & \cos T & 0 \\ -\sin c \cos T & -\sin c \sin T & \cos c \\ \cos c \cos T & \cos c \sin T & \sin c \end{bmatrix} \boldsymbol{E} \,,$$

$$\boldsymbol{D} = \mathsf{R}\boldsymbol{E} \,, \quad \boldsymbol{D}' = \mathsf{R}'\boldsymbol{E} = \mathsf{R}'\mathsf{R}^{\mathrm{T}}\boldsymbol{D} = \boldsymbol{\Omega}\boldsymbol{D} \; \forall \, \mathsf{R} \in \mathrm{SO}(3) \,, \tag{20.40}$$

$$\mathsf{R}' = \begin{bmatrix} -T' \cos T & -T' \sin T & 0 \\ +T' \sin c \sin T & -T' \sin c \cos T & 0 \\ -T' \cos c \sin T & +T \sin c \cos T & 0 \end{bmatrix} \,,$$

$$\boldsymbol{\Omega} := \mathsf{R}'\mathsf{R}^{\mathrm{T}} = \begin{bmatrix} 0 & +T' \sin c & -T' \cos c \\ -T' \sin c & 0 & 0 \\ +T' \cos c & 0 & 0 \end{bmatrix} \,,$$

$$\kappa_{\mathrm{g}} = +T' \sin c \,, \quad \kappa_{\mathrm{n}} = -T' \cos c \,, \quad \tau_{\mathrm{g}} = 0 \,,$$

$$\|\dot{\boldsymbol{X}}\| = \frac{dS}{dT} = \frac{A_1 \cos c}{\sqrt{1 - E^2 \sin^2 c}} \,,$$

$$T' := \frac{\mathrm{d}T}{\mathrm{d}S} = \frac{\sqrt{1 - E^2 \sin^2 c}}{A_1 \cos c}$$

$$\Rightarrow \tag{20.41}$$

$$\kappa_{\mathrm{g}} = +\frac{\sqrt{1 - E^2 \sin^2 c}}{A_1} \tan c = \text{const} \,,$$

$$\kappa_{\mathrm{n}} = -\frac{\sqrt{1 - E^2 \sin^2 c}}{A_1} = \text{const} \,,$$

$$\tau_{\mathrm{g}} = 0 \,.$$

End of Example.

Obviously, the parallel circle $B = $ const is a *geodesic circle*. Note that for a sphere $\mathbb{S}^2_R$ great circles are *geodesics*, but small circles are *geodesic circles*. Following the curvature measure representation in Corollary 20.1, we can characterize *geodesics* and *geodesic circles* by differential equations.

Corollary 20.2 (Geodesics, geodesic circles).

A curve $C(S)$ is a *geodesic* if and only if

$$U''^M + \{^M_{KL}\}U'^K U'^L = 0 .$$
(20.42)

A curve $C(S)$ is a *geodesic circle* if and only if

$$U'''^M + G_{KL}U''^K U''^L U'^M + 3\{^M_{KL}\}U''^K U'^L + 2G_{KL}\{^L_{PQ}\}U''^K U'^P U'^Q U'^M +$$

$$+(\{^M_{KL}\},_P + \{^Q_{KL}\}\{^M_{QP}\})U'^K U'^L U'^P + G_{KL}\{^K_{PQ}\}\{^L_{ST}\}U'^P U'^Q U'^S U'^T U'^M = 0 .$$
(20.43)

End of Corollary.

In the *tangent space* $\{T_U\mathbb{E}^2_{A_1,A_2}, G_{KL}\}$, which is spanned by the two tangent vectors $\boldsymbol{G}_1$ and $\boldsymbol{G}_2$ ($\boldsymbol{G}_1$ and $\boldsymbol{G}_2$ are neither orthogonal nor normalized), $\boldsymbol{C}_1$ and $\boldsymbol{C}_2$ (orthonormal *Cartan frame*), or $\boldsymbol{D}_1$ and $\boldsymbol{D}_2$ (orthonormal *Darboux frame*) at the point $U_0 = \{U_0^1, U_0^2\}$ (e. g. $\{L_0, B_0\}$ or $\{U_0, V_0\}$), we define (*Riemann*) *polar coordinates* and *normal coordinates* by (20.44), in particular, referring to (20.45) called $\{$"eastern"/"right"/"horizontal"$\}$ and (20.46) called $\{$"northern"/"up"/"vertical"$\}$.

$$x = r\cos\alpha , \quad y = r\sin\alpha ,$$
(20.44)

$$\frac{\frac{\partial \boldsymbol{X}}{\partial L}}{\|\frac{\partial \boldsymbol{X}}{\partial L}\|} = \frac{\boldsymbol{G}_1}{\|\boldsymbol{G}_1\|} =: \boldsymbol{C}_1 ,$$
(20.45)

$$\frac{\frac{\partial \boldsymbol{X}}{\partial B}}{\|\frac{\partial \boldsymbol{X}}{\partial B}\|} = \frac{\boldsymbol{G}_2}{\|\boldsymbol{G}_2\|} =: \boldsymbol{C}_2 .$$
(20.46)

The polar coordinate α is called "*East azimuth*" (ninety degrees *minus* "North azimuth" or minus ninety degrees plus "South azimuth"/"astronomical azimuth") while r characterizes the Euclidean distance of a point in $\{T_U\mathbb{E}^2_{A_1,A_2}, G_{KL}\}$ with respect to the origin $\{U^1, U^2\}$. Figure 20.3 illustrates the tangent space $\{T_{U_0}\mathbb{E}^2_{A_1,A_2}, G_{KL}\}$ at the point $U_0 = \{U_0^1, U_0^2\}$. Furthermore, Figure 20.3 illustrates the Cartan two-leg $\{\boldsymbol{C}_1(\text{East}), \boldsymbol{C}_2(\text{North})\}$. In contrast, Table 20.1 summarizes the various definitions of polar and normal coordinates with respect to alternative azimuth definitions.

Table 20.1. Various definitions of (Riemann) polar and normal coordinates.

orthonormal two-leg (Cartan two-leg)	azimuth	(Riemann) polar/normal coordinates
East, North : $\quad \frac{\boldsymbol{X}_L}{\|\boldsymbol{X}_L\|} = \boldsymbol{C}_1 , \; \frac{\boldsymbol{X}_B}{\|\boldsymbol{X}_B\|} = \boldsymbol{C}_2$	East azimuth	$x = r\cos\alpha , \; y = r\sin\alpha$
North, East : $\quad \frac{\boldsymbol{X}_B}{\|\boldsymbol{X}_B\|} = \boldsymbol{C}_1^* , \; \frac{\boldsymbol{X}_L}{\|\boldsymbol{X}_L\|} = \boldsymbol{C}_2^*$	North azimuth (left oriented), $\alpha^* = 90° - \alpha$	$x^* = r\cos\alpha^* = r\sin\alpha ,$ $y^* = r\sin\alpha^* = r\cos\alpha$
South, East : $\quad -\frac{\boldsymbol{X}_B}{\|\boldsymbol{X}_B\|} = \boldsymbol{C}_1^{**} , \; \frac{\boldsymbol{X}_L}{\|\boldsymbol{X}_L\|} = \boldsymbol{C}_2^{**}$	South azimuth (right oriented), $\alpha^{**} = 90° + \alpha$	$x^{**} = r\cos\alpha^{**} = -r\sin\alpha ,$ $y^{**} = r\sin\alpha^{**} = r\cos\alpha$

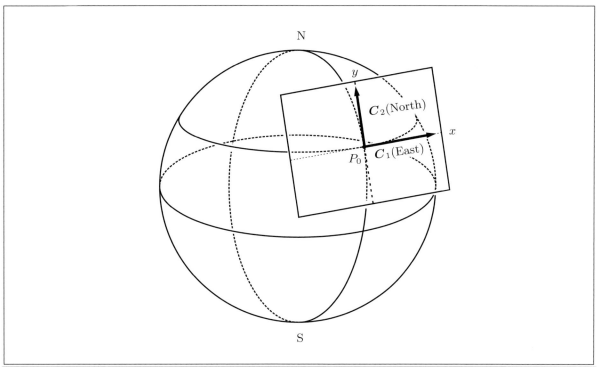

Fig. 20.3. Oblique tangential plane $T_{U_0}\mathbb{E}^2_{A_1,A_2}$, Cartan frame C_1(East) and C_2(North) at point $P_0(U_0)$.

Let us discuss how to relate the polar or normal tangential coordinates $\{\alpha, r\}$ to those coordinates which parameterize $\mathbb{E}^2_{A_1,A_2}$, here {longitude L, latitude B} or {meta-longitude U, meta-latitude V}, respectively. At first, let us identify the curve $C : [0, \infty] \to \{\mathbb{E}^2_{A_1,A_2}, G_{KL}\}$ with a *geodesic* defined by $\{\kappa_g = 0, (20.42)\}$. Preparatory is Corollary 20.3.

Corollary 20.3 (Geodesic Darboux frame $\{D_1, D_2, D_3\}$, Gauss frame $\{G_1, G_2, G_3\}$ in $\mathbb{E}^2_{A_1,A_2}$).

$$G_{1K}U'^K = \sqrt{G_{11}}\cos\alpha , \ G_{2K}U'^K = \sqrt{G_{22}}\cos\beta$$
$$\forall$$
$$\begin{cases} \cos\alpha := \langle D_1 \,|\, G_1 \rangle / (\|D_1\|\|G_1\|) , \\ \cos\beta := \langle D_1 \,|\, G_2 \rangle / (\|D_1\|\|G_2\|) , \\ \cos(\alpha+\beta) = G_{12}/(\sqrt{G_{11}}\sqrt{G_{22}}) . \end{cases} \tag{20.47}$$

If $G_{12} = 0$, then (20.48) holds.

$$L' := \frac{\mathrm{d}L}{\mathrm{d}S} = \frac{\cos\alpha}{\sqrt{G_{11}}} , \ B' := \frac{\mathrm{d}B}{\mathrm{d}S} = \frac{\sin\alpha}{\sqrt{G_{22}}} . \tag{20.48}$$

End of Corollary.

The proof is straightforward from the definitions of the angles α and β and the representation of the Darboux one-leg D_1 according to (20.11). We have to interpret the results (20.47) and (20.48) as follows: The East azimuth α can be related to L' or B', respectively, either by cosine or sinus normalized by the roots of metric coefficients $\sqrt{G_{11}}$ or $\sqrt{G_{22}}$, respectively. Finally, we define the polar coordinate r as the length S of the *geodesic* starting from the point $P_0(L_0, B_0)$ and leading to the point $P(L, B)$ or from (U_0, V_0) to (U, V), respectively.

Definition 20.4 (Riemann mapping).

The mapping $\{L, B\} \mapsto \{\alpha, r\}$ with respect to the initial point $\{L_0, B_0\}$ is denoted as *Riemann* (polar/normal coordinates) if (20.49) and (20.50) hold.

$$\alpha = \arccos \sqrt{G_{11}} L_0' = \arcsin \sqrt{G_{22}} B_0' \, , \ r = S \, , \tag{20.49}$$

$$x = S\sqrt{G_{11}} L_0' \, , \ y = S\sqrt{G_{22}} B_0' \, . \tag{20.50}$$

Alternatively, the mapping $\{U, V\} \mapsto \{\alpha, r\}$ with respect to the initial point $\{U_0, V_0\}$ is called *Riemann* (polar/normal coordinates) if (20.51) and (20.52) hold.

$$\alpha = \arccos(G_{1K} U_0'^{K} / \sqrt{G_{11}}) \, , \ r = S \, , \tag{20.51}$$

$$x = SG_{1K} U_0'^{K} / \sqrt{G_{11}} \, , \ y = S\sqrt{1 - G_{1K} G_{1L} U_0'^{K} U_0'^{L} / G_{11}} \, . \tag{20.52}$$

End of Definition.

20-2 Lagrange portrait, Hamilton portrait, Lie series, Clairaut constant

Lagrange and Hamilton portrait of a geodesic, Legendre series, Hamilton equations, initial value and boundary value problem, Riemann polar and normal coordinates, Lie series, Clairaut constant, the case of the ellipsoid-of-revolution.

In order to *materialize* the definition of *(Riemann) polar coordinates and normal coordinates*, in particular, Definition 20.4 and formulae (20.49)–(20.52), we have to solve the system of second order ordinary differential equations (20.42) of a geodesic in $\mathbb{E}^2_{A_1, A_2}$. We refer to (20.42) as the geodesic equations in the *Lagrange portrait*: they can be derived from a *stationary Lagrangean functional* of the arc length (20.53) with S_A and S_B as *fixed boundaries*.

$$\delta \int_{S_A=0}^{S_B} dS = 0 \, . \tag{20.53}$$

Alternatively, the geodesic equations as a system of *two second order ordinary differential equations* can be transformed into a system of *four first order ordinary differential equations* subject to the *Hamilton portrait* of a geodesic, for example, following E. Grafarend and R. J. You (1995) as a sample reference. The four first order ordinary differential equations can be reduced to three in case of rotational symmetry, for instance, for an ellipsoid-of-revolution in terms of $\{L, B, \alpha\}$ in phase space. Note that for both systems, we here present to you the first solutions in terms of Legendre series and the second solutions in terms of Hamilton equations, both for the initial value problem and for the boundary value problem.

20-21 Lagrange portrait of a geodesic: Legendre series, initial/boundary values

The ellipsoid-of-revolution $\mathbb{E}^2_{A_1, A_2}$ is an *analytic manifold*. Therefore, in this context the following *Taylor expansion* exists:

$$U^A(S) = U_0^A + SU_0'^A + \tfrac{1}{2!} S^2 U_0''^A + \lim_{n\to\infty} \sum_{m=3}^{n} \tfrac{1}{n!} S^m U_0^{(m)A} \, . \tag{20.54}$$

Question: "But how to *effectively* compute the higher derivatives of $U^A(S)$ with respect to an initial point U_0^A and subject to the differential equations which govern a *geodesic*, a submanifold in $\mathbb{E}^2_{A_1, A_2}$?" Answer: "A proper answer is given by the *Legendre recurrence* ("Legendre series") of $U^{(m)A}$ in terms of U'^A summarized in Box 20.3."

Box 20.3 (The Legendre recurrence of $U^{(m)A}$ in terms of U'^A, index set $A_1, A_2, \ldots, A_{m-1}, A_m \in \{1, 2\}$ (Legendre 1806)).

$$U''^A = -\{{}^A_{A_1 A_2}\} U'^{A_1} U'^{A_2} \quad \text{(geodesic, (20.42))} , \tag{20.55}$$

$$U'''^A =$$

$$= -\{{}^A_{A_1 A_2}\}_{,A_3} U'^{A_3} U'^{A_2} U'^{A_1} + 2\{{}^A_{A_1 A_2}\}\{{}^{A_1}_{A_3 A_4}\} U'^{A_2} U'^{A_3} U'^{A_4} ,$$

$$U^{(4)A} =$$

$$= -\{{}^A_{A_1 A_2}\}_{,A_3 A_4} U'^{A_4} U'^{A_3} U'^{A_2} U'^{A_1} + 3\{{}^A_{A_1 A_2}\}_{,A_3} \{{}^{A_3}_{A_4 A_5}\} U'^{A_1} U'^{A_2} U'^{A_4} U'^{A_5} \tag{20.56}$$

$$+\{{}^A_{A_1 A_2}\}_{,A_3} \{{}^{A_1}_{A_4 A_5}\}^A U'^{A_2} U'^{A_3} U'^{A_4} U'^{A_5} + 6\{{}^A_{A_1 A_2}\}\{{}^{A_1}_{A_3 A_4}\}\{{}^{A_3}_{A_5 A_6}\} U'^{A_2} U'^{A_4} U'^{A_5} U'^{A_6} ,$$

etc.

If we replace $U^{(m)A}$ in the Taylor expansion (20.54) by means of the Legendre recurrence of Box 20.3, we have solved *the initial value problem* of the geodesic (20.42) in terms of power series $U_0'^{A_1}$, $U_0'^{A_1} U_0'^{A_2}$, $U_0'^{A_1} U_0'^{A_2} U_0'^{A_3}$ etc. generating *an exponential map*, in particular

$$U^A(S) =$$

$$= U_0^A + S U_0'^A + S^2 A^A_{A_1, A_2} U_0'^{A_1} U_0'^{A_2} + \cdots + \lim_{n \to \infty} \sum_{m=3}^n S^m A^A_{A_1 A_2 \ldots A_m} U_0'^{A_1} U_0'^{A_2} \ldots U_0'^{A_m} . \tag{20.57}$$

With respect to the *first chart*, part of the minimal atlas of $\mathbb{E}^2_{A_1, A_2}$, namely the orthogonal coordinates $\{L, B\}$, we transform $U_0'^A$ via

$$L_0' = U_0^{1'} = \frac{x}{S\sqrt{G_{11}}} , \quad B_0' = U_0^{2'} = \frac{y}{S\sqrt{G_{22}}} \tag{20.58}$$

into (Riemann) normal coordinates $\{x, y\}$ (inverse Riemann mapping, inverse Riemann cha-cha-cha). Accordingly, we succeed to represent the Taylor series (20.54) in terms of (Riemann) normal coordinates $\{x, y\}$, in particular, $x^1 := x/\sqrt{G_{11}}$ and $x^2 := y/\sqrt{G_{22}}$

$$L(S_B) - L(S_A = 0) = \lim_{n \to \infty} \sum_{m=1}^n A^L_{A_1 \ldots A_m} x^{A_1} \ldots x^{A_m} ,$$

$$\tag{20.59}$$

$$B(S_B) - B(S_A = 0) = \lim_{n \to \infty} \sum_{m=1}^n A^B_{A_1 \ldots A_m} x^{A_1} \ldots x^{A_m} .$$

By *standard series inversion* of the homogeneous two-dimensional polynomial (20.59), we have solved *the boundary value problem* for given values $\{L_A, B_A\} := \{L(S_A = 0), B(S_A = 0)\}$ and $\{L_B, B_B\} := \{L(S_B), B(S_B)\}$ coordinating the points $P_A = P(L_A, B_A)$ and $P_B = P(L_B, B_B)$, respectively, particularly in the form of the polynomial for

$$x^A = \lim_{n \to \infty} \sum_{m=1}^n A^A_{A_1 \ldots A_m} [U^{A_1}(S_B) - U^{A_1}(S_A)] \ldots [U^{A_m}(S_B) - U^{A_m}(S_A)] . \tag{20.60}$$

The *Lagrange portrait* of a geodesic is based upon *Legendre series* up to order five in terms of series $\{U^1, U^2\} = \{L, B\}$, power series $\{S^0, S^1, \ldots, S^n\}$ in terms of *distance functions*. The initial values $\{L_0, B_0, L_0', B_0'\}$ constitute the *initial value problem*. In contrast, in the boundary value problem, the homogeneous polynomial in terms of $\{L_A - L_B, B_A - B_B\}$ as power series is given, while the Riemann Cartesian coordinates $\{x, y\} = \{x^1, x^2\}$ are completely unknown.

20-22 Hamilton portrait of a geodesic: Hamilton equations, initial/boundary values

The *Hamilton portrait* of a *geodesic*, here a submanifold in $\mathbb{E}^2_{A_1, A_2}$, is based upon the generalized momenta given by (20.61), in particular, for an orthogonal set of coordinates $\{L, B\}$ given by (20.62).

$$P_K := G_{KL} U'^L = \begin{cases} \sqrt{G_{11}} \cos \alpha \\ \sqrt{G_{22}} \cos \beta \end{cases}, \tag{20.61}$$

$$P_1 = G_{11} L' = \sqrt{G_{11}} \cos \alpha = N(B) \cos B \cos \alpha ,$$
$$P_2 = G_{22} B' = \sqrt{G_{22}} \sin \alpha M(B) \sin \alpha . \tag{20.62}$$

The *Hamilton equations* of a *geodesic* as a system of four first order ordinary differential equations can be written as (20.63) for a *Hamilton function* (20.64) which is produced by the *Legendre transformation* of the *Lagrange function* $2\mathcal{L}^2 := U'^K G_{KL} U'^L$.

$$\frac{dU^K}{dS} = G^{KL} P_L = \frac{\partial H^2}{\partial P_K} ,$$
$$\frac{dP_K}{dS} = -\frac{1}{2} \frac{\partial G^{AB}}{\partial U^K} P_A P_B = -\frac{\partial H^2}{\partial U^K} , \tag{20.63}$$

$$H^2 := P_K \frac{dU^K}{dS} - \mathcal{L}^2 = \frac{1}{2} G^{KL} P_K P_L . \tag{20.64}$$

First, let us assume that the differentiable manifold $\mathbb{E}^2_{A_1, A_2}$ is partially covered by a set of orthogonal coordinates $\{L, B\}$, especially in the sense of $G_{12} = 0$: the Hamilton equations are firstly specified towards (20.65), (20.66), and (20.67).

$$L' = G^{11} P_1 = \cos \alpha / \sqrt{G_{11}} ,$$
$$B' = G^{22} P_2 = \sin \alpha / \sqrt{G_{22}} , \tag{20.65}$$

$$P_1' = P_L' = -\frac{1}{2}(G_{,L}^{11} P_1^2 + G_{,L}^{22} P_2^2) = -\frac{1}{2}(\cos^2 \alpha G_{11} G_{,L}^{11} + \sin^2 \alpha G_{22} G_{,L}^{22}) , \tag{20.66}$$

$$P_2' = P_B' = -\frac{1}{2}(G_{,B}^{11} P_1^2 + G_{,B}^{22} P_2^2) = -\frac{1}{2}(\cos^2 \alpha G_{11} G_{,B}^{11} + \sin^2 \alpha G_{22} G_{,B}^{22}) . \tag{20.67}$$

If we compare (20.67) and (20.62), differentiated by $(\sqrt{G_{22}} \sin \alpha)'$, $G_{,B}^{11} = G_{11,B}^{-1} = -G_{11}^{-2} G_{11,B}$, and $G_{,B}^{22} = G_{22,B}^{-1} = -G_{22}^{-2} G_{22,B}$, we are led to (20.68). Of course, the same result would have been achieved by the comparison of (20.66) and (20.62), differentiated by $(\sqrt{G_{11}} \cos \alpha)'$.

$$\alpha' = \frac{1}{\sqrt{G_{11} G_{22}}} \left(-\frac{\partial \sqrt{G_{22}}}{\partial L} \sin \alpha + \frac{\partial \sqrt{G_{11}}}{\partial B} \cos \alpha \right) . \tag{20.68}$$

Second, we specify the metric tensor G_{KL} by Box 20.1, 1st chart, in terms of orthogonal coordinates $\{L, B\}$ covering partially $\mathbb{E}^2_{A_1, A_2}$. Obviously, $P'_L = 0$ (which holds for arbitrary surfaces-of-revolution) generates the *conservation of angular momentum* $P_1 = N(B) \cos B \cos \alpha = A$, where the constant A is the *A. C. Clairaut* constant.

$$L' = \frac{\cos \alpha}{N(B) \cos B} \, , \quad B' = \frac{\sin \alpha}{M(B)} \, , \tag{20.69}$$

$$P'_1 = P'_L = [N(B) \cos B \cos \alpha]' = 0 \, , \tag{20.70}$$
$$P'_2 = P'_B = \frac{1}{2} \left[\cos^2 \alpha \frac{\mathrm{d}}{\mathrm{d}B} \ln(N^2(B) \cos^2 B) + \sin^2 \alpha \frac{\mathrm{d}}{\mathrm{d}B} \ln(M^2(B)) \right] \, ,$$

$$\alpha' = -\frac{\tan B}{N(B)} \cos \alpha \, . \tag{20.71}$$

The ellipsoid-of-revolution $\mathbb{E}^2_{A_1, A_2}$ is an *analytic manifold*. Thus, there exists the Taylor expansion in phase space $\{L, B, P_L, P_B\}$ or $\{L, B, \alpha\}$, respectively, namely

$$L(S) = L_0 + SL'_0 + \frac{1}{2!} S^2 L''_0 + \lim_{n \to \infty} \sum_{m=3}^{n} \frac{1}{m!} S^m L_0^{(m)} \, ,$$

$$B(S) = B_0 + SB'_0 + \frac{1}{2!} S^2 B''_0 + \lim_{n \to \infty} \sum_{m=3}^{n} \frac{1}{m!} S^m B_0^{(m)} \, , \tag{20.72}$$

$$\alpha(S) = \alpha_0 + S\alpha'_0 + \frac{1}{2!} S^2 \alpha''_0 + \lim_{n \to \infty} \sum_{m=3}^{n} \frac{1}{m!} S^m \alpha_0^{(m)} \, .$$

Question.

Question: "But how to *effectively* compute the higher derivatives of $\{L_0^{(m)}, B_0^{(m)}, \alpha_0^{(m)}\}$ with respect to an initial point $\{L_0, B_0, \alpha_0\}$ and subject to the differential equations which govern a *geodesic* in the Hamilton portrait, a submanifold in $\mathbb{E}^2_{A_1, A_2}$?" Answer: "A proper answer is immediately given by the *Lie recurrence* ("Lie series") of $\{L_0^{(m)}, B_0^{(m)}, \alpha_0^{(m)}\}$ in terms of $x = r \cos \alpha_0 = S\sqrt{G_{11}} L'_0$ and $y = r \sin \alpha_0 = S\sqrt{G_{22}} B'_0$ summarized in Box 20.4."

If we replace $\{L_0^{(m)}, B_0^{(m)}, \alpha_0^{(m)}\}$ in the Taylor expansion (20.72) by means of the *Lie recurrence* of Box 20.4, we have solved the *initial value problem* of the geodesic (20.69) and (20.71) in the Hamilton portrait in terms of power series x, y, x^2, xy, y^2 etc., in particular

$$L = L_0 + [10]x + [11]xy + [12]xy^2 + [30]x^3 + O_{4L} \, , \tag{20.73}$$

$$B = B_0 + [01]y + [20]x^2 + [02]y^2 + [03]y^3 + O_{4B} \, , \tag{20.74}$$

$$\alpha = \alpha_0 + [10]_\alpha x + [11]_\alpha xy + [12]_\alpha xy^2 + [30]_\alpha x^3 + O_{4\alpha} \, . \tag{20.75}$$

The coefficients $[\mu\nu]$ are given in Box 20.5. Solving the initial value problem for $\mathbb{E}^2_{A_1,A_2}$ with semiaxes A_1 and A_2 of Earth dimension up to an accuracy of $l := L - L_0 = 0''.0003$, $b := B - B_0 = 0''.0002$, and $\alpha - \alpha_0 = 0''.001$, we are limited to distances up to $100\,\text{km}$ for series expansion *up to order five* (H. Boltz 1942).

Box 20.4 (The Lie recurrence ("Lie series") of $L^{(m)}, B^{(m)}$ in terms of $x = r\cos\alpha_0 = SN(B_0)\cos B_0 L_0'$ and $y = r\sin\alpha_0 = SM(B_0)B_0'$, $E^2 = (A_1^2 - A_2^2)/A_1^2$, $N_0 = N(B_0)$, $M_0 = M(B_0)$).

$$L' = \frac{\cos\alpha}{N(B)\cos B}, \quad B' = \frac{\sin\alpha}{M(B)}, \quad \alpha' = -\frac{\tan B}{N(B)}\cos\alpha, \tag{20.76}$$

$$L'' = \frac{-\sin\alpha}{N(B)\cos(B)}\alpha' + \cos\alpha\left(\frac{d}{dB}\frac{1}{N(B)\cos B}\right)B' = \frac{2\sin\alpha\cos\alpha\sin B}{N^2(B)\cos^2 B},$$

$$B'' = \frac{\cos\alpha}{M(B)}\alpha' + \sin\alpha\left(\frac{d}{dB}\frac{1}{M(B)}\right)B' = -\frac{\cos^2\alpha\tan B + 3E^2\sin^2\alpha\tan B}{N^2(1-E^2)^2}, \tag{20.77}$$

$$\alpha'' = \frac{\tan B\sin\alpha}{N(B)}\alpha' - \cos\alpha\left(\frac{d}{dB}\frac{\tan B}{N(B)}\right)B' = -\frac{2\tan^2 B\sin\alpha\cos\alpha}{N^2(B)} - \frac{\sin\alpha\cos\alpha}{N(B)M(B)}$$

etc.

$$SL_0' = \frac{x}{N_0\cos B_0}, \quad SB_0' = \frac{y}{M_0}, \quad S\alpha_0' = \frac{\tan B_0}{N_0}x, \tag{20.78}$$

$$SL_0'' = \frac{2\sin B_0}{N_0^2\cos^2 B_0}xy,$$

$$SB_0'' = -\frac{\tan B_0}{N_0^2(1-E^2)^2}x^2 - \frac{3E^2\tan B_0}{N_0^2(1-E^2)^2}y^2, \tag{20.79}$$

$$S\alpha_0'' = -\left(\frac{2\tan^2 B_0}{N_0^2} + \frac{1}{N_0 M_0}\right)xy,$$

$$E^2 = \frac{A_1^2 - A_2^2}{A_1^2}, \quad N_0 = N(B_0), \quad M_0 = M(B_0) \tag{20.80}$$

etc.

By series inversion of the *homogeneous two-dimensional polynomial* (20.72), $l := L - L_0 = l(x,y)$ and $b := B - B_0 = b(x,y)$, we solved the *boundary value problem* for given values $\{L(S_A = 0), B(S_A = 0)\}$ and $\{L(S_B), B(S_B)\}$, in particular, in the form of the polynomials (20.81) and (20.82). The coefficients $(\mu\nu)$ are given in Box 20.6. Numerical examples for both the initial and the boundary value problem can be found in A. Schoedlbauer (1981b).

$$x = (10)l + (11)lb + (12)lb^2 + (30)l^3 + O_{4x}, \tag{20.81}$$

$$y = (01)b + (20)l^2 + (02)b^2 + (03)b^3 + O_{4y}. \tag{20.82}$$

Box 20.5 (Coefficients of the solution of the initial value problem with respect to the Hamilton portrait $\eta_0^2 := E^2/(1-E^2)\cos^2 B_0$, $V_0^2 := 1 + \eta_0^2$, $t_0 := \tan B_0$).

$$[10] = \frac{1}{N_0 \cos B_0} \;, \quad [01] = \frac{V_0^2}{N_0} \;, \quad [02] = -\frac{3V_0^2\eta_0^2 t_0}{2N_0^2} \;, \quad [20] = -\frac{V_0^2 t_0}{2N_0^2} \;, \quad [11] = \frac{t_0}{N_0^2 \cos B_0} \;, \quad (20.83)$$

$$[12] = \frac{1 + 3t_0^2 + \eta_0^2}{3N_0^3 \cos B_0} \;, \quad [21] = -\frac{V_0^2(1 + 3t_0^2 + \eta_0^2 - 9\eta_0^2 t_0^2)}{6N_0^3} \;, \quad (20.84)$$

$$[03] = -\frac{V_0^2 \eta_0^2 (1 - t_0^2 + \eta_0^2 - 5\eta_0^2 t_0^2)}{2N_0^3} \;, \quad [30] = -\frac{t_0^2}{3N_0^3 \cos B_0} \;, \quad (20.85)$$

$$[13] = \frac{t_0(2 + 3t_0^2 + \eta_0^2 - \eta_0^4)}{3N_0^4 \cos B_0} \;, \quad [31] = -\frac{t_0(1 + 3t_0^2 + \eta_0^2)}{3N_0^4 \cos B_0} \;, \quad (20.86)$$

$$[40] = \frac{V_0^2 t_0(1 + 3t_0^2 + \eta_0^2 - 9\eta_0^2 t_0^2)}{24N_0^4} \;, \quad (20.87)$$

$$[22] = \frac{-V_0^2 t_0(4 + 6t_0^2 - 13\eta_0^2 - 9\eta_0^2 t_0^2 - 17\eta_0^4 + 45\eta_0^4 t_0^2)}{12N_0^4} \;, \quad (20.88)$$

$$[04] = \frac{V_0^2 \eta_0^2 t_0(12 + 69\eta_0^2 - 45\eta_0^2 t_0^2 + 57\eta_0^4 - 105\eta_0^4 t_0^2)}{24N_0^4} \;, \quad (20.89)$$

$$[50] = \frac{t_0^2(1 + 3t_0^2 + \eta_0^2)}{15N_0^5 \cos B_0} \;, \quad (20.90)$$

$$[41] = \frac{V_0^2}{120N_0^5}[1 + 30t_0^2 + 45t_0^4 + \eta_0^2(2 - 72t_0^2 - 90t_0^4) + \eta_0^4(1 - 102t_0^2 + 225t_0^4)] \;, \quad (20.91)$$

$$[32] = \frac{1 + 20t_0^2 + 30t_0^4 + \eta_0^2(2 + 13t_0^2) + \eta_0^4(1 - 7t_0^2)}{15N_0^5 \cos B_0} \;, \quad (20.92)$$

$$[23] = \frac{V_0^2}{60N_0^5}[-4 - 30t_0^2(1 + t_0^2) + 9\eta_0^2(1 + 2t_0^2 + 5t_0^4) +$$
$$+2\eta_0^4(15 - 177t_0^2) + \eta_0^6(17 - 402t_0^2 + 525t_0^4)] \;, \quad (20.93)$$

$$[14] = \frac{1}{15N_0^5 \cos B_0}[2 + 15t_0^2 + 15t_0^4 + 3\eta_0^2(1 + 2t_0^2) - 3\eta_0^4 t_0^2 - \eta_0^6(1 - 6t_0^2)] \;, \quad (20.94)$$

$$[05] = \frac{V_0^2 \eta_0^2}{40N_0^5}[4 - 4t_0^2 + \eta_0^2(27 - 142t_0^2 + 15t_0^4) +$$
$$+2\eta_0^4(21 - 226t_0^2 + 105t_0^4) + \eta_0^6(19 - 314t_0^2 + 315t_0^4)] \;, \quad (20.95)$$

$$[10]_\alpha = -\frac{t_0}{N_0} \;, \quad [11]_\alpha = -\frac{1 + 2t_0^2 + \eta_0^2}{2N_0^2} \;, \quad [30]_\alpha = \frac{t_0(1 + 2t_0^2)}{6N_0^3} \;,$$

$$(20.96)$$

$$[12]_\alpha = -\frac{t_0(5 + 6t_0^2)}{6N_0^3} \;, \quad [31]_\alpha = \frac{1 + 20t_0^2 + 24t_0^4}{24N_0^4} \;, \quad [13]_\alpha = -\frac{5 + 28t_0^2 + 24t_0^4}{24N_0^4} \;.$$

Box 20.6 (Coefficients of the solution of the boundary value problem with respect to the Hamilton portrait $\eta_0^2 := E^2/(1-E^2)\cos^2 B_0$, $V_0^2 := 1 + \eta_0^2$, $t_0 := \tan B_0$).

$$(10) = N_0 \cos B_0 \,, \quad (01) = \frac{1}{V_0^2} N_0 \,, \tag{20.97}$$

$$(02) = \frac{3\eta_0^2 t_0}{2V_0^4} N_0 \,, \quad (20) = \frac{t_0}{2} N_0 \cos^2 B_0 \,, \tag{20.98}$$

$$(11) = -\frac{t_0}{V_0^2} N_0 \cos B_0 \,, \tag{20.99}$$

$$(12) = -\frac{2 + 2\eta_0^2 + 9\eta_0^2 t_0^2}{6V_0^4} N_0 \cos B_0 \,, \quad (21) = \frac{1 - 3t_0^2 + \eta_0^2}{6V_0^2} N_0 \cos^2 B_0 \,, \tag{20.100}$$

$$(03) = \frac{\eta_0^2(1 - t_0^2 + \eta_0^2 + 4\eta_0^2 t_0^2)}{2V_0^6} N_0 \,, \quad (30) = -\frac{t_0^2}{6} N_0 \cos^3 B_0 \,, \tag{20.101}$$

$$(13) = -\frac{\eta_0^2 t_0(7 - 3t_0^2 + 7\eta_0^2 + 12\eta_0^2 t_0^2)}{6V_0^6} N_0 \cos B_0 \,, \quad (31) = -\frac{t_0(1 - t_0^2 + \eta_0^2)}{6V_0^2} N_0 \cos^3 B_0 \,, \tag{20.102}$$

$$(40) = \frac{t_0(1 - t_0^2 + \eta_0^2)}{24} N_0 \cos^4 B_0 \,, \tag{20.103}$$

$$(22) = \frac{-t_0(4 + 3\eta_0^2 + 9\eta_0^2 t_0^2 - \eta_0^4)}{12V_0^4} N_0 \cos^2 B_0 \,, \tag{20.104}$$

$$(04) = \frac{-\eta_0^2 t_0(4 + 17\eta_0^2 - 9\eta_0^2 t_0^2 + 13\eta_0^4 + 76\eta_0^4 t_0^2)}{8V_0^8} N_0 \,, \tag{20.105}$$

$$(50) = -\frac{t_0^2(3 - t_0^2 + 3\eta_0^2)}{120} N_0 \cos^5 B_0 \,, \tag{20.106}$$

$$(41) = \frac{N_0 \cos^4 B_0}{360V_0^2}[7 - 50t_0^2 + 15t_0^4 + 2\eta_0^2(7 - 37t_0^2) + \eta_0^4(7 - 24t_0^2)] \,, \tag{20.107}$$

$$(32) = -\frac{N_0 \cos^3 B_0}{180V_0^4}[8 - 40t_0^2 + \eta_0^2(16 - 31t_0^2 - 45t_0^4) + \eta_0^4(8 + 9t_0^2)] \,, \tag{20.108}$$

$$(23) = \frac{N_0 \cos^2 B_0}{180V_0^6}[-8 - \eta_0^2(7 + 174t_0^2 - 45t_0^4) + \\ + 2\eta_0^4(5 - 83t_0^2 - 90t_0^4) + 3\eta_0^6(3 + 2t_0^2)] \,, \tag{20.109}$$

$$(14) = -\frac{N_0 \cos B_0}{360V_0^8}[8 + 4\eta_0^2(28 - 69t_0^2) + \\ + \eta_0^4(200 - 507t_0^2 + 405t_0^4) + 3\eta_0^6(32 - 77t_0^2 - 1140t_0^4)] \,, \tag{20.110}$$

$$(05) = \frac{N_0\eta_0^2}{40V_0^{10}}[-4 + 4t_0^2 + \eta_0^2(3 - 98t_0^2 + 15t_0^4) + \\ + 2\eta_0^4(9 - 179t_0^2 + 90t_0^4) + \eta_0^6(11 - 256t_0^2 - 1320t_0^4)] \,. \tag{20.111}$$

Important!

For the ellipsoid-of-revolution, we present the solution of the *initial value problem* given $\{L_0, B_0, \alpha_0\}$ relating $x = r\cos\alpha_0 = SN(B_0)\cos B_0 L'_0$, $y = r\sin\alpha_0 = SM(B_0)B'_0$. Box 20.5 contains the coefficients $[\mu\nu]$ up to order five based upon the Lie recurrence. In contrast, we add the solution of the *boundary value problem* $x(l,b)$ and $y(l,b)$ in terms of the coefficients $(\mu\nu)$ up to order five in Box 20.6. This approximation is accurate to the order smaller than $0''.0003$ for $l := L - L_0$, $0''.0002$ for $b := B - B_0$, and $0''.001$ for $\alpha - \alpha_0$ for distances up to $100\,\mathrm{km}$ (Boltz approximation).

20-3 Soldner coordinates: geodetic parallel coordinates

Soldner coordinates: geodetic parallel coordinates. Geodetic projection, geodetic field of geodesics, and meridian convergence. Soldner coordinates as elements of an ellipsoid-of-revolution.

In this section, "geodetic parallel coordinates" as developed by J. H. Soldner (astronomer and geodesist: 1776–1833) are reviewed. His basic paper is entitled J. H. Soldner, Theorie der Landervermessung 1810, Herausgeber J. Frischauf, Oswalds Klassiker der exakten Wissenschaften, No. 184, Leipzig 1911. It formed the basis of local surveys in many countries like Bavaria, Württemberg, Baden, Hessen, and Prussia. The basis of "geodetic parallel coordinates" competing with C. F. Gauss's Gauss–Krueger coordinates is the notion of a "geodetic field of geodesics" in the following sense.

Important!

If a field of geodesics *and* its orthogonal trajectories are used as coordinate lines, then we can use its arc length of geodesics as coordinates. In reverse, if we express in a coordinate system the line element as $\mathrm{d}s^2 = \mathrm{d}u^2 + g(u,v)\mathrm{d}v^2$, in consequence, the u coordinate line is a geodesic.

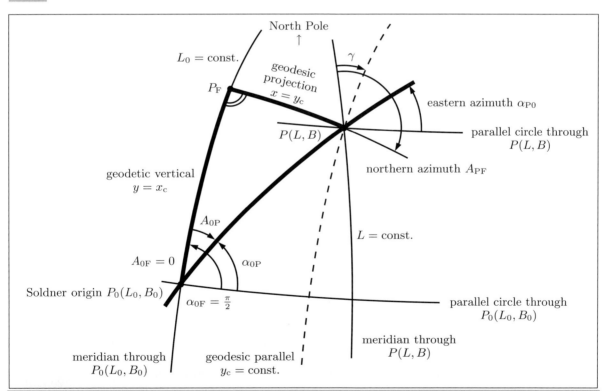

Fig. 20.4. Soldner coordinates, geodetic parallels, conventional coordinate system $\{x_\mathrm{c}, y_\mathrm{c}\}$, systematic Soldner coordinates $\{y,x\} = \{x_\mathrm{c}, y_\mathrm{c}\}$, footpoint P_F of a geodetic projection, conventional (northern) azimuth A, systematic (eastern) azimuth α.

The construction principle is as follows. A point $P_0(L_0, B_0)$ is defined as the *origin* of a *local reference system*. The x_c axis of the coordinate system, which is called "Hochwert" (Soldner wording), agrees with the chosen *reference meridian* L_0. A point $P(L, B)$ is described by *geodetic parallel coordinates* as follows. Through the point P, we compute the *unique geodetic line*, which cuts the meridian L_0 at a right angle to produce the *footprint point* P_F. The length of the geodesic P–P_F is chosen as the y_c coordinate, which is called "Rechtswert" (Soldner wording). The angle which is in between the local meridian of the point P *and* the *geodetic parallel* through the point P is called *meridian convergence* $\gamma = A_{PF} - \pi/2$. The angle γ is fixed as the northern part of the meridian, lefthand-oriented positive. We always say that the azimuth of the coordinate line $y_c = $ const. is the angle γ. Most notable, the geodetic parallel $y_c = $ const. is *not* a geodesic. Compare with Fig. 20.4.

Important!

> The y_c lines produce the *geodesic field*. In contrast, the geodetic parallels are the orthogonal trajectories $ds^2 = E(x_c, y_c)dx_c^2 + dy_c^2 = dx^2 + G(x, y)dy^2$.

20-31 First problem of Soldner coordinates: geodetic parallel coordinates, input $\{L_0, B_0, x_c = y, y_c = x\}$ versus output $\{L, B, \gamma \text{ (meridian convergence)}\}$

For the problem of given coordinates $\{L_0, B_0, x_c = y, y_c = x\}$ and unknown coordinates $\{L, B, \gamma\}$, we use the standard method of Legendre series $u = s \cos\alpha_{0P}$ and $v = s \sin\alpha_{0P}$. In Box 20.7, the details are collected.

Box 20.7 (The problem of given coordinates $\{L_0, B_0, x_c = y, y_c = x\}$ and unknown coordinates $\{L, B, \gamma\}$).

First, we rewrite (20.73)–(20.75) in terms of $u = s \cos\alpha_{0P}$ and $v = s \sin\alpha_{0P}$ and obtain (the coefficients $[\mu\nu]$ are collected in Box 20.5 and have to be computed at the point B_0)

$$
\begin{aligned}
B_P = B_0 &+ [01]v & \text{versus} \quad L_P = L_0 &+ [10]u & \text{versus} \quad \alpha_{P0} = \alpha_{0P} &+ [10]_\alpha u \\
&+ [20]u^2 & &+ [11]uv & &+ [11]_\alpha uv \\
&+ [02]v^2 & &+ [12]uv^2 & &+ [12]_\alpha uv^2 \\
&+ [21]u^2v & &+ [30]u^3 & &+ [30]_\alpha u^3 \\
&+ [03]v^3 & &+ [31]u^3v & &+ [31]_\alpha u^3v \quad (20.112) \\
&+ [40]u^4 & &+ [13]uv^3 & &+ [13]_\alpha uv^3 \\
&+ [22]u^2v^2 & &+ \ldots & &+ \ldots \\
&+ [04]v^4 & & & & \\
&+ \ldots & & & &
\end{aligned}
$$

First step: determine L_F, B_F, α_{F0}, starting point P_0, $s = y = x_c$, $x = y_c = 0$, $\alpha_{F0} = \pi/2$ given.

As a result, we compute $\{u = 0, v = y\}$ and obtain the series (the coefficients $[\mu\nu]$ are collected in Box 20.5 and have to be computed at the point B_0)

$$B_F - B_0 = [01]y + [02]y^2 + [03]y^3 + [04]y^4 + \cdots \quad \text{(a polynomial in } y\text{)} ,$$
$$L_F - L_0 = 0 , \quad \alpha_{F0} - \alpha_{0F} = 0 , \tag{20.113}$$

$$B_F = \bar{B}_0 + (B_F - B_0) , \quad L_F = L_0 , \quad \alpha_{F0} = \alpha_{0F} = \pi/2 . \tag{20.114}$$

Continuation of Box.

Second step: determine L, B, γ, starting point P_{F}, $s = x = y_{\mathrm{c}}$, $y = x_{\mathrm{c}} = 0$, $\alpha_{\mathrm{FP}} = \alpha_{\mathrm{F0}} + 3\pi/2 = 0$ given.
As a result, we compute $u = x, v = 0$, which leads to the series
(the coefficients $[\mu\nu]$ are collected in Box 20.5 and have to be computed at the point B_{F})

$$B - B_{\mathrm{F}} = [20]x^2 + [40]x^4 + \cdots \quad \text{(an even polynomial in } x\text{)},$$

$$L - L_{\mathrm{F}} = [10]x + [30]x^3 + [50]x^5 + \cdots \quad \text{(an odd polynomial in } x\text{)}, \tag{20.115}$$

$$\alpha_{\mathrm{PF}} = \gamma = [10]_\alpha x + [30]_\alpha x^3 + \cdots \pm \pi \quad \text{(an odd polynomial in } x\text{)},$$

$$B = B_{\mathrm{F}} + (B - B_{\mathrm{F}}),$$
$$L = L_{\mathrm{F}} + (L - L_{\mathrm{F}}). \tag{20.116}$$

An alternative procedure is the following one-step method: set up a Taylor series for the coefficients $[\mu\nu]$ in (20.115), which have to be computed at the point B_{F}, with B_0 as the point of expansion, i.e.

$$[\mu\nu]_{\mathrm{F}} = [\mu\nu]_0 + \left.\frac{\partial[\mu\nu]}{\partial B}\right|_0 (B_{\mathrm{F}} - B_0) + \frac{1}{2!}\left.\frac{\partial^2[\mu\nu]}{\partial B^2}\right|_0 (B_{\mathrm{F}} - B_0)^2 + \mathrm{hit}\,([\mu\nu]), \tag{20.117}$$

$$[\mu\nu]_{\alpha\,\mathrm{F}} = [\mu\nu]_{\alpha\,0} + \left.\frac{\partial[\mu\nu]_\alpha}{\partial B}\right|_0 (B_{\mathrm{F}} - B_0) + \frac{1}{2!}\left.\frac{\partial^2[\mu\nu]_\alpha}{\partial B^2}\right|_0 (B_{\mathrm{F}} - B_0)^2 + \mathrm{hit}\,([\mu\nu]_\alpha). \tag{20.118}$$

The terms $B_{\mathrm{F}} - B_0$, $(B_{\mathrm{F}} - B_0)^2$ etc. are computed from the first step, for instance,

$$B_{\mathrm{F}} - B_0 = [01]_{\,0}\, y + [02]_{\,0}\, y^2 + [03]_{\,0}\, y^3 + [04]_{\,0}\, y^4 + \cdots,$$
$$\left(B_{\mathrm{F}} - B_0\right)^2 = \left([01]_{\,0}\, y + [02]_{\,0}\, y^2 + [03]_{\,0}\, y^3 + [04]_{\,0}\, y^4 + \cdots\right)^2 \quad \text{etc.}, \tag{20.119}$$

leading to

$$[\mu\nu]_{\mathrm{F}} = [\mu\nu]_0 + \left.\frac{\partial[\mu\nu]}{\partial B}\right|_0 \left([01]_{\,0}\, y + [02]_{\,0}\, y^2 + [03]_{\,0}\, y^3 + [04]_{\,0}\, y^4 + \cdots\right) +$$
$$+ \frac{1}{2!}\left.\frac{\partial^2[\mu\nu]}{\partial B^2}\right|_0 \left([01]_{\,0}\, y + [02]_{\,0}\, y^2 + [03]_{\,0}\, y^3 + [04]_{\,0}\, y^4 + \cdots\right)^2$$
$$+ \cdots + \tag{20.120}$$
$$+ \frac{1}{n!}\left.\frac{\partial^n[\mu\nu]}{\partial B^n}\right|_0 \left([01]_{\,0}\, y + [02]_{\,0}\, y^2 + [03]_{\,0}\, y^3 + [04]_{\,0}\, y^4 + \cdots\right)^n,$$

$$[\mu\nu]_{\alpha\,\mathrm{F}} = [\mu\nu]_{\alpha\,0} + \left.\frac{\partial[\mu\nu]_\alpha}{\partial B}\right|_0 \left([01]_{\alpha\,0}\, y + [02]_{\alpha\,0}\, y^2 + [03]_{\alpha\,0}\, y^3 + [04]_{\alpha\,0}\, y^4 + \cdots\right) +$$
$$+ \frac{1}{2!}\left.\frac{\partial^2[\mu\nu]_\alpha}{\partial B^2}\right|_0 \left([01]_{\alpha\,0}\, y + [02]_{\alpha\,0}\, y^2 + [03]_{\alpha\,0}\, y^3 + [04]_{\alpha\,0}\, y^4 + \cdots\right)^2$$
$$+ \cdots + \tag{20.121}$$
$$+ \frac{1}{n!}\left.\frac{\partial^n[\mu\nu]_\alpha}{\partial B^n}\right|_0 \left([01]_{\alpha\,0}\, y + [02]_{\alpha\,0}\, y^2 + [03]_{\alpha\,0}\, y^3 + [04]_{\alpha\,0}\, y^4 + \cdots\right)^n.$$

The final step consists of substituting (20.120) and (20.121) into (20.115), and summing up (20.115) and (20.114). The finally resulting representation for the ellipsoidal coordinates $\{L, B\}$ of point P, given the Soldner coordinates $\{x, y\}$ with respect to the point $P_0(L_0, B_0)$, is of the form (20.73)–(20.75) with Soldner coefficients $[\mu\nu]_L^S$ and $[\mu\nu]_B^S$:

$$L = L_{\mathrm{P}} = L_0 + \sum_{\mu=0}^{\infty} \sum_{\nu=0}^{\infty} [\mu\nu]_L^S x^\mu y^\nu \ ,$$

$$B = B_{\mathrm{P}} = B_0 + \sum_{\mu=0}^{\infty} \sum_{\nu=0}^{\infty} [\mu\nu]_B^S x^\mu y^\nu \ , \tag{20.122}$$

$$\gamma = \alpha_{\mathrm{PF}} = \sum_{\mu=0}^{\infty} \sum_{\nu=0}^{\infty} [\mu\nu]_\alpha^S x^\mu y^\nu \ .$$

As an alternative, the numerical computation of $\{L, B, \gamma\}$ is done using the two-step-method combined with a numerical integration of (20.69) and (20.71), for example, by classical Runge–Kutta techniques. First integrate (20.69) and (20.71) numerically with initial values (20.123) in order to obtain (20.124), then integrate (20.69) and (20.71) a second time with initial values (20.125) with the result for (20.126):

$$\{L_0 = L_{P_0}, B_0 = B_{P_0}, s = x_{\mathrm{c}} = y, \alpha_0 = \alpha_{0F} = \pi/2\} \ , \tag{20.123}$$

$$\{L_{\mathrm{F}}, B_{\mathrm{F}}, \alpha_{F0} = \alpha_{0F}\} \ , \tag{20.124}$$

$$\{L_0 = L_F, B_0 = B_F, s = y_{\mathrm{c}} = x, \alpha_0 = \alpha_{FP} = 0\} \ , \tag{20.125}$$

$$\{L = L_{\mathrm{P}}, B = B_{\mathrm{P}}, \gamma = \alpha_{\mathrm{PF}}\} \ . \tag{20.126}$$

A numerical example is given in Table 20.2.

Table 20.2. First problem of Soldner coordinates: input $\{L_0, B_0, x, y\}$ versus output $\{L, B, \gamma = \alpha_{\mathrm{PF}}\}$.

Reference ellipsoid, Bessel:

$A_1 = 6\,377\,397.155\,\mathrm{m}, \ E^2 = 0.006\,674\,372\,20.$

Soldner origin, Tübingen observatory:

$B_0 = 48°31'15".723\,4\,\mathrm{N}, \ L_0 = 9°3'7".144\,5\,\mathrm{E}.$

Soldner coordinates:

$y = 29\,682.228\,\mathrm{m}, \ x = 2\,469.517\,\mathrm{m}.$

Ellipsoidal coordinates, footpoint P_{F}:

$B_{\mathrm{F}} = 48°47'16".741\,0, \ L_{\mathrm{F}} = 9°3'7".144\,5.$

Ellipsoidal coordinates, point P:

$B = B_{\mathrm{P}} = 48°47'16".723\,4, \ L - L_{\mathrm{P}} = 9°5'8".145\,0.$

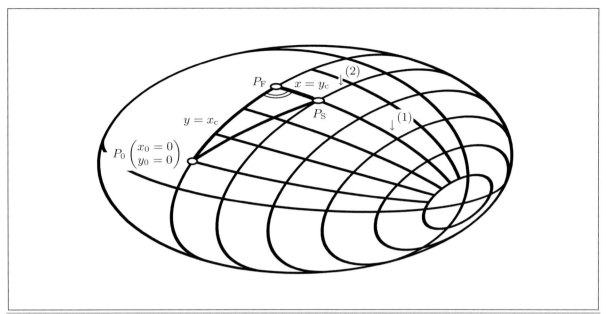

Fig. 20.5. Soldner coordinates as elements of an ellipsoid-of-revolution. $y = $ const. (1), $x = $ const. (2).

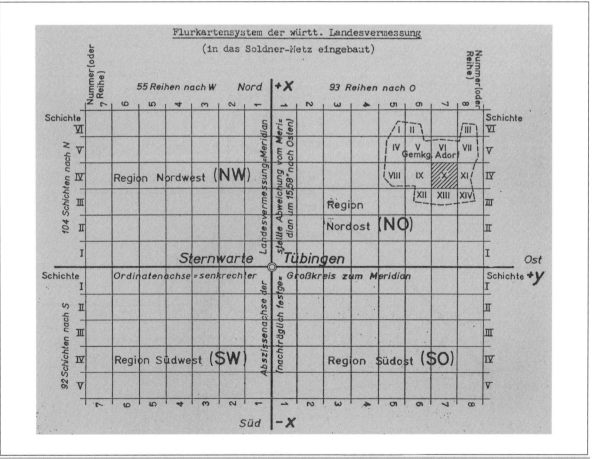

Fig. 20.6. An example of a Soldner map, centered at the Tübingen Observatory, Germany, original map. $x \equiv y_c$, $y \equiv y_c$.

20-32 Second problem of Soldner coordinates: geodetic parallel coordinates, input $\{L, B, L_0, B_0\}$ versus output $\{x = y_c, y = x_c\}$

The inverse problem to derive the Soldner coordinates $\{x = y_c, y = x_c\}$ from the given values of $\{L, B, L_0, B_0\}$ is again solved by series inversion, the result of which is provided by (20.127) and (20.128), with $(\mu\nu)_x^S$ and $(\mu\nu)_y^S$ as Soldner inverse series coefficients.

$$x = \sum_{\mu=0}^{\infty} \sum_{\nu=0}^{\infty} (\mu\nu)_x^S (B - B_0)^\mu (L - L_0)^\nu \,, \tag{20.127}$$

$$y = \sum_{\mu=0}^{\infty} \sum_{\nu=0}^{\infty} (\mu\nu)_y^S (B - B_0)^\mu (L - L_0)^\nu \,. \tag{20.128}$$

20-4 Fermi coordinates: oblique geodetic parallel coordinates

Fermi coordinates: oblique geodetic parallel coordinates. Geodetic projection. The two-step solution and the two-step solution equations.

E. Fermi (1901–1954), Italian–American physicist, developed special geodetic coordinates which might be called "oblique geodetic parallel coordinates". They are generated as outlined in Box 20.8. Compare with Fig. 20.7.

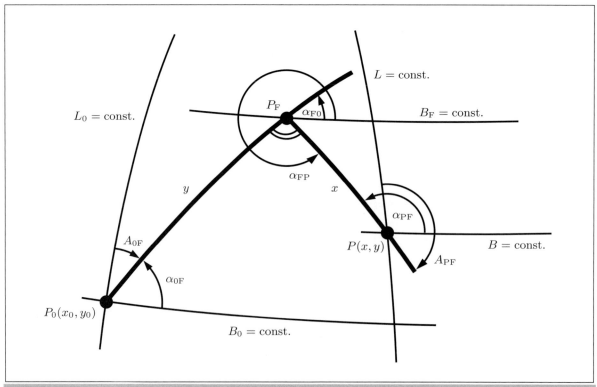

Fig. 20.7. Fermi coordinates $\{x, y\}$. $\alpha_{FP} = \alpha_{F0} + 3\pi/2$.

Box 20.8 (Oblique geodetic parallel coordinates, Fermi coordinates).

Choose an *origin* $P_0(L_0, B_0)$ of an arbitrary coordinate system and a second point $P(L, B)$ moving on the ellipsoid-of-revolution. Choose a *fixed reference point* $P_F(L_F \neq L_0, B_F)$. The first coordinate axis is generated as the *geodetic projection* of the point P onto the point P_F. The second coordinate axis can be arbitrarily chosen, for instance, as a second geodetic projection of the point P_F onto the point P_0, which meets at the point P_F at right angles. Such a pair of coordinates $\{x, y\}$ establishes two geodesics, which are easily computed once P_0 and P_F are fixed. Here, we only present the two-step solution. We note in passing that the one-step solution operates as described in Section 20-3 dealing with Soldner coordinates.

First step:

determine L_F, B_F, $\alpha_{F0} \neq \alpha_{0F} \neq \pi/2$, starting point P_0, $s = y = x_c$, $x = y_c = 0$, $\alpha_{0F} \neq \pi/2$ given.

As a result, $u = y \cos \alpha_{0F} =: u_0 \neq 0$ and $v = y \sin \alpha_{0F} =: v_0 \neq y$ lead to the general series (20.112),

just replacing B_P, L_P, α_{P0}, and α_{0P} by B_F, L_F, α_{F0}, and α_{0F}, respectively.

Second step:

determine B_F, L_F, $\alpha_{FP} \neq \gamma$, starting point P_F, $s = x = y_c$, $y = x_c = 0$, $\alpha_{FP} = \alpha_{F0} + 3\pi/2$ given.

$$
\begin{aligned}
L &= L_F + [10]_F u_F + [11]_F u_F v_F + [12]_F u_F v_F^2 + \cdots \\
&= L_0 + [10]_0 u_0 + [11]_0 u_0 v_0 + [12]_0 u_0 v_0^2 + \cdots \\
&\quad + [10]_F u_F + [11]_F u_F v_F + [12]_F u_F v_F^2 + \cdots,
\end{aligned}
\tag{20.129}
$$

$$
\begin{aligned}
B &= B_F + [01]_F v_F + [20]_F u_F^2 + [02]_F v_F^2 + [21]_F u_F^2 v_F + \cdots \\
&= B_0 + [01]_0 v_0 + [20]_0 u_0^2 + [02]_0 v_0^2 + [21]_0 u_0^2 v_0 + \cdots \\
&\quad + [01]_F v_F + [20]_F u_F^2 + [02]_F v_F^2 + [21]_F u_F^2 v_F + \cdots,
\end{aligned}
\tag{20.130}
$$

$$
\begin{aligned}
\alpha_{PF} &= \alpha_{FP} + [10]_{\alpha F} u_F + [11]_{\alpha F} u_F v_F + [12]_{\alpha F} u_F v_F^2 + \cdots \\
&= \alpha_{F0} + [10]_{\alpha F} u_F + [11]_{\alpha F} u_F v_F + [12]_{\alpha F} u_F v_F^2 + \cdots + \tfrac{3\pi}{2} \\
&= \alpha_{0F} + [10]_{\alpha 0} u_0 + [11]_{\alpha 0} u_0 v_0 + [12]_{\alpha 0} u_0 v_0^2 + \cdots + \tfrac{3\pi}{2} \\
&\quad + [10]_{\alpha F} u_F + [11]_{\alpha F} u_F v_F + [12]_{\alpha F} u_F v_F^2 + \cdots.
\end{aligned}
\tag{20.131}
$$

The coefficients $[\mu\nu]_0$ and $[\mu\nu]_F$, respectively, have to be computed at the point B_0 and B_F, respectively.

A numerical example is given in Tables 20.3 and 20.4.

Table 20.3. Fermi coordinates: input $\{L_0, B_0, x, y\}$ versus output $\{L, B\}$. Part one.

Reference ellipsoid, WGS84 (GRS80):

$A_1 = 6\,378\,137$ m, $E^2 = 0.006\,694\,380\,02$.

Fermi origin:

$B_0 = 48°$ N, $L_0 = 9°$ E.

Fermi coordinates:

$y = 234\,123.034$ m, $x = 65\,356.124$ m.

Initial azimuth α_{0F}:

$\alpha_{0F} = 50°48'13''.512\,4$.

Table 20.4. Fermi coordinates: input L_0, B_0, x, y versus output L, B . Part two.

Ellipsoidal coordinates, footpoint P_F:

$B_F = 49°36'49".5658, \; L_F = 11°2'50".4690$.

Azimuth α_{F0}:

$\alpha_{F0} = 49°15'46".4706$.

Azimuth α_{FP}:

$\alpha_{FP} = 319°15'46".4706$.

Azimuth α_{PF}:

$\alpha_{PF} = 318°44'47".4562$.

Ellipsoidal coordinates, point P:

$B = B_P = 49°13'41".7797, \; L = L_P = 11°43'38".0925$.

Length of the geodesic $s = P_0 - P$ as a result from the corresponding boundary value problem:

$s = 243\,070.157 \text{ m}$.

Azimuth of s in P_0:

$\alpha_{0P} = 35°12'9".6294$.

Azimuth of s in P:

$\alpha_{P0} = 33°9'22".3342$.

20-5 Deformation analysis: Riemann, Soldner, Gauss–Krueger coordinates

Riemann coordinates, Soldner coordinates, and Gauss–Krueger coordinates. Deformation analysis: Cauchy–Green matrix, Jacoby matrix, principal distortions.

In the following, the metric dS^2 of $\mathbb{E}^2_{A_1, A_2}$ is to be compared with the metric ds^2 of the chart established by *Riemann normal coordinates* $\{x, y\}$, namely

$$dS^2 = G_{KL}(U^M)dU^K dU^L \quad \text{versus} \quad ds^2 = g_{kl}(u^m)du^k du^l \; , \qquad (20.132)$$

$$dS^2 = N^2(B)\cos^2 B dL^2 + M^2(B)dB^2 \quad \text{versus} \quad ds^2 = dx^2 + dy^2 = r^2 d\alpha^2 + dr^2 \; , \quad (20.133)$$

subject to the mapping equations $x(l, b)$ $(\to (20.81))$ and $y(l, b)$ $(\to (20.82))$. By *pullback*, we derive the *left Cauchy-Green tensor* C_l with the *right metric matrix* $\mathsf{G}_r = \mathsf{I}_2$ and the *left Jacobian matrix* J_l, namely

$$\mathsf{C}_l = \mathsf{J}_l^T \mathsf{G}_r \mathsf{J}_l \quad \text{or} \quad c_{KL} = \frac{\partial x^i}{\partial U^K}\delta_{ij}\frac{\partial x^j}{\partial U^L} \; , \qquad (20.134)$$

$$\mathsf{J}_l = \begin{bmatrix} \frac{\partial x}{\partial L} & \frac{\partial x}{\partial B} \\ \frac{\partial y}{\partial L} & \frac{\partial y}{\partial B} \end{bmatrix} =$$

$$= \begin{bmatrix} (10) + (11)b + (12)b^2 + 3(30)l^2 + O_{3x} & (11)l + 2(12)lb + O_{3x} \\ 2(20)l + O_{3y} & (01) + 2(02)b1 + 3(03)b^2 + O_{3y} \end{bmatrix} \qquad (20.135)$$

The matrix $[c_{KL}]$ is given in Box 20.9. Let us canonically compare the two positive-definite, symmetric matrices left metric matrix $\mathsf{G}_l = \{G_{KL}\}$ and left Cauchy-Green matrix $\mathsf{C}_l = \{c_{KL}\}$, namely by the *simultaneous diagonalization* (20.136).

$$\mathsf{F}_l^{\mathsf{T}}\mathsf{C}_l\mathsf{F}_l = \operatorname{diag}(A_1^2, A_2^2) \quad \text{versus} \quad \mathsf{F}_l^{\mathsf{T}}\mathsf{G}_l\mathsf{F}_l = \mathsf{I}_2 \ . \tag{20.136}$$

By means of the left Frobenius matrix F_l, the left Cauchy–Green matrix C_l is transformed into a diagonal matrix, $((A_1, A_2)$ are the eigenvalues). In contrast, the left metric matrix G_l is transformed into a unit matrix I_2. The problem to simultaneously diagonalize the two symmetric matrices $\{\mathsf{C}_l, \mathsf{G}_l\}$, where G_l is positive-definite, is equivalent to the *general eigenvalue–eigenvector problem*. Here, we are left with *eigenvalues* $\{A_1, A_2 = 1\}$ (typical for (Riemann) polar/normal coordinates), which are given by Box 20.10. Finally, we compute the *maximum angular distortion* (20.137) in Box 20.11.

$$\omega := 2\arcsin\frac{A_1 - A_2}{A_1 + A_2} \ . \tag{20.137}$$

Box 20.9 (Left Cauchy–Green deformation tensor, Riemann coordinates).

$$c_{11} =$$

$$= \left(1 - \frac{2t_0}{V_0^2}b + \frac{3t_0^2 - 2 - 2\eta_0^2 - 9\eta_0^2 t_0^2}{3V_0^4}b^2 - \frac{\cos^2 B_0 t_0}{3}bl^2 + \frac{t_0(2 - \eta_0^2(5 + 12t_0^2) - \eta_0^4(7 + 12t_0^2))}{3V_0^6}b^3 + \right.$$

$$+ \frac{V_0^2 t_0^2 \cos^4 B_0}{12}l^4 - \frac{(14 - 60t_0^2 + \eta_0^2(28 - 63t_0^2) + \eta_0^4(14 - 3t_0^2))\cos^2 B_0}{90V_0^4}b^2 l^2 +$$

$$\left. + \frac{b^4}{60V_0^8}\left[4 - \eta_0^2(24 - 292t_0^2 + 60t_0^4) - \eta_0^4(60 - 369t_0^2 - 240t_0^4) - \eta_0^6(32 - 77t_0^2 - 1140t_0^4)\right]\right) \times$$

$$\times N_0^2 \cos^2 B_0 \ ,$$

$$c_{12} = c_{21} =$$

$$= \left(-\frac{1}{3V_0^2}bl + \frac{t(2 - 3\eta_0^2 - 5\eta_0^4)}{6V_0^6}b^2 l + \frac{\cos^2 B_0 t_0}{6}l^3 + \right.$$

$$+ \frac{4 - 2\eta_0^2(17 - 81t_0^2) - \eta_0^4(80 - 39t_0^2) - \eta_0^6(42 + 123t_0^2)}{90V_0^8}b^3 l +$$

$$\left. + \frac{(4 - 10t_0^2 + \eta_0^2(8 + 17t_0^2) + \eta_0^4(4 + 27t_0^2))\cos^2 B_0}{90V_0^4}bl^3\right) \times \tag{20.138}$$

$$\times N_0^2 \cos^2 B_0 \ ,$$

$$c_{22} =$$

$$= \left(\frac{1}{V_0^4} + \frac{6\eta_0^2 t_0}{V_0^6}b + \frac{3(1 - t_0^2 + \eta_0^2 + 7\eta_0^2 t_0^2)}{V_0^8}b^2 + \frac{\cos^2 B_0}{3V_0^2}l^2 - \frac{4\eta_0^2 t_0(1 + 2\eta_0^2 + \eta_0^4(1 + 10t_0^2))}{V_0^{10}}b^3 + \right.$$

$$+ \frac{4\eta_0^2 t_0 \cos^2 B_0}{3V_0^4}bl^2 + \frac{(3 - 5t_0^2 + \eta_0^2(6 - 11t_0^2) + 3\eta_0^4(1 - 2t_0^2))\cos^4 B_0}{45V_0^4}l^4 +$$

$$+ \frac{4(2 + \eta_0^2(13 - 9t_0^2) + \eta_0^4(20 + 27t_0^2) + 9\eta_0^6(1 + 4t_0^2))\cos^2 B_0}{45V_0^8}b^2 l^2 +$$

$$\left. + \frac{b^4}{V_0^{12}}\left[\eta_0^2(t_0^2 - 1 + \eta_0^2(3 - 41t_0^2 + 6t_0^4) + \eta_0^4(9 - 127t_0^2 + 54t_0^4) + \eta_0^6(5 - 85t_0^2 - 522t_0^4))\right]\right) \times$$

$$\times N_0^2 \ .$$

Box 20.10 (General eigenvalue problem (Λ_1, Λ_2), Riemann coordinates).

$$\Lambda_1 = 1 + \frac{1}{6V_0^2}b^2 + \frac{V_0^2 \cos^2 B_0}{6}l^2 - \frac{t_0(1+2\eta_0^2)\cos^2 B_0}{6}bl^2 +$$

$$+ \frac{\eta_0^2 t_0(1 - \eta_0^2(71 - 72t_0^2) - 72\eta_0^4(1 - 4t_0^2))}{6V_0^6}b^3 +$$

$$+ \frac{(7 - 5t_0^2 + \eta_0^2(14 - 29t_0^2) + \eta_0^4(7 - 24t_0^2))\cos^4 B_0}{360}l^4 -$$

$$- \frac{(1 + \eta_0^2(8 - 7t_0^2) + 13\eta_0^4(1 - t_0^2) + 6\eta_0^6(1 - t_0^2))\cos^2 B_0}{60V_0^4}b^2l^2 +$$

$$+ \frac{b^4}{360V_0^8}\Big[7 + 2\eta_0^2(19 - 12t_0^2) + \eta_0^4(55 + 1107t_0^2 - 1080t_0^4) +$$

$$+ \eta_0^6(24 - 2109t_0^2 + 7560t_0^4) - 3240\eta_0^8 t_0^2(1 + 4t_0^2)\Big] ,$$

$$\Lambda_2 = 1 .$$

(20.139)

Box 20.11 (Maximum angular distortion (ω), Riemann coordinates).

$$\omega = \frac{1}{6V_0^2}b^2 + \frac{V_0^2 \cos^2 B_0}{6}l^2 - \frac{t_0(1+2\eta_0^2)\cos^2 B_0}{6}bl^2 +$$

$$+ \frac{\eta_0^2 t_0(1 - \eta_0^2(71 - 72t_0^2) - 72\eta_0^4(1 + 4t_0^2))}{6V_0^6}b^3 +$$

$$+ \frac{(2 - 5t_0^2 + \eta_0^2(4 - 29t_0^2) + \eta_0^4(2 - 24t_0^2))\cos^4 B_0}{360}l^4 -$$

$$- \frac{(8 + \eta_0^2(34 - 21t_0^2) + \eta_0^4(44 - 39t_0^2) + 18\eta_0^6(1 - t_0^2))\cos^2 B_0}{180V_0^4}b^2l^2 +$$

$$+ \frac{b^4}{360V_0^8}\Big[2 + 4\eta_0^2(7 - 6t_0^2) + \eta_0^4(50 + 1107t_0^2 - 1080t_0^4) +$$

$$+ \eta_0^6(24 - 2109t_0^2 + 7560t_0^4) - 3240\eta_0^8 t_0^2(1 + 4t_0^2)\Big] .$$

(20.140)

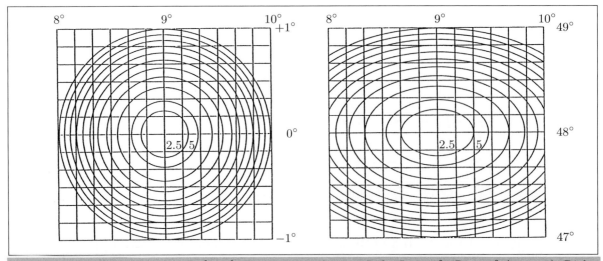

Fig. 20.8. Principal distortion Λ_1 [ppm], Riemann coordinates. Left: $L_0 = 9°$, $B_0 = 0°$ (equator). Right: $L_0 = 9°$, $B_0 = 48°$.

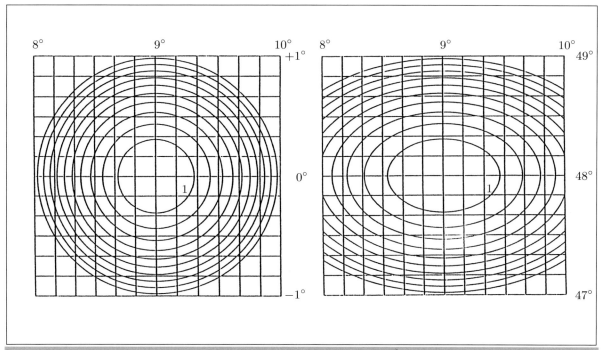

Fig. 20.9. Maximum angular distortion ω [$\circ$], Riemann coordinates. Left: $L_0 = 9°$, $B_0 = 0°$ (equator). Right: $L_0 = 9°$, $B_0 = 48°$.

Figures 20.8–20.10 illustrate over a $2° \times 2°$ grid the principal distortion $\Lambda_1(x, y)$ as well as the maximum angular distortion ω for $L_0 = 9°$ and (i) $B_0 = 0°$, (ii) $B_0 = 48°$, and (iii) $B_0 = 70°$, respectively. Figure 20.11 illustrates the local Riemann mapping and the distortion ellipses with axes $\{\Lambda_1, \Lambda_2 = 1\}$.

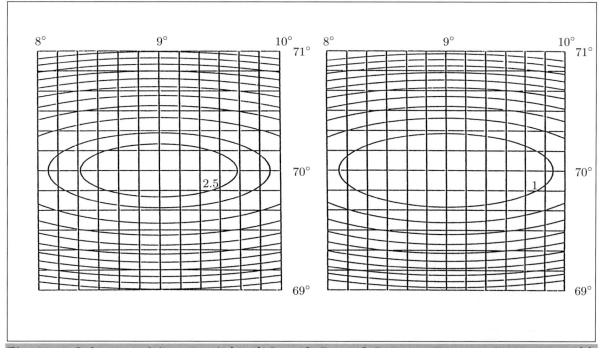

Fig. 20.10. Left: principal distortion Λ_1 [ppm], $L_0 = 9°$, $B_0 = 70°$. Right: maximum angular distortion ω [$\circ$], $L_0 = 9°$, $B_0 = 70°$.

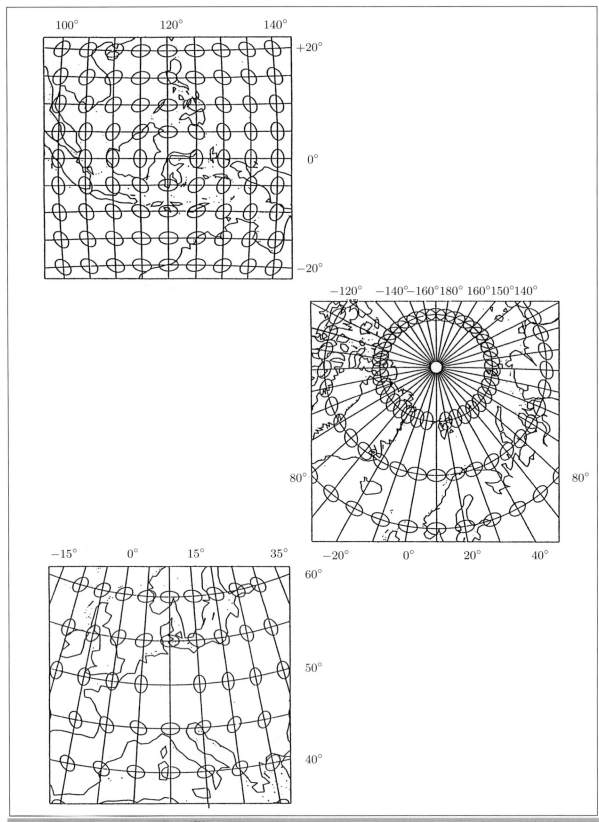

Fig. 20.11. Riemann mapping, (Riemann) normal coordinates, distortion ellipses with axes $\Lambda_1, \Lambda_2 - 1$. Top: $L_0 = 120°$, $B_0 = 0°$. Middle: $L_0 = 10°$, $B_0 = 80°$. Bottom: $L_0 = 10°$, $B_0 = 50°$.

For a local representation of the ellipsoidal surface, namely the reference figure of the Earth, conformal Gauss-Krueger and parallel Soldner coordinates are the most popular. Therefore, they are compared with (Riemann) polar/normal coordinates. As a *measure* of the *total deformation energy* (total distortion energy), according to G. B. Airy (1861), we introduce (20.142) once we map the area over the *symmetric strip* (20.141).

$$[l_{\mathrm{W}} = -l_{\mathrm{E}}, +l_{\mathrm{E}} = -l_{\mathrm{W}}] \times [B_{\mathrm{S}} = B_0 + b_{\mathrm{S}}, B_{\mathrm{N}} = B_0 + b_{\mathrm{N}}] \,. \tag{20.141}$$

The indices W, E, S, and N refer to "West", "East", "South", and "North". The area of the symmetric strip is provided by (20.143). Results characterizing the total deformation energy are collected in Corollary 20.5.

$$I_{\mathrm{A}} :=$$

$$:= \frac{1}{2S_{\mathbb{E}^2_{A_1,A_2}}} \int\limits_{l_{\mathrm{W}}}^{l_{\mathrm{E}}} \mathrm{d}l \int\limits_{B_{\mathrm{S}}}^{B_{\mathrm{N}}} \mathrm{d}B \frac{\cos B \{(A_1 - 1)^2 + (A_2 - 1)^2\}}{(1 - E^2 \sin^2 B)^2} \,, \tag{20.142}$$

$$S_{\mathbb{E}^2_{A_1,A_2}} = \int \mathrm{d}S_{\mathbb{E}^2_{A_1,A_2}} :=$$

$$:= A_1^2 (1 - E^2) \int\limits_{-l_{\mathrm{E}}}^{l_{\mathrm{E}}} \mathrm{d}l \int\limits_{B_{\mathrm{S}}}^{B_{\mathrm{N}}} \mathrm{d}B \frac{\cos B}{(1 - E^2 \sin^2 B)^2} = \tag{20.143}$$

$$= 2A_1^2 (1 - E^2) l_{\mathrm{E}} [\sin B_{\mathrm{N}} + \frac{2}{3} E^2 \sin^3 B_{\mathrm{N}} - (\sin B_{\mathrm{S}} + \frac{2}{3} E^2 \sin^3 B_{\mathrm{S}}) + \mathrm{O}(E^4)] \,.$$

Corollary 20.5 ($\mathbb{E}^2_{A_1,A_2}$, the total deformation energy, the Airy measure, Gauss–Krueger coordinates, Soldner coordinates, and Riemann coordinates).

The principal distortions read as follows ($l := L - L_0$, $b := B - B_0$).

Conformal coordinates of type Gauss–Krueger:

$$A_1 = A_2 = 1 + \frac{\cos^2 B}{2} \left(1 + \frac{E^2}{1 - E^2}\right) l^2 + \mathrm{O}_{\mathrm{GK}}(l^4) \,. \tag{20.144}$$

Parallel coordinates of type Soldner:

$$A_1 = 1 + \frac{\cos^2 B}{2} \left(1 + \frac{E^2}{1 - E^2} \cos^2 B\right) l^2 + \mathrm{O}_{\mathrm{S}}(l^4) \,, \tag{20.145}$$

$$A_2 = 1 \,.$$

Normal coordinates of type Riemann:

$$A_1 = 1 + \frac{\cos^2 B_0}{6} \frac{1 - E^2 \sin^2 B_0}{1 - E^2} l^2 + \frac{1}{6} \frac{1 - E^2}{1 - E^2 \sin^2 B_0} b^2 + \mathrm{O}_{\mathrm{R}}(l^3, b^3) \,, \tag{20.146}$$

$$A_2 = 1 \,.$$

The total deformation energy (total distortion energy, total distance distortion) over the symmetric strip $[-l_E, +l_E] \times [B_S = B_0 + b_S, B_N = B_0 + b_N]$, $L_0 = (L_W + L_E)/2$, and $B_0 = (B_S + B_N)/2$ are given as follows.

Conformal coordinates of type Gauss–Krueger:

$$I_{AGK} = \frac{1}{20} \frac{l_E^4}{(1 - E^2)^2} \times$$

$$\times \left(1 - \frac{2}{3} \frac{\sin^3 B_N - \sin^3 B_S}{\sin B_N - \sin B_S} + \frac{1}{5} \frac{\sin^5 B_N - \sin^5 B_S}{\sin B_N - \sin B_S}\right) \times$$

$$\times \left(1 - \frac{2}{3} E^2 \frac{\sin^3 B_N - \sin^3 B_S}{\sin B_N - \sin B_S} + O_2(E^4)\right) + \tag{20.147}$$

$$+ O_{GK}(l_E^6) \,,$$

$$I_{AGK}(B_N = -B_S) = \frac{1}{20} \frac{l_E^4}{(1 - E^2)^2} \times$$

$$\times \left(1 - \frac{2}{3} \sin^2 B_N (1 + E^2) + \sin^4 B_N \left(\frac{4}{9} E^2 + \frac{1}{5}\right) - \frac{2}{15} E^2 \sin^6 B_N + O_2(E^4)\right) + \tag{20.148}$$

$$+ O_{GK}(l_E^6) \,.$$

Parallel coordinates of type Soldner:

$$I_{AS} = \frac{1}{2} I_{AGK} + O_S(l_E^6) \,. \tag{20.149}$$

Normal coordinates of type Riemann:

$$I_{AR} = \frac{1}{2 \cos B_0} \left[\frac{A}{5} l_E^4 + \frac{B}{9} \frac{b_N^3 - b_S^3}{b_N - b_S} l_E^2 + \frac{C}{5} \frac{b_N^5 - b_S^5}{b_N - b_S} +\right.$$

$$\left. + \frac{D}{10} (b_N + b_S) l_E^4 + \frac{P}{12} \frac{b_N^4 - b_S^4}{b_N - b_S} l_E^2 + \frac{Q}{6} \frac{b_N^6 - b_S^6}{b_N - b_S}\right] \times \tag{20.150}$$

$$\times \left[1 - 2E^2 \sin^2 B_0 + \sin B_0 (1 - 2E^2 \sin^2 B_0 - 4E^2 \cos^2 B_0)(b_N + b_S)/2 \cos B_0\right] +$$

$$+ O_R(6) \,,$$

$$I_{AR}(b_N = -b_S) = \frac{\cos^4 B_0}{360} \left(1 + 2E^2 \cos^2 B_0 + O(E^4)\right) l_E^4 +$$

$$+ \frac{\cos^2 B_0}{324} \left(1 + O(E^4)\right) b_N^2 l_E^2 + \frac{\left(1 - 2E^2 \cos^2 B_0 + O(E^4)\right)}{360} b_N^4 + \tag{20.151}$$

$$+ O_R(6) \,.$$

(A, B, C, D, P, Q are given by (20.173)).

End of Corollary.

Proof (Gauss–Krueger, $\Lambda_1 = \Lambda_2 = \Lambda$, $l := L - L_0$).

$$I_{\text{AGK}} = \frac{1}{S_{\mathbb{E}^2_{A_1,A_2}}} \int dS_{\mathbb{E}^2_{A_1,A_2}} (\Lambda - 1)^2 = \frac{A_1^2(1 - E^2)}{S_{\mathbb{E}^2_{A_1,A_2}}} \int\limits_{-l_E}^{+l_E} dl \int\limits_{B_S}^{B_N} dB \frac{\cos B(\Lambda - 1)^2}{(1 - E^2 \sin^2 B^2)^2} \; , \quad (20.152)$$

$$\Lambda - 1 = \frac{1}{2} \cos^2 B \frac{1 - E^2 \sin^2 B}{1 - E^2} l^2 +$$

$$+ O_{\text{GK}}(l^4) \; ,$$

$$(\Lambda - 1)^2 = \frac{1}{4} \cos^4 B \frac{(1 - E^2 \sin^2 B)^2}{(1 - E^2)^2} l^4 + \quad (20.153)$$

$$+ O_{\text{GK}}(l^6) \; ,$$

$$\int dS_{\mathbb{E}^2_{A_1,A_2}} (\Lambda - 1)^2 = \frac{1}{4} \frac{A_1^2}{1 - E^2} \int\limits_{-l_E}^{+l_E} dl \left[l^4 \int\limits_{B_S}^{B_N} dB \cos^5 B + O_{\text{GK}}(l^6) \right] =$$

$$= \frac{1}{10} \frac{A_1^2}{1 - E^2} l_E^5 \left[\sin B_N - \sin B_S - \frac{2}{3}(\sin^3 B_N - \sin^3 B_S) + \frac{1}{5}(\sin^5 B_N - \sin^5 B_S) \right] + \quad (20.154)$$

$$+ O_{\text{GK}}(l_E^7) \; ,$$

$$\frac{1}{S_{E^2_{A_1,A_2}}} = \frac{1}{2A_1^2(1 - E^2)} l_E^{-1} \left[\sin B_N - \sin B_S + \frac{2}{3} E^2 (\sin^3 B_N - \sin^3 B_S) + O_1(E^4) \right]^{-1} =$$

$$\quad (20.155)$$

$$= \frac{1}{2A_1^2(1 - E^2)} l_E^{-1} (\sin B_N - \sin B_S)^{-1} \left[1 - \frac{2}{3} E^2 \frac{\sin^3 B_N - \sin^3 B_S}{\sin B_N - \sin B_S} + O_2(E^4) \right] \; ,$$

$$I_{\text{AGK}} = \frac{1}{20} \frac{l_E^4}{(1 - E^2)^2} \left(1 - \frac{2}{3} \frac{\sin^3 B_N - \sin^3 B_S}{\sin B_N - \sin B_S} + \frac{1}{5} \frac{\sin^5 B_N - \sin^5 B_S}{\sin B_N - \sin B_S} \right) \times$$

$$\times \left[1 - \frac{2}{3} E^2 \frac{\sin^3 B_N - \sin^3 B_S}{\sin B_N - \sin B_S} + O_2(E^4) \right] + \quad (20.156)$$

$$+ O_{\text{GK}}(l_E^6) \; ,$$

$$I_{\text{AGK}}(B_N = -B_S) = \frac{1}{20} \frac{l_E^4}{(1 - E^2)^2} \times$$

$$\times \left[1 - \frac{2}{3} \sin^2 B_N(1 + E^2) + \sin^4 B_N \left(\frac{4}{9} E^2 + \frac{1}{5} \right) - \frac{2}{15} E^2 \sin^6 B_N + O_2(E^4) \right] + \quad (20.157)$$

$$+ O_{\text{GK}}(l_E^6) \; .$$

End of Proof.

Proof (Soldner, $l := L - L_0$).

$$\Lambda_1 = 1 + \frac{\cos^2 B}{2}\frac{1 - E^2 \sin^2 B}{1 - E^2}l^2 + \mathrm{O_S}(l^4)\,, \quad \Lambda_2 = 1\,, \tag{20.158}$$

$$I_{\mathrm{AS}} = \frac{1}{2S_{\mathbb{E}^2_{A_1,A_2}}}\int \mathrm{d}S_{\mathbb{E}^2_{A_1,A_2}}(\Lambda_1 - 1)^2 = \frac{A_1^2(1-E^2)}{2S_{\mathbb{E}^2_{A_1,A_2}}}\int\limits_{-l_{\mathrm{E}}}^{+l_{\mathrm{E}}}\mathrm{d}l\int\limits_{B_{\mathrm{S}}}^{B_{\mathrm{N}}}\mathrm{d}B\,\frac{\cos B(\Lambda_1 - 1)^2}{(1 - E^2 \sin^2 B)^2}\,, \tag{20.159}$$

$$\Lambda_1 - 1 = \frac{\cos^2 B}{2}\frac{1 - E^2 \sin^2 B}{1 - E^2}l^2 + \mathrm{O_S}(l^4)\,,$$

$$(\Lambda_1 - 1)^2 = \frac{1}{4}\cos^4 B\frac{(1 - E^2 \sin^2 B)^2}{(1 - E^2)^2}l^4 + \mathrm{O_S}(l^6)\,, \tag{20.160}$$

$$\int \mathrm{d}S_{\mathbb{E}^2_{A_1,A_2}}(\Lambda_1 - 1)^2 = \frac{1}{4}\frac{A_1^2}{1 - E^2}\int\limits_{-l_{\mathrm{E}}}^{+l_{\mathrm{E}}}\mathrm{d}l\left[l^4\int\limits_{B_{\mathrm{S}}}^{B_{\mathrm{N}}}\mathrm{d}B\,\cos^5 B + \mathrm{O_S}(l^6)\right] =$$

$$= \frac{1}{10}\frac{A_1^2}{1 - E^2}l_{\mathrm{E}}^5\left[\sin B_{\mathrm{N}} - \sin B_{\mathrm{S}} - \frac{2}{3}(\sin^3 B_{\mathrm{N}} - \sin^3 B_{\mathrm{S}}) + \frac{1}{5}(\sin^5 B_{\mathrm{N}} - \sin^5 B_{\mathrm{S}})\right] + \tag{20.161}$$

$$+\,\mathrm{O_S}(l_{\mathrm{E}}^7)\,,$$

$$\frac{1}{2S_{\mathbb{E}^2_{A_1,A_2}}} = \frac{1}{4A_1^2(1 - E^2)}l_{\mathrm{E}}^{-1}(\sin B_{\mathrm{N}} - \sin B_{\mathrm{S}})^{-1}\times$$

$$\times\left[1 - \frac{2}{3}E^2\frac{\sin^3 B_{\mathrm{N}} - \sin^3 B_{\mathrm{S}}}{\sin B_{\mathrm{N}} - \sin B_{\mathrm{S}}} + \mathrm{O}_2(E^4)\right]\,, \tag{20.162}$$

$$I_{\mathrm{AS}} = \frac{1}{40}\frac{l_{\mathrm{E}}^4}{(1 - E^2)^2}\times$$

$$\times\left(1 - \frac{2}{3}\frac{\sin^3 B_{\mathrm{N}} - \sin^3 B_{\mathrm{S}}}{\sin B_{\mathrm{N}} - \sin B_{\mathrm{S}}} + \frac{1}{5}\frac{\sin^5 B_{\mathrm{N}} - \sin^5 B_{\mathrm{S}}}{\sin B_{\mathrm{N}} - \sin B_{\mathrm{S}}}\right)\times$$

$$\times\left[1 - \frac{2}{3}E^2\frac{\sin^3 B_{\mathrm{N}} - \sin^3 B_{\mathrm{S}}}{\sin B_{\mathrm{N}} - \sin B_{\mathrm{S}}} + \mathrm{O}_2(E^4)\right] + \tag{20.163}$$

$$+\,\mathrm{O_S}(l_{\mathrm{E}}^6)\,,$$

$$I_{\mathrm{AS}}(B_{\mathrm{N}} = -B_{\mathrm{S}}) = \frac{1}{40}\frac{l_{\mathrm{E}}^4}{(1 - E^2)^2}\times$$

$$\times\left[1 - \frac{2}{3}\sin^2 B_{\mathrm{N}}(1 + E^2) + \sin^4 B_{\mathrm{N}}\left(\frac{4}{9}E^2 + \frac{1}{5}\right) - \frac{2}{15}E^2\sin^6 B_{\mathrm{N}} + \mathrm{O}_2(E^4)\right] + \tag{20.164}$$

$$+\,\mathrm{O_S}(l_{\mathrm{E}}^6)\,.$$

Fnd of Proof.

Proof (Riemann, $l := L - L_0$, $b := B - B_0$).

$$\Lambda_1 = 1 + \frac{\cos^2 B_0}{6} \frac{1 - E^2 \sin^2 B_0}{1 - E^2} l^2 + \frac{1}{6} \frac{1 - E^2}{1 - E^2 \sin^2 B_0} b^2 + O_R(l^3, b^3) , \quad \Lambda_2 = 1 , \qquad (20.165)$$

$$I_{AR} = \frac{1}{2 S_{\mathbb{E}^2_{A_1, A_2}}} \int dS_{\mathbb{E}^2_{A_1, A_2}} (\Lambda_1 - 1)^2 = \frac{A_1^2 (1 - E^2)}{2 S_{\mathbb{E}^2_{A_1, A_2}}} \int\limits_{-l_E}^{+l_E} dl \int\limits_{B_S}^{B_N} dB \frac{\cos B (\Lambda_1 - 1)^2}{(1 - E^2 \sin^2 B)^2} , \qquad (20.166)$$

$$\Lambda_1 - 1 = \frac{\cos^2 B_0}{6} \frac{1 - E^2 \sin^2 B_0}{1 - E^2} l^2 + \frac{1}{6} \frac{1 - E^2}{1 - E^2 \sin^2 B_0} b^2 + O_R(l^3, b^3) ,$$

$$(20.167)$$

$$(\Lambda_1 - 1)^2 = \frac{\cos^4 B_0}{36} \frac{(1 - E^2 \sin^2 B_0)^2}{(1 - E^2)^2} l^4 + \frac{1}{36} \frac{(1 - E^2)^2}{(1 - E^2 \sin^2 B_0)^2} b^4 + \frac{\cos^2 B_0}{18} l^2 b^2 + O_R(l^5, b^5) ,$$

$$\cos B = \cos(B_0 + b) = \cos B_0 - \sin B_0 b + O_C(b^2) ,$$

$$\left(1 - E^2 \sin^2 B^2\right)^{-2} = 1 + 2 E^2 \sin^2 B + O(E^4) =$$

$$= \left[1 + 2 E^2 \sin B_0 + 4 E^2 \sin B_0 \cos B_0 + O(E^4)\right] b + O(b^2) ,$$

$$\cos B (1 - E^2 \sin^2 B)^{-2} = \qquad (20.168)$$

$$= \cos B_0 - \sin B_0 b + 2 E^2 \sin^2 B_0 \cos B_0 - 2 E^2 \sin^3 B_0 b + 4 E^2 \sin B_0 \cos^2 B_0 b + O(b^2) ,$$

$$B_S = B_0 + b_S , \quad B_N = B_0 + b_N ,$$

$$S_{\mathbb{E}^2_{A_1, A_2}} = A_1^2 (1 - E^2) \int\limits_{-l_E}^{+l_E} dl \int\limits_{b_S}^{b_N} db (\cos B_0 + 2 E^2 \sin^2 B_0 \cos B_0) +$$

$$(20.169)$$

$$+ A_1^2 (1 - E^2) \int\limits_{-l_E}^{+l_E} dl \int\limits_{b_S}^{b_N} db \left[-\sin B_0 - 2 E^2 \sin^3 B_0 + 4 E^2 \sin B_0 \cos^2 B_0 + o(E^4) \right] b + O(b^2) ,$$

$$S_{\mathbb{E}^2_{A_1, A_2}} = A_1^2 (1 - E^2) l_E \Big(2 (\cos B_0 + 2 E^2 \sin^2 B_0 \cos B_0)(b_N - b_S) +$$

$$(20.170)$$

$$+ \left[-\sin B_0 - 2 E^2 \sin^3 B_0 + 4 E^2 \sin B_0 \cos^2 B_0 + O(E^4) \right] (b_N^2 - b_S^2) + O(b_N^3 - b_S^3) \Big) ,$$

$$I_{AR} = \frac{A_1^2 (1 - E^2)}{2 S_{\mathbb{E}^2_{A_1, A_2}}} \int\limits_{-l_E}^{+l_E} dl \int\limits_{b_S}^{b_N} db \bigg[\cos B_0 + 2 E^2 \sin^2 B_0 \cos B_0 +$$

$$+ \left[-\sin B_0 - 2 E^2 \sin^3 B_0 + 4 E^2 \sin B_0 \cos^2 B_0 \cos^2 B_0 + O(E^4) \right] b + O(b^2) \bigg] \times \qquad (20.171)$$

$$\times \left[\frac{\cos^4 B_0}{36} \frac{(1 - E^2 \sin^2 B_0)^2}{(1 - E^2)^2} l^4 + \frac{\cos^2 B_0}{18} l^2 b^2 + \frac{1}{36} \frac{(1 - E^2)^2}{(1 - E^2 \sin^2 B_0)^2} b^4 + O_R(l^5, b^5) \right] ,$$

$$I_{AR} = \frac{A_1^2(1-E^2)}{2S_{\mathbb{E}^2_{A_1,A_2}}} \int\limits_{-l_E}^{+l_E} dl \int\limits_{b_S}^{b_N} db \big[Al^4 + Bl^2b^2 + Cb^4 + Dl^4b + Pl^2b^3 + Qb^5 + \mathrm{O_R}(6) \big] \, , \quad (20.172)$$

$$A := \big[\cos B_0 + 2E^2 \sin^2 B_0 \cos B_0 + \mathrm{O}(E^4) \big] \frac{\cos^4 B_0}{36} \frac{(1-E^2 \sin^2 B_0)^2}{(1-E^2)^2} \, ,$$

$$B := \big[\cos B_0 + 2E^2 \sin^2 B_0 \cos B_0 + \mathrm{O}(E^4) \big] \frac{\cos^2 B_0}{18} \, ,$$

$$C := \big[\cos B_0 + 2E^2 \sin^2 B_0 \cos B_0 + \mathrm{O}(E^4) \big] \frac{1}{36} \frac{(1-E^2)^2}{(1-E^2 \sin^2 B_0)^2} \, ,$$

$$(20.173)$$

$$D := \frac{-\sin B_0 \cos^4 B_0}{36} \big[3 + 2E^2(3 - \cos^2 B_0) \big] + \mathrm{O}(E^4) \, ,$$

$$P := \frac{-\sin B_0 \cos^2 B_0}{9} \big[1 + 2E^2(1 - 2\cos^2 B_0) \big] + \mathrm{O}(E^4) \, ,$$

$$Q := \frac{-\sin B_0}{36} \big[1 + 2E^2(1 - 5\cos^2 B_0) \big] + \mathrm{O}(E^4) \, ,$$

$$I_{AR} = \frac{A_1^2(1-E^2)}{S_{\mathbb{E}^2_{A_1,A_2}}} \Big[\frac{A}{5} l_E^5 (b_N - b_S) + \frac{B}{9} l_E^3 (b_N^3 - b_S^3) +$$

$$+ \frac{C}{5} l_E (b_N^5 - b_S^5) + \frac{D}{10} l_E^5 (b_N^2 - b_S^2) + \frac{P}{12} l_E^3 (b_N^4 - b_S^4) + \frac{Q}{6} l_E (b_N^6 - b_S^6) + \mathrm{O}(8) \Big] \, ,$$

$$(20.174)$$

$$\frac{1}{S_{\mathbb{E}^2_{A_1,A_2}}} =$$

$$= \frac{1 - 2E^2 \sin^2 B_0 + + \sin B_0(1 - 2E^2 \sin^2 B_0 - 4E^2 \cos^2 B_0)(b_N + b_S)/2 \cos B_0 + \mathrm{O}(b^2)}{2A_1^2(1-E^2) \cos B_0 l_E (b_N - b_S)} \, ,$$

$$(20.175)$$

$$I_{AR} = \frac{1}{2\cos B_0} \Big[\frac{A}{5} l_E^4 + \frac{B}{9} \frac{b_N^3 - b_S^3}{b_N - b_S} l_E^2 + \frac{C}{5} \frac{b_N^5 - b_S^5}{b_N - b_S} +$$

$$+ \frac{D}{10}(b_N + b_S) l_E^4 + \frac{P}{12} P \frac{b_N^4 - b_S^4}{b_N - b_S} l_E^2 + \frac{Q}{6} \frac{b_N^6 - b_S^6}{b_N - b_S} \Big] \times$$

$$(20.176)$$

$$\times \frac{1 - 2E^2 \sin^2 B_0 + \sin B_0(1 - 2E^2 \sin^2 B_0 - 4E^2 \cos^2 B_0)(b_N + b_S)}{2\cos B_0} + \mathrm{O_R}(6) \, ,$$

$$I_{AR}(b_N = -b_S) = \frac{\cos^4 B_0}{360}(1 + 2E^2 \cos^2 B_0 + \mathrm{O}(E^4)) l_E^4 +$$

$$+ \frac{\cos^2 B_0}{324}(1 + \mathrm{O}(E^4)) b_N^2 l_E^2 + \frac{(1 - 2E^2 \cos^2 B_0 + \mathrm{O}(E^4))}{360} b_N^4 + \mathrm{O_R}(6) \, .$$

$$(20.177)$$

End of Proof.

As an example, we compare in Table 20.5 the total deformation energy for conformal Gauss–Krueger coordinates, parallel Soldner coordinates, and normal Riemann coordinates. Obviously, in the range of application, normal Riemann coordinates generate the *smallest global distortion*, followed by parallel Soldner coordinates (factor half compared to conformal Gauss–Krueger coordinates), and conformal Gauss-Krueger coordinates.

Table 20.5. Total deformation energy. $L_0 = 9°$. Three cases: (i) $B_0 = 0°$, (ii) $B_0 = 48°$, (iii) $B_0 = 70°$. $l_E = 2°$, $b_N = -b_S = 2°$. I_{AGK}, I_{AS}, I_{AR}.

$L_0 = 9°$	I_{AR} ("normal Riemann")	I_{AGK} ("conformal Gauss–Krueger")	I_{AS} ("parallel Soldner")
$B_0 = 0°$	1.283×10^{-8}	7.518×10^{-8}	3.759×10^{-8}
$B_0 = 48°$	6.983×10^{-9}	1.503×10^{-8}	7.517×10^{-9}
$B_0 = 70°$	4.710×10^{-10}	1.048×10^{-9}	5.239×10^{-10}

Here, we take reference to G. B. Airy's definition of "balance of erros". Conformal geodesics were treated by E. Beltrami (1869), A. Fialkow (1939), A. M. Legendre (1806), J. A. Schouten (1954), W. A. Vogel (1970, 1973), J. Weingarten (1861), and K. Yano (1970, 1940a,b). H. Boltz (1943) presented formulae and tables for the normal computation of Gauss–Krueger coordinates which we use here. For the optimal Universal Transverse Mercator Projection, we refer to E. W. Grafarend (1995), the Riemann coordinates and its deformation analysis to E. W. Grafarend and R. Syffus (1995). The special Newton form of a geodesic in Maupertuis' gauge on the sphere and the ellipsoid-of-revolution is presented in Appendix E to which we refer. H. Lichtenegger (1987) presented his theory of three boundary problems and one initial problem. Here, we present only two out of four.

21 Datum problems

Analysis versus synthesis, Cartesian approach versus curvilinear approach. Local reference system versus global reference system. Datum parameters, collinearities, error propagation. Partial least squares, ridge type regression (Tychonov regularization), truncated or total least squares. Gauss–Krueger coordinates and UTM coordinates. Stochastic pseudo-observations and variance–covariance matrix, dispersion matrix.

The evolutionary process of $(2 + 1)$-dimensional geodesy separating horizontal control and vertical control towards three-dimensional geodesy, namely enforced by satellite global positioning systems ("global problem solver": GPS), confronts us with the problem of *curvilinear geodetic datum transformations* of the following type. In a local two-dimensional geodetic network ellipsoidal longitude and ellipsoidal latitude (equivalent: Gauss–Krueger coordinates, UTM coordinates) are available in a local geodetic reference system. From a global three-dimensional geodetic network, namely for a few identical points, ellipsoidal longitude, ellipsoidal latitude, and ellipsoidal height are known in a global geodetic reference system. Here, we aim at the analysis of datum parameters (seven parameter global conformal group $C_7(3)$: translation, rotation, scale). We set up curvilinear linearized pseudo-observational equations for given ellipsoidal longitude and ellipsoidal latitude, with incremental parameters of translation (three parameters), rotation (three parameters), and scale (one parameter), and with incremental form parameters of the ellipsoid-of-revolution (two parameters: semi-major axis, squared eccentricity). In particular, we investigate the rank deficiencies in the curvilinear datum transformation Jacobi matrix (collinearities). A strict collinearity between the incremental datum parameter t_z and the incremental semi-major axis δA has been identified. For geodetic networks of regional extension, we experienced also configurations close to a collinearity.

Section 21-1.

A regression system close to a collinearity (near linear dependence) experiences damaging effects on the ordinary least squares estimator, as small changes in the Jacobi matrix or in the vector of pseudo-observations may result in unproportionally large changes in the solution. Three methods of overcoming the problem of collinearity are currently used: (i) partial least squares, for example, P. J. Young (1994), (ii) ridge type regression (Tychonov regularization), for example, A. K. Saleh and B. M. G. Kibria (1993), and (iii) truncated or total least squares, for example, R. D. Fierro and J. R. Bunch (1994). Indeed, in the analysis of the curvilinear datum transformation Jacobi matrix for regional geodetic networks, we identified a spectral condition number (the ratio of the largest and smallest eigenvalue) of the order of 10^9. For some reasons, we have accordingly chosen partial least squares in analyzing the datum parameters from horizontal control, exclusively. Our results are presented in Section 21-1.

Section 21-2.

In contrast, Section 21-2 is devoted to the synthesis of ellipsoidal longitude and ellipsoidal latitude known in the local reference system and to be transformed into the global reference system or, equivalently into Gauss–Krueger coordinates or UTM coordinates from local to global reference. A real data example is given.

Section 21-3.

Finally, Section 21-3 three reviews the error propagation in the analysis and synthesis of a curvilinear datum transformation.

21-1 Analysis of a datum problem

Datum transformation, datum parameters (translation, rotation, scale). General conformal group, special orthogonal group. Jacobi matrix, chain Jacobi matrix.

Analysis is understood as the determination of *datum parameters* between two sets of curvilinear coordinates of identical points which cover $\mathbb{R}^3$ equipped with an Euclidean metric. The datum parameters like translation $\boldsymbol{t}$, rotation R, and scale $1 + s$ characterize a *datum transformation* ("Kartenwechsel") which leaves (as a passive transformation) angles and distance ratios equivariant. In its linear variant, they are the parameters of the conformal group $C_7(3)$, the seven-parameter transformation in $\mathbb{R}^3$. The general conformal group $C_{10}(3)$, in contrast, includes three parameters more, as outlined in E. Grafarend and G. Kampmann (1996), for instance. As soon as we cover $\mathbb{R}^3$ by Cartesian coordinates (say $\{x\ y\ z\}$ in the local reference system versus $\{X\ Y\ Z\}$ in the global reference system), we arrive at the forward transformation (21.1) of datum type versus the backward transformation (21.3) of datum type. Actually, (21.1) and (21.3) are datum transformations of $\mathbb{R}^4$ covered by homogeneous coordinates. All datum transformations are written in the sequence (i) rotation, (ii) scale, and (iii) translation, what has to be mentioned since the transformation elements are non-commutative. The formulae (21.1)–(21.5) constitute Box 21.1. We here note that SO(3) abbreviates the manifold of the three-dimensional *special orthogonal group*, namely $\mathrm{SO}(3) := \{\mathrm{R} \,|\, \mathrm{R}^{\mathrm{T}}\mathrm{R} = \mathrm{I}_3 \quad \mathrm{R} = +1\}$, where I_3 is the three-dimensional unit matrix.

Box 21.1 (The conformal group $C_7(3)$, the forward and the backward transformation, Cartesian coordinates $\{x\ y\ z\}$ of local type versus Cartesian coordinates $\{X\ Y\ Z\}$ of global type covering $\mathbb{R}^3$ equipped with an Euclidean metric).

Forward transformation:

$$\begin{bmatrix} x \\ y \\ z \end{bmatrix} = (1+s)\mathrm{R}\begin{bmatrix} X \\ Y \\ Z \end{bmatrix} + \begin{bmatrix} t_x \\ t_y \\ t_z \end{bmatrix} \qquad \begin{bmatrix} t_x \\ t_y \\ t_z \end{bmatrix} =: \boldsymbol{t} \in \mathbb{R}^3 \tag{21.1}$$

$$s \in \mathbb{R}^+ \qquad \mathrm{R} \in \mathrm{SO}(3)$$

$$\begin{bmatrix} x \\ y \\ z \\ 1 \end{bmatrix} = \left[\begin{array}{c|c} (1+s)\mathrm{R} & \boldsymbol{t} \\ \hline 0 & 1 \end{array}\right]\begin{bmatrix} X \\ Y \\ Z \\ 1 \end{bmatrix} \tag{21.2}$$

Backward transformation:

$$\begin{bmatrix} X \\ Y \\ Z \end{bmatrix} = (1+s^*)\mathrm{R}^*\begin{bmatrix} x \\ y \\ z \end{bmatrix} + \begin{bmatrix} t_x^* \\ t_y^* \\ t_z^* \end{bmatrix} \qquad \begin{bmatrix} t_x^* \\ t_y^* \\ t_z^* \end{bmatrix} =: \boldsymbol{t}^* \in \mathbb{R}^3 \tag{21.3}$$

$$s^* \in \mathbb{R}^+ \qquad \mathrm{R}^* \in \mathrm{SO}(3)$$

$$\begin{bmatrix} X \\ Y \\ Z \\ 1 \end{bmatrix} = \left[\begin{array}{c|c} (1+s^*)\mathrm{R}^* & \boldsymbol{t}^* \\ \hline 0 & 1 \end{array}\right]\begin{bmatrix} x \\ y \\ z \\ 1 \end{bmatrix} \tag{21.4}$$

"Backward–forward":

$$(1+s^*)\mathrm{R}^* = (1+s)^{-1}\mathrm{R}^{\mathrm{T}} \qquad \boldsymbol{t}^* = -(1+s)^{-1}\mathrm{R}^{\mathrm{T}}\boldsymbol{t} \tag{21.5}$$

Since the datum parameters of a geodetic datum transformation are usually close to the identity, Box 21.2 is a collection of the rotation matrix close to the identity, parameterized by *Cardan angles* into the forward and backward transformations (21.8) and (21.9) of that type, finally by (21.10) given as a system of linear equations with seven parameters (three for translation, three for rotation, and one for scale) as unknowns.

Box 21.2 (The conformal group $C_7(3)$, forward and backward transformation close to the identity).

$$R = R_1(\alpha)R_2(\beta)R_3(\gamma) \in SO(3) \ , \tag{21.6}$$

$$R_1(\alpha) = R_1(0) + R_1'(0)\alpha + O_1(\alpha^2) \ , \quad R_1'(0) = \begin{bmatrix} 0 & 0 & 0 \\ 0 & 0 & 1 \\ 0 & -1 & 0 \end{bmatrix} \ ,$$

$$R_2(\beta) = R_2(0) + R_2'(0)\beta + O_2(\beta^2) \ , \quad R_2'(0) = \begin{bmatrix} 0 & 0 & -1 \\ 0 & 0 & 0 \\ 1 & 0 & 0 \end{bmatrix} \ , \tag{21.7}$$

$$R_3(\gamma) = R_3(0) + R_3'(0)\gamma + O_3(\gamma^2) \ , \quad R_3'(0) = \begin{bmatrix} 0 & 1 & 0 \\ -1 & 0 & 0 \\ 0 & 0 & 0 \end{bmatrix} \ .$$

Forward transformation close to the identity:

$$x = X + t_x - Z\beta + Y\gamma + Xs + O_{2x} \ ,$$
$$y = Y + t_y + Z\alpha - X\gamma + Ys + O_{2y} \ , \tag{21.8}$$
$$z = Z + t_z - Y\alpha + X\beta + Zs + O_{2z} \ .$$

Backward transformation close to the identity:

$$X = x - t_x - y\gamma + z\beta - xs + O_{2X} \ ,$$
$$Y = y - t_y - z\alpha + x\gamma - ys + O_{2Y} \ , \tag{21.9}$$
$$Z = z - t_z - x\beta + y\alpha - zs + O_{2Z} \ .$$

"Backward–forward":

$$\begin{bmatrix} x - X \\ y - Y \\ z - Z \end{bmatrix} = \begin{bmatrix} 1 & 0 & 0 & 0 & -Z & Y & X \\ 0 & 1 & 0 & Z & 0 & -X & Y \\ 0 & 0 & 1 & -Y & X & 0 & Z \end{bmatrix} \begin{bmatrix} t_x \\ t_y \\ t_z \\ \alpha \\ \beta \\ \gamma \\ s \end{bmatrix} \ . \tag{21.10}$$

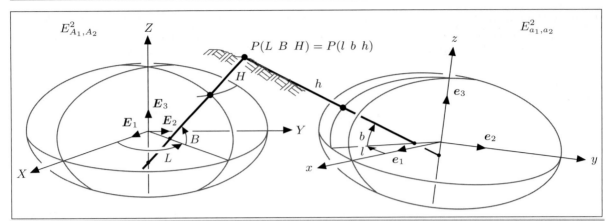

Fig. 21.1. Gauss surface normal coordinates, curvilinear geodetic datum transformations, "global" $L\ B\ H$ versus "local" $l\ b\ h$.

Since the "legal" geodetic coordinates relate to longitude l and L, latitude b and B, and height $h(l, b)$ and $H(L, B)$ of a topographic point as an element of the Earth's two-dimensional surface and with respect to an ellipsoid-of-revolution $\mathbb{E}^2_{a_1\ a_2}$ and $\mathbb{E}^2_{A_1\ A_2}$ of local type and global type, Box 21.3 summarizes the standard transformations of ellipsoidal coordinates $\{l, b, h\}$ and $\{L, B, H\}$ into Cartesian coordinates $\{x, y, z\}$ and $\{X, Y, Z\}$ and *vice versa*, namely by means of (21.13)–(21.17). In addition, we introduce the curvature radii $\{n(b), m(b)\}$ and $\{N(B), M(B)\}$ of $\mathbb{E}^2_{a_1\ a_2}$ and $\mathbb{E}^2_{A_1\ A_2}$. While the direct transformation (21.11) is given in a closed form, the inverse transformation (21.12) (following E. Grafarend and P. Lohse (1991)) is a complicated closed form, in particular, for $b(x, y, z)$, $B(X, Y, Z)$ and $h(x, y, z)$, $H(X, Y, Z)$. The height functions $h(l, b)$ and $H(L, B)$ have been computed with respect to a set of functions which are orthonormal with respect to an ellipsoid-of-revolution by E. Grafarend and J. Engels (1992). Note that one set of ellipsoidal coordinates $\{l, b, h\}$ and, $\{L, B, H\}$ does not cover the Earth's surface completely, namely due to the pole singularity of these coordinates.

$$\{l, b, h\} \mapsto \{x, y, z\}\ ,\quad \{L, B, H\} \mapsto \{X, Y, Z\}\ , \tag{21.11}$$

$$\{x, y, z\} \mapsto \{l, b, h\}\ ,\quad \{X, Y, Z\} \mapsto \{L, B, H\}\ . \tag{21.12}$$

The main idea of the applied datum transformation of curvilinear coordinates as being outlined in Box 21.3 is the following. In the global coordinate system, ellipsoidal coordinates $\{L, B, H\}$ are available, for example, from a survey by means of the Global Positioning System (GPS, GLONASS, PRARE), but in the local coordinate system, only ellipsoidal longitude, ellipsoidal latitude are accessible. It has to be emphasized that, due to the older separation of "horizontal control" and "vertical control", the ellipsoidal height $h(l, b)$ in the local coordinate system is not available. It is for this reason that, by means of (21.18) and (21.19) , we have only formulated the curvilinear datum transformation close to the identity for ellipsoidal longitude, ellipsoidal latitude $\{l, b\}$ in the local reference system as a function of $\{L, B, H\}$ and $\{t_x, t_y, t_z, \alpha, \beta, \gamma, s\}$, respectively. A closed form expression was derived only for ellipsoidal longitude l. The Taylor series expansion up to second order in terms of datum parameters of type translation, rotation, and scale as well as of variation of semi-major axis δA and of squared first eccentricity δE^2 is outlined by (21.20). The detailed results of the linearization are collected in Box 21.4. In particular, we end up with the linear system of first order $\boldsymbol{y} = \mathsf{A}\boldsymbol{x} + \text{hit}$, where hit means "higher order terms", introducing $l - L, b - B$ as the given vector $\boldsymbol{y}$ and $\{t_x, t_y, t_z, \alpha, \beta, \gamma, s, \delta A, \delta E^2\}$ as the unknown vector $\boldsymbol{x}$. The Jacobi matrix A is rigorously computed by the chain rule $\mathsf{J}_{23}\mathsf{J}_1$, where the Jacobi matrix J_1 contains the derivatives of Cartesian coordinates with respect to the datum parameters as well as ellipsoidal form parameters. In contrast, J_{23} includes the relevant 2×3 submatrix $\partial(l, b)/\partial(x, y, z)$ at the Taylor point of the general 3×3 matrix $\partial(l, b, h)/\partial(x, y, z)$. Indeed, the complicated derivative matrix $\partial(l, b, h)/\partial(x, y, z)$ is computed from its simple inverse $\partial(x, y, z)/\partial(l, b, h)$. (21.28) is the closed form representation of the Jacobi matrix A given as a function of $\{L, B, H\}$ of the global curvilinear coordinate system!

Box 21.3 (Curvilinear coordinate conformal transformation (datum transformation) extended by ellipsoid parameters, ellipsoid-of-revolution $\mathbb{E}^2_{A_1,A_2}$ versus $\mathbb{E}^2_{a_1,a_2}$, ellipsoidal coordinates $\{l, b, h\}$ of local type versus ellipsoidal coordinates $\{L, B, H\}$ of global type).

$A_1, a_1 \in \mathbb{R}^+$ semi-major axes; $E := \sqrt{(A_1^2 - A_2^2)/A_1^2}$, $e = \sqrt{(a_1^2 - a_2^2)/a_1^2}$ relative eccentricities.

Direct transformation $\{l, b, h\} \mapsto \{x, y, z\}$:

$$x = \left[\frac{a_1}{\sqrt{1 - e^2 \sin^2 b}} + h(l, b)\right] \cos b \cos l = [n(b) + h(l, b)] \cos b \cos l \,,$$

$$y = \left[\frac{a_1}{\sqrt{1 - e^2 \sin^2 b}} + h(l, b)\right] \cos b \sin l = [n(b) + h(l, b)] \cos b \sin l \,,$$

$$z = \left[\frac{a_1(1 - e^2)}{\sqrt{1 - e^2 \sin^2 b}} + h(l, b)\right] \sin b \quad = [(1 - e^2)n(b) + h(l, b)] \sin b \,,$$

(21.13)

$$n(b) := \frac{a_1}{\sqrt{1 - e^2 \sin^2 b}} \,, \quad m(b) := \frac{a_1(1 - e^2)}{\sqrt{(1 - e^2 \sin^2 b)^3}} \,.$$

(21.14)

Direct transformation $\{L, B, H\} \mapsto \{X, Y, Z\}$:

$$X = \left[\frac{A_1}{\sqrt{1 - E^2 \sin^2 B}} + H(L, B)\right] \cos B \cos L = [N(B) + H(L, B)] \cos B \cos L \,,$$

$$Y = \left[\frac{A_1}{\sqrt{1 - E^2 \sin^2 B}} + H(L, B)\right] \cos B \sin L = [N(B) + H(L, B)] \cos B \sin L \,,$$

(21.15)

$$Z = \left[\frac{A_1(1 - E^2)}{\sqrt{1 - E^2 \sin^2 B}} + H(L, B)\right] \sin B \quad = [(1 - E^2)N(B) + H(L, B)] \sin B \,,$$

$$N(B) := \frac{A_1}{\sqrt{1 - E^2 \sin^2 B}} \,, \quad M(B) := \frac{A_1(1 - E^2)}{\sqrt{(1 - E^2 \sin^2 B)^3}} \,.$$

(21.16)

Inverse transformation $\{x, y, z\} \mapsto \{l, b, h\}$:

$$l = \arctan(y/x) + \left(-\frac{1}{2}\operatorname{sgn} y - \frac{1}{2}\operatorname{sgn} y \operatorname{sgn} x + 1\right)\pi$$

$$\in \{\mathbb{R} \,|\, 0 \leq l < 2\pi\} \,,$$

$$b = \arctan\left[\frac{z}{\sqrt{x^2 + y^2}} \frac{a_1(1 - e^2 \sin^2 b)^{-1/2} + h(l, b)}{a_1(1 - e^2)(1 - e^2 \sin^2 b)^{-1/2} + h(l, b)}\right] =$$

(21.17)

$$= \arctan\left[\frac{z}{\sqrt{x^2 + y^2}} \frac{n(b) + h(l, b)}{(1 - e^2)n(b) + h(l, b)}\right]$$

$$\in \{\mathbb{R} \,|\, -\pi/2 < b < +\pi/2\} \,.$$

$$h(l, b) = h(x, y, z) \quad \text{(F. W. Grafarend and P. Lohse (1991))}.$$

Continuation of Box.

Curvilinear coordinate conformal transformation close to the identity
(datum transformation).

1st variant:

$$l =$$

$$= \arctan \left(Y + t_y + Z\alpha - X\gamma + Ys \right) / \left(X + t_x - Z\beta + Y\gamma + Xs \right) , \tag{21.18}$$

$$b = b(X, Y, Z, t_x, t_y, t_z, \alpha, \beta, \gamma, s) .$$

2nd variant:

$$\tan l =$$

$$= \Big([A_1/\sqrt{1 - E^2 \sin^2 B} + H(L, B)] \cos B \sin L + t_y$$

$$+ [A_1(1 - E^2)/\sqrt{1 - E^2 \sin^2 B} + H(L, B)] \sin B\alpha$$

$$- [A_1/\sqrt{1 - E^2 \sin^2 B} + H(L, B)] \cos B \sin L\gamma$$

$$+ [A_1/\sqrt{1 - E^2 \sin^2 B} + H(L, B)] \cos B \sin Ls \Big)$$

$$/ \Big([A_1/\sqrt{1 - E^2 \sin^2 B} + H(L, B)] \cos B \cos L + t_x \tag{21.19}$$

$$- [A_1(1 - E^2)/\sqrt{1 - E^2 \sin^2 B} + H(L, B)] \sin B\beta$$

$$+ [A_1/\sqrt{1 - E^2 \sin^2 B} + H(L, B)] \cos B \sin L\gamma$$

$$+ [A_1/\sqrt{1 - E^2 \sin^2 B} + H(L, B)] \cos B \cos Ls \Big) ,$$

$$l = l(L, B, H(L, B), t_x, t_y, t_z, \alpha, \beta, \gamma, s, A_1, E^2) ,$$

$$b = b(L, B, H(L, B), t_x, t_y, t_z, \alpha, \beta, \gamma, s, A_1, E^2) .$$

Taylor expansion:

$$l =$$

$$= L + l_{t_x} t_x + l_{t_y} t_y + l_{t_z} t_z + l_\alpha \alpha + l_\beta \beta + l_\gamma \gamma + l_s s + l_A \delta A + l_E \delta E^2 ,$$

$$b = \tag{21.20}$$

$$= B + b_{t_x} t_x + b_{t_y} t_y + b_{t_z} t_z + b_\alpha \alpha + b_\beta \beta + b_\gamma \gamma + b_s s + b_A \delta A + b_E \delta E^2 ,$$

subject to

$$l_{t_x} := \frac{\partial l}{\partial t_x} (t_x = t_y = t_z = 0, \alpha = \beta = \gamma = 0, s = 0, A_1 = a_1, E^2 = e^2) \tag{21.21}$$

etc.

and

$$\delta A := A_1 - a_1 , \quad \delta E^2 := E^2 - e^2 , \quad \delta E = (2E)^{-1} \delta E^2 . \tag{21.22}$$

Box 21.4 (Curvilinear conformal coordinate transformation (datum transformation) extended by ellipsoid parameters close to identity, Jacobi matrix).

$$
\boldsymbol{y} = \begin{bmatrix} l - L \\ b - B \end{bmatrix} = \mathsf{A} \begin{bmatrix} t_x \\ t_y \\ t_z \\ \alpha \\ \beta \\ \gamma \\ s \\ \delta A \\ \delta E^2 \end{bmatrix} + O_{2lb} = \mathsf{A}\boldsymbol{x} + O_{2x} \ . \tag{21.23}
$$

Jacobi matrix:

$$
\mathsf{A} := \left\{ \frac{\partial(l,b)}{\partial(t_x,t_y,t_z,\alpha,\beta,\gamma,s,A_1,E^2)} \right\}_{\text{taylor}} \in \mathbb{R}^{2\times 9} \ , \tag{21.24}
$$

$$
\text{taylor} := \{t_x = t_y = t_z = \alpha = \beta = \gamma = s = 0, A_1 = a_1, E^2 = e^2\} \ .
$$

Chain Jacobi matrix:

$$
\mathsf{A} := \mathsf{J}_{23}\mathsf{J}_1 = \left\{ \frac{\partial(l,b)}{\partial(x,y,z)} \right\}_{\text{taylor}} \left\{ \frac{\partial(x,y,z)}{\partial(t_x,t_y,t_z,\alpha,\beta,\gamma,s,A_1,E^2)} \right\}_{\text{taylor}} \ , \tag{21.25}
$$

$$
\left\{ \frac{\partial(x,y,z)}{\partial(l,b,h)} \right\}^{-1} = \left\{ \frac{\partial(l,b,h)}{\partial(x,y,z)} \right\} \ ,
$$

$$
\mathsf{J}^2 := \left\{ \frac{\partial(x,y,z)}{\partial(l,b,h)} \right\} = \begin{bmatrix} \partial x/\partial l & \partial x/\partial b & \partial x/\partial h \\ \partial y/\partial l & \partial y/\partial b & \partial y/\partial h \\ \partial z/\partial l & \partial z/\partial b & \partial z/\partial h \end{bmatrix} \ , \quad \mathsf{J}_2 := \left(\mathsf{J}^2\right)^{-1} = \left\{ \frac{\partial(l,b,h)}{\partial(x,y,z)} \right\} \ , \tag{21.26}
$$

$$
\mathsf{J}^2 = \begin{bmatrix} -(n+h)\cos b\sin l & -(m+h)\sin b\cos l & \cos b\cos l \\ (n+h)\cos b\cos l & -(m+h)\sin b\sin l & \cos b\sin l \\ 0 & (m+h)\cos b & \sin b \end{bmatrix} , \quad \mathsf{J}_2 = \begin{bmatrix} -\frac{\sin l}{(n+h)\cos b} & \frac{\cos l}{(n+h)\cos b} & 0 \\ -\frac{\cos l\sin b}{m+h} & -\frac{\sin l\sin b}{m+h} & \frac{\cos b}{m+h} \\ \cos b\cos l & \cos b\sin l & \sin b \end{bmatrix} , \tag{21.27}
$$

$$
\mathsf{J}_{23} := \left\{ \frac{\partial(l,b)}{\partial(x,y,z)} \right\}_{\text{taylor}} = \begin{bmatrix} -\frac{\sin L}{(N+H)\cos B} & +\frac{\cos L}{(N+H)\cos B} & 0 \\ -\frac{\cos L\sin B}{M+H} & -\frac{\sin L\sin B}{M+H} & \frac{\cos B}{M+H} \end{bmatrix} \in \mathbb{R}^{2\times 3} \ ,
$$

$$
\mathsf{J}_1 := \left\{ \frac{\partial(x,y,z)}{\partial(t_x,t_y,t_z,\alpha,\beta,\gamma,s,A_1,E^2)} \right\}_{\text{taylor}} = \tag{21.28}
$$

$$
= \begin{bmatrix} 1 & 0 & 0 & 0 & -Z & Y & X & \frac{N\cos B\cos L}{A_1} & \frac{M\cos B\sin^2 B\cos L}{2(1-E^2)} \\ 0 & 1 & 0 & Z & 0 & -X & Y & \frac{N\cos B\sin L}{A_1} & \frac{M\cos B\sin^2 B\sin L}{2(1-E^2)} \\ 0 & 0 & 1 & -Y & X & 0 & Z & \frac{N\sin B(1-E^2)}{A_1} & \frac{M\sin^3 B - 2N\sin B}{2} \end{bmatrix} \in \mathbb{R}^{3\times 9} \ .
$$

Various remarks with respect to the rank and the stability of the Jacobi matrix A have to be made. First, the Jacobi matrix A indicates that columns seven and eight, namely $\boldsymbol{a}_7$ and $\boldsymbol{a}_8$, are linearly dependent, in particular, $\mathsf{A}\boldsymbol{a}_8 = \boldsymbol{a}_7$. Obviously, the incremental scale s and the incremental semi-major axis δA cannot be determined independently. Since we cannot discriminate s and δA, we may consult data files of the global and local reference ellipsoid in order to fix the values $\delta A := A - a$ as well as $\delta E^2 := E^2 - e^2$ and remove by $b - B - (a_{28}\delta A + a_{29}\delta E^2)$ the quantities $\{\delta A, \delta E^2\}$ from the analysis. Note that only the differences in latitude are influenced by $\{s, \delta A, \delta E^2\}$, respectively. Second, for a geodetic network for which both $\{l, b\}$ and $\{L, B, H\}$ are available, the extension in latitude $\{b, B\}$ may lead to an instability within the Jacobi matrix A as we have experienced. In the analysis of column $\boldsymbol{a}_3$ (acting on translation component t_z) and column $\boldsymbol{a}_7$ (acting on scale s), $\sin B$ is approximately discriminating the two columns. In the above-quoted network configuration, $\boldsymbol{a}_3$ and $\boldsymbol{a}_7$ are nearly linearly dependent. The following rationale has accordingly been chosen, following the method of partial least squares (P. J. Young (1994)), for instance.

Important!

The system (21.23) of linear equations $\boldsymbol{y} = \mathsf{A}\boldsymbol{x} + \text{hit}$ is characterized by two pseudo-observed ellipsoidal longitude and ellipsoidal latitude differences. In case we have access to the ellipsoidal form parameter variation δA and δE^2, we are left with two equations for seven datum parameters per station point. Obviously, in order to determine the seven datum parameters, we need at least four station points with the data $\{l, b\}$ and $\{L, B, H\}$ available. But, in general, we are left with an adjustment problem when the data are accessible at four or more station points. Since ellipsoidal longitude and ellipsoidal latitude $\{l, b\}$ are elements of $\mathbb{E}^2_{a_1, a_2}$, the distance between the adjusted points and the given data points $\{l, b\}$ has to be minimized. The distance along a geodesic as outlined in Box 21.5 originates from a series expansion of the minimal geodesic on $\mathbb{E}^2_{a_1, a_2}$. A zero order approximation is the Euclidean distance known as the method of least squares. (For a review of robust distance functions for those pseudo-observations given on a circle $\mathbb{S}^1_r$ or on a sphere $\mathbb{S}^2_r$ of radius r, we refer to Y. M. Chan and X. He (1993)). Here, we have chosen the zero order approximation, the method of least squares $\|\boldsymbol{y} - \mathsf{A}\boldsymbol{x}\|^2 = \min$ leading to $\hat{\boldsymbol{x}} = (\mathsf{A}^{\mathsf{T}}\mathsf{P}\mathsf{A})^{-1}\mathsf{A}^{\mathsf{T}}\mathsf{P}\boldsymbol{y}$ as the best approximation of the datum parameters.

In the first step of the partial least squares solution, we have given the prior information of the rotation parameters as well as the scale parameter a large weight. Accordingly, we have solved for the translation parameters $\boldsymbol{x}_1 := [t_x, t_y, t_z]$ exclusively. The second step is split up into a forward and a backward one. First, we remove the data $\hat{\boldsymbol{x}}_1$ (translation parameters) of best approximation from the reduced pseudo-observed data, namely $\boldsymbol{y} - (\boldsymbol{a}_8\delta A - \boldsymbol{a}_9\delta E^2)$. Second, in the 2nd partial least squares solution, we associate to the prior information of the scale parameter a large weight. Indeed, we compute the rotation parameters $\boldsymbol{x}_2 := [\alpha, \beta, \gamma]$ exclusively from γ. The third step is split up again into a forward and a backward one. First, we remove the data $\hat{\boldsymbol{x}}_2$ (rotation parameters) of best approximation from the reduced pseudo-observed data, namely $\boldsymbol{y}_1 - \mathsf{A}_2\hat{\boldsymbol{x}}_2 =: \boldsymbol{y}_2$. Second, in the 3rd partial least squares solution, we finally compute the scale parameter $\boldsymbol{x}_3 = s$ exclusively from $\boldsymbol{y}_2$.

Box 21.5 (The distance between the points l, b and l_0, b_0 on $\mathbb{E}^2_{a_1, a_2}$ along a minimal geodesic).

$$
s^2 = \frac{a_1^2}{1 - e^2 \sin^2 b_0}\left[(b - b_0)^2 \frac{(1 - e^2)^2}{(1 - e^2 \sin^2 b_0)^2} + (l - l_0)^2 \cos^2 b_0 + \right.
$$

$$
+ 3(b - b_0)^3 \frac{(1 - e^2)^2 \cos b_0 \sin b_0}{(1 - e^2 \sin^2 b_0)^3} - (b - b_0)(l - l_0)^2 \frac{(1 - e^2)\cos b_0 \sin b_0}{1 - e^2 \sin^2 b_0} +
$$

$$
+ (b - b_0)^4 \frac{(1 - e^2)e^2[4 - 8\sin^2 b_0 + e^2 \sin^2 b_0(25 - 21\sin^2 b_0)]}{4(1 - e^2 \sin^2 b_0)^4} -
$$

$$
- (b - b_0)^2(l - l_0)^2 \frac{(1 - e^2)\cos^2 b_0(2 + 7e^2 \sin^2 b_0)}{6(1 - e^2 \sin^2 b_0)^2} - (l - l_0)^4 \frac{\cos^2 b_0 \sin^2 b_0}{12}\left.\right] + O(l^5, b^5) \, .
$$

(21.29)

			$l\,[°]$	$b\,[°]$	$L\,[°]$	$B\,[°]$	$H\,[m]$	$l-L\,['']$	$a_{28}\delta A+$ $+a_{29}\delta E^2$ $['']$	$b-B-$ $-a_{28}\delta A+$ $+a_{29}\delta E^2$ $['']$
1	0080	FRIEDRICHSKOOG	8.83971094	54.03082844	8.83867346	54.02922780	40.68	3.7349	-2.0470	7.8093
2	0090	WILHELMSHAVEN	8.14499399	53.51627163	8.14406386	53.51473059	46.29	3.3485	-2.0587	7.6065
3	0100	HAMBURG	9.98504636	53.55284548	9.98382969	53.55129680	93.49	4.3800	-2.0579	7.6332
4	0120	PILSUM	7.06140395	53.48529231	7.06064275	53.48375630	40.69	2.7403	-2.0594	7.5891
5	0121	Z-PILSUM	7.06159494	53.48535029	7.06083372	53.48381426	40.49	2.7404	-2.0594	7.5891
6	0130	BREMEN	8.80467149	53.07310638	8.80363900	53.07161511	65.83	3.7170	-2.0683	7.4369
7	0140	0-HOHENBUENSTORF	10.47774276	53.05213785	10.47645177	53.05065050	149.86	4.6476	-2.0687	7.4232
8	0141	1-HOHENBUENSTORF	10.47759020	53.05211734	10.47629924	53.05062999	152.29	4.6475	-2.0687	7.4232
9	0150	LATHEN	7.31835171	52.85570063	7.31754684	52.85423549	52.39	2.8975	-2.0728	7.3473
10	0161	1-FLADDERLOHSN	8.10138243	52.57088270	8.10045598	52.56944831	84.05	3.3352	-2.0785	7.2423
11	0170	BENTHEIM	7.04636382	52.30667538	7.04559715	52.30527264	75.38	2.7600	-2.0837	7.1335
12	0180	HANNOVER	9.74745372	52.37262889	9.74627739	52.37121615	120.25	4.2348	-2.0824	7.1683
13	0200	BRAUNSCHWEIG	10.46187383	52.29875798	10.46058928	52.29735451	137.03	4.6244	-2.0838	7.1363
14	0210	KOETERBERG	9.32549890	51.85709904	9.32438718	51.85574388	537.88	4.0022	-2.0918	6.9704
15	0220	SOESTWARTE	8.05583341	51.73579178	8.05490915	51.73445181	195.16	3.3273	-2.0941	6.9180
16	0230	ODERBRUECK	10.58209949	51.78296334	10.58079995	51.78161869	925.01	4.6783	-2.0930	6.9338
17	0280	KNUELL	9.42326622	50.91770097	9.42215045	50.91645424	689.72	4.0168	-2.1075	6.5957
18	0290	STEGSKOPF	8.03085144	50.71570294	8.02993792	50.71447390	555.84	3.2887	-2.1106	6.5352
19	0330	KOBLENZ	7.56928987	50.36015617	7.56844470	50.35896466	156.21	3.0426	-2.1159	6.4054
20	0340	PREMICH	10.00672693	50.31844766	10.00553192	50.31727074	469.86	4.3021	-2.1164	6.3533
21	0350	KLOPPENHEIM	8.73094320	50.22093592	8.72993266	50.21976312	222.44	3.6379	-2.1179	6.3399
22	0360	COBURG	10.99414188	50.26545845	10.99280510	50.26429415	502.44	4.8124	-2.1172	6.3086
23	0370	0-WIESBADEN	8.22757167	50.08929277	8.22663682	50.08813287	240.12	3.3655	-2.1196	6.2953
24	0380	HOESBACH	9.21232692	50.02581248	9.21125229	50.02466309	298.87	3.8687	-2.1205	6.2583
25	0410	NIEDERWEILER	7.29654706	49.91593987	7.29574695	49.91479387	526.50	2.8804	-2.1218	6.2474
26	0420	WUERZBURG	9.90608076	49.78352281	9.90491003	49.78240502	398.47	4.2146	-2.1236	6.1476
27	0430	HOHENTHAN	12.38484929	49.76738233	12.38331460	49.76628738	817.97	5.5249	-2.1236	6.0654

Fig. 21.2. Pseudo-observations, part one. Courtesy of Landesvermessungsamt Baden–Württemberg. 1st: point counter. 2nd: station number. 3rd: station name. 4th: $l\,[°]$. 5th: $b\,[°]$. 6th: $L\,[°]$. 7th: $B\,[°]$. 8th: $H\,[m]$. 9th: $l-L\,['']$. 10th: lateral variation $['']$ due to δA, δE^2. 11th: $b-B-$ (column10) $['']$.

			l [°]	b [°]	L [°]	B [°]	H [m]	$l-L$ ["]	lat. var. ["]	$b-B-$(col 10) ["]
28	0440	GROSSBOCKENHEIM	8.16015531	49.59810421	8.15923675	49.59699745	349.74	3.3068	-2.1259	6.1102
29	0450	ALGERSDORF	11.40859654	49.57327390	11.40721139	49.57219331	556.72	4.9865	-2.1261	6.0162
30	0460	KEWELSBERG	6.50033497	49.46746977	6.49964808	49.46636999	488.86	2.4728	-2.1274	6.0866
31	0470	KATZENBUCKEL	9.04967481	49.46374566	9.04863213	49.46265847	562.57	3.7536	-2.1274	6.0413
32	0480	DILLENBERG	10.78353740	49.45410763	10.78224537	49.45303542	477.82	4.6513	-2.1276	5.9875
33	0490	HOECHERBERG	7.26688662	49.39848305	7.26609242	49.39739208	592.77	2.8591	-2.1282	6.0557
34	0500	GERABRONN	9.92417693	49.25906044	9.92301202	49.25800319	525.91	4.1937	-2.1298	5.9359
35	0510	0-WETTZELL	12.87968741	49.14595402	12.87809615	49.14493798	661.12	5.7285	-2.1310	5.7887
36	0511	1-WETTZELL	12.88010165	49.14468550	12.87851044	49.14366961	661.19	5.7284	-2.1310	5.7882
37	0520	EICHLBERG	11.71389413	49.08811385	11.71247549	49.08709150	613.23	5.1071	-2.1316	5.8121
38	0530	KARLSRUHE	8.48649743	48.99795511	8.48554043	48.99691551	325.92	3.3452	-2.1327	5.8753
39	0540	GAMMERSFELD	11.05955113	48.80886467	11.05823173	48.80786734	586.73	4.7499	-2.1346	5.7250
40	0550	HAHNENBUEHL	9.08535449	48.78784982	9.08431743	48.78683877	567.86	3.7334	-2.1348	5.7746
41	0560	OBERKOCHEN	10.08741297	48.79318110	10.08623219	48.79217803	815.14	4.2508	-2.1347	5.7457
42	0562	2-OBERKOCHEN	10.08930834	48.79206055	10.08812732	48.79105761	781.92	4.2517	-2.1347	5.7453
43	0570	ALTREICHENAU	13.72726796	48.76817483	13.72556458	48.76721078	863.88	6.1322	-2.1349	5.6055
44	0580	HAID	12.80172624	48.66930452	12.80015679	48.66834235	461.62	5.6500	-2.1360	5.5998
45	0590	SCHWEITENKIRCHEN	11.60830958	48.50791207	11.60691832	48.50695549	568.61	5.0085	-2.1375	5.5812
46	0600	SCHUTTERWALD	7.89457050	48.44652456	7.89370568	48.44554102	198.54	3.1134	-2.1382	5.6789
47	0610	RAICHBERG	8.99655866	48.30177907	8.99554449	48.30082229	970.75	3.6510	-2.1392	5.5836
48	0620	WUERDING	13.35630361	48.36019749	13.35466304	48.35927757	363.74	5.9061	-2.1389	5.4506
49	0640	MUENCHEN	11.59267280	48.14231119	11.59129051	48.14139708	580.60	4.9763	-2.1407	5.4315
50	0650	SCHELLENBERG	9.59903456	48.05093752	9.59794033	48.05001440	680.88	3.9393	-2.1414	5.4647
51	0670	BELCHEN	7.83782984	47.81886951	7.83698176	47.81795487	1389.61	3.0531	-2.1430	5.4357
52	0690	HOHENPEISSENBERG	11.01624485	47.80205752	11.01495396	47.80117688	1027.88	4.6472	-2.1433	5.3136
53	0700	DETTENDORF	11.93994342	47.80569078	11.93851781	47.80481937	744.73	5.1322	-2.1433	5.2804
54	0710	ROSSFELD	13.09738640	47.62879150	13.09580365	47.62795310	1583.25	5.6979	-2.1443	5.1626
55	0921	1-BERLIN-W	13.31389330	52.49071768	13.31216472	52.48931302	108.41	6.2229	-2.1801	7.1369
56	1000	LEIPZIG	12.41485046	51.31366543	12.41327911	51.31239133	274.84	5.6569	-2.1013	6.6880
57	1030	OLBERSDORF	14.77796449	50.87021939	14.77604292	50.86901842	376.29	6.9176	-2.0083	6.4318
58	1040	HOHENST-E	12.72474145	50.81306034	12.72313194	50.81184778	524.21	5.7942	-2.1092	6.4744
59	1060	ALTENBERG	13.73529234	50.75285380	13.73353115	50.75165642	939.00	6.3403	-2.1099	6.4205
60	1080	GASSENREUTH	12.06058998	50.34003910	12.05908665	50.33887826	662.34	5.4120	-2.1161	6.2951

Fig. 21.3. Pseudo-observations, part two. Courtesy of Landesvermessungsamt Baden–Württemberg. 1st: point counter. 2nd: station number. 3rd: station name. 4th: l [°]. 5th: b [°]. 6th: L [°]. 7th: B [°]. 8th: H [m]. 9th: $l-L$ ["]. 10th: lateral variation ["] due to δA, δF^2. 11th: $b-B-$(column 10) ["].

A real case study is the following. For the analytic part, i.e. the determination of local to global transformation parameters, we have chosen 60 points from the German and European GPS reference network (DREF 91, EUREF), where global coordinates $\{X, Y, Z\}$ given with respect to the GRS 80 datum have been determined. The same points are equipped with official local coordinates $\{x, y, z\}$ in the Gauss–Krueger system given with respect to the Bessel ellipsoid and Potsdam datum. In the above Figs. 21.2 and 21.3, we have listed these data being transformed into ellipsoidal coordinates $\{L, B, H\}$ and $\{l, b, h\}$, respectively. (Ellipsoidal longitude l and ellipsoidal latitude b of local type ("Rauenberg", Bessel ellipsoid-of-revolution, $a_1 = 6377397.155\,\mathrm{m}$, $e^2 = 0.006674372$) versus ellipsoidal longitude L and ellipsoidal latitude B, ellipsoidal height H of global type (DREF ITRF 91, GRS 80, $A_1 = 6378137\,\mathrm{m}$, $E^2 = 0.0066943800229$) of German 1st order stations.) No heights are available in the local system. Column 10 shows the impact of the incremental semi-major axis δA and squared first eccentricity δE^2 on the pseudo-latitudes in the partial least squares process. The estimated transformation parameters according to the partial least squares process as described before are presented in Table 21.1.

Table 21.1. Analysis of datum parameters of type translation, rotation, and scale. This analysis is based upon the pseudo-observations presented in Figs. 21.2 and 21.3.

$$t_x = -610.144\,\mathrm{m}\,, \quad \alpha = -1''.0396$$

$$t_y = -21.658\ \mathrm{m}\,, \quad \beta = -0''.1859$$

$$t_z = -421.401\,\mathrm{m}\,, \quad \gamma = -1''.2712$$

$$s = -0.519\,485 \times 10^{-6}$$

$$\sqrt{\|\mathbf{y} - \mathsf{A}\hat{\mathbf{x}}\|^2/(n-7)} \text{ (longitude)} = 0''.03117710\,, \quad \sqrt{\|\mathbf{y} - \mathsf{A}\hat{\mathbf{x}}\|^2/(n-7)} \text{ (latitude)} = 0''.02640297$$

$$\mathsf{A} =$$

$$= \begin{bmatrix} -\dfrac{\sin L}{(N+H)\cos B} & \dfrac{\cos L}{(N+H)\cos B} & 0 & \dfrac{[(1-E^2)N+H]\sin B\cos L}{(N+H)\cos B} & \dfrac{[(1-E^2)N+H]\sin B\sin L}{(N+H)\cos B} \\[2ex] -\dfrac{\sin B\cos L}{M+H} & -\dfrac{\sin B\sin L}{M+H} & \dfrac{\cos B}{M+H} & \dfrac{[N^2(E^2-1)-MH]\sin L}{M(M+H)} & \dfrac{[N^2(1-E^2)+MH]\cos L}{M(M+H)} \end{bmatrix} \cdots$$

$$\cdots \begin{bmatrix} -1 & 0 & 0 & 0 \\[2ex] 0 & -\dfrac{E^2N\sin B\cos B}{M+H} & -\dfrac{E^2N\sin B\cos B}{A_1(M+H)} & -\dfrac{[E^2M\sin^2 B+2(1-E^2)N]\sin B\cos B}{2(M+H)(1-E^2)} \end{bmatrix}$$

$$\in \mathbb{R}^{2\times 9}\,.$$

(21.30)

21-2 Synthesis of a datum problem

Datum transformation, datum parameters (translation, rotation, scale). Synthesis matrix, local heights. Jacobi matrix, chain Jacobi matrix.

Synthesis is understood as the determination of curvilinear global coordinates of points from curvilinear local coordinates based upon given datum parameters. Since the datum parameters are close to the identity, Box 21.6 collects the formulae for the computation of ellipsoidal longitude L and ellipsoidal latitude B sought for in the global reference system from known ellipsoidal longitude l and ellipsoidal latitude b and given datum parameters $\{t_x t_y, t_z, \alpha, \beta, \gamma, s, \delta a, \delta e^2\}$ by a Taylor series expansion. In particular, Box 21.7 highlights the computation of the synthesis matrix B. Since in the local reference system pseudo-observations of ellipsoidal heights $h(l, b)$ are not available, in general, we here study by Box 21.8 the impact of local heights on the synthesis matrix B, namely by the decomposition $\mathsf{B} = \mathsf{B}_0 + h\mathsf{B}_1$.

Box 21.6 (Inverse coordinate conformal transformation (datum transformation) extended by ellipsoid parameters close to the identity).

Inverse transformation $\{X, Y, Z\} \mapsto \{L, B, H\}$:

$$L = \arctan(Y/X) + \left(-\frac{1}{2}\operatorname{sgn} Y - \frac{1}{2}\operatorname{sgn} Y \operatorname{sgn} X + 1\right)\pi \in \{\mathbb{R} \mid 0 \leq L < 2\pi\} \ ,$$

$$B = \arctan\left[\frac{Z}{\sqrt{X^2 + Y^2}}\frac{A_1(1 - E^2\sin^2 B)^{-1/2} + H(L, B)}{A_1(1 - E^2)(1 - E^2\sin^2 B)^{-1/2} + H(L, B)}\right] = $$

$$= \arctan\left[\frac{Z}{\sqrt{X^2 + Y^2}}\frac{N(B) + H(L, B)}{(1 - E^2)N(B) + H(L, B)}\right] \in \{\mathbb{R} \mid -\pi/2 < B < +\pi/2\} \ .$$

(21.31)

Inverse curvilinear coordinate conformal transformation close to the identity
(datum transformation).

1st variant:

$$L = \arctan\left(y - t_y - z\alpha + x\gamma - ys\right)/\left(x - t_z + z\beta - y\gamma - xs\right) \ ,$$

$$B = B(x, y, z, t_x, t_y, t_z, \alpha, \beta, \gamma, s) \ .$$

(21.32)

2nd variant:

$$\tan L =$$

$$= \Big(\left[a_1/\sqrt{1 - e^2\sin^2 b} + h(l, b)\right]\cos b \sin l - t_y \quad - \left[a_1(1 - e^2)/\sqrt{1 - e^2\sin^2 b} + h(l, b)\right]\sin b\alpha$$

$$+ \left[a_1/\sqrt{1 - e^2\sin^2 b} + h(l, b)\right]\cos b \sin l\gamma \quad - \left[a_1/\sqrt{1 - e^2\sin^2 b} + h(l, b)\right]\cos b \sin ls\Big)$$

$$/\Big(\left[a_1/\sqrt{1 - e^2\sin^2 b} + h(l, b)\right]\cos b \cos l - t_x + \left[a_1(1 - e^2)/\sqrt{1 - e^2\sin^2 b} + h(l, b)\right]\sin b\beta$$

$$- \left[a_1/\sqrt{1 - e^2\sin^2 b} + h(l, b)\right]\cos b \sin l\gamma \quad - \left[a_1/\sqrt{1 - e^2\sin^2 b} + h(l, b)\right]\cos b \cos ls\Big) \ ,$$

(21.33)

$$L = L(l, b, h(l, b), t_x, t_y, t_z, \alpha, \beta, \gamma, s, a_1, e^2) \ , \quad B = B(l, b, h(l, b), t_x, t_y, t_z, \alpha, \beta, \gamma, s, a_1, e^2) \ .$$

Taylor expansion:

$$L = l + L_{t_x}t_x + L_{t_y}t_y + L_{t_z}t_z + L_\alpha\alpha + L_\beta\beta + L_\gamma\gamma + L_s s + L_a\delta a + L_e\delta e^2 \ ,$$

$$B = b + B_{t_x}t_x + B_{t_y}t_y + B_{t_z}t_z + B_\alpha\alpha + B_\beta\beta + B_\gamma\gamma + B_s s + B_a\delta a + b_e\delta e^2 \ ,$$

(21.34)

subject to

$$L_{t_x} := \frac{\partial L}{\partial t_x}(t_x = t_y = t_z = 0, \alpha = \beta = \gamma = 0, s = 0, a_1 = A_1, e^2 = E^2)$$

(21.35)

etc.

and

$$\delta a := a_1 - A_1 = -\delta A \ , \quad \delta e^2 := e^2 - E^2 = -\delta E^2 \ , \quad \delta o = (2o)^{-1}\delta o^2 \ .$$

(21.36)

Box 21.7 (Inverse curvilinear conformal coordinate transformation (datum transformation) extended by ellipsoid parameters close to identity, Jacobi matrix).

$$
\begin{bmatrix} L \\ B \end{bmatrix} = \begin{bmatrix} l \\ b \end{bmatrix} + \mathsf{B} \begin{bmatrix} t_x \\ t_y \\ t_z \\ \alpha \\ \beta \\ \gamma \\ s \\ \delta a \\ \delta e^2 \end{bmatrix} + \mathrm{O}_{2LB} = \begin{bmatrix} l \\ b \end{bmatrix} + \mathsf{B}\boldsymbol{x} + \mathrm{O}_{2LB} \ . \tag{21.37}
$$

Jacobi matrix:

$$
\mathsf{B} := \left\{ \frac{\partial(L,B)}{\partial(t_x, t_y, t_z, \alpha, \beta, \gamma, s, a_1, e^2)} \right\}_{\text{taylor}} \in \mathbb{R}^{2 \times 9} \ , \tag{21.38}
$$
$$
\text{taylor} := \{ t_x = t_y = t_z = \alpha = \beta = \gamma = s = 0, a_1 = A_1, e^2 = E^2 \} \ .
$$

Chain Jacobi matrix:

$$
\mathsf{B} := \mathsf{J}'_{23} \mathsf{J}'_1 = \left\{ \frac{\partial(L,B)}{\partial(X,Y,Z)} \right\}_{\text{taylor}} \left\{ \frac{\partial(X,Y,Z)}{\partial(t_x, t_y, t_z, \alpha, \beta, \gamma, s, a_1, e^2)} \right\}_{\text{taylor}} \ , \tag{21.39}
$$

$$
\left\{ \frac{\partial(X,Y,Z)}{\partial(L,B,H)} \right\}^{-1} = \left\{ \frac{\partial(L,B,H)}{\partial(X,Y,Z)} \right\} \ , \tag{21.40}
$$

$$
\mathsf{J}'^2 := \left\{ \frac{\partial(X,Y,Z)}{\partial(L,B,H)} \right\} = \begin{bmatrix} \partial X/\partial L & \partial X/\partial B & \partial X/\partial H \\ \partial Y/\partial L & \partial Y/\partial B & \partial Y/\partial H \\ \partial Z/\partial L & \partial Z/\partial B & \partial Z/\partial H \end{bmatrix} \ , \quad \mathsf{J}'_2 := \left(\mathsf{J}'^2 \right)^{-1} = \left\{ \frac{\partial(L,B,H)}{\partial(X,Y,Z)} \right\} \ ,
$$

$$
\mathsf{J}'^2 =
$$
$$
\mathsf{J}'_2 =
$$
$$
\begin{bmatrix} -(N+H)\cos B \sin L & -(M+H)\sin B \cos L & \cos B \cos L \\ (N+H)\cos B \cos L & -(M+H)\sin B \sin L & \cos B \sin L \\ 0 & (M+H)\cos B & \sin B \end{bmatrix}, \quad \begin{bmatrix} -\frac{\sin L}{(N+H)\cos B} & \frac{\cos L}{(N+H)\cos B} & 0 \\ -\frac{\cos L \sin B}{M+H} & -\frac{\sin L \sin B}{M+H} & \frac{\cos B}{M+H} \\ \cos B \cos L & \cos B \sin L & \sin B \end{bmatrix}, \tag{21.41}
$$

$$
\mathsf{J}'_{23} = \left\{ \frac{\partial(L,B)}{\partial(X,Y,Z)} \right\}_{\text{taylor}} = \begin{bmatrix} -\frac{\sin l}{(n+h)\cos b} & +\frac{\cos l}{(n+h)\cos b} & 0 \\ -\frac{\cos l \sin b}{m+h} & -\frac{\sin l \sin b}{m+h} & \frac{\cos b}{m+h} \end{bmatrix} \in \mathbb{R}^{2 \times 3} \ ,
$$

$$
\mathsf{J}'_1 = \left\{ \frac{\partial(X,Y,Z)}{\partial(t_x, t_y, t_z, \alpha, \beta, \gamma, s, a_1, e^2)} \right\}_{\text{taylor}} = \tag{21.42}
$$

$$
= \begin{bmatrix} -1 & 0 & 0 & 0 & z & -y & -x & \frac{n \cos b \cos l}{a_1} & \frac{m \cos b \sin^2 b \cos l}{2(1-e^2)} \\ 0 & -1 & 0 & -z & 0 & x & -y & \frac{n \cos b \sin l}{a_1} & \frac{m \cos b \sin^2 b \sin l}{2(1-e^2)} \\ 0 & 0 & -1 & y & -x & 0 & -z & \frac{n \sin b (1-e^2)}{a_1} & \frac{m \sin^3 b - 2n \sin b}{2} \end{bmatrix} \in \mathbb{R}^{3 \times 9} \ .
$$

Box 21.8 (Decomposition of the synthesis matrix $B = B_0 + hB_1 + O_2(h^2)$.).

$$B =$$

$$= \begin{bmatrix} \frac{\sin l}{(n+h)\cos b} & -\frac{\cos l}{(n+h)\cos b} & 0 & -\frac{[(1-e^2)n+h]\sin b\cos l}{(n+h)\cos b} & -\frac{[(1-e^2)n+h]\sin b\sin l}{(n+h)\cos b} \\ \frac{\sin b\cos l}{m+h} & \frac{\sin b\sin l}{m+h} & -\frac{\cos b}{m+h} & \frac{[n^2(1-e^2)+mh]\sin l}{m(m+h)} & -\frac{[n^2(1-e^2)+mh]\cos l}{m(m+h)} \end{bmatrix} \cdots$$

$$\cdots \begin{bmatrix} 1 & 0 & 0 & 0 \\ 0 & \frac{e^2 n\sin b\cos b}{m+h} & -\frac{e^2 n\sin b\cos b}{a_1(m+h)} & -\frac{[e^2 m\sin^2 b + 2(1-e^2)n]\sin b\cos b}{2(m+h)(1-e^2)} \end{bmatrix} \in \mathbb{R}^{2\times 9},$$

(21.43)

$$B_0 :=$$

$$= \begin{bmatrix} \frac{\sin l}{n\cos b} & -\frac{\cos l}{n\cos b} & 0 & -(1-e^2)\tan b\cos l & -(1-e^2)\tan b\sin l \\ \frac{\sin b\cos l}{m} & \frac{\sin b\sin l}{m} & -\frac{\cos b}{m} & \frac{[n^2(1-e^2)]\sin l}{m^2} & -\frac{[n^2(1-e^2)]\cos l}{m^2} \end{bmatrix} \cdots$$

$$\cdots \begin{bmatrix} 1 & 0 & 0 & 0 \\ 0 & \frac{e^2 n\sin b\cos b}{m} & -\frac{e^2 n\sin b\cos b}{a_1 m} & -\frac{e^2 m\sin^3 b\cos b + 2(1-e^2)n\sin b\cos b}{2m(1-e^2)} \end{bmatrix},$$

(21.44)

$$B_1 :=$$

$$= \begin{bmatrix} -\frac{\sin l}{n^2\cos b} & \frac{\cos l}{n^2\cos b} & 0 & -\frac{e^2}{n}\tan b\cos l & -\frac{e^2}{n}\tan b\sin l \\ -\frac{\sin b\cos l}{m^2} & -\frac{\sin b\sin l}{m^2} & \frac{\cos b}{m^2} & \frac{[m^2 - n^2(1-e^2)]\sin l}{m^3} & \frac{[n^2(1-e^2)-m^2]\cos l}{m^3} \end{bmatrix} \cdots$$

$$\cdots \begin{bmatrix} 0 & 0 & 0 & 0 \\ 0 & -\frac{e^2 n\sin b\cos b}{m^2} & \frac{e^2 n\sin b\cos b}{a_1 m^2} & \frac{[e^2 m\sin^2 b + 2(1-e^2)n]\sin b\cos b}{2m^2(1-e^2)} \end{bmatrix}.$$

(21.45)

Example 21.1 numerically illustrates that for the computation of global $\{L, B\}$ from local $\{l, b\}$ and from datum parameters $\{t_x, t_y, t_z, \alpha, \beta, \gamma, s, \delta a, \delta e^2\}$ we can neglect hB_1. Accordingly, the synthesis of $\{L, B\}$ is performed by $[L, B]^T = [l, b]^T + B_0[t_x, t_y, t_z, \alpha, \beta, \gamma, s, \delta a, \delta e^2]^T$.

Example 21.1 (Synthesis of $\{L, B\}$ from $\{l, b\}$ and from datum parameters $\{t_x, t_y, t_z, \alpha, \beta, \gamma, s, \delta A, \delta E^2\}$, simulation of impact of local height h).

$$a_1 := 6\,377\,397.155\,[\text{m}], \quad e^2 := 0.006\,674\,372, \quad A_1 := 6\,378\,137\,[\text{m}], \quad E^2 := 0.006\,694\,380\,022\,9.$$

$$l = 14°, \quad b = 40°, \quad h = 1000\,\text{m (assumed)}$$
$$\{t_x, t_y, t_z, \alpha, \beta, \gamma, s, \delta A, \delta E^2\} \text{ from first section.}$$

$L_0 = L(l, b)$	$L = L(l, b, h)$
$13°59'54''.269\,6$	$13°59'54''.270\,4$
$B_0 = B(l, b)$	$B = B(l, b, h)$
$40°00'00''.069\,44$	$40°00'00''.069\,42$

End of Example.

21-3 Error propagation in analysis and synthesis of a datum problem

Nonlinear error propagation. Dispersion, dispersion transformation, dispersion matrix, variance–covariance matrix. Stochastic pseudo-observations.

First, in the *analysis* of the datum parameters from given $\{l, b\}$ and $\{L, B, H\}$ *pseudo-observations* by means of a best approximation (21.46) subject to (21.47), via *nonlinear error propagation*, we have to consider the *variance–covariance matrix/dispersion matrix* of the datum parameters $x = [t_x, t_y, t_z, \alpha, \beta, \gamma, s]$ functionally related to the variance–covariance matrix/dispersion matrix of the pseudo-observations $\{l, b\}$ and $\{L, B, H\}$. The stochastic pseudo-observations $\{l, b\}$ and $\{L, B, H\}$ enter via the relative data vector (21.48) and via the analysis matrix (21.49). Accordingly, we expand $\hat{\boldsymbol{x}}(l, b, L, B, H)$ into the dispersion (21.50) as outlined in Box 21.9. If a *prior dispersion matrix* of the form parameters $\{A_1, E^2, a_1, e^2\}$ is available, it could also be implemented.

Second, the *synthesis* of global ellipsoidal coordinates $\{L, B\}$ from given local ellipsoidal coordinates $\{l, b\}$ and datum parameters/ellipsoidal form parameters $\{t_x, t_y, t_z, \alpha, \beta, \gamma, s, \delta a, \delta e^2\}$, we again experience the impact of the dispersion matrix of $\{l, b\}$ as well as of $\{t_x, \ldots, \delta e^2\}$ via nonlinear error propagation. The random character of the pseudo-observations $\{l, b\}$ enters firstly linearly and secondly nonlinearly via $B_0(l, b)$, while the stochastic *a posteriori* parameters $\{\hat{t}_x, \ldots, \delta\hat{e}^2\}$ enter linearly. Box 21.10 reviews the expansion $[L, B](l, b, \hat{t}_x, \ldots, \delta\hat{e}^2)$ towards the dispersion (21.51).

$$\hat{\boldsymbol{x}} = \left(\mathsf{A}^\mathrm{T}\mathsf{P}\mathsf{A}\right)^{-1}\mathsf{A}^\mathrm{T}\mathsf{P}\boldsymbol{y} =: \mathsf{L}\boldsymbol{y} \,, \tag{21.46}$$

$$\mathsf{L} = \left(\mathsf{A}^\mathrm{T}\mathsf{P}\mathsf{A}\right)^{-1}\mathsf{A}^\mathrm{T}\mathsf{P} \,, \tag{21.47}$$

$$\boldsymbol{y} := [l - L, b - B]^\mathrm{T} \,, \tag{21.48}$$

$$\mathsf{A} = \mathsf{A}(L, B, H) \,, \tag{21.49}$$

$$\mathsf{D}(\hat{\boldsymbol{x}}) = \mathsf{M}_k \mathsf{D}_{kl}(l, b, L, B, H)\mathsf{M}_l^\mathrm{T} \,, \tag{21.50}$$

$$\mathsf{D}(L, B) = \mathsf{K}_\mu \mathsf{D}_{\mu\nu}(l, b, \hat{t}_x, \ldots, \delta\hat{e}^2)\mathsf{K}_\nu \,. \tag{21.51}$$

Box 21.9 (Error propagation with respect to analysis $\hat{\boldsymbol{x}}(l, b, L, B, H)$).

$$\mathrm{d}\hat{\boldsymbol{x}} = \mathrm{d}\mathsf{L}\boldsymbol{y} + \mathsf{L}\mathrm{d}\boldsymbol{y} \,. \tag{21.52}$$

$$\mathrm{d}\mathsf{L} =$$
$$= -(\mathsf{A}^\mathrm{T}\mathsf{A})\mathrm{d}\mathsf{A}^\mathrm{T}\mathsf{A}(\mathsf{A}^\mathrm{T}\mathsf{A})^{-1}\mathsf{A}^\mathrm{T} + (\mathsf{A}^\mathrm{T}\mathsf{A})^{-1}\mathrm{d}\mathsf{A}^\mathrm{T} - (\mathsf{A}^\mathrm{T}\mathsf{A})^{-1}\mathsf{A}^\mathrm{T}\mathrm{d}\mathsf{A}(\mathsf{A}^\mathrm{T}\mathsf{A})^{-1}\mathsf{A}^\mathrm{T} = \tag{21.53}$$
$$= -\mathsf{N}\mathrm{d}\mathsf{A}^\mathrm{T}\mathsf{A}\mathsf{L} - \mathsf{L}\mathrm{d}\mathsf{A}\mathsf{L} + \mathsf{N}^{-1}\mathrm{d}\mathsf{A}^\mathrm{T}$$

subject to

$$\mathsf{N} := \mathsf{A}^\mathrm{T}\mathsf{A} \,, \quad \mathsf{P} = \mathsf{I} \,. \tag{21.54}$$

$$\mathrm{d}\hat{\boldsymbol{x}} = -\mathsf{N}\mathrm{d}\mathsf{A}^\mathrm{T}\mathsf{A}\hat{\boldsymbol{x}} - \mathsf{L}\mathrm{d}\mathsf{A}\hat{\boldsymbol{x}} + \mathsf{N}^{-1}\mathrm{d}\mathsf{A}^\mathrm{T}\boldsymbol{y} + \mathsf{L}\mathrm{d}\boldsymbol{y} \,,$$

$$\mathrm{vec}\,\mathrm{d}\hat{\boldsymbol{x}} = \left[-(\mathsf{A}\hat{\boldsymbol{x}})^\mathrm{T} \otimes \mathsf{N}\right]\mathrm{vec}\,\mathrm{d}\mathsf{A}^\mathrm{T} - \left[(\hat{\boldsymbol{x}})^\mathrm{T} \otimes \mathsf{L}\right]\mathrm{vec}\,\mathrm{d}\mathsf{A} + \left[(\boldsymbol{y})^\mathrm{T} \otimes \mathsf{N}^{-1}\right]\mathrm{vec}\,\mathrm{d}\mathsf{A}^\mathrm{T} + \mathsf{L}\,\mathrm{vec}\,\mathrm{d}\boldsymbol{y} \tag{21.55}$$

subject to

(decomposition of a double Cayley product by a Kronecker–Zehfuss product)

$$\mathrm{vec}\,(\mathsf{A}\mathsf{B}\mathsf{C}) = (\mathsf{C}^\mathrm{T} \otimes \mathsf{A})\,\mathrm{vec}\,\mathsf{B} \,. \tag{21.56}$$

Continuation of Box.

$$\text{vec}\, d\hat{\boldsymbol{x}} = \mathsf{Q}\, \text{vec}\, d\mathsf{A} + \mathsf{R}\, \text{vec}\, d\boldsymbol{y} \tag{21.57}$$

subject to

$$\mathsf{Q} := \big[-(\mathsf{A}\hat{\boldsymbol{x}})^{\mathrm{T}} \otimes \mathsf{N} \big] \mathsf{I}_{7,2} + \big[(\boldsymbol{y})^{\mathrm{T}} \otimes \mathsf{N}^{-1} \big] \mathsf{I}_{7,2} - \big[(\hat{\boldsymbol{x}})^{\mathrm{T}} \otimes \mathsf{L} \big] \, ,$$

$$\mathsf{R} := \mathsf{L} \, . \tag{21.58}$$

$$\text{vec}\, d\hat{\boldsymbol{x}} = \mathsf{Q}\, \text{vec} \left(\frac{\partial \mathsf{A}}{\partial p_k} dp_k \right) + \mathsf{R}\, \text{vec} \left(\frac{\partial \boldsymbol{y}}{\partial p_k} dp_k \right) \tag{21.59}$$

with respect to the parameters $p_k := \{l, b, L, B, H\}$.

Dispersion transformation:

$$\mathsf{D}(\hat{\boldsymbol{x}}) = \mathsf{M}_k \mathsf{D}(p_k, p_l) \mathsf{M}_l^{\mathrm{T}} \tag{21.60}$$

subject to

$$\mathsf{M}_k := \mathsf{Q} \frac{\partial \mathsf{A}}{\partial p_k} + \mathsf{R} \frac{\partial \boldsymbol{y}}{\partial p_k} \tag{21.61}$$

with

$$\{k, l\} \in \{1, 2, 3, 4, 5\} \, . \tag{21.62}$$

Box 21.10 (Error propagation with respect to synthesis $[L, B](l, b, \hat{t}_x, \ldots, \delta\hat{e}^2)$).

Parameters:

$$L_i := [L, B] \, , \quad l_i := [l, b] \, , \quad t_p := [t_x, t_y, t_z, \alpha, \beta, \gamma, s, \delta a, \delta e^2] \, . \tag{21.63}$$

Error propagation:

$$\frac{\partial L_i}{\partial l_j} = \delta_{ij} + \frac{\partial b_{ip}^0}{\partial l_j} t_p \, , \quad \frac{\partial L_i}{\partial t_p} = b_{ip}^0 \, , \tag{21.64}$$

$$\mathsf{D}(L_i, L_j) = \big(\delta_{im} + b_{ip,m}^0 t_p \big) \mathsf{D}(l_m, l_n) \big(\delta_{jn} + b_{jq,n}^0 t_q \big) + b_{ip}^0 \mathsf{D}(t_p, t_q) b_{jq}^0 \, . \tag{21.65}$$

(Summation convention over repeated indices, $\{i, j, m, n\} \in \{1, 2\}$ and $\{p, q\} \in \{1, \ldots, 9\}$.)

21-4 Gauss–Krueger/UTM coordinates: from a local to a global datum

Transformation of conformal coordinates of type Gauss–Krueger or UTM from a local datum (regional, National, European) to a global datum (WGS 84).

A key problem of contemporary geodetic positioning is the transformation of *mega data sets* of conformal coordinates of type Gauss–Krueger or UTM from a *local datum*, also called regional, National, European etc., to a *global datum*, for instance, the *World Geodetic System 1984* (*WGS 84*) with reference to M. J. Boyle (1987). As an example, let us refer to the mega data sets of more than 150 Mio. Gauss–Krueger conformal coordinates of Germany, where the West German conformal coordinates relate to the *Bessel reference ellipsoid*, while the East German conformal coordinates relate to the *Krassowsky reference ellipsoid*. Thanks to the satellite Global Positioning System ("global problem solver": GPS) and advanced computer software, geodetic positions are given as conformal coordinates of type Gauss–Krueger or UTM relating to the reference ellipsoid of the World Geodetic System (WGS 84). In connection with a chart, GPS-derived conformal coordinates of type Gauss–Krueger or UTM can only be used when they are transformed from the global datum to the local datum, which the chart is based upon. Or *vice versa*: the conformal coordinates of type Gauss–Krueger or UTM which are presented in a chart of local datum have to be transformed to the global datum. The transformation of conformal coordinates (Gauss–Krueger or UTM) from a local datum to a global one and *vice versa* is the objective of our contribution.

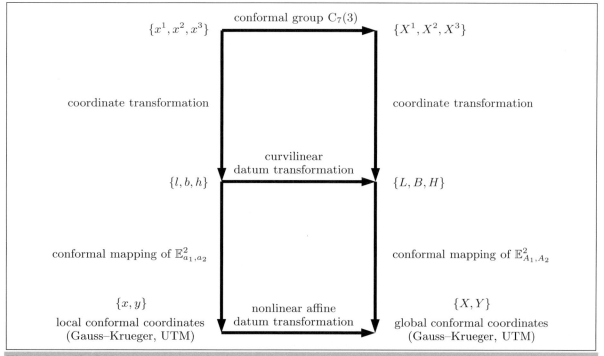

Fig. 21.4. The basic commutative diagram, rectangular, curvilinear, and conformal datum transformation, with three parameters of translation, three parameters of rotation, and one scale parameter, local reference ellipsoid-of-revolution $\mathbb{E}^2_{a_1,a_2}$, global reference ellipsoid-of-revolution $\mathbb{E}^2_{A_1,A_2}$.

As outlined by means of the commutative diagram of Fig. 21.4, Cartesian coordinates $\{X^1, X^2, X^3\}$ (capital letters: global datum) are first transformed into Cartesian coordinates $\{x^1, x^2, x^3\}$ (small letters: local datum) by means of the conformal group $C_7(3)$. Notably the conformal group $C_7(3)$ (seven parameters in a three-dimensional Weitzenböck space: three parameters for translation, three parameters for rotation, one scale parameter) leaves angles and distance ratios invariant (equivariant, form invariant). Such a datum transformation is called a *rectangular datum transformation*.

Second, in contrast, surface normal ellipsoidal coordinates of type ellipsoidal longitude, ellipsoidal latitude, ellipsoidal height $\{L, B, H\}$ and $\{l, b, h\}$ replace the Cartesian coordinates as user coordinates. $\{L, B, H\}$ refer to a global datum, in particular, to a global reference ellipsoid-of-revolution $\mathbb{E}^2_{A_1, A_2}$ (semi-major axis A_1, semi-minor axis A_2, relative eccentricity squared $E^2 := (A_1^2 - A_2^2)/A_1^2$), while $\{l, b, h\}$ refer to a local reference ellipsoid-of-revolution $\mathbb{E}^2_{a_1, a_2}$ (semi-major axis a_1, semi-minor axis a_2, relative eccentricity squared $e^2 := (a_1^2 - a_2^2)/a_1^2$. Accordingly, a transformation of ellipsoidal coordinates from $\{l, b, h\}$ to $\{L, B, H\}$ or *vice versa* is called a *curvilinear datum transformation*, i. e. close to the identity, expressed as a linear function represented by three parameters $\{t_x, t_y, t_z\}$ of translation, three parameters $\{\alpha, \beta, \gamma\}$ of rotation, and one scale parameter s. Such a curvilinear datum transformation (user oriented) has been investigated by A. Leick and B. H. W. van Gelder (1975), T. Soler (1976), R. Schreiber (1991), E. Grafarend, F. Krumm, and F. Okeke (1995) as well as F. Okeke (1997). Third, the target of our contribution is the datum transformation of conformal coordinates $\{X, Y\}$ of type Gauss–Krueger or UTM from a local datum to global one and *vice versa*. The first subsection is devoted to the derivation of the direct equations of the datum transformation $\{x, y\} \mapsto \{X, Y\}$, where we take advantage of computer-aided bivariate polynomial inversion pioneered by H. Glasmacher and K. Krack (1984), G. Joos and K. Joerg (1991), and E. Grafarend, T. Krarup and R. Syffus (1996). The second subsection collects the inverse equations $\{X, Y\} \mapsto \{x, y\}$ of a datum transformation of conformal coordinates of type Gauss–Krueger or UTM. Both transformations, direct and inverse equations, respectively, amount to bivariate polynomials with coefficients which depend on the datum transformation parameters $\{t_x, t_y, t_z, \alpha, \beta, \gamma, s\}$ and the change of the form parameter $\delta E^2 := E^2 - e^2$. Some remarks to our notation have to be made. We already mentioned that all quantities which refer to a global datum are written in capital letters, while those with reference to a local datum are written in small letters.

21-41 Direct transformation of local conformal into global conformal coordinates

The problem of generating Gauss–Krueger or UTM conformal coordinates $\{X, Y\}$ of a global reference ellipsoid-of-revolution $\mathbb{E}^2_{A_1, A_2}$ in terms of Gauss–Krueger or UTM conformal coordinates $\{x, y\}$ of a local (regional, National, European) reference ellipsoid-of-revolution $\mathbb{E}^2_{a_1, a_2}$ under a curvilinear datum transformation is solved here by a three-step-procedure according to the commutative diagram of Fig. 21.5. The first step is a representation of global conformal coordinates of type Gauss–Krueger or UTM in terms of conformal bivariate polynomials $X(L-L_0, B-B_0)$ and $Y(L-L_0, B-B_0)$ with respect to surface normal ellipsoidal longitude/ellipsoidal latitude increments $\{L-L_0, B-B_0\}$. The second step is divided into two sub-steps. First we transform the global ellipsoidal longitude/ellipsoidal latitude increments into local ellipsoidal longitude/ellipsoidal latitude increments $\{l - l_0, b - b_0\}$ by means of a curvilinear datum transformation (i. e. a linear function of the three parameters of translation, three parameters of rotation, and one scale parameter). Second, we implement the transformation $\{L-L_0, B-B_0\} \mapsto \{l-l_0, b-b_0\}$ into the representation $\{X(l-l_0, b-b_0), Y(l-l_0, b-b_0)\}$, in particular, including the curvilinear datum transformation for polynomial coefficients, too. Finally, the third step is split into two sub-steps. First, we repeat the representation of local conformal coordinates of type Gauss–Krueger or UTM in terms of conformal bivariate polynomials $\{x(l - l_0, b - b_0), y(l - l_0, b - b_0)\}$ with respect to surface normal ellipsoidal longitude/ellipsoidal latitude increments $\{l - l_0, b - b_0\}$, namely in order to construct the inverse polynomials $\{l - l_0(x, y), b - b_0(x, y)\}$ by bivariate series inversion. Second, we transfer the bivariate polynomial representation $\{l - l_0(x, y), b - b_0(x, y)\}$ to the power series $\{X(l - l_0), Y(b - b_0)\}$ in order to achieve the final bivariate general polynomials $\{X(x, y), Y(x, y)\}$.

21-411 The first step: conformal coordinates in a global datum

Conformal coordinates of an ellipsoid-of-revolution $\mathbb{E}^2_{A_1, A_2}$ in a global frame of reference (semi-major axis A_1, semi-minor axis A_2, relative eccentricity squared $E^2 := (A_1^2 - A_2^2)/A_1^2$) of type Gauss–Krueger or UTM are generated by a polynomial representation in terms of surface normal ellipsoidal longitude L and ellipsoidal latitude B with respect to an evaluation point $[L_0, B_0]$: see Box 21.11 in connection with Boxes 15.4 and 15.5.

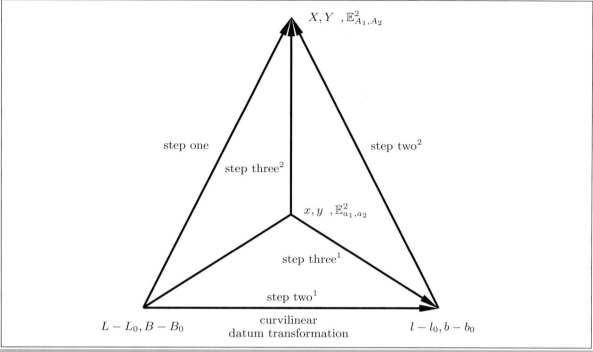

Fig. 21.5. The basic commutative diagram, genesis of the transformation of Cartesian conformal coordinates x, y in a local reference system to Cartesian conformal coordinates X, Y in a global reference system by means of a curvilinear datum transformation, three-step-procedure.

L_0 is also called the *ellipsoidal longitude* of the meridian of reference. While the coefficient Y_{00} represents the arc length of the meridian L_0 of reference, the coefficients are generated by solving the vector-valued boundary value problem of the Korn–Lichtenstein equations of conformal mapping subject to the integrability conditions, the vector-valued Laplace–Beltrami equations. The constraint of the vector-valued boundary value problem is the equidistant mapping of the meridian L_0 of reference and outlined by E. Grafarend (1995) and E. Grafarend and R. Syffus (1997).

Box 21.11 (Polynomial representation of conformal coordinates of type Gauss–Krueger or UTM, ellipsoid-of-revolution $\mathbb{E}^2_{A_1, A_2}$ (Easting X, Northing Y), surface normal ellipsoidal longitude L and ellipsoidal latitude B, evaluation point L_0, B_0, optimal factor of conformality $\rho = 0.999\,578$ (UTM) for a strip $[-l_{\mathrm{E}}, +l_{\mathrm{E}}] \times [B_{\mathrm{S}}, B_{\mathrm{N}}] = [-3.5°, +3.5°] \times [80°\mathrm{S}, 84°\mathrm{N}]$; coefficients are given in Boxes 15.4 and 15.5).

$$X =$$

$$= \rho \Big[X_{10}(L - L_0) + X_{11}(L - L_0)(B - B_0) + X_{30}(L - L_0)^3 + X_{12}(L - L_0)(B - B_0)^2 +$$

$$+ X_{31}(L - L_0)^3(B - B_0) + X_{13}(L - L_0)(B - B_0)^3 \Big] +$$

$$+ O(5) ,$$

$$(21.66)$$

$$Y =$$

$$= \rho \Big[Y_{00} + Y_{01}(B - B_0) + Y_{20}(L - L_0)^2 + Y_{02}(B - B_0)^2 + Y_{21}(L - L_0)^2(B - B_0) +$$

$$+ Y_{03}(B - B_0)^3 + Y_{40}(L - L_0)^4 + Y_{22}(L - L_0)^2(B - B_0)^2 + Y_{04}(B - B_0)^4 \Big] +$$

$$+ O(5) .$$

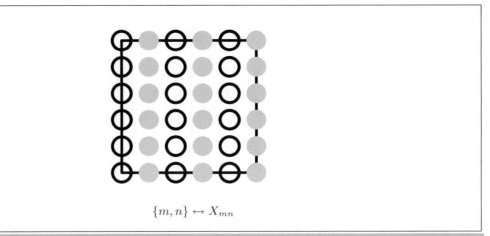

$$\{m, n\} \leftrightarrow X_{mn}$$

Fig. 21.6. Polynomial diagram, the polynomial representation of the conformal coordinate Easting X in a global frame of reference of type Gauss–Krueger or UTM, the solid dots illustrate non-zero monomials, the open circles zero monomials, according to D. Cox, J. Little and D. O'Shea (1996, pp. 433–443).

21-412 The second step: curvilinear datum transformation

With reference to E. Grafarend, F. Krumm and F. Okeke (1995, pp. 344–348), let us introduce the curvilinear datum transformation from a local datum (regional, National, European) to a global datum close to the identity reviewed in Box 21.12. Note that l and b refer to surface normal ellipsoidal longitude/ellipsoidal latitude of an ellipsoid-of-revolution $\mathbb{E}^2_{a_1, a_2}$ in a local frame of reference (semi-major axis a_1, semi-minor axis a_2, relative eccentricity squared $e^2 := (a_1^2 - a_2^2)/a_1^2$). The datum transformation close to the identity is expressed as a linear function of the datum transformation parameters, three parameters of translation $\{t_x, t_y, t_z\}$, three parameters of rotation $\{\alpha, \beta, \gamma\}$ and one scale parameter s. Note that the datum transformation in longitude close to the identity is independent of the translation parameter t_z and of the scale parameter s. In contrast, the datum transformation in latitude close to the identity does not depend on the rotation parameter γ (z axis rotation). The coefficients $\{s_{10}, \ldots, s_{27}\}$ are given in Box 21.13. The synthesis matrix S of a datum transformation close to the identity depends on the ellipsoidal height h in a local frame of reference. Only in a few cases local ellipsoidal height information is available. Within the matrix decomposition $\mathsf{S}(h) = \mathsf{S}_0 + h\mathsf{S}_1 + \mathrm{O}(h^2)$, we study the influence of local ellipsoidal height h.

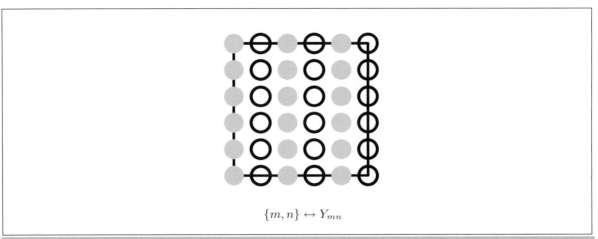

$$\{m, n\} \leftrightarrow Y_{mn}$$

Fig. 21.7. Polynomial diagram, the polynomial representation of the conformal coordinate Northing Y in a global frame of reference of type Gauss–Krueger or UTM, the solid dots illustrate non-zero monomials, the open circles zero monomials, according to D. Cox, J. Little and D. O'Shea (1996, pp. 433–443).

Box 21.12 (Curvilinear datum transformation, synthesis close to the identity).

$$L =$$

$$= l + s_{10} + s_{11}t_x + s_{12}t_y + s_{13}t_z + s_{14}\alpha + s_{15}\beta + s_{16}\gamma + s_{17}s = l + \delta L ,$$

$$\delta L :=$$

$$= s_{10} + s_{11}t_x + s_{12}t_y + s_{13}t_z + s_{14}\alpha + s_{15}\beta + s_{16}\gamma + s_{17}s ,$$

$$(21.67)$$

$$B =$$

$$= b + s_{20} + s_{21}t_x + s_{22}t_y + s_{23}t_z + s_{24}\alpha + s_{25}\beta + s_{26}\gamma + s_{27}s = b + \delta B ,$$

$$\delta B :=$$

$$= s_{20} + s_{21}t_x + s_{22}t_y + s_{23}t_z + s_{24}\alpha + s_{25}\beta + s_{26}\gamma + s_{27}s .$$

Box 21.13 (Synthesis matrix S of a curvilinear datum transformation close to the identity, decomposition $S(h) = S_0 + hS_1 + O(h^2)$).

$$n(b) := a_1/(1 - e^2 \sin^2 b)^{1/2},$$

$$m(b) := a_1(1 - e^2)/(1 - e^2 \sin^2 b)^{3/2},$$

$$p(b) := [A_1(1 - e^2 \sin^2 b)^{1/2}]/a_1(1 - E^2 \sin^2 b)^{1/2},$$

$$q(b) := [A_1(1 - E^2)(1 - e^2 \sin^2 b)^{3/2}]/a_1(1 - e^2)(1 - E^2 \sin^2 b)^{3/2}.$$

$$s_{10} = 0 , \quad s_{13} = 0 , \quad s_{17} = 0 ,$$

$$s_{11} =$$

$$= + \sin l/[(pn + h) \cos b] =$$

$$= + \sin l/[pn \cos b] - h \sin l/[p^2 n^2 \cos b] + O(h^2) ,$$

$$s_{12} =$$

$$= - \cos l/[(pn + h) \cos b] =$$

$$= - \cos l/[pn \cos b] + h \cos l/[p^2 n^2 \cos b] + O(h^2) ,$$

$$s_{14} =$$

$$(21.68)$$

$$= - \tan b \cos l[n(1 - e^2) + h]/(pn + h) =$$

$$= - \tan b \cos l(1 - e^2)/p + h \tan b \cos l(1 - e^2 - p)/(np^2) + O(h^2) ,$$

$$s_{15} =$$

$$= - \tan b \sin l[n(1 - e^2) + h]/(pn + h) =$$

$$= - \tan b \sin l(1 - e^2)/p + h \tan b \sin l(1 - e^2 - p)/(np^2) + O(h^2) ,$$

$$s_{16} =$$

$$= (n + h)/(pn + h) = 1/p + h(p - 1)/(p^2 n) + O(h^2) .$$

> **Continuation of Box.**
>
> $$s_{26} = 0 \, ,$$
>
> $$s_{20} =$$
>
> $$= -\cos b \sin b (qE^2 - e^2) n/(qm + h) =$$
>
> $$= -\cos b \sin b (qE^2 - e^2) n/qm + h \cos b \sin b (qE^2 - e^2) n/(qm)^2 + \mathrm{O}(h^2) \, ,$$
>
> $$s_{21} =$$
>
> $$= \sin b \cos l/(qm + h) =$$
>
> $$= \sin b \cos l/(qm) - h \sin b \cos l/(qm)^2 + \mathrm{O}(h^2) \, ,$$
>
> $$s_{22} =$$
>
> $$= \sin b \sin l/(qm + h) =$$
>
> $$= \sin b \sin l/(qm) - h \sin b \sin l/(qm)^2 + \mathrm{O}(h^2) \, ,$$
>
> $$s_{23} =$$
>
> $$= -\cos b/(qm + h) =$$ (21.69)
>
> $$= -\cos b/(qm) + h \cos b/(qm)^2 + \mathrm{O}(h^2) \, ,$$
>
> $$s_{24} =$$
>
> $$= +\sin l (a_1^2 + hn)/[n(qm + h)] =$$
>
> $$= +a_1^2 \sin l/(nqm) + h \sin l (nqm - a_1^2)/(nq^2 m^2) + \mathrm{O}(h^2) \, ,$$
>
> $$s_{25} =$$
>
> $$= -\cos l (a_1^2 + hn)/[n(qm + h)] =$$
>
> $$= -a_1^2 \cos l/(nqm) - h \cos l (nqm - a_1^2)/(nq^2 m^2) + \mathrm{O}(h^2) \, ,$$
>
> $$s_{27} =$$
>
> $$= e^2 \sin b \cos b n/(qm + h) =$$
>
> $$= e^2 \sin b \cos b n/(qm) - h e^2 \sin b \cos b n/(qm)^2 + \mathrm{O}(h^2) \, .$$

As soon as we implement the curvilinear datum transformation close to the identity (21.67) into the global representation (21.66) of conformal coordinates, we are led to the first version of global conformal coordinates (21.70) of Box 21.14 in terms of local ellipsoidal coordinates: see Box 21.15 and Box 21.16. Note that the coefficients $\{U_{MN}, V_{MN}\}$ in Box 21.15 and Box 21.16 depend on the coefficients $\{X_{MN}, Y_{MN}\}$ which, in turn, are given in terms of the parameters A_1, E^2, B_0. Accordingly, all the coefficients are transformed under the change of the form parameter E^2 from the global system of reference to the local one. The evaluation points $\{L_0, B_0\}$ and $\{l_0, b_0\}$ have been chosen to be identical in order to conserve the meridians-of-reference in both reference systems as well as the parallel-of-reference. Box 21.17 is a collection of Taylor expansions of order one of conformal series coefficients under a variation of form parameter δE^2 listed separately in Box 21.18 and Box 21.19. In consequence, upon a replacement of global coefficients by local coefficients of Box 21.18 in the coefficients of Box 21.19 and Box 21.15 , we derive the second version of global conformal coordinates (21.70) in terms of local ellipsoidal coordinates with the coefficients of Box 21.20 and Box 21.21.

Box 21.14 (Polynomial representation of conformal coordinates of type Gauss–Krueger or UTM after a curvilinear datum transformation).

$$X(l - l_0, b - b_0) =$$

$$= \rho \Big[U_{00} + U_{10}(l - l_0) + U_{01}(b - b_0) + U_{20}(l - l_0)^2 + U_{11}(l - l_0)(b - b_0) + U_{02}(b - b_0)^2 +$$

$$+ U_{30}(l - l_0)^3 + U_{21}(l - l_0)^2(b - b_0) + U_{12}(l - l_0)(b - b_0)^2 + U_{03}(b - b_0)^3 \Big] +$$

$$+ O(4) \, ,$$

$$Y(l - l_0, b - b_0) =$$

$$= \rho \Big[V_{00} + V_{10}(l - l_0) + V_{01}(b - b_0) + V_{20}(l - l_0)^2 + V_{11}(l - l_0)(b - b_0) + V_{02}(b - b_0)^2 +$$

$$+ V_{30}(l - l_0)^3 + V_{21}(l - l_0)^2(b - b_0) + V_{12}(l - l_0)(b - b_0)^2 + V_{03}(b - b_0)^3 \Big] +$$

$$+ O(4) \, .$$

$$(21.70)$$

Box 21.15 (Polynomial coefficients of a conformal series of type Gauss–Krueger or UTM after a curvilinear datum transformation, Easting $X(l - l_0, b - b_0)$, first version).

$$U_{00} = X_{10}\delta L + X_{11}\delta L \delta B \, ,$$
$$U_{10} = X_{10} + X_{11}\delta B \, , \qquad U_{01} = X_{11}\delta L \, ,$$
$$U_{20} = 3X_{30}\delta L \, , \qquad\qquad U_{11} = X_{11} + 2X_{12}\delta B \, , \quad U_{02} = X_{12}\delta L \, ,$$
$$U_{30} = X_{30} + X_{31}\delta B \, , \qquad U_{21} = 3X_{31}\delta L \, , \qquad\qquad U_{12} = X_{12} + 3X_{13}\delta B \, , \quad U_{03} = X_{13}\delta L \, .$$

$$(21.71)$$

Box 21.16 (Polynomial coefficients of a conformal series of type Gauss–Krueger or UTM after a curvilinear datum transformation, Northing $Y(l - l_0, b - b_0)$, first version).

$$V_{00} = Y_{00} + Y_{01}\delta L + Y_{20}\delta L^2 + Y_{02}\delta B^2 \, ,$$
$$V_{10} = 2Y_{20}\delta L \, , \qquad\qquad V_{01} = Y_{01} + 2Y_{02}\delta B \, ,$$
$$V_{20} = Y_{20} + Y_{21}\delta B \, , \qquad V_{11} = 2Y_{21}\delta L \, , \qquad\qquad V_{02} = Y_{02} + 3Y_{03}\delta B \, ,$$
$$V_{30} = 4Y_{40}\delta L \, , \qquad\qquad V_{21} = Y_{21} + 2Y_{22}\delta B \, , \quad V_{12} = 2Y_{22}\delta L \, , \qquad\qquad V_{03} = Y_{03} + 4Y_{04}\delta B \, .$$

$$(21.72)$$

Box 21.17 (Transformation of the coefficients of a conformal series under a curvilinear datum transformation, including the change of form parameter by δE^2, transformation close to the identity).

$$X_{10} = \frac{A_1}{a_1} \left(x_{10} + x_{10,e^2}\delta E^2 \right) \, , \quad X_{12} = \frac{A_1}{a_1} \left(x_{12} + x_{12,e^2}\delta E^2 \right) \, ,$$
$$X_{11} = \frac{A_1}{a_1} \left(x_{11} + x_{11,e^2}\delta E^2 \right) \, , \quad X_{30} = \frac{A_1}{a_1} \left(x_{30} + x_{30,e^2}\delta E^2 \right) \, ,$$

$$(21.73)$$

$$Y_{00} = \frac{A_1}{a_1} \left(y_{00} + y_{00,e^2}\delta E^2 + y_{00,e^2e^2}\frac{(\delta E^2)^2}{2} \right) \, ,$$
$$Y_{01} = \frac{A_1}{a_1} \left(y_{01} + y_{01,e^2}\delta E^2 \right) \, , \quad Y_{02} = \frac{A_1}{a_1} \left(y_{02} + y_{02,e^2}\delta E^2 \right) \, ,$$
$$Y_{20} = \frac{A_1}{a_1} \left(y_{20} + y_{20,e^2}\delta E^2 \right) \, , \quad Y_{21} = \frac{A_1}{a_1} \left(y_{21} + y_{21,e^2}\delta E^2 \right) \, ,$$
$$Y_{03} = \frac{A_1}{a_1} \left(y_{03} + y_{03,e^2}\delta E^2 \right) \, .$$

$$(21.74)$$

Box 21.18 (Transformation of the coefficients of a conformal series under a curvilinear datum transformation, including the change of form parameter by δE^2, East components X_{MN}).

$$x_{10,e^2} =$$

$$= \frac{a_1 \cos b_0 \sin^2 b_0}{2(1 - e^2 \sin^2 b_0)^{3/2}} \, ,$$

$$x_{11,e^2} =$$

$$= \frac{a_1 \sin b_0 (2 - 3 \sin^2 b_0 + e^2 \sin^2 b_0)}{2(1 - e^2 \sin^2 b_0)^{5/2}} \, ,$$

$$x_{12,e^2} =$$

$$= \frac{a_1 \cos b_0 (2 - 9 \sin^2 b_0 + 11 e^2 \sin^2 b_0 - 6 e^2 \sin^4 b_0 + 2 e^4 \sin^4 b_0)}{4(1 - e^2 \sin^2 b_0)^{7/2}} \, ,$$

$$x_{30,e^2} =$$

$$= \frac{a_1 \cos b_0 (2 - 3 \sin^2 b_0 - 3 e^2 \sin^2 b_0 + 6 e^2 \sin^4 b_0 - e^2 \sin^6 b_0 - e^4 \sin^6 b_0)}{12(1 - e^2)^2 (1 - e^2 \sin^2 b_0)^{3/2}} \, .$$

(21.75)

Box 21.19 (Transformation of the coefficients of a conformal series under a curvilinear datum transformation, including the change of form parameter by δE^2, North components Y_{MN}).

$$y_{00,e^2} =$$

$$= -\frac{a_1}{4} \left(b_0 + \frac{3}{2} \sin 2b_0 + \frac{3}{8} e^2 \left[b_0 + 2 \sin 2b_0 - \frac{5}{4} \sin 4b_0 \right] + \right.$$

$$\left. + \frac{15}{64} e^4 \left[b_0 + \frac{9}{4} \sin 2b_0 - \frac{9}{4} \sin 4b_0 + \frac{7}{12} \sin 6b_0 \right] \right) + O(e^6) \, ,$$

$$y_{00,e^2 e^2} =$$

$$= -\frac{3a_1}{32} \left(b_0 + 2 \sin 2b_0 - \frac{5}{4} \sin 4b_0 \right) + O(e^2) \, ,$$

$$y_{01,e^2} =$$

$$= -\frac{a_1 (2 - 3 \sin^2 b_0 + e^2 \sin^2 b_0)}{2(1 - e^2 \sin^2 b_0)^{5/2}} \, ,$$

$$y_{20,e^2} =$$

$$= \frac{a_1 \cos b_0 \sin^3 b_0}{4(1 - e^2 \sin^2 b_0)^{3/2}} \, ,$$

(21.76)

$$y_{02,e^2} =$$

$$= \frac{3a_1 \cos b_0 \sin b_0 (2 - 4 e^2 + 3 e^2 \sin^2 b_0 - e^4 \sin^2 b_0)}{4(1 - e^2 \sin^2 b_0)^{7/2}} \, ,$$

$$y_{21,e^2} =$$

$$= \frac{a_1 \sin^2 b_0 (3 - 4 \sin^2 b_0 + e^2 \sin^4 b_0)}{4(1 - e^2 \sin^2 b_0)^{5/2}} \, ,$$

$$y_{03,e^2} =$$

$$= \frac{a_1 [2 - 4 e^2 - \sin^2 b_0 (4 - 29 e^2 + 27 e^4) - 2 e^2 \sin^4 b_0 (11 - 18 e^2 + 2 e^4) - 3 e^4 \sin^6 b_0 (3 - e^2)]}{4(1 - e^2 \sin^2 b_0)^{9/2}} \, .$$

Box 21.20 (Polynomial coefficients of a conformal series of Box 21.11 of type Gauss–Krueger or UTM after a curvilinear transformation, Easting $X(l - l_0, b - b_0, \rho, t_x, t_y, t_z, \alpha, \beta, \gamma, s, \delta E^2)$, second version).

$$U_{00} = \frac{A_1}{a_1} \left[\left(x_{10} + x_{10,e^2} \delta E^2 \right) \delta L + x_{11} \delta L \delta B \right] ,$$

$$U_{10} = \frac{A_1}{a_1} \left(x_{10} + x_{10,e^2} \delta E^2 + x_{11} \delta B \right) ,$$

$$U_{01} = \frac{A_1}{a_1} x_{11} \delta L ,$$

$$U_{20} = 3 \frac{A_1}{a_1} x_{30} \delta L ,$$

$$U_{11} = \frac{A_1}{a_1} \left(x_{11} + x_{11,e^2} \delta E^2 + 2 x_{12} \delta B \right) ,$$

$$U_{02} = \frac{A_1}{a_1} x_{12} \delta L ,$$

$$U_{30} = \frac{A_1}{a_1} \left(x_{30} + x_{30,e^2} \delta E^2 + x_{31} \delta B \right) ,$$

$$U_{21} = 3 \frac{A_1}{a_1} x_{31} \delta L ,$$

$$U_{12} = \frac{A_1}{a_1} \left(x_{12} + x_{12,e^2} \delta E^2 + 3 x_{13} \delta B \right) ,$$

$$U_{03} = \frac{A_1}{a_1} x_{13} \delta L .$$

(21.77)

Box 21.21 (Polynomial coefficients of a conformal series of Box 21.11 of type Gauss–Krueger or UTM after a curvilinear transformation, Northing $Y(l - l_0, b - b_0, \rho, t_x, t_y, t_z, \alpha, \beta, \gamma, s, \delta E^2)$, second version).

$$V_{00} = \frac{A_1}{a_1} \left[y_{00} + y_{00,e^2} \delta E^2 + y_{00,e^2 e^2} \frac{(\delta E^2)^2}{2} + (y_{01} + y_{01,e^2} \delta E^2) \delta B + y_{20} \delta L^2 + y_{02} \delta B^2 \right] ,$$

$$V_{10} = 2 \frac{A_1}{a_1} y_{20} \delta L ,$$

$$V_{01} = \frac{A_1}{a_1} \left(y_{01} + y_{10,e^2} \delta E^2 + 2 y_{02} \delta B \right) ,$$

$$V_{20} = \frac{A_1}{a_1} \left(y_{20} + y_{20,e^2} \delta E^2 + y_{21} \delta B \right) ,$$

$$V_{11} = 2 \frac{A_1}{a_1} y_{21} \delta L ,$$

$$V_{02} = \frac{A_1}{a_1} \left(y_{02} + y_{02,e^2} \delta E^2 + 3 y_{03} \delta B \right) ,$$

(21.78)

$$V_{30} = 4 \frac{A_1}{a_1} y_{40} \delta L ,$$

$$V_{21} = \frac{A_1}{a_1} \left(y_{21} + y_{21,e^2} \delta E^2 + 2 y_{22} \delta B \right) ,$$

$$V_{12} = 2 \frac{A_1}{a_1} y_{22} \delta L ,$$

$$V_{03} = \frac{A_1}{a_1} \left(y_{03} + y_{30,e^2} \delta E^2 + 4 y_{04} \delta B \right) .$$

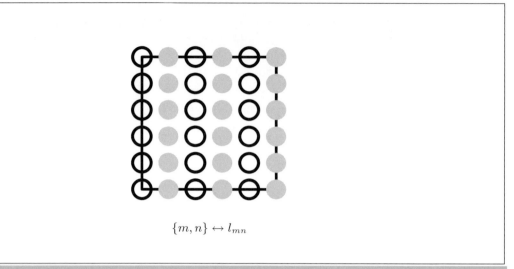

$$\{m, n\} \leftrightarrow l_{mn}$$

Fig. 21.8. Polynomial diagram, the polynomial representation of longitude increments $l - l_0$, conformal coordinates $\{x, y\}$ of type Gauss–Krueger or UTM, the solid dots illustrate non-zero monomials, the open circles zero monomials, according to D. Cox, J. Little and D. O'Shea (1996, pp. 433–443).

21-413 The third step: global conformal coordinates in a local datum

By means of the bivariate series (21.70) of Box 21.14 subject to the coefficients of Boxes 21.20 and 21.21, we have succeeded to express global conformal coordinates of type Gauss–Krueger or UTM in terms of local ellipsoidal coordinates, namely the eastern coordinate $X(l - l_0, b - b_0, \rho, t_x, t_y, t_z, \alpha, \beta, \gamma, s, \delta E^2)$ and the northern coordinate $Y(l - l_0, b - b_0, \rho, t_x, t_y, t_z, \alpha, \beta, \gamma, s, \delta E^2)$. Only the transformation (the third step) from local ellipsoidal coordinates $\{l - l_0, b - b_0\}$ is left. Such a transformation is achieved by a bivariate series inversion outlined in E. Grafarend, T. Krarup and R. Syffus (1996), namely of type (21.66). In Box 21.14, we have reviewed the result of a bivariate series inversion of conformal type subject to the coefficients given by Boxes 21.23 and 21.24 as well as illustrated by Figs. 21.8 and 21.9.

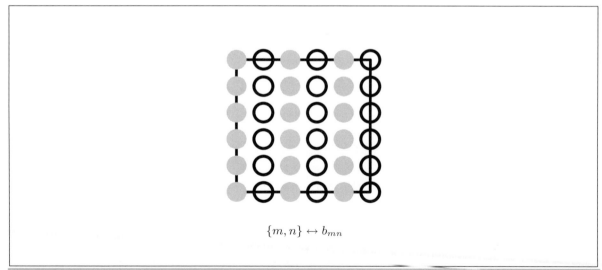

$$\{m, n\} \leftrightarrow b_{mn}$$

Fig. 21.9. Polynomial diagram, the polynomial representation of latitude increments $b - b_0$, conformal coordinates $\{x, y\}$ of type Gauss–Krueger or UTM, the solid dots illustrate non-zero monomials, the open circles zero monomials, according to D. Cox, J. Little and D. O'Shea (1996, pp. 433–443).

Box 21.22 (Series inversion of a local system of conformal coordinates of type Gauss–Krueger or UTM into ellipsoidal coordinates).

$$l - l_0 =$$

$$= l_{10}\frac{x}{\rho} + l_{11}\frac{x}{\rho}\left(\frac{y}{\rho} - y_{00}\right) + l_{30}\left(\frac{x}{\rho}\right)^3 + l_{12}\frac{x}{\rho}\left(\frac{y}{\rho} - y_{00}\right)^2 + \mathrm{O}(4) ,$$

$$b - b_0 =$$

$$= b_{01}\left(\frac{y}{\rho} - y_{00}\right) + b_{20}\left(\frac{x}{\rho}\right)^2 + b_{02}\left(\frac{y}{\rho} - y_{00}\right)^2 + b_{21}\left(\frac{x}{\rho}\right)^2\left(\frac{y}{\rho} - y_{00}\right) + b_{03}\left(\frac{y}{\rho} - y_{00}\right)^3 + \mathrm{O}(4) .$$

(21.79)

Box 21.23 (Coefficients of a bivariate series inversion of conformal type, the ellipsoidal longitude increments $l - l_0$).

$$l_{10} = \frac{(1 - e^2 \sin^2 b_0)^{1/2}}{a_1 \cos b_0} ,$$

$$l_{11} = \frac{\tan b_0 (1 - e^2 \sin^2 b_0)}{a_1^2 \cos b_0} ,$$

$$l_{30} = \frac{-(1 - e^2 \sin^2 b_0)^{3/2}(1 + \sin^2 b_0 - 3e^2 \sin^2 b_0 + e^2 \sin^4 b_0)}{6a_1^3(1 - e^2)\cos^3 b_0} ,$$

$$l_{12} = \frac{+(1 - e^2 \sin^2 b_0)^{3/2}(1 + \sin^2 b_0 - 3e^2 \sin^2 b_0 + e^2 \sin^4 b_0)}{2a_1^3(1 - e^2)\cos^3 b_0} .$$

(21.80)

Box 21.24 (Coefficients of a bivariate series inversion of conformal type, the ellipsoidal latitude increments $b - b_0$).

$$b_{01} = \frac{(1 - e^2 \sin^2 b_0)^{3/2}}{a_1(1 - e^2)} ,$$

$$b_{20} = \frac{-\tan b_0 (1 - e^2 \sin^2 b_0)^2}{2a_1^2(1 - e^2)} ,$$

$$b_{02} = \frac{-3e^2 \cos b_0 \sin b_0 (1 - e^2 \sin^2 b_0)^2}{2a_1^2(1 - e^2)^2} ,$$

$$b_{21} = \frac{-(1 - e^2 \sin^2 b_0)^{5/2}(1 - 5e^2 \sin^2 b_0 + 4e^2 \sin^4 b_0)}{2a_1^3(1 - e^2)^2 \cos^2 b_0} ,$$

$$b_{03} = \frac{-e^2(1 - e^2 \sin^2 b_0)^{5/2}(1 - 2\sin^2 b_0 - 5e^2 \sin^2 b_0 + 6e^2 \sin^4 b_0)}{2a_1^3(1 - e^2)^3} .$$

(21.81)

As soon as we implement (21.79) into (21.70), we gain the final transformation of Cartesian conformal coordinates $\{x, y\}$ in a local frame of reference into Cartesian conformal coordinates $\{X, Y\}$ in a global frame of reference. Indeed, via the coefficients of the bivariate polynomials (21.82) of Box 21.25 listed in Boxes 21.26 and 21.27, the final transformation depends on the parameters of a curvilinear datum transformation, namely three parameters $\{t_x, t_y, t_z\}$ of translation, three parameters $\{\alpha, \beta, \gamma\}$ of rotation, one scale parameter s and the change of the form parameter $\delta E^2 = E^2 - e^2$. The bivariate polynomial representation (21.82) is given up to order three due to the limited space in printing the lengthy coefficients in Boxes 21.26 and 21.27. Indeed, the final transformation $X(x, y, \rho, t_x, t_y, t_z, \alpha, \beta, \gamma, s, \delta E^2)$, $Y(x, y, \rho, t_x, t_y, t_z, \alpha, \beta, \gamma, s, \delta E^2)$ highlights the key result of a datum transformation from local conformal coordinates of type Gauss–Krueger or UTM to global conformal coordinates of type Gauss–Krueger or UTM. We therefore summarize the results as follows.

Box 21.25 (Polynomial representation of global conformal coordinates X, Y in terms of local conformal coordinates x, y due to a curvilinear datum transformation, Gauss–Krueger conformal mapping or UTM, polynomial degree three, Easting X, x, Northing Y, y).

$$X(x, y, \rho, t_x, t_y, t_z, \alpha, \beta, \gamma, s, \delta E^2) =$$

$$= \rho \left[\bar{x}_{00} + \bar{x}_{10} \frac{x}{\rho} + \bar{x}_{01} \left(\frac{y}{\rho} - y_{00} \right) + \bar{x}_{20} \left(\frac{x}{\rho} \right)^2 + \bar{x}_{11} \frac{x}{\rho} \left(\frac{y}{\rho} - y_{00} \right) + \bar{x}_{02} \left(\frac{y}{\rho} - y_{00} \right)^2 + \right.$$

$$\left. + \bar{x}_{30} \left(\frac{x}{\rho} \right)^3 + \bar{x}_{21} \left(\frac{x}{\rho} \right)^2 \left(\frac{y}{\rho} - y_{00} \right) + \bar{x}_{12} \frac{x}{\rho} \left(\frac{y}{\rho} - y_{00} \right)^2 + \bar{x}_{03} \left(\frac{y}{\rho} - y_{00} \right)^3 \right] + \mathrm{O}(4) ,$$

$$(21.82)$$

$$Y(x, y, \rho, t_x, t_y, t_z, \alpha, \beta, \gamma, s, \delta E^2) =$$

$$= \rho \left[\bar{y}_{00} + \bar{y}_{10} \frac{x}{\rho} + \bar{y}_{01} \left(\frac{y}{\rho} - y_{00} \right) + \bar{y}_{20} \left(\frac{x}{\rho} \right)^2 + \bar{y}_{11} \frac{x}{\rho} \left(\frac{y}{\rho} - y_{00} \right) + \bar{y}_{02} \left(\frac{y}{\rho} - y_{00} \right)^2 + \right.$$

$$\left. + \bar{y}_{30} \left(\frac{x}{\rho} \right)^3 + \bar{y}_{21} \left(\frac{x}{\rho} \right)^2 \left(\frac{y}{\rho} - y_{00} \right) + \bar{y}_{12} \frac{x}{\rho} \left(\frac{y}{\rho} - y_{00} \right)^2 + \bar{y}_{03} \left(\frac{y}{\rho} - y_{00} \right)^3 \right] + \mathrm{O}(4) .$$

Box 21.26 (The polynomial coefficients, the East components, namely $\bar{x}_{mn}$, the datum transformation of conformal coordinates).

$$\bar{x}_{00} = U_{00} ,$$

$$\bar{x}_{10} = U_{10} l_{10} ,$$

$$\bar{x}_{01} = U_{01} b_{01} ,$$

$$\bar{x}_{20} = U_{01} b_{20} + U_{20} l_{10}^2 ,$$

$$\bar{x}_{11} = U_{10} l_{11} + U_{11} l_{10} b_{01} , \bar{x}_{02} = U_{01} b_{02} + U_{02} b_{01}^2 ,$$

$$\bar{x}_{30} = U_{10} l_{30} + U_{11} l_{10} b_{20} + U_{30} l_{10}^3 ,$$

$$\bar{x}_{21} = U_{01} b_{21} + 2 U_{20} l_{10} l_{11} + 2 U_{02} b_{01} b_{20} + U_{21} b_{01} l_{10}^2 ,$$

$$\bar{x}_{12} = U_{10} l_{12} + U_{11} (l_{10} b_{02} + l_{11} b_{01}) + U_{12} l_{10} b_{01}^2 ,$$

$$\bar{x}_{03} = U_{01} b_{03} + 2 U_{02} b_{01} b_{02} + U_{03} b_{01}^3 .$$

$$(21.83)$$

Box 21.27 (The polynomial coefficients, the North components, namely $\bar{y}_{mn}$, the datum transformation of conformal coordinates).

$$\bar{y}_{00} = V_{00} ,$$

$$\bar{y}_{10} = V_{10} l_{10} ,$$

$$\bar{y}_{01} = V_{01} b_{01} ,$$

$$\bar{y}_{20} = V_{01} b_{20} + V_{20} l_{10}^2 ,$$

$$\bar{y}_{11} = V_{10} l_{11} + V_{11} l_{10} b_{01} , \bar{y}_{02} = V_{01} b_{02} + V_{02} b_{01}^2 ,$$

$$\bar{y}_{30} = V_{10} l_{30} + V_{11} l_{10} b_{20} + V_{30} l_{10}^3 ,$$

$$\bar{y}_{21} = V_{01} b_{21} + 2 V_{20} l_{10} l_{11} + 2 V_{02} b_{01} b_{20} + V_{21} b_{01} l_{10}^2 ,$$

$$\bar{y}_{12} = V_{10} l_{12} + V_{11} (l_{10} b_{02} + l_{11} b_{01}) + V_{12} l_{10} b_{01}^2 ,$$

$$\bar{y}_{03} = V_{01} b_{03} + 2 V_{02} b_{01} b_{02} + V_{03} b_{01}^3 .$$

$$(21.84)$$

Lemma 21.1 (Local conformal coordinates are transformed into global conformal coordinates of type Gauss–Krueger or UTM).

Let there be given conformal coordinates $\{x, y\}$ of type Gauss–Krueger or UTM of a local reference ellipsoid-of-revolution $\mathbb{E}^2_{a_1, a_2}$. Then, under a curvilinear datum transformation (21.67) and (21.66) represented by three parameters $\{t_x, t_y, t_z\}$ of translation, three parameters $\{\alpha, \beta, \gamma\}$ of rotation, and one scale parameter s, the conformal coordinates $\{X, Y\}$ of type Gauss–Krueger or UTM of a global reference ellipsoid-of-revolution $\mathbb{E}^2_{A_1, A_2}$ are represented by the bivariate polynomial

$$
X = X(x, y, \rho, t_x, t_y, t_z, \alpha, \beta, \gamma, s, \delta E^2) = \rho \sum_{m=0, n=0, m+n=N}^{\infty} \bar{x}_{mn} \left(\frac{x}{\rho}\right)^m \left(\frac{y}{\rho} - y_{00}\right)^n,
$$

$$
(21.85)
$$

$$
Y = Y(x, y, \rho, t_x, t_y, t_z, \alpha, \beta, \gamma, s, \delta E^2) = \rho \sum_{m=0, n=0, m+n=N}^{\infty} \bar{y}_{mn} \left(\frac{x}{\rho}\right)^m \left(\frac{y}{\rho} - y_{00}\right)^n.
$$

X and Y are given by (21.82) in Box 21.25 up to order three. The coefficients $\bar{x}_{mn}$ and $\bar{y}_{mn}$, which are given in Box 21.26 and Box 21.27, are product sums of the coefficients U_{MN} and V_{MN} of Box 21.20 and Box 21.21 and of the coefficients l_{mn} and b_{mn} of Box 21.23 and Box 21.24. y_{00} indicates the arc length of the meridian-of-reference l_0 in the interval $[0, b_0]$.

End of Lemma.

21-42 Inverse transformation of global conformal into local conformal coordinates

The software attached to a satellite Global Positioning System (GPS) allows the direct conversion of global ellipsoidal coordinates $\{L - L_0, B - B_0\}$ into global conformal coordinates $\{X, Y\}$ of type Gauss–Krueger or UTM, namely with reference to a global reference ellipsoid-of-revolution $\mathbb{E}^2_{A_1, A_2}$, i.e. WGS 84. In order to locate an observer with first hand information of global conformal coordinates $\{X, Y\}$ of type Gauss–Krueger or UTM in a Gauss–Krueger or UTM chart given in a local reference system (regional, National, European), we are left with the problem of transforming global conformal coordinates $\{X, Y\}$ into local conformal coordinates of type Gauss–Krueger or UTM, the chart coordinates. The problem is solved by the inverse representation of the bivariate polynomials $\{X(x, y), Y(x, y)\}$: such bivariate polynomials are inverted by means of the GKS algorithm presented by E. Grafarend, T. Krarup and R. Syffus (1996). Box 21.28 contains the inverse bivariate polynomials $\{x(X, Y), y(X, Y)\}$ with respect to the coefficients of Box 21.29 and Box 21.30, where the datum parameters are included in the coefficients $\{x^{MN}, y^{MN}\}$.

Important!

The direct and inverse equations for a datum transformation of conformal coordinates of type Gauss–Krueger or UTM from a local datum (regional, National, European) to a global datum (i.e. WGS 84) are given in terms of a bivariate polynomial representation. The polynomial coefficients depend on the datum transformation parameters, namely three parameters of translation, three parameters of rotation, one scale parameter, and one form parameter change, in total eight parameters. The form parameter change accounts for the variation of the eccentricity of the reference ellipsoid-of-revolution under the change from one geodetic datum to another one, namely from local to global or *vice versa*. The equations generating the transformation of local conformal coordinates of type Gauss–Krueger or UTM to global conformal coordinates of the same type enable us to transform mega data sets stored in data bases or in charts from the local datum (the datum of the data base, the datum of the chart) to the global datum (the datum of satellite derived coordinates by means of the Global Positioning System, i.e. WGS 84) and *vice versa*.

Box 21.28 (Polynomial representation of local conformal coordinates x, y in terms of global conformal coordinates X, Y due to a curvilinear datum transformation, Gauss–Krueger conformal mapping or UTM, polynomial degree three, Easting X, x, Northing Y, y).

$$x(X, Y, \rho, t_x, t_y, t_z, \alpha, \beta, \gamma, s, \delta E^2) =$$

$$= \rho \left[x^{10} \left(\frac{X}{\rho} - \bar{x}_{00} \right) + x^{01} \left(\frac{Y}{\rho} - \bar{y}_{00} \right) + x^{20} \left(\frac{X}{\rho} - \bar{x}_{00} \right)^2 + x^{11} \left(\frac{X}{\rho} - \bar{x}_{00} \right) \left(\frac{Y}{\rho} - \bar{y}_{00} \right) + \right.$$

$$+ x^{02} \left(\frac{Y}{\rho} - \bar{y}_{00} \right)^2 + x^{30} \left(\frac{X}{\rho} - \bar{x}_{00} \right)^3 + x^{21} \left(\frac{X}{\rho} - \bar{x}_{00} \right)^2 \left(\frac{Y}{\rho} - \bar{y}_{00} \right) +$$

$$\left. + x^{12} \left(\frac{X}{\rho} - \bar{x}_{00} \right) \left(\frac{Y}{\rho} - \bar{y}_{00} \right)^2 + x^{03} \left(\frac{Y}{\rho} - \bar{y}_{00} \right)^3 \right] +$$

$$+ O(4) ,$$

$$y(X, Y, \rho, t_x, t_y, t_z, \alpha, \beta, \gamma, s, \delta E^2) = \tag{21.86}$$

$$= \rho \left[y_{00} + y^{10} \left(\frac{X}{\rho} - \bar{x}_{00} \right) + y^{01} \left(\frac{Y}{\rho} - \bar{y}_{00} \right) + y^{20} \left(\frac{X}{\rho} - \bar{x}_{00} \right)^2 + y^{11} \left(\frac{X}{\rho} - \bar{x}_{00} \right) \left(\frac{Y}{\rho} - \bar{y}_{00} \right) + \right.$$

$$+ y^{02} \left(\frac{Y}{\rho} - \bar{y}_{00} \right)^2 + y^{30} \left(\frac{X}{\rho} - \bar{x}_{00} \right)^3 + y^{21} \left(\frac{X}{\rho} - \bar{x}_{00} \right)^2 \left(\frac{Y}{\rho} - \bar{y}_{00} \right) +$$

$$\left. + y^{12} \left(\frac{X}{\rho} - \bar{x}_{00} \right) \left(\frac{Y}{\rho} - \bar{y}_{00} \right)^2 + y^{03} \left(\frac{Y}{\rho} - \bar{y}_{00} \right)^3 \right] +$$

$$+ O(4) .$$

Box 21.29 (The polynomial coefficients, the East components, namely x^{MN}, the datum transformation of conformal coordinates).

$$\bar{x}_{00} = \left[\delta L \frac{a_1 \cos b_0}{(1 - e^2 \sin^2 b_0)^{1/2}} + \delta E^2 \delta L \frac{a_1 \cos b_0 \sin^2 b_0}{2(1 - e^2 \sin^2 b_0)^{3/2}} - \delta L \delta B \frac{a_1 (1 - e^2) \sin b_0}{(1 - e^2 \sin^2 b_0)^{3/2}} \right] \frac{A_1}{a_1} ,$$

$$x^{10} = \left[1 - \delta E^2 \frac{\sin^2 b_0}{2(1 - e^2 \sin^2 b_0)} + \delta B \frac{(1 - e^2) \tan b_0}{(1 - e^2 \sin^2 b_0)} \right] \frac{a_1}{A_1} ,$$

$$x^{01} = \left[\delta L \sin b_0 \right] \frac{a_1}{A_1} ,$$

$$x^{20} = \left[-\delta L \frac{\cos b_0 (1 - e^2 \sin^2 b_0)^{3/2}}{2a_1 (1 - e^2)} \right] \left(\frac{a_1}{A_1} \right)^2 , \tag{21.87}$$

$$x^{02} = \left[+\delta L \frac{\cos b_0 (1 - e^2 \sin^2 b_0)^{3/2}}{2a_1 (1 - e^2)} \right] \left(\frac{a_1}{A_1} \right)^2 ,$$

$$x^{11} = \left[\delta B \frac{1 + e^2 \sin^2 b_0 - 2e^2 \sin^4 b_0}{a_1 \cos^2 b_0 (1 - e^2 \sin^2 b_0)^{1/2}} - \delta E^2 \frac{\cos b_0 \sin b_0}{a_1 (1 - e^2)(1 - e^2 \sin^2 b_0)^{1/2}} \right] \left(\frac{a_1}{A_1} \right)^2 .$$

Continuation of Box.

$$x^{30} = \left[\delta B \frac{\tan b_0 [1 + 3e^2 - \sin^2 b_0(4 + 11e^2 - 3e^4) + 2e^2 \sin^4 b_0(7 - e^2) - 4e^4 \sin^6 b_0]}{6a_1^2(1 - e^2)\cos^2 b_0} - \right.$$

$$\left. - \delta E^2 \frac{1 - 2\sin^2 b_0(2 - e^2) + e^2 \sin^4 b_0}{6a_1^2(1 - e^2)^2} \right] \left(\frac{a_1}{A_1} \right)^3 ,$$

$$x^{21} = \left[+ \delta L \frac{\sin b_0(1 - e^2 \sin^2 b_0)^2(1 + 3e^2 - 4e^2 \sin^2 b_0)}{2a_1^2(1 - e^2)^2} \right] \left(\frac{a_1}{A_1} \right)^3 ,$$

$$x^{12} = \left[\delta B \frac{\tan b_0 [2 + 3e^2 - e^2 \sin^2 b_0(11 + e^2) + e^2 \sin^4 b_0(4 + 5e^2) - 2e^4 \sin^6 b_0]}{2a_1^2(1 - e^2)\cos^2 b_0} - \right.$$

$$\left. - \delta E^2 \frac{1 - 2\sin^2 b_0 + e^2 \sin^4 b_0}{2a_1^2(1 - e^2)^2} \right] \left(\frac{a_1}{A_1} \right)^3 ,$$

$$x^{03} = \left[- \delta L \frac{\sin b_0(1 - e^2 \sin^2 b_0)^2(1 + 3e^2 - 4e^2 \sin^2 b_0)}{6a_1^2(1 - e^2)^2} \right] \left(\frac{a_1}{A_1} \right)^3 .$$

(21.88)

Box 21.30 (The polynomial coefficients, the North components, namely y^{MN}, the datum transformation of conformal coordinates).

$$\bar{y}_{00} = \left[y_{00} + y_{00,e^2} \delta E^2 + y_{00,e^2 e^2} \frac{(\delta E^2)^2}{2} + \delta B \frac{a_1(1 - e^2)}{(1 - e^2 \sin^2 b_0)^{3/2}} - \right.$$

$$\left. - \delta B \delta E^2 \frac{a_1(2 - 3\sin^2 b_0 + e^2 \sin^2 b_0)}{2(1 - e^2 \sin^2 b_0)^{5/2}} + \delta L^2 \frac{a_1 \cos b_0 \sin b_0}{2(1 - e^2 \sin^2 b_0)^{1/2}} + \delta B^2 \frac{3a_1 e^2(1 - e^2)\cos b_0 \sin b_0}{2(1 - e^2 \sin^2 b_0)^{5/2}} \right] \frac{A_1}{a_1} ,$$

$$y^{10} = \left[- \delta L \sin b_0 \right] \frac{a_1}{A_1} , \quad y^{01} = \left[1 + \delta E^2 \frac{2 - 3\sin^2 b_0 + e^2 \sin^2 b_0}{2(1 - e^2)(1 - e^2 \sin^2 b_0)} - \delta B \frac{3e^2 \cos b_0 \sin b_0}{(1 - e^2 \sin^2 b_0)} \right] \frac{a_1}{A_1} ,$$

$$y^{20} = - \left[\delta B \frac{1 - 2\sin^2 b_0 - 3e^2 \sin^2 b_0 + 4e^2 \sin^4 b_0}{2a_1 \cos^2 b_0(1 - e^2 \sin^2 b_0)^{1/2}} + \delta E^2 \frac{\cos b_0 \sin b_0}{2a_1(1 - e^2)(1 - e^2 \sin^2 b_0)^{1/2}} \right] \left(\frac{a_1}{A_1} \right)^2 ,$$

$$y^{11} = - \left[\delta L \frac{\cos b_0(1 - e^2 \sin^2 b_0)^{3/2}}{a_1(1 - e^2)} \right] \left(\frac{a_1}{A_1} \right)^2 ,$$

$$y^{02} = - \left[\delta B \frac{3e^2(1 - 2\sin^2 b_0 + e^2 \sin^2 b_0)}{2a_1(1 - e^2)(1 - e^2 \sin^2 b_0)^{1/2}} + \delta E^2 \frac{3\cos b_0 \sin b_0}{2a_1(1 - e^2)(1 - e^2 \sin^2 b_0)^{1/2}} \right] \left(\frac{a_1}{A_1} \right)^2 ,$$

(21.89)

$$y^{30} = - \left[\delta L \frac{\sin b_0(1 - e^2 \sin^2 b_0)^2(1 + 3e^2 - 4e^2 \sin^2 b_0)}{6a_1^2(1 - e^2)^1} \right] \left(\frac{a_1}{A_1} \right)^3 ,$$

$$y^{21} = \left[\delta B \frac{\tan b_0 [2 + 5e^2 - 3e^2 \sin^2 b_0(5 + e^2) + 3e^2 \sin^4 b_0(2 + 3e^2) - 4e^4 \sin^6 b_0]}{2a_1^2(1 - e^2)\cos^2 b_0} - \right.$$

$$\left. - \delta E^2 \frac{1 - 2\sin^2 b_0 + e^2 \sin^4 b_0}{2a_1^2(1 - e^2)^2} \right] \left(\frac{a_1}{A_1} \right)^3 ,$$

$$y^{12} = + \left[\delta L \frac{\sin b_0(1 - e^2 \sin^2 b_0)^2(1 + 3e^2 - 4e^2 \sin^2 b_0)}{2a_1^2(1 - e^2)^1} \right] \left(\frac{a_1}{A_1} \right)^3 ,$$

$$y^{03} = \left[\delta B \frac{e^2 \cos b_0 \sin b_0[4 - 3e^2 - 2e^2 \sin^2 b_0 + e^4 \sin^2 b_0]}{2a_1^2(1 - c^2)^2} - \delta E^2 \frac{1 - 2\sin^2 b_0 + e^2 \sin^4 b_0}{2a_1^2(1 - e^2)^2} \right] \left(\frac{a_1}{A_1} \right)^3 .$$

21-43 Numerical results

Here, we depart from the polynomial representation of the global conformal coordinates $\{X, Y\}$ in terms of local conformal coordinates $\{x, y\}$ due to a curvilinear datum transformation and its inverse by Box 21.31 and Box 21.32. In our case studies, we concentrate on the State of Baden–Württemberg. The transformation of 50 BWREF points from a global to a local datum and *vice versa* has been computed. Table 21.2 summarizes those datum transformation parameters that are available to us.

Box 21.31 (Polynomial representation of the global conformal coordinates X, Y in terms of local conformal coordinates x, y due to a curvilinear datum transformation, Gauss–Krueger conformal mapping or UTM, polynomial degree three, Easting X, x, Northing Y, y).

$$X = X(x, y, \rho, t_x, t_y, t_z, \alpha, \beta, \gamma, s, A_1, E^2, a_1, e^2) =$$

$$= \rho \left[\bar{x}_{00} + \bar{x}_{10} \frac{x}{\rho} + \bar{x}_{01} \left(\frac{y}{\rho} - y_{00} \right) + \bar{x}_{20} \left(\frac{x}{\rho} \right)^2 + \bar{x}_{11} \frac{x}{\rho} \left(\frac{y}{\rho} - y_{00} \right) + \bar{x}_{02} \left(\frac{y}{\rho} - y_{00} \right)^2 + \right.$$

$$\left. + \bar{x}_{30} \left(\frac{x}{\rho} \right)^3 + \bar{x}_{21} \left(\frac{x}{\rho} \right)^2 \left(\frac{y}{\rho} - y_{00} \right) + \bar{x}_{12} \frac{x}{\rho} \left(\frac{y}{\rho} - y_{00} \right)^2 + \bar{x}_{03} \left(\frac{y}{\rho} - y_{00} \right)^3 \right] + O(4) ,$$

$$(21.90)$$

$$Y = Y(x, y, \rho, t_x, t_y, t_z, \alpha, \beta, \gamma, s, , A_1, E^2, a_1, e^2) =$$

$$= \rho \left[\bar{y}_{00} + \bar{y}_{10} \frac{x}{\rho} + \bar{y}_{01} \left(\frac{y}{\rho} - y_{00} \right) + \bar{y}_{20} \left(\frac{x}{\rho} \right)^2 + \bar{y}_{11} \frac{x}{\rho} \left(\frac{y}{\rho} - y_{00} \right) + \bar{y}_{02} \left(\frac{y}{\rho} - y_{00} \right)^2 + \right.$$

$$\left. + \bar{y}_{30} \left(\frac{x}{\rho} \right)^3 + \bar{y}_{21} \left(\frac{x}{\rho} \right)^2 \left(\frac{y}{\rho} - y_{00} \right) + \bar{y}_{12} \frac{x}{\rho} \left(\frac{y}{\rho} - y_{00} \right)^2 + \bar{y}_{03} \left(\frac{y}{\rho} - y_{00} \right)^3 \right] + O(4) .$$

Box 21.32 (Polynomial representation of the local conformal coordinates x, y in terms of global conformal coordinates X, Y due to a curvilinear datum transformation, Gauss–Krueger conformal mapping or UTM, polynomial degree three, Easting X, x, Northing Y, y).

$$x = x(X, Y, \rho, t_x, t_y, t_z, \alpha, \beta, \gamma, s, A_1, E^2, a_1, e^2) =$$

$$= \rho \left[x^{10} \left(\frac{X}{\rho} - \bar{x}_{00} \right) + x^{01} \left(\frac{Y}{\rho} - \bar{y}_{00} \right) + x^{20} \left(\frac{X}{\rho} - \bar{x}_{00} \right)^2 + x^{11} \left(\frac{X}{\rho} - \bar{x}_{00} \right) \left(\frac{Y}{\rho} - \bar{y}_{00} \right) + \right.$$

$$+ x^{02} \left(\frac{Y}{\rho} - \bar{y}_{00} \right)^2 + x^{30} \left(\frac{X}{\rho} - \bar{x}_{00} \right)^3 + x^{21} \left(\frac{X}{\rho} - \bar{x}_{00} \right)^2 \left(\frac{Y}{\rho} - \bar{y}_{00} \right) +$$

$$\left. + x^{12} \left(\frac{X}{\rho} - \bar{x}_{00} \right) \left(\frac{Y}{\rho} - \bar{y}_{00} \right)^2 + x^{03} \left(\frac{Y}{\rho} - \bar{y}_{00} \right)^3 \right] + O(4) ,$$

$$(21.91)$$

$$y = y(X, Y, \rho, t_x, t_y, t_z, \alpha, \beta, \gamma, s, A_1, E^2, a_1, e^2) =$$

$$= \rho \left[y_{00} + y^{10} \left(\frac{X}{\rho} - \bar{x}_{00} \right) + y^{01} \left(\frac{Y}{\rho} - \bar{y}_{00} \right) + y^{20} \left(\frac{X}{\rho} - \bar{x}_{00} \right)^2 + y^{11} \left(\frac{X}{\rho} - \bar{x}_{00} \right) \left(\frac{Y}{\rho} - \bar{y}_{00} \right) + \right.$$

$$+ y^{02} \left(\frac{Y}{\rho} - \bar{y}_{00} \right)^2 + y^{30} \left(\frac{X}{\rho} - \bar{x}_{00} \right)^3 + y^{21} \left(\frac{X}{\rho} - \bar{x}_{00} \right)^2 \left(\frac{Y}{\rho} - \bar{y}_{00} \right) +$$

$$\left. + y^{12} \left(\frac{X}{\rho} - \bar{x}_{00} \right) \left(\frac{Y}{\rho} - \bar{y}_{00} \right)^2 + y^{03} \left(\frac{Y}{\rho} - \bar{y}_{00} \right)^3 \right] + O(4) .$$

Table 21.2. Datum transformation. Datum parameters global (WGS 84) to local (BW).

$$t_x = 592.271 \, \text{m} , \quad t_y = 76.286 \, \text{m} ,$$

$$t_z = 407.335 \, \text{m}$$

$$\alpha = -1.092843'' , \quad \beta = -0.097832'' ,$$

$$\gamma = 1.604106''$$

$$s = 8.537829 \, \text{ppm}$$

$$a_1 = 6377397.155 \, \text{m} , \quad A_1 = 6378137 \, \text{m}$$

$$e^2 = 0.006674372231 , \quad E^2 = 0.00669437999$$

For space reasons, we review the results for only ten points, both for the forward and backward transformations. Table 21.3 and Table 21.4 represent the differences between the Gauss–Krueger conformal coordinates $\{X, Y\}$ and those computed ones $\{X(\text{trans}), Y(\text{trans})\}$. Indeed, the differences of the Easting were larger than those of the Northing. We have to mention that all transformation parameters were based on those data of "Deutsches Hauptdreiecksnetz" (DHDN). Accordingly, the accuracy of the transformation cannot be better than a few centimeters. For a more detailed analysis, we have chosen five points (Katzenbuckel, Gerabronn, Karlsruhe, Stuttgart, Oberkochen) whose Gauss–Krueger conformal coordinates as well as ellipsoidal heights are given in Table 21.5. Table 21.6 and Table 21.7 summarize those polynomial coefficients given in Box 21.31 and Box 21.32 representing (21.92) and (21.93), respectively. Note that $\bar{x}_{10}x/\rho$ is denoted as $\bar{a}_{10}$, $\bar{x}_{10}(x/\rho - y_{00})$ is denoted as $\bar{a}_{01}$ etc. From those tables, we conclude that there are only three terms larger than a centimeter. Accordingly, with such results, we can reduce the computational efforts by 30%. Indeed, we need only the coefficients $\{a_{10}, a_{01}, b_{00}, b_{10}, b_{01}\}$ and $\{\bar{a}_{00}, \bar{a}_{10}, \bar{a}_{01}, \bar{b}_{00}, \bar{b}_{10}, \bar{b}_{01}\}$, respectively. For fast less accurate computations, we can disregard the coefficients a_{01} and $\bar{a}_{01}$. The value of such a term is smaller than 10 cm. Obviously, just for mapping purposes this accuracy is sufficient: it is an advantage when you have to compute datum transformations of conformal coordinates for mega data sets.

$$X =$$
$$= X(x, y, \rho, t_x, t_y, t_z, \alpha, \beta, \gamma, s, A_1, E^2, a_1, e^2) ,$$
$$Y =$$
$$= Y(x, y, \rho, t_x, t_y, t_z, \alpha, \beta, \gamma, s, A_1, E^2, a_1, e^2) ,$$
(21.92)

$$x =$$
$$= x(X, Y, \rho, t_x, t_y, t_z, \alpha, \beta, \gamma, s, A_1, E^2, a_1, e^2) ,$$
$$y =$$
$$= y(X, Y, \rho, t_x, t_y, t_z, \alpha, \beta, \gamma, s, A_1, E^2, a_1, e^2) .$$
(21.93)

Finally, we repeat all computations by replacing the "global" reference system of type WGS 84 by the new World Geodetic Datum 2000, E. Grafarend and A. Ardalan (1999). Table 21.8 reviews the best estimates of type semi-major axis A_1, semir-minor axis A_2 and linear eccentricity $\epsilon = \sqrt{A_1^2 - A_2^2}$ both for the tide-free geoid-of-reference and for the zero-frequency tide geoid-of-reference. The related data of transformation of type UTM $\{X_{84}, Y_{84}\}$ versus $\{X_{2000}, Y_{2000}\}$, originating from a reference system of Bessel type, are collected in Table 21.9 and Table 21.10. Indeed, they document variations of the order of a few decimeter!

Table 21.3. Difference between Gauss–Krueger conformal coordinates X and X(trans): Easting.

point	X [m]	X(trans) [m]	dX [mm]
6324	3558357.7304	3558357.7333	−2 9
6417	3473105.6664	3473105.6701	−3 7
6520	3503525.3824	3503525.3858	−3 4
6725	3567188.4423	3567188.4454	−3 1
6922	3529538.2613	3529538.2647	−3 4
7016	3462353.7891	3462353.7930	−3 9
7220	3506195.9031	3506195.9068	−3 7
7226	3579947.1053	3579947.1084	−3 1
7316	3462442.3184	3462442.3224	−4 0
7324	3556797.2523	3556797.2556	−3 3

Table 21.4. Difference between Gauss–Krueger conformal coordinates Y and Y(trans): Northing.

point	Y [m]	Y(trans) [m]	dY [mm]
6324	5502059.3111	5502059.3116	−0 5
6417	5488314.3903	5488314.3904	−0 1
6520	5481082.8905	5481082.8908	−0 3
6725	5458730.7146	5458730.7152	−0 6
6922	5437066.5236	5437066.5240	−0 4
7016	5429412.0806	5429412.0807	−0 1
7220	5405925.8183	5405925.8187	−0 4
7226	5406962.3048	5406962.3055	−0 7
7316	5386837.0856	5386837.0857	−0 1
7324	5387475.3472	5387475.3477	−0 5

Table 21.5. Some selected BW points, Gauss–Krueger conformal coordinates, Easting x and Northing y, ellipsoidal height h, name of the point.

point	x [m]	y [m]	h [m]	name
6520	3503600.491	5480643.197	514.164	Katzenbuckel
6725	3567263.651	5458291.202	477.449	Gerabronn
7016	3462429.201	5428972.406	277.644	Karlsruhe
7220	3506271.260	5405486.180	519.481	Stuttgart
7226	3580022.573	5406522.794	734.318	Oberkochen

Table 21.6. Transformation from a local to a global reference system, polynomial coefficients.

point 6520	$\bar{a}_{00}$	-75.66265	$\bar{b}_{00}$	5485113.89500
	$\bar{a}_{10}$	3601.00945	$\bar{b}_{10}$	-0.04992
	$\bar{a}_{01}$	-0.05588	$\bar{b}_{01}$	-4030.98709
	$\bar{a}_{20}$	-0.00001	$\bar{b}_{20}$	0.00002
	$\bar{a}_{11}$	-0.00013	$\bar{b}_{11}$	0.00003
	$\bar{a}_{02}$	0.00002	$\bar{b}_{02}$	0.00004
	$\bar{a}_{30}$	0.00000	$\bar{b}_{30}$	0.00000
	$\bar{a}_{21}$	0.00000	$\bar{b}_{21}$	0.00000
	$\bar{a}_{12}$	0.00000	$\bar{b}_{12}$	0.00000
	$\bar{a}_{03}$	0.00000	$\bar{b}_{03}$	0.00000
point 6725	$\bar{a}_{00}$	-84.78176	$\bar{b}_{00}$	5462873.73554
	$\bar{a}_{10}$	67273.28296	$\bar{b}_{10}$	-1.03759
	$\bar{a}_{01}$	-0.06389	$\bar{b}_{01}$	-4142.01666
	$\bar{a}_{20}$	-0.00471	$\bar{b}_{20}$	0.00589
	$\bar{a}_{11}$	-0.00234	$\bar{b}_{11}$	0.00058
	$\bar{a}_{02}$	0.00002	$\bar{b}_{02}$	0.00004
	$\bar{a}_{30}$	-0.00011	$\bar{b}_{30}$	-0.00002
	$\bar{a}_{21}$	0.00000	$\bar{b}_{21}$	-0.00002
	$\bar{a}_{12}$	0.00000	$\bar{b}_{12}$	0.00000
	$\bar{a}_{03}$	0.00000	$\bar{b}_{03}$	0.00000
point 7016	$\bar{a}_{00}$	-69.99003	$\bar{b}_{00}$	5429511.99128
	$\bar{a}_{10}$	-37576.15475	$\bar{b}_{10}$	0.47341
	$\bar{a}_{01}$	-0.00126	$\bar{b}_{01}$	-100.33634
	$\bar{a}_{20}$	-0.00121	$\bar{b}_{20}$	0.00177
	$\bar{a}_{11}$	0.00003	$\bar{b}_{11}$	0.00000
	$\bar{a}_{02}$	0.00000	$\bar{b}_{02}$	0.00000
	$\bar{a}_{30}$	0.00002	$\bar{b}_{30}$	0.00000
	$\bar{a}_{21}$	0.00000	$\bar{b}_{21}$	0.00000
	$\bar{a}_{12}$	0.00000	$\bar{b}_{12}$	0.00000
	$\bar{a}_{03}$	0.00000	$\bar{b}_{03}$	0.00000
point 7220	$\bar{a}_{00}$	-76.25150	$\bar{b}_{00}$	5407273.53640
	$\bar{a}_{10}$	6272.14936	$\bar{b}_{10}$	-0.08549
	$\bar{a}_{01}$	-0.01837	$\bar{b}_{01}$	-1347.65541
	$\bar{a}_{20}$	-0.00004	$\bar{b}_{20}$	0.00005
	$\bar{a}_{11}$	-0.00007	$\bar{b}_{11}$	0.00002
	$\bar{a}_{02}$	0.00000	$\bar{b}_{02}$	0.00000
	$\bar{a}_{30}$	0.00000	$\bar{b}_{30}$	0.00000
	$\bar{a}_{21}$	0.00000	$\bar{b}_{21}$	0.00000
	$\bar{a}_{12}$	0.00000	$\bar{b}_{12}$	0.00000
	$\bar{a}_{03}$	0.00000	$\bar{b}_{03}$	0.00000
point 7226	$\bar{a}_{00}$	-86.67361	$\bar{b}_{00}$	5407274.38804
	$\bar{a}_{10}$	80033.90922	$\bar{b}_{10}$	-1.23993
	$\bar{a}_{01}$	-0.00482	$\bar{b}_{01}$	-310.92454
	$\bar{a}_{20}$	-0.00682	$\bar{b}_{20}$	0.00777
	$\bar{a}_{11}$	-0.00020	$\bar{b}_{11}$	0.00005
	$\bar{a}_{02}$	0.00000	$\bar{b}_{02}$	0.00000
	$\bar{a}_{30}$	-0.00018	$\bar{b}_{30}$	-0.00003
	$\bar{a}_{21}$	0.00000	$\bar{b}_{21}$	0.00000
	$\bar{a}_{12}$	0.00000	$\bar{b}_{12}$	0.00000
	$\bar{a}_{03}$	0.00000	$\bar{b}_{03}$	0.00000

Table 21.7. Transformation from a global to a local reference system, polynomial coefficients.

point 6520	a_{00}	0.00000	b_{00}	5484673.72823
	a_{10}	3600.53002	b_{10}	0.04992
	a_{01}	0.05588	b_{01}	-4030.54839
	a_{20}	0.00001	b_{20}	-0.00002
	a_{11}	0.00013	b_{11}	-0.00003
	a_{02}	0.00002	b_{02}	-0.00004
	a_{30}	0.00000	b_{30}	0.00000
	a_{21}	0.00000	b_{21}	0.00000
	a_{12}	0.00000	b_{12}	0.00000
	a_{03}	0.00000	b_{03}	0.00000
point 6725	a_{00}	0.00000	b_{00}	5462432.75066
	a_{10}	67263.59513	b_{10}	1.03753
	a_{01}	0.06390	b_{01}	-4142.55231
	a_{20}	0.00471	b_{20}	-0.00589
	a_{11}	0.00234	b_{11}	-0.00058
	a_{02}	-0.00002	b_{02}	-0.00004
	a_{30}	0.00011	b_{30}	0.00002
	a_{21}	0.00000	b_{21}	0.00002
	a_{12}	0.00000	b_{12}	0.00000
	a_{03}	0.00000	b_{03}	0.00000
point 7016	a_{00}	0.00000	b_{00}	5429072.73102
	a_{10}	-37570.86126	b_{10}	-0.47340
	a_{01}	0.00126	b_{01}	-99.89921
	a_{20}	0.00121	b_{20}	-0.00177
	a_{11}	-0.00003	b_{11}	0.00000
	a_{02}	0.00000	b_{02}	0.00000
	a_{30}	-0.00002	b_{30}	0.00000
	a_{21}	0.00000	b_{21}	0.00000
	a_{12}	0.00000	b_{12}	0.00000
	a_{03}	0.00000	b_{03}	0.00000
point 7220	a_{00}	0.00000	b_{00}	5406833.68349
	a_{10}	6271.26892	b_{10}	0.08549
	a_{01}	0.01837	b_{01}	-1347.56586
	a_{20}	0.00004	b_{20}	-0.00005
	a_{11}	0.00007	b_{11}	-0.00002
	a_{02}	0.00000	b_{02}	0.00000
	a_{30}	0.00000	b_{30}	0.00000
	a_{21}	0.00000	b_{21}	0.00000
	a_{12}	0.00000	b_{12}	0.00000
	a_{03}	0.00000	b_{03}	0.00000
point 7226	a_{00}	0.00000	b_{00}	5406833.68349
	a_{10}	80022.44576	b_{10}	1.23987
	a_{01}	0.00483	b_{01}	-312.04750
	a_{20}	0.00682	b_{20}	-0.00777
	a_{11}	0.00020	b_{11}	-0.00005
	a_{02}	0.00000	b_{02}	0.00000
	a_{30}	0.00018	b_{30}	0.00003
	a_{21}	0.00000	b_{21}	0.00000
	a_{12}	0.00000	b_{12}	0.00000
	a_{03}	0.00000	b_{03}	0.00000

Table 21.8. World Geodetic Datum 2000 (WGS 2000), E. Grafarend and A. Ardalan (1999).

"tide-free"		
A_1 [m]	A_2 [m]	ϵ [m]
6378136.572	6356751.920	521853.58

"zero-frequency"		
A_1 [m]	A_2 [m]	ϵ [m]
6378136.602	6356751.860	521854.674

Table 21.9. Transformation from conformal coordinates of type Gauss–Krueger (Bessel reference ellipsoid) to conformal coordinates of type Gauss–Krueger (WGS 84 and WGS 2000, "tide-free geoid").

point	X_{84} [m]	Y_{84} [m]	$X_{2000,\mathrm{tf}}$ [m]	$Y_{2000,\mathrm{tf}}$ [m]
6324	3558357.6411	5502059,3874	3558357.6374	5502059.0228
6417	3473105.6872	5488314.2616	3473105.6989	5488313.8979
6520	3503525.2908	5481082.8581	3503525.2906	5481082.4948
6725	3567188.4302	5458730.6878	3567188.4259	5458730.3259
6922	3529538.2090	5437066.5340	3529538.2071	5437066.1735
7016	3462353.8528	5429412.1301	3462353.8552	5429411.7702
7220	3506195.8794	5405925.7956	3506195.8790	5405925.4371
7226	3579947.2236	5406962.2314	3579947.2185	5406961.8728
7316	3462442.3282	5386837.1695	3462442.3306	5386836.8123
7324	3556797.3245	5387475.3087	3556797.3209	5387474.9514

Table 21.10. Transformation from conformal coordinates of type Gauss–Krueger (Bessel reference ellipsoid) to conformal coordinates of type Gauss–Krueger (WGS 84 and WGS 2000, "zero-frequency tide geoid").

point	X_{84} [m]	Y_{84} [m]	$X_{2000,\mathrm{zf}}$ [m]	$Y_{2000,\mathrm{zf}}$ [m]
6324	3558357.6411	5502059.3874	3558357.6372	5502059.0320
6417	3473105.6872	5488314.2616	3473105.6990	5488313.9072
6520	3503525.2908	5481082.8581	3503525.2906	5481082.5041
6725	3567188.4302	5458730.6878	3567188.4256	5458730.3353
6922	3529538.2090	5437066.5340	3529538.2070	5437066.1830
7016	3462353.8528	5429412.1301	3462353.8553	5429411.7796
7220	3506195.8794	5405925.7956	3506195.8790	5405925.4467
7226	3579947.2236	5406962.2314	3579947.2182	5406961.8824
7316	3462442.3282	5386837.1695	3462442.3308	5386836.8219
7324	3556797.3245	5387475.3087	3556797.3207	5387474.9610

Note that for our numerical computations, we took advantage of E. Grafarend (1995), E. Grafarend and R. Syffus (1998e), D. Friedrich (1998), and E. Grafarend and A. Ardalan (1999).

21-5 Mercator coordinates: from a global to a local datum

Transformation of conformal coordinates of type Mercator from a global datum (WGS 84) to a local datum (regional, National, European).

The equations which govern the datum transformation in the extended form of parameters of the *Universal Mercator Projection* (UMP) are discussed here.

Section 21-51.

In Section 21-51, the basic equations are reviewed: compare with Box 21.33 and Table 21.11.

Section 21-52.

In Section 21-52, a numerical exampls is presented: compare with Tables 21.12–21.19.

21-51 Datum transformation extended by form parameters of the UMP

Let us refer to Definition 21.2 as the universal Mercator projection of the ellipsoid-of-revolution $\mathbb{E}^2_{a_1,a_2}$ in local coordinates, namely ellipsoidal coordinates in a local datum.

Definition 21.2 (Universal Mercator projection, local coordinates).

A conformal transformation of ellipsoidal coordinates of type "surface normal" ellipsoidal longitude λ and "surface normal" ellipsoidal latitude φ into Cartesian coordinates $\{x, y\}$ with respect to a local ellipsoid-of-revolution $\mathbb{E}^2_{a_1,a_2}$ is called a *universal Mercator projection* if (21.94) holds, where a_1 denotes the semi-major axis, a_2 the semi-minor axis, and $e = \sqrt{1 - a_2^2/a_1^2}$ the relative eccentricity of $\mathbb{E}^2_{a_1,a_2}$.

$$x = a_1 \lambda \ ,$$

$$y = a_1 \ln \left[\tan \left(\frac{\pi}{4} + \frac{\varphi}{2} \right) \left(\frac{1 - e \sin \varphi}{1 + e \sin \varphi} \right)^{e/2} \right] \ . \tag{21.94}$$

End of Definition.

In order to transform Mercator coordinates which are given in a global datum with respect to the ellipsoid-of-revolution $\mathbb{E}^2_{A_1,A_2}$ into Mercator coordinates which are given in a local datum with respect to the ellipsoid-of-revolution $\mathbb{E}^2_{a_1,a_2}$, we take advantage of the Taylor expansion of second order, namely

$$a_1 = A_1 + \delta a \ , \quad e = E + \delta e \ ,$$

$$\lambda = \Lambda + \delta \Lambda \ , \quad \varphi = \Phi + \delta \Phi \tag{21.95}$$

so that

$$x(\lambda, a_1) = x(\Lambda + \delta\Lambda, \Phi + \delta\Phi, A_1 + \delta A) = A_1 \Lambda + \Lambda \delta a + A_1 \delta \Lambda + \delta A \delta \Lambda + \mathrm{O}_{3x} \ ,$$

$$x := x_0 + x_1 + x_2 + x_3 + \mathrm{O}_{3x} \tag{21.96}$$

and

$$y(\varphi, a_1, e) =$$

$$= y(\Phi + \delta\Phi, A_1 + \delta a, E + \delta e) =$$

$$= A_1 \ln\left[\tan\left(\frac{\pi}{4} + \frac{\Phi}{2}\right)\left(\frac{1 - E\sin\Phi}{1 + E\sin\Phi}\right)^{E/2}\right] +$$

$$+ \ln\left[\tan\left(\frac{\pi}{4} + \frac{\Phi}{2}\right)\left(\frac{1 - E\sin\Phi}{1 + E\sin\Phi}\right)^{E/2}\right]\delta a + A_1\left[\frac{1}{2}\ln\left(\frac{1 - E\sin\Phi}{1 + E\sin\Phi}\right) - \frac{E\sin\Phi}{1 - E^2\sin^2\Phi}\right]\delta e +$$

$$+ A_1\left[\left(\frac{1 - E^2}{\cos\Phi}\right)\left(\frac{1}{1 - E^2\sin^2\Phi}\right)\right]\delta\Phi - A_1\left[\frac{\sin\Phi}{(1 - E^2\sin^2\Phi)^2}\right](\delta e)^2 + \tag{21.97}$$

$$+ \frac{1}{2}A_1\left[\left(\frac{1 - E^2}{\cos\Phi}\right)\left(\frac{1 + 2E^2 - 3E^2\sin^2\Phi}{(1 - E^2\sin^2\Phi)^2}\right)\tan\Phi\right](\delta\Phi)^2 +$$

$$+ \left[\frac{1}{2}\ln\left(\frac{1 - E\sin\Phi}{1 + E\sin\Phi}\right) - \left(\frac{E\sin\Phi}{1 - E^2\sin^2\Phi}\right)\right]\delta a\,\delta e -$$

$$- 2A_1 E\left[\frac{\cos\Phi}{(1 - E^2\sin^2\Phi)^2}\right]\delta\Phi\,\delta e + \left[\left(\frac{1 - E^2}{\cos\Phi}\right)\left(\frac{1}{1 - E^2\sin^2\Phi}\right)\right]\delta\Phi\,\delta a +$$

$$+ O_{3y},$$

$$y :=$$

$$:= y_0 + y_1 + y_2 + y_3 + y_4 + y_5 + y_6 + y_7 + y_8 + \tag{21.98}$$

$$+ O_{3y}.$$

δa, δe and $\delta\Lambda$, $\delta\Phi$, respectively, as increments, account for the variation of the semi-major axis $a_1 - A_1$, the variation of the relative eccentricity $e - E$, the variation of the ellipsoidal longitude $\lambda - \Lambda$, and the variation in the ellipsoidal latitude $\varphi - \Phi$ under a geodetic datum transformation, namely, the conformal group C(3), subject to a variation of the form parameters $\{a_1 \to A_1, e \to E\}$ from $\mathbb{E}^2_{a_1,a_2}$ to $\mathbb{E}^2_{A_1,A_2}$. Here, we refer to the curvilinear datum transformation, namely $\{\lambda \to \Lambda, \varphi \to \Phi\}$. As soon as we implement the curvilinear datum transformation extended by the ellipsoidal form parameters $\{\delta a, \delta e^2 = 2e\delta e\}$ in (21.97), we arrive at the linear representation of local coordinates of the universal Mercator projection as a function of global coordinates and extended datum parameters of Box 21.33. Note that the algorithmic version of the datum transformation of UMP coordinates is given by Table 21.11. Assume that we have measured the ellipsoidal coordinates of a point by means of $\{\Lambda, \Phi, H\}$, for instance, by satellite positioning technology of type GPS, GLONASS, or other. First, for the synthesis of the design matrix A, we need the global ellipsoidal height. Second, we have to get information of the variation of the seven datum parameters and the two form parameters, namely about the basic data which established a local and a global UMP chart. Finally, by means of (21.99) and (21.100), we are able to compute Easting $x(\Lambda)$ and Northing $y(\Phi)$, namely local UMP coordinates from global ellipsoidal coordinates $\{\Lambda, \Phi, H\}$.

Table 21.11. Algorithm for computing coordinates of the universal Mercator projection as a function of global coordinates (GPS, GLONASS) and extended datum parameters

Step one.

Collect global coordinates of type Λ, Φ, H by means of GPS, GLONASS, or other satellite positioning system.

Step two.

Collect the elements of a curvilinear datum transformation, namely three translation parameters t_x, t_y, t_z , three rotation parameters α, β, γ , one scale parameter s, and two ellipsoidal form parameters $\delta a, \delta e^2 = 2e\delta e$.

Step three.

Compute $x(\Lambda)$ by means of (21.99) as Easting ("Rechswert").

Step four.

Compute $y(\Phi)$ by means of (21.100) as Northing ("Hochwert").

Box 21.33 (Local coordinates of the universal Mercator projection as a function of global coordinates and extended datum parameters).

$$x(\Lambda) = A_1 \Lambda + \Lambda \delta a + A_1 (a_{11} t_x + a_{12} t_y + a_{14} \alpha + a_{15} \beta - \gamma) + \delta a \delta \Lambda \,, \tag{21.99}$$

$$y(\Phi) = A_1 \ln \left[\tan \left(\frac{\pi}{4} + \frac{\Phi}{2} \right) \left(\frac{1 - E \sin \Phi}{1 + E \sin \Phi} \right)^{E/2} \right] +$$

$$+ A_1 \left[\left(\frac{1 - E^2}{\cos \Phi} \right) \left(\frac{1}{1 - E^2 \sin^2 \Phi} \right) \right] [a_{21} t_x + a_{22} t_y + a_{23} t_z + a_{24} \alpha + a_{25} \beta + a_{27} s] +$$

$$+ \left[\ln \tan \left(\frac{\pi}{4} + \frac{\Phi}{2} \right) \left(\frac{1 - E \sin \Phi}{1 + E \sin \Phi} \right)^{E/2} + A_1 \left[\left(\frac{1 - E^2}{\cos \Phi} \right) \left(\frac{1}{1 - E^2 \sin^2 \Phi} \right) \right] a_{28} \right] \delta a + \tag{21.100}$$

$$+ \left(A_1 \left[\frac{1}{2} \ln \left(\frac{1 - E \sin \Phi}{1 + E \sin \Phi} \right) - \frac{E \sin \Phi}{1 - E^2 \sin^2 \Phi} \right] + 2 E A_1 \left[\left(\frac{1 - E^2}{\cos \Phi} \right) \left(\frac{1}{1 - E^2 \sin^2 \Phi} \right) \right] a_{29} \right) \delta e +$$

$$+ O_{2y} \,.$$

21-52 Numerical results

In order to test the algorithm for computing coordinates of the universal Mercator projection as a function of global coordinates (GPS, GLONASS) and extended datum parameters, we present some numerical examples. Special emphasis is on the estimation of the order of magnitude of the nonlinear terms in (21.96) and (21.97). Let us begin with a set of extended datum parameters as given in Table 21.12. These represent the transformation of global curvilinear coordinates given in Table 21.14 into local curvilinear coordinates as given in Table 21.13. By means of (21.94), we have computed Easting and Northing for the five points given in Table 21.15. In contrast, by means of Table 21.17, we have computed the terms $\{x_0, x_1, x_2, x_3\}$ as well as $\{y_0, y_1, y_2, y_3, y_4, y_5, y_6, y_7, y_8\}$ of (21.96) and (21.97), which sum up to Easting and Northing in Table 21.17. Obviously, the bilinear term $x_3 := \delta a \delta \Lambda$ accounts for approximately 2 cm, while the terms y_5 (quadratic in δe^2) 0.2 cm, y_6 (quadratic in $\delta \Phi^2$) 0.4 cm, y_7 (bilinear in $\delta a \delta e$) 0.8 cm, and finally y_8 (bilinear in $\delta \Phi \delta e$) 5 cm. As a computational test, we have compared the difference between Easting and Northing in local coordinates and global coordinates (columns 4 and 5 of Table 21.17), namely in the submillimeter range. If we neglect the quadratic-bilinear terms of (21.96) and (21.97), respectively, we document errors of the order of 40 cm according to Table 21.18 and Table 21.19.

Table 21.12. Datum parameters.

$$t_x = -584.911\,\mathrm{m}\,, \quad t_y = -66.121\,\mathrm{m}\,,$$

$$t_z = -402.257\,\mathrm{m}$$

$$\alpha = 0".0038\,, \quad \beta = 0''.0013\,,$$

$$\gamma = -2".3960$$

$$s = -10.11 \times 10^{-6}\,\mathrm{ppm}$$

$$a_1 = 6377397.155\,\mathrm{m}\,, \quad A_1 = 6378137.000\,\mathrm{m}$$

$$e^2 = 0.0066743700\,, \quad E^2 = 0.00669438$$

Table 21.13. Local coordinates.

		λ			φ		h
point	(deg	min	sec)	(deg	min	sec)	(m)
1	6	0	0.0000	12	0	0.0000	1500.000
2	15	0	0.0000	6	0	0.0000	1000.000
3	14	0	0.0000	1	0	0.0000	500.000
4	6	0	0.0000	15	0	0.0000	1400.000
5	12	0	0.0000	9	0	0.0000	200.000

Table 21.14. Global coordinates.

		Λ			Φ		H
point	(deg	min	sec)	(deg	min	sec)	(m)
1	5	59	57.7487	12	0	9.6854	1486.898
2	14	59	54.7547	6	0	11.4852	946.327
3	13	59	55.1020	1	0	12.8335	415.134
4	5	59	57.7485	15	0	8.7493	1401.750
5	11	59	55.7340	9	0	10.6083	167.849

Table 21.15. Easting and Northing. Computed from (21.96) and (21.97).

point	Easting (m)	Northing (m)
1	667839.46837	1336701.67669
2	1669598.67093	664614.06647
3	1558292.09287	110569.36672
4	667839.46837	1677985.89579
5	1335678.93674	999245.35612

Table 21.16. Values (in m) of the terms of (21.96) and (21.97).

terms	point 1	point 2	point 3	point 4	point 5
x_0	667847.32987	1669630.16630	1558321.41475	667847.32369	1335701.97592
x_1	−77.46831	−193.67215	−180.76066	−77.46831	−154.93747
x_2	69.61488	162.19559	151.45635	69.62106	131.91359
x_3	−0.00807	−0.01881	−0.01756	−0.00807	−0.01530
y_0	1337134.43138	665032.56636	110974.20953	1678425.85952	999671.26343
y_1	−155.10363	−77.14180	−12.87267	−194.69242	−115.95890
y_2	26.56589	13.35831	2.23700	33.07201	19.98880
y_3	−304.22496	−354.73999	−394.24321	−278.34340	−329.95309
y_4	−0.01987	−0.00999	−0.00167	−0.02474	−0.01495
y_5	0.00153	0.00105	0.00021	0.00160	0.00136
y_6	−0.00308	−0.00154	−0.00025	−0.00383	−0.00231
y_7	−0.00586	−0.00707	−0.00794	−0.00523	−0.00649
y_8	0.03528	0.04114	0.04573	0.03228	0.03827

Table 21.17. Easting and Northing. Computed from (21.96) and (21.97) and their differences δE and δN computed from Table 21.15.

point	Easting (m)	Northing (m)	δE (m)	δN (m)
1	667839.46837	1336701.67668	0.00000	0.00001
2	1669598.67093	664614.06646	0.00000	0.00001
3	1558292.09287	110569.36671	0.00000	0.00001
4	667839.46837	1677985.89579	0.00000	0.00000
5	1335678.93674	999245.35611	0.00000	0.00001

Table 21.18. Easting and Northing. Computed from (21.96) and (21.97) without the second-order terms and their differences δE and δN computed from Table 21.15.

point	Easting (m)	Northing (m)	δE (m)	δN (m)
1	667839.47645	1336701.66858	−0.00808	0.00801
2	1669598.68975	664614.04287	−0.01882	0.02360
3	1558292.11044	110569.33064	−0.01757	0.03608
4	667839.47645	1677985.89572	−0.00808	0.00007
5	1335678.95205	999245.34023	−0.01531	0.01589

Table 21.19. Easting and Northing. Computed from (21.99) and (21.100) and their differences δE and δN computed from Table 21.15.

point	Easting (m)	Northing (m)	δE (m)	δN (m)
1	667839.47033	1336701.69887	−0.00196	−0.02218
2	1669598.66261	664614.09336	+0.00832	−0.02689
3	1558292.08521	110569.39362	+0.00766	−0.02690
4	667839.47117	1677985.91764	−0.00280	−0.02185
5	1335678.93078	999245.38223	+0.00596	−0.02611

The chapters of the Appendix that follow may supply the readers with additional interesting information. In particular, further details regarding elliptic integrals, Korn–Lichtenstein equations, and geodesics.

A Law and order

Relation preserving maps. Symmetric relations, asymmetric relations, and antisymmetry. The Cartesian product and the Venn diagram. Euler circles, power sets, partitioning or fibering.

A-1 Law and order: Cartesian product, power sets

In daily life, we make comparisons of type ... is higher than ... is smarter than ... is faster than .... Indeed, we are dealing with *order*. Mathematically speaking, *order is a binary relation* of type

$$
\begin{aligned}
xR_1y: \; x < y \;\text{(smaller than)}\,, &\quad xR_3y: \; x > y \;\text{(larger than, } R_3 = R_1^{-1})\,, \\
xR_2y: \; x \le y \;\text{(smaller-equal)}\,, &\quad xR_4y: \; x \ge y \;\text{(larger-equal, } R_4 = R_2^{-1})\,.
\end{aligned}
\tag{A.1}
$$

Example A.1 (Real numbers).

$$x \text{ and } y, \text{ for instance, can be real numbers: } x, y \in \mathbb{R}.$$

End of Example.

Question: "What is a relation?" Answer: "Let us explain the term *relation* in the frame of the following example."

Example A.2 (Cartesian product).

Define the left set $A = \{a_1, a_2, a_3\}$ as the set of balls in a left basket, and $\{a_1, a_2, a_3\} = \{\text{red,green,blue}\}$. In contrast, the right set $B = \{b_1, b_2\}$ as the set of balls in a right basket, and $\{b_1, b_2\} = \{\text{yellow,pink}\}$. Sequentically, we take a ball from the left basket as well as from the right basket such that we are led to the combinations

$$
A \times B = \Big\{ (a_1, b_1), (a_1, b_2), (a_2, b_1), (a_2, b_2), (a_3, b_1), (a_3, b_2) \Big\},
\tag{A.2}
$$

$$
A \times B = \Big\{ (\text{red,yellow}), (\text{red,pink}), (\text{green,yellow}), (\text{green,pink}), (\text{blue,yellow}), (\text{green,pink}) \Big\}.
\tag{A.3}
$$

End of Example.

Definition A.1 (Reflexive partial order).

Let M be a non-empty set. The binary relation R_2 on M is called *reflexive partial order* if for all $x, y, z \in M$ the following three conditions are fulfilled:

$$
\begin{array}{lll}
\text{(i)} & x \le x & \text{(reflexivity)}\,, \\
\text{(ii)} & \text{if } x \le y \text{ and } y \le x, \text{ then } x = y & \text{(antisymmetry)}\,, \\
\text{(iii)} & \text{if } x \le y \text{ and } y \le z, \text{ then } x \le z & \text{(transitivity)}\,.
\end{array}
\tag{A.4}
$$

End of Definition.

Obviously, in a reflexive relation R, any element $x \in M$ is in relation R to itself. But a relation is not symmetric if at least one element $x \in M$ is in relation to an element $y \in M$, which in turn is not in relation to x. If xRy, but by no means yRx applies, we speak of an *asymmetric relation*. This notion should not be confused with *antisymmetry*: if xRy and yRx applies for all $x, y \in M$, then $x = y$ is implied. And R is *transitive* in M if, for all $x, y, z \in M$, xRy and yRz implies xRz. Now we are prepared for to introduce more strictly the method of a *Cartesian product*.

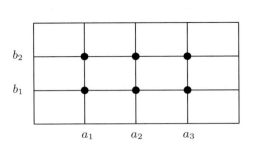

Fig. A.1. Cartesian product, Cartesian coordinate system.

The elements of the *Cartesian product* can be illustrated as point set if A and B are bounded subsets of $\mathbb{R}$. Actually, we consider the ordered pair (a, b) as (x, y) coordinate of the point $P(a, b)$ in a *Cartesian coordinate system* such that all ordered pairs of $A \times B$ are represented as points within a rectangle. Example A.2 and Fig. A.1 have indeed prepared the following definition.

Definition A.2 (Cartesian product).

The Cartesian product $A \times B$ of arbitrary sets A and B is the set of all ordered pairs (a, b) whose left element is $a \in A$ and whose right element is $b \in B$. Symbolically, we write

$$A \times B := \{(a, b) \mid a \in A, b \in B\} . \tag{A.5}$$

End of Definition.

For the Cartesian product, alternative notions are *direct product, product set, cross set, pair set,* or *union set*.

Exercise A.1 (Cartesian product).

The set of theoretical operations $\cap, \cup, \Delta, \setminus$, and $\times$, we shortly call *intersection, union, symmetric difference, difference,* and *Cartesian product*. They are illustrated by Figs. A.2–A.8. With respect to these operations, draw the Cartesian product $A \times B$ of the following sets A and B:

$$
\begin{aligned}
&\text{(i)} && A := \{x \in \mathbb{N} \mid x \in [1; 3] \vee x = 4\} , \\
&&& B := \{y \in \mathbb{N} \mid y \in [1; 2] \vee y = 3\} , \\
&\text{(ii)} && A := \{1, 2, 3\} , \\
&&& B := \{y \in \mathbb{N} \mid y \in [1; 2[\cup \{3\}\} , \\
&\text{(iii)} && A := [1; 2] \cup\,]3; 4[, \\
&&& B := [0; 1] \cup [3; 4[.
\end{aligned}
\tag{A.6}
$$

Here, we have applied the definitions of a closed, left and right open intervals:

$$
\begin{aligned}
[x; y] &:= x \leq \bullet \leq y , \\
]x; y] &:= x < \bullet \leq y , \\
[x; y[&:= x \leq \bullet < y , \\
]x; y[&:= x < \bullet < y .
\end{aligned}
\tag{A.7}
$$

End of Exercise.

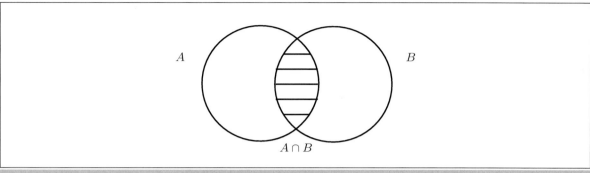

Fig. A.2. Venn diagram/Euler circles $A \cap B$: the intersection $A \cap B$ of two sets A and B is the set of all elements which are elements of the set A *and* the set B: $A \cap B := \{x \mid x \in A \wedge x \in B\}$.

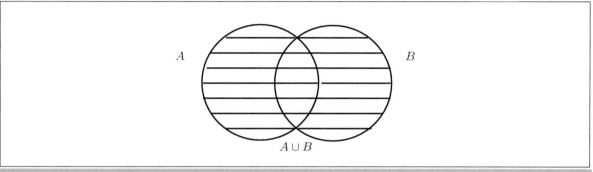

Fig. A.3. Venn diagram/Euler circles $A \cup B$: the union $A \cup B$ of two sets A and B is the set af all elements which are in the set A *or* the set B: $A \cup B := \{x \mid x \in A \vee x \in B\}$.

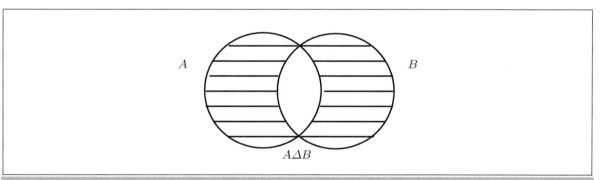

Fig. A.4. Venn diagram/Euler circles: the symmetric difference $A\Delta B$ of two sets A and B is the set af all elements which are *either* in set A *or* in set B: $A\Delta B := \{x \mid x \in A \dot\vee x \in B\}$.

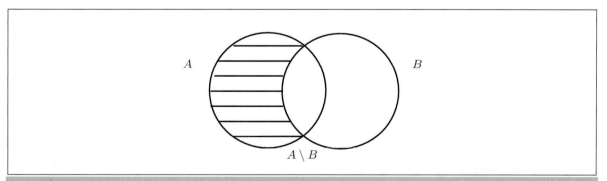

Fig. A.5. Venn diagram/Euler circles: the difference set $A \setminus B$ of two sets A and B is the set af all elements of A which are *not* in B: $A \setminus B := \{x \mid x \in A \wedge x \notin B\}$.

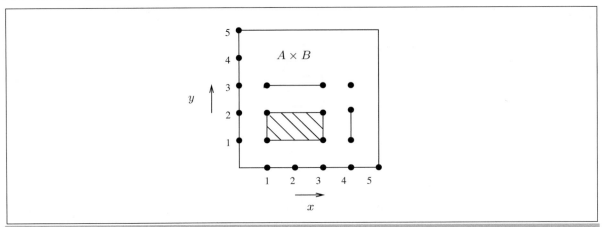

Fig. A.6. Cartesian product $A \times B$; $A := \{x \in \mathbb{N} \mid x \in [1;3] \lor x = 4\}$ and $B := \{y \in \mathbb{N} \mid y \in [1;2] \lor y = 3\}$.

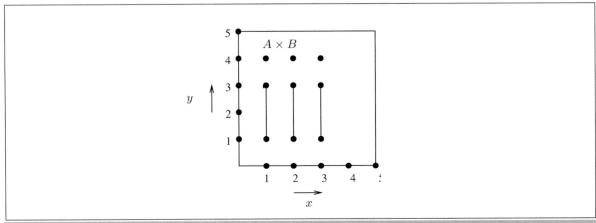

Fig. A.7. Cartesian product $A \times B$; $A := \{1,2,3\}$ and $B := [1;2[\cup\{3\}$.

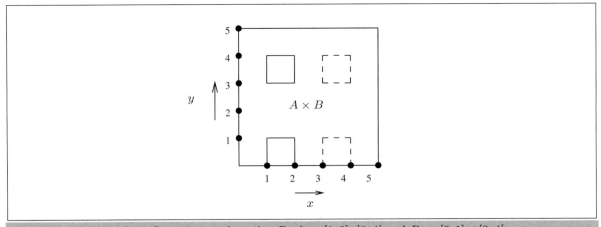

Fig. A.8. Cartesian product $A \times B$; $A := [1;2]\cup]3;4[$ and $B := [0;1] \cup [3;4[$.

In order to interpret the Cartesian product $A \times B$ as a set of third kind, we have to understand better the *power set* $P(A)$ of a set, the intersection and union of *set systems*, and the partitioning of a set system into subsets called *fibering*.

Example A.3 (Power set).

The *power set* as the set of all subsets of a set A may be demonstrated by the set $A = \{1, 2, 3\}$, whose complete list of subsets read

$$M_1 = \emptyset \,,$$

$$M_2 = \{1\}, M_3 = \{2\}, M_4 = \{3\} \,,$$

$$M_5 = \{1, 2\}, M_6 = \{2, 3\}, M_7 = \{3, 1\} \,, \tag{A.8}$$

$$M_8 = \{1, 2, 3\} \,,$$

namely built on 8 elements:

$$\mathrm{power}(M) = \{\emptyset, \{1\}, \{2\}, \{3\}, \{12\}, \{13\}, \{23\}, \{123\}\} \,. \tag{A.9}$$

End of Example.

Exercise A.2 (Power set).

If n is the number of elements of a set A for which we write $|A| = n$, then

$$|\mathrm{power}(A)| = 2^n \,, \tag{A.10}$$

namely the power set of A has exactly 2^n elements. The result motivates the name *power set*. The proof is based on complete induction:

$$\begin{array}{ll} A & \emptyset \; 1 \; 2 \; 3 \; 4 \; 5 \; 6 \; \ldots \; n \\ \mathrm{power}(A) & 1 \; 2 \; 4 \; 8 \; 16 \; 32 \; 64 \; \ldots \; 2^n \end{array} \,. \tag{A.11}$$

End of Exercise.

Example A.3 and Exercise A.2 have already used the following definition.

Definition A.3 (Power set).

The power set of a set A, shortly written $\mathrm{power}(A)$, is by definition the set of all subsets M of A:

$$\mathrm{power}(A) := \{M \mid M \subseteq A\} \tag{A.12}$$

End of Definition.

$\mathrm{power}(A)$ is a set sytem whose elements are just all subsets of A. If A is a set of *first kind*, $\mathrm{power}(A)$ is a set of *second kind*. Inclusions of the above type can be illustrated by *Hasse diagrams*, also called *order diagrams* (H. Hasse 1896–1979). In such a diagram, two sets M_1 and M_2 are identified by two points and are connected by a straight line if the *lower* set M_2 is a subset of M_1 or $M_2 \subseteq M_1$. In this way, a set M is contained in any set which is *above* of M, illustrated by an upward line.

Example A.4 (Hasse diagram).

For the set $A = \{1, 2, 3\}$, $|A| = 3$:

$$\mathrm{power}(A) = \{\emptyset, \{1\}, \{2\}, \{3\},$$

$$\{12\}, \{13\}, \{23\}, \{123\}\} \,, \tag{A.13}$$

$$|\mathrm{power}(A)| = 8 \,.$$

End of Example.

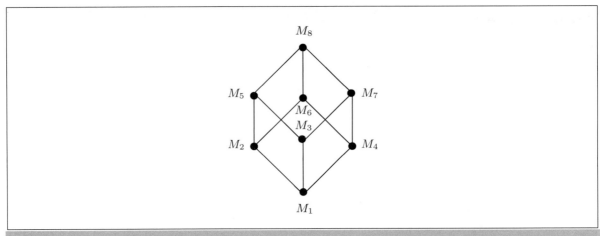

Fig. A.9. Hasse diagram for power(A), $|A| = 3$, $|\text{power}(A)| = 8$.

A-2 Law and order: Fibering

For a set system (which we experienced by power(A), for instance) $\mathbb{M} = \{A_1, A_2, ..., A_n\}$, we call $\cap\, \mathbb{M} = \cap_{i=1}^{n} A_i$ and $\cup\, \mathbb{M} = \cup_{i=1}^{n} A_i$ intersection *and* union of the set system. The inverse operation of the union of a set system, namely the *partitioning* or *fibering* of a set system into specific subsets is given by the following definition.

Definition A.4 (Fibering).

A set system $\mathbb{M} = \{M_1, M_2, \ldots, M_n\}$ $(n \in \mathbb{N}^*)$ of subsets $M_1, \ldots, M_n$ is called a *partitioning* or a *fibering* of $\mathbb{M}$ if and only if

(i) $M_i \neq \emptyset$ for any $i \in \{1, 2, ..., n\}$,

(ii) $M_i \cap M_j = \emptyset$ for any $i, j \in \{1, 2, ..., n\}$,

(iii) $M = M_1 \cup M_2 \cup \ldots \cup M_n = \cup\mathbb{M} = \cup_{i=1}^{n} M_i$

holds. These subsets of M, the elements of $\mathbb{M}$, are called *fibres* of M or of $\mathbb{M}$, respectively.

End of Definition.

In other words, $M_1, \ldots, M_n$ are *non-empty subsets* of M, their paired *intersection* $M_i \cap M_j$ is the *empty set* and their ordered *union* is M again.

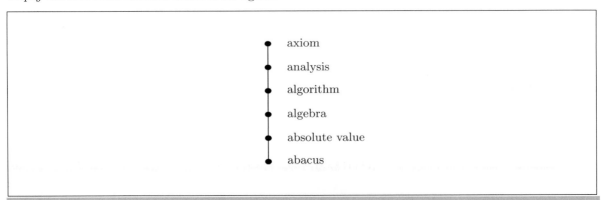

Fig. A.10. Hasse diagram, lexicographic order.

Example A.5 (Fibering).

Let $M = \mathbb{N}^*$, $M_1 = \{1\}$, $M_2 = \mathbb{P}$ (set of prime numbers) *and* $M_3 = \{x \mid ab = x$ for any $a \in \mathbb{P}$ and for $b \in \mathbb{N}^* \setminus \{1\}\}$ the set of compound numbers. Then

$$\mathbb{M} = \{M_1, M_2, M_3\} \tag{A.14}$$

is a *partitioning* or a *fibering* of M. For instance,

$$M_1 = \{1\} \,,$$

$$M_2 = \{2, 3, 5, 7, 11, 13, 17, \ldots\} \,, \tag{A.15}$$

$$M_3 = \{4, 6, 8, 9, 10, 12, 14, \ldots\}$$

fulfills

(i) $M_0 \neq \emptyset$,

(ii) $M_i \cap M_j = \emptyset \quad (i, j \in \{1, 2, 3\}, i \neq j)$, $\tag{A.16}$

(iii) $M = M_1 \cup M_2 \cup M_3 = \mathbb{N}^*$.

End of Example.

Exercise A.3 (Fibering).

Find *three* fibres of the set of

$$\mathbb{M} = \{M_0, M_1, M_2\} \ . \tag{A.17}$$

Answer:

$$M_0 := 3\mathbb{Z} := \{x \mid 3y = x \text{ and } y \in \mathbb{Z}\} =$$

$$= \{\ldots, -6, -3, 0, 3, 6, \ldots\} \,,$$

$$M_1 := 3\mathbb{Z} + 1 := \{x \mid 3y + 1 = x \text{ and } y \in \mathbb{Z}\} =$$

$$= \{\ldots, -5, -2, 1, 4, 7, \ldots\} \,, \tag{A.18}$$

$$M_2 := 3\mathbb{Z} + 2 := \{x \mid 3y + 2 = x \text{ and } y \in \mathbb{Z}\} =$$

$$= \{\ldots, -4, -1, 2, 5, 8, \ldots\} \ .$$

End of Exercise.

Example A.6 (Hasse diagram, lexicographic order).

The *Hasse diagram* of a *lexicographic order* in a set of words which begin with the initial letter "a" is called a chain. That is, all elements are ordered along a half line or line; there are no bifurcations: compare with Fig. A.10.

End of Example.

Example A.7 (Inverse relation).

The *smaller-equal-relation* $\leq$ is in the space of real numbers a reflexive partial order. Its *inverse relation* is the *inverse relation* $\geq$, again a reflexive partial order.

End of Example.

Definition A.5 (Irreflexive partial order).

Let M be a non-empty set. The binary relation R_1 on M is called *irreflexive partial order* if for all $x, y, z \in M$ the two conditions

$$
\begin{aligned}
&\text{(i)} \quad -x < x \ (x < x \text{ is not true}) \\
&\qquad \text{(irreflexivity) ,} \\
&\text{(ii)} \ \text{if } x < y \text{ and } y < 2, \text{ then } x < 2 \\
&\qquad \text{(transitivity)}
\end{aligned}
\tag{A.19}
$$

are fulfilled.

End of Definition.

B The inverse of a multivariate homogeneous polynomial

Univariate, bivariate, and multivariate polynomials and their inversion formulae. Cayley multiplication and Kronecker–Zehfuss product. Triangular matrix.

For inversion problems of map projections like the computation of conformal coordinates of type *Gauss–Krueger* or *Universal Transverse Mercator Projection* (UTM), or alternative coordinates of type *Riemann* or *Soldner–Fermi*, we may take advantage of the inversion of (i) an *univariate* homogeneous polynomial outlined in Section B-1, (ii) a *bivariate* homogeneous polynomial outlined in Section B-2, or (iii) a *trivariate*, in general, multivariate homogeneous polynomial of degree n, which is discussed in Section B-3.

Technical aside. Note that on the basis of an algorithm that is outlined in R. Koenig and K. H. Weise (1951, p. 465–466, 501–511), H. Glasmacher and K. Krack (1964, degree 6) as well as G. Joos and K. Jörg (1991, degree 5) have developed *symbolic computer manipulation software* for the inversion of a bivariate homogeneous polynomial.

Furthermore, solutions for the inversion of a univariate homogeneous polynomial are already tabulated in M. Abramowitz and J. A. Stegun (1965, p. 16, degree 7). However, we follow here E. Grafarend, T. Krarup, and R. Syffus (1996), where the inversion of a general multivariate homogeneous polynomial of degree n suited for symbolic computer manipulation is presented. For the mathematical foundation of the GKS algorithm, we refer to H. Bass, E. H. Cornell, and D. Wright (1962).

B-1 Inversion of a univariate homogeneous polynomial of degree n

Assume the univariate homogeneous polynomial of degree n, namely $y(x)$ of Box B.1, to be given and find the inverse univariate homogeneous polynomial of degree n, namely $x(y)$, i.e. from the set of coefficients $\{a_{11}, a_{12}, \ldots, a_{1n-1}, a_{1n}\}$, by the algorithm that is outlined in Box B.1, find the set of coefficients $\{b_{11}, b_{12}, \ldots, b_{1n-1}, b_{1n}\}$.

Box B.1 (Algorithm for the construction of an inverse univariate homogeneous polynomial of degree n).

$$y(x) = a_{11}x + a_{12}x^2 + \cdots + a_{1n-1}x^{n-1} + a_{1n}x^n ,$$
$$x(y) = b_{11}x + b_{12}x^2 + \cdots + b_{1n-1}x^{n-1} + b_{1n}x^n . \tag{B.1}$$

GKS algorithm: given $\{a_{11}, a_{12}, \ldots, a_{1n-1}, a_{1n}\}$, find $\{b_{11}, b_{12}, \ldots, b_{1n-1}, b_{1n}\}$.

Forward substitution:

$$x = b_{11}y + b_{12}y^2 + \cdots + b_{1n-1}y^{n-1} + b_{1n}y^n + \beta'_{1n+1} ,$$
$$x^2 = b_{22}y^2 + b_{23}y^3 + \cdots + b_{2n-1}y^{n-1} + b_{2n}y^n + \beta'_{2n+1} , \tag{B.2}$$

$$x^{n-1} = b_{n-1n-1}y^{n-1} + b_{n-1n}y^n + \beta'_{n-1n+1} ,$$
$$x^n = b_{nn}y^n + \beta'_{nn+1} , \tag{B.3}$$

$$\begin{bmatrix} y \\ y^2 \\ \cdot \\ y^n \end{bmatrix} = \begin{bmatrix} a_{11} & a_{12} & \cdots & a_{1n} \\ 0 & a_{22} & \ldots & a_{2n} \\ \cdot & \cdot & \ldots & \cdot \\ 0 & 0 & \ldots & a_{nn} \end{bmatrix} \begin{bmatrix} x \\ x^2 \\ \cdot \\ x^n \end{bmatrix} + \alpha'_n , \tag{B.4}$$

Continuation of Box.

$$\text{subject to}$$

$$a_{22} = a_{11}^2 \,,$$
$$a_{23} = 2a_{11}a_{12} \,,$$
$$a_{24} = 2a_{11}a_{13} + a_{12}^2 \,,$$
$$a_{25} = 2a_{11}a_{14} + 2a_{12}a_{13} \,,$$
$$\text{etc.}$$

$$a_{33} = a_{11}^3 \,,$$
$$a_{34} = 3a_{11}^2 a_{12} \,,$$
$$a_{35} = 3a_{11}^2 a_{13} + 3a_{11}a_{12}^2 \,,$$
$$\text{etc.}$$

$$a_{44} = a_{11}^4 \,,$$
$$a_{45} = 4a_{11}^3 a_{12} \,,$$
$$\text{etc.}$$

$$\text{(B.5)}$$

$$\nabla_A = \begin{bmatrix} a_{11} & a_{12} & a_{13} & \cdots & a_{1n} \\ 0 & a_{11}^2 & 2a_{11}a_{12} & \cdots & a_{2n} \\ . & . & \cdots & \cdots & . \\ 0 & 0 & \cdots & a_{n-1n-1} & a_{n-1n} \\ 0 & 0 & \cdots & 0 & a_{nn} \end{bmatrix} \,, \tag{B.6}$$

$$\nabla_B = \begin{bmatrix} b_{11} & b_{12} & \cdots & b_{1n-1} & b_{1n} \\ 0 & b_{22} & \cdots & b_{2n-1} & b_{2n} \\ . & . & \cdots & . & . \\ 0 & 0 & \cdots & b_{n-1n-1} & b_{n-1n} \\ 0 & 0 & \cdots & 0 & b_{nn} \end{bmatrix} \,. \tag{B.7}$$

Consult Box B.4 for the general representation of a_{mn}.

Backward substitution:

$$\nabla_A \nabla_B = I$$
$$\Leftrightarrow$$
$$\text{(i)} \quad b_{11}a_{11} = 1$$
$$\Rightarrow$$
$$b_{11} = a_{11}^{-1} \,,$$

$$\text{(ii)} \quad b_{11}a_{12} + b_{12}a_{22} = 0$$
$$\Rightarrow$$
$$b_{12} = -b_{11}a_{12}a_{22}^{-1} = -a_{11}^{-3}a_{12} \,,$$

$$\text{(iii)} \quad b_{11}a_{13} + b_{12}a_{23} + b_{13}a_{33} = 0$$
$$\Rightarrow$$
$$b_{13} = -\left(b_{11}a_{13} + b_{12}a_{23}\right)a_{33}^{-1} = a_{11}^{-1}\left(a_{12}a_{22}^{-1}a_{23} - a_{13}\right)a_{33}^{-1} \,,$$

$$\text{(iv)} \quad b_{11}a_{14} + b_{12}a_{24} + b_{13}a_{34} + b_{14}a_{44} = 0$$
$$\Rightarrow$$
$$b_{14} = -\left(b_{11}a_{14} + b_{12}a_{24} + b_{13}a_{34}\right)a_{44}^{-1} = a_{11}^{-1}\left[a_{12}a_{22}^{-1}\left(a_{24} - a_{23}a_{33}^{-1}a_{34}\right) + a_{13}a_{33}^{-1}a_{34} - a_{14}\right]a_{44}^{-1} \,.$$

$$\text{(B.8)}$$

Consult Box B.5 for the general representation of b_{1n}.

Notable for the GKS algorithm is the following. In the first step or the forward substitution, a set of equations $\{x, x^2, \ldots, x^{n-1}, x^n\}$ is constructed by substituting $x(y)$ into the powers $x, x^2, \ldots, x^{n-1}, x^n$, finally written into a matrix equation. The upper triangular matrix ∇_A is gained by a multinomial expansion as indicated. In contrast, the second step or the backward substitution is based upon the upper triangular matrix $\nabla_B := \nabla_A^{-1}$, the inversion of ∇_A. Its first row contains the unknown coefficients we are looking for: $\{b_{11}, b_{12}, \ldots, b_{1n-1}, b_{1n}\}$ The construction of ∇_A as well as ∇_A^{-1} can be based on symbolic computer manipulation. The algebraic manipulation becomes more concrete when we pay attention to Examples B.1 and B.2. The first example aims at the inversion of an univariate homogeneous polynomial of degree $n = 2$, namely $y(x) = a_{11}x + a_{12}x^2 \rightarrow x(y) = b_{11}y + b_{12}y^2$. The GKS algorithm determines the set of coefficients $\{b_{11}, b_{12}\}$ from the two given coefficients a_{11} and a_{12}. In contrast, the second example focuses on the inversion of an univariate homogeneous polynomial of degree $n = 3$, namely $y(x) = a_{11}x + a_{12}x^2 + a_{13}x^3 \rightarrow x(y) = b_{11}y + b_{12}y^2 + b_{13}y^3$. The GKS algorithm determines the set of coefficients $\{b_{11}, b_{12}, b_{13}\}$ from the three given coefficients a_{11}, a_{12}, and a_{13} .

Example B.1 (Inversion of an univariate homogeneous polynomial of degree $n = 2$).

Assume the univariate homogeneous polynomial $y(x) = a_{11}x + a_{12}x^2$ to be given and find the inverse univariate homogeneous quadratic polynomial $x(y) = b_{11}y + b_{12}y^2$ by the GKS algorithm.

1st step:

$$
\begin{aligned}
x(y) &= b_{11}y + b_{12}y^2 = b_{11}a_{11}x + \left(b_{11}a_{12} + b_{12}a_{11}^2\right)x^2 + \beta'_{13} , \\
x^2(y) &= b_{22}y^2 + \beta'_{23} = b_{22}a_{11}^2 x^2 + \beta'_{23} .
\end{aligned}
\tag{B.9}
$$

2nd step (forward substitution):

$$
\begin{bmatrix} x \\ x^2 \end{bmatrix} = \begin{bmatrix} b_{11} & b_{12} \\ 0 & b_{22} \end{bmatrix} \begin{bmatrix} a_{11} & a_{12} \\ 0 & a_{22} \end{bmatrix} \begin{bmatrix} x \\ x^2 \end{bmatrix} + \begin{bmatrix} \beta'_{13} \\ \beta'_{23} \end{bmatrix} = \begin{bmatrix} b_{11}a_{11} & b_{11}a_{12} + b_{12}a_{22} \\ 0 & b_{22}a_{22} \end{bmatrix} \begin{bmatrix} x \\ x^2 \end{bmatrix} + \begin{bmatrix} \beta'_{13} \\ \beta'_{23} \end{bmatrix} , \tag{B.10}
$$

subject to $a_{22} = a_{11}^2$.

Both the matrices $A := \nabla_A$ and $B := \nabla_B$ are upper triangular such that

$$
\nabla_A \nabla_B = I_2 \Leftrightarrow \nabla_B = \nabla_A^{-1} . \tag{B.11}
$$

3rd step (backward substitution):

$$
\nabla_B = \nabla_A^{-1} = (a_{11}a_{22})^{-1} \begin{bmatrix} a_{22} & -a_{12} \\ 0 & a_{11} \end{bmatrix} = \begin{bmatrix} a_{11}^{-1} & -a_{11}^{-3}a_{12} \\ 0 & a_{11}^{-2} \end{bmatrix} \Rightarrow b_{11} = a_{11}^{-1} , \ b_{12} = -a_{11}^{-3}a_{12} , \tag{B.12}
$$

or

$$
\begin{aligned}
b_{22}a_{22} &= b_{22}a_{11}^2 = 1 \Rightarrow b_{22} = a_{22}^{-1} = a_{11}^{-2} , \quad b_{11}a_{11} = 1 \Rightarrow b_{11} = a_{11}^{-1} , \\
b_{11}a_{12} + b_{12}a_{22} &= 0 \Rightarrow b_{12} = -a_{22}^{-1}a_{12}b_{11} = -a_{11}^{-3}a_{12} ,
\end{aligned}
\tag{B.13}
$$

$$
x(y) = a_{11}^{-1}y - a_{11}^{-3}a_{12}y^2 . \tag{B.14}
$$

End of Example.

Example B.2 (Inversion of an univariate homogeneous polynomial of degree $n = 3$).

Assume the univariate homogeneous polynomial $y(x) = a_{11}x + a_{12}x^2 + a_{13}x^3$ to be given and find the inverse univariate homogeneous quadratic polynomial $x(y) = b_{11}y + b_{12}y^2 + b_{13}y^3$ by the GKS algorithm.

1st step:

$$
\begin{aligned}
x(y) &= b_{11}y + b_{12}y^2 + b_{13}y^3 \\
&= b_{11}a_{11}x + \left(b_{11}a_{12} + b_{12}a_{11}^2\right)x^2 + \left(b_{11}a_{13} + 2b_{12}a_{11}a_{12} + b_{13}a_{11}^3\right)x^3 + \beta_{14}' , \\
x^2(y) &= b_{22}y^2 + b_{23}y^3 + \beta_{24}' \\
&= b_{22}a_{11}^2 x^2 + \left(2b_{12}a_{11}a_{12} + b_{23}a_{11}^3\right)x^3 + \beta_{24}' , \\
x^3(y) &= b_{33}y^3 + \beta_{34}' \\
&= b_{33}a_{11}^3 x^3 + \beta_{34}' .
\end{aligned}
\tag{B.15}
$$

2nd step (forward substitution):

$$
\begin{bmatrix} x \\ x^2 \\ x^3 \end{bmatrix} =
\begin{bmatrix} b_{11} & b_{12} & b_{13} \\ 0 & b_{22} & b_{23} \\ 0 & 0 & b_{33} \end{bmatrix}
\begin{bmatrix} a_{11} & a_{12} & a_{13} \\ 0 & a_{22} & a_{23} \\ 0 & 0 & a_{33} \end{bmatrix}
\begin{bmatrix} x \\ x^2 \\ x^3 \end{bmatrix} +
\begin{bmatrix} \beta_{14}' \\ \beta_{24}' \\ \beta_{34}' \end{bmatrix} ,
\tag{B.16}
$$

subject to $a_{22} = a_{11}^2$, $a_{23} = 2a_{11}a_{12}$, and $a_{33} = a_{11}^3$.

Both the matrices $\mathsf{A} := \triangledown_A$ and $\mathsf{B} := \triangledown_B$ are upper triangular such that

$$
\triangledown_A \triangledown_B = \mathsf{I}_3 \Leftrightarrow \triangledown_B = \triangledown_A^{-1} .
\tag{B.17}
$$

3rd step (backward substitution):

$$
\triangledown_B = \triangledown_A^{-1} =
\begin{bmatrix} a_{11} & a_{12} & a_{13} \\ 0 & a_{11}^2 & 2a_{11}a_{12} \\ 0 & 0 & a_{11}^3 \end{bmatrix}^{-1} =
\begin{bmatrix} a_{11}^{-1} & -a_{11}^{-3}a_{12} & a_{11}^{-4}(2a_{11}^{-1}a_{12}^2 - a_{13}) \\ 0 & a_{11}^{-2} & -2a_{11}^{-4}a_{12} \\ 0 & 0 & a_{11}^{-3} \end{bmatrix} \Rightarrow
\tag{B.18}
$$

$$
\Rightarrow b_{11} = a_{11}^{-1} , \quad b_{12} = -a_{11}^{-3}a_{12} , \quad b_{13} = a_{11}^{-4}(2a_{11}^{-1}a_{12}^2 - a_{13}) ,
$$

or

$$
\begin{aligned}
b_{33}a_{33} &= b_{33}a_{11}^3 = 1 \Rightarrow b_{33} = a_{33}^{-1} = a_{11}^{-3} , \quad b_{22}a_{22} = b_{22}a_{11}^2 = 1 \Rightarrow b_{22} = a_{22}^{-1} = a_{11}^{-2} , \\
b_{22}a_{23} &+ b_{23}a_{33} = 2a_{11}^{-1}a_{12} + b_{23}a_{11}^3 = 0 \Rightarrow b_{23} = -2a_{11}^{-4}a_{12} , \\
b_{11}a_{11} &= 1 \Rightarrow b_{11} = a_{11}^{-1} , \quad b_{11}a_{12} + b_{12}a_{22} = a_{11}^{-1}a_{12} + b_{12}a_{11}^2 = 0 \Rightarrow b_{12} = -a_{11}^{-3}a_{12} , \\
b_{11}a_{13} &+ b_{12}a_{23} + b_{13}a_{33} = a_{11}^{-1}a_{13} - a_{11}^{-3}a_{12}a_{23} + b_{13}a_{11}^3 = 0 \Rightarrow b_{13} = a_{11}^{-4}(2a_{11}^{-1}a_{12}^2 - a_{13}) ,
\end{aligned}
\tag{B.19}
$$

$$
x(y) = a_{11}^{-1}y - a_{11}^{-3}a_{12}y^2 + a_{11}^{-4}(2a_{11}^{-1}a_{12}^2 - a_{13})y^3 .
\tag{B.20}
$$

End of Example.

B-2 Inversion of a bivariate homogeneous polynomial of degree n

Assume the bivariate homogeneous polynomial of degree n, namely $\boldsymbol{y}(\boldsymbol{x})$ or $\{y_1(x_1, x_2), y_2(x_1, x_2)\}$ of Box B.2, to be given and find the inverse bivariate homogeneous polynomial of degree n, namely $\boldsymbol{x}(\boldsymbol{y})$ or $\{x_1(y_1, y_2), x_2(y_1, y_2)\}$, i.e. from the set of coefficients $\{\mathsf{A}_{11}, \mathsf{A}_{12}, \ldots, \mathsf{A}_{1n-1}, \mathsf{A}_{1n}\}$, by the algorithm that is outlined in Box B.2, find the set of coefficients $\{\mathsf{B}_{11}, \mathsf{B}_{12}, \ldots, \mathsf{B}_{1n-1}, \mathsf{B}_{1n}\}$.

Box B.2 (Algorithm for the construction of an inverse bivariate homogeneous polynomial of degree n).

$$\boldsymbol{y}(\boldsymbol{x}) := \begin{bmatrix} y_1 \\ y_2 \end{bmatrix} =$$

$$= \mathsf{A}_{11} \begin{bmatrix} x_1 \\ x_2 \end{bmatrix} + \mathsf{A}_{12} \begin{bmatrix} x_1 \\ x_2 \end{bmatrix} \otimes \begin{bmatrix} x_1 \\ x_2 \end{bmatrix} + \cdots + \mathsf{A}_{1n-1} \underbrace{\begin{bmatrix} x_1 \\ x_2 \end{bmatrix} \otimes \cdots \otimes \begin{bmatrix} x_1 \\ x_2 \end{bmatrix}}_{n-1 \text{ times}} + \mathsf{A}_{1n} \underbrace{\begin{bmatrix} x_1 \\ x_2 \end{bmatrix} \otimes \cdots \otimes \begin{bmatrix} x_1 \\ x_2 \end{bmatrix}}_{n \text{ times}} = \quad \text{(B.21)}$$

$$= \sum_{k=1}^{n} \mathsf{A}_{1k} \begin{bmatrix} x_1 \\ x_2 \end{bmatrix}^{[k]},$$

$$\boldsymbol{x}(\boldsymbol{y}) := \begin{bmatrix} x_1 \\ x_2 \end{bmatrix} =$$

$$= \mathsf{B}_{11} \begin{bmatrix} y_1 \\ y_2 \end{bmatrix} + \mathsf{B}_{12} \begin{bmatrix} y_1 \\ y_2 \end{bmatrix} \otimes \begin{bmatrix} y_1 \\ y_2 \end{bmatrix} + \cdots + \mathsf{B}_{1n-1} \underbrace{\begin{bmatrix} y_1 \\ y_2 \end{bmatrix} \otimes \cdots \otimes \begin{bmatrix} y_1 \\ y_2 \end{bmatrix}}_{n-1 \text{ times}} + \mathsf{B}_{1n} \underbrace{\begin{bmatrix} y_1 \\ y_2 \end{bmatrix} \otimes \cdots \otimes \begin{bmatrix} y_1 \\ y_2 \end{bmatrix}}_{n \text{ times}} = \quad \text{(B.22)}$$

$$= \sum_{k=1}^{n} \mathsf{B}_{1k} \begin{bmatrix} y_1 \\ y_2 \end{bmatrix}^{[k]}.$$

GKS algorithm: given $\{\mathsf{A}_{11}, \mathsf{A}_{12}, \ldots, \mathsf{A}_{1n-1}, \mathsf{A}_{1n}\}$, find $\{\mathsf{B}_{11}, \mathsf{B}_{12}, \ldots, \mathsf{B}_{1n-1}, \mathsf{B}_{1n}\}$.

1st polynomial:

$$\begin{bmatrix} x_1 \\ x_2 \end{bmatrix} = \mathsf{B}_{11} \sum_{k=1}^{n} \mathsf{A}_{1k} \begin{bmatrix} x_1 \\ x_2 \end{bmatrix}^{[k]} + \mathsf{B}_{12} \left(\sum_{k_1=1}^{n} \mathsf{A}_{1k_1} \begin{bmatrix} x_1 \\ x_2 \end{bmatrix}^{[k_1]} \right) \otimes \left(\sum_{k_2=1}^{n} \mathsf{A}_{1k_2} \begin{bmatrix} x_1 \\ x_2 \end{bmatrix}^{[k_2]} \right) + \cdots +$$

$$+ \mathsf{B}_{1n-1} \left(\sum_{k_1=1}^{n} \mathsf{A}_{1k_1} \begin{bmatrix} x_1 \\ x_2 \end{bmatrix}^{[k_1]} \right) \otimes \cdots \otimes \left(\sum_{k_{n-1}=1}^{n} \mathsf{A}_{1k_{n-1}} \begin{bmatrix} x_1 \\ x_2 \end{bmatrix}^{[k_{n-1}]} \right) +$$

$$+ \mathsf{B}_{1n} \left(\sum_{k_1=1}^{n} \mathsf{A}_{1k_1} \begin{bmatrix} x_1 \\ x_2 \end{bmatrix}^{[k_1]} \right) \otimes \cdots \otimes \left(\sum_{k_n=1}^{n} \mathsf{A}_{1k_n} \begin{bmatrix} x_1 \\ x_2 \end{bmatrix}^{[k_n]} \right) =$$

$$\quad \text{(B.23)}$$

$$= \mathsf{B}_{11}\mathsf{A}_{11} \begin{bmatrix} x_1 \\ x_2 \end{bmatrix} + (\mathsf{B}_{11}\mathsf{A}_{12} + \mathsf{B}_{12}\mathsf{A}_{11} \otimes \mathsf{A}_{11}) \begin{bmatrix} x_1 \\ x_2 \end{bmatrix}^{[2]} + \cdots +$$

$$+ (\mathsf{B}_{11}\mathsf{A}_{1n-1} + \cdots + \mathsf{B}_{1n-1}\mathsf{A}_{n-1n-1}) \begin{bmatrix} x_1 \\ x_2 \end{bmatrix}^{[n-1]} +$$

$$+ (\mathsf{B}_{11}\mathsf{A}_{1n} + \cdots + \mathsf{B}_{1n}\mathsf{A}_{nn}) \begin{bmatrix} x_1 \\ x_2 \end{bmatrix}^{[n]} +$$

$$+ \beta'_{1n+1} .$$

Continuation of Box.

2nd polynomial:

$$
\begin{bmatrix} x_1 \\ x_2 \end{bmatrix}^{[2]} = B_{22} \begin{bmatrix} y_1 \\ y_2 \end{bmatrix}^{[2]} + B_{23} \begin{bmatrix} y_1 \\ y_2 \end{bmatrix}^{[3]} + \cdots + B_{2n-1} \begin{bmatrix} y_1 \\ y_2 \end{bmatrix}^{[n-1]} + B_{2n} \begin{bmatrix} y_1 \\ y_2 \end{bmatrix}^{[n]} + \beta'_{2n+1} =
$$

$$
= B_{22}\,(A_{11} \otimes A_{11}) \begin{bmatrix} x_1 \\ x_2 \end{bmatrix}^{[2]} + (B_{22}A_{23} + B_{23}A_{33}) \begin{bmatrix} x_1 \\ x_2 \end{bmatrix}^{[3]} + \cdots + (B_{22}A_{2n} + \cdots + B_{2n}A_{nn}) \begin{bmatrix} x_1 \\ x_2 \end{bmatrix}^{[n]} + \qquad (B.24)
$$

$$
+ \beta'_{2n+1}\ .
$$

nth polynomial:

$$
\begin{bmatrix} x_1 \\ x_2 \end{bmatrix}^{[n]} = B_{nn}A_{nn} \begin{bmatrix} x_1 \\ x_2 \end{bmatrix}^{[n]} + \beta'_{nn+1}\ . \qquad (B.25)
$$

(According to E. Grafarend and B. Schaffrin (1993) or W. H. Steeb (1991).)

Forward substitution:

$$
\begin{bmatrix} \begin{bmatrix} x_1 \\ x_2 \end{bmatrix}^{[1]} \\ \begin{bmatrix} x_1 \\ x_2 \end{bmatrix}^{[2]} \\ \cdot \\ \begin{bmatrix} x_1 \\ x_2 \end{bmatrix}^{[n]} \end{bmatrix} = \begin{bmatrix} B_{11} & B_{12} & \dots & B_{1n} \\ 0 & B_{22} & \dots & B_{2n} \\ \cdot & \cdot & \dots & \cdot \\ 0 & 0 & \dots & B_{nn} \end{bmatrix} \begin{bmatrix} A_{11} & A_{12} & \dots & A_{1n} \\ 0 & A_{22} & \dots & A_{2n} \\ \cdot & \cdot & \dots & \cdot \\ 0 & 0 & \dots & A_{nn} \end{bmatrix} = \begin{bmatrix} \begin{bmatrix} x_1 \\ x_2 \end{bmatrix}^{[1]} \\ \begin{bmatrix} x_1 \\ x_2 \end{bmatrix}^{[2]} \\ \cdot \\ \begin{bmatrix} x_1 \\ x_2 \end{bmatrix}^{[n]} \end{bmatrix} + \begin{bmatrix} \beta'_{1n+1} \\ \beta'_{2n+1} \\ \cdot \\ \beta'_{nn+1} \end{bmatrix}, \qquad (B.26)
$$

subject to

$$
A_{22} = A_{11} \otimes A_{11}\ ,\quad A_{23} = A_{11} \otimes A_{12} + A_{12} \otimes A_{11}\ ,
$$

$$
A_{2n} = \sum_{i=1}^{n-1} A_{1i} \otimes A_{1n-i}\ ;
$$

$$
A_{33} = A_{11}^{[3]}\ ,
$$

$$
A_{3n} = \sum_{i=1}^{n-2} A_{1i} \otimes \sum_{j=1}^{n-i-1} A_{1j} \otimes A_{1n-i-j}\ ;
$$

$$
A_{44} = A_{11}^{[4]}\ ,
$$

$$
A_{4n} = \sum_{i=1}^{n-3} A_{1i} \otimes \left[\sum_{j=1}^{n-i-2} A_{1j} \otimes \left(\sum_{k=1}^{n-i-j-1} A_{1k} \otimes A_{1n-i-j-k} \right) \right]\ ;
$$

$$
A_{5n} = \sum_{i=1}^{n-4} A_{1i} \otimes \left[\sum_{j=1}^{n-i-3} A_{1j} \otimes \left(\sum_{k=1}^{n-i-j-2} A_{1k} \otimes \left[\sum_{l=1}^{n-i-j-k-1} A_{1l} \otimes A_{1n-i-j-k-l} \right] \right) \right]\ .
$$

$$
(B.27)
$$

(Consult Box B.4 for the general representation of A_{mn}.)

Continuation of Box.

Backward substitution:

$$\nabla_A \nabla_B = I$$

$$\Leftrightarrow$$

$$\begin{bmatrix} B_{11} & B_{12} & & B_{1n-1} & B_{1n} \\ 0 & B_{22} & & B_{2n-1} & B_{2n} \\ & & & & \\ 0 & 0 & & B_{n-1n-1} & B_{n-1n} \\ 0 & 0 & & 0 & B_{nn} \end{bmatrix} \begin{bmatrix} A_{11} & A_{12} & & A_{1n-1} & A_{1n} \\ 0 & A_{22} & & A_{2n-1} & A_{2n} \\ & & & & \\ 0 & 0 & & A_{n-1n-1} & A_{n-1n} \\ 0 & 0 & & 0 & A_{nn} \end{bmatrix} = I \qquad (B.28)$$

(i) $B_{11}A_{11} = I_2 \Rightarrow B_{11} = A_{11}^{-1}$;

(ii) $B_{12}A_{12} + B_{12}A_{22} = 0 \Rightarrow B_{12} = -B_{11}A_{12}A_{22}^{-1} = -A_{11}^{-1}A_{12}A_{22}^{-1[2]}$;

(iii) $B_{11}A_{13} + B_{12}A_{23} + B_{13}A_{33} = 0 \Rightarrow B_{13} = -(B_{11}A_{13} + B_{12}A_{23})A_{33}^{-1} =$
$$= A_{11}^{-1}\left(A_{12}A_{22}^{-1}A_{23} - A_{13}\right)A_{33}^{-1} ;$$ (B.29)

(iv) $B_{11}A_{14} + B_{12}A_{24} + B_{13}A_{34} + B_{14}A_{44} = 0 \Rightarrow B_{14} = -(B_{11}A_{14} + B_{12}A_{24} + B_{13}A_{34})A_{44}^{-1} =$
$$= A_{11}^{-1}\left[A_{12}A_{22}^{-1}\left(A_{24} - A_{23}A_{33}^{-1}A_{34}\right) + A_{13}A_{33}^{-1}A_{34} - A_{14}\right]A_{44}^{-1}$$

(Consult Box B.5 for the general representation of B_{1n}.)

Notable for the GKS algorithm is the following. In the first step or the forward substitution, a set of equations for (B.30) with respect to the *Kronecker–Zehfuss product* is constructed by substituting (B.31) into (B.32) into the powers of (B.33) set up in matrix equations for the first polynomial, the second polynomial, and finally the nth polynomial:

$$\begin{bmatrix} x_1 \\ x_2 \end{bmatrix} \quad \begin{bmatrix} x_1 \\ x_2 \end{bmatrix}^{[2]} \quad \begin{bmatrix} x_1 \\ x_2 \end{bmatrix}^{[n-1]} \begin{bmatrix} x_1 \\ x_2 \end{bmatrix}^{[n]} \qquad (B.30)$$

$$\boldsymbol{y}(\boldsymbol{x}) := \begin{bmatrix} y_1 \\ y_2 \end{bmatrix} = A_{11}\begin{bmatrix} x_1 \\ x_2 \end{bmatrix} + A_{12}\begin{bmatrix} x_1 \\ x_2 \end{bmatrix} \otimes \begin{bmatrix} x_1 \\ x_2 \end{bmatrix} + \cdots \qquad (B.31)$$

$$\boldsymbol{x}(\boldsymbol{y}) := \begin{bmatrix} x_1 \\ x_2 \end{bmatrix} = B_{11}\begin{bmatrix} y_1 \\ y_2 \end{bmatrix} + B_{12}\begin{bmatrix} y_1 \\ y_2 \end{bmatrix} \otimes \begin{bmatrix} y_1 \\ y_2 \end{bmatrix} + \cdots \qquad (B.32)$$

$$\begin{bmatrix} x_1 \\ x_2 \end{bmatrix} \quad \begin{bmatrix} x_1 \\ x_2 \end{bmatrix}^{[n]} = \begin{bmatrix} x_1 \\ x_2 \end{bmatrix} \underbrace{\otimes \cdots \otimes}_{n \text{ times}} \begin{bmatrix} x_1 \\ x_2 \end{bmatrix} \qquad (B.33)$$

Throughout, we particularly take advantage of the fundamental Kronecker–Zehfuss product rule $(AB) \otimes (BD) = (A \otimes B)(C \otimes D)$, i.e. its reduction to the Cayley product of two matrices. A heavy computation of the matrices $\{A_{22}\ A_{23}\quad A_{33}\ A_{34}\quad\}$ is taken over by the general representation of $A_{mn} \forall m < n$ in Box B.4. Finally, the upper triangular matrix ∇_A is gained such that the backward substitution can be started: $\nabla_B := \nabla_A^{-1}$ is constructed. Note that it is very helpful that only its first row is needed, which contains the unknown coefficients $\{B_{11}\ B_{12}\quad B_{1n-1}\ B_{1n}\}$, summarized in Box B.5. Furthermore, note that a symbolic computer manipulation of Box B.5 is available from the author.

The elaborate algebraic manipulation becomes more clear when we consider Examples B.3 and B.4. The first example illustrates our intention to invert a vector-valued bivariate homogeneous polynomial of degree $n = 2$, namely $\boldsymbol{y}(\boldsymbol{x}) = \mathsf{A}_{11}\boldsymbol{x} + \mathsf{A}_{12}\boldsymbol{x} \otimes \boldsymbol{x} \rightarrow \boldsymbol{x}(\boldsymbol{y}) = \mathsf{B}_{11}\boldsymbol{y} + \mathsf{B}_{12}\boldsymbol{y} \otimes \boldsymbol{y}$. The Kronecker–Zehfuss product of column arrays is explicitly given. The GKS algorithm determines the set of matrices $\{\mathsf{B}_{11}, \mathsf{B}_{12}\}$ from the two given matrices A_{11} and A_{12}. In contrast, the second example introduces the problem of inversion of a vector-valued bivariate homogeneous polynomial of degree $n = 3$, namely $\boldsymbol{y}(\boldsymbol{x}) = \mathsf{A}_{11}\boldsymbol{x} + \mathsf{A}_{12}\boldsymbol{x} \otimes \boldsymbol{x} + \mathsf{A}_{13}\boldsymbol{x} \otimes \boldsymbol{x} \otimes \boldsymbol{x} \rightarrow \boldsymbol{x}(\boldsymbol{y}) = \mathsf{B}_{11}\boldsymbol{y} + \mathsf{B}_{12}\boldsymbol{y} \otimes \boldsymbol{y} + \mathsf{B}_{13}\boldsymbol{y} \otimes \boldsymbol{y} \otimes \boldsymbol{y}$. An explicit representation of the Kronecker–Zehfuss product of double and triple column arrays is presented. Again, the GKS algorithm determines the set of matrices $\{\mathsf{B}_{11}, \mathsf{B}_{12}, \mathsf{B}_{13}\}$ from the three given matrices A_{11}, A_{12}, and A_{13}.

Example B.3 (Inversion of a bivariate homogeneous polynomial of degeree $n = 2$).

Assume the bivariate homogeneous polynomial $\boldsymbol{y}(\boldsymbol{x}) = \mathsf{A}_{11}\boldsymbol{x} + \mathsf{A}_{12}\boldsymbol{x} \otimes \boldsymbol{x}$ to be given and find the inverse bivariate homogeneous polynomial $\boldsymbol{x}(\boldsymbol{y}) = \mathsf{B}_{11}\boldsymbol{y} + \mathsf{B}_{12}\boldsymbol{y} \otimes \boldsymbol{y}$ by the GKS algorithm.

Basic equations:

$$\boldsymbol{y}(\boldsymbol{x}) = \begin{bmatrix} y_1 \\ y_2 \end{bmatrix} = \mathsf{A}_{11} \begin{bmatrix} x_1 \\ x_2 \end{bmatrix} + \mathsf{A}_{12} \begin{bmatrix} x_1 \\ x_2 \end{bmatrix} \otimes \begin{bmatrix} x_1 \\ x_2 \end{bmatrix} \tag{B.34}$$

or

$$\begin{aligned} y_1 &= a_{11}^{11}x_1 + a_{11}^{12}x_2 + a_{12}^{11}x_1^2 + \left(a_{12}^{12} + a_{12}^{13}\right)x_1x_2 + a_{12}^{14}x_2^2 \,, \\ y_2 &= a_{11}^{21}x_1 + a_{11}^{22}x_2 + a_{12}^{21}x_1^2 + \left(a_{12}^{22} + a_{12}^{23}\right)x_1x_2 + a_{12}^{24}x_2^2 \,, \end{aligned} \tag{B.35}$$

$$\begin{bmatrix} x_1 \\ x_2 \end{bmatrix} \otimes \begin{bmatrix} x_1 \\ x_2 \end{bmatrix} = \begin{bmatrix} x_1 \\ x_2 \end{bmatrix}^{[2]} = \begin{bmatrix} x_1^2 \\ x_1x_2 \\ x_2x_1 \\ x_2^2 \end{bmatrix} \tag{B.36}$$

(Kronecker–Zehfuss product);

$$\boldsymbol{x}(\boldsymbol{y}) = \begin{bmatrix} x_1 \\ x_2 \end{bmatrix} = \mathsf{B}_{11} \begin{bmatrix} y_1 \\ y_2 \end{bmatrix} + \mathsf{B}_{12} \begin{bmatrix} y_1 \\ y_2 \end{bmatrix} \otimes \begin{bmatrix} y_1 \\ y_2 \end{bmatrix} \tag{B.37}$$

or

$$\begin{aligned} x_1 &= b_{11}^{11}y_1 + b_{11}^{12}y_2 + b_{12}^{11}y_1^2 + \left(b_{12}^{12} + b_{12}^{13}\right)y_1y_2 + b_{12}^{14}y_2^2 \,, \\ x_2 &= b_{11}^{21}y_1 + b_{11}^{22}y_2 + b_{12}^{21}y_1^2 + \left(b_{12}^{22} + b_{12}^{23}\right)y_1y_2 + b_{12}^{24}y_2^2 \,, \end{aligned} \tag{B.38}$$

$$\begin{bmatrix} y_1 \\ y_2 \end{bmatrix} \otimes \begin{bmatrix} y_1 \\ y_2 \end{bmatrix} = \begin{bmatrix} y_1 \\ y_2 \end{bmatrix}^{[2]} = \begin{bmatrix} y_1^2 \\ y_1y_2 \\ y_2y_1 \\ y_2^2 \end{bmatrix} \tag{B.39}$$

(Kronecker–Zehfuss product).

1st polynomial:

$$\begin{bmatrix} x_1 \\ x_2 \end{bmatrix} = B_{11}\left(A_{11}\begin{bmatrix} x_1 \\ x_2 \end{bmatrix} + A_{12}\begin{bmatrix} x_1 \\ x_2 \end{bmatrix} \otimes \begin{bmatrix} x_1 \\ x_2 \end{bmatrix} \right) +$$

$$+ B_{12}\left(A_{11}\begin{bmatrix} x_1 \\ x_2 \end{bmatrix} + A_{12}\begin{bmatrix} x_1 \\ x_2 \end{bmatrix} \otimes \begin{bmatrix} x_1 \\ x_2 \end{bmatrix} \right) \otimes \left(A_{11}\begin{bmatrix} x_1 \\ x_2 \end{bmatrix} + A_{12}\begin{bmatrix} x_1 \\ x_2 \end{bmatrix} \otimes \begin{bmatrix} x_1 \\ x_2 \end{bmatrix} \right) = \tag{B.40}$$

$$= B_{11}A_{11}\begin{bmatrix} x_1 \\ x_2 \end{bmatrix} + (B_{11}A_{12} + B_{12}A_{11}\otimes A_{11})\begin{bmatrix} x_1 \\ x_2 \end{bmatrix}^{[2]} + \beta'_{13}.$$

According to E. Grafarend and B. Schaffrin (1993) or W. H. Steeb (1991). Note that we here have used $(AC) \otimes (BD) = (A \otimes B)(C \otimes D)$, i. e.

$$\left(A_{11}\begin{bmatrix} x_1 \\ x_2 \end{bmatrix} \right) \otimes \left(A_{11}\begin{bmatrix} x_1 \\ x_2 \end{bmatrix} \right) = (A_{11} \otimes A_{11})\left(\begin{bmatrix} x_1 \\ x_2 \end{bmatrix} \otimes \begin{bmatrix} x_1 \\ x_2 \end{bmatrix} \right). \tag{B.41}$$

2nd polynomial:

$$\begin{bmatrix} x_1 \\ x_2 \end{bmatrix} \otimes \begin{bmatrix} x_1 \\ x_2 \end{bmatrix} = B_{11}\begin{bmatrix} y_1 \\ y_2 \end{bmatrix} \otimes \begin{bmatrix} y_1 \\ y_2 \end{bmatrix} + \beta'_{23} = B_{22}(A_{11} \otimes A_{11})\left(\begin{bmatrix} x_1 \\ x_2 \end{bmatrix} \otimes \begin{bmatrix} x_1 \\ x_2 \end{bmatrix} \right) + \beta'_{23}. \tag{B.42}$$

Forward substitution:

$$\begin{bmatrix} \begin{bmatrix} x_1 \\ x_2 \end{bmatrix} \\ \begin{bmatrix} x_1 \\ x_2 \end{bmatrix} \otimes \begin{bmatrix} x_1 \\ x_2 \end{bmatrix} \end{bmatrix} = \begin{bmatrix} B_{11} & B_{12} \\ 0 & B_{22} \end{bmatrix}\begin{bmatrix} A_{11} & A_{12} \\ 0 & A_{22} \end{bmatrix}\begin{bmatrix} \begin{bmatrix} x_1 \\ x_2 \end{bmatrix} \\ \begin{bmatrix} x_1 \\ x_2 \end{bmatrix} \otimes \begin{bmatrix} x_1 \\ x_2 \end{bmatrix} \end{bmatrix} + \begin{bmatrix} \beta'_{13} \\ \beta'_{23} \end{bmatrix}, \tag{B.43}$$

subject to $A_{22} = A_{11} \otimes A_{11}$. Note that the matrices $A := \triangledown_A$ and $B := \triangledown_B$ are upper triangular such that $\triangledown_A \triangledown_B = I_2 \Leftrightarrow \triangledown_B = \triangledown_A^{-1}$.

Backward substitution:

$$\triangledown_A \triangledown_B = I_6 \Leftrightarrow \begin{bmatrix} B_{11} & B_{12} \\ 0 & B_{22} \end{bmatrix}\begin{bmatrix} A_{11} & A_{12} \\ 0 & A_{22} \end{bmatrix} = I_6 ; \tag{B.44}$$

$$\begin{aligned} &\text{(i)} \quad B_{11}A_{11} = I_2 \Rightarrow B_{11} = A_{11}^{-1}, \\ &\text{(ii)} \quad B_{11}A_{12} + B_{12}A_{22} = 0 \Rightarrow B_{12} = -B_{11}A_{12}A_{22}^{-1} = -A_{11}^{-1}A_{12}\left(A_{11}^{-1} \otimes A_{11}^{-1} \right). \end{aligned} \tag{B.45}$$

First, we have used $(A \otimes B)^{-1} = A^{-1} \otimes B^{-1}$ for two invertible square matrices A and B. Second, we have used the standard solution of a system of upper triangular matrix equations. For the inverse polynomial representation, only the elements of the first row of the matrix $B := \triangledown_B$ are of interest. An explicit write-up is

$$\begin{bmatrix} x_1 \\ x_2 \end{bmatrix} = A_{11}^{-1}\begin{bmatrix} y_1 \\ y_2 \end{bmatrix} - A_{11}^{-1}A_{12}\left(A_{11}^{-1} \otimes A_{11}^{-1} \right)\begin{bmatrix} y_1 \\ y_2 \end{bmatrix}^{[2]}. \tag{B.46}$$

End of Example.

Example B.4 (Inversion of a bivariate homogeneous polynomial of degree $n = 3$).

Assume the bivariate homogeneous polynomial $\boldsymbol{y}(\boldsymbol{x}) = \mathsf{A}_{11}\boldsymbol{x} + \mathsf{A}_{12}\boldsymbol{x} \otimes \boldsymbol{x} + \mathsf{A}_{13}\boldsymbol{x} \otimes \boldsymbol{x} \otimes \boldsymbol{x}$ to be given and find the inverse bivariate homogeneous polynomial $\boldsymbol{x}(\boldsymbol{y}) = \mathsf{B}_{11}\boldsymbol{y} + \mathsf{B}_{12}\boldsymbol{y} \otimes \boldsymbol{y} + \mathsf{B}_{13}\boldsymbol{y} \otimes \boldsymbol{y} \otimes \boldsymbol{y}$ by the GKS algorithm.

Basic equations:

$$\boldsymbol{y}(\boldsymbol{x}) = \begin{bmatrix} y_1 \\ y_2 \end{bmatrix} =$$

$$= \mathsf{A}_{11} \begin{bmatrix} x_1 \\ x_2 \end{bmatrix} + \mathsf{A}_{12} \begin{bmatrix} x_1 \\ x_2 \end{bmatrix} \otimes \begin{bmatrix} x_1 \\ x_2 \end{bmatrix} + \mathsf{A}_{13} \begin{bmatrix} x_1 \\ x_2 \end{bmatrix} \otimes \begin{bmatrix} x_1 \\ x_2 \end{bmatrix} \otimes \begin{bmatrix} x_1 \\ x_2 \end{bmatrix}$$

(B.47)

$$\begin{bmatrix} x_1 \\ x_2 \end{bmatrix} \otimes \begin{bmatrix} x_1 \\ x_2 \end{bmatrix} \otimes \begin{bmatrix} x_1 \\ x_2 \end{bmatrix} = \begin{bmatrix} x_1 \\ x_2 \end{bmatrix}^{[3]} = \begin{bmatrix} x_1^3 \\ x_1^2 x_2 \\ x_1^2 x_2 \\ x_1 x_2^2 \\ x_2 x_1^2 \\ x_2^2 x_1 \\ x_2^2 x_1 \\ x_2^3 \end{bmatrix} \in \mathbb{R}^{8 \times 1}$$

(B.48)

(triple Kronecker–Zehfuss product);

$$\boldsymbol{x}(\boldsymbol{y}) = \begin{bmatrix} x_1 \\ x_2 \end{bmatrix} =$$

$$= \mathsf{B}_{11} \begin{bmatrix} y_1 \\ y_2 \end{bmatrix} + \mathsf{B}_{12} \begin{bmatrix} y_1 \\ y_2 \end{bmatrix} \otimes \begin{bmatrix} y_1 \\ y_2 \end{bmatrix} + \mathsf{B}_{13} \begin{bmatrix} y_1 \\ y_2 \end{bmatrix} \otimes \begin{bmatrix} y_1 \\ y_2 \end{bmatrix} \otimes \begin{bmatrix} y_1 \\ y_2 \end{bmatrix}$$

(B.49)

$$\begin{bmatrix} y_1 \\ y_2 \end{bmatrix} \otimes \begin{bmatrix} y_1 \\ y_2 \end{bmatrix} \otimes \begin{bmatrix} y_1 \\ y_2 \end{bmatrix} = \begin{bmatrix} y_1 \\ y_2 \end{bmatrix}^{[3]} = \begin{bmatrix} y_1^3 \\ y_1^2 y_2 \\ y_1^2 y_2 \\ y_1 y_2^2 \\ y_2 y_1^2 \\ y_2^2 y_1 \\ y_2^2 y_1 \\ y_2^3 \end{bmatrix} \in \mathbb{R}^{8 \times 1}$$

(B.50)

(triple Kronecker–Zehfuss product).

1st polynomial:

$$\begin{bmatrix} x_1 \\ x_2 \end{bmatrix} = \mathsf{B}_{11} \sum_{k=1}^{3} \mathsf{A}_{1k} \begin{bmatrix} x_1 \\ x_2 \end{bmatrix}^{[k]} +$$

$$+\mathsf{B}_{12} \left(\sum_{k_1=1}^{3} \mathsf{A}_{1k_1} \begin{bmatrix} x_1 \\ x_2 \end{bmatrix}^{[k_1]} \right) \otimes \left(\sum_{k_2=1}^{3} \mathsf{A}_{1k_2} \begin{bmatrix} x_1 \\ x_2 \end{bmatrix}^{[k_2]} \right) +$$

$$+\mathsf{B}_{13} \left(\sum_{k_1=1}^{3} \mathsf{A}_{1k_1} \begin{bmatrix} x_1 \\ x_2 \end{bmatrix}^{[k_1]} \right) \otimes \left(\sum_{k_2=1}^{3} \mathsf{A}_{1k_2} \begin{bmatrix} x_1 \\ x_2 \end{bmatrix}^{[k_2]} \right) \otimes \left(\sum_{k_3=1}^{3} \mathsf{A}_{1k_3} \begin{bmatrix} x_1 \\ x_2 \end{bmatrix}^{[k_3]} \right) = \qquad (\text{B.51})$$

$$= \mathsf{B}_{11}\mathsf{A}_{11} \begin{bmatrix} x_1 \\ x_2 \end{bmatrix} + (\mathsf{B}_{11}\mathsf{A}_{12} + \mathsf{B}_{12}\mathsf{A}_{11} \otimes \mathsf{A}_{11}) \begin{bmatrix} x_1 \\ x_2 \end{bmatrix}^{[2]} +$$

$$+ [\mathsf{B}_{11}\mathsf{A}_{13} + \mathsf{B}_{12}(\mathsf{A}_{11} \otimes \mathsf{A}_{12} + \mathsf{A}_{12} \otimes \mathsf{A}_{11}) + \mathsf{B}_{13}(\mathsf{A}_{11} \otimes \mathsf{A}_{11} \otimes \mathsf{A}_{11})] \begin{bmatrix} x_1 \\ x_2 \end{bmatrix}^{[3]} + \beta'_{14} \,.$$

2nd polynomial:

$$\begin{bmatrix} x_1 \\ x_2 \end{bmatrix} \otimes \begin{bmatrix} x_1 \\ x_2 \end{bmatrix} = \begin{bmatrix} x_1 \\ x_2 \end{bmatrix}^{[2]} = \mathsf{B}_{22} \begin{bmatrix} y_1 \\ y_2 \end{bmatrix}^{[2]} + \mathsf{B}_{23} \begin{bmatrix} y_1 \\ y_2 \end{bmatrix}^{[3]} + \beta'_{24} =$$

$$\qquad (\text{B.52})$$

$$= \mathsf{B}_{22}(\mathsf{A}_{11} \otimes \mathsf{A}_{11}) \begin{bmatrix} x_1 \\ x_2 \end{bmatrix}^{[2]} + [\mathsf{B}_{22}(\mathsf{A}_{11} \otimes \mathsf{A}_{12} + \mathsf{A}_{12} \otimes \mathsf{A}_{11}) + \mathsf{B}_{23}(\mathsf{A}_{11} \otimes \mathsf{A}_{11} \otimes \mathsf{A}_{11})] \begin{bmatrix} x_1 \\ x_2 \end{bmatrix}^{[3]} + \beta'_{24} \,.$$

3rd polynomial:

$$\begin{bmatrix} x_1 \\ x_2 \end{bmatrix} \otimes \begin{bmatrix} x_1 \\ x_2 \end{bmatrix} \otimes \begin{bmatrix} x_1 \\ x_2 \end{bmatrix} = \begin{bmatrix} x_1 \\ x_2 \end{bmatrix}^{[3]} = \mathsf{B}_{33}(\mathsf{A}_{11} \otimes \mathsf{A}_{11} \otimes \mathsf{A}_{11}) \begin{bmatrix} x_1 \\ x_2 \end{bmatrix}^{[3]} + \beta'_{34} \,. \qquad (\text{B.53})$$

Forward substitution:

$$\begin{bmatrix} \begin{bmatrix} x_1 \\ x_2 \end{bmatrix}^{[1]} \\ \begin{bmatrix} x_1 \\ x_2 \end{bmatrix}^{[2]} \\ \begin{bmatrix} x_1 \\ x_2 \end{bmatrix}^{[3]} \end{bmatrix} = \begin{bmatrix} \mathsf{B}_{11} & \mathsf{B}_{12} & \mathsf{B}_{13} \\ 0 & \mathsf{B}_{22} & \mathsf{B}_{23} \\ 0 & 0 & \mathsf{B}_{33} \end{bmatrix} \begin{bmatrix} \mathsf{A}_{11} & \mathsf{A}_{12} & \mathsf{A}_{13} \\ 0 & \mathsf{A}_{22} & \mathsf{A}_{23} \\ 0 & 0 & \mathsf{A}_{33} \end{bmatrix} \begin{bmatrix} \begin{bmatrix} x_1 \\ x_2 \end{bmatrix}^{[1]} \\ \begin{bmatrix} x_1 \\ x_2 \end{bmatrix}^{[2]} \\ \begin{bmatrix} x_1 \\ x_2 \end{bmatrix}^{[3]} \end{bmatrix} + \begin{bmatrix} \beta'_{14} \\ \beta'_{24} \\ \beta'_{34} \end{bmatrix} , \qquad (\text{B.54})$$

subject to

$$\mathsf{A}_{22} = \mathsf{A}_{11} \otimes \mathsf{A}_{11} \,, \quad \mathsf{A}_{23} = \mathsf{A}_{11} \otimes \mathsf{A}_{12} + \mathsf{A}_{12} \otimes \mathsf{A}_{11} \,, \quad \mathsf{A}_{33} = \mathsf{A}_{11} \otimes \mathsf{A}_{11} \otimes \mathsf{A}_{11} \,. \qquad (\text{B.55})$$

Both the matrices $\mathsf{A} := \triangledown_A$ and $\mathsf{B} := \triangledown_B$ are upper triangular such that

$$\triangledown_A \triangledown_B = \mathsf{I}_{14} \Leftrightarrow \triangledown_B = \triangledown_A^{-1}. \qquad (\text{B.56})$$

Backward substitution:

$$\nabla_A \nabla_B = \mathsf{I}_{14} \Leftrightarrow \begin{bmatrix} \mathsf{B}_{11} & \mathsf{B}_{12} & \mathsf{B}_{13} \\ 0 & \mathsf{B}_{22} & \mathsf{B}_{23} \\ 0 & 0 & \mathsf{B}_{33} \end{bmatrix} \begin{bmatrix} \mathsf{A}_{11} & \mathsf{A}_{12} & \mathsf{A}_{13} \\ 0 & \mathsf{A}_{22} & \mathsf{A}_{23} \\ 0 & 0 & \mathsf{A}_{33} \end{bmatrix} = \mathsf{I}_{14} ; \tag{B.57}$$

$$\text{(i)} \quad \mathsf{B}_{11}\mathsf{A}_{11} = \mathsf{I}_2$$
$$\Rightarrow$$
$$\mathsf{B}_{11} = \mathsf{A}_{11}^{-1} ;$$

$$\text{(ii)} \quad \mathsf{B}_{12}\mathsf{A}_{12} + \mathsf{B}_{12}\mathsf{A}_{22} = 0$$
$$\Rightarrow$$
$$\mathsf{B}_{12} = -\mathsf{B}_{11}\mathsf{A}_{12}\mathsf{A}_{22}^{-1} = -\mathsf{A}_{11}^{-1}\mathsf{A}_{12}\mathsf{A}_{22}^{-1\,[2]} ; \tag{B.58}$$

$$\text{(iii)} \quad \mathsf{B}_{11}\mathsf{A}_{13} + \mathsf{B}_{12}\mathsf{A}_{23} + \mathsf{B}_{13}\mathsf{A}_{33} = 0$$
$$\Rightarrow$$
$$\mathsf{B}_{13} = -\left(\mathsf{B}_{11}\mathsf{A}_{13} + \mathsf{B}_{12}\mathsf{A}_{23}\right)\mathsf{A}_{33}^{-1} =$$
$$= -\mathsf{A}_{11}^{-1}\mathsf{A}_{13}\mathsf{A}_{11}^{-1\,[3]} + \mathsf{A}_{11}^{-1}\mathsf{A}_{12}\mathsf{A}_{11}^{-1\,[2]}\left(\mathsf{A}_{11} \otimes \mathsf{A}_{12} + \mathsf{A}_{12} \otimes \mathsf{A}_{11}\right)\mathsf{A}_{11}^{-1\,[3]}$$
$$= \mathsf{A}_{11}^{-1}\left[-\mathsf{A}_{13} + \mathsf{A}_{12}\left(\mathsf{A}_{11}^{-1} \otimes \mathsf{A}_{11}^{-1}\right)\left(\mathsf{A}_{11} \otimes \mathsf{A}_{12} + \mathsf{A}_{12} \otimes \mathsf{A}_{11}\right)\right]\left(\mathsf{A}_{11}^{-1} \otimes \mathsf{A}_{11}^{-1} \otimes \mathsf{A}_{11}^{-1}\right)$$

First, we have used $(A \otimes B)^{-1} = A^{-1} \otimes B^{-1}$ for two invertible square matrices A and B, secondly we have used the standard solution of a system of upper triangular matrix equations. For the inverse polynomial representation, only the elements of the first row of the matrix $\mathsf{B} := \nabla_B$ are of interest. An explicit write-up is

$$\begin{bmatrix} x_1 \\ x_2 \end{bmatrix} = \mathsf{A}_{11}^{-1}\begin{bmatrix} y_1 \\ y_2 \end{bmatrix} - \mathsf{A}_{11}^{-1}\mathsf{A}_{12}\left(\mathsf{A}_{11}^{-1} \otimes \mathsf{A}_{11}^{-1}\right)\begin{bmatrix} y_1 \\ y_2 \end{bmatrix}^{[2]} -$$
$$-\mathsf{A}_{11}^{-1}\left[\mathsf{A}_{13} - \mathsf{A}_{12}\left(\mathsf{A}_{11}^{-1} \otimes \mathsf{A}_{11}^{-1}\right)\left(\mathsf{A}_{11} \otimes \mathsf{A}_{12} + \mathsf{A}_{12} \otimes \mathsf{A}_{11}\right)\right]\left(\mathsf{A}_{11}^{-1} \otimes \mathsf{A}_{11}^{-1} \otimes \mathsf{A}_{11}^{-1}\right)\begin{bmatrix} y_1 \\ y_2 \end{bmatrix}^{[3]} \tag{B.59}$$

End of Example.

B-3 Inversion of a multivariate homogeneous polynomial of degree n

Assume a multivariate homogeneous polynomial of degree n, namely $\boldsymbol{y}(\boldsymbol{x})$, to be given and find the inverse multivariate homogeneous polynomial of degree n, namely $\boldsymbol{x}(\boldsymbol{y})$, and

$$\boldsymbol{y}(\boldsymbol{x}) = \mathsf{A}_{11}\boldsymbol{x} + \mathsf{A}_{12}\boldsymbol{x}^{[2]} + \cdots + \mathsf{A}_{1n-1}\boldsymbol{x}^{[n-1]} + \mathsf{A}_{1n}\boldsymbol{x}^{[n]} = \sum_{k=1}^{n} \mathsf{A}_{1k}\boldsymbol{x}^{[k]} \; \forall \; \boldsymbol{x} \in \mathbb{R}^p$$

$$\boldsymbol{x}(\boldsymbol{y}) = \mathsf{B}_{11}\boldsymbol{y} + \mathsf{B}_{12}\boldsymbol{y}^{[2]} + \cdots + \mathsf{B}_{1n-1}\boldsymbol{y}^{[n-1]} + \mathsf{B}_{1n}\boldsymbol{y}^{[n]} = \sum_{k=1}^{n} \mathsf{B}_{1k}\boldsymbol{y}^{[k]} \; \forall \; \boldsymbol{y} \in \mathbb{R}^p \tag{B.60}$$

This defines the general problem of homogeneous polynomial series inversion. It reduces to construct the matrices $\{\mathsf{B}_{11}\ \mathsf{B}_{12} \quad \mathsf{B}_{1n-1}\ \mathsf{B}_{1n}\}$ from the given matrices $\{\mathsf{A}_{11}\ \mathsf{A}_{12} \quad \mathsf{A}_{1n-1}\ \mathsf{A}_{1n}\}$ by the GKS algorithm as outlined in Box B.3.

Box B.3 (Algorithm for the construction of a multivariate homogeneous polynomial of degree n).

1st polynomial:

$$\boldsymbol{x} = \sum_{k_1=1}^{n} \mathsf{B}_{1k_1} \left(\sum_{k_2=1}^{n} \mathsf{A}_{1k_2} \boldsymbol{x}^{[k_2]} \right)^{[k_1]} + \beta'_{1n+1} . \tag{B.61}$$

2nd polynomial:

$$\boldsymbol{x}^{[2]} = \sum_{k=2}^{n} \mathsf{B}_{2k} \boldsymbol{y}^{[k]} + \beta'_{2n+1} = \sum_{k_1=2}^{n} \mathsf{B}_{2k_1} \left(\sum_{k_2=1}^{n} \mathsf{A}_{1k_2} \boldsymbol{x}^{[k_2]} \right)^{[k_1]} + \beta'_{2n+1} . \tag{B.62}$$

nth polynomial:

$$\boldsymbol{x}^{[n]} = \mathsf{B}_{nn} \boldsymbol{y}^{[n]} + \beta'_{nn+1} = \mathsf{B}_{nn} \mathsf{A}_{nn} \boldsymbol{x}^{[n]} + \beta'_{nn+1} . \tag{B.63}$$

Again taking advantage of the basic product rule between Cayley and Kronecker–Zehfuss multiplication, i.e. $(AC) \otimes (BD) = (A \otimes B)(C \otimes D)$, we arrive at the following results.

Forward substitution:

$$\begin{bmatrix} \boldsymbol{x} \\ \boldsymbol{x}^{[2]} \\ \cdot \\ \boldsymbol{x}^{[n-1]} \\ \boldsymbol{x}^{[n-1]} \end{bmatrix} = \nabla_B \nabla_A \begin{bmatrix} \boldsymbol{x} \\ \boldsymbol{x}^{[2]} \\ \cdot \\ \boldsymbol{x}^{[n-1]} \\ \boldsymbol{x}^{[n-1]} \end{bmatrix} . \tag{B.64}$$

(Note that ∇_A is given by Box B.4.)

Backward substitution:

$$\nabla_A \nabla_B = \mathsf{I}$$

$$\Rightarrow \tag{B.65}$$

see first row B_{1n} given in Box B.5.

First, in the forward substitution, we construct a set of equations for $\{\boldsymbol{x}, \boldsymbol{x}^{[2]}, \ldots, \boldsymbol{x}^{[n-1]}, \boldsymbol{x}^{[2]}\}$, where the column array powers are to be understood with respect to the Kronecker–Zehfuss product. Again, advantage is taken from the basic product rule $(AC) \otimes (BD) = (A \otimes B)(C \otimes D)$, also called "reduction of the Kronecker–Zehfuss product to the Cayley product", by replacing such an identity into the power series equations, the upper triangular matrix ∇_A as well as its inverse $\nabla_B := \nabla_A^{-1}$. Second, while Box B.4 collects the input matrices A_{mn}, which built up ∇_A, by means of Box B.5, we compute its inverse ∇_B, namely its first row $\{\mathsf{B}_{11}, \mathsf{B}_{12}, \ldots, \mathsf{B}_{1n-1}, \mathsf{B}_{1n}\}$, the one being only needed. A symbolic computer manipulation software package of Box B.5 is available from the author, in particular, for the backward substitution step.

Box B.4 (Inversion of a multivariate homogeneous polynomial of degree n: the upper triangular matrix $A := \nabla_A$).

Recurrence relation:

$$A_{mn} = \sum_{i=1}^{n-(m-1)} A_{1i} \otimes A_{m-1\,n-i} \;\; \forall \, m \le n \; . \tag{B.66}$$

Inversion relations (A_{11}, A_{1n} given):

$$A_{22} = A_{11}^{[2]} \; , \;\; A_{2n} = \sum_{i=1}^{n-1} A_{1i} \otimes A_{1\,n-i} \; ,$$

$$A_{33} = A_{11}^{[3]} \; , \;\; A_{3n} = \sum_{i=1}^{n-2} A_{1i} \otimes \sum_{j=1}^{n-i-1} A_{1j} \otimes A_{1\,n-i-j} \; ,$$

$$A_{44} = A_{11}^{[4]} \; , \;\; A_{4n} = \sum_{i=1}^{n-3} A_{1i} \otimes \sum_{j=1}^{n-i-2} A_{1j} \otimes \sum_{k=1}^{n-i-j-1} A_{1k} \otimes A_{1\,n-i-j-k} \; ,$$

$$A_{55} = A_{11}^{[5]} \; , \;\; A_{5n} = \sum_{i=1}^{n-4} A_{1i} \otimes \sum_{j=1}^{n-i-3} A_{1j} \otimes \sum_{k=1}^{n-i-j-2} A_{1k} \otimes \sum_{l=1}^{n-i-j-k-1} A_{1l} \otimes A_{1\,n-i-j-k-l} \; , \tag{B.67}$$

$$\ldots$$

$$A_{mn} = \sum_{k_1=1}^{n-(m-1)} A_{1k_1} \otimes \sum_{k_2=1}^{n-k_1-(m-2)} A_{1k_2} \otimes \sum_{k_3=1}^{n-k_1-k_2-(m-3)} A_{1k_3} \otimes \cdots \otimes$$

$$\otimes \cdots \otimes \sum_{k_{m-1}=1}^{n-k_1-k_2-\cdots-k_{m-2}} A_{1k_{m-1}} \otimes A_{1\,n-k_1-k_2-\cdots-k_{m-2}-k_{m-1}} \; .$$

Box B.5 (Inversion of a multivariate homogeneous polynomial of degree n: 1st row of the triangular matrix $B := \nabla_B$).

Recurrence relation:

$$B_{1n} = \left(- \sum_{i=1}^{n-1} B_{1i} A_{in} \right) A_{nn}^{-1} \;\; \forall \, n \ge 2 \; , \quad \text{subject to} \quad A_{nn}^{-1} = \left(A_{11}^{[n]} \right)^{-1} = \left(A_{11}^{-1} \right)^{[n]} \; . \tag{B.68}$$

Inversion relations (A_{11}, A_{1n} given):

$$B_{11} = +A_{11}^{-1} \; , \;\; B_{12} = -A_{11}^{-1} A_{12} A_{22}^{-1} \; , \;\; B_{13} = +A_{11}^{-1} \left[A_{12} A_{22}^{-1} A_{23} - A_{13} \right] A_{33}^{-1} \; ,$$

$$B_{14} = +A_{11}^{-1} \left[A_{12} A_{22}^{-1} \left(A_{24} - A_{23} A_{33}^{-1} A_{34} \right) + A_{13} A_{33}^{-1} A_{34} - A_{14} \right] A_{44}^{-1} \; ,$$

$$B_{15} = +A_{11}^{-1} \left[A_{12} A_{22}^{-1} \left(A_{25} - A_{24} A_{44}^{-1} A_{45} - A_{23} A_{33}^{-1} \left(A_{35} - A_{34} A_{44}^{-1} A_{45} \right) \right) + \right.$$

$$\left. + A_{13} A_{33}^{-1} \left(A_{35} - A_{34} A_{44}^{-1} A_{45} \right) + A_{14} A_{44}^{-1} A_{45} - A_{15} \right] A_{55}^{-1} \; ,$$

$$\ldots$$

$$B_{1n} = +A_{11}^{-1} \left[- A_{1n} + \sum_{i=2}^{n-1} A_{1i} A_{ii}^{-1} A_{in} - \sum_{i=2}^{n-2} A_{1i} A_{ii}^{-1} A_{i\,n-1} A_{n-1\,n-1}^{-1} A_{n-1\,n} - \right.$$

$$\left. - \sum_{i=2}^{n-3} A_{1i} A_{ii}^{-1} A_{i\,n-2} A_{n-2\,n-2}^{-1} \left(A_{n-2\,n} - A_{n-2\,n-1} A_{n-1\,n-1}^{1} A_{n-1\,n} \right) - \cdots \right] \; \forall \, n > 2 \; . \tag{B.69}$$

C Elliptic integrals

Elliptic kernel, elliptic modulus, elliptic functions, and elliptic integrals. Differential equations of elliptic functions. Sinus amplitudinis, cosinus amplitudinis, and delta amplitudinis.

We experience *elliptic integrals* when we are trying to compute the length of a meridian arc or the length of a geodesic of an ellipsoid-of-revolution. Here, we begin with an interesting example from circular trigonometry, which is leading us to the notion of *elliptic integrals of the first kind* as well as *elliptic functions*.

C-1 Introductory example

Example C.1 (Elliptic functions).

$$u := \arcsin x \Rightarrow \begin{bmatrix} u' = \dfrac{1}{\sqrt{1-x^2}} \,, \\ u'^2 = \dfrac{1}{1-x^2} \,, \end{bmatrix} \quad u := F(x) := \int_0^x \frac{dx'}{\sqrt{1-x'^2}} = \arcsin x \,. \tag{C.1}$$

End of Example.

C-2 Elliptic kernel, elliptic modulus, elliptic functions, elliptic integrals

Such a well-known formula from circular trigonometry expresses the inverse function of $\sin x$ as an integral. The integral kernel is $1/\sqrt{1-x^2}$. As soon as we switch to elliptic trigonometry, the following "elliptic kernel" appears:

$$\frac{1}{\sqrt{1-x^2}} \frac{1}{\sqrt{1-k^2x^2}} \quad (0 \le k^2 \le 1) \,. \tag{C.2}$$

We here note that the factor k is called the *elliptic modulus* and that the function $x = \operatorname{am}(u,k)$ is called *Jacobian amplitude* within

$$u := F(x,k) := \int_0^x \frac{dx'}{\sqrt{1-x'^2}\sqrt{1-k^2x'^2}} \,, \quad x = F^{-1}(u,k) = \operatorname{am}(u,k) \,. \tag{C.3}$$

By means of the elementary substitution $x = \sin \Phi$, the elliptic integral of the first kind is reduced to the following normal trigonometric form:

$$u := F(\phi,k) := \int_0^\phi \frac{d\phi'}{\sqrt{1-k^2\sin^2\phi'}} \,, \quad \phi = F^{-1}(u,k) = \operatorname{am}(u,k) \,. \tag{C.4}$$

We here additionally note that C. G. J. Jacobi (1804–1851) and N. H. Abel (1802–1829) had the bright idea to replace Legendre's elliptic integral of the first kind by its inverse function. The inverse function is the "simplest elliptic function": see Definition C.1.

Definition C.1 (Elliptic functions).

$$\sin \phi = \sin \operatorname{am}(u,k) =: \operatorname{sn}(u,k) \,,$$
$$\cos \phi = \cos \operatorname{am}(u,k) =: \operatorname{cn}(u,k) \,, \tag{C.5}$$

$$\sqrt{1-k^2\sin^2\phi} = \sqrt{1-k^2\sin^2\operatorname{am}(u,k)} =: \operatorname{dn}(u,k) \,, \tag{C.6}$$

to be read "sinus amplitudinis", "cosinus amplitudinis", and "delta amplitudinis".

End of Definition.

These functions are doubly periodic generalizations of the circular trigonometric functions satisfying

$$\operatorname{sn}(u,0) = \sin u \,,$$
$$\operatorname{cn}(u,0) = \cos u \,, \qquad\qquad (C.7)$$
$$\operatorname{dn}(u,0) = 1 \,.$$

$\operatorname{sn}(u,k)$, $\operatorname{cn}(u,k)$, and $\operatorname{dn}(u,k)$ may also be defined as solutions of the differential equations (C.9), (C.11), and (C.13) of first order or as solutions of the differential equations (C.15), (C.17), and (C.19) of second order : see Lemma C.2 and Lemma C.3.

Lemma C.2 (Differential equations of elliptic functions).

The elliptic functions $\operatorname{sn}(u,k)$, $\operatorname{cn}(u,k)$, and $\operatorname{dn}(u,k)$ satisfy the trigonometric and the algebraic differential equations of first order that follow:

$$x = \operatorname{sn} u \,, \quad \operatorname{sn}'u = \frac{\mathrm{dsn}\, u}{\mathrm{d}u} = \operatorname{cn} u \operatorname{dn} u \,, \qquad\qquad (C.8)$$

$$\left(\frac{\mathrm{d}x}{\mathrm{d}u}\right)^2 = \left(1 - x^2\right)\left(1 - k^2 x^2\right) \,; \qquad\qquad (C.9)$$

$$y = \operatorname{cn} u \,, \quad \operatorname{cn}'u = \frac{\mathrm{dcn}\, u}{\mathrm{d}u} = -\operatorname{sn} u \operatorname{dn} u \,, \qquad\qquad (C.10)$$

$$\left(\frac{\mathrm{d}y}{\mathrm{d}u}\right)^2 = \left(1 - y^2\right)\left(1 - k^2 + k^2 y^2\right) \,; \qquad\qquad (C.11)$$

$$z = \operatorname{dn} u \,, \quad \operatorname{dn}'u = \frac{\mathrm{ddn}\, u}{\mathrm{d}u} = -k^2 \operatorname{sn} u \operatorname{cn} u \,, \qquad\qquad (C.12)$$

$$\left(\frac{\mathrm{d}z}{\mathrm{d}u}\right)^2 = \left(1 - z^2\right)\left[z^2 - \left(1 - k^2\right)\right] \,. \qquad\qquad (C.13)$$

End of Lemma.

Lemma C.3 (Differential equations of elliptic functions).

The elliptic functions $\operatorname{sn}(u,k)$, $\operatorname{cn}(u,k)$, and $\operatorname{dn}(u,k)$ satisfy the algebraic differential equations of second order that follow:

$$x = \operatorname{sn} u \,, \qquad\qquad (C.14)$$

$$x'' = \frac{\mathrm{d}^2 x}{\mathrm{d}u^2} = -(1 + k^2)x + 2k^2 x^3 \,; \qquad\qquad (C.15)$$

$$y = \operatorname{cn} u \,, \qquad\qquad (C.16)$$

$$y'' = \frac{\mathrm{d}^2 y}{\mathrm{d}u^2} = -\left(1 - 2k^2\right) y - 2k^2 y^3 \,; \qquad\qquad (C.17)$$

$$z = \operatorname{dn} u \,, \qquad\qquad (C.18)$$

$$z'' = \frac{\mathrm{d}^2 z}{\mathrm{d}u^2} = +\left(2 - k^2\right) z - 2z^3 \,. \qquad\qquad (C.19)$$

End of Lemma.

Proof (Proof of formulae (C.8) and (C.9)).

$$x := \operatorname{sn} u \, , \ \sqrt{1 - x^2} = \operatorname{cn} u \, , \ \sqrt{1 - k^2 x^2} = \operatorname{dn} u \, , \tag{C.20}$$

$$\frac{\mathrm{d}x}{\mathrm{d}u} = \frac{1}{\mathrm{d}u/\mathrm{d}x} = \sqrt{(1 - x^2)(1 - k^2 x^2)} \, ,$$

$$\frac{\mathrm{d}x}{\mathrm{d}u} = \operatorname{cn} u \, \operatorname{dn} u \quad \mathrm{q. \, e. \, d.} \tag{C.21}$$

$$\left(\frac{\mathrm{d}x}{\mathrm{d}u} \right)^2 = \left(1 - x^2 \right) \left(1 - k^2 x^2 \right) \quad \mathrm{q. \, e. \, d.}$$

End of Proof (Proof of formulae (C.8) and (C.9)).

Proof (Proof of formulae (C.10) and (C.11)).

$$y := \operatorname{cn} u = \sqrt{1 - x^2} \, , \ \operatorname{sn} u = \sqrt{1 - y^2} = x \, , \ \operatorname{dn} u = \sqrt{1 - k^2 (1 - y^2)} = \sqrt{1 - k^2 x^2} \, , \tag{C.22}$$

$$\frac{\mathrm{d}y}{\mathrm{d}u} = \frac{\mathrm{d}\sqrt{1 - x^2}}{\mathrm{d}u} = -\frac{x}{\sqrt{1 - x^2}} \frac{\mathrm{d}x}{\mathrm{d}u} \, ,$$

$$\frac{\mathrm{d}y}{\mathrm{d}u} = -x \sqrt{1 - k^2 x^2} = -\operatorname{sn} u \, \operatorname{dn} u \quad \mathrm{q. \, e. \, d.} \tag{C.23}$$

$$\left(\frac{\mathrm{d}y}{\mathrm{d}u} \right)^2 = \left(1 - y^2 \right) \left(1 - k^2 + k^2 y^2 \right) \quad \mathrm{q. \, e. \, d.}$$

End of Proof (Proof of formulae (C.10) and (C.11)).

Proof (Proof of formulae (C.12) and (C.13)).

$$z := \operatorname{dn} u = \sqrt{1 - k^2 x^2} \, , \ \operatorname{sn} u = \frac{\sqrt{1 - z^2}}{k} = x \, , \ \operatorname{cn} u = \frac{\sqrt{k^2 - (1 - z^2)}}{k} = \sqrt{1 - x^2} \, , \tag{C.24}$$

$$\frac{\mathrm{d}z}{\mathrm{d}u} = \frac{\mathrm{d}\sqrt{1 - k^2 x^2}}{\mathrm{d}u} = -\frac{k^2 x}{\sqrt{1 - k^2 x^2}} \frac{\mathrm{d}x}{\mathrm{d}u} \, ,$$

$$\frac{\mathrm{d}z}{\mathrm{d}u} = -k^2 \operatorname{sn} u \, \operatorname{cn} u \quad \mathrm{q. \, e. \, d.}$$

$$\tag{C.25}$$

$$\frac{\mathrm{d}z}{\mathrm{d}u} = -k^2 x \sqrt{1 - x^2} = -\sqrt{1 - z^2} \sqrt{k^2 - (1 - z^2)} \, ,$$

$$\left(\frac{\mathrm{d}z}{\mathrm{d}u} \right)^2 = \left(1 - z^2 \right) \left[z^2 - \left(1 - k^2 \right) \right] \quad \mathrm{q. \, e. \, d.}$$

End of Proof (Proof of formulae (C.12) and (C.13)).

$$x' = \text{sn}'u = \text{cn}\,u\,\text{dn}\,u$$

$$\Rightarrow$$

$$x'' = \text{sn}''u = \text{cn}'u\,\text{dn}\,u + \text{cn}\,u\,\text{dn}'u\,,$$

$$x'' = -\text{sn}\,u\,\text{dn}^2u - k^2\text{sn}\,u\,\text{cn}^2u\,,$$

$$x'' = -x\left(1 - k^2x^2\right) - k^2x\left(1 - x^2\right) = -x + k^2x^3 - k^2x + k^2x^3\,,$$

$$x'' = -x\left(1 + k^2\right) + 2k^2x^3 \quad \text{q.e.d.}$$

(C.26)

$$y' = \text{cn}'u = -\text{sn}\,u\,\text{dn}\,u$$

$$\Rightarrow$$

$$y'' = \text{cn}''u = -\text{sn}'u\,\text{dn}\,u - \text{sn}\,u\,\text{dn}'u\,,$$

$$y'' = -\text{cn}\,u\,\text{dn}^2u + k^2\text{sn}^2u\,\text{cn}\,u\,,$$

$$y'' = -y\left[1 - k^2\left(1 - y^2\right)\right] + k^2y\left(1 - y^2\right) = -y + k^2y\left(1 - y^2\right) + k^2y\left(1 - y^2\right)\,,$$

$$y'' = -y + 2k^2y\left(1 - y^2\right) = -y + 2k^2y - 2k^2y^3\,,$$

$$y'' = -y\left(1 - 2k^2\right) - 2k^2y^3 \quad \text{q.e.d.}$$

(C.27)

$$z' = \text{dn}'u = -k^2\text{sn}\,u\,\text{cn}\,u$$

$$\Rightarrow$$

$$z'' = -k^2\text{sn}'u\,\text{cn}\,u - k^2\text{sn}\,u\,\text{cn}'u\,,$$

$$z'' = -k^2\text{cn}^2u\,\text{dn}\,u + k^2\text{sn}^2u\,\text{dn}\,u\,,$$

$$z'' = -\left[k^2 - \left(1 - z^2\right)\right]z + \left(1 - z^2\right)z = -k^2z + 2z\left(1 - z^2\right)\,,$$

$$z'' = +\left(2 - k^2\right)z - 2z^3 \quad \text{q.e.d.}$$

(C.28)

The standard elliptic functions satisfy special identities collected in Corollary C.4, where $1 - k^2$ is the *complementary elliptic modulus*. Similarly, we summarize addition formulae of elliptic functions in Corollary C.5.

Corollary C.4 (Special identities).

$$\text{sn}^2 u + \text{cn}^2 u = 1 \ , \ k^2 \text{sn}^2 u + \text{dn}^2 u = 1 \ ,$$

$$k^2 \text{cn}^2 u + \left(1 - k^2\right) = \text{dn}^2 u \ , \ \text{cn}^2 u + \left(1 - k^2\right) \text{sn}^2 u = \text{dn}^2 u \ .$$

(C.29)

End of Corollary.

Corollary C.5 (Addition formulae).

$$\text{sn}(u + v) = \frac{\text{sn}\,u \ \text{cn}\,v \ \text{dn}\,v + \text{sn}\,v \ \text{sn}\,u \ \text{dn}\,u}{1 - k^2 \text{sn}^2 u \ \text{sn}^2 v} \ ,$$

$$\text{cn}(u + v) = \frac{\text{cn}\,u \ \text{cn}\,v - \text{sn}\,u \ \text{sn}\,v \ \text{dn}\,u \ \text{dn}\,v}{1 - k^2 \text{sn}^2 u \ \text{sn}^2 v} \ ,$$

(C.30)

$$\text{dn}(u + v) = \frac{\text{dn}\,u \ \text{dn}\,v - k^2 \text{sn}\,u \ \text{sn}\,v \ \text{cn}\,u \ \text{cn}\,v}{1 - k^2 \text{sn}^2 u \ \text{sn}^2 v} \ .$$

End of Corollary.

Series expansions of "sinus amplitudinis", "cosinus amplitudinis", and "delta amplitudinis" are useful in Mathematical Cartography. In the following Corollary C.6, series expansions of "sinus amplitudinis", "cosinus amplitudinis", and "delta amplitudinis" are summarized.

Corollary C.6 (Series expansions of elliptic functions).

$$\text{sn}\,(u, k) = u - \frac{1}{6}\left(1 + k^2\right) u^3 + \frac{1}{120}\left(1 + 14k^2 + k^4\right) u^5 + \text{O}(u^7) \ ,$$

$$\text{cn}\,(u, k) = 1 - \frac{1}{2}u^2 + \frac{1}{24}\left(1 - 4k^2\right) u^4 - \frac{1}{720}\left(1 + 44k^2 + 16k^4\right) u^6 + \text{O}(u^8) \ ,$$

(C.31)

$$\text{dn}\,(u, k) = 1 - \frac{1}{2}k^2 u^2 + \frac{1}{24}\left(4k^2 + k^4\right) u^4 - \frac{1}{720}\left(16k^2 + 44k^3 + k^6\right) u^6 + \text{O}(u^8) \ .$$

End of Corollary.

In analysis, we have been made familiar with integrals of type $\int R\left(x, \sqrt{ax^2 + 2bx + c}\right)dx$, whose kernel is a square root of a polynomial up to degree two like $\int dx/x$, $\int dx/(1 + x^2)$, or $\int dx/\sqrt{1 - x^2}$. Those integrals can be integrated, for example, to $\ln x$, $\arctan x$, or $\arcsin x$. Alternatively, if the integral is of the form $\int R\left(x, \sqrt{a_4 x^4 + a_3 x^3 + a_2 x^2 + a_1 x + a_0}\right)dx$ ($a_3 \neq 0$, $a_4 \neq 0$ admitted), we arrive at elliptic integrals whose kernel is a square root of a polynomial up to degree four, with distinct modal points. Definition C.7 is a collective summary of elliptic integrals of the first, second, and third kinds given in both polynomial form and trigonometric form. In general, integrals of the following form are needed:

$$I = \int R\left(x, \sqrt{a_J x^J + a_{J-1} x^{J-1} + \ldots + a_1 x + a_0}\right) dx \quad (J \geq 5) \ .$$

(C.32)

Definition C.7 (Elliptic integrals of the first, second, and third kinds).

The following integrals $F(x, k)$, $E(x, k)$, and $\pi(x, k)$ are called *normal elliptic integrals* of the first, the second, and the third kind:

$$F(x, k) := \int_0^x \frac{\mathrm{d}x'}{\sqrt{\left(1 - x'^2\right)\left(1 - k^2 x'^2\right)}} \ ,$$

$$E(x, k) := \int_0^x \frac{\sqrt{1 - k^2 x'^2}\,\mathrm{d}x'}{\sqrt{\left(1 - x'^2\right)}} \ , \tag{C.33}$$

$$\pi(x, k) := \int_0^x \frac{\mathrm{d}x'}{\left(1 + n x'^2\right)\sqrt{\left(1 - x'^2\right)\left(1 - k^2 x'^2\right)}} \ .$$

The following integrals $F(\phi, k)$, $E(\phi, k)$, and $\pi(\phi, k)$ are called *trigonometric elliptic integrals* of the first, the second, and the third kind:

$$F(\phi, k) := \int_0^\phi \frac{\mathrm{d}\phi'}{\sqrt{1 - k^2 \sin^2 \phi'}} \ ,$$

$$E(\phi, k) := \int_0^\phi \sqrt{1 - k^2 \sin^2 \phi'}\,\mathrm{d}\phi' = \int_0^u \mathrm{dn}^2(u', k)\,\mathrm{d}u' \ , \tag{C.34}$$

$$\pi(\phi, k) := \int_0^\phi \frac{\mathrm{d}\phi'}{1 + n\sin^2 \phi' \sqrt{1 + k^2 \sin^2 \phi'}} \ .$$

Note that for $\phi = \pi/2$, i.e. $F(\pi/2, k)$, $E(\pi/2, k)$, and $\pi(\pi/2, k)$, these trigonometric elliptic integrals are called *complete*.

End of Definition.

For the numerical analysis of elliptic functions, the periodicity of $\mathrm{sn}\,u$, $\mathrm{cn}\,u$, and $\mathrm{dn}\,u$ is of focal interest. Lemma C.8 defines the periodic properties of $\mathrm{sn}\,u$, $\mathrm{cn}\,u$, and $\mathrm{dn}\,u$ more precisely. Note that a proof can be based upon the addition theorem of elliptic functions.

Lemma C.8 (Periodicity of elliptic functions: C. G. J. Jacobi).

The elliptic functions $\mathrm{sn}\,u$, $\mathrm{cn}\,u$, and $\mathrm{dn}\,u$ are doubly periodic in the sense of ($n, m \in \{1, 2, 3, 4, 5 \ldots\}$)

$$\mathrm{sn}\,(u + 4mK + 2niK') = \mathrm{sn}\,u \ ,$$

$$\mathrm{cn}\,[u + 4mK + 2n(K + iK')] = \mathrm{cn}\,u \ , \tag{C.35}$$

$$\mathrm{dn}\,(u + 2mK + 4niK') = \mathrm{dn}\,u \ ,$$

subject to

$$K = \int_0^{\pi/2} \frac{\mathrm{d}\phi}{\sqrt{1 - k^2 \sin^2 \phi}} \ , \quad K' = \int_0^{\pi/2} \frac{\mathrm{d}\phi}{\sqrt{1 - (1 - k^2)\sin^2 \phi}} \ . \tag{C.36}$$

End of Lemma.

These notes on elliptic functions and elliptic integrals are by no means sufficient to get some insight into these special functions. For numerical calculations based upon the *Landen transformation*, we refer to R. Bulirsch (1965a,b,1969a,b), M. Gerstl (1984), and A. Kneser (1928). General references are M. Abramowitz and J. A. Stegun (1972), R. Ayoub (1984), R. Bellman (1961), R. Cooke (1994), D. Dumont (1981), R. Fricke (1913), P. M. Porter (1989), W. H. Press et al. (1992), M. Rosen (1981), A. Schett (1977), E. I. Slavutin (1973), L. A. Sorokina (1983), J. Spanier and K. B. Oldham (1987), J. Stillwell (1989), F. Toelke (1967), and F. Tricomi (1948).

D Korn–Lichtenstein and d'Alembert–Euler equations

Conformal mapping, Korn–Lichtenstein equations and d'Alembert–Euler (Cauchy–Riemann) equations. Polynomial solutions. Conformeomorphism, condition of conformality.

D-1 Korn–Lichtenstein equations

Our starting point for the construction of a *conformal diffeomorphism* (in short *conformeomorphism*) is provided by (D.1) of a left two-dimensional Riemann manifold $\{\mathbb{M}^2, \mathsf{G}_l\}$ subject to the left metric $\mathsf{G}_l = \{G_{MN}\}$ parameterized by the left coordinates $\{U, V\} = \{U^1, U^2\}$ and a right two-dimensional Euclidean manifold $\{\mathbb{M}^2, \mathsf{G}_r = \mathsf{I}_2\}$ subject to the right metric $\mathsf{G}_r = \mathsf{I}_2 = \mathrm{diag}\,[1, 1]$ or $\{g_{\mu\nu}\} = \{\delta_{\mu\nu}\}$ parameterized by the right coordinates $\{x, y\} = \{x^1, x^2\}$ of Cartesian type.

$$\begin{bmatrix} \mathrm{d}x \\ \mathrm{d}y \end{bmatrix} = \mathsf{J}_l \begin{bmatrix} \mathrm{d}U \\ \mathrm{d}V \end{bmatrix} , \quad \mathsf{J}_l = \begin{bmatrix} x_U & x_V \\ y_U & y_V \end{bmatrix} . \tag{D.1}$$

J_l constitutes the left Jacobi matrix of first partial derivatives, which is related to the right Jacobi matrix J_r by means of the duality $\mathsf{J}_l \mathsf{J}_r = \mathsf{I}_2$ or $\mathsf{J}_l = \mathsf{J}_r^{-1}$ and $\mathsf{J}_r = \mathsf{J}_l^{-1}$ with $\mathsf{J}_l \in \mathbb{R}^{2 \times 2}$ and $\mathsf{J}_r \in \mathbb{R}^{2 \times 2}$. Let us compare the left and the right symmetric metric forms, the squared infinitesimal arc lengths (D.2) subject to the summation convention over repeated indices which run here from one to two, i.e. $\{M, N, \mu, \nu\} \in \{1, 2\}$. (D.3) constitutes the right Cauchy–Green deformation tensor which has to be constraint to the condition of conformality (D.4).

$$\mathrm{d}S^2 = \mathrm{d}U^M G_{MN} \mathrm{d}U^N = \mathrm{d}x^\mu \frac{\partial U^M}{\partial x^\mu} G_{MN} \frac{\partial U^N}{\partial x^\nu} = \qquad \text{versus} \qquad \begin{aligned} \mathrm{d}s^2 &= \mathrm{d}x^2 + \mathrm{d}y^2 = \\ &= \mathrm{d}x^\mu \delta_{\mu\nu} \mathrm{d}x^\nu , \end{aligned} \tag{D.2}$$
$$= \mathrm{d}x^\mu C_{\mu\nu} \mathrm{d}x^\nu$$

$$C_{\mu\nu} := \frac{\partial U^M}{\partial x^\mu} G_{MN} \frac{\partial U^N}{\partial x^\nu} \quad \text{or} \quad \mathsf{C}_r := \mathsf{J}_r^{\mathrm{T}} \mathsf{G}_l \mathsf{J}_r , \tag{D.3}$$

$$C_{\mu\nu} = \lambda^2 \delta_{\mu\nu} \quad \text{or} \quad \mathsf{C}_r = \lambda^2 \mathsf{I}_2 . \tag{D.4}$$

Consequently, the left symmetric metric form (D.5) enjoys a particular structure which is called *conformally flat*. The factor of conformality λ^2 is generated by the simultaneous diagonalization of the matrix pair $\{\mathsf{C}_r, \mathsf{G}_r\}$, namely from the general eigenvalue–eigenvector problem $(\mathsf{C}_r - \lambda^2 \mathsf{G}_r)\mathbf{F}_r = 0$, the characteristic equation $|\mathsf{C}_r - \lambda^2 \mathsf{G}_r| = 0$ subject to $\mathsf{G}_r = \mathsf{I}_2$ and the canonical conformality postulate $\lambda_1^2 = \lambda_2^2 = \lambda^2$. The condition of conformality transforms the right Cauchy–Green deformation tensor into (D.6).

$$\mathrm{d}S^2 = \lambda^2 (\mathrm{d}x^2 + \mathrm{d}y^2) , \tag{D.5}$$

$$\mathsf{C}_r = \mathsf{J}_r^{\mathrm{T}} \mathsf{G}_l \mathsf{J}_r = \lambda^2 \mathsf{I}_2 \quad \text{or} \quad \mathsf{C}_r^{-1} = \mathsf{J}_l \mathsf{G}_l^{-1} \mathsf{J}_l^{\mathrm{T}} = \frac{1}{\lambda^2} \mathsf{I}_2 . \tag{D.6}$$

(D.6) can be interpreted as an orthogonality condition of the rows of the left Jacobi matrix with respect to the inverse left metric matrix G_l^{-1}. G_l^{-1}-orthogonality of the rows of the left Jacobi matrix J_l implies (D.7). (D.7) be derived from (D.8), namely with respect to the permutation symbol (D.9).

$$\mathrm{d}x = \star\mathrm{d}y \,, \tag{D.7}$$

$$\mathrm{d}x = x_U\mathrm{d}U + x_V\mathrm{d}V = x_I\mathrm{d}U^I \,,$$

$$x_1 := \frac{\partial x}{\partial U} \,, \quad x_2 := \frac{\partial x}{\partial V} \,,$$

$$\star\mathrm{d}y := e_{IJ}\sqrt{\det[\mathsf{G}_l]}\mathsf{G}^{JK}y_K\mathrm{d}U^I \ \forall \ \{I,J,K\} \in \{1,2\} \,, \tag{D.8}$$

$$y_1 := \frac{\partial y}{\partial U} \,, \quad y_2 := \frac{\partial y}{\partial V} \,,$$

$$e_{IJ} = \begin{cases} +1 & \text{even permutation of the indices} \\ -1 & \text{odd permutation of the indices} \\ 0 & \text{otherwise} \end{cases} , \tag{D.9}$$

$$\mathrm{d}x := x_I\mathrm{d}U^I =$$

$$= e_{IJ}\sqrt{\det[\mathsf{G}_l]}\mathsf{G}^{JK}y_K\mathrm{d}U^I = \star\mathrm{d}y \Leftrightarrow \frac{\partial x}{\partial U^I} = e_{IJ}\sqrt{\det[\mathsf{G}_l]}\mathsf{G}^{JK}\frac{\partial y}{\partial U^K} \,. \tag{D.10}$$

Indeed, we have to take advantage of the *Hodge star operator*, which generalizes the cross product on $\mathbb{R}^3$. Lemma D.1 outlines the definition of the Hodge star operator of a one-differential form. It may be a surprise that (D.10) constitutes the Korn–Lichtenstein equations of a conformal mapping parameterized by $\{x(U,V), y(U,V)\}$. Note the representation (D.11) of the inverse left metric matrix in order to derive the third version of the Korn–Lichtenstein equations in Lemma D.1.

$$\mathsf{G}_l = G_{IJ} = \begin{bmatrix} G_{11} & G_{12} \\ G_{21} & G_{22} \end{bmatrix}$$

$$(\text{subject to } G_{21} = G_{12})$$

$$\Leftrightarrow \tag{D.11}$$

$$\mathsf{G}_l^{-1} = G^{JK} = \begin{bmatrix} G^{11} & G^{12} \\ G^{21} & G^{22} \end{bmatrix} = \frac{1}{\det[\mathsf{G}_l]}\begin{bmatrix} G_{22} & -G_{12} \\ -G_{12} & G_{11} \end{bmatrix} \,.$$

Lemma D.1 (Conformeomorphism, $\mathbb{M}_l^2 \mapsto \mathbb{M}_r^2 = \mathbb{E}_r^2$, Korn–Lichtenstein equations).

Equivalent formulations of the Korn–Lichtenstein equations which produce a conformal mapping $\mathbb{M}_l^2 \mapsto \mathbb{M}_r^2 = \mathbb{E}_r^2$ are the following:

$$\mathrm{d}x = \star\mathrm{d}y \,, \tag{D.12}$$

$$\frac{\partial x}{\partial U^I} = e_{IJ}\sqrt{\det[\mathsf{G}_l]}\mathsf{G}^{JK}\frac{\partial y}{\partial U^K} \,, \tag{D.13}$$

$$x_U = \frac{1}{\sqrt{\det[\mathsf{G}_l]}}\left(-G_{12}y_U + G_{11}y_V\right) \,,$$

$$x_V = \frac{1}{\sqrt{\det[\mathsf{G}_l]}}\left(-G_{22}y_U + G_{12}y_V\right) \,. \tag{D.14}$$

End of Lemma

Generalizations to a conformeomorphism of higher order, namely $\mathbb{M}_l^3 \mapsto \mathbb{M}_r^3 = \mathbb{E}_r^3$, lead to the Zund equations, and its generalizations $\mathbb{M}_l^n \mapsto \mathbb{M}_r^n = \mathbb{E}_r^n$ are referred to E. Grafarend and R. Syffus (1998d).

D-2 D'Alembert–Euler (Cauchy–Riemann) equations

Once an *isometric coordinate system* of a two-dimensional *Riemann manifold* $\{\mathbb{M}^2, \lambda^{-2}\delta_{\mu\nu}\}$ ("surface")
has been established, an *alternative isometric coordinate system* can be constructed by solving a
boundary value problem of the *d'Alembert–Euler equations* (*Cauchy–Riemann equations*) subject to
the integrability conditions of harmonicity type. Here, we are going to construct fundamental solutions
of these basic equations governing conformal mapping.

> **Lemma D.2 (Fundamental solution of the d'Alembert–Euler equations subject to integrability conditions
> of harmonicity, polynomial representation).**

A fundamental solution of the d'Alembert–Euler equations (Cauchy–Riemann equations) subject to
the integrability conditions of harmonicity type is

$$x = \alpha_0 + \alpha_1 q + \beta_1 p +$$

$$+ \sum_{r=2}^{n} \alpha_r \sum_{s=0}^{[r/2]} (-1)^s \binom{r}{2s} q^{r-2s} p^{2s} + \sum_{r=2}^{N} \beta_r \sum_{s=1}^{[(r+1)/2]} (-1)^{s+1} \binom{r}{2s-1} q^{r-2s+1} p^{2s-1}, \tag{D.15}$$

$$y = \beta_0 + \beta_1 q + \alpha_1 p +$$

$$+ \sum_{r=2}^{n} \beta_r \sum_{s=0}^{[r/2]} (-1)^s \binom{r}{2s} q^{r-2s} p^{2s} + \sum_{r=2}^{N} \alpha_r \sum_{s=1}^{[(r+1)/2]} (-1)^{s+1} \binom{r}{2s-1} q^{r-2s+1} p^{2s-1}. \tag{D.16}$$

> **End of Lemma.**

> **Proof.**

Once the fundamental solution of the d'Alembert-Euler equations is based on the function space of
homogeneous polynomials

$$P_r(q,p) = \sum_{\alpha+\beta=r} c_{\alpha\beta} q^\alpha p^\beta \ (0 \le \alpha \le r), \tag{D.17}$$

the vectorial Laplace-Beltrami equation has to be fulfilled. $(r-1)$ constraints are given for $(r+1)$
coefficients such that for any $r \ge 2$ two linear independent harmonic polynomials exist which we are
going to construct.

$$P_r(q,p) = c_{r,0}q^r + c_{r-1,1}q^{r-1}p + c_{r-2,2}q^{r-2}p^2 + c_{r-3,3}q^{r-3}p^3 + c_{r-4,4}q^{r-4}p^4 + \cdots$$

$$\cdots + c_{3,r-3}q^3 p^{r-3} + c_{2,r-2}q^2 p^{r-2} + c_{1,r-1}qp^{r-1} + c_{0,r}p^r, \tag{D.18}$$

$$\Delta P_r(q,p) = (\tfrac{\partial^2}{\partial q^2} + \tfrac{\partial^2}{\partial p^2})P_r(q,p) =$$

$$= r(r-1)c_{r,0}q^{r-2} + (r-2)(r-3)c_{r-2,2}q^{r-4}p^2 + (r-4)(r-5)c_{r-4,4}q^{r-6}p^4 + \cdots$$

$$\cdots + 2c_{r-2,2}q^{r-2} + 4 \cdot 3c_{r-4,4}q^{r-4}p^4 + \cdots \tag{D.19}$$

$$\cdots + (r-1)(r-2)c_{r-1,1}q^{r-3}p + (r-3)(r-4)c_{r-3,3}q^{r-5}p^3 + \cdots$$

$$\cdots + 3 \cdot 2c_{r-3,3}q^{r-3}p + \cdots .$$

Obviously, the recurrence relation (D.20) connects the coefficients of even second index with each other, similarly the coefficients of odd second index according to the following set of coefficient pairs: $c_{r,0}|c_{r-2,2}$, $c_{r-1,1}|c_{r-3,3}$, $c_{r-2,2}|c_{r-4,4}$, $c_{r-3,3}|c_{r-5,5}$ etc.

$$c_{k,r-k}k(k-1) + c_{k-2,r-k+2}(r-k+2)(r-k+1) = 0$$
$$\Leftrightarrow$$
$$c_{k-2,r-k+2} = -\frac{c_{k,r-k}k(k-1)}{(r-k+2)(r-k+1)}. \tag{D.20}$$

Those harmonic polynomials with coefficients of even second index are denoted by $P_{r,1}$, while alternatively those with coefficients of odd second index are denoted by $P_{r,2}$. Once we chose $c_{r,0} = 1$, the recurrence relation leads to (D.21) or (D.22). In contrast, once we choose $c_{r-1,1} = r$ we are led to (D.23) or (D.24).

$$P_{r,1}(q,p) = q^r - \frac{r(r-1)}{2}q^{r-2}p^2 + \frac{(r-3)(r-2)(r-1)}{4*3*2}q^{r-4}p^4 + \cdots, \tag{D.21}$$

$$P_{r,1}(q,p) = \sum_{s=0}^{[\frac{r}{2}]}(-1)^s \binom{r}{2s}q^{r-2s}p^{2s}, \tag{D.22}$$

$$P_{r,2}(q,p) = rq^{r-1}p - \frac{(r-2)(r-1)r}{3*2}q^{r-3}p^3 + \cdots, \tag{D.23}$$

$$P_{r,2}(q,p) = \sum_{s=1}^{[\frac{r+1}{2}]}(-1)^{s+1} \binom{r}{2s-1}q^{r-(2s-1)}p^{2s-1}. \tag{D.24}$$

$[\frac{r}{2}]$ denotes the largest natural number $\leq \frac{r}{2}$, $[(r+1)/2]$ the largest natural number $\leq \frac{r+1}{2}$, respectively. In summarizing, the general solution of the first Laplace–Beltrami equation is given by (D.25).

$$x = \alpha_0 + \alpha_1 q + \beta_1 p + \sum_{r=2}^{N}\alpha_r \sum_{s=0}^{[\frac{r}{2}]}(-1)^s \binom{r}{2s}q^{r-2s}p^{2s} + \sum_{r=2}^{N}\beta_r \sum_{s=1}^{[\frac{r+1}{2}]}(-1)^{s+1}\binom{r}{2s-1}q^{r-2s+1}p^{2s-1}. \tag{D.25}$$

Next, we implement the terms $x = P_{r,1}(q,p)$ and $x = P_{r,2}(q,p)$ in the d'Alembert–Euler equations (Cauchy–Riemann equations) (D.26). Obviously, there hold the polynomial relations (D.27).

$$\frac{\partial P_{r,1}}{\partial q} = \sum_{s=0}^{[\frac{r}{2}]}(-1)^s(r-2s)\binom{r}{2s}q^{r-2s-1}p^{2s} = \sum_{s=0}^{[\frac{r+1}{2}-1]}(-1)^s(r-2s)\binom{r}{2s}q^{r-2s-1}p^{2s},$$

$$\frac{\partial P_{r,1}}{\partial p} = \sum_{s=0}^{[\frac{r}{2}]}(-1)^s 2s\binom{r}{2s}q^{r-2s}p^{2s-1} = \sum_{s=1}^{[\frac{r}{2}]}(-1)^s 2s\binom{r}{2s}q^{r-2s}p^{2s-1},$$

$$\frac{\partial P_{r,2}}{\partial q} = \sum_{s=1}^{[\frac{r+1}{2}]}(-1)^{s+1}(r-2s+1)\binom{r}{2s-1}q^{r-2s}p^{2s-1} = \sum_{s=1}^{[\frac{r}{2}]}(-1)^{s+1}2s\binom{r}{2s}q^{r-2s}p^{2s-1}, \tag{D.26}$$

$$\frac{\partial P_{r,2}}{\partial p} = \sum_{s=1}^{[\frac{r+1}{2}]}(-1)^{s+1}(2s-1)\binom{r}{2s-1}q^{r-2s+1}p^{2s-2} = \sum_{s'=0}^{[\frac{r+1}{2}]-1}(-1)^{s'}(r-2s')\binom{r}{2s'}q^{r-2s'-1}p^{2s'},$$

$$\frac{\partial P_{r,1}}{\partial q} = \frac{\partial P_{r,2}}{\partial p}, \quad \frac{\partial P_{r,1}}{\partial p} = -\frac{\partial P_{r,2}}{\partial q}. \tag{D.27}$$

The general solution of the d'Alembert–Euler equations (Cauchy–Riemann equations) subject to the integrability conditions of the harmonicity type thus can be represented by

$$x = \alpha_0 + \alpha_1 q + \beta_1 p + \alpha_2(q^2 - p^2) + 2\beta_2 qp +$$

$$+ \sum_{r=3}^{N} \alpha_r \sum_{s=0}^{[\frac{r}{2}]} (-1)^s \binom{r}{2s} q^{r-2s} p^{2s} +$$

$$+ \sum_{r=3}^{N} \beta_r \sum_{s=1}^{[\frac{r+1}{2}]} (-1)^{s+1} \binom{r}{2s-1} q^{r-2s+1} p^{2s-1} ,$$

(D.28)

$$y = \beta_0 - \beta_1 q + \alpha_1 p - \beta_2(q^2 - p^2) + 2\alpha_2 qp -$$

$$- \sum_{r=3}^{N} \beta_r \sum_{s=0}^{[\frac{r}{2}]} (-1)^s \binom{r}{2s} q^{r-2s} p^{2s} +$$

$$+ \sum_{r=3}^{N} \alpha_r \sum_{s=1}^{[\frac{r+1}{2}]} (-1)^{s+1} \binom{r}{2s-1} q^{r-2s+1} p^{2s-1} .$$

(D.29)

It should be mentioned that the fundamental solution is *not* in the class of separation of variables, namely of type $f(q)g(p)$. Accordingly, we present an alternative fundamental solution of the d'Alembert–Euler equations (Cauchy–Riemann equations) subject to the integrability conditions of the harmonicity type now in the class of *separation of variables*.

End of Proof.

The fundamental solution (D.15), (D.16), (D.28), and (D.29) of the equations which govern conformal mapping of type isometric cha–cha–cha can be interpreted as following. In matrix notation, namely based upon the Kronecker–Zehfuss product, we write

$$\begin{bmatrix} x \\ y \end{bmatrix} = \begin{bmatrix} \alpha_0 \\ \beta_0 \end{bmatrix} + (\alpha_1 \mathsf{I} + \beta_1 \mathsf{A}) \begin{bmatrix} q \\ p \end{bmatrix} +$$

$$+ \left[\operatorname{vec} \begin{bmatrix} \alpha_2 & \beta_2 \\ \beta_2 & -\alpha_2 \end{bmatrix}, \operatorname{vec} \begin{bmatrix} -\beta_2 & \alpha_2 \\ \alpha_2 & -\beta_2 \end{bmatrix} \right]' \begin{bmatrix} q \\ p \end{bmatrix} \otimes \begin{bmatrix} q \\ p \end{bmatrix} + \mathsf{O}_3 ,$$

(D.30)

where we identify the transformation group of motion (translation (α_0, β_0), rotation β_1, in total three parameters), the transformation group of dilatation (one parameter α_1) and the special-conformal transformation (two parameters (α_2, β_2)), actually the six-parameter $O(2, 2)$ sub-algebra of the infinite dimensional conformal algebra $C(\infty)$ in two dimensions $\{q, p\} \in \mathbb{R}^2$. We here note in passing that the "small rotation parameter" β_1 operates on the antisymmetric matrix (D.31), while the matrices $\{\mathsf{H}^1, \mathsf{H}^2\}$, which generate the special-conformal transformation are traceless and symmetric, a property being enjoyed by all coefficient matrices of conformal transformations of higher order. A more detailed information is D. G. Boulware, L. S. Brown and R. D. Peccei (1970) and S. Ferrara, A. F. Grillo and R. Gatto (1972).

$$\mathsf{A} := \begin{bmatrix} 0 & 1 \\ -1 & 0 \end{bmatrix} ,$$

(D.31)

$$\mathsf{H}^1 := \begin{bmatrix} \alpha_2 & \beta_2 \\ \beta_2 & -\alpha_2 \end{bmatrix} , \quad \mathsf{H}^2 := \begin{bmatrix} -\beta_2 & \alpha_2 \\ \alpha_2 & \beta_2 \end{bmatrix} .$$

Lemma D.3 (Fundamental solution of the d'Alembert-Euler equations subject to integrability conditions of harmonicity, separation of variables).

A fundamental solution of the d'Alembert–Euler equations (Cauchy–Riemann equations) subject to the integrability conditions of harmonicity type in the class of *separation of variables* is

$$x(q,p) = x_0 + \sum_{m=1}^{M} [A_m \exp(mq) + C_m \exp(-mq)] \cos mp +$$
$$+ \sum_{m=1}^{M} [B_m \exp(mq) + D_m \exp(-mq)] \sin mp , \tag{D.32}$$

$$y(q,p) = y_0 + \sum_{m=1}^{M} [B_m \exp(mq) - D_m \exp(-mq)] \cos mp +$$
$$+ \sum_{m=1}^{M} [-A_m \exp(mq) + C_m \exp(-mq)] \sin mp . \tag{D.33}$$

End of Lemma.

Proof.

By *separation-of-variables*, namely by using $x(q,p) = f(q)g(p)$ and $y(q,p) = F(q)G(p)$, the vectorial Laplace–Beltrami equation leads to (D.34) and to similar equations for $F(q)$ and $G(p)$.

$$\frac{f''}{f} + \frac{g''}{g} = 0 \Rightarrow \frac{f''}{f} = -\frac{g''}{g} =: m^2 \Rightarrow f'' = m^2 f$$
$$\Rightarrow \tag{D.34}$$
$$f = \bar{c}_m \exp(mq) + \bar{d}_m \exp(-mq) \Rightarrow g'' = -m^2 g \Rightarrow g = \bar{a}_m \cos mp + \bar{b}_m \sin mp .$$

Superposition of base functions gives the setups (D.35) and (D.36). The d'Alembert–Euler equations (Cauchy–Riemann equations) $x_p = y_q$ and $x_q = -y_p$ then are specified by (D.37) and (D.38).

$$x(q,p) = \sum_{m=1}^{M} \Big[A_m \exp(mq) \cos mp + B_m \exp(mq) \sin mp +$$
$$+ C_m \exp(-mq) \cos mp + D_m \exp(-mq) \sin mp \Big] + x_0 , \tag{D.35}$$

$$y(q,p) = \sum_{m'=1}^{M} \Big[A'_{m'} \exp(m'q) \cos m'p + B'_{m'} \exp(m'q) \sin m'p +$$
$$+ C'_{m'} \exp(-m'q) \cos m'p + D'_{m'} \exp(-m'q) \sin m'p \Big] + y_0 , \tag{D.36}$$

$$x_q = \sum_{m=1}^{M} \Big[mA_m \exp(mq) \cos mp + mB_m \exp(mq) \sin mp -$$
$$- mC_m \exp(-mq) \cos mp - mD_m \exp(-mq) \sin mp \Big] , \tag{D.37}$$

$$y_p = \sum_{m'=1}^{M} \Big[- m' A'_{m'} \exp(m'q) \sin m'p + m' B'_{m'} \exp(m'q) \cos m'p -$$
$$- m' C'_{m'} \exp(-m'q) \sin m'p + m' D'_{m'} \exp(-m'q) \cos m'p \Big] \tag{D.38}$$
$$\Leftrightarrow$$
$$A'_{m'} = B_m , \quad B'_{m'} = -A_m , \quad C'_{m'} = -D_m , \quad D'_{m'} = C_m .$$

End of Proof.

Lemma D.4 (Fundamental solution of the d'Alembert-Euler equations subject to integrability conditions of harmonicity, separation of variables).

A fundamental solution of the d'Alembert-Euler equations (Cauchy-Riemann equations) subject to the integrability conditions of harmonicity type in the class of *separation of variables* is

$$x(q,p) =$$

$$= x_0 + \sum_{m=1}^{M} [I_m \cosh mq + K_m \sinh mq] \cos mp + \sum_{m=1}^{M} [J_m \cosh mq + L_m \sinh mq] \sin mp , \tag{D.39}$$

$$y(q,p) =$$

$$= y_0 + \sum_{m=1}^{M} [L_m \cosh mq + J_m \sinh mq] \cos mp + \sum_{m=1}^{M} [-K_m \cosh mq - I_m \sinh mq] \sin mp . \tag{D.40}$$

End of Lemma.

Proof.

By *separation-of-variables*, namely by using $x(q,p) = f(q)g(p)$ and $y(q,p) = F(q)G(p)$, the vectorial Laplace–Beltrami equation leads to (D.34) and to similar equations for $F(q)$ and $G(p)$.

$$\frac{f''}{f} + \frac{g''}{g} = 0 \Rightarrow \frac{f''}{f} = -\frac{g''}{g} =: m^2 \Rightarrow f'' = m^2 f$$

$$\Rightarrow \tag{D.41}$$

$$f = \bar{k}_m \cosh mq + \bar{l}_m \sinh mq \Rightarrow g'' = -m^2 g \Rightarrow g = \bar{i}_m \cos mp + \bar{j}_m \sin mp .$$

Superposition of base functions gives the setups (D.42) and (D.43). The d'Alembert–Euler equations (Cauchy–Riemann equations) $x_p = y_q$ and $x_q = -y_p$ then are specified by (D.44) and (D.45).

$$x(q,p) = \sum_{m=1}^{M} \Big[I_m \cosh mq \cos mp + J_m \cosh mq \sin mp +$$

$$+ K_m \sinh mq \cos mp + L_m \sinh mq \sin mp \Big] + x_0 , \tag{D.42}$$

$$y(q,p) = \sum_{m'=1}^{M} \Big[I'_{m'} \cosh m'q \cos m'p + J'_{m'} \cosh m'q \sin m'p +$$

$$+ K'_{m'} \sinh m'q \cos m'p + L'_{m'} \sinh m'q \sin m'p \Big] + y_0 , \tag{D.43}$$

$$x_q = \sum_{m=1}^{M} \Big[m I_m \sinh mq \cos mp + m J_m \sinh mq \sin mp +$$

$$+ m K_m \cosh mq \cos mp + m L_m \cosh mq \sin mp \Big] , \tag{D.44}$$

$$y_p = \sum_{m'=1}^{M} \Big[-m' I'_{m'} \cosh m'q \sin m'p + m' J'_{m'} \cosh m'q \cos m'p -$$

$$- m' K'_{m'} \sinh m'q \sin m'p + m' L'_{m'} \sinh m'q \cos m'p \Big] \tag{D.45}$$

$$\Leftrightarrow$$

$$I'_{m'} = L_m , \quad J'_{m'} = -K_m , \quad K'_{m'} = J_m , \quad L'_{m'} = -I_m .$$

End of Proof.

Example D.1 ($\mathbb{S}_r^2$, transverse Mercator projection).

As an example, let us construct the transverse Mercator projection *locally* for the sphere $\mathbb{S}_r^2$ based on the fundamental solution (D.15), (D.16), (D.28), and (D.29) of the d'Alembert–Euler equations (Cauchy–Riemann equations). Let us depart from the equidistant mapping of the L_0 meta-equator, namely the *boundary condition*

$$x = x\{q(L = L_0, B), p(L = L_0, B)\} = rB \ , \quad y = y\{q(L = L_0, B), p(L = L_0, B)\} = 0 \ . \quad \text{(D.46)}$$

There remains the task to express the boundary conditions in the function space (D.15) and (D.16). There are two ways in solving this problem.

First choice.

A Taylor series expansion of $B(Q)$ around $B_0(Q_0)$ leads *directly* to

$$B = \arcsin(\tanh Q) \ ,$$

$$B = B_0 + \Delta B = B_0 + b \ \forall \ b := \Delta B \ ,$$

$$Q = Q_0 + \Delta Q = Q_0 + q \ \forall \ q := \Delta Q$$

$$\Rightarrow$$

$$b = d_1 q + d_2 q^2 + \sum_{r=3}^{N \to \infty} d_r q^r \quad \text{(D.47)}$$

$$\forall$$

$$d_1 = \frac{\mathrm{d}B}{\mathrm{d}Q}(Q_0) \ , \quad d_2 = \frac{1}{2!} \frac{\mathrm{d}^2 B}{\mathrm{d}Q^2}(Q_0) \ , \quad \cdots \ , \quad d_r = \frac{1}{r!} \frac{\mathrm{d}^r B}{\mathrm{d}Q^r}(Q_0) \ .$$

The standard coefficients are given by

$$d_1 = \cos B_0 \ ,$$

$$d_2 = -\frac{1}{2} \cos^2 B_0 \tan B_0 \ ,$$

$$d_3 = -\frac{1}{6} \cos^3 B_0 [1 - \tan^2 B_0] \ , \quad \text{(D.48)}$$

$$d_4 = \frac{1}{24} \cos^4 B_0 \tan B_0 [5 - \tan^2 B_0] \ ,$$

$$d_5 = \frac{1}{120} \cos^5 B_0 [5 - 18 \tan^2 B_0 + \tan^4 B_0] \ .$$

Second choice.

The first or *remove step* is materialized by a *Taylor series expansion* of $Q(B)$ around $Q_0(B_0)$ and leads *directly* to

$$Q = \operatorname{artanh}(\sin B) = \ln \tan \left(\frac{\pi}{4} + \frac{B}{2} \right) \ , \quad \text{(D.49)}$$

$$Q = Q_0 + \Delta Q = Q_0 + q \quad \forall \ q := \Delta Q \ ,$$

$$B = B_0 + \Delta B = B_0 + b \quad \forall \ b := \Delta B$$

$$\Rightarrow$$

$$q = c_1 b + c_2 b^2 + \sum_{r=3}^{\infty} c_r b^r \quad \text{(D.50)}$$

$$\forall$$

$$c_1 = \frac{\mathrm{d}Q}{\mathrm{d}B}(B_0) \ , \quad c_2 = \frac{1}{2!} \frac{\mathrm{d}^2 Q}{\mathrm{d}B^2}(B_0) \ , \quad \cdots \ , \quad c_r = \frac{1}{r!} \frac{\mathrm{d}^r Q}{\mathrm{d}B^r}(B_0) \ .$$

Consequently, the second or *restore step* is based upon *standard series inversion* (D.51) according to (D.52) and (D.53).

$$b = d_1 q + d_2 q^2 + \sum_{r=3}^{\infty} d_r q^r \,, \tag{D.51}$$

$$c_1 = \frac{1}{\cos B_0} \,, \quad c_2 = \frac{1}{2} \frac{\tan B_0}{\cos B_0} \,,$$

$$c_3 = \frac{1}{6} \frac{1}{\cos B_0} [1 + 2\tan^2 B_0] \,, \quad c_4 = \frac{1}{24} \frac{\tan B_0}{\cos B_0} [5 \tan^2 B_0] \,, \tag{D.52}$$

$$c_5 = \frac{1}{120} \frac{1}{\cos B_0} [5 + 28\tan^2 B_0 + 24\tan^4 B_0] \,,$$

$$q = \sum_{r=1}^{N\to\infty} c_r b^r \Leftrightarrow b = \sum_{r=1}^{N\to\infty} d_r q^r \,,$$

$$d_1 = \frac{1}{c_1} \,, \quad d_2 = -\frac{c_2}{c_1^3} \,, \quad d_3 = \frac{2c_2^2}{c_1^5} - \frac{c_3}{c_1^4} \,, \quad d_4 = -\frac{5c_2^3}{c_1^7} + \frac{5c_2 c_3}{c_1^5} - \frac{c_4}{c_1^5} \,. \tag{D.53}$$

Thus we are led to the series representation of the boundary condition, namely

$$x\{q, p = 0\} = rB_0 + rb = rB_0 + r\sum_{r=1}^{N\to\infty} d_r q^r = \alpha_0 + \alpha_1 q + \alpha_2 q^2 + \sum_{r=3}^{N\to\infty} \alpha_r q^r \,, \tag{D.54}$$

$$y\{q, p = 0\} = \beta_0 - \beta_1 q - \beta_2 q^2 - \sum_{r=3}^{N\to\infty} \beta_r q^r = 0 \tag{D.55}$$

$$\Leftrightarrow$$

$$\alpha_0 = rB_0 \,, \quad \alpha_1 = rd_1 \,, \quad \alpha_2 = rd_2 \,, \quad \cdots \,, \quad \alpha_r = rd_r \,; \quad \beta_0 = \beta_1 = \beta_2 = \cdots = \beta_r = 0 \,. \tag{D.56}$$

This leads to the local representation of the transverse Mercator projection in terms of the incremental isometric longitude/latitude p/q, namely

$$x(q, p) = rB_0 + r\cos B_0 q - \frac{1}{2} r\cos^2 B_0 \tan B_0 (q^2 - p^2) + \cdots \,,$$
$$\tag{D.57}$$
$$y(q, p) = r\cos B_0 p - r\cos^2 B_0 \tan B_0 qp - \cdots \,.$$

Once we *remove* the incremental isometric longitude/latitude p/q, respectively, in favor incremental longitude/latitude l/b, respectively, with respect to the *first chart*, we gain

$$x(q, p) = rB_0 + r\cos B_0 \sum_{r=1}^{N\to\infty} c_r b^r - \frac{1}{2} r\cos^2 B_0 \tan B_0 \left(\sum_{r,s=1}^{N\to\infty} c_r c_s b^r b^s - l^2 \right) - \cdots \,,$$
$$\tag{D.58}$$
$$y(q, p) = r\cos B_0 l - r\cos^2 B_0 \tan B_0 \sum_{r=1}^{N\to\infty} c_r b^r l - \cdots \,.$$

How does the local representation of the transverse Mercator projection reflect its representation in closed form? We only have to apply a *Taylor series expansion* around $\{B_0, L_0\}$ or around $\{Q_0, L_0\}$ to arrive at (D.58).

End of Example.

Example D.2 ($\mathbb{S}_r^2$, transverse Mercator projection).

As an example, let us construct the transverse Mercator projection *locally* for the sphere $\mathbb{S}_r^2$ based on the fundamental solution of the d'Alembert–Euler equations (Cauchy–Riemann equations) in terms of separation of variables. Let us depart from the equidistant mapping of the L_0 meta-equator, namely the *boundary condition*

$$x = x\{q(L = L_0, B), p(L = L_0, B)\} = rB \ , \quad y = y\{q(L = L_0, B), p(L = L_0, B)\} = 0 \ . \quad \text{(D.59)}$$

There remains the task to express the boundary conditions in the function space. Here, we depart from (D.60). A Taylor series expansion of $B(Q)$ for $Q \geq 0$ (northern hemisphere), $u := \exp Q$, leads to (D.61).

$$B(Q) = 2 \arctan \exp Q - \frac{\pi}{2} \ , \quad \text{(D.60)}$$

$$\arctan u = \frac{\pi}{2} - \frac{1}{u} + \frac{1}{3u^3} - \frac{1}{5u^5} + \sum_{r=3}^{N \to \infty} (-1)^{r+1} \frac{1}{2r+1} \frac{1}{u^{2r+1}} =$$
$$= \frac{\pi}{2} + \sum_{r=0}^{N \to \infty} (-1)^{r+1} \frac{\exp[-(2r+1)Q]}{2r+1} \ . \quad \text{(D.61)}$$

We are thus led to the series representation of the boundary condition, for example, to (D.62) and (D.63), where we have elimineated the coefficients $\{A_m, B_m\}$ by the postulate $q \to \infty$, $\{x, y\}$ finite.

$$x\{q, p = 0\} = x_0 + \sum_{m=1}^{M \to \infty} C_m \exp(-mq) = r \left(\frac{\pi}{2} - 2 \sum_{n=0}^{N \to \infty} (-1)^{n+1} \frac{\exp[-(2n+1)Q]}{2n+1} \right) \ , \quad \text{(D.62)}$$

$$y\{q, p = 0\} = y_0 + \sum_{m=1}^{M \to \infty} D_m \exp(-mq) = 0 \ . \quad \text{(D.63)}$$

Once we compare the coefficients, we find (D.64). Finally, we find the local representation of the transverse Mercator projection in terms of isometric longitude and latitude p/q, namely (D.65), and in terms of longitude and latitude $\{l = L - L_0, B\}$, namely (D.66).

$$x_0 = r\frac{\pi}{2} \ , \quad y_0 = 0 \ ,$$
$$C_{2n} = 0 \ , \quad C_{2n+1} = r\frac{2(-1)^{n+1}}{2n+1} \ , \quad D_m = 0 \ , \quad \text{(D.64)}$$

$$x(q, p) = r\frac{\pi}{2} + 2r \sum_{n=0}^{N \to \infty} (-1)^{n+1} \frac{1}{2n+1} \exp[-(2n+1)q] \cos(2n+1)p \ ,$$
$$y(q, p) = 2r \sum_{n=0}^{N \to \infty} (-1)^n \frac{1}{2n+1} \exp[-(2n+1)q] \sin(2n+1)p \ , \quad \text{(D.65)}$$

$$x(B, l) = r\frac{\pi}{2} + 2r \sum_{n=0}^{N \to \infty} (-1)^{n+1} \frac{1}{2n+1} \frac{\cos(2n+1)l}{[\tan(\frac{\pi}{4} + \frac{B}{2})]^{2n+1}} \ ,$$
$$y(B, l) = 2r \sum_{n=0}^{N \to \infty} (-1)^n \frac{1}{2n+1} \frac{\sin(2n+1)l}{[\tan(\frac{\pi}{4} + \frac{B}{2})]^{2n+1}} \ . \quad \text{(D.66)}$$

End of Example.

Example D.3 (Optimal universal transverse Mercator projection, biaxial ellipsoid $\mathbb{E}^2_{a,b}$).

First, let us here establish the boundary condition for the universal transverse Mercator projection modulo an unknown dilatation factor. Second, with respect to the d'Alembert–Euler equations (Cauchy–Riemann equations), let us here solve the firstly formulated boundary value problem. Finally, the unknown dilatation factor is optimally determined by an optimization of the total distance distortion measure (Airy optimum) *or* of the total areal distortion.

The boundary condition.

With reference to the two examples (transverse Mercator projection of the sphere $\mathbb{S}^2_r$), we *generalize* the equidistant mapping of the L_0 meta-equator, namely the *boundary condition* (D.67), by an unknown *dilatation factor* ϱ, subject to later optimization. We begin with the solution to the problem to express the boundary condition in the function space.

$$x = x\{q(L = L_0, B), p(L = L_0, B)\} = \varrho a(1 - e^2) \int_0^B \frac{\mathrm{d}B}{(1 - e^2 \sin^2 B)^{3/2}} \, , \tag{D.67}$$

$$y = y\{q(L = L_0, B), p(L = L_0, B)\} = 0 \, .$$

Since the *boundary condition* (D.67) is given in terms of surface normal ellipsoidal latitude B, in the *first step*, we have to introduce ellipsoidal isometric latitude $Q(B): [-\pi/2, +\pi/2] \mapsto [0, \pm\infty]$ by (D.68), where $e^2 = (a^2 - b^2)/a^2 = 1 - (b^2/a^2)$ is the first numerical eccentricity.

$$Q = \ln\left[\tan\left(\frac{\pi}{4} + \frac{B}{2}\right)\left(\frac{1 - e\sin B}{1 + e\sin B}\right)^{e/2}\right] = \ln\tan\left(\frac{\pi}{4} + \frac{B}{2}\right) - \frac{e}{2}\ln\frac{1 + e\sin B}{1 - e\sin B} = \tag{D.68}$$

$$= \operatorname{artanh}(\sin B) - e\operatorname{artanh} e\sin B \, .$$

In the *second step*, we set up a *uniformly convergent series expansion* of the integral transformation according to (D.69) by means of the *recurrence relation* (D.70). The coefficients a_r are collected in (D.71). They are given as polynomials in $1, e^2, e^4$ etc. and $\cos B, \cos(2B), \cos(3B)$ etc.

$$x = \varrho \sum_{r=0}^{\infty} a_r q^r \; \forall \; x_0 := x(L_0, B_0) \, , \quad a_r = \frac{1}{r!}\frac{\mathrm{d}^r x}{\mathrm{d}Q^r}(Q_0(B_0)) \, , \quad q := Q - Q_0 = Q - Q(B_0) \, , \tag{D.69}$$

$$\frac{\mathrm{d}^r x}{\mathrm{d}Q^r} = \frac{\mathrm{d}}{\mathrm{d}B}\left(\frac{\mathrm{d}^{r-1}x}{\mathrm{d}Q^{r-1}}\right)\frac{\mathrm{d}B}{\mathrm{d}Q} \; \forall \; \frac{\mathrm{d}B}{\mathrm{d}Q} = \cos B\frac{1 - e^2\sin^2 B}{1 - e^2} \, , \tag{D.70}$$

$$a_0 = x_0 = a(1 - e^2)\int_0^B \frac{\mathrm{d}B}{(1 - e^2\sin^2 B)^{3/2}} \, , \quad a_1 = \frac{a\cos B}{\sqrt{1 - e^2\sin^2 B}} \, , \quad a_2 = \frac{-a\cos B\sin B}{2\sqrt{1 - e^2\sin^2 B}} \, ,$$

$$a_3 = \frac{-a\cos B\left(1 - 2\sin^2 B + e^2\sin^4 B\right)}{6\left(1 - e^2\right)\sqrt{1 - e^2\sin^2 B}} \, , \quad a_4 = \frac{a\cos B\sin B}{24(1 - e^2)^2\sqrt{1 - e^2\sin^2 B}} \times$$

$$\times[5 - 6\sin^2 B - e^2(1 + 6\sin^2 B - 9\sin^4 B) + e^4\sin^4 B(3 - 4\sin^2 B)] \, , \tag{D.71}$$

$$a_5 = \frac{-a\cos B}{120(1 - e^2)^3\sqrt{1 - e^2\sin^2 B}} \times$$

$$\times[-5 + 28\sin^2 B - 24\sin^4 B + e^2(1 + 16\sin^2 B - 86\sin^4 B + 72\sin^6 B) +$$

$$+ e^4\sin^4 B(-26 + 100\sin^2 B - 77\sin^4 B) + e^6\sin^6 B(12 - 39\sin^2 B + 28\sin^4 B)] \, .$$

Solution of the boundary value problem.

Let us consider the boundary value problem for the d'Alembert–Euler equations (Cauchy–Riemann equations) subject to the integrability conditions of harmonicity type, namely (D.72) and (D.73), in the function space of polynomial type (D.74) and (D.75). Once we compare the boundary conditions in the base $\{q^{r-2s}p^{2s}, q^{r-2s+1}p^{2s-1}\}$, we are led to (D.76).

$$x_p = y_q , \quad x_q = -y_p ,$$

$$\Delta x = x_{pp} + x_{qq} = 0 , \quad \Delta y = y_{pp} + y_{qq} = 0 , \tag{D.72}$$

$$x(0,q) = \sum_{r=0}^{\infty} \varrho a_r q^r, \ y(0,q) = 0 , \tag{D.73}$$

$$x(p,q) = \alpha_0 + \alpha_1 q + \beta_1 p + \alpha_2 (q^2 - p^2) + 2\beta_2 qp +$$

$$+ \sum_{r=3}^{N} \alpha_r \sum_{s=0}^{[r/2]} (-1)^s \binom{r}{2s} q^{r-2s} p^{2s} + \sum_{r=3}^{N} \beta_r \sum_{s=1}^{[(r+1)/2]} (-1)^{s+1} \binom{r}{2s-1} q^{r-2s+1} p^{2s-1} , \tag{D.74}$$

$$y(p,q) = \beta_0 - \beta_1 q + \alpha_1 p - \beta_2 (q^2 - p^2) + 2\alpha_2 qp -$$

$$- \sum_{r=3}^{N} \beta_r \sum_{s=0}^{[r/2]} (-1)^s \binom{r}{2s} q^{r-2s} p^{2s} + \sum_{r=3}^{N} \alpha_r \sum_{s=1}^{[(r+1)/2]} (-1)^{s+1} \binom{r}{2s-1} q^{r-2s+1} p^{2s-1} , \tag{D.75}$$

$$\alpha_r = \varrho a_r(B_0), \ \beta_r = 0 \ \forall \ r = 0, \ldots, \infty . \tag{D.76}$$

End of Example.

Corollary D.5 ($\mathbb{E}^2_{a,b}$, universal transverse Mercator projection modulo unknown dilatation parameter, c:c:cha–cha–cha).

The solution of the boundary value problem, where the L_0 meta-equator is modulo a dilatation parameter ϱ equidistantly mapped, is given by (D.77), (D.78), and the a_r of Table D.1.

$$x(p,q) = \varrho \left[a_0 + a_1 q + a_2 (q^2 - p^2) + \sum_{r=3}^{N} a_r \sum_{s=0}^{[r/2]} (-1)^s \binom{r}{2s} q^{r-2s} p^{2s} \right] , \tag{D.77}$$

$$y(p,q) = \varrho \left[a_1 p + 2a_2 qp + \sum_{r=3}^{N} a_r \sum_{s=1}^{[(r+1)/2]} (-1)^{s+1} \binom{r}{2s-1} q^{r-2s} p^{2s-1} \right] . \tag{D.78}$$

End of Corollary.

In practice, points on the biaxial ellipsoid are given in the chart of surface normal coordinates of type {longitude L, latitude B}, but *not* in the chart of isometric coordinates $\{p, q\}$. Thus, similarly to the second example "second choice", we take advantage of the *Taylor series expansion* of $Q(B)$ around $Q_0(B_0)$ as given by (D.79) according to *standard series inversion*. Additionally, implementing the *remove step* $b \mapsto q(b)$, we finally arrive at Corollary D.6.

$$q = \sum_{r=1}^{\infty} c_r b^r \Leftrightarrow b = \sum_{r=1}^{\infty} d_r q^r . \tag{D.79}$$

Corollary D.6 ($\mathbb{E}^2_{a,b}$, universal transverse Mercator projection modulo unknown dilatation parameter, c:c:cha–cha–cha).

The solution of the boundary value problem, where the L_0 meta-equator is modulo a dilatation parameter ϱ equidistantly mapped, is given by (D.80), (D.81), and the a_r of Tables D.2 and D.3.

$$x(l,b) = \varrho\Big[a_0 + a_{10}b + a_{20}b^2 + a_{02}l^2 + a_{30}b^3 + a_{12}bl^2 + a_{40}b^4 +$$
$$+a_{22}b^2l^2 + a_{04}l^4 + a_{50}b^5 + a_{32}b^3l^2 + a_{14}bl^4 + o_x(b^6,l^6)\Big]\,, \tag{D.80}$$

$$y(l,b) = \varrho\Big[a_{01}l + a_{11}bl + a_{21}b^2l + a_{03}l^3 + a_{31}b^3l + a_{13}bl^3 +$$
$$+a_{41}b^4l + a_{23}b^2l^3 + a_{05}l^5 + o_y(b^6,l^6)\Big]\,. \tag{D.81}$$

End of Corollary.

Table D.1. $\mathbb{E}^2_{a,b}$, the transverse Mercator projection, the series expansion $q(b) = \sum_{r=1}^N c_r b^r$ for $B = B_0$, the coefficients c_r are given up to $N = 5$, up to $N = 10$ available from the author.

$$c_1 = Q'(B) = \frac{1-e^2}{\cos B \left(1 - e^2 \sin^2 B\right)}\,,$$

$$c_2 = \frac{1}{2!}Q''(B) = \frac{\sin B \left[1 + e^2 \left(1 - 3\sin^2 B\right) + e^4 \left(-2 + 3\sin^2 B\right)\right]}{2\cos^2 B \left(1 - e^2 \sin^2 B\right)^2}\,,$$

$$c_3 = \frac{1}{3!}Q'''(B) = \frac{1}{6\cos^3 B \left(1 - e^2 \sin^2 B\right)^3}\left[1 + \sin^2 B - e^2 \left(-1 + 5\sin^2 B + 2\sin^4 B\right) - \right.$$

$$\left. -e^4 \left(2 - 10\sin^2 B + 11\sin^4 B - 9\sin^6 B\right) - e^6 \sin^2 B \left(6 - 13\sin^2 B + 9\sin^4 B\right)\right]\,,$$

$$c_4 = \frac{1}{4!}Q^{IV}(B) = \frac{\sin B}{24\cos^4 B \left(1 - e^2 \sin^2 B\right)^4} \times$$

$$\times[5 + \sin^2 B + e^2 \left(-1 - 18\sin^2 B - 5\sin^4 B\right) + e^4 \left(20 - 63\sin^2 B + 96\sin^4 B - 17\sin^6 B\right) +$$

$$+e^6 \left(-24 + 104\sin^2 B - 159\sin^4 B + 82\sin^6 B - 27\sin^8 B\right) +$$

$$+e^8 \sin^2 B \left(-24 + 68\sin^2 B - 65\sin^4 B + 27\sin^6 B\right)]\,,$$

$$c_5 = \frac{1}{5!}Q^V(B) = \frac{1}{120\cos^5 B \left(1 - e^2 \sin^2 B\right)^5} \times$$

$$\times[5 + 18\sin^2 B + \sin^4 B - e^2 \left(1 + 22\sin^2 B + 93\sin^4 B + 4\sin^6 B\right) -$$

$$-e^4 \left(-20 + 136\sin^2 B - 338\sin^4 B + 68\sin^6 B - 86\sin^8 B\right) -$$

$$-e^6 \left(24 - 380\sin^2 B + 1226\sin^4 B - 1588\sin^6 B + 1178\sin^8 B - 220\sin^{10} B\right) -$$

$$-e^8 \sin^2 B \left(240 - 1100\sin^2 B + 1936\sin^4 B - 1633\sin^6 B + 518\sin^8 B - 81\sin^{10} B\right) -$$

$$-e^{10} \sin^4 B \left(120 - 420\sin^2 B + 541\sin^4 B - 298\sin^6 B + 81\sin^8 B\right)]\,.$$

Table D.2. $\mathbb{E}^2_{a,b}$, the transverse Mercator projection, the series expansions $x(l,b)$ and $y(l,b)$ based upon $x(q) = \sum_{r=0}^{\infty} a_r q^r$. Part one.

$$q(b) = \sum_{r_1=1} c_{r_1} b^{r_1} ,$$

$$q^2(b) = \sum_{r_1,r_2} c_{r_1} c_{r_2} b^{r_1} b^{r_2} = \sum_{r=2}^{\infty} (cc)_r b^r ,$$

$$q^3(b) = \sum_{r_1,r_2,r_3} c_{r_1} c_{r_2} c_{r_3} b^{r_1} b^{r_2} b^{r_3} = \sum_{r=3}^{\infty} (ccc)_r b^r$$

etc.

$$a_{10}(b) = \frac{a \left(1 - e^2\right)}{\left(1 - e^2 \sin^2 b\right)^{\frac{3}{2}}} ,$$

$$a_{01}(b) = \frac{a \cos b}{\sqrt{1 - e^2 \sin^2 b}} ,$$

$$a_{11}(b) = \frac{-a \left(1 - e^2\right) \sin b}{\left(1 - e^2 \sin^2 b\right)^{\frac{3}{2}}} ,$$

$$a_{02}(b) = \frac{a \cos b \sin b}{2 \sqrt{1 - e^2 \sin^2 b}} ,$$

$$a_{03}(b) = \frac{a \cos b \left(1 - 2 \sin^2 b + e^2 \sin^4 b\right)}{6 \left(1 - e^2\right) \sqrt{1 - e^2 \sin^2 b}} ,$$

$$a_{12}(b) = \frac{a \left(1 - 2 \sin^2 b + e^2 \sin^4 b\right)}{2 \left(1 - e^2 \sin^2 b\right)^{\frac{3}{2}}} ,$$

$$a_{13}(b) = \frac{a \sin b}{6 \left(1 - e^2\right) \left(1 - e^2 \sin^2 b\right)^{\frac{3}{2}}} \times$$
$$\times \left[-5 + 6 \sin^2 b + e^2 \left(1 + 6 \sin^2 b - 9 \sin^4 b\right) + e^4 \sin^4 b \left(-3 + 4 \sin^2 b\right)\right] ,$$

$$a_{04}(b) = \frac{a \cos b \sin b}{24 \left(1 - e^2\right)^2 \sqrt{1 - e^2 \sin^2 b}} \times$$
$$\times \left[5 - 6 \sin^2 b - e^2 \left(1 + 6 \sin^2 b - 9 \sin^4 b\right) + e^4 \sin^4 b \left(3 - 4 \sin^2 b\right)\right] ,$$

$$a_{14}(b) = \frac{a}{24 \left(1 - e^2\right)^2 \left(1 - e^2 \sin^2 b\right)^{\frac{3}{2}}} \times$$
$$\times [5 - 28 \sin^2 b + 24 \sin^4 b + e^2 \left(-1 - 16 \sin^2 b + 86 \sin^4 b - 72 \sin^6 b\right) +$$
$$+ e^4 \sin^4 b \left(26 - 100 \sin^2 b + 77 \sin^4 b\right) + e^6 \sin^6 b \left(-12 + 39 \sin^2 b - 28 \sin^4 b\right)] ,$$

$$a_{05}(b) = \frac{-a \cos b}{120 \left(1 - e^2\right)^3 \sqrt{1 - e^2 \sin^2 b}} \times$$
$$\times [-5 + 28 \sin^2 b - 24 \sin^4 b + e^2 \left(1 + 16 \sin^2 b - 86 \sin^4 b + 72 \sin^6 b\right) +$$
$$+ e^4 \sin^4 b \left(-26 + 100 \sin^2 b - 77 \sin^4 b\right) + e^6 \sin^6 b \left(12 - 39 \sin^2 b + 28 \sin^4 b\right)] .$$

Table D.3. $\mathbb{E}^2_{a,b}$, the transverse Mercator projection, the series expansions $x(l,b)$ and $y(l,b)$ based upon $x(q) = \sum_{r=0}^{\infty} a_r q^r$. Part two.

$$a_{20}(b) = \frac{3\, a\, e^2\, \left(1 - e^2\right)\, \cos b\, \sin b}{2\left(1 - e^2 \sin^2 b\right)^{\frac{5}{2}}} \; ,$$

$$a_{30}(b) = \frac{-a\, e^2\, \left(1 - e^2\right)}{2\left(1 - e^2 \sin^2 b\right)^{\frac{7}{2}}} \times$$
$$\times \left[-1 + 2\sin^2 b + e^2 \sin^2 b \left(-4 + 3\sin^2 b\right)\right] \; ,$$

$$a_{21}(b) = \frac{-a\, \left(1 - e^2\right)\, \cos b\, \left(1 + 2\, e^2 \sin^2 b\right)}{2\left(1 - e^2 \sin^2 b\right)^{\frac{5}{2}}} \; ,$$

$$a_{40}(b) = \frac{-a\, e^2\, \left(1 - e^2\right)\, \sin b}{8\, \cos b\, \left(1 - e^2 \sin^2 b\right)^{\frac{9}{2}}} \times$$
$$\times \left[4 + e^2 \left(-15 + 22\sin^2 b\right) + e^4 \sin^2 b \left(-20 + 9\sin^2 b\right)\right] \; ,$$

$$a_{31}(b) = \frac{-a\, \left(1 - e^2\right)\, \sin b}{6\left(1 - e^2 \sin^2 b\right)^{\frac{7}{2}}} \times$$
$$\times \left[-1 + e^2 \left(9 - 10\sin^2 b\right) + e^4 \sin^2 b \left(6 - 4\sin^2 b\right)\right] \; ,$$

$$a_{22}(b) = \frac{a\, \cos b\, \sin b\, \left(-4 + e^2 \left(3 + 2\sin^2 b\right) - e^4 \sin^4 b\right)}{4\left(1 - e^2 \sin^2 b\right)^{\frac{5}{2}}} \; ,$$

$$a_{50}(b) = \frac{a\, e^2\, \left(1 - e^2\right)}{40\, \cos^2 b\, \left(1 - e^2 \sin^2 b\right)^{\frac{11}{2}}} \times$$
$$\times [-4 + 8\sin^2 b + e^2 \left(15 - 128\sin^2 b + 116\sin^4 b\right) +$$
$$+ e^4 \sin^2 b \left(180 - 362\sin^2 b + 164\sin^4 b\right) + e^6 \sin^4 b \left(120 - 136\sin^2 b + 27\sin^4 b\right)] \; ,$$

$$a_{41}(b) = \frac{-a\, \left(1 - e^2\right)}{24\, \cos b\, \left(1 - e^2 \sin^2 b\right)^{\frac{9}{2}}} \times$$
$$\times [-1 + e^2 \left(9 - 36\sin^2 b\right) + e^4 \sin^2 b \left(72 - 60\sin^2 b\right) +$$
$$+ e^6 \sin^4 b \left(24 - 8\sin^2 b\right)] \; ,$$

$$a_{32}(b) = \frac{a}{12\left(1 - e^2 \sin^2 b\right)^{\frac{7}{2}}} \times$$
$$\times [-4 + 8\sin^2 b + e^2 \left(3 - 16\sin^2 b + 4\sin^4 b\right) +$$
$$+ e^4 \sin^2 b \left(12 - 10\sin^2 b + 4\sin^4 b\right) - e^6 \sin^8 b] \; ,$$

$$a_{23}(b) = \frac{a\, \cos b}{12\left(1 - e^2\right)\left(1 - e^2 \sin^2 b\right)^{\frac{5}{2}}} \times$$
$$\times [-5 + 18\sin^2 b + e^2 \left(1 + 8\sin^2 b - 45\sin^4 b\right) +$$
$$+ e^4 \sin^2 b \left(2 - 15\sin^2 b + 46\sin^4 b\right) + e^6 \sin^6 b \left(6 - 16\sin^2 b\right)] \; .$$

E Geodesics

Geodetic curvature, geodetic torsion, geodesics. Lagrangean portrait, Hamiltonian portrait. Maupertuis gauge, Universal Lambert Projection, Universal Stereographic Projection, dynamic time.

Three topics are presented here.

Section E-1.

In Section E-1, we review the presentation of *geodetic curvature*, *geodetic torsion*, and *normal curvatures* of a submanifold on a two-dimensional Riemann manifold.

Section E-2.

In Section E-2, in some detail, we review the Darboux equations. Relatively unknown is the derivation of the differential equations of third order of a geodesic circle.

Section E-3.

In Section E-3, we concentrate on the Newton form of a geodesic in Maupertuis gauge on the sphere and the ellipsoid-of-revolution.

E-1 Geodetic curvature, geodetic torsion, and normal curvature

For the proof of Corollary 20.1, we depart from the *Darboux equations* (20.11). According to (20.11), the first *Darboux vector* $\boldsymbol{D}_1$ is represented in terms of the tangent vectors $\{\boldsymbol{G}_1, \boldsymbol{G}_2\}$, thus enjoying the derivative (E.1). In contrast, (E.2) expresses the derivative of the third Darboux vector $\boldsymbol{D}_3 = \boldsymbol{G}_3$, which coincides with the *surface normal vector*.

$$\boldsymbol{D}_1' = \boldsymbol{G}_{K,L} U'^K U'^L + \boldsymbol{G}_K U''^K , \tag{E.1}$$

$$\boldsymbol{D}_3' = \boldsymbol{G}_3' = \boldsymbol{G}_{3,L} U'^L . \tag{E.2}$$

The *Gauss–Weingarten* derivational equations govern surface geometry, in particular

$$\boldsymbol{G}_{K,L} = \{^M_{KL}\}\boldsymbol{G}_M + H_{KL}\boldsymbol{G}_3 \quad \text{(C. F. Gauss 1827)} ,$$

$$\boldsymbol{G}_{3,L} = -H_{LM} G^{MK} \boldsymbol{G}_K \quad \text{(J. Weingarten 1861)} . \tag{E.3}$$

Once being implemented into $\boldsymbol{D}_1'$ and $\boldsymbol{D}_3'$, we gain (E.4), to be confronted with (E.5).

$$\boldsymbol{D}_1' = \{^M_{KL}\} U'^K U'^L \boldsymbol{G}_M + U''^K \boldsymbol{G}_K + H_{KL} U'^K U'^L \boldsymbol{D}_3 ,$$

$$\boldsymbol{D}_3' = -H_{LM} G^{MK} \boldsymbol{G}_K , \tag{E.4}$$

$$\boldsymbol{D}_1' = +\kappa_g \boldsymbol{D}_2 + \kappa_n \boldsymbol{D}_3 ,$$

$$\boldsymbol{D}_3' = -\kappa_n \boldsymbol{D}_1 - \tau_g \boldsymbol{D}_2 . \tag{E.5}$$

Proof (Corollary 20.1).

First statement:

$$(\{^M_{KL}\}U'^K U'^L + U''^M)\boldsymbol{G}_M = \kappa_{\mathrm{g}}\boldsymbol{D}_2$$
$$\Longrightarrow$$
$$\kappa^2_{\mathrm{g}} = \langle \kappa_{\mathrm{g}}\boldsymbol{D}_2 | \kappa_{\mathrm{g}}\boldsymbol{D}_2 \rangle =$$
$$= (\{^{M_1}_{K_1 L_1}\}U'^{K_1} U'^{L_1} + U''^{M_1})G_{M_1 M_2}(\{^{M_2}_{K_2 L_2}\}U'^{K_2} U'^{L_2} + U''^{M_2})$$
$$\Longrightarrow$$
$$(20.28)\ .$$

$$\text{(E.6)}$$

Second statement:

$$\Longrightarrow$$
$$\kappa_{\mathrm{n}} = H_{KL}U'^K U'^L$$
$$\Longrightarrow$$
$$(20.29)\ .$$

$$\text{(E.7)}$$

Third statement:

$$\boldsymbol{D}''_1 = \kappa'_{\mathrm{g}}\boldsymbol{D}_2 + \kappa_{\mathrm{g}}\boldsymbol{D}'_2 + \kappa'_{\mathrm{n}}\boldsymbol{D}_3 + \kappa_{\mathrm{n}}\boldsymbol{D}'_3\ ,$$
$$\boldsymbol{D}'_2 = -\kappa_{\mathrm{g}}\boldsymbol{D}_1 + \tau_{\mathrm{g}}\boldsymbol{D}_3$$
$$\Longrightarrow$$
$$\boldsymbol{D}''_1 = -(\kappa^2_{\mathrm{g}} + \kappa^2_{\mathrm{n}})\boldsymbol{D}_1 + (\kappa'_{\mathrm{g}} - \kappa_{\mathrm{n}}\tau_{\mathrm{g}})\boldsymbol{D}_2 + (\kappa_{\mathrm{g}}\tau_{\mathrm{g}} + \kappa'_{\mathrm{n}})\boldsymbol{D}_3\ ,$$

$$\text{(E.8)}$$

$$\boldsymbol{D}_1 = \boldsymbol{G}_K U'^K\ ,$$
$$\boldsymbol{D}'_1 = \boldsymbol{G}_{K,L}U'^K U'^L + \boldsymbol{G}_K U''^K\ ,$$
$$\boldsymbol{D}''_1 = \boldsymbol{G}_{K,LM}U'^K U'^L U'^M + \boldsymbol{G}_{K,L}(U''^K U'^L + U'^K U''^L) + \boldsymbol{G}_{K,L}U''^K U'^L + \boldsymbol{G}_K U'''^K\ ,$$

$$\text{(E.9)}$$

$$\boldsymbol{G}_{K,LM} =$$
$$= \{^N_{KL}\}_{,M}\boldsymbol{G}_N + \{^N_{KL}\}\boldsymbol{G}_{N,M} + H_{KL,M}\boldsymbol{G}_3 + H_{KL}\boldsymbol{G}_{3,M}\ ,$$
$$\boldsymbol{G}_{K,LM} =$$
$$= -H_{KL}H_{MM_2}G^{M_2 K_2}\boldsymbol{G}_{K_2} +$$
$$+ \{^{N_1}_{KL}\}\{^{M_2}_{N_1 M}\}\boldsymbol{G}_{M_2} + \{^{N_1}_{KL}\}H_{N_1,M}\boldsymbol{G}_3 + \{^N_{KL}\}_{,M}\boldsymbol{G}_N + H_{KL,M}\boldsymbol{G}_3\ ,$$

$$\text{(E.10)}$$

$$\boldsymbol{D}''_1 =$$
$$= [U'''^M + (2U''^K U'^L + U'^K U''^L)\{^M_{KL}\} + U'^K U'^L U'^{M_2}(\{^M_{KL}\}_{,M_2} + \{^{M_1}_{KL}\}\{^M_{M_1 M_2}\}) -$$
$$- H_{KL}H_{M_1 M_2}G^{M_1 M}]\boldsymbol{G}_M + [(2U''^K U'^L + U'^K U''^L)H_{KL} +$$
$$+ U'^K U'^L U'^{M_2}(\{^{M_1}_{KL}\}H_{M_1 M_2} + H_{KL,M_2})]\boldsymbol{G}_3\ .$$

$$\text{(E.11)}$$

$\{{}^M_{KL}\} = \{{}^M_{LK}\}$, "symmetry" of Christoffel symbols leads to

$$D''_1 =$$

$$= [U'''{}^M + 3U''{}^K U'{}^L \{{}^M_{KL}\} + U'{}^K U'{}^L U'{}^N(\{{}^M_{KL}\}_{,N} + \{{}^Q_{KL}\}\{{}^M_{NQ}\} - H_{KL}H_{NQ}G^{QM})]\boldsymbol{G}_M + \quad \text{(E.12)}$$

$$+ [3U''{}^K U'{}^L H_{KL} + U'{}^K U'{}^L U'{}^N(\{{}^Q_{KL}\}H_{NQ} + H_{KL,N})]\boldsymbol{D}_3 .$$

τ_g then is obtained as

$$\kappa_g \tau_g + \kappa'_n = 3U''{}^K U'{}^L H_{KL} + U'{}^K U'{}^L U'{}^N(\{{}^Q_{KL}\}H_{NQ} + H_{KL,N}) ,$$

$$\kappa'_n = H_{KL,N}U'{}^K U'{}^L U'{}^N + 2H_{KL}U''{}^K U'{}^L$$

$$\Longrightarrow$$

$$\kappa_g \tau_g = H_{KL}U''{}^K U'{}^L + \{{}^Q_{KL}\}H_{NQ}U'{}^K U'{}^L U'{}^N \qquad \text{(E.13)}$$

$$\Longrightarrow$$

$$\tau_g = \frac{H_{KL}U''{}^K U'{}^L + \{{}^Q_{KL}\}H_{NQ}U'{}^K U'{}^L U'{}^N}{\sqrt{G_{M_1 M_2}(\{{}^{M_1}_{K_1 L_1}\} + U''{}^{M_1})(\{{}^{M_2}_{K_2 L_2}\} + U''{}^{M_2})}} \Longrightarrow (20.30) .$$

End of Proof.

E-2 The differential equations of third order of a geodesic circle

For the proof of Corollary 20.2, we depart from the *Darboux derivational equations* under the postulate of *geodesic circle* $\kappa_g = \text{const.}$, $\kappa_n = \text{const.}$, and $\tau_g = 0$, namely

$$\boldsymbol{D}''_1 = -(\kappa_g^2 + \kappa_n^2)\boldsymbol{D}_1 = -(\kappa_g^2 + \kappa_n^2)U'{}^M \boldsymbol{G}_M . \qquad \text{(E.14)}$$

Proof (Corollary 20.2).

$$\boldsymbol{D}''_1 = [U'''{}^M + 3U''{}^K U'{}^L\{{}^M_{KL}\} + U'{}^K U'{}^L U'{}^N(\{{}^M_{KL}\}_{,N} + \{{}^Q_{KL}\}\{{}^M_{NQ}\} - H_{KL}H_{NQ}G^{QM})]\boldsymbol{G}_M$$

$$\Longrightarrow \qquad \text{(E.15)}$$

$$(\kappa_g^2 + \kappa_n^2)U'{}^M = U'''{}^M + 3U''{}^K U'{}^L + U'{}^K U'{}^L U'{}^N(\{{}^M_{KL}\}_{,N} + \{{}^Q_{KL}\}\{{}^M_{NQ}\} - H_{KL}H_{NQ}G^{QM}) ,$$

$$\kappa_g^2 + \kappa_n^2 = G_{M_1 M_2}(U'{}^{K_1}U'{}^{L_1}\{{}^{M_1}_{K_1 L_1}\} + U''{}^{M_1})(U'{}^{K_2}U'{}^{L_2}\{{}^{M_2}_{K_2 L_2}\} + U''{}^{M_2}) +$$

$$+ H_{K_1 L_1}H_{K_2 L_2}U'{}^{K_1}U'{}^{K_2}U'{}^{L_1}U'{}^{L_2}$$

$$\Longrightarrow$$

$$U'''{}^M + 3U''{}^K U'{}^L\{{}^M_{KL}\} + U'{}^K U'{}^L U'{}^N(\{{}^M_{KL}\}_{,N} + \{{}^Q_{KL}\}\{{}^M_{NQ}\}) - U'{}^K U'{}^L U'{}^N H_{KL}H_{NQ}G^{QM} +$$

$$+ G_{M_1 M_2}(U'{}^{K_1}U'{}^{L_1}\{{}^{M_1}_{K_1 L_1}\} + U''{}^{M_1})(U'{}^{K_2}U'{}^{L_2}\{{}^{M_2}_{K_2 L_2}\}U''{}^{M_2})U'{}^M + \quad \text{(E.16)}$$

$$+ H_{K_1 L_1}H_{K_2 L_2}U'{}^{K_1}U'{}^{K_2}U'{}^{L_1}U'{}^{L_2}U'{}^M = 0$$

$$\Leftrightarrow$$

$$U'''{}^M + G_{KL}U''{}^K U''{}^L U'{}^M + 3\{{}^M_{KL}\}U''{}^K U'{}^L + 2G_{KL}\{{}^L_{PQ}\}U''{}^K U'{}^L U'{}^P U'{}^M +$$

$$+ (\{{}^M_{KL}\}_{,P} + \{{}^Q_{KL}\}\{{}^M_{QP}\})U'{}^K U'{}^L U'{}^P + G_{KL}\{{}^K_{PQ}\}\{{}^L_{ST}\}U'{}^P U'{}^Q U'{}^S U'{}^T U'{}^M = 0 .$$

Compare our result with W. O. Vogel (1970 p. 642, formula 3).

End of Proof.

E-3 The Newton form of a geodesic in Maupertuis gauge (sphere, ellipsoid-of-revolution)

Geodesics, in particular *minimal geodesics,* are of focal geodetic interest. In terms of Riemann normal coordinates, they are used in map projections to establish *azimuthal maps* – maps on a local tangential plane $T_p\mathbb{M}$ – which are geodetic with respect to the point p of evaluation. Straight lines in a Riemann map (plane chart) are the shortest on the surface, for example the Earth, with respect to the point p of evaluation. In geodetic navigation – aerial navigation, space navigation – minimal geodesies are applied to connect points on the Earth surface or in space. In both applications, initial value problems, boundary value problems as well as their mixed forms play the dominant role in solving the differential equations of a geodesic. (See E. Lichtenegger (1987) for a review of the four fundamental geodesic problems.)

Section E-31.

The differential equations of a geodesic can be written either as a system of two differential equations of second order (Lagrange portrait) or as a system of four differential equations of first order (Hamilton portrait) as long as we refer to two-dimensional surfaces. The Lagrange portrait and the Hamilton portrait of a geodesic is presented in the first section.

Section E-32.

Recently, H. Goenner, E. Grafarend and R. J. You (1994) have shown that the Newton law balancing inertial forces, in particular accelerations, and acting forces, in particular, those forces which are being derived from a potential, can be interpreted as a set of three geodesics in a three-dimensional Riemann manifold. In this case, the three-dimensional Riemann manifold is parameterized by conformal coordinates (isometric coordinates), in short, the three-dimensional Riemann manifold is said to be *conformally flat.* In addition, the factor of conformality represents as a Maupertuis gauge the potential of conservative forces. In turn, we try in the second section to express geodesics firstly by conformal coordinates (isometric coordinates) and secondly by Maupertuis gauge, in particular, aiming at a representation of geodesics in their Newton form (Newton portrait).

Section E-33, Section E-34.

The program to express the minimal geodesies as the Newton law is realized for geodesics in the sphere $\mathbb{S}^2_R$ in the third section and in the ellipsoid-of-revolution $\mathbb{E}^2_{A_1,A_2}$ in the fourth section. Extensive numerical examples in the form of computer graphics are given.

Section E-35, Section E-36.

We refer to Section E-35 for a review of Maupertuis gauged geodesics parameterized in normal coordinates with respect to a local tangent plane. We refer to Section E-36 for a review of Lie series, Maupertuis gauged geodesics, and the Hamilton portrait.

E-31 The Lagrange portrait and the Hamilton portrait of a geodesic

Let there be given a two-dimensional Riemann manifold $\{\mathbb{M}^2, q_{\mu\nu}\}$ with the standard metric G given by $\mathsf{G} - \{g_{\mu\nu}\} \in \mathbb{R}^{2\times2}$, symmetric and positive-definite, shortly a *surface.* For geodetic applications, we shall assume the following important properties.

Important!

$\{\mathbb{M}^2, g_{\mu\nu}\}$: orientable, star-shaped, second order Hölder continuous, compact.

Thus, we have excluded corners, edges, and self-intersections. $\{\mathbb{M}^2, g_{\mu\nu}\}$ is totally covered by a set of charts which form by their union a complete atlas. Any chart is an open subset of a two-dimensional Euclidean manifold $\mathbb{E}^2 := \{\mathbb{R}^2, \delta_{\mu\nu}\}$ with standard canonical metric $\mathsf{I} = \{\delta_{\mu\nu}\}$, the unit matrix. Indeed, we shall deal only with topographic surfaces which are assumed to be topologically similar to the sphere. Thus, a minimal atlas of $\{\mathbb{M}^2, g_{\mu\nu}\}$ is established by two charts, for example, as described by J. Engels and E. Grafarend (1995), based upon the following coordinates.

Important!

Quasi-spherical coordinates. Meta-quasi-spherical coordinates.

According to a standard theorem, for example, S. S. Chern (1955) applied to two-dimensional Riemann manifolds, conformal (isothermal, isometric) coordinates $\{q^1, q^2\}$ always exist. They establish a conformal diffeomorphism $\{\mathbb{M}^2, g_{\mu\nu}\} \to \{\mathbb{R}^2, \delta_{\mu\nu}\}$ which is angle preserving. In the following, we shall adopt $\{\mathbb{M}^2, g_{\mu\nu}\}$ to be *conformally flat*, in particular $g_{\mu\nu} = \lambda^2(q^1, q^2)\delta_{\mu\nu}$, where $\lambda^2(q^1, q^2)$ is the factor of conformality. The infinitesimal distance between two points $(q, q + dq)$ in $\{\mathbb{M}^2, \lambda^2\delta_{\mu\nu}\}$ can correspondingly be represented by (E.17) and (E.18). Obviously, dq^μ is an element of the tangent space $T_q\mathbb{M}^2$ at q, while $g_{\mu\nu}dq^\nu$ is an element of the cotangent space $^*T_q\mathbb{M}^2$, which is the dual space of $T_q\mathbb{M}^2 := \{\mathbb{R}^2, \mathsf{G}_q\}$.

$$ds^2 = dq^\mu g_{\mu\nu} dq^\nu \ \forall \ g_{\mu\nu} = \lambda^2(q^1, q^2)\delta_{\mu\nu} \ ,$$

$$dq^\mu \in T_q\mathbb{M}^2 \ , \quad g_{\mu\nu}dq^\nu \in^* T_q\mathbb{M}^2 \ ,$$

(E.17)

$$ds^2 = \lambda^2(q^1, q^2)\left[(dq^1)^2 + (dq^2)^2\right] \ .$$

(E.18)

E-311 Lagrange portrait

The Lagrange portrait of the parameterized curve, called *minimal geodesic* $q^\mu(s)$, a one-dimensional submanifold in the two-dimensional Riemann manifold, is provided by (E.19) and (E.20). The functional (E.19) subject to (E.20) is minimal if the following hold: (α) zero first variation (zero Fréchet derivative) and (β) positive second variation (positive second Fréchet derivative).

$$\int L\left(q, \frac{dq}{d\tau}\right)d\tau = \min \quad \text{or} \quad \int L^2\left(q, \frac{dq}{d\tau}\right)d\tau = \min \quad \text{(fixed boundary points)}, \quad \text{(E.19)}$$

$$2L^2\left(q, \frac{dq}{d\tau}\right) := \frac{ds^2}{d\tau^2} = \frac{dq^\mu}{d\tau}g_{\mu\nu}\frac{dq^\nu}{d\tau} \ \forall \ g_{\mu\nu} = \lambda^2(q^1, q^2)\delta_{\mu\nu} \ ,$$

$$2L^2\left(q, \frac{dq}{d\tau}\right) := \lambda^2(q^1, q^2)\left[\left(\frac{dq^1}{d\tau}\right)^2 + \left(\frac{dq^2}{d\tau}\right)^2\right] \ .$$

(E.20)

Under these necessary and sufficient conditions, a minimal geodesic (E.21) as a self-adjoint system of two differential equations of second order with respect to the parameter arc length s and the Christoffel symbols $[\mu\nu, \lambda]$ of the first kind is being derived. (Only with respect to Christoffel symbols of the first kind the system of differential equations (E.21) is self-adjoint: the alternative representation in terms of Christoffel symbols of the second kind with the second derivative $\mathrm{d}^2 q^\mu / \mathrm{d}s^2$ as the leading term is not, thus cannot be derived from a variational principle.)

$$g_{\mu\nu} \frac{\mathrm{d}^2 q^\nu}{\mathrm{d}s^2} + [\mu\nu, \lambda] \frac{\mathrm{d}q^\nu}{\mathrm{d}s} \frac{\mathrm{d}q^\lambda}{\mathrm{d}s} = 0 \; \forall \; g_{\mu\nu} = \lambda^2(q^1, q^2)\delta_{\mu\nu} \, ,$$

$$\lambda^2 \frac{\mathrm{d}^2 q^\mu}{\mathrm{d}s^2} + (\partial_\nu \lambda^2) \frac{\mathrm{d}q^\nu}{\mathrm{d}s} \frac{\mathrm{d}q^\mu}{\mathrm{d}s} - \frac{1}{2\lambda^2} \partial_\mu \lambda^2 = 0 \, . \tag{E.21}$$

E-312 Hamilton portrait

In contrast, the Hamilton portrait of a minimal geodesic $q^\mu(s)$ in $\{\mathbb{M}^2, \lambda^2 \delta_{\mu\nu}\}$ is based upon the generalized momentum, the generalized velocity field $g_{\mu\nu} \mathrm{d}q^\nu / \mathrm{d}\tau$, being an element of the cotangent space ${}^* T_q \mathbb{M}^2$ at point q, namely

$$p_\mu := \frac{\partial L^2}{\partial \frac{\mathrm{d}q^\mu}{\mathrm{d}\tau}} = g_{\mu\nu} \frac{\mathrm{d}q^\nu}{\mathrm{d}\tau} \in {}^* T_q \mathbb{M}^2 \; \forall \; g_{\mu\nu} = \lambda^2(q^1, q^2)\delta_{\mu\nu} \, ,$$

$$p_\mu = \lambda^2(q^1, q^2) \frac{\mathrm{d}q^\mu}{\mathrm{d}\tau} \in {}^* T_q \mathbb{M}^2 \, . \tag{E.22}$$

By means of the Legendre transformation, the dual of the Lagrangean $L^2(q, \mathrm{d}q/\mathrm{d}\tau)$, in particular the Hamiltonian $H^2(q, p)$, is established according to (E.23) with respect to the four-dimensional phase space $\{p_\mu, q^\mu\} \in {}^* T\mathbb{M}^2$, where ${}^* T\mathbb{M}^2$ is the union of cotangent spaces ${}^* T_q \mathbb{M}^2$ at points q, in particular ${}^* T\mathbb{M}^2 \cup_q {}^* T_q \mathbb{M}^2_q$, supported by a symplectic metric (skewquadratic form) outlined elsewhere. The Hamilton variational principle is formulated by (E.24) and (E.25).

$$H^2 := p_\mu \frac{\mathrm{d}q^\mu}{\mathrm{d}\tau} - L^2 = \frac{1}{2} g^{\mu\nu} p_\mu p_\nu \; \forall \; g_{\mu\nu} = \lambda^2(q^1, q^2)\delta_{\mu\nu} \, ,$$

$$H^2 = \frac{1}{2\lambda^2(q^1, q^2)} (p_1^2 + p_2^2) \, , \tag{E.23}$$

$$\int \left(p_\mu \frac{\mathrm{d}q^\mu}{\mathrm{d}\tau} - H^2 \right) \mathrm{d}\tau = \min \quad \text{(fixed boundary points in phase space)} \, , \tag{E.24}$$

$$2H^2(q, p) = g^{\mu\nu} p_\mu p_\nu \; \forall \; g_{\mu\nu} = \lambda^2(q^1, q^2)\delta_{\mu\nu} \, ,$$

$$2H^2(q, p) = \frac{1}{\lambda^2(q^1, q^2)} (p_1^2, p_2^2) \, . \tag{E.25}$$

The functional (E.24) subject to (E.25) is minimal if the following holds: (α) zero first variation (zero Fréchet derivative) and (β) positive second variation (positive second Fréchet derivative). Under these necessary and sufficient conditions, a minimal geodesic in phase space is given by (E.26) and (E.27).

$$\frac{dq^{\mu}}{d\tau} = \frac{\partial H^2}{\partial p_{\mu}} = g^{\mu\nu} p_{\nu} \ \forall \ g^{\mu\nu} = \frac{1}{\lambda^2(q^1, q^2)} \delta_{\mu\nu} \ ,$$

$$\frac{dp_{\mu}}{d\tau} = -\frac{\partial H^2}{\partial q^{\mu}} = -\frac{1}{2}\frac{\partial g^{\alpha\beta}}{\partial q^{\mu}} p_{\alpha} p_{\beta} \ \forall \ g^{\mu\nu} = \frac{1}{\lambda^2(q^1, q^2)} \delta_{\mu\nu} \ ,$$

(E.26)

$$\frac{dq^{\mu}}{d\tau} = \frac{\partial H^2}{\partial p_{\mu}} = \frac{1}{\lambda^2(q^1, q^2)} p_{\mu} \ ,$$

$$\frac{dp_{\mu}}{d\tau} = -\frac{(p_1^2 + p_2^2)}{2}\frac{\partial}{\partial q^{\mu}}\frac{1}{\lambda^2(q^1, q^2)} \ .$$

(E.27)

Note that the Hamilton equations as a system of four differential equations of first order in the variable $\{q(\tau), p(\tau)\}$ do not appear in the form we are used to from mechanics, a result caused by the effect that no dynamical time has been introduced. (E.27) can also be directly derived by (E.21) as soon as the system of two differential equations of second order in the variable $q^{\mu}(\tau)$ is transformed by means of (E.22) into a system of four differential equations of first order in the "state variable" $\{q(\tau), p(\tau)\}$. Furthermore, note that a condensed form of the Hamilton equations is achieved as soon as we introduce the antisymmetric metric tensor (symplectic tensor) (E.28).

$$\Omega :=$$

$$:= \{\omega^{ij}\} = \begin{bmatrix} 0_{\mu\nu} & +\delta_{\mu\nu} \\ -\delta_{\mu\nu} & 0_{\mu\nu} \end{bmatrix} = \begin{bmatrix} 0 & +I_2 \\ -I_2 & 0 \end{bmatrix} \ ,$$

$$\Omega^{-1} :=$$

$$:= \{\omega_{ij}\} = \begin{bmatrix} 0_{\mu\nu} & -\delta_{\mu\nu} \\ +\delta_{\mu\nu} & 0_{\mu\nu} \end{bmatrix} = \begin{bmatrix} 0 & -I_2 \\ +I_2 & 0 \end{bmatrix} \ .$$

(E.28)

In terms of the four-vector (E.29) (state vector) being an element of the phase space, the Hamilton equations in their contravariant and covariant form, respectively, are given by (E.30). In particular, the phase space is equipped with the metric (E.31).

$$y := \begin{bmatrix} q^1 \\ q^2 \\ p_1 \\ p_2 \end{bmatrix} \ ,$$

(E.29)

$$\frac{dy^i}{d\tau} - \omega^{ij}\frac{\partial H^2}{\partial y^j} = 0 \ \forall \ i, j \in \{1, 2, 3, 4\} \ ,$$

$$\omega^{ij}\frac{dy^i}{d\tau} - \frac{\partial H^2}{\partial y^j} = 0 \ \forall \ i, j \in \{1, 2, 3, 4\} \ ,$$

(E.30)

$$\omega_2 :=$$

$$:= \frac{1}{2}\omega_{ij} dy^i \wedge dy^j =$$

$$= dp_{\mu} \wedge dq^{\mu} = dp_1 \wedge dq^1 + dp_2 \wedge dq^2 \ .$$

(E.31)

E-32 The Maupertuis gauge and the Newton portrait of a geodesic

How can we gauge dynamic time into the system of differential equations for a geodesic? According to Maupertuis (1744) elaborated by C. G. J. Jacobi (1866), let us represent the arc length ds according to (E.32) as the product of the factor of conformality λ^2 and the time differential dt or, equivalently, according to (E.33), identifying the factor of conformality as the kinetic energy (E.34). Therefore, we introduce dynamic time into the one-dimensional submanifold $q(\tau) \to q(t)$, the minimal geodesic in the two-dimensional Riemann manifold.

$$ds := \lambda^2(q^1, q^2)dt , \tag{E.32}$$

$$\lambda^4(q^1, q^2) :=$$

$$:= \frac{ds^2}{dt^2} = \frac{dq^\mu}{dt} g_{\mu\nu} \frac{dq^\nu}{dt} \ \forall \ g_{\mu\nu} = \lambda^2(q^1, q^2)\delta_{\mu\nu} ,$$

$$\lambda^2(q^1, q^2) :=$$

$$:= \frac{ds}{dt} = \left(\frac{dq^1}{dt}\right)^2 + \left(\frac{dq^2}{dt}\right)^2 , \tag{E.33}$$

$$2T := \left(\frac{dq^1}{dt}\right)^2 + \left(\frac{dq^2}{dt}\right)^2 . \tag{E.34}$$

E-321 Lagrange portrait

The Laqrangean $L^2(q, dq/d\tau)$ is transformed into the Lagrangean $L^2(q, dq/dt)$ subject to the metric $g_{\mu\nu} = \lambda^2(q^1, q^2)\delta_{\mu\nu}$ and $\lambda^2 = (\dot{q}^1)^2 + (\dot{q}^2)^2$, where the dot differentiation is defined with respect to the dynamic time t. We here note in passing that the prime differentiation is defined with respect to the affine parameter arc length s.

$$L^2(q(t), \dot{q}(t)) = \frac{1}{2}\delta_{\mu\nu}\dot{q}^\mu\dot{q}^\nu + \frac{1}{2}\lambda^2(q^1, q^2) , \tag{E.35}$$

$$q'^\mu = \frac{1}{\lambda^2(q^1, q^2)}\dot{q}^\mu ,$$
$$\lambda^2 q''^\mu = \lambda^{-2}\ddot{q}^\mu - \lambda^{-4}(\partial_\nu\lambda^2)\dot{q}^\nu\dot{q}^\mu . \tag{E.36}$$

Implementing (E.36) into (E.21) or directly deriving from (E.32), we are led to Corollary E.1. Obviously, thanks to the Maupertuis gauge of a geodesic, in particular, by introducing dynamic time t, we have found the Newton form of a geodesic, an extremely elegant form we shall apply furtheron.

Corollary E.1 (Newton portrait of a Maupertuis gauged geodesic).

$$\ddot{q}^\mu - \partial_\mu\frac{\lambda^2}{2} = 0 . \tag{E.37}$$

End of Corollary.

E-322 Hamilton portrait

It should not be too surprising that by introducing the Maupertuis gauge, in particular dynamic time t, we are led to familiar Hamiltonian equations. The Hamiltonian $H^2(q(\tau), p(\tau))$ is transformed into the Hamiltonian $H^2(q(t), p(t))$ subject to the metric $g_{\mu\nu} = \lambda^2(q^1, q^2)\delta_{\mu\nu}$ and $\lambda^2 = (\dot{q}^1)^2 + (\dot{q}^2)^2$.

$$H^2(q(t), p(t)) =$$

$$= \frac{1}{2}\delta^{\mu\nu}p_\mu p_\nu - \frac{1}{2}\lambda^2(q^1, q^2) .$$

(E.38)

(E.38) used as the input to minimize the functional leads to Corollary E.2.

Corollary E.2 (Newton portrait of a Maupertuis gauged geodesic in phase space).

$$\frac{dq^\mu}{dt} = \frac{\partial H^2}{\partial p_\mu} = \delta^{\mu\nu}p_\nu ,$$

(E.39)

$$\frac{dp_\mu}{dt} = -\frac{\partial H^2}{\partial q^\mu} = \frac{1}{2}\frac{\partial}{\partial q^\mu}\lambda^2(q^1, q^2) .$$

(E.40)

End of Corollary.

E-33 A geodesic as a submanifold of the sphere (conformal coordinates)

Let us represent the differential equations of a geodesic as a submanifold of the sphere $\mathbb{M}^2 := \mathbb{S}_R^2$ as functions $q^\mu(t)$ as well as the Lagrangean $L^2(q(t), \dot{q}(t))$ and the Hamiltonian $H^2(q(t), p(t))$ with respect to conformal coordinates (isothermal, isometric) of the most important applicable map projections listed below. Compare with Figs. E.1–E.4. The related computer-graphical illustrations of Maupertuis gauged geodesics document the elegance of those submanifolds in $\mathbb{S}_R^2$ parameterized in dynamic time.

> **Important!**
>
> (i) The Universal Polar Stereographic Projection (UPS) (central perspective projection from the South Pole to a tangent plane $T_{np}\mathbb{M}^2$ at the North Pole), (ii) the Universal Mercator Projection (UM) (conformal diffeomorphism of the sphere onto a circular cylinder $\mathbb{S}_R^1 \times \mathbb{R}$ where the circle $\mathbb{S}_R^1$ is chosen as the equator of the sphere), (iii) the Universal Transverse Mercator Projection (UTM) (conformal diffeomorphism of the sphere onto a circular cylinder $\mathbb{S}_R^1 \times \mathbb{R}$ where the circle $\mathbb{S}_R^1$ is chosen as a definite meridian of the sphere), and (iv) the Universal Conic Lambert Projection (UC) (conformal diffeomorphism of the sphere onto a circular cone $\{\boldsymbol{x} \in \mathbb{R}^3 \,|\, (x^2 + y^2)/a^2 - z^2/b^2 = 0, a^2 + b^2 = R^2\}$ where the circle $\mathbb{S}_a^1$ is chosen as a definite parallel circle (line-of-contact) and b as the definite distance of the circle $\mathbb{S}_a^1$ from the equatorial plane).

Boxes E.1–E.3 contain (i) the conformal coordinates as a function of spherical longitude and spherical latitude, especially the factor of conformality, (ii) the Lagrangean version versus the Hamiltonian version of a geodesic in $\mathbb{S}_R^2$ in terms of conformal coordinates (isometric coordinates) and Maupertuis gauge, and (iii) the differential equations of a geodesic in $\mathbb{S}_R^2$ in terms of conformal coordinates (isometric coordinates) and Maupertuis gauge: Lagrange portrait, two differential equations of second order, Hamilton portrait, four differential equations of first order.

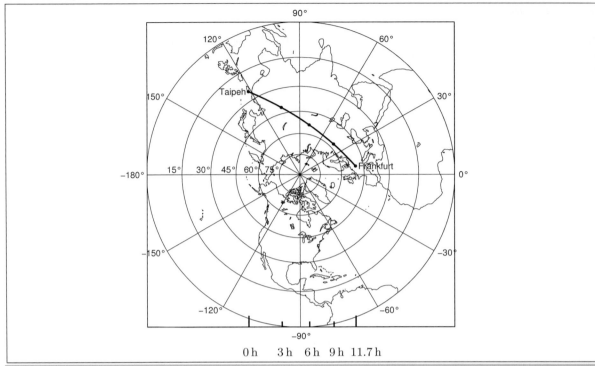

Fig. E.1. Maupertuis gauged geodesic on $\mathbb{S}^2_R$, conformal coordinates generated by the universal stereographic projection, solution of the boundary value problem, departing point Frankfurt, arrival point Taipeh, time given for constant speed movement with respect to the arrival point.

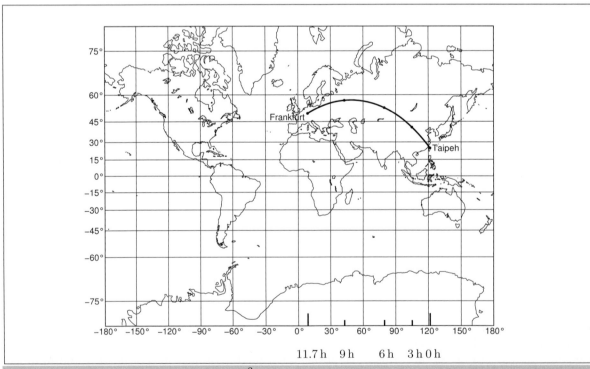

Fig. E.2. Maupertuis gauged geodesic on $\mathbb{S}^2_R$, conformal coordinates generated by the universal Mercator projection, solution of the boundary value problem, departing point Frankfurt, arrival point Taipeh, time given for constant speed movement with respect to the arrival point.

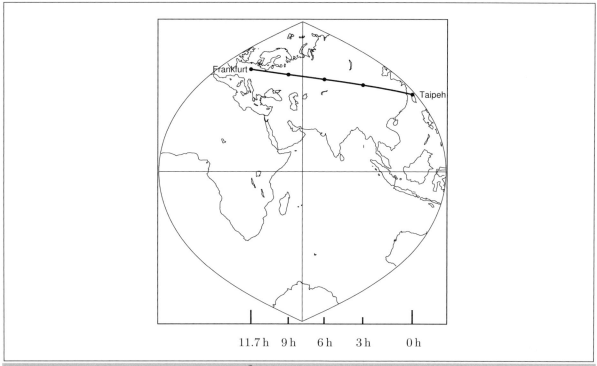

Fig. E.3. Maupertuis gauged geodesic on $\mathbb{S}_R^2$, conformal coordinates generated by the universal transverse Mercator projection, solution of the boundary value problem, departing point Frankfurt, arrival point Taipeh, time given for constant speed movement with respect to the arrival point.

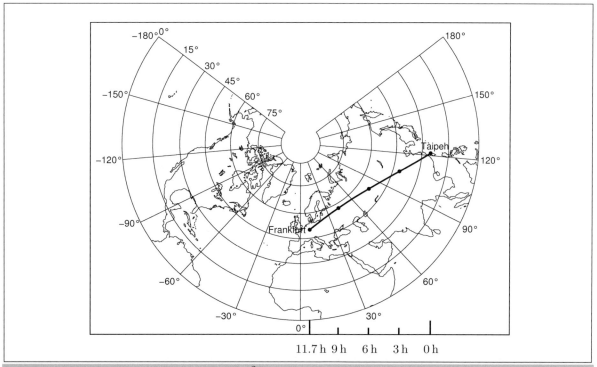

Fig. E.4. Maupertuis gauged geodesic on $\mathbb{S}_R^2$, conformal coordinates generated by the universal Lambert projection, solution of the boundary value problem, departing point Frankfurt, arrival point Taipeh, time given for constant speed movement with respect to the arrival point.

Box E.1 (The conformal coordinates as functions of spherical longitude and spherical latitude and the factor of conformality).

(i) Universal Polar Stereographic Projection (UPS):

$$q^1 = 2R\tan\left(\frac{\pi}{4} - \frac{\Phi}{2}\right)\cos\Lambda \ , \quad q^2 = 2R\tan\left(\frac{\pi}{4} - \frac{\Phi}{2}\right)\sin\Lambda \ , \tag{E.41}$$

$$\lambda = \cos^2\left(\frac{\pi}{4} - \frac{\Phi}{2}\right) = \frac{4R^2}{4R^2 + (q^1)^2 + (q^2)^2} \ , \tag{E.42}$$

$$ds^2 = \frac{16R^4}{(4R^2 + (q^1)^2 + (q^2)^2)^2}\left[(dq^1)^2 + (dq^2)^2\right] \ . \tag{E.43}$$

(ii) Universal Mercator Projection (UM):

$$q^1 = R\Lambda \ , \quad q^2 = R\ln\tan\left(\frac{\pi}{4} + \frac{\Phi}{2}\right) = R\,\text{artanh}\,(\sin\Phi) \ , \tag{E.44}$$

$$\lambda = \cos\Phi = \frac{1}{\cosh q^2/R} \ , \tag{E.45}$$

$$ds^2 = \frac{1}{\cosh^2 q^2/R}\left[(dq^1)^2 + (dq^2)^2\right] \ . \tag{E.46}$$

(iii) Universal Transverse Mercator Projection (UTM):

$$q^1 = R\arctan\frac{\tan\Phi}{\cos(\Lambda - \Lambda_0)} \ , \quad q^2 = R\,\text{artanh}\,(-\cos\Phi\sin(\Lambda - \Lambda_0)) \ , \tag{E.47}$$

$$\lambda = \cos\left[\arcsin(-\cos\Phi\sin(\Lambda - \Lambda_0))\right] \ , \tag{E.48}$$

$$\cos\Phi = \cos\frac{q^1}{R}1 + \tan^2\frac{q^1}{R}\tanh^2\frac{q^2}{R} \ , \quad \sin(\Lambda - \Lambda_0) = -\tanh\frac{q^2}{R}/\cos\Phi \ , \tag{E.49}$$

$$\lambda = \cos\left[\arcsin\left(\tanh\frac{q^2}{R}\right)\right] \ , \tag{E.50}$$

$$ds^2 = \cos^2\left[\arcsin\left(\tanh\frac{q^2}{R}\right)\right]\left[(dq^1)^2 + (dq^2)^2\right] \ . \tag{E.51}$$

(iv) Universal Conic Projection (UC):

$$q^1 = r\cos\alpha \ , \quad q^2 = r\sin\alpha \ , \quad \alpha = n\Lambda \ , \quad r = c\left(\tan\left(\frac{\pi}{4} - \frac{\Phi}{2}\right)\right)^n \ , \tag{E.52}$$

$$\lambda = \frac{r\cos\Phi}{cn\left(\tan\left(\frac{\pi}{4} - \frac{\Phi}{2}\right)\right)^n} \ . \tag{E.53}$$

Continuation of Box.

Variant one (UC). Equidistant map of the parallel circle Φ_0/line-of-contact:

$$n = \sin \Phi_0 \; , \quad c = R \frac{\cot \Phi_0}{\left(\tan\left(\frac{\pi}{4} - \frac{\Phi}{2}\right)\right)^n} \; , \tag{E.54}$$

$$\lambda = \frac{\cos \Phi}{\cos \Phi_0} \left(\frac{\tan\left(\frac{\pi}{4} - \frac{\Phi_0}{2}\right)}{\tan\left(\frac{\pi}{4} - \frac{\Phi}{2}\right)}\right)^n = \frac{1}{n} \sin \arctan\left(\frac{1}{c}\sqrt{(q^1)^2 + (q^2)^2}\right)^{1/n} \; , \tag{E.55}$$

$$ds^2 = \frac{1}{n^2} \sin^2 \arctan\left(\frac{1}{c}\sqrt{(q^1)^2 + (q^2)^2}\right)^{1/n} \left[(dq^1)^2 + (dq^2)^2\right] \; . \tag{E.56}$$

Variant two (UC). Equidistant map of two parallel circles Φ_1/Φ_2 (Lambert conformal projection):

$$n = \frac{\ln \frac{\cos \Phi_1}{\cos \Phi_2}}{\ln \frac{\tan\left(\frac{\pi}{4} - \frac{\Phi_1}{2}\right)}{\tan\left(\frac{\pi}{4} - \frac{\Phi_2}{2}\right)}} \; , \quad c = \frac{R \cos \Phi_1}{n \left(\tan\left(\frac{\pi}{4} - \frac{\Phi_1}{2}\right)\right)^n} \; , \tag{E.57}$$

$$\lambda = \frac{\cos \Phi}{\cos \Phi_2} \left(\frac{\tan\left(\frac{\pi}{4} - \frac{\Phi_2}{2}\right)}{\tan\left(\frac{\pi}{4} - \frac{\Phi}{2}\right)}\right)^n = \frac{1}{n} \sin \arctan\left(\frac{1}{c}\sqrt{(q^1)^2 + (q^2)^2}\right)^{1/n} \; , \tag{E.58}$$

$$ds^2 = \frac{1}{n^2} \sin^2 \arctan\left(\frac{1}{c}\sqrt{(q^1)^2 + (q^2)^2}\right)^{1/n} \left[(dq^1)^2 + (dq^2)^2\right] \; . \tag{E.59}$$

Box E.2 (The Lagrangean version versus the Hamiltonian version of a geodesic in $\mathbb{S}_R^2$ in terms of conformal coordinates (isometric coordinates) and Maupertuis gauge).

(i) Universal Polar Stereographic Projection (UPS):

$$
\begin{aligned}
L^2(q(t), \dot{q}(t)) &= \frac{1}{2}\delta_{\mu\nu}\dot{q}^\mu\dot{q}^\nu + \frac{8R^4}{(4R^2 + (q^1)^2 + (q^2)^2)^2} \; , \\
H^2(q(t), p(t)) &= \frac{1}{2}(p_1^2 + p_2^2) - \frac{8R^4}{(4R^2 + (q^1)^2 + (q^2)^2)^2} \; .
\end{aligned}
\tag{E.60}
$$

(ii) Universal Mercator Projection (UM):

$$L^2(q(t), \dot{q}(t)) = \frac{1}{2}\delta_{\mu\nu}\dot{q}^\mu\dot{q}^\nu + \frac{1}{2\cosh^2(q^2/R)} \; , \quad H^2(q(t), p(t)) = \frac{1}{2}(p_1^2 + p_2^2) - \frac{1}{2\cosh^2(q^2/R)} \; . \tag{E.61}$$

(iii) Universal Transverse Mercator Projection (UTM):

$$
\begin{aligned}
L^2(q(t), \dot{q}(t)) &= \frac{1}{2}\delta_{\mu\nu}\dot{q}^\mu\dot{q}^\nu + \frac{1}{2}\cos^2 \arcsin[\tanh(q^2/R)] \; , \\
H^2(q(t), p(t)) &= \frac{1}{2}(p_1^2 + p_2^2) - \frac{1}{2}\cos^2 \arcsin[\tanh(q^2/R)] \; .
\end{aligned}
\tag{E.62}
$$

(iv) Universal Conic Projection (UC):

$$
\begin{aligned}
L^2(q(t), \dot{q}(t)) &= \frac{1}{2}\delta_{\mu\nu}\dot{q}^\mu\dot{q}^\nu + \frac{1}{2n}\sin^2 \arctan\left(\frac{1}{c}\sqrt{(q^1)^2 + (q^2)^2}\right)^{1/n} \; , \\
H^2(q(t), p(t)) &= \frac{1}{2}(p_1^2 + p_2^2) - \frac{1}{2n}\sin^2 \arctan\left(\frac{1}{c}\sqrt{(q^1)^2 + (q^2)^2}\right)^{1/n} \; .
\end{aligned}
\tag{E.63}
$$

Box E.3 (The differential equations of a geodesic in $\mathbb{S}^2_R$ in terms of conformal coordinates (isometric coordinates) and Maupertuis gauge: Lagrange portrait, two differential equations of second order, Hamilton portrait, four differential equations of first order).

(i) Universal Polar Stereographic Projection (UPS):

$$\ddot{q}^\mu + \frac{32R^4}{(4R^2 + (q^1)^2 + (q^2)^2)^3} q^\mu = 0 \ \forall \ \mu = 1, 2 \ ,$$

$$\dot{q}^\mu = \delta^{\mu\nu} p_\nu \ \forall \ \mu, \nu = 1, 2$$

(summation convention) ,

$$\dot{p}_\mu = -\frac{32R^4}{(4R^2 + (q^1)^2 + (q^2)^2)^3} q^\mu \ \forall \ \mu = 1, 2 \ .$$

(E.64)

(ii) Universal Mercator Projection (UM):

$$\ddot{q}^1 = 0 \ , \ \ \ddot{q}^1 + \frac{1}{R} \frac{\sinh(q^2/R)}{\cosh^3(q^2/R)} = 0 \ ,$$

$$\dot{q}^1 = p_1 \ , \ \ \dot{q}^2 = p_2 \ ,$$

$$\dot{p}_1 = 0 \ (p_1 \ \text{cyclic}) \ , \ \ p_1 = \text{const.} \ ,$$

$$\dot{p}_2 = -\frac{1}{R} \frac{\sinh(q^2/R)}{\cosh^3(q^2/R)} \ .$$

(E.65)

(iii) Universal Transverse Mercator Projection (UTM):

$$\ddot{q}^1 = 0 \ ,$$

$$\ddot{q}^2 + \frac{1}{2R} \sin\left(2\arcsin\left(\tanh\frac{q^2}{R}\right)\right) \sqrt{1 - \tanh^2 \frac{q^2}{R}} = 0 \ ,$$

$$\dot{q}_1 = p_1 \ , \ \ \dot{q}^2 = p_2 \ ,$$

$$\dot{p}_1 = 0 \ (p_1 \ \text{cyclic}) \ , \ \ p_1 = \text{const.} \ ,$$

$$\dot{p}_2 = -\frac{1}{2R} \sin\left(2\arcsin\left(\tanh\frac{q^2}{R}\right)\right) \sqrt{1 - \tanh^2 \frac{q^2}{R}} \ .$$

(E.66)

(iv) Universal Conic Projection (UC) (variant one and variant two):

$$\ddot{q}^\mu - \frac{1}{2c^{1/n}n^3} \sin\left(2\arcsin\left(\frac{1}{c}\sqrt{(q^1)^2 + (q^2)^2}\right)^{1/n}\right) \times$$

$$\times \left(1 + \left(\frac{1}{c}\sqrt{(q^1)^2 + (q^2)^2}\right)^{2/n}\right)^{-1} \left[(q^1)^2 + (q^2)^2\right]^{(1-n)/2n} q^\mu = 0 \ ,$$

$$\dot{q}_1 = p_1 \ , \ \ \dot{q}_2 = p_2 \ ,$$

(E.67)

$$\dot{p}_\mu = \frac{1}{2c^{1/n}n^3} \sin\left(2\arctan\left(\frac{1}{c}\sqrt{(q^1)^2 + (q^2)^2}\right)^{1/n}\right) \times$$

$$\times \left(1 + \left(\frac{1}{c}\sqrt{(q^1)^2 + (q^2)^2}\right)^{2/n}\right)^{-1} \left[(q^1)^2 + (q^2)^2\right]^{(1-n)/2n} q^\mu \ .$$

E-34 A geodesic as a submanifold of the ellispoid-of-revolution (conformal coordinates)

Let us represent the differential equations of a geodesic as a submanifold of the ellispoid-of-revolution $\mathbb{M}^2 := \mathbb{E}^2_{A,B}$ as functions $q^\mu(t)$ as well as the Lagrangean L^2 and the Hamiltonian H^2 with respect to conformal coordinates (isothermal, isometric) of the most important applicable map projections listed below. Compare with Fig. E.5, which illustrates the Maupertuis gauged geodesic $\mathbb{E}^2_{A,B}$ generated by the Universal Transverse Mercator projection (UTM) with the dilatation factor $\rho = 0.999\,578$ (E. Grafarend 1994). Compare with Boxes E.4–E.6, which collect the central relations.

Important!

(i) The Universal Polar Stereographic Projection (UPS) (central perspective projection from the South Pole to a tangent plane $T_{\mathrm{np}}\mathbb{M}^2$ at the North Pole), (ii) the Universal Mercator Projection (UM) (conformal diffeomorphism of the ellipsoid-of-revolution onto a circular cylinder $\mathbb{S}^1_A \times \mathbb{R}$ where the circle $\mathbb{S}^1_A$ is chosen as the equator of the ellipsoid-of-revolution), (iii) the Universal Transverse Mercator Projection (UTM) (conformal diffeomorphism of the ellipsoid-of-revolution onto a elliptic cylinder (Λ_0 meta-equator), where the Λ_0 ellipse is chosen as the reference meridian of the ellipsoid-of-revolution), and (iv) the Universal Conic Lambert Projection (UC) (conformal diffeomorphism of the ellipsoid-of-revolution onto a circular cone $\{\boldsymbol{x} \in \mathbb{R}^3 \,|\, (x^2 + y^2)/a^2 - z^2/b^2 = 0,\ a^2 + b^2 = \text{const.}\}$, where the circle $\mathbb{S}^1_a$ with $a = A\cos\Phi/(1 - e^2\sin^2\Phi)^{-1/2}$ is chosen as a definite parallel circle (line-of-contact) and b as the latitude dependent definite distance of the circle $\mathbb{S}^1_a$ from the equatorial plane, and $e^2 = (A^2 - B^2)/A^2 = 1 - (B^2/A^2))$.

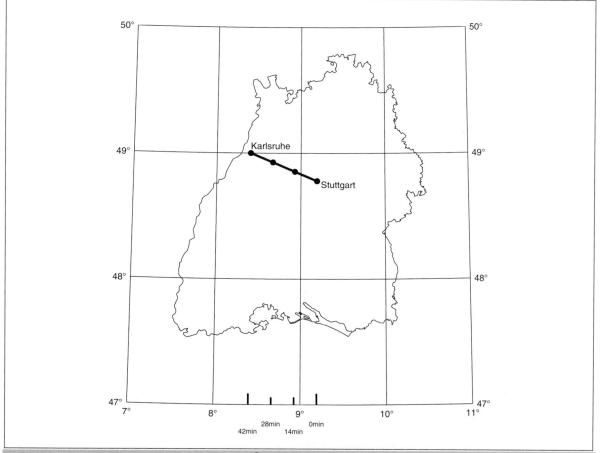

Fig. E.5. Maupertuis gauged geodesic on $\mathbb{E}^2_{A,B}$, conformal coordinates generated by the universal transverse Mercator projection, solution of the boundary value problem, departing point Stuttgart, arrival point Karlsruhe, time given for constant speed movement (70 km/h) with respect to the arrival point.

Box E.4 (The conformal coordinates as functions of ellipsoidal longitude and ellipsoidal latitude and the factor of conformality).

(i) Universal Polar Stereographic Projection (UPS):

$$q^1 = f(\Phi)\cos\Lambda \ , \quad q^2 = f(\Phi)\sin\Lambda \ , \tag{E.68}$$

$$f(\Phi) := \frac{2A}{\sqrt{1-e^2}}\left(\frac{1-e}{1+e}\right)^{\frac{e}{2}}\left(\frac{1+e\sin\Phi}{1-e\sin\Phi}\right)^{\frac{e}{2}}\tan\left(\frac{\pi}{4}-\frac{\Phi}{2}\right) \ , \tag{E.69}$$

$$\lambda = \frac{A\cos\Phi}{\sqrt{1-e^2\sin^2\Phi}}\frac{1}{f(\Phi)} = \frac{A\cos f^{-1}\left(\sqrt{(q^1)^2+(q^2)^2}\right)}{\sqrt{1-e^2\sin^2 f^{-1}\left(\sqrt{(q^1)^2+(q^2)^2}\right)}}\frac{1}{\sqrt{(q^1)^2+(q^2)^2}} \ , \tag{E.70}$$

$$ds^2 = \frac{A^2\cos^2 f^{-1}\left(\sqrt{(q^1)^2+(q^2)^2}\right)}{1-e^2\sin^2 f^{-1}\left(\sqrt{(q^1)^2+(q^2)^2}\right)}\frac{1}{(q^1)^2+(q^2)^2}\left[(dq^1)^2+(dq^2)^2\right] \ . \tag{E.71}$$

($\lambda(q^1,q^2)$ from series inversion of $q^1(\Lambda,\Phi)$ and $q^2(\Lambda,\Phi)$ leading to $\Lambda(q^1,q^2)$ and $\Phi(q^1,q^2)$. See J. P. Snyder (1987), pp. 162.)

(ii) Universal Mercator Projection (UM):

$$q^1 = A\Lambda \ , \quad q^2 = A f(\Phi) \ , \tag{E.72}$$

$$f(\Phi) := \ln\tan\left(\frac{\pi}{4}+\frac{\Phi}{2}\right)\left(\frac{1-e\sin\Phi}{1+e\sin\Phi}\right)^{\frac{e}{2}} \ , \tag{E.73}$$

$$\lambda = \frac{\sqrt{1-e^2\sin^2\Phi}}{\cos\Phi} = \frac{\sqrt{1-e^2\sin^2 f^{-1}(q^2/A)}}{\cos f^{-1}(q^2/A)} \ , \tag{E.74}$$

$$ds^2 = \frac{1-e^2\sin^2 f^{-1}(q^2/A)}{\cos^2 f^{-1}(q^2/A)}\left[(dq^1)^2+(dq^2)^2\right] \ . \tag{E.75}$$

($\lambda(q^2)$ from series inversion of $q^2(\Phi)$ leading to $\Phi(q^2)$. See J. P. Snyder (1987), pp. 45.)

(iii) Universal Transverse Mercator Projection (UTM):

$$q^1 = \rho[a_0 + a_{10}b + a_{20}b^2 + a_{02}l^2 + a_{30}b^3 + a_{12}bl^2 + a_{40}b^4 + a_{22}b^2l^2 + a_{04}l^4 + a_{50}b^5 +$$
$$+ a_{32}b^3l^2 + a_{14}bl^4 + O_1(b^6,l^6)] \quad \text{(northern)} \ ,$$

$$\tag{E.76}$$

$$q^2 = \rho[a_{01}l + a_{11}bl + a_{21}b^2l + a_{03}l^3 + a_{31}b^3l + a_{13}bl^3 + a_{41}b^4l +$$
$$+ a_{23}b^2l^3 + a_{05}l^5 + O_2(b^6,l^6)] \quad \text{(eastern)} \ .$$

(Valid $\forall\, b := \Phi - \Phi_0, \ l := \Lambda - \Lambda_0, \ \rho := 0.999\,578$ (dilatation factor). Coefficients see E. Grafarend (1994), Table 3.3.)

Continuation of Box.

$$\frac{1}{\lambda^2} = \rho^2[1 + c_{02}l^2 + c_{12}bl^2 + c_{04}l^4 + c_{22}b^2l^2 + c_{14}bl^4 + c_{32}b^3l^2 + O_{\lambda^2}(b^6, l^6)] \, ,$$

$$\lambda^2 = \frac{1}{\rho^2}[1 + d_{02}(q^2)^2 + d_{12}q^1(q^2)^2 + d_{22}(q^1)^2(q^2)^2 + \tag{E.77}$$

$$+d_{32}(q^1)^3(q^2)^2 + d_{04}(q^2)^4 + d_{14}q^1(q^2)^4 + O_{\lambda^2}((q^1)^6, (q^2)^6)] \, ,$$

$$ds^2 = \lambda^2(q^1, q^2)\left[(dq^1)^2 + (dq^2)^2\right] \, . \tag{E.78}$$

($\lambda^2(q^1, q^2)$ from series inversion of $q^1(l, b)$ and $q^2(l, b)$ leading to $l(q^1, q^2)$ and $b(q^1, q^2)$.)

The coefficients $c_{\mu\nu}$ and $d_{\mu\nu}$ are defined as follows
($\eta_0^2 = \frac{e^2}{1-e^2}\cos^2\Phi_0$ and $t = \tan\Phi_0$):

$$c_{02} = (1 + \eta_0^2)\cos^2\Phi_0 \, ,$$

$$c_{12} = -2t_0(1 + 2\eta_0^2)\cos^2\Phi_0 \, ,$$

$$c_{04} = \frac{1}{12}(8 - 4t_0^2 + 20\,\eta_0^2 - 28\,\eta_0^2t_0^2 + 16\eta_0^4 - 48\eta_0^4t_0^2 - 24\eta_0^6t_0^2 + 4\eta_0^6)\cos^4\Phi_0 \, ,$$

$$c_{22} = (t_0^2 - 1 - 2\,\eta_0^2 + 6\,\eta_0^2t_0^2)\cos^2\Phi_0 \, , \tag{E.79}$$

$$c_{14} = \frac{1}{3}(-7 + 2t_0^2 - 44\eta_0^2 + 28\,\eta_0^2t_0^2 - 56\eta_0^4 + 72\eta_0^4t_0^2 + 48\eta_0^6t_0^2 - 22\eta_0^6)t_0\cos^4\Phi_0 \, ,$$

$$c_{32} = \frac{2}{3}(2 + 10\eta_0^2 + 6\,\eta_0^2t_0^2)t_0\cos^2\Phi_0 \, ,$$

$$d_{02} = \frac{-2(1 - e^2\sin^2\Phi_0)^2}{A^2(1-e^2)} \, ,$$

$$d_{12} = \frac{4(1 - e^2\sin^2\Phi_0)^{5/2}}{A^3(1-e^2)^2}\sin\Phi_0\cos\Phi_0 \, ,$$

$$d_{22} = \frac{-2e^2(1 - e^2\sin^2\Phi_0)^3}{A^4(1-e^2)^3} \times$$

$$\times[-1 + 2\sin^2\Phi_0 + e^2\sin^2\Phi_0(6 - 7\sin^2\Phi_0)] \, ,$$

$$d_{32} = \frac{-4e^2(1 - e^2\sin^2\Phi_0)^{7/2}}{3A^5(1-e^2)^4} \times \tag{E.80}$$

$$\times[2 + e^2(9 - 22\sin^2\Phi_0) + e^4\sin^2\Phi_0(-24 + 35\sin^2\Phi_0)]\sin\Phi_0\cos_0 \, ,$$

$$d_{04} = \frac{3(1 - e^2\sin^2\Phi_0)^4}{A^4(1-e^2)^2} - \frac{(1 - e^2\sin^2\Phi_0)^3}{12A^4(1-e^2)^3} \times$$

$$\times[1 + e^2(3 - 5\sin^2\Phi_0) + e^4\sin^2\Phi_0(-27 + 28\sin^2\Phi_0)] \, ,$$

$$d_{14} = \frac{-12(1 - e^2\sin^2\Phi_0)^{9/2}}{A^5(1-e^2)^3}\sin\Phi_0\cos\Phi_0 + \frac{2e^2(1 - e^2\sin^2\Phi_0)^{7/2}}{A^5(1-e^2)^4} \times$$

$$\times[2 + e^2(9 - 19\sin^2\Phi_0) + e^4\sin^2\Phi_0(-27 + 35\sin^2\Phi_0)]\sin\Phi_0\cos\Phi_0 \, .$$

Continuation of Box.

(iv) Universal Conic Projection (UC):

$$q^1 = r \cos \alpha \; , \;\; q^2 = r \sin \alpha \; , \tag{E.81}$$

$$\alpha = n \, \Lambda \; , \;\; r = c \left(\tan \left(\tfrac{\pi}{4} - \tfrac{\Phi}{2} \right) \left(\tfrac{1 + e \sin \Phi}{1 - e \sin \Phi} \right)^{\frac{e}{2}} \right)^n \; , \tag{E.82}$$

$$\lambda = \frac{A \cos \Phi}{c \, n \sqrt{1 - e^2 \sin^2 \Phi}} \left(\tan \left(\tfrac{\pi}{4} - \tfrac{\Phi}{2} \right) \left(\tfrac{1 + e \sin \Phi}{1 - e \sin \Phi} \right)^{\frac{e}{2}} \right)^{-n} =$$

$$= \frac{A \cos r^{-1} \left(\sqrt{(q^1)^2 + (q^2)^2} \right)}{c \, n \sqrt{1 - e^2 \sin^2 r^{-1} \left(\sqrt{(q^1)^2 + (q^2)^2} \right)}} \times$$

$$\times \left(\tan \left(\tfrac{\pi}{4} - \frac{r^{-1} \left(\sqrt{(q^1)^2 + (q^2)^2} \right)}{2} \right) \left(\frac{1 + e \sin r^{-1} \left(\sqrt{(q^1)^2 + (q^2)^2} \right)}{1 - e \sin r^{-1} \left(\sqrt{(q^1)^2 + (q^2)^2} \right)} \right)^{\frac{e}{2}} \right)^{-n} . \tag{E.83}$$

Variant one (UC). Equidistant map of the parallel circle Φ_0/line-of-contact:

$$n = \sin \Phi_0 \; , \tag{E.84}$$

$$c = \frac{A \cot \Phi_0}{\sqrt{1 - e^2 \sin^2 \Phi_0}} \left(\tan \left(\tfrac{\pi}{4} - \tfrac{\Phi_0}{2} \right) \left(\tfrac{1 + e \sin \Phi_0}{1 - e \sin \Phi_0} \right)^{\frac{e}{2}} \right)^{-n} , \tag{E.85}$$

$$\mathrm{d}s^2 = \frac{A^2 \cos^2 r^{-1} \left((q^1)^2 + (q^2)^2 \right)}{c^2 n^2 \left(1 - e^2 \sin^2 r^{-1} \left(\sqrt{(q^1)^2 + (q^2)^2} \right) \right)} \times$$

$$\times \left(\tan^2 \left(\tfrac{\pi}{4} - \frac{r^{-1} \left(\sqrt{(q^1)^2 + (q^2)^2} \right)}{2} \right) \left(\frac{1 + e \sin r^{-1} \left(\sqrt{(q^1)^2 + (q^2)^2} \right)}{1 - e \sin r^{-1} \left(\sqrt{(q^1)^2 + (q^2)^2} \right)} \right)^{\frac{e}{2}} \right)^{-n} \left[(q^1)^2 + (q^2)^2 \right] . \tag{E.86}$$

Variant two (UC). Equidistant map of two parallel circles Φ_1/Φ_2 (Lambert conformal projection):

$$n = \frac{\ln \left(\cos \Phi_1 \sqrt{1 - e^2 \sin^2 \Phi_1} \right) - \ln \left(\cos \Phi_2 \sqrt{1 - e^2 \sin^2 \Phi_2} \right)}{\ln \tan \left(\tfrac{\pi}{4} - \tfrac{\Phi_1}{2} \right) \left(\tfrac{1 + e \sin \Phi_1}{1 - e \sin \Phi_1} \right)^{\frac{e}{2}} - \ln \tan \left(\tfrac{\pi}{4} - \tfrac{\Phi_2}{2} \right) \left(\tfrac{1 + e \sin \Phi_2}{1 - e \sin \Phi_2} \right)^{\frac{e}{2}}} \; , \tag{E.87}$$

$$c = \frac{A \cos \Phi_1}{n \sqrt{1 - e^2 \sin^2 \Phi_1}} \left(\tan \left(\tfrac{\pi}{4} - \tfrac{\Phi_1}{2} \right) \left(\tfrac{1 + e \sin \Phi_1}{1 - e \sin \Phi_1} \right)^{\frac{e}{2}} \right)^{-n} , \tag{E.88}$$

$$\mathrm{d}s^2 = \frac{A^2 \cos^2 r^{-1} \left((q^1)^2 + (q^2)^2 \right)}{c^2 n^2 \left(1 - e^2 \sin^2 r^{-1} \left(\sqrt{(q^1)^2 + (q^2)^2} \right) \right)} \times$$

$$\times \left(\tan^2 \left(\tfrac{\pi}{4} - \frac{r^{-1} \left(\sqrt{(q^1)^2 + (q^2)^2} \right)}{2} \right) \left(\frac{1 + e \sin r^{-1} \left(\sqrt{(q^1)^2 + (q^2)^2} \right)}{1 - e \sin r^{-1} \left(\sqrt{(q^1)^2 + (q^2)^2} \right)} \right)^{e} \right)^{-n} \left[(q^1)^2 + (q^2)^2 \right] . \tag{E.89}$$

($\lambda(q^1, q^2)$ from series inversion of $q^1(\Lambda, \Phi)$ and $q^2(\Lambda, \Phi)$ leading to $\Lambda(q^1, q^2)$ and $\Phi(q^1, q^2)$.
See J. P. Snyder (1987), pp. 109.)

Box E.5 (The Lagrangean version versus the Hamiltonian version of a geodesic in $\mathbb{E}^2_{A,B}$ in terms of conformal coordinates (isometric coordinates) and Maupertuis gauge).

(i) Universal Polar Stereographic Projection (UPS):

$$L^2(q(t), \dot{q}(t)) = \tfrac{1}{2}\left[(\dot{q}^1)^2 + (\dot{q}^2)^2\right] + \frac{A^2 \cos^2 f^{-1}\left(\sqrt{(q^1)^2 + (q^2)^2}\right)}{2\left(1 - e^2 \sin^2 f^1\left(\sqrt{(q^1)^2 + (q^2)^2}\right)\right)} \frac{1}{(q^1)^2 + (q^2)^2}, \tag{E.90}$$

$$H^2(q(t), p(t)) = \tfrac{1}{2}\left(p_1^2 + p_2^2\right) - \frac{A^2 \cos^2 f^{-1}\left(\sqrt{(q^1)^2 + (q^2)^2}\right)}{2\left(1 - e^2 \sin^2 f^1\left(\sqrt{(q^1)^2 + (q^2)^2}\right)\right)} \frac{1}{(q^1)^2 + (q^2)^2}. \tag{E.91}$$

(ii) Universal Mercator Projection (UM):

$$L^2(q(t), \dot{q}(t)) = \tfrac{1}{2}\left[(\dot{q}^1)^2 + (\dot{q}^2)^2\right] + \frac{1 - e^2 \sin^2 f^1(q^2/A)}{2\cos^2 f^{-1}(q^2/A)}, \tag{E.92}$$

$$H^2(q(t), p(t)) = \tfrac{1}{2}\left(p_1^2 + p_2^2\right) - \frac{1 - e^2 \sin^2 f^{-1}(q^2/A)}{2\cos^2 f^1(q^2/A)}. \tag{E.93}$$

(iii) Universal Transverse Mercator Projection (UTM):

$$L^2(q(t), \dot{q}(t)) = \tfrac{1}{2}\left[(\dot{q}^1)^2 + (\dot{q}^2)^2\right] + \frac{1}{2\rho^2}(1 + d_{02}(q^2)^2 + d_{12}q^1(q^2)^2 + d_{22}(q^1)^2(q^2)^2 +$$
$$+ d_{32}(q^1)^3(q^2)^2 + d_{04}(q^2)^4 + d_{14}q^1(q^2)^4 + O_{\lambda^2}((q^1)^6, (q^2)^6)), \tag{E.94}$$

$$H^2(q(t), p(t)) = \tfrac{1}{2}\left(p_1^2 + p_2^2\right) - \frac{1}{2\rho^2}(1 + d_{02}(q^2)^2 + d_{12}q^1(q^2)^2 + d_{22}(q^1)^2(q^2)^2 +$$
$$+ d_{32}(q^1)^3(q^2)^2 + d_{04}(q^2)^4 + d_{14}q^1(q^2)^4 + O_{\lambda^2}((q^1)^6, (q^2)^6)). \tag{E.95}$$

(iv) Universal Conic Projection (UC) (variant one and variant two):

$$L^2(q(t), \dot{q}(t)) = \tfrac{1}{2}\left[(\dot{q}^1)^2 + (\dot{q}^2)^2\right] + \frac{A^2 \cos^2 r^{-1}\left((q^1)^2 + (q^2)^2\right)}{2c^2 n^2\left(1 - e^2 \sin^2 r^1\left(\sqrt{(q^1)^2 + (q^2)^2}\right)\right)} \times$$

$$\times \left(\tan^2\left(\frac{\pi}{4} - \frac{r^{-1}\left(\sqrt{(q^1)^2 + (q^2)^2}\right)}{2}\right)\left(\frac{1 + e \sin r^{-1}\left(\sqrt{(q^1)^2 + (q^2)^2}\right)}{1 - e \sin r^{-1}\left(\sqrt{(q^1)^2 + (q^2)^2}\right)}\right)^e\right)^{-n}, \tag{E.96}$$

$$H^2(q(t), p(t)) = \tfrac{1}{2}\left(p_1^2 + p_2^2\right) - \frac{A^2 \cos^2 r^{-1}\left((q^1)^2 + (q^2)^2\right)}{2c^2 n^2\left(1 - e^2 \sin^2 r^1\left(\sqrt{(q^1)^2 + (q^2)^2}\right)\right)} \times$$

$$\times \left(\tan^2\left(\frac{\pi}{4} - \frac{r^{-1}\left(\sqrt{(q^1)^2 + (q^2)^2}\right)}{2}\right)\left(\frac{1 + e \sin r^{-1}\left(\sqrt{(q^1)^2 + (q^2)^2}\right)}{1 - e \sin r^{-1}\left(\sqrt{(q^1)^2 + (q^2)^2}\right)}\right)^e\right)^{-n}. \tag{E.97}$$

Box E.6 (The differential equations of a geodesic in $\mathbb{E}^2_{A,B}$ in terms of conformal coordinates (isometric coordinates) and Maupertuis gauge: Lagrange portrait, two differential equations of second order, Hamilton portrait, four differential equations of first order).

(i) Universal Polar Stereographic Projection (UPS):

$$\ddot{q}^\mu + \frac{A^2}{(q^1)^2+(q^2)^2} \frac{(1-e^2)\cos f^{-1}\left(\sqrt{(q^1)^2+(q^2)^2}\right)\sin f^{-1}\left(\sqrt{(q^1)^2+(q^2)^2}\right)}{\left(1-e^2\sin^2 f^{-1}\left(\sqrt{(q^1)^2+(q^2)^2}\right)\right)^2} \times$$

$$\times \frac{\partial f^{-1}\left(\sqrt{(q^1)^2+(q^2)^2}\right)}{\partial q^\mu} - \frac{A^2}{\left[(q^1)^2+(q^2)^2\right]^2}\frac{\cos^2 f^{-1}\left(\sqrt{(q^1)^2+(q^2)^2}\right)}{1-e^2\sin^2 f^{-1}\left(\sqrt{(q^1)^2+(q^2)^2}\right)}q^\mu = 0 \tag{E.98}$$

$$\forall\, \mu = 1,2\,,$$

$$\dot{q}^\mu = \delta^{\mu\nu}p_\nu \ \forall\, \mu,\nu = 1,2\,, \tag{E.99}$$

$$\dot{p}_\mu = -\frac{A^2}{(q^1)^2+(q^2)^2}\frac{(1-e^2)\cos f^{-1}\left(\sqrt{(q^1)^2+(q^2)^2}\right)\sin f^{-1}\left(\sqrt{(q^1)^2+(q^2)^2}\right)}{\left(1-e^2\sin^2 f^{-1}\left(\sqrt{(q^1)^2+(q^2)^2}\right)\right)^2} \times$$

$$\times \frac{\partial f^{-1}\left(\sqrt{(q^1)^2+(q^2)^2}\right)}{\partial q^\mu} + \frac{A^2}{\left[(q^1)^2+(q^2)^2\right]^2}\frac{\cos^2 f^{-1}\left(\sqrt{(q^1)^2+(q^2)^2}\right)}{1-e^2\sin^2 f^{-1}\left(\sqrt{(q^1)^2+(q^2)^2}\right)}q^\mu \tag{E.100}$$

$$\forall\, \mu = 1,2\,.$$

(ii) Universal Mercator Projection (UM):

$$\ddot{q}^1 = 0\,,\quad \ddot{q}^2 - \frac{(1-e^2)\sin f^{-1}(q^2/A)}{\cos^3 f^{-1}(q^2/A)}\frac{\partial f^{-1}(q^2/A)}{\partial q^2} = 0\,, \tag{E.101}$$

$$\dot{q}^\mu = \delta^{\mu\nu}p_\nu\,, \tag{E.102}$$

$$\dot{p}_1 = 0 \ (p_1 \text{ cyclic})\,,\quad p_1 = \text{const.}\,, \tag{E.103}$$

$$\dot{p}_2 = \frac{(1-e^2)\sin f^{-1}(q^2/A)}{\cos^3 f^{-1}(q^2/A)}\frac{\partial f^{-1}(q^2/A)}{\partial q^2}\,. \tag{E.104}$$

(iii) Universal Transverse Mercator Projection (UTM):

$$\ddot{q}^1 - \frac{1}{2\rho^2}[d_{12}(q^2)^2 + 2d_{22}q^1(q^2)^2 + 3d_{32}(q^1)^2(q^2)^2 + d_{14}(q^2)^4 + O_{\lambda^2}((q^1)^5,(q^2)^6)] = 0\,,$$
$$\ddot{q}^2 - \frac{1}{2\rho^2}[2d_{02}q^2 + 2d_{12}q^1q^2 + 2d_{22}(q^1)^2q^2 + 2d_{32}(q^1)^3q^2 + 4d_{04}(q^2)^3 + \tag{E.105}$$
$$+ 4d_{14}q^1(q^2)^3 + O_{\lambda^2}((q^1)^6,(q^2)^5)] = 0\,,$$

$$\dot{q}^\mu = \delta^{\mu\nu}p_\nu\,, \tag{E.106}$$

$$\dot{p}_1 = \frac{1}{2\rho^2}[d_{12}(q^2)^2 + 2d_{22}q^1(q^2)^2 + 3d_{32}(q^1)^2(q^2)^2 + d_{14}(q^2)^4 + O_{\lambda^2}((q^1)^5,(q^2)^6)]\,,$$
$$\dot{p}_2 = \frac{1}{2\rho^2}[2d_{02}q^2 + 2d_{12}q^1q^2 + 2d_{22}(q^1)^2q^2 + 2d_{32}(q^1)^3q^2 + 4d_{04}(q^2)^3 + \tag{E.107}$$
$$+ 4d_{14}q^1(q^2)^3 + O_{\lambda^2}((q^1)^6,(q^2)^5)]\,.$$

Continuation of Box.

(iv) Universal Conic Projection (UC) (variant one and variant two):

$$
\ddot{q}_\mu - \left[\frac{A^2(1-e^2)\cos^2 r^{-1}\left(\sqrt{(q^1)^2+(q^2)^2}\right)\left(1+\sin r^{-1}\left(\sqrt{(q^1)^2+(q^2)^2}\right)\right)^{-1}}{2c^2 n\left(1-e^2\sin^2 r^{-1}\left(\sqrt{(q^1)^2+(q^2)^2}\right)\right)\left(1-e\sin r^{-1}\left(\sqrt{(q^1)^2+(q^2)^2}\right)\right)^2} \times\right.
$$

$$
\times \left(\tan\left(\frac{\pi}{4} - \frac{r^{-1}\left(\sqrt{(q^1)^2+(q^2)^2}\right)}{2}\right)\left(\frac{1+e\sin r^{-1}\left(\sqrt{(q^1)^2+(q^2)^2}\right)}{1-e\sin r^{-1}\left(\sqrt{(q^1)^2+(q^2)^2}\right)}\right)^{\frac{e}{2}}\right)^{-2n-1} -
$$

$$
- \frac{A^2(1-e^2)\cos r^{-1}\left(\sqrt{(q^1)^2+(q^2)^2}\right)\sin r^{-1}\left(\sqrt{(q^1)^2+(q^2)^2}\right)}{c^2 n^2\left(1-e^2\sin^2 r^{-1}\left(\sqrt{(q^1)^2+(q^2)^2}\right)\right)} \times \qquad (E.108)
$$

$$
\left. \times \left(\tan\left(\frac{\pi}{4} - \frac{r^{-1}\left(\sqrt{(q^1)^2+(q^2)^2}\right)}{2}\right)\left(\frac{1+e\sin r^{-1}\left(\sqrt{(q^1)^2+(q^2)^2}\right)}{1-e\sin r^{-1}\left(\sqrt{(q^1)^2+(q^2)^2}\right)}\right)^{\frac{e}{2}}\right)^{-2n}\right] \times
$$

$$
\times \frac{\partial r^{-1}\left(\sqrt{(q^1)^2+(q^2)^2}\right)}{\partial q^\mu} = 0 \, ,
$$

$$
\dot{q}^\mu = \delta^{\mu\nu}p_\nu \, , \qquad (E.109)
$$

$$
\dot{p}_\mu = \left[\frac{A^2(1-e^2)\cos^2 r^{-1}\left(\sqrt{(q^1)^2+(q^2)^2}\right)\left(1+\sin r^{-1}\left(\sqrt{(q^1)^2+(q^2)^2}\right)\right)^{-1}}{2c^2 n\left(1-e^2\sin^2 r^{-1}\left(\sqrt{(q^1)^2+(q^2)^2}\right)\right)\left(1-e\sin r^{-1}\left(\sqrt{(q^1)^2+(q^2)^2}\right)\right)^2} \times\right.
$$

$$
\times \left(\tan\left(\frac{\pi}{4} - \frac{r^{-1}\left(\sqrt{(q^1)^2+(q^2)^2}\right)}{2}\right)\left(\frac{1+e\sin r^{-1}\left(\sqrt{(q^1)^2+(q^2)^2}\right)}{1-e\sin r^{-1}\left(\sqrt{(q^1)^2+(q^2)^2}\right)}\right)^{\frac{e}{2}}\right)^{-2n-1} -
$$

$$
- \frac{A^2(1-e^2)\cos r^{-1}\left(\sqrt{(q^1)^2+(q^2)^2}\right)\sin r^{-1}\left(\sqrt{(q^1)^2+(q^2)^2}\right)}{c^2 n^2\left(1-e^2\sin^2 r^{-1}\left(\sqrt{(q^1)^2+(q^2)^2}\right)\right)^2} \times \qquad (E.110)
$$

$$
\left. \times \left(\tan\left(\frac{\pi}{4} - \frac{r^{-1}\left(\sqrt{(q^1)^2+(q^2)^2}\right)}{2}\right)\left(\frac{1+e\sin r^{-1}\left(\sqrt{(q^1)^2+(q^2)^2}\right)}{1-e\sin r^{-1}\left(\sqrt{(q^1)^2+(q^2)^2}\right)}\right)^{\frac{e}{2}}\right)^{-2n}\right] \times
$$

$$
\times \frac{\partial r^{-1}\left(\sqrt{(q^1)^2+(q^2)^2}\right)}{\partial q^\mu} \, .
$$

E-35 Maupertuis gauged geodesics (normal coordinates, local tangent plane)

The unit tangent vector (Darboux one-leg) $\boldsymbol{d}$ at a point $\boldsymbol{q}$ of a geodesic can be represented in the local base $\{\boldsymbol{g}_1, \boldsymbol{g}_2\}$, the local tangent vectors (Gauss two-leg), which spans the tangent space $T_q\mathbb{M}^2$ at a point $\boldsymbol{q}$, namely by (E.111), where $\boldsymbol{x}'\left(q^1, q^2\right)$ denotes the immersion $\{\mathbb{M}^2, q_{\mu\nu}\} \to \{\mathbb{R}^3, \delta_{ij}\}$. Note that from now on, we assume $\{\boldsymbol{g}_1, \boldsymbol{g}_2\}$ to be an orthogonal, but not normalized Gauss two-leg which can be materialized by orthogonal coordinates $\{q^1, q^2\}$.

$$
\boldsymbol{d} = \boldsymbol{x}'(q^1, q^2) = \frac{\partial \boldsymbol{x}}{\partial q^\mu}q'^\mu = \boldsymbol{g}_1 q'^1 + \boldsymbol{g}_2 q'^2 \, . \qquad (E.111)
$$

Alternatively, we can represent $\boldsymbol{d}$ by polar coordinates as follows: by means of the scalar products (inner products) (E.112), we introduce the azimuth α with respect to the first Gauss two-leg $\boldsymbol{g}_1$. Inserting (E.111) into (E.112) and differentiating (E.112), we obtain (E.113) or, in terms of conformal coordinates, we obtain (E.114). The parameter change $s \to t$, (E.112) implemented, leads to (E.115).

$$\frac{\langle \boldsymbol{d}, \boldsymbol{g}_1 \rangle}{\|\boldsymbol{d}\|\|\boldsymbol{g}_1\|} := \cos \alpha \,,$$

$$\frac{\langle \boldsymbol{d}, \boldsymbol{g}_2 \rangle}{\|\boldsymbol{d}\|\|\boldsymbol{g}_2\|} := \sin \alpha \,,$$

(E.112)

(i)

$$\langle \boldsymbol{d}, \boldsymbol{g}_1 \rangle = g_{11} q'^1 = \|\boldsymbol{d}\|\|\boldsymbol{g}_1\| \cos \alpha = \sqrt{g_{11}} \cos \alpha \,,$$

(ii)

$$\langle \boldsymbol{d}, \boldsymbol{g}_2 \rangle = g_{22} q'^2 = \|\boldsymbol{d}\|\|\boldsymbol{g}_2\| \sin \alpha = \sqrt{g_{22}} \sin \alpha \,,$$

(E.113)

(iii)

$$\frac{\mathrm{d}\alpha}{\mathrm{d}s} = \alpha'(s) = \frac{1}{\sqrt{g_{11} g_{22}}} \left(-\frac{\partial \sqrt{g_{22}}}{\partial q^1} \sin \alpha + \frac{\partial \sqrt{g_{11}}}{\partial q^2} \cos \alpha \right) \cos \alpha \,,$$

$$q'^1 = \frac{\cos \alpha}{\lambda} \,,$$

$$q'^2 = \frac{\sin \alpha}{\lambda} \,,$$

$$\alpha' = \frac{1}{\lambda^2} \left(-\frac{\partial \lambda}{\partial q^1} \sin \alpha + \frac{\partial \lambda}{\partial q^2} \cos \alpha \right) \,,$$

(E.114)

$$\frac{\mathrm{d}q^1}{\mathrm{d}t} = \frac{\mathrm{d}q^1}{\mathrm{d}s} \frac{\mathrm{d}s}{\mathrm{d}t} = \lambda \cos \alpha \,,$$

$$\frac{\mathrm{d}q^2}{\mathrm{d}t} = \frac{\mathrm{d}q^2}{\mathrm{d}s} \frac{\mathrm{d}s}{\mathrm{d}t} = \lambda \sin \alpha \,,$$

$$\frac{\mathrm{d}\alpha}{\mathrm{d}t} = \frac{\mathrm{d}\alpha}{\mathrm{d}s} \frac{\mathrm{d}s}{\mathrm{d}t} = \frac{\partial \lambda}{\partial q^2} \cos \alpha - \frac{\partial \lambda}{\partial q^1} \sin \alpha \,.$$

(E.115)

Similarly, the generalized momenta $p_\mu \in^* T_q \mathbb{M}^2$ are represented by (E.116), leading to the Hamilton equations of a geodesic in terms of polar coordinates, namely to (E.117), and these are solved by means of Lie series in Section E-36.

$$p_1 = \lambda \cos \alpha \,, \quad p_2 = \lambda \sin \alpha \,,$$

(E.116)

$$\dot{q}^1 = \lambda \cos \alpha \,, \quad \dot{q}^2 = \lambda \sin \alpha \,,$$

$$\dot{p}_1 = \frac{1}{2} \frac{\partial \lambda^2}{\partial q^1} \,, \quad \dot{p}_2 = \frac{1}{2} \frac{\partial \lambda^2}{\partial q^2} \,.$$

(E.117)

Finally, as local polar/normal coordinates, we take advantage of Definition E.3.

Definition E.3 (Local polar/normal coordinates).

In the tangent plane $T_q\mathbb{M}^2$, we introduce local polar/normal coordinates by (E.118), where t is the parameter of the Maupertuis gauged geodesic and α is the azimuth of the geodesic passing the point $q \in \{\mathbb{M}^2, g_{\mu\nu}\}$.

$$u^1 = u := t\cos\alpha \,,$$

$$u^2 = u := t\sin\alpha \,. \tag{E.118}$$

End of Definition.

E-36 Maupertuis gauged geodesics (Lie series, Hamilton portrait)

$\mathbb{M}^2 := \mathbb{E}^2_{A,B}$ and the geodesic "Maupertuis gauged" in their Hamilton form are analytic. Accordingly, we can solve (E.113) by the Taylor expansion (E.119).

$$q^1 = q_0^1 + \frac{dq^1}{dt}(t=0)t + \frac{1}{2!}\frac{d^2q^1}{dt^2}(t=0)t^2 + \lim_{n\to\infty}\sum_{m=3}^{n}\frac{1}{m!}\frac{d^{(m)}q^1}{dt^m}(t=0)t^m \,,$$

$$q^2 = q_0^2 + \frac{dq^2}{dt}(t=0)t + \frac{1}{2!}\frac{d^2q^2}{dt^2}(t=0)t^2 + \lim_{n\to\infty}\sum_{m=3}^{n}\frac{1}{m!}\frac{d^{(m)}q^2}{dt^m}(t=0)t^m \,. \tag{E.119}$$

Important!

By means of (E.113) and (E.118), we here are able to take advantage of the Lie recurrence ("Lie series") which is summarized in the following Box E.7 in order to formulate the solution of the *initial value problem* ("erste geodätische Hauptaufgabe"). By standard series inversion of the homogeneous polynomial $q^\mu - q_0^\mu$, we have finally solved the *boundary value problem* ("zweite geodätische Hauptaufgabe") in terms of local polar/normal coordinates.

$$q^\mu = q_0^\mu + \lambda_0 u^\mu + a_{\mu\gamma}u^\mu u^\gamma + \lim_{n\to\infty}\sum_{m=3}^{n}a^\mu_{\mu_1\dots\mu_m}u^{\mu_1}\dots u^{\mu_m} \,, \tag{E.120}$$

$$u^\mu = b_0(q^\mu - q_0^\mu) + b_{\mu\gamma}(q^\mu - q_0^\mu)(q^\gamma - q_0^\gamma) + \lim_{n\to\infty}\sum_{m=3}^{n}b^\mu_{\mu_1\dots\mu_m}(q^{\mu_1} - q_0^{\mu_1})\dots(q^{\mu_m} - q_0^{\mu_m}) \,. \tag{E.121}$$

By means of the Lie series to solve (E.113), which is generated by the Universal Transverse Mercator Projection (UTM), we are led to (E.122) and (E.123), with $\Delta q^1 := q^1 - q_0^1$ and $\Delta q^2 := q^2 - q_0^2$. The coefficients $f_{\mu\nu}$, $g_{\mu\nu}$, $r_{\mu\nu}$, $h_{\mu\nu}$, and $k_{\mu\nu}$ are listed in Box E.8.

$$q^1 = q_0^1 + f_{10}u + f_{20}u^2 + f_{02}v^2 + f_{30}u^3 + f_{03}v^3 + f_{21}u^2v + f_{12}uv^2 + \cdots \,,$$

$$q^2 = q_0^2 + g_{01}v + g_{20}u^2 + g_{02}v^2 + g_{30}u^3 + g_{03}v^3 + g_{21}u^2v + g_{12}uv^2 + \cdots \,, \tag{E.122}$$

$$\alpha = \alpha_0 + r_{10}u + r_{01}v + r_{20}u^2 + r_{02}v^2 + r_{11}uv + \cdots \,,$$

$$u = t\cos\alpha_0 =$$

$$= h_{10}\Delta q^1 + h_{20}(\Delta q^1)^2 + h_{02}(\Delta q^2)^2 + h_{30}(\Delta q^1)^3 +$$

$$+h_{03}(\Delta q^2)^3 + h_{21}(\Delta q^1)^2\Delta q^2 + h_{12}\Delta q^1(\Delta q^2)^2 + \cdots \,,$$

$$v = t\sin\alpha_0 = \tag{E.123}$$

$$= k_{01}\Delta q^2 + k_{20}(\Delta q^1)^2 + k_{02}(\Delta q^2)^2 + k_{30}(\Delta q^1)^3 +$$

$$+k_{03}(\Delta q^2)^3 + k_{21}(\Delta q^1)^2\Delta q^2 + k_{12}\Delta q^1(\Delta q^2)^2 + \cdots \,.$$

Box E.7 (Lie recurrence up to the third derivatives, all quantities are to be taken at Taylor point $t = 0$, higher derivatives are available from the authors).

$$\dot{q}^1 = \lambda \cos \alpha \, ,$$

$$\dot{q}^1 t = \lambda u = \lambda u^1 \, ,$$

$$\dot{q}^2 = \lambda \sin \alpha \, ,$$

$$\dot{q}^2 t = \lambda v = \lambda u^2 \, ,$$

$$\dot{\alpha} = \lambda_{.2} \cos \alpha - \lambda_{.1} \sin \alpha \, ,$$

$$\dot{\alpha} t = \lambda_{.2} u - \lambda_{.1} v \, ,$$

$$\ddot{q}^1 = \dot{\lambda} \cos \alpha - \lambda \dot{\alpha} \sin \alpha = \lambda \lambda_{.1}(\cos^2 \alpha + \sin^2 \alpha) \, ,$$

$$\ddot{q}^1 t^2 = \lambda \lambda_{.1}(u^2 + v^2) \, ,$$

$$\ddot{q}^2 = \dot{\lambda} \sin \alpha + \lambda \dot{\alpha} \cos \alpha = \lambda \lambda_{.2}(\cos^2 \alpha + \sin^2 \alpha) \, ,$$

$$\ddot{q}^2 t^2 = \lambda \lambda_{.2}(u^2 + v^2) \, ,$$

$$\ddot{\alpha} = \dot{\lambda}_{.2} \cos \alpha - \dot{\lambda}_{.1} \sin \alpha - \lambda_{.2} \dot{\alpha} \sin \alpha - \lambda_{.1} \dot{\alpha} \sin \alpha \, ,$$

$$\ddot{\alpha} t^2 = (\lambda \lambda_{.12} - \lambda_{.1} \lambda_{.2})u^2 - (\lambda \lambda_{.12} - \lambda_{.1} \lambda_{.2})v^2 + (\lambda \lambda_{.22} - \lambda \dot{\lambda}_{.11} + (\lambda_{.1})^2 - (\lambda_{.2})^2)uv \, ,$$

$$\dddot{q}^1 = \dot{\lambda} \lambda_{.1}(\cos^2 \alpha + \sin^2 \alpha) + \lambda \dot{\lambda}_{.1}(\cos^2 \alpha + \sin^2 \alpha) \, ,$$

(E.124)

$$\dddot{q}^1 t^3 = (\lambda(\lambda_{.1})^2 + \lambda^2 \lambda_{.11})u^3 + (\lambda \lambda_{.1} \lambda_{.2} +$$

$$+ \lambda^2 \lambda_{.12})u^2 v + (\lambda(\lambda_{.1})^2 + \lambda^2 \lambda_{.11})uv^2 + (\lambda \lambda_{.1} \lambda_{.2} + \lambda^2 \lambda_{.12})v^3 \, ,$$

$$\dddot{q}^2 = \dot{\lambda} \lambda_{.2}(\cos^2 \alpha + \sin^2 \alpha) + \lambda \dot{\lambda}_{.2}(\cos^2 \alpha + \sin^2 \alpha) \, ,$$

$$\dddot{q}^2 t^3 = (\lambda(\lambda_{.2})^2 + \lambda^2 \lambda_{.22})v^3 + (\lambda \lambda_{.1} \lambda_{.2} + \lambda^2 \lambda_{.12})uv^2 +$$

$$+ (\lambda(\lambda_{.2})^2 + \lambda^2 \lambda_{.22})u^2 v + (\lambda \lambda_{.1} \lambda_{.2} + \lambda^2 \lambda_{.12})u^3 \, ,$$

$$\dddot{\alpha} = -4(\lambda \lambda_{.12} - \lambda_{.1} \lambda_{.2})\dot{\alpha} \cos \alpha \sin \alpha +$$

$$+ (\lambda \lambda_{.22} - \lambda \lambda_{.11} + (\lambda_{.1})^2 - (\lambda_{.2})^2)(\cos^2 \alpha - \sin^2 \alpha)\dot{\alpha} +$$

$$+ (\cos^2 \alpha - \sin^2 \alpha)\tfrac{d}{dt}(\lambda \lambda_{.12} - \lambda_{.1} \lambda_{.2}) +$$

$$+ \cos \alpha \sin \alpha \tfrac{d}{dt}((\lambda \lambda_{.22} - \lambda \lambda_{.11} + (\lambda_{.1})^2 - (\lambda_{.2})^2)) \, ,$$

$$\dddot{\alpha} t^3 = (\lambda \lambda_{.2} \lambda_{.22} - 2\lambda \lambda_{.2} \lambda_{.11} + (\lambda_{.1})^2 \lambda_{.2} - (\lambda_{.2})^3 + \lambda^2 \lambda_{.121})u^3 +$$

$$+ (5\lambda_{.1}(\lambda_{.2})^2 - (\lambda_{.1})^3 - 6\lambda \lambda_{.2} \lambda_{.12} + 2\lambda \lambda_{.1} \lambda_{.11} - \lambda \lambda_{.1} \lambda_{.22} + 2\lambda^2 \lambda_{.122} - \lambda^2 \lambda_{.111})u^2 v +$$

$$+ (-5(\lambda_{.1})^2 \lambda_{.2} - (\lambda_{.2})^3 + 6\lambda \lambda_{.1} \lambda_{.12} - 2\lambda \lambda_{.2} \lambda_{.22} + \lambda \lambda_{.2} \lambda_{.11} - 2\lambda^2 \lambda_{.112} + \lambda^2 \lambda_{.222})uv^2 +$$

$$+ (-\lambda \lambda_{.1} \lambda_{.11} + 2\lambda \lambda_{.1} \lambda_{.22} + (\lambda_{.1})^3 - \lambda_{.1}(\lambda_{.2})^2 - \lambda^2 \lambda_{.122})v^3 \, .$$

Note that $\lambda_{.\mu}$, $\lambda_{.\mu\nu}$, and $\lambda_{.\mu\nu\gamma}$ denote the first derivative with respect to q^μ, the second derivative with respect to q^μ and q^ν, and the third derivative with respect to q^μ, q^ν, and q^γ.

Box E.8 (The coefficients of the Lie series up to the third order, all quantities are to be taken af Taylor point $t = 0$).

$$f_{10} = \lambda = \frac{1}{\rho} \left(1 + \frac{d_{02}}{2}(q^2)^2 + \frac{d_{12}}{2}q^1(q^2)^2 \right) ,$$

$$f_{20} = \frac{1}{4\rho^2} \left(d_{12}(q^2)^2 + 2d_{22}q^1(q^2)^2 \right) ,$$

$$f_{30} = \frac{1}{12\rho^3} \left(2d_{22}(q^2)^2 + 6d_{32}q^1(q^2)^2 \right) ,$$

$$f_{03} = \frac{1}{12\rho^3} \left(2d_{12}q^2 + (d_{12}d_{02} + 4d_{14})(q^2)^3 + 4d_{22}q^1q^2 + 6d_{32}(q^1)^2q^2 \right) ,$$

$$f_{02} = f_{20} , \quad f_{21} = f_{03} , \quad f_{12} = f_{30} ,$$

(E.125)

$$g_{01} = \lambda = \frac{1}{\rho} \left(1 + \frac{d_{02}}{2}(q^2)^2 + \frac{d_{12}}{2}q^1(q^2)^2 \right) ,$$

$$g_{20} = \frac{1}{4\rho^2} \left(2d_{02}q^2 + 2d_{12}q^1q^2 + 2d_{22}(q^1)^2q^2 + 4d_{04}(q^1)^2 \right) ,$$

$$g_{30} = \frac{1}{12\rho^3} \left(2d_{12}q^2 + (d_{12}d_{02} + 4d_{14})(q^2)^3 + 4d_{22}q^1q^2 + 6d_{32}(q^1)^2q^2 \right) ,$$

$$g_{03} = \frac{1}{12\rho^3} \left(2d_{02} + 2d_{12}q^1 + 2d_{22}(q^1)^2 + (d_{02}^2 + 12d_{04})(q^2)^2 + 2d_{32}(q^1)^3 + \right.$$
$$\left. + (2d_{12}d_{02} + 12d_{14})q^1(q^2)^3 \right) ,$$

$$g_{02} = g_{20} , \quad g_{21} = g_{03} , \quad g_{12} = g_{30} ,$$

(E.126)

$$r_{10} = \frac{1}{\rho} \left(d_{02}q^2 + d_{12}q^1q^2 + d_{22}(q^1)^2q^2 + \left(2d_{04} - \frac{1}{2}d_{02}^2 \right)(q^2)^3 \right) ,$$

$$r_{01} = -\frac{1}{\rho} \left(\frac{1}{2}d_{12}(q^2)^2 + d_{22}q^1(q^2)^2 \right) ,$$

$$r_{20} = \frac{1}{2\rho^2} \left(d_{12}q^2 + 2d_{22}q^1q^2 + 3d_{32}(q^1)^2q^2 + \left(2d_{14} - \frac{1}{2}d_{02}d_{12} \right)(q^2)^3 \right) ,$$

$$r_{11} = \frac{1}{2\rho^2} \left(d_{02} + d_{12}q^1 + d_{22}(q^1)^2 + (6d_{04} - d_{22})(q^2)^2 + d_{32}(q^1)^3 + \right.$$
$$\left. + (2d_{02}d_{12} + 6d_{14} - 2d_{32})q^1(q^2)^2 \right) ,$$

$$r_{02} = r_{20} ,$$

(E.127)

$$h_{10} = \frac{1}{f_{10}} , \quad h_{20} = -\frac{f_{20}}{(f_{10})^3} ,$$

$$h_{02} = -\frac{f_{02}}{f_{10}(g_{01})^2} , \quad h_{30} = \frac{2(f_{20})^2}{(f_{10})^5} - \frac{f_{30}}{(f_{10})^4} ,$$

$$h_{03} = \frac{2f_{02}g_{02}}{f_{10}(g_{01})^4} - \frac{f_{03}}{f_{10}(g_{01})^3} , \quad h_{21} = \frac{2f_{02}g_{02}}{(f_{10})^3(g_{01})^2} - \frac{f_{21}}{(f_{10})^3g_{01}} ,$$

$$h_{12} = \frac{2f_{20}g_{02}}{(f_{10})^3(g_{01})^2} - \frac{f_{12}}{(f_{10})^2(g_{01})^2} ,$$

(E.128)

$$k_{01} = \frac{1}{g_{01}} , \quad k_{02} = -\frac{g_{02}}{(g_{01})^3} ,$$

$$k_{20} = -\frac{g_{20}}{g_{01}(f_{10})^2} , \quad k_{03} = \frac{2(g_{20})^2}{(g_{01})^5} - \frac{g_{03}}{(g_{01})^4} ,$$

$$k_{30} = \frac{2f_{20}g_{20}}{(f_{10})^4g_{01}} - \frac{g_{30}}{g_{01}(f_{10})^3} , \quad k_{21} = \frac{2g_{02}g_{20}}{(f_{10})^2(g_{01})^3} - \frac{g_{21}}{(f_{10})^2(g_{01})^2} ,$$

$$k_{12} = \frac{2f_{02}g_{20}}{(f_{10})^2(g_{01})^3} - \frac{g_{12}}{f_{10}(g_{01})^3} .$$

(E.129)

Last but not least, let us here additionally note that Appendix E-3 is based upon the contribution of E. Grafarend and R. J. You (1995).

F Mixed cylindric map projections

Mixed cylindric map projections of the ellipsoid-of-revolution. Lambert projection and Sanson–Flamsteed projection, generalized Lambert projection and generalized Sanson–Flamsteed projection. The mapping equations and the conditions. Vertical coordinates and horizontal coordinates, horizontal weighted mean and vertical weighted mean. Deformation analysis of vertically/horizontally averaged equiareal cylindric mappings. Cauchy–Green deformation tensor and principal stretches.

The *mixed spherical map projections* of equiareal, cylindric type to be considered here are based upon the *Lambert projection* and the sinusoidal *Sanson–Flamsteed projection*. These cylindric and pseudo-cylindrical map projections of the sphere are generalized to the ellipsoid-of-revolution (biaxial ellipsoid). They are used in consequence by two lemmas to generate a horizontal (a vertical) weighted mean of equiareal cylindric map projections of the *ellipsoid-of-revolution*. Its left–right deformation analysis via further results leads to the left–right principal stretches/eigenvalues and left–right eigenvectors/eigenspace as well as the maximal left–right angular distortion for these new mixed cylindric map projections of ellipsoidal type. Detailed illustrations document the cartographic synergy of mixed cylindric map projections.

Mixed cylindric map projections of the sphere are very popular projections, for instance, those equiareal cylindric projections named after H. C. Foucaut (1862) or A. M. Nell (1890) and E. Hammer (1900). The variants are the horizontal or vertical weighted means of the Lambert equiareal projections and the equiareal pseudo-cylindrical mapping of sinusoidal type which is also known as Sanson–Flamsteed projection of the sphere. In this chapter, let us exclusively aim at mixed cylindric map projections of the ellipsoid-of-revolution.

Section F-1.

In Section F-1, we derive the conditions which must be fulfilled for a pseudo-cylindrical equiareal map projection of a biaxial ellipsoid and the general form of ellipsoidal equiareal pseudo-cylindrical mapping equations of Box F.2.

Section F-2.

In Section F-2, we start from the setup of the ellipsoidal generalized Lambert projection and the ellipsoidal generalized Sanson–Flamsteed projection, which are both equiareal. By two lemmas, we shall present the vertical–horizontal mean of the generalized Lambert projection and the generalized Sanson–Flamsteed projection of the biaxial ellipsoid.

Section F-3.

The *deformation analysis* of vertically and horizontally averaged equiareal cylindric mappings is the topic of Section F-3. We especially compute the left–right principal stretches, their corresponding eigenvectors/eigenspace, and the maximal left angular distortion.

The following references are appropriate. The equiareal cylindric map projection of the sphere is addressed to J. H. Lambert (1772), while the equiareal pseudo-cylindrical map projection of the sphere, namely of sinusoidal type, to N. Sanson (1650), J. Cossin (1570), and J. Flamsteed (1692). Equally weighted vertical and horizontal components of Lambert and Sanson–Flamsteed map projections of the sphere have been presented by H. C. Foucaut (1862), A. M. Nell (1890), and E. Hammer (1900). The theory of weighted means of general map projections of the sphere has been critically reviewed by W. R. Tobler (1973).

F-1 Pseudo-cylindrical mapping: biaxial ellipsoid onto plane

First, we refer to a chart of the *biaxial ellipsoid* $\mathbb{E}^2_{A,B}$ (ellipsoid-of-revolution, spheroid) with semi-major axis A and semi-minor axis B based upon local coordinates of type {longitude Λ, latitude Φ} as surface coordinates summarized in Box F.1. Second, we set up pseudo-cylindrical mapping equations of the biaxial ellipsoid $\mathbb{E}^2_{A,B}$ onto the Euclidean plane $\{\mathbb{R}^2, \delta_{\mu\nu}\}$ in terms of surface normal ellipsoidal {longitude Λ, latitude Φ}, in particular

$$x = x(\Lambda, \Phi) = A\Lambda \frac{\cos\Phi}{\sqrt{1 - E^2 \sin^2\Phi}} g(\Phi) \,,$$

(F.1)

$$y = y(\Phi) = f(\Phi) \,,$$

$$\boldsymbol{x} := \{\boldsymbol{x} \in \mathbb{R}^2 | ax + by + c = 0\} \,.$$

(F.2)

The structure of the pseudo-cylindrical mapping equations with unknown functions $f(\Phi)$ and $g(\Phi)$ is motivated by the postulate that for $g(\Phi) = 1$ parallel circles of $\mathbb{E}^2_{A,B}$ should be mapped equidistantly onto $\{\mathbb{R}^2, \delta_{\mu\nu}\}$. ($[\delta_{\mu\nu}]$ denotes the unit matrix, all indices run from one to two.) Note that for zero relative eccentricity $E = 0$, we arrive at the pseudo-cylindrical mapping equations of the sphere $\mathbb{S}^2_R$.

Box F.1 (Chart of the biaxial ellipsoid $\mathbb{E}^2_{A,B}$).

$$\boldsymbol{X} \in \mathbb{E}^2_{A,B} := \left\{ \boldsymbol{X} \in \mathbb{R}^3 \,\middle|\, \frac{X^2 + Y^2}{A^2} + \frac{Z^2}{B^2} = 1, A \in \mathbb{R}^+, B \in \mathbb{R}^+, A > B \right\} \,.$$

(F.3)

Chart (surface normal longitude Λ, surface normal latitude Φ):

$$\boldsymbol{X}(\Lambda, \Phi) =$$

$$= e_1 \frac{A\cos\Phi\cos\Lambda}{\sqrt{1 - E^2\sin^2\Phi}} + e_2 \frac{A\cos\Phi\sin\Lambda}{\sqrt{1 - E^2\sin^2\Phi}} + e_3 \frac{A(1 - E^2)\sin\Phi}{\sqrt{1 - E^2\sin^2\Phi}}$$

(F.4)

$$(0 < \Lambda < 2\pi, -\pi/2 < \Phi < +\pi/2) \,,$$

$$\Lambda(\boldsymbol{X}) =$$

$$= \arctan\frac{Y}{X} + 180° \left[1 - \tfrac{1}{2}\mathrm{sgn}(Y) - \tfrac{1}{2}\mathrm{sgn}(Y)\mathrm{sgn}(X)\right] \,,$$

(F.5)

$$\Phi(\boldsymbol{X}) =$$

$$= \arctan\frac{1}{1 - E^2} \frac{Z}{\sqrt{X^2 + Y^2}} \,.$$

(F.6)

Relative eccentricity:

$$E^2 := (A^2 - B^2)/A^2 = 1 - B^2/A^2 \,.$$

(F.7)

Next, by means of a Corollary F.1, we want to show that in the class of pseudo-cylindrical mapping equations given by (F.1) only equiareal map projections are possible. At the same, we correct a printing error in formula (1.9) of E. Grafarend and A. Heidenreich (1995 p. 166).

Corollary F.1 (Pseudo-cylindrical mapping equations).

In the class of pseudo-cylindrical mapping equations of type (F.1), only equiareal map projections are possible restricting the unknown functions to (F.8). Conformal map projections are not in the class of pseudo-cylindrical mapping equations.

$$g = Mf'^{-1} \quad \text{or} \quad f' = Mg^{-1} \, . \tag{F.8}$$

End of Corollary.

For the proof, we are going to construct the left Cauchy–Green deformation tensor according to E. Grafarend (1995 pp. 432–436) and solve its general eigenvalue problem with respect to the metric matrix of the biaxial ellipsoid $\mathbb{E}^2_{A,B}$, i.e. in order to compute the left principal stretches $\{\Lambda_1, \Lambda_2\}$. The tests $\Lambda_1 \Lambda_2 = 1$ for an *equiareal mapping* and $\Lambda_1 = \Lambda_2$ for a *conformal mapping* according to E. Grafarend (1995 p. 449) are performed.

Proof.

The infinitesimal distance ds between two points $\boldsymbol{x}$ and $\boldsymbol{x} + d\boldsymbol{x}$ both elements of the plane $\{\mathbb{R}^2, \delta_{\mu\nu}\}$ is pullback transformed into the $\{\Lambda, \Phi\}$ coordinate representation, in particular

$$ds^2 = dx^2 + dy^2 = \delta_{\mu\nu} dx^\mu dx^\nu =$$

$$= \delta_{\mu\nu} \frac{\partial x^\mu}{\partial U^A} \frac{\partial x^\nu}{\partial U^B} dU^A dU^B \ \forall \ U^1 = \Lambda, U^2 = \Phi \, , \tag{F.9}$$

$$ds^2 = c_{AB} dU^A dU^B \ \forall \ c_{AB} := \delta_{\mu\nu} \frac{\partial x^\mu}{\partial U^A} \frac{\partial x^\nu}{\partial U^B} \, .$$

Throughout, we apply the summation convention over repeated indices. For example, we here apply $a_\mu b_\mu = a_1 b_1 + a_2 b_2$. In addition, we adopt the symbols provided by (F.10) as the symbols for the *meridional radius of curvature* $M(\Phi)$ and the *normal radius of curvature* $N(\Phi)$, respectively.

$$M(\Phi) := \frac{A(1 - E^2)}{(1 - E^2 \sin^2 \Phi)^{3/2}} \, , \quad N(\Phi) := \frac{A}{(1 - E^2 \sin^2 \Phi)^{1/2}} \, . \tag{F.10}$$

Indeed, $M(\Phi)$ is the radius of curvature of the meridian, the coordinate line $\Lambda = \text{const.}$, *but* $N(\Phi)$ as the transverse radius of curvature of a curve formed by the intersection of the normal or transverse plane $\mathbb{P}^2 \perp T_x \mathbb{E}^2_{A,B}$ which is normal to the tangent space $T_x \mathbb{E}^2_{A,B}$ of the biaxial ellipsoid $\mathbb{E}^2_{A,B}$ and is perpendicular to the meridian at a point $\{\Lambda, \Phi\} \in \mathbb{E}^2_{A,B}$. In contrast, $N(\Phi) \cos \Phi = L(\Phi)$ is the radius of curvature of the parallel circle, the coordinate line $\Phi = \text{const.}$ The principal curvature radii of the biaxial ellipsoid $\mathbb{E}^2_{A,B}$ are $\{M(\Phi), N(\Phi)\}$.

$$x = x(\Lambda, \Phi) = [N(\Phi) \cos \Phi] \Lambda g(\Phi) = L(\Phi) \Lambda g(\Phi) \, ,$$

$$y = y(\Phi) = f(\Phi) \, . \tag{F.11}$$

The left Cauchy–Green deformation tensor c_{AB} is generated by (F.12). $(x_\Lambda, y_\Lambda, x_\Phi, y_\Phi)$ denote the partials of (x, y) with respect to (Λ, Φ) so that we are led to (F.13).

$$c_{11} = x_\Lambda^2 + y_\Lambda^2 \, , \quad c_{12} = x_\Lambda x_\Phi + y_\Lambda y_\Phi \, , \quad c_{21} = c_{12} \, , \quad c_{22} = x_\Phi^2 + y_\Phi^2 \, , \tag{F.12}$$

$$\{c_{AB}\} = \begin{bmatrix} L^2 g^2 & \Lambda L g(L'g + Lg') \\ \Lambda L g(L'g + Lg') & \Lambda^2 (L'g + Lg')^2 + f'^2 \end{bmatrix} \, . \tag{F.13}$$

The infinitesimal distance dS between two points X and $X+dX$, both elements of the biaxial ellipsoid $\mathbb{E}^2_{A,B}$, represented in terms of the first chart is

$$dS^2 = G_{AB}dU^A dU^B$$

$$\forall$$

$$\{G_{AB}\} = \begin{bmatrix} N^2(\Phi)\cos^2\Phi & 0 \\ 0 & M^2(\Phi) \end{bmatrix} = \begin{bmatrix} L^2 & 0 \\ 0 & M^2 \end{bmatrix} .$$

(F.14)

Simultaneous diagonalization of the two matrices $\{c_{AB}\}$ and $\{G_{AB}\}$ being positive–definite leads to the general eigenvalue problem (F.15), leading to the left principal stretches (F.16), solved by (F.17), subject to (F.18).

$$|c_{AB} - \Lambda_S^2 G_{AB}| = 0 = \begin{vmatrix} L^2 g^2 - \Lambda_S^2 L^2 & \Lambda Lg(L'g + Lg') \\ \Lambda Lg(L'g + Lg') & \Lambda^2(L'g + Lg')^2 + f'^2 - \Lambda_S^2 M^2 \end{vmatrix} ,$$

(F.15)

$$\Lambda_S^4 - \Lambda_S^2 \left[\Lambda^2 \frac{(L'g + Lg')^2}{M^2} + \frac{f'^2}{M^2} + g^2 \right] + \frac{g^2 f'^2}{M^2} = 0 ,$$

(F.16)

$$\Lambda_1 = \sqrt{\frac{1}{2}(a + \sqrt{a^2 - 4b})} , \quad \Lambda_2 = \sqrt{\frac{1}{2}(a - \sqrt{a^2 - 4b})} ,$$

(F.17)

$$a := \Lambda^2(L'g + Lg')^2/M^2 + f'^2/M^2 + g^2 ,$$

$$b := g^2 f'^2/M^2 .$$

(F.18)

The postulate of equiareal mapping $\Lambda_1\Lambda_2 = 1$ leads to (F.19) being equivalent to $b = 1$ or $g = Mf'^{-1}$ or $g^2 f'^2/M^2 = 1$. Obviously, the postulate of a conformal mapping $\Lambda_1 = \Lambda_2$ cannot be fulfilled since $a^2 = 4b$ leads to a nonlinear functional $F(g'^4, g'^2, g^4, g^2, f'^4, f'^2)$.

$$\frac{1}{2}(a + \sqrt{a^2 - 4b})\frac{1}{2}(a - \sqrt{a^2 - 4b}) = 1$$

(F.19)

End of Proof.

In summarizing, we are led to the *equiareal pseudo-cylindrical mapping equations* of Box F.2.

Box F.2 (Equiareal pseudo-cylindrical mapping).

$$x = \frac{A^2(1 - E^2)\cos\Phi}{(1 - E^2\sin^2\Phi)^2} \Lambda \frac{1}{f'(\Phi)} = M(\Phi)N(\Phi)\cos\Phi\,\Lambda\frac{1}{f'(\Phi)} ,$$

(F.20)

$$y = f(\Phi) .$$

F-2 Mixed equiareal cylindric mapping: biaxial ellipsoid onto plane

The variants of mixed equiareal mappings of the biaxial ellipsoid onto the plane are generated by the weighted mean of the normal equiareal cylindric mapping (for the sphere $\mathbb{S}^2_R$ this is the Lambert equiareal projection) and the equiareal pseudo-cylindrical mapping of sinusoidal type (for the sphere $\mathbb{S}^2_R$ this is the Sanson–Flamsteed equiareal projection). The separate mappings are beforehand collected in Corollary F.2.

Corollary F.2 (Ellipsoidal equiareal cylindric projection of normal type).

The ellipsoidal equiareal cylindric projection of normal type (generalized Lambert projection) in the class of cylindric projections in terms of surface normal longitude Λ and latitude Φ of $\mathbb{E}^2_{A,B}$ is represented by (F.21) mapping the circular equator equidistantly.

$$x = A\Lambda ,$$

$$
y = \frac{A(1-E^2)}{4E}\left[\ln\frac{1+E\sin\Phi}{1-E\sin\Phi} + \frac{2E\sin\Phi}{1-E^2\sin^2\Phi}\right] =
$$

$$
= \frac{A(1-E^2)}{2E}\left[\operatorname{artanh}(E\sin\Phi) + \frac{E\sin\Phi}{1-E^2\sin^2\Phi}\right] .
$$

(F.21)

End of Corollary.

Proof.

For the proof, let us depart from the setup $x = A\Lambda$ and $y = f(\Phi)$ as special case of (F.11) with $L(\Phi) = 1$ and $g(\Phi) = 1$ such that the equator is mapped equidistantly. The left Cauchy–Green deformation tensor c_{AB} is generated by $c_{11} = A^2$, $c_{12} = 0$, and $c_{22} = f'^2(\Phi)$ such that the left principal stretches, following (F.12)–(F.14), amount to (F.22), which leads by partial integration (decomposition into fractions) subject to the condition $f(\Phi = 0) = 0$ directly to (F.21).

$$
\Lambda_1 = \frac{A}{N(\Phi)\cos\Phi} , \quad \Lambda_2 = \frac{f'(\Phi)}{M(\Phi)} ,
$$

$$
\Lambda_1\Lambda_2 = 1
$$

$$
\Leftrightarrow
$$

(F.22)

$$
\mathrm{d}f = A^{-1}N(\Phi)M(\Phi)\cos\Phi\,\mathrm{d}\Phi ,
$$

$$
\mathrm{d}f = A(1-E^2)\frac{\cos\Phi}{(1-E^2\sin^2\Phi)^2}\mathrm{d}\Phi .
$$

End of Proof.

Corollary F.3 (Ellipsoidal equiareal projection of pseudo-sinusoidal type).

The ellipsoidal equiareal projection of pseudo-sinusoidal type (generalized Sanson–Flamsteed projection) in the class of pseudo-cylindrical projections is represented in terms of surface normal longitude Λ and latitude Φ of $\mathbb{E}^2_{A,B}$ by (F.23) mapping a parallel circle equidistantly.

$$
x = \frac{A\cos\Phi}{\sqrt{1-E^2\sin^2\Phi}}\Lambda ,
$$

$$
y \approx A(1-E^2)\left[\Phi + \frac{3}{8}E^2(2\Phi - \sin 2\Phi) + \mathrm{O}(E^4)\right] \approx
$$

(F.23)

$$
\approx A\left[\Phi - \frac{1}{8}E^2(2\Phi + 3\sin 2\Phi) + \mathrm{O}(E^4)\right] .
$$

End of Corollary.

Proof.

For the proof, let us depart from the setup (F.24), which leads under the postulate of an equiareal mapping via (F.8) of Corollary F.1 to (F.25), expressing the arc length of the meridian, namely in terms of the standard elliptic integral of second kind, here instead by series expansion.

$$x = x(\Lambda, \Phi) =$$

$$= \frac{A \cos \Phi}{\sqrt{1 - E^2 \sin^2 \Phi}} \Lambda = [N(\Phi) \cos \Phi] \Lambda = L(\Phi) \Lambda , \tag{F.24}$$

$$y = f(\Phi) ,$$

$$g(\Phi) = 1 , \quad f'(\Phi) = M(\Phi) ,$$

$$f(\Phi) = \int_0^\Phi M(\Phi^*) \mathrm{d}\Phi^* = \int_0^\Phi \frac{A(1 - E^2) \mathrm{d}\Phi^*}{(1 - E^2 \sin^2 \Phi^*)^{3/2}} . \tag{F.25}$$

Note the integral kernel expansion as powers of E^2 is *uniformly convergent*.

$$f(\Phi) \approx A(1 - E^2) \times$$

$$\times \left[\Phi + \frac{3}{8} E^2 (2\Phi - \sin 2\Phi) + \frac{15}{256} E^4 (12\Phi - 8\sin 2\Phi + \sin 4\Phi) + \mathrm{O}(E^6) \right] . \tag{F.26}$$

End of Proof.

Note that the coordinate lines $\Lambda = \text{const.}$, the *meridians*, are mapped close to a sinusoidal arc as can been seen by Φ in (F.23). Here, we find the reason for our term "pseudo-sinusoidal". This is in contrast to the coordintae lines $\Phi = \text{const.}$: the *parallel circles* are mapped onto straight lines parallel to the x axis represented by $x = c_1 \Lambda$ and $y = c_2$, where c_1, c_2 are constants.

The *first variant* of mixed equiareal mapping of the biaxial ellipsoid onto the plane is generated as *weighted mean* of the *vertical coordinates* $\{y_{\mathrm{gL}}, y_{\mathrm{gSF}}\}$ with respect to the *generalized Lambert coordinate* y_{gL} ($\rightarrow$ (F.21)) and the *generalized Sanson–Flamsteed coordinate* y_{gSF} ($\rightarrow$ (F.23)). In contrast, the *horizontal coordinate* x is constructed from the postulate of equiareal pseudo-cylindrical mapping equations of the type (F.20).

Definition F.4 (Equiareal mapping of pseudo-cylindrical type: vertical coordinate mean).

An equiareal mapping of pseudo-cylindrical type of the biaxial ellipsoid is called *vertical coordinate mean* if (F.27) holds where α and β are weight coefficients.

$$x = x(\Lambda, \Phi) =$$

$$= \frac{A^2 (1 - E^2) \cos \Phi}{(1 - E^2 \sin^2 \Phi)^2} \Lambda \frac{1}{f'(\Phi)} ,$$

$$y = y(\Phi) = \tag{F.27}$$

$$= \frac{\alpha y_{\mathrm{gL}} + \beta y_{\mathrm{gSF}}}{\alpha + \beta} =: f(\Phi) .$$

End of Definition.

The unknown function $f(\Phi)$ is constructed as follows. Obviously, the *vertical coordinate mean* $y(\Phi)$, i.e. (F.28), is the basis which generates the *horizontal coordinate* $x(\Lambda, \Phi)$, namely (F.29), subject to (F.30) and (F.31).

$$f(\Phi) \approx \frac{A(1-E^2)}{\alpha+\beta} \times$$

$$\times \left(\beta \left[\Phi + \frac{3}{8} E^2 (2\Phi - \sin 2\Phi) + \frac{15}{256} E^4 (12\Phi - 8\sin 2\Phi + \sin 4\Phi) + \mathrm{O}(E^6) \right] + \right. \tag{F.28}$$

$$\left. + \frac{\alpha}{4E} \left[\ln \frac{1 + E\sin\Phi}{1 - E\sin\Phi} + \frac{2E\sin\Phi}{1 - E^2 \sin^2\Phi} \right] \right),$$

$$f'(\Phi) = \frac{1}{\alpha+\beta}(\alpha y'_{gL} + \beta y'_{gSF}), \quad \frac{1}{f'(\Phi)} = \frac{\alpha+\beta}{\alpha y'_{gL} + \beta y'_{gSF}}, \tag{F.29}$$

$$y'_{gL}(\Phi) = A(1-E^2)\frac{\cos\Phi}{(1 - E^2 \sin^2\Phi)^2} = A^{-1}N(\Phi)M(\Phi)\cos\Phi,$$

$$\tag{F.30}$$

$$y'_{gSF}(\Phi) = \frac{A(1-E^2)}{(1 - E^2 \sin^2\Phi)^{3/2}} = M(\Phi),$$

$$x = x(\Lambda, \Phi) = N(\Phi)M(\Phi)\cos\Phi\Lambda\frac{1}{f'(\Phi)} =$$

$$\tag{F.31}$$

$$= \frac{N(\Phi)M(\Phi)\cos\Phi\Lambda(\alpha+\beta)}{\alpha A^{-1}N(\Phi)M(\Phi)\cos\Phi + \beta M(\Phi)} = \frac{(\alpha+\beta)A\cos\Phi}{\alpha\cos\Phi + \beta\sqrt{1 - E^2 \sin^2\Phi}}\Lambda.$$

Thus, we have proven Lemma F.5.

Lemma F.5 (Vertical mean of the generalized Lambert projection and of the generalized Sanson–Flamsteed projection of the biaxial ellipsoid $\mathbb{E}^2_{A,B}$).

The vertical mean of the generalized Lambert projection and of the generalized Sanson–Flamsteed projection of the biaxial ellipsoid leads to the equiareal mapping of pseudo-cylindrical type represented by (F.32).

$$x = x(\Lambda, \Phi) = \frac{(\alpha+\beta)A\cos\Phi}{\alpha\cos\Phi + \beta\sqrt{1 - E^2 \sin^2\Phi}}\Lambda,$$

$$y = y(\Phi) \approx \frac{A(1-E^2)}{\alpha+\beta} \times$$

$$\tag{F.32}$$

$$\times \left(\beta \left[\Phi + \frac{3}{8} E^2 (2\Phi - \sin 2\Phi) + \frac{15}{256} E^4 (12\Phi - 8\sin 2\Phi + \sin 4\Phi) + \mathrm{O}(E^6) \right] + \right.$$

$$\left. + \frac{\alpha}{4E} \left[\ln \frac{1 + E\sin\Phi}{1 - E\sin\Phi} + \frac{2E\sin\Phi}{1 - E^2 \sin^2\Phi} \right] \right).$$

End of Lemma.

The *second variant* of mixed equiareal mapping of the biaxial ellipsoid onto the plane is generated as weighted mean of the *horizontal coordinates* $\{x_{\mathrm{gL}}, x_{\mathrm{gSF}}\}$ with respect to the generalized Lambert coordinate x_{gL} ($\rightarrow$ (F.21)) and the generalized Sanson–Flamsteed coordinate x_{gSF} ($\rightarrow$ (F.23)). In contrast, the *vertical coordinate* y is constructed from the postulate of equiareal pseudo-cylindrical mapping equations of the type (F.20).

Definition F.6 (Equiareal mapping of pseudo-cylindrical type: horizontal coordinate mean).

An equiareal mapping of pseudo-cylindrical type of the biaxial ellipsoid is called *horizontal coordinate mean* if (F.33) holds where α and β are weight coefficients.

$$x = x(\Lambda, \Phi) =$$

$$= \frac{\alpha x_{\mathrm{gL}} + \beta x_{\mathrm{gSF}}}{\alpha + \beta} = \frac{A^2(1 - E^2)\cos\Phi}{(1 - E^2 \sin^2\Phi)^2} \Lambda \frac{1}{f'(\Phi)} , \tag{F.33}$$

$$y = y(\Phi) =: f(\Phi) .$$

End of Definition.

This time, in constructing the unknown function $f(\Phi)$, the horizontal coordinate mean $x(\Lambda, \Phi)$, i.e. (F.34), is the basis generating the vertical coordinate $y(\Phi)$, i.e. (F.35)–(F.38).

$$\frac{\alpha x_{\mathrm{gL}} + \beta x_{\mathrm{gSF}}}{\alpha + \beta} = \frac{A^2(1 - E^2)\cos\Phi}{(1 - E^2 \sin^2\Phi)^2} \frac{\Lambda}{f'(\Phi)} , \tag{F.34}$$

$$x_{\mathrm{gL}} = A\Lambda , \quad x_{\mathrm{gSF}} = \frac{A\cos\Phi\Lambda}{\sqrt{1 - E^2 \sin^2\Phi}} , \tag{F.35}$$

$$f'(\Phi) = \frac{A^2(1 - E^2)\cos\Phi\Lambda}{(1 - E^2 \sin^2\Phi)^2} \frac{\alpha + \beta}{\alpha x_{\mathrm{gL}} + \beta x_{\mathrm{gSF}}} ,$$

$$f'(\Phi) = \frac{A^2(1 - E^2)\cos\Phi\,\Lambda}{(1 - E^2 \sin^2\Phi)^2} \frac{\alpha + \beta}{\alpha A\Lambda + \beta(A\cos\Phi)\Lambda/\sqrt{1 - E^2 \sin^2\Phi}} = \tag{F.36}$$

$$= \frac{(\alpha + \beta)A(1 - E^2)\cos\Phi(1 - E^2 \sin^2\Phi)^{-3/2}}{\alpha\sqrt{1 - E^2 \sin^2\Phi} + \beta\cos\Phi} ,$$

$$f(\Phi = 0) = 0$$

$$\Leftrightarrow$$

$$f(\Phi) = \int_0^\Phi f'(\Phi^*)\mathrm{d}\Phi^* , \tag{F.37}$$

$$y(\Phi) = A(1 - E^2)(\alpha + \beta) \int_0^\Phi \frac{\cos\Phi^*(1 - E^2 \sin^2\Phi^*)^{-3/2}}{\alpha\sqrt{1 - E^2 \sin^2\Phi^*} + \beta\cos\Phi^*} \mathrm{d}\Phi^* . \tag{F.38}$$

In detail, we use

$$(1 - E^2 \sin^2 \Phi)^{-3/2} \approx$$

$$\approx 1 + \frac{3}{2} E^2 \sin^2 \Phi + \frac{15}{8} E^4 \sin^4 \Phi + \mathrm{O}(E^6) \,,$$

$$\sqrt{1 - E^2 \sin^2 \Phi} \approx$$

$$\approx 1 - \frac{1}{2} E^2 \sin^2 \Phi - \frac{1}{8} E^4 \sin^4 \Phi + \mathrm{O}(E^6) \,,$$

(F.39)

$$\alpha \sqrt{1 - E^2 \sin^2 \Phi} + \beta \cos \Phi \approx$$

$$\approx \alpha + \beta \cos \Phi - \frac{\alpha}{2} E^2 \sin^2 \Phi - \frac{\alpha}{8} E^4 \sin^4 \Phi + \mathrm{O}(E^6) \,,$$

$$(\alpha \sqrt{1 - E^2 \sin^2 \Phi} + \beta \cos \Phi)^{-1} \approx$$

$$\approx \frac{1}{\alpha + \beta \cos \Phi} + \frac{\alpha E^2 \sin^2 \Phi}{2(\alpha + \beta \cos \Phi)^2} + \frac{\alpha E^4 (3\alpha + \beta \cos \Phi) \sin^4 \Phi}{8(\alpha + \beta \cos \Phi)^3} + \mathrm{O}(E^6) \,,$$

(F.40)

$$\frac{\cos \Phi (1 - E^2 \sin^2 \Phi)^{-3/2}}{\alpha \sqrt{1 - E^2 \sin^2 \Phi} + \beta \cos \Phi} \approx$$

$$\approx \frac{\cos \Phi}{\alpha + \beta \cos \Phi} \times$$

$$\times \left[1 + \frac{E^2 (4\alpha + 3\beta \cos \Phi) \sin^2 \Phi}{2(\alpha + \beta \cos \Phi)} + \right.$$

$$\left. + \frac{E^4 (24\alpha^2 + 37\alpha\beta \cos \Phi + 15\beta^2 \cos^2 \Phi) \sin^4 \Phi}{8(\alpha + \beta \cos \Phi)^2} + \mathrm{O}(E^6) \right] \,,$$

(F.41)

$$\int_0^\Phi \frac{\cos \Phi^* (1 - E^2 \sin^2 \Phi^*)^{-3/2}}{\alpha \sqrt{1 - E^2 \sin^2 \Phi^*} + \beta \cos \Phi^*} \mathrm{d}\Phi^* \approx$$

$$\approx \frac{2\alpha}{\beta \sqrt{\alpha^2 - \beta^2}} \arctan \frac{(\beta - \alpha) \tan \Phi/2}{\sqrt{\alpha^2 - \beta^2}} \left[1 + E^2 \left(1 - \frac{\alpha^2}{2\beta^2} \right) \right] +$$

$$+ \frac{\Phi}{\beta} \left[1 + E^2 \left(\frac{3}{4} - \frac{\alpha^2}{2\beta^2} \right) \right] +$$

(F.42)

$$+ E^2 \frac{(\alpha - \beta \cos \Phi)(2\alpha + 3\beta \cos \Phi) \sin \Phi}{4\beta^2 (\alpha + \beta \cos \Phi)} + \mathrm{O}(E^4) \,.$$

For the term-wise integration in (F.42) by MATHEMATICA, uniform convergence of the kernel has been necessary. In summarizing, we have proven Lemma F.7.

Lemma F.7 (Horizontal mean of the generalized Lambert projection and of the generalized Sanson–Flamsteed projection of the biaxial ellipsoid $\mathbb{E}^2_{A,B}$).

The horizontal mean of the generalized Lambert projection and of the generalized Sanson–Flamsteed projection of the biaxial ellipsoid leads to the equiareal mapping of pseudo-cylindrical type represented by (F.43).

$$
x = x(\Lambda, \Phi) = \frac{A\,\Lambda}{\alpha + \beta}\left(\alpha + \frac{\beta \cos \Phi}{\sqrt{1 - E^2 \sin^2 \Phi}}\right). \tag{F.43}
$$

$$
\alpha > \beta:
$$

$$
y = y(\Phi) \approx A(1 - E^2)(\alpha + \beta)\times
$$

$$
\times\left(\frac{2\alpha}{\beta\sqrt{\alpha^2 - \beta^2}}\arctan\frac{(\beta - \alpha)\tan\Phi/2}{\sqrt{\alpha^2 - \beta^2}}\left[1 + E^2(1 - \frac{\alpha^2}{2\beta^2})\right]+\right.
$$

$$
+\frac{\Phi}{\beta}\left[1 + E^2\left(\frac{3}{4} - \frac{\alpha^2}{2\beta^2}\right)\right]+
$$

$$
\left.+E^2\frac{(\alpha - \beta\cos\Phi)(2\alpha + 3\beta\cos\Phi)\sin\Phi}{4\beta^2(\alpha + \beta\cos\Phi)} + \mathrm{O}(E^4)\right). \tag{F.44}
$$

$$
\alpha < \beta:
$$

$$
y = y(\Phi) \approx A(1 - E^2)(\alpha + \beta)\times
$$

$$
\times\left(-\frac{\alpha}{\beta\sqrt{\beta^2 - \alpha^2}}\ln\frac{\beta + \alpha\cos\Phi + \sqrt{\beta^2 - \alpha^2}\sin\Phi}{\alpha + \beta\cos\Phi}\left[1 + E^2\left(1 - \frac{\alpha^2}{2\beta^2}\right)\right]+\right.
$$

$$
+\frac{\Phi}{\beta}\left[1 + E^2\left(\frac{3}{4} - \frac{\alpha^2}{2\beta^2}\right)\right]+
$$

$$
\left.+E^2\frac{(\alpha - \beta\cos\Phi)(2\alpha + 3\beta\cos\Phi)\sin\Phi}{4\beta^2(\alpha + \beta\cos\Phi)} + \mathrm{O}(E^4)\right). \tag{F.45}
$$

$$
\alpha = \beta = 1:
$$

$$
x = x(\Lambda, \Phi) \approx \frac{1}{2}A\,\Lambda\frac{\sqrt{1 - E^2\sin^2\Phi} + \cos\Phi}{\sqrt{1 - E^2\sin^2\Phi}},
$$

$$
y = y(\Phi) \approx 2A(1 - E^2)\times \tag{F.46}
$$

$$
\times\left[\Phi\left(1 + \frac{E^2}{4}\right) - \tan\frac{\Phi}{2}\left(1 + \frac{E^2}{2}\right) + E^2\frac{1 - \cos\Phi}{4(1 + \cos\Phi)}(2 + 3\cos\Phi)\sin\Phi + \mathrm{O}(E^2)\right].
$$

End of Lemma.

F-3 Deformation analysis of vertically/horizontally averaged equiareal cylindric mappings

The deformation analysis of vertically and horizontally averaged equiareal cylindric mappings based upon E. Grafarend (1995) will inform us about the minimal and maximal scale distortions, also called *left* and *right principal stretches*, as well as the maximal angular distortion. Indeed, we collect in Corollary F.8 the representation of left principal stretches $\{\Lambda_1, \Lambda_2\}$ in terms of the invariant tr $\left[\mathsf{C}_l \mathsf{C}_l^{-1}\right]$, both for the vertical as well as the horizontal mean of mixed equiareal cylindric mappings of the biaxial ellipsoid $\mathbb{E}_{A,B}^2$. In Boxes F.3 and F.4, we present the items of *left Cauchy–Green deformation tensor* based upon (F.32) or (F.43)–(F.46). Corollary F.9, in contrast, reviews the representation of maximal *left angular distortion* in terms of the sum and the difference of the *left principal stretches*. The two left eigenvectors of the left Cauchy–Green deformation are by Corollary F.10 computed in the basis of images of tangent vectors $\{\boldsymbol{G}_1, \boldsymbol{G}_2\}$ which locally span $T\mathbb{E}_{A,B}^2$, the tangent space at $\{\Lambda, \Phi\}$ of $\mathbb{E}_{A,B}^2$. In order to analyse the distortion measures in the chart $(x, y) \in \{\mathbb{R}^2, \delta_{\alpha\beta}\}$ the coordinates of the *right Cauchy–Green deformation tensor* for the vertical-horizontal mixed equiareal cylindric mapping (F.32) and (F.43) in Boxes F.5 and F.6 have been computed. The right principal stretches are inversely related to the left principal stretches. The right eigenvectors of the right Cauchy–Green deformation tensor of the vertical–horizontal mean of mixed equiareal cylindric mapping of $\mathbb{E}_{A,B}^2$ are given in Corollary F.11, in particular, the orientation of the right principal stretches.

> **Corollary F.8 (Left principal stretches of the vertical–horizontal mean of mixed equiareal cylindric mapping of the biaxial ellipsoid $\mathbb{E}_{A,B}^2$).**

The left principal stretches $\{\Lambda_1, \Lambda_2\}$ of the coordinates c_{AB} of the left Cauchy–Green deformation tensor normalized with respect to the coordinates $G_{A,B}$ of the left metric tensor are represented by (F.47) if the mapping equations (F.32) for the vertical mean and (F.43)–(F.46) for the horizontal mean of mixed equiareal mappings of the biaxial ellipsoid apply.

$$\Lambda_1 = +\sqrt{\frac{1}{2}\left[\left(\text{tr}\left[\mathsf{C}_l\mathsf{G}_l^{-1}\right]\right) + \sqrt{\left(\text{tr}\left[\mathsf{C}_l\mathsf{G}_l^{-1}\right]\right)^2 - 4}\right]}\,,$$

$$\Lambda_2 = +\sqrt{\frac{1}{2}\left[\left(\text{tr}\left[\mathsf{C}_l\mathsf{G}_l^{-1}\right]\right) - \sqrt{\left(\text{tr}\left[\mathsf{C}_l\mathsf{G}_l^{-1}\right]\right)^2 - 4}\right]}\,.$$

(F.47)

First, for the vertical mean holds (F.48).

$$\text{tr}\left[\mathsf{C}_l\mathsf{G}_l^{-1}\right] = \frac{c_{11}}{G_{11}} + \frac{c_{22}}{G_{22}} =$$

$$= \frac{A^4(\alpha+\beta)^4[A^2\Lambda^2\beta^2\sin^2\Phi + (\alpha L + \beta A)^2] + (\alpha L + \beta A)^6}{A^2(\alpha+\beta)^2(\alpha L + \beta A)^4}\,.$$

(F.48)

Second, for the horizontal mean holds (F.49).

$$\text{tr}\left[\mathsf{C}_l\mathsf{G}_l^{-1}\right] = \frac{c_{11}}{G_{11}} + \frac{c_{22}}{G_{22}} =$$

$$= \frac{(\alpha A + \beta L)^4 + \Lambda^2\beta^2\sin^2\Phi L^2(\alpha A + \beta L)^2 + L^4(\alpha+\beta)^4}{(\alpha+\beta)^2 L^2(\alpha A + \beta L)^2}\,.$$

(F.49)

End of Corollary.

Proof (of (F.47)).

For the proof of (F.47), we depart from the general eigenvalue problem (F.15), whose characteristic equation is solved under the postulate of equal area $\Lambda_1\Lambda_2 = 1$, which is equivalent to $\det\left[\mathsf{C}_l\mathsf{G}_l^{-1}\right] = 1$.

$$\left|c_{AB} - \Lambda_S^2 G_{AB}\right| = 0$$

$$\Rightarrow \tag{F.50}$$

$$\Lambda_S^4 - \Lambda_S^2 \mathrm{tr}\left[\mathsf{C}_l\mathsf{G}_l^{-1}\right] + \det\left[\mathsf{C}_l\mathsf{G}_l^{-1}\right] = 0\ ,$$

$$\Lambda_{1,2}^2 = \frac{1}{2}\left[\left(\mathrm{tr}\left[\mathsf{C}_l\mathsf{G}_l^{-1}\right]\right) \pm \sqrt{\left(\mathrm{tr}\left[\mathsf{C}_l\mathsf{G}_l^{-1}\right]\right)^2 - 4\left(\det\left[\mathsf{C}_l\mathsf{G}_l^{-1}\right]\right)}\right]$$

$$\Rightarrow \tag{F.51}$$

$$\det\left[\mathsf{C}_l\mathsf{G}_l^{-1}\right] = 1\ .$$

End of Proof.

Proof (of (F.48)).

For the proof of (F.48), we depart from the vertical mixed equiareal cylindric mapping (F.32), compute the Jacobi matrix of partial derivatives of $\{x(\Lambda,\Phi),\ y(\Phi)\}$ and constitute the coordinates of the left Cauchy–Green deformation tensor c_{AB} as well as the matrix $\mathsf{C}_l\mathsf{G}_l^{-1}$ implementing (F.14) and finally the trace $\mathrm{tr}\left[\mathsf{C}_l\mathsf{G}_l^{-1}\right]$.

$$x(\Lambda,\Phi) = \frac{(\alpha+\beta)A\cos\Phi}{\alpha\cos\Phi + \beta\sqrt{1 - E^2\sin^2\Phi}}\Lambda\ ,$$

$$y(\Phi) = \frac{A(1-E^2)}{\alpha+\beta}\left(\beta\int_0^\Phi \frac{d\Phi^*}{(1 - E^2\sin^2\Phi^*)^{3/2}} + \frac{\alpha}{4E}\left[\ln\frac{1 + E\sin\Phi}{1 - E\sin\Phi} + \frac{2E\sin\Phi}{1 - E^2\sin^2\Phi}\right]\right)\ , \tag{F.52}$$

$$x_\Lambda = \frac{A(\alpha+\beta)L}{\alpha L + \beta A}\ ,\quad y_\Lambda = 0\ ,\quad x_\Phi = -\frac{A^2(\alpha+\beta)\beta\sin\Phi M\Lambda}{(\alpha L + \beta A)^2}\ ,\quad y_\Phi = \frac{M(\alpha L + \beta A)}{A(\alpha+\beta)}\ , \tag{F.53}$$

$$c_{11} = x_\Lambda^2 + y_\Lambda^2 = \frac{A^2(\alpha+\beta)^2 L^2}{(\alpha L + \beta A)^2}\ ,$$

$$c_{12} = x_\Lambda x_\Phi + y_\Lambda y_\Phi = -\frac{A^3(\alpha+\beta)^2\beta\sin\Phi L M\Lambda}{(\alpha L + \beta A)^3}\ , \tag{F.54}$$

$$c_{22} = M^2\left[\frac{A^4(\alpha+\beta)^2\beta^2\sin^2\Phi\Lambda^2}{(\alpha L + \beta A)^4} + \frac{(\alpha L + \beta A)^2}{A^2(\alpha+\beta)^2}\right]\ ,$$

$$\det\left[c_{AB}\right] = \det\left[G_{AB}\right] = \frac{A^4(1-E^2)^2\cos^2\Phi}{(1 - E^2\sin^2\Phi)^4}\ , \tag{F.55}$$

$$\mathrm{tr}\left[\mathsf{C}_l\mathsf{G}_l^{-1}\right] = \frac{c_{11}}{G_{11}} + \frac{c_{22}}{G_{22}} = \frac{A^2(\alpha+\beta)^2}{(\alpha L + \beta A)^2} + \frac{A^6(\alpha+\beta)^4\beta^2\Lambda^2\sin^2\Phi + (\alpha L + \beta A)^6}{A^2(\alpha+\beta)^2(\alpha L + \beta A)^4}\ .$$

End of Proof.

Proof (of (F.49)).

For the proof of (F.49), we depart from the horizontal mixed equiareal cylindric mapping (F.43)–(F.46), compute the Jacobi matrix of partial derivatives of $\{x(\Lambda, \Phi), y(\Phi)\}$ and constitute the coordinates of the left Cauchy–Green deformation tensor c_{AB} as well as the matrix $\mathsf{C}_l\mathsf{G}_l^{-1}$ implementing (F.14) and finally the trace $\operatorname{tr}\left[\mathsf{C}_l\mathsf{G}_l^{-1}\right]$.

$$x(\Lambda, \Phi) = \frac{A\Lambda}{\alpha + \beta}\left(\alpha + \frac{\beta \cos \Phi}{\sqrt{1 - E^2 \sin^2 \Phi}}\right) ,$$

$$y(\Phi) = A(1 - E^2)(\alpha + \beta)\int_0^{\Phi} \frac{\cos \Phi^*(1 - E^2 \sin^2 \Phi^*)^{-3/2}}{\alpha\sqrt{1 - E^2 \sin^2 \Phi^*} + \beta \cos \Phi^*}\mathrm{d}\Phi^* ,$$

(F.56)

$$x_\Lambda = \frac{\alpha A + \beta L}{\alpha + \beta} , \quad y_\Lambda = 0 , \quad x_\Phi = -\frac{\beta \sin \Phi M\Lambda}{\alpha + \beta} , \quad y_\Phi = \frac{(\alpha + \beta)LM}{\alpha A + \beta L} ,$$

(F.57)

$$c_{11} = x_\Lambda^2 + y_\Lambda^2 = \frac{(\alpha A + \beta L)^2}{(\alpha + \beta)^2} , \quad c_{12} = x_\Lambda x_\Phi + y_\Lambda y_\Phi = -\frac{\beta \sin \Phi M(\alpha A + \beta L)\Lambda}{(\alpha + \beta)^2} ,$$

$$c_{22} = x_\Phi^2 + y_\Phi^2 = M^2\left[\frac{\beta^2 \sin^2 \Phi\Lambda^2}{(\alpha + \beta)^2} + \frac{(\alpha + \beta)^2 L^2}{(\alpha A + \beta L)^2}\right] ,$$

(F.58)

$$\det[c_{AB}] = \det[G_{AB}] = \frac{A^4(1 - E^2)^2 \cos^2 \Phi}{(1 - E^2 \sin^2 \Phi)^4} ,$$

$$\operatorname{tr}\left[\mathsf{C}_l\mathsf{G}_l^{-1}\right] = \frac{c_{11}}{G_{11}} + \frac{c_{22}}{G_{22}} = \frac{(\alpha A + \beta L)^2}{(\alpha + \beta)^2 L^2} + \frac{\beta^2\Lambda^2 \sin^2 \Phi(\alpha A + \beta L)^2 + (\alpha + \beta)^4 L^2}{(\alpha + \beta)^2(\alpha A + \beta L)^2} .$$

(F.59)

End of Proof.

Box F.3 (Left Cauchy–Green deformation tensor (vertical mixed equiareal cylindric mapping)).

The coordinates of the left Cauchy–Green deformation tensor
for the vertical mixed equiareal cylindric mapping:

$$c_{11} = \frac{A^2(\alpha + \beta)^2 L^2}{(\alpha L + \beta A)^2} , \quad c_{12} = -\frac{A^3(\alpha + \beta)^2\beta \sin \Phi LM\Lambda}{(\alpha L + \beta A)^3} ,$$

$$c_{22} = M^2\left[\frac{A^6(\alpha + \beta)^4\beta^2 \sin^2 \Phi\Lambda^2 + (\alpha L + \beta A)^6}{A^2(\alpha + \beta)^2(\alpha L + \beta A)^4}\right] .$$

(F.60)

Box F.4 (Left Cauchy–Green deformation tensor (horizontal mixed equiareal cylindric mapping)).

The coordinates of the left Cauchy–Green deformation tensor
for the horizontal mixed equiareal cylindric mapping:

$$c_{11} = \frac{(\alpha A + \beta L)^2}{(\alpha + \beta)^2} , \quad c_{12} = -\frac{\beta \sin \Phi M(\alpha A + \beta L)\Lambda}{(\alpha + \beta)^2} ,$$

$$c_{22} = M^2\left[\frac{\beta^2\Lambda^2 \sin^2 \Phi(\alpha A + \beta L)^2 + (\alpha + \beta)^4 L^2}{(\alpha + \beta)^2(\alpha A + \beta L)^2}\right] .$$

(F.61)

Corollary F.9 (Maximal left angular distortion of the vertical–horizontal mean of mixed equiareal cylindric mapping of the biaxial ellipsoid $\mathbb{E}^2_{A,B}$).

The right maximal angular distortion Ω generated by the mapping equations (F.32) for the vertical mean and (F.43)–(F.46) for the horizontal mean of mixed equiareal mappings of the biaxial ellipsoid is represented by (F.62) subject to (F.63), where $\operatorname{tr}\left[\mathsf{C}_l\mathsf{G}_l^{-1}\right]$ is given by (F.48) for the vertical mean of (F.32) and by (F.49) for the horizontal mean of (F.43)–(F.46).

$$\Omega = \arcsin\frac{\Lambda_1 - \Lambda_2}{\Lambda_1 + \Lambda_2} \, , \tag{F.62}$$

$$\Lambda_1 - \Lambda_2 = \sqrt{\left(\operatorname{tr}\left[\mathsf{C}_l\mathsf{G}_l^{-1}\right]\right) - 2} \, , \quad \Lambda_1 + \Lambda_2 = \sqrt{\left(\operatorname{tr}\left[\mathsf{C}_l\mathsf{G}_l^{-1}\right]\right) + 2} \, . \tag{F.63}$$

End of Corollary.

Proof.

For the proof of (F.63) just compute $(\Lambda_1 - \Lambda_2)^2$ and $(\Lambda_1 + \Lambda_2)^2$ under the postulate of an equiareal mapping $\Lambda_1\Lambda_2 = 1$, namely by means of (F.48).

$$
\begin{aligned}
(\Lambda_1 - \Lambda_2)^2 &= \Lambda_1^2 + \Lambda_2^2 - 2\Lambda_1\Lambda_2 = \Lambda_1^2 + \Lambda_2^2 - 2 = \operatorname{tr}\left[\mathsf{C}_l\mathsf{G}_l^{-1}\right] - 2 \, , \\
(\Lambda_1 + \Lambda_2)^2 &= \Lambda_1^2 + \Lambda_2^2 + 2\Lambda_1\Lambda_2 = \Lambda_1^2 + \Lambda_2^2 + 2 = \operatorname{tr}\left[\mathsf{C}_l\mathsf{G}_l^{-1}\right] + 2 \, .
\end{aligned}
\tag{F.64}
$$

End of Proof.

Corollary F.10 (The left eigenvectors of the left Cauchy–Green deformation tensor).

The left eigenvectors of the left Cauchy–Green deformation tensor normalized with respect to the left metric tensor can be represented with respect to the basis $\{\boldsymbol{G}_1, \boldsymbol{G}_2\}$ which spans the local tangent space $T\mathbb{E}^2_{A,B}$

End of Corollary.

The infinitesimal distance dS between two points $\boldsymbol{X}$ and $\boldsymbol{X}+\mathrm{d}\boldsymbol{X}$ both elements of the biaxial ellipsoid $\mathbb{E}^2_{A,B}$, see (F.65), is *push–forward* transformed into the $\{x, y\}$ coordinate representation, in particular, into (F.66) and (F.67).

$$
\begin{aligned}
\boldsymbol{f}_{l1} &= \frac{-(c_{12} - \Lambda_1^2 G_{12})\boldsymbol{G}_2 + (c_{22} - \Lambda_1^2 G_{22})\boldsymbol{G}_2}{\sqrt{G_{22}(c_{12} - \Lambda_1^2 G_{12})^2 + G_{11}(c_{22} - \Lambda_1^2 G_{22})^2 - 2G_{12}(c_{12} - \Lambda_1^2 G_{12})(c_{22} - \Lambda_1^2 G_{22})}} \, , \\
\boldsymbol{f}_{l2} &= \frac{-(c_{12} - \Lambda_2^2 G_{12})\boldsymbol{G}_1 + (c_{11} - \Lambda_2^2 G_{11})\boldsymbol{G}_2}{\sqrt{G_{11}(c_{12} - \Lambda_2^2 G_{12})^2 + G_{22}(c_{11} - \Lambda_2^2 G_{11})^2 - 2G_{12}(c_{12} - \Lambda_2^2 G_{12})(c_{11} - \Lambda_2^2 G_{11})}} \, ,
\end{aligned}
\tag{F.65}
$$

$$
\begin{aligned}
\mathrm{d}S^2 &= G_{AB}\mathrm{d}U^A\mathrm{d}U^B = G_{AB}\frac{\partial U^A}{\partial u^\alpha}\frac{\partial U^B}{\partial u^\beta}\mathrm{d}u^\alpha\mathrm{d}u^\beta \\
&\forall\, u^1 = x \, , \quad u^2 = y \, ,
\end{aligned}
\tag{F.66}
$$

$$
\begin{aligned}
\mathrm{d}S^2 &= C_{\alpha\beta}\mathrm{d}u^\alpha\mathrm{d}u^\beta \\
&\forall\, C_{\alpha\beta} := G_{AB}\frac{\partial U^A}{\partial u^\alpha}\frac{\partial U^B}{\partial u^\beta} \, .
\end{aligned}
\tag{F.67}
$$

The right Jacobi matrix $[\partial U^A/\partial u^\alpha] =: \mathsf{J}_r$ is the inverse of the left Jacobi matrix $[\partial u^\mu/\partial U^A] =: \mathsf{J}_l$, i.e. $\mathsf{J}_r = \mathsf{J}_l^{-1}$. Since we have already computed J_l, we are left with the problem of calculating

$$\mathsf{J}_r = \mathsf{J}_l^{-1} = (x_\Lambda y_\Phi)^{-1} \begin{bmatrix} y_\Phi & -x_\Phi \\ 0 & x_\Lambda \end{bmatrix} = \begin{bmatrix} \Lambda_x & \Lambda_y \\ \Phi_x & \Phi_y \end{bmatrix} , \tag{F.68}$$

$$\Lambda_x = x_\Lambda^{-1} , \quad \Lambda_y = -x_\Phi (x_\Lambda y_\Phi)^{-1} , \quad \Phi_x = 0 , \quad \Phi_y = y_\Phi^{-1} , \tag{F.69}$$

$$C_{11} = G_{11}\Lambda_x^2 + G_{22}\Phi_x^2 ,$$

$$C_{12} = G_{11}\Lambda_x\Lambda_y + G_{22}\Phi_x\Phi_y , \tag{F.70}$$

$$C_{22} = G_{11}\Lambda_y^2 + G_{22}\Phi_y^2 ,$$

$$C_{11} = \frac{N^2 \cos^2 \Phi}{x_\Lambda^2} ,$$

$$C_{12} = -\frac{N^2 \cos^2 \Phi x_\Phi}{x_\Lambda^2 y_\Phi} , \tag{F.71}$$

$$C_{22} = \frac{N^2 \cos^2 \Phi x_\Phi^2}{x_\Lambda^2 y_\Phi^2} + \frac{M^2}{y_\Phi^2} .$$

The coordinates of the right Cauchy–Green deformation tensor for the vertical as well as the horizontal mixed equiareal cylindric mapping of (F.32) and (F.43)–(F.46) are collected in Box F.5 and Box F.6. The results enable us to compute the right eigenvectors given by Corollary F.11. Indeed, they are needed to orientate by $\tan\varphi = C_{12}/(\lambda_2^2 - C_{11})$ the right principal stretches, once we relate $\Lambda_1 = \lambda_1^{-1}$ such that $\lambda_1 = \Lambda_1^{-1}$ and $\lambda_2 = \Lambda_2^{-1}$ holds. There exists a right analogue ω of the left maximal angular distortion Ω, namely of (F.62) , as soon as we replace left principal stretches by right ones.

Box F.5 (The coordinates of the right Cauchy–Green deformation tensor for the vertical mixed equiareal cylindric mapping of (F.32)).

The coordinates of the right Cauchy–Green deformation tensor for the vertical mixed equiareal cylindric mapping of (F.32) are provided by

$$C_{11} = \frac{(\alpha L + \beta A)^2}{A^2(\alpha + \beta)^2} , \quad C_{12} = \frac{A\beta \sin \Phi}{\alpha L + \beta A} ,$$

$$C_{22} = \frac{A^2(\alpha + \beta)^2}{(\alpha L + \beta A)^2} \left(1 + \frac{(A^2\beta^2 \sin^2 \Phi)}{(\alpha L + \beta A)^2} \right) . \tag{F.72}$$

Box F.6 (The coordinates of the right Cauchy–Green deformation tensor for the horizontal mixed equiareal cylindric mapping of (F.43)–(F.46)).

The coordinates of the right Cauchy–Green deformation tensor for the horizontal mixed equiareal cylindric mapping of (F.43)–(F.46) are provided by

$$C_{11} = \frac{(\alpha + \beta)^2 L^2}{(\alpha A + \beta L)^2} , \quad C_{12} = \frac{\beta \sin \Phi L\Lambda}{\alpha A + \beta L} ,$$

$$C_{22} = \frac{1}{(\alpha + \beta)^2} \left(\beta^2 \sin^2 \Phi\Lambda^2 + \frac{(\alpha A + \beta L)^2}{L^2} \right) . \tag{F.73}$$

Corollary F.11 (The right eigenvectors of the right Cauchy–Green deformation tensor).

The right eigenvectors of the right Cauchy–Green deformation tensor normalized with respect to the right metric tensor $G_r = I_2$ can be represented by

$$\boldsymbol{f}_{r1} = \frac{-C_{12}\boldsymbol{e}_2 + (C_{22} - \lambda_1^2)\boldsymbol{e}_1}{\sqrt{C_{12}^2 + (C_{22} - \lambda_1^2)^2}} \ , \quad \boldsymbol{f}_{r2} = \frac{-C_{12}\boldsymbol{e}_1 + (C_{11} - \lambda_2^2)\boldsymbol{e}_2}{\sqrt{C_{12}^2 + (C_{11} - \lambda_2^2)^2}} \ , \tag{F.74}$$

namely with respect to the orthonormal basis $\{\boldsymbol{e}_1, \boldsymbol{e}_2\}$ which spans $\{\mathbb{R}^2, \delta_{\alpha\beta}\}$. The coordinates of the eigenvectors $\{\boldsymbol{f}_{r1}, \boldsymbol{f}_{r2}\}$ generate the orthonormal matrix

$$\mathsf{F}_r = \begin{bmatrix} \cos\varphi & \sin\varphi \\ -\sin\varphi & \cos\varphi \end{bmatrix} = [\boldsymbol{f}_{r1}, \boldsymbol{f}_{r2}] \ , \tag{F.75}$$

such that

$$\tan\varphi = \frac{C_{12}}{-(C_{11} - \lambda_2^2)} = \frac{2C_{12}}{C_{22} - C_{11} - \sqrt{(C_{11} - C_{22})^2 + (2C_{12})^2}} \ , \quad \cot\varphi = \frac{2C_{12}}{C_{11} - C_{22}} \ . \tag{F.76}$$

End of Corollary.

As a visualization for the derived pseudo-cylindrical mappings of the biaxial ellipsoid onto the plane, the following Figs. F.1–F.10 here are given for different weight parameters α and β including their Tissot indicatrices.

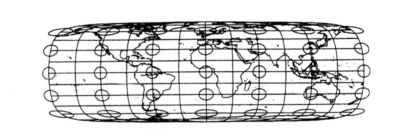

Fig. F.1. Vertical weighted mean of the generalized Lambert projection and the generalized Sanson–Flamsteed projection of the biaxial ellipsoid $\mathbb{E}_{A,B}^2$, squared relative eccentricity $E^2 = 0.1$, weight parameters $\alpha = 1$, $\beta = 0.1$.

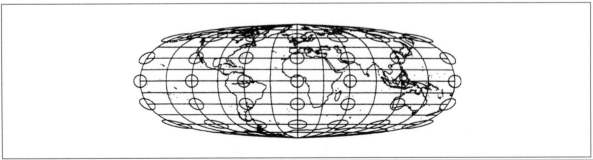

Fig. F.2. Vertical weighted mean of the generalized Lambert projection and the generalized Sanson–Flamsteed projection of the biaxial ellipsoid $\mathbb{E}_{A,B}^2$, squared relative eccentricity $E^2 = 0.1$, weight parameters $\alpha = 1$, $\beta = 0.5$.

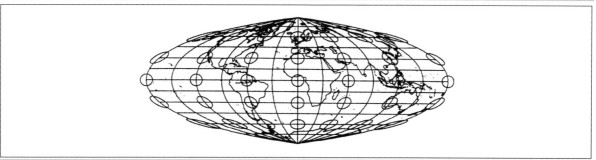

Fig. F.3. Vertical weighted mean of the generalized Lambert projection and the generalized Sanson–Flamsteed projection of the biaxial ellipsoid $\mathbb{E}^2_{A,B}$ (generalized Foucaut projection), squared relative eccentricity $E^2 = 0.1$, weight parameters $\alpha = \beta = 1$.

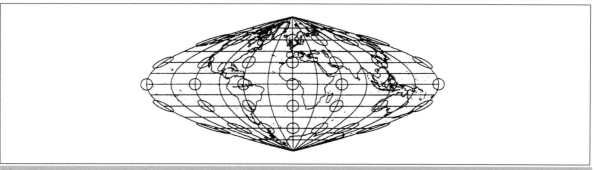

Fig. F.4. Vertical weighted mean of the generalized Lambert projection and the generalized Sanson–Flamsteed projection of the biaxial ellipsoid $\mathbb{E}^2_{A,B}$, squared relative eccentricity $E^2 = 0.1$, weight parameters $\alpha = 0.5$, $\beta = 1$.

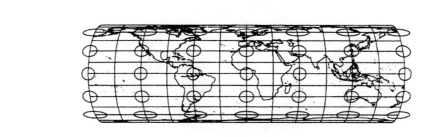

Fig. F.5. Vertical weighted mean of the generalized Lambert projection and the generalized Sanson–Flamsteed projection of the biaxial ellipsoid $\mathbb{E}^2_{A,B}$, squared relative eccentricity $E^2 = 0.1$, weight parameters $\alpha = 0.1$, $\beta = 1$.

Fig. F.6. Horizontal weighted mean of the generalized Lambert projection and the generalized Sanson–Flamsteed projection of the biaxial ellipsoid $\mathbb{E}^2_{A,B}$, squared relative eccentricity $E^2 = 0.1$, weight parameters $\alpha = 1$, $\beta = 0.1$.

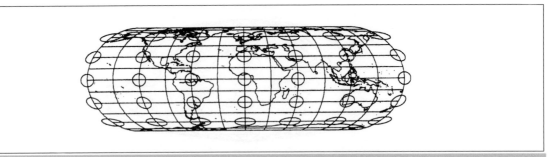

Fig. F.7. Horizontal weighted mean of the generalized Lambert projection and the generalized Sanson–Flamsteed projection of the biaxial ellipsoid $\mathbb{E}^2_{A,B}$, squared relative eccentricity $E^2 = 0.1$, weight parameters $\alpha = 1$, $\beta = 0.5$.

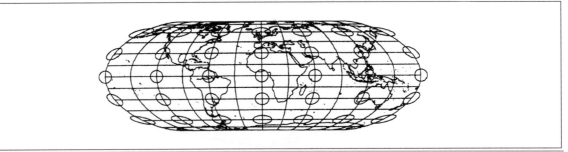

Fig. F.8. Horizontal weighted mean of the generalized Lambert projection and the generalized Sanson–Flamsteed projection of the biaxial ellipsoid $\mathbb{E}^2_{A,B}$ (generalized Nell–Hammer projection), squared relative eccentricity $E^2 = 0.1$, weight parameters $\alpha = \beta = 1$.

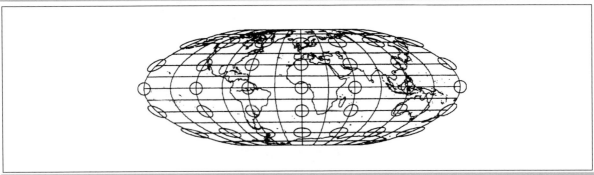

Fig. F.9. Horizontal weighted mean of the generalized Lambert projection and the generalized Sanson–Flamsteed projection of the biaxial ellipsoid $\mathbb{E}^2_{A,B}$, squared relative eccentricity $E^2 = 0.1$, weight parameters $\alpha = 0.5$, $\beta = 1$.

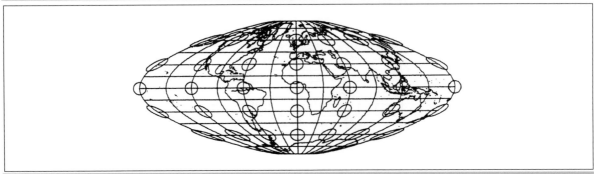

Fig. F.10. Horizontal weighted mean of the generalized Lambert projection and the generalized Sanson–Flamsteed projection of the biaxial ellipsoid $\mathbb{E}^2_{A,B}$, squared relative eccentricity $E^2 = 0.1$, weight parameters $\alpha = 0.1$, $\beta = 1$.

Note that our results are based upon the contributions of J. Cossin (1570), J. Flamsteed (1692), H. C. Foucaut (1862), E. Grafarend (1995), E. Grafarend and A. Heidenreich (1995), E. Grafarend and R. Syffus (1997c), E. Hammer (1900), J. H. Lambert (1772), A. M. Nell (1890), N. Sanson (1675) and M. R. Tobler (1973).

G Generalized Mollweide projection

> Generalized Mollweide projection of the ellipsoid-of-revolution, the standard Mollweide projection, the generalized Mollweide projection, general equiareal pseudo-cylindrical mapping equations.

The *standard Mollweide projection* of the sphere $\mathbb{S}^2_R$ which is of type *equiareal pseudo-cylindrical* is *generalized* to the *biaxial ellipsoid* $\mathbb{E}^2_{A_1\,A_2}$.

Important!

Within the class of pseudo-cylindrical mapping equations (G.8) of $\mathbb{E}^2_{A_1\,A_2}$ (semi-major axis A_1, semi-minor axis A_2) it is shown by solving the general eigenvalue problem (Tissot analysis) that only equiareal mappings, no conformal mappings exist. The mapping equations (G.20), which generalize those from $\mathbb{S}^2_R$ to $\mathbb{E}^2_{A_1\,A_2}$, lead under the equiareal postulate to a generalized Kepler equation (G.39), which is solved by Newton iteration, for instance, see Table G.1. Two variants of the ellipsoidal Mollweide projection, in particular, (G.35) and (G.36) versus (G.37) and (G.38), are presented, which guarantee that parallel circles (coordinate lines of constant ellipsoidal latitude) are mapped onto straight lines in the plane, while meridians (coordinate lines of constant ellipsoidal longitude) are mapped onto ellipses of variable axes. A theorem collects the basic results. Computer graphical examples illustrate the first pseudo-cylindrical map projection of $\mathbb{E}^2_{A_1\,A_2}$ of type *generalized Mollweide*.

With advent of artificial satellites of the Earth measuring precisely its size and shape, the ellipsoidal reference figure becomes more and more obvious. In order to present an equiareal map of a biaxial reference ellipsoid (which is of central importance for a graphical representation of environmental data, for example, from *remote sensing*) the popular Mollweide projection of the sphere $\mathbb{S}^2_R$ of radius R has to be generalized into an equiareal pseudo-cylindrical projection of the ellipsoid-of-revolution or spheroid $\mathbb{E}^2_{A_1\,A_2}$ with the semi-major axis A_1 and with the semi-minor axis A_2, for example, according to the *Geodetic Reference System* 1980.

> **Section G-1.**

In order to generalize the standard Mollweide projection of $\mathbb{S}^2_R$ towards $\mathbb{E}^2_{A_1\,A_2}$ (compare with J. P. Snyder (1977, 1979)), Section G-1 is a setup of *general pseudo-cylindrical mapping equations* of class (G.8) which allow only equiareal, but no conformal projections. (We met the same situation for $\mathbb{S}^2_R$.) By solving a general eigenvalue–eigenvector problem, the principal distortions as well as their directions (eigenvectors) are computed, in particular, on the basis of the metric of the plane represented by *pullback* in terms of the left Cauchy–Green deformation tensor and of the metric of the spheroid.

> **Section G-2.**

In Section G-2, we specialize the *ellipsoidal Mollweide projection* by an "Ansatz" (G.20), leading to the problem to solve a generalized Kepler equation (G.39), for example, by Newton iteration, see Table G.1. The basic results are collected in three corollaries and one theorem.

> **Section G-3.**

Finally, completing the preceding Sections G-1 and G-2, Section G-3 presents computer graphics of the generalized Mollweide projection of $\mathbb{E}^2_{A_1\,A_2}$.

G-1 The pseudo-cylindrical mapping of the biaxial ellipsoid onto the plane

First, we construct a *minimal atlas* of the *biaxial ellipsoid* $\mathbb{E}^2_{A_1\,A_2}$ ("ellipsoid-of-revolution", "spheroid") based on local coordinates of type {longitude, latitude} and {meta-longitude, meta-latitude}.

$$\boldsymbol{X} \in \mathbb{E}^2_{A_1, A_2} :=$$

$$:= \left\{ \boldsymbol{X} \in \mathbb{R}^3 \,\middle|\, \frac{X^2 + Y^2}{A_1^2} + \frac{Z^2}{A_2^2} = 1, A_1 \in \mathbb{R}^+, A_2 \in \mathbb{R}^+, A_1 > A_2 \right\} , \tag{G.1}$$

$$\boldsymbol{X}(L, B) =$$

$$= \boldsymbol{e}_1 \frac{A_1 \cos \Phi \cos \Lambda}{\sqrt{1 - E^2 \sin^2 \Phi}} + \boldsymbol{e}_2 \frac{A_1 \cos \Phi \sin \Lambda}{\sqrt{1 - E^2 \sin^2 \Phi}} + \boldsymbol{e}_3 \frac{A_1 (1 - E^2) \sin \Phi}{\sqrt{1 - E^2 \sin^2 \Phi}} . \tag{G.2}$$

For surface normal ellipsoidal {longitude Λ, latitude Φ}, we choose the open set $0 < \Lambda < 2\pi$ and $-\pi/2 < \Phi < +\pi/2$ in order to avoid any *coordinate singularity* once we endow the manifold $\mathbb{E}^2_{A_1, A_2}$ with a differentiable structure. Indeed, (G.2) covers all points of $\mathbb{E}^2_{A_1, A_2}$ except the meridians $\Lambda = 0$ and $\Lambda = \pi$, respectively, as well as the poles $\Phi = \pm\pi/2$. E denotes the relative eccentricity of the biaxial ellipsoid $\mathbb{E}^2_{A_1, A_2}$ defined by $E^2 := (A_1^2 - A_2^2)/A_1^2$. In order to guarantee *bijectivity* of the mapping $\{X, Y, Z\} \mapsto \{\Lambda, \Phi\}$, we use

$$\Lambda(\boldsymbol{X}) = \arctan \frac{Y}{X} + \left[-\frac{1}{2} \operatorname{sgn}(Y) - \frac{1}{2} \operatorname{sgn}(Y) \operatorname{sgn}(X) + 1 \right] 180° , \tag{G.3}$$

$$\Phi(\boldsymbol{X}) = \arctan \frac{1}{\sqrt{1 - E^2}} \frac{Z}{\sqrt{X^2 + Y^2}} . $$

In order to establish the second set of local coordinates, in particular {meta-longitude, meta-latitude}, we transform the orthonormal triad $\{\boldsymbol{e}_1, \boldsymbol{e}_2, \boldsymbol{e}_3\}$, which spans $\mathbb{E}^3 = \{\mathbb{R}^3, \delta_{ij}\}$ where δ_{ij} is the Kronecker symbol for a unit matrix $(i, j = 1, 2, 3)$, into the transverse orthonormal triad $\{\boldsymbol{e}_{1'}, \boldsymbol{e}_{2'}, \boldsymbol{e}_{3'}\}$ by means of the rotation matrices $\mathsf{R}_3(\Lambda_0)$ and $\mathsf{R}_1(\pi/2)$.

$$[\boldsymbol{e}_{1'}, \boldsymbol{e}_{2'}, \boldsymbol{e}_{3'}] = [\boldsymbol{e}_1, \boldsymbol{e}_2, \boldsymbol{e}_3] \mathsf{R}_3^{\mathrm{T}}(\Lambda_0) \mathsf{R}_1^{\mathrm{T}}(\pi/2) , \tag{G.4}$$

$$\boldsymbol{X}' \in \mathbb{E}^2_{A_1, A_2} :=$$

$$:= \left\{ \boldsymbol{X}' \in \mathbb{R}^3 \,\middle|\, \frac{X'^2 + Z'^2}{A_1^2} + \frac{Y'^2}{A_2^2} = 1, A_1 \in \mathbb{R}^+, A_2 \in \mathbb{R}^+, A_1 > A_2 \right\} , \tag{G.5}$$

$$\boldsymbol{X}'(\alpha, \beta) =$$

$$= \boldsymbol{e}_{1'} \frac{A_1 \cos \alpha \cos \beta}{\sqrt{1 - E^2 \sin^2 \alpha \cos^2 \beta}} + \boldsymbol{e}_{2'} \frac{A_1 (1 - E^2) \sin \alpha \cos \beta}{\sqrt{1 - E^2 \sin^2 \alpha \cos^2 \beta}} + \boldsymbol{e}_{3'} \frac{A_1 \sin \beta}{\sqrt{1 - E^2 \sin^2 \alpha \cos^2 \beta}} . \tag{G.6}$$

For surface normal ellipsoidal {meta-longitude α, meta-latitude β}, we choose the open set $0 < \alpha < 2\pi$ and $-\pi/2 < \beta < +\pi/2$ in order to avoid any *coordinate singularity* once we differentiate $\boldsymbol{X}'(\alpha, \beta)$ with respect to meta-longitude and meta-latitude. In order to ensure *bijectivity* of the mapping $\{X', Y', Z'\} \mapsto \{\alpha, \beta\}$, we use

$$\alpha(\boldsymbol{X}') = \arctan \frac{1}{1 - E^2} \frac{Y'}{X'} + \left[-\frac{1}{2} \operatorname{sgn}(Y') - \frac{1}{2} \operatorname{sgn}(Y') \operatorname{sgn}(X') \right] 180° , \tag{G.7}$$

$$\beta(\boldsymbol{X}') = \arctan(1 - E^2) \frac{Z'}{\sqrt{(1 - E^2)X'^2 + Y'^2}} . $$

The union of the two charts $\boldsymbol{X}(L, B) \cup \boldsymbol{X}'(\alpha, \beta)$ covers the entire biaxial ellipsoid $\mathbb{E}^2_{A_1, A_2}$, thus generating a *minimal atlas*. (Surfaces which are topological similar to the sphere, for example, the biaxial ellipsoid, are uniquely described by a minimal atlas of two charts. An alternative example is the torus whose minimal atlas is generated by *three charts*.)

Second, we set up *pseudo-cylindrical mapping equations* of the biaxial ellipsoid $\mathbb{E}^2_{A_1,A_2}$ onto the plane $\mathbb{R}^2$ in terms of surface normal ellipsoidal {longitude Λ, latitude Φ}, in particular

$$x = x(\Lambda, \Phi) = A_1 \Lambda \frac{\cos \Phi}{\sqrt{1 - E^2 \sin^2 \Phi}} g(\Phi) \,,$$

$$y = y(\Phi) = A_1 f(\Phi) \,,$$

$$\boldsymbol{x} := \left\{ \boldsymbol{x} \in \mathbb{R}^2 \,\middle|\, ax + by + c = 0 \right\} \,.$$

(G.8)

The structure of the pseudo-cylindrical mapping equations with unknown functions $f(\Phi)$ and $g(\Phi)$ is motivated by the postulate that for $g(\Phi) = 1$ parallel circles of $\mathbb{E}^2_{A_1,A_2}$ should be mapped equidistantly onto $\mathbb{R}^2$. Note that for zero relative eccentricity $E = 0$, we arrive at the pseudo-cylindrical mapping equations of the sphere $\mathbb{S}^2_R$.

Third, by means of Corollary G.1, we want to show that in the class of pseudo-cylindrical mapping equations (G.8), only equiareal map projections are possible.

Corollary G.1 (Pseudo-cylindrical mapping equations).

In the class of the pseudo-cylindrical mapping equations of type (G.8), only equiareal map projections are possible restricting the unknown functions to (G.9). Note that conformal map projections are not in the class of pseudo-cylindrical mapping equations.

$$f' = g^{-1} \quad \text{or} \quad g = f'^{-1} \,.$$

(G.9)

End of Corollary.

For the proof, we are going to construct the left Cauchy–Green deformation tensor and solve its general eigenvalue problem in order to compute the principal distortions $\{\Lambda_1, \Lambda_2\}$. The tests $\Lambda_1 \Lambda_2 = 1$ for equiareal and $\Lambda_1 = \Lambda_2$ for conformality are performed.

Proof.

The infinitesimal distance $\mathrm{d}s$ between two points $\boldsymbol{x}$ and $\boldsymbol{x} + \mathrm{d}\boldsymbol{x}$, both elements of the plane $\{\mathbb{R}^2, \delta_{\mu\nu}\}$ is by pullback transformed into a $\{\Lambda, \Phi\}$ coordinate representation, in particular

$$\mathrm{d}s^2 = \mathrm{d}x^2 + \mathrm{d}y^2 = \delta_{\mu\nu} \mathrm{d}x^\mu \mathrm{d}x^\nu = \delta_{\mu\nu} \frac{\partial x^\mu}{\partial U^A} \frac{\partial x^\nu}{\partial U^B} \mathrm{d}U^A \mathrm{d}U^B \ \forall \ U^1 = \Lambda, \ U^2 = \Phi \,,$$

$$\mathrm{d}s^2 = c_{AB} \mathrm{d}U^A \mathrm{d}U^B \ \forall \ c_{AB} := \delta_{\mu\nu} \frac{\partial x^\mu}{\partial U^A} \frac{\partial x^\nu}{\partial U^B} \,.$$

(G.10)

Throughout, we apply the summation convention over repeated indices, for example, $a_\mu b_\mu = a_1 b_1 + a_2 b_2$ and $a_i b_i = a_1 b_1 + a_2 b_2 + a_3 b_3$. In addition, we adopt the notation (G.11), the symbols for the meridional radius of curvature $M(\Phi)$ and the normal radius of curvature $N(\Phi)$, respectively.

$$M(\Phi) := \frac{A_1 (1 - E^2)}{(1 - E^2 \sin^2 \Phi)^{3/2}} \,,$$

$$N(\Phi) := \frac{A_1}{(1 - E^2 \sin^2 \Phi)^{1/2}} \,.$$

(G.11)

The normal or transverse radius of curvature of $\mathbb{E}^2_{A_1,A_2}$ is the curvature radius of a curve formed by the intersection of the normal or transverse plane $\mathbb{R}^2 \perp T_x \mathbb{E}^2_{A_1,A_2}$ which is normal to the tangent space $T_x \mathbb{E}^2_{A_1,A_2}$ of the ellipsoid $\mathbb{E}^2_{A_1,A_2}$ and is perpendicular to the meridian at a point $\boldsymbol{X} \in \mathbb{E}^2_{A_1,A_2}$.

$$x = x(\Lambda, \Phi) = N(\Phi) \Lambda \cos \Phi g(\Phi) = A_1 \Lambda P(\Phi) g(\Phi) \,,$$

$$y = y(\Phi) = A_1 f(\Phi) \,.$$

(G.12)

Thus, the left Cauchy–Green deformation tensor c_{AB} is generated by

$$c_{11} = \left(\frac{\partial x}{\partial \Lambda}\right)^2 + \left(\frac{\partial y}{\partial \Lambda}\right)^2 = A_1^2 P^2 g^2 \ ,$$

$$c_{12} = c_{21} = \frac{\partial x}{\partial \Lambda}\frac{\partial x}{\partial \Phi} + \frac{\partial y}{\partial \Lambda}\frac{\partial y}{\partial \Phi} = A_1^2 \Lambda P g (P'g + Pg') \ , \tag{G.13}$$

$$c_{22} = \left(\frac{\partial x}{\partial \Phi}\right)^2 + \left(\frac{\partial y}{\partial \Phi}\right)^2 = A_1^2 \Lambda^2 (P'g + Pg')^2 + A_1^2 f'^2 \ ,$$

$$\{c_{AB}\} = A_1^2 \begin{bmatrix} P^2 g^2 & \Lambda P P' g^2 + \Lambda P^2 g g' \\ \Lambda P P' g^2 + \Lambda P^2 g g' & \Lambda^2 P'^2 g^2 + \Lambda^2 P^2 g'^2 + 2\Lambda^2 P P' g g' + f'^2 \end{bmatrix} \ . \tag{G.14}$$

The infinitesimal distance dS between two points X and $X+dX$, both elements of the biaxial ellipsoid $\mathbb{E}^2_{A_1,A_2}$, is in terms of the first chart represented by (G.15). Simultaneous diagonalization of the two matrices $\{c_{AB}\}$ and $\{G_{AB}\}$ being positive–definite leads to the general eigenvalue problem (G.16) with the eigenvalues (principal distortions) $\{\Lambda_1, \Lambda_2\}$ given by (G.17).

$$dS^2 = G_{AB}dU^A dU^B \ \forall \ \{G_{AB}\} = \begin{bmatrix} N^2(\Phi)\cos^2\Phi & 0 \\ 0 & M^2(\Phi) \end{bmatrix} = A_1^2 \begin{bmatrix} P^2 & 0 \\ 0 & M^2 \end{bmatrix} \ , \tag{G.15}$$

$$\left| c_{AB} - \Lambda_S^2 G_{AB} \right| = 0 \ , \tag{G.16}$$

$$\Lambda_S^4 [\Lambda^2 (P'g + Pg')^2 + f'^2 + g^2 M^2] + g^2 f'^2 = 0$$

$$\Leftrightarrow$$

$$\Lambda_S^2 = \frac{1}{2}(\Lambda^2 P'^2 g^2 + \Lambda^2 P^2 g'^2 + 2\Lambda P P' g g' + f'^2 + g^2 M^2) \pm$$

$$\pm \sqrt{\frac{1}{4}(\Lambda^2 P'^2 g^2 + \Lambda P P' g g' + f'^2 + g^2 M^2)^2 + 1} \tag{G.17}$$

$$\Leftrightarrow$$

$$\Lambda_S^2 = a \pm b \Leftrightarrow \Lambda_1^2 = a + b \ , \ \ \Lambda_2^2 = a - b \ .$$

The postulate of an equiareal mapping $\Lambda_1^2 \Lambda_2^2 = 1$ leads to $(a+b)(a-b) = a^2 - b^2 = 1$ or $gf'^2 = 1$ or $g = f'^{-1}$ or $g^{-1} = f'$. Obviously, the postulate of a conformal mapping $\Lambda_1^2 = \Lambda_2^2$ cannot be fulfilled since $a + b \neq a - b$, $b \neq 0$ holds, in general.

End of Proof.

In summarizing, we are led to the *equiareal pseudo-cylindrical mapping equations* (G.18) with the *principal distortions* (G.19).

$$x = A_1 \Lambda \frac{\cos\Phi}{\sqrt{1 - E^2 \sin^2\Phi}}\frac{1}{f'(\Phi)} = N(\Phi)\Lambda\cos\Phi\frac{1}{f'(\Phi)} = A_1^2\Lambda\frac{\cos\Phi}{\sqrt{1 - E^2\sin^2\Phi}}\frac{1}{\frac{dy}{d\Phi}} \ , \tag{G.18}$$

$$y = A_1 f(\Phi) \ ,$$

$$\Lambda_S^2 = \frac{1}{2}(\Lambda^2 P'^2 + Q^2 + 1) \pm \sqrt{\frac{1}{4}(\Lambda^2 P'^2 + Q^2 + 1) + 1}$$

$$\forall \ \ P'(\Phi) = -\frac{\sin\Phi}{\sqrt{1 - E^2\sin^2\Phi}} + \frac{E^2\sin\Phi\cos^2\Phi}{(1 - E^2\sin^2\Phi)^{3/2}} \ . \tag{G.19}$$

G-2 The generalized Mollweide projections for the biaxial ellipsoid

The characteristics of the spherical Mollweide projection within the class of pseudo-cylindrical mappings are as follows. The graticule parallel circles are mapped on parallel *straight lines*, while meridians on *ellipses*. We shall keep these properties for the ellipsoidal Mollweide projection by the "Ansatz"

$$x(\Lambda, \Phi) = a\Lambda \cos t, \; y(\Phi) = b \sin t . \tag{G.20}$$

The polar coordinate t can be interpreted as the *reduced latitude* of the ellipse $x^2/a^2(\Lambda) + y^2/b^2 = 1$, where $a(\Lambda) = a\Lambda$ holds. Note that for longitude $\Lambda = 0$, $a(\Lambda) = 0$ follows. Therefore, the central *Greenwich meridian* $\Lambda = 0$ is mapped onto a straight line. Of course, any other central meridian could have been chosen alternatively. Again, we compute the left Cauchy–Green deformation tensor, this time represented in terms of $t(\Phi)$ as follows.

$$\frac{\partial x}{\partial \Lambda} = a \cos t(\Phi) , \quad \frac{\partial x}{\partial \Phi} = -a\Lambda \sin t(\Phi)\frac{dt}{d\Phi} , \quad \frac{\partial y}{\partial \Lambda} = 0 , \quad \frac{\partial y}{\partial \Phi} = b \cos t(\Phi)\frac{dt}{d\Phi} , \tag{G.21}$$

$$\text{(G.13), (G.21)}$$
$$\Rightarrow$$

$$\{c_{AB}\} = \begin{bmatrix} a^2 \cos^2 t(\Phi) & -a^2\Lambda \sin t(\Phi) \cos t(\Phi)\frac{dt}{d\Phi} \\ -a^2\Lambda \sin t(\Phi) \cos t(\Phi)\frac{dt}{d\Phi} & a^2\Lambda^2 \sin^2 t(\Phi) \left(\frac{dt}{d\Phi}\right)^2 + b^2 \cos^2 t(\Phi) \left(\frac{dt}{d\Phi}\right)^2 \end{bmatrix} , \tag{G.22}$$

$$\text{(G.14), (G.15), (G.22)} \Rightarrow \left| c_{AB} - \Lambda_S^2 G_{AB} \right| = 0$$
$$\Leftrightarrow$$

$$\begin{vmatrix} a^2 \cos^2 t - \Lambda_S^2 G_{11} & -a^2\Lambda \sin t \cos t \frac{dt}{d\Phi} \\ -a^2 \sin t \cos t \frac{dt}{d\Phi} & (a^2\Lambda^2 \sin^2 t + b^2 \cos^2 t)\left(\frac{dt}{d\Phi}\right)^2 - \Lambda_S^2 G_{22} \end{vmatrix} = 0$$
$$\Leftrightarrow$$

$$\Lambda_S^4 G_{11}G_{22} - \Lambda_S^2[G_{11}(a^2\Lambda^2 \sin^2 t + b^2 \cos^2 t)\left(\frac{dt}{d\Phi}\right)^2 + G_{22}a^2 \cos^2 t] + \tag{G.23}$$

$$+a^2 \cos^2 t(a^2\Lambda^2 \sin^2 t + b^2 \cos^2 t)\left(\frac{dt}{d\Phi}\right)^2 - a^4\Lambda^2 \sin^2 t \cos^2 t \left(\frac{dt}{d\Phi}\right)^2 = 0$$
$$\Leftrightarrow$$

$$\Lambda_S^4 - \Lambda_S^2 \frac{G_{11}(a^2\Lambda^2 \sin^2 t + b^2 \cos^2 t)\left(\frac{dt}{d\Phi}\right)^2}{G_{11}G_{22}} + \frac{G_{22}a^2 \cos^2 t}{G_{11}G_{22}} +$$

$$+\frac{a^2 \cos^2 t(a^2\Lambda^2 \sin^2 t + b^2 \cos^2 t)\left(\frac{dt}{d\Phi}\right)^2}{G_{11}G_{22}} - \frac{a^4\Lambda^2 \sin^2 t \cos^2 t \left(\frac{dt}{d\Phi}\right)^2}{G_{11}G_{22}} = 0 ,$$

$$\Lambda_S^2 = \frac{1}{2}\frac{G_{11}(a^2\Lambda^2 \sin^2 t + b^2 \cos^2 t)\left(\frac{dt}{d\Phi}\right)^2}{G_{11}G_{22}} + \frac{G_{22}a^2 \cos^2 t}{G_{11}G_{22}} \pm$$

$$\pm \left[\frac{1}{4}\left(\frac{G_{11}(a^2\Lambda^2 \sin^2 t + b^2 \cos^2 t)\left(\frac{dt}{d\Phi}\right)^2}{G_{11}G_{22}} + \frac{G_{22}a^2 \cos^2 t}{G_{11}G_{22}}\right)^2 - \right.$$

$$\left. -\frac{a^2 \cos^2 t(a^2\Lambda^2 \sin^2 t + b^2 \cos^2 t)}{G_{11}G_{22}} - \frac{a^4\Lambda^2 \sin^2 t \cos^2 t}{G_{11}G_{22}}\left(\frac{dt}{d\Phi}\right)^2 \right]^{1/2} \tag{G.24}$$

$$= \mu \pm \sqrt{\mu^2 - \gamma} ,$$

$$\Lambda_1^2 \Lambda_2^2 = 1 \Leftrightarrow \mu^2 - (\mu^2 - \gamma) = 1 \Leftrightarrow \gamma = 1$$

$$\Leftrightarrow$$

$$[a^2 \cos^2 t(a^2 \Lambda^2 \sin^2 t + b^2 \cos^2 t) - a^4 \Lambda^2 \sin^2 t \cos^2 t] \left(\frac{dt}{d\Phi}\right)^2 = G_{11} G_{22} \qquad (G.25)$$

$$\Leftrightarrow$$

$$ab \cos^2 t \frac{dt}{d\Phi} = \sqrt{G_{11} G_{22}} = M(\Phi) N(\Phi) \cos \Phi = A_1^2 (1 - E^2) \frac{\cos \Phi}{(1 - E^2 \sin^2 \Phi)^2} \ .$$

Let us collect the previous results in Corollary G.2.

Corollary G.2 (Generalized Kepler equation, generalized Mollweide projection for the biaxial ellipsoid).

Under the equiareal pseudo-cylindrical mapping equations (G.26), where parallel circles are mapped onto straight lines and meridians are mapped onto ellipses $x^2/a^2(\Lambda) + y^2/b^2 = 1$ with $a(\Lambda) = a\Lambda$, the *generalized Kepler equation* (G.27) has to be solved.

$$x(\Lambda, \Phi) = a\Lambda \cos t(\Phi) \ , \quad y(\Phi) = b \sin t(\Phi) \ , \qquad (G.26)$$

$$ab \cos^2 t \frac{dt}{d\Phi} = A_1^2 (1 - E^2) \frac{\cos \Phi}{(1 - E^2 \sin^2 \Phi)^2} \ . \qquad (G.27)$$

End of Corollary.

Next, let us integrate the *generalized Kepler equation.*

$$\frac{dt}{d\Phi} = \frac{A_1^2 (1 - E^2)}{ab} \frac{1}{\cos^2 t(\Phi)} \frac{\cos \Phi}{(1 - E^2 \sin^2 \Phi)^2} \ , \quad \cos^2 t(\Phi) \, dt = \frac{A_1^2 (1 - E^2)}{ab} \frac{\cos \Phi}{(1 - E^2 \sin^2 \Phi)^2} \, d\Phi. \ (G.28)$$

The *forward substitution* $x = E \sin \Phi$ leads to (G.29).

$$\int_0^t \cos^2 t \, dt = \frac{A_1^2 (1 - E^2)}{Eab} \int_0^x \frac{1}{(1 - x^2)^2} \, dx$$

$$\Leftrightarrow \qquad\qquad\qquad (G.29)$$

$$\frac{1}{2} t + \frac{1}{4} \sin 2t = \frac{A_1^2 (1 - E^2)}{abE} \left[\frac{x}{2(1 - x^2)} + \operatorname{ar\,tanh} x \right] \ .$$

The *backward substitution* finally leads to the integrated generalized Kepler equations (G.30).

$$\frac{1}{2} t + \frac{1}{4} \sin 2t =$$

$$= \frac{A_1^2}{ab} \frac{1 - E^2}{E} \left[\frac{E \sin \Phi}{2(1 - E^2 \sin^2 \Phi)} + \frac{1}{2} \operatorname{ar\,tanh} (E \sin \Phi) \right] \ . \qquad (G.30)$$

Due to $\operatorname{ar\,tanh} (E \sin \Phi) = \frac{1}{2} \ln \frac{1 + E \sin \Phi}{1 - E \sin \Phi}$, we alternatively obtain (G.31).

$$2t + \sin 2t =$$

$$= \frac{A_1^2}{ab} \frac{1 - E^2}{E} \left[\ln \frac{1 + E \sin \Phi}{1 - E \sin \Phi} + \frac{2E \sin \Phi}{1 - E^2 \sin^2 \Phi} \right] \ . \qquad (G.31)$$

Note that the total area of the biaxial ellipsoid $\mathbb{E}^2_{A_1,A_2}$ (for example, E. Grafarend and J. Engels (1992)) is represented by (G.32). This result can be used to determine the unknown ellipsoid axes a and b according to the *Mollweide gauge*. In case of the sphere $\mathbb{S}^2_R$, C. B. Mollweide (1805) has proposed that the *half–sphere* $-\pi/2 \le \Lambda \le +\pi/2$ should be mapped equiareally onto a *circle* in the plane $\mathbb{P}^2$. For the biaxial ellipsoid $\mathbb{E}^2_{A_1,A_2}$, a corresponding postulate would define the *half–ellipsoid* $-\pi/2 \le \Lambda \le +\pi/2$ to be mapped equiareally onto an *ellipse*, in particular (G.33).

$$S_{\mathbb{E}^2_{A_1,A_2}} = \pi A_1^2 \frac{1-E^2}{E}\left[\ln\frac{1+E}{1-E} + \frac{2E}{1-E^2}\right],\tag{G.32}$$

$$a(\Lambda = \pi/2) = a\pi/2\,,\ \pi a(\Lambda = \pi/2)b = \frac{1}{2}S_{\mathbb{E}^2_{A_1,A_2}}$$
$$\Leftrightarrow$$
$$\frac{1}{2}\pi^2 ab = \frac{1}{2}S_{\mathbb{E}^2_{A_1,A_2}}\tag{G.33}$$
$$\Leftrightarrow$$
$$\pi^2 ab = S_{\mathbb{E}^2_{A_1,A_2}}$$
$$\Leftrightarrow$$

$$\pi ab = A_1^2 \frac{1-E^2}{E}\left[\ln\frac{1+E}{1-E} + \frac{2E}{1-E^2}\right].\tag{G.34}$$

Corollary G.3 (Generalized Mollweide gauge for the biaxial ellipsoid).

Under the postulate that the half–ellipsoid $-\pi/2 \le \Lambda \le +\pi/2$ to be mapped equiareally onto an ellipse in $\mathbb{P}^2$, the generalized Mollweide gauge (G.34) holds. The result (G.34) approaches the original Mollweide gauge once we set the relative eccentricity $E = 0$.

End of Corollary.

There are two variants of interest in order to determine the axes a and b of the ellipse that is defined by $x^2/a^2(\Lambda) + y^2/b^2 = 0$. (i) Variant one is being motived by the original Mollweide projection for the sphere $\mathbb{S}^2_R$. Accordingly, we define (G.35) and (G.36). If we set $E = 0$, the *spherical Mollweide projection* is derived. (ii) An alternative variant is the postulate of an equidistant mapping of the equator of $\mathbb{E}^2_{A_1,A_2}$ according to (G.37) and (G.38).

$$b := A_1\sqrt{2}\,,\tag{G.35}$$

$$(\text{G.34}),\ (\text{G.35}) \Leftrightarrow a = \frac{A_1(1-E^2)}{\pi E\sqrt{2}}\left[\ln\frac{1+E}{1-E} + \frac{2E}{1-E^2}\right];\tag{G.36}$$

$$x(\Phi = 0) = 2\pi a := 2\pi A_1\,,\ y(\Phi = 0) = 0$$
$$\Leftrightarrow\tag{G.37}$$
$$a := A_1\,,$$

$$(\text{G.34}),\ (\text{G.37}) \Leftrightarrow b = \frac{A_1(1-E^2)}{\pi E}\left[\ln\frac{1+E}{1-E} + \frac{2E}{1-E^2}\right].\tag{G.38}$$

As soon as we implement the axe a and the axe b in the *generalized Kepler equation* (G.31), we gain its final form (G.39). We here note the symmetry of the right–side representation. Table G.1 is a *Newton iteration solution* (see W. Toernig (1979)), for instance) of the generalized Kepler equation for the biaxial ellipsoid $\mathbb{E}^2_{A_1,A_2}$ for the axes A_1 and A_2 as well as the relative eccentricity E according to the Geodetic Reference System 1980 (Bulletin Geodesique 58 (1984) pp. 388–398).

$$2t + \sin 2t = \pi \frac{\ln \frac{1+E\sin\Phi}{1-E\sin\Phi} + \frac{2E\sin\Phi}{1-E^2\sin^2\Phi}}{\ln \frac{1+E}{1-E} + \frac{2E}{1-E^2}} \ . \tag{G.39}$$

Table G.1. Newton iterative solution of the generalized Kepler equation. Parameters: $A_1 = 6\,378\,137\,\mathrm{m}$, $A_2 = 6\,356\,752.314\,1\,\mathrm{m}$, $E^2 = 0.006\,694\,380\,002\,290$.

Φ	$t[\mathrm{rad}]$	$t[°]$
$90°$	1.56673055580	89.767053190
$80°$	1.23781233200	70.921424475
$70°$	1.03751932140	59.445479975
$60°$	0.86517291851	49.570758193
$50°$	0.70717579230	40.518189428
$40°$	0.55790136480	31.965394499
$30°$	0.41428104971	23.736556358
$20°$	0.27434065788	15.718562294
$10°$	0.13663879358	7.828826413
$0°$	0.00000000000	0.000000000

Before we present examples of the ellipsoidal Mollweide projection, we briefly summarize the basic results in Theorem G.4.

Theorem G.4 (Generalized Mollweide projection of the biaxial ellipsoid $\mathbb{E}^2_{A_1,A_2}$).

In the class of pseudo-cylindrical mappings of the biaxial ellipsoid $\mathbb{E}^2_{A_1,A_2}$, the equations (G.40) generate an equiareal mapping.

$$x(\Lambda, \Phi) = a\Lambda \cos t(\Phi), \ y(\Phi) = b \sin t(\Phi) \ . \tag{G.40}$$

For a given ellipsoidal latitude Φ, the reduced latitude t is a solution of the transcendental equation (G.41), which for relative eccentricity $E = 0$ coincides with the Kepler equation.

$$2t + \sin 2t = \pi \frac{\ln \frac{1+E\sin\Phi}{1-E\sin\Phi} + \frac{2E\sin\Phi}{1-E^2\sin^2\Phi}}{\ln \frac{1+E}{1-E} + \frac{2E}{1-E^2}} \ . \tag{G.41}$$

There are two variants for the gauge of the axes a and b of the ellipse $x^2/a^2(\Lambda) + y^2/b^2 = 1$, $a(\Lambda) = a\Lambda$, in particular, $b := A_1\sqrt{2}$, $a = $ (G.36) (variant one) and $a := A_1$, $b = $ (G.38) (variant two). The principal distortions Λ_1 and Λ_2 of the generalized Mollweide projection are given by (G.24) inserting $dt/d\Phi$ according to (G.25) and $t(\Phi)$, solution of (G.41), which are the eigenvalues of the general eigenvalue problem $(c_{AB} - \Lambda^2 G_{AB})E^{BC} = 0$. The principal distortions are plotted along the eigenvectors which constitute the eigenvector matrix $\{E^{BC}\}$

The coordinate lines $\Phi = $ const. ($t = $ const.) called *parallel circles* are mapped onto straight lines $y = $ const. while the coordinate lines $\Lambda = $ const. called *meridians* are mapped onto ellipses $x^2/a^2(\Lambda) + y^2/b^2 = 1$, $a(\Lambda) = a\Lambda$, a straight line for $\Lambda = 0$ (Greenwich meridian), in particular. For relative eccentricity $E = 0$, the generalized Mollweide projection of the biaxial ellipsoid $\mathbb{E}^2_{A_1,A_2}$, variant one, coincides with the standard Mollweide projection of the sphere $\mathbb{S}^2_R$.

End of Theorem.

G-3 Examples

The following six examples illustrate by computer graphics the generalized Mollweide projection of the biaxial ellipsoid.

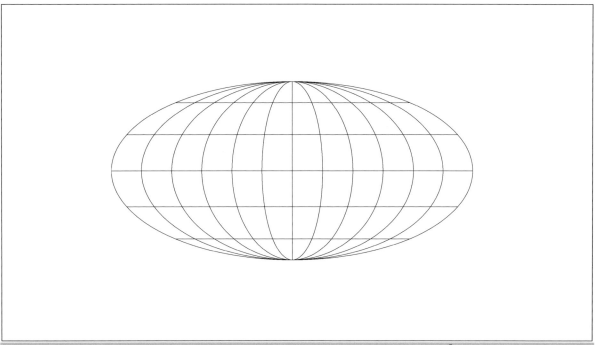

Fig. G.1. Standard Mollweide projection of the sphere $\mathbb{S}^2_R$.

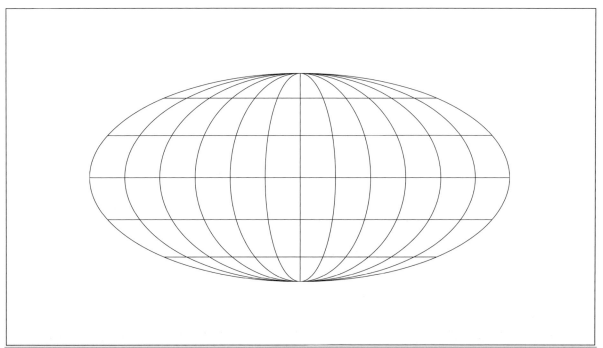

Fig. G.2. Generalized Mollweide projection of the biaxial ellipsoid $\mathbb{E}^2_{A_1, A_2}$, E^2 (Geodetic Reference System 1980), variant one.

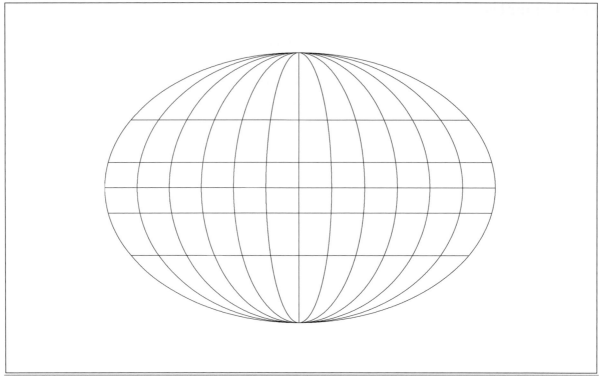

Fig. G.3. Generalized Mollweide projection of the biaxial ellipsoid $\mathbb{E}^2_{A_1,A_2}$, $E^2 = 0\,7$, variant one.

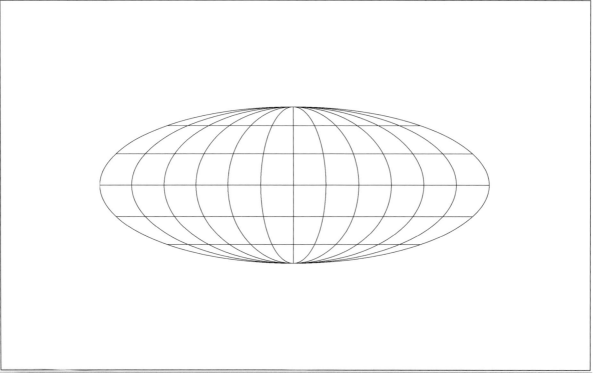

Fig. G.4. Generalized Mollweide projection of the biaxial ellipsoid $\mathbb{E}^2_{A_1,A_2}$, E^2 (Geodetic Reference System 1980), variant two.

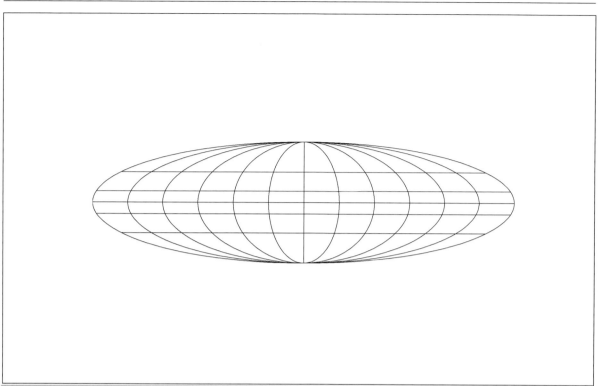

Fig. G.5. Generalized Mollweide projection of the biaxial ellipsoid $\mathbb{E}^2_{A_1,A_2}$, $E^2 = 0\,7$, variant two.

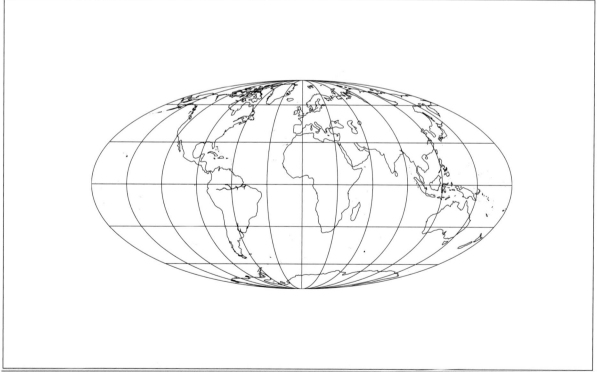

Fig. G.6. Generalized Mollweide projection of the Earth (biaxial ellipsoid $\mathbb{E}^2_{A_1,A_2}$, E^2 (Geodetic Reference System 1980)), continental contour lines.

Aside.

With our students of Stuttgart University has been the vote that the Mollweide projection was in the list of the two most popular map projections.

Note that the original contribution mapping the sphere is due to C. B. Mollweide (1805). Here, we use the works of E. Grafarend and J. Engels (1992), E. Grafarend and A. Heidenreich (1995), J. P. Snyder (1977, 1979) for mapping the ellipsoid-of-revolution. The numerical treatment is based upon the work of W. Toernig (1979).

H Generalized Hammer projection

Generalized Hammer projection of the ellipsoid-of-revolution: azimuthal, transverse, rescaled equiareal. Mapping equations. Univariate series inversion.

The *classical Hammer projection* of the *sphere*, which is azimuthal, transverse rescaled equiareal, is *generalized* to the *ellipsoid-of-revolution*. Its first constituent, the azimuthal transverse equiareal projection of the biaxial ellipsoid, is derived giving the equations for an equiareal transverse azimuthal projection. The second constituent, the equiareal mapping of the biaxial ellipsoid with respect to a transverse frame of reference and a change of scale, is reviewed. Then considered results give collections of the general mapping equations generating the ellipsoidal Hammer projection, which finally lead to a world map.

Aside.
One of the most widely used equiareal map projection is the *Hammer projection* of the *sphere* (Hammer 1892). It maps parallel circles and meridians of the sphere onto algebraic curves of fourth order; its limit line ($\Lambda = \pm\pi$) is an ellipse with respect to the gauge $c_1 = 2$, $c_2 = 1$, $c_3 = 1/2$, and $c_4 = 1$ as illustrated by Fig. H.1 with respect to the mapping equations (H.87), (H.88), and (H.96), respectively. Many celestial bodies like the Earth are pronounced *ellipsoidal*. It is therefore our target to generalize the spherical Hammer projection to an ellipsoid-of-revolution to which we refer as a biaxial ellipsoid.

Section H-1, Section H-3, Section H-4.

The first constituent of the Hammer projection is the transverse equiareal projection onto a tangent plane, namely of azimuthal type. Section H-1 outlines accordingly the introduction of a transverse reference frame for a biaxial ellipsoid. In particular, formulae in (H.12) are derived which constitute the transformation from surface normal ellipsoidal longitude/latitude $\{\Lambda^*, \Phi^*\}$ to surface normal ellipsoidal meta-longitude/meta-latitude $\{A^*, B^*\}$ defined in the ellipsoidal transverse frame of reference. In consequence, the transverse equiareal mapping of a biaxial ellipsoid onto a transverse tangent plane is given by corollaries in terms of ellipsoidal meta-longitude/ meta-latitude. Finally, the elaborate transformation $\{\Lambda^*, \Phi^*\} \rightarrow \{A^*, B^*\}$ in these transverse equiareal mapping equations is performed, leading to the final result of Lemma H.3 and Corollary H.4. Mathematical details are collected in Sections H-3 and H-4

Section H-2, Section H-5.

The second constituent of the Hammer projection is an alternative equiareal mapping of the biaxial ellipsoid with respect to a transverse frame of reference and a change of scale outlined in Section H-2. As a reference, we firstly present an equiareal mapping from a left biaxial ellipsoid to a right biaxial ellipsoid. We secondly give the explicit form of the mapping equations generating an equiareal map from a left biaxial ellipsoid to a right biaxial ellipsoid with respect to a transverse frame of reference and a change of scale. The general mapping equations $x = c_1 x^*(\Lambda^*, \Phi^*)$ and $y = c_2 y^*(\Lambda^*, \Phi^*)$ are specified by means of the differential equations $d\Lambda^*/d\Lambda = c_3$ and $d\Phi^*/d\Phi = c_4(1 - E^2 \sin^2 \Phi^*)^2 \cos \Phi/[(1 - E^2 \sin^2 \Phi)^2 \cos \Phi^*]$ subject to the initial data $\{\Lambda^*(\Lambda = 0) = 0, \Phi^*(\Phi = 0) = 0\}$ of left and right longitude/latitude. In order to guarantee an equiareal mapping, the constants of gauge have to fulfill $c_1 c_2 c_3 c_4 = 1$. Termwise integration of the differential equation $d\Phi^*/d\Phi$ leads to the problem to determine $\sin \Phi^*$ from a homogeneous polynomial equation which is solved by univariate series inversion in Section H-5. Third, the general equations of the ellipsoidal Hammer projection are presented in Lemma H.5 and made more specific by Corollary H.6, namely with respect to the Hammer gauge $\{c_1 = 2, c_2 = 1, c_3 = 1/2, c_4 = 1\}$ which guarantees a map of the total left biaxial ellipsoid onto one chart. Finally, Fig. H.3 presents a world map of the ellipsoidal Hammer projection for a relative eccentricity $E \neq 0$.

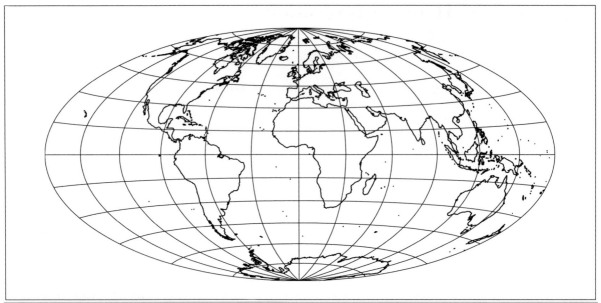

Fig. H.1. Hammer projection of the sphere $\mathbb{S}_R^1$.

Due to length restriction of the manuscript, we had to exclude the deformation/distortion analysis of the ellipsoidal Hammer projection. Another investigation of how parallel circles and meridians are mapped onto transverse tangent plane of the biaxial ellipsoid has to be performed. For an up-to-date reference of the Hammer projection of the sphere under a general gauge $\{c_1, c_2, c_3, c_4\}$, we refer to J. Hoschek (1984) and K. H. Wagner (1962).

H-1 The transverse equiareal projection of the biaxial ellipsoid

The first constituent of the Hammer projection is the transverse equiareal projection, now being developed for the ellipsoid-of-revolution. First, we set up the transverse reference frame which leads to an elliptic meta-equator and a circular zero meta-longitude meta-meridian. In particular, we derive the mapping equations from surface normal ellipsoidal longitude/latitude to surface normal ellipsoidal meta-longitude/meta-latitude. Second, we derive the differential equation for a transverse equiareal projection of the biaxial ellipsoid and find its integral in terms of meta-longitude/meta-latitude. Third, we express the mapping equations which generate a transverse equiareal diffeomorphism in terms of ellipsoidal longitude/latitude. Mathematical details are presented in Sections H-3 and H-4.

H-11 The transverse reference frame

First, let us orientate a set of orthonormal base vectors $\{E_1, E_2, E_3\}$ along the principal axes of (H.1). Against this frame of reference $\{E_1, E_2, E_3; 0\}$ consisting of the orthonormal base vectors $\{E_1, E_2, E_3\}$ and the origin 0, we introduce the oblique frame of reference $\{E_{1'}, E_{2'}, E_{3'}; 0\}$ by means of (H.2).

$$\mathbb{E}^2_{A_1, A_2} := \left\{ \boldsymbol{X} \in \mathbb{R}^3 \,\big|\, (X^2 + Y^2)/A_1^2 + Z^2/A_2^2 = 1, \mathbb{R}^+ \ni A_1 > A_2 \in \mathbb{R}^+ \right\} , \tag{H.1}$$

$$\begin{bmatrix} \boldsymbol{E}_{1'} \\ \boldsymbol{E}_{2'} \\ \boldsymbol{E}_{3'} \end{bmatrix} = \mathsf{R}_1(I)\mathsf{R}_3(\Omega) \begin{bmatrix} \boldsymbol{E}_1 \\ \boldsymbol{E}_2 \\ \boldsymbol{E}_3 \end{bmatrix} . \tag{H.2}$$

The rotation around the 3 axis, we have denoted by Ω, the right ascension of the ascending node, while the rotation around the intermediate 1 axis by I, the inclination. R_1 and R_3, respectively, are orthonormal matrices such that (H.3) holds.

$$
R_1(I)R_3(\Omega) = \begin{bmatrix} \cos\Omega & \sin\Omega & 0 \\ -\sin\Omega\cos I & +\cos\Omega\cos I & \sin I \\ +\sin\Omega\sin I & -\cos\Omega\sin I & \cos I \end{bmatrix} \in \mathbb{R}^{3\times 3} \ . \tag{H.3}
$$

Accordingly, (H.4) is a representation of the placement vector $\boldsymbol{X}$ in the orthonormal bases $\{\boldsymbol{E}_1, \boldsymbol{E}_2, \boldsymbol{E}_3\}$ and $\{\boldsymbol{E}_{1'}, \boldsymbol{E}_{2'}, \boldsymbol{E}_{3'}\}$, respectively. We aim at a transverse orientation of the oblique frame of reference $\{\boldsymbol{E}_{1'}, \boldsymbol{E}_{2'}, \boldsymbol{E}_{3'}; 0\}$, which is characterized by a base vector $\boldsymbol{E}_{3'}$ in the equatorial plane $\mathbb{P}^2(X, Y)$ and the base vectors $\{\boldsymbol{E}_{1'}, \boldsymbol{E}_{2'}\}$ in the rotated plane $\mathbb{P}^2(-Y, -Z)$. Such an orientation of the transverse frame of reference $\{\boldsymbol{E}_{1'}, \boldsymbol{E}_{2'}, \boldsymbol{E}_{3'}; 0\}$ is achieved choosing the inclination $I = 270°$ ($\cos I = 0, \sin I = -1$), for instance, namely (H.5) or (H.6).

$$
\boldsymbol{X} = \sum_{i=1}^{3} \boldsymbol{E}_i X^i = \sum_{i'=1}^{3} \boldsymbol{E}_{i'} X^{i'} \ , \tag{H.4}
$$

$$
\boldsymbol{E}_{1'} = \boldsymbol{E}_1\cos\Omega + \boldsymbol{E}_2\sin\Omega \ , \quad \boldsymbol{E}_{2'} = -\boldsymbol{E}_3 \ , \quad \boldsymbol{E}_{3'} = -\boldsymbol{E}_1\sin\Omega + \boldsymbol{E}_2\cos\Omega \ , \tag{H.5}
$$

$$
X' = X\cos\Omega + Y\sin\Omega \ , \quad Y' = -Z \ , \quad Z' = -X\sin\Omega + Y\cos\Omega \ . \tag{H.6}
$$

Example H.1 (An example: $\Omega = 270°$ ($\cos\Omega = 0, \sin\Omega = -1$)).

As an example, let us choose $\Omega = 270°$ ($\cos\Omega = 0, \sin\Omega = -1$) so that we obtain (H.7) or (H.8), identified as western, southern, and Greenwich.

$$
\boldsymbol{E}_{1'} = -\boldsymbol{E}_2 \ , \quad \boldsymbol{E}_{2'} = -\boldsymbol{E}_3 \ , \quad \boldsymbol{E}_{3'} = \boldsymbol{E}_1 \ , \tag{H.7}
$$

$$
X' = -Y \ , \quad Y' = -Z \ , \quad Z' = X \ . \tag{H.8}
$$

Indeed, the meta-equator is elliptic in the plane $\mathbb{P}^2(X, Y)$ directed towards Greenwich.

End of Example.

The example may motivate the exotic choice of $\Omega = 270°$, $I = 270°$. Fig. H.2 illustrates the special transverse frame of reference $\{\boldsymbol{E}_{1'}, \boldsymbol{E}_{2'}, \boldsymbol{E}_{3'}; 0\}$.

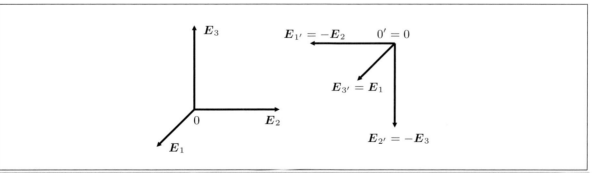

Fig. H.2. The frame of reference $\boldsymbol{E}_1, \boldsymbol{E}_2, \boldsymbol{E}_3; 0$ and the transverse frame of reference $\boldsymbol{E}_{1'}, \boldsymbol{E}_{2'}, \boldsymbol{E}_{3'}; 0$ for the special choice $\Omega = 270°$, $I = 270°$.

While (H.1) is a representation of the biaxial ellipsoid $\mathbb{E}^2_{A_1\,A_2}$ of semi-major axis A_1 and semi-minor axis A_2 in terms of $\{X, Y, Z\}$ coordinates along the orthonormal basis $\{\boldsymbol{E}_1, \boldsymbol{E}_2, \boldsymbol{E}_3\}$, (H.9) is the analogous representation of $\mathbb{E}^2_{A_1\,A_2}$ in terms of $\{X', Y', Z'\}$ along the transverse orthonormal basis $\{\boldsymbol{E}_{1'}, \boldsymbol{E}_{2'}, \boldsymbol{E}_{3'}\}$.

$$\mathbb{E}^2_{A_1\,A_2} = \left\{ \boldsymbol{X} \in \mathbb{R}^3 \,\middle|\, (X'^2 + Z'^2)/A_1^2 + Y'^2/A_2^2 = 1, \mathbb{R}^+ \ni A_1 > A_2 \in \mathbb{R}^+ \right\}\,. \tag{H.9}$$

The meta-equator $X'^2/A_1^2 + Y'^2/A_2^2 = 1$ is elliptic, while the meta-meridian $X'^2 + Z'^2 = A_1^2$ is circular. These properties of the meta-equator and the meta-meridian motivate the introduction of surface normal ellipsoidal meta-longitude/meta-latitude $\{A^*, B^*\}$, namely in order to parameterize $\mathbb{E}^2_{A_1\,A_2}$ according to (H.10) in contrast to (H.11) with respect to surface normal ellipsoidal longitude/latitude $\{\Lambda^*, \Phi^*\}$ and with respect to relative eccentricity $E^2 = (A_1^2 - A_2^2)/A_1^2$.

$$X' = \frac{A_1\sqrt{1 - E^2}}{\sqrt{1 - E^2\cos^2 A^*}}\cos B^* \cos A^*\,,\quad Y' = \frac{A_1\sqrt{1 - E^2}}{\sqrt{1 - E^2\cos^2 A^*}}\cos B^* \sin A^*\,,$$

$$Z' = A_1\sin B^*\,, \tag{H.10}$$

$$X = \frac{A_1\cos\Phi^*\cos\Lambda^*}{\sqrt{1 - E^2\sin^2\Phi^*}}\,,\quad Y = \frac{A_1\cos\Phi^*\sin\Lambda^*}{\sqrt{1 - E^2\sin^2\Phi^*}}\,,\quad Z = \frac{A_1(1 - E^2)\sin\Phi^*}{\sqrt{1 - E^2\sin^2\Phi^*}}\,. \tag{H.11}$$

The third equation of (H.10) as well as the second equation of (H.10) divided by the first equation of (H.10) subject to (H.8) lead to the transformation $\{\Lambda^*, \Phi^*\} \to \{A^*, B^*\}$, namely $\tan A^* = Y'/X'$, $\sin B^* = Z'/A'$ and

$$\tan A^* = \frac{-(1 - E^2)\tan\Phi^*}{\cos(\Lambda^* - \Omega)}\,,\quad \sin B^* = \frac{\cos\Phi^*\sin(\Lambda^* - \Omega)}{\sqrt{1 - E^2\sin^2\Phi^*}}\,. \tag{H.12}$$

Later on, we have to use $\sin A^*$, $\cos A^*$, $\cos B^*$, which is derived from (H.12) to coincide with

$$\cos A^* = \frac{1}{\sqrt{1 + \tan^2 A^*}}\,,$$

$$\sin A^* = \frac{\tan A^*}{\sqrt{1 + \tan^2 A^*}}\,, \tag{H.13}$$

$$\cos A^* = \frac{\cos(\Lambda^* - \Omega)}{\sqrt{\cos^2(\Lambda^* - \Omega) + (1 - E^2)^2\tan^2\Phi^*}}\,,$$

$$\sin A^* = \frac{-(1 - E^2)\tan\Phi^*}{\sqrt{\cos^2(\Lambda^* - \Omega) + (1 - E^2)^2\tan^2\Phi^*}}\,, \tag{H.14}$$

$$\cos B^* = \frac{\sqrt{1 - E^2\sin^2\Phi^* - \cos^2\Phi^*\sin^2(\Lambda^* - \Omega)}}{\sqrt{1 - E^2\sin^2\Phi^*}}\,. \tag{H.15}$$

For the special choice $\Omega = 270°$, the transformation $\{\Lambda^*, \Phi^*\} \to \{A^*, B^*\}$ is given by

$$\tan A^* = \frac{(1 - E^2)\tan\Phi^*}{\sin\Lambda^*}\,,\quad \sin B^* = \frac{\cos\Phi^*\cos\Lambda^*}{\sqrt{1 - E^2\sin^2\Phi^*}}\,. \tag{H.16}$$

H-12 The equiareal mapping of the biaxial ellipsoid onto a transverse tangent plane

We are going to construct the equiareal mapping of the biaxial ellipsoid $\mathbb{E}^2_{A_1,A_2}$ onto the transverse tangent plane normal to $\boldsymbol{E}_{3'}$ which is parameterized either by Cartesian coordinates $\{x^*, y^*\}$ or by polar coordinates $\{\alpha, r\}$ related by $x^* = r\cos\alpha$, $y^* = r\sin\alpha$. The mapping $\{A^*, B^*\}$ is transverse azimuthal by means of $\{\alpha = A^* + \pi, r = r(A^*, B^*)\}$, namely

$$x^* = -r(A^*, B^*)\cos A^* \,,$$

$$y^* = -r(A^*, B^*)\sin A^* \,. \tag{H.17}$$

Meta-longitude A^* coincides with the polar coordinate α, the western azimuth in case of $\Omega = 270°$, in the transverse plane; the radius r is an unknown function $r(A^*, B^*)$ of meta-longitude A^* and meta-latitude B^* which has to be determined. In order to derive the unknown function $r(A^*, B^*)$, we calculate the left Cauchy–Green deformation tensor $\mathsf{C}_l := \{c_{KL}\}$, i.e. the infinitesimal distance between two points in the plane covered by $\{\alpha, r\}$, namely

$$\mathrm{d}s^2 = g_{kl}\mathrm{d}u^k\mathrm{d}u^l = g_{11}\mathrm{d}\alpha^2 + g_{22}\mathrm{d}r^2 = r^2\mathrm{d}\alpha^2 + \mathrm{d}r^2 \,, \tag{H.18}$$

$$\mathrm{d}s^2 = g_{kl}\frac{\partial u^k}{\partial U^K}\frac{\partial u^l}{\partial U^L}\mathrm{d}U^K\mathrm{d}U^L = c_{KL}\mathrm{d}U^K\mathrm{d}U^L = c_{11}\mathrm{d}A^{*^2} + 2c_{12}\mathrm{d}A^*\mathrm{d}B^* + c_{22}\mathrm{d}B^{*^2} \,, \tag{H.19}$$

$$c_{11} = r^2\left(\frac{\partial\alpha}{\partial A^*}\right)^2 + \left(\frac{\partial r}{\partial A^*}\right)^2 \,,$$

$$c_{12} = r^2\frac{\partial\alpha}{\partial A^*}\frac{\partial\alpha}{\partial B^*} + \frac{\partial r}{\partial A^*}\frac{\partial r}{\partial B^*} \,, \tag{H.20}$$

$$c_{22} = r^2\left(\frac{\partial\alpha}{\partial B^*}\right)^2 + \left(\frac{\partial r}{\partial B^*}\right)^2 \,,$$

$$\frac{\partial\alpha}{\partial A^*} = 1 \,, \quad \frac{\partial r}{\partial A^*} = r_{A^*} \,,$$

$$\frac{\partial\alpha}{\partial B^*} = 0 \,, \quad \frac{\partial r}{\partial B^*} = r_{B^*} \,, \tag{H.21}$$

$$\mathsf{C}_l := \{c_{KL}\} = \begin{bmatrix} r^2 + r_{A^*}^2 & r_{A^*}r_{B^*} \\ r_{A^*}r_{B^*} & r_{B^*}^2 \end{bmatrix} \,. \tag{H.22}$$

Corollary H.1 (Equiareal mapping of the biaxial ellipsoid onto the transverse tangent plane).

The mapping of the biaxial ellipsoid onto the transverse tangent plane normal to $\boldsymbol{E}_{3'}$ is equiareal if (H.23) and (H.24) hold with respect to the left Cauchy–Green deformation tensor $\mathsf{C}_l = \{c_{KL}\}$ of type (H.22) and the left metric tensor $\mathsf{G}_l = \{G_{KL}\}$ of $\mathbb{E}^2_{A_1,A_2}$.

$$\det\left[\mathsf{C}_l\mathsf{G}_l^{-1}\right] = 1 \,, \tag{H.23}$$

$$rr_{B^*} = \frac{1}{2}r_{B^*}^2 = -\sqrt{\det\left[G_{KL}\right]} \,. \tag{H.24}$$

End of Corollary.

For the proof of (H.23), we refer to E. Grafarend (1995).

Proof.

(H.24) follows directly from (H.23) and (H.22). The negative sign has been chosen in order to guarantee that the orientation of $\mathbb{E}^2_{A_1,A_2}$ is conserved. Traditionally, the polar distance $\Delta^* := \pi/2 - B^*$ is chosen for (H.24) generating the positive sign within

$$rr_{\Delta^*} = +\sqrt{\det[G_{KL}]} \,. \tag{H.25}$$

Furthermore, we compute $G_{KL}(A^*, B^*)$, namely

$$G_{11} = \langle \boldsymbol{X}_{A^*} \mid \boldsymbol{X}_{A^*} \rangle =$$

$$= \frac{A_1^2(1 - E^2)}{(1 - E^2 \cos^2 A^*)^3}(1 - 2E^2 \cos^2 A^* + E^4 \cos^2 A^*) \cos^2 B^* \,,$$

$$G_{12} = \langle \boldsymbol{X}_{A^*} \mid \boldsymbol{X}_{B^*} \rangle =$$

$$= \frac{A_1^2 E^2 (1 - E^2) \sin A^* \cos A^* \sin B^* \cos B^*}{(1 - E^2 \cos^2 A^*)^2} \,,$$

$$G_{22} = \langle \boldsymbol{X}_{B^*} \mid \boldsymbol{X}_{B^*} \rangle = \tag{H.26}$$

$$= \frac{A_1^2(1 - E^2 + E^2 \cos^2 B^* \sin^2 A^*)}{1 - E^2 \cos^2 A^*} \,,$$

$$\det[G_{KL}] = \frac{A_1^4(1 - E^2)^2 \cos^2 B^*}{(1 - E^2 \cos^2 A^*)^3} \times$$

$$\times \left[1 - E^2 \cos^2 A^* + \frac{E^2 \cos^2 B^* \sin^2 A^*}{1 - E^2} \right] \,,$$

and subsequently

$$r^2 = -2 \int\limits_{\pi/2}^{B^*} \sqrt{\det[G_{KL}](A^*, B^*)} dB^* \,, \tag{H.27}$$

for $r(B^* = \pi/2) = 0 = r(\Delta^* = 0)$

$$r^2 = -\frac{2A_1^2(1 - E^2)}{(1 - E^2 \cos^2 A^*)^{3/2}} \times$$

$$\times \int\limits_{\pi/2}^{B^*} dB^* \cos B^* \sqrt{1 - E^2 \cos^2 A^* + \frac{E^2 \cos^2 B^* \sin^2 A^*}{1 - E^2}} \,. \tag{H.28}$$

End of Proof.

Corollary H.2 (Equiareal mapping of the biaxial ellipsoid onto the transverse tangent plane, special case $\Omega = I = 3\pi/2$).

The mapping of the biaxial ellipsoid onto the transverse tangent plane normal to $\boldsymbol{E}_{3'}$ is equiareal if

(i)
$$\alpha = A^* + \pi \, ,$$

(H.29)

(ii)
$$r = \frac{A_1(1 - E^2)^{1/4}}{(1 - E^2 \cos^2 A^*)^{3/4}} \times$$

$$\times \left[- \sin B^* \sqrt{(1 - E^2)(1 - E^2 \cos^2 A^*) + E^2 \cos^2 B^* \sin^2 A^*} - \right.$$

$$- \frac{1 - 2E^2 \cos^2 A^* + E^4 \cos^2 A^*}{E \sin A^*} \arcsin \frac{E \sin B^* \sin A^*}{\sqrt{1 - 2E^2 \cos^2 A^* + E^4 \cos^2 A^*}} +$$

$$+ \sqrt{(1 - E^2)(1 - E^2 \cos^2 A^*)} +$$

$$\left. + \frac{1 - 2E^2 \cos^2 A^* + E^4 \cos^2 A^*}{E \sin A^*} \arcsin \frac{E \sin A^*}{\sqrt{1 - 2E^2 \cos^2 A^* + E^4 \cos^2 A^*}} \right]^{1/2} \, ,$$

(H.30)

(iii)
$$r = 2A_1 \sin \frac{\Delta^*}{2} =$$

$$= A_1 \sqrt{2} \sqrt{1 - \cos \Delta^*} = A_1 \sqrt{2} \sqrt{1 - \sin B^*}$$

(H.31)

$$(\text{if } E = 0) \, .$$

End of Corollary.

The proof for the integrals is presented in Section H-3.

H-13 The equiareal mapping in terms of ellipsoidal longitude, ellipsoidal latitude

We implement the transformation $\{\Lambda^*, \Phi^*\} \to \{A^*, B^*\}$ into the mapping equations (H.29)–(H.31) which generate an equiareal mapping onto the transverse tangential plane according to (H.21) and (H.16), respectively, in particular (H.14). Let us decompose (H.30) term-wise, namely

$$r^2(A^*, B^*) =$$

$$= \frac{A_1^2 \sqrt{1 - E^2}}{(1 - E^2 \cos^2 A^*)^{3/2}} (t_1 + t_2 + t_3 + t_4) \, .$$

(H.32)

The elaborate computation of the factor and the four terms t_1, t_2, t_3 and t_4 as functions of $\{\Lambda^*, \Phi^*\}$ has been performed in Section H-4. Here, the result is presented in form of Lemma H.3 and Corollary H.4.

Lemma H.3 (Equiareal mapping of the biaxial ellipsoid onto the transverse plane).

The mapping of the biaxial ellipsoid onto the transverse tangent plane normal to $\boldsymbol{E}_{3'}$ is equiareal if

(i)

$$\alpha = \arctan \frac{(1 - E^2)\tan \Phi^*}{-\cos(\Lambda^* - \Omega)} \,,$$

$$r = A_1 \frac{\sqrt{\cos^2(\Lambda^* - \Omega)\cos^2 \Phi^* + (1 - E^2)^2 \sin^2 \Phi^*}}{(\cos^2(\Lambda^* - \Omega)\cos^2 \Phi^* + (1 - E^2)\sin^2 \Phi^*)^{3/4}} \sqrt{t_1^* + t_2^* + t_3^* + t_4^*}$$

(H.33)

(in polar coordinates) ,

subject to

$$t_1^* = -\frac{\sin(\Lambda^* - \Omega)\cos \Phi^*}{1 - E^2 \sin^2 \Phi^*}\sqrt{\cos^2(\Lambda^* - \Omega)\cos^2 \Phi^* + (1 - E^2)\sin^2 \Phi^*} \,,$$

$$t_2^* = -\frac{1 - \sin^2(\Lambda^* - \Omega)\cos^2 \Phi^*}{E \sin \Phi^*}\arcsin \frac{E \sin(\Lambda^* - \Omega)\sin \Phi^* \cos \Phi^*}{\sqrt{(1 - E^2 \sin^2 \Phi^*)(1 - \sin^2(\Lambda^* - \Omega)\cos^2 \Phi^*)}} \,,$$

(H.34)

$$t_3^* = \sqrt{\cos^2(\Lambda^* - \Omega)\cos^2 \Phi^* + (1 - E^2)\sin^2 \Phi^*} \,,$$

$$t_4^* = \frac{1 - \sin^2(\Lambda^* - \Omega)\cos^2 \Phi^*}{E \sin \Phi^*}\arcsin \frac{E \sin \Phi^*}{\sqrt{1 - \sin^2(\Lambda^* - \Omega)\cos^2 \Phi^*}} \,,$$

(ii)

$$\cos \alpha = \frac{-\cos(\Lambda^* - \Omega)}{\sqrt{\cos^2(\Lambda^* - \Omega) + (1 - E^2)^2 \tan^2 \Phi^*}} \,,$$

$$\sin \alpha = \frac{(1 - E^2)\tan \Phi^*}{\sqrt{\cos^2(\Lambda^* - \Omega) + (1 - E^2)^2 \tan^2 \Phi^*}}$$

(H.35)

(in Cartesian coordinates $x^* = r \cos \alpha$ and $y^* = r \sin \alpha$) ,

(iii)

$$\alpha = \lim_{E \to 0} \alpha(E) = \arctan \frac{\tan \Phi^*}{-\cos(\Lambda^* - \Omega)} \,,$$

$$r = \lim_{E \to 0} r(E) = A_1 \sqrt{2}\sqrt{1 - \sin(\Lambda^* - \Omega)\cos \Phi^*}$$

(H.36)

(if $E = 0$) .

End of Lemma.

Corollary H.4 (Equiareal mapping of the biaxial ellipsoid onto the transverse tangent plane, special case $\Omega = 3\pi/2$).

The mapping of the biaxial ellipsoid onto the transverse tangent plane normal to $\boldsymbol{E}_{3'}$ is equiareal if

(i)

$$\alpha = \arctan \frac{(1 - E^2)\tan \Phi^*}{\sin \Lambda^*} \,,$$

$$r = A_1 \frac{\sqrt{\sin^2 \Lambda^* \cos^2 \Phi^* + (1 - E^2)^2 \sin^2 \Phi^*}}{(\sin^2 \Lambda^* \cos^2 \Phi^* + (1 - E^2)\sin^2 \Phi^*)^{3/4}} \sqrt{t_1^* + t_2^* + t_3^* + t_4^*}$$

(H.37)

(in polar coordinates) ,

subject to

$$t_1^* = -\frac{\cos \Lambda^* \cos \Phi^*}{1 - E^2 \sin^2 \Phi^*} \sqrt{\sin^2 \Lambda^* \cos^2 \Phi^* + (1 - E^2)\sin^2 \Phi^*} \,,$$

$$t_2^* = -\frac{1 - \cos^2 \Lambda^* \cos^2 \Phi^*}{E \sin \Phi^*} \arcsin \frac{E \cos \Lambda^* \sin \Phi^* \cos \Phi^*}{\sqrt{(1 - E^2 \sin^2 \Phi^*)(1 - \cos^2 \Lambda^* \cos^2 \Phi^*)}} \,,$$

(H.38)

$$t_3^* = \sqrt{\sin^2 \Lambda^* \cos^2 \Phi^* + (1 - E^2)\sin^2 \Phi^*} \,,$$

$$t_4^* = \frac{1 - \cos^2 \Lambda^* \cos^2 \Phi^*}{E \sin \Phi^*} \arcsin \frac{E \sin \Phi^*}{\sqrt{1 - \cos^2 \Lambda^* \cos^2 \Phi^*}} \,,$$

(ii)

$$\cos \alpha = \frac{\sin \Lambda^*}{\sqrt{\sin^2 \Lambda^* + (1 - E^2)^2 \tan^2 \Phi^*}} \,, \quad \sin \alpha = \frac{(1 - E^2)\tan \Phi^*}{\sqrt{\sin^2 \Lambda^* + (1 - E^2)^2 \tan^2 \Phi^*}}$$

(H.39)

(in Cartesian coordinates $x^* = r \cos \alpha$ and $y^* = r \sin \alpha$) ,

(iii)

$$\alpha = \lim_{E \to 0} \alpha(E) = \arctan \frac{\tan \Phi^*}{\sin \Lambda^*} \,, \quad r = \lim_{E \to 0} r(E) = A_1 \sqrt{2}\sqrt{1 - \cos \Lambda^* \cos \Phi^*}$$

(H.40)

(if $E = 0$) .

End of Corollary.

H-2 The ellipsoidal Hammer projection

The second constituent of the Hammer projection is a proper change of scale of the transverse equiareal projection which conserves the local area.

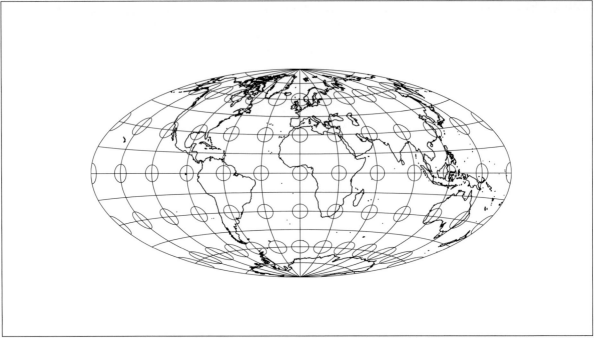

Fig. H.3. The ellipsoidal Hammer projection, squared relative eccentricity $E^2 = 0.1$.

Section H-21.

As a starting point, we set up in Section H-21 the equations of an equiareal mapping from a left biaxial ellipsoid to a right biaxial ellipsoid in order to be motivated for the structure of a change of scale.

Section H-22.

Section H-22 introduces in detail the rescaled equations $x = c_1 x^*(\Lambda^*, \Phi^*)$ $y = c_2 y^*(\Lambda^*, \Phi^*)$ of a transverse equiareal projection with respect to a right biaxial ellipsoid $\mathbb{E}^2_{A_1, A_2}$. Surface normal ellipsoidal longitude $\Lambda^*(\Lambda, \Phi; c_3, c_4)$ and latitude $\Phi^*(\Lambda, \Phi; c_3, c_4)$ of the right biaxial ellipsoid are in consequence related to surface normal ellipsoidal longitude/latitude $\{\Lambda, \Phi\}$ of the left biaxial ellipsoid, particularly postulating $d\Lambda^*/d\Lambda = c_3$, $d\Phi^*/d\Phi = c_4(1 - E^2 \sin^2 \Phi^*)^2 \cos \Phi / [(1 - E^2 \sin^2 \Phi)^2 \cos \Phi^*]$, where the scale constants $\{c_1, c_2, c_3, c_4\}$ are chosen in such a way to guarantee an areomorphism by $c_1 c_2 c_3 c_4 = 1$. The final form of the mapping equations generating the *ellipsoidal Hammer projection* is achieved by the inversion of an odd homogeneous polynomial equation for $\sin \Phi^*$ outlined in Section H-5.

H-21 The equiareal mapping from a left biaxial ellipsoid to a right biaxial ellipsoid

The equiareal mapping of a left biaxial ellipsoid $\mathbb{E}^2_{A_1, A_2}$ to a right biaxial ellipsoid $\mathbb{E}^2_{A_{1*}, A_{2*}}$ which is outlined by Box H.1 is of preparatory nature for the following section. We assume that pointwise the surface normal ellipsoidal longitude $\{\Lambda, \Lambda^*\}$ of types left and right coincide, but the function which relates surface normal ellipsoidal latitude from the left to the right, namely $\Phi^*(\Phi)$, is unknown. Based upon the structure of the mapping equations (H.41), the postulate of an equiareal mapping (H.23), in particular $\det [C_l G_l^{-1}] = 1$, leads to the left Cauchy–Green deformation tensor (H.42) with respect to the left metric tensor (H.43) of $\mathbb{E}^2_{A_1, A_2}$. The equivalence of $\det [C_l G_l^{-1}]$ with $\det [C_l] = \det [G_l]$ leads to the differential equation (H.44) for the unknown function $\Phi^*(\Phi)$. For the identities of (H.45) and (H.46), we have used only the positive preserving diffeomorphism $[d\Lambda^*, d\Phi^*]^T = J[d\Lambda, d\Phi]^T, |J| > 0$, namely a positive determinant of the Jacobi matrix J. Left and right integration of (H.46) with respect to the condition $\Phi^*(\Phi = 0) = 0$ leads finally to the mapping equations in (H.47) of equiareal type from a left biaxial ellipsoid $\mathbb{E}^2_{A_1, A_2}$ to a right biaxial ellipsoid $\mathbb{E}^2_{A_{1*}, A_{2*}}$.

Box H.1 (Equiareal mapping from a left biaxial ellipsoid to a right biaxial ellipsoid).

$$\Lambda^* = \Lambda^*(\Lambda) , \quad \Phi^* = \Phi^*(\Phi) , \tag{H.41}$$

$$\mathbf{C}_l = \begin{bmatrix} G^*_{11} \Lambda^{*2}_\Lambda & 0 \\ 0 & G^*_{22} \Phi^{*2}_\Phi \end{bmatrix} = \begin{bmatrix} \frac{A^2_{1*} \cos^2 \Phi^*}{1 - E^2_* \sin^2 \Phi^*} \Lambda^{*2}_\Lambda & 0 \\ 0 & \frac{A^2_{1*}(1 - E^2_*)^2}{(1 - E^2_* \sin^2 \Phi^*)^3} \Phi^{*2}_\Phi \end{bmatrix} , \tag{H.42}$$

$$\mathbf{G}_l = \begin{bmatrix} \frac{A^2_1 \cos^2 \Phi}{1 - E^2 \sin^2 \Phi} & 0 \\ 0 & \frac{A^2_1(1 - E^2)^2}{(1 - E^2 \sin^2 \Phi)^3} \end{bmatrix} , \tag{H.43}$$

$$\det[\mathbf{C}_l] = \det[\mathbf{G}_l] \Leftrightarrow \frac{A^4_{1*}(1 - E^2_*)^2 \cos^2 \Phi^*}{(1 - E^2_* \sin^2 \Phi^*)^4} \Phi^{*2}_\Phi = \frac{A^4_1(1 - E^2)^2 \cos^2 \Phi}{(1 - E^2 \sin^2 \Phi)^4} , \tag{H.44}$$

$$\Phi^*_\Phi = \frac{d\Phi^*}{d\Phi} = \frac{(1 - E^2_* \sin^2 \Phi^*)^2}{(1 - E^2 \sin^2 \Phi)^2} \frac{\cos \Phi}{\cos \Phi^*} \frac{A^2_1(1 - E^2)}{A^2_{1*}(1 - E^2_*)} , \tag{H.45}$$

$$A^2_{1*}(1 - E^2_*) \left[\frac{\operatorname{ar\,tanh}(E_* \sin \Phi^*)}{2E_*} + \frac{\sin \Phi^*}{2(1 - E^2_* \sin^2 \Phi^*)} \right] =$$
$$= A^2_1(1 - E^2) \left[\frac{\operatorname{ar\,tanh}(E \sin \Phi)}{2E} + \frac{\sin \Phi}{2(1 - E^2 \sin^2 \Phi)} \right] . \tag{H.46}$$

H-22 The explicit form of the mapping equations generating an equiareal map

We here consider the explicit form of the mapping equations generating an equiareal map from a left biaxial ellipsoid to a right biaxial ellipsoid with respect to a transverse frame of reference and a change of scale. Let us begin with the setup (H.47) of Box H.2 of general mapping equations $x = c_1 x^*(\Lambda^*, \Phi^*)$ and $y = c_2 y^*(\Lambda^*, \Phi^*)$ in a transverse frame of reference and under a change of scale with respect to gauge constants $\{c_1, c_2\}$. Those coordinates $\{\Lambda^*, \Phi^*\}$ which characterize a point on the right ellipsoid-of-revolution are related via (H.48) $\{\Lambda^*(\Lambda), \Phi^*(\Phi)\}$ to the coordinates $\{\Lambda, \Phi\}$ characteristic for a point on the left ellipsoid-of-revolution. Note that right ellipsoidal longitude/latitude depend only on left ellipsoidal longitude/latitude. In addition, we assume a coincidence between the semi-major axis A_{1*} and A_1, respectively, the semi-minor axis A_{2*} and A_2 between right and left $\mathbb{E}^2_{A_{1*}, A_{2*}}$ and $\mathbb{E}^2_{A_1, A_2}$, respectively, expressed by (H.49). Next, by (H.50) and (H.51), we subscribe the differential relations $d\Lambda^*/d\Lambda = c_3$ subject to $\Lambda^*(\Lambda = 0) = 0$ and $d\Phi^*/d\Phi = c_4(1 - E^2 \sin^2 \Phi^*)^2 \cos \Phi/[(1 - E^2 \sin^2 \Phi)^2 \cos \Phi^*]$ subject to $\Phi^*(\Phi = 0) = 0$ with respect to gauge constants $\{c_3, c_4\}$. (H.45) has motivated (H.51). The detailed computation of the left Cauchy–Green tensor via (H.52), (H.53), (H.54), and (H.55) leads us to the postulate (H.56) of an equiareal mapping. Indeed, we take advantage via (H.56), (H.57), and (H.58) of the fact that $\{x^*(\Lambda^*, \Phi^*), y^*(\Lambda^*, \Phi^*)\}$ is already an equiareal mapping. Thus, we may consider the transformation $\{x^*, y^*\} \to \{x, y\}$ as a change from one equiareal chart to another equiareal chart (a:a: cha–cha–cha). (H.59) is a representation of the postulate of an areomorphism which leads by subscribing $d\Lambda^*/d\Lambda = c_3$ to the explicit form of $d\Phi^*/d\Phi$ of type (H.52), too. Indeed, we do not have to postulate (H.51)! In order to guarantee an equiareal mapping $\{\Lambda, \Phi\} \to \{x, y\}$, the gauge constants have to fulfill (H.60), namely $c_1 c_2 c_3 c_4 = 1$.

Box H.2 (The equiareal mapping from a left biaxial ellipsoid to a right biaxial ellipsoid with respect to a transverse frame of reference and a change of scale (the Hammer projection of the ellipsoid-of-revolution)).

$$x = c_1 x^*(\Lambda^*, \Phi^*) \ , \quad y = c_2 y^*(\Lambda^*, \Phi^*) \ , \tag{H.47}$$

$$\Lambda^*(\Lambda) \ , \quad \Phi^*(\Phi) \ , \tag{H.48}$$

subject to

$$A_{1*} = A_1 \ , \quad A_{2*} = A_2, \ E^* = E \ , \tag{H.49}$$

$$\Lambda_\Lambda^* = \frac{\mathrm{d}\Lambda^*}{\mathrm{d}\Lambda} = c_3 \ , \tag{H.50}$$

$$\Phi_\Phi^* = \frac{\mathrm{d}\Phi^*}{\mathrm{d}\Phi} = c_4 \frac{(1 - E^2 \sin^2 \Phi^*)^2}{(1 - E^2 \sin^2 \Phi)^2} \frac{\cos \Phi}{\cos \Phi^*} \ . \tag{H.51}$$

The left Cauchy–Green deformation tensor:

$$x_\Lambda = \frac{\partial x}{\partial \Lambda} = c_1 \frac{\partial x^*}{\partial \Lambda^*} \frac{\mathrm{d}\Lambda^*}{\mathrm{d}\Lambda} = c_1 \Lambda_\Lambda^* x_{\Lambda^*}^* \ , \quad x_\Phi = \frac{\partial x}{\partial \Phi} = c_1 \frac{\partial x^*}{\partial \Phi^*} \frac{\mathrm{d}\Phi^*}{\mathrm{d}\Phi} = c_1 \Phi_\Phi^* x_{\Phi^*}^* \ ,$$

$$y_\Lambda = \frac{\partial y}{\partial \Lambda} = c_2 \frac{\partial y^*}{\partial \Lambda^*} \frac{\mathrm{d}\Lambda^*}{\mathrm{d}\Lambda} = c_2 \Lambda_\Lambda^* y_{\Lambda^*}^* \ , \quad y_\Phi = \frac{\partial y}{\partial \Phi} = c_2 \frac{\partial y^*}{\partial \Phi^*} \frac{\mathrm{d}\Phi^*}{\mathrm{d}\Phi} = c_2 \Phi_\Phi^* y_{\Phi^*}^* \ , \tag{H.52}$$

$$\mathrm{d}s^2 = \mathrm{d}x^2 + \mathrm{d}y^2 =$$

$$= (x_\Lambda^2 + y_\Lambda^2)\mathrm{d}\Lambda^2 + 2(x_\Lambda x_\Phi + y_\Lambda y_\Phi)\mathrm{d}\Lambda\mathrm{d}\Phi + (x_\Phi^2 + y_\Phi^2)\mathrm{d}\Phi^2 = \sum_{A,B=1}^{2} c_{AB}\mathrm{d}U^A\mathrm{d}U^B \ , \tag{H.53}$$

$$c_{11} = \Lambda_\Lambda^{*2}(c_1^2 x_{\Lambda^*}^{*2} + c_2^2 y_{\Lambda^*}^{*2}) \ , \quad c_{12} = \Lambda_\Lambda^* \Phi_\Phi^*(c_1^2 x_{\Lambda^*}^* x_{\Phi^*}^* + c_2^2 y_{\Lambda^*}^* y_{\Phi^*}^*) \ , \quad c_{22} = \Phi_\Phi^{*2}(c_1^2 x_{\Phi^*}^{*2} + c_2^2 y_{\Phi^*}^{*2}) \ , \tag{H.54}$$

$$\sqrt{\det[\mathsf{C}_l]} = \sqrt{c_{11}c_{22} - c_{12}^2} = c_1 c_2 \Lambda_\Lambda^* \Phi_\Phi^* (x_{\Lambda^*}^* y_{\Phi^*}^* - x_{\Phi^*}^* y_{\Lambda^*}^*) \ . \tag{H.55}$$

The postulate of an equiareal mapping:

$$\sqrt{\det[\mathsf{C}_l]} = \sqrt{\det[\mathsf{G}_l]} \Leftrightarrow c_1 c_2 \Lambda_\Lambda^* \Phi_\Phi^* (x_{\Lambda^*}^* y_{\Phi^*}^* - x_{\Phi^*}^* y_{\Lambda^*}^*) = \frac{A^2(1 - E^2) \cos \Phi}{(1 - E^2 \sin^2 \Phi)^2} \ , \tag{H.56}$$

$$x_{\Lambda^*}^* y_{\Phi^*}^* - x_{\Phi^*}^* y_{\Lambda^*}^* = \sqrt{\det[\mathsf{C}_l^*]} = \sqrt{\det[\mathsf{G}_l^*]} \ , \tag{H.57}$$

$$x_{\Lambda^*}^* y_{\Phi^*}^* - x_{\Phi^*}^* y_{\Lambda^*}^* = \frac{A_1^2(1 - E^2) \cos \Phi^*}{(1 - E^2 \sin^2 \Phi^*)^2} \ , \tag{H.58}$$

$$\frac{\sqrt{\det[\mathsf{C}_l]}}{\sqrt{\det[\mathsf{G}_l]}} = 1 \Leftrightarrow \Lambda_\Lambda^* = c_3 c_1 c_2 \Lambda_\Lambda^* \Phi_\Phi^* \frac{(1 - E^2 \sin^2 \Phi)^2}{(1 - E^2 \sin^2 \Phi^*)^2} \frac{\cos \Phi^*}{\cos \Phi} = 1$$

$$\Leftrightarrow$$

$$c_1 c_2 c_3 \Phi_\Phi^* \frac{(1 - E^2 \sin^2 \Phi)^2}{(1 - E^2 \sin^2 \Phi^*)^2} \frac{\cos \Phi^*}{\cos \Phi} = 1 \ , \quad \Phi_\Phi^* = c_4 \frac{(1 - E^2 \sin^2 \Phi^*)^2}{(1 - E^2 \sin^2 \Phi)^2} \frac{\cos \Phi}{\cos \Phi^*} \tag{H.59}$$

$$\Leftrightarrow$$

$$c_1 c_2 c_3 c_4 = 1 \ . \tag{H.60}$$

Box H.3 outlines the explicit solutions of the differential equations (H.50) and (H.51) transformed into (H.61) and (H.62) and being subjected to the initial values $\Lambda^*(\Lambda = 0) = 0$ and $\Phi^*(\Phi = 0) = 0$ as given by (H.63). Left and right integration of (H.61) and (H.62) with respect to the initial data (H.63) lead us directly to the solutions (H.64) and (H.65). For zero relative eccentricity, $E = 0$, by (H.66), we arrive at the spherical solution $\sin \Phi^* = c_4 \sin \Phi$. But for the ellipsoidal case, a series expansion of (H.65) in even powers of E, namely E^0, E^2, E^4 etc., represents $\sin \Phi^*$ as a homogeneous polynomial of odd degree which has to be inverted as outlined in Section H-3. As a result, we gain $\sin \Phi^*(\sin \Phi; c_4)$ of type (H.67).

Box H.3 (The mapping equations of equiareal type from a left biaxial ellipsoid to a right biaxial ellipsoid with respect to a transverse frame of reference and a change of scale (the Hammer projection of $\mathbb{E}^2_{A_1, A_2}$)).

$$\mathrm{d}\Lambda^* = c_3 \mathrm{d}\Lambda , \tag{H.61}$$

$$\frac{\cos \Phi^*}{(1 - E^2 \sin^2 \Phi^*)^2} \mathrm{d}\Phi^* = c_4 \frac{\cos \Phi}{(1 - E^2 \sin^2 \Phi)^2} \mathrm{d}\Phi , \tag{H.62}$$

subject to

$$\Lambda^*(\Lambda = 0) = 0 , \quad \Phi^*(\Phi = 0) = 0 , \tag{H.63}$$

$$\Lambda^* = c_3 \Lambda , \tag{H.64}$$

$$\frac{\operatorname{ar tanh}(E \sin \Phi^*)}{2E} + \frac{\sin \Phi^*}{2(1 - E^2 \sin^2 \Phi^*)} =$$
$$= c_4 \left[\frac{\operatorname{ar tanh}(E \sin \Phi)}{2E} + \frac{\sin \Phi}{2(1 - E^2 \sin^2 \Phi)} \right] . \tag{H.65}$$

If $E = 0$, then

$$\sin \Phi^* = c_4 \sin \Phi . \tag{H.66}$$

If $E \neq 0$, then

$$\sin \Phi^* =$$
$$= c_4 \sin \Phi \left[1 + \frac{2}{3} E^2 \sin^2 \Phi (1 - c_4^2) + \frac{1}{15} E^4 \sin^4 \Phi (9 - 20 c_4^2 + 11 c_4^4) + \mathrm{O}(E^6) \right] . \tag{H.67}$$

As prepared by Box H.2 and Box H.3, we can finally present by Lemma H.5 the equiareal mapping of the biaxial ellipsoid with respect to a transverse frame of reference and a change of scale, in short the *Hammer projection of the biaxial ellipsoid*. In particular, the transfer of the four characteristic terms $\{t_1^*, t_2^*, t_3^*, t_4^*\}$, namely (H.34), being functions of $\Lambda^* = c_3 \Lambda$ and $\sin \Phi^*(\sin \Phi; c_4)$, has to be made. While case (i) of Corollary H.6 highlights the general ellipsoidal Hammer projection, case (ii) is its specific form for zero relative eccentricity, $E = 0$, namely its spherical counterpart. For the choice $c_1 = 2$, $c_2 = 1$, $c_3 = 1/2$, $c_4 = 1$ of case (iii), we receive by means of (H.89)–(H.92) the ellipsoidal mapping equations of special equiareal projection in the Hammer gauge. In contrast, case (iv) specializes, for $E = 0$, (H.96) to the spherical mapping equations in the Hammer gauge, indeed the original Hammer mapping equations (Hammer 1892). Various alternative variants of the ellipsoidal mapping equations of equiareal type can be chosen, for different gauge constants $\{c_1, c_2, c_3, c_4\}$ as long as they fulfill $c_1 c_2 c_3 c_4 = 1$. In particular, they refer to a pointwise map of the North Pole or not or to other criteria.

Lemma H.5 (The equiareal mapping of the biaxial ellipsoid with respect to a transverse frame of reference and a change of scale (the Hammer projection of $\mathbb{E}^2_{A_1,A_2}$)).

The mapping of the right biaxial ellipsoid $\mathbb{E}^2_{A_1,A_2}$ with respect to left biaxial ellipsoid $\mathbb{E}^2_{A_{1*},A_{2*}}$ subject to $A_{1*} = A_1$, $A_{2*} = A_2$ onto the transverse tangent plane normal to $\boldsymbol{E}_{3'}$ and with respect to a change of scale is equiareal if

$$x = c_1 r(\Lambda,\Phi;c_3,c_4) \cos\alpha(\Lambda,\Phi;c_3,c_4) ,\tag{H.68}$$

$$y = c_2 r(\Lambda,\Phi;c_3,c_4) \sin\alpha(\Lambda,\Phi;c_3,c_4) ,\tag{H.69}$$

subject to

$$\cos\alpha(\Lambda,\Phi;c_3,c_4) =$$
$$= \frac{-\cos[\Lambda^*(\Lambda;c_3) - \Omega]}{\sqrt{\cos^2[\Lambda^*(\Lambda;c_3) - \Omega] + (1-E^2)^2 \tan^2[\Phi^*(\Phi;c_4)]}} ,\tag{H.70}$$

$$\sin\alpha(\Lambda,\Phi;c_3,c_4) =$$
$$= \frac{(1-E^2)\tan[\Phi^*(\Phi;c_4)]}{\sqrt{\cos^2[\Lambda^*(\Lambda;c_3) - \Omega] + (1-E^2)^2 \tan^2[\Phi^*(\Phi;c_4)]}} ,\tag{H.71}$$

$$r^2(\Lambda,\Phi;c_3,c_4) =$$
$$= A_1^2 \Big(\cos^2[\Lambda^*(\Lambda;c_3) - \Omega] \cos^2[\Phi^*(\Phi;c_4)] + (1-E^2)^2 \sin^2[\Phi^*(\Phi;c_4)] \Big)$$
$$\Big/ \Big[\Big(\cos^2[\Lambda^*(\Lambda;c_3) - \Omega] \cos^2[\Phi^*(\Phi;c_4)] + (1-E^2)^2 \sin^2[\Phi^*(\Phi;c_4)] \Big)^{3/2} \Big] \times$$
$$\times \Big[t_1^*(\Lambda,\Phi;c_3,c_4) + t_2^*(\Lambda,\Phi;c_3,c_4) + t_3^*(\Lambda,\Phi;c_3,c_4) + t_4^*(\Lambda,\Phi;c_3,c_4) \Big] ,\tag{H.72}$$

$$\Lambda^* = c_3\Lambda ,\tag{H.73}$$

$$\sin\Phi^* =$$
$$= c_4\sin\Phi\Big[1 + \frac{2}{3}E^2\sin^2\Phi(1-c_4^2) + \frac{1}{15}E^4\sin^4\Phi(9 - 20c_4^2 + 11c_4^4) + \mathrm{O}(E^6)\Big] ,\tag{H.74}$$

$$c_1 c_2 c_3 c_4 = 1 .\tag{H.75}$$

End of Lemma.

Corollary H.6 (The equiareal mapping of the biaxial ellipsoid with respect to a transverse frame of reference and a change of scale, special case $\Omega = 3\pi/2$ (the Hammer projection of $\mathbb{E}^2_{A_1,A_2}$)).

(i)

The mapping of the right biaxial ellipsoid $\mathbb{E}^2_{A_1,A_2}$ with respect to left biaxial ellipsoid $\mathbb{E}^2_{A_{1*},A_{2*}}$ subject to $A_{1*} = A_1$, $A_{2*} = A_2$ onto the transverse tangent plane specialized by $\Omega = 3\pi/2$ being normal to $\boldsymbol{E}_{3'}$ and with respect to a change of scale is equiareal if

$$x = c_1 r(\Lambda,\Phi;c_3,c_4) \cos\alpha(\Lambda,\Phi;c_3,c_4) ,$$
$$y = c_2 r(\Lambda,\Phi;c_3,c_4) \sin\alpha(\Lambda,\Phi;c_3,c_4) ,\tag{H.76}$$

subject to

$$\cos\alpha(\Lambda,\Phi;c_3,c_4) = \frac{\sin[\Lambda^*(\Lambda;c_3)]}{\sqrt{\sin^2[\Lambda^*(\Lambda;c_3)] + (1-E^2)^2\tan^2[\Phi^*(\Phi;c_4)]}} \; , \qquad (\text{H.77})$$

$$\sin\alpha(\Lambda,\Phi;c_3,c_4) = \frac{(1-E^2)\tan[\Phi^*(\Phi;c_4)]}{\sqrt{\sin^2[\Lambda^*(\Lambda;c_3)] + (1-E^2)^2\tan^2[\Phi^*(\Phi;c_4)]}} \; , \qquad (\text{H.78})$$

$$r^2(\Lambda,\Phi;c_3,c_4) = A_1^2\Big(\sin^2[\Lambda^*(\Lambda;c_3)]\cos^2[\Phi^*(\Phi;c_4)] + (1-E^2)^2\sin^2[\Phi^*(\Phi;c_4)]\Big)$$
$$/\Big[(\sin^2[\Lambda^*(\Lambda;c_3)]\cos^2[\Phi^*(\Phi;c_4)] + (1-E^2)^2\sin^2[\Phi^*(\Phi;c_4)])^{3/2}\Big] \times \qquad (\text{H.79})$$
$$\times\Big[t_1^*(\Lambda,\Phi;c_3,c_4) + t_2^*(\Lambda,\Phi;c_3,c_4) + t_3^*(\Lambda,\Phi;c_3,c_4) + t_4^*(\Lambda,\Phi;c_3,c_4)\Big] \; ,$$

$$\Lambda^* = c_3\Lambda \; , \qquad (\text{H.80})$$

$$\sin\Phi^* = c_4\sin\Phi\Big[1 + \frac{2}{3}E^2\sin^2\Phi(1-c_4^2) + \frac{1}{15}E^4\sin^4\Phi(9 - 20c_4^2 + 11c_4^4) + \mathrm{O}(E^6)\Big] \; , \qquad (\text{H.81})$$

$$c_1c_2c_3c_4 = 1 \; . \qquad (\text{H.82})$$

(ii)

If the relative eccentricity vanishes, $E = 0$, then we arrive at the Hammer projection of the sphere $\mathbb{S}^2_{A_1}$, namely

$$x = c_1 r(\Lambda,\Phi;c_3,c_4)\cos\alpha(\Lambda,\Phi;c_3,c_4) \; ,$$
$$y = c_2 r(\Lambda,\Phi;c_3,c_4)\sin\alpha(\Lambda,\Phi;c_3,c_4) \; , \qquad (\text{H.83})$$

subject to

$$\cos\alpha(\Lambda,\Phi;c_3,c_4) = \frac{\sqrt{1 - c_4^2\sin^2\Phi}\,\sin c_3\Lambda}{\sqrt{1 - (1 - c_4^2\sin^2\Phi)\cos^2 c_3\Lambda}} \; , \qquad (\text{H.84})$$

$$\sin\alpha(\Lambda,\Phi;c_3,c_4) = \frac{c_4\sin\Phi}{\sqrt{1 - (1 - c_4^2\sin^2\Phi)\cos^2 c_3\Lambda}} \; , \qquad (\text{H.85})$$

$$r = A_1\sqrt{2}\sqrt{1 - \sqrt{1 - c_4^2\sin^2\Phi}\cos c_3\Lambda} \; , \qquad (\text{H.86})$$

$$x = c_1 A_1\sqrt{2}\,\frac{\sqrt{1 - c_4^2\sin^2\Phi}\,\sin c_3\Lambda}{\sqrt{1 + \sqrt{1 - c_4^2\sin^2\Phi}\cos c_3\Lambda}} \; , \qquad (\text{H.87})$$

$$y = c_2 A_1\sqrt{2}\,\frac{c_4\sin\Phi}{\sqrt{1 + \sqrt{1 - c_4^2\sin^2\Phi}\cos c_3\Lambda}} \; . \qquad (\text{H.88})$$

(iii)

If we choose $c_1 = 2$, $c_2 = 1$, $c_3 = 1/2$, and $c_4 = 1$ which fulfills $c_1 c_2 c_3 c_4 = 1$ (Hammer's choice), then the mapping of the right biaxial ellipsoid $\mathbb{E}^2_{A_1\,A_2}$ with respect to left biaxial ellipsoid $\mathbb{E}^2_{A_{1*}\,A_{2*}}$ subject to $A_{1*} = A_1$, $A_{2*} = A_2$ onto the transverse tangent plane being normal to $\boldsymbol{E}_{3'}$ and rescaled, namely of equiareal type, reduces to

$$x = 2r(\Lambda, \Phi)\cos\alpha(\Lambda, \Phi)\,, \quad y = r(\Lambda, \Phi)\sin\alpha(\Lambda, \Phi)\,, \tag{H.89}$$

$$\cos\alpha(\Lambda, \Phi) = \frac{\sin\Lambda/2}{\sqrt{\sin^2\Lambda/2 + (1 - E^2)^2 \tan^2\Phi}}\,,$$
$$\sin\alpha(\Lambda, \Phi) = \frac{(1 - E^2)\tan\Phi}{\sqrt{\sin^2\Lambda/2 + (1 - E^2)^2 \tan^2\Phi}}\,, \tag{H.90}$$

$$r^2(\Lambda, \Phi) = A_1^2 \frac{\sin^2\Lambda/2\cos^2\Phi + (1 - E^2)^2 \sin^2\Phi}{(\sin^2\Lambda/2\cos^2\Phi + (1 - E^2)\sin^2\Phi)^{3\ 2}}(t_1^* + t_2^* + t_3^* + t_4^*)\,, \tag{H.91}$$

$$t_1^* = -\frac{\cos\Lambda/2\cos\Phi}{1 - E^2\sin^2\Phi}\sqrt{\sin^2\Lambda/2\cos^2\Phi + (1 - E^2)\sin^2\Phi}\,,$$
$$t_2^* = -\frac{1 - \cos^2\Lambda/2\cos^2\Phi}{E\sin\Phi}\arcsin\frac{E\sin\Phi\cos\Phi\cos\Lambda/2}{\sqrt{(1 - E^2\sin^2\Phi)(1 - \cos^2\Lambda/2\cos^2\Phi)}}\,,$$
$$t_3^* = \sqrt{\sin^2\Lambda/2\cos^2\Phi + (1 - E^2)\sin^2\Phi}\,,$$
$$t_4^* = \frac{1 - \cos^2\Lambda/2\cos^2\Phi}{E\sin\Phi}\arcsin\frac{E\sin\Phi}{\sqrt{1 - \cos^2\Lambda/2\cos^2\Phi}}\,. \tag{H.92}$$

(iv)

If the relative eccentricity vanishes, $E = 0$, then we arrive at the special Hammer projection of the sphere $\mathbb{S}^2_{A_1}$ subject to $c_1 = 2$, $c_2 = 1$, $c_3 = 1/2$, and $c_4 = 1$, namely

$$x = 2r(\Lambda, \Phi)\cos\alpha(\Lambda, \Phi)\,, \quad y = r(\Lambda, \Phi)\sin\alpha(\Lambda, \Phi)\,, \tag{H.93}$$

$$\cos\alpha(\Lambda, \Phi) = \frac{\cos\Phi\sin\Lambda/2}{\sqrt{1 - \cos^2\Phi\cos^2\Lambda/2}}\,, \quad \sin\alpha(\Lambda, \Phi) = \frac{\sin\Phi}{\sqrt{1 - \cos^2\Phi\cos^2\Lambda/2}}\,, \tag{H.94}$$

$$r = A\sqrt{2}\sqrt{1 - \cos\Phi\cos\Lambda/2}\,, \tag{H.95}$$

$$x = 2A_1\sqrt{2}\frac{\cos\Phi\sin\Lambda/2}{\sqrt{1 + \cos\Phi\cos\Lambda/2}}\,, \quad y = A_1\sqrt{2}\frac{\sin\Phi}{\sqrt{1 + \cos\Phi\cos\Lambda/2}}\,. \tag{H.96}$$

End of Corollary.

As a visualization for the derived mapping equations for the *ellipsoidal Hammer projection*, at the beginning of this section, Fig. H.3 is given including the *Tissot indicatrices*.

H-3 An integration formula

(H.28) may be written as an integration formula. The following relations (H.97)–(H.100) specify this integration formula. If the relative eccentricity approaches $E = 0$, then the radial coordinate specializes to (H.101) according to the L'Hôpital Rule (H.102) and (H.103).

$$\int \sin x \sqrt{1 + p^2 \sin^2 x}\, dx = -\frac{1}{2} \cos x \sqrt{1 + p^2 \sin^2 x} - \frac{1+p^2}{2p} \arcsin \frac{p \cos x}{\sqrt{1+p^2}} , \tag{H.97}$$

subject to

$$p^2 := \frac{1}{1 - E^2} \frac{E^2 \sin^2 A^*}{1 - E^2 \cos^2 A^*} , \tag{H.98}$$

$$r^2 = \frac{A_1^2 (1 - E^2)}{1 - E^2 \cos^2 A^*} \times$$

$$\times \left[- \cos \Delta^* \sqrt{1 + \frac{E^2 \sin^2 A^* \sin^2 \Delta^*}{(1 - E^2)(1 - E^2 \cos^2 A^*)}} - \frac{\sqrt{1 - E^2}\sqrt{1 - E^2 \cos^2 A^*}}{E \sin A^*} \times \right.$$

$$\times \left(1 + \frac{E^2 \sin^2 A^*}{(1 - E^2)(1 - E^2 \cos^2 A^*)}\right) \times$$

$$\left. \times \arcsin \left(\frac{E \sin A^*}{\sqrt{1 - E^2}\sqrt{1 - E^2 \cos^2 A^*}} \frac{\cos \Delta^*}{\sqrt{1 + \frac{E^2 \sin^2 A^*}{(1-E^2)(1-E^2 \cos^2 A^*)}}} \right) \right]_0^{\Delta^*} , \tag{H.99}$$

$$r^2 = \frac{A_1^2 \sqrt{1 - E^2}}{(1 - E^2 \cos^2 A^*)^{3/2}} \times$$

$$\times \left[- \cos \Delta^* \sqrt{(1 - E^2)(1 - E^2 \cos^2 A^* + E^2 \sin^2 \Delta^* \sin^2 A^*)} - \right.$$

$$- \frac{1 - 2E^2 \cos^2 A^* + E^4 \cos^2 A^*}{E \sin A^*} \arcsin \frac{E \cos \Delta^* \sin A^*}{\sqrt{1 - 2E^2 \cos^2 A^* + E^4 \cos^2 A^*}} + \tag{H.100}$$

$$+ \sqrt{(1 - E^2)(1 - E^2 \cos^2 A^*)} +$$

$$\left. + \frac{1 - 2E^2 \cos^2 A^* + E^4 \cos^2 A^*}{E \sin A^*} \arcsin \frac{E \sin A^*}{\sqrt{1 - 2E^2 \cos^2 A^* + E^4 \cos^2 A^*}} \right] ,$$

$$r = A_1 \sqrt{2}\sqrt{1 - \cos \Delta^*} = A_1 \sqrt{2}\sqrt{1 - \sin B^*} , \tag{H.101}$$

$$\lim_{x \to 0} \frac{\arcsin x}{x} = \lim_{x \to 0} \frac{a}{\sqrt{1 - a^2 x^2}} = a , \tag{H.102}$$

$$\lim_{E \to 0} \frac{\arcsin(E \sin A^* \cos \Delta^*)}{E \sin A^*} = \cos \Delta^* ,$$

$$\lim_{E \to 0} \frac{\arcsin(E \sin A^*)}{E \sin A^*} = 1 . \tag{H.103}$$

H-4 The transformation of the radial function $r(A^*, B^*)$ into $r(\Lambda^*, \Phi^*)$

In this section, the transformation of the radial function $r(A^*, B^*)$ into $r(\Lambda^*, \Phi^*)$ is presented. The following relations (H.104)–(H.112) specify this transformation.

First factor (see (H.14), (H.30)):

$$\frac{A_1^2 \sqrt{1 - E^2}}{(1 - E^2 \cos^2 A^*)^{3/2}} = \frac{A_1^2}{(1 - E^2)} \frac{(\cos^2(\Lambda^* - \Omega) \cos^2 \Phi^* + (1 - E^2)^2 \sin^2 \Phi^*)^{3/2}}{(\cos^2(\Lambda^* - \Omega) \cos^2 \Phi^* + (1 - E^2) \sin^2 \Phi^*)^{3/2}} . \tag{H.104}$$

First term (see (H.12),(H.14), (H.15), (H.30)):

$$t_1 = t_1(A^*, B^*) := -\sin B^* \sqrt{(1 - E^2)(1 - E^2 \cos^2 A^*) + E^2 \cos^2 B^* \sin^2 A^*} , \tag{H.105}$$

$$t_1 = t_1(\Lambda^*, \Phi^*) := -\frac{\sin(\Lambda^* - \Omega) \cos \Phi^* (1 - E^2)}{1 - E^2 \sin^2 \Phi^*} \times$$
$$\times \frac{\sqrt{\cos^2(\Lambda^* - \Omega) \cos^2 \Phi^* + (1 - E^2) \sin^2 \Phi^*}}{\sqrt{\cos^2(\Lambda^* - \Omega) \cos^2 \Phi^* + (1 - E^2)^2 \sin^2 \Phi^*}} . \tag{H.106}$$

Second term (see (H.12),(H.14), (H.15), (H.30)):

$$t_2 = t_2(A^*, B^*) := -\frac{1 - 2E^2 \cos^2 A^* + E^4 \cos^2 A^*}{E \sin A^*} \times$$
$$\times \arcsin \frac{E \sin B^* \sin A^*}{\sqrt{1 - 2E^2 \cos^2 A^* + E^4 \cos^2 A^*}} , \tag{H.107}$$

$$t_2 = t_2(\Lambda^*, \Phi^*) := -\frac{1 - E^2}{E \sin \Phi^*} \frac{1 - \sin^2(\Lambda^* - \Omega) \cos^2 \Phi^*}{\sqrt{\cos^2(\Lambda^* - \Omega) \cos^2 \Phi^* + (1 - E^2)^2 \sin^2 \Phi^*}} \times$$
$$\times \arcsin \frac{E \sin(\Lambda^* - \Omega) \sin \Phi^* \cos \Phi^*}{\sqrt{(1 - E^2 \sin^2 \Phi^*)(1 - \sin^2(\Lambda^* - \Omega) \cos^2 \Phi^*)}} . \tag{H.108}$$

Third term (see (H.14), (H.30)):

$$t_3 = t_3(A^*, B^*) := \sqrt{1 - E^2} \sqrt{1 - E^2 \cos^2 A^*} , \tag{H.109}$$

$$t_3 = t_3(\Lambda^*, \Phi^*) := (1 - E^2) \frac{\sqrt{\cos^2(\Lambda^* - \Omega) \cos^2 \Phi^* + (1 - E^2) \sin^2 \Phi^*}}{\sqrt{\cos^2(\Lambda^* - \Omega) \cos^2 \Phi^* + (1 - E^2)^2 \sin^2 \Phi^*}} . \tag{H.110}$$

Fourth term (see (H.14), (H.30)):

$$t_4 = t_4(A^*, B^*) := \frac{1 - 2E^2 \cos^2 A^* + E^4 \cos^2 A^*}{E \sin A^*} \arcsin \frac{E \sin A^*}{\sqrt{1 - 2E^2 \cos^2 A^* + E^4 \cos^2 A^*}} \,, \quad \text{(H.111)}$$

$$t_4 = t_4(\Lambda^*, \Phi^*) := \frac{1 - E^2}{E \sin \Phi^*} \times$$

$$\times \frac{1 - \sin^2(\Lambda^* - \Omega) \cos^2 \Phi^*}{\sqrt{\cos^2(\Lambda^* - \Omega) \cos^2 \Phi^* + (1 - E^2)^2 \sin^2 \Phi^*}} \arcsin \frac{E \sin \Phi^*}{\sqrt{1 - \sin^2(\Lambda^* - \Omega) \cos^2 \Phi^*}} \,. \quad \text{(H.112)}$$

H-5 The inverse of a special univariate homogeneous polynomial

In order to solve (H.65) for $\sin \Phi^*$, we proceed to present the series expansions of $\operatorname{ar\,tanh} x$ in (H.113) and of $(1 + x)^{-1}$ in (H.114) (compare with M. Abramowitz and J. A. Stegun 1965). Those series expansions is applied to the two terms (H.115) and (H.116) which appear in (H.65). In particular, we recognize the homogeneous polynomial form of (H.117) as soon as we substitute $x := \sin \Phi^*$ by (H.118) and y by (H.119). The inverse of the univariate homogeneous polynomial (H.120) represented by (H.121) is computed up to degree five. Forward and backward substitution amount to (H.129) reevaluated by means of the final solution to the inversion by (H.131).

$$\operatorname{ar\,tanh} x = x + \frac{x^3}{3} + \frac{x^5}{5} + \frac{x^7}{7} + \mathrm{O}(x^9) \quad (|x| < 1) \,, \quad \text{(H.113)}$$

$$(1 + x)^{-1} = 1 - x + x^2 - x^3 + \mathrm{O}(x^4) \quad (-1 < x < 1) \,, \quad \text{(H.114)}$$

$$\frac{1}{2E} \operatorname{ar\,tanh}(E \sin \Phi^*) = \frac{1}{2} \sin \Phi^* + \frac{1}{6} E^2 \sin^3 \Phi^* + \frac{1}{10} E^4 \sin^5 \Phi^* + \mathrm{O}(E^6) \,, \quad \text{(H.115)}$$

$$\frac{\sin \Phi^*}{2(1 - E^2 \sin^2 \Phi^*)} = \frac{1}{2} \sin \Phi^* + \frac{1}{2} E^2 \sin^3 \Phi^* + \frac{1}{2} E^4 \sin^5 \Phi^* + \mathrm{O}(E^6) \,, \quad \text{(H.116)}$$

$$\frac{1}{2E} \operatorname{ar\,tanh}(E \sin \Phi^*) + \frac{\sin \Phi^*}{2(1 - E^2 \sin^2 \Phi^*)} = \sin \Phi^* + \frac{2}{3} E^2 \sin^3 \Phi^* + \frac{3}{5} E^4 \sin^5 \Phi^* + \mathrm{O}(E^6) \,. \quad \text{(H.117)}$$

(H.65) can now be written as univariate special homogeneous polynomial of degree n, namely

$$x := \sin \Phi^* \,, \quad \text{(H.118)}$$

$$y := c_4 \left[\frac{1}{2E} \operatorname{ar\,tanh}(E \sin \Phi) + \frac{\sin \Phi}{2(1 - E^2 \sin^2 \Phi)} \right] =$$

$$= c_4 \left[\sin \Phi + \frac{2}{3} E^2 \sin^3 \Phi + \frac{3}{5} E^4 \sin^5 \Phi + \mathrm{O}(E^6) \right] \,, \quad \text{(H.119)}$$

$$y(x) = a_{11} x + a_{13} x^3 + a_{15} x^5 + \cdots + a_{1n} x^n \quad (n \text{ odd}) \,, \quad \text{(H.120)}$$

$$x(y) = b_{11} y + b_{13} y^3 + b_{15} y^5 + \cdots + b_{1n} y^n \quad (n \text{ odd}) \,, \quad \text{(H.121)}$$

subject to

$$a_{11} = 1 \,. \quad \text{(H.122)}$$

Following E. Grafarend et al. (1996), we can immediately formulate the series expansion with respect to an upper triangular matrix R_A truncated up to degree five according to (H.123) subject to (H.124) as a forward substitution.

$$\begin{bmatrix} y \\ y^3 \\ y^5 \end{bmatrix} = \begin{bmatrix} 1 & a_{13} & a_{15} \\ 0 & 1 & a_{35} \\ 0 & 0 & 1 \end{bmatrix} \begin{bmatrix} x \\ x^3 \\ x^5 \end{bmatrix} + r = R_A \begin{bmatrix} x \\ x^3 \\ x^5 \end{bmatrix} + r \ , \tag{H.123}$$

$$a_{13} = \frac{2}{3} E^2 \ , \quad a_{15} = \frac{3}{5} E^4 \ , \quad a_{35} = 3a_{13} = 2E^2 \ . \tag{H.124}$$

In contrast, the backward substitution leads to (H.125) or (H.126).

$$\begin{bmatrix} x \\ x^3 \\ x^5 \end{bmatrix} = \begin{bmatrix} 1 & b_{13} & b_{15} \\ 0 & 1 & b_{35} \\ 0 & 0 & 1 \end{bmatrix} \begin{bmatrix} y \\ y^3 \\ y^5 \end{bmatrix} + s = R_B \begin{bmatrix} y \\ y^3 \\ y^5 \end{bmatrix} + s \ , \tag{H.125}$$

$$R_A R_B = I_3 \ , \quad \begin{bmatrix} 1 & b_{13} & b_{15} \\ 0 & 1 & b_{35} \\ 0 & 0 & 1 \end{bmatrix} \begin{bmatrix} 1 & a_{13} & a_{15} \\ 0 & 1 & a_{35} \\ 0 & 0 & 1 \end{bmatrix} = \begin{bmatrix} 1 & 0 & 0 \\ 0 & 1 & 0 \\ 0 & 0 & 1 \end{bmatrix} \ . \tag{H.126}$$

Finally, we obtain (H.127) and thus (H.128).

$$b_{13} = -a_{13} \ , \quad b_{15} = 3a_{13}^2 - a_{15} \ , \quad b_{35} = -3a_{13} \ , \tag{H.127}$$

$$b_{13} = -\frac{2}{3} E^2 \ , \quad b_{15} = \frac{11}{15} E^4 \ . \tag{H.128}$$

Using the first row of R_B, we arrive at (H.129).

$$x = y + b_{13} y^3 + b_{15} y^5 + O(y^7) \ . \tag{H.129}$$

The powers of y are computed according to (H.130), leading to (H.131).

$$y = c_4 \left[\sin \Phi + \frac{2}{3} E^2 \sin^3 \Phi + \frac{3}{5} E^4 \sin^5 \Phi + O(E^6) \right] \ ,$$

$$y^3 = c_4^3 \left[\sin^3 \Phi + 2E^2 \sin^5 \Phi + \frac{47}{15} E^4 \sin^7 \Phi + O(E^6) \right] \ , \tag{H.130}$$

$$y^5 = c_4^5 \left[\sin^5 \Phi + \frac{10}{3} E^2 \sin^7 \Phi + \frac{67}{9} E^4 \sin^9 \Phi + O(E^6) \right] \ ,$$

$$\sin \Phi^* = c_4 \sin \Phi \left[1 + \frac{2}{3} E^2 \sin^2 \Phi (1 - c_4^2) + \frac{1}{15} E^4 \sin^4 \Phi (9 - 20c_4^2 + 11c_4^4) + O(E^6) \right] \ . \tag{H.131}$$

With our students of Stuttgart University has been the vote that the Hammer projection was in the list of the two most popular map projections.

Note that the original contribution "mapping the sphere" is due to E. Hammer (1892). Here, we use the works of E. Grafarend and R. Syffus (1997e), J. Hoschek (1984), and K. H. Wagner (1962). We used M. Abramowitz and J. A. Stegun (1965) as well as E. Grafarend, T. Krarup, and R. Syffus for the mathematical details. Special reference is also E. Grafarend (1995).

I Mercator projection and polycylindric projection

Optimal Mercator projection and optimal polycylindric projection of conformal type. Case study Indonesia. Universal Transverse Mercator Projection (UTM), Universal Polycylindric Projection (UPC).

As a conformal mapping of the sphere $\mathbb{S}^2_R$ or as a conformal mapping of the ellipsoid-of-revolution $\mathbb{E}^2_{A_1, A_2}$, the Mercator projection maps the equator equidistantly while the transverse Mercator projection maps the transverse meta-equator, the meridian-of-reference, with equidistance. Accordingly, the Mercator projection is very well suited to geographic regions which extend East–West along the equator. In contrast, the transverse Mercator projection is appropriate for those regions which have a South–North extension. Like the optimal transverse Mercator projection, which is also known as the *Universal Transverse Mercator Projection* (UTM) and which maps the meridian-of-reference Λ_0 with an optimal dilatation factor $\rho = 0.999\,578$ with respect to the World Geodetic Reference System (WGS 84) and a strip $[\Lambda_0 - \Lambda_W, \Lambda_0 - \Lambda_E] \times [\Phi_S, \Phi_N] = [-3.5°, +3.5°] \times [-80°, +84°]$, we construct an optimal dilatation factor ρ for the optimal Mercator projection, summarized as the *Universal Mercator Projection* (UM), and an optimal dilatation factor ρ_0 for the optimal polycylindric projection for various strip widths which maps parallel circles Φ_0 equidistantly except for a dilatation factor ρ_0 summarized as the *Universal Polycylindric Projection* (UPC). It turns out that the optimal dilatation factors are independent of the longitudinal extension of the strip and depend only on the latitude Φ_0 of the parallel circle-of-reference and the southern and northern extension, namely the latitudes Φ_S and Φ_N, of the strip. For instance, for a strip $[\Phi_S, \Phi_N] = [-1.5°, +1.5°]$ along the equator $\Phi_0 = 0$, the optimal Mercator projection with respect to WGS 84 is characterized by an optimal dilatation factor $\rho = 0.999\,887$ (strip width $3°$). For other strip widths and different choices of the parallel circle-of-reference Φ_0, precise optimal dilatation factors are given. Finally, the UPC for the geographic region of Indonesia is presented as an example.

> **Important!**
>
> The Mercator projection of the sphere $\mathbb{S}^2_R$ or of the ellipsoid-of-revolution $\mathbb{E}^2_{A_1, A_2}$ is, amongst conformality, characterized by the equidistant mapping of the equator. In contrast, the transverse Mercator projection is conformal and maps the transverse meta-equator, the meridian-of-reference, equidistantly. Accordingly, the Mercator projection is very well suited to regions which extend East–West along the equator, while the transverse Mercator projection fits well to those regions which have a South–North extension. For geographic regions which are centered along lines neither equatorial, parallel circles, nor meridians, the oblique Mercator projection according to J. Engels and E. Grafarend (1995) is the conformal mapping which has to be preferred.

A typical feature of the Universal Transverse Mercator Projection (UTM) is the equidistant mapping of the central meridian of a zone except for a dilatation factor ρ which is determined by an optimality criterion. As outlined in E. Grafarend (1995), the Airy criterion of a minimal average distortion over the zone leads to an optimal value of the dilatation factor ρ depending on the strip width. An Airy optimal dilatation factor ρ, in addition, sets the average areal distortion over the zone to zero, which is quite a welcome result of an optimal map projection. Here we aim at a similar result for the Universal Mercator Projection (UM) and for the Universal Polycylindric Projection (UPC): the classical Mercator projection is designed Airy optimal for a finite zone along the equator. The equator is equidistantly mapped except for an Airy optimal dilatation factor. In particular, we analyze the Airy optimal dilatation factor as a function of the strip width. The UM strip is bounded by a southern as well as a northern parallel circle. While UM with an Airy optimal dilatation factor is well suited for geographic regions along the equator, the Airy optimal UPC has its merits for those territories which extend along a parallel circle – as a case study, Indonesia has been chosen. For such a conformal projection, a chosen parallel circle is equidistantly mapped except for a dilatation factor ρ_0 which is designed Airy optimal for a zone bounded by a southern as well as a northern parallel circle. For both types of optimal mapping, namely UM and UPC, the Airy criterion of a minimal average distortion over the zone produces zero average areal distortion, too.

Section I-1.

In detail, Section I-1 focuses on the optimal Mercator projection of the ellipsoid-of-revolution $\mathbb{E}^2_{A_1,A_2}$ with respect to the WGS 84. Figure I.1 displays the Airy optimal dilatation factor as a function of the strip width, while Table I.1 lists various optimal dilatation factors for the given strip widths $3°$, $6°$, $12°$, $20°$.

Section I-2.

In contrast, Section I-2 presents the optimal polycylindric projection of conformal type. Figure I.2 displays various Airy optimal dilatation factors for given parallel circles-of-reference parameterized by the ellipsoidal latitude Φ_0 and the strip width $\Phi_N - \Phi_S$ of northern and southern boundaries. Tables I.2 and I.3 are detailed lists of various optimal dilatation factors in different zones sorted by the strip widths of $3°$ and $6°$. As a detailed example, the optimal UPC for the geographic region of Indonesia is presented as a case-study.

Particular reference is made to G. B. Airy (1861) for the Airy optimality criterion, to Snyder (1987) with respect to the Mercator projection, to J. Engels and E. Grafarend (1995) with respect to the oblique Mercator Projection, and to E. Grafarend (1995) for a review of the Tissot distortion analysis of a map projection and for the optimal transverse Mercator projection.

I-1 The optimal Mercator projection (UM)

Here we present three definitions which relate to the generalized Mercator projection, the Airy optimal generalized Mercator projection (UM) and finally the generalized Mercator projection of least total areal distortion. Three lemmas and one corollary describe in detail the optimal Mercator projection which is finally illustrated by one table, one figure and two examples with respect to WGS 84.

Definition I.1 (Generalized Mercator projection, mapping equations).

The conformal mapping of the ellipsoid-of-revolution (I.1) with semi-major axis A_1, semi-minor axis A_2, and relative eccentricity squared $E^2 := (A_1^2 - A_2^2)/A_1^2$ onto the developed circular cylinder $\mathbb{C}^2_{\rho A_1}$ of radius ρA_1 is called a *generalized Mercator projection* if the equator of $\mathbb{E}^2_{A_1,A_2}$ is mapped equidistantly except for a dilatation factor ρ such that the mapping equations (I.2) hold with respect to surface normal coordinates (longitude Λ, latitude Φ) which parameterize $\mathbb{E}^2_{A_1,A_2}$.

$$\boldsymbol{X} \in \mathbb{E}^2_{A_1,A_2} :=$$

$$:= \left\{ \boldsymbol{X} \in \mathbb{R}^3 \left| \frac{X^2 + Y^2}{A_1^2} + \frac{Z^2}{A_2^2} = 1,\, A_1 \in \mathbb{R}^+,\, A_2 \in \mathbb{R}^+,\, A_1 > A_2 \right. \right\},$$

(I.1)

$$x = \rho A_1 (\Lambda - \Lambda_0),$$

$$y = \rho A_1 \ln \left(\tan \left(\frac{\pi}{4} + \frac{\Phi}{2} \right) \left(\frac{1 - E \sin \Phi}{1 + E \sin \Phi} \right)^{E/2} \right).$$

(I.2)

Λ_0 is called the surface normal longitude-of-reference. The plane covered by the chart $\{x, y\}$, Cartesian coordinates, with an Euclidean metric, namely $\{\mathbb{R}^2, \delta_{kl}\}$ (Kronecker delta, unit matrix) is generated by developing the circular cylinder $\mathbb{C}^2_{\rho A_1}$ of radius ρA_1.

End of Definition.

Lemma I.2 (Generalized Mercator projection, principal stretches).

With respect to the left Tissot distortion measure represented by the matrix $\mathbf{C}_l \mathbf{G}_l^{-1}$ of the left Cauchy–Green deformation tensor $\mathbf{C}_l = \mathbf{J}_l^T \mathbf{G}_r \mathbf{J}_l$ multiplied by the inverse of the left metric tensor $\mathbf{G}_l$, the matrix of the metric tensor of $\mathbb{E}^2_{A_1, A_2}$, the left principal stretches of the generalized Mercator projection are given by

$$\Lambda_1 = \Lambda_2 = \rho \frac{\sqrt{1 - E^2 \sin^2 \Phi}}{\cos \Phi} \; . \tag{I.3}$$

The eigenvalues $\{\Lambda_1, \Lambda_2\}$ cover the eigenspace of the left Tissot matrix $\mathbf{C}_l \mathbf{G}_l^{-1}$. Due to conformality, they are identical, $\Lambda_1 = \Lambda_2 = \Lambda_S$. $\mathbf{J}_l$ denotes the left Jacobi map $(dx, dy) \mapsto (d\Lambda, d\Phi)$, $\mathbf{G}_r$ the matrix of the right metric tensor of the plane generated by developing the circular cylinder $\mathbb{C}^2_{\rho A_1}$ of radius ρA_1, namely the unit matrix $\mathbf{G}_r = \mathbf{I}_2$.

End of Lemma.

Definition I.3 (Generalized Mercator projection, Airy optimum).

The generalized Mercator projection of the ellipsoid-of-revolution $\mathbb{E}^2_{A_1, A_2}$ onto the developed circular cylinder $\mathbb{C}^2_{\rho A_1}$ of radius ρA_1 is called *Airy optimal* if the deviation from an isometry (I.4) in terms of the left principal stretches $\{\Lambda_1, \Lambda_2\}$ averaged over a mapping area of interest, namely the surface integral (I.5), is minimal with respect to the unknown dilatation factor ρ.

$$\frac{(\Lambda_1 - 1)^2 + (\Lambda_2 - 1)^2}{2} \; , \tag{I.4}$$

$$J_{lA} := \frac{1}{2S} \int_{\text{area}} \left[(\Lambda_1 - 1)^2 + (\Lambda_2 - 1)^2 \right] dS = \min_{\rho} \; . \tag{I.5}$$

End of Definition.

The infinitesimal surface element of $\mathbb{E}^2_{A_1, A_2}$ is represented by the expression $\sqrt{\det[\mathbf{G}_l]} d\Lambda d\Phi$, namely by (I.6). In contrast, for the equatorial strip $[\Lambda_W, \Lambda_E] \times [\Phi_S, \Phi_N]$ between a longitudinal extension $(\Lambda_0 - \Delta\Lambda, \Lambda_0 + \Delta\Lambda)$ and a latitudinal extension (Φ_S, Φ_N), the finite area is computed by (I.7); the subscripts S, N, E, and W denote South, North, East, and West. The actual computation up to the fourth order in relative eccentricity, namely $O(E^4)$, is performed by an uniform convergent series expansion of $(1 - x)^{-2}$ for $|x| < 1$ and a term-wise integration, namely interchanging summation and integration. Note that along the surface normal longitude of reference Λ_0 the strip has been chosen symmetrically such that (I.8), i.e. $\Lambda_E - \Lambda_W = \Lambda_0 + \Delta\Lambda - (\Lambda_0 - \Delta\Lambda) = 2\Delta\Lambda$, holds.

The areal element of $\mathbb{E}^2_{A_1, A_2}$ is provided by

$$dS = \frac{A_1^2 (1 - E^2) \cos \Phi}{(1 - E^2 \sin^2 \Phi)} d\Lambda d\Phi \; , \tag{I.6}$$

$$S := \text{area} \left\{ \mathbb{E}^2_{A_1, A_2} \Big|_{\substack{\Lambda_W = \Lambda_0 - \Delta\Lambda \le \Lambda \le \Lambda_\varepsilon = \Lambda_0 - \Delta\Lambda \\ \Phi_S \le \Phi \le \Phi_N}} \right\} = \int_{\Lambda_W}^{\Lambda_E} d\Lambda \int_{\Phi_S}^{\Phi_N} \frac{A_1^2 (1 - E^2) \cos \Phi}{(1 - E^2 \sin^2 \Phi)^2} d\Phi =$$

$$= A_1^2 (1 - E) \int_{\Lambda_W}^{\Lambda_E} d\Lambda \int_{\Phi_S}^{\Phi_N} \cos \Phi [1 + 2E^2 \sin^2 \Phi + O(E^4)] d\Phi = \tag{I.7}$$

$$= 2A_1^2 (1 - E^2) \Delta\Lambda [\sin \Phi_N + \frac{2}{3} E^2 \sin^3 \Phi_N - (\sin \Phi_S + \frac{2}{3} E^2 \sin^3 \Phi_S) + O(E^4)] \; ,$$

$$\Lambda_E - \Lambda_W = 2\Delta\Lambda \; . \tag{I.8}$$

In case of the generalized Mercator projection, the left *Airy distortion energy* J_{lA} is the quadratic form in terms of the dilatation factor ρ, in particular

$$J_{lA}(\rho) = c_0 - 2c_1\rho + c_2\rho^2 \,, \tag{I.9}$$

such that

$$c_0 = 1 \,,$$

$$c_1 = \left[(\Phi_N - \Phi_S)\left(1 + \frac{3}{4}E^2 + \frac{45}{64}E^4\right) - \frac{3}{4}\left(1 + \frac{15}{16}E^2\right)E^2\left(\cos\Phi_N\sin\Phi_N - \cos\Phi_S\cos\Phi_S\right) - \right.$$

$$\left. - \frac{15}{32}E^4\left(\cos\Phi_N\sin^3\Phi_N - \cos\Phi_S\sin^3\Phi_S\right)\right] /$$

$$\left[\sin\Phi_N + \frac{2}{3}E^2\sin^3\Phi_N + \frac{3}{5}E^4\sin^5\Phi_N - \sin\Phi_S - \frac{2}{3}E^2\sin^3\Phi_S - \frac{3}{5}E^4\sin^5\Phi_S\right] + O(E^6) \,, \tag{I.10}$$

$$c_2 = \left[(1 + E^2 + E^4)\ln\left(\tan\left(\frac{\pi}{4} + \frac{\Phi_N}{2}\right) / \tan\left(\frac{\pi}{4} + \frac{\Phi_S}{2}\right)\right) - E^2(1 + E^2)(\sin\Phi_N - \sin\Phi_S) - \right.$$

$$\left. - \frac{1}{3}E^3(\sin^3\Phi_N - \sin^3\Phi_S)\right] /$$

$$\left[\sin\Phi_N + \frac{2}{3}E^2\sin^3\Phi_N + \frac{3}{5}E^4\sin^5\Phi_N - \sin\Phi_S - \frac{2}{3}E^2\sin^3\Phi_S - \frac{3}{5}E^4\sin^5\Phi_S\right] + O(E^6)$$

hold.

Constitutional elements of the left Airy distortion energy are

$$J_{lA} = \frac{1}{S}\int_S (\Lambda_S - 1)^2 dS = \frac{1}{S}\int_S (\Lambda_S^2 - 2\Lambda_S + 1)dS = 1 - \frac{2}{S}\int_S \Lambda_S dS + \frac{1}{S}\int_S \Lambda_S^2 dS \,, \tag{I.11}$$

$$\int_S \Lambda_S dS = \rho\int_S \frac{\sqrt{1 - E^2\sin^2\Phi}}{\cos\Phi}\frac{A_1^2(1 - E^2)\cos\Phi}{(1 - E^2\sin^2\Phi)^2}d\Lambda d\Phi \,, \tag{I.12}$$

$$\int_S \Lambda_S^2 dS = \rho^2\int_S \frac{1 - E^2\sin^2\Phi}{\cos^2\Phi}\frac{A_1^2(1 - E^2)\cos\Phi}{(1 - E^2\sin^2\Phi)^2}d\Lambda d\Phi \,, \tag{I.13}$$

$$J_{lA}(\rho) = c_0 - 2c_1\rho + c_2\rho^2$$

$$\Leftrightarrow$$

$$c_0 := 1$$

$$\Leftrightarrow$$

$$c_1 := \frac{1}{S}\int_S \Lambda_S dS = \frac{1}{S}A_1^2(1 - E^2)\int_S \frac{d\Lambda d\Phi}{(1 - E^2\sin^2\Phi)^{3/2}} \,,$$

$$c_2 := \frac{1}{S}\int_S \Lambda_S^2 dS = \frac{1}{S}A_1^2(1 - E^2)\int_S \frac{d\Lambda d\Phi}{\cos\Phi(1 - E^2\sin^2\Phi)} \,. \tag{I.14}$$

Furthermore, constitutional elements of the left Airy distortion energy are

$$\int_{\Lambda_W}^{\Lambda_E} \mathrm{d}\Lambda \int_{\Phi_S}^{\Phi_E} \frac{\mathrm{d}\Phi}{(1 - E^2 \sin^2 \Phi)^{3/2}} = \int_{\Phi_S}^{\Phi_N} \frac{(\Lambda_E - \Lambda_W)\mathrm{d}\Phi}{(1 - E^2 \sin^2 \Phi)^{3/2}} \,, \qquad (\mathrm{I.15})$$

$$\int \frac{\mathrm{d}x}{(1 - E^2 \sin^2 x)^{3/2}} =$$

$$= x \left(1 + \frac{3}{4}E^2 + \frac{45}{64}E^4\right) - \frac{3}{4}\cos x \sin x \left(1 + \frac{15}{16}E^2\right) E^2 - \frac{15}{32}E^4 \cos x \sin^3 x + \mathrm{O}(E^6) \qquad (\mathrm{I.16})$$

$$\Rightarrow$$

$$c_1 \text{ (see first equation of (I.10))} \,,$$

$$\int_{\Lambda_W}^{\Lambda_E} \mathrm{d}\Lambda \int_{\Phi_S}^{\Phi_E} \frac{\mathrm{d}\Phi}{\cos \Phi (1 - E^2 \sin^2 \Phi)} = \int_{\Phi_S}^{\Phi_N} \frac{(\Lambda_E - \Lambda_W)\mathrm{d}\Phi}{\cos \Phi (1 - E^2 \sin^2 \Phi)} \,, \qquad (\mathrm{I.17})$$

$$\int \frac{\mathrm{d}x}{\cos x (1 - E^2 \sin^2 x)} =$$

$$= \left(1 + E^2 + E^4\right) \ln \tan \left(\frac{\pi}{4} + \frac{x}{2}\right) - E^2 \left(1 + E^2\right) \sin x - \frac{1}{3}E^4 \sin^3 x + \mathrm{O}(E^6) \qquad (\mathrm{I.18})$$

$$\Rightarrow$$

$$c_2 \text{ (see second equation of (I.10))} \,.$$

Note that for the proof of Lemma I.4, we have collected all constitutional items in (I.10)–(I.18). Indeed, as soon as we represent the left principal stretches $\Lambda_1 = \Lambda_2 = \Lambda_S$ according to (I.3) within the left Airy distortion energy J_{lA}, in particular (I.9), we are left with the quadratic polynomial of (I.11) which constitutes the integrals of (I.12) and (I.13). First, the left principal stretch Λ_S has to be integrated over the area of interest. Second, the squared left principal stretch Λ_S^2 has to be integrated over the area of interest. In this way, we are led to the coefficients c_0, c_1, abd c_2 of type (I.14). (I.15)–(I.18) describe the involved integrals which are computed by term-wise integration of the uniformly convergent kernel series, namely by interchanging integration and summation. The integral series expansions are of the order $\mathrm{O}(E^6)$ for (I.16) and (I.18).

Lemma I.5 (Minimal Airy distortion energy).

The Airy distortion energy (I.9) is minimal if the dilatation factor amounts to $\hat{\rho} = c_1/c_2$ and

$$\hat{\rho}(\Phi_S, \Phi_N) =$$

$$= \left[(\Phi_N - \Phi_S)\left(1 + \frac{3}{4}E^2 + \frac{45}{64}E^4\right) - \frac{3}{4}\left(1 + \frac{15}{16}E^2\right) E^2 (\cos \Phi_N \sin \Phi_N - \cos \Phi_S \sin \Phi_S)\right.$$

$$\left. - \frac{15}{32}E^4 (\cos \Phi_N \sin^3 \Phi_N - \cos \Phi_S \sin^3 \Phi_S)\right] / \left[\ln \left(\tan \left(\frac{\pi}{4} + \frac{\Phi_N}{2}\right) / \tan \left(\frac{\pi}{4} + \frac{\Phi_S}{2}\right)\right) \times\right.$$

$$\left. \times \left(1 + E^2 + E^4\right) - E^2(1 + E^2)(\sin \Phi_N - \sin \Phi_S) - \frac{1}{3}E^4(\sin^3 \Phi_N - \sin^3 \Phi_S)\right] + \mathrm{O}(E^6) \,, \qquad (\mathrm{I.19})$$

$$\hat{\rho}(\Phi_S = -\Phi_N) =$$

$$= \left[\Phi_N \left(1 + \frac{3}{4}E^2 + \frac{45}{64}E^4\right) - \frac{3}{4}\left(1 + \frac{15}{16}E^2\right) E^2 \cos \Phi_N \sin \Phi_N - \frac{15}{32}E^4 \cos \Phi_N \sin^3 \Phi_N\right] /$$

$$\left[\left(1 + E^2 + E^4\right) \ln \tan \left(\frac{\pi}{4} + \frac{\Phi_N}{2}\right) - E^2(1 + E^2) \sin \Phi_N - \frac{1}{3}E^4 \sin^3 \Phi_N\right] + \mathrm{O}(E^6) \,.$$

End of Lemma.

Proof.

$\hat{\rho} = c_1/c_2$ is proven by the following procedure.

$$J_{l\mathrm{A}} = c_0 - 2c_1\rho + c_2\rho^2 = \min_{\rho} \ . \tag{I.20}$$

Necessary:

$$\frac{\mathrm{d}J_{l\mathrm{A}}}{\mathrm{d}\rho}(\rho = \hat{\rho}) = -2c_1 + c_2\hat{\rho} = 0$$

$$\Leftrightarrow \tag{I.21}$$

$$\hat{\rho} = c_1/c_2 \ .$$

Sufficient:

$$\frac{\mathrm{d}^2 J_{l\mathrm{A}}}{\mathrm{d}\rho^2}(\rho = \hat{\rho}) = 2c_2 > 0 \ . \tag{I.22}$$

(I.19) directly follows from (I.10) and $\hat{\rho} = c_1/c_2$.

End of Proof.

Before we go into numerical computations of the optimal dilatation factor $\hat{\rho}$ for the generalized Mercator projection, let us here briefly present a result for zero total areal distortion as it is outlined by E. Grafarend (1995).

Definition I.6 (Generalized Mercator projection, optimal with respect to areal distortion).

The generalized Mercator projection of the ellipsoid-of-revolution $\mathbb{E}^2_{A_1,A_2}$ onto the developed circular cylinder $\mathbb{C}^2_{\rho A_1}$ of radius ρA_1 is called *optimal with respect to areal distortion* if the deviation from an equiareal mapping $\Lambda_1\Lambda_2 - 1$ in terms of the left principal stretches (Λ_1, Λ_2) averaged over a mapping area of interest, namely the total areal distortion (I.23), is minimal with respect to the unknown dilatation factor ρ.

$$J_l := \frac{1}{S}\int_{\mathrm{S}} (\Lambda_1\Lambda_2 - 1)\mathrm{d}S = \min_{\rho} \ . \tag{I.23}$$

End of Definition.

Corollary I.7 (Generalized Mercator projection, dilatation factor).

For a generalized Mercator projection of the half-symmetric strip $[\Lambda_{\mathrm{W}} = \Lambda_0 - \Delta\Lambda, \Lambda_0 + \Delta\Lambda = \Lambda_{\mathrm{E}}] \times [\Phi_{\mathrm{S}}, \Phi_{\mathrm{N}}]$, the postulates of minimal Airy distortion energy (minimal total distance distortion) and of minimal total areal distortion lead to the same unknown dilatation factor $\hat{\rho}$ of (I.19) by first-order approximation. The total areal distortion amounts to zero.

End of Corollary.

Proof.

We start from the representation of the left principal stretches for a mapping of conformal type implemented into, firstly, $J_{l\mathrm{A}}$, secondly, J_l. The squared left principal stretches are assumed to be given by 1 plus a small quantity μ except for the dilatation factor ρ:

$$\Lambda_1^2 = \Lambda_2^2 = \Lambda_{\mathrm{S}}^2 = \rho^2(1 + \mu) \ \forall \ \mu \ll 1 \ ,$$

$$\Lambda_1 = \Lambda_2 = \Lambda_{\mathrm{S}} = \rho\left(1 + \frac{\mu}{2}\right) + \mathrm{O}(\mu^2) \ \forall \ \mu \ll 1 \ . \tag{I.24}$$

$$J_{lA}:$$

$$J_{lA} := \frac{1}{2S} \int_S [(\Lambda_1 - 1)^2 + (\Lambda_2 - 1)^2] dS \ \forall \ \Lambda_1 = \Lambda_2 = \Lambda_S \ ,$$

$$J_{lA} := \frac{1}{S} \int_S (\Lambda_S - 1)^2 dS = 1 + \frac{1}{S} \int_S \left[(1 + \mu)\rho^2 - 2\left(1 + \frac{\mu}{2}\right)\rho \right] dS + O(\mu^2) \ ,$$

(I.25)

$$J'_{lA}(\rho) = 0$$

$$\Leftrightarrow$$

$$\hat{\rho} = \int_S \left(1 + \frac{\mu}{2}\right) dS \Big/ \int_S (1 + \mu) dS + O(\mu^2)$$

$$\Leftrightarrow$$

$$\hat{\rho} = \left(1 + \frac{1}{2S} \int_S \mu dS\right) \Big/ \left(1 + \frac{1}{S} \int_S \mu dS\right) + O(\mu^2)$$

$$= 1 - \frac{1}{2S} \int_S \mu dS + O(\mu^2) \ .$$

(I.26)

$$J_l:$$

$$J_l := \frac{1}{S} \int_S (\Lambda_1 \Lambda_2 - 1) dS \ \forall \Lambda_1 = \Lambda_2 = \Lambda_S \ ,$$

$$J_l := \frac{1}{S} \int_S (\Lambda_S^2 - 1) dS \ ,$$

(I.27)

$$J_l = 0$$

$$\Leftrightarrow$$

$$\frac{1}{S} \int_S [\rho^2(1 + \mu) - 1] dS = 0$$

$$\Leftrightarrow$$

$$\rho^2(J_l = 0) = \frac{1}{1 + \frac{1}{S} \int_S \mu dS}$$

(I.28)

$$= 1 - \frac{1}{S} \int_S \mu dS + O(\mu^2)$$

$$\Leftrightarrow$$

$$\rho(J_l = 0) = 1 - \frac{1}{2S} \int_S \mu dS + O(\mu^2) \ .$$

$$\hat{\rho}(J_{lA} = \min) = \rho(J_l = 0).$$

End of Proof.

As a basis for a discussion of the Airy optimal generalized Mercator projection (UM), we refer to Table I.1 and Fig. I.1 where the Airy optimal dilatation factor $\hat{\rho}(\Phi_N)$ as a function of the strip width $2\Phi_N$ with respect to WGS 84 has been computed or plotted, respectively. Finally, we present two examples for the optimal design of the generalized Mercator projection which can be compared to those of E. Grafarend (1995) for the optimal transverse Mercator projection.

Example I.1 ($[\Phi_{\mathrm{S}} = -\Phi_{\mathrm{N}}, \Phi_{\mathrm{N}}] = [-1.5°, +1.5°]$).

For the Airy optimal generalized UM, we have chosen a strip width of 3° between $\Phi_{\mathrm{S}} = -1.5°$ southern latitude and $\Phi_{\mathrm{N}} = 1.5°$ northern latitude. Once we refer to the WGS 84, the Airy optimal dilatation factor amounts to

$$\hat{\rho} = 0.999\,887 .$$ (I.29)

End of Example.

Example I.2 ($[\Phi_{\mathrm{S}} = -\Phi_{\mathrm{N}}, \Phi_{\mathrm{N}}] = [-3°, +3°]$).

For the second example, we have chosen a strip width of 6° between $\Phi_{\mathrm{S}} = -3°$ southern latitude and $\Phi_{\mathrm{N}} = 3°$ northern latitude. Once we refer to WGS 84, the Airy optimal dilatation factor amounts to

$$\hat{\rho} = 0.999\,546 .$$ (I.30)

End of Example.

Table I.1. Airy optimal dilatation factor $\hat{\rho}$ for a symmetric strip $[\Lambda_{\mathrm{W}}, \Lambda_{\mathrm{E}}] \times [\Phi_{\mathrm{S}} = -\Phi_{\mathrm{N}}, \Phi_{\mathrm{N}}]$, $\Lambda_{\mathrm{W}} = \Lambda_0 - \Delta\Lambda$, $\Lambda_{\mathrm{E}} = \Lambda_0 + \Delta\Lambda$, generalized UM, WGS 84, $A_1 = 6\,378\,137$ m, $E^2 = 0.006\,694\,379\,990\,13$.

$\Phi_{\mathrm{N}} = 1.5°$	$\Phi_{\mathrm{N}} = 3°$	$\Phi_{\mathrm{N}} = 6°$	$\Phi_{\mathrm{N}} = 10°$
$\hat{\rho} = 0.999\,887$	$\hat{\rho} = 0.999\,546$	$\hat{\rho} = 0.998\,183$	$\hat{\rho} = 0.994\,943$

I-2 The optimal polycylindric projection of conformal type (UPC)

Here, we present two definitions which relate to the generalized polycylindric projection of conformal type and the Airy optimal generalized polycylindric projection of conformal type (UPC). Three lemmas describe in detail the optimal UPC for the ellipsoid-of-revolution, which is finally illustrated by four tables and five figures, including a detailed example for the geographic region of Indonesia. All optimal map projections refer to WGS 84.

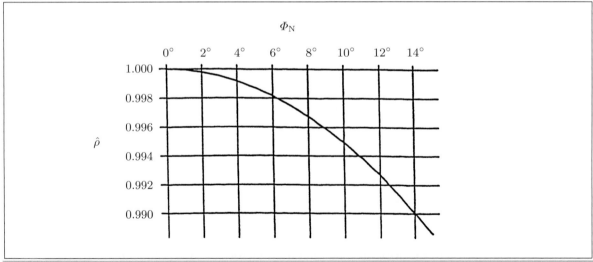

Fig. I.1. Airy optimal dilatation factor $\hat{\rho}$ for a symmetric strip $[\Lambda_{\mathrm{W}}, \Lambda_{\mathrm{E}}] \times [\Phi_{\mathrm{S}} = -\Phi_{\mathrm{N}}, \Phi_{\mathrm{N}}]$ with $\Lambda_{\mathrm{W}} = \Lambda_0 - \Delta\Lambda$ and $\Lambda_{\mathrm{E}} = \Lambda_0 + \Delta\Lambda$, generalized UM, WGS 84, $\rho(\Phi_{\mathrm{N}})$: Airy optimal dilatation factor as function of the half-strip width Φ_{N}.

Definition I.8 (Generalized polycylindric projection, mapping equations).

The conformal mapping of the ellipsoid-of-revolution (I.32) with semi-major axis A_1, semi-minor axis A_2, and relative eccentricity squared $E^2 := (A_1^2 - A_2^2)/A_1^2$ onto the developed circular cylinder $\mathbb{C}_R^2$ of radius (I.31) is called a *generalized polycylindric projection* if the parallel circle-of-reference Φ_0 is mapped equidistantly except for a dilatation factor ρ_0 such that the mapping equations (I.33) hold.

$$R = \rho_0 A_1 \cos \Phi_0 / \sqrt{1 - E^2 \sin^2 \Phi_0} \,, \tag{I.31}$$

$$\boldsymbol{X} \in \mathbb{E}_{A_1,A_2}^2 := \left\{ \boldsymbol{X} \in \mathbb{R}^3 \,\middle|\, \frac{X^2 + Y^2}{A_1^2} + \frac{Z^2}{A_2^2} = 1,\, A_1 \in \mathbb{R}^+,\, A_2 \in \mathbb{R}^+,\, A_1 > A_2 \right\} \,, \tag{I.32}$$

$$x = \rho_0 A_1 \frac{\cos \Phi_0 (\Lambda - \Lambda_0)}{\sqrt{1 - E^2 \sin^2 \Phi_0}} \,,$$

$$\tag{I.33}$$

$$y = \rho_0 A_1 \frac{\cos \Phi_0}{\sqrt{1 - E^2 \sin^2 \Phi_0}} \ln \left(\tan \left(\frac{\pi}{4} + \frac{\Phi}{2} \right) \left(\frac{1 - E \sin \Phi}{1 + E \sin \Phi} \right)^{E/2} \right) \,.$$

The plane covered by the chart (x, y). Cartesian coordinates, with an Euclidean metric, namely $\{\mathbb{R}^2, \delta_{kl}\}$ (Kronecker delta, unit matrix) is generated by developing the circular cylinder $\mathbb{C}_R^2$ of radius (I.31) with respect to the surface normal latitude Φ_0 of reference.

End of Definition.

Lemma I.9 (Generalized polycylindric projection, principal stretches).

With respect to the left Tissot distortion measure represented by the matrix $\mathsf{C}_l \mathsf{G}_l^{-1}$ of the left Cauchy–Green deformation tensor $\mathsf{C}_l = \mathsf{J}_l^{\mathrm{T}} \mathsf{G}_r \mathsf{J}_l$ multiplied by the inverse of the left metric tensor G_l, the matrix of the metric tensor of $\mathbb{E}_{A_1,A_2}^2$, the left principal stretches of the generalized polycylindric projection are given by

$$\Lambda_1 = \Lambda_2 = \rho_0 \frac{\cos \Phi_0}{\sqrt{1 - E^2 \sin^2 \Phi_0}} \frac{\sqrt{1 - E^2 \sin^2 \Phi}}{\cos \Phi} \,. \tag{I.34}$$

The eigenvalues $\{\Lambda_1, \Lambda_2\}$ cover the eigenspace of the left Tissot matrix $\mathsf{C}_l \mathsf{G}_l^{-1}$. Due to conformality, they are identical, $\Lambda_1 = \Lambda_2 = \Lambda_S$. J_l denotes the left Jacobi map $(\mathrm{d}x, \mathrm{d}y) \mapsto (\mathrm{d}\Lambda, \mathrm{d}\Phi)$, G_r the matrix of the right metric tensor of the plane generated by developing the circular cylinder $\mathbb{C}_R^2$ of radius (I.31), namely the unit matrix $\mathsf{G}_r = \mathsf{I}_2$.

End of Lemma.

Definition I.10 (Generalized polycylindric projection, Airy optimum).

The generalized polycylindric projection of conformal type of (I.33) and (I.34) of the ellipsoid-of-revolution $\mathbb{E}_{A_1,A_2}^2$ onto the developed circular cylinder $\mathbb{C}_R^2$ of radius (I.31) is called *Airy optimal* if the deviation from an isometry (I.35) in terms of the left principal stretches $\{\Lambda_1, \Lambda_2\}$, in particular (I.34), averaged over a mapping area of interest, namely the surface integral (I.36), is minimal with respect to the unknown dilatation factor ρ_0.

$$[(\Lambda_1 - 1)^2 + (\Lambda_2 - 1)^2]/2 \,, \tag{I.35}$$

$$J_{l\mathrm{A}} := \frac{1}{2S} \int_{\mathrm{area}} [(\Lambda_1 - 1)^2 + (\Lambda_2 - 1)^2] \mathrm{d}S = \min_{\rho_0} \,. \tag{I.36}$$

End of Definition.

Let us refer to the representation of the areal elements $\{dS, S\}$ of the ellipsoid-of-revolution of Section I-1. With the next step, we move on to Lemma I.11 for a representation of J_{lA} subject to $\Lambda_1 = \Lambda_2$, in particular (I.34).

Lemma I.11 (Generalized polycylindric projection, Airy distortion energy).

In case of the generalized polycylindric projection of conformal type, the left *Airy distortion energy* J_{lA} is the quadratic form in terms of the dilatation factor ρ_0, in particular

$$J_{lA}(\rho_0) = c_{00} - 2c_{01}\rho_0 + c_{02}\rho_0^2 , \qquad (I.37)$$

such that

$$c_{00} = 1 ,$$

$$c_{01} = c_1 \cos\Phi_0 / \sqrt{1 - E^2 \sin^2\Phi_0} , \qquad (I.38)$$

$$c_{02} = c_2 \cos^2\Phi_0 / (1 - E^2 \sin^2\Phi_0)$$

hold.

End of Lemma.

Once we start from the proof of Lemma I.4, the extension to the result of (I.37) and I.38) with respect to (I.34) into Lemma I.11 is straightforward.

Question. Question: "Where, with respect to the dilatation factor ρ_0, is the Airy distortion energy minimal?" Answer: "The detailed answer is given in Lemma I.12."

Lemma I.12 (Minimal Airy distortion energy).

The Airy distortion energy (I.37) is minimal if the dilatation factor amounts to $\hat\rho_0 = c_{01}/c_{02}$ and

$$\hat\rho_0(\Phi_S, \Phi_N) =$$

$$= \hat\rho_0(\Phi_S, \Phi_N) \frac{\sqrt{1 - E^2 \sin^2\Phi_0}}{\cos\Phi_0} ,$$

$$\hat\rho_0(\Phi_S = -\Phi_N) = \qquad (I.39)$$

$$= \hat\rho_0(\Phi_S = -\Phi_N) \frac{\sqrt{1 - E^2 \sin^2\Phi_0}}{\cos\Phi_0} ,$$

where c_{01} and c_{02} follow from (I.38) and (I.10), $\hat\rho_0(\Phi_S = \Phi_N)$ from (I.19) (see the first equation), and $\hat\rho_0(\Phi_S = -\Phi_N)$ from (I.19) (see the second equation), respectively.

End of Lemma.

The proof of Lemma I.12 completely follows along the lines of the proof of Lemma I.5 and is therefore not repeated here. In addition, we note zero total areal distortion over a half-symmetric strip $[\Lambda_W = \Lambda_0 - \Delta\Lambda, \Lambda_0 + \Delta\Lambda = \Lambda_E] \times [\Phi_S = \Phi_0 - \Delta\Phi, \Phi_0 + \Delta\Phi = \Phi_N]$ if the Airy optimal dilatation factor $\hat\rho_0$ of type (I.39), first equation, or of type (I.39), second equation, is implemented. Definition I.6 and Corollary I.7 apply accordingly. As a basis for a discussion of the Airy optimal UPC, let us here refer to Table I.2 and Table I.3 as well as to Fig. I.2, where the Airy optimal dilatation factor $\hat\rho_0(\Phi_0, \Phi_S = \Phi_0 - \Delta\Phi, \Phi_N = \Phi_0 + \Delta\Phi)$ as a function of the strip width $\Phi_N - \Phi_S = 2\Delta\Phi$ with respect to WGS 84 has been computed or plotted, respectively.

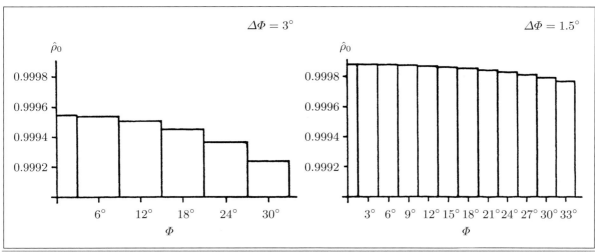

Fig. I.2. Airy optimal dilatation factor $\hat{\rho}_0$ for a symmetric strip, generalized UPC. $\hat{\rho}_0(\Phi_0, \Delta\Phi)$: Airy optimal dilatation factor $\hat{\rho}_0$ as a function of the chosen parallel circle latitude Φ_0 and strip width $\Delta\Phi$, WGS 84. Symmetric strip $[\Lambda_{\mathrm{W}} = \Lambda_0 - \Delta\Lambda, \Lambda_0 + \Delta\Lambda = \Lambda_{\mathrm{E}}] \times [\Phi_{\mathrm{S}} = \Phi_0 - \Delta\Phi, \Phi_0 + \Delta\Phi = \Phi_{\mathrm{N}}]$.

Table I.2. Airy optimal dilatation factor $\hat{\rho}_0$ for a symmetric strip, generalized UPC, WGS 84. Symmetric strip $[\Lambda_{\mathrm{W}} = \Lambda_0 - \Delta\Lambda, \Lambda_0 + \Delta\Lambda = \Lambda_{\mathrm{E}}] \times [\Phi_{\mathrm{S}} = \Phi_0 - \Delta\Phi, \Phi_0 + \Delta\Phi = \Phi_{\mathrm{N}}]$, strip width 3°, $\Delta\Phi = 1.5°$.

zone	Φ_0	Φ_{S}	Φ_{N}	$\hat{\rho}_0$	zone	Φ_0	Φ_{S}	Φ_{N}	$\hat{\rho}_0$
0	0°	−1.5°	+1.5°	0.999 887	±1	±3°	±1.5°	±4.5°	0.999 886
±2	±6°	±4.5°	±7.5°	0.999 884	±3	±9°	±7.5°	±10.5°	0.999 881
±4	±12°	±10.5°	±13.5°	0.999 876	±5	±15°	±13.5°	±16.5°	0.999 870
±6	±18°	±16.5°	±19.5°	0.999 862	±7	±21°	±19.5°	±22.5°	0.999 852
±8	±24°	±22.5°	±25.5°	0.999 840	±9	±27°	±25.5°	±28.5°	0.999 826
±10	±30°	±28.5°	±31.5°	0.999 809	±11	±33°	±31.5°	±34.5°	0.999 789
±12	±36°	±34.5°	±37.5°	0.999 764	±13	±39°	±37.5°	±40.5°	0.999 735
±14	±42°	±40.5°	±43.5°	0.999 699	±15	±45°	±43.5°	±46.5°	0.999 656
±16	±48°	±46.5°	±49.5°	0.999 602	±17	±51°	±49.5°	±52.5°	0.999 535
±18	±54°	±52.5°	±55.5°	0.999 450	±19	±57°	±55.5°	±58.5°	0.999 341
±20	±60°	±58.5°	±61.5°	0.999 197	±21	±63°	±61.5°	±64.5°	0.999 002
±22	±66°	±64.5°	±67.5°	0.998 729	±23	±69°	±67.5°	±70.5°	0.998 330
±24	±72°	±70.5°	±73.5°	0.997 714	±25	±75°	±73.5°	±76.5°	0.996 691
±26	±78°	±76.5°	±79.5°	0.994 804	±27	±81°	±79.5°	±82.5°	0.990 705
±28	±84°	±82.5°	±85.5°	0.978 842	±29	±87°	±85.5°	±88.5°	0.910 273

Table I.3. Airy optimal dilatation factor $\hat{\rho}_0$ for a symmetric strip, generalized UPC, WGS 84. Symmetric strip $[\Lambda_{\mathrm{W}} = \Lambda_0 - \Delta\Lambda, \Lambda_0 + \Delta\Lambda = \Lambda_{\mathrm{E}}] \times [\Phi_{\mathrm{S}} = \Phi_0 - \Delta\Phi, \Phi_0 + \Delta\Phi = \Phi_{\mathrm{N}}]$, strip width 6°, $\Delta\Phi = 3°$.

zone	Φ_0	Φ_{S}	Φ_{N}	$\hat{\rho}_0$	zone	Φ_0	Φ_{S}	Φ_{N}	$\hat{\rho}_0$
0	0°	−3°	+3°	0.999 546	±1	±6°	±3°	±9°	0.999 536
±2	±12°	±9°	±15°	0.999 504	±3	±18°	±15°	±21°	0.999 448
±4	±24°	±21°	±27°	0.999 362	±5	±30°	±27°	±33°	0.999 236
±6	±36°	±33°	±39°	0.999 057	±7	±42°	±39°	±45°	0.998 796
±8	±48°	±45°	±51°	0.998 407	±9	±54°	±51°	±57°	0.997 799
±10	±60°	±57°	±63°	0.996 782	±11	±66°	±63°	±69°	0.994 899
±12	±72°	±69°	±75°	0.990 804	±13	+78°	±75°	⊥81°	0.978 942
±14	±84°	±81°	±87°	0.910 374					

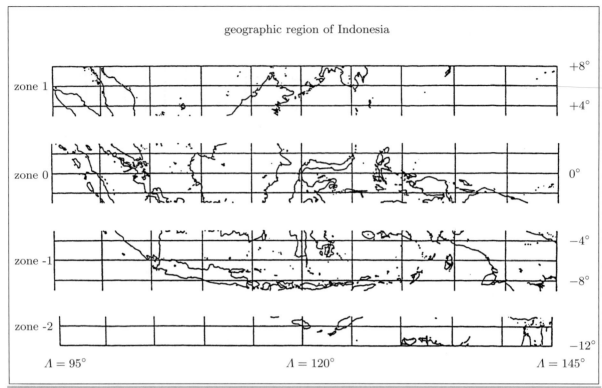

geographic region of Indonesia

zone 1   +8°

+4°

zone 0   0°

zone -1   −4°

−8°

zone -2   −12°

$\Lambda = 95°$ $\Lambda = 120°$ $\Lambda = 145°$

Fig. I.3. UPC, geographic region of Indonesia, strip $[95° < \Lambda < 145°] \times [-12° < \Phi < +8°]$ and strip width $\Phi_N - \Phi_S = 2\Delta\Phi = 6°$, the zones $0, \pm 1, -2$.

Finally, we present as an example the Airy optimal UPC for a strip system which extends to $-12°$ of southerly and $+8°$ of northerly latitude. Again this example can be considered as analogous to one given by E. Grafarend (1995) for the optimal transverse Mercator projection.

Example I.3 ($[95° < \Lambda < 145°] \times [-12° < \Phi < +8°]$).

For the Airy optimal UPC, we have chosen a strip width of $6°$ between $\Phi_S = -12°$ and $\Phi_N = 8°$ of southern and northern latitude, in particular, to match the geographic region of Indonesia. Once we refer to WGS 84, the strip system as well as the dilatation factor $\hat{\rho}_0$ per strip is illustrated by Fig. I.3, namely for the zones $0, \pm 1, -2$.

End of Example.

Furthermore, we computed by means of (I.34) the left principal stretches $\Lambda_1 = \Lambda_2 = \Lambda_S(\Phi_0, \Phi; \hat{\rho}_0)$ of each strip and plotted them in Figs. I.4–I.6. Moreover, Tables I.4 and I.5 give the latitude $\bar{\Phi}$ of each strip according to the strip width $\hat{\rho}_0$ and Φ_0 along which the mapping is equidistant. $\bar{\Phi}$ is determined by solving (I.34) for given strip width, $\hat{\rho}_0$ and Φ_0 under the condition that $\Lambda_1 = \Lambda_2 = 1$ holds. Obviously, the variation of the left principal stretch $\Lambda_1 = \Lambda_2 = \Lambda_S(\Phi_0, \Phi; \hat{\rho}_0)$ is small within the chosen strip. Alternatively, we may say that the radius of the left Tissot circle $\Lambda_1 = \Lambda_2 = \Lambda_S(\Phi_0, \Phi; \hat{\rho}_0)$ varies only for a small amount, a favourable result of the Airy optimal UPC.

Important! For regions with a East–West extension around the equator, the universal Mercator projection is Airy optimal. In contrast, the universal transverse Mercator projection is Airy optimal if the region extends North–South. The oblique Mercator projection is Airy optimal for an oblique extension of a region. The analogue statement holds for the Universal Poly-cylindric Projection (UPS), the Universal Transverse Polycylindric Projection (UTPC), and the Oblique Polycylindric Projection (OPC).

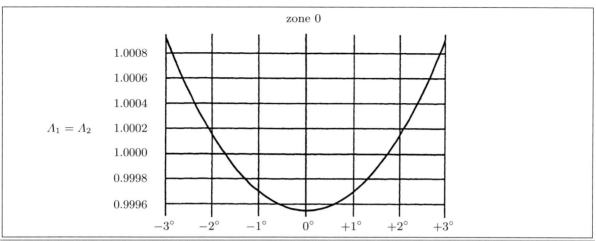

Fig. I.4. Airy optimal UPC, WGS 84, zone 0: variation of the radius of the left Tissot circle, the left principal stretches $\Lambda_1 = \Lambda_2 = \Lambda_S(\Phi_0 = 0, \Phi; \hat{\rho}_0)$.

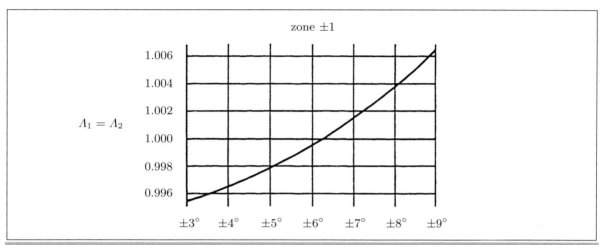

Fig. I.5. Airy optimal UPC, WGS 84, zone ± 1: variation of the radius of the left Tissot circle, the left principal stretches $\Lambda_1 = \Lambda_2 = \Lambda_S(\Phi_0 = \pm 6°, \Phi; \hat{\rho}_0)$.

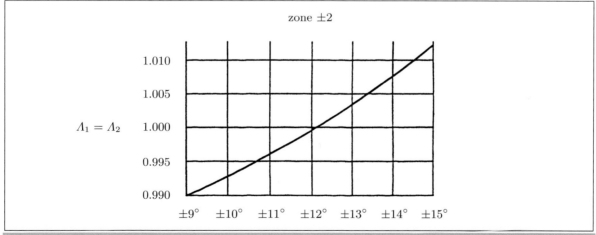

Fig. I.6. Airy optimal UPC, WGS 84, zone ± 2: variation of the radius of the left Tissot circle, the left principal stretches $\Lambda_1 = \Lambda_2 = \Lambda_S(\Phi_0 = \pm 12°, \Phi; \hat{\rho}_0)$.

Table I.4. Latitude $\bar{\Phi}$ for a symmetric strip $[\Lambda_\mathrm{W} = \Lambda_0 - \Delta\Lambda, \Lambda_0 + \Delta\Lambda = \Lambda_\mathrm{E}] \times [\Phi_\mathrm{S} = \Phi_0 - \Delta\Phi, \Phi_0 + \Delta\Phi = \Phi_\mathrm{N}]$ along which the generalized UPC is equidistant, WGS 84, strip width $3°$, $\Delta\Phi = 1.5°$.

zone	Φ_0	$\bar{\Phi}$	zone	Φ_0	$\bar{\Phi}$	zone	Φ_0	$\bar{\Phi}$	zone	Φ_0	$\bar{\Phi}$
0	$0°$	$\pm 0°51'58''$	± 1	$\pm 3°$	$\pm 3°7'23''$	± 2	$\pm 6°$	$\pm 6°3'48''$	± 3	$\pm 9°$	$\pm 9°2'36''$
± 4	$\pm 12°$	$\pm 12°2'1''$	± 5	$\pm 15°$	$\pm 15°1'41''$	± 6	$\pm 18°$	$\pm 18°1'28''$	± 7	$\pm 21°$	$\pm 21°1'20''$
± 8	$\pm 24°$	$\pm 24°1'14''$	± 9	$\pm 27°$	$\pm 27°1'11''$	± 10	$\pm 30°$	$\pm 30°1'8''$	± 11	$\pm 33°$	$\pm 33°1'7''$
± 12	$\pm 36°$	$\pm 36°1'7''$	± 13	$\pm 39°$	$\pm 39°1'8''$	± 14	$\pm 42°$	$\pm 42°1'9''$	± 15	$\pm 45°$	$\pm 45°1'11''$
± 16	$\pm 48°$	$\pm 48°1'14''$	± 17	$\pm 51°$	$\pm 51°1'18''$	± 18	$\pm 54°$	$\pm 54°1'23''$	± 19	$\pm 57°$	$\pm 57°1'28''$
± 20	$\pm 60°$	$\pm 60°1'36''$	± 21	$\pm 63°$	$\pm 63°1'45''$	± 22	$\pm 66°$	$\pm 66°1'57''$	± 23	$\pm 69°$	$\pm 69°2'12''$
± 24	$\pm 72°$	$\pm 72°2'33''$	± 25	$\pm 75°$	$\pm 75°3'3''$	± 26	$\pm 78°$	$\pm 78°3'48''$	± 27	$\pm 81°$	$\pm 81°5'4''$
± 28	$\pm 84°$	$\pm 84°7'39''$	± 29	$\pm 87°$	$\pm 87°16'10''$						

Table I.5. Latitude $\bar{\Phi}$ for a symmetric strip $[\Lambda_\mathrm{W} = \Lambda_0 - \Delta\Lambda, \Lambda_0 + \Delta\Lambda = \Lambda_\mathrm{E}] \times [\Phi_\mathrm{S} = \Phi_0 - \Delta\Phi, \Phi_0 + \Delta\Phi = \Phi_\mathrm{N}]$ along which the generalized UPC is equidistant, WGS 84, strip width $6°$, $\Delta\Phi = 3°$.

zone	Φ_0	$\bar{\Phi}$	zone	Φ_0	$\bar{\Phi}$	zone	Φ_0	$\bar{\Phi}$	zone	Φ_0	$\bar{\Phi}$
0	$0°$	$\pm 1°43'56''$	± 1	$\pm 6°$	$\pm 6°14'59''$	± 2	$\pm 12°$	$\pm 12°8'2''$	± 3	$\pm 18°$	$\pm 18°5'52''$
± 4	$\pm 24°$	$\pm 24°4'57''$	± 5	$\pm 30°$	$\pm 30°4'34''$	± 6	$\pm 36°$	$\pm 36°4'29''$	± 7	$\pm 42°$	$\pm 42°4'37''$
± 8	$\pm 48°$	$\pm 48°4'57''$	± 9	$\pm 54°$	$\pm 54°5'30''$	± 10	$\pm 60°$	$\pm 60°6'24''$	± 11	$\pm 66°$	$\pm 66°7'49''$
± 12	$\pm 72°$	$\pm 72°10'16''$	± 13	$\pm 78°$	$\pm 78°15'23''$	± 14	$\pm 84°$	$\pm 84°32'22''$			

J Gauss surface normal coordinates in geometry and gravity

Three-dimensional geodesy, minimal distance mapping, geometric heights. Reference plane, reference sphere reference ellipsoid-of-revolution, reference triaxial ellipsoid.

With the advent of artifical satellites in an Earth-bound orbit, geodesists succeeded to position points of the topographic surface $\mathbb{T}^2$ by a set of $\{X, Y, Z\} \in \mathbb{R}^3$ coordinates in a three-dimensional reference frame at the mass center of the Earth oriented along the equatorial axes at some reference epoch $t_0 \in \mathbb{R}$. In particular, global positioning systems ("global problem solver": GPS), were responsible for the materialization of *three-dimensional geodesy* in an Euclidean space. Based upon a triple $\{X, Y, Z\} \in \mathbb{R}^3$ of coordinates new concepts for converting these coordinates into heights with respect to a reference surface have been developed.

Important!

In the geometry space, the triplet $\{X, Y, Z\} \in \mathbb{T}^2$ is transformed by a geodesic projection into geometric heights with respect to (i) a *reference plane* $\mathbb{P}^2$, (ii) a *reference sphere* $\mathbb{S}_r^2$, (iii) a *reference ellipsoid-of-revolution* $\mathbb{E}_{A_1, A_2}^2$, or (iv) a *reference triaxial ellipsoid* $\mathbb{E}_{A_1, A_2, A_3}^2$.

Important!

First, the geodesic projection is performed by a straight line as the geodesic in flat geometry space. Second, the special geodesic passing the point $\{X, Y, Z\} \in \mathbb{T}$ has been chosen which has minimal distance S to the reference surface. The length of the geodesic from $\{X, Y, Z\} \in \mathbb{T}$ to $\{\hat{x}, \hat{y}, \hat{z}\} \in \mathbb{P}^2$ or $\mathbb{S}_r^2$ or $\mathbb{E}_{A_1, A_2}^2$ or $\mathbb{E}_{A_1, A_2, A_3}^2$, in short, $\{\hat{x}, \hat{y}, \hat{z}\}$ being determined by the *minimal distance mapping*, constitute the projective height in geometry space, namely of type (i) planar, (ii) spherical, (iii) ellipsoidal, or (iv) triaxial ellipsoidal.

Section J-1.

By algebraic mean, Section J-1 outlines various step procedures to establish projection heights in geometry space. By means of *minimal distance mapping*, various computational steps, either forward or backwards, are reviewed depending on the nature of the projection surface. The projection surfaces include (i) the plane, (ii) the sphere, (iv) the ellipsoid-of-revolution, and (iv) the triaxial ellipsoid.

Section J-2.

More specific, Section J-2 reviews various algorithms of computing Gauss surface normal coordinates for the case of an ellipsoid-of-revolution. The highlight is the computational algorithm by means of *Gröbner basis* and the *Buchberger algorithm* in establishing an ideal for the polynomial solution for the minimum distance mapping. From the Baltic Sea Level Project, we refer to detailed solutions of twenty-one points varying from Finland, Sweden, Lithuania, Poland, and Germany and taking reference to the World Geodetic Datum 2000 with the $\{A_1, A_2\}$ data $A_1 = 6\,378\,136.602$ m (semi-major axis) and $A_2 = 6\,656\,751.860$ m (semi-minor axis) following E. Grafarend and A. Ardalan (1999).

Section J-3.

Finally, Section J-3 presents the computation of Gauss surface normal coordinates for the case of a triaxial ellipsoid. For the Earth, we compute the position and orientation, and from parameters of the best fitting triaxial ellipsoid, we chose the *geoid* as the ideal Earth figure closest to the *mean sea level*. This important result is extended to other celestial bodies of triaxial nature, namely for Moon, Mars, Phobos, Amalthea, Io, and Mimas.

J-1 Projective heights in geometry space: from planar/spherical to ellipsoidal mapping

First, we outline how points in $\{\mathbb{R}^3, \delta_{kl}\}$ are connected by *geodesics*, namely by straight lines which are derived from a *variational principle*. The general solution of the differential equations of a geodesic, in particular, in terms of an affine parameter of its length, is represented by a linear one-dimensional manifold embedded into $\{\mathbb{R}^3, \delta_{kl}\}$. Second, based upon geodesics in $\{\mathbb{R}^3, \delta_{kl}\}$, in Fig. J.1, we introduce the *orthogonal projection* of a point P (the peak of a mountain, the top of a tower) onto a horizontal plane through P_0 by $p = \pi(P)$ generating the height $\overline{pP}$ of P with respect to the plane $\mathbb{P}_0^2$ through P_0. Alternatively, we may interpret the orthogonal projection $p = \pi(P)$ along a geodesic/straight line as a *minimal distance mapping* of P with respect to P_0 generating $p = \pi(P)$. Third, by Fig. J.2, we illustrate the minimal distance mapping of a topographic point $P \in \mathbb{T}^2$ as an element of the topographic surface (two-dimensional Riemann manifold) of the Earth along a geodesic/straight line onto a plane $\mathbb{P}^2$ which may be chosen as the horizontal plane at some reference point. In this way, we generate the orthogonal projection $p = \pi(P)$ and the length of the shortest distance $\overline{pP}$, called the geometric height of P with respect to $\mathbb{P}^2$. The choice of the height $\overline{pP}$ is very popular in photogrammetric and engineering surveying. By contrast, by Figs. J.3 and J.4, we illustrate the minimal distance mapping of a point P onto $p \in \mathbb{S}_r^2$ or $p \in \mathbb{E}_{A_1,A_2}^2$ or $p \in \mathbb{E}_{A_1,A_2,A_3}^2$ along a geodesic/straight line through P. Let us here assume that the reference surface is no longer the plane $\mathbb{P}^2$, but the sphere $\mathbb{S}_r^2$ of radius r or the ellipsoid-of-revolution $\mathbb{E}_{A_1,A_2}^2$ of semi-major axis A_1 and semi-minor axis A_2 or the triaxial ellipsoid $\mathbb{E}_{A_1,A_2,A_3}^2$ with the axes $A_1 > A_2 > A_3$. Fourth, by Box J.1, let us here outline a variant of the analytical treatment of generating geometric heights with respect to (i) the plane $\mathbb{P}^2$, (ii) the sphere $\mathbb{S}_r^2$, (iii) the ellipsoid-of-revolution $\mathbb{E}_{A_1,A_2}^2$, and (iv) the triaxial ellipsoid $\mathbb{E}_{A_1,A_2,A_3}^2$ by the minimal distance principle.

Important!

By (J.15)–(J.18), the constraint Lagrangean is defined with respect to the Euclidean distance $\|\boldsymbol{X} - \boldsymbol{x}\|^2/2$ subject to $\boldsymbol{X} \in \mathbb{T}^2$, $\boldsymbol{x} \in \mathbb{P}^2$ or $\mathbb{S}_r^2$ or $\mathbb{E}_{A_1,A_2}^2$ or $\mathbb{E}_{A_1,A_2,A_3}^2$, and the following constraint. The point $\boldsymbol{x}$ is an element of the plane $\mathbb{P}^2$, the sphere $\mathbb{S}_r^2$, the ellipsoid-of-revolution $\mathbb{E}_{A_1,A_2}^2$, or the triaxial ellipsoid $\mathbb{E}_{A_1,A_2,A_3}^2$. The constraint enters the Lagrangean by a Lagrange multiplier Λ. The routine of constraint optimization is followed by (J.9)–(J.13) The focus is on the normal equations (J.10)–(J.13), which constitute a system of algebraic equations of second degree. A solution algorithm is outlined by (J.19)–(J.23). In order to guarantee a minimal distance solution, the solution points of the nonlinear equations of normal type have to be tested with respect to the second variation, i. e. the positivity of the Hesse matrix of second derivatives with respect to the unknown coordinates $\{x^1, x^2, x^3\}$ of $p = \pi(P)$.

First, we alternatively present a second variation of the contruction of projective heights in geometry space by a minimal distance mapping of a topographic point $\boldsymbol{X} \in \mathbb{T}^2$ onto a reference surface of type (i) plane $\mathbb{P}^2$, (ii) sphere $\mathbb{S}_r^2$, (iii) ellipsoid-of-revolution $\mathbb{E}_{A_1,A_2}^2$, and (iv) triaxial ellipsoid $\mathbb{E}_{A_1,A_2,A_3}^2$, namely based upon $\|\boldsymbol{X} - \boldsymbol{x}(u,v)\| = \text{ext.}$ $\boldsymbol{x}(u,v)$ indicates a suitable parameterization of the surfaces (i)–(iv) by means of coordinates $\{u,v\}$, which constitute a chart of the Riemannian manifold (i)–(iv).

Important!

The first variation $\delta\mathcal{L}(u,v) = \delta\|\boldsymbol{X} - \boldsymbol{x}(u,v)\| = 0$ leads to the normal equations (J.15)–(J.18), which establish the orthogonality of type $\boldsymbol{X} - \boldsymbol{x}(\tilde{u}, \tilde{v}) = h\boldsymbol{n}$ and $\partial\boldsymbol{x}/\partial u^\alpha(\tilde{u}, \tilde{v}) = \boldsymbol{t}_\alpha$, namely of the normal surfaces $\boldsymbol{n}$ and the surface tangent vector $\boldsymbol{t}_\alpha$ for all $\alpha \in \{1,2\}$. In particular, projective heights for (i) the plane $\mathbb{P}^2$ by (J.5), (ii) the sphere $\mathbb{S}_r^2$ by (J.6), (iii) the ellipsoid-of-revolution $\mathbb{E}_{A_1,A_2}^2$ by (J.7), and (iv) triaxial ellipsoid $\mathbb{E}_{A_1,A_2,A_3}^2$ by (J.8). With respect to the second variation, besides (J.15)–(J.19) as the necessary condition for a minimal distance mapping, (J.19)–(J.23) establishes the sufficiency condition.

The sufficiency condition has been interpreted by the matrices of the first and second fundamental form in E. Grafarend and P. Lohse (1991). In addition, we like to refer to N. Bartelme and P. Meissl (1975), W. Benning (1974), H. Fröhlich and H. H. Hansen (1976), B. Heck (2002), M. Heikkinen (1982), M. K. Paul (1973), P. O. Penev (1978), M. Pick (1985), H. Sünkel (1976), T. Vincenty (1976, 1980), recently J. Awange and E. Grafarend (2005).

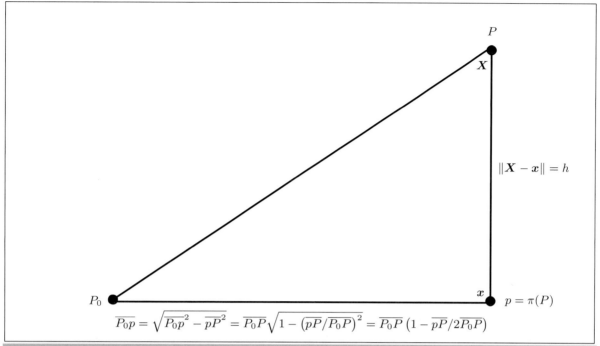

$$\overline{P_0p} = \sqrt{\overline{P_0p}^2 - \overline{pP}^2} = \overline{P_0P}\sqrt{1 - \left(\overline{pP}/\overline{P_0P}\right)^2} = \overline{P_0P}\left(1 - \overline{pP}/2\overline{P_0P}\right)$$

Fig. J.1. Projective heights in geometry space, orthogonal projection $p = \pi(P)$ of a topographic point $\boldsymbol{X} \in \mathbb{T}^2$ onto a local horizontal plane $\boldsymbol{X}_0 \in \mathbb{T}^2$, minimal distance mapping $\delta\|\boldsymbol{X} - \boldsymbol{x}\|^2 = 0$.

Indeed, for a suitable choice of surface parameters/surface coordinates of a chart, the unconstrained optimization problem may be preferable to constraint optimization, however, which we do not want to treat here.

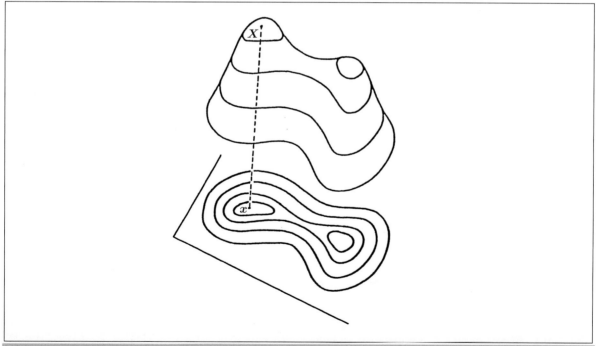

Fig. J.2. Projective heights in geometry space, minimal distance mapping with respect to a reference plane $\mathbb{P}^2$ at $\boldsymbol{X}_0 \in \mathbb{T}^2$, orthogonal projection $p = \pi(P)$.

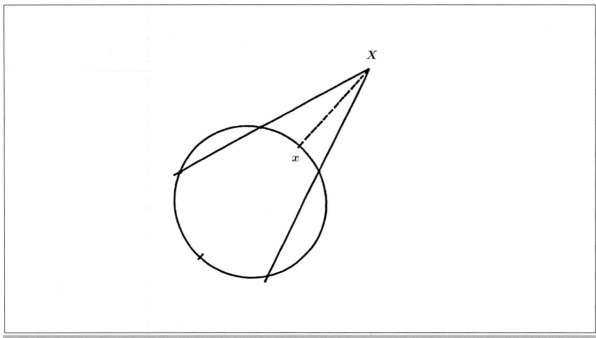

Fig. J.3. Projective heights in geometry space, minimal distance mapping with respect to a reference sphere $\mathbb{S}^2_r$, spherical heights h_S (length of the geodesic from $\boldsymbol{X} \in \mathbb{T}^2$ to $\boldsymbol{x} \in \mathbb{S}^2_r$).

Let us calculate the projection heights in geometry space, in detail, the minimal distance with respect to a reference surface of type (i) plane $\mathbb{P}^2$, (ii) sphere $\mathbb{S}^2_r$, (iii) ellipsoid-of-revolution $\mathbb{E}^2_{A_1,A_2}$, and (iv) triaxial ellipsoid $\mathbb{E}^2_{A_1,A_2,A_3}$ in Box J.2.

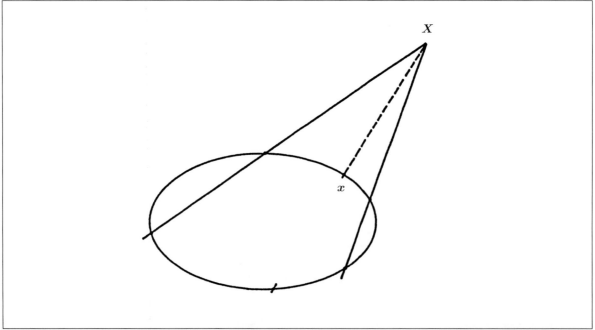

Fig. J.4. Projective heights in geometry space, minimal distance mapping with respect to a reference triaxial ellipsoid $\mathbb{E}^2_{A_1,A_2,A_3}$ or a reference ellipsoid-of-revolution $\mathbb{E}^2_{A_1,A_2}$, ellipsoidal heights h_E (length of the geodesic from $\boldsymbol{X} \in \mathbb{T}^2$ to $\boldsymbol{x} \in \mathbb{E}^2_{A_1,A_2}$ or $\mathbb{E}^2_{A_1,A_2,A_3}$).

Box J.1 (Projection heights in geometry space: minimal distance mapping).

Geodesics in geometry space:

$$\frac{d}{dt}\frac{\dot{x}^k}{\sqrt{\dot{x}^2 + \dot{y}^2 + \dot{z}^2}} = 0 \ . \tag{J.1}$$

Minimal distance mapping:

$$S := [X - x(u,v)]^2 + [Y - y(u,v)]^2 + [Z - Z(u,v)]^2 \ . \tag{J.2}$$

Reference surface:

(i)
$$\boldsymbol{x}(u,v) \in \mathbb{P}^2 \quad \text{(plane)} \ ,$$

(ii)
$$\boldsymbol{x}(u,v) \in \mathbb{S}_r^2 \quad \text{(sphere)} \ ,$$

(iii)   (J.3)
$$\boldsymbol{x}(u,v) \in \mathbb{E}_{A_1,A_2}^2 \quad \text{(ellipsoid-of-revolution)} \ ,$$

(iv)
$$\boldsymbol{x}(u,v) \in \mathbb{E}_{A_1,A_2,A_3}^2 \quad \text{(triaxial ellipsoid)} \ .$$

Heights in geometry space:

$$h := \| \boldsymbol{X} - \boldsymbol{x}(u,v) \| \ . \tag{J.4}$$

Box J.2 (Projection heights in geometry space: minimal distance mapping, formulae and relations).

Stationary functional:

(i)
$$\mathcal{L}(x^1,x^2,x^3,x^4) := \frac{1}{2}\|\boldsymbol{X} - \boldsymbol{x}\|^2 + \Lambda(a_1x + a_2y + a_3z + a_4) = \tag{J.5}$$
$$= \tfrac{1}{2}\left[(X - x^1)^2 + (Y - x^2)^2 + (Z - x^3)^2\right] + x^4(a_1x^1 + a_2x^2 + a_3x^3 + a_4) \ ,$$

(ii)
$$\mathcal{L}(x^1,x^2,x^3,x^4) := \|\boldsymbol{X} - \boldsymbol{x}\|^2 + \Lambda(x^2 + y^2 + z^2 - r^2) = \tag{J.6}$$
$$= (X - x^1)^2 + (Y - x^2)^2 + (Z - x^3)^2 + x^4[(x^1)^2 + (x^2)^2 + (x^3)^2 - r^2] \ ,$$

(iii)
$$\mathcal{L}(x^1,x^2,x^3,x^4) := \|\boldsymbol{X} - \boldsymbol{x}\|^2 + \Lambda\left[\frac{(x^1)^2 + (x^2)^2}{A_1^2} + \frac{(x^3)^2}{A_2^2} - 1\right]$$
$$\text{or} \tag{J.7}$$
$$\mathcal{L}(x^1,x^2,x^3,x^4) := (X - x^1)^2 + (Y - x^2)^2 + (Z - x^3)^2 + x^4\left(A_2^2[(x^1)^2 + (x^2)^2] + A_1^2(x^3)^2 - A_1^2A_2^2\right) \ ,$$

(iv)
$$\mathcal{L}(x^1,x^2,x^3,x^4) := \|\boldsymbol{X} - \boldsymbol{x}\|^2 + \Lambda\left[\frac{(x^1)^2}{A_1^2} + \frac{(x^2)^2}{A_2^2} + \frac{(x^3)^2}{A_3^2} - 1\right]$$
$$\text{or} \tag{J.8}$$
$$\mathcal{L}(x^1,x^2,x^3,x^4) := \|\boldsymbol{X} - \boldsymbol{x}\|^2 + x^4\left[A_2^2A_3^2x^2 + A_1^2A_3^2y^2 + A_1^2A_2^2z^2 - A_1^2A_2^2A_3^2\right] \ .$$

First variation:

$$\frac{\partial \mathcal{L}}{\partial x^\mu}(\hat{x}^\nu) = 0 \ \forall \ \mu, \nu \in \{1, 2, 3, 4\} \tag{J.9}$$

$$\Longleftrightarrow$$

(i) plane $\mathbb{P}^2$:
$$-(X - \hat{x}^1) + a_1\hat{x}^4 = 0 \ , \ \ -(Y - \hat{x}^2) + a_2\hat{x}^4 = 0 \ , \ \ -(Z - \hat{x}^3) + a_3\hat{x}^4 = 0 \ , \tag{J.10}$$
$$a_1\hat{x}^1 + a_2\hat{x}^2 + a_3\hat{x}^3 + a_4 = 0 \ ;$$

(ii) sphere $\mathbb{S}_r^2$:
$$-(X - \hat{x}^1) + \hat{x}^1\hat{x}^4 = 0 \ , \ \ -(Y - \hat{x}^2) + \hat{x}^2\hat{x}^4 = 0 \ , \ \ -(Z - \hat{x}^3) + \hat{x}^3\hat{x}^4 = 0 \ , \tag{J.11}$$
$$(\hat{x}^1)^2 + (\hat{x}^2)^2 + (\hat{x}^3)^2 - r^2 = 0 \ ;$$

(iii) ellipsoid-of-revolution $\mathbb{E}^2_{A_1, A_2}$:
$$-(X - \hat{x}^1) + A_2^2\hat{x}^1\hat{x}^4 = 0 \ , \ \ -(Y - \hat{x}^2) + A_2^2\hat{x}^2\hat{x}^4 = 0 \ , \ \ -(Z - \hat{x}^3) + A_1^2\hat{x}^3\hat{x}^4 = 0 \ , \tag{J.12}$$
$$A_2^2[(\hat{x}^1)^2 + (\hat{x}^2)^2] + A_1^2(\hat{x}^3)^2 - A_1^2A_2^2 = 0 \ ;$$

(iv) triaxial ellipsoid $\mathbb{E}^2_{A_1, A_2, A_2}$:
$$-(X - \hat{x}^1) + A_2^2A_3^2\hat{x}^1\hat{x}^4 = 0 \ , \ \ -(Y - \hat{x}^2) + A_1^2A_3^2\hat{x}^2\hat{x}^4 = 0 \ , \ \ -(Z - \hat{x}^3) + A_1^2A_2^2\hat{x}^3\hat{x}^4 = 0 \ , \tag{J.13}$$
$$A_2^2A_3^2(\hat{x}^1)^2 + A_1^2A_3^2(\hat{x}^2)^2 + A_1^2A_2^2(\hat{x}^3)^2 - A_1^2A_2^2A_3^2 = 0 \ .$$

Second variation.

The second variation decides about the solution of type "minimum" or "maximum" or "turning point".
In our case

$$\frac{1}{2}\frac{\partial^2 \mathcal{L}}{\partial x^k \partial x^l}(\hat{x}^\gamma) > 0 \tag{J.14}$$

$$\Longleftrightarrow$$

(i) plane $\mathbb{P}^2$:
$$\begin{bmatrix} +1 & 0 & 0 \\ 0 & +1 & 0 \\ 0 & 0 & +1 \end{bmatrix} > 0 \ ; \tag{J.15}$$

(ii) sphere $\mathbb{S}_r^2$:
$$\begin{bmatrix} +1 + \hat{\Lambda} & 0 & 0 \\ 0 & +1 + \hat{\Lambda} & 0 \\ 0 & 0 & +1 + \hat{\Lambda} \end{bmatrix} > 0 \ ; \tag{J.16}$$

(iii) ellipsoid-of-revolution $\mathbb{E}^2_{A_1, A_2}$:
$$\begin{bmatrix} +1 + A_2^2\hat{\Lambda} & 0 & 0 \\ 0 & +1 + A_2^2\hat{\Lambda} & 0 \\ 0 & 0 & +1 + A_1^2\hat{\Lambda} \end{bmatrix} > 0 \ ; \tag{J.17}$$

(iv) triaxial ellipsoid $\mathbb{E}^2_{A_1, A_2, A_2}$:
$$\begin{bmatrix} +1 + A_?^2A_3^2 12\hat{\Lambda} & 0 & 0 \\ 0 & +1 + A_1^2A_3^2\hat{\Lambda} & 0 \\ 0 & 0 & +1 + A_1^2A_2^2\hat{\Lambda} \end{bmatrix} > 0 \ . \tag{J.18}$$

The solution algorithm presented in Box J.3 determines in the forward step from the first variational equations (i), (ii), and (iii) the quantities $\hat{x}^1$, $\hat{x}^2$, and $\hat{x}^3$, and inserts them afterwards into (iv) as a second forward step. The backward step is organized in first solving for $\hat{x}^4$ in a polynomial equation of type linear, quadratic, or order three. Second, we have to decide whether our solution fulfills the condition of positivity of the Hesse matrix of second derivatives in order to discriminate the non-admissible solutions.

Box J.3 (Solution algorithm).

Forward step:

(i) plane:

$$(a_1^2 + a_2^2 + a_3^2)\hat{\Lambda} = a_1 X + a_2 Y + a_3 Z + a_4 \; ;$$

(J.19)

(ii) sphere:

$$(\hat{\Lambda} + 1)^2 = (X^2 + Y^2 + Z^2)/r^2$$
$$\Rightarrow$$
$$\hat{\Lambda}_- = -1 - \sqrt{X^2 + Y^2 + Z^2}/r \; ,$$
$$\hat{\Lambda}_+ = -1 + \sqrt{X^2 + Y^2 + Z^2}/r \; ;$$

(J.20)

(iii) ellipsoid-of-revolution:

$$A_1^2 A_2^2 (1 + A_1^2 \hat{\Lambda})^2 (1 + A_2^2 \hat{\Lambda})^2 - A_2 (X^2 + Y^2)(1 + A_1^2 \hat{\Lambda})^2 - A_1^2 Z^2 (1 + A_2^2 \hat{\Lambda})^2 = 0 \; ;$$

(J.21)

(iv) triaxial ellipsoid:

$$A_2^2 A_3^2 X^2 (1 + A_1^2 A_3^2 \hat{\Lambda})^2 (1 + A_1^2 A_2^2 \hat{\Lambda})^2 +$$
$$+ A_1^2 A_3^2 Y^2 (1 + A_2^2 A_3^2 \hat{\Lambda})^2 (1 + A_1^2 A_2^2 \hat{\Lambda})^2 + A_1^2 A_2^2 Z^2 (1 + A_2^2 A_3^2 \hat{\Lambda})^2 (1 + A_1^2 A_3^2 \hat{\Lambda})^2 -$$
$$- (1 + A_2^2 A_3^2 \hat{\Lambda})^2 (1 + A_1^2 A_3^2 \hat{\Lambda})^2 (1 + A_1^2 A_2^2 \hat{\Lambda})^2 A_1^2 A_2^2 A_3^2 = 0$$

(J.22)

or

$$a_6 (\hat{\Lambda}^2)^3 + a_4 (\hat{\Lambda}^2)^2 + a_2 (\hat{\Lambda}^2) + a_0 = 0$$
(cubic equation for $\hat{\Lambda}$) .

(J.23)

Backward step:

"Reset $\hat{x}^4 \sim \hat{\Lambda}$ into (i), (ii), and (iii), and solve the three equations for $\hat{x}^1$, $\hat{x}^2$, and $\hat{x}^3$, and test the condition of positivity of the Hesse matrix in order to discriminate the admissible solutions."

J-2 Gauss surface normal coordinates: case study ellipsoid-of-revolution

First, we review surface normal coordinates for the ellipsoid-of-revolution. Second, we extend the derivation to three-dimensional surface normal coordinates in terms of the forward transformation as well as of the backward transformation by means of the constraint minimum distance mapping in terms of the Buchberger algorithm.

J-21 Review of surface normal coordinates for the ellipsoid-of-revolution

The coordinates of the ellipsoid-of-revolution of type {ellipsoidal longitude, ellipsoidal latitude, ellipsoidal height} are *surface normal coordinates* in the following sense. They are founded on the famous *Gauss map* of the surface normal vector $\boldsymbol{\nu}(\boldsymbol{x})$ of the ellipsoid-of-revolution (J.24) in terms of the semi-major axis $A_1 > A_2$ and of the semi-minor axis $A_2 < A_1$. $\boldsymbol{\nu}(\boldsymbol{x})$ is also called *normal field*. Compare with Fig. J.5.

$$\mathbb{E}^2_{A_1\ A_2} := \left\{ \boldsymbol{x} \in \mathbb{R}^3 \,\middle|\, f(x\ y\ z) := \tfrac{x^2+y^2}{A_1^2} + \tfrac{z^2}{A_2^2} - 1 = 0 \ \mathbb{R}^+ \ni A_1 > A_2 \in \mathbb{R}^+ \right\} \tag{J.24}$$

Definition J.1 (Gauss map).

The spherical image of the surface norm vector $\boldsymbol{\nu}(l\ b) \in \mathbb{S}^2 := \{\boldsymbol{x} \in \mathbb{R}^3 \mid x^2 + y^2 + z^2 - 1 = 0\}$ of the ellipsoid-of-revolution $\mathbb{E}^2_{A_1\ A_2} \subset \mathbb{R}^3$ is defined as (J.25) with respect to the orthonormal left basis $\{e_1\ e_2\ e_3 \mid \mathcal{O}\}$ in the origin $\mathcal{O}$ of the ellipsoid-of-revolution.

$$\boldsymbol{\nu}(l\ b) = e_1 \cos b \cos l + e_2 \cos b \sin l + e_3 \sin b =$$

$$= [e_1\ e_2\ e_3] \begin{bmatrix} \cos b \cos l \\ \cos b \sin l \\ \sin b \end{bmatrix} \in N_{lb}\mathbb{E}^2_{A_1\ A_2} \tag{J.25}$$

$\{l\ b\}$ are called *surface normal longitude* and *surface normal latitude*, respectively. The orthonormal basis vector with respect to the origin $\mathcal{O}$ spans a three-dimensional Euclidean space.

End of Definition.

Question: "How can we find with given parameterized structure of the surface normal vector $\boldsymbol{\nu}(l\ b)$ the set of functions $\{x(l\ b)\ y(l\ b)\ z(l\ b)\}$ of the embedding of the ellipsoid-of-revolution $f(x\ y\ z) := (x^2+y^2)\ A_1^2 + z^2\ A_2^2$?" Answer: "Starting from the gradient grad of the of the gradient function $f(x\ y\ z)$, this set of functions is immediately derived as shown by the calculations that follow."

The surface normal vector of an algebraic surface ("polynom representation") has the representation (J.26), where $\|\mathrm{grad}f(x\ y\ z)\|$ is identified as the l_2 norm of Euclidean length of the gradient function $f(x\ y\ z)$. In detail, we compute (J.27).

$$\boldsymbol{\nu}(x\ y\ z) := \frac{\mathrm{grad}f(x\ y\ z)}{\|\mathrm{grad}f(x\ y\ z)\|_2} \tag{J.26}$$

$$\mathrm{grad}f(x\ y\ z) = [e_1\ e_2\ e_3] \begin{bmatrix} 2x\ A_1^2 \\ 2y\ A_1^2 \\ 2z\ A_2^2 \end{bmatrix} \tag{J.27}$$

$$\|\mathrm{grad}f(x\ y\ z)\| = \frac{2\sqrt{A_2^4(x^2+y^2) + A_1^4 z^2}}{A_1^2 A_2^2}$$

We need the relative eccentricity of the ellipsoid-of-revolution (J.28) for representing finally the surface normal vector.

$$E^2 := \frac{A_1^2 - A_2^2}{A_1^2} \quad \text{or} \quad \frac{1}{1 - E^2} =: \frac{A_1^2}{A_2^2} \tag{J.28}$$

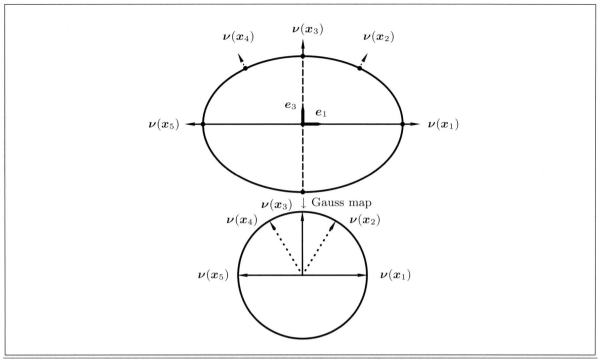

Fig. J.5. Vertical section of the ellipsoid-of-revolution $\mathbb{E}^2_{A_1,A_2} \subset \mathbb{R}^3$ and the spherical image of the surface normal vector (Gauss map) $\boldsymbol{\nu}(l,b) \in \mathbb{S}^2$, position vectors $\boldsymbol{x}_1, \ldots, \boldsymbol{x}_5$ and associated surface normal vectors $\boldsymbol{\nu}(\boldsymbol{x}_1) = \boldsymbol{\nu}_1, \ldots, \boldsymbol{\nu}(\boldsymbol{x}_5) = \boldsymbol{\nu}_5$.

Lemma J.2 ($\mathbb{E}^2_{A_1,A_2}$ surface normal vector).

Let $f(x,y,z) := (x^2+y^2)/A_1^2 + z^2/A_2^2 - 1$ be a polynomial representation of the ellipsoid-of-revolution. Then (J.29)–(J.31) are Cartesian forms of the surface normal vector.

$$\boldsymbol{\nu}(x,y,z) := \frac{\operatorname{grad} f(x,y,z)}{\|\operatorname{grad} f(x,y,z)\|_2} \ , \tag{J.29}$$

$$\boldsymbol{\nu}(x,y,z) := [\boldsymbol{e_1}, \boldsymbol{e_2}, \boldsymbol{e_3}] \begin{bmatrix} 2x/A_1^2 \\ 2y/A_1^2 \\ 2z/A_2^2 \end{bmatrix} \frac{A_1^2 A_2^2}{\sqrt{A_2^4(x^2+y^2) + a_1^4 Z^2}} \ , \tag{J.30}$$

$$\boldsymbol{\nu}(x,y,z) := [\boldsymbol{e_1}, \boldsymbol{e_2}, \boldsymbol{e_3}] \begin{bmatrix} \frac{x}{\sqrt{x^2+y^2+z^2/(1-E^2)}} \\ \frac{y}{\sqrt{x^2+y^2+z^2/(1-E^2)}} \\ \frac{z}{\sqrt{(1-E^2)^2(x^2+y^2)+z^2}} \end{bmatrix} . \tag{J.31}$$

End of Lemma.

The four steps that are outlined in Box J.4 are needed to derive the desired representation. Comparing the spherical representation and the Cartesian representation in terms of the surface normal vector (J.32) or (J.33) under the side condition (J.34), we are able to derive an *isometric embedding* $\{x(l,b), y(l,b), z(l,b)\}$ of $\mathbb{E}^2_{A_1,A_2} \subset \mathbb{R}^3$.

$$\boldsymbol{\nu}(l\ b) = \boldsymbol{\nu}(x\ y\ z) \tag{J.32}$$

$$\begin{bmatrix} \cos b \cos l \\ \cos b \sin l \\ \sin b \end{bmatrix} = \begin{bmatrix} x \\ y \\ z\ (1 - E^2) \end{bmatrix} \frac{1 - E^2}{\sqrt{(1 - E^2)^2(x^2 + y^2) + z^2}} \tag{J.33}$$

$$f(x\ y\ z) = \frac{x^2 + y^2}{A_1^2} + \frac{z^2}{A_2^2} - 1 \quad \text{or} \quad x^2 + y^2 + \frac{z^2}{1 - E^2} = A_1^2 \tag{J.34}$$

Box J.4 (Four steps towards an isometric embedding $\{x(l\ b)\ y(l\ b)\ z(l\ b)\}$ of $\mathbb{E}_{A_1,A_2}^2 \subset \mathbb{R}^3$).

The first operation. Use (J.33), the first and second equation, and add

$$\cos^2 b = (1 - E^2)^2 \frac{x^2 + y^2}{(1 - E^2)^2(x^2 + y^2) + z^2}$$
$$\Rightarrow \tag{J.35}$$
$$x^2 + y^2 = \left[x^2 + y^2 + z^2\ (1 - E^2)^2 \right] \cos^2 b$$

The second operation. Use (J.33), the third equation:

$$\sin^2 b = \frac{z^2}{(1 - E^2)^2(x^2 + y^2) + z^2}$$
$$\Rightarrow \tag{J.36}$$
$$z^2\ (1 - E^2) = (1 - E^2) \left[x^2 + y^2 + z^2\ (1 - E^2)^2 \right] \sin^2 b$$

The third operation. Replace the terms in (J.34) by (J.35) and (J.36):

$$A_1^2 = x^2 + y^2 + z^2\ (1 - E^2) = \left[x^2 + y^2 + z^2\ (1 - E^2)^2 \right](1 - E^2 \sin^2 b)$$
$$\Rightarrow$$
$$x^2 + y^2 + z^2\ (1 - E^2)^2 = \frac{A_1^2}{1 - E^2 \sin^2 b} \tag{J.37}$$
$$\Rightarrow$$
$$\sqrt{x^2 + y^2 + z^2\ (1 - E^2)^2} = A_1\ \sqrt{1 - E^2 \sin^2 b}$$

The fourth operation. Solve (J.33) for $\{x\ y\ z\}$ and replace the square root by (J.37):

$$x = \sqrt{x^2 + y^2 + z^2\ (1 - E^2)^2} \cos b \cos l$$
$$y = \sqrt{x^2 + y^2 + z^2\ (1 - E^2)^2} \cos b \sin l \tag{J.38}$$
$$z = (1 - E^2)\sqrt{x^2 + y^2 + z^2\ (1 - E^2)^2} \sin b$$

The result is summarized in Lemma J.3.

Lemma J.3 ($\mathbb{E}^2_{A_1,A_2}$ surface normal coordinates).

Let the surface normal vector $\boldsymbol{\nu}(l,b)$ of an ellipsoid-of-revolution be the spherical image (Gauss map) represented by (J.25). Then (J.39) and (J.40) hold and (J.41) is the parameter representation of the ellipsoid-of-revolution $\mathbb{E}^2_{A_1,A_2}$ in terms of *surface normal coordinates*, namely in terms of surface normal longitude and surface normal latitude.

$$\boldsymbol{\nu}(l,b) = \boldsymbol{e}_1 \frac{A_1}{\sqrt{1-E^2\sin^2 b}} \cos b \cos l + \boldsymbol{e}_2 \frac{A_1}{\sqrt{1-E^2\sin^2 b}} \cos b \sin l +$$

$$+\boldsymbol{e}_3 \frac{A_1(1-E^2)}{\sqrt{1-E^2\sin^2 b}} \sin b \ , \tag{J.39}$$

$$\boldsymbol{x}(l,b) = [\boldsymbol{e_1},\boldsymbol{e_2},\boldsymbol{e_3}] \begin{bmatrix} \cos b \cos l \\ \cos b \sin l \\ (1-E^2)\sin b \end{bmatrix} \frac{A_1}{\sqrt{1-E^2\sin^2 b}} \ , \tag{J.40}$$

$$x(l,b) = \frac{A_1 \cos b \cos l}{\sqrt{1-E^2\sin^2 b}} \ , \quad y(l,b) = \frac{A_1 \cos b \sin l}{\sqrt{1-E^2\sin^2 b}} \ ,$$

$$\tag{J.41}$$

$$z(l,b) = \frac{A_1(1-E^2)\sin b}{\sqrt{1-E^2\sin^2 b}} \ .$$

End of Lemma.

We have the proof that the normal field $\boldsymbol{\nu}(l,b)$ is not a gradient field, in consequence an anholonomic variable, to the reader. In terms of surface normal coordinates, the differential invariants $\{I, II, III\}$ take a simple form, namely

$$I \sim \mathrm{d}\boldsymbol{\mu}^2 := \langle \boldsymbol{\mu}_l \mathrm{d}l + \boldsymbol{\mu}_b \mathrm{d}b \,|\, \boldsymbol{\mu}_l \mathrm{d}l + \boldsymbol{\mu}_b \mathrm{d}b \rangle \ , \tag{J.42}$$

$$II \sim -\langle \mathrm{d}\boldsymbol{\mu} \,|\, \mathrm{d}\boldsymbol{\nu} \rangle := \langle \boldsymbol{\mu}_l \mathrm{d}l + \boldsymbol{\mu}_b \mathrm{d}b \,|\, \boldsymbol{\nu}_l \mathrm{d}l + \boldsymbol{\nu}_b \mathrm{d}b \rangle \ , \tag{J.43}$$

$$III \sim \mathrm{d}\boldsymbol{\nu}^2 := \langle \boldsymbol{\nu}_l \mathrm{d}l + \boldsymbol{\nu}_b \mathrm{d}b \,|\, \boldsymbol{\nu}_l \mathrm{d}l + \boldsymbol{\nu}_b \mathrm{d}b \rangle \ , \tag{J.44}$$

or

$$I \sim \mathrm{d}\boldsymbol{\mu}^2 = \frac{A_1^2 \cos^2 b}{1-E^2\sin^2 b} \mathrm{d}l^2 + \frac{A_1^2(1-E^2)^2}{(1-E^2\sin^2 b)^{3/2}} \mathrm{d}b^2 \ , \tag{J.45}$$

$$II \sim -\langle \mathrm{d}\boldsymbol{\mu} \,|\, \mathrm{d}\boldsymbol{\nu} \rangle = \frac{A_1^2 \cos^2 b}{\sqrt{1-E^2\sin^2 b}} \mathrm{d}l^2 + \frac{A_1(1-E^2)}{(1-E^2\sin^2 b)^{3/2}} \mathrm{d}b^2 \ , \tag{J.46}$$

$$III \sim \mathrm{d}\boldsymbol{\nu}^2 = \cos^2 b \, \mathrm{d}l^2 + \mathrm{d}b^2 \ . \tag{J.47}$$

Corollary J.4 ($\mathbb{E}^2_{A_1,A_2}$ Gauss differential invariants).

The Gauss differential invariants $\{I, II, III\}$ of the ellipsoid-of-revolution $\mathbb{E}^2_{A_1,A_2}$ are characterized by Gauss surface normal coordinates represented by (J.42), (J.43), and (J.44). Especially, the Gauss map $N\mathbb{E}^2_{A_1,A_2} \mapsto \mathbb{S}^2$ has the spherical metric III.

End of Corollary.

J-22 Buchberger algorithm of forming a constraint minimum distance mapping

The forward transformation of Gauss coordinates of an ellipsoid-of-revolution takes the form (J.48) and (J.49) illustrated by Fig. J.6. The triplet {surface normal longitude, surface normal latitude, surface normal height} describes the position of a point $\boldsymbol{X}(L, B, H)$ where the surface normal height $H(L, B)$ is a given function of longitude and latitude.

Box J.5 (Forward transformation of Gauss coordinates of an ellipsoid-of-revolution).

$$\boldsymbol{X}(L, B, H) = +\boldsymbol{e}_1 \left[\frac{A_1}{\sqrt{1 - E^2 \sin^2 B}} + H(L, B) \right] \cos B \cos L$$

$$+\boldsymbol{e}_2 \left[\frac{A_1}{\sqrt{1 - E^2 \sin^2 B}} + H(L, B) \right] \cos B \sin L \qquad \text{(J.48)}$$

$$+\boldsymbol{e}_3 \left[\frac{A_1(1 - E^2)}{\sqrt{1 - E^2 \sin^2 B}} + H(L, B) \right] \sin B ,$$

$$X = \left[\frac{A_1}{\sqrt{1 - E^2 \sin^2 B}} + H(L, B) \right] \cos B \cos L ,$$

$$Y = \left[\frac{A_1}{\sqrt{1 - E^2 \sin^2 B}} + H(L, B) \right] \cos B \sin L , \qquad \text{(J.49)}$$

$$Z = \left[\frac{A_1(1 - E^2)}{\sqrt{1 - E^2 \sin^2 B}} + H(L, B) \right] \sin B .$$

In order to solve in algorithmic form the characteristic normal equation by means of a constraint minimum distance mapping given earlier, we outline the first and second forward step of reduction, which leads us to a univariate polynomial equation of fourth order in terms of Lagrangean multipliers. As soon as we have implemented standard software to solve the fourth order equation, we continue to determine with the backward step the Cartesian coordinates $\{x_1, x_2, x_3\}$ of the point $p \in \mathbb{E}^2_{A_1,A_2}$, which has been generated by means of the *minimum distance mapping* of a point $P \in \mathbb{T}$ to $p \in \mathbb{E}^2_{A_1,A_2}$. Finally, by means of Box J.5, we convert the Cartesian coordinates $\{X, Y, Z\} \in \mathbb{T}^2$ and $\{x_1, x_2, x_3\} \in \mathbb{E}^2_{A_1,A_2}$ to Gauss ellipsoidal coordinates $\{L, B, H\}$.

Without the various forward and backward reduction steps, we could automatically generate an equivalent algorithm for solving the normal equations in a closed form by means of *Gröbner basis* and the *Buchberger algorithm* (D. Cox., J. Little, and D. O'Shea (1996), T. Becker and V. Weispfenning (1998), B. Sturmfels (1996) and R. Zippel (1993)). Let us write the *Ideal* of the polynomials in lexicographic order $x_1 > x_2 > x_3 > x_4$ (read: x_1 before x_2 before x_3 before x_4) into Box J.6. The Gröbner basis of the *Ideal* characteristic for the minimum distance mapping problem can be computed either by MATHEMATICA software or by MAPLE software.

Box J.6 (Algorithm for solving the normal equations of the constraint minimum distance mapping).

First forward step. Solve (i), (ii), and (iii) for $\{x_1, x_2, x_3\}$:

(i)
$$x_1^{\wedge}(1 + b^2 x_4^{\wedge}) = X \;\Rightarrow\; x_1^{\wedge} = \frac{X}{1 + b^2 x_4^{\wedge}}\ ,$$

(ii)
$$x_2^{\wedge}(1 + b^2 x_4^{\wedge}) = Y \;\Rightarrow\; x_2^{\wedge} = \frac{Y}{1 + b^2 x_4^{\wedge}}\ ,$$

(iii)
$$x_3^{\wedge}(1 + a^2 x_4^{\wedge}) = Z \;\Rightarrow\; x_3^{\wedge} = \frac{Z}{1 + a^2 x_4^{\wedge}}\ .$$

(J.50)

Second forward step. Substitute $\{x_1^{\wedge}, x_2^{\wedge}, x_3^{\wedge}, x_4^{\wedge}\}$:

$$x_1^{\wedge 2} + x_2^{\wedge 2} = \frac{1}{(1 + b^2 x_4^{\wedge})^2}\left(X^2 + Y^2\right)\ ,\quad x_3^{\wedge 2} = \frac{1}{(1 + a^2 x_4^{\wedge})^2} Z^2\ ,$$

$$b^2\left(x_1^{\wedge 2} + x_2^{\wedge 2}\right) + a^2 x_3^{\wedge 2} - a^2 b^2 \Leftrightarrow b^2 \frac{X^2 + Y^2}{(1 + b^2 x_4^{\wedge})^2} + a^2 \frac{Z^2}{(1 + a^2 x_4^{\wedge})^2} - a^2 b^2 = 0\ .$$

(J.51)

The characteristic quadratic equation.
Multiply the rational equation of constraint by $(1 + a^2 x_4)^2 (1 + b^2 x_4)^2$:

$$b^2(1 + a^2 x_4)^2\left(X^2 + Y^2\right) + a^2(1 + b^2 x_4)^2 Z^2 - a^2 b^2(1 + a^2 x_4)^2(1 + b^2 x_4)^2 = 0$$

$$\Leftrightarrow$$

$$(1 + 2a^2 x_4 + a^4 x_4^2)b^2\left(X^2 + Y^2\right) + (1 + 2b^2 x_4 + b^4 x_4^2)a^2 Z^2 -$$
$$- a^2 b^2(1 + 2a^2 x_4 + a^4 x_4^2)(1 + 2b^2 x_4 + b^4 x_4^2) = 0\ ,$$

(J.52)

$$-x_4^4 a^6 b^6 - 2x_4^3 a^4 b^4(a^2 + b^2) + x_4^2 a^2 b^2[a^2\left(X^2 + Y^2\right) + b^2 Z^2 - 4a^2 b^2 - a^4 - b^4] +$$
$$+ 2x_4 a^2 b^2\left(X^2 + Y^2 + Z^2\right) + b^2\left(X^2 + Y^2\right) + a^2 Z^2 - a^2 b^2 = 0\ ,$$

(J.53)

$$x_4^4 + 2x_4^3 \frac{a^2 + b^2}{a^2 b^2} + x_4^2 \frac{4a^2 b^2 + a^4 + b^4 - a^2\left(X^2 + Y^2\right) - b^2 Z^2}{a^4 b^4} -$$
$$- 2x_4 \frac{\left(X^2 + Y^2 + Z^2\right)}{a^4 b^4} - \frac{b^2\left(X^2 + Y^2\right) + a^2 Z^2 - a^2 b^2}{a^6 b^6} = 0\ .$$

(J.54)

Backward step. Substitute $\{x_1^{\wedge}(x_4^{\wedge}), x_2^{\wedge}(x_4^{\wedge}), x_3^{\wedge}(x_4^{\wedge})\}$:

$$x_1^{\wedge} = (1 + b^2 x_4^{\wedge})^{-1} X\ ,\quad x_2^{\wedge} = (1 + b^2 x_4^{\wedge})^{-1} Y\ ,\quad x_3^{\wedge} = (1 + a^2 x_4^{\wedge})^{-1} Z\ .$$

(J.55)

Test:

$$\Lambda_1 = \Lambda_2 = 1 + b^2 x_4^{\wedge} > 0,\ \Lambda_3 = 1 + a^2 x_4^{\wedge} > 0 \text{ if } \Lambda_1 = \Lambda_2 > 0 \text{ and } \Lambda_3 > 0 \text{ then end.}$$

Here, we used MATHEMATICA 2.2 for DOS 387. The executable command is "GroebnerBasis [Polynomials, Variables]" in a specified ordering. The fourteen elements of the computed Gröbner basis can be interpreted as following. The first equation is a univariate polynomial of order four in the Lagrange multiplier identical to (J.53). As soon as we substitute the admissible value x_4 into the linear equations (J.61), (J.65), and (J.69), we obtain the unknowns $\{x_1\ x_2\ x_3\} = \{x\ y\ z\}$.

Box J.7 (Closed form solution).

$$\{X, Y, Z\} \in \mathbb{T}^2, \{x_1, x_2, x_3\} \in \mathbb{E}^2_{a,a,b} \text{ to } \{L, B, H\}.$$

Pythagoras in three dimensions:

$$H := \sqrt{(X - x_1)^2 + (Y - x_2)^2 + (Z - x_3)^2} \ . \tag{J.56}$$

Convert $\{x_1, x_2, x_3\}$ and $\{X, Y, Z\}$ to $\{L, B\}$:

$$\tan L = \frac{Y - x_2}{X - x_1} = \frac{Y - y}{X - x} \ , \quad \tan B = \frac{Z - x_3}{\sqrt{(X - x_1)^2 + (Y - x_2)^2}} = \frac{Z - x_3}{\sqrt{(X - x)^2 + (Y - y)^2}} \ . \tag{J.57}$$

Box J.8 (Buchberger algorithm, Gröbner basis for solving the normal equations of the constraint minimum distance mapping).

$$\text{Ideal } I :=$$
$$:= \left[x_1 + b^2 x_1 x_4 - X, x_2 + b^2 x_2 x_4 - Y, x_3 + a^2 x_3 x_4 - Z, b^2 x_1^2 + b^2 x_2^2 - a^2 x_3^2 - a^2 b^2\right]$$

$$\text{Groebner basis } G :=$$
$$:= \left[\{x_1 + b^2 x_1 x_4 - X, x_2 + b^2 x_2 x_4 - Y, x_3 + a^2 x_3 x_4 - Z, b^2 x_1^2 + b^2 x_2^2 - a^2 x_3^2 - a^2 b^2\}\{x_1, x_2, x_3, x_4\}\right]$$

Computed Gröbner basis for the minimum distance mapping problem:

$$\begin{aligned} &a^2 b^2 x_4^4 + (2a^6 b^4 + 2a^4 b^6)x_4^3 + (a^6 b^2 + 4a^4 b^4 + a^2 b^6 - a^4 b^2 X^2 - a^4 b^2 Y^2 - a^2 b^4 Z^2)x_4^2 + \\ &+ (2a^4 b^2 + 2a^2 b^4 - 2a^2 b^2 X^2 - 2a^2 b^2 Y^2 - 2a^2 b^2 Z^2)x_4 + (a^2 b^2 - b^2 X^2 - b^2 Y^2 - a^2 Z^2) \ , \end{aligned} \tag{J.58}$$

$$\begin{aligned} &(a^4 Z - 2a^2 b^2 Z + b^4 Z)x_3 - a^6 b^6 x_4^3 - (2a^6 b^4 + a^4 b^6)x_4^2 - \\ &- (a^6 b^2 + 2a^4 b^4 - a^4 b^2 X^2 - a^4 b^2 Y^2 - a^2 b^4 Z^2)x_4 - a^2 b^4 + a^2 b^2 X^2 + a^2 b^2 Y^2 + 2a^2 b^2 Z^2 - b^4 Z^2 \ , \end{aligned} \tag{J.59}$$

$$\begin{aligned} &(2b^2 Z + b^4 x_4 Z - a^2 Z)x_3 + a^4 b^6 x_4^3 + (2a^4 b^4 + a^2 b^6)x_4^2 + \\ &+ (a^4 b^2 + 2a^2 b^4 - a^2 b^2 X^2 - a^2 b^2 Y^2 - b^4 Z^2)x_4 + a^2 b^2 - b^2 X^2 - b^2 Y^2 - 2b^2 Z^2 \ , \end{aligned} \tag{J.60}$$

$$(1 + a^2 x_4)x_3 - Z \ , \tag{J.61}$$

$$\begin{aligned} &(a^4 - 2a^2 b^2 + b^4)x_3^2 + (2a^2 b^2 Z - 2b^4 Z)x_3 - a^4 b^6 x_4^2 - \\ &- 2a^4 b^4 x_4 - a^4 b^2 + a^2 b^2 X^2 + a^2 b^2 Y^2 + b^4 Z^2 \ , \end{aligned} \tag{J.62}$$

$$\begin{aligned} &(2b^2 - a^2 + b^4 x_4)x_3^2 - a^2 Z x_3 + a^4 b^6 x_4^3 + (2a^4 b^4 + 2a^2 b^6)x_4^2 + \\ &+ (a^4 b^2 + 4a^2 b^4 - a^2 b^2 X^2 - a^2 b^2 Y^2 - b^4 Z^2)x_4 + 2a^2 b^2 - 2b^2 X^2 - 2bY^2 - 2b^2 Z^2 \ , \end{aligned} \tag{J.63}$$

$$(X^2 + Y^2)x_2 + a^2 b^4 Y x_4^2 + Y(a^2 b^2 - b^2 x_3^2 - b^2 Z x_3)x_4 + Y x_3^2 - Y^3 - Y Z x_3 - Y X^2 \ , \tag{J.64}$$

$$(1 + b^2 x_4)x_2 - Y \ , \tag{J.65}$$

$$(a^2 x_3 - b^2 x_3 + b^2 Z)x_2 - a^2 x_3 Y \ , \tag{J.66}$$

$$Y x_1 - X x_2 \ , \tag{J.67}$$

$$X x_1 + a^2 b^4 x_4^2 + (a^2 b^2 + b^2 x_3^2 - b^2 Z x_3)x_4 + x_3^2 - Z x_3 + Y x_2 - X^2 - Y^2 \ , \tag{J.68}$$

$$(1 + b^2 x_4)x_1 - X \ , \tag{J.69}$$

$$(a^2 x_3 - b^2 x_3 + b^2 Z)x_1 - a^2 X x_3 \ , \tag{J.70}$$

$$x_1^2 + a^2 b^4 x_4^2 + (2a^2 b^2 + b^2 x_3^2 - b^3 Z x_3)x_4 + 2x_3^2 - 2Z x_3 + x_2^2 - X^2 - Y^2 \ . \tag{J.71}$$

Let us adopt the World Geodetic Datum 2000 with the data "semi-major" axis $A_1 = 6\,378\,136.602\,\mathrm{m}$ and "semi-major axis" $A_2 = 6\,356\,751.860\,\mathrm{m}$ of the International Reference Ellipsoid (E. Grafarend and A. Ardalan 1999). Here, we take advantage of given Cartesian coordinates of twenty-one points of the topographic surface of the Earth presented in Table J.1. Compare with Fig. J.6.

Table J.1. Cartesian coordinates of topographic point (Baltic Sea Level Project).

station	$X\,[\mathrm{m}]$	$Y\,[\mathrm{m}]$	$Z\,[\mathrm{m}]$
Borkum (Ger)	3770667.9989	446076.4896	5107686.2085
Degerby (Fin)	2994064.9360	1112559.0570	5502241.3760
Furuögrund (Swe)	2527022.8721	981957.2890	5753940.9920
Hamina (Fin)	2795471.2067	1435427.7930	5531682.2031
Hanko (Fin)	2959210.9709	1254679.1202	5490594.4410
Helgoland (Gcr)	3706044.9443	513713.2151	5148193.4472
Helsinki (Fin)	2885137.3909	1342710.2301	5509039.1190
Kemi (Fin)	2397071.5771	1093330.3129	5789108.4470
Klagshamn (Swe)	3527585.7675	807513.8946	5234549.7020
Klaipeda (Lit)	3353590.2428	1302063.0141	5249159.4123
List/Sylt (Gcr)	3625339.9221	537853.8704	5202539.0255
Molas (Lit)	3358793.3811	1294907.4149	5247584.4010
Mäntyluoto (Fin)	2831096.7193	1113102.7637	5587165.0458
Raahe (Fin)	2494035.0244	1131370.9936	5740955.4096
Ratan (Swe)	2620087.6160	1000008.2649	5709322.5771
Spikarna (Swe)	2828573.4638	893623.7288	5627447.0693
Stockholm (Swe)	3101008.8620	1013021.0372	5462373.3830
Ustka (Pol)	3545014.3300	1073939.7720	5174949.9470
Vaasa (Fin)	2691307.2541	1063691.5238	5664806.3799
Visby(Swc)	3249304.4375	1073624.8912	5364363.0732
Ölands N. U. (Swe)	3295551.5710	1012564.9063	5348113.6687

From the algorithm of Box J.8, the first polynomial equation of fourth order of the Gröbner basis is obtained as (J.72). Numerical values are provided by Table J.2.

Table J.2. Polynomial coefficients of the univariate polynomial of order four in x_4.

point	c_0	c_1	c_2	c_3	c_4
1	$-2.3309099e+22$	$1.334253e+41$	$1.351627e+55$	$4.382358e+68$	$4.441958e+81$
2	$-1.142213e+22$	$1.3351890e+41$	$1.352005e+55$	$4.382358e+68$	$4.441958e+81$
3	$-1.720998e+22$	$1.335813e+41$	$1.352259e+55$	$4.382358e+68$	$4.441958e+81$
4	$-8.871288e+21$	$1.335264e+41$	$1.352035e+55$	$4.382358e+68$	$4.441958e+81$
5	$-1.308070e+22$	$1.335160e+41$	$1.351993e+55$	$4.382358e+68$	$4.441958e+81$
6	$-2.275210e+22$	$1.334345e+41$	$1.351665e+55$	$4.382358e+68$	$4.441958e+81$
7	$-1.272935e+22$	$1.335205e+41$	$1.352012e+55$	$4.382358e+68$	$4.441958e+81$
8	$-1.373946e+22$	$1.335906e+41$	$1.352296e+55$	$4.382358e+68$	$4.441958e+81$
9	$-1.981047e+22$	$1.334546e+41$	$1.351746e+55$	$4.382358e+68$	$4.441958e+81$
10	$-2.755981e+22$	$1.334574e+41$	$1.351758e+55$	$4.382358e+68$	$4.441958e+81$
11	$-2.330047e+22$	$1.334469e+41$	$1.351715e+55$	$4.382358e+68$	$4.441958e+81$
12	$-1.538357e+22$	$1.334580e+41$	$1.351759e+55$	$4.382358e+68$	$4.441958e+81$
13	$-1.117760e+22$	$1.335399e+41$	$1.352090e+55$	$4.382358e+68$	$4.441958e+81$
14	$-1.124559e+22$	$1.335785e+41$	$1.352246e+55$	$4.382358e+68$	$4.441958e+81$
15	$-1.200556e+22$	$1.335704e+41$	$1.352214e+55$	$4.382358e+68$	$4.441958e+81$
16	$-1.427443e+22$	$1.335496e+41$	$1.352130e+55$	$4.382358e+68$	$4.441958e+81$
17	$-1.836471e+22$	$1.335087e+41$	$1.351965e+55$	$4.382358e+68$	$4.441958e+81$
18	$-1.772332e+22$	$1.334410e+41$	$1.351690e+55$	$4.382358e+68$	$4.441958e+81$
19	$-1.012020e+22$	$1.335593e+41$	$1.352168e+55$	$4.382358e+68$	$4.441958e+81$
20	$-1.427711e+22$	$1.334856e+41$	$1.351870e+55$	$4.382358e+68$	$4.441958e+81$
21	$-1.644250e+22$	$1.334815e+41$	$1.351854e+55$	$4.382358e+68$	$4.441958e+81$

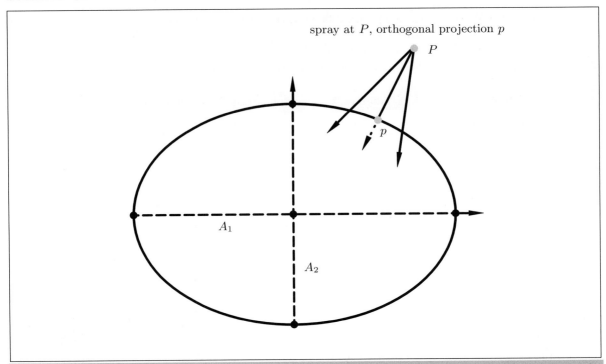

Fig. J.6. Minimum distance mapping of a point P on the Earth's topographic surface to a point p on the International Reference Ellipsoid $\mathbb{E}^2_{A_1, A_2}$.

$$c_4 x_4^4 + c_3 x_4^3 + c_2 x_4^2 + c_1 x_4 + c_0 = 0 \, ,$$

$$c_4 = a^6 b^6, \; c_3 = 2a^6 b^4 + 2a^4 b^6 \, ,$$

$$c_2 = a^6 b^2 + 4a^4 b^4 + a^2 b^6 - a^4 b^2 X^2 - a^4 b^2 Y^2 - a^2 b^4 Z^2 \, ,$$

$$c_1 = 2a^4 b^2 + 2a^2 b^4 - 2a^2 b^2 X^2 - 2a^2 b^2 Y^2 - 2a^2 b^2 Z^2 \, ,$$

$$c_0 = a^2 b^2 - b^2 X^2 - b^2 Y^2 - a^2 Z^2 \, .$$

(J.72)

J-3 Gauss surface normal coordinates: case study triaxial ellipsoid

First, we review surface normal coordinates for the triaxial ellipsoid. Second, it is our duty to review representative data for the triaxial ellipsoid for the Earth and other celestial bodies.

J-31 Review of surface normal coordinates for the triaxial ellipsoid

In case of a triaxial ellipsoid, we depart from the representation (J.73) subject to (J.74) once we use surface normal coordinates.

$$\frac{X^2}{A_1^2} + \frac{Y^2}{A_2^2} + \frac{Z^2}{A_3^2} = 1 \, , \quad \begin{bmatrix} X^2 \\ Y^2 \\ Z^2 \end{bmatrix} = \frac{A_1^2}{W} \begin{bmatrix} \cos B \cos L \\ (1 - E_{12}^2) \cos B \cos L \\ (1 - E_{13}^2) \sin B \end{bmatrix} \, ,$$

(J.73)

$$W = W(L, B) := \sqrt{1 - E_{13}^2 \sin^2 B - E_{12}^2 \cos^2 B \sin^2 L} \, .$$

(J.74)

The inverse transformation is characterized by (J.75) and (J.76) .

$$
L = \begin{cases}
\left(\arctan \dfrac{1}{1 - E_{12}^2} \dfrac{Y}{X} \right) & \text{for } X > 0 \,, \\[2ex]
\left(\arctan \dfrac{1}{1 - E_{12}^2} \dfrac{Y}{X} \right) + \pi & \text{for } X < 0 \,, \\[2ex]
(\operatorname{sgn} Y) \dfrac{\pi}{2} & \text{for } X = 0 \text{ and } Y \neq 0 \,, \\[2ex]
\text{not defined} & \text{for } X = 0 \text{ and } Y = 0 \,,
\end{cases}
\tag{J.75}
$$

$$
B = \begin{cases}
\arctan \dfrac{1 - E_{12}^2}{1 - E_{13}^2} \dfrac{Z}{\sqrt{(1 - E_{12}^2)^2 X^2 + Y^2}} & \text{for } X \neq 0 \text{ or } Y \neq 0 \,, \\[2ex]
(\operatorname{sgn} Z) \dfrac{\pi}{2} & \text{for } X = 0 \text{ and } Y = 0 \text{ and } Z \neq 0 \,, \\[2ex]
\text{not defined} & \text{for } X = 0 \text{ and } Y = 0 \text{ and } Z = 0 \,.
\end{cases}
\tag{J.76}
$$

We here note that A_1 is the semi-major axis, A_2 is the intermediate semi-major axis $A_2 < A_1$, and finally A_3 is the semi-minor axis $A_3 < A_2 < A_1$. The eccentricity of the intersection ellipses is given by (J.77) in the $\{1, 2\} = \{X, Y\}$ plane and by (J.78) in the $\{1, 3\} = \{X, Z\}$ plane.

$$
E_{12} = \sqrt{1 + A_2^2 / A_1^2} \,,
\tag{J.77}
$$

$$
E_{13} = \sqrt{1 + A_3^2 / A_1^2} \,.
\tag{J.78}
$$

Furthermore, we here point out that *elliptic heights* on top of a triaxial ellipsoid can be expressed by (J.79) subject to (J.80).

$$
X = \left[\frac{A_1}{W} + H(L, B) \right] \cos B \cos L \,,
$$

$$
Y = \left[\frac{A_1 (1 - E_{12}^2)}{W} + H(L, B) \right] \cos B \sin L \,,
\tag{J.79}
$$

$$
Z = \left[\frac{A_1 (1 - E_{13}^2)}{W} + H(L, B) \right] \sin B \,,
$$

$$
W = \sqrt{1 - E_{13}^2 \sin^2 B - E_{12}^2 \cos^2 B \sin^2 L} \,.
\tag{J.80}
$$

J-32 Position, orientation, form parameters: case study Earth

Let us here assume that we refer to the *geoid* as the equipotential surface of gravity close to the *Mean Sea Level* fitted to the triaxial ellipsoid. Actually, with respect to the biaxial ellipsoid, fitting the triaxial ellipsoid is 65% better. The difference of axes in the equatorial plane A_1–A_2 rounds up to 69 meters. With respect to the center of the best fitting triaxial ellipsoid, the mass center of the Earth is displaced approximately by 11–15 meters. The orientation of the triaxial axes with respect to the principal axes is given by

$$
A_1 = 6\,378\,173.435 \, \text{m} \quad (14°53'42'' \text{ westerly of the Greenwich meridian}) \,,
$$

$$
A_2 = 6\,378\,103.9 \, \text{m} \,, \quad A_3 = 6\,356\,754.4 \, \text{m} \,,
\tag{J.81}
$$

$$
A_1 - A_2 = 69.5 \, \text{m} \,, \quad A_1 - A_3 = 21\,419.0 \, \text{m} \,.
$$

The reciprocal polar flattening is provided by

$$\frac{1}{\alpha_{13}} := \frac{A_1}{A_1 - A_3} = 297.781\,194 \ . \tag{J.82}$$

The polar flattening is provided by

$$\alpha_{13} = 3.358\,17 \times 10^{-3} \ , \quad 1 - \alpha_{13} = 0.996\,643\,83 \ . \tag{J.83}$$

The reciprocal equatorial flattening is provided by

$$\frac{1}{\alpha_{12}} := \frac{A_1}{A_1 - A_2} = 91\,650.826 \ . \tag{J.84}$$

The equatorial flattening is provided by

$$\alpha_{12} = 1.091\,097 \times 10^{-3} \ , \quad 1 - \alpha_{12} = 0.999\,989\,089 \ . \tag{J.85}$$

Important!

The transformation of Cartesian coordinates $\{x^*, y^*, z^*\}$ in an Earth fixed equatorial reference system $f^0 = \{f_{1^0}, f_{2^0}, f_{3^0}\}$ into Cartesian coordinates $\{x, y, z\}$ in the elliptic reference system is described by the following relations.

$$\begin{bmatrix} x \\ y \\ z \end{bmatrix} = \mathsf{R}^{\mathrm{T}}(\delta\alpha, \delta\beta, \Delta\lambda) \begin{bmatrix} x' \\ y' \\ z' \end{bmatrix} + \begin{bmatrix} \Delta x \\ \Delta y \\ \Delta z \end{bmatrix} \ , \tag{J.86}$$

$$\mathsf{R}^{\mathrm{T}}(\delta\alpha, \delta\beta, \Delta\lambda) =$$

$$= \mathsf{R}_3(\Delta\lambda)\mathsf{R}_2(\delta\beta)\mathsf{R}_1(\delta\alpha) =$$

$$= \begin{bmatrix} \cos\Delta\lambda & \sin\Delta\lambda & \delta\beta\cos\Delta\lambda + \delta\alpha\sin\Delta\lambda \\ -\sin\Delta\lambda & \cos\Delta\lambda & \delta\beta\sin\Delta\lambda - \delta\alpha\cos\Delta\lambda \\ \delta\beta & -\delta\alpha & 1 \end{bmatrix} \ . \tag{J.87}$$

Important!

The numbers that follow below have been determined in the Ph. D. Thesis of B. Eitschberger (Bonn 1975). The terms $\{\Delta\lambda, \delta\alpha, \delta\beta\}$ define the orientation parameters, and the terms $\{\Delta x, \Delta y, \Delta z\}$ define the translation parameters.

$$\Delta\lambda = -14°53'42'' \ , \quad \delta\alpha = 0.16'' \ , \quad \delta\beta = 0.10'' \ , \tag{J.88}$$

$$\Delta x = -5.9\,\mathrm{cm} \ , \quad \Delta y = -2.4\,\mathrm{cm} \ , \quad \Delta z = +1.8\,\mathrm{cm} \ . \tag{J.89}$$

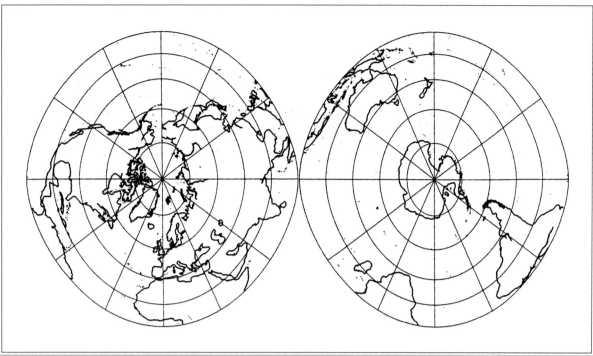

Fig. J.7. Azimuthal mapping of the triaxial ellipsoid, case study Earth.

J-33 Form parameters of a surface normal triaxial ellipsoid: Earth, Moon, Mars, Phobos, Amalthea, Io, Mimas

The following is a list of reference figures of the Earth, the Earth's moon, and other celestial bodies which are pronounced *triaxially*.

Table J.3. Form parameters of reference figures

body	axis A_1 [km]	axis $A_2 < A_1$ [km]	axis $A_3 < A_2 < A_1$ [km]	source
Earth	6 378.245	6 378.032 4	6 356.863 0	Schliephake (1955)
Earth	6 378.173 435	6 378.103 9	6 356.754 4	Eitschberger (1978) $(\Delta\Lambda = -14°53'42'')$
Moon	1 738.30	1 738.18	1 737.65	Wu (1981)
Mars	3 394.6	3 393.3	3 376.3	Wu (1981) $(\Delta\Lambda = -105°)$
Phobos	12.908	11.410	9.122	Bursa (1989)
Phobos	13.5	10.7	9.6	Bursa (1988)
Amalthea	135	82	75	Bursa (1988)
Io	1 833	1 922	1 819	Bursa (1988)
Mimas	209.1	196.1	191.9	Bursa (1988)

As an example, we illustrate by Fig. J.7 an azimuthal mapping of the triaxial ellipsoid of the Earth, which is equidistant along the meridian parameterized by polar coordinates of type (J.90) referred to the elliptic integral $\boldsymbol{E}(\cdot, \pi/2)$. Compare with the Diploma Thesis of B. Mueller (1991).

$$\alpha = \arctan\left(\sqrt{1 - E_{12}^2}\,\tan\varLambda\right)\,,\quad r = A_1\sqrt{1 - E_{12}^2\sin^2\varLambda}\;\boldsymbol{E}\left(\frac{\sqrt{E_{13}^2 - E_{12}^2\sin^2\varLambda}}{1 - E_{12}^2\sin^2\varLambda},\frac{\pi}{2}\right)\,.\quad (\text{J}.90)$$

Important!

Important references are M. Bursa (1989a,b), B. Eitschberger (1978), E. Grafarend and P. Lohse (1991), B. Heck (2002), W. Klingenberg (1982), L. P. Lee (1965), H. Merkel (1956), B. Müler (1991), G. Schliephake (1955, 1956), H. Schmehl (1927, 1930), J. P. Snyder (1985), H. Viesel (1971), J. A. Weightman (1961) and S. C. Wu (1981).

Bibliography

1. Abbas Y.A.A.H. (1993): Triple projection of a topographic surface from an external perspective center, PhD thesis, Assiut University, Assiut, Egypt 1993

2. Abdel-Latif M.S. (1985): Spherical curves of constant bearing, Dirasat 12 (1985) 85–99

3. Abdel-Latif M.S., El-Sonbaty, A., Abdel-Rahim A. (1992): On the orientation problem for non metric cameras, Proceedings of the Jordanian Conference on Civil Engineering II, 2–4 June 1992

4. Abramowitz M., Stegun J.A. (1965): Handbook of Mathematical Functions, National Bureau of Standards, Applied Mathematical Series 55, New York 1965

5. Abramowitz M., Stegun J.A. (1972): Jacobian elliptic functions and theta functions, Ch. 16 in Handbook of Mathematical Functions with Formulas, Graphs, and Mathematical Tables, 9th printing, New York Dover (1972) 567–581

6. Abrams C.W., Bowers V.L. (1973): Universal transverse Mercator grid, Department of the Army, Technical Manual TM 5-241-8, Washington D.C. 1973

7. Adams B.R. (1984): Transverse cylindrical stereographic and conical stereographic projections, The American Cartographer 11 (1984) 40–48

8. Adams O.S. (1918): Lambert projection tables for the United States: U.S. Coast and Geodetic Survey Spec. Pub. 52, 1918

9. Adams O.S. (1919): General theory of polyconic projections, Department of Commerce, U.S. Coast and Geodetic Survey, U.S. Government Printing Office, Washington 1919

10. Adams O.S. (1921): Latitude developments connected with geodesy and cartography with tables, including a table for Lambert Equal–Area Meridional projection: U.S Coast and Geodetic Survey Spec. Pub. 67, 1921

11. Adams O.S. (1925): Elliptic functions applied to conformal world maps, Dep. of Commerce, Serial Nr. 297, U.S. Government Printing Office, Washington 1925

12. Adams O.S. (1927): Tables for Albers projection: U.S. Coast and Geodetic Survey Spec. Pub. 130, 1927

13. Adams O.S., Deetz C.H. (1990): Elements of map projection, 5th ed., Special Publication 68, US Government Printing Office, Washington DC 1921 (reprint 1990)

14. Adams P.D. (1934): General theory of polyconic projections, U.S. Coast and Geodetic Survey, U.S. Government Printing Office, Washington 1934

15. Adams P.D. (1952): Conformal projections in geodesy and cartography, U.S Coast and Geodetic Survey, Special Publication No. 251, U.S. Government Printing Office, Washington 1952

16. Adolph U.-C. (1992): Neue vermittelnde Entwurfsgruppen – erzeugt über Matrizenformeln aus Mollweide–Netzen, Kartographische Nachrichten 42 (1992), 134–138

17. Aduol F.W.O. (1989): Integrierte geodätische Netzanalyse mit stochastischer Vorinformation über Schwerefeld und Referenzellipsoid, Becksche Verlagsbuchhandlung, München 1989

18. Agard S.B., Gehring F.W. (1965): Angles and quasiconformal mappings, Proc. London Math. Soc. 3 (1965) 1–21

19. Ahlfors L.V. (1939): Untersuchungen zur Theorie der konformen Abbildung und der ganzen Funktionen, Acta Soc. Sci. Fam. 1 (1939) 1–40

20. Airy G.B. (1861): Explanation of a projection by balance of errors for maps applying to a very large extent of the Earth's surface: comparison of this projection with other projections, Phil. Mag. 22 (1861) 409–442

21. Akivis M.A. (1952): Invariante Konstruktion der Geometrie einer Hyperfläche in einem konformen Raum, Matematiceskij sbornik. Moskva 31 (1952) 43–75

22. Albers H.C. (1805): Beschreibung einer neuen Kegelprojektion: Zach's Monatliche Korrespondenz zur Beförderung der Erd– und Himmelskunde, Nov., 450–459

23. Albertz J., Lehmann, H., Tauch, R. (1992): Herstellung und Gestaltung hochauflösender Satelliten–Bildkarten, Kartographische Nachrichten 42 (1992) 205–214

24. Albertz J.,Tauch R. (1991): Erfahrungen bei der Herstellung von Satelliten–Bildkarten, Sonderdruck aus: Veröffentlichungen des Zentralinstituts für Physik der Erde 118 (1991) 373–383

25. Alexander J.C. (1985): The numerics of computing geodetic ellipsoids, in: Classroom notes in applied mathematics, ed. M.S. Klamkin, Siam Review 27 (1985) 241–247

26. Almansi E. (1911): Sulle deformazioni finite di solidi elastici isotropi, Note I, Atti Accad. naz. Lincei, Re., Serie Quinta 201 (1911) 705–714

27. Alpha T.R., Gerin M. (1978): A survey on the properties and uses of selected map projections: U.S. Geol. Survey Misc. Geol. Inv. Map. I–1096, 1978

28. Altamini Z., Boucher C. (2001): The ITRS and ETRS 89 relationship: new results from ITRF 2000, in: EUREF, Dubrovnik 2001

29. Altmann S.L. (1992): Icons and symmetries, Clarendon Press, Oxford 1992

30. Amalvict M., Livieratos E. (1988): Surface mapping of a rotational reference ellipsoid onto a triaxial counterpart through strain parameters, Manuscripta Geodaetica 13 (1988) 133–138

31. American Cartographic Association (1986): Which map is best? Projections for world maps, American Congress on surveying and mapping, Falls Church, Virginia 1986

32. American Cartographic Association (1988): Choosing a world map, attributes, distortions, classes, aspects, American Congress on surveying and mapping, Bethesda, Maryland 1988

33. American Cartographic Association (1991): Matching the Map projection to the Need, American Congress on surveying and mapping, Bethesda, Maryland 1991

34. American Soc. Of Civil Engineers, American Congress on Surveying and Mapping and American Soc. for Photogrammetry and Remote Sensing (1994): Glossary of the Mapping Sciences, American Soc. of Civil Engineers, American Congress on Surveying and Mapping, American Soc. for Photogrammetry and Remote Sensing 1994

35. Andrews H.J. (1935): Note on the use of Oblique Cylindrical Orthomorphic projection: Geographical Journal 86 (1935) 446

36. Andrews H.J. (1938): An oblique Mercator projection for Europe and Asia, The Geographical Journal 92 (1938) 538ff

37. Andrews H.J. (1941): Note on the use of oblique cylindrical orthomorphic projection, The Geographical Journal 97/98 (1941) 446

38. Anserment A. (1941): Quelques charactéristiques du système de coordonnées Bonne, Schweizerische Zeitschrift für Vermessungswesen und Kulturtechnik 39 (1941) 189–192

39. Antonopoulos A. (2003): Scale effects associated to the transformation of a rotational to a triaxial ellipsoid and their connection to relativity, Artificial Satellites 38 (2003) 119–131

40. Appel K., Haken W. (1977): The solution of the Four–Color–Map problem, Scientific American 237 (1977) 108–121

41. Ardalan A.A., Safari A. (2004): Ellipsoidal terrain correction based on multi–cylindrical equal–area map projection of the reference ellipsoid, J. Geodesy (2004)

42. Aringer K. (1994): Geodätische Hauptaufgaben auf Flächen in kartesischen Koordinaten, Report C421, Deutsche Geodätische Kommission, Bayer. Akad. Wiss., München 1994

43. Armanni G. (1915): Sulle deformazioni finite di solidi elastici isotropi, Nuovo cimento 10 (1915) 427–447

44. Army Department of the (1958): Universal Transverse Mercator Grid Table for latitudes 0°–80°, International Spheroid (meters) Volume I–II, U.S. Army 1958

45. Army Department of the (1959): Universal Transverse Mercator Grid Table for latitudes 0°–80°, Bessel Spheroid (meters) Coordinates for 5–Minute intersections, U.S. Army–Headquarters 1959

46. Army Department of the (1973): Universal Transverse Mercator Grid, U.S. Army Tech. Manual TM 5–241–8.

47. Arnold G.C. (1984) : The derivation of mapping equations and distortion formulae for the satellite tracking map projections, DTIC Elect (1984) 1–63

48. Audin M. (1994): Courbes algébriques et systèmes intégrables: géodésiques des quadriques, Expo. Math. 12 (1994) 193–226

49. Awange J.L., Fukuda Y., Grafarend E.W. (2004): Exact solution of the nonlinear 7–parameter datum transformation by Groebner basis, Bollettino di Geodesia e Scienze Affini 63 (2004) 117–123

50. Axler S., Bourdon P., Ramey W. (2001): Harmonic function theory, Springer Verlag, Berlin Heidelberg New York 2001

51. Ayoub R. (1984): The lemniscate and Fagnano's contributions to elliptic integrals, Archive for History of Exact Sciences 29 (1984) 131–149

52. Baarda W. (1967): Statistical concepts in geodesy, Netherlands Geodetic Commission, Publications on Geodesy, New Series 2 (1967)

53. Baehr H.-G. (1989): Sphärische und ebene Dreiecksberechnungen mit dem Tangens des Viertelwinkels, Zeitschrift für Vermessungswesen 114 (1989) 485–493

54. Baeschlin F. (1918): Einige Entwicklungen zur Bonne'schen Kartenprojektion, Schweizerische Zeitschrift für Vermessungswesen und Kulturtechnik 16 (1918) 193–201

55. Baetslé P.-L. (1970): Optimalisation d'une représentation cartographique, Bulletin trimestriel de la Societé belge de Photogrammétrie 101 (1970) 11–26

56. Baeyer, J.J. (1862) : Das Messen auf der Sphäroidischen Erdoberfläche. Als Erläuterung meines Entwurfes zu einer mitteleuropäischen Gradmessung, Berlin 1862

57. Baily W. (1886): A map of the world on Flamsteed's projection: London, Edinburgh and Dublin Philosophical Magazine, series 5, 21 (1886) 415–416

58. Bajaj C.L., Bernardini F., Xu G. (1994): Reconstruction of surfaces and surfaces–on–surfaces from unorganized weighted points, Dept. of Computer Sciences, Purdue University, West Lafayette 1994

59. Baker J.G.P. (1986): The "dinomic" world map projection, The Cartographic Journal 23 (1986) 66–68

60. Balcerzak J. (1985): Algorithms for computation of ellipsoidal geodetic coordinates from rectangular coordinates in the Gauss–Krueger Projection by the method of expansions into power series, Geodezia i Kartografia 23 (1985) 24–37

61. Balcerzak J., Panasiuk J. (1983): Obliczanie wspólrzednych prostokatnych plaskich Gaussa–Kruegera w szerokiej strefie poludnikowej powierzchni elipsoidy, Polytechnika Warszawska, Warschau 1983

62. Baranyi J. (1968): Hungarian cartographical Studies, pages 19–31, Budapest 1968

63. Baranyi J. (1987): Konstruktion anschaulicher Erdabbildungen, Kartographische Nachrichten 37 (1987) 11–17

64. Baranyi J., Karsay F. (1972): World map projections with better shape–keeping properties, in: Hungarian Cartographical Studies, pages 13–19, Budapest 1972

65. Barber C., Cromley R., Andrle R. (1995): Evaluating alternative line simplification strategies for multiple representations of cartographic lines, Cartography and Geograph. Information Systems 22 (1995) 276–290

66. Barner M., Flohr F. (1958): Der Vierscheitelsatz und seine Verallgemeinerung, Der Mathematikunterricht 4 (1958)

67. Barrow J.D. (1983): Dimensionality, Phil. Trans. R. Soc. Lond. A 310 (1983) 337–346

68. Barsi A. (2001): Performing coordinate transformation by artificial neural network, Allgemeine Vermessungsnachrichten 108 (2001) 134–137

69. Bartel K. (1934): Malerische Perspektive, Band 1, B.G. Teubner, Leipzig / Stuttgart 1934

70. Bartelme N., Meissl P. (1975): Ein einfaches, rasches und numerisch stabiles Verfahren zur Bestimmung des kürzesten Abstandes eines Punktes von einem sphäroidischen Rotationsellipsoid, Allgemeine Vermessungsnachrichten 82 (1975) 436–439

71. Barwinsky K.-J. et al. (1992): Landesvermessung 2000, Nachrichten aus dem öffentlichen Vermessungsdienst Nordrhein–Westfalen 2 (1992) 55-84

72. Bass H., Connell E.H., Wright D. (1982): The Jacobian conjecture: reduction of degree and formal expansion of the inverse, Bulletin of the American Mathematical Society 7 (1982) 287–329

73. Batson R.M. (1973): Cartographic products from the Mariner 9 mission: Jour. Geophys. Research 78 (1973) 4424–4435

74. Batson R.M. (1976): Cartography of Mars: Am. Cartographer 3 (1976) 57–63

75. Batson R.M., Bridges H.M., Inge J.L., Isbell C., Masursky H., Strobell M.E., Tyner R.L. (1980): Mapping the Galilean satellites of Jupiter with Voyager data: Photogrammetric Engineering and Remote Sensing 46 (1980) 1303–1312

76. Battha L. (1997): Estimation of coefficients of the 2D–projective transformation with the Cauchy–function, Acta Geod. Geoph. Hung. 32 (1997) 245–248

77. Beaman W.M. (1928): Topographic mapping: U.S. Geol. Survey Bull. (1928) 788-E

78. Becker T., Weispfennig V. (1998): Gröbner bases, A computational approach to commutative algebra, Graduate text in Mathematics 141, Springer Verlag, New York 1998

79. Behnke H., Stein K. (1947/1949): Entwicklung analytischer Funktionen auf Riemannschen Flächen, Springer–Verlag, Berlin–Göttingen–Heidelberg 1947/1949

80. Behrmann W. (1909): Zur Kritik der flächentreuen Projektionen der ganzen Erde und einer Halbkugel, Sitzungsberichte der Königlich Bayerischen Akademie der Wissenschaften, Mathematisch–physikalische Klasse 13, pages 19–74, München 1909

81. Behrmann W. (1910): Die beste bekannte flächentreue Projektion der ganzen Erde, Kartographischer Monatsbericht 9, in: Dr. A. Petermanns Mitteilungen, Hrsg. P. Langhans, pages 141–144, Gotha 1910

82. Beineke D. (1991): Untersuchungen zur Robinson–Abbildung und Vorschlag einer analytischen Abbildungsvorschrift, Kartographische Nachrichten 41 (1991) 85–94

83. Beineke D. (1995): Kritik und Diskussion, Kartographische Nachrichten 45 (1995) 151–153

84. Beineke D. (2001): Verfahren zur Genauigkeitsanalyse für Altkarten, Schriftenreihe Studiengang Geodäsie und Geoinformation, Universität der Bundeswehr München, Heft 71, Neubiberg 2001

85. Beineke H.-D. (1983): Automationsgerechte Koordinatentransformation für kleinmaßstäbige Kartennetzabbildungen mit Hilfe der Tensorrechnung, Kartographische Nachrichten 33 (1983) 55–64

86. Bell S.R., Brylinski J.-L., Huckleberry A.T., Narasimhan R., Okonek C., Schumacher G., Van de Ven A., Zucker S. (1997): Complex manifolds, Springer Verlag, Berlin Heidelberg New York (2nd edition) 1997

87. Bellman R., Fan K. (1963): On systems of linear inequalities in hermitian matrix variables, Proceedings of Symposia in Pure Mathematics 7 (1963) 1–11

88. Bellman R.E. (1961): A brief introduction to theta functions, Rinehart and Winston, New York: Holt 1961

89. Beltrami E. (1866): Risoluzione del problema: riportare i punti di una superficie sopra un piano in modo che le linee geodetiche vengano rappresentate da linee rette, Annali di Matematica 7 (1866) 185–204

90. Beltrami E. (1869): Zur Theorie des Krümmungsmaßes, Mathematische Annalen 1 (1869) 575–582

91. Belykh V.N. (1988): Calculation on a computer of the complete elliptic integrals K(x) and E(x), Boundary value problems for partial differential equations, Akad. Nauk SSSR Sibirsk. Otdel., Inst. Mat., Novosibirsk, 1988, pp. 3-15

92. Benning W. (1974): Der kürzeste Abstand eines in rechtwinkligen Koordinaten gegebenen Außenpunktes vom Ellipsoid, Allgemeine Vermessungsnachrichten 81 (1974) 429–433

93. Berghaus H. (1850): Physikalischer Schul–Atlas, Gotha Verlag 1850

94. Berlyant A.M., Novakovskiy B.A. (1986): Mapping Sciences and Remote Sensing 23 (1986) 115–122

95. Berman G. (1961): The wedge product, The American Mathematical Monthly 68 (1961) 112–119

96. Bermejo M., Otero J. (2005): Minimum conformal mapping distortion according to Chebyshev's principle: a case–study over peninsular Spain, J. Geodesy 79 (2005) 124–134, DOI: 10.1007/s00190-005-0450-5

97. Bibby H.M., Reilly W.I. (1981): The use of the New Zealand map grid projection as a local survey projection, New Zealand Surveyor 8 (1981) 11–24

98. Bieberbach L. (1916): Über die Koeffizienten derjenigen Potenzreihen, welche eine schlechte Abbildung des Einheitskreises vermitteln, Sitzungsberichte der Königlich Preussischen Akademie der Wissenschaften, Verlag der Königlichen Akademie der Wissenschaften, Berlin 1916

99. Biernacki F. (1966): Theory of representation of surfaces for surveyors and cartographers, The Scientific Publications Foreign Cooperation Center of the Central Institute for Scientific, Technical and Economic Information, Warsaw 1966

100. Bills B.G. (1987): Planetary Geodesy, Reviews of Geophysics 25 (1987) 833–839

101. Bills B.G., Kiefer W.S., Jones R.L. (1987): Venus gravity: a harmonic analysis, Journal of Geophysical Research 92 (1987) 10,335–10,351

102. Bisegna P., Podio–Guidugli P. (1995): Mohr's Arbelos, Meccanica 30 (1995) 417–424

103. Bishop E. (1965): Differentiable manifolds in complex Euclidean Space, Duke Mathematical Journal 32 (1965) 1–21

104. Blaschke W. (1942): Über die Differenzialgeometrie von Gauss, in: Jahresbericht der Deutschen Mathematikervereinigung, Hrsg. E. Sperner, Leipzig und Berlin 1942

105. Blaschke W.(1929): Vorlesungen über Differentialgeometrie III, Berlin 1929

106. Blaschke W., Leichtweiß K. (1973): Elementare Differentialgeometrie, Springer, Berlin – Heidelberg – New York 1973

107. Blumenthal L M. (1953): Theory and applications of distance geometry, Clarendon Press, Oxford 1953

108. Bobenko A. (1994): Surfaces in terms of 2 by 2 matrices. Old and new integrable cases, in: Fordy A., Wood J. (eds.): Harmonic maps and integrable systems, Vieweg 1994

109. Bobenko A., Pinkall U. (1996): Discrete isothermic surfaces, J. reine angew. Math. 475 (1996) 187–208

110. Bodemueller H. (1934): Über die konforme Abbildung der Erdoberfläche mit günstiger Richtungs– und Längenreduktion für die Zwecke einer Landesvermessung, Allgemeine Vermessungsnachrichten 46 (1934) 550–566, 569–578, 585–593, 601–607

111. Boedewadt U.T. (1942): Die Fourierentwicklung des Sinus, Cosinus und der Umkehrung einer Fourierreihe, Mathematische Zeitschrift 47 (1942) 655–662

112. Boehme R. (1993): Inventory of world topographic mapping 3, Elsevier Applied Science Publishers, London–New York 1993

113. Boelesvölgyiné B.M. (1988): Vízszintes alapponthàlózatunk transzformációja a szocialista országok egységes rendszerébe, Geodézia és Kartográfia 40 (1988) 1–4

114. Böhm R. (2006): Variationen von Weltkartennetzen der Wagner-Hammer-Aitoff Entwurfsfamilie, Kartographische Nachrichten 56 (2006) 8–16

115. Boljen J. (1997): Zur Transformation von Koordinaten vom DHDN 90 in das ETRS 89, Allgemeine Vermessungsnachrichten 104 (1997) 294–300

116. Bolliger J. (1967): Die Projektionen der schweizerischen Plan– und Kartenwerke: Winterthur, Switz., Druckerei Winterthur AG, 1967

117. Boltz H. (1942): Formeln und Tafeln zur numerischen (nicht logarithmischen) Berechnung geographischer Koordinaten aus den Richtungen und Längen der Dreiecksseiten erster Ordnung. Veröffentlichungen des Geodätischen Instituts Potsdam, Neue Folge, Nr. 110

118. Boltz H. (1943): Formeln und Tafeln zur numerischen (nicht logarithmischen) Berechnung Gauss–Krueger'scher Koordinaten aus den geographischen Koordinaten, Veröffentlichungen des Geodätischen Instituts Potsdam, pages V–XVI, Frickert & Co., Potsdam 1943

119. Bomford G. (1971): Geodesy: Oxford, Eng., Clarendon Press 1971

120. Bonacker W., Anliker E. (1930): Heinrich Christian Albers, der Urheber der flächentreuen Kartenrumpfprojektion: Petermanns Geographische Mitteilungen 76 (1930) 238–240

121. Borg I., Groenen P. (1997): Modern multidimensional scaling, Springer Verlag, Berlin - Heidelberg – New York 1997

122. Borkowski K.M. (1987): Transformation of geocentric to geodetic coordinates without approximations, Astrophysics and Space Science 139 (1987) 1–4

123. Borkowski K.M. (1989): Accurate algorithms to transform geocentric to geodetic coordinates, Bulletin Géodésique 63 (1989) 50–56

124. Bormann G.E., Vozikis E. (1982): Photographische Kartenumbildung mit dem Wild–AVIOPLAN OR1, Kartographische Nachrichten 32 (1982) 201–206 + Beilage

125. Boucher C. (1980): The general theory of deformations and its applications in geodesy, Bollettino di Geodesia e Scienze Affini 39 (1980) 14–35

126. Bougainville L.A. (1756): Trait´⌐¢du calcul int´⌐¢ral, pour servir de suite ´⌐¢l'Analyse des infiniments petits de M. le marquis de l'H´⌐¢ital, Paris : H.-L. Gu´⌐¢in et L.-F. Delatour, 1754-1756, 2 vol. XXIII, 340 p. ; XXIV, 259 p.

127. Boulware D.G., Brown L.S., Peccei R.D. (1970): Deep–inelastic electroproduction and conformal symmetry, Physical Review D 2 (1970) 293–298

128. Bourguignon J.P. (1970): Transformation infinitesimal conformes fermées des variétés riemanniennes connexes complètes, C. R. Acad. Sci. Ser. A 270 (1970) 1593–1596

129. Bourguignon J.P. (1996): An introduction to geometrical variational problems, in: Lectures on geometrical variational problems, S. Nishikawa and R. Schoen (Hrsg.), pages 1–41, Springer Verlag, Berlin Heidelberg New York 1996

130. Boutoura C., Livieratos E. (1986): Strain analysis for geometric comparisons of maps, The Cartographic Journal 23 (1986) 27–34

131. Bowring B.R. (1985): The geometry of the loxodrome on the ellipsoid, The Canadian Surveyor 39 (1985) 223–230

132. Bowring B.R. (1986): The Lambert conical orthomorphic projection and computational stability, Bulletin Géodésique 60 (1986) 345–354

133. Bowring B.R. (1993): Applicable complex and unreal geodesy, Survey Review 32 (249) 145–158

134. Boyle M.J. (1987a): World Geodetic System 1984 (WGS 84), Defence Mapping Agency, Attn: PR, Building 56, U.S. Naval Observatory, Washington D.C. 1987

135. Boyle M.J. (1987b): Department of Defence World Geodetic System 1984 – It's definition and relationship with local geodetic systems, DMA Technical Report 83502.2., Washington, D.C. 1987

136. Brandenberger C. (1986): EDV–Einsatz in der Atlaskartographie, in: Digitale Technologie in der Kartographie, Wiener Symposium 1986, F. Mayer (Hrsg.), pages 105–121, Wien 1986

137. Brandenberger C. (1992): Einsatz des KIS–Systems bei der Erstellung des Kartenblockes "Erde im Sonnensystem" für den neuen Schweizer Weltatlas, Kartographische Nachrichten 42 (1992) 138–143

138. Brandstaetter G. (1967): Über den sphärischen Rückwärtsschnitt und seine Anwendung in der geodätischen Astronomie, Report C110, Deutsche Geodätische Kommission, Bayer. Akad. Wiss., München 1967

139. Brasselet J.-P. (1990): La géométrie des tracés de voies. de chemin de fer à grand vitesse, in: Geometry and Topology of Submanifolds, II, Boyom M., Morvan J.M., Verstraelen L. (Hrsg.), pages 32–49, World Scientific, Singapore 1990

140. Brauner H. (1983): Zur Theorie linearer Abbildungen, Abh. Math. Seminar Univ. Hamburg 53 (1983) 154–169

141. Bretterbauer K. (1980): Über Zentralschnitte des Rotationsellipsoides, Mitteilungen der geodätischen Institute der Technischen Universität Graz, Folge 35, Festschrift zur Emeritierung von o. Univ.-Prof. Dipl.-Ing. Dr. techn. Karl Hubeny, pages 59–67, Graz 1980

142. Bretterbauer K. (1989): Die trimetrische Projektion von Chamberlin, Kartographische Nachrichten 39 (1989) 51–55

143. Bretterbauer K. (1990): Ein Algorithmus zur massenhaften Transformation österreichischer konformer Koordinaten, Kartographische Nachrichten 40 (1990) 229–231

144. Bretterbauer K. (1991): Mathematische Lehre vom Kartenentwurf, Teil I und II, TU Wien 1991

145. Bretterbauer K. (1994): Ein Berechnungsverfahren für die Robinson–Projektion, Kartographische Nachrichten 44 (1994) 227–229

146. Bretterbauer K. (1995a): Koordinatensysteme – der rote Faden durch Geodäsie und GIS, Institutsmitteilungen des Instituts für Geodäsie der Universität Innsbruck, VIII. Int. Geodätische Woche, Heft 16, pages 1–17, Innsbruck 1995

147. Bretterbauer K. (1995b): Die Gauss–Krueger Abbildung einfach dargestellt, Österreichische Zeitschrift für Vermessung und Geoinformation 83 (1995) 146–150

148. Bretterbauer K. (2001a): Die Himmelssphäre, eben dargestellt, Sterne und Weltraum 3 (2001) 276–279

149. Bretterbauer K. (2001b): Eine Variante der trimetrischen Projektion, Kartographische Nachrichten 51 (2001) 130–132

150. Bretterbauer K. (2002a): Die runde Erde, eben dargestellt, Abbildungslehre und sphärische Kartennetzentwürfe, Geowiss. Mitt. 59, TU Wien, Wien 2002

151. Bretterbauer K. (2002b): Neue Netzentwürfe auf Basis finiter Elemente, Österreichische Zeitschrift für Vermessung und Geoinformation 90 (2002) 43–46

152. Briesemeister W. (1953): A new oblique equal–area projection, The Geographical Review 43 (1953) 260–261

153. Brill M.H. (1983): Closed–form extension of the anharmonic ratio to N–Space, Computer Vision, Graphics and Image Processing 23 (1983) 92–98

154. Britting K.A. (1971): Inertial navigation systems analysis, Wiley–Interscience, New York 1971

155. Brown B.H. (1935): Conformal and equiareal world maps, The American Mathematical Monthly 42 (1935) 212–223

156. Brown L.A. (1949): The story of maps: New York, Bonanza Books, reprint undated

157. Bruhns O.T., Xiao H., Meyers A. (1999): Self–consistent Eulerian rate type elasto–plasticity models based upon the logarithmic stress rate, Int. J. Plasticity 15 (1999) 479–520

158. Brumberg V.A., Groten E. (2001): IAU resolutions on reference systems and time scales in pratice, Astronomy and Astrophysics 367 (2001) 1070–1077

159. Brunner K. (1995): Digitale Kartographie an Arbeitsplatzrechnern, Kartographische Nachrichten 45 (1995) 63–68

160. Bruss A.R., Horn B.K.P. (1983): Passive navigation, Computer Vision, Graphics and Image Processing 21 (1983) 3–20

161. Buck H. (1997): Bezugs– und Abbildungssysteme in der Landesvermessung, Deutscher Verein für Vermessungswesen, Landesverein Baden–Württemberg, Mitteilungen 1 (1997) 27–57

162. Bugayevskiy L.M. (1994): Zur konformen Abbildung eines dreiachsigen Ellipsoids, Allgemeine Vermessungsnachrichten 101 (1994) 194–205

163. Bugayevskiy L.M., Bocharov A.Y. (1974): The use of asymmetrical conformal projections for compiling maps of extensive territories, Geodesy, Mapping and Photogrammetry 16 (1974) 177–179

164. Bugayevskiy L.M., Krasnopevtseva B.V., Shingareva K.B. (1994): Mapping of extraterrestial bodies, Allgemeine Vermessungsnachrichten 101 (1994) 194–205

165. Bugayevskiy L.M., Krasnopevtseva B.V., Shingareva K.B. (1996): Zur kartographischen Darstellung irregulärer Himmelskörper, Zeitschrift für Vermessungswesen 121 (1996) 533–540

166. Bugayevskiy L.M., Snyder J.P. (1995): Map projections. A reference manual, Taylor & Francis, London 1995

167. Bulirsch R. (1965a): Numerical calculation of elliptic integrals and elliptic functions, Numerische Mathematik 7 (1965) 78–90

168. Bulirsch R. (1965b): Numerical calculation of elliptic integrals and elliptic functions II, Numerische Mathematik 7 (1965) 353–354

169. Bulirsch R. (1969a): An extension of the Bartky–transformation to incomplete elliptic integrals of the third kind, Numerische Mathematik 13 (1969) 266–284

170. Bulirsch R. (1969b): Numerical calculation of elliptic integrals and elliptic functions. III, Numerische Mathematik 13 (1969) 305–315

171. Bulirsch R. (2001): Himmel und Erde messen, Deutscher Verein für Vermessungswesen, Mitteilungsblatt Bayern 53 (2001) 401–451

172. Bulirsch R., Gerstl M. (1983): Numerical evaluation of elliptic integrals for geodetic applications, Bollettino di Geodesia e Scienze Affini 42 (1983) 149–160

173. Bulrisch R. (1967): Numerical calculation of the sine, cosine and Fresnel integrals, Numerische Mathematik 9 (1967) 380–385

174. Burckel R. (1979): An introduction to classical complex analysis, Birkhäuser Verlag, Basel / Stuttgart 1979

175. Burger K. (1987): Bergmännisches Rißwesen–Stand und Perspektiven, Schriftenreihe Lagerstättenerfassung und -darstellung, Gebirgs–und Bodenbewegungen, Bergschäden, Ingenieurvermessung 11 (1987) 17–75

176. Burger K. et al. (1984): Isotopische Alter von pyrokiastischen Sanidinen aus Kaolin–Kohlentonsteinen als Korrelationsmarken für das mitteleuropäische Oberkarbon, Fortschr. Geol. Rheinld. u. Westf. 32 (1984) 119–150

177. Bursa M. (1970): Best–fitting tri–axial earth ellipsoid parameters derived from satellite observations, Studia geoph. et geod. 14 (1970) 1–9

178. Bursa M. (1989a): Tidal origin of tri–axiality of synchronously orbiting satellites, Bull. Astron. Inst. Czechosl. 40 (1989) 105–108

179. Bursa M. (1989b): Figure and dynamic parameters of synchronously orbiting satellites in the solar system, Bull. Astron. Inst. Czechosl. 40 (1989) 125–130

180. Bursa M. (1990a): Gravity field of satellites disintegrating at the Roche limit, Bull. Astron. Inst. Czechosl. 41 (1990) 96–103

181. Bursa M. (1990b): Estimating mean densities of saturnian tri–axial satellites, Bull. Astron. Inst. Czechosl. 41 (1990) 104–107

182. Bursa M. (1993): Distribution of gravitational potential energy within the solar system, Earth, Moon and Planets 62 (1993) 149–159

183. Bursa M. (1995): Primary and derived parameters of common relevance of astronomy, geodesy, and geodynamics, Earth, Moon and Planets 69 (1995) 51–63

184. Bursa M. (1997): Figure parameters of Ganymede, Acta Geod. Geoph. Hung. 32 (1997) 225–233

185. Bursa M. (2001): Long–term stability of geoidal geopotential from Topex/Poseidon satellite altimetry 1993–1999, Earth, Moon and Planets 84 (2001) 163–176

186. Bursa M., Bystrzycka B., Radej K., Vatrt V. (1995): Estimation of the accuracy of geopotential models, Studia geoph. et geod. 39 (1995) 365–374

187. Bursa M., Groten E., Kenyon S., Kouba J., Radej K., Vatrt V., Vojtiskova M. (2002): Earth's dimension specified by geoidal geopotential, Studia geophys. et geod. 46 (2002) 1–8

188. Bursa M., Kouba J., Müller A., Radej K., True S.A., Vatrt V., Vojtiskova M. (1999): Differences between mean sea levels for the Pacific, Atlantic and Indian Oceans from Topex/Poseidon altimetry, Studia geoph. et geod. 43 (1999) 1–6

189. Bursa M., Kouba J., Radej K., True S.A., Vatrt V., Vojtiskova M. (1999): Temporal variations in sea surface topography and dynamics of the earth's inertia ellipsoid, Studia geoph. et geod. 43 (1999) 7–19

190. Bursa M., Sima Z. (1979): Equatorial flattenings of planets: Mars, Bull. Astron. Inst. Czechosl. 30 (1979) 122–126

191. Bursa M., Sima Z. (1980): Tri–axiality of the earth, the moon and Mars, Studia geoph. et geod. 24 (1980) 211–217

192. Bursa M., Sima Z. (1984): Equatorial flattening and principal moments of inertia of the earth, Studia geoph. et geod. 28 (1984) 9–10

193. Bursa M., Sima Z. (1985a): Equatorial flattenings of planets: Venus, Bull. Astron. Inst. Czechosl. 36 (1985) 129–138

194. Bursa M., Sima Z. (1985b): Dynamic and figure parameters of Venus and Mars, Adv. Space Res. 5 (1985) 43–46

195. Bursa M., Vanysek V. (1996): Triaxiality of satellites and small bodies in the solar system, Earth, Moon and Planets 75 (1996) 95–126

196. Burstall F., Hertrich–Jeromin U., Pedit F., Pinkall U. (1997): Isothermic surfaces and curved flats, Math. Z. 225 (1997) 199–299

197. Calabi E. (1953): Isometric imbedding of complex manifolds, Annals of Mathematics 58 (1953) 1–23

198. Calapso P. (1992): Sulla superficie a linee di curvature isotherme, Rend. Circ. Mat. Palermo 17 (1992) 275–286

199. Canters F., Decleir H. (1989): The world in perspective. A directory of world map projections, J. Wiley & Sons, New York Chichester 1989

200. Cantor G. (1877): Ein Beitrag zur Mannigfaltigkeitslehre, Journal für Mathematik 84 (1877) 242–258

201. Caputo M. (1959): Conformal projection of an ellipsoid of revolution when the scale factor and ist normal derivative are assigned on a geodetic line of the ellipsoid, Journal of Geophysical Research 64 (1959) 1867–1873

202. Carathéodory C. (1912): Untersuchungen über die konformen Abbildungen von festen und veränderlichen Gebieten, Mathematische Annalen, 72. Band, Teubner Verlag, Leipzig 1912

203. Cardoso J.F., Souloumiac A. (1996): Jacobi angles for simultaneous diagonalization, SIAM Journal Matrix Analysis and Applications 17 (1996) 161–164

204. Carlson B.C. (1965): On computing elliptic integrals and functions, J. Math. and Phys. 44 (1965) 36–51

205. Carlson B.C. (1977): Elliptic integrals of the first kind, SIAM J. Math. Anal. 8 (1977) 231–242

206. Carlson B.C. (1987): A table of elliptic integrals of the second kind, Math. Comp. 49 (1987) 595–606 and S13–S17

207. Carlson B.C. (1988): A table of elliptic integrals of the third kind, Math. Comp. 51 (1988) 267–280 and S1–S5

208. Carlson B.C. (1989): A table of elliptic integrals: Cubic cases, Math. Comp. 53 (1989) 327–333

209. Carlson B.C. (1991): A table of elliptic integrals: One quadratic factor, Math. Comp. 56 (1991) 267–280

210. Carlson B.C. (1992): A table of elliptic integrals: Two quadratic factors, Math. Comp. 59 (1992) 165–180

211. Carlson B.C. (1995): Numerical computation of real or complex elliptic integrals, Numer. Algorithms 10 (1995) 13–26

212. Carlson B.C., Notis E.M. (1981): Algorithm 577. Algorithms for incomplete elliptic integrals, ACM Trans. Math. Software 7 (1981) 398–403

213. Carstensen L.W. (1986): Hypothesis testing using univariate and bivariate choropleth maps, The American Cartographer 13 (1986) 231–251

214. Carter S., West A. (1972): Tight and taut immersions, Proc. London Math. Soc. 25 (1972) 701–720

215. Castellv P. (1994): TTC – Symbolic tensor and exterior calculus, Computers in Physics, 8 (1994) 360–367

216. Castner H.W. (1990): Seeking new horizons: A perceptual approach to geographic education, McGill–Queens University Press, Montreal 1990

217. Cauchy A. (1823): Recherches sur l'équilibre et mouvement intérieur des corps solides ou fluides, élastiques ou non élastiques, Bull. Soc. Philomath (1823) 9–13,

218. Cauchy A. (1827a): De la pression ou tension dans un corps solide, Ex. de Math. 2 (1827) 42–56

219. Cauchy A. (1827b): Sur les relations qui existent dans l'état d'équilibre d'un corps solide ou fluide entre les pressions ou tensions et les forces accélératrices, Ex. de Math. 2 (1827) 108–111

220. Cauchy A. (1828): Sur les équations qui expriment les conditions d'équilibre ou les lois du mouvement intérieur d'un corps solide, élastique, ou non élastique, Ex. De Math. 3 (1828) 160–187

221. Cauchy A. (1829): Sur l'équilibre et le mouvement intérieur des corps considérés comme des masses continues, Ex. de Math. 4 (1829) 293–319

222. Cauchy A. (1850): Mémoires sur les systèmes isotropes des points matériels, Mem. Acad. Sci. 22 (1850) 615

223. Cauchy A. (1889): Oeuvres complètes, Iie série, tome VII, pp. 82–93, Gauthier–Villars et Fils, 1889

224. Cauchy A. (1890): Sur l'équilibre et le mouvement d'un système de points matériels sollicités par des forces d'attraction ou de répulsion mutuelle, Oeuvres Complètes, Iie série, tome VIII, pp. 227–252, Gauthier–Villars et Fils, 1890

225. Cayley A. (1871): On the surfaces divisible into squares by their curves of curvature, Proc. London Math. Soc. IV (1871) 8–9

226. Cazenave A., Nerem R.S. (2004): Present–day sea level change: Observations and causes, Review of Geophysics 42 (2004) 1–20

227. Cecil T.E. (1976): Taut immersions of noncompact surfaces into a Euclidean 3–space, J. Differential Geometry 11 (1976) 451–459

228. Cecil T.E., Ryan P.J. (1978): Focal sets, taut embeddings and the cyclides of Dupin, Math. Ann. 236 (1978) 177–190

229. Cecil T.E., Ryan P.J. (1980): Conformal geometry and the cyclides of Dupin, Canadian Journal of Mathematics 32 (1980) 767–783

230. Chamberlin W. (1950): The round earth on flat paper, National Geographic Society, Washington 1950

231. Chan Y.M., He X. (1993): On median–type estimators of direction for the von Mises–Fisher distribution, Biometrika 80 (1993) 869–875

232. Chatelin F. (1993): Eigenvalues of matrices, J. Wiley & Sons, New York Chichester 1993

233. Chebyshev P.L. (1962): Sur la construction des cartes géographiques, in Oevre I, Chelsea, New York (1962) 233–236, 239–247

234. Chen B.Y. (1973): Geometry of submanifolds, Marcel Dekker Inc., New York 1973

235. Chen B.Y., Deprez J., Dillen F., Verstraelen L., Vrancken L. (1988): Curves of finite type, in: Geometry and topology of submanifolds, II, Boyom M., Morvan J.-M., Verstraelen L. (Hrsg.), pages 76–110, World Scientific, Singapore 1990

236. Chen B.Y., Yano K. (1973): Special conformally flat spaces and canal hypersurfaces, Tohoku Math. Journ. 25 (1973) 177–184

237. Chen J.Y. (1980): On the geodetic problem of long distances in two different projections, Zeitschrift für Vermessungswesen 105 (1980) 256–271

238. Cheney M. (2001): A mathematical tutorial on synthetic aperture radar, Siam Review 43 (2001) 301–312

239. Chern S.S. (1955a): La géométrie des sous–variétés d'un espace Euclidien a plusieurs dimensions, in: L'Enseignement Mathématique, Fehr H., Buhl A., pages 26–46, Librarie de l'Université Georg & Cie S.A., Genève 1955

240. Chern S.S. (1955b): An elementary proof of the existence of isothermal parameters on a surface, Proc. American Math. Soc. 6 (1955) 771–782

241. Chern S.S. (1965): Minimal surfaces in an Euclidean space of N dimensions, in: Differential and combinatorial topology, Stewart S. Cairns (Hrsg.), pages 187–199, Princeton University Press, Princeton 1965

242. Chern S.S. (1967a): Complex manifolds without potential theory, D. Van Nostrand Comp., Princeton N.J. 1967

243. Chern S.S. (1967b): Studies in global geometry and analysis, Math. Ass. America, Prentice Hall, Englewood Cliffs 1967

244. Chern S.S., Hartman P., Wintner A. (1954): On isothermic coordinates, Commentarii Mathematici Helvetici 28 (1954) 301–309

245. Chern S.S: (1948): On the multiplication in the characteristic ring of a sphere bundle, Annals of Mathematics 49 (1948) 362–372

246. Chovitz B. (1952): Classification of map projections in terms of the metric tensor of second order, Boll. di Geodesia e Scienze Affini 11 (1952) 379–394

247. Chovitz B. (1954): Some applications of the classification of map projections in terms of the metric tensor of the second order, Bollettino die Geodesia e Scienze Affini 13 (1954) 47–67

248. Chovitz B. (1956): A general formula for ellipsoid–to–ellipsoid mapping, Bollettino di Geodesia e Scienze Affini 15 (1956) 1–20

249. Chovitz B. (1978): Perspective projections in terms of the metric tensor to the second order, Bollettino di Geodesia e Scienze Affini 37 (1978) 451–463

250. Chovitz B. (1979): A general theory of map projections, Bollettino di Geodesia e Scienze Affini 38 (1979) 457–479

251. Christensen A.H.J. (1992): The Chamberlin trimetic projection, Cartography and Geographic Information Systems 19 (1992) 88–100

252. Christoffel E. (1865): Über die Bestimmung der Gestalt einer krummen Oberfläche durch lokale Messungen auf derselben, J. reine angew. Math. 64 (1865) 193–209

253. Christoffel E. (1867): Über einige allgemeine Eigenschaften der Minimumsflächen, J. reine angew. Math. 67 (1867) 218–228

254. Chu M.T. (1991a): A continuous Jacobi-like approach to the simultaneous reduction of real matrices, Linear Algebra Appl. 147 (1991) 75–96

255. Chu M.T. (1991b): Least squares approximation by real normal matrices with specified spectrum, SIAM J. Matrix Anal. Appl. 12 (1991) 115–127

256. Cimbálník M. (1987): Derived geometrical constants of the geodetic reference system 1980, Studia geoph. et geod. 31 (1987) 404–406

257. Claire C.N. (1968): State plane coordinates by automatic data processing: U.S. Coast and Geodetic Survey Pub. 62–4, 1968

258. Clarke A.R., Helmert F.R. (1911): Figure of the Earth: Encyclopedia Britannica 11th ed. 8 (1911) 801–813

259. Clarke K.C., Schweizer D.M. (1991): Measuring the fractal dimension of natural surfaces using a robust fractal estimator, Cartography and Geographic Information Systems 18 (1991) 37–47

260. Claussen H. (1995): Qualitätsanforderungen an die digitale Karte aus Anwendersicht, Mitteilungen der geodätischen Institute der TU Graz, Folge 80, Graz (1995) 33–40

261. Close C. (1921): Note on a doubly–equidistant projection, Geographical Journal 57 (1921) 446–448

262. Close C. (1929): An oblique Mollweide projection of the sphere, Geographical Journal 73 (1929) 251–253

263. Close C. (1934): A doubly equidistant projection of the sphere, Geographical Journal 83 (1934) 144–145

264. Close C., Clarke A.R. (1911): Map projections: Encyclopedia Britannica, 11th ed., 17 (1911) 653–663

265. Cody W.J (1965b): Chebyshev polynomial expansions of complete elliptic integrals, Math. Comp. 19 (1965) 249–259

266. Cody W.J. (1965a): Chebyshev approximations for the complete elliptic integrals K and E, Math. Comp. 19 (1965) 105–112, for corrigenda see same journal 20 (1966) 207

267. Cogley J.G. (1984): Map projections with freely variable aspect, Eos 65 (1984) 481–482

268. Cohn H. (1967): Conformal mapping on Riemann surfaces, McGraw–Hill book company, New York 1967

269. Cole J.H. (1943): The use of the conformal sphere for the construction of map projections: Survey of Egypt paper 46, Giza (Orman) 1943

270. Colvocoresses A.P. (1969): A unified plane co–ordinate reference system, World Cartography 9 (1969) 9–65

271. Conway J.B. (1975): Functions of complex variable, Springer–Verlag, New York – Heidelberg – Berlin (1975)

272. Cooke R. (1994): Elliptic integrals and functions, in I Grattan–Guinness (ed.), Companion Encyclopedia of the History and Philosophy of the Mathematical Sciences (1994) 529–539

273. Copson E.T. (1935): Conformal representation, in: An introduction to the theory of functions of a complex variable, pages 180–204, Clarendon Press, Oxford 1935

274. Corbley K.P. (1998): Image maps created for Odra and Morava Floods of '97, GeoInformatics, September (1998) 21–25

275. Cossin J. (1570): Carte cosmographique ou universelle description du monde avec le vrai traict des vens, faict en Dieppe par Johan Cossin, marinnier en l'an 1570, ms. World map. Bibliothèque Nationale, Départment des Cartes et Pins, GE D 7896, 1570

276. Cox D., Little J., O'Shea D. (1996): Ideals, varieties and algorithms, 2nd ed., Spinger, Berlin – Heidelberg – New York 1996

277. Coxeter H.S.M. (1972): The mathematical implications of Escher's prints, in: The world of M.C. Escher, M.C. Escher and J.L. Locher (Hrsg.) pages IX–X, Abrams, New York 1972

278. Craig T. (1882): A treatise on projections, in: United States Coast and Geodetic Survey, Carlile P. Patterson (Hrsg.), pages 133–187, Government Printing Office, Washington 1882

279. Craster J.E.E. (1929): Some equal–area projections of the sphere, The Geographical Journal 74 (1929) 471–474

280. Craster J.E.E. (1938): Oblique conical orthomorphic projection for New Zealand, The Geographical Journal 92 (1938) 537–538

281. Critchfield C.L. (1989): Computation of elliptic functions, J. Math. Phys. 30 (1989) 295–297

282. Crocetto N., Russo P. (1994): Helmert's projection of a ground point onto the rotational reference ellipsoid in topocentric cartesian coordiantes, Bulletin Géodésique 69 (1994) 43–48

283. Croft S.K. (1992): Proteus: Geology, shape, and catastrophic destruction, Icarus 99 (1992) 402–419

284. Cromley R.G. (1991): Hierarchical methods of line simplification, Cartography and Geographic Information Systems 18 (1991) 125–131

285. Crowell R.H., Fox R.H. (1963): Introduction to knot theory, Ginn and Company, Boston 1963

286. Crumeyrolle A. (1990): Orthogonal and sympletic Clifford algebras, Kluwer Academic Publisher, Dordrecht - Boston - London 1990

287. Czeczor H.E. (1981): Die internationale Weltkarte (IWK) 1 : 1 000 000 für das Gebiet der Bundesrepublik Deutschland, Nachrichten aus dem Karten– und Vermessungswesen 86 (1981) 47–52

288. Dahlberg R.E. (1962): Evolution of interrupted map projects, in: Internationales Jahrbuch für Kartographie, ed. E. Imhof, C. Bertelsmann Verlag, Gütersloh 1962

289. Danielsen J. (1989): The area under the geodesic, Survey Review 30 (1989) 61–66

290. Darboux G. (1899): Sur les surfaces isothermiques, Comptes Rendus 122 (1899) 1299–1305, 1483–1487, 1538

291. Davidson E.R. (1993): Monster matrices: their eigenvalues and eigenvectors, Comput. Phys. 7 (1993) 519–522

292. Davies A. (1949): An interrupted zenithal world map, The Scottish Geographical Magazine 65 (1949) 1–7

293. Davies M.E. (1983): The shape of Io, IAU Colloquium 77, Cornell University, 1983

294. Davies M.E., Abalakin V.K., Bursa M., Lederle T., Lieske J.H., Rapp R.H., Seidelman P.K., Sinclair A.T., Teifel V.G., Tjuflin Y.S. (1986): Report of the IAU/IAG/COSPAR working group on cartographic coordinates and rotational elements of the planets and satellites; 1985, Celestial Mechanics 39 (1986) 103–113

295. Davies M.E., Abalakin V.K., Lieske J.H., Seidelman P.K., Sinclair A.T., Sinzi A.M., Smith B.A., Tjuflin Y.S. (1986): Report of the IAU working group on cartographic coordinates and rotational elements of the planets and satellites; 1982, Celestial Mechanics 29 (1983) 309–321

296. Davies M.E., Batson R.M. (1975): Surface coordinates and cartography of Mercury: Jour. Geophys. Research 80 (1975) 2417–2430

297. Davies M.E., Katayama F.Y. (1983): The control networks of Mimas and Enceladus, Icarus 53 (1983) 332–340

298. Day J.W.R. (1988): The formula for finding the ordinary latitude from the isometric latitude, Survey Review 29 (1988) 383–386

299. De Floriani L., Puppo E. (1988): Constrained Delauney triangulation for multiresolution surface description, IEEE (1988) 566–569

300. De Maupertuis P.L.M. (1744): Accord de différentes lois de la Nature. Qui avoient jusqu'ici paru incompatibles. de l'Académie Royal des Sciences de Paris 1744. Reprinted in 1965 Oeuvres IV (1744) 3–28

301. Deakin R.E. (1990): A minimum–error equal–area pseudocylindrical map projection, Cartography and Geographic Information Systems 17 (1990) 161–167

302. De Azcárraga J.A., Izquierdo J.M. (1995): Lie Groups, Lie Algebras, Cohomology and some Applications in Physics, Cambridge Monographs on Mathematical Physics 1995

303. Debenham F. (1958): The global atlas–a new view of the world from space, Simon and Schuster, New York 1958

304. Deetz C.H. (1918a). The Lambert conformal conic projection with two standard pararells, including a comparison of the Lambert projection with the Bonne and Polyconic projections: U.S. coast and Geodetic Survey Spec. Pub. 47, 1918

305. Deetz C.H. (1918b): Lambert projection tables with conversion tables: U.S. Coast and Geodetic Survey Spec. Pub.

306. Deetz C.H., Adams O.S. (1934): Elements of map projection, U.S. Coast and Geodetic Survey Special Publication No. 68, 4th ed., 1934

307. Defense Mapping Agency (1981): Glossary of mapping, charting, and geodetic terms, U.S. Government Printing Office, Washington 1981

308. Degn C. et al. (1956): Seydlitz allgemeine Erdkunde hrsg., Hannover 1956

309. Deimler W. (1914): Konforme Abbildung des ganzen Erdellipsoids auf die Kugel, Abhandlungen der Königlich Bayerischen Akademie der Wissenschaften, Mathematisch–Physikalische Klasse, 27. Abhandlung, Verlag der Königlich Bayerischen Akademie der Wissenschaften, München 1914

310. Demmel J. et al (1999): Computing the singular value decomposition with high relative accuracy, Linear Algebra and its Applications 299 (1999) 21–80

311. De Moor B., Zha H. (1991): A tree of generalizations of the ordinary singular value decomposition, Linear Algebra Appl. 147 (1991) 469–500

312. Dermanis A., Livieratos E. (1983a): Applications of deformation analysis in geodesy and geodynamics, Reviews of Geophysics and Space Physics 21 (1983) 229–238

313. Dermanis A., Livieratos E. (1983b): Applications of strain criteria in cartography, Bulletin Géodésique 57 (1983) 215–225

314. Dermanis A., Livieratos E. (1993): Dilatation, shear, rotation and energy: analysis of map projections, Bollettino di Geodesia e Science Affini 42 (1993) 53–68

315. Dermanis A., Livieratos E., Pertsinidou S. (1984): Deformation analysis of geoid to ellipsoid mapping, Quaterniones geodaesiae 4 (1984) 225–240

316. Dermott S.F. (1979): Shapes and gravitational moments of satellites and asteroids, Icarus 37 (1979) 575–586

317. Dermott S.F. (1984): Rotation and the internal structures of the major planets and their inner satellites, Phil. Trans. R. Soc. Lond. A 313 (1984) 123–139

318. Deszcz R. (1990): Examples of four–dimensional Riemannian manifolds satisfying some pseudo–symmetry curvature conditions, in: Geometry and topology of submanifolds, II, Boyom M., Morvan J.-M., Verstraelen L. (Hrsg.), World Scientific, Singapore 1990

319. Deturck D.M., Yang D. (1984): Existence of elastic deformations with prescribed principal strains and triply orthogonal systems, Duke Mathematical Journal 51 (1984) 243–260

320. Di Francesco P., Mathieu P., Sénéchal D. (1997): Conformal field theory, Springer–Verlag, New York 1997

321. Dickmann F., Zehner K. (2001): PC–basierte Kartographie und GIS–Software – Ein Produktvergleich, in: J. Dodt and W. Herzog (Eds.): Kartographisches Taschenbuch 2001, Kirschbaum Verlag, Bonn 2001, 50–64

322. Dillen F., Vrancken L. (1990). Affine differential geometry of hypersurfaces, in: Geometry and Topology of Submanifolds, II, Avignon (1990) 144–165

323. Dingeldey F. (1910): Kegelschnittsysteme, in: Repetitorium der höheren Mathematik, Vol. 2 (Geometrie), ed.: E. Pascal, Verlag B.G. Teubner, Lepzig (1910) 246–247

324. Do Carmo M.P. (1994): Differential forms and applications, Springer 1994

325. Do Carmo M., Dajczer M., Mercuri F. (1985): Compact conformally flat hypersurfaces, Transactions of the American Mathematical Society, 288 (1985) 189–203

326. Dodt J., Herzog W. (2001): Kartographisches Taschenbuch 2001, Kirschbaum Verlag, Bonn 2001

327. Dombrowski P. (1979): 150 years after Gauss' "disquisitiones generales circa superficies curvas", asterisque 62, société mathematique de France, Paris 1979

328. Donaldson S.K., Kronheimer P.B. (1990): The geometry of four–manifolds, Clarendon Press, Oxford 1990

329. Dorrer E. (1966): Direkte numerische Lösung der geodätischen Hauptaufgaben auf Rotationsflächen, Report C90, Deutsche Geodätische Kommission, Bayer. Akad. Wiss., München 1966

330. Dorrer E. : From elliptic arc length to Gauss–Krueger coordinates by analytical continuation, in: E. Grafarend, F. Krumm and V. Schwarze: Geodesy – the challenge of the 3rd Millenium, pp. 293–298, Springer Verlag, Berlin – Heidelberg – New York 2003

331. Dozier J. (1980): Improved algorithm for calculation of UTM and geodetic coordinates, NOAA Tech. Rept. NESS 81, 1980

332. Dracup J.F. (1998): A fresh look at tangent plane grids, Surveying and Land Information Systems 58 (1998) 205–221

333. Draheim H. (1986): Die Kartographie in Geschichte und Gegenwart, Kartographische Nachrichten 36 (1986) 161–172

334. Dreincourt L., Laborde J. (1932): Traité des projections des cartes géographiques: Herman et cie., Paris 1932

335. Duda F.P., Martins L.C. (1995): Compatibility conditions for the Cauchy–Green strain fields: Solutions for the plane case, Journal of Elasticity 39 (1995) 247–264

336. Dufour H.M. (1971): La projection stereographique de la sphere et de l'ellipsoide, Institut Geographique National, 1971

337. Dufour H.M. (1976): Usage systematique de la projection stereographique pour les transformations de coordonnees planes, Institut Geographique National, 1976

338. Dufour H.M. (1989a): Le système icostereographique, Institute geographique national, 1989, p. 75

339. Dufour H.M. (1989b) : La division de la sphère en éléments de meme surface et de formes voisines, avec hiérarchisation, C.R. Acad. Sci. Paris, 309 (1989) 307–310

340. Dufour H.M. (1991) : A proposal for a unique quasi–regular grid on a sphere, Manuscripta Geodaetica 16 (1991) 267–273

341. Dumitrescu V. (1968): Cartographic solution for deciphering space–photographs, Internationales Jahrbuch für Kartographie 8 (1968) 66–74

342. Dumitrescu V. (1974): Kosmographische Perspektiv–Projektionen, Allgemeine Vermessungsnachrichten 81 (1974) 142–151

343. Dumitrescu V. (1977a): Cosmographic perspective projections – The mathematical model of space–photographs, in: Studies in theoretical cartography, ed. I. Kratschmer, Deuticke, Vienna 1977

344. Dumitrescu V. (1977b): Cosmographic perspective projections, Beiträge zur theor. Kartographie, Festschrift f. E. Arnberger, Deuticke, Wien (1977) 91–106

345. Dumont D. (1981): Une approach combinatoire des fonctions elliptiques de Jacobi, Adv. Math. 41 (1981) 1–39

346. Dunkl C.F., Ramirez D.E. (1994a): Algorithm 736, Hyperelliptic intergrals and the surface measure of ellipsoids, ACM Trans. Math. Software 20 (1994) 427–435

347. Dunkl C.F., Ramirez D.E. (1994b): Computing hyperelliptic integrals for surface measure of ellipsoids, ACM Trans. Math. Software 20 (1994) 413–426

348. Dyer J.A., Snyder J.P. (1989): Minimum–error equal–area map projections, The American Cartographer 16 (1989) 39–43

349. Ecker E (1976): Über die Gauss–Krueger Abbildung, Österreichische Zeitschrift für Vermessungswesen und Photogrammetrie 65 (1976) 108–117

350. Ecker E. (1978): Conformal mappings of the earth ellipsoid, Manuscripta Geodaetica 3 (1978) 229–251

351. Ecker E. (1980): Über die inverse Gauss–Krueger–Abbildung, Österreichische Zeitschrift für Vermessungswesen und Photogrammetrie 68 (1980) 71–78

352. Eckert M. (1906): Neue Entwürfe für Erdkarten, Petermanns Geographische Mitteilungen 5 (1906) 97–109

353. Eckert–Greifendorff M. (1935): Eine neue flächentreue (azimutaloide) Erdkarte, Petermann's Mitteilungen 81 (1935) 190–192

354. Eckmann B. (1968): Continuous solutions of linear equations – some exceptional dimensions in topology, in: Battelle Rencontres, ed. C.M. DeWitt and J.A. Wheeler, W.A. Benjamin Inc., New York 1968

355. Edelman A., Elmroth E., Kagström B. (1997): A geometric approach to perturbation theory of matrices and matrix pencils. Part I: Versal deformations, SIAM J. Matrix Anal. Appl. 18 (1997) 653–692

356. Edwards H.M. (1994): Advanced calculus, a differential forms approach, Birkhäuser Verlag, Boston – Basel – Berlin 1994

357. Eels J., Lemaire, L. (1978): A report on harmonic maps, Bull. London Math. Soc. 10 (1978) 1–68

358. Egeltoft T., Stoimenov G. (1997): Map projections, Report 2017, Royal Institute of Technology, Dept. of Geodesy and Photogrammetry, Stockholm 1997

359. Egenhofer M.J. (1991): Extending SQL for graphical display, Cartography and Geographic Information Systems 18 (1991) 230–245

360. Eggert O. (1936): Die stereographische Abbildung des Erdellipsoids, Zeitschrift für Vermessungswesen 65 (1936) 153–164

361. Ehlert D. (1983a): Beziehungen zwischen Ellipsoidprametern, Sonderdruck aus: Nachrichten aus dem Karten–und Vermessungswesen, Reihe I: Originalbeiträge 91 (1983) Verlag des Instituts für Angewandte Geodäsie, Frankfurt a.M. 1983

362. Ehlert D. (1983b): Die Bessel–Helmertsche Lösung der beiden geodätischen Hauptaufgaben, Zeitschrift für Vermessungswesen 108 (1983) 495–500

363. Eisenhart L.P. (1929): Dynamical trajectories and geodesics, Annals of Mathematics, Princeton University Press, Princeton 1929

364. Eisenhart L.P. (1949): Riemannian geometry, Princeton University Press, Princeton 1949

365. Eisenhart L.P. (1961): Continuous groups of transformations, Dover Publ. New York 1961

366. Eisenlohr F. (1870): Über Flächenabbildung, Journal für reine und angewandte Mathematik 72 (1870) 143–151

367. Eitschberger B. (1978): Ein geodätiches Weltdatum aus terrestrischen und Satellitendaten, Report C 245, Deutsche Geodätische Kommision, Bayer Akad. Wiss., München 1978

368. Ekman M. (1996): The permanent problem of the permanent tide: What to do with it in geodetic reference systems? B.I.M. 125 (1996) 9508-9513

369. Embacher W. (1980): Ein Versuch zur Bestimung des gestörten Schwerevektors aus lokalen Gravimetermessungen, Zeitschrift für Vermessungswesen 105 (1980) 245–255

370. Engels J., Grafarend E.W. (1995): The oblique Mercator projection of the ellipsoid of revolution $\mathbb{E}^2_{A,B}$, J. Geodesy 70 (1995) 38–50

371. Eringen A.C. (1962): Nonlinear theory of continuous media, McGraw–Hill Book Company, New York 1962

372. Euler L. (1770): Sectio secunda de principiis motus fluidorum, Novi. Comm. Acad. Sci. Petrop 14 (1769) 270–386

373. Euler L. (1755) : Principes généraux des mouvement des fluides, Memoirs de l'Accad. des Sciences de Berlin 11 (1755) 274–315

374. Euler L. (1777a): Über die Abbildung einer Kugelfläche in die Ebene, Acta Academiae Scientiarum Petropolitanae, Petersburg 1777

375. Euler L. (1777b) : De repraesentatione superificiei sphaericae super plano, Acta Acad. Scient. Imperial. Petropolitanae pro anno, T.I. (1777) 107–132

376. Euler L. (1898): Drei Abhandlungen über Kartenprojection. Ostwald's Klassiker der exakten Naturwissenschaften, Nr. 93, Engelmann, Leipzig 1898

377. Eyton J.R. (1991): Rate–of–change maps, Cartography and Geographic Information Systems 18 (1991) 87–103

378. Faber G. (1907): Einfaches Beispiel einer stetigen nirgends differentiierbaren Funktion, Jahresbericht der deutschen Mathematiker–Vereinigung 16 (1907) 538–540

379. Fair W.G., Luke Y.L. (1967): Rational approximations to the incomplete elliptic integrals of the first and second kinds, Math. Comp. 21 (1967) 418–422

380. Fairgrieve J. (1928): A new projection, Geography (Manchester) 14 (1928) 525–526

381. Falcidiendo B., Spagnuolo M. (1991): A new method for the characterization of topographic surfaces, Int. J. Geographical Information Systems 5 (1991) 397–412

382. Fang T.-P., Piegl L.A. (1993): Delaunay triangulation using a uniform grid, IEEE Computer Graphics & Applications 5 (1993) 36–47

383. Farr T. et al. (1995): The global topography mission gains momentum, EOS, Transactions 76 (1995) 1–4

384. Fary I. (1949): Sur la courbure d'une courbe gauche faisant un nœud, Bulletin Soc. Math. France 77 (1949) 128–138

385. Fawcett C.B. (1949): A new net for a world map, Geographisches Journal 114 (1949) 68–70

386. Featherstone W. (1997): The importance of including the geoid in terrestrial survey data reduction to the geocentric datum of Australia, Australian Surveyor 6 (1997)

387. Featherstone W., Barrington T.R. (1996): A Microsoft Windows–based package to transform coordinates to the geocentric datum of Australia, Cartography 25 (1996) 81–87

388. Featherstone W., Vaníček P. (1999): The role of coordinate systems, coordinates and heights in Horizontal Datum Transformations, The Australian Surveyor 44 (1999)

389. Feeman T.G. (2000): Equal area world maps: A case study, SIAM Review 42 (2000) 109–114

390. Feeman T.G. (2002): Portraits of the Earth. A Mathematician Looks at Maps, American Mathematical Society 2002

391. Feigenbaum M.J. (1994): Riemann maps and world maps, in: Trends and perspectives in applied mathematics, ed. L. Sirovich, Springer Verlag, Berlin Heidelberg New York 1994, pp. 55–71

392. Ferrara S., Grillo A.F., Gatto R. (1972): Conformal algebra in two space-time dimensions and the Thirring model, Il Nuovo Cinmento 12A (1972) 959–968

393. Ferus D., Pedit F., Pinkall U., Sterling I. (1992): Minimal tori in S4, J. reine angew. Math. 429 (1992) 1–47

394. Fiala F. (1957): Mathematische Kartographie, VEB–Verlag Technik, Berlin 1957

395. Fialkow A. (1939): Conformal geodesics, Transactions of the American Mathematical Society 45 (1939) 443–473

396. Fierro D.R., Bunch J.R. (1994): Collinearity and total least squares, SIAM J. Matrix Anal. Appl. 15 (1994) 1167–1181

397. Finger J (1894a): Über die allgemeinsten Beziehungen zwischen den Deformationen und den zugehörigen Spannungen in aerotropen und isotropen Substanzen, Sitzber. Akad. Wiss. Wien (2a) 103 (1894) 1073–1100

398. Finger J. (1894b): Das Potential der inneren Kräfte..., Sitzber. Akad. Wiss. Wien (2a) 103 (1894) 163–200

399. Finsler P. (1918): Über Kurven und Flächen in allgemeinen Räumen, Dissertation, Universität Göttingen 1918, Nachdruck Birkhäuser Verlag, Basel 1951

400. Finsterwalder R. (1993): Die "betrachtungstreue" Azimutalprojektion, Kartographische Nachrichten 43 (1993) 234–236

401. Finzi A. (1922): Sulle varieta in rappresentazione conforme con la varieta euclidea a piu di tre dimensioni, Rend. Acc. Lincei Classe Sci Ser 5 31 (1922) 8–12

402. Firneis M.G., Firneis J. (1980): Zur symmetrischen Ableitung der Halbwinkelformeln der sphärischen Trigonometrie, Zeitschrift für Vermessungswesen 105 (1980) 271–278

403. Fite E.D., Freeman A. (1926): A book of old maps delineating American history from the earliest days down to the close of the Revolutionary War, Harvard Univ. Press, reprint 1969, Dover Publications, Cambridge 1926

404. Fitzgerald J.E. (1980): Tensorial Hencky measure of strain and strain rate for finite deformations, J. Appl. Phys. 51 (1980) 5111–5115

405. Flanders H. (1967): Differential forms, in: Studies in Mathematics, ed. S.S. Chen, Vol. 4, pp. 57–95, Prentice Hall 1967

406. Flanders H. (1970): Differential forms with applications to the Physical sciences, 4th printing, Academic Press London 1970

407. Foley T.A., Lane D.A., Nielson G.M. (1990): Interpolation of scattered data on closed surfaces, Computer Aided Geometric Design 7 (1990) 303–312

408. Forbes V.L. (1996): Archipelagic Sea Lanes: The Indonesian Case, The Indian Ocean Review June (1996) 10–14

409. Forsyth A.R. (1895): Conjugate points of geodesics on an oblate spheroid, Messenger of mathematics 25 (1895) 161–169

410. Forsyth A.R. (1918): Theory of functions of a complex variable, 3rd edition, Cambridge, England: Cambridge University Press 1918

411. Foucaut H.C. de Prépetit (1862) : Notice sur la construction de nouvelles mappemondes et de nouveaux atlas de geographie, Arras, France (1862) 5–10

412. Fox R.H. (1941): On the Lusternik–Schnirelmann category, Annals of Mathematics 42 (1941) 333–370

413. Francula N. (1971): Die vorteilhaftesten Abbildungen in der Atlaskartographie, Inaugural–Dissertation, Hohe Landwirtschaftliche Fakultät, Friedrich–Wilhelms–Universität, Bonn 1971

414. Francula N. (1980): Über die Verzerrungen in den kartographischen Abbildungen, Kartographische Nachrichten 30 (1980) 214–216

415. Francula N. (1981): Erwiderung auf die Anmerkungen zu einer Theorie kartographischer Abbildungen, Kartographische Nachrichten 31 (1981) 190–191

416. Francula N. (1985): Inverse Abbildungsfunktionen der echten kartographischen Abbildungen, in: Betrachtungen zur Kartographie: Eine Festschrift für Aloys Heupel zum 60. Geburtstag, ed. Institut für Kartographie und Topographie der Rheinischen Friedrich–Wilhelms–Universität Bonn, Kirschbaum Verlag, Bonn 1985

417. Frank A. (1940): Beiträge zur winkeltreuen Abbildung des Erdellipsoides, Zeitschrift für Vermessungswesen 65 (1940) 97–112, 145–160, 193–204

418. Franke R. (1982): Scattered data interpolation: Tests of some methods, Mathematics of Computation 38 (1982) 183–200

419. Frankich K. (1980): Mathematical cartography part one: Geographic map projections, University of Calgary, Calgary 1980

420. Frankich K. (1982): Optimization of geographic map projections for Canadian territory, PhD thesis, Simon Fraser University 1982

421. Frauendiener J., Friedrich H. (eds.) (2002): The conformal structure of space–time–geometry, analysis, numerics, Springer, Berlin – Heidelberg – New York 2002

422. Freed D.S., Uhlenbeck K.K. (1984): Instantons and four–manifolds, Springer Verlag, Berlin – Heidelberg – New York 1984

423. Freund P.G.O. (1986): Introduction to Supersymmetry, Cambridge University Press, Cambridge 1986

424. Fricke R. (1913): Elliptische Funktionen, in Encyklopädie der mathematischen Wissenschaften 2 (1913) 177–348

425. Friedmann A. (1965): Isometric embedding of Riemannian manifolds into Euclidean spaces, Reviews of Modern Physics 37 (1965) 201–203

426. Friedrich D. (1998): Krummlinige Datumstransformation – Herleitung und Vergleich unterschiedlicher Berechnungsarten, Studienarbeit Geodätisches Institut Universität Stuttgart 1998

427. Fritsch D., Walter V. (1998): Comparison of ATKIS and GDF data structures, in: Symposium on geodesy for geotechnical and structural engineering, April 20–22, 1998, Eisenstadt, Austria, publ. TU Wien, Abtlg. Ingenieurgeodäsie, Wien 1998

428. Fritsch R., Fritsch G. (1998): The four–color theorem, Springer–Verlag, New York 1998

429. Froehlich H., Hansen H.-H. (1976): Zur Lotfußpunktberechnung bei rotationsellipsoidischer Bezugsfläche, Allgemeine Vermessungsnachrichten 83 (1976) 175–178

430. Froehlich H., Krieg B., Vente S. (1996): Darstellung der Breiten– und Längenunterschiede zwischen den Systemen ETRS 89 und DHDN bzw. S42/83 für das Gebiet der Bundesrepublik Deutschland, Forum 22 (1996) 294–305

431. Froehlich H., Tenhaef M., Körner H. (2000): Geodätische Koordinatentransformationen: Ein Leitfaden, Essen 2000

432. Frolov Y.S. (1963): Method of comparative evaluation of cartographic projections, Department of Cartography, Leningrad State University, Leningrad 1963

433. Fuchs W.R. (1967): Der "kalkulierte" Zufall in der Physik, Naturwissenschaft und Medizin 20 (1967)20–29

434. Fukushima T., Ishizaki H. (1994): Numerical computation of incomplete elliptic integrals of a general form, Celestial Mech. Dynam. Astronom. 59 (1994) 237–251

435. Furtwaengler P., Wiechert E. (1906–1925): Encyklopädie der mathematischen Wissenschaften mit Einschluß ihrer Anwendungen, Band 6, 1. Teil: Geodäsie und Geophysik, B.G. Teubner Verlag, Leipzig 1906–1925

436. Gabriel R. (1979): Matrizen mit maximaler Diagonale bei unitärer Similarität , J. reine und angew. Math. 307/308 (1979) 31–52

437. Gade, K. (2005): NAVLAB – A Generic simulation and post–processing tool for navigation, Hydrographische Nachrichten 75 (2005) 4–13

438. Gaier D. (1983): Numerical methods in conformal mapping, in: H. Werner et al. (eds.), Computational Aspects of Complex Analysis, Reidel Publishing Company 1983, 51–78

439. Gall Rev. J. (1885): Use of cylindrical projections for geographical, astronomical, and scientific purposes, Scottish Geographical Magazine 1 (1885) 119–123

440. Gallot S., Hulin D., Lafontaine J. (1987): Riemannian Geometry, Springer Verlag, Berlin – Heidelberg – New York 1987

441. Gannett S.S. (1904): Geographic tables and formulas, 2nd ed., U.S. Geol. Survey Bull. (1904) 234

442. Gao Y., Lahaye F., Heroux P., Liao X., Beck N., Olynik M. (2001): Modeling and estimation of C1–P1 bias in GPS receivers, J. Geodesy 74 (2001) 621–626

443. Garabedian P.R., Spencer D.C. (1952): Complex boundary value problems, Transactions of the American Mathematical Society 73 (1952) 223–242

444. Gargiulo R., Vassallo A. (1997): La "Total Station" con algoritmi generalizzati per la soluzione dei problemi fondamentali della topografia, Bollettino della SIFET 4 (1997) 121–142

445. Gargiulo R., Vassallo A. (1998b): The spatial solution of the first fundamental geodetic problem, Survey Review 34 (1998) 405–412

446. Gargiulo R., Vassallo, A. (1998a): Complementi di geodesia geometrica topografia e cartografia nautica analitica, Genova 1998

447. Gartner G., Popp A. (1995): Kartographische Produkte für Flugpassagiere, Kartographische Nachrichten 45 (1995) 96–107

448. Garver J.B. (1988): New perspective on the world, National Geographic 12 (1988) 910–914

449. Gauss C.F. (1813): Vier Notizen über Inversion der Potenzreihen, Abhandlungen der Königl. Gesellschaft der Wiessenschaft zu Göttingen, Bd. VIII, 69–75, Göttingen 1900

450. Gauss C.F. (1816-1827): Conforme Abbildung des Sphäroids in der Ebene, Abhandlungen der Königl. Gesellschaft der Wissenschaften zu Göttingen, Bd. IX, 142–194

451. Gauss C.F. (1822): Allgemeine Auflösung der Aufgabe, die Teile einer gegebenen Fläche auf einer anderen gegebenen Fläche so abzubilden, daß die Abbildung dem Abgebildeten in den kleinsten Teilen ähnlich wird, 1822, Abhandlungen Königl. Gesellschaft der Wissenschaften zu Göttingen, Bd. IV 189–216, Göttingen 1838

452. Gauss C.F. (1827): Disquisitiones generales circa superficies curvas, Commentationes Societatis Regiae Scientiarum Gottingensis Recentioris, vol. 6, Göttingen 1827, English: 150 years after Gauss, Dombrowski, P. (ed.) Société Mathématique de France, Asterique 62 (1979) 3–81, Deutsch: Allgemeine Flächtentheorie von C.F. Gauss, Ostwald's Klassiker der Exakten Wissenschaften, Vol. 5, 5th edition, Akad. Verlagsgesellschaft, Leipzig 1921

453. Gauss C.F. (1828a): Conforme Abbildung des Sphäroids in der Ebene, Abhandlungen der Königl. Gesellschaft der Wissenschaften zu Göttingen, Nachlass (1828), Ges. Werke IX, pages 142–194, Göttingen 1903

454. Gauss C.F. (1828b): Conforme Doppelprojektion des Sphäroids auf die Kugel und die Ebene, Abhandlungen der Königl. Gesellschaft der Wissenschaften zu Göttingen, Nachlass (1828), Ges. Werke IX, pages 107–116, Göttingen 1903

455. Gauss C.F. (1828c): Conforme Übertragung des Sphäroids auf den Kegelmantel, Abhandlungen der Königl. Gesellschaft der Wissenschaften zu Göttingen, Nachlass (1828), Ges. Werke IX, pages 134–140, Göttingen 1903

456. Gauss C.F. (1828d): Stereographische Projection der Kugel auf die Ebene, Abhandlungen der Königl. Gesellschaft der Wissenschaften zu Göttingen, Nachlass (1828), Ges. Werke IX, pages 117–122, Göttingen 1903

457. Gauss C.F. (1828e): Übertragung der Kugel auf die Ebene durch Mercators Projection, Abhandlungen der Königl. Gesellschaft der Wissenschaften zu Göttingen, Nachlass (1828), Ges. Werke IX, pages 124–133, Göttingen 1903

458. Gauss C.F. (1828f): Zur Netzausgleichung, Abhandlungen der Königl. Gesellschaft der Wissenschaften zu Göttingen, Nachlass (1828), Ges. Werke IX, pages 298–183, Göttingen 1903

459. Gauss C.F. (1832): Intensitas vis magneticae terrestris ad mensuram absolutam revocata, Werke Bd. V 79–118, Göttingen 1832

460. Gauss C.F. (1838): Allgemeine Theorie des Erdmagnetismus, Werke Bd. V 119–193, Göttingen 1838

461. Gauss C.F. (1840): Allg. Lehrsätze in Beziehung auf die im verkehrten Verhältnis des Quadrates der Entfernung wirkenden Anziehungs– und Abstoßungskräfte, hg. von A. Wangerin (60 Seiten) in: Ostwald's Klassiker der exakten Wissenschaften

462. Gauss C.F. (1844): Untersuchungen über Gegenstände der höheren Geodäsie, erste Abhandlung, Abhandlungen der Königl. Gesellschaft der Wissenschaften zu Göttingen, Bd. 2 (1844), Ges. Werke IV pages 259–334, Göttingen 1880

463. Gauss C.F. (1847): Untersuchungen über Gegenstände der höheren Geodäsie, zweite Abhandlung, Abhandlungen der Königl. Gesellschaft der Wissenschaften zu Göttingen, Bd. 3 (1847), Ges. Werke IV pages 303–340, Göttingen 1880

464. Gauss C.F. (1894): Allgemeine Auflösung der Aufgabe: Die Theile einer gegebenen Fläche auf einer andern gegebenen Fläche so abzubilden, dass die Abbildung dem Abgebildeten in den kleinsten Theilen ähnlich wird, Ostwald's Klassiker der exakten Naturwissenschaften, Nr. 55, Leipzig 1894

465. Gauss C.F. (1900): Vier Notizen über Inversion der Potenzreihen (1822), Abhandlungen Königl. Gesellschaften der Wissenschaften zu Göttingen, Bd. VIII, 69–75, Göttingen 1900

466. Gerber D.E.P. (1987/88): Projektive Behandlung dreidimensionaler Netze der geometrischen Geodäsie, Siegerist Druck AG, Meisterschwanden 1998

467. Gere J.M., Weaver W. (1965): Matrix algebra for engineers, D. Van Nostrand, New York 1965

468. Gerlach C. (2003): Zur Höhensystemumstellung und Geoidberechnung in Bayern, Verlag der Bayerischen Akademie der Wissenschaften in Kommission bei Verlag C.H. Beck, München 2003 C571

469. Germain A. (1831): Mémoire sur la courbure des surfaces, Crelle's J. reine und angewandte Mathematik 7 (1831) 1–29

470. Germain A. (1865): Traité des projections des cartes géographiques, representation plane de la sphère et du sphèroide, Paris 1865

471. Gernet M. (1895) : Über Reduktion hyperelliptischer Integrale, Inagural–Dissertation, Druck von Friedrich Gutsch, Karlsruhe 1895

472. Gerstl M. (1984): Die Gauss–Kruegersche Abbildung des Erdellipsoides mit direkter Berechnung der elliptischen Integrale durch Landentransformation, Report C296, Deutsche Geodätische Kommission, Bayer. Akad. Wiss., München 1984

473. Gerstl M., Bulirsch R. (1983): Numerical evaluation of elliptic integrals for geodetic applications, Bollettino di Geodesia e Scienze Affini 42 (1983) 5–160

474. Ghitau D. (1996): Über Koordinatentransformationen in dreidimensionalen Systemen mit linearen Modellen, Zeitschrift für Vermessungswesen 121 (1996) 203–212

475. Gigas E. (1962): Die universale transversale Mercatorprojektion (UTM), Vermessungstechnische Rundschau 9 (1962) 329–334

476. Gilbarg D., Trudinger N.S. (1998): Elliptic partial differential equations of second order, Springer Verlag, Berlin – Heidelberg – New York 1998

477. Gilbert E.N. (1974): Distortion in maps, Siam Review 16 (1974) 47–62

478. Giorgi F. (1997a): Representation of heterogeneity effects in earth system modeling: Experience from land surface modeling, Reviews of Geophysics 35 (1997) 413–438

479. Giorgi F. (1997b): An approach for the representation of surface heterogeneity in land surface models, Part I: Theoretical Framework, Monthly Weather Review, American Meteorological Society, 1997

480. Glasmacher H. (1987): Die Gauss'sche Ellipsoid–Abbildung mit komplexer Arithmetik und numerischen Näherungsverfahren, Schriftenreihe Studiengang Vermessungswesen 29 (1987), Universität der Bundeswehr München, 1987

481. Glasmacher H., Krack K. (1984): Umkehrung von vollständigen Potenzreihen mit zwei Veränderlichen, in: 10 Jahre Hochschule der Bundeswehr München: Beiträge aus dem Institut für Geodäsie, ed. W. Caspary, A. Schoedlbauer und W. Welsch, Hochschule der Bundeswehr, München 1984

482. Glossary of the Mapping Sciences prepared by a joint committee of the American Society of Civil Engineers, American Congress on Surveying and Mapping, and American Society for photogrammetry and remote sensing, New York 1994

483. Goe G., van der Waerden B.L., Miller A.I. (1974): Comments on A.I. Miller's "The myth of Gauss" experiment on the Euclidean nature of physical space, Isis 65 (1974) 83–87

484. Goenner H., Grafarend E.W., You, R.J. (1994): Newton mechanics as geodesic flow on Maupertuis' manifolds: The local isometric embedding into flat spaces, Manuscripta Geodaetica 19 (1994) 339–345

485. Gold C.M. (1982): Neighbors, adjacency and theft – The Voronoi process for spatial analysis, European Conference on Geographic Information Systems (1982) 382–398

486. Goldstein M., Haussman W., Jetter, K. (1984): Best harmonic L1 approximation to subharmonic functions, J. London Math. Soc. 30 (1984) 257–264

487. Goldstein M., Haussman W., Rogge, L. (1988): On the mean value property of harmonic functions and best harmonic L1 approximation, Trans. Amer. Math. Soc. 305 (1988) 505–515

488. Golub G.H., van der Vorst H.A. (2000): Eigenvalue computation in the 20th century, J. Comput. Appl. Math. 123 (2000) 35–65

489. Golub G.H., Van Loan C.F. (1983): Matrix Computations, North Oxford Academic, Oxford (1983)

490. Goode J.P. (1925): The Homolosine projection: a new device for portraying the earth's surface entire, Assoc. Am. Geog. Annals 15 (1925) 119–125

491. Goode J.P. (1929): A new projection for the world map: The polar equal area, Annals of the Association of American Geographers 19 (1929) 157–161

492. Goussinsky B. (1951): On the classification of map projections, Empire Survey Review 11 (1951) 75–79

493. Gowdy R.H. (1995): Affine projection–tensor geometry: Lie derivatives and isometries, J. Math. Phys. 36 (1995) 1882–1907

494. Gradsteyn I.S., Ryzhik I.M. (1983): Table of Integrals, Series, and Products. Corrected and Enlarged Edition. Academic Press, New York

495. Graf F.X. (1955): Beiträge zur sphäroidischen Trigonometrie, Verlag der Bayerischen Akademie der Wissenschaften, Beck'sche Verlagsbuchhandlung, München 1955

496. Graf U. (1941): Über die Äquideformaten der flächentreuen Zylinderentwürfe, Petermanns Geographische Mitteilungen, 7/8 (1941) 281–290

497. Grafarend E.W. (1967): Bergbaubedingte Deformation und ihr Deformationstensor, Bergbauwissenschaften 14 (1967) 125–132

498. Grafarend E.W. (1969): Helmertsche Fußpunktkurve oder Mohrscher Kreis?, Allgemeine Vermessungsnachrichten 76 (1969) 239–240

499. Grafarend E.W. (1972): Hilbert–Basen zur Optimierung mehrdimensionaler Punktmannigfaltigkeiten, Zeitschrift für angewandte Mathematik und Mechanik 52 (1972) 240–241

500. Grafarend E.W. (1974): Optimization of geodetic networks, Bollettino di Geodesia e Scienze Affini 33 (1974) 351–406

501. Grafarend E.W. (1977a): Stress–strain relations in geodetic networks, Publ. Geodetic Institute, Uppsala University, No. 16, Uppsala 1977

502. Grafarend E.W. (1977b): Geodäsie: Gauss'sche oder Cartansche Flächengeometrie?, Allgemeine Vermessungsnachrichten 84 (1977) 139–150

503. Grafarend E.W. (1978): Dreidimensionale geodätische Abbildungsgleichungen und die Näherungsfigur der Erde, Zeitschrift für Vermessungswesen 103 (1978) 132–140

504. Grafarend E.W. (1979): Space–time geodesy, Boll. Geod. e Sci. Affini 38 (1979) 551–589

505. Grafarend E.W. (1981): Kommentar eines Geodäten zu einer Arbeit E.B. Christoffels. The influence of his work on mathematics and the physical sciences, eds. P.L. Butzer and F. Feher, Birkhäuser Verlag, Basel 1981, pp. 735–742

506. Grafarend E.W. (1984): Beste echte Zylinderabbildungen, Kartographische Nachrichten 34 (1984) 103–107

507. Grafarend E.W. (1992a): Four Lectures on Special and General Relativity. Lecture Notes in Earth Science. Sansò, F. and Rummel, R. (eds.), Theory of Satellite Geodesy and Gravity Field Determination 25 (1992) 115–151

508. Grafarend E.W. (1992b): The modeling of free satellite networks in spacetime. In: Proc. International workshop on global positioning systems in geosciences. Eds. S.P. Mertikas, Department of Mineral Resources Engineering, Technical University of Crete (1992) 45–66

509. Grafarend E.W. (1995): The optimal universal Mercator projection, Manuscripta Geodaetica 20 (1995) 421–468

510. Grafarend E.W. (1996): Entwerfend Festliches für Klaus Linkwitz, in: Festschrift für K. Linkwitz (eds. E. Baumann, U. Hangleiter, W. Möhlenbrink), Seiten 110–117, Schriftenreihe der Institute des Fachbereiches Vermessungswesen, Technical Report 1996.1, Stuttgart 1996

511. Grafarend E.W. (2000): Gauss'sche flächennormale Koordinaten im Geometrie– und Schwereraum, Erster Teil: Flächennormale Ellipsoidkoordinaten, Zeitschrift für Vermessungswesen 125 (2000) 136–139

512. Grafarend E.W. (2001a): Gauss surface normal coordinates in geometry and gravity space, Part 2a, Zeitschrift für Vermessungswesen 126 (2001) 373–382

513. Grafarend E.W. (2001b): Harmonic Maps, J. Geodesy 78 (2005) 594–615

514. Grafarend E.W., Ardalan A. (1997): W_0: an estimate in the Finnish Height Datum N60, epoch 1993.4 from twenty–five GPS points of the Baltic Sea Level Project. J. Geodesy 71 (1997) 674–679

515. Grafarend E.W., Ardalan A. (1999): World Geodetic Datum 2000, J. Geodesy 73 (1999) 611–623

516. Grafarend E.W., Ardalan A. (2000): The minimal distance mapping of the physical surface of the Earth onto the Somigliana–Pizetti telluroid and the corresponding quasigeoid, case study: State of Baden–Wüerttemberg, Zeitschrift für Vermessungswesen 125 (2000) 48–60

517. Grafarend E.W., Ardalan A., Kakkuri J. (2002): National height datum, the Gauss–Listing geoid level value w_0 and its time variation $\dot{w}_0$ (Baltic Sea Level Project: epochs 1990.8, 1993.8, 1997.4), J. Geodesy 76 (2002) 1–28

518. Grafarend E.W., Engels J. (1992a): A global representation of ellipsoidal heights–geoidal undulations or topographic heights – in terms of orthonormal functions, Part 1: "amplitude–modified" spherical harmonic functions, Manuscripta Geodaetica 17 (1992) 52–58

519. Grafarend E.W., Engels J. (1992b): A global representation of ellipsoidal heights–geoidal undulations or topographic heights – in terms of orthonormal functions, Part 2: "phase modified" spherical harmonic functions, Manuscripta Geodaetica 17 (1992) 59–64

520. Grafarend E.W., Heidenreich A. (1995): The generalized Mollweide projection of the biaxial ellipsoid, Bulletin Géodésique 69 (1995) 164–172

521. Grafarend E.W., Hendricks A., Gilbert A. (2000): Transformation of conformal coordinates of type Gauss–Krueger or UTM from a local datum (regional, national, European) to a global datum (WGS 84, ITRF 96) Part II: Case studies, Allgemeine Vermessungsnachrichten 107 (2000) 218–222

522. Grafarend E.W., Kampmann G. (1996): $C_{10}(3)$: The ten parameter conformal group as a datum transformation in threedimensional Euclidean space, Zeitschrift für Vermessungswesen 121 (1996) 68–77

523. Grafarend E.W, Knickmeyer E.H., Schaffrin B. (1982): Geodätische Datumtransformationen, Zeitschrift für Vermessungswesen 107 (1982) 15–25

524. Grafarend E.W., Krarup T., Syffus R. (1996): An algorithm for the inverse of a multivariate homogenous polynomial of degree n, J. Geodesy 70 (1996) 276–286

525. Grafarend E.W., Krumm F., Okeke F. (1995): Curvilinear geodetic datum transformation, Zeitschrift für Vermessungswesen 120 (1995) 334–350

526. Grafarend E.W., Lohse P. (1991): The minimal distance mapping of the topographic surface onto the (reference) ellipsoid of revolution, Manuscripta Geodaetica 16 (1991) 92–110

527. Grafarend E.W., Lohse, P., Schaffrin B. (1989): Dreidimensionaler Rückwärtsschnitt, Zeitschrift für Vermessungswesen 114 (1989) 61–67, 127–137, 172–175, 225–234, 278–287

528. Grafarend E.W., Niermann A. (1984): Beste Zylinderabbildungen, Kartographische Nachrichten 34 (1984) 103-107

529. Grafarend E.W., Okeke F. (1998): Transformation of conformal coordinates of type Mercator from a global datum (WGS 84) to a local datum (regional, national), Marine Geodesy 21 (1998) 169–180

530. Grafarend E.W., Schaffrin, B. (1976): Equivalence of estimable quantities and invariants in geodetic networks. Zeitschrift für Vermessungswesen 101 (1976) 485–491

531. Grafarend E.W., Schaffrin, B. (1982): Vectors, quaternions and spinors – a discussion of algebras underlying three–dimensional geodesy – (B. Schaffrin), Feestbunderter Gelegenheid van de 65ste Verjaardag van Professor Baarda, Deel I, ed. Geodetic Computer Centre (LGR), Delft 1982, 111–134

532. Grafarend E.W., Schaffrin B. (1989): The geometry of non–linear adjustment–The Planar trisection problem, in: Festschrift to Torben Krarup, ed. E. Kejlso, K. Poder, C.C. Tschening, Geodaetisk Institute 58 (1989) 149–172

533. Grafarend E.W., Schaffrin B. (1993): Ausgleichungsrechnung in linearen Modellen, B. I. Wissenschaftsverlag, Mannheim 1993

534. Grafarend E.W., Schwarze V. (2002): Das Global Positioning System, Physikalisches Journal 1 (2002) 39–44

535. Grafarend E.W., Shan J. (1997): Estimable quantities in projective networks, Zeitschrift für Vermessungswesen 122 (1997), 218–226, 323–333

536. Grafarend E.W., Syffus R. (1995): The oblique azimuthal projection of geodesic type for the biaxial ellipsoid: Riemann polar and normal coordinates, J. Geodesy 70 (1995) 13–37

537. Grafarend E.W., Syffus R. (1997a): Strip transformation of conformal coordinates of type Gauss–Krueger and UTM, Allgemeine Vermessungsnachrichten 104 (1997) 184–189

538. Grafarend E.W., Syffus R. (1997b): The optimal Mercator projection and the optimal polycylindric projection of conformal type – case study Indonesia – Proceedings 18th International Cartographic Conference, ed. L. Ottoson, Vol. 3, pp. 1751–1759, Stockholm 1997, Proceedings, GALOS (Geodetic Aspects of the Law of the Sea), 2nd international conference, Denpasar, Bali, Indonesia, July 1–4, 1996, pp. 183–192, Inst. of Technology, Bandung 1996

539. Grafarend E.W., Syffus R. (1997c): Mixed cylindric map projections of the ellipsoid of revolution, J. Geodesy 71 (1997) 685–694

540. Grafarend E.W., Syffus R. (1997d): Map projections of project surveying objects and architectural structures, Part 1: Projective geometry of the pneu or torus $\mathbb{T}^2_{A,B}$ with boundary, Zeitschrift für Vermessungswesen 122 (1997) 457–465

541. Grafarend E.W., Syffus R. (1997e): The Hammer projection of the ellipsoid of revolution (azimuthal, transverse, rescaled equiareal), J. Geodesy 71 (1997) 736–748

542. Grafarend E.W., Syffus R. (1997f): Map projections of project surveying objects and architectural structures, Part 2: Projective geometry of the cooling tower of the hyperboloid $\mathbb{H}^2$, Zeitschrift für Vermessungswesen 122 (1997) 560–566

543. Grafarend E.W., Syffus R. (1998a): Map projections of project surveying objects and architectural structures, Part 3: Projective geometry of the parabolic mirror or the paraboloid $\mathbb{P}^2$ with boundary, Zeitschrift für Vermessungswesen 123 (1998) 93–97

544. Grafarend E.W., Syffus R. (1998b): Map projections of project surveying objects and architectural structures, Part 4: Projective geometry of the church tower or the onion $\mathbb{Z}^2$, Zeitschrift für Vermessungswesen 123 (1998) 128–132

545. Grafarend E.W., Syffus R. (1998c): The Optimal Mercator projection and the optimal polycylindric projection of conformal type – case study Indonesia, J. Geodesy 72 (1998) 251–258

546. Grafarend E.W., Syffus R. (1998d): The solution of the Korn–Lichtenstein equations of conformal mapping: the direct generation of ellipsoidal Gauss–Krueger conformal coordinates or the Transverse Mercator Projection, J. Geodesy 72 (1998) 282–293

547. Grafarend E.W., Syffus R. (1998e): Transformation of conformal coordinates from a local datum (regional, national, European) to a global datum (WSG 84). Part I: The transformation equations, Allgemeine Vermessungsnachrichten 105 (1998) 134–141

548. Grafarend E.W., Syffus R., You R.J. (1995): Projective heights in geometry and gravity space, Allgemeine Vermessungsnachrichten 102 (1995) 382–403

549. Grafarend E.W., You R.J. (1995): The Newton form of a geodesic in Maupertuis gauge on the sphere and the biaxial ellipsoid, Zeitschrift für Vermessungswesen 120 (1995) 68–80, 509–521

550. Gravé M.D.A. (1896) : Sur la construction des Cartes géographiques, Journ. de Math. 5 (1896) 317–361

551. Green G. (1839): On the laws of reflection and refraction of light at the common surface of two non–crystallized media, Trans. Cambridge Phil. Soc. 7 (1839) 1–24

552. Green G. (1841): On the propagation of light in crystallized media, Trans. Cambridge Phil. Soc. 7 (1841) 121–140

553. Green P.J., Sibson R. (1978): Computing Dirichlet tessellations in the plane, The Computer Journal 21 (1978) 168–173

554. Greenhood D. (1964): Mapping, University of Chicago Press, Chicago 1964

555. Gretschel H. (1873): Lehrbuch der Kartenprojektionen, B.F. Voigt, Weimar 1873

556. Greuel O., Kadner H. (1990): Komplexe Funktionen und konforme Abbildungen, Teubner, Leipzig 1990

557. Gröbner W., Hofreiter N. (eds.) (1973): Integraltafel, Springer-Verlag, Wien 1973

558. Grone R., Johnson C.R., Sa E.M., Wolkowicz H. (1987): Normal matrices, Linear Algebra Appl. 87 (1987) 213–225

559. Grossmann W. (1933): Die reduzierte Länge der geodätischen Linie und ihre Anwendung bei der Berechnung rechtwinkliger Koordinaten in der Geodäsie, Zeitschrift für Vermessungswesen 16 (1933) 401–419

560. Großmann W. (1976): Geodätische Rechnungen und Abbildungen in der Landesvermessung, Verlag Konrad Wittwer, Stuttgart 1976

561. Gunning R.C., Rossi H. (1965): Analytic functions of several complex variables, Prentice–Hall Inc., Englewood Cliffs, N.J. 1965

562. Gutierrez C., Sotomayor J. (1986): Closed principal lines and bifurcation, Bol. Soc. Bras. Mat. 17 (1986) 1–19

563. Gyoerffy J. (1990): Anmerkungen zur Frage der besten echten Zylinderabbildungen, Kartographische Nachrichten 4 (1990) 140–146

564. Haag K. (1989): Automated cadastral maps as a basis for LIS in Germany, Allgemeine Vermessungsnachrichten International Edition 6 (1989) 20–25

565. Haahti H. (1965): Über konforme Differentialgeometrie und zugeordnete Verjüngungsoperatoren in Hilbert–Räumen, Annales Academiae Scientarium Fennicale A (1965) 3–20

566. Haathi H. (1960): Über konforme Abbildungen eines euklidischen Raumes in eine Riemannsche Mannigfaltigkeit, Suomalasien Tiedeakatemian, Toimituksia Annales Academiae Scientarium Fennicae Series A, Helsinki 1960

567. Haibach O. (1962): Die Anfertigung orthogonaler und plagiogonaler rißlicher Darstellungen durch Einsatz elektronischer Rechenanlagen statt der bisher rechnerischen und konstruktiven Verfahren, Sonderdruck aus Bergb.–Wiss. 9 (1962) 137–146

568. Haibach O. (1966a): Grund–und Oberflächenbestimmungen in Grund–, Seiger–, Flach– und schiefen Grundrissen, Sonderdruck aus der Zeitschrift des Deutschen Markscheider–Vereins "Mitteilungen aus dem Markscheidewesen" 3 (1966) 87–98

569. Haibach O. (1966b): Über das Leistungsvermögen und die Wechselbeziehung von Rissprojektionen und Thema, Sonderdruck aus der Zeitschrift des Deutschen Markscheider–Vereins "Mitteilungen aus dem Markscheidewesen" 4 (1966) 154–172

570. Haibach O. (1966c): Inwieweit kann ein schiefer Grundriß Winkel–, Längen–und Flächentreue besitzen, und wie ist der Gebrauch eines solchen Risses?, Bergb.–Wiss. 13 (1966) 241–246

571. Haibach O. (1967a): Der Gebrauch des Seigerrisses gezeigt an der Steigerungsgeraden bzw. Steigungslinie, Sonderdruck aus der Zeitschrift des Deutschen Markscheider–Vereins "Mitteilungen aus dem Markscheidewesen" 1 (1967) 30–43

572. Haibach O. (1967b): Genauigkeitsuntersuchungen an grundrißlichen Wert–und Kennlinien, Bergb.–Wiss. 14 (1967) 389–399

573. Haibach O., Burger K. (1982): Mathematische Grundlagen und instrumentelle Erfordernisse für Projektionszeicheneinrichtungen, die in Projektionsarten des neuzeitlichen Rißwesens eine unmittelbare (eigenhändige) Bearbeitung ermöglichen, Lippe 1982

574. Haines G.V. (1967): A Taylor expansion of the geomagnetic field in the Canadian Arctic, Publications of the Dominion Observatory, Ottawa 35 (1967) 119–140

575. Haines G.V. (1981): The modified polyconic projection, Cartographica 18 (1981) 49–58

576. Haines G.V. (1987): The inverse modified polyconic projection, Cartographica 24 (1987) 14–24

577. Hake G. (1974): Kartographie. Sammlung Göschen, de Gruyter, Berlin 1974

578. Hake G., Gruenreich D., Meng L. (2002): Kartographie, Walter de Gruyter, Berlin / New York 2002

579. Halmos F., Szádecky–Kardoss G.Y. (1967): Die einfache Bestimmung der Meridiankonvergenz bei verschiedenen Projektionen, Acta Geodaetica, Geophys. et Montanist. Acad. Sci. Hung. Tomus 2 (3–4) (1967) 351–366

580. Haltiner G.J., Williams R.T. (1980): Numerical prediction and dynamic meteorology, 2nd edition, J. Wiley & Sons, New York Chichester 1980

581. Hammer E. (1892): Über die Planisphäre von Aitow und verwandte Entwürfe, insbesondere neue flächentreue ähnlicher Art, Petermanns Geographische Mitteilungen (Gotha) 38 (1892) 85–87

582. Hammer E. (1900): Unechtzylindrische und unechtkonische flächentreue Abbildungen. Mittel zum Auftragen gegebener Bogenlängen auf gezeichneten Kreisbögen von bekannten Halbmessern, Petermanns Geographische Mitteilungen 46 (1900) 42–46

583. Hammersley J.M., Handscomb D.C. (1964): Monte Carlo methods, Methuen's Monographs on applied probability and statistics, J. Wiley & Sons, New York Chichester 1964

584. Hancock H. (1958a): Theory of elliptic functions, Dover, New York 1958

585. Hancock H. (1958b): Elliptic intergrals, Dover, New York 1958

586. Harbeck R. (1995): Erdoberflächenmodelle der Landesvermessung und ihre Anwendungsgebiete, Kartographische Nachrichten 45 (1995) 41–50

587. Harrison C.G.A. (1972): Poles of rotation, Earth and Planetary Science Letters 14 (1972) 31–38

588. Harrison R.E. (1943): The nomograph as an instrument in map making, Geographical Review 33 (1943) 655–657

589. Hasse H. (1943): Journal für die reine und angewandte Mathematik, Band 185, Walter de Gruyter, Berlin 1943

590. Hassler F.R. (1825): On the mechanical organization of a large survey, and the particular application to the survey of the coast, Amer. Philosophical Soc. Trans. 2, 385-408, 1825

591. Hauer F. (1941): Flächentreue Abbildung kleiner Bereiche des Rotationsellipsoids in der Ebene durch Systeme geringster Streckenverzerrung, Zeitschrift für Vermessungswesen 70 (1941) 194–213

592. Hauer F. (1943): Flächentreue Abbildung kleiner Bereiche des Rotationsellipsoids in der Ebene bis einschließlich Glieder 4. Ordnung, Zeitschrift für Vermessungswesen 72 (1943) 179–189

593. Hauer F. (1949): Entwicklung von Formeln zur praktischen Anwendung der flächentreuen Abbildung kleiner Bereiche des Rotationsellipsoids in die Ebene, Österreichische Zeitschrift für Vermessungswesen, Sonderheft 6 (1949) 1–31

594. Haussner R., Schering K. (eds.) (1902): Gesammelte Mathematische Werke, Mayer & Mueller, Berlin 1902

595. Hayford J.F. (1909): The figure of the earth and isostasy from measurements in the United States: U.S. Coast and Geodetic Survey 1909

596. Hayman W.K., Kershaw D., Lyons T.J. (1984): The best harmonic approximation to a continuous function, Anniversary Volume on Approximation Theory and Functional Analysis, Internat. Ser. Numer. Math. 65 (1984) 317–327

597. Hazay I. (1965): Die Bedeutung der Tissot–Indikatrix, Acta Techn. Hung. 52 (1965) 171–200

598. Hazay I. (1983): Quick zone–to–zone transformation in the Gauss–Krueger projection, Acta Geod. Geophys. et Montanist. Hung. 18 (1983) 71–81

599. Hecht H. (1995): Die elektronische Seekarte, Vermessungsingenieur 45 (1995) 104–111

600. Hecht H., Berking B., Buettgenbach G., Jonas M. (1999): Die elektronische Seekarte, Wichmann Verlag, Heidelberg 1999

601. Heck B. (2002): Rechenverfahren und Auswertemodelle der Landesvermessung, 3. Auflage (3rd new edition), Herbert Wichmann Verlag, Heidelberg 2002

602. Heck B. (2003): Rechenverfahren und Auswertemodelle der Landesvermessung, 3., neu bearbeitete und erweitete Auflage, Herbert Wichmann Verlag, Karlsruhe 2003

603. Hedrick E.R., Ingold L. (1925a): Analytic functions in three dimensions, Transactions of the American Mathematical Society 27 (1925) 551–555

604. Hedrick E.R., Ingold L. (1925b): The Beltrami equations in three dimensions, Transactions of the American Mathematical Society 27 (1925) 556–562

605. Heideman M.T., Johnson D.H., Burrus C.S. (1984): Gauss and the history of the fast Fourier transform, IEEE ASSP Magazine 1 (1984) 14–21

606. Heikkinen M. (1982): Geschlossene Formeln zur Berechnung räumlicher geodätischer Koordinaten aus rechtwinkligen Koordinaten, Zeitschrift für Vermessungswesen 107 (1982) 207–211

607. Heiskanen W.A. (1928): Ist die Erde ein dreiachsiges Ellipsoid?, Gerlands Beiträge zur Geophysik B19 (1928) 356–377

608. Heiskanen W.A. (1962): Is the Earth a triaxial ellipsoid?, J. Geophysical Research 67 (1962) 321–329

609. Heiskanen W.A., Moritz H. (1967): Physical geodesy, W.H. Freeman Publ., San Francisco 1967

610. Heitz S. (1984): Geodätische und isotherme Koordinaten auf geodätischen Bezugsflächen, Mitt. Geod. Institute, Universität Bonn, Report 66, Bonn 1984

611. Heitz S. (1985): Koordinaten auf geodätischen Bezugsflächen, Dümmler, Bonn 1985

612. Heitz S. (1988): Coordinates in geodesy, Springer, Berlin – Heidelberg – New York 1988

613. Helein F. (2002): Harmonic maps, conservation laws and moving frames, 2nd ed., Cambridge University Press, New York 2002

614. Helmert F.R. (1880): Die mathematischen und physikalischen Theorien der höheren Geodäsie, Band 1, G. Teubner, Leipzig 1880

615. Helms L.L. (1969): Introduction to potential theory, Wiley Interscience Pure and Applied Mathematics 22, New York (1969)

616. Hencky H. (1928): Über die Form des Elastizitätsgesetzes bei ideal elastischen Stoffen, Zeitschrift f. techn. Physik 9 (1928) 215–220, 457

617. Hencky H. (1929a): Welche Umstände bedingen die Verfestigung bei der bildsamen Verformung von festen isotropen Körpern?, Z. Physik 55 (1929) 145–155

618. Hencky H. (1929b): Das Superpositionsgesetz eines endlich deformierten relaxionsfähigen elastischen Kontinuums und seine Bedeutung für eine exakte Ableitung der Gleichungen für die zähe Flüssigkeit in der Eulerschen Form, Ann. Physik 5 (1929) 617–630

619. Henkel M. (1999): Conformal invariance and critical phenomena, Springer–Verlag, Berlin – Heidelberg 1999

620. Henle J.M., Kleinberg E.M. (1979): Infinitesimal Calculus, MIT, Cambridge 1979

621. Henrici P. (1986): Applied and computational complex analysis, Vol. 3, Wiley & Sons, New York – London – Sydney – Toronto 1986

622. Heppes A. (1964): Isogonale sphärische Netze, University of Science Rolando Eötvös Budapest, Sectio Mathematica, Annals 7 (1964) 41–48

623. Hertrich–Jeromin U., Hoffmann T., Pinkall U. (1999): A discrete version of the Darboux transform for isothermic surfaces; in Bobenko, A., Seiler R.: Discrete integrable geometry and physics, Oxford University Press, Oxford 1999

624. Hertrich–Jeromin U., Pedit F. (1997): Remarks on the Darboux transform of isothermic surfaces, Doc. Math. J. DMV 2 (1997) 313–333

625. Herz N. (1885): Lehrbuch der Landkartenprojektionen, Teubner, Leipzig 1885

626. Heumann C. (1941): Tables of complete elliptic integrals, Journal of Mathematics and Physics 20 (1941) 127–154

627. Higham N.J. (1984): Computing the polar decomposition – with applications, University of Manchester, Department of Mathematics, Numerical Analysis Report No. 94, Manchester 1984

628. Higham N.J. (1986): Computing the polar decomposition, SIAM J. Sci. Stat. Comput. 7 (1986) 1160–1174

629. Hildebrandt H. (1962): Die Lösung der Geodätischen Hauptaufgabe auf dem Bruns'schen Niveausphäroid mit Hilfe der Legendre'schen Reihen, Zeitschrift für Vermessungswesen 87 (1962) 299–306

630. Hill G.W. (1908): Application on Tchébychef's principle in the projection of maps, Ann. Math.10 (1908) 23–36

631. Hilliard J.A., Basoglu Ü., Muehrcke P.C. (1978): A projection handbook: Univ. Wisconsin–Madison, Cartographic Laboratory

632. Hinks A.R. (1912): Map projections: Cambridge Univ. Press, Cambridge 1912

633. Hinks A.R. (1940): Maps of the world on an oblique Mercator projection, The Geographical Journal 95 (1940) 381–383

634. Hinks A.R. (1941): More world maps on oblique Mercator projections, The Geographical Journal 97 (1941) 353–356

635. Hirsch M. (1991): Eine numerische Lösung für die Differentialgleichung der geodätischen Linie auf dem Rotationsellipsoid, Wissenschaftliche Zeitschrift der Technischen Universität Dresden 40 (1991) 145–151

636. Hirsch M.W. (1976): Differential topology, Springer–Verlag, New York 1976

637. Hirvonen R.A. (1960): New theory of the gravimetric geodesy, Publications of the Isostatic Institute of the International Association of Geodesy 32 (1960) 1–52

638. Hochstoeger F. (1995): Die Ermittlung der topographischen Abschattung von GPS–Satelliten unter Verwendung eines digitalen Geländemodells, Österreichische Zeitschrift für Vermessung und Geoinformation 83 (1995) 144–145

639. Hoellig K. (1992): B–Splines in der geometrischen Datenverarbeitung, in: Wechselwirkungen, Jahrbuch 1992, Aus Lehre und Forschung der Universität Stuttgart, Stuttgart 1992, 77–84

640. Hoermander L. (1965): The Frobenius–Nirenberg theorem, Arkiv för Matematik 5 (1965) 425–432

641. Hoermander L. (1966): An introduction to complex analysis in several variables, D. van Nostrand Company, Inc., Princeton 1966

642. Hofsommer D.J., van de Riet R.P. (1963): On the numerical calculation of elliptic integrals of the first and second kind and the elliptic functions of Jacobi, Numerische Mathematik 5 (1963) 291–302

643. Hojovec V., Jokl L. (1981): Relation between the extreme angular and areal distortion in cartographic representation, Studia geoph. et geod. 25 (1981) 132–151

644. Hooijberg M. (1997): Practical geodesy using computers, Springer Verlag, Berlin – Heidelberg – New York 1997

645. Hooijberg M. (2005): Kartenprojektion, 2. Auflage, Springer Verlag 2005

646. Hopf E. (1941): Statistik der Lösungen geodätischer Probleme vom unstabilen Typus. II., Mathematische Annalen 117 (1941) 590–608

647. Hopf H. (1948): Zur Topologie der komplexen Mannigfaltigkeiten, Studies and Essays, R. Courant Anniversary Volume (1948) 167–185

648. Hopfner E. (1933): Physikalische Geodäsie, Akademische Verlagsgesellschaft, Leipzig 1933

649. Hopfner E. (1938): Zur Berechnung des Meridianbogens, Zeitschrift für Vermessungswesen 67 (1938) 620–627

650. Hopfner F. (1940): Über die Änderung der geodätischen Kurve am Rotationsellipsoid bei einer Änderung der Ellipsoidparameter, Zeitschrift für Vermessungswesen 69 (1940) 392–402

651. Hopfner F. (1948): Lambert, Gauss, Tissot, Inaugurationsrede, Österreichische Zeitschrift für Vermessungswesen 36 (1948) 49–55

652. Horemuz M. (1999): Error calculation in Maritime delimitation between states with opposite or adjacent coasts, Marine Geodesy 22 (1999) 1–17

653. Horn B.K.P. (1987): Closed–form solution of absolute orientation using unit quaternions, J. Opt. Soc. Am. A. 4 (1987) 629–642

654. Horn B.K.P., Hilden H.M., Negahdaripour S. (1988): Closed–form solution of absolute orientation using orthonormal matrices, J. Opt. Soc. Am. A. 5 (1988) 1127–1135

655. Hoschek J. (1984): Mathematische Grundlagen der Kartographie, B.-I., Mannheim 1984

656. Hoschek J. (1992): Grundlagen der geometrischen Datenverarbeitung (2. Aufl.), Teubner Verlag, Stuttgart 1992

657. Hoschek J., Lasser D. (1989a): Baryzentrische Koordinaten, in: ibid., Grundlagen der geometrischen Datenverarbeitung, Teubner, Stuttgart 1989, 243–256

658. Hoschek J., Lasser D. (1989b): Grundlagen der geometrischen Datenverarbeitung, Teuber, Stuttgart 1989

659. Hoschek J., Lasser D. (1992): Mathematische Grundlagen der Kartographie, BI–Wissenschaftsverlag, Bibl. Institut, 2. Auflage, neu bearb. and erw. Aufl., Manheim–Wien–Zürich 1992

660. Hotine M. (1946): The orthomorphic projection of the spheroid, Empire Survey Review 8 (1946) 300–311

661. Hotine M. (1947a): The orthomorphic projection of the spheroid–II, Empire Survey Review 9 (1946) 25–35

662. Hotine M. (1947b): The orthomorphic projection of the spheroid–III, Empire Survey Review 9 (1946) 52–70

663. Hotine M. (1947c): The orthomorphic projection of the spheroid–IV, Empire Survey Review 9 (1946) 112–123

664. Hotine M. (1947d): The orthomorphic projection of the spheroid–V, Empire Survey Review 9 (1946) 157–166

665. Hotine M. (1969): Mathematical Geodesy, U.S. Department of Commerce, Washington 1969

666. Hotine M. (1991): Differential Geodesy, Springer–Verlag, Berlin – Heidelberg – New York 1991

667. Hristow W.K. (1937a): Potenzreihen zwischen den stereographischen und den geographischen Koordinaten und umgekehrt, Zeitschrift für Vermessungswesen 66 (1937) 84–89

668. Hristow W.K. (1937b): Berechnung der Koordinatendifferenzen und der Ordinatenkonvergenz aus der Länge und dem Richtungswinkel einer geodätischen Strecke für eine beliebige Fläche und ein beliebiges isothermes Koordinatensystem, Zeitschrift für Vermessungswesen 66 (1937) 171–178

669. Hristow W.K. (1938): Übergang von einer normalen winkeltreuen Kegel–Abbildung zu einer normalen flächentreuen Kegel–Abbildung und umgekehrt, Zeitschrift für Vermessungswesen 93 (1968) 693

670. Hristow W.K. (1955): Die Gauss'schen und geographischen Koordinaten auf dem Ellipsoid von Krassowsky, VEB Verlag Technik, Berlin 1955

671. Hsu M.-L. (1981): The role of projections in modern map design, Cartographica 18 (1981) 151–186

672. Hubeney K. (1980a): Festschrift zur Emeritierung von o. Univ.-Prof. Dipl.-Ing. Dr. techn. Karl Hubeney, Mitteilungen der geodätischen Institute der TU Graz, Folge 35, Graz 1980

673. Hubeny K. (1953): Isotherme Koordinatensysteme und konforme Abbildungen des Rotationsellipsoides, Sonderheft 23, Österreichische Zeitschrift für Vermessungswesen, Wien 1953

674. Hubeny K. (1980b): Eine weitere Herleitung des Theorems von Clairaut, Mitteilungen der geodätischen Institute der Technischen Universität Graz 52 (1986) 21–23

675. Huck H., Roitzsch R., Simon U., Vortisch W., Walden R., Wegner B., Wendland (1973): Beweismethoden der Differentialgeometrie im Großen, Springer Verlag, Berlin – Heidelberg – New York 1973

676. Hufnagel H. (1974): Die Peters–Projektion – eine neue und/oder aktuelle Abbildung der Erde?, Allgemeine Vermessungsnachrichten 81 (1974) 225–232

677. Hufnagel H. (1989): Ein System unecht–zylindrischer Kartennetze für Erdkarten, Kartographische Nachrichten 39 (1989) 89–96

678. Hughes D.R., Piper F.C. (1973): Projective Planes, Springer–Verlag, New York 1973

679. Hunger F. (1938): Die Überführung von Gauss–Krueger Koordinaten in das System des benachbarten Meridianstreifens, Zeitschrift für Vermessungswesen 93 (1968) 687–691

680. Id Ozone M. (1985): Non–iterative solution of the equation, Surveying and Mapping 45 (1985) 169–171

681. Ihde J. (1991): Geodätische Bezugssysteme, Vermessungstechnik 39 (1991) 13–15, 57–63

682. Ihde J. (1993): Some remarks on geodetic reference systems in Eastern Europe in preparation of a uniform European Geoid, Bulletin Géodésique 67 (1993) 81–85

683. Ihde J., Schach, H., Steinich, L. (1995): Beziehungen zwischen den geodätischen Bezussystemen Datum Rauenberg, ED 50 und System 42, Report B298, Deutsche Geodätische Kommission, Bayer. Akad. Wiss., München 1995

684. Iliffe J.C. (2000): Datums and map projections for remote sensing, GIS and surveying, Whittles Publishing, London 2000

685. Illert A. (2001): Kooperation der amtlichen Kartographie in Europa – Projekte, Partnerschaften und Produkte, in: J. Dodt and W. Herzog (Eds.): Kartographisches Taschenbuch 2001, Kirschbaum Verlag, Bonn 2001, 35–49

686. Ingwersen M. (1996): Die Berechnung Gauss'scher und geographischer Koordinaten mit Rekursionsformeln, Zeitschrift für Vermessungswesen 121 (1996) 124–132

687. International hydrografic organization (1993): A manual on technical aspects of the united nation on the law of the sea – 1982, Special Publication 51, 3rd Edition, Monaco 1993

688. Ipbuker C. (2005): A computational approach to the Robinson projection, Survey Review 38 (2005) 297–310

689. Irmisch S., Schwolow R. (1994): Erzeugung unstrukturierter Dreiecknetze mittels der Delauney–Triangulierung, Z. Flugwiss. Weltraumforsch. 18 (1994) 361–368

690. Ivory J. (1824): Solution of a geodetical problem, Philosophical Magazine and Journal 64 (1924) 35–39

691. Izotov A.A. (1959): Reference Ellipsoid and the Standard Geodetic Datum adopted in the USSR, Bulletin Géodésique 53 (1959) 1–6

692. Jackson J.E. (1987): Sphere, spheroid and projections für surveyors, BSP Professional Books, Oxford 1987

693. Jacobi C.G.J. (1826): Die Gauss'sche Methode, die Werte der Integrale näherungsweise zu finden, J. reine und angew. Math, 1 (1826) 301–308

694. Jacobi C.G.J. (1839): Note von der geodätischen Linie auf einem Ellipsoid und den verschiedenen Anwendungen einer merkwürdigen analytischen Substitution, Crelles J. 19 (1839) 309–313

695. Jacobi C.G.J. (1866a): Achtundzwanzigste Vorlesung, in: ibid., Vorlesungen über Dynamik, Georg Reimer, Berlin 1866, pp. 212–221

696. Jacobi C.G.J. (1866b): Die kürzeste Linie auf dem dreiaxigen Ellipsoid. 28. Vorlesung. Vorlesungen über Dynamik, gehalten an der Universität zu Königsberg im WS 1842–1843. Hrsg. A. Clebsch, Reimer Verlag, Berlin 1866

697. Jacobi C.G.J. (1869): Gesammelte Werke, Bände 1–8, Kgl. Preuss. Akad.Wiss., Berlin 1881, Ed. A. Clebsch Chelsea Publ. Comp., New York 1869

698. James H., Clarke A.R. (1882): On projections for map applying to a very large extend of the Earth's surface, Phil. Mag. 23 (1882) 308–312

699. Janenko N.N. (1953): Einige Fragen der Einbettungstheorie Riemannscher Metriken in Euklidische Räume, Uspechi matematiceskich nauk. Moskva 8 (1953) 21–100

700. Jank W., Kivioja L.A. (1980): Solution of the direct and inverse problems of Reference Ellipsoids by point–by–point integration using programmable pocket calculators, Surveying and Mapping 40 (1980) 325–337

701. Jennings G.A. (1994): Modern geometry with applications, Springer Verlag, Berlin – Heidelberg – New York 1994

702. Jensch G. (1970): Die Erde und ihre Darstellung im Kartenbild, Westermann, Braunschweig 1970

703. Joachimsthal F. (1890): Anwendung der Differential– und Integralrechnung auf die allgemeine Theorie der Flächen und der Linien doppelter Krümmung, Teubner, Leipzig 1890

704. Jones C.B., Bundy G.L., Ware J.M. (1995): Map generalization with a triangulated data structure, Cartography and Geograph. Information Systems 22 (1995) 317–331

705. Jones N.L., Wright S.G., Maidment D.R. (1990): Watershed delineation with triangle–based terrain models, Journal of hydraulic engineering 116 (1990) 1232–1251

706. Joos G. (1989): Pseudokonische und Pseudoazimutale Abbildungen, Selbständige Arbeit, Studiengang Vermessungswesen an der Universität Stuttgart, Stuttgart 1989

707. Joos G., Joerg K. (1991): Inversion of two bivariate power series using symbolic formula manipulation, Institute of Geodesy, University of Stuttgart, Technical Report 13, Stuttgart 1991

708. Jordan W. (1875): Zur Vergleichung der Soldner'schen rechtwinkligen sphärischen Koordinaten mit der Gauss'schen konformen Abbildung des Ellipsoids auf die Ebene, Zeitschrift für Vermessungswesen IV (1875) 27-32

709. Jordan W. (1896): Der mittlere Verzerrungsfehler, Zeitschrift für Vermessungswesen 25 (1896) 249–252

710. Jordan W., Eggert O., Kneissl M. (1959): Handbuch der Vermessungskunde, Band IV, Zweite Hälfte, J.B. Metzlersche Verlagsbuchhandlung, Stuttgart 1959

711. Junklus J.L., Turner J.D. (1978): A distortion–free map projection for analysis of satellite imagery, The Journal of the Astronautical Sciences 26 (1978) 211–234

712. Kahle H.-G. et al (2000): GPS–derived strain rate field within the boundary zones of the Eurasian, African and Arabian Plates, J. Geophys. Res. 105 (2000) 23,353–23,370

713. Kakkuri J. (1995): The Baltic Sea level project, Allgemeine Vermessungsnachrichten 102 (1995) 331–336

714. Kakkuri J. (1996): Postglacial deformation of the Fennoscandian crust, Geophysica (1996)

715. Kakkuri J., Vermeer M., Maelkki P., Boman H., Kahma K.K., Leppaeranta M. (1988): Land uplift and sea level variability spectrum using fully measured monthly means of tide gauge readings, Finnish Marine Research 256 (1988) 3–75

716. Kaltsikis C. (1980): Über bestangepaßte konforme Abbildungen, Dissertation, TU München, München 1980

717. Kanatani K. (1985): Detecting the motion of a planar surface by line and surface integrals, Computer Vision, Graphics, and Image Processing 29 (1985) 13–22

718. Kanatani K. (1987): Structure and motion from optical flow under perspective projection, Computer Vision, Graphics, and Image Processing 38 (1987) 122–146

719. Kanatani K. (1990): Group–theoretical methods in image understanding, Springer Verlag, Berlin – Heidelberg – New York 1990

720. Kao R. (1961): Geometric projections of the sphere and the spheroid, Canadian Geographer 5 (1961) 12–21

721. Karni Z., Reiner M. (1960): Measures of deformation in the strained and in the unstrained state, Bull. Res. Coun. Israel 8c volume 89, Jerusalem 1960

722. Kasner E. (1909): Natural families of trajectories conservative fields of force, Transactions of the American Mathematical Society 10 (1909) 201–203

723. Kavrajski V.V. (1958): Ausgewählte Werke, Mathematische Kartographie, Allgemeine Theorie der kartographischen Abbildungen, Kegel–und Zylinderabbildungen, ihre Anwendungen (russ.) GS VMP, Moskau 1958

724. Kaya A. (1994): An alternative formula for finding the geodetic latitude from the isometric latitude, Survey Review 253 (1994) 450–452

725. Kellogg O.D. (1912): Harmonic functions and Green's integral, Transactions of the American Mathematical Society 13 (1912) 109–132

726. Kelnhofer F. (1977): Kartennetzberechnung mittels einfacher Computerprogramme dargelegt an Beispielen abstandstreuer und flächentreuer Kegelentwürfe, in: Beiträge zur theoretischen Kartographie, Festschrift für Erik Arnberger, ed. I. Kretschmer, Franz Deuticke, Wien 1977

727. Kenney C., Laub A.J. (1991): Polar decomposition and matrix sign function condition estimates, SIAM J. Sci. Stat. Comput. 12 (1991) 488–504

728. Keuning J. (1955): The history of geographical map projections until 1600, Imago Mundi 12 (1955) 1–25

729. Killing W. (1892): Ueber die Grundlagen der Geometrie, J. Reine Angew. Math. 109 (1892) 121–186

730. Kimerling A.J. (1984): Area computation from geodetic coordinates on the spheroid, Surveying and Mapping 44 (1984) 343–351

731. Kimerling A.J., Overton W.S., White D. (1995): Statistical comparison of map projection distortions within irregular areas, Cartography and Geograph. Information Systems 22 (1995) 205–221

732. King A.C. (1988): Periodic approximations to an elliptic function, Appl. Anal. 27 (1988) 271–278

733. Kirchhoff G. (1852): Über die Gleichungen des Gleichgewichts eines elastischen Körpers bei nicht unendlich kleinen Verschiebungen seiner Teile, Sitzber. Akad. Wiss. Wien 9 (1852) 762–773

734. Kivioja L.A. (1971): Computation of geodetic direct and indirect problems by computers accumulating increments from geodetic line elements, Bulletin Géodésique 99 (1971) 55–63

735. Kline M. (1994): Projective geometry, in: From five fingers to infinity, A journey through the history of mathematics, ed. F.J. Swetz, Open Court, Chicago 1994

736. Klingatsch A. (1897): Zur ebenen rechtwinkligen Abbildung der Soldnerschen Koordinaten, Zeitschrift für Vermessungswesen 26 (1897) 431–436

737. Klingenberg W. (1973): Flächentheorie im Großen, in: ibid., Eine Vorlesung über Differentialgeometrie, Springer Verlag, Berlin – Heidelberg – New York 1973, pp. 96–113

738. Klingenberg W. (1978): Lecture on closed geodesics, Springer Verlag, Berlin – Heidelberg – New York 1978

739. Klingenberg W. (1982): Riemannian geometry, de Gruyter, Berlin – New York 1982

740. Klinghammer I., Gyöffy J. (1988): Zur Wahl der Kartennetzentwürfe für thematische Weltatlanten, in: Zum Problem der thematischen Weltatlanten, Vorträge zum Kolloquium aus Anlaß der 200–Jahr–Feier des Gothaer Verlagshauses, VEB Hermann Haack, Geographisch–Kartographische Anstalt Gotha, Gotha 1988

741. Klotz J. (1991): Eine analytische Lösung kanonischer Gleichungen der geodätischen Linie zur Transformation ellipsoidischer Flächenkoordinaten. Report C385, Deutsche Geodätische Kommission, Bayer. Akad. Wiss., München 1991

742. Klotz J. (1993): Die Transformation zwischen geographischen Koordinaten und geodätischen Polar–und Parallel–Koordinaten, Zeitschrift für Vermessungswesen 118 (1993) 217–227

743. Kneschke A. (1962a): Funktionen einer komplexen Veränderung, in: ibid., Differentialgleichungen und Randwertprobleme, Teubner, Leipzig 1962, 226–270

744. Kneschke A. (1962b): Die Differentialgleichungen der geodätischen Linien, in: ibid., Differentialgleichungen und Randwertprobleme, Teubner, Leipzig 1962, 326–337

745. Kneser A. (1928): Neue Untersuchung einer Reihe aus der Theorie der elliptischen Funktionen, Journ. f. Math. 158 (1928) 209–218

746. Knoerrer H. (1980): Geodesics on the ellipsoid, Inventions math. 59 (1980) 119–143

747. Knopp. K. (ed.)(1936): Mathematische Zeitschrift 40. Band, Julius Springer, Berlin 1936

748. Kober H. (1957): Dictionary of conformal representations, Dover Publications, 1957

749. Koch J. (1916): Die Messung der Braaker Basis 1820 und 1821 im Rahmen der Landestriangulation Dänemarks und Hannovers, Zeitschrift für Vermessungskunde 45 (1916) 11–23

750. Koch K.R. (1980): Parameterschätzung und Hypothesentests in linearen Modellen, Dümmler's Verlag, Bonn 1980

751. Koch K.R. (1982): S–transformations and projections for obtaining estimable parameters. Forty Years of Thought, Anniversary Volume on the Occasion of Professor Baarda's 65th Birthday (1982) 136–144

752. Koch K.R. (1983): Die Wahl des Datums eines trigonometrischen Netzes bei Punkteinschaltungen. Zeitschrift für Vermessungswesen 108 (1983) 104–111

753. Koch K.R. (1985): Invariante Größen bei Datumtransformationen, Vermessung, Photogrammetrie, Kulturtechnik 83 (1985) 320–322

754. Koehnlein W. (1962): Untersuchungen über große geodätische Dreiecke auf geschlossenen Rotationsflächen unter besonderer Berücksichtigung des Rotationsellipsoides, Report C51, Deutsche Geodätische Kommission, Bayer. Akad. Wiss., München 1962

755. Koenig R. (1938): Über die Umkehrung einer trigonometrischen Reihe, Berichte über die Verhandlungen der Sächs. Akademie der Wissenschaften. Math.-Nat. Klasse 90 (1938) 69–82

756. Koenig R., Weise K.H. (1951): Mathematische Grundlagen der höheren Geodäsie und Kartographie, Bd. I, Springer Verlag, Berlin – Heidelberg – New York 1951

757. Kohn J.J. (1972): Integration of complex vector fields, Bulletin of the American Mathematical Society 78 (1972) 1–11

758. Kopfermann K. (1977): Mathematische Grundstrukturen, Akademische Verlagsgesellschaft, Wiesbaden 1977

759. Koppelman W. (1959): The Rieman–Hilbert problem for finite Riemann surfaces, Communications on pure and applied mathematics 12 (1959) 13–35

760. Koppelt U., Biegel M. (1989): Spherical harmonic expansion of the continents and oceans distribution function, Studia geoph. et geod. 33 (1989) 315–321

761. Korn A. (1907): Sur les équations de elasticité, Annales de l'Ecole Normale Superieure 24 (1907) 9–75

762. Korn A. (1909): Über Minimalflächen, deren Randkurven wenig von den ebenen Kurven abweichen, Berliner Berichte, Phys. Math. Klasse, Anhang, Berlin 1909

763. Korn A. (1914): Zwei Anwendungen der Methode der sukzessiven Annäherungen, in: Mathematischen Abhandlungen Hermann Amandus Schwarz zu seinem fünfzigjährigen Doktorjubiläum, Springer Verlag, Berlin – Heidelberg – New York 1914, pp. 215–229

764. Kozák J., Jiri V. (2002): Berghaus' physikalischer Atlas: Surprising content and superior artistic images, Stud. Geophys. Geod. 46 (2002) 599–610

765. Krack K. (1980): Rechnerunterstützte Entwicklung der Legendreschen Reihen, Österreichische Zeitschrift für Vermessungswesen 68 (1980) 145–156

766. Krack K. (1981): Die Umwandlung von Gauss'schen konformen Koordinaten in geographische Koordinaten des Bezugsellipsoides auf der Grundlage des transversalen Mercatorentwurfs, Allgemeine Vermessungsnachrichten 88 (1981) 173–178

767. Krack K. (1982a): Rechnerunterstütze Ableitung der Legendreschen Reihen und Abschätzung ihrer ellipsoidischen Anteile zur Lösung der ersten geodätischen Hauptaufgabe auf Rotationsellipsoiden, Zeitschrift für Vermessungswesen 107 (1982) 118–125

768. Krack K. (1982b): Rechnerunterstützte Entwicklung der Mittelbreitenformeln und Abschätzung ihrer ellipsoiden Anteile zur Lösung der zweiten geodätischen Hauptaufgabe auf Rotationsellipsoiden, Zeitschrift für Vermessungswesen 107 (1982) 502–513

769. Krack K. (1982c): Zur direkten Berechnung der geographischen Breite aus der Meridianbogenlänge auf Rotationsellipsoiden, Allgemeine Vermessungsnachrichten 89 (1982) 122–126

770. Krack K. (1998): Ein allgemeiner Ansatz zur Lösung der ersten Grundaufgabe in der Landesvermessung mithilfe der Computeralgebra, Allgemeine Vermessungsnachrichten 105 (1998) 388–395

771. Krack K. (1999): Dreizehn Aufgaben aus der Landesvermessung im Geographischen Koordinatensystem, Schriftenreihe Studiengang Vermessungswesen, Universität der Bundeswehr München, Heft 65, Neubiberg 1999

772. Krack K. (2000): Mathematica–Programme zur Darstellung des Zusammenhangs von geographischer Breite unf Meridianbogenlänge auf Rotationsellipsoiden, in: 25 Jahre Institut für Geodäsie, Teil 1, (Caspary, W., Heister, H., Schoedlbauer, A., Welsch, W. eds) pp. 91–109, Heft 60–1, Schriftenreihe Studiengang Geodäsie und GeoInformation, Uni Bundeswehr, Neubiberg 2000

773. Krack K., Glasmacher H. (1984): Umkehrung von zwei vollständigen Potenzreihen mit zwei Veränderlichen, Heft 10 der Schriftenreihe des wiss. Studienganges Vermessungswesen an der Universität der Bundeswehr München, München 1984

774. Krack K., Schoedlbauer A. (1984): Bivariate Polynome zur genäherten Bestimmung von UTM Koordinaten (ED50) aus Gauss Krueger Koordinaten, Heft 10 der akademischen Schriftenreihe des wiss. Studienganges Vermessungswesen an der Universität der Bundeswehr München, München 1984

775. Krakiwsky E., Karimi H.A., Harris C., George J. (1987): Research into electronic maps and automatic vehicle location, Proceedings of the eigth international symposium on computer–assisted cartography, Auto Carto 8, Baltimore 1987

776. Krantz S.G., Parks H.R. (2002): The implicit function theorem–history, theory and applications, Birkhäuser, Boston 2002

777. Krauss G. et al. (1969): Die amtlichen topographischen Kartenwerke der Bundesrepublik Deutschland, Herbert Wichmann Verlag, Karlsruhe 1969

778. Kretschmer I. (1978): Irreführende Meinungen über die "Peters–Karte", Mitt. der Österreichische Geogr. Ges., Band 120, I (1978) 124–135

779. Kretschmer I. (1993): Mercators Bedeutung in der Projektionslehre (Mercatorprojektion), in: Mercator und Wandlungen der Wissenschaften im 16. und 17. Jahrhundert, M. Büttner and R. Dirven (ed.), Universitätsverlag Dr. N. Brockmeyer, Bochum 1993, pp. 151–174

780. Kretschmer I. (1994): Die Eigenschaften der "Mercatorprojektion" und ihre heutige Anwendung, in: Mercator – ein Wegbereiter neuzeitlichen Denkens, I. Hantsche (ed.), Universitätsverlag Dr. N. Brockmeyer, Bochum 1994, pp. 141–169

781. Kreyszig E. (1959): Differential geometry, University of Toronto Press, Toronto, 1959

782. Kreyszig E. (1988): Advanced engineering mathematics, 6th ed., J. Wiley & Sons, New York 1988

783. Kruecken W. (1994): Das Rätsel der Mercator–Karte 1569, Kartographische Nachrichten 44 (1994) 182–185

784. Krueger E., Roesch N. (1998): Parametersysteme auf dem dreiachsigen Ellipsoid, Kartographische Nachrichten 48 (1998) 234–237

785. Krueger L. (1883): Die geodätische Linie des Sphäroids und Untersuchung darüber, wann dieselbe aufhört, kürzeste Linie zu sein (the geodesic of the spheroid and an investigation into when the geodesic is no longer the shortest line), Inauguraldissertation an der Universität Tübingen, Schade, Berlin 1883

786. Krueger L. (1897): Zur Theorie rechtwinkliger geodätischer Koordinaten, Zeitschrift für Vermessungswesen 15 (1897) 441–453

787. Krueger L. (1903): Bemerkungen zu C.F. Gauss: Conforme Abbildungen des Sphäroids in der Ebene, C.F. Gauss, Werke, Königl. Ges. Wiss. Göttingen Bd. IX (1903) 195–204

788. Krueger L. (1912): Konforme Abbildung des Erdellipsoids in der Ebene, B.G. Teubner, Leipzig 1912

789. Krueger L. (1914): Transformation der Koordinaten bei der konformen Doppelprojektion des Erdellipsoids auf die Kugel und die Ebene, Teubner, Leipzig 1914

790. Krueger L. (1919): Formeln zur konformen Abbildung des Erdellipsoides in der Ebene. Herausgegeben von der Preußischen Landsaufnahme, Berlin, 1919

791. Krueger L. (1921): Die Formeln von C.G. Andrae, O. Schreiber, F.R. Helmert und O. Börsch für geographische Koordinaten und Untersuchung ihrer Genauigkeit, Zeitschrift für Vermessungswesen 50 (1921), 547–557

792. Krueger L. (1922): Zur stereographischen Projektion, P. Stankiewicz, Berlin 1922

793. Krueger L. (1926): Anleitung für die Truppe zum Eintragen des Gitternetzes in Karten der Maßstäbe 1:25000, 1:100000, 1:200000 und 1:300000, Deutsche Heeresvermessungsstelle, Berlin 1926

794. Krumm F., Grafarend E.W. (2002): Datum–free deformation analysis of ITRF networks, Artificial Satellites, J. Planetary Geodesy 37 (2002) 75–84

795. Krupzig E. (1983): Advanced engineering mathematics, J. Wiley & Sons, New York Chichester 1983

796. Kuehnel W. (1991): On the inner curvature of the second fundamental form, Proceedings of the 3rd Congress of Geometry, Thessaloniki (1991) 248–253

797. Kuehnel W. (2002): Differential geometry, curves–surfaces–manifolds, American Mathematical Society, Providence, Rhode Island 2002

798. Kuehnel W., Rademacher H.-B. (1995): Essential conformal fields in pseudo–Riemannian geometry, J. Math. Pures Appl. 74 (1995) 453–481

799. Kuiper N.H. (1949): On conformally–flat spaces in the large, Annals of Mathematics 50 (1949) 916–924

800. Kuiper N.H. (1950): On compact conformally Euclidean spaces of dimension > 2, Annals of Mathematics 52 (1950) 478–490

801. Kuiper N.H. (1980): Tight embeddings and maps. Submanifolds of geometrical class three in EN, in: The Chern Symposium 1979, Proceedings of the International Symposium on Differential Geometry in honor of S–.S. Chern 1979, W.-Y. Hsiang et al. (eds.), Springer, New York 1908, 97–145

802. Kulkarni R.S. (1969): Curvature structures and conformal transformations, J. Differential Geometry 4 (1969) 425–451

803. Kulkarni R.S. (1972): Conformally flat manifolds, Proceedings of the National Academy of Sciences of the United States of America 69 (1972) 2675–2676

804. Kulkarni R.S. (1988): Conformal structures and Möbius structures, in: Conformal Geometry, R.S. Kulkarni and U. Pinkall (eds.), Friedrich Vieweg & Sohn, Braunschweig / Wiesbaden 1988, pp. 1–39

805. Kulkarni R.S., Pinkall U. (eds.)(1988): Conformal geometry, Vieweg, Braunschweig / Wiesbaden 1988

806. Kumler M.P., Tobler W.R. (1991): Three world maps on a Moebius strip, Cartography and Geographic Information Systems 18 (1991) 275–276

807. Kuntz E. (1964): Die analytischen Grundlagen perspektivischer Abbildungen der Erdoberfläche aus großen Höhen, Report C69, Deutsche Geodätische Kommission, Bayer. Akad. Wiss., München 1964

808. Kuntz E. (1990): Kartennetzentwurfslehre, Wichmann, 2nd ed., Karlsruhe 1990

809. Kythe P.K. (1998): Computational conformal mapping, Birkhäuser, Boston 1998

810. Laborde Chef d'escadron (1928): La nouvelle projection du service géographique de Madagascar: Cahiers due Service géographique de Madagascar 1, Tananarive 1928

811. Lafontaine J. (1988a): Conformal geometry from the Riemannian view–point, in: Conformal Geometry, R.S. Kulkarni and U. Pinkall (eds.), Friedrich Vieweg & Sohn, Braunschweig / Wiesbaden 1988, pp. 65–92

812. Lafontaine J. (1988b): The theorem of Lelong–Ferrand and Obata, in: Conformal Geometry, R.S. Kulkarni and U. Pinkall (eds.), Friedrich Vieweg & Sohn, Braunschweig / Wiesbaden 1988, pp. 93–103

813. Lagrange J.L. (1779): Ueber Kartenprojection. Abhandlungen von Lagrange (1779) und Gau´⁻¢(1822). Herausgegeben von A. Wangerin. Ostwald's Klassiker der exakten Wissenschaften, Nr. 155, 57-101, Verlag von Wilhelm Engelmann, Leipzig.

814. Lagrange de J.L. (1781) : Sur la construction des cartes geographiques, Nouveaux Mémoires de l'Academie Royale des Sciences et Belles Lettres de Berlin 161–210, Berlin 1781

815. Lagrange de J.L. (1794): Über die Construction geographischer Karten, Ostwald's Klassiker der exakten Naturwissenschaften, Nr. 55, Engelmann, Leipzig 1894

816. Lagrange de J.L., Gauss C.F. (1894): Über Kartenprojektion, Leipzig 1894

817. Lallemand C. (1911): Sur les déformations résultant du mode de construction de la carte internationale du monde au millionième: Comtes Rendus 153 (1911) 559–567

818. Lamé M.G. (1818): Examen des différentes méthodes employées pour résoudre les problèmes de géométrie, Paris (1818) 70–72

819. Lambert J.H. (1772): Beiträge zum Gebrauche der Mathematik und deren Anwendung: Part III, section 6: Anmerkungen und Zusätze zur Entwerfung der Land–und Himmelscharten: Berlin, translated and introduced by W.R. Tobler, Univ. Michigan 1972

820. Lambert J.H. (1894): Land– und Himmelskarten, Ostwald's Klassiker der exakten Naturwissenschaften, Nr. 54, Engelmann, Leipzig 1894

821. Lancaster G.M. (1969): A characterization of certain conformally Euclidean spaces of class one, Proceedings of the American Mathematical Society 21 (1969) 623–628

822. Lancaster G.M. (1973): Canonical metrics for certain conformally Euclidean spaces of dimension three and codimension one, Duke Mathematical Journal 40 (1973) 1–8

823. Lanczos C. (1949): The variational principles of mechanics, University of Toronto Press, Toronto 1949

824. Lane E.P. (1939): Metric differential geometry of curves and surfaces, University of Chicago Press, 1939 p.189

825. Lang S. (1995): Differential and Riemannian manifolds, Springer Verlag, Berlin – Heidelberg – New York 1995

826. Langhans P. (ed.)(1935): Dr. A. Petermanns Mitteilungen, Ergänzungsband XLVIII, Heft 218–221, Justus Perthes, Gotha 1935

827. Laskowski P.H. (1989): The traditional and modern look at Tissot's indicatrix, The American Cartographer 16 (1989) 123–133

828. Laugwitz D. (1977): Differentialgeometrie, B.G. Teubner, Stuttgart 1977

829. Lawson H.B. (1977): The quantitative theory of foliations, Rhode Island 1977

830. Lee D.K. (1990): Application of theta functions for numerical evaluation of complete elliptic integrals of the first and second kinds, Comput. Phys. Comm. 60 (1990) 319–327

831. Lee L.P. (1944): The nomenclature and classification of map projections, Empire Survey Review 7 (1944) 190–200

832. Lee L.P. (1945): The transverse Mercator projection of the spheroid, Empire Survey Review 8 (1945) 142–152

833. Lee L.P. (1954): A transverse Mercator projection of the spheroid alternative to the Gauss–Krueger form, Empire Survey Review 12 (1954) 12–17

834. Lee L.P. (1962): The transverse Mercator projection of the entire spheroid, Empire Survey Review 16 (1962) 208–217

835. Lee L.P. (1963a): The transverse Mercator projection of the entire spheroid, Empire Survey Review 17 (1963) 343

836. Lee L.P. (1963b): Scale and convergence in the transverse Mercator projection of the entire spheroid, Survey Review 127 (1963) 49–51

837. Lee L.P. (1965): Some conformal projections based on elliptic functions, The Cartographical Review 55 (1965) 563–580

838. Lee L.P. (1968): Mathematical geography, The Geographical Review 58 (1968) 490–491

839. Lee L.P. (1974): A conformal projection for the map of the Pacific, New Zeeland Geographer 30 (1974) 75–77

840. Lee L.P. (1976): Conformal projections based on elliptic functions, Cartographica Monograph No 16. Univ. of Toronto Press 1976

841. Legendre A.M. (1806): Analyse des triangles tracés sur la surface d'un sphéroide. Tome VII de la 1°–série des memoires de l'Academie des Sciences, Paris 1806

842. Lehmann M. (1939): Über die Lagrangesche Projektion, Zeitschrift für Vermessungswesen 8 (1939) 329–344, 361–376, 425–432

843. Leichtweiß K. (1956): Das Problem von Cauchy in der mehrdimensionalen Differentialgeometrie, Math. Annalen 130 (1956) 442–474

844. Leichtweiß K. (1961): Zur Riemannschen Geometrie in Grassmannschen Mannigfaltigkeiten, Math. Zeitschrift 76 (1961) 334–366

845. Leichtweiß K. (1967): Zur Charakterisierung der Wendelflächen unter den vollständigen Minimalflächen, Abh. Math. Seminar Universität Hamburg, Bd. 30, Heft 1/2, pp. 36–53, Vandenhoek und Ruprecht, Göttingen 1967

846. Leick A., van Gelder B.H.W (1975): Similarity transformations and geodetic network distortions based on Doppler satellite observations. The Ohio State University, Department of Geodetic Science, Report 235, Columbus 1975

847. Leighly J.B. (1955): Aitoff and Hammer – An attempt at clarification, The Geographical Review 45 (1955) 246–249

848. Lemczyk T.Y., Yovanovich M.M. (1988): Efficient evaluation of incomplete elliptic integrals and functions, Comput. Math. Appl. 16 (1988) 747–757

849. Lense J. (1926): Über ametrische Mannigfaltigkeiten und quadratische Differentialformen mit verschwindender Diskriminante, Jahresbericht der deutschen Mathematiker–Vereinigung 35 (1926) 280–294

850. Lenzmann L. (1985): Umrechnung Gauss'scher konformer Koordinaten durch Approximationsformeln, Allgemeine Vermessungsnachrichten 92 (1985) 193–199

851. Lewis B. Sir C., Campbell Col. J.D., eds. (1951): The American Oxford atlas, Oxford Univ. Press, Oxford 1951

852. Lichtenegger H. (1972): Der Allgemeinfall kosmographischer Perspektiven, Österreichische Zeitschrift für Vermessungswesen 60 (1972) 85–90

853. Lichtenegger H. (1987): Zur numerischen Lösung geodätischer Hauptaufgaben auf dem Ellipsoid, Zeitschrift für Vermessungswesen 112 (1987) 508–515

854. Lichtenstein L. (1909–1921): Neuere Entwicklung der Potentialtheorie, konforme Abbildung, Enzyklopädie der math. Wiss. II C 3 (1909–1921) 177–377

855. Lichtenstein L. (1911): Beweis des Satzes, daß jedes hinreichend kleine, im wesentlichen stetig gekrümmte, singularitätenfreie Flächenstück auf einem Teil einer Ebene zusammenhängend und in den kleinsten Teilen ähnlich abgebildet werden kann, Preußische Akademie der Wissenschaften, Berlin 1911

856. Lichtenstein L. (1916): Zur Theorie der konformen Abbildung. Konforme Abbildung nichtanalytischer, singularitätenfreier Flächenstücke auf ebene Gebiete, Anzeiger der Akademie der Wissenschaften in Krakau 2–4 (1916) 192–217

857. Liebmann H. (1918): Die angenäherte Ermittelung harmonischer Funktionen und konformer Abbildungen, Sitzungsberichte der math.-phys. Klasse, Bayer. Akademie der Wissenschaften, München 1918, pp. 385–416

858. Lilienthal R. (1902–1927): Die auf einer Fläche gezogenen Linien, Enzyklopädie der math. Wiss. 3. Band: Geometrie, Teubner, Leipzig 1902–1927

859. Liouville J. (1850): Extension au cas de trois dimensions de la question du tracé géographique, Note VI, by G. Monge: application de l'analyse à la géométrie, 5ème édition revue corrigée par M. Liouville, Bachelier, Paris 1850

860. Livieratos E. (1987): Differential geometry treatment of a gravity field feature: the strain interpretation of the global geoid, in: Geodetic Theory and Methodology, Report 600006, University of Calgary, pp. 49–73, Calgary 1987

861. Loan C.F. van (1976): Generalizing the singular value decomposition, SIAM J. Numer. Anal. 13 (1976) 76–83

862. Loebell F. (1942): Allgemeine Theorie der Flächenabbildungen, Nachrichten a.d. Reichsvermessungsamt 18 (1942) 299–307

863. Lohse P. (1990): Dreidimensionaler Rückwärtsschnitt, Zeitschrift für Vermessungswesen 115 (1990) 162–167

864. Lomnicki A. (1956): Kartografia Matematyczna, Warszawa 1956

865. Lowell K., Gold C. (1995): Using a fuzzy surface–based cartographic representation to decrease digitizing efforts for natural phenomena, Cartography and Geograph. Information Systems 22 (1995) 222–231

866. Loxton J. (1985): The Peters phenomenon, The Cartographic Journal 22 (1985) 106–108

867. Luccio M. (2001): Telematics today, smart cars, informed drivers, GPS World Showcase 12 (2001) 28–29

868. Luke Y.L. (1968): Approximations for elliptic integrals, Math. Comp. 22 (1968) 627–634

869. Luke Y.L. (1970): Further approximations for elliptic integrals, Math. Comp. 24 (1970) 191–198

870. MacCallum M.A.H. (1983): Classifying metrics in theory and practice, in: Unified field theories on more than 4 dimensions, V. De Sabbata and E. Schmutzer (eds.), World Scientific, Singapore 1983

871. Macdonald R.R. (1968): An optimum continental projection, Cartographic Journal 5 (1968) 46–47

872. MacKay R.S. (1993): Renormalisation in area–preserving maps, World Scientific, Singapore 1993

873. Macvean D.B. (1968): Die Elementarbeit in einem Kontinuum und die Zuordnung von Spannungs– und Verzerrungstensoren, Zeitschrift für Angewandte Mathematik und Physik 19 (1968) 137–185

874. Magnus J.R., Neudecker H. (1988): Matrix differential calculus with applications in statistics and econometrics, J. Wiley & Sons, New York Chichester 1988

875. Mainwaring J. (1942): An introduction to the study of map projections, McMillan & Co., London 1942

876. Maling D.H. (1959/60): A review of some Russian map projections, Empire Survey Review 15 (1959/60) 203–215, 255–303, 294–303

877. Maling D.H. (1968): The terminology of map projecions, in: Internationales Jahrbuch für Kartographie 8 (1968) 11–64

878. Maling D.H. (1973): Projections for navigation charts, in: ibid., Coordinate Systems and Map Projections, George Philip and Son Ltd., London 1973, pp. 183–198

879. Maling D.H. (1993): Coordinate systems and map projections, 2nd edition, Pergamon Press, Oxford (1993)

880. Maling D.H. (1989): Measurements from maps: Principles and methods of cartometry, Pergamon, Oxford 1989

881. Maltman A. (1996): Geological Maps, 2nd ed., J. Wiley & Sons, New York Chichester 1996

882. Mareyen M., Becker M. (1998): On the datum realization of regional GPS networks, Allgemeine Vermessungsnachrichten 105 (1998) 396–406

883. Markuschewitsch A.I. (1955): Skizzen zur Geschichte der analytischen Funktion, Deutscher Verlag der Wissenschaften, Belin 1955

884. Marsden J.E., Hughes T.J.R. (1983): Mathematical foundations of elasticity, Prentice Hall Publ., Englewoog Cliffs 1983

885. Massey W.S. (1962): Surfaces of Gaussian curvature zero in Euclidean 3–space, Tohuku Math. J. 14 (1962) 73–79

886. Massonet D., Feigl K.L. (1998): Radar interferometry and its application to changes in the Earth's surface, Reviews of Geophysics 36 (1998) 441–500

887. Matérn B. (1986): Spatial variation, 2nd ed., Lecture notes in statistics 36, Springer–Verlag, Berlin 1986

888. Mather R.S. (1971): The analysis of the Earth's gravity field, University of New South Wales, School of Surveying, Kensington, Australia 1971

889. Mather R.S. (1973): The theory and geodetic use of some common projections, Monograph No. 1, The School of Surveying, The Univ. of New South Wales, Kensington, Australia 1973

890. Maurer H. (1935): Ebene Kugelbilder – ein Linnésches System der Kartenentwürfe, Justus Perthes, Gotha 1935

891. Mc Gehee O.C. (2000): An introduction to complex analysis, J. Wiley & Sons, New York Chichester 2000

892. McDonnell P.W. (1979): Introduction to map projections, Marcel Dekker Inc., New York and Basel 1979

893. McLachlan R. (1994): A gallery of constant–negative–curvature surfaces, The Mathematical Intelligencer 16 (1994) 31–37

894. McLachlan R.I., Segur H. (1994): A note on the motion of surfaces, Physics Letters A 194 (1994) 165–172

895. McLain D.H. (1976): Two dimensional interpolation from random data, The Computer Journal 19 (1976) 178–181

896. Mehl C. (1999): Condensed forms for skew–Hamiltonian / Hamiltonian pencils, SIAM J. Matrix Anal. Appl. 21 (1999) 454–476

897. Meichle H. (2001): Digitale Finite Element–Höhenbezugsfläche (DFHBF) für Baden–Württemberg (digital finite element height reference surface (DFHBF) for Baden–Wuerttemberg), Mitteilungen deutscher Verein für Vermessungswesen, Landesverein Baden–Württemberg, Stuttgart 2001, 23–31

898. Meissl P. (1981): Skriptum aus Ellipsoidische Geometrie nach der Vorlesung von o. Univ. Prof. Dr. Peter Meissl, TU Graz, Institut für Ellipsoidische Geodäsie, Graz 1981

899. Melluish R.K. (1931): An introduction to the mathematics of map projections, Cambridge University Press, Cambridge 1931

900. Mendlovitz M.A. (1999): More results on eigenvector saddle points and eigenpolynomials, SIAM J. Matrix Anal. Appl. 21 (1999) 593–612

901. Mercator G. (1569): Weltkarte ad usum navigatium, Duisburg 1569, repr. W. Krücken and J. Milz, Duisburg 1994

902. Merkel H. (1956): Grundzüge der Kartenprojektionslehre, Teil I und II, Report A17, Deutsche Geodätische Kommission, Bayer. Akad. Wiss., München 1956

903. Midy P. (1975): An improved calculation of the general elliptic integral of the second kind in the neighbourhood of x=0, Numer. Math 25 (1975) 99–101

904. Militärgeographische Amt (1962): Transformation von UTM– in Geographische Kordinaten und Umkehrung, Bad Godesberg 1962

905. Militärisches Geowesen (1988): World Geodetic System 1984 (WGS84), Amt für militärisches Geowesen 1988

906. Miller A.I. (1972): The myth of Gauss' experiment on the Euclidean nature of physical space, Isis 63 (1972) 345–348

907. Miller O.M. (1941): A conformal map projection for the Americas, Geographical Review 31 (1941) 100–104

908. Miller O.M. (1942): Notes on cylindrical world map projections, Geographical Review 32 (1942) 424–430

909. Miller O.M. (1953): A new conformal projection for Europe and Asia, Geographical Review 43 (1953) 405–409

910. Milnor J. (1969): A problem in cartography, The American Mathematical Monthly 6 (1969) 1101–1112

911. Milnor J. (1994): Collected papers, Vol. 1 (Geometry), Publish or Perish Inc., Houston, Texas 1994

912. Mirsky L. (1960): Symmetric gauge functions and unitary invariant norms, Quart. J. Math 11 (1960) 50–59

913. Misner C.W. (1978): Harmonic maps as models for physical theores, Physical Review D 18 (1978) 4510–4524

914. Misner C.W., Thorne K.S., Wheeler J.A. (1973): Gravitation, Freeman, New York 1973

915. Mitchell H.C., Simmons L.G. (1945): The state coordinate systems (a manual for surveyors), U.S. Coast and Geodetic Survey Spec. Pub. 235

916. Mitra S.K., Rao C.R. (1968): Simultaneous reduction of a pair of quadratic forms, Sankya A 30 (1968) 312–322

917. Mittelstaedt F.-G. (1989): Vorworte in deutschen Schulatlanten, Kartographische Nachrichten 39 (1989) 212–216

918. Mittermayer E. (1965): Formeln zur Berechnung der ellipsoidischen geographischen Endbreite für Meridianbögen beliebiger Länge, Zeitschrift für Vermessungswesen 90 (1965) 403–408

919. Mittermayer E. (1993a): Zur Integraldarstellung der geodätischen Linie auf dem Rotationsellipsoid, Zeitschrift für Vermessungswesen 118 (1993) 72–74

920. Mittermayer E. (1993b): Die Gauss'schen Koordinaten in sphärischer und ellipsoidischer Approximation / Konforme Abbildung, Zeitschrift für Vermessungswesen 118 (1993) 345–356

921. Mittermayer E. (1994): Einführung "Gauss'scher Kugelkoordinaten", Zeitschrift für Vermessungswesen 119 (1994) 24–35

922. Mittermayer E. (1996): Die sphärische konforme Meridiankonvergenz – eine räumliche Ortsfunktion, Zeitschrift für Vermessungswesen 121 (1996) 114–123

923. Mittermayer E. (1998): Krümmung und Windung der r–Linien metrischer Kugelkoordinaten (Mercator), Allgemeine Vermessungsnachrichten 105 (1998) 409–413

924. Mittermayer E. (1999): Die Gauss'schen Koordinaten als Ortsfunktionen und Funktionentheorie / Vektoranalysis, Allgemeine Vermessungsnachrichten 106 (1999) 405–416

925. Mittermayer E. (2000): Mercator–Projektion und die Kugelloxodrome, Allgemeine Vermessungsnachrichten 107 (2000) 58–67

926. Mohr O. (1928): Abhandlungen aus dem Gebiete der Technischen Mechanik, Verlag von Ernst und Sohn, Seiten 192–202, Berlin 1928

927. Mok E. (1992): A model for the transformation between satellite and terrestrial networks in Hong Kong, Survey Review 31 (1992) 344–350

928. Mollweide C.B. (1805): Mappirungskunst des Claudius Ptolemaeus, ein Beytrag zur Geschichte der Landkarten, Zach's Monatliche Korrespondenz zur Beförderung der Erd–und Himmelskunde (1805) 319–340, 504–514

929. Monastyrsky M. (1999): Doctoral dissertation, in: ibid., Riemann, topography, and physics, Birkhäuser, Boston 1999, 10–17

930. Moore J.D. (1977): Conformally flat submanifolds of Euclidean space, Math. Annals 225 (1977) 89–97

931. Moran P.A.P. (1950): Numerical integration by systematic sampling, Proceedings of the Cambridge Philosophical Society 46 (1950) 111–115

932. Morita T. (1999): Calculation of the elliptic integrals of the first and second kinds with complex modulus, Numer. Math. 82 (1999) 677–688

933. Moritz H. (1984): Geodetic reference system 1980, The Geodesist's Handbook, Bulletin Géodésique 58 (1984), 388–398

934. Moritz H. (1994): The Hamiltonian structure of refraction and of the gravity field, Manuscripta Geodaetica 20 (1994) 52–60

935. Morman K.N. Jr. (1986): The generalized strain measure with application to nonhomogeneous deformations in rubber–like solids, Journal of Applied Mechanics 53 (1986) 726–728

936. Moser J. (1980): Geometry of quadrics and spectral theory, in: Proceedings of the international symposium on differential geometry in honor of S.-S. Chern, W.Y. Hsiang et al. (eds.), Springer Verlag, Berlin – Heidelberg – New York 1980, pp. 147–188

937. Muehrcke P.C. (1986): Map use, 2nd ed., Madison, Wis.: JP Publications 1986

938. Mueller B. (1991): Kartenprojektionen des dreiachsigen Ellipsoids, Diplomarbeit Geodätisches Institut Universität Stuttgart 1991

939. Mueller F.J. (1914): Johann Georg von Soldner, der Geodät, Dissertation, Kgl. Technische Hochschule, München 1914

940. Mueller I. (1974): Global satellite triangulation and trilateration results. J. Geophys. 79 (1974) 5333–5335

941. Mueller J. (1967): Map projections in geodesy, in: International dictionary of geophysics, S.K. Runcorn et al. (eds.), Pergamon Press, Oxford 1977, pp. 910–920

942. Mundell I. (1993): Maps that shape the world, New Scientist 139 (1993) 21–23

943. Murakami M., Oki S. (1999): Realization of the Japanese geodetic datum 2000 (JGD 2000), Bulletin of the Geographical Survey Institute 45 (1999) 1–10

944. Nash J. (1956): The imbedding problem for Riemannian manifolds, Annals of Mathematics 63 (1956) 20–63

945. National Academy of Sciences (1971): North American datum: National Ocean Survey contract rept. E–53–69(N) 80 p., 7figs., 1971

946. Naumann H. (1957): Über Vektorsterne und Parallelprojektionen regulärer Polytope, Math. Zeitschrift 67 (1957) 75–82

947. Nell A.M. (1890): Äquivalente Kartenprojektionen, Petermanns Geographische Mitteilungen 36 (1890) 93–98

948. Nellis W.J., Carlson B.C. (1966): Reduction and evaluation of elliptic integrals, Math. Comp. 20 (1966) 223–231

949. Neumann L. (1923): Mathematische Geographie und Kartenentwurfslehre, Ferd. Hirt, Breslau 1923

950. Neutsch W. (1995): Koordinaten, Spektrum Adademischer Verlag, 1353 pages, Heidelberg 1995

951. Newcomb R.W. (1960): On the simultaneous diagonalization of two semidefinite matrices, Quart. J. App. Math. 19 (1960) 144–146

952. Nielson G.M. (1983): A method for interpolating scattered data based upon a minimum norm network, Mathematics of Computation 40 (1983) 253–271

953. Nielson G.M., Ramaraj R. (1987): Interpolation over a sphere based upon a minimum norm network, Computer Aided Geometric Design 4 (1987) 41–57

954. Niermann A. (1984): A comparison of various advantageous cartographic representations, Quaterniones Geodaesiae 5 (1984) 11–36, 61–104

955. Nirenberg L. (1953): The Weyl and Minkowski problems in differential geometry in the large, Communications on Pure and Applied Mathematics 6 (1953) 337–394

956. Nishikawa S. (1974): Conformally flat hypersurfaces in a Euclidean space, Tohoku Math. Journal 26 (1974) 563–572

957. Nishikawa S., Maeda Y. (1974): Conformally flat hypersurfaces in a conformally flat Riemannian manifold, Tohoku Math. Journal 26 (1974) 159–168

958. Nitzsche J.C.C. (1965): On new results in the theory of minimal surfaces, Bulletin of the Am. Math. Soc. 71 (1965) 195–270

959. Nomizu K., Rodriguez L. (1972): Umbilical submanifolds and Morse functions, Nagoya Math. Journal 48 (1972) 197–201

960. Nordenskioeld A.E. (1889): Facsimile–atlas, Dover Publications, Inc., New York 1889 (reprint 1973)

961. Obata M. (1962): Conformal transformations of compact Riemannian manifolds, Illinois Journal of Mathematics 6 (1962) 292–295

962. Obata M. (1968): The Gauss map of immersions of Riemannian manifolds in spaces of constant curvature, J. Differential Geometry 2 (1968) 217–223

963. Obata M. (1971): The conjectures on conformal transformations of Riemannian manifolds, J. Differential Geometry 6 (1971) 247–258

964. Odermatt H. (1960): Tafeln zum Projektionssystem der schweizerischen Landesvermessung, Mitteilungen aus dem Geodätischen Institut an der Eidgenössischen Technischen Hochschule in Zürich, Nr. 8, Verlag Leemann, Zürich 1960

965. Ogden R.W. (1984): Non-linear elastic deformations, J. Wiley, New York (1984)

966. O'Keefe J.A. (1953): The isoparametric method of mapping one ellipsoid on another, Transactions, American Geophysical Union 34 (1953) 869–875

967. O'Keefe J.A., Greenberg A. (1977): A note on the Van der Grinten projection of the whole earth onto a circular disc, Am. Cartographer 4 (1977) 127–132

968. Okeke F. (1997): The curvilinear datum transformation model, PhD Thesis, Geodätisches Institut Universität Stuttgart 1997

969. Oki N., Hosokawa Y., Sugimoto E., Abe Y., Taniguchi T. (1993): Portable vehicle navigation system (NV–1): Its features and operability, IEEE – IEE Vehicle Navigation & Information Systems Conference, Ottawa 1993, pp. 482–485

970. Opozda B., Verstraelen L. (1990): On a new curvature tensor in affine differential geometry, in: Geometry and topology of submanifolds, II, M. Boyom et al. (eds.), World Scientific, Singapore 1990, pp. 271–293

971. Ord–Smith R.J. (1984): Efficient geodetic calculations by microcomputer, Survey Review 27 (1984) 227–231

972. Ord–Smith R.J. (1985): Transverse Mercator projection – a simple geometrical approximation, Survey Review 28 (1985) 51–62

973. Ormeling F.J., Kraak M.J. (1990): Kartografie, Delftse Universitaire Press, Delft 1990

974. Ossermann R. (1959): Remarks on minimal surfaces, Communications on Pure and Applied Mathematics 12 (1959) 233–239

975. Ossermann R. (1980): Minimal surfaces, Gauss maps, total curvature, eigenvalue estimates, and stability, in: Proceedings of the international symposium on differential geometry in honor of S.-S. Chern, W.Y. Hsiang et al. (eds.), Springer Verlag, Berlin Heidelberg New York 1980, pp. 199–227

976. Ossermann R. (1986): A survey of minimal surfaces, Dover Publications, New York 1986

977. Osterhold M. (1993): Landesvermessung und Landinformationssysteme in den Vereinigten Staaten von Amerika, Allgemeine Vermessunsnachrichten 100 (1993) 287–295

978. Otero J. (1997): A best harmonic approximation problem arising in cartography, Atti Sem. Mat. Fis. Univ. Modena XLV (1997) 471–492

979. Otero J., Sevilla M.J. (1990): On the optimal choice of the standard parallels for a conformal conical projection, Bollettino di Geodesia e Science Affini 1 (1990) 1–14

980. Ottoson P. (2001): Geographic indexing and data management for 3D–visualisation, Dissertation, Royal Institute of Technology, Dept. of Infrastructure, Division of Geodesy and Geoinformatics, Stockholm 2001

981. Ozone M.J. (1985): Non iterative solution of the 0 equation, Surveying and Mapping 45 (1985) 169–171

982. Pachelski W. (1994): Possible users of natural (barycentric) coordinates for positioning, Schriftenreihe der Institute des Fachbereichs Vermessungswesen, Universität Stuttgart, Report Nr. 1994.2, Stuttgart 1994

983. Panteliou S.D., Dimarogonas A.D., Katz I.N. (1996): Direct and inverse interpolation for Jacobian elliptic functions, zeta function of Jacobi and complete elliptic integrals of the second kind, Comput. Math. Appl. 32 (1996) 51–57

984. Parry R.B., Perkins C.R. (1987): World mapping today, Butterworth & Co., London 1987

985. Paul M.K. (1973): A note on computation of geodetic coordinates from geocentric (Cartesian) coordinates, Bulletin Géodésique 108 (1973) 135–139

986. Pearson II F. (1977): Map projection equations, Naval Surface Weapons Center, Dahlgren 1977

987. Pearson II F. (1990): Map projections: Theory and applications, CRC Press, Boca Raton, Florida 1990

988. Pec K., Martinec Z. (1983): Expansion of Geoid heights over a triaxial Earth's ellipsoid into a spherical harmonic series, Studia geoph. et geod. 27 (1983) 217–232

989. Penev P.O. (1978): Transformation of rectangular coordinates into geographical coordinates by closed formulas, Geodesy, Mapping and Photogrammetry 20 (1978) 175–177

990. Peters A.B. (1984): Distance–related maps, The American Cartographer 11 (1984) 119–131

991. Peterson K. (1868): Ueber Curven und Flächen, A. Lang's Buchhandlung, Moskau 1868

992. Petrich M., Tobler W. (1984): The globular projection generalized, The American Cartographer 11 (1984) 101–105

993. Petrovic M. (1990): Curvature conditions on hypersurfaces of revolution, in: Geometry and topology of submanifolds, II, M. Boyom et al. (eds.), World Scientific, Singapore 1990, pp. 294–300

994. Pettengill G.H., Campbell D.B., Masursky H. (1980): The surface of Venus, Scientific American 243 (1980) 54–65

995. Pick M. (1957a): Konforme Transformation von einem Ellipsoid auf ein anderes Ellipsoid, Studia geoph. et geod. 1 (1957) 46–73

996. Pick M. (1957b): Über das Problem der Transformation eines Referenzellipsoids auf ein anderes mittels Projektion längs der Normalen, Studia geoph. et geod. 1 (1957) 372–375

997. Pick M. (1958): Zur Frage der konformen Transformation von einem Ellipsoid auf ein anderes Ellipsoid, Studia geoph. et geod. 2 (1958) 174–177

998. Pick M. (1961): Projektive Methode zur Transformation dreiachsiger Ellipsoide mit nicht parallelen Achsen, Studia geoph. et geod. 5 (1961) 191–209

999. Pick M. (1994): Transformation of the trigonometric network by the projection, in: Role of Modern Geodesy, contributions of TS ACSR, Working Groups: Global Geodesy, Building of WGS 84, Transition to NATO Standards GPS in Geodesy and Navigation, pres. at 2nd Seminar Budapest, November 22–24 1994, 13–18

1000. Pick M: (1985): Closed formulas for transformation of the Cartesian coordinate system into a system of geodetic coordinates, Studia Geoph. et Geod. 29 (1985) 112–119

1001. Pinkall U. (1988): Compact conformally flat hypersurfaces, in: Conformal geometry, R.S. Kulkarni and U. Pinkall (eds.), Friedr. Vieweg & Sohn, Braunschweig 1988, pp. 217–236

1002. Piola G. (1833): La meccanica de' corpi naturalmente estesi trattata col calcolo delle variazioni, Opusc. Mat. Fis. Di Diversi Autori. Milano. Guisti 1 (1833) 201–236

1003. Piola G. (1836): Nuova analisi per tutte le questioni della meccanica molecolare, Memorie Mat. Fis. Soc. Ital. Sci. Modena 21 (1836) 155–321

1004. Piola G. (1848): Intorno alle equazioni fondamentali del movimento di corpi qualsivogliano, considerati secondo la naturale loro forma e costituzione, Mem. Mat. Fis. Soc. Ital. Moderna 24 (1848) 1–186

1005. Pizzetti P. (1925): Höhere Geodäsie, in: Encyklopädie der mathematischen Wissenschaften, Band VI,1, P. Furtwängler and E. Wiechert (eds.), B.G. Teubner, Leipzig 1925, pp. 117–243

1006. Poder K., Hornik H. (1988): The European Datum of 1987 (ED 87), Publ. 18, Report IAG, Lisbon 1988

1007. Porter R.M. (1989): Historical development of the elliptic integral (Spanish), Congress of the Mexican Mathematical Society, Mexico City 1989, 133–156

1008. Porter W., McDonnell Jr. (1979): Introduction to map projections, Marcel Dekker, New York 1979

1009. Pottmann H. (1992): Interpolation on surfaces using minimum norm networks, Computer Aided Geometric Design 9 (1992) 51–67

1010. Pottmann H., Divivier A. (1990): Interpolation von Meßdaten auf Flächen, in: Geometrische Verfahren der graphischen Datenverarbeitung, J.L. Encarnacao et al. (eds.), Springer, Berlin 1990, pp. 104–120

1011. Pottmann H., Eck M. (1990): Modified multiquadric methods for scattered data interpolation over a sphere, Computer Aided Geometric Design 7 (1990) 313–321

1012. Pottmann H., Hagen H., Divivier A. (1991): Visualizing functions on a surface, The Journal of Visualization and Computer Animation 2 (1991) 52–58

1013. Press W.H., Flannery B.P., Teukolsky S.A., Vetterling W.T. (1988): Numerical recipes in C., Cambridge University Press, Cambridge 1988

1014. Press W.H., Flannery B.P., Teukolsky S.A., Vetterling W.T. (1992): Elliptic integrals and Jacobi elliptic functions, Numerical Recipes in FORTRAN: The Art of Scientific Computing, 2nd ed., Cambridge University Press, Cambridge, 1992, pp. 254–263

1015. Press W.H., Teukolsky S.A., Elliptic integrals, Comput. In Phys. 4 (1990) 92–96

1016. Price W.F. (1986): The new definition of the meter, Survey Review 28 (1986) 276–279

1017. Pruszko T. (1984): Dissertationes mathematicae, Panstwowe Wydawnictwo Naukowe, Warszawa 1984

1018. Rademacher H.-B. (1988): Conformal and isometric immersions of conformally flat Riemannian manifolds into spheres and Euclidean spaces, in: Conformal geometry, R.S. Kulkarni and U. Pinkall (eds.), Friedr. Vieweg & Sohn, Braunschweig 1988, pp. 191–216

1019. Rahmann G.M. (1974): Map projections, Oxford Univ. Press, Karachi 1974

1020. Raisz E. (1962): Principles of cartography, McGraw–Hill, New York 1962

1021. Rao C.R., Mitra S.K. (1971): Generalized inverse of matrices and its applications, Wiley New York (1971)

1022. Rapp R.H. (1974a): Geometric Geodesy, Volume I (Basic Principles), Ohio State University, Dep. Of Geodetic Science, Columbus 1974

1023. Rapp R.H. (1974b): Current estimates of the mean earth ellipsoid parameters, Geophys. Res. Lett. 1 (1974) 35–38

1024. Rapp R.H. (1975): Geometric Geodesy, Volume II (Advanced Techniques), Ohio State University, Dep. Of Geodetic Science, Columbus 1975

1025. Rapp R.H. (1981): Transformation of geodetic data between Reference Datums, Geometric Reference Datums, Geometric Geodesy, Volume III, Ohio State University, pages 53–57, Columbus 1981

1026. Rapp R.H. (1989): The decay of the spectrum of the gravitational potential and the topography for the Earth, Geophys. J. Int. 99 (1989) 449–455

1027. Ratner D.A. (1991): An implementation of the Robinson map projection based on cubic splines, Cartography and Geographic Information Systems 18 (1991) 104–108

1028. Reckziegel H. (1979): Completeness of curvature surfaces of an isometric immersion, J. Differential Geometry 14 (1979) 7–20

1029. Reich K. (2000): Gauss' Schüler: Studierten bei Gauss und machten Karriere. Gauss' Erfolg als Hochschullehrer (Gauss's students: studied with him and were successful. Gauss's success as a university professor), Gauss Gesellschaft E.V. Göttingen, Mitteilungen Nr. 37, pages 33–62, Göttingen 2000

1030. Reigber C., Balmino G., Moynot B. (1976): The GRIM 2 Earth gravity field model, Bayerische Akademie der Wissenschaften, Report A86, Deutsche Geodätische Kommission, Bayer. Akad. Wiss., München 1976

1031. Reigber C., Müller H., Rizos C., Bosch W., Balmino G., Moyot B. (1983): An improved GRIM3 Earth gravity model (GRIM3B), Proceedings of the International Association of Geodesy (IAG) Symposia, XVIII Gerenral Assembly, Hamburg 1983, pp. 388–415

1032. Reignier F. (1957): Les systèmes de projection et leurs applications, Institut Géographique National, Paris 1957

1033. Reilly W.I. (1973): A conformal mapping projection with minimum scale error, Survey Review 22 (1973) 57–71

1034. Reilly W.I., Bibby H.M. (1975): A conformal mapping projection with minimum scale error – part 2: Scale and convergence in projection coordinates, Survey Review 23 (1975) 79–87

1035. Reiner M. (1945): A mathematical theory of dilatancy, Amer. J. Math. 67 (1945) 350–362

1036. Reiner M. (1948): Elasticity beyond the elastic limit, Amer. J. Math. 70 (1948) 433–446

1037. Reinhardt F., Soeder H. (1980): dtv Atlas zur Mathematik, Bd. 1, München 1980

1038. Reithofer A. (1977): Koeffizististentafeln und Rechenprogramme für die Gauss–Krueger–(UTM–) Koordinaten der Ellipsoide von Bessel, Hayford, Krassowsky und des Referenzellipsoids 1967, Mitteilungen der geodätischen Institute der TU Graz, Folge 27, Graz 1977

1039. Ricci G. (1918): Sulla determinazione di varieta dotate di proprietà intrinseche date a priori – note I, Rend. Acc. Lincei Classe Sci., Ser. 5, Vol. 19, Rome 1918

1040. Richardus P., Adler R.K. (1972a): Map projections for geodesists, cartographers, and geographers, North–Holland Publ. Co., Amsterdam 1972

1041. Richardus P., Adler R.K. (1972b): Map projections, North Holland Publ. Co., London 1972

1042. Riemann B. (1851): Grundlagen für eine allgemeine Theorie der Funktionen einer veränderlichen complexen Größe, Inauguraldissertation, Göttingen 1851

1043. Riemann B. (1868): Über die Hypothesen, welche der Geometrie zugrunde liegen (Habilitationsschrift), Göttinger Abhandlungen 13 (1868) 1–20, Ges. Werke, 2. Auflage (1892) 272–287, Neue Ausgabe Springer 1921

1044. Riemann B. (1876a): Grundlagen für eine allgemeine Theorie der Functionen einer veränderlichen complexen Grösse, Gesammelte Mathematische Werke, ed. H. Weber, R. Dedekind, B.G. Teubner, Leipzig 1876, pp. 3–47

1045. Riemann B. (1876b): Theorie der Abel'schen Funktionen, Gesammelte Mathematische Werke, ed. H. Weber, R. Dedekind, B.G. Teubner, Leipzig 1876, pp. 88–144

1046. Rinner K. (1944): Allgemeine Koeffizientenbedingungen in Reihen für konforme Abbildungen des Ellipsoides in der Ebene, Zeitschrift für Vermessungswesen 73 (1944) 102–107, 232

1047. Robinson A.H. (1951): The use of deformational data in evaluating world map projections, Annals, Association of American Geographers 41 (1951) 58–74

1048. Robinson A.H. (1953): Elements of cartography, J. Wiley & Sons, New York Chichester 1953

1049. Robinson A.H. (1974): A new map projecion: Its development and characteristics, in: Internationales Jahrbuch für Kartographie, G.M. Kirschbaum und K.H. Meine (eds.), Kirschbaum–Verlag, Bonn 1974, Nachdruck Bonner Universitätsdruckerei, Bonn, pp. 145–155

1050. Robinson A.H. (1986): Which map is best? American Congress on Surveying and Mapping, Falls Church, Virginia 1986

1051. Robinson A.H. (1988): Choosing a world map, American Congress on Surveying and Mapping, Bethesda, Maryland 1988

1052. Robinson A.H., Morrison J.L., Muehrcke P.C., Kimerling A.J., Guptill S.C. (1995): Elements of cartography, 6th ed., J. Wiley & Sons, New York Chichester 1995

1053. Robinson A.H., Sale R.D., Morrison J.L. (1978): Elements of cartography, 4th ed., J. Wiley & Sons, New York Chichester 1978

1054. Robinson A.H., Sale R.D., Morrison J.L., Muehrcke P.C. (1984): Elements of cartography, J. Wiley & Sons, New York Chichester 1984

1055. Robinson A.H., Snyder J.P. (1991): Matching the map projection to the need, American Congress on Surveying and Mapping, Bethesda, Maryland 1991

1056. Rockafellar R.T. (1970): Convex analysis, Princeton University Press, Princeton, New Jersey 1970

1057. Roesch N., Kern M. (2000): Die direkte Berechnung elliptischer Integrale, Zeitschrift für Vermessungswesen 125 (2000) 209–213

1058. Rosen M. (1981): Abel's theorem on the lemniscate, Amer. Math. Monthly 88 (1981) 387–395

1059. Rosenmund M. (1903): Die Änderung des Projektionssystems der schweizerischen Landesvermessung, Haller'sche Buchdruckerei, Bern 1903

1060. Roxburgh I.W. (1992): Post Newtonian tests of quartic metric theories of gravity, Reports on Mathematical Physics 31 (1992) 171–178

1061. Royal Society (1966): Glossary of technical terms in cartography, London 1966

1062. Rubincam D.F. (1981): Latitude and longitude from Van der Grinten grid coordinates, American Cartographer 8 (1981) 177–180

1063. Rune G.A. (1954): Some formulae concerning the Transverse Mercator Projection, Bulletin Géodésique 34 (1954) 309–317

1064. Saad Y. (1992): Numerical methods for large eigenvalue problems, Manchester University Press, Manchester 1992

1065. Saalfeld A. (1991): An application of algebraic topology to an overlay problem of analytical cartography, Cartography and Geographic Information Systems 18 (1991) 23–36

1066. Sacks R. (1950): The projection of the ellipsoid, Empire Survey Revue 78 (1950) 369–375

1067. Saito T. (1970): The computation of long geodesics on the ellipsoid by non–series expanding procedure, Bulletin Géodésique 95 (1970) 341–373

1068. Sala K.L. (1989): Transformations of the Jacobian amplitude function and its calculation via the arithmetic–geometric mean, SIAM J. Math. Anal. 20 (1989) 1512–1528

1069. Saleh A.K., Kibria B.M.G. (1993): Performance of some new preliminary test ridge regression estimators and their properties. Commun. Statist. Theory Meth. 22 (1993) 2747–2764

1070. Salkowski E. (1927): Repertorium der höheren Analysis, Zweite Auflage, Zweiter Teilband, B.G. Teubner, Leipzig 1927

1071. Samelson H. (1969): Orientability of hypersurfaces in Rn, Proceedings of the American Mathematical Society 22 (1969) 301–302

1072. Sammet G. (1990): Der vermessene Planet, Gruner + Jahr, Hamburg 1990

1073. Sanchez R., Theriault Y. (1985): Des algorithmes économiques pour la projection Mercator transverse, Le Géomètre Canadien 39 (2985) 23–30

1074. Sansò F. (1972): Cartografia – Carta conforme con minime deformazioni areali, Atta della Accademia Nazionale dei Lincei, Rendiconti della Classe di Scienze fisiche, matematiche e naturali, Serie 7, Vol. 52 (1972) 197–205

1075. Sansò F. (1973): An exact solution of the roto–translation problem, Photogrammetria 29 (1973) 203–216

1076. Sansò F. (1975): Fotogrammetria – A further account of roto–translations and the use of the method of conditioned observations, Atta della Accademia Nazionale dei Lincei, Rendiconti della Classe di Scienze fisiche, matematiche e naturali, Serie 8, Vol. 60 (1976) 126-134

1077. Sansò F. (1980): Dual Relations in Geometry and Gravity Space, Zeitschrift für Vermessungswesen 105 (1980) 279–289

1078. Sanson N. (1675): Cartes generales de la géographie ancienne et nouvelle, Paris 1675

1079. Schaefer V. (1999): Quo vadis geodesia? ... Sic erit pars publica. In: Quo vadis geodesia...? Festschrift for E.W. Grafarend on the occasion of his 60th birthday. F. Krumm and V. Schwarze (Eds.), Schriftenreihe der Institute des Studiengangs Geodäsie und Geoinformatik, Technical Reports, Department of Geodesy and Geoinformatics, Report Nr. 6 (1999) 413–418

1080. Scheffers G. (1918): Flächentreue Abbildung in der Ebene, Mathematische Zeitschrift 2 (1918) 180–186

1081. Scheffers G. (1922): Anwendung der Differential– und Integral–Rechnung auf Geometrie, Zweiter Band: Einführung in die Theorie der Flächen, Walter de Gruyter & Co., Berlin und Leipzig 1922

1082. Scheffers G., Strubecker K. (1956): Wie findet und zeichnet man Gradnetze von Land– und Sternkarten?, Teubner, Stuttgart 1956

1083. Schellhammer F. (1878): Über äquivalente Abbildung, Zeitschrift für Mathematik und Physik 23 (1878) 69–83

1084. Schering E. (1857): Über die conforme Abbildung des Ellipsoides auf der Ebene, Göttingen 1857, Nachdruck in: Ges. Math. Werke von E. Schering, eds. R. Haussner und K. Schering, 1.Bd., Mayer und Müller Verlag, Berlin 1902

1085. Schett A. (1977): Recurrence formula of the taylor series expansion coefficients of the Jacobi elliptic functions, Math. Comput. 32 (1977) 1003–1005

1086. Schjerning W. (1904): Über mitabstandstreue Karten, R. Lechner, Wien 1904

1087. Schliephake G. (1956): Berechnungen auf dem dreiachsigen Erdellipsoid nach Krassowski, Vermessungstechnik 4 (1956) 7–10

1088. Schloemilch O. (1849): Die allgemeine Umkehrung gegebener Funktionen, H.W. Schmidt, Halle 1849

1089. Schmehl H. (1927): Untersuchungen über ein allgemeines Erdellipsoid, Veröffentlichungen des Preußischen Geodätischen Institutes, Neue Folge Nr. 98, Potsdam 1927

1090. Schmehl H. (1930): Geschlossene geodätische Linien auf dem Ellipsoid, Zeitschrift für Vermessungswesen 1 (1930) 1–11

1091. Schmid E. (1962): The Earth as viewed from a satellite, U.S. Department of Commerce, Coast and Geodetic Survey, Technical Bulletin No. 20, Washington 1962

1092. Schmidt H. (1937/1938): Elementare Krümmungsbetrachtungen bei konformer Abbildung, Semester–Berichte zur Pflege des Zusammenhangs von Universität und Schule 11 (1937/1938) 54–81

1093. Schmidt H. (1938): Zum Umkehrproblem bei periodischen und fastperiodischen Funktionen, Berichte über die Verhandlungen der Sächsischen Akademie der Wissenschaften, Math.–Nat. Klasse 90 (1938) 83–96

1094. Schmidt H. (1975): Ein Beitrag zur mehrfarbigen Rasterreproduktion unter besonderer Berücksichtigung großformatiger Kopierraster und einer optimalen Kombination zwischen Rasterwinklung und Rasterweiten, Dissertation, Universität Bonn, Hohe landwirtschaftliche Fakultät, Bonn 1975

1095. Schmidt H. (1999): Lösung der geodätischen Hauptaufgabe auf dem Rotationsellipsoid mittels numerischer Integration, Zeitschrift für Vermessungswesen 124 (1999) 121–128

1096. Schmidt H. (2000): Berechnung geodätischer Linien auf dem Rotationsellipsoid im Grenzbereich diametraler Endpunkte, Zeitschrift für Vermessungswesen 125 (2000) 61–64

1097. Schmidt R. (1994): Überlegungen zur Vereinheitlichung der äußeren Form der europäischen Kartenwerke, BDVI–FORUM 20 (1994) 241–249

1098. Schnaedelbach K. (1985): Zur Berechnung langer Ellipsoidsehnen und geodätischer Linien, Allgemeine Vermessungsnachrichten 92 (1985) 503–510

1099. Schneider D. (1984): Schweizerisches Projektionssystem, Bundesamt für Landestopographie, Wabern 1984

1100. Schneider U. (2004): Die Macht der Karten, Wiss. Buchgesellschaft, Primus Verlag Darmstadt, Darmstadt 2004

1101. Schnirelmann L. (1930): Über eine neue kombinatorische Invariante, Monatshefte für Mathematik und Physik 37 (1930) 131–134

1102. Schoedlbauer A. (1963): Über eine neue numerische Lösung der 1. geodätischen Hauptaufgabe auf einem Referenz–Rotationsellipsoid der Erde für Seitenlängen bis 120km, Report C58, Deutsche Geodätische Kommission, Bayer. Akad. Wiss., München 1963

1103. Schoedlbauer A. (1979): Sammlung von Rechenformeln und Rechenbeispielen zur Landesvermessung, Teil A: Lagemessung, Hochschule der Bundeswehr, München 1979

1104. Schoedlbauer A. (1980): Kugeln als Hilfsflächen bei der Lösung der beiden geodätischen Hauptaufgaben, Mitteilungen zur Emeritierung von o. Univ.-Prof. Dipl.-Ing. Dr. techn. Karl Hubeny, Mitteilungen der geodätischen Institute der Technischen Universität Graz Folge 35, Graz 1980, pp. 193–201

1105. Schoedlbauer A. (1981a): Gauss'sche konforme Abbildung von Bezugsellipsoiden in die Ebene auf der Grundlage des transversalen Mercatorentwurfs, Allgemeine Vermessungsnachrichten 88 (1981) 165–173

1106. Schoedlbauer A. (1981b): Rechenformeln und Rechenbeispiele zur Landesvermessung, Teil 1: Die geodätischen Grundaufgaben auf Bezugsellipsoiden im System der geographischen Koordinaten und die Berechnung ellipsoider Dreiecke, Herbert Wichmann Verlag, Karlsruhe 1981

1107. Schoedlbauer A. (1982a): Rechenformeln und Rechenbeispiele zur Landesvermessung, Teil 2: Geodätische Berechnungen im System der Gauss'schen konformen Abbildung eines Bezugsellipsoids unter besonderer Berücksichtigung des Gauss–Krueger– und des UTM–Koordinatensystems im Bereich der Bundesrepublik Deutschland, Herbert Wichmann Verlag, Karlsruhe 1982

1108. Schoedlbauer A. (1982b): Transformation Gauss'scher konformer Koordinaten von einem Meridianstreifen in das benachbarte unter Bezugsnahme auf strenge Formeln der querachsigen sphärischen Mercator Projektion, Allgemeine Vermessungsnachrichten 89 (1982) 18–29

1109. Schoedlbauer A. (1984): Rechenformeln und Rechenbeispiele zur Landesvermessung, Teil 3: Punkteinschaltungen im System der Gauss'schen und der geographischen Koordinaten, Herbert Wichmann Verlag, Karlsruhe 1984

1110. Schoen R. (1984): Conformal deformation of a Riemannian metric to constant scalar curvature, J. Differential Geometry 20 (1984) 479–495

1111. Schoeps D. (1964): Die Lösung der geodätischen Hauptaufgaben in der Nähe der Erdpole mit Hilfe der stereographischen Polarprojektion, Akademie–Verlag, Berlin 1964

1112. Schoppmeyer J. (1992): Farbe – Definition und Behandlung beim Übergang zur digitalen Kartographie, Kartographische Nachrichten 42 (1992) 125–134

1113. Schott C.A. (1882): A comparison of the relative value of the polyconic projection used on the coast and geodetic survey, with some other projections, Annual Report of the superintendent of the U.S. Coast and Geodetic survey showing the progress of the work during the fiscal year ending with June, 1880, Appendix No. 15, 1882, pp.287–296

1114. Schottenloher M. (1997): A mathematical introduction to conformal field theory, Springer–Verlag, Berlin 1997

1115. Schouten J.A. (1921): Über die konforme Abbildung n–dimensionaler Mannigfaltigkeiten mit quadratischer Maßbestimmung auf eine Mannigfaltigkeit mit Euklidischer Maßbestimmung, Math. Z. 11 (1921) 58–88

1116. Schouten J.A. (1954): Ricci–calculus, 2nd ed., Springer–Verlag, Berlin 1954

1117. Schouten J.A., Struik D.-J. (1938): Einführung in die neueren Methoden der Differentialgeometrie, Zweiter Band: Geometrie, P. Noordhoff N.V., Groningen–Batavia 1938

1118. Schreiber O. (1866): Theorie der Projektionsmethode der Hannoverschen Landesvermessung, Hahn'sche Buchhandlung, Hannover 1866

1119. Schreiber O. (1899): Zur konformen Doppelprojektion der Preussischen Landesaufnahme, Zeitschrift für Vermessungswesen 28 (1899) 491–502, 593–613

1120. Schreiber R. (1990): Numerische Untersuchungen zur Koordinatentransformation mit geozentrischen Datumsparamteren, in: A. Schoedlbauer (Hrsg.), Moderne Verfahren der Landesvermessung, Teil I: Global Positioning System, Beiträge zum 22. DVW–Seminar 12.–14. April 1989, Schriftenreihe Studiengang Vermessungswesen, Universität der Bundeswehr, München 1990

1121. Schreiber R. (1991): Ein klassifizierender Beitrag zur Abbildungstheorie und numerischen Genauigkeit von geodetischen Datumübergängen, Report C377, Deutsche Geodätische Kommission, Bayer. Akad. Wiss., München 1991

1122. Schroeder E. (1988): Kartenentwürfe der Erde, Teubner, Leipzig 1988

1123. Schumaker L.L. (1993): Triangulations in CAGD, IEEE Computer Graphics & Applications 13 (1993) 47–52

1124. Schwarz H.A. (1869): Über einige Abbildungsaufgaben, Journal für die reine und angewandte Mathematik 70 (1869) 105–120

1125. Schwarze V.S. (1979): Verwaltungsvorschriften für Katasterkarten, Innenministerium Baden–Württemberg, Stuttgart 1979

1126. Schwarze V.S. (1999): Soldner parallel and Fermi coordinates for the biaxial ellipsoid, private communication, Stuttgart 1999

1127. Schwarze V.S. (2002): Geodesic map projections of type Soldner and Fermi of the biaxial ellipsoid. Department of Geodetic Science, Internal Report, 24 pages, Stuttgart 2002

1128. Schwarze V.S., Hartmann, T., Leins, M. and M. Soffel (1993): Relativistic effects in satellite positioning, Manuscripta Geodaetica 18 (1993) 306–316

1129. Scotese C.R., Bambach R.K., Barton C., Van Der Voo R., Ziegler A.M. (1979): Paleozoic base maps, J. Geodesy 87 (1979) 217–277

1130. Searle S.R. (1982): Matrix algebra useful for statistics, Wiley New York 1982

1131. Semple J.G., Kneebone G.T. (1956): Algebraic projective geometry, Clarendon Press, Oxford 1956

1132. Seth B.R. (1964a): Generalized strain measure with applications to physical problems, in: Second order effects in elasticity, plasticity and fluid dynamics, Reiner, M., Abir D. (eds.), Pergamon Press, pp. 162–172, Oxford 1964

1133. Seth B.R. (1964b): Generalized strain and transition concepts for elastic–plastic deformation–creep and relaxation, IUTAM Symposium, Pergamon Press, pp. 383–389, München 1964

1134. Sevilla M.J., Malpica J.A. (1999): A minimum elastic deformation energy projection, Survey Review 35 (1999) 56–66, 109–115

1135. Shafarevich I.R. (1994): Basic algebraic geometry 1, Springer–Verlag, Berlin 1994

1136. Shebl S. A. (1995): Conformal mapping of the triaxial ellipsoid and its applications in geodesy, Doctoral thesis, Alexandria University, Alexandria 1995

1137. Shougen W., Shuqin Z. (1991): An Algorithm for Ax=(lambda)Bx with Symmetric and Positive-Definite A and B, SIAM Journal Matrix Analysis and Applications 12 (1991) 654–660

1138. Sibson R. (1978): Locally equiangular triangulations, The Computer Journal 21 (1978) 243–245

1139. Siemon K. (1935): Neue Netzentwürfe für Kurskarten von Gebieten höherer Breite, Mitteilungen des Reichsamtes für Landesaufnahme 11 (1935) 21–35

1140. Signorini A. (1930): Sulla meccanica di sistemi continui, Atti Accad. naz. Lincei, serie 6, 12 (1930) 312–316, 411–416

1141. Simo J.C., Taylor R.L. (1991): Quasi-incompressible finite elasticity in principal stretches. Continuum basis and numerical algorithms, Computer Methods in Applied Mechanics and Engineering 85 (1991) 273–310

1142. Singer I.M., Thorpe J.A. (1967): Some point set topology, in: ibid., Lecture notes on elementary topology and geometry, Scott, Foresman and Company, 1967, pp. 1–27

1143. Sjoeberg L.E. (2005a): Determination of areas on the plane, sphere and ellipsoid, Survey Reviews 2005a

1144. Sjoeberg L.E. (2005b): Precise determination of the Clairaut constant in ellipsoidal geodesy, Survey Reviews 2005b

1145. Sjoeberg L.E. (2006): New solutions to the direct and indirect geodetic problems on the ellipsoid, Zeitschrift für Geodäsie, Geoinformation und Landmanagement 131 (2006) 1–5

1146. Sjogren W.L. (1983): Planetary geodesy, Reviews of Geophysics and Space Physics 21 (1983) 528–537

1147. Skogloev E., Magnusson P., Dahlgren M. (1996): Evolution of the obliquities for ten asteroids, Planet. Space Sci. 44 (1996) 1177–1183

1148. Skorokhod A.V., Hoppensteadt and H.D. Salehi (2002): Random perturbation methods with applications in Science and Engineering, Springer Verlag, Berlin – Heidelberg – New York 2002

1149. Slavutin E.I. (1973): Euler's works on elliptic integrals (Russian), History and methodology of the natural sciences XIV: Mathematics, Moscow 1973, 181–189

1150. Slocum T.A. (1995): 1995 U.S. National Report to the International Cartographic Association, Cartography and Geographic Information Systems 22 (1995) 109–114

1151. Snedecor G.W., Cochran W.G. (1980): Statistical methods, 7th ed., Iowa State University Press, Ames 1980

1152. Snyder J.P. (1977): A comparison of pseudocylindrical map projections, American cartographer 4 (1977) 59–81

1153. Snyder J.P. (1978a): Equidistant conic map projections, Annals of the Association of American Geographers 68 (1978) 373–383

1154. Snyder J.P: (1978b): The space oblique Mercator projection, Photogrammetric Engineering and Remote Sensing 44 (1978) 585–596

1155. Snyder J.P: (1979a): Calculating map projections for the ellipsoid, The American Cartographer 6 (1979) 67–76

1156. Snyder J.P. (1979b): Projection notes, The American Cartographer 6 (1979) 81

1157. Snyder J.P. (1979c): Map projections for satellite applications, Proceedings of the American Congress on Surveying and Mapping, 39th Annual Meeting in Washington D.C., Falls Church, Virginia 1979, pp. 134–146

1158. Snyder J.P. (1981a): Map projections for satellite tracking, Photogrammetric Engineering and Remote Sensing 47 (1981) 205–213

1159. Snyder J.P. (1981b) The perspective map projection of the Earth, The American Cartographer 8 (1981) 149–160

1160. Snyder J.P. (1981c): The space oblique mercator – mathematical development, U.S. Geological Survey Bulletin 1518, United States Government Printing Office, Washington 1981

1161. Snyder J.P. (1982): Map projections used by the U.S. geological survey, Geological Survey Bulletin 1532, Second Edition, United States Government Printing Office, Washington 1982

1162. Snyder J.P. (1984a): A low–error conformal map projection for the 50 states, The American Cartographer 11 (1984) 27–39

1163. Snyder J.P. (1984b): Map–projection graphics from a personal computer, The American Cartographer 11 (1984) 132–138

1164. Snyder J.P: (1984c): Minimum–error map projections bounded by polygons, The Cartographic Journal 22 (1984) 112–120

1165. Snyder J.P. (1985a): Computer–assisted map projection research, United States Government Printing Office, Washington 1985

1166. Snyder J.P. (1985b): Conformal mapping of the triaxial ellipsoid, Survey Review 28 (1985) 130–148

1167. Snyder J.P. (1987a): Labeling projections on published maps, The American Cartographer 14 (1987) 21–27

1168. Snyder J.P. (1987b): Map projections – A working manual, U.S. Geological Survey professional paper 1395, USPOGO, Washington D.C. 1987

1169. Snyder J.P. (1988): New equal–area map projections for noncircular regions, The American Cartographer 15 (1988) 341–355

1170. Snyder J.P. (1990): The Robinson projection – A computation algorithm, Cartography and Geographic Information Systems 17 (1990) 301–305

1171. Snyder J.P. (1992): An equal–area map projection for polyhedral globes, Cartographica 29 (1992) 10–21

1172. Snyder J.P. (1994): How practical are minimum–error map projections?, Cartographic Perspectives 17 (1994) 3–9

1173. Snyder J.P. et al. (1986): Which map is best? Projections for world maps, American Congress on Surveying and Mapping, American Cartographic Association, Special Publication No. 1, Falls Church, Virginia 1986, 14 pages

1174. Snyder J.P., DeLucia A.A. (1986): An innovative world map projection, The American Cartographer 13 (1986) 165–167

1175. Snyder J.P., Steward H. (1988): Bibliography of map projections, U.S. Government Printing Office 1988

1176. Snyder J.P., Voxland P.M. (1989): An album of map projections, U.S. Geological Survey, Professional Paper 1453, 249 pages, United States Government Printing Office, Washington 1989

1177. Soldner J. (1911): Theorie der Landesvermessung (1810), Verlag von Wilhelm Engelmann, Leipzig 1911

1178. Soler, T. (1976): On differential transformations between Cartesian and curvilinear (geodetic) coordinates. The Ohio State University, Department of Geodetic Science, Report 236, Columbus 1976.

1179. Sorokina L.A. (1983): Legendre's works on the theory of elliptic integrals (Russian), Istor.-Mat. Issled. 27(1983) 163–178

1180. Soycan M. (2005): Polynomial versus similarity transformations between GPS and Turkish reference systems, Survey Review 38 (2005) 58–69

1181. Spada G. (1995): Changes in the Earth inertia tensor: The role of boundary conditions at the core–mantle interface, Geophysical Research Letters 22 (1995) 3557–3560

1182. Spallek K. (1980): Kurven und Karten, Bibliographisches Institut, Zürich 1980

1183. Spanier J., Oldham K.B. (1967): An atlas of functions, Hemisphere, Washington D.C. 1967

1184. Spanier J., Oldham K.B. (1987): The Jacobian elliptic functions, An Atlas of Functions, Washington DC: Hemisphere 1987, pp. 635–652

1185. Spata M., Froehlich H. (2001): Kartendatum–Shiftparameter, Allgemeine Vermessungsnachrichten 108 (2001) 132–133

1186. Spencer A.J. (1987): Isotropic polynomial invariants and tensor functions, in: Applications of tensor functions in solid mechanics, CISM courses and lectures 292, Springer Verlag, Berlin – Heidelberg – New York 1987, 142–186

1187. Spencer A.J., Rivlin R.S. (1958/59): The theory of matrix polynomials and its application to the mechanics of isotropic continua, Arch. Rational Mech. Anal 2 (1958/59) 309–336

1188. Spencer A.J., Rivlin R.S. (1960): Further results in the theory of matrix polynomials, Arch. Rational Mech. and Analysis 3 (1960) 214–230

1189. Spencer D.C. (1969): Overdetermined systems of linear partial differential equations, Bulletin of the American Mathematical Society 75 (1969) 179–239

1190. Spilhaus A. (1942): Maps of the whole world ocean, Geographical Review 32 (1942) 431–435

1191. Spilhaus A. (1976): New look in maps brings out patterns of plate tectonics, Smithsonian 7 (1976) 54–63

1192. Spilhaus A. (1983): World ocean maps: The proper places to interrupt, Proceedings of the American Philosophical Society 127 (1983) 50–60

1193. Spilhaus A. (1984): Plate tectonics in geoforms and jigsaws, Proceedings of the American Philosophical Society 128 (1984) 257–269

1194. Spilhaus A., Snyder J.P. (1991): World maps with natural boundaries, Cartography and Geographic Information Systems 18 (1991) 246–254

1195. Spivak M. (1979): A comprehensive introduction to differential geometry, Vol. 4, 2nd edition, Publish or Perish, Boston 1979

1196. Staude O. (1922): Fokaleigenschaften und konfokale Systeme von Flächen zweiter Ordnung, in: Repetitorium der höheren Mathematik, Vol. 2 (Geometrie), ed.: E. Pascal, Verlag B.G. Teubner, Leipzig (1922) 616–618

1197. Steeb W.H. (1991): Kronecker product of matrices and applications, B. I. Wissenschaftsverlag, Mannheim 1991

1198. Steers J.A. (1970): An introduction to the study of map projections, 15th ed., University of London Press, London 1970

1199. Stein E.M., Weiss G. (1968): Generalization of the Cauchy–Riemann equations and representations of the rotation group, American J. Math. 90 (1968) 163–196

1200. Stillwell J. (1989): Mathematics and history, New York, Berlin, Heidelberg 1989, 152–167

1201. Stooke P.J. (1991): Lunar and planetary cartographic research at the University of Western Ontario, CISM Journal ACSGC 45 (1991) 23–31

1202. Stooke P.J., Keller C.P. (1990): Map projections for non–spherical worlds, Cartographica 27 (1990) 82–100

1203. Stoughton H.W. (1984): Analysis of various algorithms to compute scale factors on the Lambert conformal conic projection, Technical paper of the American Congress on Surveying and Mapping Fall Convention, in: R.B. McEwen nad L. Starr (eds.): Research for the USGS digital cartography program, San Antonio 1984, pp. 325–333

1204. Strang van Hees G.L. (1993): Globale en lokale geodetische Systemen, Nederlandse Commissie voor Geodesie, TU Delft, Publikatie 30, Delft 1993

1205. Strasser G.L. (1957): Ellipsoidische Parameter der Erdfigur (1800–1959), Report A19, Deutsche Geodätische Kommission, Bayer. Akad. Wiss., München 1957

1206. Stroyan K. (1977): Infinitesimal analysis of curves and surfaces; in Barwise, J.: Handbook of mathematical logic, North–Holland, Amsterdam 1977

1207. Sturmfels B. (1996): Gröbner bases and convex polytopes, American Mathematical Society, Providence 1996

1208. Suenkel H. (1976): Ein nicht–iteratives Verfahren zur Transformation geodätischer Koordinaten, Öster. Zeitschrift für Vermessungswesen 64 (1976) 29–33

1209. Sun J.-G. (2000): Condition number and backward error for the generalized singular value decomposition, Siam J. Matrix Anal. Appl. 22 (2000) 323–341

1210. Swonarew K.A. (1953): Kartenentwurfslehre, VEB Verlag Technik, Berlin 1953

1211. Sylvester J., Uhlmann G. (1990): The Dirichlet to Neumann map and applications, in: D. Colton, R. Ewing and W. Rundell (eds.), Inverse problems in partial differential equations, Siam, Philadelphia 1990, pp. 101–139

1212. Synott S.P. et al. (1964): Shape of Io, Bulletin of the American Astronomical Society 16 (1984) 657

1213. Szaflarski J. (1955): Zarys kartografii, PPWK, Warszawa 1955

1214. Tarczy–Hornsch A., Hristov V.K. (1959): Tables for the Krassovsky–Ellipsoid, Akademiai Kiado, Budapest 1959

1215. Tardi P. (1952): Travaux de l'association internationale de géodésie, Tome 17, Paris 1952

1216. Taucer G. (1954): Alcune considerazioni sul teorema di Schols, Bollettino di Geodesia e Scienze Affini 13 (1954) 159–162

1217. Temme N. (1996): Special functions, J. Wiley & Sons, New York Chichester 1996

1218. Tenenblat K. (1971): On isometric immersions of Riemannian manifolds, Boletin Sociedade Brasileirade Matematica 2 (1971) 23–36

1219. Theimer J. (1933): Kartenprojektionslehre, Mont. Hochschule, Loben 1933

1220. Theissen R. (1990): Berechnung des Richtungswinkels t ohne Quadrantenabfrage, Zeitschrift für Vermessungswesen 115 (1990) 193–195

1221. Thirring W. (1988): Lehrbuch der Mathematischen Physik. Band 1: Klassische Dynamische Systeme, Springer Verlag, Berlin – Heidelberg – New York 1988

1222. Thomas P. et al. (1983a): Phoebe: Voyager 2 observations, Journal of Geophysical Research 88 (1983) 8736–8742

1223. Thomas P. et al. (1983b): Saturn's small satellites: Voyager imaging results, Journal of Geophysical Research 88 (1983) 8743–8754

1224. Thomas P.C. et al. (1994): The shape of Gaspra, Icarus 107 (1994) 23–36

1225. Thomas P.D. (1952): Conformal projections in geodesy and cartography, U.S. Coast and Geodetic Survey, Special Publication No. 251, Washington 1952

1226. Thomas P.D. (1962): Geodetic positioning of the Hawaiian Islands, Mapping 22 (1962) 89–95

1227. Thomas P.D. (1978): Conformal projections in geodesy and cartography, Spec. Publ. 251, US Gov. Printing Office, Washington DC 1978

1228. Thomas P.D. et al. (1970): Spheroidal geodesics, reference systems, & local geometry, U.S. Naval Oceanographic Office, Washington D.C. 1970

1229. Thompson M.M. (1979): Maps for America, U.S. Geol. Survey 1979

1230. Timmermann H. (1973): Koordinatenfreie Kennzeichnung von Projektionen in projektiven Räumen, Mitt. Math. Ges. Hamburg 10 (1973) 88–103

1231. Ting T.C.T. (1985): Determination of $C_{1/2}$, $C_{-1/2}$ and more general isotropic tensor functions of C, J. Elasticity 15 (1985) 319–323

1232. Tissot N.A. (1881): Mémoire sur la représentation des surfaces et les projections des cartes géographiques, Gauthier–Villars, Paris 1881

1233. Tissot N.A. (1887): Die Netzentwürfe geographischer Karten, autorisierte deutsche Bearbeitung mit einigen Zusätzen von E. Hammer, Metzler Buchh., Stuttgart 1887

1234. Tobler W.R. (1962a): A classification of map projections, Ann. Assoc. American Geographers 52 (1962) 167–175

1235. Tobler W.R. (1962b): The polar case of Hammer's projection, The Professional Geographer 14 (1962) 20–22

1236. Tobler W.R. (1963a): Geographic area and map projections, The Geographical Review 53 (1963) 59–78

1237. Tobler W.R. (1963b): Some new equal area map projections, Survey Review 17 (1963) 240–243

1238. Tobler W.R. (1964): Geographical coordinate computations: Part II, Finite map projection distortions, Technical report no. 3, ONR Task No. 389-137, Department of Geography, University of Michigan 1964

1239. Tobler W.R. (1973): The hyperelliptical and other new pseudo–cylindrical equal area map projections, Journal of Geophysical Research 78 (1973) 1753–1759

1240. Tobler W.R. (1974): Local map projections, The American Cartographer 1 (1974) 51–62

1241. Tobler W.R. (1977): Numerical approaches to map projections, in: Beiträge zur theoretischen Kartographie, Festschrift für Erik Arnberger, hg. Ingrid Kretschmer, Franz Deuticke, Wien 1977, pp.51–64

1242. Tobler W.R. (1978): A proposal for an equal area map of the entire world on Mercator's projection, The American Cartographer 5 (1978) 149–154

1243. Tobler W.R. (1986): Measuring the similarity of map projections, The American Cartographer 2 (1986) 135–139

1244. Tobler W.R. (1993): Three short papers on geographical analysis and modelling, NCGIA, Technical Report 1 (1993), University of California, Santa Barbara

1245. Todorov I.T., Mintchev M.C., Petkova V.B. (1978): Conformal invariance in quantum field theory, Scuola Normale Superiore Pisa, Classe di Scienze, Pisa 1978

1246. Toelke F. (1967): Praktische Funktionslehre, 4. Band: Elliptische Integralgruppen und Jacobische elliptische Funktionen im Komplexen, Springer–Verlag, Berlin 1967

1247. Toernig W. (1979): Numerische Mathematik für Ingenieure und Physiker, Springer Verlag Berlin – Heidelberg – New York, Band I 1979

1248. Tolstova T.I. (1969): The airy criterion as applied to Azimuthal projections, Geodesy and Aerophotography 8 (1969) 427–428

1249. Topchilov M.A. (1970): On an extension of Chebyshev's theorem to certain classes of cartographic projections, Geodesy and Aerophotography 9 (1970) 251–254

1250. Torge W. (1980): Geodätisches Datum und Datumstransformation, in: H. Pelzer (ed.), Geodätische Netze in Landes– und Ingenieurvermessung, K. Wittwer, Stuttgart 1980, pp.131–140

1251. Torge W. (1993): Von der Mitteleuropäischen Gradmessung zur Internationalen Assoziation für Geodäsie–Problemstellungen und Lösungen, Zeitschrift für Vermessungswesen 118 (1993) 595–605

1252. Torge W. (2001): Geodesy, 2nd Ed., de Gruyter, Berlin 2001

1253. Trefethen L.N. (1986): Numerical conformal mapping, Amsterdam 1986

1254. Tricomi F. (1948): Elliptische Funktionen, Akademische Verlagsgesellschaft, Leipzig 1948

1255. Tricot C. (1995): Curves and fractal dimension, Springer–Verlag, New York 1995

1256. Truesdell C. (1958): Geometric interpretation for the reciprocal deformation tensors, Quart. Appl. Math. 15 (1958) 434–435

1257. Truesdell C. (1965): The non–linear field theories of mechanics, Handbuch der Physik, Band III/3, Springer–Verlag, 1965 Truesdell C. (1966): The elements of continuum mechanics, Springer Verlag, Berlin – Heidelberg – New York 1966

1258. Truesdell C., Toupin R. (1960): The classical field theories, in: Handbuch der Physik, vol.III/I, Springer–Verlag, Berlin 1960

1259. U.S. Coast and Geodetic Survey (1882): Report of the superintendent of the U.S. Coast and Geodetic Survey June 1880: Appendix 15: A comparison of the relative value of the polyconic projection used on the Coast and Geodetic Survey, with some other projections, by C.A. Scott

1260. U.S. Coast and Geodetic Survey (1900): Tables for a polyconic projection of maps: U.S. Coast and Geodetic Survey Spec. Pub. 5, 1900

1261. U.S. Geological Survey (1964): Topographic instructions of the United States Geological Survey Book 5, Part 5B, Cartographic tables: U.S. Geol. Survey, 1964

1262. U.S. Geological Survey (1970): National atlas of the United States: U.S. Geol. Survey, 1970

1263. Uhlig F. (1973): Simultaneous block diagonalization of two real symmetric matrices, Linear Algebra Appl. 7 (1973) 281–289

1264. Uhlig F. (1976): A canonical form for a pair of real symmetric matrices that generate a nonsingurlar pencil, Linear Algebra Appl. 14 (1976) 189–210

1265. Uhlig F. (1979): A recurring theorem about pairs of quadratic forms and extensions: a survey, Linear Algebra Appl. 25 (1979) 189–210

1266. Uhlig L., Hoffmann P. (1968): Leitfaden der Navigation, Automatisierung der Navigation, transpress VEB Verlag, Berlin 1968

1267. United Nations (1963): Specifications of the international map of the world on the millionth scale v. 2, New York, United Nations

1268. Urmajew N.A. (1958): Sphäroidische Geodäsie, VEB Verlag, Berlin 1958

1269. Vakhrameyeva L.A. (1971): Conformal projections obtained from series and their application, Geodesy and Aerophotography 10 (1971) 338–340

1270. Van der Grinten A.J. (1904): Darstellung der ganzen Erdoberfläche auf einer kreisförmigen Projektionsebene, Dr. A. Petermanns Mitteilungen aus Justus Perthes' geographischer Anstalt 50 (1904) 155–159

1271. Van der Grinten A.J. (1905a): New circular projection of the whole Earth's surface, American Journal of Science 19 (1905) 357–366

1272. Van der Grinten A.J. (1905b): Zur Verebnung der ganzen Erdoberfläche, Nachtrag zu der Darstellung in Pet. Mitt. 1904: Petermanns Geographische Mitteilungen 51, 237

1273. Van der Vorst H.A., Golub G.H. (1997): 150 years old and still alive: eigenproblems, in: The state of the art in numerical analysis (eds. I.S. Duff and G.A. Watson), pages 93–119, Oxford 1997

1274. Van Dooren P. (1981): The generalized eigenstructure problem in linear system theory, IEEE Trans. Auto. Cont. AC-26 (1981) 111–128

1275. Van Roessel J.W. (1991): A new approach to plane–sweep overlay: topological structuring and line–segment classification, Cartography and Geographic Information Systems 18 (1991) 49–67

1276. Van Zandt F.K. (1976): Boundaries of the United States and the several States: U.S. Geol. Survey Prof. Paper 909, 1976

1277. Vaníček P. (1980): Tidal corrections to geodetic quantities, US Department of Commerce, National Oceanic and Atmospheric Administration, National Ocean Survey, NOAA technical report NOS 83 NGS 14, Rockville, Maryland 1980

1278. Vaníček P., Najafi-Alamdari M. (2004): Proposed new cartographic mapping for Iran, Spatial Science 49 (2004) 31–42

1279. Varga J. (1983a): A Lambert–féle szögtartó kúpvetületról, Geod. es Kartogr. Budapest 35 (1983) 25–30

1280. Varga J. (1983b): Conversions between geographical and transverse Mercator (UTM, Gauss–Krueger) grid coordinates, Periodica Polytechnica 27 (1983) 239–251

1281. Vatrt V. (1999): Methodology of testing geopotential models specified in different tide systems, Studia geoph. et geod. 43 (1999) 73–77

1282. Veblen O. (1933): Invariants of quadratic differential forms, Cambridge University Press, New York 1933

1283. Viesel H. (1971): Über einfach geschlossene geodätische Linien auf dem Ellipsoid, Archiv der Mathematik 22 (1971) 106-112

1284. Vincenty T. (1971): The meridional distance problem for desk computers, Survey Review 161 (1971) 136–140

1285. Vincenty T. (1976a): Ein Verfahren zur Bestimmung der geodätischen Höhe eines Punktes, Allgemeine Vermessungsnachrichten 83 (1976) 179

1286. Vincenty T. (1976b): Direct und inverse solutions of geodesics on the ellipsoid with applications of nested equations, Survey Review 176 (1976) 88–93, 294

1287. Vincenty T. (1976c): Determination of North American Datum of 1983–coordinates of map corners, NOS NGS 16, US Government Printing Office, Washington DC 1976

1288. Vincenty T. (1976d): Ein Verfahren zur Bestimmung der geodätischen Höhe eines Punktes, Allgemeine Vermessungsnachrichten 83 (1976) 179

1289. Vincenty T. (1980): Zur räumlich–ellipsoidischen Koordinatentransformation, Zeitschrift für Vermessungswesen 105 (1980) 519–521

1290. Vincenty T. (1985): Precise determination of the scale factor from Lambert conformal conical projection coordinates, Surveying and Mapping 45 (1985) 315–318

1291. Vincenty T. (1986a): Lambert conformal conical projection: arc–to–chord connection, Surveying and Mapping 46 (1986) 163–167

1292. Vincenty T. (1986b): Use of polynomial coefficient in conversions of coordinates on the Lambert conformal conical projection, Surveying and Mapping 46 (1986) 15–18

1293. Vincze V. (1983): Fundamental equations with general validity of real projections, Acta Geodetica et Montanistica Hung. 18 (1983) 383–401

1294. Visvalingam M., Williamson P.J. (1995): Simplification and generalization of large scale data for roads: A comparison of two filtering algorithms, Cartography and Geograph. Information Systems 22 (1995) 264–275

1295. Vogel W.O. (1970): Kreistreue Transformationen in Riemannschen Räumen, Archiv der Mathematik 21 (1970) 641–645

1296. Vogel W.O. (1973): Einige Kennzeichnungen der Homothetischen Abbildungen eines Riemannschen Raumes unter den kreistreuen Abbildungen, Manuscripta Mathematica 9 (1973) 211–228

1297. Volkov N.M. (1973): Cartographic projection of photographs of celestial bodies taken from outer space, Geodesy, Mapping and Photogrammetry 15 (1973) 87–91

1298. VonderMuehll K. (1868): Über die Abbildung von Ebenen auf Ebenen, Journal für die reine und angewandte Mathematik 69 (1868) 264–285

1299. Voosoghi B., Helali H., Sedighzadeh F. (2003): Intelligent map projection transformation, ISPRS commission IV joint workshop in challenges in geospatial analysis, integration and visualisation II, September 8–9 (2003), Stuttgart university of applied science, Stuttgart, Germany 2003

1300. Voss A. (1882): Über ein neues Princip der Abbildung krummer Oberflächen, Mathematische Annalen 19 (1882) 1–26

1301. Waalewijn A. (1986): Der Amsterdamer Pegel (NAP), Österreichische Zeitschrift für Vermessungswesen und Photogrammetrie 74 (1986) 264–270

1302. Waalewijn A. (1987): The Amsterdam Ordnance Datum (NAP), Survey Review 29 (1987) 197–204

1303. Wagner K. (1962): Kartographische Netzentwürfe, Bibliographisches Institut, Mannheim 1962

1304. Wagner K.-H. (1932): Die unechten Zylinderprojektionen, Offizin Haag–Drulin Ag, Leipzig 1932

1305. Wahba G. (1984): Surface fitting with scattered noisy data on Euclidean D–space and on the sphere, Rocky Mountain Journal of Mathematics 14 (1984) 281–299

1306. Walter W. (1974): Einführung in die Theorie der Distributionen, B.I.-Wissenschaftsverlag, Mannheim - Wien – Zürich 1974

1307. Wangerin A. (1894): Über Kartenprojektionen, Abhandlungen von J.L. Lagrange und C.F. Gauss, W. Engelmann Verlag, Leipzig 1894

1308. Wangerin A. (1921): Allgemeine Flächentheorie von C.F. Gauss, deutsche Übersetzung von Disquisitiones generales circa superficies curvas, Ostwald's Klassiker der exakten Wissenschaften Nr. 5, 5. Auflage, Akad. Verlagsgesellschaft, Leipzig 1921

1309. Ward K. (1979): Cartography in the round – The oceanographic projection, The Cartographic Journal 16 (1979) 104–116

1310. Ward M. (1960): The calculation of the complete elliptic integral of the third kind, Amer. Math. Monthly 67 (1960)

1311. Watson D.F. (1981): Computing the n-dimensional Delauney tessellation with application to Voronoi polytopes, The Computer Journal 24 (1981) 167–172

1312. Watts D. (1970): Some new map projections of the world, Geographic Journal 7 (1970) 41–46

1313. Weber H. (1867): Über ein Prinzip der Abbildung der Teile einer krummen Oberfläche auf einer Ebene, Journal für die reine und angewandte Mathematik 67 (1867) 229–247

1314. Wee C.E., Goldman R.N. (1995): Elimination and resultants, Part 1: Elimination and bivariate resultants, IEEE Computer Graphics and Applications 1 (1995) 69–77

1315. Wee C.E., Goldman R.N. (1995): Elimination and resultants, Part 2: Multivariate resultants, IEEE Computer Graphics and Applications 3 (1995) 60–69

1316. Weibel R. (1995): Map generalization in the context of digital systems, Cartography and Geograph. Information Systems 22 (1995) 259–263

1317. Weierstrass K. (1894a): Mathematische Werke Bd. I, Berlin 1894

1318. Weierstrass K. (1894b): Mathematische Werke Bd. VI, Vorlesungen über Anwendung der elliptischen Funktion, Georg Olms Verlagsbuchhandlung, Hildesheim 1894

1319. Weierstrass K. (1894c): Über die geodätischen Linien auf dem dreiaxigen Ellipsoid, Mathematische Werke Bd. I, Berlin 1894, pp. 257–266

1320. Weierstrass K. (1903a): Die Oberfläche eines dreiaxigen Ellipsoids, Vorlesungen über Anwendung der elliptischen Funktionen, Vorlesungsskript 1865, 3. Kapitel, Mathematische Werke Bd. VI, Berlin 1903, pp. 30–41

1321. Weierstrass K. (1903b): Bestimmung der geodätischen Linien auf einem Rotationsellipsoide, Mathematische Werke Bd. VI, Berlin 1903, pp. 330–344

1322. Weierstrass K. (1903c): Vorlesungen über Anwendung der elliptischen Funktionen, Vorlesungsskript 1865, 31. Kapitel, Mathematische Werke Bd. VI, Berlin 1903, pp. 345–354

1323. Weightman J.A. (1961): A projection for a triaxial ellipsoid: the generalized stereographic projection, Empire Survey Review 16 (1961) 69–78

1324. Weingarten J. (1861): Über eine Klasse auf einander abwickelbarer Flächen, Journal für reine und angewandte Mathematik 59 (1861) 382–393

1325. Weingarten J. (1883): Über die Eigenschaften des Linienelements der Flächen von konstantem Krümmungsmaß, Crelle J. 94 (1883) 181–202

1326. Weintraub S.H. (1997): Differential forms, Academic Press, San Diego 1997

1327. Wessel P., Smith W.H.F. (1991): Free software helps map and display data, EOS, Trans. American Geophys. Union 72 (99) 441–446

1328. Weyl H. (1918): Reine Infinitesimalgeometrie, Math. Z. 2 (1918) 384–411

1329. Weyl H. (1921): Zur Infinitesimalgeometrie: Einordnung der projektiven und der konformen Auffassung, Nachr. Königl. Ges. Wiss. Göttingen, Math.-Phys. Klasse, Göttingen 1921

1330. White D., Kimerling A.J., Overton W.S. (1992): Cartographic and geometric components of a global sampling design for environmental monitoring, Cartography and Geographic Information Systems 19 (1992) 5–22

1331. Wiechel H. (1879): Rationelle Gradnetzprojektionen, Civilingenieur, new series 25 (1879) 401–422

1332. Wieser M. (1995): VNS (Vehicle Navigation Systems) aus der Sicht des Geodäten, Mitteilungen der geodätischen Institute der TU Graz, Folge 80, Graz 1995, 17–24

1333. Wilkinson J.H. (1965): The algebraic eigenvalue problem, Oxford University Press, Oxford 1965

1334. Williams R., Phythian J.E. (1989): Navigating along geodesic paths on the surface of a spheroid, The Journal of Navigation 42 (1989) 129–136

1335. Wintner A. (1956): On Frenet's equations, Amer. J. Math. 78 (1956) 349–356

1336. Wloka J. (1987): Partial differential equations, Cambridge University Press, Cambridge, 1987

1337. Woestijne I. van de (1990): Minimal surfaces of the 3–dimensional Minkowski space, in: M. Boyom et al. (ed.), Geometry and topology of submanifolds, II, World Scientific, Singapore 1990, pp. 344–369

1338. Wohlrab O. (1989): Die Berechnung und graphische Darstellung von Randwertproblemen für Minimalflächen, in: H. Jürgens und D. Saupe (eds.): Visualisierung in Mathematik und Naturwissenschaften, Springer Verlag, Berlin – Heidelberg – New York 1989

1339. Wolf H. (1987): Datumsbestimmung im Bereich der deutschen Landesvermessung, Zeitschrift für Vermessungswesen 112 (1987) 406–413

1340. Wolf H. (1995): 400 Jahre Mercator – 400 Jahre Atlas, Kartographische Nachrichten 45 (1995) 146–148

1341. Wolf J.A. (1964–65): Isotropic manifolds of indefinite metric, Commentarii Mathematici Helvetici 39 (1964–65) 21–64

1342. Wolfrum O. (1984): Die Theorie der Normalschnitte und ihre Anwendungen, Technische Hochschule Darmstadt (1984) 2–17

1343. Wolkow N.M. (1969): Automatisierung und Mechanisierung in der mathematischen Kartographie, Wissenschaftliche Zeitschrift der Technischen Universität Dresden 18 (1969) 589–596

1344. Wong F.K.C. (1965): World map projections in the United States from 1940 to 1960, Syracuse University, New York 1965

1345. Wong Y.-C. (1967): Differential geometry of Grassmann manifolds, Proceedings of the National Academy of Sciences of the United States of America 57 (1967) 589–594

1346. World Geodetic System Committee (1974): The Department of Defense World Geodetic System 1972 presented by T.O. Seppelin at the Int. Symposium on problems related to the redefinition of North American Geodetic Networks, Fredericton, N.B., Canada 1974

1347. Wraight A.J., Roberts E.B. (1957): The coast and geodetic survey, 1807–1957: 150 years of history: U.S. Coast and Geodetic Survey

1348. Wray T. (1974): The seven aspects of a general map projection, Supplement 2, Canadian Cartographer 11, Monographe No.11, Cartographica, B.V. Gustell Publ., University of Toronto Press, Toronto 1974

1349. Wu S.S.C. (1978): Mars synthetic topographic mapping, Icarus 33 (1978) 417–440

1350. Wu S.S.C. (1981): A method of defining topographic datums of planetary bodies, Ann. Geophys. 37 (1981) 147–160

1351. Wuensch V. (1997): Differentialgeometrie – Kurven und Flächen, Teubner Verlagsgesellschaft, Stuttgart/Leipzig 1997

1352. Wyszecki G., Stiles W.S. (1967): Color science, concepts and methods, quantitative data and formulas, J. Wiley & Sons, New York Chichester 1967

1353. Xiao H., Bruhns O.T., Meyers A. (1997): Hypo–elasticity model based upon the logarithmic stress rate, Journal of Elasticity 47 (1997) 51–68

1354. Xiao H., Bruhns O.T., Meyers A. (1998): On objective corotational rates and their defining spin tensors, Int. J. Solids Structures 35 (1998) 4001–4014

1355. Xiao H., Bruhns O.T., Meyers A. (1999): Existence and uniqueness of the integrable–exactly hypoelastic equation and its significance to finite inelasticity, Acta Mechanica 138 (1999) 31–50

1356. Yang Q.H., Snyder J.P., Tobler W.R. (2000): Map Projection Transformation, Taylor & Francis, 2000

1357. Yano K. (1940a): Concircular geometry I. Concircular transformations, Proc. Imperial Academy (Japan) 16 (1940) 195–200

1358. Yano K. (1940b): Concircular geometry II. Integrability conditions of, Proc. Imperial Academy (Japan) 16 (1940) 354–360

1359. Yano K. (1940c): Concircular geometry III. Theory of curves, Proc. Imperial Academy (Japan) 16 (1940) 442–448

1360. Yano K. (1940d): Concircular geometry IV. Theory of subspaces, Proc. Imperial Academy (Japan) 16 (1940) 505–511

1361. Yano K. (1940e): Conformally separable quadratic differential forms, Proc. Imp. Acad. Tokyo 16 (1940) 83–86

1362. Yano K. (1942): Concircular geometry V. Einstein spaces, Proc. Imperial Academy (Japan) 18 (1942) 446–451

1363. Yano K. (1955): The theory of lie derivatives and its applications, North–Holland Publishing Co., Amsterdam 1955

1364. Yano K. (1970): On Riemannian manifolds admitting an infinitesimal conformal transformation, Math. Z. 113 (1970) 205–214

1365. Yanushaushas A.I. (1982): Three–dimensional analogues of conformal mappings (in Russian) Iz da te l'stvo Nauka, Novosibirsk 1982

1366. Yates F. (1949): Systematic sampling, Philosophical transactions of the Royal Society 241 (1949) 355–77

1367. Yeremeyev V.F., Yurkina M.J. (1969): On the orientation of the Reference Geodetic Ellipsoid, Bulletin Géodésique 91 (1969) 13–16

1368. Yoeli P. (1986): Computer executed production of a regular grid of height points from digital contours, The American Cartographer 13 (1986) 219–229

1369. Young A.E. (1920): Some investigations in the theory of map projections, Royal Geographical Soc., London 1920

1370. Young A.E. (1930): Conformal map projections, Geographical J. 76 (1930) 348–351

1371. Young P. (1994): A reformulation of the partial least squares regression algorithm, SIAM J. Sci. Comput. 15 (1994) 225–230

1372. Yurkina M.I. (1996): Gravity potential at the major vertical datum as primary geodetic constant, Studia geoph. et geod. 40 (1996) 9–13

1373. Yuzefovich Y.M. (1971): Extension of the Chebyshev–Grave theorem to a new glass of cartographic projections, Geodesy and Aerophotography 10 (1971) 155–157

1374. Zadro M., Carminelli A. (1966): Rapprezentazione conforme del geoide sull ellissoide internazionale, Bollettino di Geodesia e Scienze Affini 25 (1966) 25–36

1375. Zafindratafa G. (1990): The local structure of a 2–codimensional conformally flat submanifold in an Euclidean Space $\mathbb{R}^{n+2}$, in: M. Boyom et al. (ed.), Geometry and topology of submanifolds, II, World Scientific, Singapore 1990, pp. 386–412

1376. Zeger J. (1991): 150 Jahre Bessel Ellipsoid 1841–1991, Österreichische Zeitschrift für Vermessungswesen und Photogrammetrie 79 (1991) 337–340

1377. Zha H. (1991): The restricted singular value decomposition of matrix triplets, SIAM J. Matrix Anal. Appl. 12 (1991) 172–194

1378. Zharkov V.N., Leontjev V.V., Kozenko A.V. (1985): Models, figures, and gravitational moments of the Galilean satellites of Jupiter and icy satellites of Saturn, Icarus 61 (1985) 92–100

1379. Zilkoski D.B., Richards J.H., Young G.M. (1992): Results of the general adjustment of the North American Datum of 1988, Surveying and Mapping 52 (1992) 133–149

1380. Zippel R. (1993): Effective Polynomial Computation, Kluwer Academic, Boston 1993

1381. Zoeppritz K. (1912): Leitfaden der Kartenentwurfslehre, hrsg. von A. Bludau, Erster Teil: Die Projektionslehre, Teubner, Leipzig und Berlin 1912

1382. Zoeppritz K., Bludau A. (1912): Leitfaden der Kartenentwurfslehre, Teubner, Leipzig 1912

1383. Zund J.D. (1987): The tensorial form of the Cauchy–Riemann equations, Tensor, New Series 44 (1987) 281–290

1384. Zund J.D. (1989): Topological foundations of the Marussi–Hotine approach to geodesy, Department of Mathematical Sciences, New Mexico State University, Scientific Report No. 1, Las Cruces 1989

1385. Zund J.D. (1994a): The differential geodesy of the spherical representation, Department of Mathematical Sciences, New Mexico State University, Scientific Report No. 6, Las Cruces 1994

1386. Zund J.D. (1994b): Foundations of differential geometry, Springer Verlag, Berlin – Heidelberg – New York 1994

1387. Zund J.D., Moore W.A. (1987): Conformal geometry, Hotine's conjecture, and differential geodesy, Department of Mathematical Sciences, New Mexico State University, Scientific Report No. 1, Las Cruces 1987

Index